T0256978

At the Top of the Grand Staircase

At the Top of the Grand Staircase

LIFE OF THE PAST   James O. Farlow, editor

AT THE TOP OF THE GRAND STAIRCASE

The Late Cretaceous of Southern Utah

Edited by ALAN L. TITUS and MARK A. LOEWEN

INDIANA UNIVERSITY PRESS   Bloomington & Indianapolis

This book is a publication of

Indiana University Press
Office of Scholarly Publishing
Herman B Wells Library 350
1320 East 10th Street
Bloomington, Indiana 47405 USA

iupress.indiana.edu

Telephone orders 800-842-6796
Fax orders 812-855-7931

© 2013 by Glen Canyon Natural
History Association

All rights reserved

No part of this book may be reproduced or
utilized in any form or by any means, electronic
or mechanical, including photocopying and
recording, or by any information storage and
retrieval system, without permission in writing
from the publisher. The Association of American
University Presses' Resolution on Permissions
constitutes the only exception to this prohibition.

Manufactured in the United States of America

Library of Congress Cataloging-in-Publication Data

At the top of the grand staircase : the late
Cretaceous of Southern Utah / edited by
Alan L. Titus and Mark A. Loewen.
 pages cm. – (Life of the past)
 Includes bibliographical references and index.
 ISBN 978-0-253-00883-1 (cloth : alk.
paper)–ISBN 978-0-253-00896-1 (e-book)
1. Geology, Stratigraphic – Cretaceous.
2. Paleontology–Cretaceous. 3. Geology–Utah.
4. Animals, Fossil–Utah. 5. Grand Staircase-
Escalante National Monument (Utah) I. Titus,
Alan L. (Alan Lee), [date] editor of compilation.
II. Loewen, Mark A., editor of compilation.
 QE688.A85 2013
 551.7'7097925–dc23
                                   2013005859

1 2 3 4 5 18 17 16 15 14 13

This volume is dedicated to William Jefferson Clinton, forty-second president of the United States, who established Grand Staircase–Escalante National Monument by presidential proclamation in 1996 and indirectly spurred a new era of study on southern Utah's Cretaceous stratigraphic and paleontological record.

Proceeding eastward to the Kaiparowits the Cretaceous
presents itself in a manner which is highly significant, and
which merits careful examination.

Clarence Dutton
Tertiary History of the Grand Canyon District, 1882

Contents

Foreword

IN 1982, TOM RYER CAME TO THE UNIVERSITY OF WYO-ming to give a talk on coal geology in the Cretaceous of Utah. A few years earlier (1979), Jason Lillegraven had completed his groundbreaking *Mesozoic Mammals* book, and the gaps in our record of Cretaceous mammals were fresh in my mind. I asked Tom what was landward of these Cretaceous coals, and he replied, "Boring floodplain sediments." That did it for me. After I graduated in 1982, I drove out to the San Rafael Swell, the Henry Mountains, and the Kaiparowits Plateau, and I found fossils everywhere I went. I was most impressed by the thick and well-exposed section on the Kaiparowits Plateau. So in May 1983, I went to the Kaiparowits Plateau to begin my dissertation research.

I fully expected to have the Kaiparowits Plateau to myself. However, that same year, the remarkable collector Will Downs led Richard Cifelli (then at the Museum of Northern Arizona) to the Kaiparowits Plateau. As it turned out, we ended up camped less than a mile from each other on Horse Mountain. This was the beginning of still-continuing paleontological discoveries on the Kaiparowits Plateau. The scientific resources of the region eventually became the basis for establishing Grand Staircase–Escalante National Monument in 1996, and subsequent paleontological research on the monument has been championed by Alan Titus, to whom all of us are indebted.

As was typical of the times, the workers that had simultaneously arrived on the Kaiparowits Plateau in 1983 studied mammals; however, this volume reflects the healthy diversification of research that has taken place on the plateau over the years. This volume includes chapters on flora, invertebrates, sharks and rays, fish, frogs, salamanders, turtles, lizards, crocodylians, dinosaurs (of course!), trackways, and marine vertebrates. Also, important contributions are being made on the chronostratigraphy of the region (it is wonderful to finally have some radiometric dates!) and on other aspects of the geology and sedimentology of this extraordinary sequence.

The premise on which Grand Staircase–Escalante National Monument was established is well demonstrated in this volume, but all of us who work in this region understand that this compilation does not represent the end of scientific research here, but rather a new starting place. The monument will serve to enhance our knowledge of Earth history, and in particular the Late Cretaceous, for generations to come.

Jeffrey G. Eaton
Ogden, Utah
December 2011

DURING MAY 22–23, 2009, DOZENS OF RESEARCHERS gathered in St. George, Utah, at a conference titled "Learning from the Land: Advances in Western Interior Late Cretaceous Geology and Paleontology." I organized this conference under the sponsorship of Grand Staircase–Escalante National Monument, the Utah Museum of Natural History (now the Natural History Museum of Utah), Glen Canyon Natural History Association, the Utah Friends of Paleontology, and Grand Staircase–Escalante Partners. It was held, admittedly, largely to celebrate the success of macrovertebrate paleontology research in Grand Staircase–Escalante National Monument over the last decade. However, as the abstracts poured in, it dawned on me how diverse the topics being presented truly were – from sand volcanoes and sequence stratigraphy to mammalian paleontology. For some, the meeting represented their first foray into the region's scientific treasure trove; others presented on career-long syntheses of vast amounts of data. After digesting the talks and abstracts, it became clear that the foundation had been laid for a volume on southern Utah's Cretaceous geology and paleontology. Major contributions to vertebrate evolutionary patterns, biogeography, and biostratigraphy existed in all the unpublished data that sorely needed to see the light of day. Conversations with the helpful staff at Indiana University Press, particularly Jim Farlow and Robert Sloan, ultimately led to the present volume.

It is safe to say that 10 years earlier, very few researchers outside of the coal geology community had even heard of the Kaiparowits Plateau. Yet now, it seemed as if the place were earning itself a position of global prominence in the Cretaceous research community – perhaps even living up to claims in the National Monument's establishing proclamation that the region held "one of the best and most continuous records of Late Cretaceous terrestrial life in the world." At the very least, southern Utah had become a primary reference point for the nature and evolution of terrestrial ecosystems in southern Laramidia. As of this writing, only 10% of the estimated 250,000 hectares of Cretaceous outcrop in the Kaiparowits Basin has been intensively prospected for fossils. The percentages are even lower on the adjacent Paunsaugunt Plateau, Markagunt Plateau, and Pine Valley Mountain areas. With a minimum of 300,000 hectares of outcrop still unexplored, the region – one of the last mapped in the continental United States – also remains one of the last true frontiers in North American geology and paleontology. I hope this volume represents just the first in a series of treatises documenting the wealth of knowledge that the Cretaceous rocks of this wild and beautiful place have to offer.

Alan L. Titus
Paleontologist
Grand Staircase–Escalante National Monument
Kanab, Utah

Acknowledgments

A

WE THANK FIRST AND FOREMOST THE BUREAU OF LAND Management for their support of research on the Cretaceous of southern Utah, largely as a result of the creation of Grand Staircase-Escalante National Monument in 1996. This includes grants to partner research institutions, support of its own internal research and resource programs, and sponsorship of symposia and publications, including this volume. In particular Marietta Eaton, Carl Rountree, Jeff Jarvis, and Scott Foss need to be called out for their unfailing support.

We thank Glen Canyon Natural History Association and the Grand Staircase-Escalante Partners organizations for their continuous support of research, symposia, and publications. Many thanks are due to Chris Eaton of the former and Daisy Ballard of the latter for the huge roles they played in raising support and helping to proof the volume.

We thank the many authors who attended the 2009 symposium that led to this volume. Their hard work and follow-through with timely submission and editing of the manuscripts were truly inspiring. Along the same lines thanks must also be given to the many reviewers who took time from their busy schedules to ensure that the content is both accurate and current.

Lastly, but certainly not least, we want to recognize the editorial staff at Indiana University Press, including Bob Sloan, Jim Farlow, June Silay, and Karen Hellekson, who agreed to take on this Gordian knot and help make the editors' vision for a Utah Cretaceous volume a reality.

The editors

Contributors

L. Barry Albright III
University of North Florida, Jacksonville, FL 32224, U.S.A., lalbrigh@unf.edu

Michael A. Arthur
Department of Geoscience Pennsylvania State University, University Park, PA 16802, U.S.A., maa6@psu.edu

Richard S. Barclay
Smithsonian Institution, P.O. Box 37012, MRC 121, Washington, D.C. 20013-7012, U.S.A., barclayrs@si.edu

Samuel A. Bowring
Department of Earth, Atmospheric and Planetary Sciences, Massachusetts Institute of Technology, Cambridge, MA 01239, U.S.A., sbowring@mit.edu

Clint A. Boyd
South Dakota School of Mines and Technology, Rapid City, SD 57701, U.S.A., clintboyd@stratfit.org

Donald B. Brinkman
Royal Tyrrell Museum of Palaeontology, Drumheller, Alberta T0J 0Y0, Canada, don.brinkman@gov.ab.ca

Robert Buchwaldt
Washington University, Campus Box 1169, One Brookings Drive, St. Louis, MO 63130-4899, buckwaldt@levee.wustl.edu

Michael E. Burns
Department of Biological Sciences, University of Alberta, 11455 Saskatchewan Drive, Edmonton, Alberta, T6G 2E9, Canada, mburns@ualberta.ca

Richard L. Cifelli
Oklahoma Museum of Natural History, University of Oklahoma, 2401 Chautauqua Avenue, Norman, OK 73072, U.S.A., rlc@ou.edu

Leon P. A. M. Claessens
College of the Holy Cross, P.O. Box B, 1 College Street, Worcester, MA 01610, U.S.A., lclaesse@holycross.edu

Walter E. Dean
United States Geological Survey, Federal Center, Denver, CO 80225, U.S.A., dean@usgs.gov

Donald D. DeBlieux
Utah Geological Survey, P.O. Box 146100, Salt Lake City, UT 84114, U.S.A., dondeblieux@utah.gov

Alan L. Deino
Berkeley Geochronology Center, Berkeley, CA 94709, U.S.A., adeino@bgc.org

Jeffrey G. Eaton
Department of Geosciences, Weber State University, Ogden, UT 84408-2507, U.S.A., jeaton@weber.edu

David C. Evans
Department of Natural History (Palaeobiology), Royal Ontario Museum, 100 Queen's Park, Toronto, Ontario, M5S 2C6, Canada, davide @rom.on.ca; University of Toronto, Department of Ecology and Evolutionary Biology, 25 Willcocks Street, Toronto, Ontario, M5S 3B2, Canada, d.evans@utoronto.ca

Andrew A. Farke
Raymond M. Alf Museum of Paleontology, 1175 West Baseline Road, Claremont, CA 91711, U.S.A., afarke@webb.org

James D. Gardner
Royal Tyrrell Museum of Palaeontology, Box 7500, Drumheller, Alberta, T0J 0Y0, Canada, james.gardner@gov.ab.ca

Terry A. Gates
Department of Biology, North Carolina State University, Raleigh, NC 27695-7617, U.S.A., tagates@ncsu.edu

Michael A. Getty
Denver Museum of Nature & Science, 2001 Colorado Boulevard, Denver, CO 80205-5732, U.S.A., mike.getty@dmns.org

Gerard Gierlinski
Polish Geological Institute, Rakowiecka 4, PL-00-975, Warszawa, Poland, gierlinski@yahoo.com

David D. Gillette
Museum of Northern Arizona, Department of Geology, 3101 North Fort Valley Road, Flagstaff, AZ 86001, U.S.A., dgillette@mna.mus.az.us

Martha C. Hayden
Utah Geological Survey, P.O. Box 146100, Salt Lake City, Utah 84114-6100, U.S.A., mhayden@utah.gov

Hannah L. Hilbert-Wolf
Carleton College, Northfield, MN 55057, U.S.A., hilbertwolf@gmail.com

J. Howard Hutchison
P.O. Box 261, Escalante, UT 84726, U.S.A., Howard.Hutchison@gmail.com

Randall B. Irmis
Department of Geology and Geophysics, 115 South 1460 East, University of Utah, Salt Lake City, UT 84112-0111, U.S.A.; Natural History Museum of Utah, 300 Wakara Way, University of Utah, Salt Lake City, UT 84108, U.S.A., irmis@nhmu.utah.edu

Zubair Jinnah
School of Geosciences, University of the Witwatersrand, Private Bag 3, WITS, 2050, Johannesburg, South Africa, Zubair.jinnah@wits.ac.za

Kirk R. Johnson
National Museum of History, P.O. Box 37012, Smithsonian Institution, Washington, D.C. 20013-7012, U.S.A., kirk.johnson@si.edu

Erle G. Kauffman
Department of Geological Sciences, Indiana University, Bloomington, IN 47405, U.S.A.

Gy-Su Kim
The Webb Schools, 1175 West Baseline Road, Claremont, CA 91711, U.S.A., gkim@webb.org

James I. Kirkland
Utah Geological Survey, P.O. Box 146100, Salt Lake City, Utah 84114-6100, U.S.A., jimkirkland@utah.gov

Douglas E. Kline
Department of Earth Sciences, Denver Museum of Nature & Science, 2001 Colorado Boulevard, Denver, CO 80205-5732, U.S.A., douglas.kline@dmns.org

Michael J. Knell
Montana State University, Department of Earth Sciences, 200 Traphagen Hall, Bozeman, MT 59717-3480, U.S.A., knell.mike@gmail.com

Martin Lockley
Dinosaur Tracks Museum, University of Colorado at Denver, Denver, CO 80217, U.S.A., Martin.Lockley@UCDenver.edu

Mark A. Loewen
Department of Geology and Geophysics, 115 South 1460 East, University of Utah, Salt Lake City, UT 84112-0111, U.S.A.; Natural History Museum of Utah, 300 Wakara Way, University of Utah, Salt Lake City, UT 84108, U.S.A., mloewen@umnh.utah.edu

Eric K. Lund
Department of Biomedical Sciences, Ohio University Heritage College of Osteopathic Medicine, 228 Irvine Hall, Athens, OH 45701, U.S.A., lunde@ohio.edu

Christopher T. McGarrity
University of Toronto, Department of Ecology and Evolutionary Biology, 25 Willcocks Street, Toronto, Ontario, Canada M5S 3B2, christopher.mcgarrity@utoronto.ca

Ian M. Miller
Denver Museum of Nature & Science, 2001 Colorado Boulevard, Denver, CO 80205-5732, U.S.A., ian.miller@dmns.org

Andrew G. Neuman
Royal Tyrrell Museum of Palaeontology, Drumheller, Alberta T0J 0Y0, Canada, andrew.neuman@gov.ab.ca

Michael G. Newbrey
Royal Tyrrell Museum of Palaeontology, Drumheller, Alberta T0J 0Y0, Canada; Department of Biological Sciences, University of Alberta, Edmonton, Alberta T6G 2E9, Canada, mike.newbrey@gov.ab.ca

†Douglas J. Nichols
Denver Museum of Nature & Science, 2001 Colorado Boulevard, Denver, CO 802055732, U.S.A.

Randall L. Nydam
Midwestern University, 19555 North 59th Avenue, Glendale, AZ 85308, U.S.A., rnydam@midwestern.edu

Patrick M. O'Connor
Department of Biomedical Sciences, Ohio University Heritage College of Osteopathic Medicine, 228 Irvine Hall, Athens, OH 45701, U.S.A., oconnorp@ohio.edu

Tomáš Přikryl
Department of Palaeobiology, Geological Institute, Academy of Sciences of the Czech Republic, Rozvojová 135, CZ-165 00 Prague 6, Czech Republic, Prikryl.T@seznam.cz

Eric M. Roberts
School of Earth and Environmental Sciences, James Cook University, Townsville 4811, Queensland, Australia, eric.roberts@jcu.edu.au

Zbyněk Roček
Department of Palaeobiology, Geological Institute, Academy of Sciences of the Czech Republic, Rozvojová 135, CZ-165 00 Prague 6, Czech Republic, Rocek@gli.cas.cz

Scott D. Sampson
Denver Museum of Nature & Science, 2001 Colorado Boulevard, Denver, CO 80205-5732, U.S.A., Scott.Sampson@dmns.org

Joseph J. W. Sertich
Denver Museum of Nature & Science, 2001 Colorado Boulevard, Denver, CO 80205-5732, U.S.A., joe.sertich@dmns.org

Edward L. Simpson
Department of Physical Sciences, Indiana University of Pennsylvania, Kutztown, PA 19530, U.S.A., simpson@kutztown.edu

Leif Tapanila
Idaho State University, Department of Geosciences, Pocatello, ID 83209-8072, U.S.A.; Idaho Museum of Natural History, Division of Earth Sciences, Pocatello, ID 83209-8096, U.S.A., tapaleif@isu.edu

Sarah E. Tindall
Department of Physical Sciences, Indiana University of Pennsylvania, Kutztown, PA 19530, U.S.A., tindall@kutztown.edu

Alan L. Titus
Grand Staircase–Escalante National Monument, Kanab, Utah 84741, U.S.A., atitus@blm.gov

Matthew K. Vickaryous
Ontario Veterinary College, University of Guelph, 50 Stone Road, Guelph, Ontario, N1G 2W1, Canada, mvickary@uoguelph.ca

Jelle P. Wiersma
Department of Earth and Life Sciences, Vrije Universiteit Amsterdam, Amsterdam, The Netherlands; Department of Geology and Geophysics, 115 South 1460 East, University of Utah, Salt Lake City, UT 84112-0111, U.S.A.; Natural History Museum of Utah, 300 Wakara Way, University of Utah, Salt Lake City, UT 84108, U.S.A., jwiersma@umnh.edu

Thomas Williamson
New Mexico Museum of Natural History and Science, 1801 Mountain Road Northwest, Albuquerque, NM 87104, U.S.A., thomas.williamson@state.nm.us

Michael C. Wizevich
Department of Physics and Earth Sciences, Central Connecticut State University, New Britain, CT 06050, U.S.A., wizevichmic@ccsu.edu

Lindsay E. Zanno
North Carolina Museum of Natural Sciences, 121 West Jones Street, Raleigh, NC 27603-1368, U.S.A., lindsay.zanno@naturalsciences.org

At the Top of the Grand Staircase

1.1. Panorama of outcrops of the Paleocene-age Claron Formation exposed at Bryce Canyon National Park, Utah.

One Hundred Thirty Years of Cretaceous Research in Southern Utah

Alan L. Titus

INTRODUCTION

Southern Utah possesses a wild, stark, rugged landscape that leaves most people who experience it irrevocably and profoundly changed. Scenic wonders such as the Grand Canyon, Bryce Canyon (Fig. 1.1), Zion Canyon, The Wave, Buckskin Gulch, Cedar Breaks, and Capitol Reef are deservedly visited annually by millions of tourists from all over the globe who leave both awestruck and humbled. What is true for the tourist is even more so for the geologists and paleontologists who have worked in the canyon, mesa, plateau, and cliff outcrops of the Grand Staircase–Kaiparowits Plateau region ever since the first reports of the Powell expedition were published (e.g., Dutton, 1880; Howell, 1875). But the bedrock geology of the Grand Staircase provides not just the raw material for geomorphological agents to shape renowned scenic wonders; it also contains the fascinating saga of the evolving North American Cordilleran biosphere: a cavalcade of changing landscapes and organisms frozen in time for researchers to poke, prod, and ponder. Even though the area was first geologically mapped over 125 years ago, many basic stratigraphic and paleontological questions still remain unanswered. For vertebrate paleontologists, the region is still a frontier waiting to yield a wealth of data on the Mesozoic biosphere.

Two of the most obvious red rock massifs so sought after by the more aesthetically inclined (the Navajo and Claron formations) are interleaved with thousands of meters of subtler gray and brown strata that comprise southern Utah's Cretaceous record (Fig. 1.2). Although not as commanding to the eye, the import of this portion of the Grand Staircase story to scientists cannot be overstated: it holds a terrestrial vertebrate fossil record that is certainly the most continuous known in the southern United States, if not the world (Eaton and Cifelli, 1997). As previous scientists probed into the region's Cretaceous geology, it became clear that this portion of the Sevier Foreland Basin was not only a fastidious record keeper of the ecosphere, but also strategically placed – close enough to the orogen to be predominantly terrestrial, but far enough east to maintain at least intermittent contact with the seaway and thus contain crucial high-resolution biostratigraphic data. These conditions persisted until approximately 72 Ma, when Laramide uplifts forced the regression of the seaway into Colorado. Six and a half million years later, the great reign of nonavian dinosaurs was over, leaving a mystery whose solving has become a collective human passion.

GEOLOGICAL INVESTIGATIONS

Powell (1875:190), Howell (1875), and Gilbert (1875) were the first to document beds of Cretaceous Age exposed in what are now known as the Grand Staircase and Kaiparowits Plateau physiographic regions of southern Utah. These works provide only brief descriptions of Cretaceous stratigraphic units, mostly for mapping and summary purposes. It is clear from the geologic maps that accompany these works (Fig. 1.3) that the late 19th-century concept of the region's Cretaceous strata was essentially the same as that today. In his exhaustive monograph, Stanton (1893) concluded from available marine invertebrate fossil evidence that the lower portion of the Cretaceous sequence (Dakota and Tropic formations) was a Colorado Formation equivalent and that the thick overlying succession was terrestrial and Laramie in age.

Twenty-five years passed before any additional serious geologic studies were undertaken in the Kaiparowits region. In 1915 Herbert E. Gregory conducted his first geological reconnaissance in the Kaiparowits, with the intent to document those areas that the Powell expeditions had overlooked or only covered in a cursory way. Gregory returned to the Kaiparowits region in 1918, 1922, 1924, 1925, and 1927 to continue mapping, measuring sections, and making fossil collections. In 1921 the eminent stratigrapher and paleontologist Raymond C. Moore also started stratigraphic and paleontological studies in the region, continuing his efforts into the summer of 1923. The resulting collaborative work between these two researchers (Gregory and Moore, 1931) led to publication of the classic work that established the region's modern nomenclatural framework for Cretaceous strata. With the exception of the lowermost terrestrial portion of the section, which Gregory and Moore (1931) referred

1.2. Outcrops of the upper Campanian Kaiparowits Formation exposed at the north end of Horse Mountain, Grand Staircase–Escalante National Monument. These outcrops were among the first explored by J. Eaton and R. Cifelli in the early 1980s.

to the Dakota (?) Sandstone, the entire Cretaceous section was divided into new formations: in ascending order, they are the Tropic, Straight Cliffs, Wahweap, and Kaiparowits. Although placement of some of the contacts between these units has changed slightly, with the exception of the Dakota Formation, which has now been partly replaced with the Cedar Mountain Formation, Gregory and Moore's original names and concepts are used today essentially unchanged.

Armed with new biostratigraphic data and advances in the understanding of the Western Interior Cretaceous succession, Gregory and Moore (1931) divided the section into pre-Carlile (Dakota Formation and lower portion of the Tropic Formation), Carlile (Tropic Formation), Niobrara (Straight Cliffs Formation), early and middle Montana (Wahweap Formation), and Fruitland (Kaiparowits Formation) faunal

zones. Again, like their stratigraphic nomenclature, these age assignments still largely stand.

Meanwhile, in the Markagunt Plateau region, Richardson (1909, 1927) measured sections and described strata of Cretaceous age, primarily to determine coal potential. This same motivation would later spur many mapping studies on Cretaceous strata of the region well into the 1990s (Spieker, 1925; Cashion, 1961, 1976; Cohenour, 1963; Robison, 1966; Peterson and Horton, 1967; Waldrop and Sutton, 1967a, 1967b, 1967c; Doelling and Graham, 1972; Bowers, 1973a, 1973b, 1973c, 1975, 1981, 1983, 1991a, 1991b, 1993; Peterson and Barnum, 1973a, 1973b; Zeller, 1973a, 1973b, 1973c, 1973d, 1978, 1990a, 1990b, 1990c; Zeller and Stephens, 1973; Peterson, 1975, 1980; Zeller and Vaninetti, 1990). Gregory also continued his regional geologic studies and published a series of papers (Gregory,

Titus

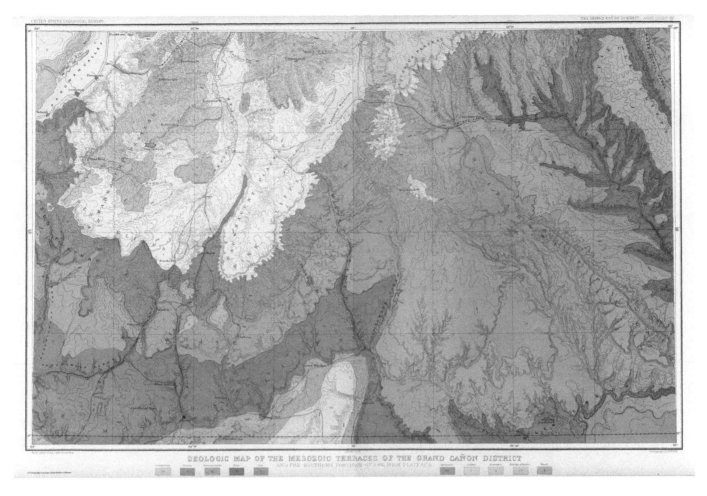

1.3. The first geologic map of the Grand Staircase region demonstrating that the originally postulated Cretaceous bedrock concepts have changed little over the last 130 years. From Dutton (1880).

1948, 1949, 1950a, 1950b, 1951) that include original stratigraphic work in Cretaceous sections using the unmodified nomenclature of his collaborative work with R. C. Moore.

In the 1960s John Lawrence and Fred "Pete" Peterson undertook detailed studies on Cretaceous strata of the Kaiparowits Plateau region as part of their respective dissertation research projects. Lawrence never finished his dissertation on Dakota and Tropic stratigraphy, but he did publish a small paper revising the definition of the base of the Tropic Formation (Lawrence, 1965). Peterson finished his dissertation in August 1969 and shortly thereafter published the bulk of it as a U.S. Geological Survey open-file report (Peterson, 1969a) and bulletin (Peterson, 1969b). In the latter work, Peterson proposed the four currently recognized members of the Straight Cliffs Formation: Tibbet Canyon, Smoky Hollow, John Henry, and Drip Tank.

During the subsequent 1970s and early 1980s much of the stratigraphic work done on the region's Cretaceous stratigraphy was connected to coal resource investigations (Vaninetti, 1978; Sargent and Hansen, 1982; Kirschbaum and McCabe, 1992). However, in the early 1980s the University of Colorado–Boulder took an interest in the regional Cretaceous framework, particularly from a cyclic stratigraphic and biochronological perspective. Dissertations and theses were written on the Dakota Formation (Gustason, 1989), Tropic Shale and nearby Black Mesa exposures of Mancos Shale (Elder, 1987; Kirkland, 1990), and the Straight Cliffs Formation (Bobb, 1991; Little, 1997). Eaton (1987, 1990, 1991), also at UC–Boulder, reviewed the entire Kaiparowits Basin and Henry Basin Cretaceous successions and subdivided the Wahweap Formation into four members. Concurrently, Patrick Goldstrand of the University of Nevada–Reno undertook studies on the Laramide portion of the section (Goldstrand, 1990, 1991, 1994; Schmitt et al., 1991; Goldstrand and Mullet, 1997). An outstanding synthesis of regional Greenhorn cyclothemic stratigraphy that resulted from this burst of activity is provided in Tibert et al. (2003). The volume of data compiled in each of these studies was comparable with the work done by Peterson (1969a), and each in its own way was a significant advance over previous work.

In the early 1990s energy development spurred another series of articles analyzing environmental facies and applying sequence stratigraphic concepts of Haq, Hardenbol, and Vail (1988) to the Straight Cliffs Formation (Shanley and McCabe, 1991, 1993, 1995; Shanley, McCabe, and Hettinger, 1992; Hettinger, McCabe, and Shanley, 1994; Hettinger, 1995). Sequence stratigraphic models developed in this period were widely cited for the next two decades. Terry Tilton also finished his dissertation on Cretaceous stratigraphy of the Paunsaugunt Plateau about this same time (Tilton, 1991). At the close of the 1990s David Ulicny applied sequence stratigraphic concepts to the Dakota Formation (Ulicny, 1999) and replaced its standard lower, middle, and upper member terminology with numbered units 1–6. Ulicny's unit terminology has been largely ignored, but his cyclic stratigraphic model for the Dakota Formation is still widely cited. In the same year, Stony Pollock finished his master's thesis on comprehensive lithofacies and petrographic study on the Capping Sandstone Member of the Wahweap Formation (Pollock, 1999) and hypothesized the Capping Sandstone Member was deposited under a radically different depositional regime than the underlying three members, in essence echoing Little's tectonic cycle hypothesis, but adding much additional data and applying the idea of Heller et al. (1988) that transverse and axial river systems in the foreland alternated back and forth, driven by the synchronous tectonic setting in the orogen. Eventually these studies were summarized by Lawton, Pollock, and Robinson (2003).

Starting in the present millennium, some of the first detailed stratigraphic investigations were done on the more poorly understood Markagunt Plateau and Pine Valley Mountain regions (Eaton et al., 2001; Laurin and Sageman, 2001; Moore and Straub, 2001; Tibert et al., 2003). Meanwhile, additional detailed studies were done on the depositional, taphonomic, and sedimentological signatures of the Straight Cliffs (Castle et al., 2004; Christensen, 2005; Allen and Johnson, 2010a, 2010b; Gallin, Johnson, and Allen, 2010), Wahweap (Simpson et al., 2008, 2009; Tindall et al., 2010), and Kaiparowits formations (Roberts, 2005, 2007) in the form of M.Sc. and Ph.D. theses or associated papers. Radiometric dates were also generated for the first time for much of the terrestrial part of the section (Biek et al., 2000; Dyman et al., 2002; Roberts et al., 2005; Jinnah et al., 2009), ending decades of speculation on the absolute ages of these units and providing the first firm basis for continental- and global-scale correlation of the terrestrial fossil faunas. Recent mapping on the Markagunt Plateau is also beginning to clear up some lingering questions on age relationships of the terrestrial upper portion of the sequence (Biek et al., 2010; Hylland, 2010). The latest tool applied to regional Cretaceous stratigraphic problems is detrital zircon analysis (Larsen et al., 2010).

In spite of these quantum increases in geological knowledge of the southern Utah region, key problems still remain to be resolved, particularly chronostratigraphic correlations into the western Markagunt and Pine Valley areas and the age relationships of all of the coarser-grained units, such as the Drip Tank Member. Facies, architectural, and faunal/floral changes that occur in penecontemporaneous sedimentary packages as one moves west from coastal settings into upper alluvial plain/alluvial fan settings also remain largely undocumented.

PALEONTOLOGICAL INVESTIGATIONS

The first mention of Cretaceous fossils from the Kaiparowits Plateau area was made by Howell (1875:271), who listed *Physa*, *Limea*, *Viviparus* spp., an indeterminate small oyster-like shell, and "fragments of large bones" from "gray arenaceous and argillaceous shales" outcropping below the Table Cliffs (an area now called The Blues). Howell referred the fossils and containing beds to the Tertiary, but it is now known that they came from the Upper Cretaceous Kaiparowits Formation. Subsequently Stanton (1893) summarized the invertebrate fauna of the Western Interior Colorado Group (Cenomanian–Santonian) and described numerous marine and brackish-water taxa collected in the Paunsaugunt Plateau and Markagunt Plateau areas by C. D. Walcott in 1882 and T. W. Stanton in 1892. Gregory and Moore (1931) reported on extensive fossil collections made in support of their Kaiparowits Plateau study. Among these are the first published identifications of Cretaceous-age terrestrial fossil vertebrate material from the region, collected from the Kaiparowits Formation in the vicinity of Canaan Peak. Identified were (with identifications made by R. S. Lull): the turtle *Baena* sp. and the dinosaurs cf. *Triceratops* sp. and *Trachodon* (represented by several elements including tibia, humerus, scapula, and an ungual).

Similar fossil material from the "Kaiparowits Formation" (now referred to the middle unit of the Wahweap Formation) of the southern Paunsaugunt Plateau was listed by Gregory (1951, with identifications by C. W. Gilmore), this time giving specific taxonomic names to three different turtles (*Basilemys* sp., *Adocus* sp., and *Baena* cf. *B. nodosa*). Fossil plants reported by Gregory (1950b) also from the "Kaiparowits Formation" (now referred to Wahweap Formation or older beds) include *Dammarites caudatus?*, *Podozamites oblongus*, *Podozamites angustifolius*, *Platanus newberryana*, *Platanus* cf. *P. primaeva*, *Betula* cf. *B. beatriciana*, *Menispermites ovalis*, *Cinnamomum* sp., and *Viburnum robustum*. Lists of fossil invertebrate species were also given for most of the Cretaceous units covered in the preceding works.

In 1969 Brigham Young University (BYU) graduate student Fred Lohrengel published his dissertation on palynology of the Kaiparowits Formation in which he assigned the Kaiparowits Formation a Hell Creek–ian (Maastrichtian) age. This work was revisited by Farabee (1991), Bowers (1972), and Nichols (1997), who concluded that the palynological evidence from the Kaiparowits Formation more strongly argued for an upper Campanian assignment. Palynological studies were also subsequently conducted on the Dakota (May and Traverse, 1973; am Ende, 1991), Straight Cliffs (Orlansky, 1971; Nichols, 1997), and Wahweap (Nichols, 1997) formations.

Possibly because of conversations with Fred Lohrengel, "Dinosaur Jim" Jensen, also from BYU, prospected the Kaiparowits Formation for one summer in 1971 and found a partial skull of the hadrosaur *Parasaurolophus* (Weishampel and Jensen, 1978). In 1977 Frank DeCourten was working at the Museum of Northern Arizona as an intern when a member of the public brought the famed paleontologist Ned Colbert some theropod dinosaur bones he had collected in the Kaiparowits Formation. Because they were Cretaceous, Ned wasn't interested, but he encouraged Frank to go and collect them, and they turned out to be an articulated partial ornithomimid (DeCourten and Russell, 1985).

During 1981–1982 Sam Webb, Jim Jensen, and Brooks Britt of BYU conducted a field campaign in the Kaiparowits Formation. They found several significant specimens, including a partial tyrannosaur (including skull elements), articulated turtles, and a partial hadrosaur skeleton. The tyrannosaur material now serves as the holotype for the newly named *Teratophoneus curriei* (Carr et al., 2011). Sam Webb, who had meant to include the Kaiparowits material in his master's thesis, never finished his degree, and most of the BYU material remains unpublished.

In 1982 University of Colorado–Boulder graduate student Jeff Eaton first visited the Kaiparowits Plateau with the objective of doing mammal biostratigraphy on the nonmarine portions of the section. Three years later he teamed up with Rich Cifelli, then the curator for the Museum of Northern Arizona, who was also working in the area, and began a 15-year-long intensive sampling and screen-washing operation through the entire Cretaceous section. Dozens of mammal species were documented (Cifelli, 1990a, 1990b, 1990c, 1990d; Cifelli and Madsen, 1986; Cifelli and Eaton, 1987; Eaton and Cifelli, 1988; Eaton, 1993, 1995, 2002, 2006a, 2006b; Cifelli and Johanson, 1994) and vast collections of other vertebrate material were amassed (Eaton, Cifelli, et al., 1999), much of which is figured in this volume for the first time. This first systematic look at the Kaiparowits Plateau's vertebrate fossil record allowed Eaton, Cifelli, and their collaborators, including J. H. Hutchison, James Kirkland, and J. Michael Parrish, to track innovation, diversity, and extinction in vertebrate ecosystems over a 25-million-year span, automatically making it one of the most continuous

Late Cretaceous terrestrial vertebrate fossil records known in the world (Eaton and Cifelli, 1997). One significant paper to result from this body of data analyzed the effects of the Cenomanian–Turonian extinction on the region's terrestrial vertebrate fauna (Eaton et al., 1997).

Later Eaton expanded the scope of his research to include outcrops in the Paunsaugunt Plateau (Eaton, 1999a, this volume, Chapter 15), Markagunt Plateau (Eaton, 2006a; Eaton, Diem, et al., 1999), and Pine Valley Mountain areas (Eaton, 1999b).

University of Colorado–Boulder Ph.D. student Will Elder began trenching and sampling the Tropic Formation in southern Utah in 1984, expanding his master's research on molluscan diversity and paleoecology of the Cenomanian–Turonian marine faunal turnover at Black Mesa, Arizona, into a general biostratigraphic and paleoecological analysis of the Cenomanian–Turonian boundary interval throughout the Western Interior (Elder, 1989). Later collaboration with Brad Sageman and Gus Gustason led to high-resolution biostratigraphically supported offshore to onshore correlation of Greenhorn cyclothemic parasequences as far west as the Markagunt Plateau (Elder, Gustason, and Sageman, 1994). The first description of a plesiosaur from the Tropic Formation was given by Gillette et al. (1999).

On September 18, 1996, Grand Staircase–Escalante National Monument was established by President William J. Clinton and kept within the Bureau of Land Management for administration. The monument proclamation calls the Cretaceous paleontology of the Kaiparowits Plateau, in part as a result of the pioneering work of Eaton and Cifelli, "world class." In 1998 an exhaustive inventory of known paleontological localities was conducted (Foster et al., 2001). Hamblin and Foster (2000) summarized the vertebrate ichnological record of Grand Staircase–Escalante National Monument, and Cobban et al. (2000) published a summary of brackish-water and marine invertebrates, tabulating data from numerous collections made from the Monument by U.S. Geological Survey mapping crews over the last 50 years.

The political act of creating Grand Staircase–Escalante National Monument eventually became a catalyst for a dramatic increase in research on the region's macrovertebrate paleontology when the monument management plan was approved in 2000. Plan decisions to inventory and research Grand Staircase–Escalante National Monument's geology and paleontology as well as to prioritize the protection of resources at risk of loss over time led to my hiring as a full-time monument paleontologist. By 2001 the formation of close partnerships between the Utah Museum of Natural History (Salt Lake City), the Utah Geological Survey (Salt Lake City), and the Museum of Northern Arizona (Flagstaff) had opened the door to intensive prospecting of the Kaiparowits

Plateau for Cretaceous macrovertebrate fossil sites in an effort to augment previously conducted microvertebrate work. Teams from all four partners documented over 150 new vertebrate fossil sites in 2001 and conducted excavations on a fully articulated gryposaur hadrosaur with skin impressions, an early middle Campanian dinosaur bonebed, and a relatively complete specimen of a new genus of chasmosaurine ceratopsid. Since 2001 the Alf Museum of Paleontology (Claremont), Midwestern University (Glendale), the Denver Museum of Nature and Science (Denver), and Montana State University (Bozeman) have also put forth major efforts in the Kaiparowits Basin, helping to document thousands of new fossil sites and amassing important new invertebrate, microvertebrate, mesovertebrate, macrovertebrate, and trace fossil specimens from both marine and terrestrial facies (Titus et al., 2005; Zanno and Sampson, 2005; Roberts and Tapanila, 2006; Albright, Gillette, and Titus, 2007a, 2007b; Gates and Sampson, 2007; Roberts, 2007; Schmeisser and Gillette, 2009; Zanno et al., 2009; Getty et al., 2010; Kirkland and DeBlieux, 2010; Sampson et al., 2010; Titus, Albright, and Barclay, 2010; Knell et al., 2011; chapters in this volume). Simpson et al. (2010) even reported on purported maniraptorian predatory digging traces from the Wahweap Formation, the first of its kind ever published. Ultimately, this research made the cover of *Time* magazine in September 2010, with the naming of two new chasmosaurine ceratopsid genera.

It is now safe to say this coordinated research is spurring a renaissance in North American Late Cretaceous paleontology, particularly in regard to biomes and the paleobiogeography of larger vertebrates of the Laramidia subcontinent (Gates et al., 2010). Work continues across the region, and new, significant fossil finds are made every field season. Considering that there are hundreds of thousands of hectares of outcrop that have yet to be surveyed even at a reconnaissance level in the Kaiparowits Basin alone, it is no exaggeration to claim the region will remain an exciting research frontier for decades to come.

REFERENCES CITED

Albright, L. B., D. D. Gillette, and A. L. Titus. 2007a. Plesiosaurs from the Upper Cretaceous (Cenomanian–Turonian) Tropic Shale of southern Utah: part 1, new records of the pliosaur *Brachauchenius lucasi*. Journal of Vertebrate Paleontology 27:31–40.

Albright, L. B., D. D. Gillette, and A. L. Titus. 2007b. Plesiosaurs from the Upper Cretaceous (Cenomanian–Turonian) Tropic Shale of southern Utah: part 2, Polycotylidae. Journal of Vertebrate Paleontology 27:41–58.

Allen, J. L., and C. L. Johnson. 2010a. Facies control on sandstone composition (and influence of statistical methods on interpretations) in the John Henry Member, Straight Cliffs Formation, Southern Utah, U.S.A. Sedimentary Geology 230:60–76.

Allen, J. L., and C. L. Johnson. 2010b. Sedimentary facies, paleoenvironments, and relative sea level changes in the John Henry Member, Cretaceous Straight Cliffs Formation, southern Utah; pp. 225–247 in S. M. Carney, D. E. Tabet, and C. L. Johnson (eds.), Geology of South-Central Utah. Utah Geological Association Guidebook 39.

am Ende, B. A. 1991. Depositional environments, palynology, and age of the Dakota Formation, south-central Utah; pp. 65–83 in J. D Nations and J. G. Eaton (eds.), Stratigraphy, Depositional Environments, and Sedimentary Tectonics of the Western Margin, Cretaceous Western Interior Seaway. Geological Society of America Special Paper 260.

Biek, R. F., G. C. Willis, M. D. Hylland, and H. H. Doelling. 2000. Geology of Zion National Park; pp. 107–137 in D. A. Sprinkel, T. C. Chidsey, and P. B. Anderson (eds.), Geology of Utah's Parks and Monuments. Utah Geological Association Publication 28.

Biek, R. F., F. Maldonado, D. W. Moore, J. J. Anderson, P. D. Rowley, V. S. Williams, D. Nealey, and E. G. Sable. 2010. Interim Geologic Map of the West Part of the Panguitch 30'×60' Quadrangle, Garfield, Iron, and Kane Counties, Utah – Year 2 Progress Report. Utah Geological Survey Open-File Report 277.

Bobb, M. C. 1991. The Calico bed, Upper Cretaceous, southern Utah; a fluvial sheet deposit in the Western Interior foreland basin and its relationship to eustasy and tectonics. M.Sc. thesis, University of Colorado, Boulder, Colorado.

Bowers, W. E. 1972. The Canaan Peak, Pine Hollow, and Wasatch formations in the Table Cliff region, Garfield County, Utah. U.S. Geological Survey Bulletin 1331-B.

Bowers, W. E. 1973a. Geologic map and coal resources of the Upper Valley quadrangle, Garfield County, Utah. U.S. Geological Survey Coal Investigations Map C-60.

Bowers, W. E. 1973b. Geologic map and coal resources of the Griffin Point quadrangle, Garfield County, Utah. U.S. Geological Survey Coal Investigations Map C-61.

Bowers, W. E. 1973c. Geologic map and coal resources of the Pine Lake quadrangle, Garfield County, Utah. U.S. Geological Survey Coal Investigations Map C-66.

Bowers, W. E. 1975. Geologic map and coal resources of the Henrieville quadrangle, Garfield and Kane counties, Utah. U.S. Geological Survey Coal Investigation Map C-74.

Bowers, W. E. 1981. Geologic map and coal deposits of the Canaan Peak quadrangle, Garfield and Kane counties, Utah. U.S. Geological Survey Coal Investigations Map C-90.

Bowers, W. E. 1983. Geologic map and coal sections of the Butler Valley quadrangle, Kane County, Utah. U.S. Geological Survey Coal Investigations Map C-95.

Bowers, W. E. 1991a. Geologic map and coal deposits of the Horse Mountain quadrangle, Kane County, Utah. U.S. Geological Survey Coal Investigations Map C-137.

Bowers, W. E. 1991b. Geologic map of the Fourmile Bench quadrangle, Kane County, Utah. U.S. Geological Survey Coal Investigations Map C-140.

Bowers, W. E. 1993. Geologic map of the Horse Flat quadrangle, Kane County, Utah. U.S. Geological Survey Coal Investigations Map C-144.

Carr, T. D., T. E. Williamson, B. B. Brooks, and K. Stadtman. 2011. Evidence for high taxonomic and morphologic tyrannosauroid diversity in the Late Cretaceous (Late Campanian) of the American Southwest and a new short-skulled tyrannosaurid from the Kaiparowits formation of Utah. Naturwissenschaften 98:241–246.

Cashion, W. B., Jr. 1961. Geology and fuels resources of the Orderville–Glendale area, Kane County, Utah. U.S. Geological Survey Coal Investigation Map C-49.

Cashion, W. B., Jr. 1976. Geologic map of the south flank of the Markagunt Plateau, northwest Kane County, Utah. U.S. Geological Survey Miscellaneous Investigations Map I-494.

Castle, J. W., F. J. Molz, S. Lu, and C. L. Dinwiddie. 2004. Sedimentology and fractal-based analysis of permeability data, John Henry Member, Straight Cliffs Formation (Upper Cretaceous), Utah, U.S.A. Journal of Sedimentary Research 74:270–284.

Christensen, A. E. 2005. Sequence stratigraphy, sedimentology and provenance of the Drip Tank Member of the Straight Cliffs Formation, Kaiparowits Formation, Southern Utah. M.S. thesis, New Mexico State University, Las Cruces, New Mexico.

Cifelli, R. L. 1990a. Cretaceous mammals of southern Utah. I. Marsupial mammals from the Kaiparowits Formation (Judithian). Journal of Vertebrate Paleontology 10:295–319.

Cifelli, R. L. 1990b. Cretaceous mammals of southern Utah. II. Marsupials and marsupial-like mammals from the Wahweap Formation (early Campanian). Journal of Vertebrate Paleontology 10:320–331.

Cifelli, R. L. 1990c. Cretaceous mammals of southern Utah. III. Therian mammals from the Turonian (early Late Cretaceous). Journal of Vertebrate Paleontology 10:332–345.

Cifelli, R. L. 1990d. Cretaceous mammals of southern Utah. IV. Eutherian mammals from the Wahweap (Aquilan) and Kaiparowits (Judithian) formations. Journal of Vertebrate Paleontology 10:346–360.

Cifelli R. L., and J. G. Eaton. 1987. Marsupial from the earliest Late Cretaceous of Western U.S. Nature 325:520–522.

Cifelli, R. L., and Z. Johanson. 1994. New marsupial from the Upper Cretaceous of Utah. Journal of Vertebrate Paleontology 14:292–295.

Cifelli, R. L., and S. K. Madsen. 1986. An Upper Cretaceous symmetrodont (Mammalia) from southern Utah. Journal of Vertebrate Paleontology 6:258–263.

Cobban, W. A., T. S. Dyman, G. L. Pollock, K. I. Takahashi, L. E. Davis, and D. B. Riggin. 2000. Inventory of dominantly marine and brackish-water fossils from Late Cretaceous rocks in and near Grand Staircase–Escalante National Monument, Utah; pp. 579–589 in D. A. Sprinkel, T. C. Chidsey, and P. B. Anderson (eds.), Geology of Utah's Parks and Monuments. Utah Geological Association Publication 28.

Cohenour, R. E. 1963. Coal Resources of Part of the Alton Area, Kane County, Utah. Utah Geological and Mineralogical Survey Report of Investigation 2, 20 p.

DeCourten, F. L., and D. A. Russell. 1985. A specimen of Ornithomimus velox (Therapoda, Ornithomimidae) from the terminal Cretaceous Kaiparowits Formation of southern Utah. Journal of Paleontology 59:1091–1099.

Doelling, H. H., and R. L. Graham. 1972. Southwestern Utah Coal Fields: Alton, Kaiparowits, Plateau and Kolob-Harmony. Utah Geological and Mineral Survey Monograph Series 1.

Dutton, C. E. 1880. Report on the Geology of the High Plateaus of Utah with Atlas. Department of the Interior. Government Printing Office, Washington, D.C., 307 p.

Dyman, T. S., W. A. Cobban, L. E. Davis, R. L. Eves, G. L. Pollock, J. D. Obradovich, A. L. Titus, K. I. Takahashi, T. C. Hester, and D. Cantu. 2002. Upper Cretaceous marine and brackish water strata at Grand Staircase–Escalante National Monument, Utah; pp. 171–198 in W. R. Lund (ed.), Field Guide to Geologic Excursions in Southwestern Utah and Adjacent Areas of Arizona and Nevada. U.S. Geological Survey Open-File Report 02–172.

Eaton, J. G. 1987. Stratigraphy, depositional environments, and age of Cretaceous mammal-bearing rocks in Utah, and systematics of the Multituberculata (Mammalia). Ph.D. dissertation, University of Colorado, Boulder, Colorado.

Eaton, J. G. 1990. Stratigraphic revision of Campanian (Upper Cretaceous) rocks in the Henry Basin, Utah. Mountain Geologist 27:27–38.

Eaton, J. G. 1991. Biostratigraphic framework for Upper Cretaceous rocks of the Kaiparowits Plateau, southern Utah; pp. 47–63 in J. D. Nations and J. G. Eaton (eds.), Stratigraphy, Depositional Environments, and Sedimentary Tectonics of the Western Margin, Cretaceous Western Interior Seaway. Geological Society of America Special Paper 260.

Eaton, J. G. 1993. Therian mammals from the Cenomanian (Upper Cretaceous) Dakota Formation, southwestern Utah. Journal of Vertebrate Paleontology 13:105–124.

Eaton, J. G. 1995. Cenomanian and Turonian (early Late Cretaceous) multituberculate mammals from southwestern Utah. Journal of Vertebrate Paleontology 15:761–784.

Eaton, J. G. 1999a. Vertebrate paleontology of the Paunsaugunt Plateau, Upper Cretaceous, southwestern Utah; pp. 335–338 in D. D. Gillette (ed.), Vertebrate Paleontology in Utah. Utah Geological Survey Miscellaneous Publication 99-1.

Eaton, J. G. 1999b. Vertebrate paleontology of the Iron Springs Formation, Upper Cretaceous, southwestern Utah; pp. 339–343 in D. D. Gillette (ed.), Vertebrate Paleontology in Utah. Utah Geological Survey Miscellaneous Publication 99-1.

Eaton, J. G. 2002. Multituberculate Mammals from the Wahweap (Campanian, Aquilan) and Kaiparowits (Campanian, Judithian) Formations, Within and Near the Grand Staircase–Escalante National Monument, Southern Utah. Utah Geological Survey Miscellaneous Publication 02-4, 66 p.

Eaton, J. G. 2006a. Late Cretaceous mammals from Cedar Canyon, southwestern Utah; pp. 373–402 in S. G. Lucas and R. M. Sullivan (eds.), Late Cretaceous Vertebrates from the Western Interior. New Mexico Museum of Natural History and Science Bulletin 35.

Eaton, J. G. 2006b. Santonian (Late Cretaceous) Mammals from the John Henry Member of the Straight Cliffs Formation, Grand Staircase–Escalante National Monument, Utah. Journal of Vertebrate Paleontology 26:446–460.

Eaton, J. G., and R. L. Cifelli. 1988. Preliminary report on the Late Cretaceous mammals of the Kaiparowits Plateau, southern Utah. University of Wyoming Contributions to Geology 26:45–55.

Eaton, J. G., and R. L. Cifelli. 1997. Cretaceous vertebrates of the Grand Staircase–Escalante National Monument; pp. 365–372 in L. M. Hill (ed.), Learning from the Land: Grand Staircase–Escalante National Monument Science Symposium Proceedings. U.S. Department of the Interior, Bureau of Land Management.

Eaton, J. G., R. L. Cifelli, J. H. Hutchison, J. I. Kirkland, and J. M. Parrish. 1999a. Cretaceous vertebrate faunas from the Kaiparowits Plateau, south central Utah; pp. 345–353 in D. D. Gillette (ed.), Vertebrate Paleontology in Utah. Utah Geological Survey Miscellaneous Publication 99-1.

Eaton, J. G., S. Diem, J. D. Archibald, C. Schierup, and H. Munk. 1999b. Vertebrate paleontology of the Upper Cretaceous rocks of the Markagunt Plateau, southwestern Utah; pp. 323–333 in D. D. Gillette (ed.), Vertebrate Paleontology in Utah. Utah Geological Survey Miscellaneous Publication 99-1.

Eaton, J. G., J. I. Kirkland, J. H. Hutchinson, R. Denton, R. C. O'Neill, and M. J. Parrish. 1997. Nonmarine extinction across the Cenomanian–Turonian (C-T) boundary, southwestern Utah, with a comparison to the Cretaceous–Tertiary (K-T) extinction event. Geological Society of America Bulletin 59:129–151.

Eaton, J. G., J. Laurin, J. I. Kirkland, N. E. Tibert, R. M. Leckie, B. B. Sageman, P. M. Goldstrand, D. W. Moore, A. W. Straub, W. A. Cobban, J. D. Dalebout. 2001. Cretaceous and early Tertiary geology of Cedar and Parowan canyons, western Markagunt Plateau, Utah; pp. 337–363 in M. C. Erskine, J. E. Faulds, J. M. Bartley, P. D. Rowley (eds.), The Geologic Transition, High Plateaus to Great Basin; a Symposium and Field Guide; the Mackin Volume. Guidebook–Pacific Section, American Association of Petroleum Geologists 78.

Elder, W. P. 1987. The paleoecology of the Cenomanian–Turonian (Cretaceous) stage boundary at Black Mesa, Arizona. Palaios 2:24–40.

Elder, W. P. 1989. Molluscan extinction patters across the Cenomanian–Turonian stage boundary in the Western Interior of the United States. Paleobiology 15:299–320.

Elder, W. P., E. R. Gustason, and B. B. Sageman. 1994. Correlation of basinal carbonate cycles to nearshore parasequences in the Late Cretaceous Greenhorn seaway, Western Interior, U.S.A. Geological Society of America Bulletin 106:892–902.

Farabee, M. J. 1991. Palynology of the upper Kaiparowits Formation (Upper Cretaceous, Campanian) in southcentral Utah. Georgia Journal of Science 49:34–35.

Foster, J. R., A. L. Titus, G. F. Winterfeld, M. C. Hayden, and A. H. Hamblin. 2001. Paleontological Survey of the Grand Staircase–Escalante National Monument, Garfield and Kane Counties, Utah. Utah Geological Survey Special Publication SS-99.

Gallin, W. N., C. L. Johnson, and J. L. Allen. 2010. Fluvial and marginal marine architecture of the John Henry Member, Straight Cliffs Formation, Kelly Grade of the Kaiparowits Plateau, south-central Utah; pp. 248–275 in S. M. Carney, D. E. Tabet, and C. L. Johnson (eds.), Geology of South-Central Utah. Utah Geological Association Guidebook 39.

Gates, T. A., and Sampson S. D. 2007. A new species of Gryposaurus (Dinosauria: Hadrosauridae) from the Upper Campanian Kaiparowits Formation of Utah. Zoological Journal of the Linnean Society of London 151:351–376.

Gates, T. A., S. D. Sampson, L. E. Zanno, E. M. Roberts, J. G. Eaton, R. L. Nydam, J. H. Hutchison, J. A. Smith, M. A. Loewen, and M. A. Getty. 2010. Biogeography of terrestrial and freshwater vertebrates from the Late Cretaceous (Campanian) Western Interior of North America. Palaeogeography, Palaeoclimatology, Palaeoecology 291:371–387.

Getty, M. A., M. A. Loewen, E. M. Roberts, A. L. Titus, and S. D. Sampson. 2010. Taphonomy of horned dinosaurs (Ornithischia: Ceratopsidae) from the late Campanian Kaiparowits Formation, Grand Staircase–Escalante National Monument, Utah; pp. 478–494 in M. J. Ryan, J. B. Chinnery-Allgeier, and D. A. Eberth (eds.),

New Perspectives on Horned Dinosaurs. Indiana University Press, Bloomington, Indiana.

Gilbert, G. K. 1875. Reports on the geology of portions of Nevada, Utah, California, and Arizona examined in the years 1871 and 1872. United States Geographical and Geological Survey West of the 100th Meridian Report 3:17–187.

Gillette, D. D., M. C. Hayden, and A. L. Titus. 1999. Occurrence and biostratigraphic framework of a plesiosaur from the Upper Cretaceous Tropic Shale of southwestern Utah; pp. 269–273 in D. D. Gillette (ed.), Vertebrate Paleontology in Utah. Utah Geological Survey Miscellaneous Publication 99–1.

Goldstrand, P. M. 1990. Stratigraphy and Paleogeography of Late Cretaceous and Paleogene Rocks of Southwest Utah. Utah Geological and Mineral Survey Miscellaneous Publication 90–2.

Goldstrand, P. M. 1991. Tectonostratigraphy, petrology, and paleogeography of Upper Cretaceous to Eocene rocks of southwest Utah. Ph.D. dissertation, University of Nevada, Reno, Nevada.

Goldstrand, P. M. 1994. Tectonic development of Upper Cretaceous to Eocene strata of southwestern Utah. Geological Society of America Bulletin 106:145–154.

Goldstrand, P. M., and D. J. Mullet. 1997. The Paleocene Grand Castle Formation–a new formation on the Markagunt Plateau of southwestern Utah; pp. 59–78 in F. Maldonado and L. D. Nealey (eds.), Geologic Transition in Southeastern Nevada, Southwestern Utah, and Northwestern Arizona. U.S. Geological Survey Bulletin 2153.

Gregory, H. E. 1948. Geology and geography of central Kane County, Utah. Geological Society of America Bulletin 59:211–248.

Gregory, H. E. 1949. Geologic and geographic reconnaissance of eastern Markagunt Plateau, Utah. Geological Society of America Bulletin 60:969–997.

Gregory, H. E. 1950a. Geology and Geography of the Zion Park Region, Utah and Arizona. U.S. Geological Survey Professional Paper 220, 200 p.

Gregory, H. E. 1950b. Geology of Eastern Iron County, Utah. Utah Geological and Mineralogical Survey Bulletin 37, 153 p.

Gregory, H. E. 1951. The Geology and Geography of the Paunsaugunt Region, Utah. U.S. Geological Survey Professional Paper 226, 116 p.

Gregory, H. E., and R. C. Moore. 1931. The Kaiparowits Region; Geographic and Geologic Reconnaissance of Parts of Utah and Arizona. U.S. Geological Survey Professional Paper 164.

Gustason, E. R. 1989. Stratigraphy and sedimentology of the Middle Cretaceous (Albian–Cenomanian) Dakota Formation, southwestern Utah. Ph.D. dissertation, University of Colorado, Boulder, Colorado.

Hamblin, A G., and J. R. Foster. 2000. Ancient animal footprints and traces in the Grand Staircase–Escalante National Monument, south-central Utah; pp. 557–568 in D. A. Sprinkel, T. C. Chidsey, and P. B. Anderson (eds.), Geology of Utah's Parks and Monuments. Utah Geological Association Publication 28.

Haq, B. U., J. Hardenbol, and P. R Vail. 1988. Mesozoic and Cenozoic chronostratigraphy and cycles of sea level change; pp. 72–108 in C. K. Wilgus, B. S. Hastings, C. G. S. C. Kendall, H. W. Posamentier, C. A. Ross, and J. C. Van Wagoner (eds.), Sea-Level Changes–An Integrated Approach. Society for Sedimentary Geology Special Publication 42.

Heller, P. L., C. L. Angevine, N. S. Winslow, and C. Paola. 1988. Two-phase stratigraphic model of foreland-basin sequences. Geology 16:501–504.

Hettinger, R. D. 1995. Sedimentological Descriptions and Depositional Interpretations, in Sequence Stratigraphic Context, of Two 300-Meter Cores from the Upper Cretaceous Straight Cliffs Formation, Kaiparowits Plateau, Kane County, Utah. U.S. Geological Survey Bulletin 2115-A.

Hettinger, R. D., P. J. McCabe, and K. W. Shanley. 1994. Detailed facies anatomy of transgressive and highstand systems tracts from the Upper Cretaceous of southern Utah, U.S.A.; pp. 235–257 in H. W. Posmentier and P. Weimer (eds.), Recent Advances in Siliciclastic Sequence Stratigraphy. American Association of Petroleum Geologists Memoir 58.

Hylland, M. D. 2010. Geologic map of the Clear Creek Mountain Quadrangle, Kane County, Utah. Utah Geological Survey Map 245.

Howell, E. E. 1875. Report on the geology of portions of Utah, Nevada, Arizona, and New Mexico. United States Geographical and Geological Survey West of the 100th Meridian Report 3:227–301.

Jinnah, Z. A., E. M. Roberts, A. D. Deino, J. S. Larsen, P. K. Link, and C. M. Fanning. 2009. New ^{40}Ar/^{39}Ar and detrital zircon U-Pb ages for the Upper Cretaceous Wahweap and Kaiparowits formations on the Kaiparowits Plateau, Utah: implications for regional correlation, provenance, and biostratigraphy. Cretaceous Research 30:287–299.

Kirkland, J. I. 1990. The Paleontology and Paleoenvironments of the Middle Cretaceous (Late Cenomanian–Middle Turonian) Greenhorn Cyclothem at Black Mesa, northeastern Arizona. Ph.D. dissertation, University of Colorado, Boulder, Colorado.

Kirkland, J. I., and D. D. DeBlieux. 2010. New basal centrosaurine ceratopsian skulls from the Wahweap Formation (Middle Campanian), Grand Staircase–Escalante National Monument, southern Utah; pp. 117–140 in M. J. Ryan, J. B. Chinnery-Allgeier, and D. A. Eberth (eds.), New Perspectives on Horned Dinosaurs. Indiana University Press, Bloomington, Indiana.

Kirschbaum, M. A., and P. J. McCabe. 1992. Controls on the accumulation of coal and on the development of anastomosed fluvial systems in the Cretaceous Dakota Formation of southern Utah. Sedimentology 39:581–598.

Knell, M. J., F. D. Jackson, A. L. Titus, L. B. Albright. 2011. A gravid fossil turtle from the Upper Cretaceous (Campanian) Kaiparowits Formation, southern Utah. Historical Biology 23:57–62.

Larsen, J. S., P. K. Link, E. M. Roberts, L. Tapanila, and C. M. Fanning. 2010. Cyclic stratigraphy of the Paleocene Pine Hollow and detrital zircon provenance of Campanian to Eocene sandstones of the Kaiparowits and Table Cliffs

basins, south-central Utah; pp. 194–224 in S. M. Carney, D. E. Tabet, and C. L. Johnson (eds.), Geology of South-Central Utah. Utah Geological Association Guidebook 39.

Laurin, J., and B. B. Sageman. 2001. Tectono-sedimentary evolution of the western margin of the Colorado Plateau during the latest Cenomanian and early Turonian; pp. 57–74 in M. C. Erskine, J. E. Faulds, J. M. Bartley, P. D. Rowley (eds.), The Geologic Transition, High Plateaus to Great Basin; a Symposium and Field Guide; the Mackin Volume. Guidebook–Pacific Section, American Association of Petroleum Geologists 78.

Lawrence, J. C. 1965. Stratigraphy of the Dakota and Tropic formations of Cretaceous age in southern Utah; pp. 71–91 in Geology and Resources of South-Central Utah. Utah Geological Society Guidebook to the Geology of Utah 19.

Lawton, T. F., S. L. Pollock, and R. A. J. Robinson. 2003. Integrating sandstone petrology and nonmarine sequence stratigraphy: application to the late Cretaceous fluvial systems of southwestern Utah, U.S.A. Journal of Sedimentary Research 73:389–406.

Little, W. W. 1997. Tectonic and eustatic controls on cyclical fluvial patterns, Upper Cretaceous strata of the Kaiparowits Basin Utah; pp. 489–504 in L. M. Hill (ed.), Learning from the Land: Grand Staircase–Escalante National Monument Science Symposium Proceedings. U.S. Department of the Interior, Bureau of Land Management.

Lohrengel, C. F., 2nd. 1969. Palynology of the Kaiparowits Formation, Garfield County, Utah. Brigham Young University Geology Studies 16:61–180.

May, F. E., and A. Traverse. 1973. Palynology of the Dakota Sandstone (middle Cretaceous) near Bryce Canyon National Park, southern Utah. Geoscience and Man 7:57–64.

Moore, D. W., and A. W. Straub. 2001. Correlations of Upper Cretaceous and Paleogene(?) rocks beneath the Claron Formation, Crow Creek, western Markagunt Plateau, southwest Utah; pp. 75–95 in M. C. Erskine, J. E. Faulds, J. M. Bartley, and P. D. Rowley (eds.), The Geologic Transition, High Plateaus to Great Basin; a Symposium and Field Guide; the Mackin Volume. Guidebook–Pacific Section, American Association of Petroleum Geologists 78.

Nichols, D. J. 1997. Palynology and ages of some Upper Cretaceous formations in the Markagunt and northwestern Kaiparowits Plateaus, southwestern Utah; pp. 81–95 in F. Maldonado and L. D. Nealey (eds.), Geologic Studies in the Basin and Range–Colorado Plateau transition zone in southeastern Nevada, southwestern Utah, and northwestern Arizona, 1995. U.S. Geological Survey Bulletin 2153-E.

Orlansky, R. 1971. Palynology of the Upper Cretaceous Straight Cliffs Sandstone, Garfield County, Utah. Utah Geological and Mineralogical Survey Bulletin 89, 57 p.

Peterson, F. 1969a. Cretaceous Sedimentation and Tectonism in the Southeastern Kaiparowits Region, Utah. U.S. Geological Survey Open-File Report 69–202.

Peterson, F. 1969b. Four New Members of the Upper Cretaceous Straight Cliffs Formation in the Southeastern Kaiparowits Region, Kane County, Utah. U.S. Geological Survey Bulletin 1274-J.

Peterson, F. 1975. Geologic map of the Sooner Bench quadrangle, Kane County, Utah. U.S. Geological Survey Miscellaneous Investigations Series Map I-874.

Peterson, F. 1980. Geologic map and coal deposits of the Big Hollow Wash quadrangle, Kane County, Utah. U.S. Geological Survey Coal Investigations Map C-84.

Peterson, F., and B. E. Barnum. 1973a. Geologic map and coal resources of the northeast quarter of the Cummings Mesa quadrangle [Navajo Point 7.5' quadrangle], Kane County, Utah. U.S. Geological Survey Coal Investigations Map C-63.

Peterson, F., and B. E. Barnum. 1973b. Geologic map and coal resources of the northwest quarter of the Cummings Mesa quadrangle [Mazuki Point 7.5' quadrangle], Kane County, Utah. U.S. Geological Survey Coal Investigations Map C-64.

Peterson, F., and G. W. Horton. 1967. Preliminary geologic map and coal deposits of the northeast quarter of the Gunsight Butte quadrangle [Sit Down Bench 7.5' quadrangle], Kane County, Utah. Utah Geological and Mineralogical Survey Map 24-F.

Pollock, S. L. 1999. Provenance, geometry, lithofacies, and age of the Upper Cretaceous Wahweap Formation, Cordilleran foreland basin, southern Utah. M.Sc. thesis, New Mexico State University, Las Cruces, New Mexico.

Powell, J. W. 1875. Exploration of the Colorado River of the West and its tributaries explored in 1869, 1870, 1871, and 1872 under the direction of the secretary of the Smithsonian Institution. Government Printing Office, Washington, D.C.

Richardson, G. B. 1909. The Harmony, Colob, and Kanab coal fields, southern Utah. U.S. Geological Survey Bulletin 341-C:379–400.

Richardson, G. B. 1927. The Upper Cretaceous section in the Colob Plateau, southwest Utah. Washington Academy of Sciences Journal 17:464–475.

Roberts, E. M. 2005. Stratigraphic, taphonomic, and paleoenvironmental analysis of the Upper Cretaceous Kaiparowits Formation, Grand Staircase–Escalante National Monument, southern Utah. Ph.D. dissertation, University of Utah, Salt Lake City, Utah.

Roberts, E. M. 2007. Facies architecture and depositional environments of the Upper Cretaceous Kaiparowits Formation, southern Utah. Sedimentary Geology 197:207–233.

Roberts, E. M., and L. Tapanila. 2006. A new social insect nest trace from the Late Cretaceous Kaiparowits Formation of southern Utah. Journal of Paleontology 80:768–774.

Roberts, E. M., A. D. Deino, and M. A. Chan. 2005. ^{40}Ar/^{39}Ar age of the Kaiparowits Formation, southern Utah, and correlation of coeval strata and faunas along the margin of the Western Interior Basin. Cretaceous Research 26:307–318.

Robison, R. A. 1966. Geology and Coal Resources of the Tropic Area, Garfield County, Utah. Utah Geological and Mineralogical Survey Special Studies 18.

Sampson, S. D., M. A. Loewen, A. Farke, E. M. Roberts, C. Forster, J. A. Smith, and A. L. Titus. 2010. New horned dinosaurs from Utah provide evidence for intracontinental dinosaur endemism. PLos One 5(9):e122292.

Sargent, K. A., and D. E. Hansen. 1982. Bedrock geologic map of the Kaiparowits coal-basin area, Utah. U.S. Geological Survey Miscellaneous Investigation Series Map I-1033-I.

Schmeisser, R. L., and D. D. Gillette. 2009. Unusual occurrence of gastroliths in a polycotylid plesiosaur from the Upper Cretaceous Tropic Shale, southern Utah. Palaios 24:453–459.

Schmitt, G. J., D. A. Jones, and P. M. Goldstrand. 1991. Braided stream deposition and provenance of the Upper Cretaceous–Paleocene(?) Canaan Peak Formation, Sevier foreland basin, southwestern Utah; pp. 27–45 in J. D. Nations and J. G. Eaton (eds.), Stratigraphy, Depositional Environments, and Sedimentary Tectonics of the Western Margin, Cretaceous Western Interior Seaway. Geological Society of America Special Paper 260.

Shanley, K. W., and P. J. McCabe. 1991. Predicting facies architecture through sequence stratigraphy; an example from the Kaiparowits Plateau, Utah. Geology 19:742–745.

Shanley, K. W., and P. J. McCabe. 1993. Alluvial architecture in a sequence stratigraphic framework: a case history from the Upper Cretaceous of southern Utah, U.S.A.; pp. 21–56 in S. Flint and I. Bryant (eds.), Quantitative Modeling of Clastic Hydrocarbon Reservoirs and Outcrop Analogues. International Association of Sedimentologists Special Publications 15. Blackwell Scientific Publications, Oxford, U.K.

Shanley, K. W., and P. J. McCabe. 1995. Sequence stratigraphy of Turonian–Santonian strata, Kaiparowits Plateau, Southern Utah, U.S.A.: implications for regional correlation and foreland basin evolution; pp. 103–136 in J. C. Van Wagoner and G. T. Bertram (eds.), Sequence Stratigraphy of Foreland Basin Deposits: Outcrop and Subsurface Examples from the Cretaceous of North America. American Association of Petroleum Geologists Memoir 64.

Shanley, K. W., P. J. McCabe, and R. D. Hettinger. 1992. Tidal influence in Cretaceous fluvial strata from Utah, U.S.A.–a key to sequence stratigraphic interpretation. Sedimentology 39:905–930.

Simpson, E. L., M. C. Wizevich, H. L. Hilbert-Wolf, S. E. Tindall, J. J. Bernard, and W. S. Simpson. 2009. An Upper Cretaceous sag pond deposit: implications for recognition of local seismicity and surface rupture along the Kaibab monocline, Utah. Geology 37:967–970.

Simpson, E. L., H. L. Hilbert-Wolf, W. S. Simpson, S. E. Tindall, J. J. Bernard, T. A. Jenesky, and M. C. Wizevich. 2008. The interaction of aeolian and fluvial processes during deposition of the Upper Cretaceous capping sandstone member, Wahweap Formation, Kaiparowits Basin, Utah, U.S.A. Palaeogeography, Palaeoclimatology, Palaeoecology 270:19–28.

Simpson, E. L., H. L. Hilbert-Wolf, M. C. Wizevich, S. E. Tindall, B. R. Fasinski, L. P. Storm, and M. D. Needle. 2010. Predatory digging behavior by dinosaurs. Geology 38:699–702.

Spieker, E. M. 1925. Geology of coal fields of Utah. United State Bureau of Mines Technical Paper 345:13–72.

Stanton, T. W. 1893. The Colorado Formation and Its Invertebrate Fauna. U.S. Geological Survey Bulletin 106.

Tibert, N. E., R. M. Leckie, J. G. Eaton, J. I. Kirkland, J. P. Colin, E. L. Leitholk, and M. E. McCormic. 2003. Recognition of relative sea-level change in Upper Cretaceous coal bearing strata–a paleoecological approach using agglutinated foraminifera and ostracods to detect key stratigraphic surfaces; pp. 263–299 in H. C. Olson and R. M. Leckie (eds.), Micropaleontological Proxies for Sea-Level Change and Stratigraphic Discontinuities. Society of Economic Paleontologists and Mineralogists Special Publication 75.

Tilton, T. L. 1991. Upper Cretaceous stratigraphy of the Southern Paunsaugunt Plateau, Kane County, Utah. Ph.D. dissertation, University of Utah, Salt Lake City, Utah.

Tindall, S. E., L. P. Storm, T. A. Jenesky, and E. L. Simpson. 2010. Growth faults in the Kaiparowits basin, Utah, pinpoint initial Laramide deformation in the western Colorado plateau. Lithosphere 2:221–231.

Titus, A. L., L. B. Albright, and R. S. Barclay. 2010. The first record of Cenomanian (Late Cretaceous) insect body fossils from the Kaiparowits Basin, northern Arizona; pp. 127–132 in M. Eaton (ed.), Learning from the Land, Grand Staircase–Escalante National Monument Science Symposium Proceedings. Grand Staircase–Escalante Partners.

Titus, A. L., J. D. Powell, E. M. Roberts, S. D. Sampson, S. L. Pollock, J. I. Kirkland, and L. B. Albright. 2005. Late Cretaceous stratigraphy, depositional environments, and macrovertebrate paleontology of the Kaiparowits Plateau, Grand Staircase–Escalante National Monument, Utah; pp. 101–128 in J. Pederson and C. M. Dehler (eds.), Interior Western United States. Geological Society of America Field Guide 6.

Ulicny, D. 1999. Sequence stratigraphy of the Dakota Formation (Cenomanian), southern Utah; interplay of eustacy and tectonics in a foreland basin. Sedimentology 46:807–836.

Vaninetti, G. E. 1978. Coal stratigraphy of the John Henry Member of the Straight Cliffs Formation, Kaiparowits Plateau, Utah. M.Sc. thesis, University of Utah, Salt Lake City, Utah.

Waldrop, H. A., and R. L. Sutton. 1967a. Preliminary geologic map and coal deposits of the northwest quarter of the Nipple Butte quadrangle [Nipple Butte 7.5' quadrangle], Kane County, Utah. Utah Geological and Mineralogical Survey Map 24-A.

Waldrop, H. A., and R. L. Sutton. 1967b. Preliminary geologic map and coal deposits of the northeast quarter of the Nipple Butte quadrangle [Tibbet Bench 7.5' quadrangle], Kane County, Utah. Utah Geological and Mineralogical Survey Map 24-B.

Waldrop, H. A., and R. L. Sutton. 1967c. Preliminary geologic map and coal deposits of the southwest quarter of the Nipple Butte quadrangle [Glen Canyon City 7.5' quadrangle], Kane County, Utah and Coconino County, Arizona. Utah Geological and Mineralogical Survey Map 24-D.

Weishampel, D. B., and J. A. Jensen. 1979. *Parasaurolophus* (Reptilia: Hadrosauridae) from Utah. Journal of Paleontology 53:1422–1427.

Zanno, L. E., and S. D. Sampson. 2005. A new oviraptorosaur (Theropoda, Maniraptora) from the Late Cretaceous (Campanian) of Utah. Journal of Vertebrate Paleontology 25:897–904.

Zanno, L. E., D. D. Gillette, L. B. Albright, and A. L. Titus, A. L. 2009. A new North American therizinosaurid and the role of herbivory in "predatory" dinosaur evolution. Proceeding of the Royal Society B: Biological Sciences 276:3505–3511.

Zeller, H. D. 1973a. Geologic map and coal resources of the Carcass Canyon quadrangle, Garfield and Kane counties, Utah. U.S. Geological Survey Coal Investigations Map C-56.

Zeller, H. D. 1973b. Geologic map and coal resources of the Canaan Creek quadrangle, Garfield County, Utah. U.S. Geological Survey Coal Investigations Map C-57.

Zeller, H. D. 1973c. Geologic map and coal resources of the Death Ridge quadrangle, Garfield and Kane counties, Utah. U.S. Geological Survey Coal Investigations Map C-58.

Zeller, H. D. 1973d. Geologic map and coal resources of the Dave Canyon quadrangle, Garfield County, Utah. U.S. Geological Survey Coal Investigations Map C-59.

Zeller, H. D. 1978. Geologic map and coal resources of the Collet Top quadrangle, Kane County, Utah. U.S. Geological Survey Coal Investigations Map C-80.

Zeller, H. D. 1990a. Geologic map and coal stratigraphy of the Needle Eye Point quadrangle, Kane County, Utah. U.S. Geological Survey Coal Investigations Map C-129.

Zeller, H. D. 1990b. Geologic map and coal stratigraphy of the East of the Navajo quadrangle, Kane County, Utah. U.S. Geological Survey Coal Investigations Map C-130.

Zeller, H. D. 1990c. Geologic map and coal stratigraphy of the Petes Cove quadrangle, Kane County, Utah. U.S. Geological Survey Coal Investigations Map C-132.

Zeller, H. D., and E. V. Stephens. 1973. Geologic map and coal resources of the Seep Flat quadrangle, Garfield and Kane counties, Utah. U.S. Geological Survey Coal Investigations Series Map C-65.

Zeller, H. D., and G. E. Vaninetti. 1990. Geologic map and coal stratigraphy of the Ship Mountain Point quadrangle and north part of the Tibbet Bench quadrangle, Kane County, Utah. U.S. Geological Survey Coal Investigations Map C-131.

Geologic Overview

Alan L. Titus, Eric M. Roberts, and L. Barry Albright III

CRETACEOUS STRATA IN SOUTHERN UTAH WERE DEPOS-
ited in the proximal portion of the Sevier Foreland Basin.
Total thickness of Cretaceous sediments probably exceeded
3000 m in the region before mid-Laramide uplift and erosion. Exposures are primarily found at the Kaiparowits Plateau and around the margins of the Markagunt and Paunsaugunt plateaus and the Pine Valley Mountain region. The
Cretaceous section is divided up into the Cedar Mountain,
Dakota, Tropic, Straight Cliffs, Wahweap, and Kaiparowits
formations east of Parowan Canyon and is contained almost
entirely within the Iron Springs Formation west. The sections are highly fossiliferous and yield one of the best records
of Late Cretaceous terrestrial ecosystem evolution known in
North America.

INTRODUCTION

The state of Utah lies within both the Cordilleran Thrust
Belt and Cordilleran Foreland Basin System (Fig. 2.1). The
boundary between these two provinces, called the Cordilleran or Wasatch Hingeline (DeCelles, 2004), roughly parallels
the east margin of the Sevier Fold and Thrust Belt. West
of the Wasatch Hingeline are the extended and dissected
remnants of thrust sheet stacks of Precambrian through early
Mesozoic sedimentary rocks. East of the Wasatch Hingeline
are thick sections of largely flat-lying Paleozoic, Mesozoic,
and Paleogene sedimentary rocks (Hintze, 1988). Cretaceous
strata crop out widely east of the Wasatch Hingeline, especially in the eastern Wasatch Plateau, Book Cliffs, Henry
Basin, La Sal-Abajo Mountains, and the southern portion of
the state (Fig. 2.2).

In southern Utah, Cretaceous strata crop out almost continuously along a 190-km-long belt from St. George at the
west end to the Fifty Mile Cliffs at the east end (Fig. 2.3).
This outcrop belt is contained almost entirely within four
prominent physiographic features: from east to west, the
Markagunt Plateau, the Paunsaugunt Plateau, the Tropic
Amphitheater, and the Kaiparowits Plateau (Doelling, 1975;
Doelling and Davis, 1989). Of these four physiographic features, the Kaiparowits Plateau (Fig. 2.3), inside Grand Staircase–Escalante National Monument, contains the largest

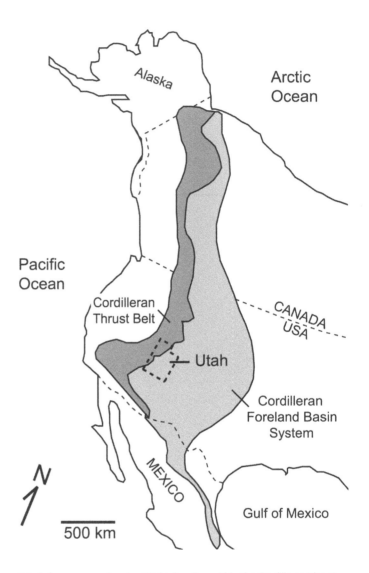

2.1. Reference map showing Utah's location within the Cordilleran Thrust
Belt and Cordilleran Foreland Basin System. Modified from DeCelles (2004).

continuous exposures of Cretaceous strata (~850,000 acres)
in the southern portion of the state (Sargent and Hansen,
1982). All Cretaceous rock on the Kaiparowits Plateau is
within the western portion of the Kaiparowits Basin, a structural low between the East Kaibab Monocline and the Circle
Cliffs uplift (Titus et al., 2005). The remaining three areas
to the west are collectively part of the Grand Staircase physiographic region. Cretaceous strata also crop out around the

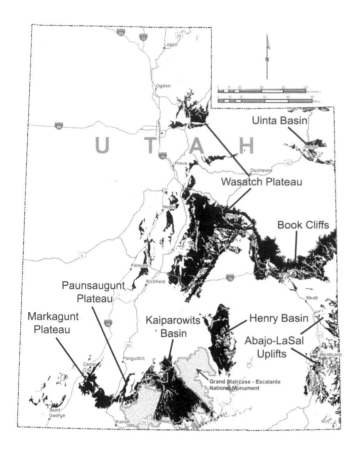

2.2. Geologic map showing Upper Cretaceous bedrock distribution (in black) in Utah.

flanks of the Pine Valley Mountains near St. George, near Gunnison Reservoir, and at Parowan Gap, west of the town of Parowan (Fig. 2.3). Forty kilometers northeast of the east edge of the Kaiparowits Plateau, vast exposures of Cretaceous strata also occur within the Henry Basin, around the flanks of the Henry Mountains (Doelling, 1975).

With the exception of those in the Pine Valley Mountains and Parowan Gap areas, most of the Cretaceous outcrops are structurally uncomplicated and are horizontal to gently dipping, except where they intersect the Hurricane Cliffs, Sevier, and Paunsaugunt fault systems, and the East Kaibab, Echo, and Waterpocket monoclines (Fig. 2.3). Both the Pausaugunt and Markagunt plateaus are extensively capped with Paleogene strata (Bowers, 1972, 1990; Biek et al., 2010), whereas the Kaiparowits Plateau had almost all of its Paleogene units removed by erosion during the Neogene.

In addition to its substantial middle and upper Campanian terrestrial stratigraphic and fossil record, the Kaiparowits Basin is regionally important for one other reason: its paleogeographic position along the western margin of the Cretaceous Western Interior Seaway. For an extended 20-million-year period, a complex interplay between eustasy, tectonics, and sedimentation rates caused alternating marine

and terrestrial deposition that has provided critical biostratigraphic constraints on the largely terrestrial Cretaceous record located to the west. In contrast, in the nearby Henry Basin, marine conditions existed almost continuously from late Cenomanian to early Campanian time (Peterson and Kirk, 1977; Eaton, 1990).

Institutional Abbreviation UMNH, Utah Museum of Natural History, Salt Lake City, Utah.

TECTONIC AND PALEOGEOGRAPHIC SETTING

Cretaceous sedimentation in southern Utah initiated in the late Early Cretaceous in response to lithospheric flexure caused by thrust loading within the Sevier Fold and Thrust Belt coupled with eastward progradation of a thick clastic wedge into the Cordilleran Foreland Basin System (DeCelles and Giles, 1996). Because pre-Barremian Lower Cretaceous strata are rare or absent in Utah (Sprinkle et al., 1999), and because underlying Jurassic units of the San Rafael Group are significantly eroded, it is likely that the entire western margin of the southern Cordilleran Foreland Basin System was uplifted between Tithonian and early Barremian time. Some authors have attributed this uplift to migration of a tectonic forebulge through the region (Currie, 1997; Willis, 1999). Many inconsistencies in age relationships between supposed pre- and postforebulge deposits (particularly the widespread absence of Kimmeridgian through Barremian strata within the proximal part of the foreland basin and the uniform age of basal Cretaceous rocks throughout the region) make it unlikely that this model is entirely correct. It is possible that southern Utah was simply a cratonic block that completely rebounded after the cessation of Late Jurassic thrusting. Thrusting during the Barremian resulted in deposition of the Cedar Mountain and San Pitch/Indianola formations in central Utah (Sprinkle et al., 1999) in a distinctly foreland basin setting. However, there is no evidence that significant movement took place within the Sevier Fold and Thrust Belt in southern Utah or southern Nevada until Albian time (Carpenter, 1989).

Late Early–Early Late Cretaceous (Albian–Turonian)

In latest Albian–early Cenomanian time the classic frontal part of the Sevier Fold and Thrust Belt began to take shape as a continuous feature along the entire Wasatch Hingeline (DeCelles, 2004). Dyman et al. (2002a) reported an age of 101.7 ± 0.42 Ma (Late Albian) for a bentonite sample collected very low in the Iron Springs Formation near Gunlock Reservoir, suggesting that flexural loading of the region

Titus, Roberts, and Albright

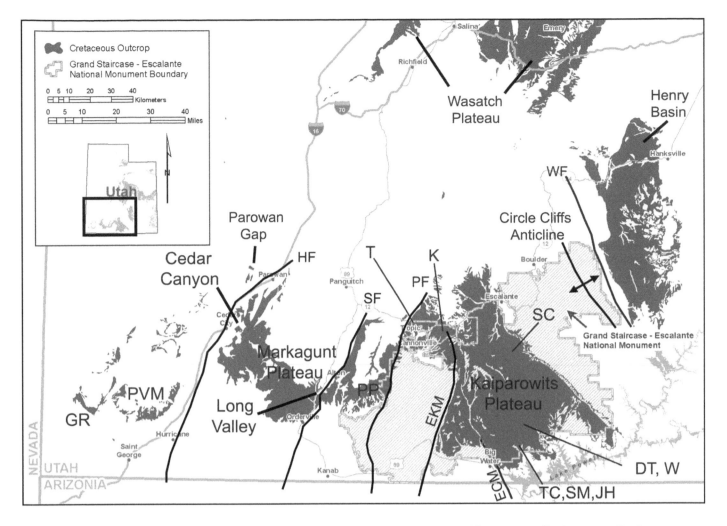

2.3. Map of south-central and southwestern Utah showing Cretaceous bedrock and the major structural features controlling its pattern. Also shown are type section locations for the Tropic (T), Straight Cliffs (SC), Wahweap (W), and Kaiparowits (K) formations, as well as the Tibbet Canyon (TC), Smoky Hollow (SM), and Drip Tank members. GR, Gunlock Reservoir; PVM, Pine Valley Mountains; HF, Hurricane Fault; SF, Sevier Fault; PP, Paunsaugunt Plateau; PF, Paunsaugunt Fault; EKM, East Kaibab Monocline; ECM, Echo Cliffs Monocline; WM, Waterpocket Monocline.

started around then. Coincident with thrust loading were the extremely high sea levels (Fig. 2.4) of the Greenhorn Eustatic Event (late Albian–middle Turonian). By late Cenomanian time, sandstone compositions in the Kaiparowits Basin indicate that granitic source areas in central Arizona were also feeding the Kaiparowits Basin (Gustason, 1989). Although DeCelles (2004) shows a magmatic arc in southern Arizona during the Cenomanian, how the tectonics of this area affected the Kaiparowits Basin is still unclear. Elder and Kirkland (1993, 1994) postulated a rifted highland margin (Mogollon Highland) inboard of the Arizona magmatic arc as the source for Kaiparowits Basin feldspathic sands.

Late Cretaceous I (Turonian–Campanian)

Between Turonian and middle Campanian time, tectonic conditions remained similar to those established during the Cenomanian. Sea levels during the Niobrara Eustatic Event were elevated almost as high as those of the Greenhorn (Hancock and Kauffman, 1979). Eastward movement of thrust sheets continued. The influx of sediment into the area from the Mogollon Highland and Sevier Fold and Thrust Belt was balanced by the creation of accommodation space and high sea levels, allowing coastal conditions to persist in the Kaiparowits Basin region for almost 15 Ma (Fig. 2.5).

Thrust system fronts continued propagating eastward in middle and late Campanian time and actually began to crosscut older Cretaceous foreland basin rocks of the Iron Springs Formation (Goldstrand, 1994; Lawton et al., 2003). Continued shortening in the Late Campanian and Maastrichtian led to large-scale partitioning of the Cordilleran Foreland Basin System through regional uplifts like the San Rafael Swell and Circle Cliffs Dome, heralding the start of the Laramide Orogeny.

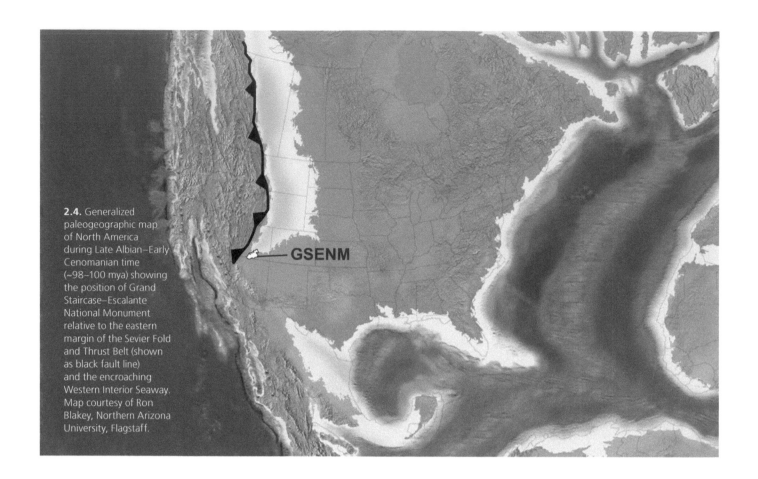

2.4. Generalized paleogeographic map of North America during Late Albian–Early Cenomanian time (~98–100 mya) showing the position of Grand Staircase–Escalante National Monument relative to the eastern margin of the Sevier Fold and Thrust Belt (shown as black fault line) and the encroaching Western Interior Seaway. Map courtesy of Ron Blakey, Northern Arizona University, Flagstaff.

GSENM

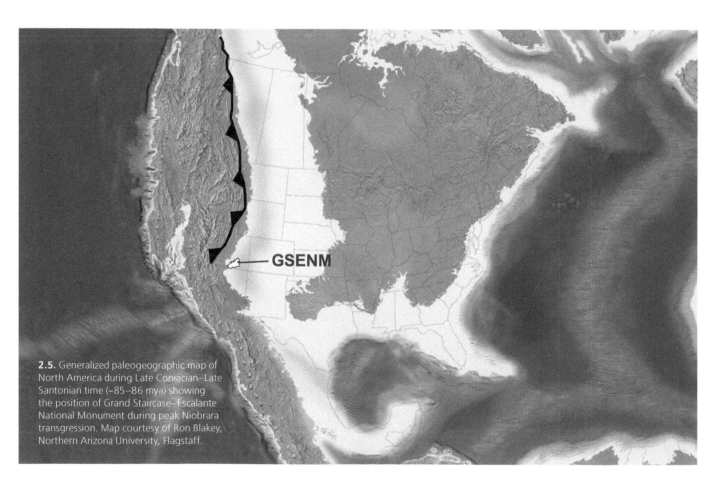

2.5. Generalized paleogeographic map of North America during Late Coniacian–Late Santonian time (~85–86 mya) showing the position of Grand Staircase–Escalante National Monument during peak Niobrara transgression. Map courtesy of Ron Blakey, Northern Arizona University, Flagstaff.

GSENM

Titus, Roberts, and Albright

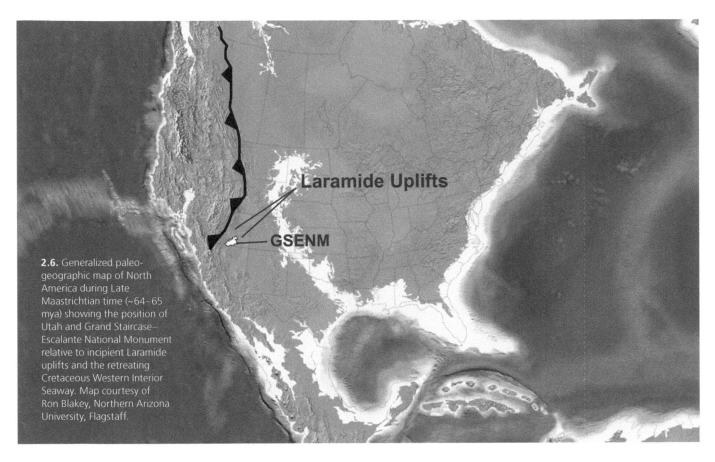

2.6. Generalized paleo-geographic map of North America during Late Maastrichtian time (~64–65 mya) showing the position of Utah and Grand Staircase–Escalante National Monument relative to incipient Laramide uplifts and the retreating Cretaceous Western Interior Seaway. Map courtesy of Ron Blakey, Northern Arizona University, Flagstaff.

Late Cretaceous II (Campanian–Maastrichtian)

Laramide phase activity along classic Sevier thrust fronts was accompanied by significant vertical offset on long-standing basement-seated structures like the East Kaibab Monocline and Waterpocket Fold (Fig. 2.6) at least as early as the middle Campanian (Tindall et al., 2010). Locally thick but laterally constrained basin fill sequences such as the Canaan Peak Formation of possible Maastrichtian age characterize this phase of Kaiparowits Basin evolution. Continued regional uplift created a regional unconformity between Cretaceous and Paleogene units that marks the top of the Mesozoic in southern Utah.

UPPER CRETACEOUS STRATIGRAPHY
OF SOUTHERN UTAH

The Cretaceous system in southern Utah is a highly heterogeneous, generally progradational foreland basin clastic succession dominated by sandstone (Little, 1997) but containing significant conglomerate, mudstone, and siltstone (Fig. 2.7). Dramatic lateral facies shifts occur up and down the Cordilleran Foreland Basin System, marking boundaries between discrete depocenters (e.g., foredeep versus wedge top) and competing influences between source areas and marine systems to the east. Pedogenic and diagenetic carbonate (siderite

and calcite), coal, and phosphate are present in minor overall amounts. Although trivial compared to the overall volume of rock, economic-grade coal deposits occur over much of the study area (Doelling and Graham, 1972).

Source areas include local intrabasinal, the Mogollon Highlands of central Arizona, the southern Utah portion of the Sevier Fold and Thrust Belt, the Delfonte Volcanic Province of southeastern California, the southern Nevada portion of the Sevier Thrust Belt, and the Sierra Nevada Magmatic Complex. Within any given formation there can be complex interplay between multiple source areas largely controlled by contemporaneous regional tectonics (Goldstrand, 1990; Eaton and Nations, 1991; Little, 1997; Lawton et al., 2003; Larsen, 2007; Roberts, 2007; Larsen et al., 2010; Jinnah and Roberts, 2011; Lawton and Bradford, 2011).

In general, the stratigraphic thickness of individual Cretaceous formations increases to the west (Fig. 2.7), toward the Sevier Fold and Thrust Belt (Eaton, 1991). One of the exceptions to this is the marine Tropic Shale, which pinches out westward toward the orogen. In spite of the general trend of unit thickening, the thickness of the entire Cretaceous section actually decreases toward the west (more than 2000 m in the Kaiparowits Basin, versus 1400 m on the west side of the Markagunt Plateau) as a result of a significant sub-Claron (Paleogene) down-cutting event that eliminated Upper and Middle Campanian strata (an estimated 1300 m of section)

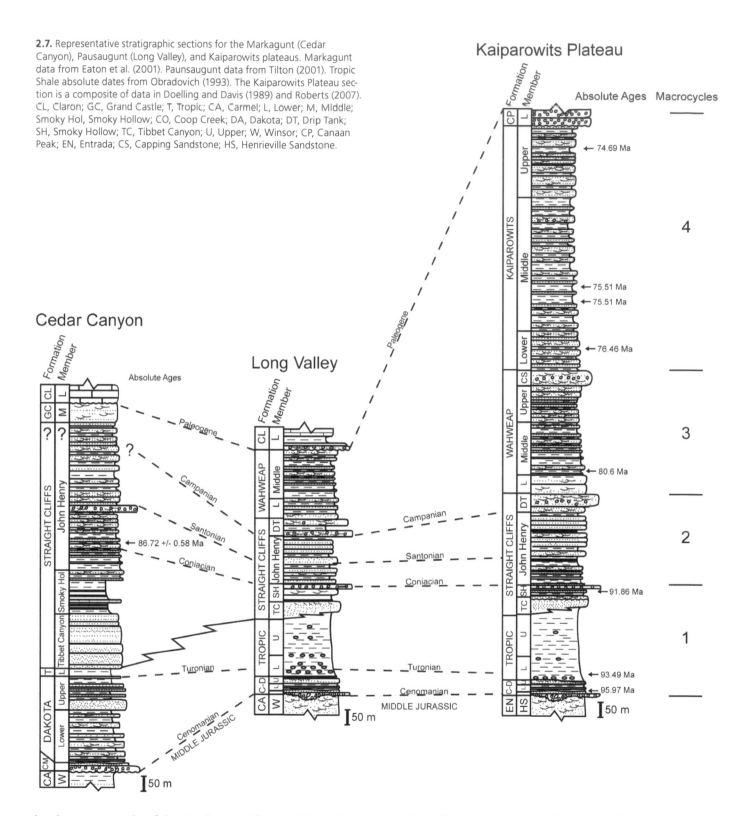

2.7. Representative stratigraphic sections for the Markagunt (Cedar Canyon), Pausaugunt (Long Valley), and Kaiparowits plateaus. Markagunt data from Eaton et al. (2001). Paunsaugunt data from Tilton (2001). Tropic Shale absolute dates from Obradovich (1993). The Kaiparowits Plateau section is a composite of data in Doelling and Davis (1989) and Roberts (2007). CL, Claron; GC, Grand Castle; T, Tropic; CA, Carmel; L, Lower; M, Middle; Smoky Hol, Smoky Hollow; CO, Coop Creek; DA, Dakota; DT, Drip Tank; SH, Smoky Hollow; TC, Tibbet Canyon; U, Upper; W, Winsor; CP, Canaan Peak; EN, Entrada; CS, Capping Sandstone; HS, Henrieville Sandstone.

by the western side of the Markagunt Plateau (Fig. 2.7). As a result, the Kaiparowits Basin contains the only significant upper Campanian sequence in the region. A relatively thick sequence of upper Campanian and Maastrichtian strata was probably present in the Henry Basin but was mostly later removed, along with the Paleogene record, by post-Laramide (mostly Neogene) erosion (Fig. 2.8).

One of the most obvious features of the region's Cretaceous record are lithologic macrocycles first pointed out by Peterson (1969a), then analyzed in detail by Little (1997). These cycles are characterized by the following generalized succession: (1) a lower sandstone-dominated fluvial channel interval; (2) a middle mudstone-dominated overbank and channel unit; (3) another sandstone-dominated fluvial

Titus, Roberts, and Albright

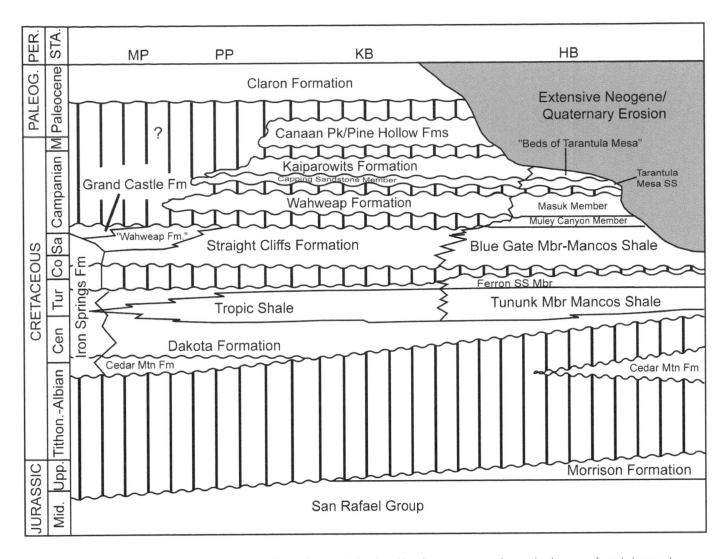

2.8. Chronostratigraphic relationships of Cretaceous units in southern Utah. Stratigraphic units are compressed upsection because of regularly spaced time axis and no vertical thickness is implied. Vertical rule indicates unconformity. PER, period; PALEOG, Paleogene; STA, stage; M, Maastrichtian; Sa, Santonian; Co, Coniacian; Tur, Turonian; Cen, Cenomanian; Tithon, Tithonian; Upp, upper; Mid, middle; MP, Markagunt Plateau; PP, Paunsaugunt Plateau; KB, Kaiparowits Basin; HB, Henry Basin.

interval; and (4) a coarse-grained amalgamated fluvial channel interval at the top. Four of these macrocycles are present in the Kaiparowits Basin area (Fig. 2.7), and the lower three can also be recognized on the Paunsaugunt and Markagunt plateaus. The lowest complete cyclic package encompasses the Dakota and Tropic formations and the lower two members of the Straight Cliffs Formation (Tibbet Canyon and Smoky Hollow). The second cycle is entirely within the upper two members of the Straight Cliffs Formation (John Henry and Drip Tank members), the third is the Wahweap Formation in its entirety, and the fourth is the Kaiparowits Formation in its entirety.

Little (1997) concluded that the coarse-grained units at the tops of these cycles were not transgressive systems tract-related incised valley fills, as hypothesized by Shanley and McCabe (1991), because they exhibit sheet-like geometry,

conformable lower contacts, and significant erosional relief on their upper surfaces. Little (1997) also concluded that these sequences are controlled by both tectonic events (thrusting in the orogen) and eustatic fluctuations, with the influence of eustasy dramatically decreasing in the upper two cycles (Wahweap and Kaiparowits formations). This decoupling of the third and fourth cycles from eustasy was primarily based on Little's assumption that the age and inferred conformable nature of the succession meant they could not be correlated with the classic Clagget (T8 is the Pakowki Formation of Canada) and Bearpaw (T9) transgressions. Radiometric dating (Jinnah et al., 2009) and magnetostratigraphic analysis (discussed below) of the Wahweap Formation exclusive of the capping sandstone member (~81–79 Ma) now show its deposition to have occurred almost precisely during the peak Clagget transgression (Jinnah and Roberts, 2011) and that

2.9. Exposures of the Cedar Mountain, Dakota, Tropic, and Straight Cliffs formations north of Big Water, Utah; south-central Kaiparowits Plateau. Cretaceous section rests with mild angular unconformity on the Jurassic Gunsight Butte Member of the Entrada Formation.

there may be a substantial diastem within the Wahweap Formation that correlates with the Judith River sea-level fall. As such, the influence of sea level on the third cycle appears to have been pronounced. The connection between the Bearpaw (T9) transgressive cycle, which started at approximately 75.5 Ma in Montana (corresponding to the SB2 horizon of Rogers and Brady, 2010 within the Judith River Formation), and deposition of the Kaiparowits Formation is slightly more difficult to establish, but it could certainly be argued, given the dates of the Bearpaw transgression, that the fourth (Kaiparowits) cycle also was influenced by eustasy.

In hindsight, it has become clear that correlation of time stratigraphic units across the Markagunt, Paunsaugunt, and Kaiparowits plateaus has been, and continues to be to some degree, hampered by dramatic changes in depositional facies and Laramide deformation on a regional scale. Only relatively recently have the actual age relationships of most of these terrestrial units been determined through detailed

mammal biostratigraphy, radiometric dating, paleomagnetic analysis, and careful petrographic analysis.

Cedar Mountain Formation

The basal beds of the Cretaceous system throughout most of southern Utah are an interval of poorly sorted massive, planar, and trough cross-stratified conglomerate deposited unconformably on Middle and Upper Jurassic strata (Fig. 2.9), in and adjacent to southeast trending incised paleovalleys (Gustason, 1989). The unit ranges in thickness from 1 to 50 m, depending on proximity to the axes of paleovalleys and distance from the Sevier Fold and Thrust Belt. Formerly referred to the basal member of the Dakota Formation (Eaton, 1991; Uličný, 1999), these beds, along with a smectitic gray mudstone exposed in the Markagunt Plateau area, are now called the Cedar Mountain Formation (Doelling et al., 2003; Hylland, 2010). High-energy fluvial channel and

Titus, Roberts, and Albright

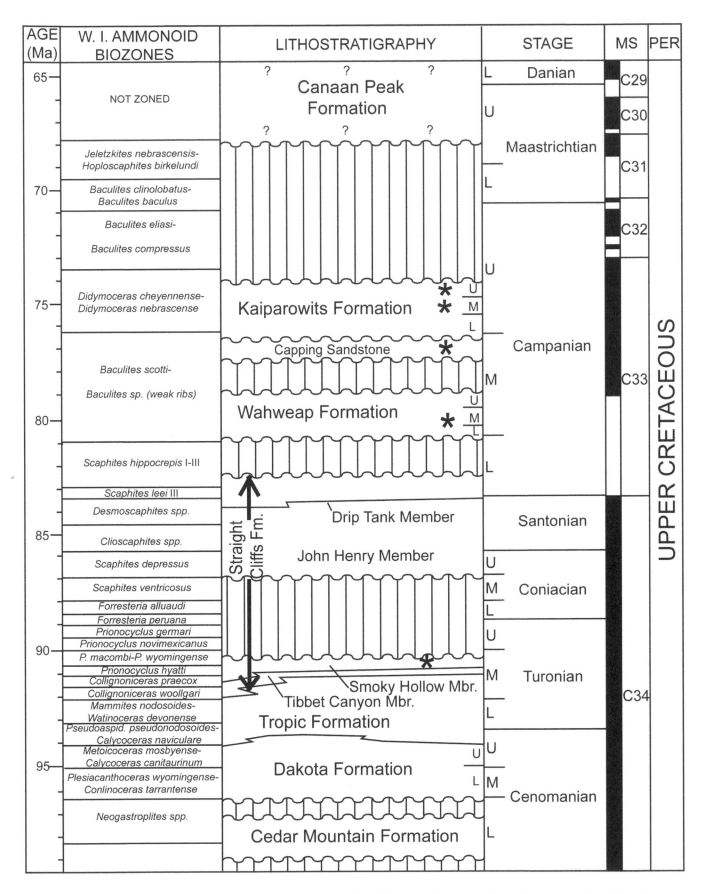

2.10. Detailed chronostratigraphic relationships of Upper Cretaceous strata of the Kaiparowits Plateau area, Grand Staircase–Escalante National Monument. *P, Prionocyclus;* L, lower; M, middle; U, upper; MS, magnetic polarity scale; Per, period. Asterisks indicate horizons yielding radiometric dates. No vertical thickness of lithostratigraphic units is implied.

associated floodplain environments are indicated (Gustason, 1989).

Clasts are mostly chert and quartzite cobbles, pebbles, and granules, but boulder-sized clasts of petrified wood (some >1 m long) and limestone are relatively common on the Markagunt and western Paunsaugunt plateaus. All identified wood samples consist of the araucarian genus *Araucarioxylon* (Gustason, 1989), which could be derived from either the Chinle or Morrison formations. The large size (>1 m) of some of the permineralized wood clasts observed in the Glendale area would suggest that derivation from the Chinle Formation is unlikely as the nearest source outcrops would have been over 140 km to the west in the Wah Wah Thrust area. A more likely source for both the wood and much of the chert and quartzite clasts is the Salt Wash Member of the Morrison Formation, remnants of which were probably widely exposed throughout the area in the early part of the Cretaceous. Current indicators such as pebble imbrication and cross-beds demonstrate unimodal southeast-directed transport (Gustason, 1989), which is normal to the Sevier Fold and Thrust Belt. There is a subtle shift in mean paleocurrent direction upsection from southeast to more east as paleovalleys filled and the rivers became less confined (Gustason, 1989).

The age of the Cedar Mountain Formation in southern Utah (Fig. 2.10) is late Albian–early Cenomanian, as established on the basis of pollen isolated from mudstone partings in the conglomerates (Doelling and Davis, 1989) and a single crystal $^{40}Ar/^{39}Ar$ date of 97.9 ± 0.5 Ma from mudstones overlying the conglomeratic interval (Biek et al., 2003; Hylland, 2010). Dyman et al. (2002a) reported an age of 101.7 ± 0.42 Ma for a bentonite sample collected from equivalent beds in the lower portion of the Iron Springs Formation near Gunlock Reservoir. A general time equivalency with the Mussentuchit Member of the Cedar Mountain Formation in the San Rafael Swell area is indicated (Kirkland and Madsen, 2007), although genetically the two formations are probably separate.

Dakota Formation

East of the western Markagunt Plateau, the interval between the lower gravelly sequence (Cedar Mountain Formation) and the gray marine beds of the Tropic Shale can be divided into two recognizable units (Fig. 2.9): a lower mudstone-dominated sequence with common amalgamated channel sequences, and an upper sandier interval with upward-coarsening sequences, hummocky cross-stratification, and marine and estuarine shell beds (Gustason, 1989). Collectively, these two units, which were formerly Peterson's (1969a) informal middle and upper members of the Dakota Formation, now

make up the entire unit as recognized in southern Utah (Doelling et al., 2003).

The Dakota Formation thickens toward the west (Gustason, 1989; Tibert et al., 2003) as a result of both proximity to source area and diachroneity of the contact (younger to the west) with the overlying Tropic Shale (Fig. 2.7). The latter unit pinches out near the west edge of the Markagunt Plateau placing the Dakota Formation in direct contact with the Tibbet Canyon Member of the Straight Cliffs Formation. In contrast, the Dakota reaches its maximum thickness in this same area, approaching 400 m (Biek et al., 2010). West of the Markagunt Plateau, equivalent strata are found in the lower part of the Iron Springs Formation (Eaton, 1999). In the Kaiparowits Basin, the Dakota Formation is relatively thin (<60 m), highly variable in thickness, and has a general thinning to the east trend. Syndepositional movement of crustal blocks within the Kaiparowits Basin area strongly controlled lateral facies variability, thickness of the formation, and diachroneity of the upper contact with the Tropic Shale (Gustason, 1989; Titus, 2002). Lithofacies associations in the Dakota Formation suggest deposition in a mix of alluvial plain, coastal plain, estuarine, and open marine depositional environments, as might be expected under the conditions of rapidly rising base level associated with the advancing Greenhorn Seaway (Kirschbaum and McCabe, 1992; Uličný, 1999).

In their pioneering work on Kaiparowits stratigraphy, Gregory and Moore (1931:95) originally defined the Dakota (?) Sandstone in the southern Utah area as all Upper Cretaceous "sandstone, coal, and shale below the lowest marine fossil-bearing beds." Gregory and Moore (1931:95) admitted that this division was "arbitrary" and probably inconsistent from section to section. The overlying yellowish sandstones with obvious marine fossils were grouped into the base of their newly defined Tropic Shale unit (Fig. 2.11).

Lawrence (1965) redefined the base of the Tropic Shale as the base of the gray homogeneous mudstone sequence and referred Gregory and Moore's lower sandstone-dominated interval to the underlying Dakota Formation (Fig. 2.11), a move that made sense from both mapping and stratigraphic perspectives. Peterson (1969a) subdivided the Dakota Formation sensu Lawrence (1965) into lower, middle, and upper members. This nomenclatural configuration stood unchanged for over 30 years, when Doelling et al. (2003) reassigned the conglomeratic beds to the Cedar Mountain Formation. Uličný (1999) replaced the informal "lower," "middle," and "upper" member terminology with numbered units 1 through 6. Subsequent workers have not adopted Uličný's terminology because they are recognizable only locally in the Kaiparowits Basin.

Gregory & Moore 1931	Lawrence 1965	Peterson 1969a		Doelling et al. 2000		Uličný 1999
Tropic Shale	Tropic Shale	Tropic Shale		Tropic Shale		Tropic Sh
	Dakota (?) Sandstone	Dakota Sandstone	Upper Member	Dakota Fm.	Upper Member	6
						5
						4
Dakota (?) Sandstone			Middle Member		Lower Member	3
						2
			Lower Member	Cedar Mtn. Formation		1

2.11. Nomenclatural history for the Cedar Mountain, Dakota, and Tropic formations in southern Utah.

The Dakota Formation was originally proposed by Meek and Hayden (1861), who designated the type section as the Missouri River bluffs in eastern Nebraska. Meek and Hayden (1856) had earlier extended the use of their informal Unit 1, which they later named the Dakota Group, into New Mexico, and it is clear that these authors presumed that the Dakota Formation was the lowest mappable Cretaceous unit throughout most of the United States Western Interior. While it might not be so obvious why a formation with a Nebraska type section should be used in Utah, nearly every study since the Dakota Group was first named has concluded that its use in the Colorado Plateau region is valid.

The Colorado Plateau "Dakota" is similar to the type Dakota, the latter of which is generally subdivided into three units: the Nishnabotna Member, and the lower and upper submembers of the Woodbury Member (Ludvigson et al., 2010). Both the Nishnabotna Member and the Cedar Mountain Formation of Doelling et al. (2003) are high-energy fluvial gravelly sequences, while the lower and upper submembers of the Woodbury Member and the Dakota Formation of Utah are low-energy fluvial and shallow marine sequences. The major difference between the two lithosomes, discounting petrographic composition and minor variations in sedimentary architecture, is the older age of the type Dakota Formation. Most of the lower two-thirds of the type section is upper Albian in age, while the upper portion is lower Cenomanian–essentially a one-substage offset from the southern Utah section, where the basal conglomeratic section is lower Cenomanian and the middle unit is middle Cenomanian (Dyman et al., 2002a). Similarities should be expected because both the Utah and Nebraska sequences were deposited in a low accommodation space setting as the Cretaceous Interior Seaway was rising and flooding the continent.

Lower Member The lower member is a variable package of sandstone, massive and laminated mudstone, claystone, with occasional bentonitic clay layers, and carbonaceous beds/coal seams. A wide variety of bed forms is present,

including lenticular (isolated channel) sandstones, contorted bedding (probably seismically induced), multistory sandstone bodies, inclined heterolithic stratification, trough cross-beds, planar cross-beds, chevron cross-beds, ripple laminations, and horizontal laminations (Uličný, 1999). Transport of sediment was largely east–southeast, away from the Sevier Fold and Thrust Belt (Gustason, 1989). The abundance of mudstone, as well as the presence of isolated channel-form sandstone bodies and other lateral accretion features, indicate deposition in meandering–anastomosing stream systems and adjacent levees, swamps, and floodplains (Gustason, 1989). Uličný (1999) hypothesized that the lower portion of the lower member was deposited just inland of a major estuarine system (Uličný's units 2 and 3A), while the upper portion was deposited in a tide dominated estuarine setting.

The middle part of the lower member yielded a $^{40}Ar/^{39}Ar$ date of 95.97 ± 0.22 Ma (Dyman et al., 2002a), placing it firmly within the middle Cenomanian (following Ogg, Agterberg, and Gradstein, 2004). Marine invertebrate fossils in the upper member suggest that the entire unit is middle Cenomanian in age (Fig. 2.10).

The lower member is abundantly fossiliferous. Most of the fossil material by volume is plant, and coal is widespread in the member. Shell beds of unionoid bivalves and viviparid-type gastropods occur at the bases of sandstone channel bodies or in ironstone concretionary bodies. Body fossils of dragonfly larvae have been found in lacustrine beds near the top of the unit associated with beautifully preserved Myrtaceae foliage in the south-central portion of the Kaiparowits Basin, just south of the Arizona–Utah border (Titus, Albright, and Barclay, 2010).

Vertebrate body fossils and traces are found in a variety of Dakota lithofacies, with larger bones, scales, and teeth frequently concentrated above scours in simple or multistory sandstone bodies, associated with intraformational conglomerates (Titus et al., 2005). Cumulatively, the lower member of the Dakota Formation in southern Utah has yielded the most diverse middle Cenomanian terrestrial vertebrate fauna known in North America. Crocodylians, turtles, and amiid fish are the most obvious taxa, indicating a paludal depositional environment, but nearly every other common Cretaceous faunal element is present, including a diverse mammal assemblage. Identifiable larger macrovertebrate remains are rare.

Excellently preserved plant macrofossils are abundant. Permineralized wood is present but uncommon. Pollen and plant macrofossils indicate the flora was almost equally dominated by mosses and ferns, araucariacean and taxodiacean conifers (sometimes found as silicified logs), and a variety of angiosperms, including species of Magnoliaceae, Myrtaceae,

Anacardiaceae, and Rosaceae (Gustason, 1989; am Ende, 1991). Collectively, the fossil fauna and flora indicate freshwater, terrestrial habitats and a relatively humid climate.

Upper Member The upper member of the Dakota has some of the lithologic character of the lower Dakota, with sandstones, carbonaceous beds, and thick mudstone sequences. However, it can be distinguished by its more yellowish color, more abundant sandstone, cyclic upward-coarsening sequences, and abundant feldspar grain content, as well as the presence of persistent, thin-bedded, hummocky, cross-stratified sand sheets (Gustason, 1989). Marine ichnotaxa such as *Ophiomorpha* and fossils of estuarine invertebrates including the bivalves *Corbula*, *Brachiodontes*, *Crassostrea*, *Flemingostrea*, and *Exogyra* are locally abundant (Cobban et al., 2000). Also locally common are shells of open marine forms like the bivalve *Inoceramus* and the ammonites *Metoicoceras* and *Dunveganoceras*. Locally, the topmost bed contains a thick accumulation of massive-shelled *Exogyra* and other bivalves such as *Pinna* that overlies a scoured thin coal seam. A complex suite of back barrier and coastal to offshore marine environments are indicated. Furthermore, the extremely complex facies architecture of estuarine, shoreface, and offshore marine environments observed in the upper member (Uličný, 1999) suggest the formation was deposited on a highly embayed coastal plain whose shoreline geometry was in part shaped by linear structural features such as the East Kaibab Monocline. Elder, Gustason, and Sageman (1994) documented the highly time-transgressive nature of the lithofacies associations (i.e., member units) westward toward the orogen. At least three transgressive–regressive parasequences can be recognized and traced over much of the Kaiparowits Basin (Uličný, 1999; Dyman et al., 2002b).

Ammonites and inoceramids have demonstrated that in the Kaiparowits Basin the upper member's lower portion is within the *Calycoceras canitaurinum* and *Dunveganoceras problematicum* ammonoid biozones (Uličný, 1999; Titus, 2002), and the upper portion in the *Dunveganoceras conditum–Metoicoceras mosbyense* ammonoid biozone. Around paleotopographic high areas (e.g., between the East Kaibab Monocline and the Echo Cliffs Monocline), the uppermost beds of the Dakota range as high as the *Vascoceras diartianum* ammonite biozone (Titus, 2002) and yield large specimens of the ammonite *Calycoceras* cf. *C. naviculare*, as well as huge numbers of the oysters *Exogyra olisiponensis* and *Exogyra kellumi* in tidal channel lags. The biostratigraphy demonstrates that the upper member is nearly equivalent to the entire Hartland Shale Member of the Greenhorn Formation (lower upper Cenomanian). Vertebrate fossils are not abundant, and where found, usually consist of localized lag accumulations of shark and pycnodont fish teeth.

Tropic Shale

The Tropic Shale was named by Gregory and Moore (1931). The type section comprises dark gray marine shale that crops out around the town of Tropic, Utah, in the Paria Amphitheater (Fig. 2.3), where it forms characteristic, sparsely vegetated gray-colored badlands. In the Kaiparowits Basin, the formation ranges in thickness from 183 to 274 m (Sargent and Hansen, 1982). The contacts with both the underlying Dakota Formation and the overlying Straight Cliffs Formation are conformable (Fig. 2.12) and defined as the change from dominantly sandstone (Dakota and Straight Cliffs formations) to dominantly mudstone (Tropic Shale). The Tropic thins to the west, pinching out near the western edge of the Markagunt Plateau, where the Tibbet Canyon Member of the Straight Cliffs Formation rests directly on the upper member of the Dakota Formation.

Lithologies are predominantly mudstone and claystone, but in the upper portion, hummocky cross-stratified and turbiditic sandstone beds become more common. One- or 2-cm-thick siltstone and fine sandstone beds containing abundant inoceramid shell debris are also common in the middle of the formation. Solid and septarian carbonate concretionary horizons are characteristic of the lower and middle parts of the formation, occurring in discrete horizons. However, uncommon concretions are also present in the upper part. In general, the abundance and stratigraphic distribution of concretions in the section increases toward the west and the paleoshoreline (Elder, 1991).

Owing to the abundance of high-resolution biostratigraphic data (ammonites and inoceramids), the Tropic has been well constrained (Fig. 2.10) in the Kaiparowits Basin as upper Cenomanian–lower Turonian (*Vascoceras diartianum–Prionocyclus hyatti* ammonite biozones). Co-occurring datable bentonite ash beds can also be correlated almost bed by bed with those in time-equivalent sections of marine strata in other parts of the Western Interior Basin (Elder, 1989; Elder, 1991; Dyman et al. 2002b; Cobban et al., 2006). In the central Markagunt Plateau where the Tropic pinches out, it is entirely early Turonian in age (Eaton et al., 2001), yielding the inoceramid *Mytiloides kossmati* and the ammonites *Watinoceras* sp. and *Fagesia catinus*. Elsewhere in Utah, Tropic equivalents have been referred to the Tununk Member of the Mancos Shale and the Allen Valley Shale of the western Wasatch (Hintze, 1988).

Tropic Shale deposition was in shallow to moderately deep-water offshore muddy shelf facies (Elder, Gustason, and Sageman, 1994). The lower two thirds of the Tropic Shale in the Kaiparowits Basin is a bluish-gray color because of its high carbonate content and encompasses 11 ammonoid biozones. The upper third is darker and noncalcareous, and

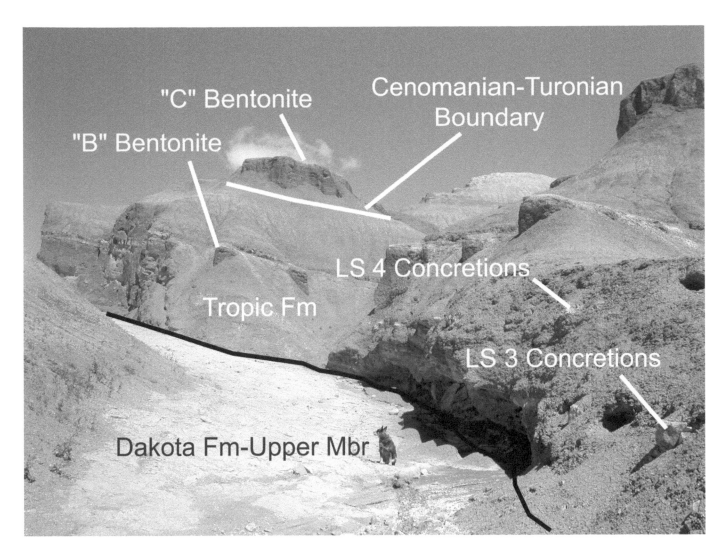

2.12. Exposures of the upper member of the Dakota Formation and lower portion of the Tropic Shale north of Big Water, Utah; south-central Kaiparowits Plateau. Here the top beds of the Dakota are within the *Vascoceras diartianum* ammonoid biozone (AB) and the contact is within the "A" bentonite of Elder (1991). Concretionary horizons LS3 and LS4 are within the *Euomphaloceras septemseriatum* AB, the "B" bentonite is within the *Neocardioceras juddi* AB, and the "C" bentonite is within the *Pseudaspidoceras flexuosum* AB. Marker bed nomenclature after Elder (1991).

it encompasses only one or two ammonoid biozones, depending on location (Dyman et al., 2002b).

Titus et al. (2005) reported the presence of cold hydrocarbon seep bioherms in the lower portion of the Tropic in the Cottonwood Canyon area (Fig. 2.13). The mounds have up to 1 m of relief and occur over syndepositional upwarps along the East Kaibab Monocline that funneled methane from underlying coal beds in the Dakota Formation (Kiel, Wiese, and Titus, 2012).

Beautifully preserved fossils are abundant throughout the Tropic (Cobban et al., 2000; Foster et al., 2001; Dyman et al., 2002b). Ammonites, inoceramid bivalves, thick- and thin-shelled oysters, and several genera of gastropods are especially common, but solitary corals, annelids (serpulids), rudistid bivalves, and other taxa are known. Three-dimensional preservation of most of the invertebrate fauna is limited to within carbonate concretions. A wide variety of small and large vertebrates, including sharks, bony fish, marine turtles, plesiosaurs, and a nearly complete theropod dinosaur have been documented (Gillette, Hayden, and Titus, 1999; Albright, Gillette, and Titus, 2007a, 2007b, this volume, Chapter 25; Schmeisser and Gillette, 2009). Vertebrate material appears to be more common in the lower portion of the lower Turonian part of the section where depositional rates were lowest, but plesiosaurs and other large vertebrates occur throughout the section.

Straight Cliffs Formation

The Straight Cliffs Formation was named the Straight Cliffs Sandstone by Gregory and Moore (1931), who described the type section from outcrops on the east face of the Kaiparowits Plateau (the Straight Cliffs of Fifty Mile Mountain), southeast of the town of Escalante (Fig. 2.3). The formation was

2.13. Specimens of the ammonite *Eumophaloceras* cf. *E. costatum* that occur in abundance around hydrocarbon cold seep limestone mounds in the lower member of the Tropic Shale, southern Kaiparowits Plateau.

further subdivided by Peterson (1969b) into, in ascending order, the Tibbet Canyon, Smoky Hollow, John Henry (Fig. 2.14), and Drip Tank members (Fig. 2.16). The type sections for all of the members are located in the south-central portion of the Kaiparowits Plateau (Fig. 2.3), where exposures are especially good. However, the formation has been mapped as far west as the west edge of the Markagunt Plateau, where it is subdivided into the Tibbet Canyon and Smoky Hollow Members, and an undifferentiated upper member (Eaton et al., 2001; Biek et al., 2010).

The Straight Cliffs Formation consists mostly of sandstone, with lesser amounts of interbedded siltstone, mudstone, and claystone. Coal, concretionary ironstone, and conglomerate are minor components (Fig. 2.7). A complex array of sedimentary structures, fossils, bedding types, architecture, and lithologies indicate a variety of marine, coastal plain, and alluvial plain environments. Fortunately, because the Straight Cliffs Formation contains the kind of large sandstone bodies that typically act as oil and gas reservoirs in the subsurface, as well as rich coal resources, it has been the subject of many detailed studies (Peterson, 1969a; Doelling

and Graham, 1972; Eaton, 1991; Hettinger, 1993, 1995; Shanley and McCabe, 1993, 1995; Little, 1997; Castle et al., 2004; Allen and Johnson, 2010a, 2010b; Gallin, Johnson, and Allen, 2010). The formation ranges in thickness from 300 to 500 m in the Kaiparowits Basin, with a definite northward thickening trend. Most of the Iron Springs Formation west of the Markagunt Plateau correlates with the Straight Cliffs Formation, and the top of the former is apparently no older than Santonian (Eaton, 1999).

Over the last 30 years differing models have been offered to explain the Straight Cliffs stratigraphic architecture. One school applied the classic sequence stratigraphic concepts of Haq, Hardenbol, and Vail (1988) and Posamentier, Jervey, and Vail (1988) to the section, hypothesizing that base-level fluctuation was largely responsible for the observed depositional systems architecture (Hettinger, 1993, 1995, 2000; Shanley and McCabe, 1993, 1995). Others (Bobb, 1991; Eaton, 1991; Little, 1997; Lawton et al., 2003) have postulated episodic thrusting and eustasy as the primary mechanism for generating accommodation space and controlling sedimentary architecture. In the latter view, high subsidence rates

Titus, Roberts, and Albright

Calico Bed · John Henry Mbr

Smoky Hollow Mbr

Tibbet Canyon Mbr

2.14. Exposures of the Tibbet Canyon, Smoky Hollow, and John Henry members of the Straight Cliffs Formation exposed on the Kelly Grade, a few kilometers east of the type sections for all three, south-central Kaiparowits Plateau. Calico bed is below a major disconformity spanning the upper Turonian and lower and middle Coniacian stages.

and accommodation in the proximal foreland basin during active thrusting tend to trap coarser sediment in axial fluvial systems, allowing mostly finer material to accumulate in the distal basin. During tectonic quiescence, a slowdown in the generation of accommodation space results in rapid proximal filling and development of a ramp geometry, leading to dispersal of coarse clastics into the distal foreland (sensu Heller et al., 1988), and the Drip Tank Member has been interpreted to have formed in this manner. Recently Gallin, Johnson, and Allen, (2010) and Allen and Johnson (2010b) have proposed models that argue multiple allogenic (e.g., regional climate, sea level, and tectonics) and autogenic (e.g., bay head delta construction, longshore drift, and barrier construction and destruction) processes played the major roles in shaping Straight Cliffs Formation architecture.

Tibbet Canyon Member The Tibbet Canyon Member consists mostly of yellowish-tan sandstone with occasional concretionary bodies, and lesser amounts of thinly bedded gray and olive-green mudstone and carbonaceous shale. The type section is in Tibbet Canyon, in the south-central portion of the Kaiparowits Plateau (Peterson, 1969b). Intrabasinal and extrabasinal conglomerates are minor components. Sandstones and conglomeratic units at the top of the member contain feldspar and nonrecycled monocrystalline quartz pebbles indicating Mogollon source areas (Peterson, 1969b). Thickness ranges between 21 and 61 m (Cobban et al., 2000) in the Kaiparowits Basin, but it thickens to 200 m on the Markagunt Plateau (Eaton et al., 2001; Biek et al., 2010).

The Tibbet Canyon Member conformably overlies the Tropic Shale east of the west side of the Markagunt Plateau. Its lower contact with the Tropic is somewhat gradational but is defined as the transition from mudstone dominated (Tropic Shale) to sandstone dominated (Tibbet Canyon

2.15. Drip Tank Canyon, the type section for the Drip Tank Member of the Straight Cliffs Formation and overlying lower, middle, and upper members of the Wahweap Formation. View is looking northwest from the south side of Drip Tank Canyon, toward Reynolds Point.

Member), or alternatively, the base of the first readily mappable sandstone bed (Peterson, 1969b). West of where the Tropic Shale pinches out, it conformably overlies the very similar sandstones of the upper member of the Dakota Formation, and the contact becomes difficult to place.

On the Paunsaugunt and Kaiparowits plateaus, the contact with the overlying Smoky Hollow Member is also relatively conformable and is placed at the transition from sand-dominated to mud-dominated lithologies. The lower half of the unit is offshore marine to estuarine in origin and exhibits tabular beds, hummocky cross-stratification, massive structureless sandstone intervals, and local dense accumulations of marine and brackish water mollusks and invertebrate bioturbation (e.g., *Ophiomorpha*).

Biostratigraphically important ammonites occur commonly in the lower portion and clearly demonstrate the diachroneity of the Tropic–Tibbet contact, which becomes younger toward the east (Fig. 2.10). In the Kaiparowits Basin, *Prionocyclus hyatti* characterizes the lower portion of the unit (i.e., middle Turonian), whereas *Collignoniceras woolgari* is found in that same interval on the west side of the Paunsaugunt Plateau. Eaton et al. (2001) reported *Collignoniceras woolgari* from the very top of the Tibbet Canyon Member

in Cedar Canyon, showing its top there is equivalent to the middle of the Tropic Shale in the Kaiparowits Basin.

A major shift in depositional architecture occurs in the middle portion of the member, where upper shoreface sandstones and estuarine shales are scoured and overlain by a highly heterogeneous sequence of trough cross-bedded sandstone, amalgamated fluvial channel deposits containing petrified wood, and localized, typically scoured overbank mudstone/carbonaceous shale bodies. Frequently the contact between the lower marine and upper fluvial sequences is marked by lag concentrations of shark teeth, turtle shell fragments, and other vertebrate fossils reworked from estuarine mudstones and redeposited into the bases of the fluvial channel sandstones. Eaton et al. (1999a) and Hettinger (2000) interpreted the upper interval as estuarine and fluviodeltaic, which is a logical facies continuum from the upper shoreface and beach deposits that occur in the lower portion of the member.

Two opposing stratigraphic interpretations of the Tibbet Canyon Member have been proposed: first, there is a significant diastem at the base of the fluviodeltaic interval in the Tibbet Canyon Member (base of the Smoky Hollow sequence) representing a major sequence boundary; or second,

Titus, Roberts, and Albright

the section is essentially conformable up to the Smoky Hollow–John Henry contact and the facies shift observed in the Tibbet Canyon Member is largely the result of allogenic coastal processes imprinted over a more gradual base level fall.

The first view was proposed by Shanley and McCabe (1991, 1995), Hettinger (1993), and Hettinger, McCabe, and Shanley (1994), who were the first to apply base level–controlled sequence stratigraphic models of Posamentier, Jervey, and Vail (1988) to the Straight Cliffs Formation. These authors emphasized the juxtaposition of Tibbet Canyon Member coastal plain and alluvial plain fluvial deposits and estuarine mudstones over the shoreface interval in the lower portion of the Tibbet as indicative of a disconformity or transgressive sequence boundary. On this basis, Shanley and McCabe (1995) designated the contact between the marine and estuarine/fluvial portions of the Tibbet Canyon members the base of their Tibbet Canyon Sequence, postulating that it marks the resumption of deposition after a base-level drop sometime during post–*Prionocyclus hyatti* time (?*Prionocyclus macombi*–lower *Prionocyclus wyomingense* ammonoid biozones). In this model the sub–Calico bed portion of the Smoky Hollow Member represents the high-stand systems tract of the Tibbet Sequence, and another major disconformity is proposed at the base of the Calico bed.

The alternative view, developed by Peterson (1969a) and reiterated by Eaton (1991), is that the sequence is essentially conformable through the top of the Calico bed and that a more significant disconformity, eliminating the upper Turonian, the lower Coniacian, and perhaps the entire middle Coniacian, is located at the John Henry–Smoky Hollow contact (the Calico bed is not always present beneath the John Henry Member).

To test these competing hypotheses, a recently discovered bentonite in the Smoky Hollow Formation, located 53.5 m above the contact with the Tibbet Canyon Member and 3.5 m below the base of the Calico bed, was collected by one of us (E.M.R.) in the Kaiparowits Basin, south of Henrieville and north of Grovenor's Arch, along Monument/County Road 420/K7450. The bentonite, which is 30 cm thick with fresh, euhedral crystals at its base, was sampled at the following GPS coordinates: WGS 84: UTM 12 0422766/4151850 (Fig. 2.16). Approximately 20 kg of sample was separated for both sanidine and zircon phenocrysts, which were independently dated both by U-Pb thermal ionization mass spectrometry analysis on zircon at the Massachusetts Institute of Technology by S. Bowring and by laser fusion $^{40}Ar/^{39}Ar$ analysis on sanidine at the Berkeley Geochronology Center by A. Deino. This is the first radiometric age date for the Straight Cliffs east of the Markagunt Plateau. The independent analyses resulted in virtually identical ages of 91.86 ± 0.34 Ma and

91.88 ± 0.70 Ma for the zircon and sanidine, respectively (Fig. 2.16). These new age data for the Smoky Hollow Member in the Kaiparowits Basin argue for a negligible age difference between the Tibbet and Smoky Hollow members, invalidating the hypothesis that a significant disconformity exists within the Tibbet Canyon Member. Moreover, the new age provides strong support to the hypothesis of Peterson (1969a) that the Tibbet Canyon Member is conformable with the overlying Smoky Hollow Member. Therefore, this interpretation is adopted here (Fig. 2.10).

Fossils are common throughout the Tibbet Canyon Member. The lower portion of the marine interval contains an abundant and highly diverse normal marine molluscan fauna, while dense accumulations of thick-shelled oysters such as *Crassostrea* are more typical of the upper shoreface sequence. Marine invertebrate trace fossils are also common throughout the lower portion. Marine and estuarine condrichthians and osteichthians are common at the top of the lower portion, having been concentrated there as lags from estuarine mudstones scoured before deposition of the upper portion. The upper portion of the member, interpreted as fluviodeltaic by Eaton et al. (1999a), contains a more diverse but essentially unstudied terrestrial vertebrate fauna that includes *Ceratodus semiplicatus*, lepisosteids, turtles, crocodylians, and rare marsupial mammals. Petrified wood and other fossil flora are also very common in the upper portion.

Smoky Hollow Member The name Smoky Hollow Member was given by Peterson (1969b) to the recessive series of carbonaceous mudstones, mudstones, coals, sandstones, and conglomerates that overlie the Tibbet Canyon Member and weather into a topographic bench. The designated type section (Peterson, 1969b) is in what is called Smoky Hollow on the 1953 edition of the United States Geological Survey Nipple Butte 15-minute quadrangle. When the area was mapped by the United States Geological Survey at the 7.5-minute level in the late 1970s, they renamed Smoky Hollow "Squaw Canyon" and named a previously unnamed tributary of Warm Creek Canyon "Smoky Hollow." Thus, the designated type section is not located in what is presently called Smoky Hollow, but rather in Squaw Canyon. In most places the unit is less than 50 m thick, but it does reach a thickness of 100 m in the northeast part of the Kaiparowits Basin (Eaton, 1991). Eaton et al. (2001) estimated a similar thickness of 107 m for the unit on the western Markagunt Plateau, where it is significantly older than the type section (*Collignoniceras woolgari* ammonoid biozone).

The lower contact with the underlying Tibbet Canyon Member is usually relatively sharp. However, channel sandstone bodies similar to those found in the underlying Tibbet Canyon Member do locally occur in the lower portion of the Smoky Hollow. Mudstones in the lower portion of

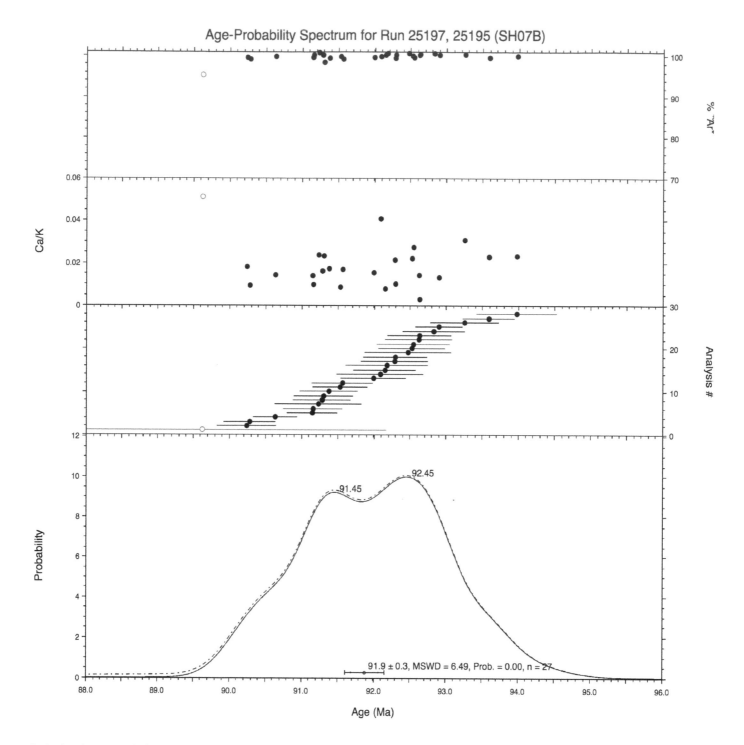

2.16. Plots showing U-Pb (facing page) and ³⁹Ar/⁴⁰Ar (above) age data for an ash sample taken from the upper portion of the Smoky Hollow Member. Note close agreement of two results.

the Smoky Hollow Member are paludal in origin and commonly carbonaceous, whereas the middle portion, which is also mudstone dominated, contains very few carbonaceous horizons, probably because of shortened floodplain lake life cycles (better drainage). Siderite concretionary masses, both bedded and lenticular, are common in the mudstones and concentrated at the bases of channel deposits.

Peterson (1969b) also coined the informal unit name Calico Bed for a mottled, leached interval of coarse-grained sandstone and pebble conglomerate of fluvial origin that usually caps the Smoky Hollow Member. The Calico bed is locally cut out or truncated by pre–John Henry erosion and nearly everywhere the Smoky Hollow–John Henry contact is sharp. Shanley and McCabe (1991) and Hettinger (1995) proposed a major sequence boundary at the base of the Calico bed and

Titus, Roberts, and Albright

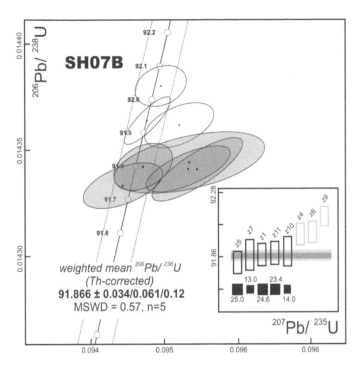

weighted mean $^{206}Pb/^{238}U$
(Th-corrected)
91.866 ± 0.034/0.061/0.12
MSWD = 0.57, n=5

considered the Smoky Hollow–John Henry contact to be a conformable transgressive systems tract–high-stand systems tract bounding surface within their Calico sequence. Given that the John Henry was deposited on considerable relief eroded into the Calico bed and underlying strata, and that there is no obvious ravinement surface or channel scour below the Calico bed, this seems unlikely (Bobb, 1991). In fact, locally, channel fill conglomerates identical to the Calico bed occur below it, and the upper portion of the Smoky Hollow Member shows gradation from mudstone and sandstone into coarser-grained facies. Peterson (1969a) and Eaton (1991) advocated the notion that a significant disconformity eliminates nearly all the upper Turonian and lower and middle Coniacian at the Smoky Hollow–John Henry contact; that interpretation is taken here. On the west side of the Paunsaugunt Plateau, over 60% of the Smoky Hollow Member is medium sandstone ·(Tilton, 2001), whereas that figure is closer to 40% over much of the Kaiparowits Basin (Peterson, 1969a). Biek et al. (2010) noted the Calico bed is thin or missing on the western Markagunt Plateau.

Previously the Smoky Hollow Member was dated entirely by the ammonite biostratigraphy of the upper and lower bounding units (i.e., middle Turonian–middle Coniacian age range). The newly dated ash bed in the upper Smoky Hollow (presented herein) yields concordant U-Pb and $^{40}Ar/^{39}Ar$ ages of ~91.86 and 91.88 Ma (Fig. 2.16), respectively, and argues for relative conformity through the entire Tibbet Canyon–Smoky Hollow interval (Fig. 2.10). Although the date would appear to be nearly 1 Ma too old relative to the Tibbet Canyon Member biostratigraphic data following Ogg, Agterberg, and Gradstein (2004), it is likely that

the Cenomanian and Turonian dates of Obradovich (1993) used by Ogg, Agterberg, and Gradstein (2004) are too young relative to results being obtained currently by using new techniques and standards (Obradovich et al., 2002). Additionally, the new $^{40}Ar/^{39}Ar$ age presented here was calculated with a new astronomically calibrated (Kuiper et al., 2008) Fish Canyon sanidine standard (28.2 Ma), which results in a ~0.5 Ma older age than if we used the uncalibrated Fish Canyon sanidine standard of 28.02 Ma. The advantage of the use of this new standard is clearly demonstrated here, in that it eliminates previous problems with discordance between $^{40}Ar/^{39}Ar$ and U-Pb ages (i.e., U-Pb yielding ~1% older ages than $^{40}Ar/^{39}Ar$). Given this correction, the new date for the Smoky Hollow Member is perfectly compatible with a middle Turonian assignment, probably within the upper portion of the *Prionocyclus hyatti* ammonoid biozone. Cumulatively, then, the evidence strongly corroborates the conformable nature of the Tibbet Canyon–Smoky Hollow succession proposed by Peterson (1969a). Furthermore, the Smoky Hollow–John Henry contact disconformity most likely spans the *Prionocyclus macombi* through *Scaphites ventricosus* ammonoid biozones (upper Turonian–middle Coniacian), as based on the presence of the upper Coniacian ammonite *Protexanites bourgeoisianum* near the base of the John Henry Member (Peterson, 1969a; Dyman et al., 2002b).

Not surprisingly, much of the missing section between the Calico bed and the John Henry Member in the Kaiparowits Basin correlates with the more seaward Ferron Sandstone (Ryer, 1991) and Juana Lopez Member of the Mancos Shale (Cobban, 1976), indicating sediment was bypassing the Kaiparowits region to be deposited in central and eastern Utah. The sub–John Henry disconformity also correlates with deposition of the Tres Hermanos Formation–Gallup Sandstone–Dilco Coal Member of the Crevasse Canyon Formation succession of western New Mexico (Molenaar et al., 2002), which records a corroborating basinward shift of coastal plain facies into the Mancos Shale depositional system.

A rich microvertebrate fauna has been documented from the Smoky Hollow Member, mostly from the middle noncarbonaceous interval (Eaton et al., 1999a). Macrovertebrate remains are less common, but three ornithopod dinosaur bonebeds, one bearing the remains of at least three individuals (UMNH VP Loc. 1266), have been discovered within Grand Staircase–Escalante National Monument in the last 15 years (Gates et al., this volume, Chapter 19). Petrified wood (primary fossils, not reworked) and 2–3-cm-diameter bone and turtle shell fragments are not uncommon in the Calico bed. The Smoky Hollow vertebrate fauna has an entirely Late Cretaceous character, with some of the Early Cretaceous holdovers like *Glyptops* and *Ceratodus* conspicuously absent (Eaton et al., 1997).

John Henry Member The John Henry Member is lithologically variable like the Smoky Hollow Member but much thicker (Fig. 2.7). Sandstone, carbonaceous and noncarbonaceous mudstone, and coal make up the bulk of the rock types. The type section, designated by Peterson (1969b), is in Squaw Canyon, to the north of the type Smoky Hollow, about 4.8 km east of John Henry Canyon. Thickness ranges between 200 and 340 m, with a definite northeast thickening trend. The lower contact with the Smoky Hollow Member is sharp and clearly incised with several meters of relief. Locally the Calico bed is cut out in the Kaiparowits Basin and the sandstones of the John Henry rest directly on the "barren" mudstones of the Smoky Hollow Member. Typical fluvial, shore-complex sedimentary structures including heterolithic epsilon-type and bidirectional cross-beds and hummocky cross-stratified sandstones characterize the John Henry Member. Coal is abundant, particularly in what are known as the Alvey (upper) and Christensen (lower) zones. Seven separate mappable shoreface sandstone bodies, labeled A to G, have been recognized in the member in its eastern exposures (Peterson, 1969a). Two marine mudstone tongues, key units for age constraint, are present in the member's eastern outcrop area (Peterson, 1969a): the informal "Lower Mudstone Tongue" at the base and the "Upper Mudstone Tongue," which grades laterally into sandstone beds "E" and "F."

Extremely detailed lithostratigraphic studies have been conducted on the John Henry Member by Vaninetti (1978), Shanley and McCabe (1991), Shanley, McCabe, and Hettinger (1992), Hettinger (1993, 1995), Hettinger, McCabe, and Shanley (1994), Allen and Johnson (2010a, 2010b), and Gallin, Johnson, and Allen, (2010). In general, environmental facies grade from open marine and shoreface on the east edge of the Kaiparowits Plateau to fluvial/overbank along the East Kaibab Monocline. Accordingly, stratigraphic architecture and lithology change radically along this same general east–west trend. In the central Kaiparowits Plateau area, which is the most heavily studied because of its estimated 28 billion short tons of recoverable coal reserves (Hettinger et al., 2000), three basic facies assemblages can be recognized in the John Henry Member (Gallin, Johnson, and Allen, 2010): (1) tidally influenced fluvial channel systems in coastal plain mires; (2) isolated distributary channels, laterally restricted channel belts, and bay head deltas in lagoons; and (3) laterally extensive channel belts and channel belt complexes in normal floodplain systems. Along the Fifty Mile Mountain, facies associations are less diverse and can be basically grouped into transgressive and regressive successions of shoreface, lagoonal, and tidally and longshore drift–influenced fluviodeltaic complexes (Allen and Johnson, 2010b).

The age of the John Henry Member is derived entirely from marine invertebrate biostratigraphic data obtained from easternmost exposures (Fig. 2.10). The "Lower Mudstone Tongue" contains the upper Coniacian index ammonite *Protexanites bourgeoisianus* (d'Orbigny), while the "Upper Mudstone Tongue" yields the upper Santonian inoceramid *Endocostea baltica* (Boehm) (Dyman et al., 2002b). Even higher in the section, Peterson (1969a) collected specimens of the late Santonian index inoceramid *Endocostea flexibaltica* (Seitz) from near the "G" sandstone. A ^{40}Ar/^{39}Ar date of 86.72 ± 0.58 Ma obtained for a euhedral biotite-bearing blue tuff bed located 270 m above the top of the Tibbet Canyon Member in Cedar Canyon (Eaton et al., 2001) is consistent with biostratigraphic data from the Kaiparowits Basin. Age-equivalent rocks in the Markagunt Plateau represent a more inland, fully alluvial plain setting than those in the Kaiparowits Basin (Eaton et al., 2001). Preliminary paleomagnetic analysis of the John Henry Member in the Kaiparowits Basin by Albright and Titus (2009) suggests that the entire member is of normal polarity and entirely within the C34n Polarity SuperChron.

The upper Coniacian–upper Santonian age of the John Henry Member correlates almost exactly with the Tres Hermanos Formation–Crevasse Canyon, lowermost Menefee Formation sequence in the San Juan Basin of New Mexico (Molenaar et al., 2002), a view first advocated by Peterson (1969a). Additionally, the John Henry Member's lower and upper marine mudstone tongues correlate precisely with the Mulatto and Satan tongues of the Mancos Shale, showing that both areas were synchronously influenced by sea level fluctuations. Both the Kaiparowits Basin and San Juan Basin Coniacian base-level records differ somewhat from that of the central Wasatch Plateau–San Rafael Swell area, where sea level (or accommodation space) rose substantially during the early Coniacian (Kennedy and Cobban, 1991). Presumably the proximity of the central Utah sections to the active Nebo Thrust Salient (DeCelles, 2004) was responsible for these differing relative sea-level curves.

Fossils are locally abundant in the John Henry Member. Marine, estuarine, and freshwater invertebrates, mostly mollusks, are all known. On the eastern side of the Kaiparowits Plateau, densely packed shell beds containing the inoceramid *Volviceramus involutus* (Sowerby) or *Crassostrea* sp. are quite common. Transgressive lags at the base of marine parasequences locally contain concentrations of bone and shark's teeth, along with abundant marine invertebrate taxa such as inoceramids and ammonites.

Macrovertebrate remains are uncommon in the eastern Kaiparowits region but increase in abundance toward the west. Terrestrial microvertebrates are locally abundant in mudstones and in lags at the base of channel sandstone bodies, particularly west of the central Kaiparowits Plateau (see Chapters 10–12, 15, and 16 in this volume). Recent discoveries of dinosaur bonebeds in the Coniacian portion of the John Henry Member exposed on the east side of the Paunsaugunt Plateau (Gates et al., this volume, Chapter 19) show that this unit will provide important data in the future. Currently the dinosaur fauna is known almost exclusively from teeth recovered during microvertebrate sampling that shows only a generic dromaeosaur–ankylosaur–hadrosaur dinosaur assemblage (Eaton et al., 1999a). Vertebrate tracksites associated with coal seams and crevasse splay–floodplain contact surfaces are not uncommon, and both large and small theropod and ornithopod dinosaur and crocodylian swim tracks are known (Foster et al., 2001).

Drip Tank Member The Drip Tank Member was named by Peterson (1969b). The type section is at the mouth of Drip Tank Canyon, in the south-central Kaiparowits Plateau (Fig. 2.15). It consists mostly of resistant, tan- to brown-colored massive, trough cross-bedded and planar cross-bedded sandstones, with locally abundant chert pebble conglomerate. The chert pebble clasts serve to differentiate the unit in many places from the overlying recessive lower sandstone unit of the Wahweap Formation in which pebble clasts are rare or absent (Fig. 2.7). Amalgamated fluvial–multistory channel architecture is dominant. Abundant log and stem casts occur above channel scours, along with regular occurrences of fragmentary dinosaur bone and other vertebrate fossils. Bed architecture shows the Drip Tank was deposited within high-energy braided, and possibly moderate-energy meandering, stream systems. Christensen's (2005) analysis of the member's composition at four locations within the Kaiparowits Basin identified both Sevier orogenic belt (quartzolithic) and Mogollon highland (quartzofeldspatholithic) sources.

In the Kaiparowits Basin the unit ranges in thickness from 43 to 150 m, with a northward thickening trend (Eaton, 1991). Because of its resistant nature, the Drip Tank locally weathers into vertical cliffs and underlies prominent benches formed in the lower portion of the Wahweap Formation. The lower contact with the John Henry Member appears somewhat gradational, and Peterson (1969b) originally included finer-grained units below the first resistant sandstone ledge in his Drip Tank. Eaton (1991) restricted the member to the resistant cliff-forming sandstone interval and also placed the ledge-forming upper portion of the type Drip Tank into the lower portion of the Wahweap Formation.

Peterson (1969a) and Eaton (1991) interpreted the John Henry–Drip Tank contact interval as relatively conformable, but others view it as a major sequence boundary (Shanley and McCabe, 1993). No radiometric dates have

been obtained from the Drip Tank, but recent palynological sampling (Christensen, 2005) suggests that, like the upper part of the John Henry Member, at least the lower portion is upper Santonian (Fig. 2.10). Preliminary results of paleomagnetic analysis by Albright and Titus at both the northern and southern ends of the Kaiparowits Plateau indicate that the middle portion of the unit is reverse polarity and falls within the lower portion of the C33r Polarity Chron. If this is correct, then the John Henry–Drip Tank contact approximates the Santonian–Campanian stage boundary and much, if not all, of the Drip Tank Member is early Campanian in age. We hypothesize that the Straight Cliffs–Wahweap contact is a disconformity and that strata representing the *Scaphites hippocrepis* II through *Baculites* sp. (weak flank ribs) ammonoid biozones are missing (Fig. 2.10). In more seaward sections in New Mexico and Montana, a major regression is recorded in the Mesaverde Group and Eagle Formation, respectively, at this same time (Hicks, Obradovich, and Tauxe, 1999). We hypothesize that during this low stand, the Kaiparowits Plateau region actually experienced sediment bypass with minimal regional incision. Radiometric dates from the overlying Wahweap Formation (Jinnah et al., 2009) demonstrate that regional sedimentation resumed at about the time of the *Baculites obtusus* ammonoid biozone (which equals the base of the Lewis Shale in the San Juan Basin).

The Drip Tank Member is the lowest unit in the Kaiparowits Basin Cretaceous succession that has not previously been recognized with certainty west of the Paunsaugunt Plateau region. In this part of the section drastic facies changes occur between isochronous strata in eastern and western sections that make lithostratigraphic correlations difficult (Eaton et al., 1999a, 2001; Eaton, 2006). However, mounting evidence indicates that at least the lower two members of the Grand Castle Formation of the Markagunt Plateau (type section in Parowan Canyon), previously thought to be Paleocene in age (Goldstrand and Mullet, 1997), correlate with the Drip Tank Member (Hunt et al., 2011), a possibility first raised by Biek et al. (2010). Biek at al. (2010) reported the presence of upper Santonian or lower Campanian palynomorphs in the middle unit of the Grand Castle Formation, and Nichols (1997) reported Santonian? pollen from an unnamed unit that overlies the Grand Castle Formation. In 2010 EMR found a dinosaur tracksite in the middle unit of the Grand Castle Formation and confirmed that the unit is Cretaceous. The most significant age data come from Eaton et al. (1999b) and Eaton (2006), who reported on an enigmatic Cretaceous mammalian assemblage recovered in Cedar Canyon (UMNH VP Loc. 11) from directly below the "white sandstone" unit of Eaton et al. (2001). Two of the pediomyid marsupials from this locality compare closely with forms from the

Upper Santonian Milk River Formation (Davis, 2007). Eaton (2006:400) concluded, in spite of its "similarity to the (Santonian) Milk River fauna," that the horizon yielding the assemblage was early Campanian in age. Eaton (this volume, Chapter 15) has since revised the age determination for the locality 11 assemblage to Wahweap age (i.e., lower middle Campanian), but the evidence is scant because of apparent biogeographic differences. Because the conglomerate shown by Eaton et al. (2001) to occur almost 200 m stratigraphically below the locality 11 horizon has been physically traced north into the lower conglomerate unit of the Grand Castle Formation (Biek et al., 2010), currently the evidence best supports an early Campanian age for the lower and middle, and possibly the upper, members of the Grand Castle Formation and equivalency with the Drip Tank Member (Hunt et al., 2011). If this is true, then the 230 m of Markagunt Plateau strata assigned by Biek et al. (2010) to the Wahweap Formation(?) is time correlative with the John Henry Member and Santonian or early Campanian in age.

Diagnostic macrofossils are not common in the Drip Tank Member in the Kaiparowits Basin, but Eaton (1991) did report fish scales and fragments of turtle shell. The unit has yielded diagnostic macrovertebrate material (including crocodylian and dinosaur specimens) in the Paunsaugunt Plateau area. Fossil wood, both silicified and limonitized, is also abundant. No vertebrate faunal lists have been generated for the Drip Tank Member, and the diagnostic material from the Paunsaugunt Plateau has not been studied. The quantities of bone seen in outcrop indicate the unit has future potential, although its tendency to weather into cliffs makes prospecting difficult.

Wahweap Formation

The Wahweap Formation was named by Gregory and Moore (1931; as the Wahweap Sandstone). The type section proposed by Gregory and Moore (1931) is here interpreted as being along the middle reach of Wahweap Creek, on the west side of the Kaiparowits Plateau (Fig. 2.3), where a thick section of resistant tan and yellowish-brown sandstones, alternating with nonresistant olive-gray mudstones, crop out between the Drip Tank Member and the overlying bluish-gray sandstones of the Kaiparowits Formation. In the Kaiparowits Basin, the formation ranges in thickness between 360 and 460 m. Over the southern portion of the Paunsaugunt Plateau, the Claron Formation (Paleocene) rests directly on the Wahweap's middle member, and the formation is less than half its Kaiparowits Basin thickness. In the northern portion of the Paunsaugunt Plateau, nearly complete sections are preserved, but poor exposures and structural complications

make them difficult to see in any one place. However, it does appear that the northern Paunsaugunt sections are thicker than those of the Kaiparowits Basin. Available biostratigraphic data make it implausible that complete sections of Wahweap are preserved on the Markagunt Plateau (Biek et al., 2011).

Eaton (1991) further divided the Wahweap Formation into four informal members (Fig. 2.15), designating the Reynolds Point area, also in the Kaiparowits Basin, as the type section for the lower three and Pardner Canyon as the type section for the uppermost because of accessibility. Contacts between the lower, middle, and upper members are all conformable but generally sharp, and they represent changes of sand/shale ratios and fluvial styles. The upper contact of the upper member with the overlying capping sandstone member represents an abrupt change in color, petrology, grain size, and fluvial style. Although there is some lateral variability in thickness, all four members can be consistently recognized throughout the Kaiparowits Basin.

Two recently dated ash beds (80.6 and 79.9 Ma) from the lower portion of the middle member (see Jinnah, this volume, Chapter 4, for a full review of the Wahweap Formation) demonstrate that the formation largely correlates with the Claggett Shale (Hicks, Obradovich, and Tauxe, 1999), and thus chronostratigraphically falls in between the Eagle and Judith River formations. Preliminary paleomagnetic data obtained by Albright and Titus indicate that both the lower and middle members are entirely reversed in their magnetic signatures and referable to the C33r Polarity Chron and that this reversal extends well up into the upper sandstone member. Jinnah et al. (2009) analyzed detrital zircons from the base of the capping sandstone member that yielded an age of 77 ± 2 Ma, indicating there is a near 2-million-year-long diastem at the upper sandstone member–capping sandstone member contact (Fig. 2.10). The upper sandstone and capping sandstone members exhibit dramatically different compositions, source areas, and sedimentary architecture (Lawton et al., 2003), and the diastem corresponds both with a major reorganization of the basin and source area systems and peak regression of the Claggett Seaway (Rogers and Brady, 2010). The base of the overlying Kaiparowits Formation has been estimated to date to 76.6 Ma (Roberts et al., this volume, Chapter 6) on the basis of dated ash beds, giving a youngest possible age of middle Campanian (76.6 Ma maximum) for the top of the Wahweap and demonstrating relatively continuous deposition through the capping sandstone member–lower Kaiparowits member interval.

Jinnah (this volume, Chapter 4) recognized 10 discrete facies associations that he grouped into channel and flood basin deposits associated with a fluvial to upper estuarine system. Overall, the Wahweap Formation was deposited under wet conditions with periods of increased aridity, particularly toward the top (Simpson et al., 2008; Jinnah, this volume, Chapter 4). The mixed influences of eustasy, thrust belt loading (in the Sevier), and initiation of Laramide tectonism in the foreland all contributed to basin subsidence and deposition of the formation (Lawton et al., 2003; Simpson et al., 2009; Tindall et al., 2010; Jinnah and Roberts, 2011). The Wahweap Formation preserves the most diverse middle Campanian terrestrial vertebrate fauna known from North America.

Lower Member The lower member is conformable with, but rests sharply on, the Drip Tank Member of the Straight Cliffs Formation. It forms alternating benches and slopes of interbedded trough cross-bedded sandstone and mudstone deposited by north- to northeast-flowing meandering rivers. Eaton (1991) reported a thickness of 65 m for the unit at Reynolds Point, the type section. Quartzofeldspathic sandstone composition, detrital zircons, and transport directions indicate Mogollon source areas were most significant (Lawton et al., 2003; Jinnah et al., 2009; Larsen et al., 2010). Intraformational conglomerates commonly overlie mudstone intervals above obvious scour surfaces. Petrified wood is common, and concentrations of large logs occur in several areas, particularly in southern outcrops. Eaton (1991) stated that the highest concentrations of logs are on channel margins, associated with crevasse splay sandstones. Other fossils are also present, with stream-worn bone, turtle shell, and gar scales especially common in vertebrate microsites associated with intraclast conglomerates concentrated as lags. Larger bone is not uncommon. One of the only true multitaxic dinosaur bonebeds known from the formation is in this member. The lower member was deposited in a low-relief, waterlogged proximal flood basin setting of a meandering river system.

Middle Member The overlying middle member is dominantly fine-grained deposits that are interbedded with trough cross-bedded and ripple laminated sandstone of quartzofeldspathic composition. It is a slope former and overlies a major bench system developed on the lower member and Drip Tank Member of the Straight Cliffs Formation. The unit is characterized by truly spectacular flood basin alluvial architecture and flood basin facies associations (Jinnah, this volume, Chapter 4). Thickness is variable, but it generally measures about 100 m; Eaton (1991) reported a thickness of 112 m at Reynolds Point, the type section. Again, the high percentage of feldspar in the sandstones suggests significant Mogollon input (Lawton et al., 2003). Intraformational conglomerate is common at the base of sandstone beds above scours. Mudstones are gray and green, and they commonly contain iron-rich carbonate concretions, blocky weathering,

slicks, and other paleosol features. Fossils are common, and the member now has the distinction of containing the highest abundance of associated macrovertebrate fossil sites for the entire formation.

Upper Member The upper member, which forms benches and ledges, is composed of quartzolithic trough cross-bedded sandstone with minor mudstone deposited by northeast-flowing meandering rivers with strong evidence of tidal influence (Jinnah and Roberts, 2011). Eaton (1991) reported a thickness of 148 m at Reynolds Point. The member typically weathers into cliffs, making it difficult to prospect for fossils, although several important macrovertebrate sites have recently been identified. Moreover, Eaton (1991) indicated the presence of large number of microvertebrate localities in the member. The quartzolithic sandstone composition indicates a shift to sources in southern Nevada and possibly southwestern Utah.

Capping Sandstone Member The capping sandstone member consists of multistoried conglomerate, pebbly sandstone, sandstone, and muddy siltstone sheets that represent braided gravel bars, channel complexes, crevasse splays, and floodplain deposits. Large siltstone and mudstone rip-up clasts (10–150 cm across) are commonly found at the bottoms of channel fill, which also contain abundant plant fragments and tree trunks with root networks. Thickness for the unit in the Kaiparowits Basin ranges 75–150 m, but on the west side of the Paunsaugunt Plateau in Hillsdale Canyon, its stratigraphic equivalent may be nearly twice as thick, with a proportional increase in clast size in the conglomerates.

Sandstone compositions are quartzose, and paleocurrents indicate deposition by east- to southeast-flowing braided rivers. The architecture, characterized by amalgamated channel deposits, has been interpreted as resulting from high-energy braided stream systems, deposited under extremely low accommodation conditions (Lawton et al., 2003). The quartzose grain compositions of the sandstones strongly contrast with the feldspatholithic compositions of the lower two sandy members, indicating a marked change in sediment source, presumably to the Sevier Fold and Thrust Belt.

The unit has not yielded much in the way of diagnostic skeletal material. A purported maniraptorian digging trace and associated prey mammal gallery-type burrows were reported from the unit by Simpson et al. (2010).

Kaiparowits Formation

The Kaiparowits Formation was named by Gregory and Moore (1931). A type area was not designated, but it is clear that those authors considered the region below the Table Cliffs as most typical. The original definition included all

2.17. Aerial view of the badlands known as The Blues, below the Table Cliffs Plateau and Powell Point (3105 m). This is the type section for the Kaiparowits Formation and is just north of the type section for the overlying Canaan Peak Formation. View is looking north across Henrieville Canyon, where cliffs of the capping sandstone member of the Wahweap Formation are visible in the lower left-hand corner. Topography is typical for outcrops of the Kaiparowits Formation.

strata between the brown sandstones of the Wahweap Formation and the pink beds of the Claron Formation. Bowers's (1972) designation of the Canaan Peak Formation fixed the upper boundary of the Kaiparowits Formation as a regional angular unconformity below coarse conglomeratic beds bearing volcanic clasts. Roberts (2007) designated the area beneath the Table Cliffs and Canaan Peak, referred to as The Blues (Fig. 2.3), as a lectostratotype section (Fig. 2.17).

The thickness of the Kaiparowits Formation at any given outcrop is locally controlled by the pre–middle Paleocene tectonic history. Regionally, the unit (along with the rest of the section) was tilted northward (dipping away from the Mogollon Rim), and locally either faulted along deep basement structures such as the Hurricane, Sevier, and Paunsaugunt faults in a general down to the east trend, or folded (e.g., Upper Valley and East Kaibab structures). As a result, the thickest sections (~865 m) are found along the axis of the Table Cliffs Syncline (Sargent and Hansen, 1982). Pre-Claron erosion almost removed Kaiparowits Formation from the entire modern Cretaceous outcrop area west of the Paunsaugunt

Fault. However, thin (60 m or less) sections can be found at the top of the Cretaceous section, immediately below the Claron Formation in the Proctor Canyon and Hillsdale Canyon areas, on the northwest side of the Paunsaugunt Plateau, where they occur over a greatly expanded section equivalent to the capping sandstone member of the Wahweap Formation.

The Kaiparowits Formation consists mostly of gray-blue to greenish-gray and gray-colored sandstones, mudstones, and siltstones, and typically weathers into classic badlands topography (Fig. 2.17). Facies and architectural analysis of the Kaiparowits Formation has provided detailed information about the formation's paleoenvironments and depositional history (Roberts et al., this volume). The Kaiparowits Formation is subdivided (following Roberts, 2007) into three informal units, upper, middle and lower, on the basis of distinct changes in alluvial architecture, channel morphology, and sandstone/mudstone ratios. The lower and upper units are characterized, respectively, by high channel/overbank ratios of 75/25 and 60/40, sheet-like major sandstones, and

Titus, Roberts, and Albright

only moderately abundant fossil preservation. In contrast, the middle unit is richly fossiliferous, contains a high proportion of lenticular channels, and has a low channel/overbank ratio of 45/55.

The fine-grained nature of the sediment and nonmarine fauna suggest deposition upon a low-relief, inland alluvial plain setting. Thick paludal deposits, large channels, and poorly developed, hydromorphic paleosols dominate the sedimentary record, and all are suggestive of a relatively wet alluvial system (Roberts, 2007), an interpretation supported by the high abundance and diversity of aquatic vertebrate, invertebrate, and plant fossils preserved within the formation.

$^{40}Ar/^{39}Ar$ analysis of four bentonites collected throughout the formation yielded a late Campanian age, between ~76.1 and 74.0 Ma (Roberts, Deino, and Chan, 2005); however, Roberts et al. (this volume) provide an additional radiometric age and justify a recalibration of their original dates (Figs. 2.7, 2.10), pushing back the formation's age by about half a million years (from ~76.6–74.5 Ma).

High-resolution stratigraphic correlation reveals that the Kaiparowits Formation is contemporaneous with some of the more important vertebrate fossil-bearing formations in the Western Interior Basin (Roberts, Deino, and Chan, 2005, this volume), providing a firm basis for addressing questions on mammal biostratigraphy, vertebrate evolution, biodiversity, and paleobiogeography (e.g., dinosaur provincialism) in the Cretaceous Western Interior Basin. The Kaiparowits Formation is considered by many to be the crown jewel of Kaiparowits Basin paleontology because of the abundance, diversity, and quality of its fossils. Previous research has conclusively demonstrated that the formation contains an outstanding record of Late Campanian terrestrial vertebrate ecosystems (Sampson et al., this volume, Chapter 28).

Canaan Peak Formation

Whether or not a significant post–Kaiparowits Formation Cretaceous stratigraphic record exists in southern Utah is still equivocal. The lower portion of the Canaan Peak Formation has been given a Cretaceous age (Bowers, 1972; Goldstrand, 1990). However, the argument for this consists primarily of it being bracketed by the Upper Campanian Kaiparowits Formation below, and the Paleocene Pine Hollow Formation above. Fossil evidence for the age of the Canaan Peak Formation consists of reworked early or middle Campanian-age palynomorphs in the lower portion of the formation (Bowers, 1972) and a distinctive early Paleocene palynomorph assemblage in the upper portion (Goldstrand, 1990). Fragmentary bone large enough to be from a dinosaur has been observed by EMR in the Canaan Peak Formation,

but it has not been collected or studied and could possibly be reworked. Because most studies have concluded it may in part be Maastrichtian in age, a brief review is given.

The Canaan Peak Formation is a relatively thin package of chert pebble and cobble conglomerate and chert lithic-rich sandstones. Lesser amounts of volcanic clasts are present. Trough cross-bedding is characteristic, and high-energy braided stream systems are the dominant depositional environment (Schmitt, Jones, and Goldstrand, 1991). Large clast sizes indicate a not-too-distant source area for some of the sediment. Goldstrand (1994) suggested mixed source areas for the lower portion of the Canaan Peak Formation, with a component coming from southeastern Nevada (source of volcanic clasts) and a more local component derived from detritus shed eastward off the Sevier Fold and Thrust Belt. Some authors have referred the upper portion of the Canaan Peak Formation to the Grand Castle Formation (Goldstrand, 1990). However, because it is now known that at least the lower two members of the Grand Castle Formation in its type area are Santonian or early Campanian in age, it would seem prudent to restrict the use of the term Grand Castle Formation from the Kaiparowits Basin.

Recent detrital zircon analysis of the Canaan Peak Formation by Larsen et al. (2010) reveals that it has a dramatically different provenance than the underlying Kaiparowits Formation. Recycled Proterozoic zircons, instead of Mesozoic magmatic arc–derived populations, dominate the assemblage. This study indicates a major change in fluvial drainage patterns and source areas between Kaiparowits and Canaan Peak time, mostly consistent with the hypothesis of Goldstrand (1990, 1994), although indicating a more important Sevier than southeastern Nevada source.

The almost total disappearance of volcanic clasts and a shift in paleocurrent directions recorded in the upper portion of the Canaan Peak (referred in the Kaiparowits Basin to the Grand Castle Formation by Goldstrand, 1990) suggest that another major source area change occurred during its deposition. The mostly likely explanation for this reorganization is Laramide tectonics-driven basin partitioning that led to isolation of the Delfonte Volcanic province from southern Utah. Units overlying the Canaan Peak all show a dramatic shift in sediment composition that indicates a quartz-rich recycled orogen source, almost certainly locally derived from the Wah Wah and Blue Mountain thrust sheets.

Jinnah et al. (2009) and Larsen et al. (2010) showed that detrital zircons in the Wahweap and Kaiparowits formations have two Cretaceous age peaks, one at around 82 Ma and another at around 100 Ma. The source for the 82 Ma population is most likely the Sierra Nevada batholith or associated volcanics, while the 100 Ma population may be from the Sierra

Nevada, or more probably from the Delfonte Volcanic field to the southeast. Only the 100 Ma zircon population continues into the Canaan Peak Formation, demonstrating the loss of the more distal source area (and supporting the Delfonte source hypothesis), probably as a result of uplift-driven basin partitioning. Regardless of the age of the Canaan Peak Formation, the significant known Cretaceous fossil record ends at the top of the underlying Kaiparowits Formation, and contact between the lower and upper members of the Canaan Peak Formation terminates the Cretaceous depositional system.

ACKNOWLEDGMENTS

We thank the support of our home organizations, Grand Staircase–Escalante National Monument, University of Queensland, and the University of Northern Florida. We are also indebted to the constructive criticism given through reviews by T. Lawton and J. Eaton. Helpful discussions were held with R. Biek and J. Kirkland of the Utah Geological Survey and M. Loewen of the Natural History Museum of Utah.

REFERENCES CITED

Albright, L. B., and A. L. Titus. 2009. Locating the Late Cretaceous Santonian–Campanian stage boundary in Grand Staircase–Escalante National Monument using magnetic polarity stratigraphy; p. 14 in A. L. Titus (ed.) Advances in Western Interior Late Cretaceous Paleontology and Geology (St. George, Utah) Abstracts with Programs.

Albright, L. B., D. D. Gillette, and A. L. Titus. 2007a. Plesiosaurs from Upper Cretaceous (Cenomanian–Turonian) Tropic Shale of southern Utah, part 1: Pliosauridae. Journal of Vertebrate Paleontology 27:31–40.

Albright, L. B., D. D. Gillette, and A. L. Titus. 2007b. Plesiosaurs from Upper Cretaceous (Cenomanian–Turonian) Tropic Shale of southern Utah, part 2: Polycotylidae. Journal of Vertebrate Paleontology 27:41–58.

Allen, J. L., and C. L. Johnson. 2010a. Facies control on sandstone composition (and influence of statistical methods on interpretations) in the John Henry Member, Straight Cliffs Formation, Southern Utah, U.S.A. Sedimentary Geology 230:60–76.

Allen, J. L., and C. L. Johnson. 2010b. Sedimentary facies, paleoenvironments, and relative sea level changes in the John Henry Member, Cretaceous Straight Cliffs Formation, Southern Utah; pp. 225–247 in S. M. Carney, D. E. Tabet, and C. L. Johnson (eds.), Geology of South-Central Utah. Utah Geological Association Guidebook 39.

am Ende, B. A. 1991. Depositional environments, palynology, and age of the Dakota Formation, south-central Utah; pp. 65–83 in J. D Nations and J. G. Eaton (eds.), Stratigraphy, Depositional Environments, and Sedimentary Tectonics of the Western Margin, Cretaceous Western Interior Seaway. Geological Society of America Special Paper 260.

Aschoff, J, and R. Steel. 2011. Anomalous clastic wedge development during the Sevier–Laramide transition, North American Cordilleran foreland basin, USA. Geological Society of America Bulletin 123:1822–1835.

Biek, R. F., J. J. Anderson, P. D. Rowley, and F. Maldonado. 2011. Interim Geologic Map of the West Part of the Panguitch 30'×60' Quadrangle, Garfield, Iron, and Kane Counties, Utah – Year 3 Progress Report. Utah Geological Survey Open-File Report 585.

Biek, R. F., G. C. Willis, M. D. Hylland, and H. H. Doelling. 2003. Geology of Zion National Park;

pp. 107–137 in D. A. Sprinkel, T. C. Chidsey, and P. B. Anderson (eds.), Geology of Utah's Parks and Monuments (2nd edition). Utah Geological Association Publication 28.

Biek, R. F., F. Maldonado, D. W. Moore, J. J. Anderson, P. D. Rowley, V. S. Williams, D. Nealey, and E. G. Sable. 2010. Interim geologic map of the west part of the Panguitch 30'×60' Quadrangle, Garfield, Iron, and Kane Counties, Utah – Year 2 Progress Report. Utah Geological Survey Open-File Report 277.

Bobb, M. C. 1991. The Calico bed, Upper Cretaceous, southern Utah; a fluvial sheet deposit in the Western Interior foreland basin and its relationship to eustacy and tectonics. M.Sc. thesis, University of Colorado, Boulder, Colorado.

Bowers, W. E. 1972. The Canaan Peak, Pine Hollow, and Wasatch Formations in the Table Cliff Region, Garfield County, Utah. U.S. Geological Survey Bulletin 1331-B.

Bowers, W. E. 1990. Geologic map of Bryce Canyon National Park and vicinity, southwestern Utah. U.S. Geological Survey Map I-2108.

Carpenter, D. G. 1989. Geology of the North Muddy Mountains, Clark County, Nevada, and regional structural synthesis; fold-thrust and Basin-Range structure in southern Nevada, Southwest Utah and Northwest Arizona. M.Sc. thesis, Oregon State University, Corvallis, Oregon.

Castle, J. W., F. J. Molz, S. Lu, and C. L. Dinwiddie. 2004. Sedimentology and fractal-based analysis of permeability data, John Henry Member, Straight Cliffs Formation (Upper Cretaceous), Utah, U.S.A. Journal of Sedimentary Research 74:270–284.

Christensen, A. E. 2005. Sequence stratigraphy, sedimentology and provenance of the Drip Tank Member of the Straight Cliffs Formation, Kaiparowits Formation, Southern Utah. M.Sc. thesis, New Mexico State University, Las Cruces, New Mexico.

Cobban, W. A. 1976. Ammonite record from the Mancos Shale of the Castle Valley–Price–Woodside area, east-central Utah. Brigham Young University Geology Studies 22:117–126.

Cobban, W. A., J. D. Obradovich, I. Walaszcyk, and K. C. McKinney. 2006. A USGS Zonal Table for the Upper Cretaceous Middle Cenomanian–Maastrichtian of the Western

Interior of the United States Based on Ammonites, Inoceramids, and Radiometric Ages. U.S. Geological Survey Open-File Report 2006-1250.

Cobban, W. A., T. S. Dyman, G. L. Pollock, K. I. Takahashi, L. E. Davis, and D. B. Riggin. 2000. Inventory of dominantly marine and brackish-water fossils from Late Cretaceous rocks in and near Grand Staircase–Escalante National Monument, Utah; pp. 579–589 in D. A. Sprinkel, T. C. Chidsey, and P. B. Anderson (eds.), Geology of Utah's Parks and Monuments. Utah Geological Association Publication 28.

Currie, B. S. 1997. Sequence stratigraphy of nonmarine Jurassic–Cretaceous rocks, central Cordilleran foreland-basin system. Geological Society of America Bulletin 109:1206–1222.

Davis, B. M. 2007. A revision of "pediomyid" marsupials from the Late Cretaceous of North America. Acta Palaeontologica Polonica 52:217–256.

DeCelles, P. G. 2004. Late Jurassic to Eocene evolution of the Cordilleran thrust belt and foreland basin system, western U.S.A. American Journal of Science 304:105–168.

DeCelles, P. G., and K. A. Giles. 1996. Foreland basin systems. Basin Research 8:105–123.

Doelling, H. H. 1975. Geology and Mineral Resources of Garfield County, Utah. Utah Geological and Mineral Survey Bulletin 107.

Doelling, H. H., and R. L. Graham. 1972. Southwestern Utah Coal Fields: Alton, Kaiparowits, Plateau and Kolob-Harmony. Utah Geological and Mineral Survey Monograph Series No.1.

Doelling, H. H., and D. D. Davis. 1989. The Geology of Kane County, Utah: Geology, Mineral Resources, Geologic Hazards. Utah Geological and Mineral Survey Bulletin 124.

Doelling, H. H., R. E. Blackett, A. H. Hamblin, J. D. Powell, and G. L. Pollock. 2003. Geology of Grand Staircase–Escalante National Monument, Utah; pp. 189–231 in D. A. Sprinkel, T. C. Chidsey, and P. B. Anderson (eds.), Geology of Utah's Parks and Monuments (2nd edition). Utah Geological Association Publication 28.

Dyman, T. S., W. A. Cobban, A. L. Titus, J. D. Obradovich, L. E. Davis, R. L. Eves, G. L. Pollock, K. I. Takahashi, and T. C. Hester. 2002a New biostratigraphic and radiometric ages for Albian–Turonian Dakota Formation and Tropic

Shale at Grand Staircase–Escalante National Monument and Iron Springs Formation near Cedar City, Parowan, and Gunlock, Utah. Geological Society of America Abstracts with Programs 34(4):A13.

Dyman, T. S., W. A. Cobban, L. E. Davis, R. L. Eves, G. L. Pollock, J. D. Obradovich, A. L. Titus, K. I. Takahashi, T. C. Hester, and D. Cantu. 2002b. Upper Cretaceous marine and brackish water strata at Grand Staircase–Escalante National Monument, Utah; pp. 171–198 in W. R. Lund (ed.), Field Guide to Geologic Excursions in Southwestern Utah and Adjacent Areas of Arizona and Nevada. U.S. Geological Survey Open-File Report 02–172.

Eaton, J. G. 1990. Stratigraphic revision of Campanian (Upper Cretaceous) rocks in the Henry Basin, Utah. Mountain Geologist 27:27–38.

Eaton, J. G. 1991. Biostratigraphic framework for Upper Cretaceous rocks of the Kaiparowits Plateau, southern Utah; pp. 47–63 in J. D. Nations and J. G. Eaton (eds.), Stratigraphy, Depositional Environments, and Sedimentary Tectonics of the Western Margin, Cretaceous Western Interior Seaway. Geological Society of America Special Paper 260.

Eaton, J. G. 1999. Vertebrate paleontology of the Iron Springs Formation, Upper Cretaceous, southwestern Utah; pp. 339–343 in D. D. Gillette (ed.), Vertebrate Paleontology in Utah. Utah Geological Survey Miscellaneous Publication 99–1.

Eaton, J. G. 2006. Late Cretaceous mammals from Cedar Canyon, southwestern Utah; pp. 373–402 in S. G. Lucas and R. M. Sullivan (eds.), Late Cretaceous Vertebrates from the Western Interior. New Mexico Museum of Natural History and Science Bulletin 35.

Eaton, J. G., and J. D. Nations. 1991. Introduction; tectonic setting along the margin of the Cretaceous Western Interior Seaway, southwestern Utah and northern Arizona; pp. 1–8 in J. D. Nations and J. G. Eaton (eds.), Stratigraphy, Depositional Environments, and Sedimentary Tectonics of the Western Margin, Cretaceous Western Interior Seaway. Geological Society of America Special Paper 260.

Eaton, J. G., R. L. Cifelli, J. H. Hutchison, J. I. Kirkland, and J. M. Parrish. 1999a. Cretaceous vertebrate faunas from the Kaiparowits Plateau, south central Utah; pp. 345–353 in D. D. Gillette (ed.), Vertebrate Paleontology in Utah. Utah Geological Survey Miscellaneous Publication 99–1.

Eaton, J. G., S. Diem, J. D. Archibald, C. Schierup, and H. Munk. 1999b. Vertebrate paleontology of the Upper Cretaceous rocks of the Markagunt Plateau, southwestern Utah; pp. 323–333 in D. D. Gillette (ed.), Vertebrate Paleontology in Utah. Utah Geological Survey Miscellaneous Publication 99–1.

Eaton, J. G., J. I. Kirkland, J. H. Hutchinson, R. Denton, R. C. O'Neill, and M. J. Parrish. 1997. Nonmarine extinction across the Cenomanian–Turonian (C-T) boundary, southwestern Utah, with a comparison to the Cretaceous–Tertiary (K-T) extinction event. Geological Society of America Bulletin 59:129–151.

Eaton, J. G., J. Laurin, J. I. Kirkland, N. E. Tibert, R. M. Leckie, B. B. Sageman, P. M. Goldstrand, D. W. Moore, A. W. Straub, W. A. Cobban,

and J. D. Dalebout. 2001. Cretaceous and early Tertiary geology of Cedar and Parowan canyons, western Markagunt Plateau, Utah; pp. 337–363 in M. C. Erskine, J. E. Faulds, J. M. Bartley, and P. D. Rowley (eds.), The Geologic Transition, High Plateaus to Great Basin; a Symposium and Field Guide; the Mackin Volume. Guidebook–Pacific Section, American Association of Petroleum Geologists 78.

Elder, W. P. 1989. Molluscan extinction patters across the Cenomanian–Turonian stage boundary in the Western Interior of the United States. Paleobiology 15:299–320.

Elder, W. P. 1991. Molluscan paleoecology and sedimentation patterns of the Cenomanian–Turonian extinction interval in the southern Colorado Plateau region; pp. 113–137 in J. D. Nations and J. G. Eaton (eds.), Stratigraphy, Depositional Environments, and Sedimentary Tectonics of the Western Margin, Cretaceous Western Interior Seaway. Geological Society of America Special Paper 260.

Elder, W. P., and J. I. Kirkland. 1993. Cretaceous paleogeography of the Colorado Plateau and adjacent areas; pp. 129–151 in M. Morales (ed.), Aspects of Mesozoic Geology and Paleontology of the Colorado Plateau. Museum of Northern Arizona Bulletin 59.

Elder, W. P., and J. I. Kirkland. 1994. Cretaceous paleogeography of the southern Western Interior region; pp. 415–440 in M. V. Caputo, J. A. Peterson, and K. J. Franczyk (eds.), Mesozoic System of the Rocky Mountain Region, U.S.A. Rocky Mountain Section, Society for Sedimentary Geology.

Elder, W. P., E. R. Gustason, and B. B. Sageman. 1994. Correlation of basinal carbonate cycles to nearshore parasequences in the Late Cretaceous Greenhorn seaway, Western Interior, U.S.A. Geological Society of America Bulletin 106:892–902.

Foster, J. R., A. L. Titus, G. F. Winterfeld, M. C. Hayden, and A. H. Hamblin. 2001. Paleontological Survey of the Grand Staircase–Escalante National Monument, Garfield and Kane Counties, Utah. Utah Geological Survey Special Publication SS-99.

Gallin, W. N., C. L. Johnson, and J. L. Allen. 2010. Fluvial and marginal marine architecture of the John Henry Member, Straight Cliffs Formation, Kelly Grade of the Kaiparowits Plateau, south-central Utah; pp. 248–275 in S. M. Carney, D. E. Tabet, and C. L. Johnson (eds.), Geology of South-Central Utah. Utah Geological Association Guidebook 39.

Gillette, D. D., M. C. Hayden, and A. L. Titus. 1999. Occurrence and biostratigraphic framework of a plesiosaur from the Upper Cretaceous Tropic Shale of southwestern Utah; pp. 269–273 in D. D. Gillette (ed.), Vertebrate Paleontology in Utah. Utah Geological Survey Miscellaneous Publication 99–1.

Goldstrand, P. M. 1990. Stratigraphy and Paleogeography of Late Cretaceous and Paleogene Rocks of Southwest Utah. Utah Geological and Mineral Survey Miscellaneous Publication 90–2.

Goldstrand, P. M. 1994. Tectonic development of Upper Cretaceous to Eocene strata of southwestern Utah. Geological Society of America Bulletin 106:145–154.

Goldstrand, P. M., and D. J. Mullet. 1997. The Paleocene Grand Castle Formation–a new formation on the Markagunt Plateau of southwestern Utah; pp. 59–78 in F. Maldonado and L. D. Nealey (eds.), Geologic Transition in Southeastern Nevada, Southwestern Utah, and Northwestern Arizona. U.S. Geological Survey Bulletin 2153.

Gregory, H. E., and R. C. Moore. 1931. The Kaiparowits Region; Geographic and Geologic Reconnaissance of Parts of Utah and Arizona. U.S. Geological Survey Professional Paper 164.

Gustason, E. R. 1989. Stratigraphy and sedimentology of the Middle Cretaceous (Albian–Cenomanian) Dakota Formation, southwestern Utah. Ph.D. dissertation, University of Colorado, Boulder, Colorado.

Hancock, J. M., and E. G. Kauffman. 1979. The great transgressions of the Late Cretaceous. Journal of the Geological Society of London 136:175–186.

Haq, B. U., J. Hardenbol, and P. R Vail. 1988. Mesozoic and Cenozoic chronostratigraphy and cycles of sea level change; pp. 72–108 in C. K. Wilgus, B. S. Hastings, C. G. S. C. Kendall, H. W. Posamentier, C. A. Ross, and J. C. Van Wagoner (eds.), Sea-Level Changes–An Integrated Approach. Society for Sedimentary Geology Special Publication 42.

Heller, P. L., C. L. Angevine, N. S. Winslow, and C. Paola. 1988. Two–phase stratigraphic model of foreland-basin sequences. Geology 16:501–504.

Hettinger, R. D. 1993. Sedimentological Descriptions and Geophysical Logs of Two 300 m Cores Collected from the Straight Cliffs Formation of the Kaiparowits Plateau, Kane County, Utah. U.S. Geological Survey Open-File Report 93–270.

Hettinger, R. D. 1995. Sedimentological Descriptions and Depositional Interpretations, in Sequence Stratigraphic Context, of Two 300-meter Cores from the Upper Cretaceous Straight Cliffs Formation, Kaiparowits Plateau, Kane County, Utah. U.S. Geological Survey Bulletin 2115-A.

Hettinger, R. D. 2000. A summary of coal distribution and geology in the Kaiparowits Plateau, Utah; pp. J1–J17 in Geologic Assessment of Coal in the Colorado Plateau: Arizona, Colorado, New Mexico, and Utah. U.S. Geological Survey Professional Paper 1625-B.

Hettinger, R. D., P. J. McCabe, and K. W. Shanley. 1994. Detailed facies anatomy of transgressive and highstand systems tracts from the Upper Cretaceous of southern Utah, U.S.A.; pp. 235–257 in H. W. Posmentier and P. Weimer (eds.), Recent Advances in Siliciclastic Sequence Stratigraphy. American Association of Petroleum Geologists Memoir 58.

Hettinger, R. D., L. N. R. Roberts, L. R. H. Biewick, and M. A. Kirschbaum. 2000. Geologic overview and resource assessment of coal in the Kaiparowits Plateau, southern Utah; pp. T1–T69 in Geologic Assessment of Coal in the Colorado Plateau: Arizona, Colorado, New Mexico, and Utah. U.S. Geological Survey Professional Paper 1625-B.

Hicks, J. F., J. D. Obradovich, and L. Tauxe. 1999. Magnetostratigraphy, isotopic age calibration and intercontinental correlation of the Red Bird section of the Pierre Shale, Niobrara County, Wyoming, U.S.A. Cretaceous Research 20:1–27.

Hintze, L. F. 1988. Geologic History of Utah. Brigham Young University Geology Studies Special Publication 7.

Hunt, G. J., R. F. Biek, D. L. DeBlieux, S. K. Madsen, A. R. C. Milner, E. M. Roberts, and J. G. Eaton. 2011. Dinosaur tracks confirm Late Cretaceous age for the lower Grand Castle Formation, Iron County, Utah: stratigraphic and tectonic implications. Abstracts with Programs, Geological Society of America, joint Cordilleran–Rocky Mountain Section Annual Meeting, Logan, Utah 43(4):16.

Hylland, M. D. 2010. Geologic map of the Clear Creek Mountain quadrangle, Kane County, Utah. Utah Geological Survey Map 245.

Jinnah, Z. A., and E. M. Roberts. 2011. Facies associations, paleoenvironments, and base-level changes in the Upper Cretaceous Wahweap Formation, Utah, U.S.A. Journal of Sedimentary Research 81:266–283.

Jinnah, Z. A., E. M. Roberts, A. D. Deino, J. S. Larsen, P. K. Link, and C. M. Fanning. 2009. New ^{40}Ar/^{39}Ar and detrital zircon U-Pb ages for the Upper Cretaceous Wahweap and Kaiparowits formations on the Kaiparowits Plateau, Utah: implications for regional correlation, provenance, and biostratigraphy. Cretaceous Research 30:287–299.

Kennedy, W. J., and W. A. Cobban. 1991. Coniacian Ammonite Faunas from the United States Western Interior. Palaeontological Association (London) Special Papers in Palaeontology 45.

Kiel, S., F. Wiese, and A. L. Titus. 2012. Shallow-water methane-seep faunas in the Cenomanian Western Interior Seaway: No evidence for onshore–offshore adaptations to deep-sea vents. Geology 40:839–842.

Kirkland, J. I., and S. K. Madsen. 2007. The Lower Cretaceous Cedar Mountain Formation, Eastern Utah: The View Up an Always Interesting Learning Curve. Utah Geological Association Publication 35.

Kirschbaum, M. A., and P. J. McCabe. 1992. Controls on the accumulation of coal and on the development of anastomosed fluvial systems in the Cretaceous Dakota Formation of southern Utah. Sedimentology 39:581–598.

Kuiper, K. F., A. Deino, F. J. Hilgen, W. Krigsman, P. R. Renne, and J. R. Wijbrans. 2008. Synchronizing rock clocks of Earth history. Science 320:500–504.

Larsen, J. S. 2007. Facies and provenance of the Pine Hollow Formation: implications for Sevier Foreland Basin evolution and the Paleocene climate of southern Utah. M.Sc. thesis, Idaho State University, Pocatello, Idaho.

Larsen, J. S., P. K. Link, E. M. Roberts, L. Tapanila, and C. M. Fanning. 2010. Cyclic stratigraphy of the Paleocene Pine Hollow and detrital zircon provenance of Campanian to Eocene sandstones of the Kaiparowits and Table Cliffs basins, south-central Utah; pp. 194–224 in S. M. Carney, D. E. Tabet, and C. L. Johnson (eds.), Geology of South-Central Utah. Utah Geological Association Guidebook 39.

Lawrence, J. C. 1965. Stratigraphy of the Dakota and Tropic formations of Cretaceous age in southern Utah; pp. 71–91 in Geology and Resources of South-Central Utah. Utah

Geological Society Guidebook to the Geology of Utah 19.

Lawton, T. F., and B. A. Bradford. 2011. Correlation and provenance of Upper Cretaceous (Campanian) fluvial strata, Utah, U.S.A., from zircon U-Pb geochronology and petrography. Journal of Sedimentary Research 82:495–512.

Lawton, T. F., S. L. Pollock, and R. A. J. Robinson. 2003. Integrating sandstone petrology and nonmarine sequence stratigraphy: application to the late Cretaceous fluvial systems of southwestern Utah, U.S.A. Journal of Sedimentary Research 73:389–406.

Little, W. W. 1997. Tectonic and eustatic controls on cyclical fluvial patterns, Upper Cretaceous strata of the Kaiparowits Basin Utah; pp. 489–504 in L. M. Hill (ed.), Learning from the Land: Grand Staircase–Escalante National Monument Science Symposium Proceedings. U.S. Department of the Interior, Bureau of Land Management.

Ludvigson, G. A., B. J. Witzke, R. M. Joeckel, R. L. Ravn, P. L. Phillips, L. A. González, and R. L. Brenner. 2010. New insights on the sequence stratigraphic architecture of the Dakota Formation in Kansas–Nebraska–Iowa from a decade of sponsored research activity. Kansas Geological Survey Bulletin 258:1–35.

Meek, F. B., and F. V. Hayden. 1856. Descriptions of new species of gastropoda and cephalopoda from the Cretaceous formations of Nebraska Territory. Proceedings of the Academy of Natural Sciences, Philadelphia 8:63–126.

Meek, F. B., and F. V. Hayden. 1861. Descriptions of new Lower Silurian, Jurassic, Cretaceous, and some Tertiary fossils collected in Nebraska Territory with some remarks on the rocks from which they were obtained. Proceedings of the Academy of Natural Sciences, Philadelphia 13:415–447.

Molenaar, C. M., W. A. Cobban, E. A. Merewether, C. L. Pillmore, D. G. Wolfe, and J. M. Holbrook. 2002. Regional stratigraphic cross sections of Cretaceous rocks from east-central Arizona to the Oklahoma Panhandle. U.S. Geological Survey Miscellaneous Field Studies Map 2382.

Nichols, D. J. 1997. Palynology and ages of some Upper Cretaceous formations in the Markagunt and northwestern Kaiparowits Plateaus, southwestern Utah; pp. 81–95 in F. Maldonado and L. D. Nealey (eds.), Geologic Studies in the Basin and Range–Colorado Plateau Transition Zone in Southeastern Nevada, Southwestern Utah, and Northwestern Arizona, 1995. U.S. Geological Survey Bulletin 2153-E.

Obradovich, J. D. 1993. A Cretaceous time scale; pp. 379–398 in W. G. E. Caldwell and E. G. Kauffman (eds.), Evolution of the Western Interior Basin. Geological Association of Canada Special Paper 39.

Obradovich, J. D., T. Matsumoto, T. Nishida, and Y. Inoue. 2002. Integrated biostratigraphic and radiometric scale on the Lower Cenomanian (Cretaceous) of Hokkaido, Japan. Proceedings of the Japan Academy, Series B, Physical and Biological Sciences 78:149–153.

Ogg, J. G., F. P. Agterberg, and F. M. Gradstein. 2004. The Cretaceous period; pp. 344–383 in F. M. Gradstein, J. G. Ogg, and A. G. Smith

(eds.), A Geologic Time Scale. Cambridge University Press, Cambridge.

Peterson, F. 1969a. Cretaceous Sedimentation and Tectonism in the Southeastern Kaiparowits Region, Utah. U.S. Geological Survey Open-File Report 69–202.

Peterson, F. 1969b. Four New Members of the Upper Cretaceous Straight Cliffs Formation in the Southeastern Kaiparowits Region, Kane County, Utah. U.S. Geological Survey Bulletin 1274-J.

Peterson, F., and A. R. Kirk. 1977. Correlation of the Cretaceous rocks in the San Juan, Black Mesa, Kaiparowits and Henry basins, southern Colorado Plateau; pp. 167–189 in J. E. Fassett (ed.), San Juan Basin III: Northwestern New Mexico. New Mexico Geological Society Guidebook, 28th Field Conference.

Pollock, S. L. 1999. Provenance, geometry, lithofacies, and age of the Upper Cretaceous Wahweap Formation, Cordilleran foreland basin, southern Utah. M.Sc. thesis, New Mexico State University, Las Cruces, New Mexico.

Posamentier, H. W., M. T. Jervey, and P. R. Vail. 1988. Eustatic controls on clastic deposition I–conceptual framework; pp. 109–124 in C. K. Wilgus, B. S. Hastings, C. G. S. C. Kendall, H. W. Posamentier, C. A. Ross, and J. C. Van Wagoner (eds.), Sea-Level Changes–An Integrated Approach. Society for Sedimentary Geology Special Publication 42.

Roberts, E. M. 2007. Facies architecture and depositional environments of the Upper Cretaceous Kaiparowits Formation, southern Utah. Sedimentary Geology 197:207–233.

Roberts, E. M., A. D. Deino, and M. A. Chan. 2005. ^{40}Ar/^{39}Ar age of the Kaiparowits Formation, southern Utah, and correlation of coeval strata and faunas along the margin of the Western Interior Basin. Cretaceous Research 26:307–318.

Rogers, R. R., and M. E. Brady. 2010. Origins of microfossil bonebeds: insights from the Upper Cretaceous Judith River Formation of north-central Montana. Paleobiology 6:80–112.

Ryer, T. A. 1991. Stratigraphy, facies, and depositional history of the Ferron Sandstone in the canyon of Muddy Creek, east-central Utah; pp. 45–54 in T. C. Chidsey Jr. (ed.), Geology of East-Central Utah. Utah Geological Association Publication 19.

Sargent, K. A., and D. E. Hansen. 1982. Bedrock geologic map of the Kaiparowits coal-basin area, Utah. U.S. Geological Survey Miscellaneous Investigation Series Map I-1033-I.

Schmeisser, R. L., and D. D. Gillette. 2009. Unusual occurrence of gastroliths in a polycotylid plesiosaur from the Upper Cretaceous Tropic Shale, southern Utah. Palaios 24:453–459.

Schmitt, G. J., D. A. Jones, and P. M. Goldstrand. 1991. Braided stream deposition and provenance of the Upper Cretaceous–Paleocene(?) Canaan Peak Formation, Sevier Foreland Basin, southwestern Utah; pp. 27–45 in J. D. Nations and J. G. Eaton (eds.), Stratigraphy, Depositional Environments, and Sedimentary Tectonics of the Western Margin, Cretaceous Western Interior Seaway. Geological Society of America Special Paper 260.

Shanley, K. W., and P. J. McCabe. 1991. Predicting facies architecture through sequence

Titus, Roberts, and Albright

stratigraphy; an example from the Kaiparowits Plateau, Utah. Geology 19:742–745.

Shanley, K. W., and P. J. McCabe. 1993. Alluvial architecture in a sequence stratigraphic framework: a case history from the Upper Cretaceous of southern Utah, U.S.A.; pp. 21–56 in S. Flint and I. Bryant (eds.), Quantitative Modeling of Clastic Hydrocarbon Reservoirs and Outcrop Analogues. International Association of Sedimentologists Special Publication 15.

Shanley, K. W., and P. J. McCabe. 1995. Sequence stratigraphy of Turonian–Santonian strata, Kaiparowits Plateau, Southern Utah, U.S.A.: implications for regional correlation and foreland basin evolution; pp. 103–136 in J. C. Van Wagoner, and G. T. Bertram (eds.), Sequence Stratigraphy of Foreland Basin Deposits, Outcrop and Subsurface Examples from the Cretaceous of North America. American Association of Petroleum Geologists Memoir 64.

Shanley, K. W., P. J. McCabe, and R. D. Hettinger. 1992. Tidal influence in Cretaceous fluvial strata from Utah, U.S.A.–a key to sequence stratigraphic interpretation. Sedimentology 39:905–930.

Simpson, E. L., M. C. Wizevich, H. L. Hilbert-Wolf, S. E. Tindall, J. J. Bernard, and W. S. Simpson. 2009. An Upper Cretaceous sag pond deposit: Implications for recognition of local seismicity and surface rupture along the Kaibab monocline, Utah. Geology 37:967–970.

Simpson, E. L., H. L. Hilbert-Wolf, W. S. Simpson, S. E. Tindall, J. J. Bernard, T. A. Jenesky, and M. C. Wizevich. 2008. The interaction of aeolian and fluvial processes during deposition of the Upper Cretaceous capping sandstone member, Wahweap Formation, Kaiparowits

Basin, Utah, U.S.A. Palaeogeography, Palaeoclimatology, Palaeoecology 270:19–28.

Simpson, E. L., H. L. Hilbert-Wolf, M. C. Wizevich, S. E. Tindall, B. R. Fasinski, L. P. Storm, and M. D. Needle. 2010. Predatory digging behavior by dinosaurs. Geology 38:699–702.

Sprinkle, D. A., M. P. Weiss, R. W. Fleming, and G. L. Waanders. 1999. Redefining the Lower Cretaceous Stratigraphy within the Central Utah Foreland Basin. Utah Geological Survey Special Study 97.

Tibert, N. E., R. M. Leckie, J. G. Eaton, J. I. Kirkland, J. P. Colin, E. L. Leithold, and M. E. McCormic. 2003. Recognition of relative sea-level change in Upper Cretaceous coal bearing strata–a paleoecological approach using agglutinated foraminifera and ostracods to detect key stratigraphic surfaces; pp. 263–299 in H. C. Olson and R. M. Leckie (eds.), Micropaleontological Proxies for Sea-Level Change and Stratigraphic Discontinuities. Society of Economic Paleontologists and Mineralogists Special Publication 75.

Tilton, T. L. 2001. Geologic Map of the Alton Quadrangle, Kane County, Utah. Utah Geological Survey Miscellaneous Publication 01–4.

Tindall, S. E., L. P. Storm, T. A. Jenesky, and E. L. Simpson. 2010. Growth faults in the Kaiparowits basin, Utah, pinpoint initial Laramide deformation in the western Colorado Plateau. Lithosphere 2:221–231.

Titus, A. L. 2002. Late Cenomanian (Late Cretaceous *Sciponoceras gracile* biozone) paleogeographic evolution of the Grand Staircase–Escalante National Monument region; implications of recent advances in high-resolution ammonoid biostratigraphy. Abstracts

with Programs, Geological Society of America, Rocky Mountain Section, 54th Annual Meeting, Cedar City, Utah 34(4):13.

Titus, A. L., L. B. Albright, and R. S. Barclay. 2010. The first record of Cenomanian (Late Cretaceous) insect body fossils from the Kaiparowits Basin, northern Arizona; pp. 127–132 in L. Hill (ed.), Learning from the Land: Grand Staircase–Escalante National Monument Science Symposium (Cedar City, Utah) Proceedings.

Titus, A. L., J. D. Powell, E. M. Roberts, S. D. Sampson, S. L. Pollock, J. I. Kirkland, and L. B. Albright. 2005. Late Cretaceous stratigraphy, depositional environments, and macrovertebrate paleontology of the Kaiparowits Plateau, Grand Staircase–Escalante National Monument, Utah; pp. 101–128 in J. Pederson and C. M. Dehler (eds.), Interior Western United States. Geological Society of America Field Guide 6.

Uličný, D. 1999. Sequence stratigraphy of the Dakota Formation (Cenomanian), southern Utah; interplay of eustacy and tectonics in a foreland basin. Sedimentology 46:807–836.

Vaninetti, G. E. 1978. Coal stratigraphy of the John Henry Member of the Straight Cliffs Formation, Kaiparowits Plateau, Utah. M.Sc. thesis, University of Utah, Salt Lake City, Utah.

Willis, G. C. 1999. The Utah thrust system; an overview; pp. 1–10 in L. E. Spangler and C. J. Allen (eds.), Geology of Northern Utah and Vicinity. Utah Geological Association Publication 27.

Accumulation of Organic Carbon–Rich Strata along the Western Margin and in the Center of the North American Western Interior Seaway during the Cenomanian–Turonian Transgression

Walter E. Dean, Erle G. Kauffman, and Michael A. Arthur

3

DURING THE CRETACEOUS PERIOD, THE NORTH AMERI-
can Western Interior Seaway occupied a rapidly subsiding
north–south-trending basin that was characterized by substantial clastic sediment input derived from uplifted volcanic terranes of the Sevier orogenic belt to the west. At
times of maximum transgression, the shoreline of the Western Interior Seaway extended as far west as western Utah
and western Arizona, where thick deposits of fine-grained
sediment rich in organic carbon (OC) were deposited. The
Tropic Shale and correlative Tununk Shale Member of the
Mancos Shale of southern Utah accumulated within a few
hundred kilometers offshore from the western margin of the
Western Interior Seaway; they record a part of the Greenhorn transgressive–regressive cycle that lasted about 4 Ma
during which the Western Interior Seaway expanded to its
maximum extent and then partially contracted. The entire
thickness of the Tropic Shale was sampled in a 280-m-long
core that was drilled and continuously cored at the base of
the Kaiparowits Plateau near the town of Escalante, Utah
(USGS #1 Escalante). The Tropic Shale, although called a
shale, is more correctly a marlstone, containing up to 60%
$CaCO_3$, with an average of 34% in the Escalante core. Biostratigraphy and bentonite beds have allowed us to correlate
peaks in concentration of $CaCO_3$ with individual limestone
beds in the Bridge Creek Limestone Member of the Greenhorn Formation in eastern Colorado and western Kansas.
The Tropic Shale in the Escalante core is characterized by
a generally upward increasing OC content, from a low value
of <1% to a high value of 3%. The Cenomanian–Turonian
(C/T) boundary is a time horizon marking a major oceanic
anoxic event. It occurs about 20 m above the base of the
Escalante core and is marked by a positive carbon isotope
excursion, typical of the C/T oceanic anoxic event around
the world. However, maximum accumulation of OC occurs
much higher in the section, long after maximum Greenhorn
transgression. Therefore, the Tropic Shale in the Escalante
core apparently has recorded both global (oceanic anoxic

event) and local events along the western shore of the Western Interior Seaway.

INTRODUCTION

The Cretaceous Period (approximately 136–65 Ma) of Earth
history offers a significant opportunity for major contributions to understanding global processes and their variations.
Cretaceous marine and terrestrial strata deposited during
this greenhouse period are extremely widespread in outcrop,
subcrop, and ocean basins. Many of the subcrop sections are
available by shallow to intermediate (up to 1000 m) continental drilling, and ocean basin sites are accessible by deep-sea
drilling. It is estimated that Cretaceous source rocks are responsible for more than 70% of the world's in place reserves
of crude oil (Tissot, 1979). Understanding the origin and
distribution of Cretaceous organic-rich sequences is tanta-
mount to understanding the origin and distribution of much
of the world's oil and gas, discovered and undiscovered. In
addition, Cretaceous strata contain major reserves of coal,
kaolinite, phosphorite, bauxite, and manganese. Accurate
prediction of the availability of such resources and an understanding of their distribution requires models based on a
comprehensive knowledge of the Cretaceous world.

The middle Cretaceous between 120 Ma and 80 Ma
(late Albian to early Campanian) is characterized by several
globally widespread episodes of organic carbon (OC) burial
in marine sequences (Arthur and Schlanger, 1979; Jenkyns,
1980, Arthur, Schlanger, and Jenkyns, 1987; Schlanger et al.,
1987; Arthur et al., 1990). These episodes represent periods of
widespread oxygen deficiency in oceanic mid- and/or deep-
water masses that have been termed oceanic anoxic events
(OAEs). One of these OAEs at the Cenomanian–Turonian
(C/T) boundary (93.4 Ma) is known as the Bonarelli Horizon
after the C/T boundary in Italy. The widespread occurrence
of OAEs in time and space within the middle Cretaceous
may imply fundamental changes in oceanic circulation or

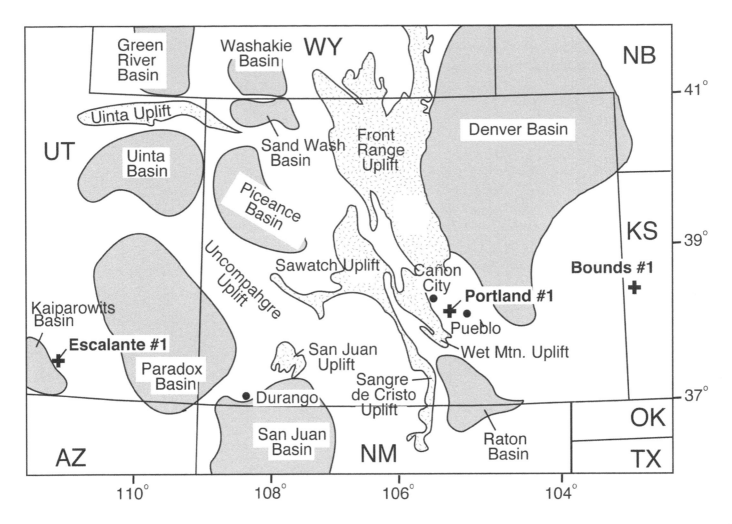

3.1. Location map showing basins and uplifts of the central Rocky Mountain region and locations of cores discussed in this chapter.

the rate and mode of delivery of organic matter to the ocean. The origins of the OAEs are not known for certain, but available data suggest that such events resulted from some combination of higher phytoplankton productivity and enhanced preservation under oxygen-depleted deep-water masses. Because of the importance of the Cretaceous to so many different disciplines of the geosciences, the Global Sedimentary Geology Program identified Cretaceous Resources, Events, and Rhythms as its first major international project (Ginsburg and Beaudoin, 1990).

During the Cretaceous Period, the North American Western Interior Seaway was a north–south arm of the east–west Tethys Sea. The Western Interior Seaway occupied a rapidly subsiding basin that was characterized by substantial clastic sediment input derived from uplifted sedimentary rocks and volcanic terranes of the Sevier orogenic belt to the west. This volcanic belt also provided abundant windblown volcanic ash that is manifested in the stratigraphic sequences of the Western Interior Seaway as thick (>3 m) to very thin beds (<1 cm) of altered volcanic ash (bentonite). The western

margin also provided coarse- to fine-grained clastic deposits, many of which are now petroleum reservoirs. At maximum transgression, the Western Interior Seaway extended in length from the Gulf of Mexico to the Arctic, and in width from Utah to Nebraska and the Dakotas. A northeast arm of the Western Interior Seaway connected to the North Atlantic.

The numerous energy-rich Cretaceous foreland basin packages of the Rocky Mountain region of North America were largely deposited under or on the margins of the once extensive Western Interior Seaway (Fig. 3.1). Because of their abundant fossil and energy resources, these Cretaceous successions of the Western Interior Seaway are among the most studied in the world, and knowledge gained from Western Interior Seaway deposits, built up over the years by industry, government, and academic scientists, has provided exceptional temporal and spatial control for working out sequence stratigraphic relationships, as well as providing extensive knowledge on the sedimentology, geochemistry, and paleobiology of these sequences. At times of maximum transgression, the shoreline of the Western Interior Seaway extended as far west as western Utah and western Arizona, where thick deposits of fine-grained sediment were deposited. At the same time, carbonate sediments relatively undiluted by fine-grained clastic sediments from the west were deposited in the deeper, central parts of the Western Interior Seaway in what are now eastern Colorado and New Mexico; the panhandles of Texas and Oklahoma; and Kansas, Nebraska, and the Dakotas. The lithified carbonate sediments are now important hydrocarbon source rocks (Clayton and Swetland, 1980). In the spirit of Cretaceous Resources, Events, and Rhythms project, the Western Interior Seaway Drilling Project was initiated in 1990 (Dean and Arthur, 1998a), which consisted of researchers from the U.S. Geological Survey (USGS), Amoco, and seven academic institutions.

The Western Interior Seaway Drilling Project proposed drilling and continuously coring a series of holes from the basin center in western Kansas and eastern and central Colorado to the western margin in Utah (Fig. 3.1). The first hole to be cored was the Amoco #1 R. K. Bounds well, Greeley County, Kansas, drilled by Amoco (Fig. 3.1; Scott et al., 1998). The second hole to be cored was the USGS #1 Portland well near Cañon City, Colorado, and the third hole was the USGS #1 Escalante well in south-central Utah (Fig. 3.1), both drilled by the USGS. All three cores are stored, and are available to the public, at the USGS Core Research Center, Lakewood, Colorado, and preliminary scientific results from these three cores are in Dean and Arthur (1998b). Data for this report, and for most of the reports in Dean and Arthur (1998b), are available online (http://hurricane.ncdc.noaa.gov/pls/paleo /contribseries.contrib).

Here we develop an understanding of the setting in which basin–margin fine-grained clastic sediments accumulated along the western margin of the Western Interior Seaway at the time of the mid-Cretaceous C/T OAE preserved in the Tropic Shale and correlative Tununk Shale Member of the Mancos Shale. We also want to compare these western clastic facies with the better-known basin-center carbonate rocks. For this comparison, we will focus on the Tropic Shale in the Escalante core in Utah and on the Bridge Creek Limestone Member of the Greenhorn Formation in the Portland core in eastern Colorado and the Bounds core in western Kansas (Fig. 3.1). We will do this by using certain geochemical "fossils" that provide clues to the depositional environmental conditions. Trends in the amount and character of OC in the strata will be related to both local sea-level changes and global events. We will examine how OC enrichment affected isotopic composition of OC, and we will also examine the geochemical cycles of other elements such as Fe, S, and certain trace elements. Another goal of this study is to see whether $CaCO_3$-rich parts of the Tropic Shale can be correlated with individual limestone beds in the Bridge Creek Limestone deposited in the center of the Western Interior Seaway in eastern Colorado and western Kansas (Cobban and Scott, 1972; Meyers, Sageman, and Hinnov, 2001). This correlation of carbonate units will be aided, when possible, by correlation of volcanic ash beds (bentonites) in the Tropic Shale with named bentonite beds in the Bridge Creek Limestone (Elder, 1985). We will place more emphasis on the Tropic Shale because much less is known about the fine-grained clastics of western margin of the Western Interior Seaway than about the basin-center carbonates.

METHODS

Concentrations of total carbon and inorganic carbon were determined by coulometric titration of CO_2 after extraction from the sediment by combustion at 950°C and acid volatilization, respectively (Engleman et al., 1985), in USGS laboratories, Denver, Colorado. Weight percentage inorganic carbon was converted to weight percentage $CaCO_3$ by dividing by 0.12, the fraction of carbon in $CaCO_3$. OC was determined as the difference between total carbon and inorganic carbon. The accuracy and precision for both total carbon and inorganic carbon, determined from hundreds of replicate standards (reagent-grade $CaCO_3$ and a sample of Tropic Shale), usually are better than 0.10 wt%.

Pyrolysis of organic matter was performed with a Rock-Eval pyrolysis instrument for comparison with similar analyses used commonly in the study of ancient black shales. The Rock-Eval pyrolysis method provides a rapid determination of the hydrocarbon (HC) type and abundance in sediments and sedimentary rocks (Espitalié et al., 1977; Peters, 1986). Programmed heating of a sample in a helium atmosphere results in generation of HC and CO_2. The HC generated are then measured with a flame ionization detector, and the CO_2 yield is monitored by a thermal conductivity detector. The yield of free or adsorbed HC is determined by heating the sample in flowing helium 250°C, and is recorded as the first characteristic peak on a pyrogram (S1; mg HC/g sample). The S1 peak is roughly proportional to the content of organic matter that can be extracted from the rock or sediment with organic solvents. The second peak is composed of pyrolytic hydrocarbons generated by thermal breakdown of kerogen as the sample is heated from 250°C to 550°C (S2, mg HC/g sample). Our data will be expressed in terms of a hydrogen index (HI) in which the S2 peak is normalized to the OC content of the sample (mg HC/g OC).

Stable carbon isotope ratios were determined by standard techniques (Pratt and Threlkeld, 1984; Dean, Arthur, and

Dean, Kauffman, and Arthur

Claypool, 1986). Powdered samples for carbon isotope determinations of OC were oven dried at 40°C and reacted with an excess of 0.5 N HCl for 24 hours to dissolve carbonate minerals. The washed residue from each sample was combusted at 1000°C under oxygen pressure in an induction furnace. The resulting CO_2 was dehydrated and purified in a high-vacuum gas-transfer system. All isotope ratios were determined by an isotope ratio mass spectrometer. Results are reported in the standard per mil (‰) δ notation relative to the Vienna Pee Dee belemnite (PDB) marine-carbonate standard, δ^{13}C ‰ = [(R_{sample}/RPDB) − 1] × 10³, where R is the ^{13}C: ^{12}C ratio. Analytical precision of these analyses is ±0.2‰.

Concentrations of total sulfur (TS) were determined with LECO combustion–infrared instrumentation (Baedecker, 1987). For inorganic geochemical analysis, powdered samples were analyzed for concentrations of 29 major, minor, and trace elements by induction-coupled argon-plasma emission spectrometry (Baedecker, 1987). As a measure of precision, 10 splits of one sample of the Tropic Shale were analyzed. The mean standard deviation and coefficient of variation (standard deviation as a percentage of the mean) for each element in these 10 splits is provided in Table 3.1.

WESTERN MARGIN

Background

High river discharge from the Sevier highland to the west of the Western Interior Seaway transported fine-grained sediments to the western margin. The river discharge also periodically created density stratification with a freshwater cap and quiescent, oxygen-deficient bottom waters resulting in preservation of laminated sediments. These fine-grained marine facies contain a valuable record of the dynamic history of structural, oceanographic, climatic, and ecologic changes in the Western Interior Seaway. Little is known, however, about the processes by which fine-grained sediments rich in OC were dispersed along the margin of the Western Interior Seaway. An understanding of these processes, and of the modes and rates of sediment accumulation in muddy offshore environments, is a prerequisite for accurate reconstruction of the history of the Western Interior Seaway.

The sediments that became the Tropic Shale and correlative Tununk Shale Member of the Mancos Shale of southern Utah accumulated within a few hundred kilometers offshore from the western margin of the Western Interior Seaway, and record a part of the Greenhorn transgressive–regressive cycle that lasted about 4 Myr, during which the Western Interior Seaway expanded to its maximum extent and then partially contracted (Kauffman, 1977, 1985; Arthur, Dean,

Table 3.1. Analyses of various elements in a sample of Cretaceous Tropic Shale from the USGS #1 Escalante well, Utah[a]

Element	Mean	SD	CV[b]
Al (%)	7.81	0.057	0.73
Ca (%)	1.81	0.032	1.75
Fe (%)	2.90	0.000	0.00
K (%)	2.14	0.052	2.41
Mg (%)	1.40	0.000	0.00
Na (%)	0.78	0.005	0.62
P (%)	0.07	0.000	0.00
Ti (%)	0.33	0.012	3.55
Mn (ppm)	100	6.9	6.90
As (ppm)	11	1.0	9.54
Ba (ppm)	347	4.8	1.39
Be (ppm)	2	0.0	0.00
Ce (ppm)	64	1.5	2.33
Co (ppm)	11	0.5	4.87
Cr (ppm)	76	0.4	0.56
Cu (ppm)	19	1.1	6.01
Ga (ppm)	18	0.7	3.98
La (ppm)	35	0.7	2.02
Li (ppm)	53	0.0	0.00
Nb (ppm)	17	0.7	4.24
Nd (ppm)	28	1.1	3.91
Ni (ppm)	15	0.7	4.79
Pb (ppm)	15	0.8	5.45
Sc (ppm)	11	0.3	2.85
Sr (ppm)	190	0.0	0.00
Th (ppm)	13	1.0	7.24
V (ppm)	111	3.2	2.85
Y (ppm)	21	0.7	3.40
Zn (ppm)	81	1.7	2.05

SD, standard deviation; CV, coefficient of variation.

[a] Mean, SD, and CV (SD as a percentage of the mean) of 10 analyses of eight major elements and 21 trace elements by ICP of a sample of Cretaceous Tropic Shale from the USGS #1 Escalante well, Utah.

[b] Average CV is 2.88.

and Claypool, 1985; Kauffman and Caldwell, 1993; Kauffman et al., 1993; Pratt et al., 1993).

The entire thickness of the Tropic Shale was sampled in the 280-m-long Escalante core that was drilled and continuously cored with better than 98% recovery at the base of the Kaiparowits Plateau near the town of Escalante, Utah (Fig. 3.1). This core collected the bottom of the Tibbet Canyon Member of the Straight Cliffs Formation, all of the marine Tropic Shale, and the top of the Dakota Sandstone (Fig. 3.2). The Escalante core provides an excellent opportunity for detailed reconstruction of processes of fine-grained sediment dispersal and accumulation of organic matter along, and just offshore from, the margin of the seaway. The Tropic Shale, although called a shale, is more correctly a marlstone, containing up to 60% $CaCO_3$, with an average of 34% in the

USGS Escalante No. 1

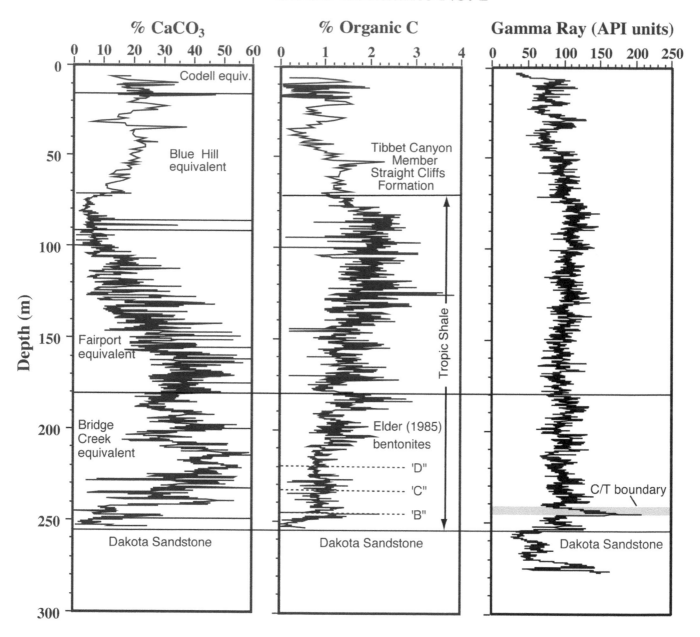

3.2. Profiles of % CaCO₃ and % OC in the USGS #1 Escalante core. Stratigraphic intervals that are equivalent in time to the Bridge Creek Limestone Member of the Greenhorn Formation, and Fairport Chalky Shale Blue Hill Shale and Codell Sandstone Members of the Carlile Shale in eastern Colorado are indicated on the profile of % CaCO₃. Also shown is the gamma ray log from the Escalante borehole.

Escalante core. The excellent biostratigraphy and abundant bentonite beds have allowed us to correlate peaks in concentration of CaCO₃ with individual limestone beds in the Bridge Creek Limestone in eastern Colorado and western Kansas.

Maximum transgression during the Greenhorn Cycle is represented in the Escalante core by the CaCO₃ maximum at about 220 m within the part of the section of Tropic Shale that is equivalent in time to the Bridge Creek Limestone Member of the Greenhorn Formation in eastern Colorado

and Kansas (Fig. 3.2). The progressive decrease in CaCO₃ content upward in the section of the Tropic Shale that is equivalent in time to the Fairport Shale in eastern Colorado and Kansas (Fig. 3.2) reflects the Greenhorn regression following the C/T high stand. The Tropic Shale in the Escalante core represents rapid deposition of fine-grained sediment in a prodeltaic environment (Leithold, 1993, 1994; Leithold and Dean, 1998). Sediment accumulation at the site of the Escalante core was both by suspension fallout from river plumes and storm-induced turbidity currents.

Sethi and Leithold (1994) recognized climatic limestone/marlstone bedding cycles in outcrop sections of the Tropic Shale and Tununk Shale in Utah that were antithetic with cycles of percentage coarse silt and sand. They interpreted the higher $CaCO_3$ intervals as representing dryer climatic intervals and the higher silt–sand intervals as representing wetter climatic intervals in response to orbital (Milankovitch) forcing. As these marine marls accumulated along the western margin of the seaway, more calcareous, basinal facies of the Greenhorn Limestone were deposited to the east in Colorado and Kansas, also in response to Milankovitch orbital forcing, with periodicities of about 20, 40, and 100 ka as based on dated bentonite beds and lithostratigraphy (Fischer, 1980; Arthur et al., 1984; Barron, Arthur, and Kauffman, 1985; Fischer, Herbert, and Silva, 1985; Kauffman, 1985; Arthur and Dean, 1991; Kauffman et al., 1993; Pratt et al., 1993; Sageman et al., 1997; Meyers, Sageman, and Hinnov, 2001). Elder, Gustason, and Sageman (1994) concluded that Milankovitch climatic forcing was the most plausible cause of cyclic sedimentation at the basin margin and as well as in the basin center.

Biostratigraphy

The Upper Cretaceous c/t boundary is a second-ranked extinction boundary, which means extinction at 75% with a loss of 30–40% of molluscan taxa. This is reflected primarily in the ammonites (class Cephalopoda) at the genus level. Bivalve groups suffered considerably lower extinction values, primarily at the species and subspecies group level. For example, one family, the Inoceramidae (class Bivalvia), lost <5% of taxa. It is one of the five major extinctions, along with three minor, ones during the Phanerozoic. It happened during or just after the global Bonarelli OAE at one of two major peaks in sea-level rise. Ammonites and inoceramids both experienced major turnovers for the first 1 My when ecospaces were depleted of competitors. Eventually, as competition increased, turnover rates increased from less than 800 ky to more than 2 My per species or subspecies.

In North America, the c/t boundary took place during a maximum transgression of the Western Interior Seaway. During this time, the sea level was the highest since the Ordovician. At maximum transgression during the Cretaceous, the Western Interior Seaway was approximately 5000 km long, connecting Arctic Canada and Alaska with the Gulf of Mexico, and approximately 1000 km wide, connecting central Utah with Nebraska and South Dakota. A northeast arm of the Western Interior Seaway originated in South Dakota, extended into southern Alberta, and connected to the North Atlantic.

Mollusks dominated the Upper Cretaceous seaway macrofauna. Ammonites and inoceramids were especially abundant, particularly in deeper-water facies dominated by shale and carbonates (Fig. 3.3). Ammonites lived in the middle and upper water column, but their dead shells settled into the dysoxic to anoxic benthic zone, where they are preserved largely intact. Their fossils are found in facies ranging from sandstone, where they are associated with diverse benthic taxa, to offshore shales and limestones where they are associated with fewer than five other molluscan species. Some of these other species, such as Lucinidae and small Ostreidae, had their own set of adaptations to oxygen-deficient environments. The inoceramids and ammonites have about the same evolutionary rates, as measured by well-dated bentonites (Obradovich, 1993). Kauffman et al. (1993) provided limited additional dates that were based on dated bentonites and on the assumption of equal evolutionary rates for the inoceramids and ammonites.

The c/t boundary faunal succession in southern Utah, based on surface collections, is the same as for the Western Interior Seaway in general: a mixture of ammonites and rapidly evolving inoceramids. However, in the Escalante core, Upper Cenomanian and Lower Turonian ammonites are very rare (a single occurrence) even though they wholly agree in age determination with the more abundant inoceramids. Therefore, we are dependent on the inoceramids for definition of the c/t boundary, with additional observations on planktic foraminifera.

The Upper Cenomanian species *Inoceramus pictus* is found in two closely spaced levels in the Escalante core at 246.0 and 246.3 m (Fig. 3.4). Above this, *Inoceramus* sp. cf. *I. pictus* occurs at 244 m. This possibly represents a zone of mixing of two c/t boundary inoceramid faunas. Above this zone of mixing, at 236 m, *Mytiloides* n. sp., flat with weakly defined, incomplete rugae, occurs. Elsewhere (e.g., Pueblo, Colorado; Red Wash, Utah) this species is limited to the basal Turonian. At Pueblo, there is a 0.5- to 1-m zone of mixing of typical Cenomanian inoceramids and those representing the basal Turonian. The possible c/t boundary in the Escalante core lies between 245 and 242 m (Fig. 3.4) at two possible positions. The lowest of these is in the peak in $\delta^{13}C$ (Fig. 3.5) indicating the global Bonarelli OAE. This is 3 m below the first appearance of a single *Whiteinella archeocretacia*, a foraminifera, which, along with early *Mytiloides* spp., signals the earliest Turonian (Fig. 3.4; Hart, 1996).

Carbon Content and Type

The Tropic Shale in the Escalante core is characterized by generally upward increasing OC content, from a low value of <1% to a high value of 3% (Fig. 3.2). This indicates that

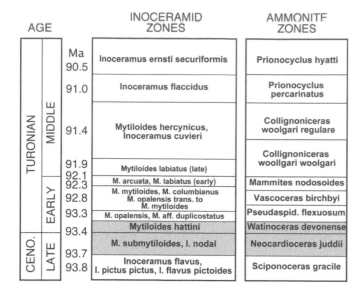

AGE				INOCERAMID ZONES	AMMONITE ZONES
TURONIAN	MIDDLE		Ma 90.5	Inoceramus ernsti securiformis	Prionocyclus hyatti
			91.0	Inoceramus flaccidus	Prionocyclus percarinatus
			91.4	Mytiloides hercynicus, Inoceramus cuvieri	Collignoniceras woolgari regulare
			91.9	Mytiloides labiatus (late)	Collignoniceras woolgari woolgari
	EARLY		92.1 92.3	M. arcuata, M. labiatus (early)	Mammites nodosoides
			92.8	M. mytiloides, M. columbianus M. opalensis trans. to M. mytiloides	Vascoceras birchbyi
			93.3	M. opalensis, M. aff. duplicostatus	Pseudaspid. flexuosum
			93.4	Mytiloides hattini	Watinoceras devonense
CENO.	LATE		93.7	M. submytiloides, I. nodal	Neocardioceras juddii
			93.8	Inoceramus flavus, I. pictus pictus, I. flavus pictoides	Sciponoceras gracile

3.3. Ammonite and inoceramid biostratigraphy at the Cenomanian–Turonian boundary in the Western Interior Basin. Biostratigraphic zones and ages are from Kauffman et al. (1993) and Obradovich (1993).

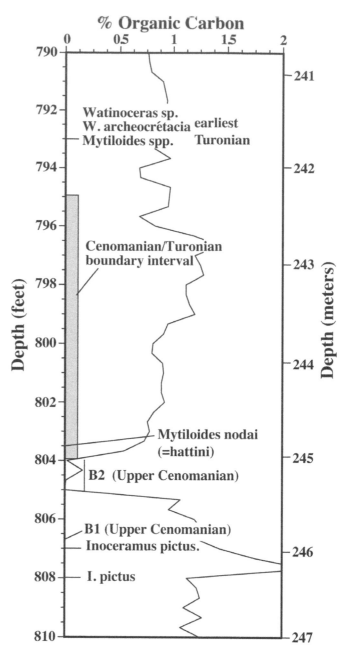

3.4. Profile of % OC and locations of key bentonite beds (B1 and B2 of Elder, 1985) and fossils that define the zone of the Cenomanian–Turonian boundary in the USGS #1 Escalante core.

the amount of organic matter that accumulated along the western margin of the Western Interior Seaway increased as the shoreline transgressed from east to west. But was this increased organic matter of marine or terrestrial origin? Rock-Eval pyrolysis provides a rapid estimate of the type, abundance, and degree of preservation (and/or source) of organic matter in sediments and sedimentary rocks (Tissot and Welte, 1984; Peters, 1986). Our data are expressed as a pyrolysis HI in milligrams of hydrocarbons (HC) per gram of OC (mg HC/g OC). This is a proxy for an atomic H/C ratio in kerogen. Terrestrial organic matter (type III) typically has HI values of <200, and well-preserved algal marine organic matter (type I and II) typically has HI values of >400. Values of HI in the Tropic Shale also generally increase upward parallel to the increase in OC (Fig. 3.5), suggesting that there was an increasing proportion of algal marine organic matter. However, the highest values of HI (200–250) are still fairly low relative to those in a good marine hydrocarbon source rock (HI >400), indicating that most of the organic matter is terrestrial (type III), which is to be expected considering the substantial influx of terrestrial clastic sediment to the Western Interior Seaway. The increase in organic matter past peak Greenhorn transgression (approximately 220 m in the Escalante core) indicates that, unlike CaCO₃ content, OC content was not coupled to the Greenhorn transgression–regression cycle (Fig. 3.2).

Isotopic Composition

An upward increase in the relative proportion of marine organic matter in the Tropic Shale in the Escalante core

is further indicated by a decrease in values of $\delta^{13}C$ of the organic matter (Fig. 3.5). Cretaceous marine OC typically has $\delta^{13}C$ values of −26‰ to −28‰, whereas Cretaceous terrestrial organic matter typically has $\delta^{13}C$ values of −25‰ to −26‰ (Arthur et al., 1985; Dean, Arthur, and Claypool, 1986). The exception is marine organic matter deposited at the C/T OAE that is up to 5‰ higher than that of either older or younger Cretaceous marine organic matter (Schlanger et al., 1987; Arthur, Dean, and Pratt, 1988; Fig. 3.5). In Cretaceous OC-rich sequences, other than C/T sequences, HI and

Dean, Kauffman, and Arthur

USGS Escalante No. 1

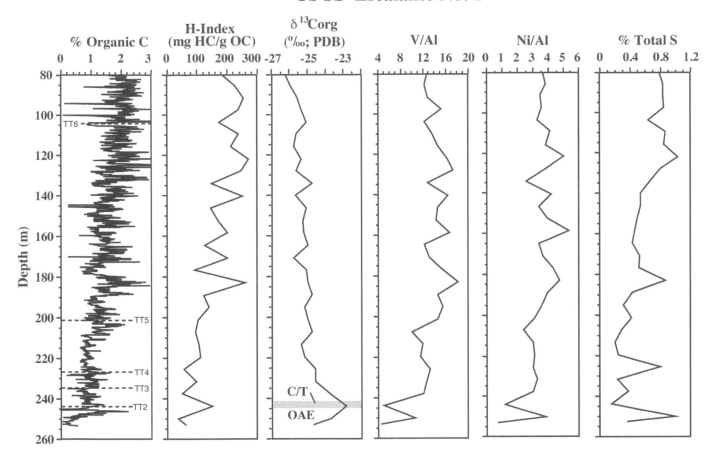

3.5. Profiles of % OC, Rock-Eval pyrolysis HI, δ¹³C of OC, ratio of vanadium to aluminum (V/Al), ratio of nickel to aluminum (Ni/Al), and % TS in the Tropic Shale in the USGS #1 Escalante core. Bentonite marker beds TT2 through TT6 of Leithold (1993, 1994) are indicated on the profile of OC. C/T, Cenomanian–Turonian boundary; OAE, oceanic anoxic event.

δ^{13}C often are negatively correlated (as in Fig. 3.4), indicating that greater hydrogen richness of organic matter is the result of a greater contribution of marine organic matter (lower values of δ^{13}C) rather than enhanced preservation of organic matter (Arthur and Dean, 1991).

With this background, we interpret the profiles in Fig. 3.5 as follows: The organic matter initially deposited in the Tropic Shale as the Cretaceous shoreline transgressed over the shoreline sands of the Dakota Group was almost entirely terrestrial, with OC values of <0.5%, HI values of <50, and δ^{13}C values of about −25‰ (above the C/T boundary). The pulse of marine organic matter production and preservation coincident with the global OAE beginning just below the C/T boundary is marked in the Escalante core by an increase in % OC, HI, and δ^{13}C (Fig. 3.5). For comparison, in the Amoco #1 Rebecca K. Bounds core in western Kansas, values of HI increase from 400 to 700 and values of δ^{13}C increase from −27‰ to −23‰ just below the C/T boundary (Dean et al., 1995; Fig. 3.6). In the Princeton University #1 core from Rock Canyon Anticline near Pueblo, Colorado (Fig. 3.1), values

of δ^{13}C increase from −28‰ to −24‰ just below the C/T boundary (Pratt, 1985; Arthur, Dean, and Pratt, 1988; Pratt et al., 1993). The lower (more negative) values of δ^{13}C in Kansas and eastern Colorado before the C/T OAE reflect the more open marine conditions at those sites.

After the C/T OAE, increasing transgression of the Greenhorn sea resulted in more marine organic matter being produced and preserved at the site of the Escalante core, as evidenced by higher values of % OC and HI, and lower (more negative) values of δ^{13}C (Fig. 3.5). At most locations throughout the world, the C/T boundary usually is marked by a peak in amount of OC as well as a positive excursion in the isotopic composition of inorganic and organic carbon. In the Escalante core, however, the peak in the amount (% OC) and marine character (higher values of HI) of the organic matter occur higher in the section. Therefore, the Tropic Shale in the Escalante core apparently has recorded both global (OAE) and local events along the western shore of the Western Interior Seaway.

Amoco R.K. Bounds No. 1

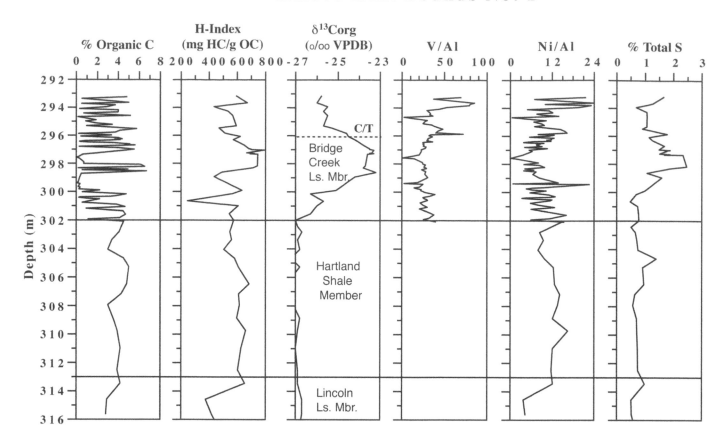

3.6. Profiles of % OC, Rock-Eval pyrolysis HI, δ¹³C of OC, V/Al ratio, Ni/Al ratio, and % TS in the Greenhorn Formation in the Amoco #1 Rebecca K. Bounds core from western Kansas (Fig. 3.1). Stratigraphic intervals of the Lincoln Limestone, Hartland Shale, and Bridge Creek Limestone members of the Greenhorn are indicated. C/T, Cenomanian–Turonian boundary.

Inorganic Geochemistry

The accumulation of more marine organic matter in the upper part of the Tropic Shale in the Escalante core also resulted in higher concentrations of sulfur (S; Fig. 3.5), likely present as pyrite (FeS₂). It also resulted in the accumulation of certain trace elements such as vanadium (V), nickel (Ni), copper (Cu), and zinc (Zn) that are commonly concentrated in oc-rich sediments where sulfate reduction has occurred (Jacobs, Emerson, and Skei, 1985; Emerson and Huested, 1991; Crusius et al., 1996). We have chosen V and Ni to represent this group, and we have normalized their concentrations to that of aluminum (Al) to factor out dilution by CaCO₃ (Fig. 3.5). These trace metals are thought to be concentrated in tetrapyrroles of algal origin, the precursors of type II and type I kerogen (Lewan and Maynard, 1982). Crude oils, particularly high-S crude oils, commonly contain high concentrations of V and Ni (Tissot and Welte, 1984). Although values of V/Al and Ni/Al increase in the more oc- and hc-rich upper part of the Tropic Shale in the Escalante

core, these values are still lower than in average shale (16 and 8.5, respectively; Table 3.1) and considerably lower than in Bridge Creek Limestone in the basin-center Bounds core in western Kansas, which contains more algal organic matter (Fig. 3.6). The Bridge Creek in the Bounds core is considerably more oc- and hc-rich than the Tropic Shale, with oc values up to 6% and hi values up to 800 (Fig. 3.6), indicating that this would be a good marine hydrocarbon source rock with type II organic matter.

Although S concentrations increase upward within the Tropic Shale along with increasing oc concentrations (Fig. 3.5), the S concentration is also elevated just below the c/t boundary coincident with increased δ¹³C values that represent a global OAE. An increase in S content also occurs just below the c/t boundary in the basin center Bounds core (Fig. 3.6). This suggests that the OAE was accompanied by increased sulfate reduction and sulfide formation, likely as pyrite, except in systems that were Fe limited.

Diagenetic relations between oc, total sulfur (TS), and total iron (TFe) can be examined in cross plots of TS versus

Dean, Kauffman, and Arthur

OC, and TS versus TFe (Fig. 3.7). The TS/OC ratio in Holocene normal marine sediments averages about 0.4 (Raiswell and Berner, 1986). These are sediments that were deposited under oxic bottom-water conditions, with sufficient labile organic matter for sulfate reduction, and anoxic sediment pore waters. The plot of TS versus OC in Tropic Shale from the Escalante core (Fig. 3.7A) shows considerable scatter with many samples having excess TS or OC values relative to the normal marine ratio of 0.4. Excess OC suggests that either sulfate was limiting (unlikely in a marine system) or that the organic matter was unreactive. The low HI values in the Tropic Shale, particularly in the lower part of the section in the Escalante core (Fig. 3.5), indicate that there was a limited supply of reactive organic matter. Excess S implies that these sediments were deposited under anoxic bottom-water conditions with iron sulfide formation in the water column (Lyons, 1997). The S/Fe ratio in pyrite is 1.15. In samples from the Escalante core, TS and TFe have a positive correlation, but the S/Fe ratio is much lower (0.467, the slope of the regression line, the solid line in Fig. 3.7B) than the 1.15 pyrite ratio (dashed line), indicating that there is excess Fe in the detrital clastic fraction not fixed in pyrite because the S from sulfate reduction was limiting. The amount of excess Fe is indicated by the positive intercept on the TFe axis–that is, at 0% TS, there is 1% TFe. In carbonate-rich environments, often there is not enough Fe from clastic material to keep pace with sulfide production, so all of the free Fe will go into the production of pyrite; the sediments will have an S/Fe ratio of 1.15, but excess sulfide escapes into the water column. Such a relation was found for the Bridge Creek Limestone in the Princeton University #1 core from Pueblo, Colorado (Dean and Arthur, 1989). The Bridge Creek was deposited in the deepest part of the Western Interior Seaway, and bottom waters were periodically dysoxic to anoxic (Pratt, 1984).

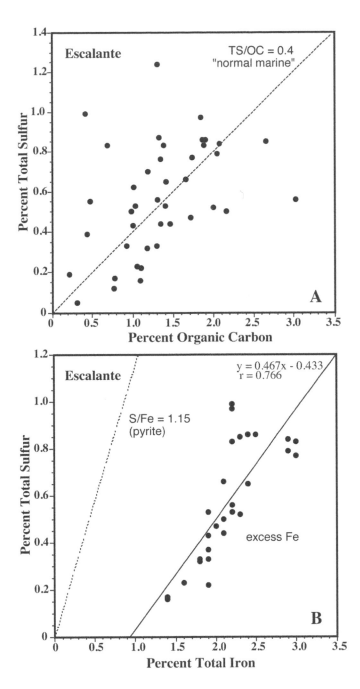

3.7. Cross plots of (A) % OC versus % TS, and (B) % TFe versus % TS in the Tropic Shale in the USGS #1 Escalante core.

CORRELATION WITH THE GREENHORN FORMATION

The Bridge Creek Limestone Member of the Greenhorn Formation is well known for its prominent limestone/marlstone cycles in outcrops in eastern Colorado and western Kansas (Fig. 3.8) and in cores (Fig. 3.9), and they have been described extensively (Hattin, 1971; Kauffman, 1977; Fischer, Herbert, and Silva, 1985; Bottjer et al., 1986; Eicher and Diner, 1989; Arthur and Dean, 1991; Pratt ct al., 1993; Sageman et al., 1997; Dean and Arthur, 1998b; Meyers, Sageman, and Hinnov, 2001). They played a historical role in the development of modern cyclostratigraphy. As pointed out by Fischer (1980), G. K. Gilbert (1895) was the first to suggest that the

bedding cycles were forced by millennial-scale orbital cyclicity, and that these cycles might be used to calibrate geologic time. Kauffman (1977) estimated bedding cycle periodicity at 60–80 ka, and later (1987), on the basis of well-dated bentonite intervals, he estimated periodicities of 20–25 ka, 40–45 ka, and 100–110 ka, which he attributed to Milankovitch orbital cyclicity. Fischer (1980) estimated periodicities of 22–27 ka, and later Fischer, Herbert, and Silva (1985) modified this estimate to 41 ka, which they attributed to an obliquity orbital

3.8. Limestone/marlstone bedding couplets in the Bridge Creek Limestone Member of the Greenhorn Formation at Pueblo reservoir, eastern Colorado. Al Fischer and Mike Arthur for scale.

Milankovitch cycle. However, age control was poor, resulting in considerable uncertainty. Elder (1985) estimated an average period of 100 ka for his limestone marker beds (Fig. 3.9) in the Bridge Creek at Pueblo, which he attributed to orbital eccentricity. Many authors cited a range of periodicities of 20 to 100 ka for the Bridge Creek bedding cycles (reviews by Schwarzacher and Fischer, 1982; Arthur et al., 1984; Pratt, 1984; Barron, Arthur, and Kauffman, 1985; Fischer, Herbert, and Silva, 1985; Bottjer et al., 1986; Arthur and Dean, 1991; Pratt et al., 1993). The most commonly cited links are changes in the Earth's orbital precession, axial obliquity, and orbital eccentricity with periodicities of about 21 ka, 41 ka, and 100 ka based on past astronomical frequencies (Berger, Loutre, and Laskar, 1992) estimated from modern orbital characteristics of the Sun–Earth system (Berger, 1984).

The Cretaceous time scale for the Western Interior was refined considerably by radiometric dates presented by Obradovich (1993). Sageman et al. (1997, 1998) and Meyers, Sageman, and Hinnov (2001) used the improved age scale for the Cenomanian and Turonian of Obradovich (1993) and continuous core coverage in the Portland core (Fig. 3.1) to quantitatively test periodicities of the Bridge Creek bedding cycles. They used several spectral methods applied to measured concentrations of $CaCO_3$ and OC, and grayscale optical densities measured on black-and-white photographs of the slabbed archive half of the Portland core. The spectra for % $CaCO_3$ showed significant peaks for all three orbital cycles (precession, obliquity, and eccentricity). The spectra for grayscale data also showed significant peaks for all three, but especially for precession and eccentricity. The spectra for % OC showed significant peaks only for obliquity.

The depositional mechanisms responsible for the Bridge Creek bedding cycles have been attributed to clastic dilution, in which climatically induced changes in the hydrologic cycle controlled the delivery of clastic material to the basin (Hattin, 1971; Arthur et al., 1984; Barron, Arthur, and Kauffman, 1985; Arthur and Dean, 1991; Pratt et al., 1993); to climate-controlled production of calcareous nanoplankton in surface waters (Eicher and Diner, 1985, 1989; Arthur and Dean, 1991); and to carbonate dissolution (Arthur and Dean, 1991). Superimposed on these primary climatically related cycles were variations in production and preservation of organic matter, burial diagenesis, and aperiodic dilution by volcanic ash. Some variations were unique to the

Dean, Kauffman, and Arthur

Western Interior Seaway and others were more widespread, such as OAEs. Climate-induced changes in the hydrologic cycle also produced fluctuations in water-column stability and bottom-water production, circulation, and oxygenation that produced changes in preservation of organic matter and carbonate. Sageman et al. (1997, 1998) and Meyers, Sageman, and Hinnov (2001) concluded that both clastic dilution and carbonate production were the main cause of the Bridge Creek bedding rhythms. Precessional cycles forced carbonate (nanofossil) production through its effect on evaporation and nutrient upwelling. Obliquity cycles had a strong influence on precipitation and runoff, and therefore on dilution by detrital clastic material. Precession and obliquity combined to produce the complex bedding pattern in the Bridge Creek. Because of the close proximity of the Escalante core to the western margin of the Western Interior Seaway, variations in $CaCO_3$ content are most likely due to clastic dilution in response to rainfall and runoff in the Sevier volcanic highlands to the west.

The part of the Tropic Shale in the Escalante core that is the time equivalent of the Bridge Creek Limestone shows considerable variation in percentage of $CaCO_3$, at a much higher frequency (Fig. 3.10) than the limestone–marlstone bedding cycles in the Bridge Creek (Figs. 3.8 and 3.9). The highest values suggest that they may be correlated with individual limestone beds in the Bridge Creek in eastern Colorado and western Kansas. The limestone marker beds of Elder (1985), selected from the marker beds of Cobban and Scott (1972), are easily correlated in outcrop because of distinctive bentonite marker beds described by Elder (1985). Zelt (1985) identified 17 bentonite layers in the Tropic and Tununk shales, which he traced throughout the region in both outcrops and in well logs. Leithold (1993, 1994) designated these bentonite beds Tr1 through Tr17 in outcrop, and the lowermost six of them have been identified in the Tropic Shale in the Escalante core (Figs. 3.5 and 3.10; Leithold and Dean, 1998). On the basis of biostratigraphy, marker beds Tr1 through Tr4 are equivalent to Elder's bentonite marker beds A, B, C, and D in the Bridge Creek (Fig. 3.10). On the basis of these bentonite marker beds, we have suggested possible correlations between peaks in percentage $CaCO_3$ in the Tropic Shale in the Escalante core and the limestone marker beds in the Bridge Creek Limestone in the USGS #1 Portland identified by Meyers, Sageman, and Hinnov (2001) (Fig. 3.10).

CONCLUSIONS

1. The 280-m-long Escalante core recovered the bottom of the Tibbet Canyon Member of the Straight Cliffs Formation, all of the marine Tropic Shale, and the top of the Dakota Sandstone.

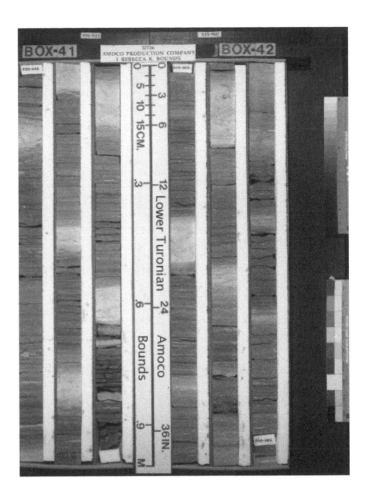

3.9. Six meters of limestone/marlstone couplets in the Bridge Creek Limestone Member of the Greenhorn Formation in the Amoco #1 Rebecca K. Bounds core from western Kansas.

2. Molluscan macrofossils across the C/T boundary are dominated by inoceramids (Bivalvia); ammonites (Cephalopoda) are extremely rare. Therefore, the C/T boundary in the Escalante core is a broadly defined zone between 245 and 242 m based on volcanic ash (bentonite) beds and sparse macrofossils.

3. The C/T boundary in the Escalante core also is marked by a positive carbon isotope excursion, as it is in many localities around the world, indicating that the Western Interior Seaway at the site of the Escalante core was responding to the global Bonarelli OAE.

4. The peak of the C/T Greenhorn transgression in the Western Interior Seaway is marked in the Escalante core by a peak in $CaCO_3$ content at about 220 m, or about 20 m above the C/T boundary. Peak OC content occurs much higher, at about 130 m, indicating that OC content was not coupled to the Greenhorn transgression but that $CaCO_3$ content was.

5. The accumulation of more marine organic matter in the upper part of the Tropic Shale in the Escalante core also resulted in the accumulation of minor and trace elements

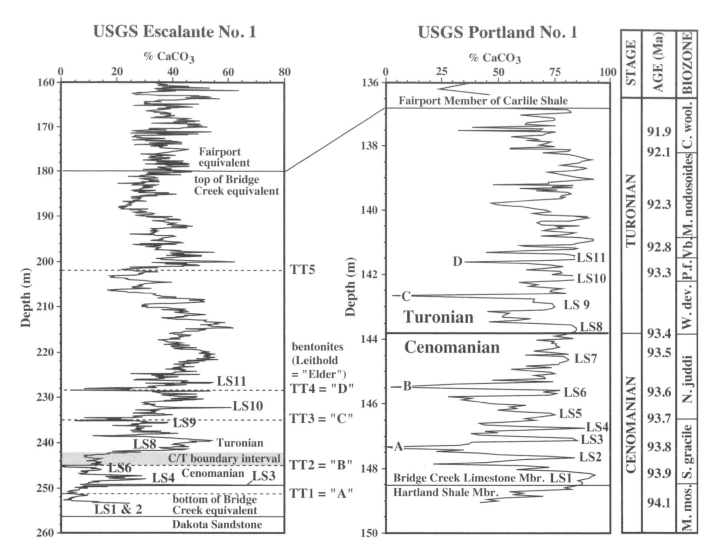

3.10. Percent CaCO$_3$ in samples of Tropic Shale in the USGS #1 Escalante core and in samples of the Greenhorn Formation in the USGS #1 Portland core showing locations of bentonite beds (A, B, C, and D beds of Elder, 1985; TT1 through TT5 beds of Leithold, 1993, 1994). Limestone marker beds (LS) are from Elder (1985) selected from marker beds of Cobban and Scott (1972). Ages are the ages of biozone boundaries interpolated between radiometric ages of bentonites A, B, C, and D by Obradovich (1993) and Kauffman et al. (1993). Modified from Meyers, Sageman, and Hinnov (2001).

such as S, V, Ni, Cu, and Zn that are commonly concentrated in oc-rich sediments where sulfate reduction has occurred.

6. On the basis of biostratigraphy and distinctive bentonite marker beds, peaks in CaCO$_3$ content may be correlated to individual limestone beds in the Bridge Creek Limestone Member of the Greenhorn Formation in eastern Colorado and western Kansas. The prominent cyclicity of the Bridge Creek Limestone beds and interbedded marlstones has been variously attributed to Milankovitch cycles in the Earth's orbital precession, axial obliquity, and/or eccentricity. The Greenhorn marlstone beds also have been attributed to dilution by detrital clastic material, CaCO$_3$ production, and/or CaCO$_3$ dissolution. The cyclic variations in CaCO$_3$ in the Tropic Shale are most likely due to clastic dilution in response to rainfall and runoff in the Sevier volcanic highlands to the west of the Western Interior Seaway.

ACKNOWLEDGMENTS

We thank M. Lewan, B. Sageman, and D. Sawyer for careful reviews, and J. Honey for providing more than 2000 carbon analyses. This research was supported by the Global Change and Climate History Program of the U.S. Geological Survey.

Dean, Kauffman, and Arthur

Arthur, M. A., and W. E. Dean. 1991. An holistic geochemical approach to cyclomania–examples from Cretaceous pelagic limestone sequences; pp. 126–166 in G. Einsele, W. Ricken, and A. Seilacher (eds.), Cycles and Events in Stratigraphy. Springer-Verlag, New York.

Arthur, M. A., and S. O. Schlanger. 1979. Cretaceous "oceanic anoxic events" as causal factors in development of reef-reservoired giant oil fields. American Association of Petroleum Geologists Bulletin 63:870–885.

Arthur, M. A., W. E. Dean, and G. E. Claypool. 1985. Anomalous ^{13}C enrichment in modern marine organic carbon. Nature 315:216–218.

Arthur, M. A., W. E. Dean, and L. M. Pratt. 1988. Geochemical and climatic effects of increased organic carbon burial at the Cenomanian/Turonian boundary. Nature 335:714–717.

Arthur, M. A., S. O. Schlanger, and H. C. Jenkyns. 1987. The Cenomanian–Turonian oceanic anoxic event, II. Paleoceanographic controls on organic matter production and preservation; pp. 401–420 in J. Brooks and A. Fleet (eds.), Marine Petroleum Source Rocks. Geological Society of London Special Publication 2.

Arthur, M. A., H.-J. Brumsack, H. C. Jenkyns, and S. O. Schlanger. 1990. Stratigraphy, geochemistry, and paleoceanography of organic carbon-rich Cretaceous sequences; pp. 75–119 in R. N. Ginsburg and B. Beaudoin (eds.), Cretaceous, Resources, Events, and Rhythms: Background and Plans for Research. Kluwer Academic Publishers, Dordrecht.

Arthur, M. A., W. E. Dean, D. A. Bottjer, and P. A. Scholle. 1984. Rhythmic bedding in Mesozoic–Cenozoic pelagic carbonate sequences; the primary and diagenetic origin of Milankovitch-like cycles; pp. 191–222 in A. Berger, J. Imbrie, J. D. Hays, G. Kukla, and B. Saltzman (eds.), Milankovitch and Climate, Pt. 1. Reidel, Amsterdam.

Arthur, M. A., W. E. Dean, R. M. Pollastro, and P. A. Scholle. 1985. Comparative geochemical and mineralogical studies of two cyclic transgressive limestone units, Cretaceous Western Interior Basin, U.S.; pp. 16–27 in L. M. Pratt, E. G. Kauffman, and F. B. Zelt (eds.), Fine-Grained Deposits and Biofacies of the Cretaceous Western Interior Seaway: Evidence of Cyclic Sedimentary Processes. Society of Economic Paleontologists and Mineralogists, Field Trip Guidebook 4, 1985 Midyear Meeting, Golden, Colorado.

Baedecker, P. A., ed. 1987. Geochemical Methods of Analysis. U.S. Geological Survey Bulletin 1770.

Barron, E. J., M. A. Arthur, and E. G. Kauffman. 1985. Cretaceous rhythmic bedding sequences: a plausible link between orbital variations and climate. Earth and Planetary Science Letters 72:327–340.

Berger, A. 1984. Accuracy and frequency stability of the Earth's orbital elements during the Quaternary; pp. 3–40 in A. Berger, J. Imbrie, J. D. Hays, G. Kukla, and B. Saltzman (eds.), Milankovitch and Climate, Pt. 1. Reidel, Amsterdam.

Berger, A., M. F. Loutre, and J. Laskar. 1992. Stability of the astronomical frequencies over the Earth's history for paleoclimate studies. Science 255:560–566.

Bottjer, D. J., M. A. Arthur, W. E. Dean, D. E. Hattin, and C. E. Savrda. 1986. Rhythmic bedding produced in Cretaceous pelagic carbonate environments: sensitive recorder of climatic cycles. Paleoceanography 1:467–481.

Clayton, J. L., and P. J. Swetland. 1980. Petroleum generation and migration in Denver Basin. American Association of Petroleum Geologists Bulletin 64:1613–1633.

Cobban, W. A., and R. W. Scott. 1972. Stratigraphy and ammonite fauna of the Graneros Shale and Greenhorn Limestone near Pueblo, Colorado. U.S. Geological Survey Professional Paper 645.

Crusius, J., S. Calvert, T. Pedersen, and D. Sage. 1996. Rhenium and molybdenum enrichments in sediments as indicators of oxic, suboxic, and sulfidic conditions of deposition. Earth Planetary Science Letters 145:65–78.

Dean, W. E., and M. A. Arthur. 1989. Iron–sulfur–carbon relationships in organic-carbon-rich sequences I: Cretaceous Western Interior Seaway. American Journal of Science 289:708–743.

Dean, W. E., and M. A. Arthur. 1998a. Cretaceous Western Interior Seaway drilling project: an overview; pp. 1–10 in W. E. Dean and M. A. Arthur (eds.), Stratigraphy and Paleoenvironments of the Cretaceous Western Interior Seaway. Society for Sedimentary Geology Concepts in Sedimentology and Paleontology 6.

Dean, W. E., and M. A. Arthur, eds. 1998b. Stratigraphy and Paleoenvironments of the Cretaceous Western Interior Seaway. Society for Sedimentary Geology Concepts in Sedimentology and Paleontology 6.

Dean, W. E., M. A. Arthur, and G. E. Claypool. 1986. Depletion of ^{13}C in Cretaceous marine organic matter–source, diagenetic, or environmental signal? Marine Geology 70:119–157.

Dean, W. E., M. A. Arthur, B. B. Sageman, and M. D. Lewan. 1995. Core Descriptions and Preliminary Geochemical Data for the Amoco Production Company Rebecca K. Bounds #1 Well, Greeley County, Kansas. U.S. Geological Survey Open-File Report 95–209.

Eicher, D. L., and R. Diner. 1985. Foraminifera as indicators of water mass in the Cretaceous Greenhorn Sea, Western Interior; pp. 60–71 in L. M. Pratt, E. G. Kauffman, and F. B. Zelt (eds.), Fine-Grained Deposits and Biofacies of the Cretaceous Western Interior Seaway: Evidence of Cyclic Sedimentary Processes. Society of Economic Paleontologists and Mineralogists, Field Trip Guidebook 4, 1985 Midyear Meeting, Golden, Colorado.

Eicher, D. L., and R. Diner. 1989. Origin of the Cretaceous Bridge Creek cycles in the Western Interior, United States. Palaeogeography, Palaeoclimatology, and Palaeoecology 74:127–146.

Elder, W. P. 1985. Biotic patterns across the Cenomanian–Turonian extinction boundary near Pueblo, Colorado; pp. 157–169 in L. M. Pratt, E. G. Kauffman, and F. B. Zelt (eds.), Fine-Grained Deposits and Biofacies of the Cretaceous Western Interior Seaway: Evidence of Cyclic Sedimentary Processes. Society of Economic Paleontologists and Mineralogists, Field Trip Guidebook 4, 1985 Midyear Meeting, Golden, Colorado.

Elder, W. P., E. R. Gustason, and B. B. Sageman. 1994. Correlation of basinal carbonate cycles to nearshore parasequences in the Late Cretaceous Greenhorn seaway, Western Interior, U.S.A. Geological Society of America Bulletin 106:892–902.

Emerson, S. R., and S. S. Huested. 1991. Ocean anoxia and concentrations of molybdenum and vanadium in seawater. Marine Chemistry 34:177–196.

Engleman, E. E., L. L. Jackson, D. R. Norton, and A. G. Fischer. 1985. Determination of carbonate carbon in geological materials by coulometric titration. Chemical Geology 53:125–128.

Espitalié, J., M. Madec, B. Tissot, J. J. Menning, and P. Leplat. 1977. Source rock characterization method for petroleum exploration. Offshore Technology Conference, Houston, Texas, Paper 2935.

Fischer, A. G. 1980. Gilbert-bedding rhythms and geochronology; pp. 93–104 in E. I. Yochelson (ed.), The Scientific Ideas of G. K. Gilbert. Geological Society of America Special Paper 183.

Fischer, A. G., T. Herbert, and I. P. Silva. 1985. Carbonate bedding cycles in Cretaceous pelagic and hemipelagic sequences; pp. 1–10 in L. M. Pratt, E. G. Kauffman, and F. B. Zelt (eds.), Fine-Grained Deposits and Biofacies of the Cretaceous Western Interior Seaway: Evidence of Cyclic Sedimentary Processes. Society of Economic Paleontologists and Mineralogists, Field Trip Guidebook 4, 1985 Midyear Meeting, Golden, Colorado.

Gilbert, G. K. 1895. Sedimentary measurement of geologic time. Journal of Geology 3:121–127.

Ginsburg, R. N., and B. Beaudoin, eds. 1990. Cretaceous Resources, Events and Rhythms: Background and Plans for Research. Kluwer Academic Publishers, Dordrecht.

Hart, M. B. 1996. Recovery of the food chain after the Late Cenomanian extinction event; pp. 265–277 in M. B. Hart (ed.), Biotic Recovery from Mass Extinction Events. Geological Society of London Special Publication 102.

Hattin, D. E. 1971. Widespread, synchronously deposited, burrow-mottled limestone beds in Greenhorn Limestone (Upper Cretaceous) of Kansas and central Colorado. American Association of Petroleum Geologists Bulletin 55:412–431.

Jacobs, L., S. Emerson, and J. Skei. 1985. Partitioning and transport of metals across the O_2/H_2S interface in a permanently anoxic basin, Framvaren Fjord, Norway. Geochimica et Cosmochimca Acta 49:1433–1444.

Jenkyns, H. C. 1980. Cretaceous anoxic events–from continents to oceans. Journal of the Geological Society of London 137:171–188.

Kauffman, E. G. 1977. Geological and biological overview: Western Interior Cretaceous Basin; pp. 75–99 in E. G. Kauffman (ed.), Cretaceous Facies, Faunas and Paleoenvironments across the Western Interior Basin. Mountain Geologist 14.

Kauffman, E. G. 1985. Cretaceous evolution of the Western Interior Basin of the United States; pp. iv–xii in L. M. Pratt, E. G. Kauffman, and F. B. Zelt (eds.), Fine-Grained Deposits and Biofacies of the Cretaceous Western Interior Seaway: Evidence of Cyclic Sedimentary Processes. Society of Economic Paleontologists and Mineralogists, Field Trip Guidebook 4, 1985 Midyear Meeting, Golden, Colorado.

Kauffman, E. G. 1987. High-resolution event stratigraphy: concepts, methods, and Cretaceous examples; pp. 2–34 in E. G. Kauffman, B. B. Sageman, E. R. Gustason, and W. P. Elder (eds.), High-Resolution Event Stratigraphy, Greenhorn Cyclothem (Cretaceous: Cenomanian–Turonian), Western Interior Basin of Colorado and Utah. Geological Society of America, Rocky Mountain Section, 1987 Regional Meeting, Field Trip Guidebook.

Kauffman, E. G., and W. G. E. Caldwell. 1993. The Western Interior Basin in time and space; pp. 1–30 in W. G. E. Caldwell and E. G. Kauffman (eds.), Evolution of the Western Interior Basin. Geological Association of Canada Special Paper 39.

Kauffman, E. G., B. B. Sageman, J. I. Kirkland, W. P. Elder, P. J. Harries, and T. Villamil. 1993. Molluscan biostratigraphy of the Cretaceous Western Interior Basin, North America; pp. 397–434 in W. G. E. Caldwell and E. G. Kauffman (eds.), Evolution of the Western Interior Basin. Geological Association of Canada Special Paper 39.

Leithold, E. L. 1993. Preservation of laminated shale in ancient clinoforms: comparison to modern subaqueous deltas. Geology 21:359–362.

Leithold, E. L. 1994. Stratigraphical architecture at the muddy margin of the Cretaceous Western Interior Seaway, southern Utah. Sedimentology 41:521–542.

Leithold, E. L., and W. E. Dean. 1998. Depositional processes and carbon burial on a Turonian prodelta at the margin of the Western Interior Seaway; pp. 189–200 in W. E. Dean and M. A. Arthur (eds.), Stratigraphy and Paleoenvironments of the Cretaceous Western Interior Seaway. Society for Sedimentary Geology Concepts in Sedimentology and Paleontology 6.

Lewan, M. D., and J. B. Maynard. 1982. Factors controlling enrichment of vanadium and nickelin the bitumen of organic sedimentary rocks. Geochimica et Cosmochimica Acta 46:2547–2560.

Lyons, T. W. 1997. Sulfur isotopic trends and pathways of iron sulfide formation in upper Holocene sediments of the anoxic Black Sea. Geochimica et Cosmochimica Acta 61:3367–3382.

Meyers, S. R., B. B. Sageman, and L. A. Hinnov. 2001. Integrated quantitative stratigraphy of the Cenomanian–Turonian Bridge Creek Limestone Member using evolutive harmonic analysis and stratigraphic modeling. Journal of Sedimentary Research 71:628–644.

Obradovich, J. 1993. A Cretaceous time scale; pp. 379–396 in W. G. E. Caldwell and E. G. Kauffman (eds.), Evolution of the Western Interior Basin. Geological Association of Canada Special Paper 39.

Peters, K. E. 1986. Guidelines for evaluating petroleum source rock using programmed pyrolysis. American Association of Petroleum Geologists Bulletin 70:318–329.

Pratt, L. M. 1984. Influence of paleoenvironmental factors on preservation of organic matter in Middle Cretaceous Greenhorn Formation, Pueblo, Colorado. American Association of Petroleum Geologists Bulletin 68:1146–1159.

Pratt, L. M. 1985. Isotopic studies of organic matter and carbonate in rocks of the Greenhorn marine cycle; pp. 38–48 in L. M. Pratt, E. G. Kauffman, and F. B. Zelt (eds.), Fine-Grained Deposits and Biofacies of the Cretaceous Western Interior Seaway: Evidence of Cyclic Sedimentary Processes. Society of Economic Paleontologists and Mineralogists, Field Trip Guidebook 4, 1985 Midyear Meeting, Golden, Colorado.

Pratt, L. M., and C. N. Threlkeld. 1984. Stratigraphic significance of $^{13}C/^{12}C$ ratios in mid-Cretaceous rocks of the Western Interior, U.S.A.; pp. 305–312 in D. F. Stott and D. J. Glass (eds.), Mesozoic of Middle North America. Canadian Society of Petroleum Geology Memoir 9.

Pratt, L. M., M. A. Arthur, W. E. Dean, and P. A. Scholle. 1993. Paleo-oceanographic cycles and events during the Late Cretaceous in the Western Interior Seaway of North America; pp. 333–354 in W. G. E. Caldwell and E. G. Kauffman (eds.), Evolution of the Western Interior Basin. Geological Association of Canada Special Paper 39.

Raiswell, R., and R. A. Berner. 1986. Pyrite and organic matter in Phanerozoic normal marine shales. Geochimica et Cosmochimica Acta 50:1967–1976.

Sageman, B. B., J. Rich, M. A. Arthur, G. E. Birchfield, and W. E. Dean. 1997. Evidence for Milankovitch periodicities in Cenomanian–Turonian lithologic and geochemical cycles. Journal of Sedimentary Research 67:286–301.

Sageman, B. B., J. Rich, M. A. Arthur, W. E. Dean, C. E. Savrda, and T. J. Bralower. 1998. Multiple Milankovitch cycles; pp. 153–171 in W. E. Dean and M. A. Arthur (eds.), Stratigraphy and Paleoenvironments of the Cretaceous Western Interior Seaway. Society for Sedimentary Geology Concepts in Sedimentology and Paleontology 6.

Schlanger, S. O., M. A. Arthur, H. C. Jenkyns, and P. A. Scholle. 1987. The Cenomanian–Turonian oceanic anoxic event, I. Stratigraphy and distribution of organic carbon–rich beds and the marine ^{13}C excursion; pp. 371–399 in J. Brooks and A. Fleet (eds.), Marine Petroleum Source Rocks. Geological Society of London Special Publication 26.

Schwarzacher, W., and A. G. Fischer. 1982. Limestone–shale bedding and perturbations of the earth's orbit; pp. 72–95 in G. Einsele and A. Seilacher (eds.), Cyclic and Event Stratification. Springer-Verlag, New York.

Scott, R. W., M. J. Evetts, P. C. Franks, J. A. Bergen, and J. A. Stein. 1998. Timing of mid-Cretaceous relative sea level changes in the Western Interior: Amoco #1 Bounds core; pp. 11–34 in W. E. Dean and M. A. Arthur (eds.), Stratigraphy and Paleoenvironments of the Cretaceous Western Interior Seaway. Society for Sedimentary Geology Concepts in Sedimentology and Paleontology 6.

Sethi, P. S., and E. L. Leithold. 1994. Climatic cyclicity and terrigenous sediment influx to the early Turonian Greenhorn Sea, southern Utah. Journal of Sedimentary Research 64:26–39.

Tissot, B. P. 1979. Effects on prolific petroleum source rocks and major coal deposits caused by sea-level change. Nature 77:463–465.

Tissot, B. P., and D. H. Welte. 1984. Petroleum Formation and Occurrence (2nd edition). Springer-Verlag, New York.

Zelt, F. B. 1985. Natural gamma-ray spectrometry, lithofacies, and depositional environments of selected Upper Cretaceous marine mudrocks, western United States including Tropic Shale and Tununk Member of Mancos Shale. Ph.D. dissertation, Princeton University, Princeton, New Jersey.

Tectonic and Sedimentary Controls, Age, and Correlation of the Upper Cretaceous Wahweap Formation, Southern Utah

4

Zubair Jinnah

THE WAHWEAP FORMATION WAS DEPOSITED BY FLUVIAL to estuarine systems in the central part of the Western Interior Basin, between the Sevier fold-and-thrust belt to the west and the Western Interior Seaway to the east, during the Late Cretaceous. Today, the Wahweap Formation is exposed on the Markagunt, Paunsaugunt, and Kaiparowits plateaus in southern Utah. Equivalent rocks, assigned to the Masuk and Tarantula Mesa formations, also crop out in the Henry Mountains in south-central Utah. After the initial, preliminary description of the Wahweap Formation by Gregory and Moore (1931), little follow-up work had been conducted in the formation until two decades ago, when a renaissance of geologic and paleontologic investigations by multiple workers led to great improvements in understanding the vertebrate faunas, depositional environments, sequence stratigraphy, and depositional age of the formation. This chapter summarizes the geologic work performed in the Wahweap Formation and its seaward equivalent strata in the Henry Mountains to date, as well as presenting new data and interpretations on the sedimentology and regional lithostratigraphic, chronostratigraphic, and sequence stratigraphic correlations.

A marked variation in alluvial style and architecture is observed in the lower portion of the Wahweap Formation between outcrop in the proximal Markagunt and Paunsaugunt plateaus (mudrock dominated with isolated channel bodies) compared to correlative distal exposures in the Kaiparowits Plateau (parts of the stratigraphy dominated by amalgamated channel bodies). The Masuk Formation in the Henry Mountains, which correlates to the lower part of the Wahweap Formation, bears remarkable similarity to the Wahweap Formation on the Kaiparowits Plateau, although preserving coal beds in its lower portion. These differences in alluvial architecture between the outcrops proximal to the orogenic belt (Markagunt and Paunsaugunt plateaus) and more distal (Kaiparowits Plateau and Henry Basin) is explained in terms of different allogenic controls (tectonism proximally versus eustasy distally) in these two separate regions.

The uppermost portion of the Wahweap Formation, called the capping sandstone member (Eaton, 1991), has more consistent alluvial architecture and thickness across

its proximal–distal extent. The Tarantula Mesa Formation, a more distal correlate of the capping sandstone, has similar alluvial architecture and thickness, and together, these coarse-grained units appear to represent a period of gravel progradation into the Western Interior foredeep during a period of low accommodation space associated with thrust quiescence in the Sevier belt.

INTRODUCTION

Western Interior Basin

The Western Interior Basin in North America extends from Alberta in the north to the Gulf of Mexico in the south and is the best-studied retroarc foreland basin in the world (Kauffman, 1977; Weimer, 1986; Dickinson et al., 1988; Kauffman and Caldwell, 1993; Van Wagoner, 1995; DeCelles and Currie, 1996; Catuneanu, Sweet, and Miall, 2000; Miall et al., 2008). Subsidence and subsequent sedimentation in the Western Interior Basin began as a result of eastward vergent thrusting (Sevier Orogeny) associated with subduction of the Farallon Plate beneath the North American Plate during the Late Jurassic, with widespread continental deposition of the Morrison Formation (DeCelles, 2004; Miall et al., 2008). Deposition extended until the Eocene, by which time basement-cored uplifts (Laramide Orogeny) had effectively partitioned the basin into a number of small, isolated depocenters (Goldstrand, 1994; DeCelles, 2004; Lawton, 2008). Although sedimentation transpired across the basin for over 100 Ma, extensive unconformities (commonly interpreted to record the passing of the forebulge) have removed much of the Upper Jurassic to Early Cretaceous strata in many parts of the basin, particularly across much of southern Utah. Because of this, and the exceptional exposure of Upper Cretaceous layers across much of the basin (e.g., San Juan Basin of northwestern New Mexico; Book Cliffs of central Utah; the disturbed belt and Judith Breaks of north and central Montana; Dinosaur Provincial Park of Alberta), Upper Cretaceous strata of western North America are among the best-studied sedimentary successions on Earth in terms of their geology and paleontology. This is also because Upper

Cretaceous rocks of the Western Interior Basin incorporate fluvial, marginal marine, and fully marine deposits, many of which are fossiliferous, and present unique opportunities to study and test models related to such diverse fields as vertebrate biogeography (Rowe et al., 1992; Lehman, 1997; Brinkman, Ryan, and Eberth, 1998; Sampson, 2009; Gates et al., 2010), sequence stratigraphy and fluvial architecture (Shanley and McCabe, 1991, 1995; Van Wagoner, 1995; Yoshida, Willis, and Miall, 1996; McLaurin and Steel, 2000, 2007; Adams and Bhattacharya, 2005), and foreland basin development (Fouch et al., 1983; Lawton, 1986; Heller and Paola, 1989; Catuneanu, Sweet, and Miall, 2000; DeCelles, 2004; Horton, Constenius, and DeCelles, 2004).

In the northern part of the Western Interior Basin (Montana and Wyoming), the study of interfingering terrestrial and marine deposits (Gill and Cobban, 1973), together with a high-resolution biostratigraphic framework (from ammonites in marine deposits), has revealed that the Western Interior Seaway underwent four transgressive–regressive cycles during the Late Cretaceous (Kauffman, 1977). The transgressive phases of these cycles, termed the Greenhorn (T6), Niobrara (T7), Claggett (T8), and Bearpaw (T9) cycles, are represented by westward-thinning tongues of marine shales that are preserved between contemporaneous continental sedimentary rocks. The Greenhorn transgression peaked at 92.8 Ma (Hancock and Kauffman, 1979) whereas the Niobrara cycle had several episodes of sea-level rise and fall that have been biostratigraphically determined to be diachronous, occurring between 89.3 and 87.6 Ma (Merewether, Cobban, and Obradovich, 2007). The Claggett transgression is widely correlated to a global eustatic sea level rise at ~80 Ma and is more tightly constrained to ~79.6 Ma in the Two Medicine Formation in Montana (Rogers, 1998). The ultimate Bearpaw transgression is diachronous, dated between 74.5 and 72 Ma in the northern Western Interior Basin, but has been shown to have peaked much earlier to the south in the San Juan and Kaiparowits basins, around 75.7 Ma (Molenaar et al., 2002; Roberts, 2007).

In the central part of the Western Interior Basin (Book Cliffs in central Utah), well-preserved fluvial–marine transitions are exposed in continuous cliff faces perpendicular to depositional strike. This style of preservation has allowed important sequence stratigraphic surfaces in the marine realm to be traced into the terrestrial deposits and was key to the development of nonmarine sequence stratigraphy (Van Wagoner, 1995; Robinson and Slingerland, 1998; Miall and Arush, 2001; McLaurin and Steel, 2007).

In southern Utah, Upper Cretaceous strata are present across a similar proximal–distal transect of the basin, although they are preserved in a number of structurally

distinct regions (the Markagunt, Paunsaugunt, and Kaiparowits plateaus and the Henry Mountains; Fig. 4.1). Detailed geological and paleontological work has only begun relatively recently in southern Utah, with the Kaiparowits Formation being the focus of most of the paleontological work, while geological work has been more evenly spread over the Straight Cliffs, Wahweap, and Kaiparowits formations (Shanley and McCabe, 1991, 1995; Lawton, Pollock, and Robinson, 2003; Roberts, Deino, and Chan, 2005; Roberts, 2007; Jinnah et al., 2009b). The Upper Cretaceous sequence remains best known from the Kaiparowits Plateau, where exposures are virtually continuous, whereas the Markagunt and Paunsaugunt plateaus are both more densely vegetated and thus more poorly understood, although new mapping by the Utah Geological Survey is addressing this. Poor correlation of formations and members across the three plateaus has hindered stratigraphic work and impeded attempts to fully understand development of the foreland basin and controls on sedimentation in southern Utah, and represents a major goal of this study and other researchers.

In this chapter, depositional environments, age, and taphonomy are reviewed, coupled with detailed characterization of thickness, facies, and alluvial architecture across a coeval sequence of inland (Wahweap Formation) to seaward (Masuk Formation) Upper Cretaceous alluvial to coastal plain strata in southern Utah (Fig. 4.1). This study is intended to provide depositional context for these fossil-rich deposits and test recent concepts and models of alluvial sequence stratigraphy and allogenic controls on sedimentation in the Western Interior Basin.

Kaiparowits Plateau

The first work to deal with the stratigraphic and structural features of the Kaiparowits Plateau was done by Gregory and Moore (1931), who coined the term Wahweap Sandstone for a 380-m-thick series of sandstone and shale characterized by a lack of coal beds and a paucity of fossils. Although Gregory and Moore (1931) did not subdivide the Wahweap, they did note that the upper third of the unit was dominated by cliff-forming sandstone. Peterson and Kirk (1977) studied Upper Cretaceous successions from parts of New Mexico, Utah, and Arizona and placed them into transgressive–regressive cycles. They correlated the Wahweap Formation in the Kaiparowits region with the Masuk Formation in the Henry Mountains region, both of which were assigned a Lower Campanian age. Eaton (1990) revised the stratigraphy of the Henry Mountains region, providing additional evidence for correlation between the Wahweap and Masuk formations.

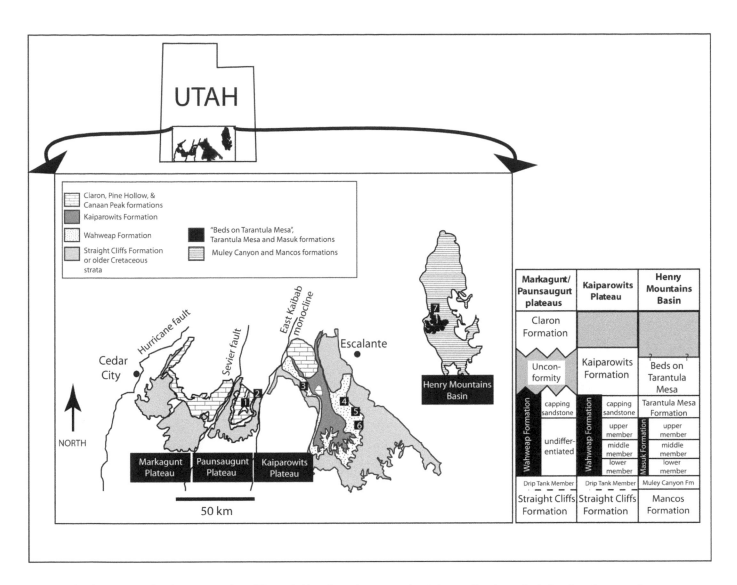

4.1. Map showing outcrop of Wahweap, Masuk, and Tarantula Mesa formations across the study area. Numbers refer to location of measured sections in Fig. 4.4.

Detailed bio- and lithostratigraphic work (Eaton, 1991) resulted in the subdivision of the Wahweap Formation on the Kaiparowits Plateau into four informal members. The lower member was defined as a series of interbedded mudstones and sandstones capped by a ~5-m-thick sandstone body. The middle member was defined as a mudrock-dominated unit characterized by recessive weathering, and the upper member was named for an increasingly sandstone-rich succession with interbedded mudstone. Together, the lower, middle, and upper members were found by Eaton (1991) to be 315 m thick at their type section at Reynolds Point. The capping sandstone member, which lies above the upper member, was measured by Eaton (1991) at its type section at Pardner Canyon and found to be 94 m, bringing the total composite thickness of the Wahweap Formation to 409 m. Eaton (1991) also undertook sedimentological study of the Wahweap

Formation and postulated a meandering river paleoenvironment for the lower, middle, and upper members and a braided river environment for the capping sandstone.

In contrast to Eaton (1991), who defined informal members for the Wahweap Formation on the Kaiparowits Plateau, Doelling (1997) divided the Wahweap into a lower and upper member, with the lower member consisting of interbedded sandstone, siltstone, and shale, whereas the upper member (which correlates with Eaton's capping sandstone) consists of orange to tan fine- to medium-grained sandstone. While Doelling's scheme does not provide as much resolution as that of Eaton on the Kaiparowits Plateau, it has the advantage of being applicable across the extent of the Wahweap Formation from the Markagunt to the Kaiparowits Plateau. In this chapter, the more commonly used subdivision of Eaton (1991) is adopted, with the thick sandstone/conglomerate unit at the top of the Wahweap Formation being referred to as the capping sandstone.

Following Eaton's (1991) stratigraphic work, a number of sequence stratigraphic and basin development studies have been undertaken on the Upper Cretaceous strata of the Kaiparowits Plateau (Shanley and McCabe, 1995; Little, 1997; Lawton, Pollock, and Robinson, 2003). These are discussed briefly here. Shanley and McCabe (1991, 1995) developed a model for nonmarine sequence stratigraphy through the study of contemporaneous terrestrial and marginal marine facies in the Straight Cliffs Formation. By studying alluvial architecture in the nonmarine facies and determining relative sea level through correlation with marginal marine facies, they determined that fluvial sequences in the Straight Cliffs Formation could be divided into amalgamated fluvial (regressive), isolated fluvial (transgressive), tidally influenced (high stand), alluvial plain, and coal-bearing facies tracts. The amalgamated fluvial facies tract, which overlies the sequence boundary, represents amalgamated channel bodies deposited in low-accommodation space settings immediately after regression.

Little (1997) provided a general sequence stratigraphic framework for the entire Upper Cretaceous sequence of the Kaiparowits Plateau that was based on initiation of tectonic subsidence and the generation of accommodation space. In his model, sequence boundaries are overlain by high-accommodation space, mud-rich deposits, and amalgamated facies cap stratigraphic sequences.

Lawton, Pollock, and Robinson (2003) also provided a sequence stratigraphic framework for rocks of the Drip Tank, Wahweap, and Kaiparowits formations, more similar in principle to the scheme that Shanley and McCabe (1995) used for the Straight Cliffs Formation. Lawton, Pollock, and Robinson (2003) placed the Drip Tank Member of the Straight Cliffs Formation and the lower member of the Wahweap Formation into the amalgamated fluvial facies tract, the middle member of the Wahweap Formation into the isolated fluvial facies tract, and the capping sandstone member of the Wahweap Formation into the amalgamated fluvial facies tract of the overlying sequence. They thus placed a sequence boundary between the upper and capping sandstone members of the Wahweap Formation.

In addition to their sequence stratigraphic work, Lawton, Pollock, and Robinson (2003) provided the first detailed interpretation of the basinal context in which the Wahweap Formation was deposited. They regarded the lower, middle, and upper members to have been deposited in a foredeep setting, whereas the capping sandstone was deposited in a wedge-top setting marked by the basin partitioning into different depocenters on the Markagunt, Paunsaugunt, and Kaiparowits plateaus. They presented two lines of evidence for the shift from a foredeep to a wedge-top depozone: first, both paleocurrent indicators and sandstone composition show that the lower, middle, and upper members had a source area in the Mogollan highland and Cordilleran magmatic arc to the south and southwest, whereas the capping sandstone had a source area in the Sevier fold-and-thrust belt to the west. Second, they found that the capping sandstone was thicker on the west side of the East Kaibab Monocline (Markagunt and Paunsaugunt plateaus) than on the east (Kaiparowits Plateau). They interpreted this to mean that syndepositional normal movement (possibly from sediment and tectonic loading on the hanging wall) along the East Kaibab Monocline partitioned the foreland basin at this time into separate depocenters on the Markagunt and Paunsaugunt plateaus (with high accommodation space) and the Kaiparowits Plateau (with low accommodation space). More recent work by Simpson et al. (2009) and Hilbert-Wolf et al. (2009) has documented the formation of seismites and fault-related sag ponds in the upper member and capping sandstone around the East Kaibab Monocline that are strongly suggestive of syndepositional movement along this structure during deposition of the capping sandstone.

Detailed local studies of the structural and sedimentological setting of the capping sandstone have been undertaken by Pollock (1999), Simpson et al. (2008), and Hilbert-Wolf et al. (2009). Pollock (1999) looked at alluvial architecture and thickness trends, as well as sandstone petrography, whereas Simpson et al. (2008) did a detailed study of sedimentary structures in the capping sandstone and concluded that the preservation of eolian stratification in subaerially exposed braid bars indicated general aridity at the time of deposition. Hilbert-Wolf et al. (2009) documented the preservation of seismites in the capping sandstone on the Kaiparowits Plateau close to the East Kaibab Monocline and attributed their development to either local normal faulting or regional seismicity in the Sevier fold-and-thrust belt.

The Wahweap Formation appears to be on the cusp of yielding a paleontological bonanza. Paleontological investigations in the formation only go back only as far as the 1980s, when Jeff Eaton and Rich Cifelli first recognized the wealth and significance of microvertebrates, especially mammals in the Wahweap Formation (Eaton and Cifelli, 1988; Cifelli, 1990; Eaton, 2002). They documented productive microvertebrate localities in the lower and middle members (Cifelli, 1990) as well as the upper member (Eaton, 1991), but not the capping sandstone. Only in the last decade has the large vertebrate potential in the Wahweap Formation begun to be assessed and realized. Work by the Utah Geological Survey and Utah Museum of Natural History crews has documented significant new dinosaur discoveries (Sampson et al., 2002; Kirkland and DeBlieux, 2005, 2010; DeBlieux et

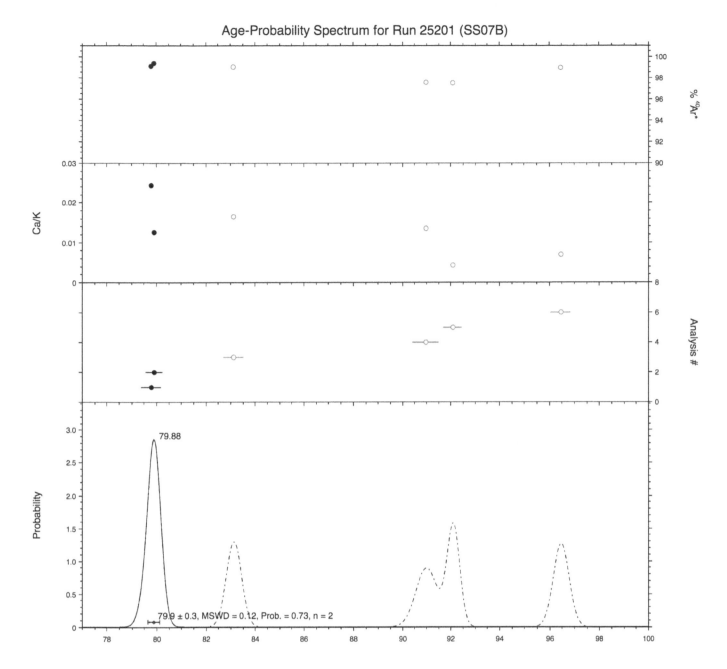

4.2. New age date from a reworked ash horizon in the middle member (54 m above the Drip Tank–Wahweap contact).

al., 2009; O'Connor et al., 2009), primarily within the lower and middle members, but also in the upper member. Alongside the wealth of mammal fossils and emerging dinosaur discoveries are a wealth of other small vertebrates, including turtles, crocodiles, and fish, but also an important record of continental invertebrates and plant macrofossils (Kirkland, Lewis, and Estep, 1998; Foster et al., 2001).

AGE OF THE WAHWEAP FORMATION

Until recently, the depositional age of the Wahweap Formation was constrained by comparison with mammalian faunas from the northern part of the Western Interior Basin (North

American Land Mammal Ages). The Wahweap mammalian fauna was found to belong to an earlier time than common Judithian faunas, and the Kaiparowits Basin thus gained recognition as recording an important pre-Judithian locality. Later work in the underlying Straight Cliffs Formation (particularly the John Henry Member) recorded definite Aquilan faunas (Eaton, 2006) that have contributed to a sense of uncertainty about the placement of the Wahweap in terms of its North American Land Mammal Age. This question remains to be resolved.

Radiometric dating of detrital zircons using U-Pb (Larsen, 2007; Dickinson and Gehrels, 2008) and devitrified ash (bentonite)-hosted sanidine (Ar-Ar; Jinnah et al., 2009b) from

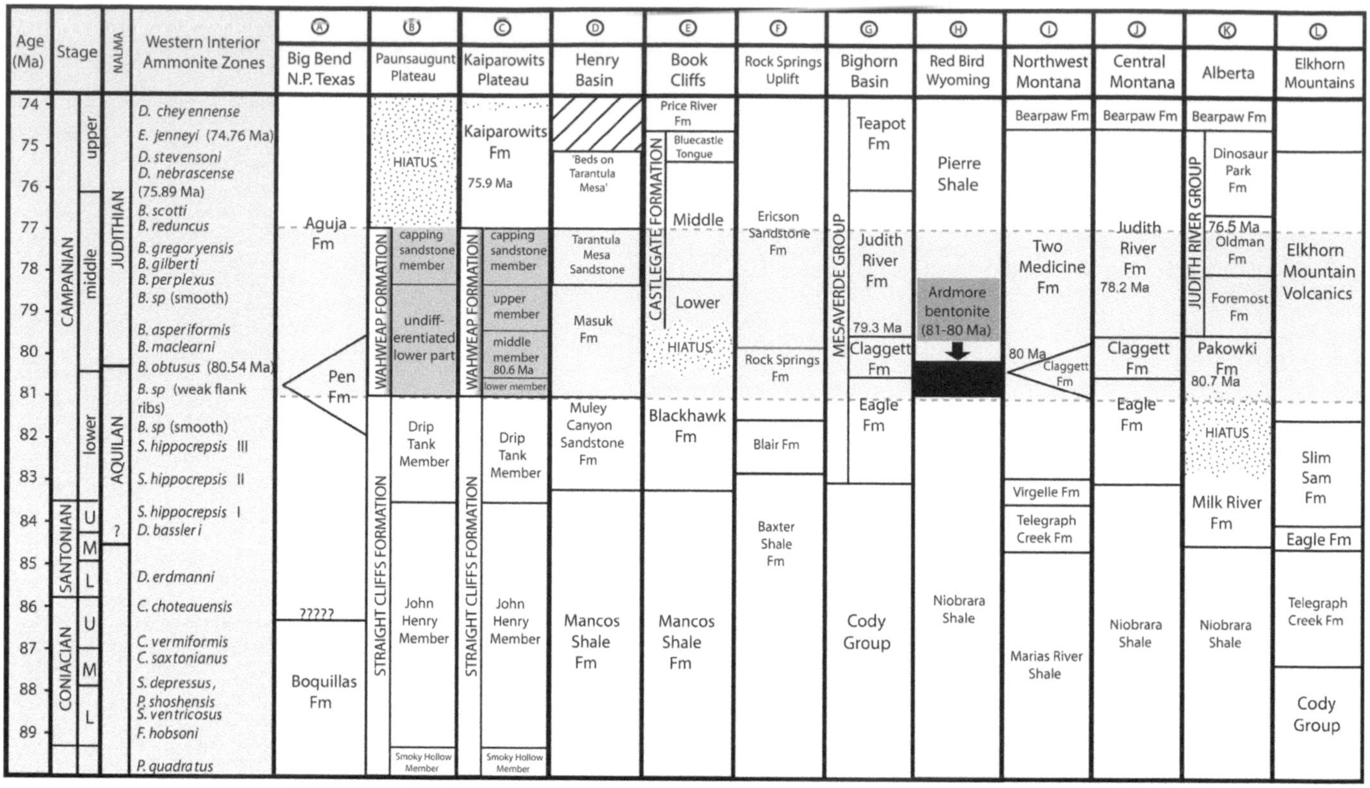

4.3. Revised correlation diagram for the Wahweap Formation (after Jinnah et al., 2009b). Ages for other successions are from various sources.

the Kaiparowits Plateau has helped to provide clarity on age and regional stratigraphic correlations for the Wahweap Formation. Detrital zircon dating in the formation has resulted in more informed provenance and paleofluvial drainage reconstructions across the basin; it has also provided improved age constraints on the otherwise undatable capping sandstone member, which hosts a zircon dated at 77 ± 2 Ma (Jinnah et al., 2009b).

$^{40}Ar/^{39}Ar$ dating was conducted on sanidine phenocrysts from a bentonite bed in the lower part of the middle member. Jinnah et al. (2009b) reported the age of this bentonite horizon as 80.1 ± 0.3 Ma. $^{40}Ar/^{39}Ar$ dating is a method in which ages are determined for the bentonite bed in question, as well as for a highly calibrated standard of known age (Renne et al., 1998). The 28.02 Ma Fish Canyon Tuff is currently the most widely used $^{40}Ar/^{39}Ar$ standard (Renne et al., 1994, 1998). However, Kuiper et al. (2008) recently recalibrated the Fish Canyon Tuff to 28.201 Ma. Ar-Ar dates that used the Fish Canyon tuff as a standard, including the date from the Wahweap Formation (Jinnah et al., 2009b), therefore also require revision. Recalculation of the age of the tuff in the middle member of the Wahweap Formation yields an age of 80.6 Ma. Additionally, a second ash bed has recently been dated. This ash bed occurs approximately 10 m above the 80.6 Ma ash bed (~60-m level) in the same Star

Seep field area. Although this new ash was exceptionally fine grained and partially contaminated, it did yield two young sanidine crystals that provide a concordant age of 79.9 ± 0.3 Ma using the recalibrated Fish Canyon Tuff Standard (Fig. 4.2). On the bases of these two ash beds from the lower part of the middle member and from dated ash beds from the base of the Kaiparowits Formation (76.5 Ma; revised from Roberts, Deino, and Chan, 2005) and Smoky Hollow Member of the Straight Cliffs Formation (approximately 91.9 Ma; revised from O'Connor et al., 2009), the lower contact of the Wahweap Formation is estimated to be approximately 81 Ma, and the upper contact is found to be approximately 77 Ma (Fig. 4.3).

Age constraint of the rocks assigned to the Wahweap Formation on the Markagunt and Paunsaugunt plateaus is less precise as a result of ambiguous biostratigraphy, poor lithostratigraphic correlation, and a lack of preserved ash beds for radiometric dating. In Cedar Canyon on the Markagunt Plateau, rocks equivalent to the Straight Cliffs Formation have been identified, which Eaton et al. (2001) dated to 86.72 ± 0.58 Ma. Several workers have regarded a conglomerate unit within this succession as the Drip Tank Sandstone (Moore and Straub, 2001; Lawton, Pollock, and Robinson, 2003; Sable and Hereford, 2004). The mudrock-rich strata that overlie this conglomerate are regarded as Santonian to

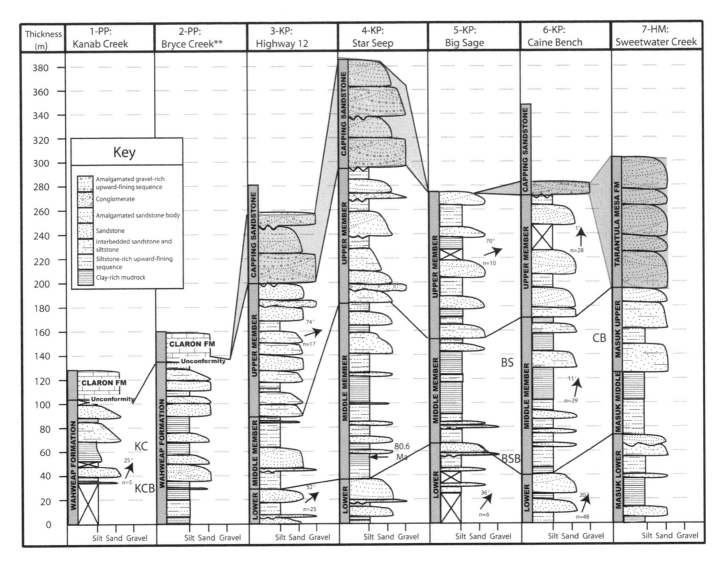

4.4. Stratigraphic sections and correlation through the Wahweap, Masuk, and Tarantula Mesa formations. MP, Markagunt Plateau; PP, Paunsaugunt Plateau; KP, Kaiparowits Plateau; HM, Henry Mountains. Section marked with an asterisk was measured partially by Tilton (1991) and partially by Pollock (1999). Section with a double asterisk was measured by Pollock (1999).

earliest Campanian (Nichols, 1997; Lawton, Pollock, and Robinson, 2003) and are thus probably older than the type Wahweap Formation. The rocks of Cedar Canyon are not dealt with further in this study.

On the eastern border of the Markagunt Plateau, rocks assigned to the Wahweap Formation by Pollock (1999) crop out along Highway 14. These rocks are dissimilar to the type Wahweap on the Kaiparowits Plateau and are significantly thinner. However, they are lithologically similar to the Wahweap Formation on the Paunsaugunt Plateau.

On the Paunsaugunt Plateau, there is a lack of clear age constraint on the Wahweap Formation. Preliminary biostratigraphic work on the Paunsaugunt Plateau by Eaton (1993) found faunal assemblages in the highest Cretaceous beds there that are similar to those of the Kaiparowits Formation on the Kaiparowits Plateau; however, the basis for these

correlations still remains equivocal. Complex facies changes and the dramatic thinning of the Late Cretaceous section across the Paunsaugunt Fault continue to make detailed correlations with the Kaiparowits Plateau difficult.

METHODS

Eleven stratigraphic sections were measured through the Wahweap Formation, and one section was measured through the Masuk Formation (Figs. 4.4, 4.5). Additionally, sections measured by others on the Markagunt and Paunsaugunt plateaus and Henry Mountains region have been added to the newly measured sections. Detailed sedimentological analysis was undertaken as the sections were measured. The results of this work are briefly summarized in Tables 4.1 and 4.2, but the bulk of this work has been presented elsewhere (Jinnah

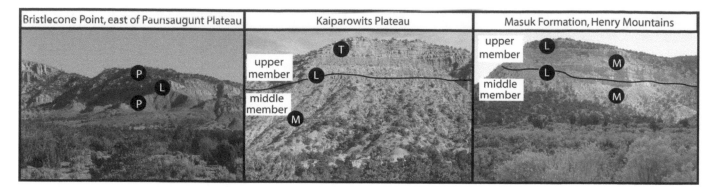

| Bristlecone Point, east of Paunsaugunt Plateau | Kaiparowits Plateau | Masuk Formation, Henry Mountains |

4.5. Dominant facies associations of the Wahweap Formation across the Paunsaugunt and Kaiparowits plateaus, and the Masuk Formation in the Henry Mountains. P, pedogenic mudrock; M, mudrock nonpedogenic; L, lenticular sandstone; T, tabular sandstone.

and Roberts, 2011). Stratigraphic architecture and inferred accommodation space were then compared across the Markagunt, Paunsaugunt, and Kaiparowits plateaus.

RESULTS

Wahweap Formation West of East Kaibab Monocline

On the Markagunt Plateau along Highway 89, the thickness of the lower, mudrock-rich part of the Wahweap has been measured at approximately 165 m by Tilton (1991). The sequence is dominated by red, yellow, and gray mudrock, in which nonpedogenic and pedogenic units can be distinguished. Nonpedogenic mudrock has a color of 5PB 5/1 blueish gray or 10Y 8/1 blueish white and may have a massive sedimentary texture or else contain weakly developed planar bedding. Pedogenic mudrock has a variable color (e.g., 5RP 4/2 grayish purple, 5R 4/4 weak red, 10YR 4/2 dark yellowish orange, 5R 8/8 red) and contains slickenside structures, root mottles (5Y 6/6 yellow), and horizontal stratification. Nonpedogenic and pedogenic mudrock typically occur cyclically through the lower part of the Wahweap Formation, with a nonpedogenic unit between 0.4 and 2 m thick at the base of the cycle and a pedogenic unit of between 1 and 2 m at the top. The cyclical interbedded nature of nonpedogenic and pedogenic mudrock therefore suggests that floodplain sequences on the Markagunt Plateau are cumulative paleosols.

Sandstone bodies have a lenticular geometry with erosive bases and sharp or gradational upper contacts. They lack internal erosive surfaces and can thus be classified as single story, and are typically less than 2 m thick and up to 20 m wide. Trough cross-stratification, ripple lamination, and planar bedding all occur within channel sandstones. The largest trough cross-stratification foresets are approximately 0.6 m thick. Although lateral accretion macroform structures are not obvious, the overall fluvial architecture is most consistent

with low-gradient channel systems with stable banks (i.e., meandering to anastomosing style streams). On the Markagunt and Paunsaugunt plateaus, there are no heterolithic stratification sets or any other sedimentary structures indicative of tidal influence within sandstone bodies.

The capping sandstone unit on the Markagunt Plateau is approximately 70 m thick (Pollock, 1999). The fluvial architecture within the capping sandstone is more amalgamated than in the lower part of the Wahweap Formation, and there is minimal preservation of overbank fines. Gravel bed forms and major and minor tabular sandstone bodies form the bulk of the capping sandstone. Intraformational and extraformational conglomerates lenses occur throughout. Overall, these features indicate a high-gradient, most likely braided fluvial system.

The Wahweap Formation on the Paunsaugunt Plateau is similar to that on the Markagunt Plateau. The lower part has a thickness of approximately 100 m and consists of a mudrock-dominated succession with a high degree of pedogenesis, including slickensides, root mottling, and horizonization. Thin sandstone bodies occur throughout the lower part of the succession, characterized by both lenticular and sheet-like geometries. Like the Markagunt Plateau, sandstone bodies on the Paunsaugunt Plateau lack internal erosive surfaces (although their basal surfaces may be erosive). Upper surfaces of sandstone bodies are sharp or gradational.

On the Paunsaugunt Plateau, outcrop of the capping sandstone is poor: in many places the entire unit has been eroded, and the Claron Formation unconformably overlies the lower part of the Wahweap Formation. Where the capping sandstone is preserved, it has been interpreted as paleovalley fill to account for its inconsistent thickness (Pollock, 1999).

Between the eastern edge of the Paunsaugunt Plateau at the Paunsaugunt fault and the western edge of the Kaiparowits Plateau at the East Kaibab monocline, there is a moderate amount of exposure of the lower, mudrock-dominated part of the Wahweap Formation along several hills. These outcrops (best observed at Bristlecone Point) constitute pedogenic mudrock and siltstones that closely resemble the lower Wahweap Formation on the Paunsaugunt Plateau. At this locality, the lower part of the Wahweap Formation is overlain unconformably by the Claron Formation.

Wahweap Formation on the Kaiparowits Plateau

Summaries of detailed facies and architectural analysis of the Wahweap Formation are given in Tables 4.1 and 4.2, which form the basis for comparison with new sedimentological data I present here for the Wahweap Formation on the Markagunt and Paunsaugunt plateaus, and the downdip correlative Masuk Formation. A more comprehensive review of facies and architecture of the Wahweap of the Kaiparowits Plateau is provided by Jinnah and Roberts (2011), with a summary here. Compared to the relatively thin exposures on the Markagunt and Paunsaugunt plateaus, the Wahweap Formation attains its maximum thickness of ~400 m on the Kaiparowits Plateau. Several well-developed, laterally extensive amalgamated sandstone bodies occur on the Kaiparowits Plateau and form the basis for subdividing the Wahweap Formation into the lower, middle, upper and capping sandstone members. Additionally, the Wahweap Formation on the Kaiparowits Plateau is more sand rich close to the East Kaibab Monocline in the west and more mud rich toward the east. In contrast to the Markagunt and Paunsaugunt plateaus, vertical stacking patterns of a number of stratigraphic sections on the Kaiparowits Plateau show distinct cyclicity in alluvial architecture. For example, laterally extensive amalgamated sandstone bodies occur in the lower, upper, and capping sandstone members, whereas aggradational mudrock occurs in the lower part of the lower member and the entire middle member. Mudrock units typically display features indicative of subaqueous deposition at their base and there is a transition to features more consistent with subaerial exposure and incipient pedogenesis toward the top. Minor occurrence of calcrete nodules in mudrocks (and both calcrete and mudrock fragments as rip-up clasts in intraformational conglomerates) indicate that the environment of deposition was seasonally arid. Sandstone bodies (both lenticular and sheet geometry) are either single story or multistory and are typically larger than on the Markagunt or Paunsaugunt plateaus. On the Kaiparowits Plateau, large sandstone bodies may extend for tens to hundreds of meters (or more) and be tens of meters thick. Several facies associations indicative of tidal influence (Table 4.2) have been documented in the middle and upper members.

The lower three members of the Wahweap Formation on the Kaiparowits Plateau attain a thickness of between 200 m in the north and 280 m in the south. The capping sandstone is at its thickest to the north of the Kaiparowits Plateau, where it attains a thickness of more than 140 m. Across most of the southern part of the Kaiparowits Plateau, the capping sandstone was extensively eroded during the Neogene, but where the complete unit is seen to occur, its thickness is found to be between 60 m (close to Death Ridge) and 90 m (at Star Seep on Smoky Mountain Road; Fig. 4.6).

Masuk and Tarantula Mesa Formations in the Henry Mountains Region

The Masuk Formation in the Henry Mountains region is regarded as equivalent to the lower part of the Wahweap Formation, whereas the Tarantula Mesa Formation is equivalent to the capping sandstone (Eaton, 1990). Like the lower part of the Wahweap on the Kaiparowits Plateau, the Masuk Formation is divided into a lower, middle, and upper member with virtually identical distinguishing characteristics. Together, the three members of the Masuk Formation attain a thickness of 194 m. Overall, the Masuk Formation contains more mudrock than the Wahweap Formation, and in this succession, mudrock typically possesses well-developed laminations and a drab color. There is little evidence of subaerial exposure, pedogenesis, or seasonal aridity. Additionally, unlike the Wahweap Formation, there is a dominance of lenticular over tabular sandstone body geometry in the sandstone-rich lower and upper members. Sandstone bodies vary from meters to hundreds of meters in lateral extent. The Tarantula Mesa Formation, which is 100 to 125 m thick as measured by Smith (1983), is similar to the capping sandstone member in both architecture and thickness, but the capping sandstone generally contains beds of both intraformational and extraformational conglomerate that are not well developed in the Tarantula Mesa Formation.

DISCUSSION

Interpretations of Accommodation Space

Interpretations of accommodation space in the Markagunt, Paunsaugunt, Kaiparowits, and Henry Mountains basins are made on the basis of alluvial architecture and total stratigraphic thickness (Table 4.3). On the Markagunt and Paunsaugunt plateaus, the isolated nature of channels in the lower part of the Wahweap Formation, and their entombment in thick pedogenic mudrock sequences, coupled with their lack of significant internal erosional discontinuities, suggests that

Table 4.1. Lithofacies of the Wahweap Formation

Lithofacies	Description	Grain size/description	Color	Features
Gmi	Matrix supported trough cross-bedded or massive intraformational conglomerate	*Clasts*–Intraformational siltstone and mudstone, shell, bone and wood, 2–200 mm, rounded to angular, poorly sorted. *Matrix*–Medium-coarse sand.	*Clasts*–Medium light gray (10Y 6/2). *Matrix*–Grayish orange pink (10R 8/2) to reddish brown (10R 5/4).	Occurs in scour fills, at the base of sandstone-dominated units, or overlies erosional surfaces within amalgamated sandstone bodies.
Gce	Clast supported trough cross-bedded extraformational conglomerate	*Clasts*–Typically sub-rounded to well rounded chert or limestone, 2–100mm with rare wood or bone. *Matrix*–Medium-coarse sand.	*Matrix*–Pale yellow (5Y 8/3) to moderate orange pink (5Y 8/4).	Typically erode into mudstone or siltstone, and grade upward into sandstone. Forms units up to several meters thick.
Gci	Clast supported intraformational conglomerate	*Clasts*–Intraformational siltstone and mudstone, fragmentary shell, bone and wood, 2–150 mm, rounded to sub-rounded. *Matrix*–Medium-coarse sand.	*Clasts*–Medium light gray (10Y 6/2). *Matrix*–Pale yellow (5Y 8/3) to moderate orange pink (5Y 8/4).	Forms localised units overlying erosional surfaces within amalgamated sandstone bodies.
Gs	Clast supported shell conglomerate	*Clasts*–Fragmentary to complete shells (gastropods and bivalves; 90%). Fragmentary vertebrate remains. Granule sized intraformational siltstone. *Matrix*–Medium-coarse sand.	*Matrix*–Pale yellow (5Y 8/3) to moderate orange pink (5Y 8/4).	Forms localised units, up to 1 m thick, which contain 90% shell material (dominantly fragmentary), as well as minor vertebrate bone and intraformational mudrock.
Gc	Clast supported calcrete nodule conglomerate	*Clasts*–Entire or fragmentary calcrete nodules, 20–150 mm. *Matrix*–Medium sand.	*Clasts*–White (8N) to very pale orange (10YR 8/3) or strong brown (5YR 4/6). *Matrix*–Pale yellow (5Y 8/3) to moderate orange pink (5Y 8/4).	Forms thin but laterally extensive units. Often overlies deeply incisive contacts within thick sandstone bodies.
Sh	Horizontally laminated sandstone	Very fine to medium sand, minor silt.	Very pale orange (10YR 8/3).	Mud drapes on cosets. Plant impressions on bedding plane.
Shl	Horizontally laminated sandstone with parting lineation	Medium sand.	Pale yellow (5Y 8/3).	Parting lineation.
Sr	Rippled sandstone	Very fine to medium sand.	Dusky red (5R 3/4) to pale yellow (5Y 8/3).	Localised facies, associated with Sh and Fg. Sr often grades upward into Fg.
St	Trough cross-bedded sandstone	Fine to very coarse sand. Rare granule- to pebble-sized clasts.	Pale yellow (5Y 7/4) to pale red (5R 6/2).	Often overlies Gc, Gm or Sm. Often has fragmentary plant or vertebrate remains.
Sp	Planar cross-bedded sandstone	Medium sand.	Pale yellow (5Y 8/3) to moderate orange pink (5Y 8/4), greyish orange pink (10R 8/2).	Occurs in association with St or Sm, or finely interbedded with Fg. Sometimes has wood with *Teredolites* boring or other plant material with sulfurous rims.
Sm	Massive to weakly laminated sandstone	Fine to coarse sand. Rare granule- to cobble-sized siltstone clasts.	Pale yellow (5Y 7/4) to greyish orange pink (10R 8/2).	Often has erosive base and sometimes has well cemented sulfide nodules.
Fp	Red/yellow pedogenic mudstone and siltstone	Clay, silt, minor sand.	Light brown (5YR 5/8) to pale reddish brown (10R 5/4).	Oriented nodules, rhizocretions and wood fragments. Weakly developed horizontal laminations.
Fg	Gray laminated claystone and siltstone	Clay, silt.	Grayish red purple (5RP 4/2) to medium light gray (10Y 6/2).	Well developed horizontal laminations occur. Contains nodules (goethite?) and invertebrates (molluscs and crabs), vertebrates and plant matter. Often has gypsum veins.
Fc	Carbonaceous mudstone	Clay, silt.	Dark gray, weak red (5R 4/2) or dark yellowish brown (10YR 4/2).	Localised beds associated with Fg. High organic content, including wood fragments with sulfurous rims. Has gypsum veins and weakly developed horizontal laminations.
Fcr	Calcrete nodule siltstone	Silt with minor sand.	5Y 8/2 pale greenish yellow (matrix) 8N white (nodules).	Dense accumulations of calcium carbonate nodules up to 100 mm in diameter.
Fb	Bentonitic claystone	Clay with minor silt, sand.	Pale yellow (5Y 8/3) to blueish gray (5PB 5/1).	Phenocrysts of biotite and feldspar, concentrated near the base. Upper contact gradational with overlying Fg. Distinctive popcorn weathering due to clay content.
C	Coal	–	Black.	Occurs as isolated lenses, less than 0.5 m wide and 10 cm thick, in association with Fg and Fc.

Jinnah

Table 4.2. Facies associations of the Wahweap Formation

Facies association	Description	Lithofacies[a]	Diagnostic features	Fossils	Interpretation	Occurrence[b]
FA1	Single-story lenticular and stone	Gci, Gmi, Sh, St, Sp, Sr	Erosive base, no internal erosive surfaces, lenticular geometry	Fragmentary bone, wood	Channel fill	MK, PN, KP, middle KP, upper HM
FA2	Multistory lenticular sandstone	Gci, Gmi, St, Sh, Sp	Erosive base, internal erosive surfaces, lenticular geometry	Shell conglomerate, fragmentary bone, wood	Amalgamated channel fill deposits	KP, middle KP upper HM
FA3	Major tabular sandstone body	Gci, Gmi, Gc, Gs St, Sp, Sh, Shl, Sr, Sm, Fcr, C	Erosive/sharp base, internal erosive surfaces, tabular geometry	Shell conglomerate, fragmentary bone, wood	Amalgamated channel fill deposits	KP, lower KP, upper KP, capping sandstone
FA4	Gravel bed form	Gce, Gci	Erosive/sharp base, elongate parallel to paleocurrent	Fragmentary vertebrate remains (extremely rare)	Braid bars	KP, lower KP, capping sandstone
FA5	Low-angle heterolithic cross-strata	Sp, Fg	Erosive base, interlaminated sand, silt and clay at cm scale, laminations inclined relative to basal surface	Wood	Inclined heterolithic stratification	KP, middle KP, upper HM
FA6	Minor tabular sandstone body	Sh, Sr	Sharp base, no internal erosive surfaces, tabular geometry, fining upward	None	Crevasse channel	MK, PN, KP, lower KP, middle KP, capping sandstone, HM
FA7	Lenticular interlaminated sandstone and mudrock	Sh, St, Fg	Erosive base, lenticular geometry, sand and silt interlaminated parallel to basal surface	Fragmentary bone, wood	Tidally-influenced channel	KP, middle KP, upper HM
FA8	Inclined interbedded sandstone and mudrock	Sh, Fg, Fp	Sharp base, interlaminated sand, silt and clay at cm to m scale, fining upward	Abundant fragmentary or complete vertebrate, invertebrate and plant remains	Proximal overbank deposits–levee, splay	KP, lower KP, middle KP, upper HM
FA9	Laterally extensive mudrock	Fg, Fc, Fp, Fb, Fcr, C	Sharp or gradational base, abundant planar bedded mudrock	Fragmentary or complete vertebrate, invertebrate and plant remains	Distal overbank deposits–back swamp, pond, lake, and soil	KP, middle HM
FA10	Lenticular mudrock	Sm, Fg	Erosive base, lenticular geometry, abundant mudrock	None	Muddy channel	KP, middle HM
FA11	Pedogenic mudrock	Fp, Fg	Slickenside structures, root mottles, preserved soil horizons	None	Cumulative paleosol	MK, PN

[a] See Table 4.1 for a listing of lithofacies.

[b] MK, Markagunt Plateau; PN, Paunsaugunt Plateau; KP, Kaiparowits Plateau; HM, Henry Mountains.

deposition took place in a relatively high accommodation space setting. Cumulative paleosols, such as those described here from the Markagunt and Paunsaugunt plateaus, indicate conditions where sedimentation occurs at a constant rate (Kraus, 1999). On the Paunsaugunt Plateau, the preserved thickness of the succession is only 100 m (although the composite thickness may be greater), but the alluvial architecture suggests a high accommodation space setting. Because the lower part of the Wahweap Formation is mostly overlain unconformably by the Paleocene–Eocene Claron Formation in this area, the thickness is most likely accounted for by erosion between the Campanian and the Paleocene. Santucci and Kirkland (2010) suggest that this erosion is related to uplift along the Paunsaugunt fault during the latest Cretaceous to Early Tertiary.

The capping sandstone, which contains numerous erosive surfaces, intraformational conglomerate lenses, and very little preserved overbank material, was deposited under conditions of much less accommodation space than the lower part of the Wahweap Formation.

On the Kaiparowits Plateau, accommodation space varied during deposition of the lower, middle, and upper members. Deposition of the lower member took place in a low accommodation space setting, and the middle member was deposited in a high accommodation space setting. The basal part of the upper member was deposited under high accommodation space settings, although accommodation space gradually declined as the member was deposited, and the amalgamated sandstones at the top of the upper member suggests that deposition at that time occurred in a low accommodation space setting. Alluvial architecture of the capping sandstone on the Kaiparowits Plateau is mostly similar to that on the Markagunt and Paunsaugunt plateaus and indicates low accommodation space.

In the Henry Mountains, the lower, middle, and upper members of the Masuk Formation closely mirror the

4.6. Stratigraphy of the Wahweap Formation on the Kaiparowits Plateau. (A) Star Seep, southern Kaiparowits Plateau. (B) Upper Valley along Utah Highway 12, northern Kaiparowits Plateau.

Wahweap Formation in terms of the position of amalgamated sandstone units, and these members were deposited under similar accommodation space settings to their equivalents on the Kaiparowits Plateau. The Tarantula Mesa Formation, like the capping sandstone, has numerous erosive surfaces and a lack of overbank material; it is similarly interpreted to have been deposited in a low accommodation space setting.

Environmental Variation from the Markagunt Plateau to the Henry Mountains

Distinct differences in sedimentology are recognized between the lower part of the Wahweap Formation on the Markagunt and Paunsaugunt plateaus and the Kaiparowits Plateau and Henry Mountains. The rivers that deposited the lower part of the Wahweap Formation on the Markagunt and Paunsaugunt plateaus had small channels (as evidenced by the size of lenticular sandstones) with confined flow. The sediment that accumulated on the floodplains of these rivers

was well drained and vegetated, and underwent pedogenesis to a significant degree, resulting in the development of features such as slickensides, root mottling, and horizonization. Kraus (1999) documented the relationship between sedimentation rate and pedogenesis. Cumulative soil horizons such as those described here from the Markagunt and Paunsaugunt plateaus are typical of scenarios where floodplains receive constant sedimentation and there is little erosion. Pedogenesis in these conditions is interrupted by sedimentation on the floodplain, resulting in interbedded pedogenic and nonpedogenic horizons.

There was little change in environment throughout deposition of the succession on the Markagunt and Paunsaugunt plateaus, resulting in a fairly uniform lithostratigraphy. The lack of such features as heterolithic stratification and brackish-water trace fossils suggest a fully fluvial environment.

On the Kaiparowits Plateau, larger channels with more interconnectivity developed, resulting in the deposition of amalgamated sheet sandstones. Overbank deposition

Table 4.3. Summary of thickness (T) and accommodation space (AS) conditions across the extent of the Wahweap, Masuk, and Tarantula Mesa formations

	Paunsaugunt Plateau			Kaiparowits Plateau			Henry Mountains		
Wahweap unit	T	AS	Formation/member	T	AS	Formation/member	T	AS
Capping sandstone	75 ma	Low	Wahweap/capping sandstone	90 m	Low	Tarantula Mesa Formation	125 m[b]	Low
Lower part	100 m	High	Wahweap/upper	280 m	High, decreasing	Masuk/upper	194 m	High, decreasing
			Wahweap/middle		High, increasing	Masuk/middle		High, increasing
			Wahweap/lower		Low, increasing	Masuk/lower		Low, increasing

[a] Thickness measured by Tilton (1991).

[b] Thickness measured by Smith (1983).

Jinnah

occurred abundantly, especially in the lower and middle members. Sediment accumulations on floodplains underwent incipient pedogenesis, as evidenced by the presence of slickensides and root mottling, but well-defined soil horizons did not develop. Soil development was likely interrupted by rapid and continuous overbank deposition or else by large standing bodies of water developing on low-lying floodplains. Heterolithic stratification with regular sand–mud couplets at the top of the middle and base of the upper member record a marine incursion and the development of estuarine conditions in the Kaiparowits Plateau.

In the more distal Masuk Formation, large channels persisted. Floodplain sediment accumulations that bear the remains of aquatic invertebrates and have no evidence of pedogenesis suggest a constantly waterlogged floodplain with large standing bodies of water in a low-lying coastal plain. Although heterolithic stratification is abundant throughout the section, the invertebrate fossils reported by Eaton (1990) from the Masuk Formation suggest a dominantly freshwater setting with periodic incursions of brackish water.

In terms of fossil preservation throughout the stratigraphic section, several trends are notable. On the Markagunt and Paunsaugunt plateaus, fossils are most commonly preserved as individual elements from a variety of taxa within fossiliferous horizons (microsites) (Eaton, 1999). Sites with associated or articulated material are uncommon within the Wahweap Formation in these areas. On the Kaiparowits Plateau, where the Wahweap Formation is very well exposed and has been more extensively prospected, fossil microsites that yield a higher number of taxa represented by isolated or fragmentary remains are most common in the sandstone-rich lower and upper members (Cifelli, 1990; Eaton, 1991). Sites with associated or articulated remains are much less common than microsites (DeBlieux et al. 2009); the mudrock-rich middle member is more likely to contain such sites. Jinnah et al. (2009a) document a site with associated hadrosaur remains from the middle member.

These trends in fossil preservation are related to the environment of deposition and are ultimately controlled by the accommodation space available at time of deposition. Large amalgamated channel bodies of the lower and upper members typically contain much material that has been extensively reworked, resulting in preservation of isolated elements, whereas rapid floodplain sedimentation in the middle member (and parts of the lower member) was more conducive to preserving associated fossil material.

Sequence Controls

There is a change in fluvial architecture and sandstone: mudstone ratio in the lower part of the Wahweap Formation across the East Kaibab Monocline/Paunsaugunt fault. To the west, mudrock-dominated strata with a high degree of pedogenesis and a mostly uniform vertical stacking pattern occur, whereas to the east, the Wahweap and Masuk formations display an overall upward-coarsening pattern. On the Markagunt and Paunsaugunt plateaus, the dominant architectural style of isolated lenticular sandstone bodies encased in mudrock suggests that high-accommodation space conditions prevailed throughout deposition whereas on the Kaiparowits Plateau and in the Masuk Formation the variation in fluvial architecture suggests that there was fluctuation between high- and low-accommodation space in these regions. The poor correlation of high- and low-accommodation space deposits across the East Kaibab Monocline suggests that different mechanisms were responsible for generating and influencing accommodation space in these discrete regions during deposition of the lower part of the Wahweap Formation. Because the western deposits were proximal to the Sevier fold-and-thrust belt and far from the contemporaneous shoreline, it is most likely that tectonic forces were the controlling mechanism for accommodation space generation in this part of the basin. For the deposits of the Kaiparowits Plateau and the Henry Mountains, which were proximal to the contemporaneous shoreline, eustatic variations in the Western Interior Seaway would have also had an influence on sedimentation. As such, the Markagunt and Paunsaugunt plateaus host the deposits of proximal river systems (sensu Catuneanu, 2006), whereas the Kaiparowits Plateau and Henry Mountains Basin host distal deposits.

By the time of capping sandstone and Tarantula Mesa deposition, eastward migration of the fold-and-thrust belt, together with a regional drop in base level, eliminated eustatic influence as a factor in deposition and alluvial architecture. The deposition of terrestrial strata (Tarantula Mesa Formation) across the former shoreline, with little evidence of tidal influence, in the Henry Mountains Basin suggests that the Western Interior Seaway had retreated significantly by this time.

Basin Development

Within foreland systems, four depocenters – wedge top, foredeep (foreland basin), forebulge, and back bulge – have been identified (DeCelles and Giles, 1996; Catuneanu, 2004). Each of these depocenters is characterized by different patterns of lateral and vertical stratigraphic architecture, as well as by varying accommodation space exemplified in the concept of reciprocal stratigraphies, where high accommodation space deposits in the foredeep are contemporaneous with low accommodation space deposits in the forebulge and vice versa (Heller et al., 1988; Catuneanu, 2004). The most

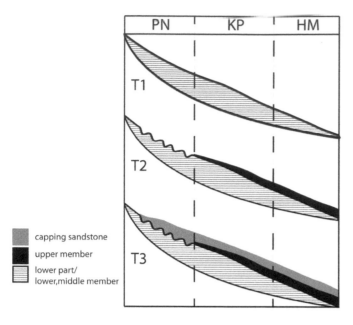

4.7. Diagram showing deposition of different members across the study area through time (T1, T2, and T3). PN, Paunsaugunt Plateau, KP, Kaiparowits Plateau, HM, Henry Mountains.

proximal depocenter, the wedge top, is characterized by piggyback basins that develop on active thrust sheets (Ori and Friend, 1984), coarse clastic wedges, discontinuous facies, and syndepositional deformation of sediments by tectonic activity. The foredeep is typically the zone of maximum subsidence and rapid generation of accommodation space due to crustal flexure as a response to tectonic loading in the adjacent thrust belt. Periods of intense thrust activity therefore drive near concurrent accommodation space generation in the foredeep. In the forebulge, deposition is controlled by the underfilled versus overfilled nature of the basin and the rate of sediment supply (Burbank et al. 1989; Catuneanu, 2004).

Studies of the sedimentary record that focus on identifying different depocenters and the transitions between these depocenters have important implications for understanding foreland basin evolution and controls on the sedimentary record and can also aid in local and regional correlation by explaining proximal–distal facies variations (Ori and Friend, 1984; Lawton and Trexler, 1991; DeCelles, 2004). The alluvial architecture of the lower part of the Wahweap Formation on the Markagunt and Paunsaugunt plateaus suggests that deposition took place in a relatively high accommodation space setting with no discernible changes in facies architecture within this interval. However, the maximum preserved thickness of this unit is about 165 m along Highway 89–less than the lower, middle, and upper members on the Kaiparowits Plateau (280 m), where the alluvial architecture suggests that accommodation space was more variable. It is therefore likely that there was deep incision along the lower

Wahweap-capping sandstone contact on the Markagunt and Paunsaugunt plateaus.

In the study of Lawton, Pollock, and Robinson (2003), the capping sandstone is interpreted to form in a wedge top depocenter. Evidence cited for this includes the drastic change in alluvial architecture (and hence accommodation space), sandstone petrology (from quartzolithic/subfeldspathic to quartzose; Pollock, 1999), and inconsistent thickness of the capping sandstone across the Paunsaugunt and Kaiparowits plateaus.

The capping sandstone is overlain unconformably by the Paleocene–Eocene Claron Formation on the Markagunt and Paunsaugunt plateaus. In many places, the capping sandstone is not present at all, and the Claron Formation lies directly on the lower part of the Wahweap Formation. In some places, the entire Wahweap Formation is missing and the Claron overlies the Drip Tank Sandstone. On the southern Kaiparowits Plateau, the capping sandstone is undergoing modern-day erosion across most of its extent, but the total thickness has been measured at between 60 and 90 m. The consistency of both the thickness and the fluvial architecture of the capping sandstone and equivalent Tarantula Mesa Formation across large structural features such as the Sevier and Paunsaugunt faults and the East Kaibab monocline, which separate distinct proximal–distal facies in the lower part of the Wahweap Formation, show that the capping sandstone forms an extensive conglomeratic sheet that was deposited across the boundaries of several structural basins (Fig. 4.7). The presence of seismites close to the East Kaibab monocline (Hilbert-Wolf et al., 2009) shows that some syndepositional tectonic activity occurred along this structure, but the similar thickness of the capping sandstone on either side of the structure, together with the remarkable continuity of lithofacies, indicates that the Markagunt, Paunsaugunt, and Kaiparowits basins were not fully partitioned at the time of deposition, and syndepositional fault movement did not disrupt sedimentation patterns in the wedge top.

Alternatively, the continuity of the capping sandstone across the Sevier fault and East Kaibab monocline could indicate that deposition took place in a foredeep rather than wedge-top setting, during a time of thrust quiescence and low-accommodation space in the foredeep. There has been debate about the exact conditions required for progradation of coarse-grained units into the distal parts of a foreland basin. In the two-phase model of Heller et al. (1988), gravel deposits accumulate in the proximal reaches of the foredeep syntectonically, when uplift of thrust sheets and simultaneous subsidence in the foredeep provide a source area and abundant accommodation space respectively. Progradation of coarse-grained sediment into the distal reaches of the foredeep occurs during periods of thrust quiescence when accommodation space in the proximal foredeep has been reduced and transverse drainages are set up. Burbank et al. (1989) later recognized that other factors such as rate of sediment supply also affect the distance that coarse-grained units prograde. Although syndepositional movement along large faults in the foredeep may affect the progradation distance of coarse-grained units, this does not appear to be the case in the capping sandstone–Tarantula Mesa system, which is largely uniform in thickness across the Sevier fault, Paunsaugunt fault, and East Kaibab monocline.

The great distance of progradation of coarse-grained detritus in the capping sandstone–Tarantula Mesa system suggests that deposition took place during a phase of thrust-belt quiescence, which is in keeping with the interpretation of a low accommodation space depositional setting for these rocks. In the Sevier fold-and-thrust belt in central Utah (close to the source of the capping sandstone), thrusting on the Paxton thrust ended in the Early Campanian, whereas thrusting on the Gunnison thrust began in the Late Campanian (DeCelles and Coogan, 2006). A period of thrust quiescence during capping sandstone and Tarantula Mesa deposition in the mid-Campanian is therefore reasonable.

CONCLUSIONS

Study of the Wahweap Formation across the Markagunt, Paunsaugunt, and Kaiparowits plateaus, as well as the Masuk and Tarantula Mesa formations, suggests that variable accommodation space was generated in these different regions during deposition of the lower part of the succession. The lower part of the Wahweap Formation on the Markagunt and Paunsaugunt plateaus is characterized by the deposits of small channels in a high accommodation space setting. On the Kaiparowits Plateau, accommodation space varies from low in the lower member to high in the middle member, and back to low in the upper member. The Masuk Formation, also divided into lower, middle, and upper members, mirrors the pattern in the Kaiparowits Plateau. The variation in accommodation space suggests that different controlling mechanisms operated proximally and distally: tectonic activity in the Sevier fold-and-thrust belt was the driving force of sequences in the Markagunt and Paunsaugunt plateaus, whereas a combination of eustasy and tectonics drove sequence development in the Kaiparowits Plateau and Henry Mountains Basin.

During deposition of the capping sandstone and Tarantula Mesa Formation, withdrawal of the Western Interior Seaway and the development of fully terrestrial strata across the study area indicate that tectonic forces exerted the major control on the sequence. The proximal–distal continuity of facies in the capping sandstone and Tarantula Mesa Formation, together with their consistent thickness, suggests that deposition of these strata took place in the foredeep depocenter as a result of the distal progradation of coarse sediments during a period of thrust quiescence.

ACKNOWLEDGMENTS

Funding for this project came in part from the BHP Billiton Academic Fellowship. I thank A. Titus, D. Powell, and M. Eaton from the BLM and J. Rechsteiner from the Forest Service for assistance with research permits. A. Deino (Berkeley Geochronology Center) ran sample SS07B for new age dates. Thanks are due to E. Roberts for much assistance and useful discussions, as well as to M. Getty and the UMNH field crew for their camaraderie in the field. This chapter was greatly improved by reviews from J. Kirkland and J. Eaton.

REFERENCES CITED

Adams, M. M., and J. P. Bhattacharya. 2005. No change in fluvial style across a sequence boundary, Cretaceous Blackhawk and Castlegate formations of central Utah, U.S.A. Journal of Sedimentary Research 75:1038–1051.

Brinkman, D. B., M. J. Ryan, and D. A. Eberth. 1998. The paleogeographic and stratigraphic distribution of ceratopsids (Ornithischia) in the Upper Judith River Group of Western Canada. Palaios 13:160–169.

Burbank, D. W., R. A. Beck, R. G. H. Raynolds, R. Hobbs, and R. A. K. Tahirkheli. 1989. Thrusting and gravel progradation in foreland basins: a test of post-thrusting gravel dispersal. Geology 16:1143–1146.

Catuneanu, O. 2004. Retroarc foreland systems–evolution through time. Journal of African Earth Sciences 38:225–242.

Catuneanu, O. 2006. Principles of Sequence Stratigraphy. Elsevier, Amsterdam.

Catuneanu, O., A. R. Sweet, and A. D. Miall. 2000. Reciprocal stratigraphy of the Campanian–Paleocene Western Interior of North America. Sedimentary Geology 134:235–255.

Cifelli, R. L. 1990. Cretaceous mammals of southern Utah. II. Marsupials and marsupial-like mammals from the Wahweap Formation (Early Campanian). Journal of Vertebrate Paleontology 10:320–331.

DeBlieux, D. D., J. I. Kirkland, J. G. Eaton, A. L. Titus, M. A. Getty, S. D. Sampson, and M. C. Hayden. 2009. Vertebrate paleontology of the Lower–Middle Campanian Wahweap Formation in Grand Staircase–Escalante National Monument (GSENM); p. 17 in Advances in Western Interior Late Cretaceous Paleontology and Geology, Abstracts with Program.

DeCelles, P. G. 2004. Late Jurassic to Eocene evolution of the Cordilleran thrust belt and foreland basin system, western U.S.A. American Journal of Science 304:105–168.

DeCelles, P. G., and J. C. Coogan. 2006. Regional structure and kinematic history of the Sevier fold-and-thrust belt, central Utah. Geological Society of America Bulletin 118:841–864.

DeCelles, P. G., and B. S. Currie. 1996. Long-term sediment accumulation in the Middle Jurassic–early Eocene Cordilleran retroarc foreland-basin system. Geology 24:591–594.

DeCelles, P. G., and K. N. Giles. 1996. Foreland basin systems. Basin Research 8:105–123.

Dickinson, W. R., and G. E. Gehrels. 2008. Sediment delivery to the Cordilleran foreland basin: insights from U-Pb ages from detrital zircons in Upper Jurassic and Cretaceous strata of the Colorado Plateau. American Journal of Science 308:1041–1082.

Dickinson, W. R., M. A. Klute, M. J. Hayes, S. U. Janecke, E. R. Lundin, M. A. McKittrik, and M. D. Olivares. 1988. Paleogeographic and paleotectonic setting of Laramide sedimentary basins in the central Rocky Mountain region. Geological Society of America Bulletin 100:1023–1039.

Doelling, H. H. 1997. Interim geologic map of the Smoky Mountain 30'×60' quadrangle, Kane and San Juan Counties, Utah and Coconino County, Arizona. Utah Geological Survey Open-File Report 359.

Eaton, J. G. 1990. Stratigraphic revision of Campanian (Upper Cretaceous) rocks in the Henry Basin, Utah. Mountain Geologist 27:27–38.

Eaton, J. G. 1991. Biostratigraphic framework for the Upper Cretaceous rocks of the Kaiparowits Plateau, southern Utah; pp. 47–63 in J. D. Nations and J. G. Eaton (eds.), Stratigraphy, Depositional Environments, and Sedimentary Tectonics of the Western Margin, Cretaceous Western Interior Seaway. Geological Society of America Special Paper 260.

Eaton, J. G. 1993. Mammalian paleontology and correlation of the uppermost Cretaceous rocks of the Paunsaugunt Plateau; pp. 163–180 in M. Morales (ed.), Aspects of Mesozoic Geology and Paleontology of the Colorado Plateau. Museum of Northern Arizona Bulletin 59.

Eaton, J. G. 1999. Vertebrate paleontology of the Paunsaugunt Plateau, Upper Cretaceous, southwestern Utah; pp. 335–339 in D. D. Gillette (ed.), Vertebrate Paleontology in Utah. Utah Geological Survey Miscellaneous Publication 99–1.

Eaton, J. G. 2002. Multituberculate Mammals from the Wahweap (Campanian, Aquilan) and Kaiparowits (Campanian, Judithian) Formations, Within and Near Grand Staircase–Escalante National Monument, Southern Utah. Utah Geological Survey Miscellaneous Publication 02–4.

Eaton, J. G. 2006. Santonian (Late Cretaceous) mammals from the John Henry Member of the Straight Cliffs Formation, Grand Staircase–Escalante National Monument, Utah. Journal of Vertebrate Paleontology 26:446–460.

Eaton, J. G., and R. L. Cifelli. 1988. Preliminary Report on Late Cretaceous Mammals of the Kaiparowits Plateau, Southern Utah. University of Wyoming Contributions to Geology 26:45–55.

Eaton, J. G., J. Laurin, J. I. Kirkland, N. E. Tibert, R. M. Leckie, B. B. Sageman, P. M. Goldstrand, D. W. Moore, A. W. Straub, W. A. Cobban, and J. D. Dalebout. 2001. Cretaceous and Early Tertiary geology of Cedar and Parowan canyons, western Markagunt Plateau, Utah: Utah Geological Association Field Trip Road Log; pp. 337–363 in M. C. Erskine, J. E. Faulds, J. M. Bartley, P. D. Rowley (eds.), The Geologic Transition, High Plateaus to Great Basin; a Symposium and Field Guide; the Mackin Volume. Guidebook–Pacific Section, American Association of Petroleum Geologists 78.

Foster, J. R., A. L. Titus, G. F. Winterfield, M. C. Hayden, and A. H. Hamblin. 2001. Paleontological survey of the Grand Staircase–Escalante National Monument, Garfield and Kane counties, Utah. Utah Geological Survey Special Study 99.

Fouch, T. D., T. F. Lawton, J. D. Nichols, W. B. Cashion, and W. A. Cobban. 1983. Patterns and timing of synorogenic sedimentation in Upper Cretaceous rocks of central and northwestern Utah; pp. 305–336 in M. W. Reynolds and E. D. Dolly (eds.), Mesozoic Paleogeography of the West-Central United States. Society of Economic Paleontologists and Mineralogists, Rocky Mountain Section.

Gates, T. A., S. D. Sampson, L. E. Zanno, E. M. Roberts, J. G. Eaton, R. L. Nydam, J. H. Hutchison, J. A. Smith, M. A. Loewen, and M. A. Getty. 2010. Biogeography of terrestrial and freshwater vertebrates from the Late Cretaceous (Campanian) Western Interior of North America. Palaeogeography, Palaeoclimatology, Palaeoecology 291:371–387.

Gill, J. R., and W. A. Cobban. 1973. Stratigraphy and Geologic History of the Montana Group and Equivalent Rocks, Montana, Wyoming, and North and South Dakota. U.S. Geological Survey Professional Paper 776.

Goldstrand, P. M. 1994. Tectonic development of Upper Cretaceous to Eocene strata of southwestern Utah. Geological Society of America Bulletin 106:145–154.

Gregory, H. E., and R. C. Moore. 1931. The Kaiparowits Region. U.S. Geological Survey Professional Paper 164.

Hancock, J. M., and E. G. Kauffman. 1979. The great transgressions of the Late Cretaceous. Journal of the Geological Society of London 136:175–186.

Heller, P. L., and C. Paola. 1989. The paradox of Lower Cretaceous gravels and the initiation of thrusting in the Sevier orogenic belt, United States Western Interior. Geological Society of America Bulletin 101:864–875.

Heller, P. L., C. L. Angevine, N. S. Winslow, and C. Paola. 1988. Two-phase stratigraphic model of foreland-basin sequences. Geology 16:501–504.

Hilbert-Wolf, H. L., E. L. Simpson, W. S. Simpson, S. E. Tindall, and M. C. Wizevich. 2009. Insights into syndepositional fault movements in a foreland basin; trends in seismites of the Upper Cretaceous, Wahweap Formation, Kaiparowits Basin, Utah, U.S.A. Basin Research 21:856–871.

Horton, B. K., K. N. Constenius, and P. G. DeCelles. 2004. Tectonic control on coarse-grained foreland-basin sequences: an example from the Cordilleran foreland basin, Utah. Geology 32:637–640.

Jinnah, Z. A., and E. M. Roberts. 2011. Facies associations, paleoenvironments and base level changes in the Upper Cretaceous Wahweap Formation. Journal of Sedimentary Research 81:266–283.

Jinnah, Z. A., M. A. Getty, and T. A. Gates. 2009a. Taphonomy of a hadrosaur bonebed from the Upper Cretaceous Wahweap Formation, Grand Staircase–Escalante National Monument, Utah; p. 28 in Advances in Western Interior Late Cretaceous Paleontology and Geology, St. George, Utah, Abstracts with Program.

Jinnah, Z. A., E. M. Roberts, A. L. Deino, J. S. Larsen, P. K. Link, and C. M. Fanning. 2009b. New ^{40}Ar-^{39}Ar and detrital zircon U-Pb ages for the Upper Cretaceous Wahweap and Kaiparowits formations on the Kaiparowits Plateau, Utah: implications for regional correlation, provenance and biostratigraphy. Cretaceous Research 30:287–299.

Kauffman, E. G. 1977. Geological and biological overview: Western Interior Cretaceous Basin. Mountain Geologist 14:75–99.

Kauffman, E. G., and W. G. E. Caldwell. 1993. The Western Interior Basin in space and time; pp. 1–30 in W. G. E. Caldwell and E. G. Kauffman (eds.), Evolution of the Western Interior Basin. Geological Association of Canada Special Paper 39.

Kirkland, J. I., and D. DeBlieux. 2005. Dinosaur remains from the lower to middle Campanian Wahweap Formation at Grand Staircase–Escalante National Monument, southern Utah. Journal of Vertebrate Paleontology 25(3, Supplement):78A.

Kirkland, J. I., and D. DeBlieux. 2010. New basal centrosaurine ceratopsian skulls from the Wahweap Formation (Middle Campanian), Grand Staircase–Escalante National Monument, southern Utah; pp. 117–140 in M. J. Ryan, B. J. Chinnery-Allgeier, D. A. Eberth, and P. Dodson (eds.), New Perspectives on Horned Dinosaurs. Indiana University Press, Bloomington, Indiana.

Kirkland, J. I., S. G. Lucas, and J. W. Estep. 1998. Cretaceous dinosaurs of the Colorado Plateau; pp. 79–89 in S. G. Lucas, J. I. Kirkland, and J. W. Estep (eds.), Lower and Middle Cretaceous Terrestrial Ecosystems. New Mexico Museum of Natural History and Science Bulletin 14.

Kraus, M. J. 1999. Paleosols in clastic sedimentary rocks: their geologic applications. Earth Science Reviews 47:41–70.

Kuiper, K. F., A. Deino, F. J. Hilgen, W. Krijgsman, R. Renne, and J. R. Wijbrans. 2008. Synchronizing rock clocks of Earth history. Science 320:500–504.

Larsen, J. S. 2007. Facies and provenance of the Pine Hollow Formation: implications for Sevier foreland basin evolution and the Paleocene climate of southern Utah. M.Sc. thesis, Idaho State University, Pocatello, Idaho.

Lawton, T. F. 1986. Fluvial systems of the Upper Cretaceous Mesaverde Formation, Central Utah: a record of transition from thin-skinned to thick-skinned deformation in the foreland region; pp. 423–442 in J. A. Peterson (ed.), Paleotectonics and Sedimentation in the Rocky Mountain Region. American Association of Petroleum Geologists Memoir 41.

Lawton, T. F. 2008. Laramide sedimentary basins; pp. 429–450 in A. D. Miall (ed.), The Sedimentary Basins of the United States and Canada. Sedimentary Basins of the World 5.

Lawton, T. F., and J. H. Trexler. 1991. Piggyback basin in the Sevier orogenic belt, Utah: implications for development of the thrust wedge. Geology 19:827–830.

Lawton, T. F., S. L. Pollock, and R. A. J. Robinson. 2003. Integrating sandstone petrology and nonmarine sequence stratigraphy: application to the Late Cretaceous fluvial systems of southwestern Utah, U.S.A. Journal of Sedimentary Research 73:389–406.

Lehman, T. M. 1997. Late Campanian dinosaur biogeography in the western interior of North America; pp. 223–240 in D. L. Wolberg, E. Stump, and G. D. Rosenberg (eds.), Dinofest International. Academy of Natural Sciences, Philadelphia.

Little, W. W. 1997. Tectonic and eustatic controls on cyclical fluvial patterns, Upper Cretaceous strata of the Kaiparowits Basin, Utah; pp. 489–504 in L. M. Hill (ed.), Learning from the Land: Grand Staircase–Escalante National Monument Science Symposium Proceedings. U.S. Department of the Interior, Bureau of Land Management.

McLaurin, B. T., and R. J. Steel. 2000. Fourth order nonmarine to marine sequences, middle Castlegate Formation, Book Cliffs, Utah. Geology 28:359–362.

McLaurin, B. T., and R. J. Steel. 2007. Architecture and origin of an amalgamated fluvial sheet sand, lower Castlegate Formation, Book Cliffs, Utah. Sedimentary Geology 197:291–311.

Merewether, E. A., W. A. Cobban, and J. D. Obradovich. 2007. Regional disconformities in Turonian and Coniacian (Upper Cretaceous) strata in Colorado, Wyoming, and adjoining states–biochronological evidence. Rocky Mountain Geology 42:95–122.

Miall, A. D., and M. Arush. 2001. The Castlegate Sandstone of the Book Cliffs, Utah: sequence stratigraphy, paleogeography, and tectonic controls. Journal of Sedimentary Research 71:537–548.

Miall, A. D., O. Catuneanu, B. K. Vakarelov, and R. Post. 2008. The Western Interior Basin; pp. 329–362 in A. D. Miall (ed.), The Sedimentary Basins of the United States and Canada. Sedimentary Basins of the World 5.

Molenaar, C. M., W. A. Cobban, E. A. Merewether, C. L. Pillmore, D. G. Wolfe, and J. M. Holbrook. 2002. Regional stratigraphic cross sections of Cretaceous rocks from east-central Arizona to the Oklahoma panhandle. U.S. Geological Survey Miscellaneous Field Studies Map MF-2382.

Moore, D. W., and A. W. Straub. 2001. Correlation of Upper Cretaceous and Paleogene (?) rocks beneath the Claron Formation, Crow Creek, western Markagunt Plateau, southwest Utah; pp. 75–95 in M. C. Erskine, J. E. Faulds, J. M. Bartley, and P. D. Rowley (eds.), The Geologic Transition, High Plateaus to Great Basin: A Symposium and Field Guide; the Mackin Volume. Guidebook–Pacific Section, American Association of Petroleum Geologists 78.

Nichols, D. J. 1997. Palynology and ages of some Upper Cretaceous formations in the Markagunt and northwestern Kaiparowits plateaus, southwestern Utah; pp. 81–95 in F. Maldonado and L. D. Nealey (eds.), Geologic Studies in the Basin and Range–Colorado Plateau Transition Zone in Southeastern Nevada, Southwestern Utah, and Northwestern Arizona, 1995. U.S. Geological Survey Bulletin 2153-E.

O'Connor, P. M., E. M. Roberts, L. Tapanila, A. Deino, T. Hieronymus, and Z. Jinnah. 2009. Reconnaissance paleontology in the Straight Cliffs and Wahweap formations (Turonian–Campanian), Grand Staircase–Escalante National Monument, Utah; p. 40 in Advances in Western Interior Late Cretaceous Paleontology and Geology, St. George, Utah, Abstracts with Program.

Ori, G. G., and P. F. Friend. 1984. Sedimentary basins formed and carried on active thrust sheets. Geology 12:475–478.

Peterson, F., and A. R. Kirk. 1977. Correlation of Cretaceous rocks in the San Juan, Black Mesa, Kaiparowits and Henry basins, southern Colorado Plateau. New Mexico Geological Society Field Guide 28:167–178.

Pollock, S. L. 1999. Provenance, geometry, lithofacies and age of the Upper Cretaceous Wahweap Formation, Cordilleran foreland basin, southern Utah. M.Sc. thesis, New Mexico State University, Las Cruces, New Mexico.

Renne, P. R., C. C. Swisher, A. L. Deino, D. B. Karner, T. L. Owens, and D. J. DePaolo. 1998. Intercalibration of standards, absolute ages and uncertainties in $^{40}Ar/^{39}Ar$ dating. Chemical Geology 145:117–152.

Renne, P. R., A. L. Deino, R. C. Walter, B. D. Turrin, C. C. Swisher, T. A. Becker, G. H. Curtis, W. D. Sharp, and A. R. Jaouni. 1994. Intercalibration of astronomical and radioisotopic time. Geology 22:783–786.

Roberts, E. M. 2007. Facies architecture and depositional environments of the Upper Cretaceous Kaiparowits Formation, southern Utah. Sedimentary Geology 197:207–233.

Roberts, E. M., A. L. Deino, and M. A. Chan. 2005. $^{40}Ar/^{39}Ar$ age of the Kaiparowits Formation, southern Utah, and correlation of contemporaneous Campanian strata and vertebrate faunas along the margin of the Western Interior Basin. Cretaceous Research 26:307–318.

Robinson, R. A. J., and R. L. Slingerland. 1998. Grain-size trends, basin subsidence and sediment supply in the Campanian Castlegate Sandstone and equivalent conglomerates of central Utah. Basin Research 10:109–127.

Rogers, R. R. 1998. Sequence analysis of the Upper Cretaceous Two Medicine and Judith River formations, Montana: nonmarine response to the Claggett and Bearpaw marine cycles. Journal of Sedimentary Research 68:615–631.

Rowe, T., R. L. Cifelli, T. M. Lehman, and A. Weil. 1992. The Campanian Terlingua local fauna, with a summary of other vertebrates from the Aguja Formation, Trans-Pecos, Texas. Journal of Vertebrate Paleontology 12:472–493.

Sable, E. G., and R. Hereford. 2004. Geologic map of the Kanab 30'×60' quadrangle, Utah and Arizona. U.S. Geological Survey Map I-2655.

Sampson, S. D. 2009. Dinosaur Odyssey. University of California Press, Berkeley, California.

Sampson, S. D., M. A. Loewen, T. A. Gates, L. E. Zanno, and J. I. Kirkland. 2002. New evidence of dinosaurs and other vertebrates from the Upper Cretaceous Wahweap and Kaiparowits formations, Grand Staircase–Escalante National Monument, southern Utah; p. 5 in Geological Society of America, Rocky Mountain Section 54th Annual Meeting, Abstracts with Programs 34.

Santucci, V. L., and J. I. Kirkland. 2010. An overview of National Park Service paleontological resources from the parks and monuments in Utah; pp. 589–623 in D. A. Sprinkel, T. C. Chidsey Jr., and P. B. Anderson (eds.), Geology of Utah's Parks and Monuments (3rd edition). Utah Geological Association Publication 28.

Shanley, K. W., and P. J. McCabe. 1991. Predicting facies architecture through sequence stratigraphy–an example from the Kaiparowits Plateau, Utah. Geology 19:742–745.

Shanley, K. W., and P. J. McCabe. 1995. Sequence stratigraphy of Turonian–Santonian strata, Kaiparowits Plateau, southern Utah, U.S.A.: implications for regional correlation and foreland basin evolution; pp. 103–136 in J. C. Van Wagoner and G. T. Bertram (eds.), Sequence Stratigraphy of Foreland Basin Deposits: Outcrop and Subsurface Examples from the Cretaceous of North America. American Association of Petroleum Geologists Memoir 64.

Simpson, E. L., M. C. Wizevich, H. L. Hilbert-Wolf, S. E. Tindall, J. J. Bernard, and W. S. Simpson. 2009. An Upper Cretaceous sag pond deposit: implications for recognition of local seismicity and surface rupture along the Kaibab monocline, Utah. Geology 37:967–970.

Simpson, E. L., H. L. Hilbert-Wolf, W. S. Simpson, S. E. Tindall, J. J. Bernard, T. A. Jenesky, and M. C. Wizevich. 2008. The interaction of aeolian and fluvial processes during deposition of the Upper Cretaceous capping sandstone member, Wahweap Formation, Kaiparowits Basin, Utah, U.S.A. Palaeogeography, Palaeoclimatology, Palaeoecology 270:19–28.

Smith, C. 1983. Geology, depositional environments and coal resources of the Mt. Pennell 2 NW Quadrangle, Garfield County, Utah. Brigham Young University Geological Studies 30:145–169.

Tilton, T. L. 1991. Upper Cretaceous stratigraphy of the Southern Paunsaugunt Plateau, Kane County, Utah. Ph.D. dissertation, University of Utah, Salt Lake City, Utah.

Van Wagoner, J. C. 1995. Sequence stratigraphy and marine to nonmarine facies architecture of foreland basin strata, Book Cliffs, Utah, U.S.A.; pp. 137–223 in J. C. Van Wagoner and G. T. Bertram (eds.), Sequence Stratigraphy of Foreland Basin Deposits: Outcrop and Subsurface Examples from the Cretaceous of North America. American Association of Petroleum Geologists Memoir 64.

Weimer, R. J. 1986. Relationship of unconformities, tectonics, and sea level changes in the Cretaceous of the Western Interior, United States; pp. 397–442 in J. A. Peterson (ed.), Paleotectonics and Sedimentation in the Rocky Mountain Region, United States. American Association of Petroleum Geologists Memoir 41.

Yoshida, S., A. Willis, and A. D. Miall. 1996. Tectonic control of nested sequence architecture in the Castlegate Sandstone (Upper Cretaceous), Book Cliffs, Utah. Journal of Sedimentary Research 66:737–748.

Implications of the Internal Plumbing of a Late Cretaceous Sand Volcano: Grand Staircase–Escalante National Monument, Utah

Edward L. Simpson, Hannah L. Hilbert-Wolf, Michael C. Wizevich, and Sarah E. Tindall

A VERTICAL, CROSS-SECTIONAL EXPOSURE THROUGH A well-preserved sand blow and the associated volcano was discovered 1.8 m above the base of the Upper Cretaceous upper member of the Wahweap Formation. Preserved features within the feeder conduit (pipe) of the sand volcano facilitate reconstruction of the vertical fluid flow and interpretation of liquefied sand flow generated by local seismogenic faulting. In the lower reaches of the conduit, a dilation fracture crosscuts a low-permeability, fine-grained sandstone seal. Above this, the conduit widens and the edges become more diffuse in the overlying higher-permeability sandstone, and conversely, the pipe contracts in diameter as it passes through lower-permeability sandstones. This change in character of the pipe, in addition to the structureless sandstone adjacent to the pipe, indicates that lateral flow to the conduit was greater in the high-permeability zones. Within the pipe, subvertical stringers of granules 30 cm below the vent indicate that fluid flow of the liquefied sediment was of sufficient velocity to move granules. Medium to fine sand, elutriated from the sediment in the conduit, forms the subarial sand volcano cone. The internally massive surface volcano is slightly asymmetrical, measuring ~120 cm in apparent diameter and ~20 cm in height and is onlapped by the lee face of a fluvial dune.

Modern sand volcanoes often develop proximal to faults as a result of high-magnitude seismic events. This ancient sand volcano is located in close proximity to a series of normal faults and, along with other preserved syntectonic deposits in the Wahweap Formation, indicates that the fault slip history includes intense, high-magnitude seismic activity.

INTRODUCTION

Sand blows are pressurized, liquefied sand–water mixtures that eject sediment onto the Earth's surface, forming a sand volcano. A seismic origin for sand blows was established when their distribution was correlated with the 1811–1812 earthquake activity along the New Madrid fault system (Obermeier, 1996; Montenat et al., 2007). In addition, nonseismic sand volcanoes can be produced by relatively low-energy sand boils. Such sand volcanoes have been documented along man-made levees paralleling the Mississippi River, under which water seeps during floods and is expelled from fractures as sand boils (Li et al., 1996). Another occurrence of nonseismic sand volcanoes has been documented in Fremont Valley, California, where sediment-laden runoff was intercepted by a preexisting ground fissure and later discharged down slope as sand boils onto the land surface (Holzer and Clark, 1993).

Sand volcanoes and their associated subsurface deformation are commonly used to reconstruct paleoseimogenic activity (Obermeier, 1996). There is potential, then, for preserved, nonseismic features to be misinterpreted as earthquake-induced liquefaction features. Li et al. (1996) have established criteria to differentiate between lower energy, nonseismic sand boils, and high-energy, earthquake-induced sand blows.

The recognition and interpretation of seismically induced sand volcanoes in the rock record is challenging because the exposed surface volcano is prone to significant erosion (Obermeier, 1996; Montenat et al., 2007). Obermeier (1996) identified only large sand blows in association with the 1811–1812 New Madrid meizoseismal zone, even though eyewitnesses reported smaller blows as well. Clearly, preservation depends on factors such as the size of the volcano and time before burial, so it is likely that the subaerial sand volcanoes are rarely preserved. Thus, venting activity must often be interpreted from cross-sectional exposures of the subsurface (internal) plumbing system (Montenat, 1980). A conceptual model of the internal conduits has been developed from the study of modern and ancient sand blows (Obermeier, 1996; Lui and Li, 2001; Maurya et al., 2006; Pringle et al., 2007). Montenat et al. (2007) proposed a series of criteria to permit sand blow identification in the rock record when sand volcanoes are not preserved.

Recent workers have established the syndepositional seismic nature of normal faults along the Cockscomb, in the Upper Cretaceous Wahweap Formation, in Grand Staircase–Escalante National Monument, Utah (Fig. 5.1; Hilbert-Wolf

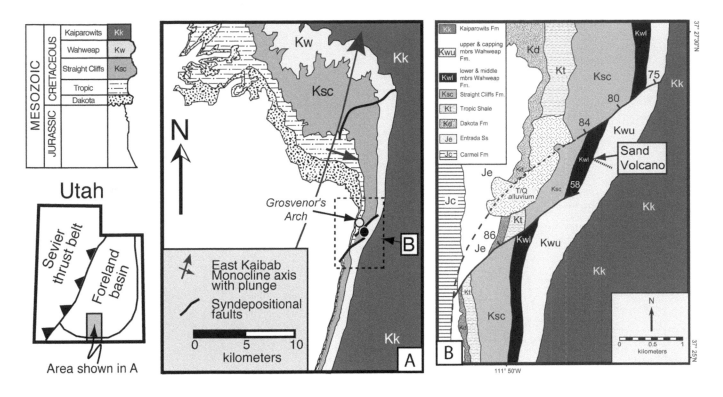

5.1. Stratigraphy, tectonic setting, and geologic map of the study area. The preserved sand volcano is located along the Cockscomb in Grand Staircase–Escalante National Monument. Geologic map A modified from Sargent and Hansen (1982). Geologic map B modified from Tindall et al. (2010).

et al., 2009; Simpson et al., 2009; Tindall et al., 2010). This chapter identifies and interprets a well-preserved sand volcano associated with these faults and documents: (1) the effects that subtle variations in permeability and seals/caps have on fluid flow; (2) the development of fractured beds at the base of the conduit; and (3) the variability in internal architecture of an individual sand volcano, which will aid in the future identification of these features in cross section.

GEOLOGICAL SETTING

Within Grand Staircase–Escalante National Monument, the Upper Cretaceous Wahweap Formation conformably overlies the Straight Cliffs Formation and underlies the Kaiparowits Formation (Fig. 5.1; Peterson, 1969; Eaton, 1991; Little, 1997; Lawton, Pollack, and Robinson, 2003). Eaton (1991) informally subdivided the Wahweap Formation, from oldest to youngest, into the lower, middle, upper, and capping sandstone members. In the Cockscomb area of Grand Staircase–Escalante National Monument, we placed the base of the upper member at the first laterally extensive, slope-forming sandstone bed. The upper member is distinguished from the capping sandstone member by a change in lithology, from a sublithic arenite to a quartz arenite, which is visually distinguishable as a shift in color from tan to white

(Eaton, 1991; Eaton and Nations, 1991; Pollock, 1999; Lawton, Pollack, and Robinson, 2003).

Fluvial deposits characterize the upper member (Little, 1997; Lawton, Pollack, and Robinson, 2003; Wizevich et al., 2008; Jinnah and Roberts, 2009). $^{40}Ar/^{39}Ar$ radiometric dating of tuffs in the Wahweap and Kaiparowits formations (Roberts, Deino, and Chan, 2005; Jinnah et al., 2009) and microvertebrate biostratigraphy (Eaton, 1991, 2002) establish a Late Cretaceous Campanian age for the upper member of the Wahweap Formation.

Seismogenically generated soft-sediment deformation occurs throughout the upper and capping member sandstones of the Wahweap Formation, attesting to the syndepositional nature of the normal faults (Fig. 5.2; Hilbert-Wolf et al., 2009). Varying degrees of deformation in the sandstones reflect the amount, intensity, and duration of seismic shaking, as well as the depth of the sediment at the time of shaking. Deformation types include convoluted bedding, sand intrusions and volcanoes, and completely obliterated massive layers that can be traced laterally for hundreds of meters (Hilbert-Wolf et al., 2009). In addition, surface rupturing events at the contact between the upper and capping sandstone members indicate that seismic movement occurred on one of the local normal faults, resulting in the development of a sag pond (Simpson et al., 2009).

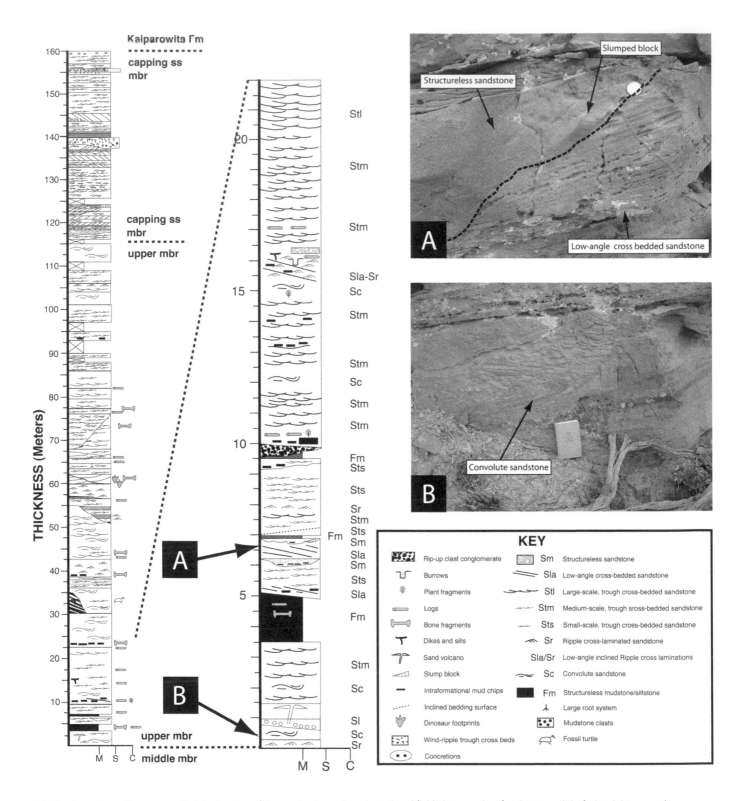

5.2. Stratigraphic section measured at the location of the sand volcano (see Fig. 5.1) and field photographs of sedimentary lithofacies. (A) Low-angle, cross-bedded sandstone crosscut by an erosive-based, massive sandstone. Note the preservation of a slump block of low-angle sandstone on the channel margin (dashed line). Coin is 2.5 cm wide. (B) Convolute sandstone bed approximately 2 m above the contact between the middle and upper members. Field notebook is 12.0 cm wide.

DESCRIPTION

A cross section through a sand blow is preserved approximately 1.8 m above the contact between the middle and upper members of the Wahweap Formation (Fig. 5.2). Lithofacies associated with (but not involved with) the sand blow include low-angle, cross-stratified sandstone, medium-scale trough cross-beds (Fig. 5.2), erosive-based, massive sandstones, some

Simpson et al.

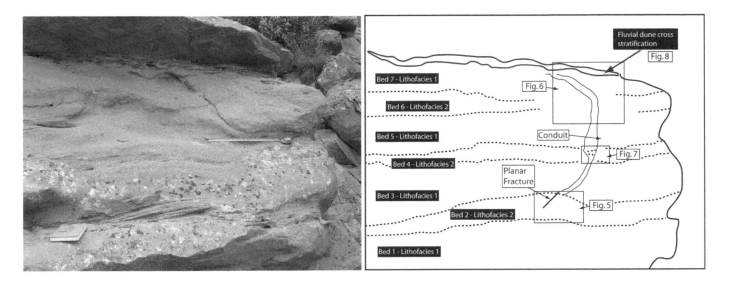

5.3. Field photograph and line drawing of cross section through the sand blow in the upper member of the Wahweap Formation. Outcrop tilts away from the plane of the photograph. Boxes indicate beds and lithofacies discussed in the text. Dashed lines indicate boundaries between lithofacies.

with slump blocks (Fig. 5.2A), low-angle cross-bedded sandstone (Fig. 5.2A), and low-angle inversely graded lamination.

Above the first soft-sediment deformed horizon encountered in the measured section through the upper member (Fig. 5.2), seven beds are crosscut by the sand blow conduit system that leads vertically to the preserved sand volcano that caps bed 7 (Figs. 5.3 and 5.4). These beds consist of two vertically alternating lithofacies. Lithofacies 1, which is found in beds 1, 3, 5, and 7, consists of medium-grained, massive sandstone that ranges up to 25 cm in thickness. Vertically within lithofacies 1, the intensity of liquefaction and fluidization increases, as does the number of beds with thick pebble layers. Lithofacies 2, which composes beds 2, 4, and 6, contains low-angle, cross-stratified, fine-grained sandstone, often with mudstone rip-up clasts, that parallel stratification. Thickness ranges from 15 to 25 cm.

The conduit appears to have developed from a white, planar fracture, which crosscuts and dips approximately 50° with respect to the low-angle stratification of bed 2 (Fig. 5.5). At the base of bed 3, the bleached planar fracture expands in width upward. The expansion of the conduit is constricted through the lithofacies 2 of beds 4 and 6, but the conduit widens and its edges become more diffuse in beds 3, 5, and 7 (Fig. 5.4). Internally, conduit fill contains poorly developed, near-vertical stratification that parallels the conduit margin (Fig. 5.6). Vertical stratification is developed as a subtle variation in grain size. Deformation in the form of folds and convolute bedding extends horizontally outward from the conduit, and the width of the deformed zones increases vertically (Figs. 5.3 and 5.4). Flame features that deform the contacts between beds are better developed near the conduit, and their amplitude increases vertically. The flames are associated with the

thickening of beds 3, 5, and 7 (Fig. 5.6). In the conduit, very coarse to granule-size sediment from bed 5 collapsed below the original level of bed 4 (Fig. 5.7).

In cross section, the sand volcano is cone shaped, with low-angle flanks, and has an internally massive structure (Fig. 5.6). The overall geometry is slightly asymmetrical, measuring ~120 cm in apparent diameter and ~20 cm in maximum height. From the base to the top, the volcano is graded from medium to fine sand. The shortest flank of the volcano onlaps or drapes the older toe sets of fluvial dune cross-bedding (Fig. 5.8). The same set of cross-stratification ultimately buried the sand volcano. Both the flanks along with the older cross-stratification are slightly deformed as a result of fluid flow through this upper sediment associated with sand volcano construction. The younger cross-stratification covering the sand volcano is undeformed. On the basis of two-dimensional measurements and the assumption of a conical geometry, the volume of sand expelled by the sand blow was approximately 0.08 m^3.

INTERPRETATION

The sand volcano in the upper member of the Wahweap Formation is interpreted as a seismogenic fluidization feature induced by intense earthquake activity along local normal faults. A seismogenic origin is supported by both the surface expression of the sand volcano and the internal architecture of the feeder conduit.

The lithofacies associated with the sand volcano are consistent with a braided stream setting that experienced high-discharge events. The upper member has been interpreted as a braided stream system with low sinuosity (Little, 1997;

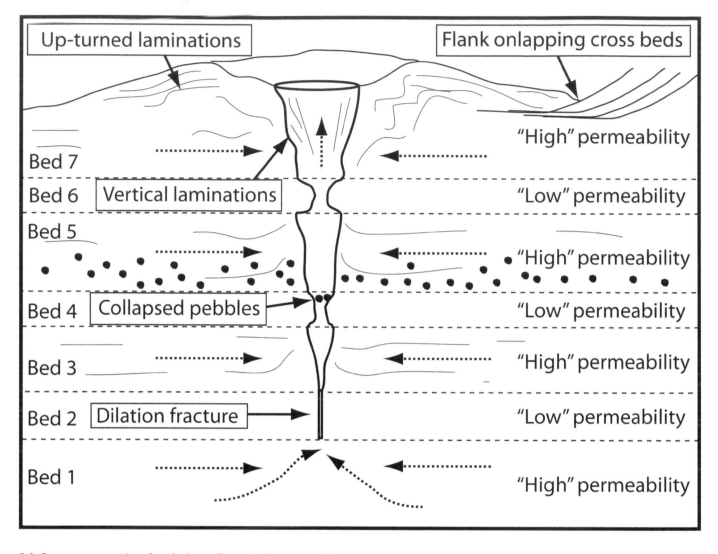

Up-turned laminations

Flank onlapping cross beds

Bed 7 ....▶ ↑ ◀............ "High" permeability

Bed 6 | Vertical laminations | "Low" permeability

Bed 5 ....▶ ◀............ "High" permeability

Bed 4 | Collapsed pebbles | → "Low" permeability

Bed 3 ....▶ ◀............ "High" permeability

Bed 2 | Dilation fracture | → "Low" permeability

Bed 1 ....▶ ◀............ "High" permeability

5.4. Cartoon reconstruction of sand volcano, illustrating the various relationships discussed in the text. This cartoon does not match up with Fig. 5.3 because the outcrop is at an angle, and this reconstruction is done without distortion. Dashed lines indicate flow lines.

Lawton, Pollack, and Robinson, 2003; Wizevich et al., 2008; Jinnah and Roberts, 2009). Lithofacies 1 and 2 reflect deposition of through cross-bedding in bars, specifically bar cores. The trough cross-beds were subsequently modified by liquefaction and fluidization, and by smaller bar top bed forms. The recognition of erosive-based, massive sandstones, some with slump blocks, horizontally stratified sandstone, and low-angle, inversely graded lamination in the lower part of the upper member of this section, indicates that the ephemeral braid system present at the time of the sand blow may have been drier than the stream system that is preserved at stratigraphically higher levels. The low-angle, inversely graded lamination in the upper member is characteristic of wind-ripple stratification (Hunter, 1977; Fryberger and Schenk, 1988). Complex reworking of bar tops is interpreted as related to more semiarid to arid conditions (Bullard and Livingstone, 2002). Similarly, in the capping sandstone member periods of greater aridity are also recognized (Simpson et al., 2008).

The preservation of sand volcano cones is rare because such surface features often succumb to rapid erosion (Montenat, 1980; Montenat et al., 2007). However, in the New Madrid Seismic Zone, 200-year-old sand volcanoes are still recognizable (Lui and Li, 2001; Obermeier, Pond, and Olsen, 2001; Hough et al., 2005; Cox et al., 2007). The fortuitous preservation of the upper member sand volcano cone is probably the result of rapid burial by prograding dunes (trough cross-beds, lithofacies 1) deposited on the down-dropped hanging wall of a nearby normal fault, and may have been enhanced by a period of prolonged exposure during this drier interval. However, the most remarkable aspect of the upper member sand volcano is the exposure (beneath the cone) of the feeder conduit system and its relationships within units of differing permeability.

The 1.2-m diameter of the upper member sand volcano is consistent with other sand volcanoes reported in the literature (Rajendran et al., 2001, 2002; Audemard et al., 2005;

Simpson et al.

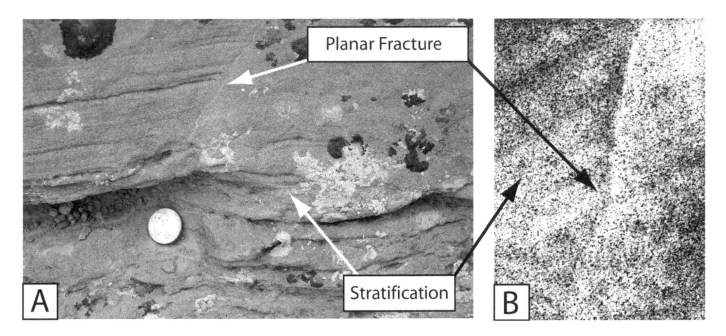

5.5. Planar fracture feature. (A) Fracture feature crosscutting bed 2; note the loss of stratification on the top of the footwall of the fracture. Coin is 2.5 cm wide. (B) Scan of thin section of the fracture feature; note that the stratification is well developed and is not offset. Thin section is 10 by 15 mm. Location is shown in Fig. 5.3.

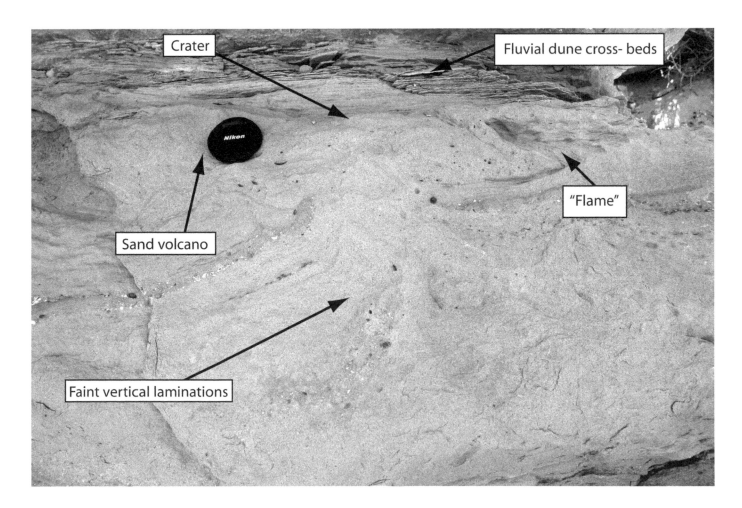

5.6. Field photograph of sand blow upper conduit and sand volcano. Note the flame developed at the end of the conduit and the faint, near-vertical laminations within the conduit. Coin is 2.5 cm wide. Location is shown in Fig. 5.3.

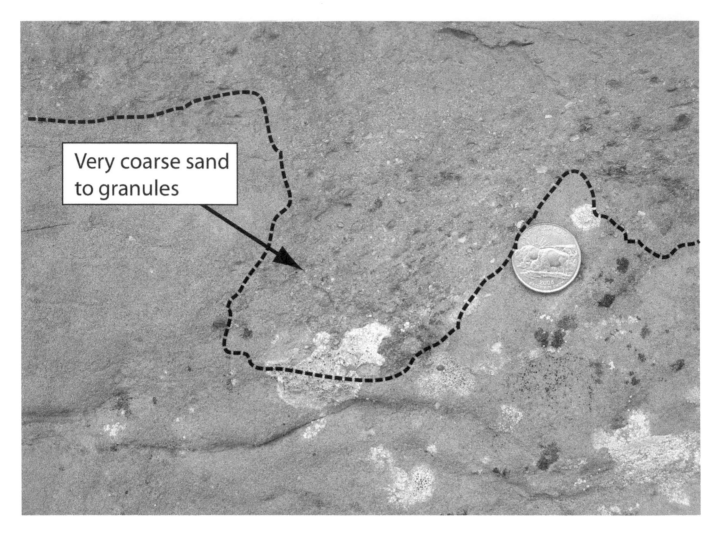

5.7. Field photograph of sand blow conduit. Note the collapse of very coarse to granule-size material from the overlying bed into the conduit. Coin is 2.5 cm wide. Location is shown in Fig. 5.3. Dashed line highlights the base of the very coarse sand to granules.

Maurya et al., 2006; Jonk, Hurst, and Cronn, 2007; Montenat et al., 2007). The typical anatomy of a sand volcano edifice displays stratification and a notable reduction of grain size from the base to the top, reflecting a reduction in the fluid flow velocity and expulsion rate as fluids rupture through a low-permeability siltstone or mudstone seal (Saucier, 1989; Obermeier, Pond, and Olsen, 2001; Jonk, Hurst, and Cronn, 2007). Similar grain-size variation exists in the upper member sand volcano cone.

The upper member sand volcano was deformed by a fluidization event. The deformation may be related to continued compaction or to collapse of the fluid source beds after initial rapid fluidization and development of the sand volcano.

Samaila et al. (2006) show steep-angle flanks of sand volcanoes. Jonk, Hurst, and Cronn (2007) examined bedding plane exposures of Carboniferous three-dimensional preserved sand volcanoes and plotted width versus height of sand volcanoes, which revealed a linear relationship. The

upper-member sand volcano plots slightly outside the envelope, being of lower slope angle per diameter. New Madrid sand volcanoes have a lower height-to-width ratio (Obermeier, 1996) than the Wahweap sand volcano. These variations may be related to the grain size controlling the angle of repose or to the medium (air or water) in which the sand volcano was extruded. No definitive work has examined this relationship.

In the absence of a preserved sand volcano, Montenat et al. (2007) proposed criteria to recognize a sand blow in cross section. These include: (1) an asymmetric, upturned sandy bed; (2) preservation of primary features in this upturned bed; (3) overturned bed crumbled or partly liquefied/convolute; (4) expulsion of fine-grained sediment; (5) surface flow; and (6) fragments of sand beds injected up the conduit. All of the above criteria are recognized in the upper member sand blow, strongly supporting a seismogenic origin. Application of these criteria to the capping member shows that they are useful when the sand volcano is not preserved (Fig. 5.9). Li

Simpson et al.

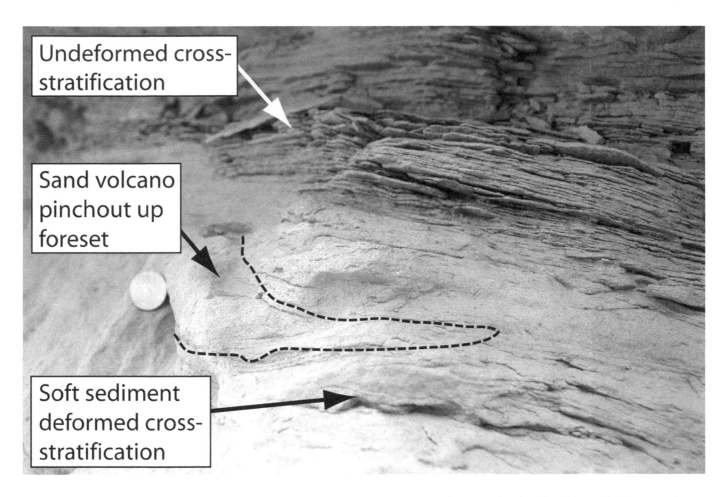

5.8. Strike view of foresets. Dip direction is to the top of the photograph. Note the reduction in deformation after the onlap of the sand volcano onto the foreset of the cross-stratification set. Coin is 2.5 cm wide.

et al. (1996) developed criteria to distinguish sand blows from sand boils in more modern settings over a limited geographic range, but the application of these criteria to the rock record is hindered by the absence of three-dimensional exposure and other limits imposed by the rock record.

The fracture feature (Figs. 5.4 and 5.5) is most likely dilational in origin. During dilation, the fracture was a high-permeability conduit that contributed significant fluid flow to the sand blow from the liquefied sediment in bed 1 (Bernard, Eichhubl, and Aydin, 2002; Fossen et al., 2007). The grains in the walls of the fracture were scrubbed of clay coatings, permitting differential diagenesis; quartz cement highlighted the fracture white. Vertical variations in conduit diameter are a reflection of subtle differences in permeability. Slight variations in compaction of sediment occur between lithofacies 2 (beds 2, 4, and 6) and lithofacies 1 (beds 1, 3, 5, and 7). Relatively low-permeability zones associated with low-angle stratification (beds 2, 4, and 6) are reflected in narrow diameters and allowed overpressuring and limited vertical fluid flow, whereas wider diameters are linked to relatively high-permeability, structureless sandstones (beds 3, 5, and 7). The

change in character of the pipe and the presence of structureless sandstone near the pipe indicate that lateral flow to the conduit was greater in the high-permeability zones (beds 3, 5, and 7). These contributed the largest volume of fluids and also probably the highest fluid flow velocities near the conduit. Consequently, the primary sedimentary features of these beds experienced the most destruction and deformation. Mount (1993) recognized fluidization halos around pipes fed by lateral flow and noted slight variations in clay content, which acted to develop temporary seals to vertical fluid flow. Obermeier, Pond, and Olsen (2001) showed that sill development can occur at permeability boundaries and that the piping often thins upward. Within the upper member pipe, a winnowed concentration of pebbles 30 cm below the vent indicates that vertical fluid flow was of sufficient velocity to elutriate fine sand, which later formed the surface sand volcano. The velocity was not great enough to eject very coarse to granular sediment. The duration of the flow is unknown. Modern sand volcanoes have been reported to extrude fluids up to 3 weeks after their formation (Rajendran et al., 2001, 2002; Audemard et al., 2005; Maurya et al., 2006).

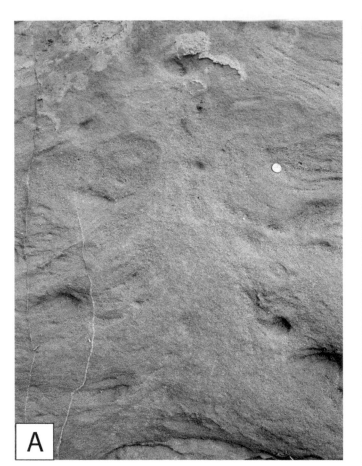
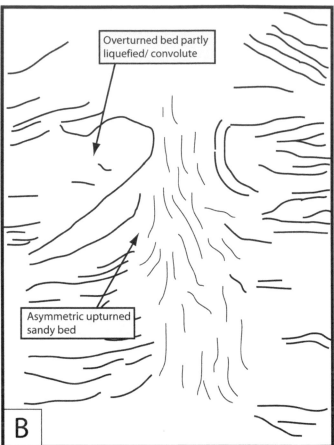

5.9. Preserved sand blow in the capping sandstone member of the Wahweap Formation. (A) Field photograph. Scale is 2.5 cm wide. (B) Interpretive line drawing of the sand blow in (A).

Single conduits can host multiple fluid flow events (Saucier, 1989) or multiphase ejection (Pringle et al., 2007). The most commonly studied sand blows possess a thick permeability seal composed of siltstones or mudstones and are probably single flow events (Audemard and Santis, 1991; Wesnousky and Leffler, 1994; Obermeier, 1996). The upper member sand contrasts with these sand blows because of the absence of a thick unit of fine, low-permeability material, like those typical of overbank deposits.

DISCUSSION

The upper member outcrop represents a unique cross section through a sand blow that yields important insights into the complexities of fluid flow and its reaction to slight differences in permeability. Often such conduits do not show variation in width where they cross different compositional layers (Obermeier, Pond, and Olsen, 2001; Thakkar and Goyal, 2004). This sand volcano and associated fluid flow event were likely generated by a single seismic event because of its shallow depth and the disruption of the seal. Fluidization occurred at

shallow depths in the absence of a thick permeability barrier (cf. Obermeier, Pond, and Olsen, 2001).

Castilla and Audemard (2007) used a worldwide database to argue that blow diameter versus epicentral distance follows an inverse power law, such that larger-diameter sand volcanoes are restricted to areas proximal to the epicenter, whereas smaller-diameter sand volcanoes occur at much greater distances, up to ~450 km from the seismic source. The upper member sand volcano is less than 1 m in diameter. Sand volcanoes of this diameter can occur over a wide range of epicentral distances. Thus, smaller-diameter sand volcanoes are consistent with lower-magnitude earthquakes with a small epicentral distance (Castilla and Audemard, 2007). Another common technique is to plot the distance of the most distal liquefaction versus magnitude (Ambraseys, 1988; Olsen, Green, and Obermeier, 2005). This technique is problematic for our Cretaceous example because the distance from the epicenter to the most distal liquefaction is unknown. Hilbert-Wolf et al. (2009) examined the capping sandstone and were able to isolate the impact of movement of the local normal faults on sedimentation, suggesting that

Simpson et al.

these faults were seismically active into capping sandstone member deposition. All studies of seismically generated liquefaction are consistent with seismogenic movement on these Cretaceous local faults.

SUMMARY

A preserved cross section through a sand blow in the Upper Cretaceous upper member of the Wahweap Formation demonstrates that slight variations in permeability controlled the morphology of the conduits and associated deformation. The sand blow is associated with active seismogenic movements on synsedimentary faults.

ACKNOWLEDGMENTS

We are grateful for the cooperation of the Grand Staircase–Escalante National Monument staff, particularly A. Titus and D. Powell. Acknowledgment is made to the donors of the American Chemical Society Petroleum Research Fund for their support of this research. In addition, Kutztown University Research Committee, Pennsylvania State System of Higher Education Professional Development Committee, Central Connecticut State University School of Arts and Sciences Dean's Initiative, a Connecticut State University Research Grant, and Carleton College's Robert J. Kolenkow and Robert A. Reitz Undergraduate Research grant provided funding. We thank D. Loope for his thoughtful review.

REFERENCES CITED

Ambraseys, N. N. 1988. Engineering seismology. Earthquake Engineering and Structural Dynamics 17:1–105.

Audemard, F. A., and E. de Santis. 1991. Survey of liquefaction structures induced by recent moderate earthquakes. Bulletin of the International Association of Engineering Geology 44:1–16.

Audemard, F. A., J. G. Gómez, H. J. Tavera, and G. N. Orihuela. 2005. Soil liquefaction during the Arequipa Mw 8.4, June 23, 2001, earthquake, southern coastal Peru. Engineering Geology 78:237–255.

Bernard, X. D., P. Eichhubl, and A. Aydin. 2002. Dilation bands: a new form of localized failure in granular material. Geophysical Research Letters 29:1–4.

Bullard, J. E., and I. Livingstone. 2002. Interactions between aeolian and fluvial systems in dryland environments. Area 34:8–16.

Castilla, R. A., and F. A. Audemard. 2007. Sand blows as a potential tool for magnitude estimation of pre-instrumental earthquakes. Journal of Seismology 11:473–487.

Cox, R. T., A. A. Hill, D. Larsen, T. Holzer, S. L. Forman, T. Noce, C. Gardner, and J. Morat. 2007. Seismotectonic implications of sand blows in the southern Mississippi Embayment. Engineering Geology 89:278–299.

Eaton, J. G. 1991. Biostratigraphic framework for the Upper Cretaceous rocks of the Kaiparowits Plateau, southern Utah; pp. 47–63 in J. D. Nations and J. G. Eaton (eds.), Stratigraphy, Depositional Environments, and Sedimentary Tectonics of the Western Margin, Cretaceous Western Interior Seaway. Geological Society of America Special Publication 260.

Eaton, J. G. 2002. Multituberculate Mammals from the Wahweap (Campanian, Aquilan) and Kaiparowits (Campanian, Judithian) Formations, Within and Near the Grand Staircase–Escalante National Monument, Southern Utah. Utah Geological Survey Miscellaneous Publication 02-4.

Eaton, J. G., and J. D. Nations. 1991. Introduction: tectonic setting along the margin of the Cretaceous Western Interior Seaway, southwestern Utah and North Arizona; pp.

1–8 in J. D. Nations and J. G. Eaton (eds.), Stratigraphy, Depositional Environments, and Sedimentary Tectonics of the Western Margin, Cretaceous Western Interior Seaway. Geological Society of America Special Publication 260.

Fossen, H., R. A. Schultz, Z. K. Shipton, and K. Mair. 2007. Deformation bands in sandstone: a review. Journal of the Geological Society, London 164:755–769.

Fryberger, S. G., and C. J. Schenk. 1988. Pin stripe lamination: a distinctive feature in modern and ancient eolian sediments. Sedimentary Geology 55:1–15.

Hilbert-Wolf, H. L., E. L. Simpson, W. S. Simpson, S. E. Tindall, and M. C. Wizevich. 2009. Insights into syndepositional fault movement in a foreland basin; trends in seismites of the Upper Cretaceous, Wahweap Formation, Kaiparowits Basin, Utah. Basin Research 21:856–871.

Holzer, T. L., and M. M. Clark. 1993. Sand boils without earthquakes. Geology 21:873–876.

Hough, S. E., R. Bilham, K. Mueller, W. Stephenson, R. Williams, and J. Odum. 2005. Wagon loads of sand blows in White County, Illinois. Seismological Research Letters 76:373–386.

Hunter, R. E. 1977. Basic types of stratification in small eolian dunes. Sedimentology 24:361–387.

Jinnah, Z. A., and E. M. Roberts. 2009. Facies associations and paleoenvironment of the Upper Cretaceous Wahweap Formation: implications for sequence stratigraphy; p. 29 in Advances in Western Interior Late Cretaceous Paleontology and Geology Abstracts with Programs.

Jinnah, Z. A., E. M. Roberts, A. L. Deino, J. S. Larsen, P. K. Link, and C. M. Fanning. 2009. New ^{40}Ar-^{39}Ar and detrital zircon U-Pb ages for the Upper Cretaceous Wahweap and Kaiparowits Formations on the Kaiparowits Plateau, Utah: implications of regional correlation, provenance, and biostratigraphy. Cretaceous Research 30:287–299.

Jonk, R., A. Hurst, and B. T. Cronn. 2007. Variations in sediment extrusion in basin-floor, slope and delta-front settings: sand volcanoes and extruded sand sheets from the Namurian of County Clare, Ireland; pp. 221–226 in A.

Hurst and J. Cartwright (eds.), Sand Injectites: Implications for Hydrocarbon Exploration and Production. American Association of Petroleum Geologists Memoir 87.

Lawton, T. F., S. L. Pollack, and R. A. J. Robinson. 2003. Integrating sandstone petrology and nonmarine sequence stratigraphy: applications to the Late Cretaceous fluvial systems of southwestern Utah, U.S.A. Journal of Sedimentary Research 73:389–406.

Li, Y., J. Craven, E. S. Schweig, and S. F. Obermeier. 1996. Sand boils induced by the 1993 Mississippi River flood: could they one day be misinterpreted as earthquake-induced liquefaction? Geology 24:171–174.

Little, W. M. 1997. Tectonic and eustatic controls on cyclic fluvial patterns, Upper Cretaceous strata of the Kaiparowits Basin, Utah; pp. 489–504 in L. M. Hill (ed.), Learning from the Land: Grand Staircase–Escalante National Monument Science Symposium Proceedings. U.S. Department of the Interior, Bureau of Land Management.

Lui, L., and Y. Li. 2001. Identification of liquefaction and deformation features using ground penetrating radar in the New Madrid seismic zone, U.S.A. Journal of Applied Geophysics 47:199–215.

Maurya, D. M., B. Goyal, A. K. Patidar, N. Mulchandani, M. G. Thakkar, and L. S. Chamyal. 2006. Ground penetrating radar imaging of two large sand blow craters related to the 2001 Bhuj earthquake, Kachchh, western India. Journal of Applied Geophysics 60:142–152.

Montenat, C. 1980. Relations entre déformations synsédimentaires et paléoséismicité dans le Messinien de San Miguel de Salinas (Cordillères bétiques orientales, Espagne). Bulletin Société Géologique, France 22:501–509.

Montenat, C., P. Barrier, O. d'Estevou, and C. Hibsch. 2007. Seismites: an attempt at critical analysis and classification. Sedimentary Geology 196:5–30.

Mount, J. F. 1993. Formation of fluidization pipes during liquefaction: examples from the Uratanna Formation (Lower Cambrian), South Australia. Sedimentology 40:1027–1037.

Obermeier, S. F. 1996. Using liquefaction features for paeoseismic analysis; pp. 331–396 in J. P. McCalpin (ed.), Paleoseismology. Academic Press, London.

Obermeier, S. F., E. C. Pond, and S. M. Olsen. 2001. Paleoliquefaction Studies in Continental Settings: Geological and Geotechnical Factors in Interpretation and Back-analysis. U.S. Geological Survey Open-File Report 01–29.

Olsen, S. M., R. A. Green, and S. F. Obermeier. 2005. Revised magnitude-bound relation for the Wabash Valley seismic zone of the central United States. Seismological Research Letters 76:756–771.

Peterson, F. 1969. Cretaceous sedimentation and tectonism in the southeastern Kaiparowits region, Utah. Ph.D. dissertation, Stanford University, Stanford, California.

Pollock, S. L. 1999. Provenance, geometry, lithofaces and age of the Upper Cretaceous Wahweap Formation, Cordilleran Foreland Basin, Southern Utah. M.S. thesis, New Mexico State University, Las Cruces, New Mexico.

Pringle, J. K., A. R. Westerman, D. A. Stanbrook, D. I. Tatum, and A. R. Gardiner. 2007. Sand volcanoes of the Carboniferous Ross Formation, County Clare, Western Ireland: 3-D internal sedimentary structure and formation; pp. 227–231 in A. Hurst and J. Cartwright (eds.), Sand Injectites: Implications for Hydrocarbon Exploration and Production. American Association of Petroleum Geologists Memoir 87.

Rajendran, K., C. P. Rajendran, M. Thakkar, and R. K. Gartia. 2002. Sand blows from the 2001 Bhuj earthquake reveal clues on past seismicity. Current Science 83:603–610.

Rajendran, K., C. P. Rajendran, M. Thakkar, and M. Tuttle. 2001. The 2001 Kutch (Bhuj) earthquake: coseismic surface features and their significance. Current Science 80:1397–1405.

Roberts, E. M., A. L. Deino, and M. A. Chan. 2005. ^{40}Ar/^{39}Ar age of the Kaiparowits Formation, southern Utah, and correlation of contemporaneous Campanian strata and vertebrate faunas along the margin of the Western Interior Basin. Cretaceous Research 26:307–318.

Samaila, N. K., M. B. Abubakar, E. F. Dike, and C. Obaje. 2006. Description of soft-sediment deformation structures in the Cretaceous Bima Sandstone from the Yola Arm, Upper Benue Trough, Northeastern Nigeria. Journal of African Earth Sciences 44:66–74.

Sargent, K. A., and D. E. Hansen. 1982. Bedrock geologic map of the Kaiparowits coal-basin area, Utah. U.S. Geological Survey Map I-1033-I.

Saucier, R. T. 1989. Evidence for episodic sand-blow activity during the 1811–1812 New Madrid (Missouri) earthquake series. Geology 17:103–106.

Simpson, E. L., M. C. Wizevich, H. L. Hilbert-Wolf, S. E. Tindall, J. J. Bernard, and W. S. Simpson. 2009. An upper Cretaceous sag pond deposit: implications for the recognition of local seismicity and surface rupture along the Kaibab monocline, Utah. Geology 37:967–970.

Simpson, E. L., H. L. Wolf, W. S. Simpson, S. E. Tindall, J. J. Bernard, T. A. Jenesky, and M. C. Wizevich. 2008. The interaction of aeolian and fluvial processes during deposition of the late Cretaceous capping sandstone member, Wahweap Formation, Kaiparowits Basin, Utah, U.S.A. Palaeogeography, Palaeoclimatology, Palaeoecology 270:19–28.

Thakkar, M. G., and B. Goyal. 2004. On the relation between magnitude and liquefaction dimension at the epicentral zone of 2001 Bhuj earthquake. Current Science 87:811–817.

Tindall, S. E., L. P. Storm, T. A. Jenesky, and E. L. Simpson. 2010. Growth faults in the Kaiparowits basin, Utah, pinpoint initial Laramide deformation in the western Colorado Plateau. Lithosphere 2:221–231.

Wesnousky, S. G., and L. M. Leffler. 1994. A Search for Paleoliquefaction and Evidence Bearing on Recurrence Behavior of the Great 1811–12 New Madrid Earthquakes. U.S. Geological Survey Professional Paper 1538-H.

Wizevich, M. C., E. L. Simpson, S. E. Tindall, J. Bernard, H. L. Wolf, W. S. Simpson, L. Storm, and S. Paese. 2008. Tectonic control on the alternating fluvial style in the Late Cretaceous upper member of the Wahweap Formation, southern Utah. Geological Society of America Abstracts with Programs 40(1):40.

Simpson et al.

The Kaiparowits Formation: A Remarkable Record of Late Cretaceous Terrestrial Environments, Ecosystems, and Evolution in Western North America

6

Eric M. Roberts, Scott D. Sampson, Alan L. Deino,
Samuel A. Bowring, and Robert Buchwaldt

UPDATED SEDIMENTOLOGICAL AND PALEONTOLOGICAL data support earlier assertions that the Kaiparowits Formation was deposited in a wet alluvial to coastal plain setting with an abundance of large river channels and perennial ponds, lakes, and wetlands. A synthesis of available geochronological data from contemporaneous Upper Cretaceous continental sedimentary units was compiled, and many ages were recalibrated on the basis of new standards to provide the most up-to-date correlations of coeval strata and associated faunas across the Western Interior Basin. Recalibration of Kaiparowits Formation ash beds demonstrates that the formation is approximately half a million years older than previously suggested, deposited ~76.6–74.5 Ma. In addition, a new ash bed (bentonite) from the Horse Mountain area, collected in the lower portion of the middle unit of the Kaiparowits Formation (~190-m level), was radiometrically dated by both $^{40}Ar/^{39}Ar$ and U-Pb techniques, resulting in similar ages of 75.97 ± 0.18 Ma and 76.26 ± 0.05 Ma. Importantly, both ages are consistent with dated ash beds sitting above and below this level. Measured sections from throughout the outcrop expanse of the Kaiparowits Formation are correlated by an updated tephrostratigraphy, and key vertebrate fossil sites from throughout the formation are precisely tied into this stratigraphy. Updated geochronology and stratigraphy reveals that many of the most richly fossiliferous intervals across the Western Interior Basin are constrained to extremely narrow temporal intervals. The term *taphozone* was coined to describe broad geographic zones of exceptional and elevated continental fossil preservation within narrow temporal windows. Taphozones imply the existence of regional or basin-scale controls on fossil preservation. The Kaiparowits–Dinosaur Park–upper Two Medicine taphozone is assigned to the widespread interval of middle to late Campanian strata in the Western Interior Basin defined by the Kaiparowits, Dinosaur Park, and upper Two Medicine formations. Preliminary analysis of this phenomenon suggests that synchronous deposition of large volumes of volcanic ash across the basin during this time may be the primary driver of elevated fossil preservation in the Kaiparowits–Dinosaur Park–upper Two Medicine taphozone.

INTRODUCTION

The Kaiparowits Formation provides a truly exceptional window into Late Cretaceous terrestrial ecosystems of the Western Interior of North America. A wealth of attention has recently been devoted to the formation, resulting in numerous new and important paleontological discoveries and greatly improved temporal and paleoenvironmental contexts. The renaissance of research in the Kaiparowits Formation over the last 10–20 years has been prompted by a realization of the untapped richness of continental vertebrate, invertebrate, paleobotanical, and trace fossils entombed within the myriad of wet, subhumid channel, overbank, and lacustrine depositional environments that make up this sequence. Arguably, the greatest attribute of the formation, from a paleontological viewpoint, is the outstanding potential for high-resolution evolutionary and ecological investigations of its preserved flora and fauna. The unusually thick (860 m) succession was deposited in a remarkably short ~2 Myr interval, recording one of the fastest sediment accumulation rates of any richly fossiliferous continental sedimentary sequence in the world (Roberts, 2005). The significance of this fact is greatly enhanced by the voluminous and high-quality preservation of most major groups of continental fossils, including aquatic and terrestrial vertebrate skeletal accumulations, macro- and microinvertebrate shelled organisms, plant macro- and microfossils (leaves, wood, palynomorphs), and continental trace fossil assemblages from multiple substrates. Moreover, recent high-precision radioisotopic dating in the formation (and in coeval units) reveals that the Kaiparowits Formation is nearly identical in age (Late Campanian) to many of the other most important fossiliferous Campanian formations or intervals across the Western Interior Basin, including the Dinosaur Park Formation and portions of the Judith River, Two Medicine, Fruitland, and possibly Aguja formations (Roberts, Deino, and Chan, 2005). This fact has been recognized by various workers as a critical opportunity for high-resolution comparisons of closely coeval floras and faunas across the more than 2000-km north–south-trending corridor of fossiliferous strata, which is well exposed from Alberta to Mexico. Sampson et al. (2004, 2010) and Gates et al. (2010) used this improved chronostratigraphic control to test long-debated hypotheses concerning the distribution patterns and provinciality of subcontinental faunas in the Western Interior Basin. Sampson et al. (2004, 2010) argued that provincialism hypotheses put forward a decade earlier by Lehman (1997) could now be revisited because many of the criticisms associated with temporofaunal correlations of these faunas have been resolved as a result of advances in radioisotopic dating and greatly expanded fossil collections from critical areas (i.e., Kaiparowits Formation). In particular, studies focused on dinosaurs, such as those of Gates and Sampson (2007), Gates et al. (2010), and Sampson et al. (2010), recognize patterns of provinciality among large dinosaurs up and down the western margin of the Western Interior Seaway, indicating that many groups were characterized by smaller-than-expected geographic ranges, given the size of many dinosaur species, lending support to Lehman's (1997) provincialism hypothesis, yet differing in complexity of biogeographic patterns. Moreover, enhanced fossil collection and stratigraphic control enabled the documentation of rapid evolutionary turnover between large dinosaur taxa and evidence of overlapping ranges between certain ceratopsian and hadrosaur taxa, a phenomenon not previously observed in the Western Interior Basin (Ryan and Evans, 2005; Gates and Sampson, 2007; Sampson et al., 2010). Gates et al. (2010) synthesized data on all major vertebrate groups in the target formation to provide a more holistic review of paleogeographic patterns within this temporal interval. Their results reveal a more pronounced latitudinally controlled biogeographic pattern than even Lehman (1997) and Sampson et al. (2004) realized and strongly supports the idea of discrete centers of endemism among many vertebrate groups across the basin.

The Kaiparowits Formation also reveals important information relating to the evolution of Late Cretaceous landscapes (paleoenvironments and paleogeography) and regional tectonics during a time of major orogenic and basinal transitions in the Western Interior. The goal of this chapter is to provide updated paleoenvironmental, paleoclimatic, and geochronological context for the remarkable floras and faunas preserved within the Kaiparowits Formation. Moreover, updated regional stratigraphic correlations with other portions of the Western Interior Basin are outlined, along with presentation of a new radioisotopic date from the Kaiparowits Formation and recalibration of older dates from the Kaiparowits and coeval units in the Western Interior

Basin. Finally, a discussion of the taphonomic significance of widespread instances of exceptional fossil preservation across narrow temporal windows, including the Kaiparowits time frame, is given.

METHODS

An overview of the regional geology and previous work conducted in the Kaiparowits Formation and other Upper Cretaceous units in the Kaiparowits Basin is presented by Titus et al. (this volume, Chapter 2) and thus will not be discussed in great detail here.

Here we report on the results of research conducted in the Kaiparowits Formation over the last 9 years. Data have been collected from 10 primary locations along the length of the Kaiparowits Plateau that provide information for most outcrop areas with significant exposures (Fig. 6.1). Detailed stratigraphic sections have been measured at each locality with a Jacob staff and Brunton compass; in many cases, multiple sections were measured in each area. However, Roberts and colleagues (Roberts, Deino, and Chan, 2005; Roberts, 2007; Roberts, Tapanila, and Mijal, 2008) provide a detailed review of the various methods used to study the geology of the Kaiparowits Formation.

Ar/Ar Methods

The sample was disaggregated with warm tap water and repeated agitation, then sieved through 20, 40, 60, and 100 mesh screens. Feldspar was separated from the remaining coarse fraction with a Franz isodynamic separator. Then sanidine was isolated from plagioclase by density separations via heavy liquids.

The largest (>300 µm) sanidine concentrates were placed into wells in an aluminum disk in preparation for irradiation. The arrangement consisted of 12 wells (0.80 inches deep by 0.130 inches in diameter) in a 0.260-square-inch-diameter circle, with four standards at the cardinal points and unknowns in the remaining positions (Best et al., 1995). After additional protective packaging, the samples were irradiated for 90 hours in the Cd-lined, in-core CLICIT facility of the Oregon State University TRIGA reactor. Sanidine from the Fish Canyon Tuff of Colorado was used as a mineral standard, with a reference age of 28.201 ± 0.046 Ma (Kuiper et al., 2008).

$^{40}Ar/^{39}Ar$ extractions were performed at the Berkeley Geochronology Center; a focused CO_2 laser was used to fuse and rapidly liberate trapped argon from individual sanidine crystals.

Gasses were scrubbed with SAES getters for several minutes to remove impurities (CO, CO_2, N_2, O_2, and H_2),

Roberts et al.

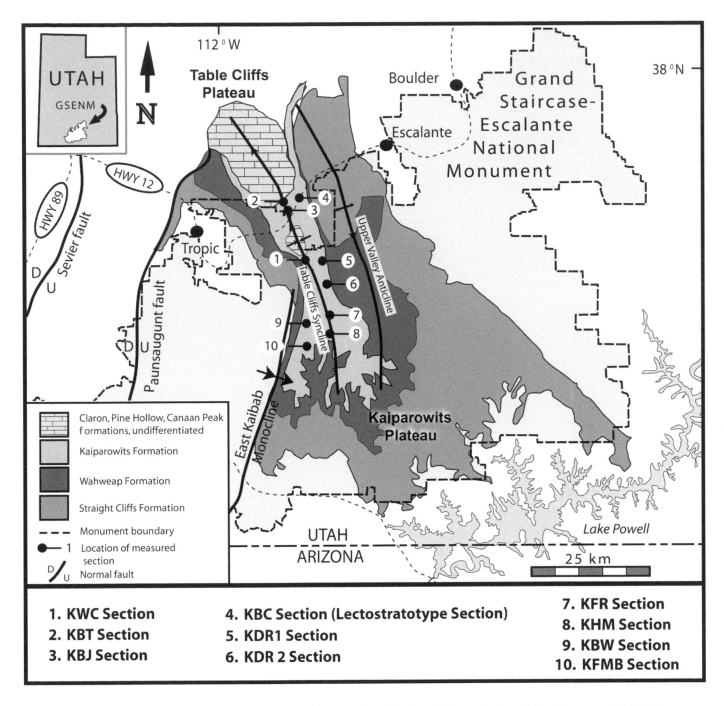

6.1. Outcrop extent and measured section locations in the Kaiparowits Formation within Grand Staircase–Escalante National Monument, Utah. KBC, Kaiparowits Blues Ceratopsian; KBJ, Kaiparowits Blues Jodi's; KBT, Kaiparowits Blues Tyrannosaur; KBW, Kaiparowits Blues Wash; KDR-1, Kaiparowits Death Ridge 1; KDR-2, Kaiparowits Death Ridge 2; KFMB, Kaiparowits Four Mile Bench; KFR, Kaiparowits Fossil Ridge; KHM, Kaiparowits Horse Mountain; KWC, Kaiparowits Wahweap Creek.

followed immediately by measurement of the purified noble gases for five argon isotopes on a MAP 215–50 mass spectrometer for approximately 30 minutes.

U-Pb Methods

Heavy minerals were separated via standard heavy liquid and magnetic techniques. Zircon grains were hand-selected for analysis from the least magnetic fraction on the basis of the absence of cracks, inclusions, and surface contamination. In order to minimize the effects of Pb loss, the grains were subjected to a version of the thermal annealing and acid leaching (also known as chemical abrasion technique of Mattinson (2005) before isotope dilution thermal ionization mass spectrometry analyses using the mixed ^{205}Pb-^{233}U-^{235}U EARTHTIME tracer solution. Details of zircon

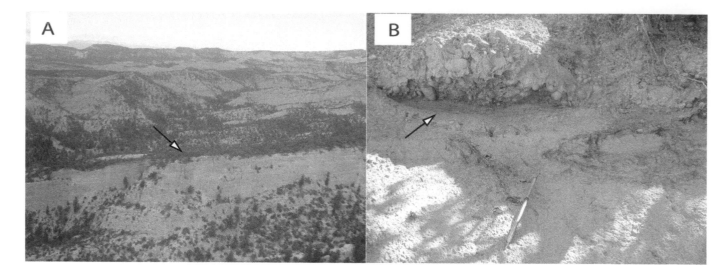

6.2. Photographs showing the stratigraphic relationships between the Kaiparowits Formation, (A) the underlying Wahweap Formation, and (B) the overlying Canaan Peak Formation. Arrows point to contacts.

pretreatment, dissolution, and U and Pb chemical extraction procedures are described in Ramezani et al. (2007).

U and Pb isotopic measurements were performed on a VG Sector-54 multicollector thermal ionization mass spectrometer at the Massachusetts Institute of Technology. Pb and U were loaded together on a single Re filament in a silica gel/phosphoric acid mixture (Gerstenberger and Hasse, 1997). Pb isotopes were measured by peak hopping with a single Daly photomultiplier detector, and U isotopic measurements were made in static mode with multiple Faraday collectors. Details of fractionation and blank corrections are given in Table 6.3. Data reduction, date calculation, and the generation of plots used the algorithms of McLean et al. (2011) and the data reduction plotting program Redux (Bowring et al., 2011). U-Pb errors on analyses from this study are reported in the following manner, unless otherwise noted: ±X/Y/Z, where X is the internal error in absence of all systematic errors, Y includes the tracer calibration error, and Z includes both tracer calibration and decay constant errors of Jaffey et al. (1971).

Stratigraphy and Field Relations

The Kaiparowits Formation is well exposed throughout the Kaiparowits Plateau, reaching up to 860 m thick. The Kaiparowits Formation is largely conformable, albeit erosionally in places, on the Wahweap Formation and is overlain with angular unconformity by conglomeratic facies of the Canaan Peak Formation (Fig. 6.2). Roberts (2007) designated a lectostratotype or principal reference section for the Kaiparowits Formation in The Blues area (KBC section) and informally subdivided the formation into three units, lower, middle, and upper. Ten detailed sections have been measured, and here we present schematic sections for each of the most significant outcrop areas of the formation (Fig. 6.1). Stratigraphic correlation of the Kaiparowits Formation is possible through gross recognition of the lower, middle, and upper units. Higher resolution stratigraphic correlation is achieved through recognition of discrete volcanic marker horizons (bentonites). A correlation chart has been constructed to illustrate the stratigraphic relationships between each of these areas and to help decipher the distribution of fossil localities within the formation (Fig. 6.3).

Chronostratigraphy and Geochronology

Although some early estimates of the Kaiparowits Formation based on palynology (Lohrengel, 1969) and dinosaurs (DeCourten and Russell, 1985) suggested a Maastrichtian age, the mammalian biostratigraphy developed for the Kaiparowits Formation by Jeff Eaton and Rich Cifelli (Eaton and Cifelli, 1988; Cifelli, 1990a, 1990b, 1990c; Eaton, 2002) clearly demonstrated a Campanian age for the fauna. Reevaluation of the palynology by Bowers (1990) and Eaton (1991) further supported a Campanian age. In 2005 the first radioisotopic dates from the formation were reported, providing more precise constraints on the mammalian biostratigraphy and revealing a remarkably short temporal duration for the 860-m-thick formation, deposited ~76.1–74.0 Ma (Roberts, Deino, and Chan, 2005). This preliminary geochronologic study yielded concordant laser-fusion $^{40}Ar/^{39}Ar$ dates on sanidine from four different bentonite horizons dispersed throughout the formation that obey stratigraphic superposition. However, a recent astronomical recalibration of the age of the

Roberts et al.

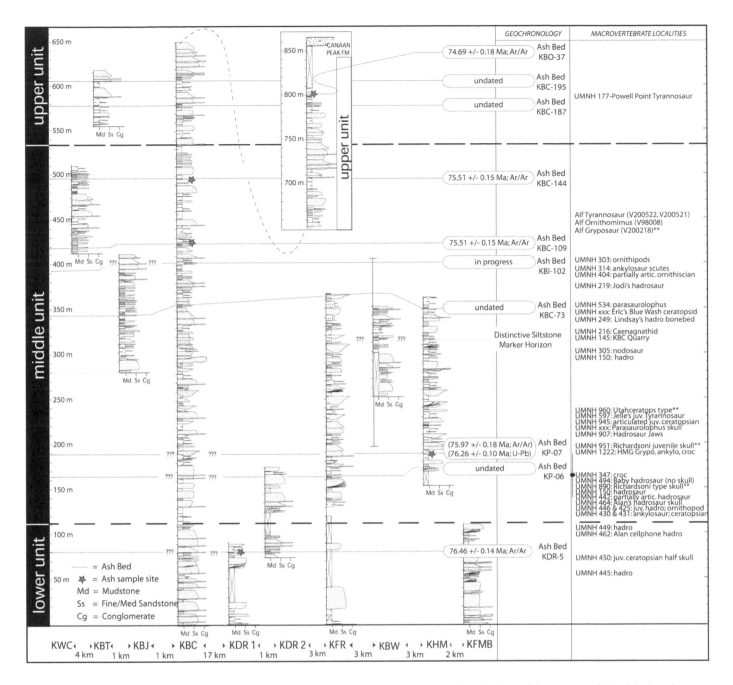

6.3. Correlated measured sections throughout the Kaiparowits Formation. Ten sections are correlated on the basis of the presence of 12 ash beds and one regionally extensive calcareous siltstone (lacustrine). Major vertebrate macrofossil localities are precisely tied to the updated chronostratigraphy and lithostratigraphy. Location of each section is shown in Fig. 6.1. Four of the ⁴⁰Ar/³⁹Ar ages (KDR-5, KBC-109, KBC-144, KBO-37) are recalibrated from Roberts, Deino, and Chan (2005), and one new bentonite (KP-07) was dated by both ⁴⁰Ar/³⁹Ar and U-Pb techniques in this study.

⁴⁰Ar/³⁹Ar standard used in these studies (sanidine from the Fish Canyon Tuff of Colorado) has led to a revision of the reference age from 28.02 Ma to 28.201 ± 0.046 Ma (Kuiper et al., 2008). A significant outcome of the recalibration of the Fish Canyon Tuff sanidine is the subsequent improvement in the correspondence between ⁴⁰Ar/³⁹Ar and U-Pb dates. This has been a long-standing problem in geochronology (Kuiper et al., 2008), highlighted by ~1% discordance between these two dating systems. The recalibrated Fish Canyon Tuff sanidine eruption age of Kuiper et al. (2008) reportedly increases

the concordance between ⁴⁰Ar/³⁹Ar and U-Pb ages; however, few case studies have tested this. Renne et al. (2010) recently suggested an age for the Fish Canyon Tuff sanidine of 28.305 Ma but later modified this value to 28.294 Ma (Renne et al., 2011). Most recently, Meyers et al. (2012) developed an intercalibrated astrochronologic and radioisotopic timescale for the Cenomanian–Turonian boundary using Ar-Ar, U-Pb geochronology, and astrochronology. They concluded that an eruption age for the Fish Canyon Tuff sanidine of 28.201 Ma (sensu Kuiper et al., 2008) is supported by their

Table 6.1. Summary of isotopic age data and recalibration results for dated Upper Campanian ash beds across the Western Interior Basin

Sitew	Sample no.	Stratigraphy	Location	Dating system and facility	Mineral
Kaiparowits Plateau–Kaiparowits and Wahweap formations[a]	KBO-37	Kaiparowits–790 m	Blues	40Ar/39Ar; BGC	S
	KBC-109	Kaiparowits–420 m	Blues	40Ar/39Ar; BGC	S
	KBC-144	Kaiparowits–490 m	Blues	40Ar/39Ar; BGC	S
	KP-7	Kaiparowits–190 m	Horse Mnt	40Ar/39Ar; BGC	S
	KP-7	Kaiparowits–190 m	Horse Mnt	U-Pb TIMS; MIT	Z
	KDR-5	Kaiparowits 80 m	Death Ridge	40Ar/39Ar; BGC	S
	SS07B	Wahweap ~60 m	Star Seep	40Ar/39Ar; BGC	S
	CF05B	Wahweap ~50 m	Star Seep	40Ar/39Ar; BGC	S
Fruitland–Kirtland formations[b]	Ash J	KL	Hunter Wash	40Ar/39Ar; USGS	S
	Ash H	KL	Hunter Wash	40Ar/39Ar; USGS	S
	Ash 4	KL	Hunter Wash	40Ar/39Ar; USGS	S
	Ash2	Base KL	Hunter Wash	40Ar/39Ar; USGS	S
	DEP	Lower FR	Dog Eye Pond	40Ar/39Ar; USGS	S
	Huerfanito	LEWIS Sh	SW San Juan	40Ar/39Ar; USGS	S
Big Bend National Park–Aguja and Javelina formations[c]	Bryer	Aguja u. shale	Big Bend	U-Pb SHRIMP-RG; Stanford/USGS	Z
	Befus	Aguja u. shale	Big Bend	U-Pb SHRIMP-RG; Stanford/USGS	Z
	Lehman	Javelina	Big Bend	U-Pb	M
Two Medicine Formation[d]	TM-6	10 m below top	Blacktail Cr	40Ar/39Ar; BGC	P
	TM-6	10 m below top	Blacktail Cr	40Ar/39Ar; BGC	B
	TM-4	480 m above base	Type Area	40Ar/39Ar; BGC	S
	TM-4	480 m above base	Type Area	40Ar/39Ar; BGC	P
	WOFS/U5	108 m above base	Shields Crossing	40Ar/39Ar; BGC	P
	WOFS/U5	108 m above base	Shields Crossing	40Ar/39Ar; BGC	B
	TM-3	105 m above base	Choteau	40Ar/39Ar; BGC	P
	TM-3	105 m above base	Choteau	40Ar/39Ar; BGC	B
	RT/TM-7	105 m above base	Choteau	40Ar/39Ar; BGC	P
	RT/TM-7	105 m above base	Choteau	40Ar/39Ar; BGC	B
	Foreman	~265 m; mid	Hadro Hill	40Ar/39Ar; BGC	S
	Foreman	~265 m; mid	Hadro Hill	40Ar/39Ar; BGC	P
Judith River Formation[e]	84MG8-3-4	38 m above 7-16-1	Kennedy Coulee	40Ar/39Ar; BGC	S
	84MG8-3-4	38 m above 7-16-1	Kennedy Coulee	40Ar/39Ar; BGC	B
	84MG7-16-1	Near base	Kennedy Coulee	40Ar/39Ar; BGC	S
	84MG7-16-1	Near base	Kennedy Coulee	40Ar/39Ar; BGC	B
Dinosaur Provincial Park, Oldman, Dinosaur Park, and Bearpaw formations[f]	DPP-7	11 m above base	Bearpaw Fm	40Ar/39Ar; BGC	S
	Plateau Tuff	25 m below top	Dinosaur Park Fm	40Ar/39Ar; BGC	S
	Top Oldman	4 m below top	Oldman Fm	40Ar/39Ar; BGC	S

P, plagioclase; B, biotite; S, sanidine; Z, zircon; FCT, Fish Canyon tuff; BTQD, Braintree quartz diorite; TCR, Taylor Creek rhyolite; BGC, Berkeley Geochronology Center; NA, not applicable.

[a] Modified from Roberts, Deino, and Chan (2005), Jinnah et al. (2009), and Jinnah (this volume).

[b] Modified from Fassett and Steiner (1997).

[c] Modified from Lehman et al. (2006), Breyer et al. (2007), and Befus et al. (2008).

[d] Modified from Rogers, Swisher, and Horner (1993) and Foreman et al. (2008).

[e] Modified from Goodwin and Deino (1989).

[f] Modified from Eberth and Hamblin (1993) and Eberth and Dieno (1992).

intercalibration. As such, part of this chapter focuses on (1) recalculating old Kaiparowits ^{40}Ar/^{39}Ar dates using the new Fish Canyon Tuff sanidine date, and (2) dating a new Kaiparowits ash (KP-07) to compare ^{40}Ar/^{39}Ar and U-Pb dates. This fits within the greater goal of the Kaiparowits Basin Project, which is to continue to improve and refine the geochronology of the Kaiparowits Formation through identification and dating of new ash beds and through improved sampling and

Legacy age	Legacy λ	Legacy standard (t)	New λ	New standard (t)	Converted age or new age	Relative change
74.21 ± 0.18 Ma	5.543E-10/y	FCT; 28.02 Ma	5.463E-10/y	FCT: 28.201 Ma	74.69 ± 0.18 Ma	0.65%
75.02 ± 0.15 Ma	5.543E-10/y	FCT; 28.02 Ma	5.463E-10/y	FCT: 28.201 Ma	75.51 ± 0.15 Ma	0.65%
75.02 ± 0.15 Ma	5.543E-10/y	FCT; 28.02 Ma	5.463E-10/y	FCT: 28.201 Ma	75.51 ± 0.15 Ma	0.65%
NA	NA	NA	5.463E-10/y	FCT: 28.201 Ma	75.97 ± 0.18 Ma	NA
NA					76.26 ± 0.05 Ma	NA
75.96 ± 0.14 Ma	5.543E-10/y	FCT; 28.02 Ma	5.463E-10/y	FCT: 28.201 Ma	76.46 ± 0.14 Ma	0.65%
NA	NA	NA	5.463E-10/y	FCT: 28.201 Ma	79.9 ± 0.3 Ma	NA
80.1 ± 0.15 Ma	5.54E-10	FCT; 28.02 Ma	5.463E-10/y	FCT: 28.201 Ma	80.63 ± 0.15 Ma	0.65%
73.04 ± 0.25 Ma	4.962E-10/y	TCR; 28.32 Ma	5.463E-10/y	FCT: 28.201 Ma	72.66 ± 0.25 Ma	−0.53%
73.37 ± 0.18 Ma	4.962E-10/y	TCR; 28.32 Ma	5.463E-10/y	FCT: 28.201 Ma	72.98 ± 0.18 Ma	−0.53%
74.11 ± 0.62 Ma	4.962E-10/y	TCR; 28.32 Ma	5.463E-10/y	FCT: 28.201 Ma	73.72 ± 0.62 Ma	−0.53%
74.56 ± 0.13 Ma	4.962E-10/y	TCR; 28.32 Ma	5.463E-10/y	FCT: 28.201 Ma	74.17 ± 0.13 Ma	−0.53%
75.56 ± 0.41 Ma	4.962E-10/y	TCR; 28.32 Ma	5.463E-10/y	FCT: 28.201 Ma	75.16 ± 0.41 Ma	−0.53%
75.76 ± 0.34 Ma	4.962E-10/y	TCR; 28.32 Ma	5.463E-10/y	FCT: 28.201 Ma	75.36 ± 0.34 Ma	−0.53%
72.6 ± 1.5 Ma	NA	NA	?	R33 BTQD: 419 Ma	NA	NA
76.9 ± 1.2 Ma	NA	NA	?	R33 BTQD: 419 Ma	NA	NA
69.0 ± 0.9 Ma	NA	NA	?	?	NA	NA
74.08 ± 0.1 Ma	NA	NA	NA	NA	NA	NA
73.56 ± 1.14 Ma	NA	NA	NA	NA	NA	NA
74.07 ± 0.72 Ma	5.543E-10/y	FCT; 27.84 Ma	5.463E-10/y	FCT: 28.201 Ma	75.03 ± 0. Ma	1.28%
74.27 ± 0.15 Ma	NA	NA	NA	NA	NA	NA
79.60 ± 0.09 Ma	NA	NA	NA	NA	NA	NA
78.26 ± 0.14 Ma	NA	NA	NA	NA	NA	NA
79.77 ± 0.11 Ma	NA	NA	NA	NA	NA	NA
79.71 ± 0.03 Ma	NA	NA	NA	NA	NA	NA
80.0 ± 0.114 Ma	NA	NA	NA	NA	NA	NA
80.04 ± 0.07 Ma	NA	NA	NA	NA	NA	NA
77.52 ± 0.19 Ma	5.543E-10/y	FCT; 28.02 Ma	5.463E-10/y	FCT: 28.201 Ma	78.03 ± 0.19 Ma	0.65%
75.8 ± 0.7 Ma	NA	NA	NA	NA	NA	NA
78.2 ± 0.2 Ma	5.543E-10/y	FCT; 28.02 Ma	5.463E-10/y	FCT: 28.201 Ma	78.71 ± 0. Ma	0.65%
78.2 ± 0.2 Ma	5.543E-10/y	NA	NA	NA	NA	NA
78.5 ± 0.2 Ma	5.543E-10/y	FCT; 28.02 Ma	5.463E-10/y	FCT: 28.201 Ma	79.02 ± 0. Ma	0.65%
79.5 ± 0.2 Ma	5.543E-10/y	NA	NA	NA	NA	NA
74.98 ± 0.24 Ma	5.543E-10/y	FCT; 28.02 Ma	5.463E-10/y	FCT: 28.201 Ma	75.47 ± 0.24 Ma	0.65%
75.3 ± 0.3 Ma	5.543E-10/y	FCT; 28.02 Ma	5.463E-10/y	FCT: 28.201 Ma	75.79 ± 0.24 Ma	0.65%
76.5 ± 0.5 Ma	5.543E-10/y	FCT; 28.02 Ma	5.463E-10/y	FCT: 28.201 Ma	77.00 ± 0.24 Ma	0.65%

multiple dating techniques of new and previously dated bentonite beds.

Recalculation of Previous $^{40}Ar/^{39}Ar$ Ages Based on New Fish Canyon Tuff Standard

Utilizing the new Fish Canyon Tuff sanidine date of Kuiper et al. (2008), the $^{40}Ar/^{39}Ar$ ages of Kaiparowits bentonites KDR-5 (75.96 ± 0.14 Ma), KBC-109 (75.02 ± 0.15 Ma), KBC-144 (75.02 ± 0.15 Ma), and KBO-37 (74.21 ± 0.18 Ma) all become roughly half a million years older, recalculated to 76.46 ± 0.14 Ma, 75.51 ± 0.15 Ma, 75.51 ± 0.15 Ma, and 74.69 ± 0.18 Ma, respectively (Table 6.1; all uncertainty quoted at 1σ internal). Thus, utilizing the average rock accumulation rate of 41 cm/ka for the 710 m of strata bracketed between bentonites KDR-5 and KBO-37 and linearly extrapolating, the age of the

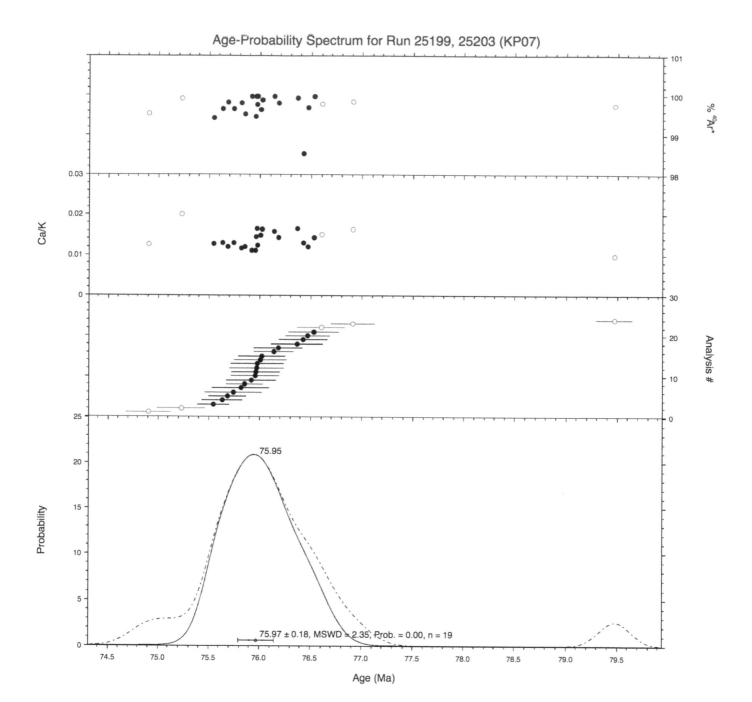

6.4. Age probability diagram for KP-07 $^{40}Ar/^{39}Ar$ age determination on sanidine.

formation is still estimated to span 2.1 Myr (sensu Roberts, Deino, and Chan, 2005), deposited approximately between 76.6 to 74.5 Ma. Recalculated $^{40}Ar/^{39}Ar$ ages for the Kaiparowits Formation, along with recalculated $^{40}Ar/^{39}Ar$ ages (on sanidine) from stratigraphically relevant late Campanian ash beds from across the Western Interior Basin, are presented in Table 6.1. All updated $^{40}Ar/^{39}Ar$ ages on sanidine were converted by the EARTHTIME Ar tool (http://www .earth-time.org/ar-ar.html). Only those ash beds in which

the legacy decay constant and fluence monitor (old age standard) are known (i.e., published in the literature) have been included. Many of the older radioisotopic ages from coeval units are based on older standards, resulting in varying degrees of relative change in dates. Many of the radioisotopic dates presented in Table 6.1 are based on minerals other than sanidine or used different dating systems than $^{40}Ar/^{39}Ar$. Thus they are not relevant to age recalculation, but they have been included for comparison.

Roberts et al.

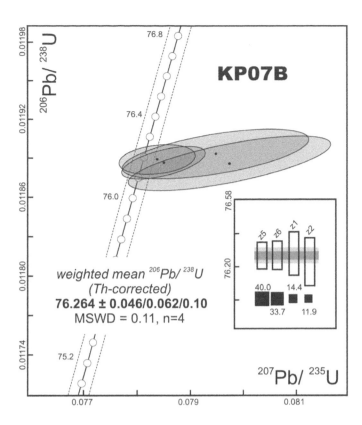

6.5. Age probability diagram for KP-07 U-Pb dilution thermal ionization mass spectrometry age determination on zircon.

Bentonite Stratigraphy in the Kaiparowits Formation

Since 2007, three new bentonite horizons (KP-06, KP-07, KBI-118) have been identified in the Kaiparowits Formation and placed within the detailed stratigraphic framework of the formation (Fig. 6.3). The new bentonite horizons are useful not only for their potential to be dated and to refine the chronostratigraphy, but also for their utility as marker beds for local stratigraphic and faunal correlation. Of the three new bentonites identified, KP-06 and KP-07 were identified in the lower portion of the middle unit (177-m and 190-m levels, respectively) in the Horse Mountain area, whereas KBI-118 was identified in the upper portion of the middle member (400-m level) in The Blues area, ~250 m to the south of the type section. All three ash beds can be traced across their respective outcrop areas, particularly KP-07 (B. Albright, pers. comm.). New measured sections and bentonite correlations in isolated sections, such as Blues Wash and south of Canaan Peak along Wahweap Creek (Fig. 6.3), permit greatly improved stratigraphic control and high-precision correlation of important fossil localities (Figs. 6.1 and 6.3).

^{40}Ar/^{39}Ar and U-Pb Dating of New Bentonite Sample (KP-07)

In order to fill one of the largest geochronologic gaps in the Kaiparowits Formation that exist between the top of the lower unit (80-m level) and the upper part of the middle unit (420-m level), bentonite KP-07 was selected for radiometric dating. Moreover, in an effort to more fully explore the possibility for high-resolution dating in the Kaiparowits Formation, dating of this ash bed was performed as part of a pilot study for assessing the age correspondence between ^{40}Ar/^{39}Ar dates on sanidine and U-Pb dates on zircon in Kaiparowits ash beds. Bentonite sample KP-07 was collected near the base of Horse Mountain on a prominent bench just below a small rise. The bentonite displays classic popcorn weathering and is easily recognizable and laterally traceable. It is light yellowish-green in color and up to 25 cm thick. It was collected at GPS coordinates (WGS 84) 37°26′10.88″N/111°41′46.94″W.

METHODS

A deep pit was excavated into a fresh hill slope exposure so we could obtain a fresh sample KP-07. We hand-sorted ~15 kg of the freshest sample from near the base of the bentonite to obtain the largest crystals. Phenocryst separations and Ar/Ar dating (on sanidine) were conducted following the same methodology as outlined in Roberts, Deino, and Chan (2005). The only departure was the use of the new Fish Canyon Tuff sanidine reference age of 28.201 ± 0.046 Ma (Kuiper et al., 2008) in the ^{40}Ar/^{39}Ar age calculation. Zircons were isolated from the heavy mineral separates and analyzed at the Massachusetts Institute of Technology by chemical abrasion thermal ionization mass spectrometry U-Pb dating techniques (sensu Mattinson, 2005).

RESULTS

Summary ^{40}Ar/^{39}Ar and U-Pb analytical data for KP-07 are presented in Tables 6.2 and 6.3. The sanidine yields a ^{40}Ar/^{39}Ar age of 75.97 ± 0.18 Ma (±1σ internal) (Fig. 6.4). The same sample yielded abundant clear zircons. We selected six clear pinkish 150–200-μm idiomorphic zircon grains with sharp edges and no macroscopic inclusions. Four analyses yielded a cluster on the Concordia diagram (Fig. 6.5). Two analyses revealed elevated contents of common Pb and larger error ellipses, whereas the other two had higher radiogenic-to-common ratios. Together they yield a weighted mean date

Table 6.2. Summary of analytical data for $^{40}Ar/^{39}Ar$ analysis on sanidine from bentonite sample KP-07 from the Kaiparowits Formation

Lab ID no.	J (×10⁻³) Value	±1σ	³⁹Ar Mol (× 10⁻¹⁴)	Ca/K Value	±1σ	⁴⁰Ar*/³⁹Ar Value	±1σ	Age (Ma) Value	±1σ	w/±J Value	±1σ	MSWD	Prob.
KP07:													
25199-02	26.363	0.060	13.18	0.0141062	0.0006118	1.6363	0.0046	76.19006	0.2107313	0.27	0.00		
25199-05	26.363	0.060	21.68	0.018932	0.0003783	1.6253	0.0034	75.6919	0.154505	0.23	0.00		
25199-07	26.363	0.060	14.08	0.0128138	0.0006929	1.6242	0.0038	75.63881	0.1726468	0.24	0.00		
25199-08	26.363	0.060	42.38	0.0110605	0.0002991	1.6312	0.0047	75.96127	0.2151482	0.27	0.00		
25199-09	26.363	0.060	29.61	0.0126731	0.0005712	1.6222	0.0030	75.5489	0.1358489	0.22	0.00		
25199-10	26.363	0.060	10.14	0.011954	0.000745	1.6425	0.0041	76.47366	0.187109	0.25	0.00		
25199-12	26.363	0.060	12.65	0.0119535	0.0008252	1.6289	0.0034	75.85392	0.1563572	0.23	0.00		
25203-26	26.176	0.060	28.87	0.012913	0.0010813	1.6434	0.0051	75.98569	0.231818	0.29	0.00		
25203-27	26.176	0.060	22.90	0.0127694	0.0014056	1.6532	0.0048	76.4263	0.2185403	0.28	0.00		
25203-28	26.176	0.060	11.88	0.0156729	0.0021458	1.6470	0.0037	76.14712	0.1657546	0.24	0.00		
25203-30	26.176	0.060	11.86	0.0161544	0.0020935	1.6444	0.0045	76.02836	0.204334	0.27	0.00		
25203-36	26.176	0.060	10.28	0.0109978	0.0025575	1.6420	0.0049	75.92297	0.2228852	0.28	0.00		
25203-37	26.176	0.060	8.82	0.0147072	0.0027392	1.6440	0.0052	76.0101	0.2370023	0.29	0.00		
25203-38	26.176	0.060	20.55	0.0140967	0.0013192	1.6556	0.0048	76.53582	0.2150136	0.28	0.00		
25203-39	26.176	0.060	8.52	0.011599	0.0028807	1.6399	0.0056	75.82362	0.2519305	0.30	0.00		
25203-42	26.176	0.060	15.81	0.0144119	0.0015231	1.6431	0.0046	75.96899	0.2099075	0.27	0.00		
25203-44	26.176	0.060	11.07	0.0163304	0.0021376	1.6519	0.0050	76.37022	0.2250828	0.28	0.00		
25203-47	26.176	0.060	13.30	0.0163813	0.0018477	1.6433	0.0052	75.978	0.2367024	0.29	0.00		
25203-48	26.176	0.060	9.28	0.0128526	0.0026514	1.6382	0.0056	75.7501	0.254031	0.31	0.00		
****				**0.0121522**	**0.0001737**	**1.6373**	**0.0010**	**75.971491**	**0.0049913**	**0.18**	**0.00**	**2.34**	**0.00**
Omitted, 1.5 nMADs from median age:													
25199-01	26.363	0.060	16.06	0.0125484	0.0005882	1.6081	0.0043	74.90661	0.1983166	0.26	0.00		
25199-03	26.363	0.060	13.88	0.0149484	0.0009583	1.6454	0.0046	76.60597	0.2081628	0.27	0.00		
25203-32	26.176	0.060	9.48	0.0200025	0.0029353	1.6268	0.0047	75.23455	0.2108959	0.27	0.00		
25203-35	26.176	0.060	11.80	0.0163047	0.0023595	1.6639	0.0043	76.91223	0.1956708	0.26	0.00		
Omitted, Ca/K >1:													
25199-04	26.363	0.060	1.07	4.71488	0.0293463	1.6387	0.0115	76.30033	0.523748	0.55	0.00		
25199-11	26.363	0.060	1.25	5.563566	0.0506044	8.1532	0.0202	351.2484	0.7910486	1.07	0.00		
25203-29	26.176	0.060	1.58	6.173433	0.0465825	1.6902	0.0132	78.10244	0.5968914	0.62	0.00		
25203-31	26.176	0.060	0.28	6.609179	0.1567269	1.7503	0.0554	80.81636	2.501571	2.51	0.00		
25203-33	26.176	0.060	0.10	8.454077	0.3976608	2.0480	0.1435	94.21049	6.430504	6.43	0.00		
25203-34	26.176	0.060	1.02	6.160251	0.075284	1.6331	0.0174	75.51904	0.78625	0.80	0.00		
25203-40	26.176	0.060	1.84	6.748619	0.063482	1.6696	0.0113	77.16961	0.5121121	0.54	0.00		
25203-41	26.176	0.060	0.96	4.172736	0.052381	1.6664	0.0184	77.02676	0.8324836	0.85	0.00		
25203-45	26.176	0.060	0.53	11.04448	0.1353716	1.5710	0.0332	72.70489	1.507746	1.52	0.00		
25203-46	26.176	0.060	1.34	3.704773	0.0493322	1.6410	0.0154	75.87733	0.6975218	0.72	0.00		
25203-49	26.176	0.060	0.41	8.339255	0.1472765	1.6314	0.0393	75.44303	1.777942	1.79	0.00		
Omitted arbitrarily:													
25199-06	26.363	0.060	19.93	0.0096543	0.0004019	1.7085	0.0034	79.47917	0.1546748	0.24	0.00		

Roberts et al.

Table 6.3. Details of fractionation and blank corrections for KP07B zircon (chemical abrasion thermal ionization mass spectrometry)[a]

Assessment	Element	Fraction				
		z1	z2	z5	z6	z7
Composition	Th/U[b]	1.24	1.27	0.79	0.38	0.50
	Pb*[c] (pg)	13.3	14.4	15.1	5.8	2.8
	Pbc[d] (pg)	2.06	2.58	1.10	0.34	0.33
	Pb*/Pbc[e]	6.4	5.6	13.7	16.9	8.6
Isotopic ratio	$^{206}Pb/^{204}Pb$[f]	339	295	777	1061	530
	$^{208}Pb/^{206}Pb$[f]	0.396	0.407	0.254	0.122	0.160
	$^{206}Pb/^{238}U$[g]	0.01190	0.01190	0.01190	0.01190	0.01280
	$\pm2\sigma$ (%)	0.162	0.178	0.097	0.106	0.148
	$^{207}Pb/^{235}U$[g]	0.079	0.080	0.078	0.078	0.085
	$\pm2\sigma$ (%)	2.21	2.39	0.91	1.00	1.61
	$^{207}Pb/^{206}Pb$[g]	0.048309	0.048497	0.047644	0.047718	0.047949
	$\pm2\sigma$ (%)	2.120	2.273	0.890	0.941	1.501
Date (Ma)	$^{206}Pb/^{238}U$[h]	76.277	76.229	76.267	76.267	81.986
	$\pm2\sigma$ (abs)	0.122	0.135	0.073	0.080	0.120
	$^{207}Pb/^{235}U$[h]	77.47	77.71	76.43	76.55	82.48
	$\pm2\sigma$ (abs)	1.65	1.79	0.67	0.74	1.27
	$^{207}Pb/^{206}Pb$[h]	114.4	123.6	81.6	85.3	96.8
	$\pm2\sigma$ (abs)	50.0	53.5	21.1	22.3	35.5
	Correlation coefficient	0.58	0.70	0.25	0.59	0.73

[a] Blank composition: $^{206}Pb/^{204}Pb$ = 18.24 ± 0.21; $^{207}Pb/^{204}Pb$ = 15.34 ± 0.16; $^{208}Pb/^{204}Pb$ = 37.35 ± 0.20.

[b] Th contents calculated from radiogenic ^{208}Pb and the $^{207}Pb/^{206}Pb$ date of the sample, assuming concordance between U-Th and Pb systems.

[c] Total mass of radiogenic Pb.

[d] Total mass of common Pb.

[e] Ratio of radiogenic Pb (including ^{208}Pb) to common Pb.

[f] Measured ratio corrected for fractionation and spike contribution only.

[g] Measured ratios corrected for fractionation, tracer, blank, and initial common Pb.

[h] Isotopic dates calculated using the decay constants λ238 = 1.55125E-10 and λ235 = 9.8485E-10 (Jaffey et al., 1971).

of 76.264 ± 0.046/0.062/0.10 (MSWD [mean square of weighted deviation] = 0.11) (Fig. 6.5). However, the two most precise analyses control the weighted mean (80%) and result in a low MSWD (Fig. 6.5). Two additional analyses are distinctly older, with $^{206}Pb/^{238}U$ dates of approximately 82 Ma, and reflect either incorporation during eruption of magma or a background detrital component (Table 6.3). These new data result in a notable improvement in general correspondence between $^{40}Ar/^{39}Ar$ (Fish Canyon Tuff sanidine = 28.201 Ma) and U-Pb dates (see Kuiper et al., 2008, for a full discussion). More importantly, the new KP-07 dates, when combined with the recalculated $^{40}Ar/^{39}Ar$ dates of both underlying ash bed KDR-5 (76.45 ± 0.14 Ma) and overlying ash bed KBC-109 (75.51 ± 0.15 Ma), obey stratigraphic superposition (Table 6.1; Fig. 6.3).

Sedimentology

Roberts (2007) carried out a detailed alluvial architecture analysis of the Kaiparowits Formation, identifying 14 lithofacies and nine facies associations. The facies associations include the following: intraformational conglomerate (FA1);

mollusk shell conglomerate (FA2); major tabular sandstone (FA3); major lenticular sandstone (FA4); minor tabular and lenticular sandstone (FA5); finely laminated, calcareous siltstone (FA6); inclined heterolithic sandstone and mudstone (FA7); sandy mudstone (FA8); and carbonaceous mudstone (FA9). Each of these facies associations are briefly reviewed and illustrated here.

Intraformational (FA1) and Mollusk Shell (FA2) Conglomerates

Intraformational conglomerates are a common facies association in the Kaiparowits Formation (Fig. 6.6), most typically associated with the major lenticular (FA4) and major sheet (FA3) sandstone facies associations. They range in thickness from a few centimeters to several meters and are dominated by conglomerate and subordinate sandstone facies. Clast composition is overwhelmingly intraformationally derived as a result of fluvial incision and accumulation of lithic clasts, large mudstone blocks, and bone, wood, and shell fragments. They are interpreted to represent high-energy fluvial flow conditions and associated depositional events in channels.

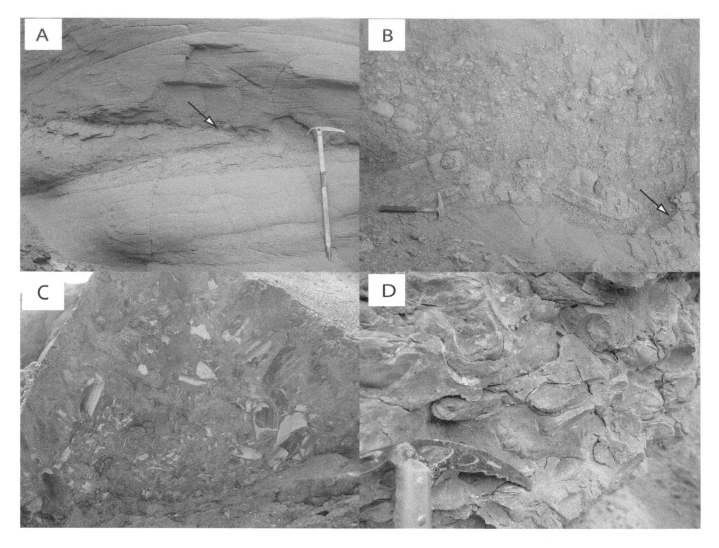

6.6. Photographs of facies associations 1 and 2 (FA1, FA2). (A) Thin intraformation conglomerate (FA1) developed at the erosive contact (arrow) between two stacked major tabular sandstone bodies (FA3) in the upper member (ice axe for scale); (B) well-developed intraformational conglomerate (FA1), also located at erosional contact (arrow) between two major tabular sandstone bodies (FA3) in the upper member (rock hammer for scale); (C) thin mollusk shell conglomerate (FA2) found at the base of a major lenticular sandstone in the lower member (photo is ~90 cm wide); (D) 2-m-thick mollusk shell conglomerate (FA2) in middle member (clam calamity bed).

They may also represent thalweg deposits of fluvial channels associated with normal flow conditions. These units are most common at the base of channels or along fourth- or fifth-order surfaces that delineate stacked, multistory channels. Intraformational conglomerates, along with sandy mudstones (FA8), represent a very common location for vertebrate and invertebrate microsite fossil concentrations. FA8-sourced microsites are considered to be allochthonous in nature, as opposed to FA8 microsites, which are considered autochthonous in nature (Roberts, 2007).

Major Tabular (FA3) and Lenticular (FA4) Sandstones

Thick sandstone units in the Kaiparowits can be divided into tabular (FA3) and lenticular (FA4) units on the basis of their alluvial architecture and stacking patterns (Fig. 6.7). Both are characterized by sand bodies greater than 1.5–2 m thick, but typically exceeding 5 m, composed of a variety of sandstone and conglomerate lithofacies typified by St, Sp, Sh, Gmm, Gsm, Se, and Tr (see Roberts, 2007, for lithofacies descriptions). The Kaiparowits Formation appears to be mudrock dominated, but on closer inspection, sandstone is nearly subequal in the middle portion of the formation and dominant in the upper and lower parts. Additionally, sandstones of the Kaiparowits Formation are dominated by feldspatholithic petrofacies with high proportions of chert lithic and volcanic lithic grains (Fig. 6.8), with low to moderate amounts of matrix and cement that leads to rapid surficial weathering. A closer inspection reveals that most of the tree-covered ridges present in the badlands are actually FA3 or FA4 units.

Roberts et al.

6.7. Outcrop photographs of FA3 and FA4. (A, B) Major lenticular sandstones (FA4) in the middle (A) and lower (B) members (arrows show bases of sandstones); (C) stacked major tabular sandstone bodies in the middle member (arrows show base of sandstone bodies); (D) major lenticular sandstone with well-exposed lateral accretion elements south of Canaan Peak; (E) lateral accretion elements in major tabular sandstone (FA3) at Four-Mile Bench; (F) stacked major sheet sandstones (FA3); contacts shown by arrows) in the upper member in the Blues.

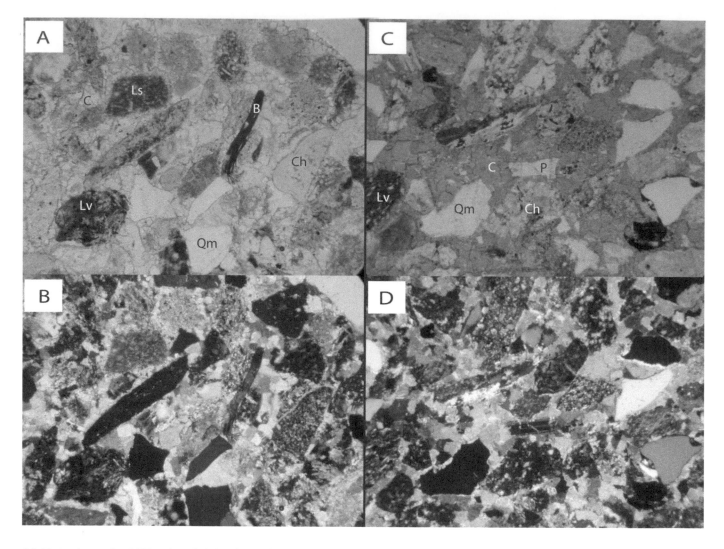

6.8. Photomicrographs of FA3 and FA4 (4×). (A, B) Typical major tabular sandstone (FA3) with monocrystalline quartz (Qm), chert (Ch), biotite (B), sedimentary (Ls) and volcanic lithic (Lv) grains and coarse blocky calcite cement (C) (A, plane polarized light [ppl]; B, cross polarized light [cpl], 4×). (C, D) Typical major lenticular sandstone (FA4) with monocrystalline quartz, lithic volcanics, chert, plagioclase (P), and coarse blocky calcite cement (stained with Alarizin Red for calcite) (C, ppl; D, cpl).

Moreover, the grain size of FA3 and FA4 units is remarkably fine, dominated by fine- to very fine-grained sandstone. In many cases, trough cross-bedding is well preserved, along with abundant macroform architectural elements, including lateral accretion. Both facies associations are characterized by the presence of basal fourth- or fifth-order surfaces and typified by single-story units; however, FA3 may also commonly be multistoried, particularly in the lower and upper units of the formation. Major tabular sandstones are differentiated from major lenticular sandstones by the following features: geometry; width/thickness ratio (>15 for FA3, <15 for FA4; sensu Friend, Slater, and Williams, 1979); and presence or absence of lateral accretion (common in FA3, rare in FA4). Both preserve abundant fossils, particularly carbonized logs, bivalve shell beds, and vertebrate bones. FA3 sand bodies are interpreted to reflect sinuous fluvial channel deposits,

suggestive of meandering- or perhaps anastomosing-style rivers. Channel reconstructions presented in Roberts (2007) indicate that typical bankfull widths ranged 20–80 m, whereas bankfull depths may have been between 3 and 10 m. These ratios are most consistent with suspended-load, meandering rivers with stable banks.

The lenticular geometry of FA3 units, coupled with the common observation of multiple FA3 bodies at the same stratigraphic level, is most consistent with rapid avulsion, as commonly associated with anastomosing style channels (Nadon, 1994). Other such diagnostic features of anastomosing rivers as in situ, upright trees in levee deposits, encasement of sand bodies in thick overbank fines, and presence of abundant crevasse splay deposits (FA5) further support this interpretation.

Roberts et al.

6.9. Photographs of FA5, FA8, and FA9. (A) Outcrop image of Horse Mountain area showing heterogeneous accumulation of FA minor tabular sandstones (FA5) interbedded with sandy (FA8) and carbonaceous mudstones (FA9) above a major lenticular sandstone body (FA4); (B) close-up view of a sandy mudstone (FA8) showing massive bedding and small calcium carbonate nodules (arrow) (scale bar in cm); (C) photomicrograph of a sandy mudstone (FA8) with large mollusk shell fragments (arrows) (scale bar in cm); (D) field photograph of carbonaceous mudstone (FA9) with a turtle carapace and several mollusks highlighted by the arrows (scale in cm).

Minor Tabular and Lenticular Sandstone (FA5)

Minor tabular and lenticular sandstone bodies are abundant throughout the Kaiparowits Formation, represented by 0.2–2-m-thick units packaged between thick overbank sequences (FA8, FA9) (Fig. 6.9). Whereas minor tabular sandstones have width/thickness ratios greater than 15, minor lenticular sandstones are recognized by having ratios less than 15. FA5 is commonly bioturbated, including preservation of the new social insect nesting trace fossils, *Socialites tumulus* (Roberts and Tapanila, 2006). It also preserves plant macrofossils, invertebrates, and rare vertebrate material. On the basis of their architecture and their sedimentary structures, minor tabular and lenticular sandstones are interpreted to represent crevasse splay and crevasse channel deposits, respectively (Roberts, 2007).

Finely Laminated, Calcareous Siltstone (FA6)

This rare facies association is diagnosed by laterally extensive beds of finely laminated, calcareous siltstone that range 10–250 cm thick. They are very rare but are easy to correlate over many kilometers. Insect burrows, along with aquatic mollusks preserved in moldic form, and locally abundant plant macrofossils (e.g., ferns) are present within this facies association. Additionally, unusual domal structures have also been observed at several locations within FA6, which may be produced by cyanobacterial mats. This facies association is consistent with deposition in large, shallow lakes.

Inclined Heterolithic Sandstone and Mudstone (FA7)

An important but rare facies association in the Kaiparowits Formation is FA7, which is characterized by repeated beds of inclined sandstone and mudstone. FA7 is restricted to a narrow interval between 160 and 190 m. Flaser, wavy, and lenticular bedding is common, and trough cross-beds in this facies association are commonly clay or carbon lined; the brackish-water trace fossil *Teredolites* has been observed in several places (Tapanila and Roberts, this volume). Considered together, this facies association is interpreted as inclined heterolithic stratification, which is typically interpreted as point bar lateral accretion in a tidally influenced channel. Although FA7 units have been sampled, no evidence of marine or brackish-water microfossils or vertebrates have been found (J. Eaton, pers. comm.). However, glauconite, an authigenic mineral commonly associated with shallow marine environments, has been observed in a number of samples. Moreover, slightly higher up in section and not associated with FA7 is a remarkable shell bed (FA2) known as the clam calamity bed (Roberts, Tapanila, and Mijal, 2008) that preserves a death assemblage of millions of unionoid bivalves. The bivalves are all considered to be fluvial (perhaps brackish); however, many of the recently dead, gaping shells preserve skeletons of the encrusting brackish-water bryozoan *Conopeum* sp. (Roberts, Tapanila, and Mijal, 2008; Tapanila and Roberts, this volume). Certainly the preponderance of data suggests that brackish-water conditions did exist in Kaiparowits river systems for short durations, but that for most of the Kaiparowits time, fully alluvial conditions dominated.

Sandy Mudstone (FA8) and Carbonaceous Mudstone (FA9)

Both FA8 and FA9 are interested as overbank facies associations characterized by fine-grained facies (Fig. 6.9). Both are typically richly fossiliferous and dominated by freshwater aquatic invertebrates and plants (Miller et al., this volume; Tapanila and Roberts, this volume). FA8 is coarser grained, typified by clay with a subequal proportion of silt and sand and blocky to platy weathering in fresh sequences, whereas FA9 is more rich in clay, with lesser amounts of silt and sand, with platy to fissile fresh weathering characteristics. FA9 is most easily identified by its dark gray to black color, associated with its much higher concentration of plant remains, including amber and coal blebs and stringers. FA8 is more typically green in color, with fewer plant fragments and much more abundant aquatic invertebrate and vertebrate microfossils, along with the rare well-preserved terrestrial vertebrate skeletal concentration. In some instances, such features as slickensides, rootlets, small calcium carbonate nodules, and drab color banding are present in FA8.

FA8 is interpreted to represent a range of overbank environments dominated by perennial ponds and lakes. The lenticular nature of many FA8 and FA9 units suggests that channel-fill oxbow-style lakes were common. Weakly developed paleosols, indicative of waterlogged soil conditions, are relatively common. FA9 is more diagnostic of marsh or swamp conditions, presumably associated low-lying areas (back-swamp depocenter) adjacent to major channel belts. Sedimentation rates on the floodplain were likely too high (up to 57 cm/ka) for thick coals to develop. The abundance of FA8 and FA9 across the Kaiparowits Formation, but particularly within the middle unit, suggests a relatively wet, low-lying alluvial system characterized by large, slow-moving channels and well-vegetated, stable overbank sequences.

Depositional Environments and Paleoclimate

Synthesized sedimentological, paleontological, and paleobotanical data give a clearer understanding of the paleoenvironmental and paleoclimatic conditions that persisted during the deposition of the Kaiparowits Formation. Previous workers have debated whether the formation was deposited within an inland, fully alluvial depositional setting (Eaton, 1991; Little, 1995) or an upper delta plain system (Lohrengel, 1969). The results of Roberts (2007) and other workers support both suggestions but demonstrate that the majority of the formation was accumulated under fully alluvial conditions, inland of contemporaneous shorelines. The exception where shorelines appear to have been closer to the Kaiparowits Plateau during the late Campanian is a discrete, tidally influenced to estuarine interval within the lower half of the middle unit (~120–320 m) (Roberts, 2007; Tapanila and Roberts, this volume).

The Kaiparowits Formation was dominated by an array of continental depositional environments, with large, deep fluvial channels supported by stable banks. Rivers flowed generally west across the alluvial–coastal plain, draining into the retreating Western Interior Seaway. The floodplains to these large channels were dominated by long-lived perennial ponds, wetlands, and lakes. The nearly ubiquitous nature of large aquatic mollusks, coupled with the abundance of aquatic vertebrates (Gates et al., 2010) and plants (Miller et al., this volume) in many overbank units, testifies to the wet nature of this alluvial system and the persistence of standing water deposits. Evidence of some pedogenesis in many sandy mudstones (FA8), including weak soil horizon development, incipient caliche (calcium carbonate nodules), ped structures, and slicken sides, all indicate mild seasonal variation in precipitation and water levels (Roberts, 2007). Generally, paleosol features in the Kaiparowits Formation are indicative of waterlogged soils with high water table; however, some horizons are much better developed, consistent with longer exposure and more intense pedogenesis. These better-developed paleosols are interpreted to represent more distal floodplain facies, rather than intervals of increased aridity. Lohrengel (1969) first proposed that the Gulf Coast region of the United States today would be a good modern analog for the Kaiparowits Formation. This assertion has been well supported by data collected here and by numerous workers over the last decade.

Taphonomy

The high diversity and abundance of fossil preservation in the Kaiparowits Formation have been topics of interest and speculation for many years. Comparisons have been made between the Kaiparowits Formation and coeval formations like the Two Medicine and Dinosaur Park; however, at the formation scale, the Kaiparowits is also unique in many respects. Various workers have investigated aspects of the vertebrate and invertebrate taphonomy of the Kaiparowits Formation, including Roberts, Tapanila, and Mijal (2008), Lund et al. (2009), Getty et al. (2010), and Tapanila and Roberts (this volume). A full review of this work will not be presented here; however, it is worth summarizing the key taphonomic modes and characteristics of the formation.

The most notable observation is the high abundance and diversity of aquatic vertebrates (Gates et al., 2010), invertebrates (Tapanila and Roberts, this volume), and plants (Lohrengel, 1969; Miller et al., this volume) within the formation. This pattern is generally consistent with sedimentological evidence for abundant channels and wetlands. With regard to vertebrate trends, Getty et al. (2010) observed that few bonebeds exist within the Kaiparowits Formation, which is in contrast to many coeval formations to the north like the Dinosaur Park, Two Medicine, and Judith River formations, where large dinosaur bonebeds are commonplace. Instead of bonebeds, large vertebrate remains in the Kaiparowits Formation are dominantly recovered as partial to fully articulated skeletons of single individuals in large channel sandstones.

Eberth and Currie (2005) associate many of the bonebeds in the Dinosaur Park Formation with repeated large-scale flooding events. Although there is evidence in the Kaiparowits for such large-scale catastrophic flooding events, as demonstrated by the clam calamity bed in the middle unit (Roberts, Tapanila, and Mijal, 2008), they were clearly rare events. Another striking observation about the Kaiparowits taphonomy is the general rarity of articulated or associated vertebrate skeletons in overbank-hosted mudstone facies. The few that do exist tend to show highly elevated levels of bone modification, including insect traces, tooth traces, trample marks, and abundant evidence of wet-rot-style bone weathering (Roberts, 2005; Getty et al., 2010). This is most likely attributable to rapid decomposition and breakdown of organic material on the floodplain due to a humid, wet climate, and hence decreased preservation potential and quality of these types of fossil sites in general. Another interesting phenomenon observed in many of these sites is the preservation of skin impressions with many articulated skeletons (Lund et al., 2009), something that has not been reported to this extent anywhere else in the world. Lund et al. (2009) attributed this to a combination of rapid burial events in the Kaiparowits Formation and the particularly robust nature of ceratopsian and especially hadrosaur skin.

Microsites are nearly ubiquitous in the Kaiparowits Formation, and like many other formations through time and space, microsites are divided into two types: fluvial and pond. Unlike many formations, where fluvial microsites are dominant, the Kaiparowits Formation is represented by more pond microsites than fluvial microsites. Fluvial microsites are typically dominated by time-averaged mixtures of allochthonous to para-autochthonous terrestrial and aquatic vertebrates, with abundant large fluvial bivalves and carbonized logs. Pond microsites on the other hand, are dominated by less time-averaged assemblages of autochthonous low-energy aquatic mollusks, vertebrates (turtles, crocodiles, fish), and plants.

Regional Stratigraphic Correlations
and the Concept of Taphozones

Recalibration of older isotopic dates from the Kaiparowits Formation and coeval units in the Western Interior Basin, coupled with the addition of new isotopic ages from the Kaiparowits, Wahweap, and Aguja formations, provides an important opportunity to improve intrabasinal correlations of richly fossiliferous continental sequences in the Western Interior Basin (Fig. 6.10). These recalibrated $^{40}Ar/^{39}Ar$ dates, new U-Pb, and legacy K/Ar ages are synthesized with biostratigraphic data where necessary to construct an up-to-date stratigraphic correlation chart (Table 6.1; Figs. 6.10, 6.11) for comparing Upper Cretaceous continental fossil-bearing units along the western margin of the Western Interior Basin.

Many of the most richly fossiliferous, terrestrial vertebrate-, invertebrate-, and paleobotanical-bearing formations in the Western Interior Basin are closely contemporaneous with the Kaiparowits Formation (Fig. 6.11). The temporal interval between ~76.6 and 74.5 Ma that the Kaiparowits

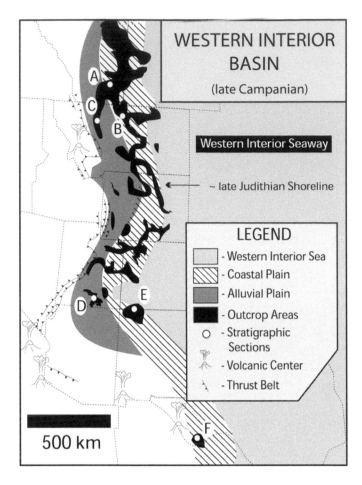

WESTERN INTERIOR
BASIN
(late Campanian)

Western Interior Seaway

~ late Judithian Shoreline

LEGEND

- Western Interior Sea
- Coastal Plain
- Alluvial Plain
- Outcrop Areas
○ - Stratigraphic Sections
- Volcanic Center
- Thrust Belt

500 km

6.10. Map of the Late Campanian continental strata in Western Interior Basin and location of principal vertebrate fossil-bearing units (modified from Roberts, Deino, and Chan, 2005).

Formation occupies, and approximately half a million years above and below this (~77–74 Ma), represents a remarkable basin-wide occurrence of richly fossiliferous terrestrial units with a high concentration of volcanic ashes that permits precise correlations. We refer to this narrow interval of time between ~77 and 74 Ma in Laramidia/Western Interior Basin as the Kaiparowits–Dinosaur Park–upper Two Medicine taphozone because it represents one of the single most extensive, richly fossiliferous windows into a Mesozoic terrestrial ecosystem in the world. The term *taphozone* is used here not in the biostratigraphic sense but rather in the taphonomic sense to indicate a broad geographic area of exceptional and elevated preservation of multiple fossil groups across a narrow temporal window.

The Kaiparowits–Dinosaur Park–upper Two Medicine taphozone includes the following formations: Dinosaur Park Formation and the most fossiliferous upper portions of the Judith River and Two Medicine formations to the north (Goodwin and Deino, 1989; Eberth and Hamblin, 1993; Rogers, Swisher, and Horner, 1993); the Fruitland Formation and the lowest portion of the Kirtland Formation (Fassett and

Steiner, 1997) in New Mexico, although the primary fossil-producing intervals of both units are slightly younger than the Kaiparowits Formation; and to the far south in Texas, the upper shale member of the Aguja Formation, which at least partially correlates to the Kaiparowits Formation and partially falls within the Kaiparowits–Dinosaur Park–upper Two Medicine taphozone (Rowe et al., 1992; Lehman, 1997; Sankey, 2001; McDowell, Lehman, and Connelly, 2004; Lehman et al., 2006; Breyer et al., 2007; Befus et al., 2008) (Figs. 6.10 and 6.11).

The Kaiparowits Formation and the Kaiparowits–Dinosaur Park–upper Two Medicine taphozone are both coincident with what has been called the zenith of dinosaur diversity in the Late Cretaceous (Sloan, 1976; Dodson, 1983; Dodson and Tatarinov, 1990). It is little wonder that this time period has been associated with the peak of dinosaur diversity: it represents a period of time up and down Laramidia and the Western Interior Basin with highly elevated preservation potential for not only fossil vertebrates, but also invertebrates and plants. In reality, the Kaiparowits–Dinosaur Park–upper Two Medicine taphozone represents a potentially unique and extraordinary opportunity to analyze a virtual snapshot of a Mesozoic terrestrial ecosystem, including patterns of evolution and extinction, within a narrow, well-dated time slice of time at a continental scale.

Although the concept of taphozones will not be developed in further detail here, the overarching taphonomic controls on preservation in taphozones like the Kaiparowits–Dinosaur Park–upper Two Medicine taphozone are necessarily regional in scale and of great interest. For instance, what links exceptional terrestrial fossil preservation (i.e., high abundance and multiple fossil groups – vertebrates, invertebrates, plants) across widespread geographic regions during remarkably short temporal intervals? A general review of multiple factors that have an influence on fossil preservation, including tectonics and subsidence, sediment supply and accumulation rates, paleoenvironments, and paleoclimates, in the Campanian Western Interior Basin reveals very little correspondence between any of the main formations. Interestingly, the only thing that seems to link each of these units is their temporal correspondence and their abundance of well-dated volcanic ash beds, perhaps indicating a link between widespread magmatism and preservation. Understanding basin- or regional-scale tectonic controls on continental taphonomy holds potential for better addressing problems associated with the fidelity of the continental fossil record and hence has implications for improving the accuracy associated evolutionary and paleoecological studies. Although this question cannot be adequately answered here, it should be a topic of future studies.

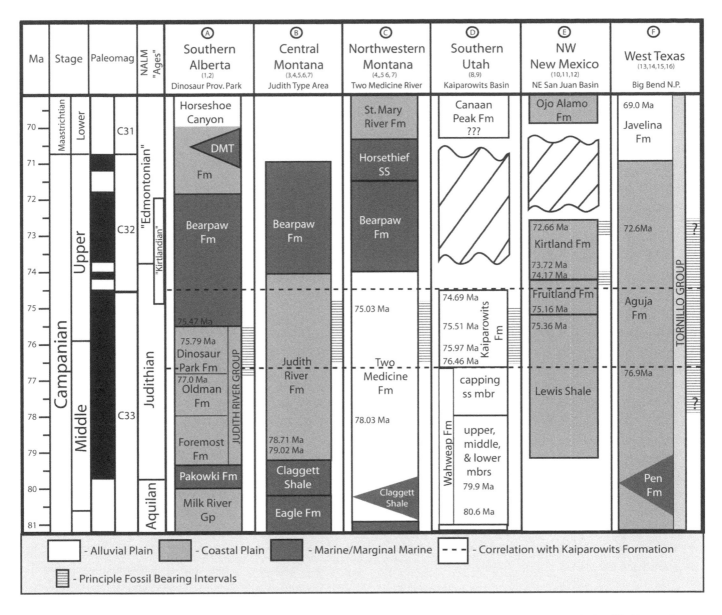

6.11. Updated correlation chart of principal vertebrate-bearing continental deposits of Campanian age in the Western Interior Basin (modified from Roberts, Deino, and Chan, 2005).

REGIONAL TECTONICS AND SEDIMENTARY PROVENANCE

The exceptionally high sediment accumulation rates documented for the Kaiparowits Formation, particularly in the middle to upper units (up to 576 m/Ma), are unparalleled in the Cretaceous Western Interior foreland basin. They indicate a period of extremely rapid subsidence and generation of accommodation space in the foreland. This led to the low stream gradients where anastomosing channels and overbank wetlands could develop. High sediment supply, coupled with high precipitation (as documented by Miller et al., this volume), prohibited development of thick coals. Together with changing paleocurrent patterns recorded throughout

the Kaiparowits Formation, these features have been interpreted to be the result of initiation of Laramide uplifts in the foreland basin (Roberts, 2007; Lawton and Bradford, 2011).

Sandstones of the Kaiparowits Formation are immature both texturally and compositionally. Above the basal 50 m of the Kaiparowits Formation, feldspathic litharenites (feldspatholithic petrofacies of Lawton, Pollock, and Robinson, 2003; Lawton and Bradford, 2011) dominate, with a very high proportion of angular chert grains and volcanic rock fragments. Gay and Bronson (1991) also noted the remarkably high concentrations of heavy minerals in the formation, particularly magnetite. Together, these petrographic features are taken to be a reflection of a volcanic-dominated provenance, which Lawton and Bradford (2011) interpreted as coming from the Jurassic Cretaceous border rift system of southern Arizona. Correlation with coeval units in central Utah is more difficult as a result of the paucity of radioisotopic geochronological control, variable tectonic influences, and the concentration of high-order sequences recognized in the Book Cliffs. However, on the basis of new temporal data by Lawton and Bradford (2011), it appears that the Kaiparowits Formation correlates to the upper Castlegate Sandstone in Price Canyon and the upper Neslen Formation, Farrer Formation, and perhaps part of the Tuscher Formation near Green River. Identification and dating of in situ ash beds in the Book Cliffs will ultimately be important for testing these correlations.

CONCLUSIONS

Nearly a decade of stratigraphic and sedimentologic investigations into the Kaiparowits Formation in close collaboration with vertebrate, invertebrate, and paleobotanical researchers has produced a high level of understanding of the age, paleoenvironments, paleoclimate, paleogeography, and paleoecology of the Kaiparowits Formation. The Kaiparowits represents a wet, humid alluvial system dominated by large channels and abundant wetland swamps, ponds, and lakes. It preserves an unbelievably abundant and diverse range of fossils, including continental and aquatic vertebrates. Terrestrial and aquatic invertebrates and a range of aquatic and terrestrial plant macrofossils, spores, and palynomorphs are also highly abundant within the formation. At least 12 different volcanic ash beds have been discovered and correlated across the Kaiparowits Formation, and five of these have been dated and recalibrated to provide a high-resolution chronostratigraphy. The 860-m-thick Kaiparowits Formation was deposited roughly 76.6–74.5 Ma, indicating an incredibly rapid sediment accumulation rate of up to 570 m/Ma, which is nearly an order of magnitude higher than most other contemporaneous continental sedimentary units in the Western Interior Basin from Texas to Alberta.

The temporal interval roughly associated with the Kaiparowits Formation is all the more important because it is well represented up and down the Laramidia/Western Interior Basin by equally other fossiliferous units, including the Aguja and Fruitland formations to the south and the Two Medicine, Judith River, and Dinosaur Park formations to the North. Moreover, this narrow window of time is perhaps the single best dated interval of time across the Laramidia/Western Interior Basin, with multiple dated (and yet to be dated) volcanic ash beds in each of these formations. We coin the term *taphozone* to describe narrow temporal windows of exceptionally fossiliferous continental deposits, and we consider the Kaiparowits Formation and coeval units to make up the Kaiparowits–Dinosaur Park–upper Two Medicine taphozone. Taphozones such as this represent unprecedented opportunities to study the mode and tempo of vertebrate, invertebrate, and plant evolution and ecology in ancient continental ecosystems.

ACKNOWLEDGMENTS

Fieldwork in the Kaiparowits Formation was supported in part by the Bureau of Land Management–Grand Staircase–Escalante National Monument (JSA055008) and the National Science Foundation (EAR 0745454, 0819953). U-Pb geochronology at the Massachusetts Institute of Technology was supported by the National Science Foundation (EAR 0643158). We are particularly grateful to A. Titus for his strong support and his role in facilitating and encouraging research in the Grand Staircase–Escalante National Monument. D. Powell and C. Shelton also deserve thanks for their parts in supporting this research and helping with permits. M. Getty, B. Gates, M. Loewen, L. Tapanila, Z. Jinnah, J. Smith, L. Zanno, J. Weirsma, R. Barclay, B. Foreman, P. O'Connor, T. Heironymous, K. Johnson, I. Miller, and many others have provided invaluable collaboration and critical insights into the inner workings and soul of the Kaiparowits Formation. We thank D. Lofgren for providing measured sections and fossil locality details for Alf museum sites in the Kaiparowits. J. Eaton also provided much critical help, insight, and feedback about the Kaiparowits Formation over the years. Finally, we thank S. Roberts for his field assistance and access to the Upper Valley camp.

REFERENCES CITED

Befus, K. S., R. E. Hanson, T. M. Lehman, and W. R. Griffin. 2008. Cretaceous basaltic phreatomagnetic volcanism in West Texas: Maar complex at Pena Mountain, Big Bend National Park. Journal of Volcanology and Geothermal Research 173:245–264.

Best, M. G., E. H. Christiansen, A. L. Deino, C. S. Gromme, and D. G. Tingey. 1995. Correlation and emplacement of a large, zoned, discontinuously exposed ash-flow sheet: ^{40}Ar/^{39}Ar chronology, paleomagnetism, and petrology of the Pahrangat Formation, Nevada. Journal of Geophysical Research 100(B12):24593–24609.

Bowers, W. E. 1990. Geologic maps of Bryce Canyon National Park and vicinity, southwestern Utah. U.S. Geological Survey Map I-2108.

Bowring, J. F., N. M. McLean, and S. A. Bowring. 2011. Engineering cyber infrastructure for U-Pb geochronology: Tripoli and U-Pb_Redux. Geochemistry, Geophysics, Geosystems 12:Q0AA19.

Breyer, J. A., A. B. Busbey, R. E. Hanson, W. R. Griffin, U. S. Hargrove, S. C. Bergman, and K. S. Befus. 2007. Evidence for Late Cretaceous

Roberts et al.

volcanism in Trans-Pecos Texas. Journal of Geology 115:243–251.

Cifelli, R. L. 1990a. Cretaceous mammals of southern Utah. I. Marsupials from the Kaiparowits Formation (Judithian). Journal of Vertebrate Paleonotology 10:295–319.

Cifelli, R. L. 1990b. Cretaceous mammals of southern Utah. II. Marsupials and marsupial-like mammals from the Wahweap Formation (early Campanian). Journal of Vertebrate Paleonotology 10:320–345.

Cifelli, R. L. 1990c. Cretaceous mammals of southern Utah. IV. Eutherian mammals from the Wahweap (Aquilan) and Kaiparowits (Judithian) formations. Journal of Vertebrate Paleontology 10:346–360.

DeCourten, F. L., and D. A. Russell. 1985. A specimen of Ornithomimus velox (Theropoda, Ornithomimidae) from the terminal Cretaceous Kaiparowits Formation of southern Utah. Journal of Paleontology 59:1091–1099.

Dodson, P. J. 1983. A faunal review of the Judith River (Oldman) Formation, Dinosaur Provincial Park, Alberta. Mosasaur 1:89–118.

Dodson, P. J., and L. P. Tatarinov. 1990. Dinosaur extinction; pp. 55–62 in D. B. Weishample, P. J. Dodson, and H. Osmólska (eds.), The Dinosauria. University of California Press, Berkeley, California.

Eaton, J. G. 1991. Biostratigraphic framework for the Upper Cretaceous rocks of the Kaiparowits Plateau, southern Utah; pp. 47–61 in J. D. Nations and J. G. Eaton (eds.), Stratigraphy, Depositional Environments, and Sedimentary Tectonics of the Western Margin, Cretaceous Western Interior Seaway. Geological Society of America Special Paper 260.

Eaton, J. G. 2002. Multituberculate Mammals from the Wahweap (Campanian, Aquilan) and Kaiparowits (Campanian, Judithian) Formations, Within and Near Grand Staircase–Escalante National Monument, Southern Utah. Utah Geological Survey Miscellaneous Publication 02–4.

Eaton, J. G., and R. L. Cifelli. 1988. Preliminary report on Late Cretaceous mammals of the Kaiparowits Plateau, southern Utah. University of Wyoming Contributions to Geology 26:45–55.

Eberth, D. A., and P. J. Currie. 2005. Vertebrate taphonomy and taphonomic modes; pp. 453–477 in P. J. Currie, and E. B. Koppelhus (eds.), Dinosaur Provincial Park: A Spectacular Ecosystem Revealed. Indiana University Press, Bloomington, Indiana.

Eberth, D. A., and A. L. Deino. 1992. A geochronology of the nonmarine Judith River Formation of southern Alberta. SEPM Theme Meeting, Mesozoic of the Western Interior, 24–25.

Eberth, D. A., and A. P. Hamblin. 1993. Tectonic, stratigraphic and sedimentologic significance of a regional discontinuity in the upper Judith River Group (Belly River wedge) of southern Alberta, Saskachewan, and northern Montana. Canadian Journal of Earth Sciences 30:174–200.

Fassett, J. E., and M. B. Steiner. 1997. Precise age of C33N–C32R magnetic-polarity reversal, San Juan Basin, New Mexico and Colorado. New Mexico Geological Society Guidebook 48:239–247.

Foreman, B. Z., R. R. Rogers, A. L. Deino, K. R. Wirth, J. T. Thole. 2008. Geochemical characterization of bentonite beds in the Two Medicine Formation (Campanian, Montana), including a new $^{40}Ar/^{39}Ar$ age. Cretaceous Research 29:373–385.

Friend, P. F., M. J. Slater, and R. C. Williams. 1979. Vertical and lateral building of river sandstone bodies, Ebro Basin, Spain. Journal of the Geological Society of London 146:39–46.

Gates, T. A., and S. D. Sampson. 2007. A new species of Gryposaurus (Dinosauria: Hadrosauridae) from the Upper Campanian Kaiparowits Formation of Utah. Zoological Journal of the Linnean Society of London 151:351–376.

Gates, T. A., S. D. Sampson, L. E. Zanno, E. M. Roberts, J. G. Eaton, R. L. Nydam, J. H. Hutchison, J. A. Smith, M. A. Loewen, and M. A. Getty. 2010. Biogeography of terrestrial and freshwater vertebrates from the Late Cretaceous (Campanian) Western Interior of North America. Palaeogeography, Palaeoclimatology, Palaeoecology 291:371–387.

Gay, S. P., and W. H. Bronson. 1991. Syngenetic magnetic anomaly sources: three examples. Geophysics 56:902–913.

Gerstenberger, H., and G. Hasse. 1997. A highly effective emitter substance for mass spectrometric Pb isotope ratio determinations. Chemical Geology 136:309–312.

Getty, M. A., M. A. Loewen, E. M. Roberts, A. L. Titus, and S. D. Sampson. 2010. Taphonomy of horned dinosaurs (Ornithischia: Ceratopsidae) from the late Campanian Kaiparowits Formation, Grand Staircase–Escalante National Monument, Utah; pp. 478–494 in M. J. Ryan, B. J. Chinnery-Allgeier, and D. A. Eberth (eds.), New Perspectives on Horned Dinosaurs: The Royal Tyrrell Museum Ceratopsian Symposium. Indiana University Press, Bloomington, Indiana.

Goodwin, M. B., and A. L. Deino. 1989. The first radiometric ages from the Judith River Formation (Upper Cretaceous), Hill County, Montana. Canadian Journal of Earth Sciences 26:1384–1391.

Jaffey, A. H., K. F. Flynn, L. E. Glendenin, W. C. Bentley, and A. M. Essling. 1971. Precision measurements of half-lives and specific activities of 235 U and 138 U. Physical Review C 4:1889–1906.

Jinnah, Z. A., E. M. Roberts, A. L. Deino, J. S. Larsen, P. K. Link, and C. M. Fanning. 2009. New $^{40}Ar/^{39}Ar$ and detrital zircon U-Pb ages for the Upper Cretaceous Wahweap and Kaiparowits formations in the Kaiparowits Plateau, Utah: implications for regional correlation, provenance, and biostratigraphy. Cretaceous Research 30:287–299.

Kuiper, K. F., A. Deino, F. J. Hilgen, W. Krijgsman, P. R. Renne, and J. R. Wijbrans. 2008. Synchronizing rock clocks of earth history. Science 320:500–504.

Lawton, T. F., and B. W. Bradford. 2011. Correlation and provenance of Upper Cretaceous (Campanian) fluvial strata in the Cordilleran Foreland Basin, Utah, U.S.A., from zircon U-Pb geochronology and petrography. Journal of Sedimentary Research 81:495–512.

Lawton, T. F., S. L. Pollock, and R. A. J. Robinson. 2003. Integrating sandstone petrology and nonmarine sequence stratigraphy: application to the Late Cretaceoous fluvial systems of southwestern Utah, U.S.A. Journal of Sedimentary Research 73:389–406.

Lehman, T. M. 1997. Late Campanian dinosaur biogeography in the western interior of North America; pp. 223–240 in D. L. Wolberg, E. Stump, and G. D. Rosenberg (eds.), Dinofest International. Philadelphia Academy of Natural Sciences, Philadelphia, Pennsylvania.

Lehman, T. M., F. W. McDowell, and J. N. Connelly. 2006. First isotopic (U-Pb) age for the Late Cretaceous Alamosaurus vertebrate fauna of west Texas, and its significance as a link between two faunal provinces. Journal of Vertebrate Paleontology 26:922–928.

Little, W. W. 1995. The influence of tectonics and eustacy on alluvial architecture, Middle Coniacian through Campanian strata of the Kaiparowits Basin, Utah. Ph.D. dissertation, University of Colorado, Boulder, Colorado.

Lohrengel, C. F. 1969. Palynology of the Kaiparowits Formation, Garfield County, Utah. Brigham Young University Geology Studies 6:61–180.

Lund, E. K., M. A. Loewen, E. M. Roberts, S. D. Sampson, and M. A. Getty. 2009. Dinosaur skin impressions in the Upper Cretaceous (Late Campanian) Kaiparowits Formation, Southern Utah: insights into preservation mode; p. 32 in Advances in Western Interior Late Cretaceous Paleontology and Geology, St. George, Utah, Abstracts with Program.

Mattinson, J. M. 2005. Zircon U-Pb chemical abrasion ("CA-TIMS") method: combined annealing and multi-step partial dissolution analysis for improved precision and accuracy of zircon ages. Chemical Geology 220:47–66.

McDowell, F. W., T. M. Lehman, and J. N. Connelly. 2004. A U-Pb age for the Late Cretaceous Alamosaurus vertebrate fauna of west Texas. Geological Society of America Abstracts with Programs 36(4):6.

McLean, N. M., J. F. Bowring, and S. A. Bowring. 2011. An algorithm for U-Pb isotope dilution data reduction and uncertainty propagation. Geochemistry Geophysics Geosystems 12, Q0AA18.

Meyers, S. R., S. E. Siewert, B. S. Singer, B. B. Sageman, D. J. Condon, J. D. Obradovich, B. R. Jicha, and D. A. Sawyer. 2012. Intercalibration of radioisotopic and astrochronologic time scales for the Cenomanian–Turonian boundary interval, Western Interior Basin, U.S.A. Geology 40:7–10.

Nadon, G. C. 1994. The genesis and recognition of anastomosed fluvial deposits: data from the St. Mary River Formation, southwestern Alberta, Canada. Journal of Sedimentary Research B64:451–463.

Ramezani, J., S. A. Bowring, M. W. Martin, D. J. Lehrmann, P. Montgomery, P. Enos, J. L. Payne, M. J. Orchard, H. Wang, and J. Wei. 2007. Timing the recovery from the end-Permian extinction; geochronologic and biostratigraphic constraints from south China. Geology 37:137–138.

Renne, P. R., R. Mundil, G. Balco, K. Min, and K. R. Ludwig. 2010. Joint determination of ^{40}K decay constants and $^{40}Ar*/^{40}K$ for the Fish Canyon sanidine $^{40}Ar/^{39}Ar$ standard, and improved accuracy for Ar/Ar geo-chronology. Geochimica et Cosmochimica Acta 74:5349–5367.

Renne, P. R., G. Balco, K. Ludwig, R. Mundil, and K. Min. 2011. Response to the comment by W. H. Schwarz et al. on "Joint determination of 40K decay constants and ^{40}Ar/^{40}K for the Fish Canyon sanidine standard, and improved accuracy for ^{40}Ar/^{39}Ar geochronology' by P. R. Renne et al. (2010)." Geochimica et Cosmochimica Acta 75:5097–5100.

Roberts, E. M. 2005. Stratigraphic, taphonomic and paleoenvironmental analysis of the Upper Cretaceous Kaiparowits Formation, Grand Staircase–Escalante National Monument, Utah. Ph.D. dissertation, University of Utah, Salt Lake City, Utah.

Roberts, E. M. 2007. Facies architecture and depositional environments of the Upper Cretaceous Kaiparowits Formation, southern Utah. Sedimentary Geology 197:207–233.

Roberts, E. M., and M. A. Chan. 2010. Variations in iron oxide, iron sulfide and carbonate concretions and their distributions in fluvio-deltaic and neashore sandstones: Cretaceous examples from the Kaiparowits Plateau, Utah and San Juan Basin, New Mexico; pp. 151–177 in S. M. Carney, D. E. Tabet, and C. L. Johnson (eds.), Geology of South-Central Utah. Utah Geological Association Publication 39.

Roberts, E. M., and L. Tapanila. 2006. A new social insect nest trace from the Late Cretaceous Kaiparowits Formation of southern Utah. Journal of Paleontology 80:768–774.

Roberts, E. M., A. L. Deino, and M. A. Chan. 2005. ^{40}Ar/^{39}Ar age of the Kaiparowits Formation, southern Utah, and correlation of coeval strata and faunas along the margin of the Western Interior Basin. Cretaceous Research 26:307–318.

Roberts, E. M., L. Tapanila, and B. Mijal. 2008. Taphonomy and sedimentology of storm-generated continental shell beds: a case example from the Cretaceous Western Interior Basin. Journal of Geology 116:462–479.

Rogers, R. R., C. C. Swisher, and J. R. Horner. 1993. ^{40}Ar/^{39}Ar age and correlation of the nonmarine Two Medicine Formation (Upper Cretaceous), northwestern Montana, U.S.A. Canadian Journal of Earth Sciences 30:1066–1075.

Rowe, T., R. L. Cifelli, T. M. Lehman, and A. Weil. 1992. The Campanian Terlingua local fauna, with a summary of other vertebrates from the Aguja Formation, Trans-Pecos, Texas. Journal of Vertebrate Paleontology 12:472–493.

Ryan, M. J., and D. C. Evans. 2005. Ornithischian dinosaurs; pp. 312–348 in P. J. Currie and E. B. Koppelhus (eds.), Dinosaur Provincial Park: A Spectacular Ancient Ecosystem Revealed. Indiana University Press, Bloomington, Indiana.

Sampson, S., M. Loewen, E. Roberts, J. Smith, L. Zanno, and T. Gates. 2004. Provincialism in Late Cretaceous terrestrial faunas: new evidence from the Campanian Kaiparowits Formation of Utah. Journal of Vertebrate Paleontology 24:108A.

Sampson, S. D., M. A. Loewen, A. Farke, E. M. Roberts, C. Forster, J. A. Smith, and A. L. Titus. 2010. New horned dinosaurs from Utah provide evidence for intracontinental dinosaur endemism. PLoS One (9):e122292.

Sankey, J. T. 2001. Late Campanian southern dinosaurs, Aguja Formation, Big Bend, Texas. Journal of Paleontology 75:208–215.

Sloan, R. E. 1976. The ecology of dinosaur extinction; pp. 134–155 in C. S. Churcher (ed.), Essays on Palaeontology in Honour of Loris Shano Russell. University of Toronto Press, Toronto.

Roberts et al.

A Late Campanian Flora from the Kaiparowits Formation, Southern Utah, and a Brief Overview of the Widely Sampled but Little-Known Campanian Vegetation of the Western Interior of North America

7

Ian M. Miller, Kirk R. Johnson, Douglas E. Kline, †Douglas J. Nichols, and Richard S. Barclay

FOSSIL-BEARING TERRESTRIAL STRATA OF CAMPANIAN age are widespread in the Western Interior Basin of North America and contain some of the world's best known and most diverse dinosaurian faunas. More than 30 Campanian megafloras have been found from Texas to the Arctic, but our understanding of the vegetation they represent is poor because it is based on outdated 19th- and early 20th-century collections and studies. Nonetheless, these megafloral assemblages provide paleoclimate estimates that correlate with latitude and vegetation patterns that follow geographic barriers. Recent work on newly discovered sites since the early 1980s is largely unpublished or is focused on describing a few taxa or the nonangiosperm portion of the flora. We have discovered and extensively sampled 10 new megafloral sites from the upper Campanian Kaiparowits Formation (~76–74 Ma) in the exposure known as The Blues near Escalante, Utah. Results from a single, precisely dated (~75.69–75.72 Ma) site suggest that the flora of the Kaiparowits is exceptionally diverse (87 morphotypes) and dominated by angiosperms (62 dicot leaf morphotypes) that grew in a subtropical, megathermal, and relatively wet climate (mean annual temperature ~20°C and mean annual precipitation ~1.8 m). The palynoflora of the formation is correspondingly diverse and contains several paleoenvironmental indicators that suggest the formation was deposited in a slow-moving freshwater environment. Overall, the Kaiparowits flora has similarities to other Western Interior Campanian floras and some Maastrichtian floras; it hosts a diverse aquatic component mirroring the high abundance and diversity of aquatic vertebrate and invertebrate fossils in the formation. This element of the flora corroborates the sedimentological interpretation that the Kaiparowits Formation was deposited in a lowland floodplain with extensive ponding.

INTRODUCTION

Deciphering the history of flora, vegetation, and climate through the Late Cretaceous is fundamental to understanding the evolution and ecological radiation of angiosperms, the ecology of diverse dinosaurian faunas, and the dynamics and evolution of greenhouse climate systems. The Western Interior Basin of North America, a contiguous, continent-scale basin, which lay east of the Western Cordillera and stretched from the Gulf of Mexico to the Arctic Sea, was an active depocenter during the Late Cretaceous (Roberts and Kirschbaum, 1995; Miall et al., 2008). As such, it provides a natural laboratory to study Late Cretaceous ecosystems because the basin contains widespread formations with extensive outcrops that preserve abundant fossil plants (Crabtree, 1987a; Wolfe and Upchurch, 1987; Upchurch and Wolfe, 1993; Aulenback, 2009) and animals (Weishampel et al., 2004) over a more than 30° latitudinal range. During the Late Cretaceous, the terrestrial ecosystem of the Western Interior Basin is most completely preserved in the fluviodeltaic deposits of Campanian age (83.5 to 70.7 Ma in Gradstein, Ogg, and Smith, 2004) on the western margin of the basin (Jinnah et al., 2009). These deposits offer the best opportunity to form a Late Cretaceous continent-scale synthesis of flora, fauna, and climate.

Despite an abundance of fossil floras (Fig. 7.1), our knowledge of vegetation in the Western Interior Basin during the Campanian comes mostly from syntheses of plant megafossil publications that are typically more than 80 years old (Wolfe and Upchurch, 1987; McClammer and Crabtree, 1989; Upchurch and Wolfe, 1993). These megafloras have poor collection data, and their taxonomy is outdated. Furthermore, the age of these floras are known only to stage or substage. More recent megafloral assemblage work is unpublished, consisting of theses (Crabtree, 1987b; Van Boskirk, 1998) or of newly collected but unanalyzed sites. Though exceptionally

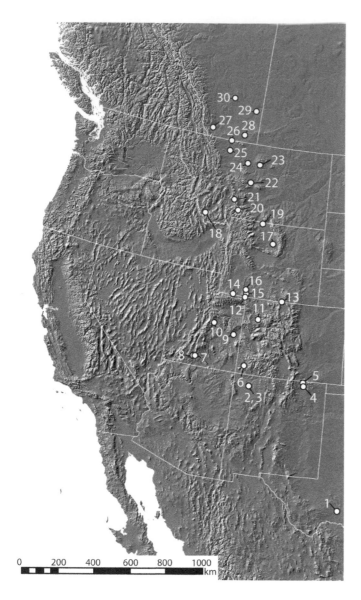

7.1. Digital elevation model of western North America from Mexico to Canada showing locations of the 30 Campanian floras (white circles) from the Western Interior Basin listed in Table 7.1.

based on compilations of megafloral data that have used estimates of the position of the Late Cretaceous North American pole that are obsolete. In general, however, published paleolatitudes for Western Interior Basin megafloras have changed by less than 5°N latitude and have moved north relative to previous estimates. Paleolatitude positions in Table 7.1 are based on a recent estimate of the Late Cretaceous North American pole position.) In short, the present understanding of Campanian vegetation and climate in North America, particularly where it has been based on megafloral assemblages in outdated publications, must be viewed in light of the quality or type of primary data.

In this chapter, we provide a megafloral and palynological characterization, correlation, and paleoclimatic analysis of a new, systematically collected flora from a well-documented depositional setting in the upper Campanian Kaiparowits Formation (~76–74 Ma) in southern Utah. By means of depositional rates and a suite of radiometric ages, we are able to give the most precise age yet for any Campanian flora in North America. We also give a summary of Campanian climate and vegetation in the Western Interior Basin to provide context for the new Kaiparowits flora. Though our current analysis is preliminary and based presently on a single locality, we have discovered 10 productive megafloral sites in the Kaiparowits Formation. Once extensively collected and analyzed, these sites, plus the one described herein, will provide a foundation for an updated synthesis of Campanian vegetation and terrestrial climate in the Western Interior Basin.

CAMPANIAN CLIMATE AND VEGETATION IN THE WESTERN INTERIOR BASIN

Campanian Climate

The Late Cretaceous is characterized by a sustained greenhouse climate. It is best documented in marine isotope records, which show warm to extremely warm sea-surface (~25–35+°C at low latitudes) and bottom-water temperatures (~10–15+°C at low latitudes) (Frakes, 1999; Pearson et al., 2001; Huber, Norris, and MacLeod, 2002; Norris et al., 2002; Schouten et al., 2003; Forster et al., 2007). Though short-wavelength temperature variability is observed in marine records (Price and Hart, 2002; Wilson, Norris, and Cooper, 2002; Miller et al., 2003), overall paleoclimate is thought to be generally equable, showing a broad warming trend toward the Turonian–Coniacian, slight cooling into the Campanian, followed by increased cooling in the Maastrichtian (Huber, Norris, and MacLeod, 2002), and a sharp warming just before the end of the Maastrichtian (Huber and Watkins, 1992; Barrera and Savin, 1999; Olsson, Wright, and Miller, 2001; Wilf, Johnson, and Huber, 2003).

useful for piecing together aspects of Campanian vegetation, the most current published studies focus on single plant taxa (Stockey and Rothwell, 2003; Stockey, Rothwell, and Johnson, 2007), partial floras (Aulenback, 2009), or palynology (Nichols and Sweet, 1993; Bramen and Koppelhus, 2003). The exception is the newly published, in-situ, ash-bed flora at Big Cedar Ridge in Wyoming, which provides a snapshot of predominantly herbaceous vegetation on the Campanian landscape (Wing et al., 2012). Knowledge of Campanian terrestrial climate is derived from the same historic floral collections (Wolfe and Upchurch, 1987) or is based on faunal assemblages from limited stratigraphic ranges (Amiot et al., 2004) or from model results (Donnadieu et al., 2006; Sewall et al., 2007). An additional complexity is that reconstructions of continent-scale Campanian vegetation and climate are

Miller et al.

Table 7.1. Campanian megafloras from the North American Western Interior Basin

Flora/formation[a]	Latitude (°N)[b]	Longitude (°E)[b]	State/province	Paleolatitude (°N)[c]	Substage[d]
1 Aguja	30.5	−101.75	Texas	36.9, 33.8	Campanian (undivided)
2 Fruitland	36.5	−108	New Mexico	44.2	Upper Campanian
3 Kirtland	36.5	−108	New Mexico	44.2	Upper Campanian
4 Vermejo	36.8	−104.5	New Mexico	43.6	Upper Campanian
5 Trinidad Sandstone	37	−104.5	New Mexico	43.8	Upper Campanian
6 Prince River	37.5	−108.5	Colorado	45.3	Upper Campanian
7 Kaiparowits[e]	37.6	−111.85	Utah	46.2	Upper Campanian
8 Wahweap[e]	37.6	−111.95	Utah	46.3, 43.1	Middle Campanian
9 Neslen[e]	39	−109.5	Utah	47.0, 43.8	Middle Campanian
10 Blackhawk	39.5	−111	Utah	47.8, 44.7	Lower Campanian
11 Mesa Verde (Iles)	40	−108	Colorado	47.6	Upper Campanian
12 Rock Springs[e]	40.9	−109	Wyoming	48.7, 45.5	Lower to middle Campanian
13 Medicine Bow	41	−106.5	Wyoming	48.8, 45.0	Campanian (undivided)
14 Adaville	41	−110	Wyoming	49.0, 45.8	Lower Campanian
15 Almond (Point of Rocks)	41	−109	Wyoming	48.8, 45.6	Middle Campanian
16 Almond[e]	41.75	−109.2	Wyoming	49.5, 46.3	Middle Campanian
17 Big Cedar Ridge	44	−107.7	Wyoming	51.3	Upper Campanian
18 Beaverhead	45	−113	Montana	53.5, 50.3	Lower Campanian
19 Eagle (Van Boskirk, 1998)	45	−108.9	Wyoming	52.5, 49.3	Lower Campanian
20 Cokedale	45.5	−110.5	Montana	53.4, 50.2	Lower Campanian
21 Maudlow	46	−111	Montana	53.9, 50.8	Lower Campanian
22 Eagle (historical site)	47	−110	Montana	54.6, 51.5	Lower Campanian
23 Judith River	48	−109.5	Montana	55.4	Upper Campanian
24 Claggett	48	−110.5	Montana	55.7, 52.5	Middle Campanian
25 Two Medicine	48.5	−112	Montana	56.5, 53.4	Lower Campanian
26 Milk River	49	−112	Alberta	57.0, 53.8	Lower Campanian
27 Allison	49.5	−113.75	Alberta	57.9, 54.7	Lower Campanian
28 Oldman	49.5	−110.75	Alberta	57.1, 54.0	Middle Campanian
29 Dinosaur Park[e]	50.75	−110.5	Alberta	58.2, 55.1	Middle to upper Campanian
30 Horseshoe Canyon	51.25	−112.4	Alberta	59.2	Upper Campanian

[a] To compile the list of floras and formations, we used the following sources and the references cited therein: Crabtree (1987a), Wolfe and Upchurch (1987), McClammer and Crabtree (1989), Upchurch and Wolfe (1993), Van Boskirk (1998), Aulenback (2009), Wing et al. (2012) and *The Compendium Index of North American Mesozoic and Cenozoic Type Fossil Plants at the Yale Peabody Museum*.

[b] Latitude and longitude were rounded to the nearest quarter degree unless precise measurements were available.

[c] Mean paleolatitudes were calculated from Torsvik et al. (2008) using the 20 Ma running mean global apparent polar wander path in 10 Ma increments. All floras are restored using the North American 70 Ma pole estimate for comparison to the Kaiparowits flora. For floras that fall in the early and middle Campanian, we also give their restored position using the North American 80 Ma pole estimate using the format "70 Ma estimate, 80 Ma estimate." Depending on the longitude of the flora, there is an approximately 3° change in the paleolatitudinal position between the 70 Ma and 80 Ma estimates for the North American pole.

[d] We give the Campanian substage for each flora according to the North American Western Interior Ammonite Zonation of Cobban et al. (2006). The age of each flora is based on the temporal range of the formations in which they occur following the Western Interior Basin correlation charts of Jinnah et al. (2009), Roberts, Deino, and Chan (2005), and Roberts and Kirschbaum (1995) or the approximate stratigraphic position assigned to them by McClammer and Crabtree (1989).

[e] Unpublished floras known only from collections at the Denver Museum of Nature & Science.

The Late Cretaceous terrestrial record of climate is less well known. The primary direct proxy for terrestrial paleoclimate during the Late Cretaceous is dicotyledonous angiosperm leaf physiognomy where, for a given fossil flora, the presence or absence of teeth is correlated to mean annual temperature (MAT) (Wolfe, 1979; Wing and Greenwood, 1993; Wilf, 1997; Miller, Brandon, and Hickey, 2006) and leaf size is correlated to mean annual precipitation (MAP) (Wilf et al., 1998). In the Western Interior Basin, the leaf-based temperature trends broadly reflect those in the marine realm, but they lack detail in terms of precise age estimates, rigorous statistical evaluation (including standardization of estimates using accurate meridional temperature profiles), and uniformity in temporal distribution (whole stages may be represented by a single flora). Recent syntheses have focused on the beginning and the end of the Late Cretaceous. In particular, Miller, Brandon and Hickey (2006) showed that meridianal profiles of Albian–Cenomanian MAT were higher than those previously estimated because new approximations of the Cretaceous long-normal paleomagnetic pole (Housen et al., 2003) significantly changed the paleolatitude of the floras used to create the meridianal MAT gradient. Similarly, Johnson (2002) and Wilf, Johnson and Huber (2003) demonstrated that late Maastrichtian terrestrial MAT records were far more complex than previously thought, and mirror much more closely temperature records known from the marine realm. These recent syntheses highlight the need for a reexamination of the terrestrial record of temperature through the Late Cretaceous.

Almost all of what we know about Campanian MAT and MAP in the Western Interior Basin comes from the megafloral record. These data indicate a slight cooling leading into and through the stage but overall generally equable temperatures characterized by low-latitudinal MAT gradients (0.3°C/°N) (Wolfe and Upchurch, 1987) and near-polar MAT values of approximately 5–8°C (Parrish and Spicer, 1988). On the basis of megafloral assemblages, Upchurch and Wolfe (1993) located the 20°C MAT isotherm around 50 to 55°N and argued for little or no freezing at 60°N. These estimates agree well with oxygen isotope records of temperature derived from faunal remains, which indicate a latitudinal MAT gradient of 0.4°C/°N and suggest that freezing temperatures were not reached until about 80°N (Amiot et al., 2004).

Three latitudinal bands of relative MAP occur in the Western Interior Basin during the Campanian (Wolfe and Upchurch, 1987). Fossil wood and leaves indicate a belt of low rainfall south of about 50°N, high rainfall from 50 to 60°N, and moderate to high rainfall north of approximately 60°N (Wolfe and Upchurch, 1987; Upchurch and Wolfe, 1993). In the low-rainfall belt, precipitation seems to be distributed throughout the year with infrequent drought years (Falcon-Lang, 2003), whereas north of 60°N, precipitation appears seasonal (Upchurch and Wolfe, 1993). However, recent geochemical work has found evidence that a strong monsoon was active in the Sevier foreland (Fricke, Foreman, and Sewall, 2010) in the belt of low and evenly distributed precipitation identified by Upchurch and Wolfe (1993). Nonetheless, at least in a broad sense, Campanian megafloral assemblages in the Western Interior Basin give mean annual temperature and precipitation estimates that correlate with latitude.

Late Cretaceous Vegetation

The number and spatial distribution of fossil floras from the Late Cretaceous in North America is remarkable (Fig. 7.1, Table 7.1 for the Campanian) compared to those of other continents. However, with few exceptions (Johnson, 2002; Wing et al., 2012), precise age estimates for these floras and statistical evaluation of floral data sets are lacking. Moreover, taphonomic analyses of transported floras in fluvio-deltaic depositional settings – the most common preservation state of megafossil floras –, particularly in terms of facies control on the distribution and composition of floral elements, have not been conducted. The best continent-scale records of vegetation are derived from the study of palynomorphs. They indicate an explosion of angiosperm species in the late Early Cretaceous at low paleolatitudes with a corresponding taxonomic demise of other major plant clades, followed by a steady increase in angiosperm diversity in more northerly latitudes through the Late Cretaceous (Lidgard and Crane, 1990; Lupia, Lidgard, and Crane, 1999). Most megafloral syntheses show the same trend (Upchurch and Wolfe, 1993; Wing and Boucher, 1998); however, recent systematic collections of exceptionally well-preserved floras in ash-bed deposits demonstrate that syntheses based on megafloral collections from non–ash bed sites elucidate only part of the ecosystem. The ash-fall taphonomic window suggests that though angiosperms diversified through the Late Cretaceous, they occupied a limited ecospace relative to present (Wing, Hickey and Swisher, 1993; Wing et al., 2012). This is supported by analysis of fossil wood sites that suggest that cupressoid gymnosperms were the dominant overstory tree in some stable coastal plain settings in the late Campanian (Davies-Vollum et al., 2011). It is clear from these studies that taphonomy imposes a significant bias on our understanding of the evolution and ecology of vegetation in the Late Cretaceous.

Campanian Vegetation

The palynological record in the Western Interior Basin during the Campanian offers the broadest view of regional vegetation. Most notably, North America was characterized by three different palynological provinces (Herngreen et al., 1996). The *Normapolles* province occupied eastern North America and stretched through Greenland to Europe and western Asia. The Western Cordillera, however, was characterized by two floral provinces: the *Aquilapollenites* province on the eastern slope, and the continental margin province along the west coast (Frederiksen, 1988). The *Aquilapollenites* province was the more prominent of the two and stretched through the Western Interior Basin up into Alaska and across into eastern Asia; the continental margin province was confined to the North American west coast. These palynological records indicate that geographic barriers like the Western Interior Seaway and mountains of the Western Cordillera served to separate broad floral provinces (Nichols and Sweet, 1993, and others note some latitudinal divisions with these provinces). On the other hand, large latitudinal expanses, like that from the southern Western Interior Basin

Miller et al.

to Alaska, did not segment floral provinces, suggesting that latitude-coupled climate barriers either did not exist or did not play an important role in continent-scale floral patterns.

Campanian megafloral assemblages in the Western Interior Basin show that the floral diversity, though dominated by angiosperms, contained a varied and sometimes archaic nonangiosperm component. Among the nonangiosperms were bryophytes, lycopsids, sphenophytes, ferns, and a number of gymnosperm clades. Of the nongymnosperms, ferns are the most abundant, particularly in northern floras, where they comprised as much as 25% of species diversity (Upchurch and Wolfe, 1993; Wing et al., 2012). The gymnosperms included conifers, cycads, bennettites, ginkgophytes, and seed ferns. Of these groups, conifers were the most diverse; they were particularly common in backswamp deposits, where their abundance often exceeded that of the angiosperms (Crabtree, 1987a; Davies-Vollum et al., 2011). The conifers of this age were primarily taxodiaceous; strap-leaved araucarians were uncommon but appear to have been widespread (Upchurch and Wolfe, 1993). The extinct family Cheirolepidiaceae was present but rare (Watson, 1988; Upchurch and Wolfe, 1993; Lupia, Lidgard, and Crane, 1999). Several conifer genera of unknown affinity were also present and locally abundant. These include *Moriconia* and *Protophyllocladus*. Cycads and ginkgophytes showed low diversity at low latitudes but an increase in both diversity and abundance in northern floras (Crabtree, 1987a; Upchurch and Wolfe, 1993). Notable among the ginkgophytes is the report of the archaic genus *Sphenobaiera* from Alberta (Crabtree, 1987a). Another archaic group, the bennettites, appears to become extinct during the Campanian. The youngest North American records are from the early Campanian Allison (Berry, 1929) (Fig. 7.1, #27) and Adaville (Upchurch and Wolfe, 1993) (Fig. 7.1, #14) floras in the Western Interior Basin and the early Campanian beds of Vancouver Island (Stockey and Rothwell, 2003). Most recently, Aulenback (2009) notes possible bennettitalean seed cones from the late Campanian to Maastrichtian Horseshoe Canyon flora (Fig. 7.1, #30). Finally, seed ferns are reported from the Campanian of Alaska (Hollick, 1930), but these fossils may be misidentified specimens of *Protophyllocladus*.

Angiosperms were the most diverse clade in Campanian floras, where they typically comprised 60–90% of floodplain species diversity in areas where 20 or more megafloral species were identified (Upchurch and Wolfe, 1993; Wing, Hickey, and Swisher, 1993). Within the angiosperms, dicots show the highest diversity. Platanoids were exceptionally diverse (Crabtree, 1987a) when compared to the extant diversity of the group. Water plants (e.g., *Quereuxia, Brasenites, Nelumbites, Nymphaeites*) were also diverse and geographically

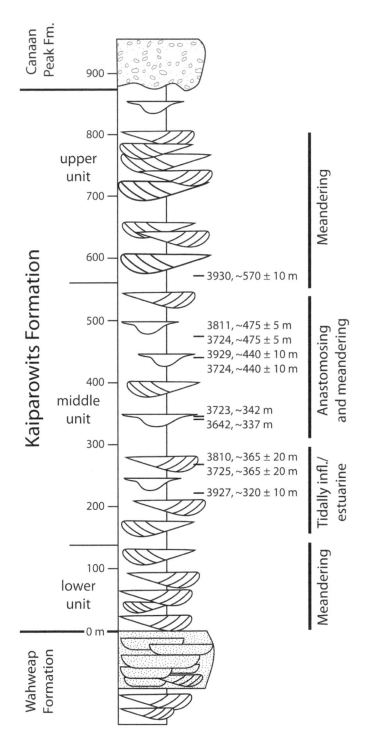

7.2. Representative stratigraphic column for the Kaiparowits Formation redrawn from Roberts (2007) showing the location of 10 productive megafloras. The stratigraphic level of each flora was provided by Eric Roberts (pers. comm., 2011). All floras, except one (DMNH Loc. 3930), occur in the middle unit, which exhibits tidally influenced/estuarine and anastomosing/meandering channel architecture.

widespread; not surprisingly, they are found mostly in paludal facies (Crabtree, 1987a). Several major evolutionary events occurred in monocots, including the first widespread occurrence of palm megafossils in the Western Interior Basin

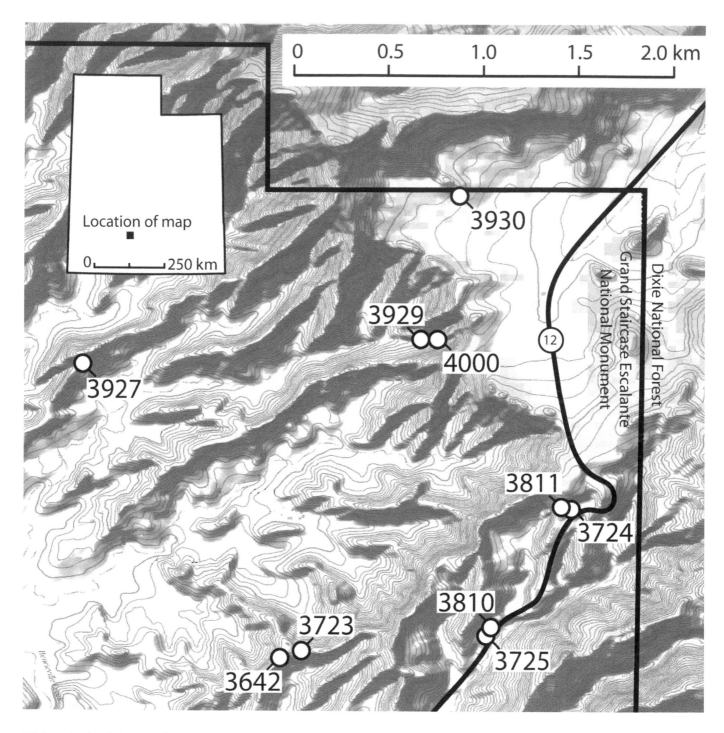

7.3. Map showing the location of 10 productive fossil plant megafloras in the section known as The Blues north of Highway 12. DMNH Loc. 3642 is flora #7 in Fig. 7.1. Inset map showing Utah with the location of the enlarged map showing floral localities.

during the early Campanian (Crabtree, 1987b). Pinnate forms radiated through the Campanian, reaching a northern extent of about 55°N paleolatitude (Crabtree, 1987b). By the late Campanian, palms dominated the angiosperm portion of the forest floor cover in some floodplain ecosystems (Wing et al., 2012). Gingers (Zingiberales) also first appear in the Campanian based on the presence of seeds (Friis, 1988). New and

eventually common aquatic monocots originate, including *Cobbania*. (Stockey, Rothwell, and Johnson, 2007).

The sequence of Campanian megafloral assemblages in the Western Interior Basin represents the longest and most complete sample for this time period in the world. However, major questions remain concerning floral evolution and vegetational dynamics in the region. For example, what are

Miller et al.

the sample-size and facies-normalized latitudinal gradients of megafloral and palynofloral diversity, major plant clade abundances, and leaf-based food web proxies like insect damage diversity and abundance? Was there a latitudinal gradient in leaf deciduousness? Were Campanian angiosperm forests closed canopy and multistratal, or open canopy and fundamentally ruderal? What known and new taxa are herbaceous? Are any of the plant ecological patterns correlated with dinosaurian compositional or diversity gradients? As noted above, previous authors have attempted to answer some of these questions, but with the exception of new analyses on exceptionally preserved floras like Wing et al. (2012) their techniques were qualitative and were based on the same historic collections with outdated taxonomy. The application of several new paleoecological proxies involving leaf size, venation networks/density, and petiole width to leaf area (Royer et al., 2007; Boyce, 2008; Blonder et al., 2011) to new systematic megafloral collections from the Western Interior Basin may provide quantitative answers to these questions.

FLORA OF THE KAIPAROWITS FORMATION

Stratigraphy and Depositional Setting

The Kaiparowits Basin, located in south-central Utah in the Grand Staircase–Escalante National Monument, contains ~2 km of nearly continuous Upper Cretaceous alluvial plain, marginal marine, and marine sediments, which accumulated in the foreland of the Sevier Fold and Thrust Belt (DeCelles, 2004; Titus et al., 2005). The late Campanian Kaiparowits Formation (76.1–74.0 Ma) (Roberts, Deino and Chan, 2005) occurs close to the top of the sequence and comprises nearly half of the section. The unusual thickness (~860 m) of the Kaiparowits Formation is most likely due to increased Laramide flexural and/or dynamic subsidence (DeCelles, 2004) combined with climatic and eustatic

controls (Roberts, 2007). Much of the Kaiparowits Formation exhibits exceptional preservation and an excellent record of vertebrate (Sampson et al., 2004; Titus et al., 2005), invertebrate (Roberts, Tapanila and Mijal, 2008), trace (Roberts and Tapanila, 2006), and palynomorph fossils (Lohrengel, 1969). Though plant compression–impression fossils have been reported throughout the formation (Roberts, 2005, 2007), little work has been done to systematically prospect, quarry, and analyze plants from the Kaiparowits. Given the thickness, duration, and existing stratigraphic framework, including multiple, stratigraphically spaced radiometric ages, the Kaiparowits offers a remarkable opportunity to investigate local floral change across landscapes (i.e., plants tied to facies associations) and through time during the first part of the late Campanian.

Roberts (2007) divided the Kaiparowits Formation into three units (lower, middle, and upper) on the basis of lithostratigraphy and fluvial architecture (Fig. 7.2; see also Roberts, 2007:fig. 10). He found that the formation alternated from fluvial to partially tidally influenced estuarine conditions and was characterized by thick floodplain deposits composed of abundant pond, overbank, suspended load channel, and waterlogged paleosol beds (Roberts, 2007). Though the entire formation preserves fossils, the best preservation occurs in the middle unit (Roberts, 2007, and references therein). Reconnaissance by the Denver Museum of Nature & Science (DMNH) in 2002, 2003, and 2008–2010 led to the discovery of 10 new fossil plant sites and several additional productive palynomorph horizons covering a range of facies associations at multiple levels in the formation. (Fig. 7.3 shows the approximate location of the new megafloral sites in the section called The Blues.) We have prepared and analyzed the megaflora from only one of these sites (DMNH Loc. 3642) for diversity and paleoclimatic estimates. In addition, we have reanalyzed a 40-year-old pollen collection and study from the section (Lohrengel, 1969) and have added several

Table 7.2. Morphotype summary by major taxonomic category for DMNH Loc. 3642

Major taxonomic category	No. of morphotypes	No. of specimens	% Morphotypes	% Specimens
Equisetopsida (*Equisetum* sp.)	1	1	1.1	0.1
Polypodiopsida	8	26	10.5	3.8
Pinopsida				
Cycadales	2	4	2.3	0.6
Coniferales	4	31	4.6	4.6
Angiospermophytina				
Liliopsida	4	43	4.6	6.3
Magnoliopsida				
Reproductive organs	6	8	6.9	1.2
Leaves	62	566	70.1	83.4
Count for vegetative morphotypes	80	671		
Count for all morphotypes	87	679		

Table 7.3. Voucher specimen data for fossil plant morphotypes from DMNH Loc. 3642[a]

Bin[b]	Morphotype no.[c]	No. of specimens[d]	Margin[e]	Leaf size[f]						
				Lepto	Nano	Micro	Noto	Meso	Macro	Mega
1	KP1	2	–	–	–	–	–	–	–	–
2	KP2	1	–	–	–	–	–	–	–	–
2	KP3	1	–	–	–	–	–	–	–	–
2	KP4	1	–	–	–	–	–	–	–	–
2	KP5	1	–	–	–	–	–	–	–	–
3	KP6	2	–	–	–	–	–	–	–	–
6	KP7	1	–	–	–	–	–	–	–	–
7	KP8	15	–	–	–	–	–	–	–	–
7	KP9	1	–	–	–	–	–	–	–	–
7	KP10	2	–	–	–	–	–	–	–	–
7	KP11	3	–	–	–	–	–	–	–	–
7	KP12	1	–	–	–	–	–	–	–	–
7	KP13	2	–	–	–	–	–	–	–	–
7	KP14	1	–	–	–	–	–	–	–	–
7	KP15	1	–	–	–	–	–	–	–	–
8	KP16	2	–	–	–	–	–	–	–	–
8	KP17	2	–	–	–	–	–	–	–	–
10	KP18	2	–	–	–	–	–	–	–	–
10	KP19	12	–	–	–	–	–	–	–	–
10	KP20	10	–	–	–	–	–	–	–	–
10	KP21	7	–	–	–	–	–	–	–	–
11	KP22	10	–	–	–	–	–	–	–	–
11	KP23	1	–	–	–	–	–	–	–	–
11	KP24	31	1	0.000	0.000	0.333	0.333	0.333	0.000	0.000
11	KP25	1	–	–	–	–	–	–	–	–
12	KP26	1	1	0.000	1.000	0.000	0.000	0.000	0.000	0.000
13	KP27	5	1	0.000	0.000	0.000	0.000	1.000	0.000	0.000
13	KP28	1	1	0.000	0.000	0.000	1.000	0.000	0.000	0.000
13	KP29	1	0	–	–	–	–	–	–	–
14	KP30	2	1	0.000	0.000	0.000	0.000	1.000	0.000	0.000
15	KP31	18	0	0.000	0.000	0.000	0.500	0.500	0.000	0.000
15	KP32	2	0	0.000	0.000	0.000	0.500	0.500	0.000	0.000
15	KP33	2	0	0.000	0.000	0.000	1.000	0.000	0.000	0.000
16	KP34	1	0	0.000	0.000	0.000	0.000	1.000	0.000	0.000
16	KP35	1	0	0.000	0.000	0.000	1.000	0.000	0.000	0.000
17	KP36	40	0	0.000	0.000	0.333	0.333	0.333	0.000	0.000
17	KP37	4	0	0.000	0.000	0.000	1.000	0.000	0.000	0.000
17	KP38	6	0	0.000	0.000	0.000	1.000	0.000	0.000	0.000
22	KP39	2	1	0.000	0.000	0.000	0.000	1.000	0.000	0.000
22	KP40	3	1	0.000	0.000	0.000	0.000	1.000	0.000	0.000
22	KP41	1	1	0.000	0.000	0.000	0.000	1.000	0.000	0.000
22	KP42	1	1	0.000	0.000	0.000	1.000	0.000	0.000	0.000
22	KP43	8	1	0.000	0.000	0.000	1.000	0.000	0.000	0.000
22	KP44	11	1	0.000	0.000	0.000	1.000	0.000	0.000	0.000
23	KP45	42	1	0.000	0.000	0.333	0.333	0.333	0.000	0.000
24	KP46	26	1	0.000	0.000	0.333	0.333	0.333	0.000	0.000
24	KP47	10	1	0.000	0.000	0.000	0.500	0.500	0.000	0.000
24	KP48	4	1	0.000	0.000	0.333	0.333	0.333	0.000	0.000
24	KP49	3	1	0.000	0.000	0.000	0.500	0.500	0.000	0.000
28	KP50	1	1	0.000	0.000	1.000	0.000	0.000	0.000	0.000
29	KP51	6	1	0.000	0.000	0.333	0.333	0.333	0.000	0.000
29	KP52	2	1	–	–	–	–	–	–	–
29	KP53	9	1	0.000	0.000	0.000	0.500	0.500	0.000	0.000

Miller et al.

Bin[b]	Morphotype no.[c]	No. of specimens[d]	Margin[e]	Leaf size[f]						
				Lepto	Nano	Micro	Noto	Meso	Macro	Mega
29	KP54	1	1	–	–	–	–	–	–	–
30	KP55	5	1	0.000	0.000	1.000	0.000	0.000	0.000	0.000
30	KP56	1	1	–	–	–	–	–	–	–
30	KP57	5	1	0.000	0.000	0.500	0.500	0.000	0.000	0.000
30	KP58	2	1	0.000	0.000	0.000	1.000	0.000	0.000	0.000
30	KP59	1	1	0.000	0.000	0.000	0.000	1.000	0.000	0.000
30	KP60	1	1	0.000	0.000	0.000	1.000	0.000	0.000	0.000
31	KP61	6	0	0.000	0.000	0.000	0.500	0.500	0.000	0.000
31	KP62	3	0	0.000	0.000	0.000	0.000	1.000	0.000	0.000
31	KP63	2	0	0.000	0.000	0.000	1.000	0.000	0.000	0.000
31	KP64	4	1	0.000	0.000	0.000	1.000	0.000	0.000	0.000
31	KP65	79	0	0.000	0.000	0.000	0.000	1.000	0.000	0.000
33	KP66	4	0	0.000	0.000	0.333	0.333	0.333	0.000	0.000
33	KP67	84	0	0.000	0.000	0.333	0.333	0.333	0.000	0.000
33	KP68	109	0	0.000	0.000	0.333	0.333	0.333	0.000	0.000
33	KP69	5	0	0.000	0.000	0.500	0.500	0.000	0.000	0.000
33	KP70	4	0	0.000	0.000	0.000	0.000	1.000	0.000	0.000
33	KP71	1	0	0.000	0.000	0.000	0.000	1.000	0.000	0.000
33	KP72	1	0	0.000	0.000	0.000	1.000	0.000	0.000	0.000
35	KP73	2	1	0.000	0.000	0.000	0.000	1.000	0.000	0.000
35	KP74	2	1	0.000	0.000	0.000	0.000	1.000	0.000	0.000
35	KP75	3	1	0.000	0.000	0.000	0.000	1.000	0.000	0.000
35	KP76	1	1	0.000	0.000	0.000	0.000	1.000	0.000	0.000
36	KP77	1	1	0.000	0.000	1.000	0.000	0.000	0.000	0.000
37	KP78	12	1	0.000	0.000	0.000	0.500	0.500	0.000	0.000
38	KP79	3	1	0.000	0.000	0.000	1.000	0.000	0.000	0.000
39	KP80	2	1	0.000	0.000	0.000	0.500	0.500	0.000	0.000
39	KP81	2	1	0.000	0.000	0.500	0.500	0.000	0.000	0.000
40	KP82	1	1	0.000	0.000	1.000	0.000	0.000	0.000	0.000
40	KP83	2	1	0.000	1.000	0.000	0.000	0.000	0.000	0.000
41	KP84	1	–	–	–	–	–	–	–	–
41	KP85	1	1	0.000	1.000	0.000	0.000	0.000	0.000	0.000
41	KP86	1	0	0.000	0.000	1.000	0.000	0.000	0.000	0.000
41	KP87	1	1	0.000	0.000	1.000	0.000	0.000	0.000	0.000

[a] Although their size and margin data are listed, morphotypes KP24 (monocot *Cobbania* sp.) and KP26 (dicot *Brasenites* sp.) are aquatic plants and are not included in the calculation of paleoclimate estimates. A dash indicates it was impossible to determine margin type or leaf size, or where they are not applicable.

[b] "Bin" refers to the morphological bin system used by the Paleobotany Project at DMNS (paleobotanyproject.org). 1, Winged reproductive structures. 2, Nonwinged reproductive structures. 3, Reproductive axes. 4, Indeterminate but distinct plant fragments. 5, Lycopsids. 6, Sphenopsids. 7, Filicales. 8, Cycadaleans. 9, Ginkgoaleans. 10, Coniferales. 11, Monocot leaves. 12–41, Dicot angiosperm leaf categories.

[c] The morphotype number is a combination of an abbreviation for the Kaiparowits Formation (KP) and a sequential listing of the number of morphotypes from the formation.

[d] The number of specimens in each morphotype.

[e] Margin type: 0, Toothed. 1, Entire.

[f] See Ellis et al. (2009) for definitions of leaf sizes.

new palynomorph samples that correspond to megafloral sites. This chapter is based on the single megafloral site and the pollen reanalysis. Future work will focus on preparing and analyzing additional in-house megafossil collections, increasing the number of well-quarried sites from other facies associations and from the lower and upper units, and systematically resampling the entire formation for fossil pollen.

The megafloral site, DMNH Loc. 3642, is preserved in a relatively thin (~30 cm) sandy siltstone to silty sandstone horizon in a sequence of inclined heterolithic sandstone, siltstone, and mudstone beds assigned to facies association 7 (FA7) of Roberts (2007). In the formation, FA7 occurs within the lower and middle units and is interpreted as a point-bar accretion deposit that formed under fluctuating, possibly

7.4. Eight fern morphotypes found in the Kaiparowits Formation. (A) KP10, DMNH 33384; (B) KP11, DMNH 33385; (C) KP14, DMNH 33389; (D) KP8, DMNH 33387; (E) KP9, DMNH 33383; (F) KP13, DMNH 33388; (G) KP12, DMNH 33386; (H) KP15, DMNH 33390. Scale bars = 1 cm.

tidal, currents (Roberts, 2007). The site is located ~337 m above the base of the formation and sits ~85 m below an ash bed dated at 75.51 Ma (^{40}Ar/^{39}Ar), and ~115 m above and ash bed dated at 75.97 Ma (E. Roberts, pers. comm., 2011). We used a depositional rate of ~41 cm/ka to estimate the age of the flora from DMNH Loc. 3642 to be ~75.69–75.72 Ma (Roberts et al. chapter 6).

Megafloral Diversity and Composition

The Kaiparowits flora was collected by standard bench-quarrying techniques. The current voucher collection (Tables 7.2, 7.3) was built from a combination of field-census collecting and cherry-picking (a field collection process focused on collecting only new or exceptionally preserved specimens).

Miller et al.

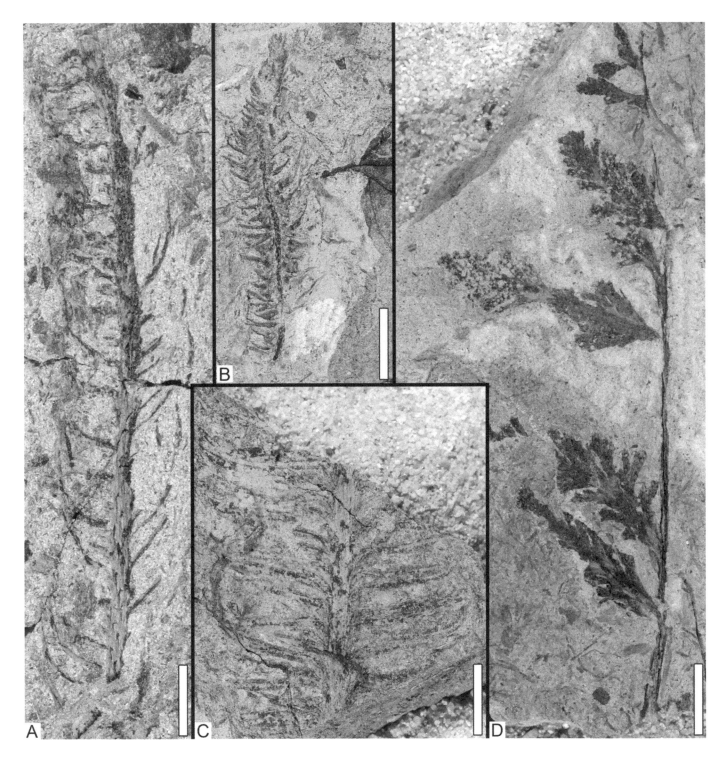

7.5. Four conifer foliage types from site DMNH Loc. 3642 in the Kaiparowits Formation. (A) KP21, DMNH 30350; (B) KP19, DMNH 30351; (C) KP20, DMNH 33391; (D) KP18, DMNH 33392. Scale bars = 1 cm; specimens shown at the same scale.

Rather than attempting to produce normalized alpha diversity curves in advance of a well-documented morphotype list from multiple localities in the formation, we focus on describing the overall diversity and composition of the single site (DMNH Loc. 3642). We also present select aquatic plant specimens from the other, not as yet fully processed, collections from the formation.

We sorted the flora into plant macrofossil morphotypes following the concept described by Johnson (1989). Independent of the Linnaean system, this method uses the morphological characters of disassociated plant organs (e.g., leaves, fruits, stems) to circumscribe discrete operational taxonomic units before a formal taxonomy is erected. Such morphotypes are based on multiple well-preserved specimens and

7.6. Thirteen representative leaf morphotypes from site DMNH Loc. 3642 in the Kaiparowits Formation, exclusive of the platanoid leaf morphotypes (see fig. 7.7). We list distinguishing characteristics for each morphotype. (A) KP45, DMNH 33453, is the second most common nonplatanoid morphotype in the flora (42 voucher specimens). It exhibits the highest leaf rank among morphotypes from the site (see Hickey and Wolfe 1975 and references therein for leaf ranks); (B) KP57, DMNH 33452 is most similar to KP55 but shows a pronounced asymmetric leaf base, which is also slightly embayed; (C) KP75, DMNH 33454, is distinguished by its palmate/pinnate primary venation (varies from specimen to specimen) combined with an entire margin and strong, brochidodromous, agrophic veins and well-organized higher-order venation; (D) KP55, DMNH 33450, is the only morphotype with trifoliate leaf attachment; (E) KP49, DMNH 33399, is distinguished by its numerous and prominent intersecondary veins; (F) KP85, DMNH 33455, is the only morphotype that appears herbaceous, on the basis of overall morphology and leaf architecture; (G) KP73, DMNH 33394, is distinguished by its festooned brochidodromous secondary veins; (H) KP66, DMNH 33398, is distinguished by its consistently small size and relatively large teeth confined to the distal half of the leaf; (I) KP77, DMNH 33396, is distinguished by its low leaf rank and asymmetric leaf shape; (J) KP59, DMNH 33451, is distinguished by its decurrent leaf base with crowded basal secondary veins and its disorganized tertiary venation; (K) KP82, DMNH 33397, is the only morphotype with a nearly orbicular leaf shape, decurrent leaf base (shown on counterpart), and a slightly emarginate leaf apex; (L) KP81, DMNH 33395, is distinguished by a deeply cordate leaf base and prominent agrophic veins; (M) KP80, DMNH 33393, is distinguished by an orbicular leaf shape and large teeth (teeth are visible in the upper left-hand corner of the figured specimen). Scale bar in the lower right-hand corner is for all specimens and equals 1 cm.

7.7. Nine platanoid leaf morphotypes from site DMNH Loc. 3642 in the Kaiparowits Formation. All of the platanoid morphotypes are represented by four or more voucher specimens (range, 4–109 specimens; Table 7.3), with the exception of KP71 (B), which is represented by a single specimen. We list distinguishing characteristics for each morphotype; (A) KP78, DMNH 30352, is distinguished by its reduced teeth, which are sometimes recessed, its broad lobes, and its markedly decurrent leaf base; (B) KP71, DMNH 30353, is distinguished by the straight course of its secondary veins, which originate from the midvein opposite to one another, compound agrophic veins, and small, rounded, and infrequent teeth; (C) KP36, DMNH 30354, is distinguished by its pinnate primary vein architecture, well-impressed and strictly opposite percurrent tertiary venation, and relatively large teeth; (D) KP69, DMNH 30355, is distinguished by irregular secondary vein course, disorganized tertiary venation, and numerous small and distinct teeth; (E) KP37, DMNH 30356, is distinguished by its large teeth, which are approaching the size of lobes; (F) KP38, DMNH 30357, is most similar to KP69 (D), but has fewer and larger teeth and better-organized percurrent tertiary venation; (G) KP70, DMNH 30358, is distinguished by its consistently obovate leaf shape, with large, rounded teeth restricted to the distal half of the leaf, and well-organized opposite percurrent tertiary venation; (H) KP67, DMNH 30359, is most similar to KP36 (C) but exhibits lobes, smaller teeth, less well-impressed venation, and more widely spaced and less well-organized percurrent tertiary venation; (I) KP68, DMNH 30360, is the most common platanoid morphotype in the collection (109 voucher specimens). It is most similar to KP71 (B). However, it is distinguished by a tendency for having a slightly obovate leaf shape and for having actinodromous venation, a typically naked base, secondary veins that consistently curve toward the apex, and well-organized opposite percurrent tertiary venation. Scale bar in the lower right-hand corner is for all specimens and equals 1 cm.

closely approximate biological species, allowing for paleo-ecological and paleoclimate analysis.

We identified 87 morphotypes comprising equisetopsid, fern, cycad, conifer, and angiosperm stem, seed, and leaf remains from site DMNH Loc. 3642 (Table 7.2). The overall composition of the nondicot leaf taxa is similar to other Late Cretaceous floras from the Western Interior Basin and consists of eight ferns (Fig. 7.4), two possible cycads, four cupressaceous/taxodiaceous conifers (Fig. 7.5), and four monocots, which included two palms and two aquatic species (see Fig. 7.9). A third pinnate palm morphotype was found on a loose surface block in the lower part of the Kaiparowits section.

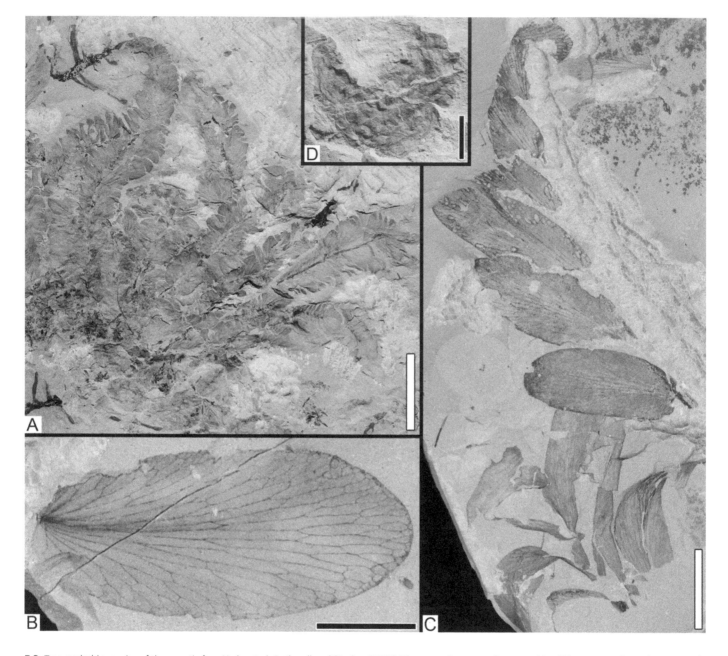

7.8. Two probable species of the aquatic fern *Hydropteris* Rothwell and Stockey (1994). These specimens are from pond localities separate from the censused channel deposit locality. (A) *Hydropteris* sp., KP88, DMNH 30368, DMNH Loc. 3723; scale bar = 1 cm; (B) *Hydropteris* cf. *H. pinnata*, KP89, DMNH 30370, DMNH Loc. 3725; scale bar = 1 cm; (C) *Hydropteris* cf. *H. pinnata*, KP89, DMNH 30369, DMNH Loc. 3725; pinna with anastomosing venation; scale bar = 0.5 cm; (D) *Hydropteris* cf. *H. pinnata*, KP89, DMNH 30371, DMNH Loc. 3725; sporocarp; scale bar = 0.5 cm.

Notably absent are the *Dammarites* cone scales that are often associated with the *Araucaria*-like foliage genus *Elatides* (Fig. 7.5).

The dicot morphotypes overwhelmingly dominate the flora both in terms of the number of species present (62 morphotypes) and the number of megafossils recovered (~70% and ~83%, respectively; Tables 7.2, 7.3) (Fig. 7.6 shows representative dicot leaf morphotypes). Among the dicots, the most dominant morphotypes belong to the platanoids. They comprise nine leaf morphotypes (~15% of the dicot morphotypes) and nearly 50% of the dicot leaf specimens found

(Fig. 7.7). Throughout the Western Interior Basin, platanoids are common in Upper Cretaceous fluvial, and in particular channel-deposit, floras (Crabtree, 1987a). For example, the Hell Creek Formation contains more than 30 platanoid morphotypes mostly concentrated in channel-deposit settings (Johnson, 2002). In that formation, platanoids comprise ~10% of the species diversity. Ponded-water deposits in the Kaiparowits have yielded a different flora with few to no platanoids. Like the Hell Creek Formation, our initial observation is that in the Kaiparowits Formation platanoids are dominant in channel settings but become markedly scarcer

Miller et al.

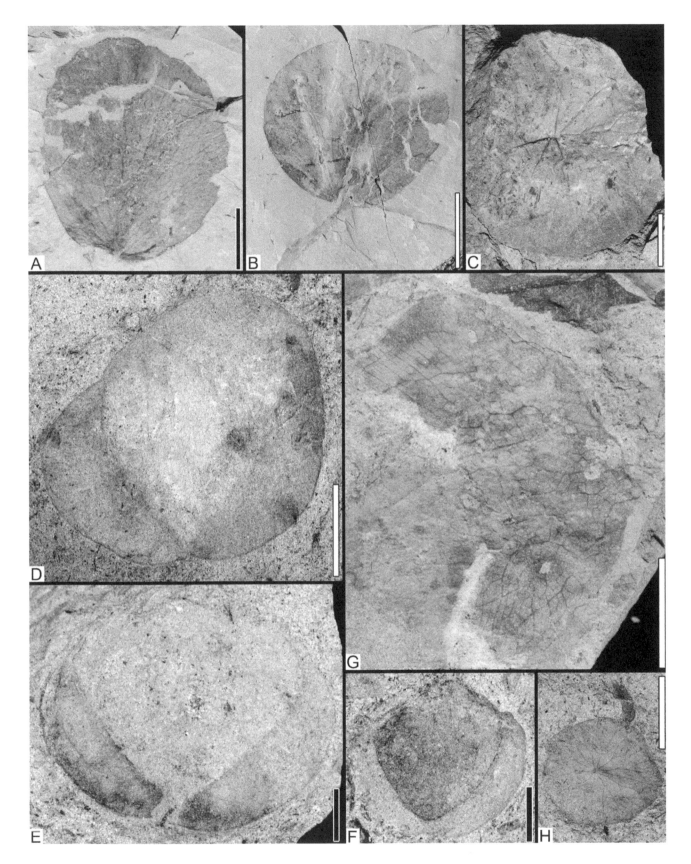

7.9. Aquatic angiosperms from the Kaiparowits Formation. (A) *Quereuxia* sp., KP90, DMNH 30366, DMNH Loc. 3723; scale bar = 0.5 cm; (B) *Quereuxia* sp., KP90, DMNH 30367, DMNH Loc. 3723; scale bar = 1 cm; (C) *Nelumbites* sp. KP91, DMNH 30372, DMNH Loc. 3811; scale bar = 1 cm; (D–F) possible new species of the extinct aquatic angiosperm *Cobbania*. (D) KP24, DMNH 30362, DMNH Loc. 3642, scale bar = to 1 cm; (E) KP74, DMNH 30363, DMNH Loc. 3642, scale bar = 0.5 cm; (F) KP24, DMNH 30364, DMNH Loc. 3642, scale bar = 0.5 cm; (G) *Cobbania* cf. *C. corrugata,* KP26, DMNH 30365, DMNH Loc. 3642, scale bar = 1 cm; (H) *Brasenites* sp., KP25, DMNH 30361, DMNH Loc. 3642, scale bar = 0.5 cm.

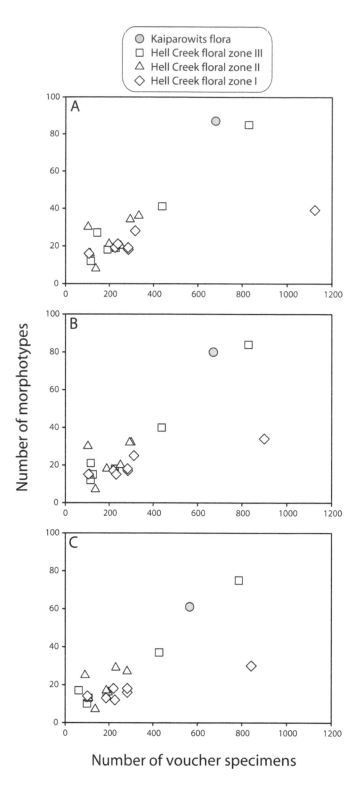

7.10. Diversity and voucher specimen counts for Kaiparowits DMNH Loc. 3642 and Hell Creek channel floras (*n* = 20). Hell Creek Formation floral data are from Johnson (2002) and Wilf and Johnson (2004) and are divided into floral zones after Johnson (2002). (A) All morphotypes, including seeds and flowers; (B) all leaf morphotypes; (C) only dicot leaf morphotypes.

both in terms of species abundance and number of specimens as quiet-water deposits are approached.

Platanoid leaf architecture is remarkably consistent throughout the Late Cretaceous, and some genera, like *Credneria*, have ranges of more than 30 million years. Platanoid species, however, have much shorter durations; this may be due to the problem of overspeciation of floras in historical treatments. Though the formations are separated by ~10 million years and ~10° of latitude, we note the leaf architectural similarity between some Kaiparowits and Hell Creek platanoids. The apparent niche conservatism of the platanoid clade and the presence of long-lived platanoid species indicate some level of stasis in Late Cretaceous biomes across the Western Interior Basin.

Aquatic plants show a similar pattern to platanoids through the Late Cretaceous. Their niche, however, is fundamentally more consistent because water temperature fluctuates less than air temperature, and as long as freshwater bodies are present, they are not susceptible to precipitation patterns. The abundance of aquatic plant morphotypes in the Kaiparowits is high and matches that of the most diverse aquatic plant assemblages known from the Late Cretaceous. Kaiparowits morphotypes include two aquatic ferns (*Hydropteris* cf. *H. pinnata* and *Hydropteris* sp.) and five aquatic angiosperms, including *Quereuxia* sp., *Cobbania* cf. *C. corrugata*, cf. *Cobbania*, *Brasenites* sp., and *Nelumbites* sp. (Figs. 7.8, 7.9). The censused quarry yielded only *Cobbania* cf. *C. corrugata* (one specimen) (Fig. 7.9G), cf. *Cobbania* (32 specimens) (Fig. 7.9D–F), and *Brasenites* sp. (one specimen) (Fig. 7.9H). The *Hydropteris* species and *Quereuxia* were found in ponded-water deposits. The flora of one of these sites (DMNH Loc. 3723), which also contains an abundance of well-preserved invertebrates, is composed almost entirely of *Quereuxia*, where whole plants appear to be preserved. The only other plant found at DMNH Loc. 3723 is *Hydropteris* sp. (Fig. 7.5A). *Hydropteris* cf. *H. pinnata* is relatively abundant at a second site (DMNH Loc. 3725), which appears to be a pond that was periodically inundated by higher-energy water. This site is also characterized by a diverse dicot flora that appears to be compositionally very different from that of site presented herein (DMNH Loc. 3642). Finally, the specimen of *Nelumbites* sp. (Fig. 7.9C) occurs in a channel deposit in a thin, trough-cross-bedded, fine- to medium-grained sandstone interbed (DMNH Loc. 3811). The high diversity of aquatic plants in the Kaiparowits mirrors the similarly high diversity of aquatic vertebrates and invertebrates found in the formation and corroborates the findings of Roberts (2007) that the Kaiparowits accumulated in a floodplain depositional setting dominated by large low-gradient rivers and large bodies of ponded water.

Miller et al.

The dicot leaf morphotype count from DMNH Loc. 3642 is among the most diverse for a single locality in the Late Cretaceous. Though more than 75 Late Cretaceous floras are known from the Western Interior Basin (see *The Compendium Index of North American Mesozoic and Cenozoic Type Fossil Plants*, http://www.peabody.yale.edu/collections /pb/eCI/), few have been analyzed and published using modern techniques, and of these, none are Campanian. The only systematically collected floras of Late Cretaceous age that give site counts of dicot leaf morphotypes and depositional classification of those sites are from the late Maastrichtian Hell Creek flora in the Williston Basin (Johnson, 2002). Despite the temporal and geographic differences, we compare the Kaiparowits data to that of the Hell Creek Formation.

We compare the diversity of the Kaiparowits flora (DMNH Loc. 3642 only) to 20 channel floras from the Hell Creek Formation that are each represented by more than 100 voucher specimens. In advance of a strict comparative census collection from the Kaiparowits, we plot the number of voucher specimens versus the morphotype diversity for each site (Fig. 7.10). Though the approach is not quantitative given the varying collection styles used to generate the data, which specimens were chosen to be collected in the field, and the lack of data normalization, the plots provide a qualitative comparison that supports the finding that the Kaiparowits flora is among the most diverse in the Late Cretaceous.

Leaf-Based Estimates of Paleoclimate

Two relationships have been widely used to utilize the fossil leaf physiognomy of woody dicotyledonous angiosperms to estimate terrestrial paleoclimate. These are the correlation between the percentage of smooth-margined leaf species at a site and local mean annual temperature (MAT) (Wolfe, 1971, 1978, 1979: Wing and Greenwood, 1993; Wilf, 1997; Miller, Brandon, and Hickey, 2006; Greenwood, 2007) and the correlation between the average size of leaf species and local MAP (Wilf et al., 1998, 1999). The validity of both techniques has recently been questioned in terms of the degree to which physiognomic characters are primarily phylogenetic signals rather than predictors of climate (Wilf, 2008; Little, Kembel, and Wilf, 2010). However, recent global data sets have supported the utility of leaf margin estimates of MAT, although with larger uncertainties (Peppe et al., 2011). Fossil leaf margin estimates of MAT have also been shown to mirror deep-sea temperature records (Wilf, Johnson, and Huber, 2003; Wing et al., 2005), correlate with latitude (the primary variable in terrestrial low-elevation surface temperature) (Wolfe and Upchurch, 1987; Miller, Brandon, and Hickey,

2006), and follow food web dynamics that are indirectly related to air temperature (Currano et al., 2008). Moreover, Little, Kembel, and Wilf (2010) acknowledge that leaf-based proxies for paleotemperature are at least reliable for regional comparative analysis. For this study, we used the Miller, Brandon and Hickey (2006) regression and error estimate based on North and Central American floras, which includes an estimate of extrabinomial error, to determine MAT from DMNH Loc. 3642. For MAP, we used the regression provided by Wilf et al. (1998) and followed the leaf size definitions given by Ellis et al. (2009). The fossil data for our estimates are provided in Table 7.3.

We used 60 and 55 dicot leaf morphotypes to estimate MAT and MAP, respectively. One dicot leaf morphotype has the morphology and architecture of an aquatic plant (Fig. 7.10H) and another is distinctly herbaceous (Fig. 7.6F), so they were excluded from our paleoclimate analysis. Several additional morphotypes did not preserve enough of the leaf to accurately estimate leaf size. We found that the percentage of entire margined dicot morphotypes (P) is 0.65, which gives a MAT of 20.16 ± 2.33°C (±1 SE). The leaf size of the Kaiparowits dicot morphotypes covers a range from nanophyll to mesophyll, with the majority of specimens falling in the notophyll (39.5%) and mesophyll (38.6%) categories. We estimate a MAP of 1.78 m (1.24–2.55 m) (±1 SE). The MAT and MAP results from DMNH Loc. 3642 indicate a paleoclimate that was most similar to the modern climate on the Gulf Coast of North America between Texas and Florida (see Global Historical Climatology Network, version 2, 1697 to present, ftp.ncdc.noaa.gov) (Peterson and Vose, 2001). In the context of the Campanian paleoclimate trends described by Wolfe and Upchurch (1987), our MAT estimate from the Kaiparowits flora agrees well with their interpretations, as the site is located slightly south of their 20°C isotherm. In terms of MAP, however, the Kaiparowits falls in their belt of low precipitation, yet gives a MAP estimate of approximately ~2 m. Average leaf size of fossil floras is often influenced by taphonomy and collecting style. The Wolfe and Upchurch (1987) MAP pattern in the Western Interior Basin is based on historical fossil collections, which may be significantly biased as a result of these factors.

Notably, the dicot leaf estimates of paleoclimate from the Kaiparowits agree well with the findings of Roberts (2007), who argued that the sedimentology of the formation indicated that it was deposited in a relatively wet, subhumid environment. Future work on the paleoclimate of the Kaiparowits flora will focus on estimating relative differences in paleoclimatic estimates as a result of taphonomic factors resulting from variation in depositional settings (Burnham et al., 2001; Greenwood, 2005) in the formation and placing the

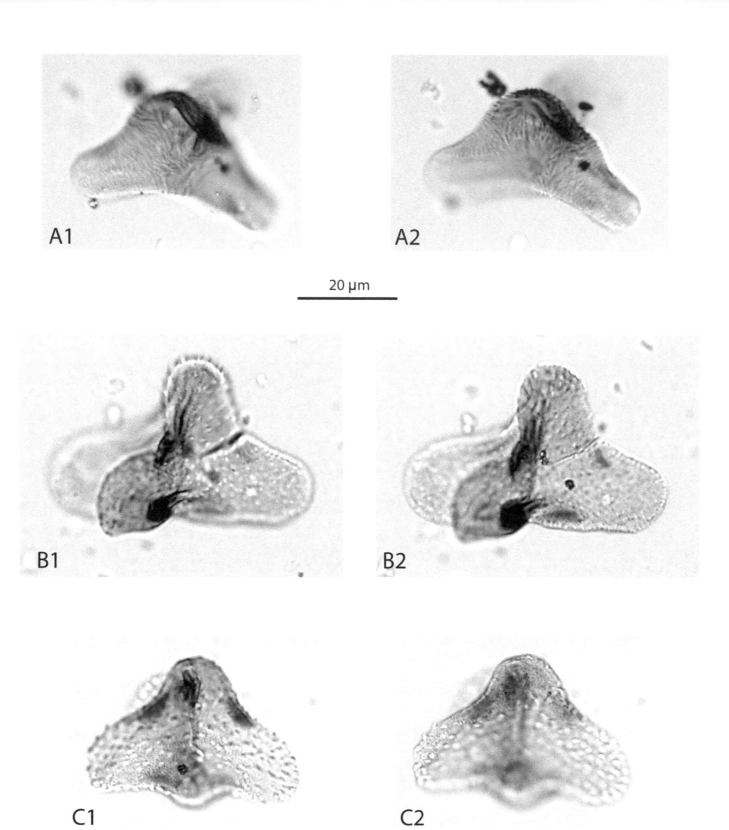

7.11. (A1, A2) *Aquilapollenites pyriformis* (two levels of focus on a single specimen) CFL-K-212A (138.4, 17.7 [paired numbers in parentheses are stage coordinates]); (B1, B2) *Aquilapollenites quadrilobus* (two levels of focus on a single specimen) CFL-K-212A (141.8, 6.8); (C1, C2) *Aquilapollenites turbidus* (two levels of focus on a single specimen) CFL-K-160A (131.9, 5.0).

Miller et al.

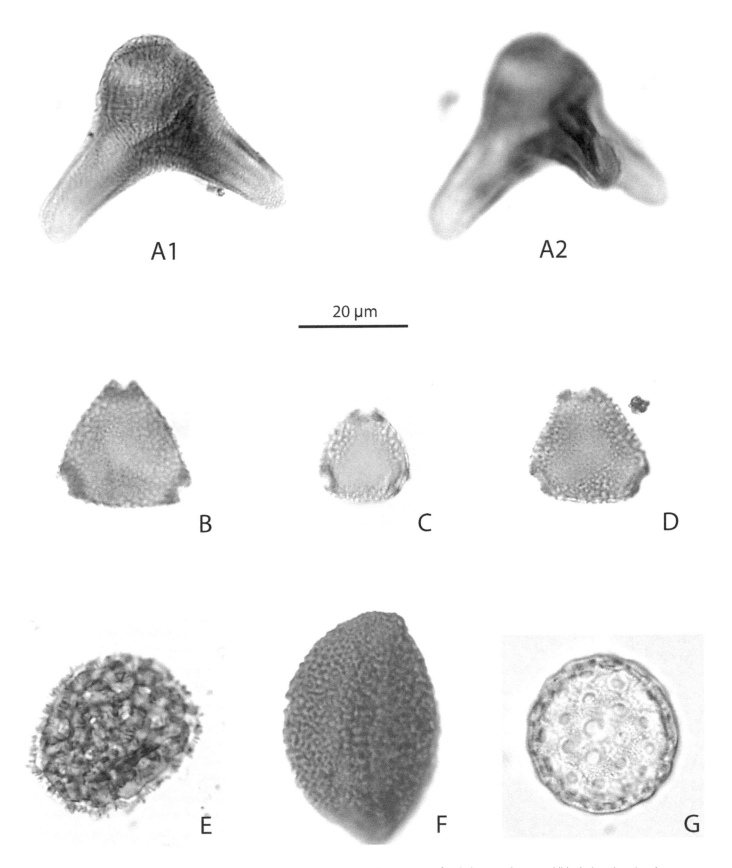

7.12. (A1, A2) *Aquilapollenites striatus* (two levels of focus on a single specimen) above CFL, 200 feet below conglomerate (slide designation given by Lohrengel, 1969) (132.8, 10.0 [paired numbers in parentheses are stage coordinates]); (B) *Tschudypollis* spp. Above CFL, 200 feet below conglomerate (134.6,6.3); (C) *Tschudypollis* spp. CFL-K-170C (117.3, 17.6); (D) *Tschudypollis* spp. Above CFL, 200 feet below conglomerate (136.6,6.4); (E) *Erdtmanipollis* sp. CFL-K-212A (128.3, 7.5); (F) *Arecipites* sp. DBP-2009–037 (113.8, 7.3); (G) *Chenopodipollis* sp. CFL-K-160A (123.5, 14.0).

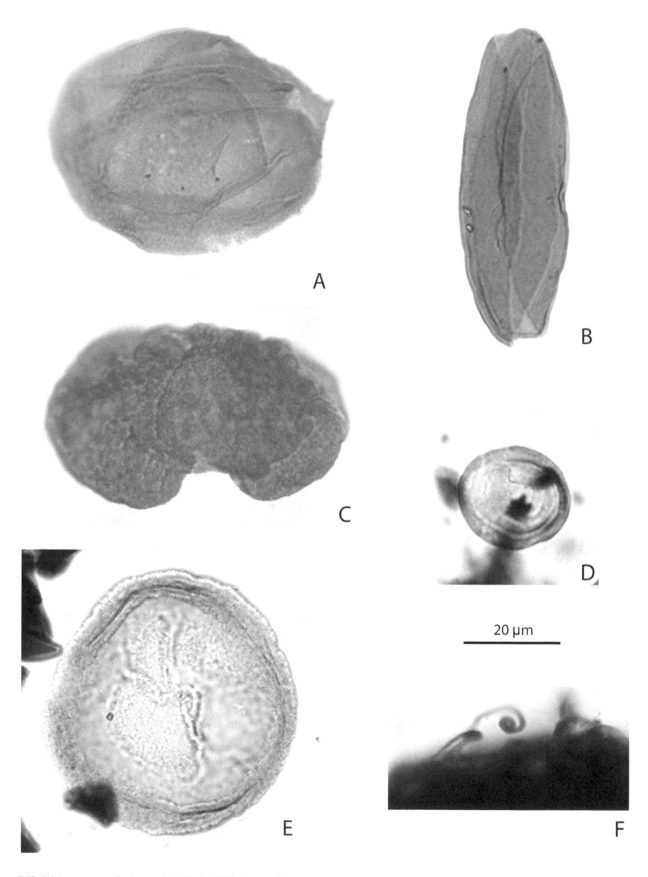

7.13. (A) *Inaperturopollenites* sp. DBP-2009–037 (121.4, 6.7 [paired numbers in parentheses are stage coordinates]); (B) *Cycadopites* sp. DBP-2009-037 (111.2, 8.4); (C) *Rugubivesiculites* sp. DBP-2009-037 (126.7, 5.1); (D) *Classopollis* sp. CFL-K-332B (129.0, 6.3); (E) *Araucariacites* sp. CFL-K-344W (138.1, 17.8); (F) *Azolla circinnata* (close-up view of edge of specimen showing glochidia) CFL-K-256A (126.5, 5.4).

Miller et al.

Kaiparowits paleoclimate estimates in an updated regional context at the scale of the Western Interior Basin.

Palynofloral Diversity and Composition

Before the present study, six publications addressed the palynology of the Kaiparowits Formation (Lohrengel, 1969; Bowers, 1972; Eaton, 1991; Nichols and Sweet, 1993; Nichols, 1997). The earliest and most comprehensive of these was by Lohrengel (1969), and in this chapter, we primarily review that work. The other studies amount to a few additions to the Kaiparowits palynofloral list (Bowers, 1972; Farabee, 1991) or are focused on updating the age determination of the formation from Maastrichtian to Campanian (Eaton, 1991; Nichols and Sweet, 1993; Nichols, 1997).

Lohrengel (1969) identified 80 palynomorph species from the formation. These include two species of freshwater algal cysts, 26 species of spores (including bryophytes and ferns), 20 species of gymnosperm pollen, and 32 species of angiosperm pollen. However, 13 of these identifications were tentative, and 36 of his species were identified only to the level of genus. On the basis of our review of Lohrengel's descriptions and photomicrographs, we question some of his identifications, even at the generic level. Admittedly, the published reproductions of Lohrengel's photomicrographs are of poor quality, so in many cases it is difficult to see distinguishing palynomorph characteristics.

Lohrengel's (1969) work indicates that the Kaiparowits palynoflora is composed of roughly twice as many species of pollen (including both gymnosperms and angiosperms) as spores. Lohrengel reported that the palynoflora was dominated by gymnosperms, despite the greater diversity of angiosperms, and found that five gymnosperm species, three angiosperm species, and one fern species each comprised more than 10% of specimens present in at least one of the 15 stratigraphic levels he analyzed within the formation.

Of the most commonly occurring gymnosperm palynomorphs, four are inaperturate pollen (three of these species are assigned to the genus *Inaperturopollenites*—by far the most abundant group—and one to the genus *Araucariacites*), one is bisaccate and assigned to the genus *Pityosporites*, and one is monosulcate and assigned to the genus *Entylissa*. Among the angiosperms, the most commonly occurring species belong to the genus *Tricolpopollenites* (one species) and "*Proteacidites*" (two species) and the most common spore belongs to the fern genus *Laevigatosporites*.

Bowers (1972) added seven species, including *Aquilapollenites*, *Classopollis*, *Ephedra*, *Erdtmanipollis*, *Eucommiidites*, *Kuylisporites*, and *Rugubivesiculites*. In our review of Lohrengel's (1969) material, we observed and here report *Classopollis* sp. (Fig. 7.13D), *Erdtmanipollis* sp. (Fig. 7.12E),

Rugubivesiculites sp. (Fig. 7.13C), and *Ephedra* sp. (not illustrated). We did not find any examples of species in *Eucommiidites* or *Kuylisporites*. We note that the species *Aquilapollenites* cf. *A. senonicus* reported by Bowers (1972) is more likely *Aquilapollenites quadrilobus*, based on our review of Lohrengel's (1969) material.

The present palynological study includes two parts. First, we reexamined most of Lohrengel's (1969) slides. We discuss the biostratigraphically important palynomorphs and present several additional species that Lohrengel did not report. Second, we present a short summary of our initial results from new palynological samples collected from the Kaiparowits Formation in conjunction with our recent megafloral study. Lohrengel's slides are currently held in the Department of Earth Sciences at the Denver Museum of Nature & Science.

Lohrengel (1969) reported two angiosperm pollen genera that have biostratigraphic importance: *Aquilapollenites* and "*Proteacidites*." The latter genus was renamed *Tschudypollis*. We hereafter informally reassign all "*Proteacidites*" identifications in Lohrengel (1969) as *Tschudypollis*. (Nichols, 2002) A more in-depth taxonomic treatment, which will formally recombine these species from the Kaiparowits, is needed but is beyond the scope of this study. *Tschudypollis* ranges from the Coniacian to the Maastrichtian in the Western Interior Basin (Nichols, 1994; Nichols and Johnson, 2002). Species of *Aquilapollenites*, however, have shorter temporal ranges and are the most biostratigraphically useful palynomorphs during the Campanian (Tschudy and Leopold, 1971; Nichols and Sweet, 1993). Utilizing recent studies of *Aquilapollenites*, we have reassigned Lohrengel's species in the genus to produce a more accurate assessment of this palynoflora.

Lohrengel (1969) described and illustrated four species of *Tschudypollis*. He reported the genus in all 15 samples he analyzed from the Kaiparowits Formation. He regarded two of these species, *Tschudypollis retusus* Anderson and *T. thalmannii* Anderson, as common (more than 10% of the grains present) at the level 262 m above the base of his measured section. Until a reevaluation of these species in *Tschudypollis* is conducted, we assign them to *Tschudypollis* sp. Specimens of this group are illustrated in Fig. 7.12B–D.

Lohrengel (1969) also reported four species of *Aquilapollenites* from the Kaiparowits. In our review, we identified one additional *Aquilapollenites* species. These five species more or less segregate by stratigraphic level within the formation, thus indicating their potential value as guide fossils. The first of these, "*Aquilapollenites* n. sp. A," is restricted to the basal 1.5 m of the formation. No slides were available from this level, and we found no occurrence of this species in the remainder of the section. Eaton (1991:58) stated that this species is "almost identical to *A. trialatus*." However, Lohrengel described it having a gemmate exine, whereas it actually has a reticulate exine. Our comparison of Lohrengel's specimen (1969:pl. 10, fig. 13) to the literature (Tschudy and Leopold, 1971:pl.4, figs. 4A, B; Nichols and Sweet, 1993:pl.1, fig. 46) indicates that Lohrengel's "*Aquilapollenites* n. sp. A" is *A. trialatus* var. *variabilis* Tschudy and Leopold (1971). This species is characteristic of the upper Campanian and lower Maastrichtian (Tschudy and Leopold, 1971; Nichols, 1994).

Between 137.2 and 381.0 m above the base of the formation, Lohrengel (1969) reported "*Aquilapollenites* cf. *A. delicatus.*" Eaton (1991:58) found this species to be "almost identical to *A. turbidus*," but he could not see the spines on the equatorial projections in Lohrengel's plate 10, fig. 10. Our photomicrographs of this species (Fig. 7.11, C1, C2) show these spines, and we agree with Eaton that this species is actually *A. turbidus* Tschudy and Leopold. This species has a range of upper Campanian to Maastrichtian (Tschudy and Leopold, 1971).

Lohrengel (1969) also reported *Aquilapollenites pyriformis* Norton from the same stratigraphic interval(Fig. 7.11, A1, A2). Though the complete stratigraphic range of this species has not been established, it has been reported previously in the Maastrichtian (Nichols, 2002).

Lohrengel (1969) recorded "*Aquilapollenites* n. sp. B" from one sample located 262.1 m above the base of the formation. Lohrengel's species (Fig. 7.11, B1, B2) appears to be *A. quadrilobus* Rouse. The spines on the major pole of *A. quadrilobus* are diagnostic, and they differentiate this species from the very similar *A. senonicus*, which has a reticulate polar sculpture. The spines on the distal pole of *A. quadrilobus* from the Kaiparowits are small, measuring ~1 μm in length. The pattern of these spines, at a lower magnification, could easily be interpreted as a reticulum, leading to the misidentification as *A. senonicus*. For this reason, we suspect that Bowers's (1972) report of "*Aquilapollenites* cf. *A. senonicus*" in the formation is more likely *A. quadrilobus*. *Aquilapollenites quadrilobus* is illustrated in Fig. 7.11, B1, B2. This species has a range of late Campanian to Maastrichtian (Tschudy and Leopold, 1971).

We also identified *A. striatus* Funkhouser (=*Striatocorpus striatus* of Farabee, Vezey, and Skarvala, 1991) from Lohrengel's (1969) sample located 61 m below the contact with the overlying Canaan Peak Formation. Lohrengel did not report this species, but his slide is labeled "Kaiparowits? above CFL, 200 feet below conglomerate." No definitive occurrences of *A. striatus* have been observed below this level. Farabee, Vezey, and Skarvala (1991:pl. 2, figs. 1–5), however, reported "*Striatocorpus* cf. *striatus*" from the formation and illustrated three specimens. They noted the dimensions of these specimens as intermediate between *A. pyriformis* Norton (="*Striatocorpus*" *pyriformis*) and *A. striatus*. Unfortunately,

the stratigraphic position of these specimens was not given. From our experience with the Kaiparowits palynomorphs, we would reassign most of the *Aquilapollenites* cf. *A. striatus* in Farabee, Vezey, and Skarvala (1991:pl. 2, figs. 1, 2, 4, and 5) to *A. pyriformis*. The specimen in fig. 3 of Farabee, Vezey, and Skarvala, however, appears to be intermediate between *A. pyriformis* and *A. striatus* in that it has more of the less complex striations characteristic of *A. pyriformis*, while its form more closely resembles that of *A. striatus*. The major pole is more pronounced and the apices of the equatorial projections are more rounded than those typical of *A. pyriformis*. Nonetheless, we argue that the *A. striatus* figured here (Fig. 7.12, A1, A2) is the first report of the species from the Kaiparowits Formation. *Aquilapollenites striatus* has a range of late Campanian to lower Maastrichtian.

Of the five species of *Aquilapollenites* present in the Kaiparowits Formation, two have particular stratigraphic significance in age determination and correlation. *A. trialatus* var. *variabilis* first appears in the lower Campanian and ranges into the lower part of the upper Campanian (Tschudy and Leopold, 1971). *A. striatus* first appears near the upper part of the upper Campanian and ranges into the Maastrichtian. On the basis of the bracketing provided by these species of *Aquilapollenites*, the age of the Kaiparowits Formation is middle to late Campanian, which agrees with radiometric dates determined by Roberts, Deino and Chan (2005)

We also identified several palynomorphs not reported by Lohrengel (1969). These are: *Azolla circinata* (Fig. 7.13F), *Chenopodipollis* sp. (Fig. 7.12G), *Classopollis* sp. (Fig. 7.13D), *Erdtmanipollis* sp. (Fig. 7.12E), and the acritarch *Pseudoschizaea* sp. (not illustrated).

Our analysis of sediment samples from the megafloral localities resulted in good recovery of palynomorphs from fine-grained pond environment samples, but poor recovery from channel and point bar deposits, including the megafloral census site (DMNH Loc. 3642), where the matrix is sandy. One of the pond sites (DMNH Loc. 3723) that gave good recovery is approximately 7 m stratigraphically above the census megaflora reported here. Our preliminary results from this site indicate ranges and relative abundances of species consistent with those given by Lohrengel (1969). Selected species of gymnosperm pollen from this site are illustrated in Fig. 7.13A–C. Further analysis of the Kaiparowits palynomorphs, including an updated sampling campaign, is planned.

Palynofloral Paleoenvironmental Indicators

Two species of the water fern *Azolla* occur in the Kaiparowits Formation. One (*A. cretacea*) was reported by Lohrengel (1969), and we found the second (*A. circinata*) widely distributed within the formation. Modern *Azolla* inhabits slow-moving freshwater but is not found in stagnant swamp environments.

Roberts (2007) argued for a zone of tidally influenced deposition between 160 and 190 m above the base of the formation. He based his interpretation on bedding indicators and the presence of *Teredolites*, a marine- to brackish-water trace fossil. In our review of Lohrengel's (1969) material, including a sample from the 167.6-m mark, we did not find dinoflagellate cysts that would support either marine- or brackish-water conditions. Further palynological sampling at specific *Teredolites* horizons would be a prime target to look for palynomorph evidence of brackish water.

CONCLUSIONS

Initial analysis of the fossil plant remains from the Kaiparowits Formation indicates the presence of a diverse flora, which includes angiosperms as well as equisetopsids, ferns, cycads, and conifers. Dicot leaf morphotypes are the most diverse element of the flora, and this component is dominated by platanoid taxa—a trend common in Late Cretaceous floras—that are in general very similar to those found in younger floras in the Western Interior Basin. The diversity of dicot morphotypes from the single analyzed site (DMNH Loc. 3642) is as high as the most diverse floras found in channel settings of Late Cretaceous age. Angiosperm pollen is also diverse and includes multiple biostratigraphically important species in the genus *Aquilapollenites*. Importantly, these species are used as middle to late Campanian indicators, which is herein corroborated by the recent radiometric dates from the formation. Aquatic angiosperms are also diverse and include

multiple mega- and palynofloral species mirroring the high diversity of aquatic vertebrates and invertebrates found in the formation. Leaf physiognomic analysis of the dicot portion of the megaflora showed that the vegetation grew in a subtropical and relatively wet climate similar to modern-day Gulf Coast temperatures and precipitation, which agrees with the sedimentological interpretations of Roberts (2007). Though Campanian mega- and palynofloral assemblages in the Western Interior Basin provide a remarkable opportunity to greatly expand our knowledge of the evolution of Late Cretaceous vegetation, our interpretations of dinosaur ecology, and our understanding of greenhouse climates, an update to syntheses based on historic collections of fossil plants is sorely needed.

ACKNOWLEDGMENTS

This project was funded by NSF-EAR 0745495 to K. Johnson and I. Miller. We thank A. Titus, E. Roberts, S. Roberts, S. Sampson, M. Getty, M. Loewen, L. Miller, C. DeForest, S. and C. Russell, J. Downs, J. VandenBrooks, B. Allen, M. Gorman, and M. Poltenovage for help in the field. We thank the Leaf Whackers volunteer group at Denver Museum of Nature & Science for help preparing and curating the megafossil specimens. We also thank E. Roberts for stratigraphic advice and the loan of Lohrengel's study slides; A. Titus for help with permits, field advice, and hospitality in Kanab; S. Russell for help morphotyping the megafloral specimens and help with background literature; and the Utah Department of Transportation for access to the right of way on Highway 12 to collect fossils. Reviews by L. J. Hickey and an anonymous reviewer greatly improved the chapter.

REFERENCES CITED

Amiot, R., C. Lecuyer, E. Buffetaut, F. Fluteau, S. Legendre, and F. Martineau. 2004. Latitudinal temperature gradient during the Cretaceous upper Campanian–middle Maastrichtian: δ18O record of continental vertebrates. Earth and Planetary Science Letters 226:255–272.

Aulenback, K. R. 2009. Identification Guide to the Fossil Plants of the Horseshoe Canyon Formation of Drumheller, Alberta. University of Calgary Press, Calgary.

Barrera, E., and S. M. Savin. 1999. Evolution of late Campanian–Maastrichtian marine climates and oceans; pp. 245–282 in E. Barrera and C. C. Johnson (eds.), Evolution of the Cretaceous Ocean-Climate System. Geological Society of America Special Paper 332.

Berry, E. W. 1929. The Allison flora. Canada National Museum Bulletin 58:66–72.

Blonder, B., C. Violle, L. P. Bentley, and B. J. Enquist. 2011. Venation networks and the origin of the leaf economics spectrum. Ecology Letters 14:91–100.

Bowers, W. E. 1972. The Canaan Peak, Pine Hallow, and Wasatch Formations in the Table Cliff Region, Garfield County, Utah. U.S. Geological Survey Survey Bulletin 1331-B.

Boyce, C. K. 2009. Seeing the forest with the leaves – clues to canopy placement from leaf fossil size and venation characteristics. Geobiology 7:192–199.

Bramen, D. R., and E. B. Koppelhus. 2005. Campanian palynomorphs; pp. 101–131 in P. J. Currie and E. B. Koppelhus (eds.), Dinosaur Provincial Park: A Spectacular Ecosystem Revealed. Indiana University Press, Bloomington, Indiana.

Burnham, R. J., N. Pitman, K. R. Johnson, and P. Wilf. 2001. Habitat-related error in estimating temperatures from leaf margins in a humid tropical forest. American Journal of Botany 88:1096–1102.

Cobban, W. A., I. Walaszczyk, J. D. Obradovich, and K. C. McKinney. 2006. A USGS Zonal Table for the Upper Cretaceous Middle

Cenomanian–Maastrichtian of the Western Interior of the United States Based on Ammonites, Inoceramids, and Radiometric Ages. U.S. Geological Survey Open-File Report 2006–1250.

Crabtree, D. R. 1987a. Angiosperms of the northern Rocky Mountains: Albian to Campanian (Cretaceous) megafossil floras. Annals of the Missouri Botanical Garden 74:707–747.

Crabtree, D. R. 1987b. The early Campanian flora of the Two Medicine Formation, northcentral Montana. Ph.D. dissertation, University of Montana, Missoula, Montana.

Currano, E. D., P. Wilf, S. L. Wing, C. C. Labandeira, E. C. Lovelock, and D. L. Royer. 2008. Sharply increased insect herbivory during the Paleocene–Eocene thermal maximum. Proceedings of the National Academy of Sciences of the United States of America 105:1960–1964.

Davies-Vollum, K. S., Boucher, L. D., Hudson, P., and Proskurowski, A. Y. 2011. A Late Cretaceous coniferous woodland from the San Juan Basin, New Mexico. Palaios 26:89–98.

DeCelles, P. G. 2004. Late Jurassic to Eocene evolution of the Cordilleran thrust belt and foreland basin system, western U.S.A. American Journal of Science 304:105–168.

Donnadieu, Y., R. Pierrehumbert, R. Jacob, and F. Fluteau. 2006. Modeling the primary control of paleogeography on Cretaceous climate. Earth and Planetary Science Letters 248:426–437.

Eaton, J. G. 1991. Biostratigraphic framework for the Upper Cretaceous rocks of the Kaiparowits Plateau, southern Utah; pp. 47–63 in J. D. Nations, and J. G. Eaton (eds.), Stratigraphy, Depositional Environments, and Sedimentary Tectonics of the Western Margin, Cretaceous Western Interior Seaway. Geological Society of America Special Paper 260.

Ellis, B., D. C. Daly, L. J. Hickey, K. R. Johnson, J. D. Mitchell, P. Wilf, and S. L. Wing. 2009. Manual of Leaf Architecture. Cornell University Press, Ithaca, New York.

Falcon-Lang, H. J. 2003. Growth interruptions in silicified conifer woods from the Upper Cretaceous Two Medicine Formation, Montana, USA: implications for paleoclimate and dinosaur paleoecology. Palaeogeography, Palaeoclimatology, Palaeoecology 199:299–314.

Farabee, M. J. 1991. Palynology of the Upper Kaiparowits Formation (Upper Cretaceous, Campanian) in southcentral Utah [abstract]. Georgia Journal of Science 4:34.

Farabee, M. J., E. L. Vezey, and J. J. Skarvala. 1991. Systematics of the genus *Striatocorpus* (Krutzsch, 1970). Palynology 15:81–90.

Forster, A., S. Schouten, M. Baas, and J. S. S. Damste. 2007. Mid-Cretaceous (Albian–Santonian) sea surface temperature record of the tropical Atlantic Ocean. Geology 35:919–922.

Frakes, L. A. 1999. Estimating the global thermal state from Cretaceous sea surface and continental temperature data; pp. 49–58 in E. Barrera and C. C. Johnson (eds.), Evolution of the Cretaceous Ocean-Climate System. Geological Society of America Special Paper 332.

Frederiksen, N. O. 1988. Tectonic and paleogeografic setting of a new latest Cretaceous floristic province in North America. Palaios 2:533–542.

Fricke, H. C., B. Z. Foreman, and J. O. Sewall. 2010. Integrated climate model–oxygen isotope evidence for a North American monsoon during the Late Cretaceous. Earth and Planetary Science Letters 289:11–21.

Friis, E. M. 1988. *Spirematospermum chandlerae* sp. nov., an extinct species of Zingiberaceae from the North American Cretaceous. Tertiary Research 9:7–12.

Gradstein, F. M., J. G. Ogg, and A. G. Smith. 2004. A Geologic Time Scale. Cambridge University Press, Cambridge.

Greenwood, D. R. 2005. Leaf margin analysis: taphonomic constraints. Palaios 20:498–505.

Greenwood, D. R. 2007. Fossil angiosperm leaves and climate: from Wolfe and Dilcher to Burnham and Wilf. Courier Forschungsinstitut Senckenberg 258:95–108.

Herngreen, G. F. W, M Kedves, L. V. Rovnina, and S. B. Smirnova. 1996. Cretaceous palynofloral provinces: a review; pp. 1157–1188 in J. Jansonius and D. C. McGregor (eds.), Palynology: Principles and Applications, Volume 3. American Association of Stratigraphic Palynologists Foundation. Dallas, Texas.

Hickey, L. J., and J. A. Wolfe. 1975. The bases of angiosperm phylogeny: vegetative morphology. Annals of the Missouri Botanical Garden 62:538–589.

Hollick, A. 1930. The Upper Cretaceous Flora of Alaska. U.S. Geological Survey Professional Paper 159.

Housen, B. A., M. E. Beck Jr., R. F. Burmester, T. Fawcett, P. Petro, R. Sargent, K. Addis, K. Curtis, J. Ladd, N. Liner, B. Molitor, T. Montgomery, I. Mynatt, B. Palmer, D. Tucker, and I. White. 2003. Paleomagnetism of the Mount Stuart batholith revisited again: what has been learned since 1972? American Journal of Science 303:263–299.

Huber, B. T., and D. K. Watkins. 1992. Biogeography of Campanian–Maastrichtian calcareous plankton in the region of the Southern Ocean; pp. 31–60 in J. P. Kennett and D. A. Warnke (eds.), The Antarctic Paleoenvironment: A Perspective on Global Change, Volume 2. American Geophysical Union Antarctic Research 56.

Huber, B. T., R. D. Norris, and K. G. MacLeod. 2002. Deep-sea paleotemperature record of extreme warmth during the Cretaceous. Geology 30:123–126.

Jinnah, Z. A., E. M. Roberts, A. L. Deino, J. S. Larson, P. K. Link, and C. M. Fanning. 2009. New ^{40}Ar-^{39}Ar and detrital zircon U-Pb ages for the Upper Cretaceous Wahweap and Kaiparowits formations on the Kaiparowits Plateau, Utah: implications for regional correlation, provenance, and biostratigraphy. Cretaceous Research 30:287–299.

Johnson, K. R. 1989. High-resolution megafloral biostratigraphy spanning the Cretaceous–Tertiary boundary in the northern Great Plains. Ph.D. dissertation, Yale University, New Haven, Connecticut.

Johnson, K. R. 2002. Megaflora of the Hell Creek and lower Fort Union Formations in the western Dakotas: vegetational response to climate change, the Cretaceous–Tertiary boundary event, and rapid marine transgression; pp. 329–392 in J. H. Hartman, K. R. Johnson, and D. J. Nichols (eds.), The Hell Creek Formation and the Cretaceous–Tertiary Boundary in the North Great Plains: An Integrated Continental Record of the End of the Cretacous. Geological Society of America Special Paper 361.

Lidgard, S., and P. R. Crane. 1990. Angiosperm diversification and Cretaceous floristic trends: a comparison of palynomorphs and leaf macrofloras. Paleobiology 16:77–93.

Little, S. A., S. W. Kembel, and P. Wilf. 2010. Paleotemperature proxies from leaf fossils reinterpreted in light of evolutionary history. PloS One 5:e15161.

Lohrengel, C. F. 1969. Palynology of the Kaiparowits Formation, Garfield County, Utah. Brigham Young University Geology Studies 16:61–180.

Lupia, R., S. Lidgard, and P. R. Crane. 1999. Comparing palynological abundance and diversity: implications for biotic replacement during the Cretaceous angiosperm radiation. Paleobiology 25:305–340.

McClammer, J. U. J., and D. R. Crabtree. 1989. Post-Barremian (Early Cretaceous) to Paleocene paleobotanical collections in the Western Interior of North America. Review of Paleobotany and Palynology 57:221–232.

Miall, A. D., O. Catuneanu, B. K. Vakarelov, and R. Post. 2008. The Western Interior Basin; pp. 329–358 in A. D. Miall (ed.), The Sedimentary Basins of the United States and Canada. Elsevier, Amsterdam.

Miller, I. M., M. T. Brandon, and L. J. Hickey. 2006. Using leaf margin analysis to estimate the mid-Cretaceous (Albian) paleolatitude of the Baja BC block. Earth and Planetary Science Letters 245:95–114.

Miller, K. G., P. J. Sugarman, J. V. Browning, M. A. Kominz, J. C. Hernandez, R. K. Olsson, J. D. Wright, M. D. Feigenson, and W. V. Sickel. 2003. Late Cretaceous chronology of large, rapid sea-level changes: glacioeustasy during the greenhouse world. Geology 31:585–588.

Nichols, D. J. 1994. A revised palynostratigraphic zonation of the nonmarine Upper Cretaceous, Rocky Mountain region, United States; pp. 503–521 in M. V. Caputo, J. A. Peterson, and K. J. Franczyk (eds.), Mesozoic Systems of the Rocky Mountain region, U.S.A. Rocky Mountain Section of the Society for Sedimentary Geology.

Nichols, D. J. 1997. Palynology and Ages of Some Upper Cretaceous Formations in the Markagunt and Northwestern Kaiparowits Plateaus, Southwestern Utah. U.S. Geological Survey Bulletin 2153-E.

Nichols, D. J. 2002. Palynology and palynostratigraphy of the Hell Creek Formation in North Dakota–a microfossil record of plants at the end of Cretaceous time; pp. 391–454 in J. H. Hartman, K. R. Johnson, and D. J. Nichols (eds.), The Hell Creek Formation and the Cretaceous–Tertiary Boundary in the North Great Plains: An Integrated Continental Record of the End of the Cretacous. Geological Society of America Special Paper 361.

Nichols, D. J., and K. R. Johnson. 2002. Palynology and microstratigraphy of Cretaceous–Tertiary boundary sections in southwestern North Dakota; pp. 95–143 in J. H. Hartman, K. R. Johnson, and D. J. Nichols (eds.), The Hell Creek Formation and the Cretaceous–Tertiary Boundary in the North Great Plains: An Integrated Continental Record of the End of the Cretacous. Geological Society of America Special Paper 361.

Nichols, D. J., and A. R. Sweet. 1993. Biostratigraphy of Upper Cretaceous nonmarine palynofloras in a north–south transect of the Western Interior Basin; pp. 539–584 in W. G. E. Caldwell and E. G. Kauffman (eds.), Evolution of the Western Interior Basin. Geological Association of Canada Special Paper 39.

Norris, R. D., K. L. Bice, E. A. Magno, and P. A. Wilson. 2002. Jiggling the tropical thermostat in the Cretaceous hothouse. Geology 30:299–302.

Olsson, R. K., J. D. Wright, and K. G. Miller. 2001. Paleobiogeography of *Pseudotextularia elegans* during the latest Maastrichtian global warming event. Journal of Foraminiferal Research 31:275–282.

Miller et al.

Parrish, J. T., and R. A. Spicer. 1988. Late
Cretaceous terrestrial vegetation: a near-polar
temperature curve. Geology 16:22–25.

Pearson, P. N., P. W. Ditchfield, J. Singano, K. G.
Harcourt-Brown, C. J. Nicholas, R. K. Olsson,
N. J.Shackleton, and M. A. Hall. 2001. Warm
tropical sea surface temperatures in the
Late Cretaceous and Eocene epochs. Nature
413:481–487.

Peppe, D. J., D. L. Royer, B. Cariglino, S. Y. Oliver,
S. Newman, E. Leight, G. Enikolopov, M.
Fernandez-Burgos, F. Herrera, J. M. Adams,
E. Correa, E. D. Currano, J. M. Erickson, L. F.
Hinojosa, J. W. Hoganson, A. Iglesias, C. A.
Jaramillo, K. R. Johnson, G. J. Jordan, N. J. B.
Kraft, E. C. Lovelock, C. H. Lusk, U. Niinements,
J. Penuelas, G. Rapson, S. L. Wing, and I. J.
Wright. 2011. Sensitivity of leaf size and shape
to climate: global patterns and paleoclimatic
applications. New Phytologist 190:724–739.

Peterson, T. C., and R. S. Vose. 1997. An
overview of the Global Historical Climatology
Network temperature database. American
Meteorological Society Bulletin 78:2837–2849.

Price, G. D., and M. B. Hart. 2002. Isotopic
evidence for Early to mid-Cretaceous
ocean temperature variability. Marine
Micropaleontology 46:45–58.

Roberts, E. M. 2005. Stratigraphic, taphonomic
and paleoenvironmental analysis of the Upper
Cretaceous Kaiparowits Formation, Grand
Staircase–Escalante National Monument,
southern Utah. Ph.D. dissertation, University of
Utah, Salt Lake City, Utah.

Roberts, E. M. 2007. Facies architecture and
depositional environments of the Upper
Cretaceous Kaiparowits Formation, southern
Utah. Sedimentary Geology 197:207–233.

Roberts, E. M., and L. Tapanila. 2006. A new
social insect nest from the Upper Cretaceous
Kaiparowits Formation of southern Utah.
Journal of Paleontology 80:768–774.

Roberts, E. M., A. D. Deino, and M. A. Chan.
2005. ⁴⁰Ar/³⁹Ar age of the Kaiparowits
Formation, southern Utah, and correlation of
coeval strata and faunas along the margin of
the Western Interior Basin. Cretaceous Research
26:307–318.

Roberts, E. M., L. Tapanila, and B. Mijal. 2008.
Taphonomy and sedimentology of storm-
generated continental shell beds: a case
example from the Cretaceous Western Interior
Basin. Journal of Geology 116:462–479.

Roberts, L. N. R., and M. A. Kirschbaum. 1995.
Paleogeography of the Late Cretaceous of the
Western Interior of middle North America–coal
distribution and sediment accumulation.
U.S. Geological Survey Professional Paper
1561:1–65.

Rothwell, G. W., and R. A. Stockey. 1994. The
role of Hydropteris pinnata gen. et sp. nov. in
reconstructing the cladistics of heterosporous
ferns. American Journal of Botany 81:479–492.

Royer, D. L., L. Sack, P. Wilf, C. H. Lusk, G. J.
Jordan, U. Niinements, I. J. Wright, M. Westoby,
B. Cariglino, P. D. Cooley, A. D. Cutter, K. R.
Johnson, C. C. Labandeira, A. T. Moles, M. B.
Palmer, and F. Valladares. 2007. Fossil leaf
economics quantified: calibration, Eocene
case study, and implications. Paleobiology
33:574–589.

Sampson, S. D., M. A. Loewen, E. M. Roberts,
J. A. Smith, L. E. Zanno, and T. A. Gates. 2004.
Provincialism in Late Cretaceous terrestrial
faunas: new evidence from the Campanian
Kaiparowits Formation. Journal of Vertebrate
Paleontology 24:108A.

Schouten, S., E. C. Hopmans, A. Forster, Y. V.
Breugel, M. M. M. Kuypers, and J. S. S. Damste.
2003. Extremely high sea-surface temperatures
at low latitudes during the middle Cretaceous
as revealed by archaeal membrane lipids.
Geology 31:1069–1072.

Sewall, J. O., R. S. W. van de Wal, K. van der
Zwan, C. van Ooosterhout, H. A. Dijkstra, and
C. R. Scotese. 2007. Climate model boundary
conditions for four Cretaceous time slices.
Climate of the Past Discussions 3:791–810.

Stockey, R. A., and G. W. Rothwell. 2003.
Anatomically preserved Williamsonia
(Williamsoniaceae): evidence for Bennettitalean
reproduction in the Late Cretaceous of Western
North America. International Journal of Plant
Sciences 164:251–262.

Stockey, R. A., G. W. Rothwell, and K. R. Johnson.
2007. Cobbania corrugata gen. et comb. nov.
(Araceae): a floating aquatic monocot from the
Upper Cretaceous of western North America.
American Journal of Botany 94:609–624.

Titus, A. L., J. D. Powell, E. M. Roberts, S. D.
Sampson, S. L. Pollock, J. I. Kirkland, and
L. B. Albright. 2005. Late Cretaceous
stratigraphy, depositional environments,
and macrovertebrate paleontology of the
Kaiparowits Plateau, Grand Staircase–Escalante
National Monument, Utah; pp. 1–27 in J.
Pederson and C. M. Dehler (eds.), Interior
Western United States. Geological Society of
America Field Guide 6.

Torsvik, T. H., R. D. Muller, R. Van der Voo, B.
Steinberger, and C. Gaina. 2008. Global
plate motion frames: toward a unified model.
Reviews of Geophysics 46:RG3004, 1–44.

Tschudy, B. D., and E. B. Leopold. 1971.
Aquilapollenites (Rouse) Funkhouser–selected
Rocky Mountain taxa and their stratigraphic
ranges; pp. 113–167 in R. Kosanke and A. T.
Cross (eds.), Symposium on Palynology of the
Late Cretaceous and Early Tertiary. Geological
Society of America Special Paper 127.

Upchurch, G. R., and J. A. Wolfe. 1993.
Cretaceous vegetation of the Western Interior
and adjacent regions of North America;
pp. 243–281 in W. G. E. Caldwell and E. G.
Kauffman (eds.), Evolution of the Western
Interior Basin. Geological Association of
Canada Special Paper 39.

Van Boskirk, M. C. 1998. The flora of the
Eagle Formation and its significance for
Late Cretaceous floristic evolution. Ph.D.
dissertation, Yale University, New Haven,
Connecticut.

Watson, J. 1988. The Cheirolepidiaceae; pp.
382–447 in C. B. Beck (ed.), Origin and
Early Evolution of Gymnosperms. Columbia
University Press, New York.

Weishampel, D. B., P. M. Barrett, R. A. Coria,
J. Loeuff, X. Xu, X. Zhau, A. Shani, E. M. P.
Gomani, and C. R. Noto. 2004. Dinosaur
distribution; pp. 517–606 in D. B. Weishampel,
P. Dodson, and H. Osmólska (eds.), The
Dinosauria. University of California Press,
Berkeley, California.

Wilf, P. 1997. When are leaves good
thermometers? A new case for leaf margin
analysis. Paleobiology 23:373–390.

Wilf, P. 2008. Fossil angiosperm leaves:
paleobotany's difficult children prove
themselves. Paleontological Society Papers
14:319–333.

Wilf, P., and K. R. Johnson. 2004. Land plant
extinction at the end of the Cretaceous: a
quantitative analysis of the North Dakota
megaflora record. Paleobiology 30:347–368.

Wilf, P., K. R. Johnson, and B. T. Huber. 2003.
Correlated terrestrial and marine evidence for
global climate changes before mass extinction
at the Cretaceous–Paleogene boundary.
Proceedings of the National Academy of
Sciences of the United States of America
100:599–604.

Wilf, P., S. L. Wing, D. R. Greenwood, and
C. L. Greenwood. 1998. Using fossil leaves
as paleoprecipitation indicators: an Eocene
example. Geology 26:203–206.

Wilf, P., S. L. Wing, D. R. Greenwood, and
C. L. Greenwood. 1999. Using fossil leaves
as paleoprecipitation indicators: an Eocene
example: reply. Geology 27:92.

Wilson, P. A., R. D. Norris, and M. J. Cooper. 2002.
Testing the Cretaceous greenhouse hypothesis
using glassy foraminiferal calcite from the core
of the Turonian tropics on Demerara Rise.
Geology 30:607–610.

Wing, S. L., and L. D. Boucher. 1998. Ecological
aspects of the Cretaceous flowering plant
radiation. Annual Reviews in Earth and
Planetary Sciences 26:379–421.

Wing, S. L., and D. R. Greenwood. 1993.
Fossils and fossil climate: the case for
equable continental interiors in the Eocene.
Philosophical Transactions of the Royal Society
of London B 341:243–252.

Wing, S. L., L. J. Hickey, and C. C. Swisher.
1993. Implications of an exceptional fossil
flora for Late Cretaceous vegetation. Science
363:342–344.

Wing, S. L., G. J. Harrington, F. A. Smith, J. L.
Bloch, D. M. Boyer, and K. H. Freeman. 2005.
Transient floral change and rapid global
warming at the Paleocene–Eocene boundary.
Science 310:993–996.

Wing, S. L., C. A. Stromberg, L. J. Hickey, F.
Tiver, B. Willis, R. J. Burnham, and A. K.
Behrensmeyer. 2012. Floral and environmental
gradients on a Late Cretaceous landscape.
Ecological Monographs, 82(1):23–47.

Wolfe, J. A. 1971. Tertiary climatic fluctuations
and methods of analysis of Tertiary floras.
Palaeogeography, Palaeoclimatology,
Palaeoecology 9:27–57.

Wolfe, J. A. 1978. A paleobotanical interpretation
of Tertiary climates in the North Hemisphere.
American Scientist 66:694–703.

Wolfe, J. A. 1979. Temperature parameters
of humid to mesic forests of eastern Asia
and relation to forests of other regions of
the northern hemisphere and Australasia.
U.S. Geological Survey Professional Paper
1106:1–37.

Wolfe, J. A., and G. R. Upchurch. 1987. North
American nonmarine climates and vegetation
during the Late Cretaceous. Palaeogeography,
Palaeoclimatology, Palaeoecology 61:33–77.

Continental Invertebrates and Trace Fossils from the Campanian Kaiparowits Formation, Utah

8

Leif Tapanila and Eric M. Roberts

A SURVEY OF OVER 50 LOCALITIES FOR INVERTEBRATE fossils and their traces in the mudstones and sandstones of the Kaiparowits Formation, which spans 1.8 million years of Late Cretaceous (Campanian) time, demonstrates that it is one of the most prolific units in the Western Interior. Pulmonates, caenogastropods, and freshwater bivalves dominate the invertebrate fossil record both in number and diversity, and these are accompanied by ostracodes and a unique occurrence of bryozoan. Trace fossils such as the type specimens of *Socialites* nests and *Osteocallis* bone scrapings strongly suggest the activity of insects despite their absence in the body fossil record. At least 35 different aquatic and terrestrial gastropod and 13 freshwater bivalve morphotypes described from the formation support other independent evidence that the Campanian of southern Utah had a warm, humid climate with perennial aquatic environments. The highest-quality autochthonous preservation, including primary aragonite shell, occurs in facies interpreted as overbank ponds, lakes, and marshes. The greatest local diversity of 20 aquatic mollusk and one brackish-water bryozoan taxa is reported from a thick shell conglomerate that is interpreted as a mass-mortality storm deposit. Maximum diversity of mollusks in the middle of the formation appears to coincide with an increase in facies recording wetter fluvial environments, but future collecting efforts are required to isolate this relationship from taphonomic or sampling biases that may contribute to this preliminary signal.

INTRODUCTION

The Kaiparowits Formation in southern Utah (Fig. 8.1) is a major center for new discoveries in Cretaceous vertebrate paleontology, driven largely by prospecting and excavation efforts for archosaurs (Sampson et al., 2004; Wiersma, Hutchison, and Gates, 2004; Titus et al., 2005; Zanno and Sampson, 2006; Gates and Sampson, 2007). Although they have received scant attention by comparison, the invertebrate and trace fossil records are also proving to be exceptionally abundant, well preserved, and diverse throughout the formation. Gregory and Moore (1931), and later Gregory (1950, 1951) recorded a total of 11 gastropod and six bivalve taxa from the

Kaiparowits Formation as a basis for biostratigraphic comparisons, and they concluded that the Kaiparowits Formation molluscan fauna resembled that of the Fruitland Formation of New Mexico in both age and general composition. Nearly 70 years later, the partial temporal correlation of the Fruitland and Kaiparowits formations was independently confirmed by radiometric dating of bentonites by Fassett and Steiner (1997) and by Roberts, Deino, and Chan (2005) and Roberts et al. (this volume, Chapter 6), respectively. The focus of our research is describing the total preserved biodiversity of the Kaiparowits Formation and identifying the interplay between environment and taphonomy in the preservation of continental fossils. This chapter furthers this goal by providing a summary of known invertebrate and trace fossil diversity, along with the first systematic descriptions of continental mollusks for the Kaiparowits Formation.

GEOLOGIC SETTING

The Kaiparowits Formation (Fig. 8.2) represents the thickest unit (860 m) of a >2000-m-thick package of Upper Cretaceous strata deposited within the southern Cordilleran foreland basin, as a result of the combined influences of loading to the west and southwest in the Sevier Fold and Thrust Belt and Mogollon Highlands, respectively. Sediment of the Kaiparowits Formation was shed off each of these uplifts and deposited within the foredeep depocenter as the Western Interior Seaway retreated from Utah. A wet, humid continental setting is inferred (Roberts et al., this volume, Chapter 6), dominated by large, slow-moving channel systems with stable banks and abundant overbank pond and wetland environments (Fig. 8.3). The Kaiparowits Formation is subdivided into three informal members, termed the lower, middle, and upper units, which reflect major changes in alluvial architecture through the succession. The lower and upper units represent channel-dominated successions, whereas the middle unit is dominated by overbank pond and lake mudstones. Maximum transgression of the Bearpaw Sea into Utah is recorded within the lower part of the middle unit, and recognized by an increase in tidally influenced facies, brackish-marine trace fossils (e.g., *Teredolites*), and

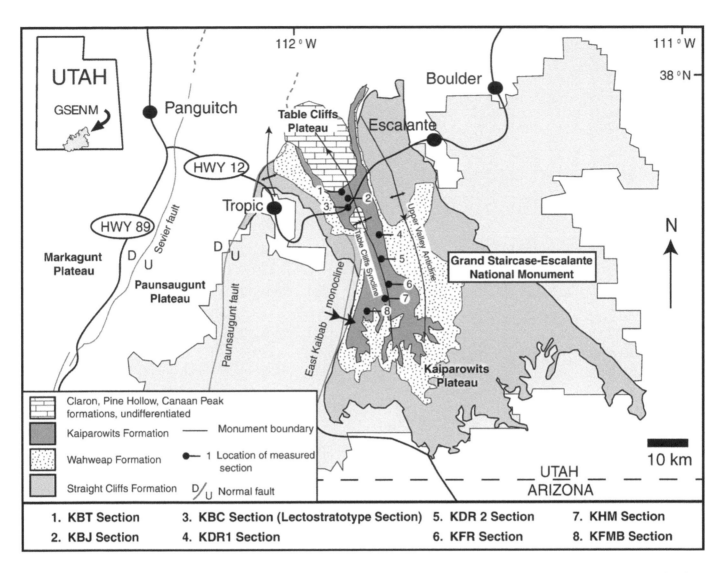

8.1. Map of Kaiparowits Formation outcrops and measured sections within the Grand Staircase–Escalante National Monument, southern Utah. Modified from Roberts (2007).

a unique occurrence of the brackish-water bryozoan *Conopeum* (Roberts, 2007; Roberts, Tapanila, and Mijal, 2008). Above this interval (~300-m level), no tidally influenced facies or brackish-water fauna are known in the formation. The Kaiparowits Formation contains a wealth of vertebrate fossils, plant macrofossils, and invertebrate body and trace fossils. Over one dozen ash beds have also been identified in the Kaiparowits Formation, five of which have been dated radiometrically, providing a high-resolution temporal (Late Campanian) framework (Fig. 8.2).

METHODS

Field investigation of the sedimentology and invertebrate paleontology of the Kaiparowits Formation was initiated in the summer of 2001. Collecting continued on an annual basis through 2009. Sampling protocol is variable among

localities, from census and surface collecting to grab samples. For this reason, the total diversity at any particular locality (alpha diversity) may be subject to sampling bias. In 2009, census collection of freshly exposed outcrop has improved the quality and thoroughness of collecting efforts, though we regard most localities to be undersampled, especially for thin-shelled and small taxa. Fifty-three mollusk localities span most of the 856-m Kaiparowits Formation, although the majority is concentrated in the mudstone-rich middle unit (Fig. 8.2). Because of the volatile nature of systematic taxonomy among continental mollusks (e.g., polyphyly; Jablonski, Finarelli, and Roy, 2006), a conservative approach is used here. Fifty-two unique taxa of gastropod and bivalve are recognized in this chapter, and each is figured. Presence/ absence data for each taxon is provided per locality.

Institutional Abbreviations IMNH, John A. White Repository at Idaho Museum of Natural History, Pocatello,

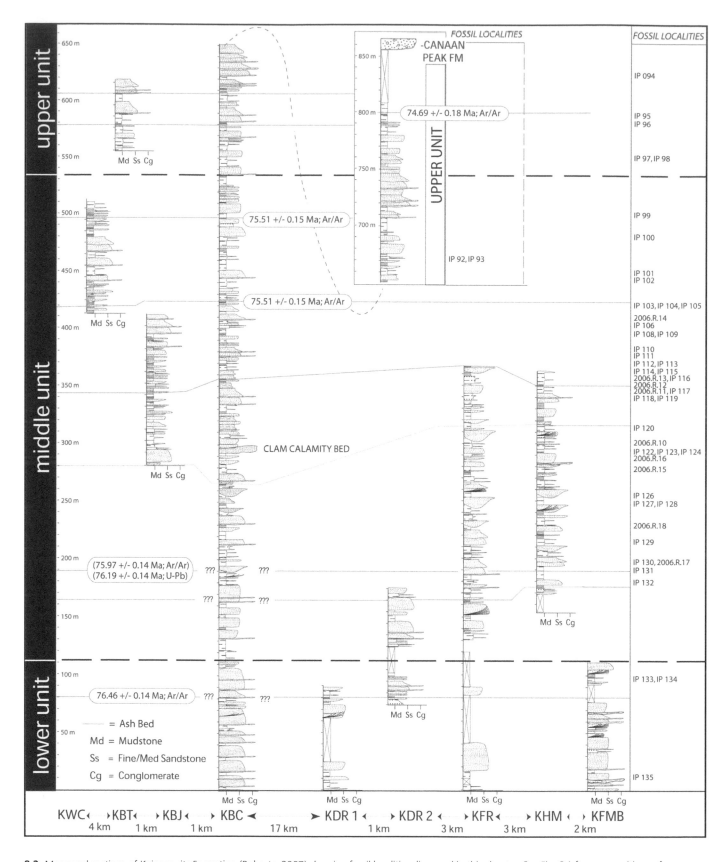

8.2. Measured sections of Kaiparowits Formation (Roberts, 2007) showing fossil localities discussed in this chapter. See Fig. 8.1 for map positions of measured sections.

Tapanila and Roberts

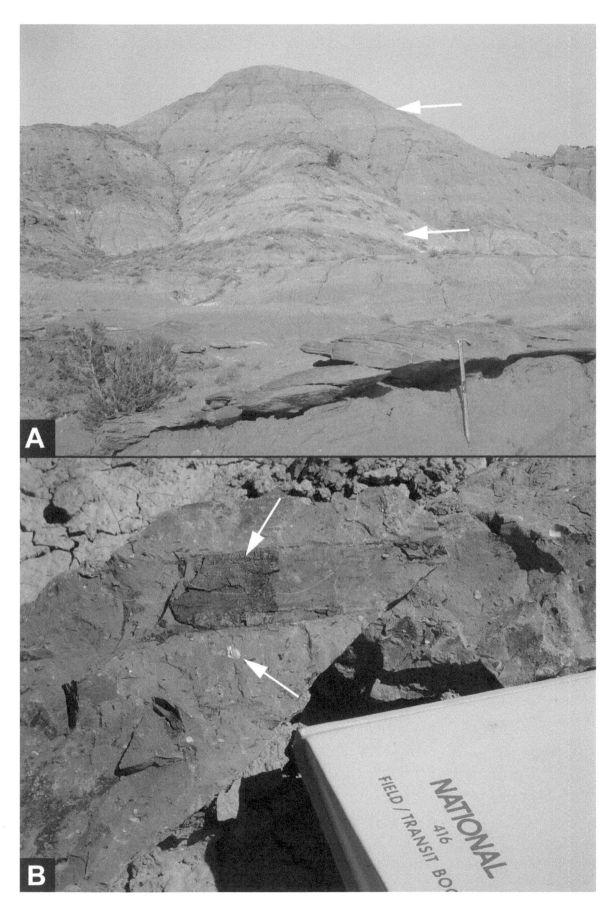

8.3. Typical mollusk-dominated wetland facies. (A) A 25-m-thick interval of overbank fines (between the two arrows); (B) close-up view of paludal mudstone with coalified wood (top arrow) and mollusk shell fragments (bottom arrow).

Idaho; UMNH, Utah Museum of Natural History, Salt Lake City, Utah.

FOSSIL INVERTEBRATES

Gastropods are the most diverse of the Kaiparowits invertebrates, from the aquatic gill-breathing prosobranchs to the lung-breathing pulmonates of primarily aquatic (basommatophoran) and terrestrial (stylommatophoran) affinity. Thirty-six taxa are recognized (Figs. 8.4, 8.5). Freshwater bivalves (Fig. 8.6) are assigned herein to 16 taxa (or morphotypes), which likely overestimates the number of species, owing to the convergent morphologies that unrelated taxa adopt when exposed to similar environmental conditions (Ortmann, 1920; Watters, 1994). Table 8.1 summarizes, by locality, the presence or absence of each molluscan taxon described below.

Class GASTROPODA Cuvier, 1797
Subclass PULMONATA Cuvier, 1817
Order STYLOMMATOPHORA Schmidt, 1855
Family DISCIDAE (Thiele, 1931)
Genus *DISCUS* Fitzinger, 1833
DISCUS sp.

Description Low spired to planispiral shell with three or more whorls (Fig. 8.5X). Specimens flattened, with thin shells and ornamented with very fine transverse, curved ridges. Aperture obscured. Measurements (mm): width 22.

Family HELICIDAE Rafinesque, 1815
Genus "*HELIX*" Linnaeus, 1758
"*HELIX*" sp.

Description Heliciform shell of medium size and thickness with prominent growth lines (Fig. 8.5U,V). Five whorls are acutely convex with a subtle suture, especially toward apex. Aperture does not extend below last whorl. Measurements (mm): height 11, width 15, aperture height 6.

Family POLYGYRIDAE Pilsbry, 1895
Genus *POLYGYRA* Say, 1818
POLYGYRA sp.

Description Small- to medium-size heliciform shell. Shell is thin and ornamented by very fine, low-relief spiral lines (Fig. 8.5S,T). Four to five whorls that are highly convex. First two to three whorls have smooth suture, which becomes distinct toward aperture. Final volution increases rapidly in size, and aperture is wide and extends below final whorl. Measurements (mm): height 7.5, width 14, aperture height 6.5.

Family INDETERMINATE
Stylommatophoran sp. 1

Description Heliciform shell with slightly raised spire. Convex whorls number five; the first two are smooth, becoming ornamented with fine transverse, curved ridges (Fig. 8.5Y). Suture is distinct. Base of last whorl is flat, but aperture of specimen is obscured. Measurements (mm): height 2, width 6.

Stylommatophoran sp. 2

Description Small, thin, smooth shell with low spire shape. Four whorls are convex with strong angulation at base (Fig. 8.5Q,R). Sutures are slightly indented. Aperture is wide, nearly rhombic, and forms flat bottom surface of final whorl. Measurements (mm): height 4.5, width 7, aperture height 2.5.

Order BASOMMATOPHORA Schmidt, 1855
Family ACROLOXIDAE Thiele, 1931
Genus *ACROLOXUS* Beck, 1838
ACROLOXUS sp.

Description Dextral, patelliform, low-relief shell. Aperture nearly circular (Fig. 8.4T). Apex forms curved, acutely conical shape, but shell apex not preserved well enough to determine whether dimple is present or absent. Radial plications and fine growth bands ornament thin shell. Measurements (mm): aperture dimensions 4.9 × 4.8, shell height 1.4.

Family PLANORBIDAE Rafinesque, 1815
Genus *PLANORBIS* Müller, 1774
PLANORBIS (BATHYOMPHALUS)
CHACOENSIS Stanton, 1917

Description Planispiral shell with four whorls. Umbilicus perforate (Fig. 8.5M,N). Aperture deeply curved at base and flat on top. Fine growth lines ornament the thin shell. Measurements (mm): height 3.7, width 7.4, aperture height 3.7.

Genus *BULINUS* Müller, 1781
BULINUS? SUBELONGATUS Meek
and Hayden in Meek, 1876

Description Sinistral, olive-shaped shell with elevated spire (Fig. 8.4Q). Seven whorls are weakly convex and have slight sutures. Shell is thin and contains fine growth lines. Aperture elongate terminates at base of last whorl. Measurements (mm): height 18, width 8, aperture height 11.

Table 8.1. Presence or absence of molluscan taxon by locality

Locality	Section	Height (m)	Facies	Discus sp.	Helix'	Polygyra	Stylomma. sp. 1	Stylomma. sp. 2	Acroloxus	Planorbis (B.) chacoensis	Physa copei	Physa globosa	Physa felixi	Bulinus? subelongatus	Neritina baueri	Neritina cf. N. baptista	V. cf V. paludinaeformis	Viviparus sp. 1	Viviparus sp. 2	Viviparus sp. 3	Viviparus sp. 4	Viviparus sp. 5	Viviparus sp. 6	Viviparus sp. 7	Viviparus retusus	Tulotomops thompsoni	Campeloma sp.	Lioplacodes subtortuosa	Lioplacodes nebrascensis	Lioplacodes sp. 1	Lioplacodes sp. 2	Lioplacodes sp. 3	Caenogastropoda sp. 1	Caenogastropoda sp. 2	Caenogastropoda sp. 3	Caenogastropoda sp. 4	Ampullariid sp. 1	Ampullariid sp. 2	Bivalve sp. 1	Bivalve sp. 2	Bivalve sp. 3	Plesielliptio sp. 1	Plesielliptio sp. 2	Plesielliptio sp. 3	Proparreysia sp. 1	Proparreysia sp. 2	Bivalve sp. 4	Anodonta	Bivalve sp. 5	Bivalve sp. 6	"Unio" reesidei	Bivalve sp. 7	Sphaerium	Bivalve sp. 8	wood	nacre	bone	coprolite	eggshell	operculum	SUM
IP-092	KBC	668	8				1					1				1														1	1	1																															5
IP-093	KBC	668	8												1					1																					1																						3
IP-094	KBC	600-650								1			1																																		1															3
IP-095	KBT	590	3.1									1								1											1	1																										1	1		1			5
IP-096	KBT	588	8																											1	1	1																									1	1	1		1		4
IP-097	KBT	550	8										1																	1	1	1	1																									1					5
IP-098	KBT	550	8																											1		1	1																									1					3
IP-099	KBC	499	8		1															1																																											1
IP-100	KBC	480																												1																																1
IP-101	KWC	450	3.1																	1									1	1								1																				1				4
IP-102	KBC	445				1												1	1												1											1										1											6
IP-103	KBC	420	8																	1		1																					1	1	1																	5
IP-104	KBC	420	8							1			1																																								1								1		3
IP-105	BICS	420	9					1																																																						1
2006.R.14	KBC	410										1							1																				1	1	1		1		1						1						1	1					8
IP-106	BICS	404	8						1				1			1														1								1	1															1									7
IP-107	S.Can	400-700																																					1		1						1														3
IP-108	KBC	400	8																																								1	1																		2
IP-109	KBC	400	8										1			1															1																															3
IP-110	KBC	385	8	1												1									1						1																															3
IP-111	BICS	380	8						1				1			1	1																																													1	4
IP-112	BICS	370	9				1					1							1							1	1		1	1					1													1											1	1			8
IP-113	BICS	370	8				1			1	1																	1	1																							1					1	1		1	6
IP-114	KBJ	365-385	4.1																																																										0
IP-115	KBJ	365-385	4.1									1																		1																											1	1			4
2006.R.13	KBC	360																		1		1	1																	1																				4
IP-116	BICS	360	9			1							1									1		1																																					1	4
2006.R.12	KBC	355																1	1	1	1																				1																				5
2006.R.11	KBC	345	8											1						1	1		1	1								1																					1			1					7
IP-117	BICS	343	8	1	1							1		1	1	1												1			1																					1							1	1	7

Family PHYSIDAE Fitzinger, 1833
Genus *PHYSA* Draparnaud, 1801
PHYSA COPEI White, 1877

Description Sinistral, oblong shell with short conical spire of five whorls (Fig. 8.4P). Medium- to large-size shell is thin with fine growth lines. Final whorl is two thirds of shell height. Aperture elongate terminates at base of last whorl. Measurements (mm): height 33, width 18, aperture height 23.

PHYSA GLOBOSA Yen, 1954

Description Very small, sinistral, bulbous shell with elevated, scalariform spire (Fig. 8.4S). Fine growth lines ornament the thin shell. Five whorls are slightly convex. Aperture is obscured in specimen. Measurements (mm): height 3.5, width 2.5, aperture height 2.

PHYSA FELIXI White, 1880

Description Very large, sinistral, semispherical shell with flat to paucispiral apex (Fig. 8.4O). Shell thin and ornamented with fine growth lines and crenulations near aperture. Four whorls terminate with elongate aperture. Measurements (mm): height 50, width 40, aperture height 47.

PHYSA sp.

Description Sinistral, bulbous, paucispiral shell with three whorls (Fig. 8.4R). Shell is thin with fine growth lines and crenulated near apex. Elongate aperture extends below final whorl. Measurements (mm): height 17, width 12, aperture height 14.5.

Subclass CAENOGASTROPODA Cox, 1959
Family NERITIDAE Rafinesque, 1815
Genus *NERITINA* Rafinesque, 1815
NERITINA BAUERI Stanton, 1917

Description Large, ovoid shell with flat spire (Fig. 8.5Z,AA). Two to three volutions rapidly expand to form large elongate, oval aperture. Spiral lines faintly impressed on outer whorl of steinkern specimen. Measurements (mm): height 26, width 25, aperture height 25.

NERITINA cf. *N. BAPTISTA* White, 1883

Description Small, dextral, cap-shaped shell (Fig. 8.4U,V). Two to three whorls form flat spiral that enlarges rapidly to oval aperture. Apex lies close to aperture and is not prominent. Concentric growth bands ornament thin shell. Measurements (mm): aperture dimensions 7.2 × 5.0, shell height 4.4.

Family VIVIPARIDAE Gray, 1847
Genus *VIVIPARUS* Montfort, 1810
VIVIPARUS cf. *V. PALUDINAEFORMIS* Hall, 1845

Description Shell of medium size and thin (Fig. 8.4K). Whorls count five and are weakly convex, with a curved lower margin and slight indentation of suture. Spire has higher angle compared to *Viviparus* sp. 2. Fine growth lines on whorl are parallel to shell axis. Measurements (mm): height 27, width 18, aperture height 15.

VIVIPARUS RETUSUS Meek and Hayden, 1856

Description Subconical, medium- to small-size shell and thin (Fig. 8.4A,B). Low-angle spire with weakly convex whorls and rounded lower margin forms slightly impressed suture. Fine growth lines. Measurements (mm): height 15, width 15, aperture height 7.

VIVIPARUS sp. 1

Description Bulbous shell is very large (Fig. 8.4G,H). Whorls count four or five and are weakly convex, with a curved lower margin giving a suture having a slight indentation. Fine growth lines on shell are parallel to the axis. Specimen is flattened. Measurements (mm): height 50, width 38, aperture height 27.

VIVIPARUS sp. 2

Description Shell of medium size and thin (Fig. 8.4C,D). Short, wide angle spire. Five whorls are flat to very weakly convex, with an angular lower margin that produces a slightly indented suture. Aperture extends below base of final whorl. Shorter and flatter than *Viviparus* cf. *V. paludinaeformis*. Measurements (mm): height 15, width 18, aperture height 7.

VIVIPARUS sp. 3

Description Shell of medium size, calcitized (Fig. 8.4J). Moderate spire angle, between *Viviparus* sp. 2 and *V.* cf. *V. paludinaeformis*. Five whorls are moderately convex, with curved lower margin that produces an indented suture. Faint growth lines. Measurements (mm): height 19, width 18, aperture height 10.

VIVIPARUS sp. 4

Description Shell of medium size, calcitized (Fig. 8.4I). High-angle spire with seven whorls that are moderately convex with indented suture. Similar to *Viviparus* sp. 3. Specimen incomplete. Measurements (mm): height 19, width 16, height of second-to-last whorl 12.

VIVIPARUS sp. 5

Description Medium shell, preserved as steinkern (Fig. 8.4E,F). Wide spire angle forms four or five whorls that are scalariform. Whorl is convex at top and flattens below to form a strong angle at base. Suture not preserved. Impression of growth lines on last whorl. Measurements (mm): height 19, width 21, aperture height 11.

VIVIPARUS sp. 6

Description Shell large and thin (Fig. 8.4M). Whorls numbering five form high spire. Upper and lower part of whorl form sharp angle with flat middle section. Aperture large and extends below base of last whorl. Parallel growth lines angle away from base of whorl. Measurements (mm): height 33, width 33, aperture height 18.

VIVIPARUS sp. 7

Description Small shell of four whorls, preserved as steinkern (Fig. 8.4L). Moderate spire angle of flat to slightly convex whorls form modest angle at base. Suture not preserved. Aperture wide and extends below lowest whorl. Measurements (mm): height 15, width 13, aperture height 8.

Tapanila and Roberts

8.4. Gastropods from the Kaiparowits Formation, southern Utah. (A,B) *Viviparus retusus;* (A) oblique apical view; (B) ventral view (IMNH IP 118/1); (C,D) *Viviparus* sp. 2; (C) apical view; (D) ventral view (IMNH IP 111/1); (E,F) *Viviparus* sp. 5; (E) apical view (IMNH 2006.R.16/21); (F) ventral view (IMNH 2006.R.16/11); (G,H) *Viviparus* sp. 1; (G) dorsal view; (H) ventral view (IMNH 2006.R.17/1); (I) *Viviparus* sp. 4, ventral view (IMNH IP 131/1); (J) *Viviparus* sp. 3, dorsal view (IMNH IP 93/2); (K) *Viviparus paludinaeformis,* ventral view (IMNH IP 115/1); (L) *Viviparus* sp. 7, ventral view (IMNH 2006.R.16/12); (M) *Viviparus* sp. 6, dorsal view (IMNH 2006.R.16/13); (N) *Tulotomops thompsoni,* dorsal view (IMNH IP 123/3); (O) *Physa felixi,* Dorsal view (IMNH IP 116/2); (P) *Physa copei,* dorsal view (IMNH IP 113/1); (Q) *Bulinus? subelongatus,* dorsal view (IMNH 2006.R.11/1); (R) *Physa* sp., ventral view (IMNH IP 112/3); (S) *Physa globosa,* dorsal view (IMNH IP 92/4); (T) *Acroloxus,* apical view (IMNH IP 105/1); (U,V) *Neritina* cf. *N. baptista;* (U) lateral view; and (V) apical view (IMNH 2006.R.16/14).

Genus TULOTOMOPS Wenz, 1939
TULOTOMOPS THOMPSONI (White, 1876)

Description Shell large and very thick forms conical spire (Fig. 8.4N). Five whorls are flat to slightly convex with a strongly angled base to form a distinct suture. Shell is ornamented by two spiral bands and knobs that increase in prominence on lower whorls. Measurements (mm): height 34, width 34, aperture height 14.

Genus CAMPELOMA Rafinesque, 1819
CAMPELOMA sp.

Description Small- to medium-size, globose shell with moderately high spire (Fig. 8.5J). Specimen has thick calcitized shell with sinuous growth lines. Six to seven whorls are convex at base and flatten near top to form rounded step and pronounced suture. Lower whorls proportionally larger than whorls of the spire. Measurements (mm): height 19, width 11, aperture height 10.

Genus LIOPLACODES Meek, 1864
LIOPLACODES SUBTORTUOSA
Meek and Hayden in Meek, 1876

Description High spire, carinate shell of medium size and seven whorls (Fig. 8.5G,L). Shell thick with sinuous growth lines. Carination located low on uppermost whorls and migrates to a central position on lowest whorls. Suture is pronounced. Aperture elongate with round base and sinuous peripheral margin that extends below final whorl. Measurements (mm): height 23, width 15, aperture height 15.

LIOPLACODES NEBRASCENSIS
Meek and Hayden in Meek, 1876

Description Globular, high-spire shell of medium size (Fig. 8.5H,I). Shell is moderately thick with prominent, sinuous growth lines. Seven whorls are moderately convex and form distinct suture, but less so than in *Campeloma* sp. Oval aperture extends below final whorl and has sinuous peripheral margin. Measurements (mm): height 18, width 11, aperture height 10.

LIOPLACODES sp. 1

Description Medium spire shell with rounded conical shape (Fig. 8.5K). Seven convex whorls form distinct suture. Thin shell has sinuous growth lines and pair of spiral lines. Measurements (mm): height 15, width 6, aperture height 5.

LIOPLACODES sp. 2

Description Small shell with moderate spire and bulbous conical shape (Fig. 8.5F). Five whorls are convex with indented suture. Thin shell has sinuous aperture and pair of spiral lines. Measurements (mm): height 5, width 4, aperture height 3.

LIOPLACODES sp. 3

Description Small shell with moderate spire and bulbous conical shape (Fig. 8.5E). Six whorls are convex with slight suture. Thin shell has pair of spiral lines. Shell is proportionally larger in final whorl than *Lioplacodes* sp. 2. Measurements (mm): height 9, width 6.5, aperture height 5.

Family AMPULLARIIDAE Gray, 1824
Ampullariid indet. sp. 1

Description Ovoid, paucispiral shell of five whorls with smooth surface (Fig. 8.5W). Aperture elongate with convex periphery that does not extend below last whorl. Thin shell is tan in color. Measurements (mm): height 9, width 9, aperture height 5.5.

Ampullariid indet. sp. 2

Description Small, ovoid, paucispiral shell with five whorls (Fig. 8.5O,P). Thin shell has convex whorls, with the final whorl much wider than previous. Large oval aperture extends up to the shoulder of the previous whorl, and extends below the final whorl. Measurements (mm): height 7.5, width 9.5, aperture height 5.

8.5. Gastropods and trace fossils from the Kaiparowits Formation, southern Utah. (A) Caenogastropoda indet. sp. 3, dorsal view (IMNH IP 102/1); (B) Caenogastropoda indet. sp. 4, ventral view (IMNH IP 133/1); (C) Caenogastropoda indet. sp. 1, ventral view (IMNH IP 117/2); (D) Caenogastropoda indet. sp. 2, ventral view (IMNH IP 112/1); (E) *Lioplacodes* sp. 3, dorsal view (IMNH IP 92/2); (F) *Lioplacodes* sp. 2, ventral view (IMNH IP 110/1); (G) *Lioplacodes subtortuosa*, ventral view (IMNH IP 123/2); (H,I) *Lioplacodes nebrascensis;* (H) ventral; (I) dorsal view (IMNH IP 97/1); (J) *Campeloma* sp., dorsal view, last whorl broken (IMNH IP 96/1); (K) *Lioplacodes* sp. 1, ventral view (IMNH IP 92/3); (L) *Lioplacodes subtortuosa,* dorsal view (IMNH IP 112/2); (M,N) *Planorbis (B.) chacoensis;* (M) ventral and (N) apical view (IMNH IP 106/4); (O,P) Ampullariid sp. 2; (O) ventral view (IMNH IP 106/2); (P) apical view (IMNH IP 106/3); (Q,R) Stylommatophora sp. 2; (Q) ventral view; (R) apical view (IMNH IP 92/1); (S,T) *Polygyra;* (S) ventral view; (T) apical view (IMNH IP 117/1); (U,V) *"Helix;"* (U) ventral view; (V) apical view (IMNH IP 123/1); (W) Ampullariid sp. 1, ventral view (IMNH IP 106/1); (X) *Discus* sp., apical view (IMNH IP 119/1); (Y) Stylommatophora sp. 1, apical view (IMNH IP 116/1); (Z,AA) *Neritina baueri;* (Z) dorsal view; (AA) ventral view (IMNH IP 93/1); (BB) *Teredolites* casts (IMNH IP 114/1); (CC) vertical burrow cast (IMNH 2006.R.15/1).

A-D, F, M-T, W, Y

3 mm

E, G-L, U-V

X, Z, AA

10 mm

BB-CC

8.6. Bivalves from the Kaiparowits Formation, southern Utah. (A,B) *Plesielliptio* sp. 1; (A) left external valve (IMNH 2006.R.16.1); (B) right side of steinkern (IMNH 2006.R.16.2); (C,D) bivalve indet. sp. 1; (C) left external valve (IMNH 2006.R.16.4); (D) right side of steinkern (IMNH 2006.R.16/15); (E,F) bivalve indet. sp. 2; (E) left external valve (IMNH 2006.R.16/16); (F) left side of steinkern (IMNH 2006.R.16/17); (G,H) *Anodonta;* (G) right external valve of juvenile (IMNH 2006.R.16.9); (H) right side of steinkern (IMNH 2006.R.16.8); (I,J) *Plesielliptio* sp. 2; (I) right external valve; with (J) dorsal view of ornament (IMNH 2006.R.16/18); (K,L) bivalve indet. sp. 3; (K) left side of steinkern (IMNH 2006.R.16.5); (L) right external valve (IMNH 2006.R.16.6); (M) bivalve indet. sp. 6, Left external valve (IMNH IP 107/1); (N) bivalve indet. sp. 5, left external valve (IMNH IP 125/1); (O) *Proparreysia* sp. 1, right external valve (IMNH 2006.R.16/19); (P) *Proparreysia* sp. 2, right external valve (IMNH IP 132/1); (Q) bivalve indet. sp. 4, left side of steinkern (IMNH IP 94/1); (R) *"Unio" reesidei,* right external valve (IMNH IP 135/1); (S) *Plesielliptio* sp. 3, right external valve (IMNH IP 140/1, float specimen); (T) *Sphaerium,* right external valve (IMNH IP 117/3); (U) bivalve indet. sp. 8, left external valve (IMNH IP 134/1); (V) bivalve indet. sp. 7, external impression of very large right valve (IMNH IP 124/1).

FAMILY INDETERMINATE
Caenogastropoda indet. sp. 1

Description Small, thin shell with high spire (Fig. 8.5C). Whorls count five and are highly convex, giving shell a rounded outline. Suture is moderately depressed. Elongate aperture extends below base of last whorl. Measurements (mm): height 8, width 5, aperture height 3.5.

Caenogastropoda indet. sp. 2

Description Small, high-spire shell with seven whorls that are weakly convex. Specimen has thin calcitized shell and moderate suture (Fig. 8.5D). Aperture elongate and extends below final whorl. Measurements (mm): height 10.5, width 5, aperture height 4.

Caenogastropoda indet. sp. 3

Description Very small high spire, conical shell with seven whorls (Fig. 8.5A). Slight convexity of whorls produces minor suture. Thin, delicate shell. Measurements (mm): height 4, width 2, height of last whorl 1.5.

Caenogastropoda indet. sp. 4

Description Very high-spire shell of five whorls that are slightly convex and produce moderate suture (Fig. 8.5B). Specimen has calcified shell. Aperture elongate and short relative to final whorl. Measurements (mm): height 8.5, width 4, aperture height 3.5.

Class BIVALVIA Linnaeus, 1758
Genus ANODONTA Lamarck, 1799
ANODONTA sp.

Description Steinkern preserved only (Fig. 8.6G,H). No notch for dentition. Rectangular to elongate oval outline with umbones at midline. Minor concavity of ventral margin. Measurements of steinkern (mm): length 66.4, height 30.3, half width 12.1.

Genus PLESIELLIPTIO Russell, 1934
PLESIELLIPTIO sp. 1

Description Very large, lanceolate form with thick shell (Fig. 8.6A,B). Deep concave groove from umbo to posteroventral end. Umbones located near anterior end, similar to Bivalve indet. sp. 2. Concave ventral margin. Plicate ornament on anterodorsal surface. Steinkern has deep recess for dentition, forms hook shape at anterior end. Wedge-shaped umbo is at 50° angle relative to the ventral pallial line. Measurements (mm): length 126.7, height 54.8, width 38.1. Resembles *Plesielliptio* type 1 of Roberts, Tapanila, and Mijal (2008).

PLESIELLIPTIO sp. 2

Description Small, circular to oval form with moderate to thin shell (Fig. 8.6I,J). Convex ventral surface. Umbo projects above dorsal margin and located just anterior of midline. Two lines extend posteroventrally from ringed umbo. Steinkern has wedge-shaped umbo that has 60° angle relative to the ventral pallial line. Measurements (mm): length 31.1, height 28.3, width 9.

PLESIELLIPTIO sp. 3

Description Small size, thin shell, with elongate oval to rectangular outline (Fig. 8.6S). Slight umbo positioned closer to midline than to anterior end. Two lines extend ventroposteriorly from ringed umbo. Measurements (mm): length 39.8, height 18, half width 3.2.

Genus PROPARREYSIA Pilsbry, 1921
PROPARREYSIA sp. 1

Description Triangular shell with umbones at midline (Fig. 8.6O). Convex ventral margin. Chevron ornamentation extends ventrally from umbo, and strong angulation at posterior end. Narrower in form compared to *Proparreysia* sp. 2. Measurements (mm): length 39.4, height 36.1, width 21.5.

PROPARREYSIA sp. 2

Description Triangular shell with umbones at midline (Fig. 8.6P). Convex ventral margin. Chevron ornamentation extends ventrally from umbo and becomes concentric banding near ventral margin. Fold at posterior end. Wider than *Proparreysia* sp. 1. Measurements (mm): length 42.5, height 33.3, half width 13.3.

Genus SPHAERIUM Scopoli, 1777
SPHAERIUM sp.

Description Very small, thin shell with circular outline. Umbones prominent and located at midline. Concentric growth lines prominent. Measurements (mm): length 6.6, height 7.1, half width 2.1.

Genus *UNIO* Philipsson in Retzius, 1788
"UNIO" REESIDEI Stanton, 1917

Description Medium size, thin shell, with oval to broadly triangular outline (Fig. 8.6R). Umbones prominent and positioned just anterior of midline. Plication originates from posterodorsal margin and shifts to concentric growth bands anteriorly. Steinkern shows small indentation for dentition and relatively narrow width, in contrast to the prominent hook and inflated form observed in the steinkern of Bivalve indet. sp. 4. Measurements (mm): length 53.8, height 37.4, half width 17.

Bivalve indet. sp. 1

Description Large, oval to quadrate form with very thick shell (Fig. 8.6C,D). Posterior narrows to form stout termination with angulation at posteroventral margin. Ventral surface is slightly concave. Umbones of low relief are positioned very close to the anterior end of the shell. Steinkern has deep recess for dentition. Wedge-shaped umbo is at 40° angle relative to the ventral pallial line. Measurements (mm): length 69.6, height 50.4, width 39.6.

Bivalve indet. sp. 2

Description Large, moderately thick shell that is oval (Fig. 8.6E,F). Slight angulation at posteroventral margin. Ventral surface slightly convex. Umbones set close to anterior end, but not as far as in Bivalve indet. sp. 1. Steinkern has deep recess for dentition. Wedge-shaped umbo is at 50° angle relative to the ventral pallial line. Measurements (mm): length 69.5, height 44.0, width 27.5.

Bivalve indet. sp. 3

Description Medium-sized oval to quadrate form with moderate to thin shell. Ventral surface is convex (Fig. 8.6K,L). Umbones nearly at anterior end. Two lines extend from umbo to posterior end. Steinkern has deep recess for dentition. Wedge-shaped umbo is at 60° angle relative to the ventral pallial line. Measurements (mm): length 47, height 38.4, half width 19.2.

Bivalve indet. sp. 4

Description Steinkern preserved only (Fig. 8.6Q). Inflated form with broadly triangular to oval outline. Umbones located just anterior of midline. Deep sulcus for dentition at anterior end produces a distinctive hook shape. Posterior ventral margin convex with ruffled commissure. Measurements of steinkern (mm): length 56, height 44.8, width 30.7.

Bivalve indet. sp. 5

Description Medium size, moderately thick shell with oval to comma-shaped outline (Fig. 8.6N). Umbones at extreme anterior end. Strongly convex dorsoposterior and concave ventroposterior. Concentric growth bands deflect in sigmoidal pattern from umbo to ventral margin. Measurements (mm): length 65.6, height 40.4, width 22.5.

Bivalve indet. sp. 6

Description Medium-size, moderately thick shell, with oval outline (Fig. 8.6M). Umbones form prominent beak and are positioned halfway between anterior end and midline. Dorsal margin extends above umbo in posterior direction. Ventral margin slightly convex. Measurements (mm): length 56.9, height 35.3, width 35.8.

Bivalve indet. sp. 7

Description Very large shell with broadly oval outline (Fig. 8.6V). Straight dorsal margin and broadly convex ventral margin. Posterior margin is convex and tapers slightly. Concentric growth rings are prominent. Specimens preserved as poor-quality steinkerns and one very large external impression. Measurements (mm): length 122.3, height 94, half width 33.3.

Bivalve indet. sp. 8

Description Small, thin shell with elongate outline (Fig. 8.6U). Flat to slightly convex ventral margin. Umbones raised and just anterior of midline. Concentric growth bands. Measurements (mm): length 15.5, height 10.1, half width 2.

NONMOLLUSCAN INVERTEBRATES

Bryozoans

One locality, referred to as the clam calamity bed, preserves an abundant monotaxic occurrence of bryozoans in the middle part of the Kaiparowits Formation (Fig. 8.7A). The genus is tentatively identified as *Conopeum* and is found encrusting the inner and outer shells of unionoid bivalves and large gastropods. Roberts, Tapanila, and Mijal (2008) described the clam calamity outcrop in detail and interpreted the accumulation of shells as the result of intense storm (possibly cyclonic) and flood deposition. The co-occurrence

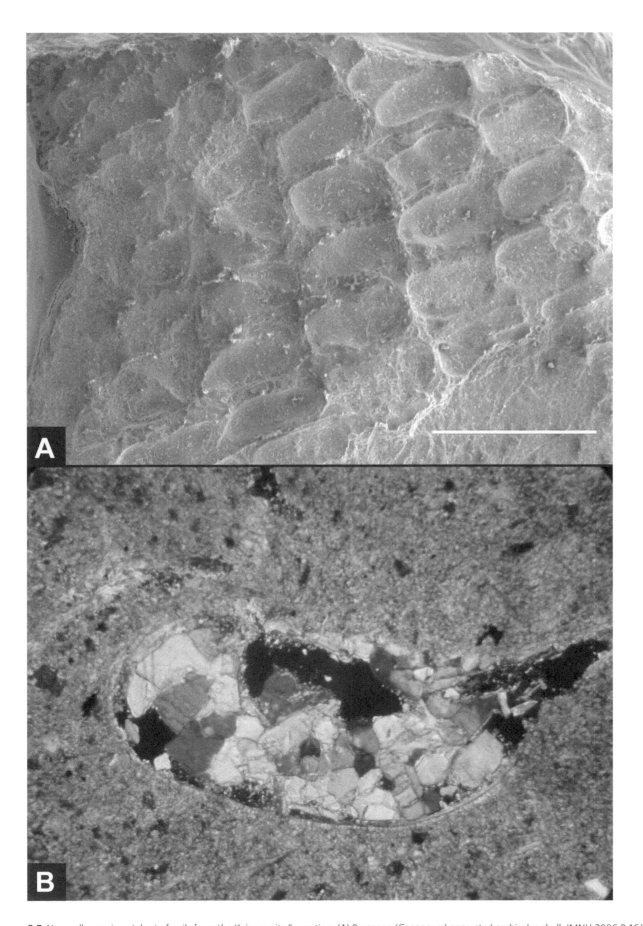

8.7. Nonmolluscan invertebrate fossils from the Kaiparowits Formation. (A) Bryozoan (*Conopeum*) encrusted on bivalve shell, IMNH 2006.R.16/20; scale bar = 1 mm; (B) Ostracode (2.5 mm across) in thin section from the silty mudstone of FA8 (of Roberts, 2007), 315 m level of KBC section.

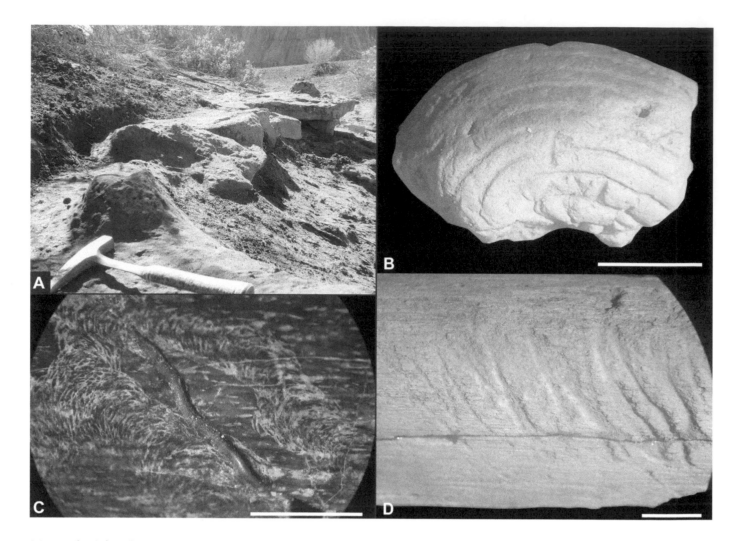

8.8. Trace fossils from the Kaiparowits Formation. (A) Sandstone ledge preserving multiple *Socialites tumulus* social insect nests, 490-m level of middle unit; (B) coprolite with striae, locality IMNH IP 93, 668-m level of upper unit; (C) *Osteocallis mandibulus* on dinosaur bone, UMNH VP-16028, 315-m level of middle unit; (D) bite traces on dinosaur bone, near base of middle unit. (B,D) Scale bars = 5 mm.

of brackish-tolerant *Conopeum* on freshwater bivalve shells demonstrates the proximity of the Kaiparowits fluvial system to the margin of the Bearpaw Sea, and this observation helps identify maximum transgression during the deposition of the lower middle unit.

Ostracodes

Brouwers and Eaton (2000) briefly mentioned the occurrence of aquatic ostracodes in the Kaiparowits Formation, but they did not provide additional details beyond their mere presence in the formation (Fig. 8.7B).

ICHNOLOGY

The Kaiparowits is the type formation for two trace fossil ichnotaxa (*Socialites tumulus* and *Osteocallis mandibulus*) that preserve behaviors of animals (i.e., insects) not currently

represented in the body fossil record. In addition, ichnology studies are providing important constraints on the paleoenvironmental setting of the Kaiparowits fluvial system, particularly as it relates to the shoreline of the Western Interior Seaway.

Social Insect Nests

Nine volcano-shaped domes, *Socialites tumulus*, were described by Roberts and Tapanila (2006) from 25 m² of a laterally continuous sandstone bed, located in the upper middle unit (475 m from the base) of the Kaiparowits Formation (Fig. 8.8A). Detailed sectioning of the structures revealed two types of burrow morphologies within and on top of the domes. Narrow cylindrical galleries are the dominant burrow type, and interconnect to less frequent, J-shaped, wide chambers. All burrows are filled with passive overlying sediment, suggesting that the cavities were maintained open; however,

Tapanila and Roberts

the burrows are unlined. Sedimentary context demonstrates that the colonized sandstone bed was capped by fine-grained sediments, likely deposited from a crevasse splay that immediately covered the burrow system. Several of the domes display evidence for recovery of the burrow system after the splay event. By comparison to other continental mound-shaped traces, *Socialites* was interpreted as a social insect nest. The lack of sedimentary linings on the burrow walls is most consistent with the activity of ants; however, termites or another social insect cannot be ruled out. The occurrence of social insect nests demonstrates that the Kaiparowits floodplain supported construction of long-term invertebrate dwelling structures, but it also indicates that a shallow water table and occasional flooding were considerable hazards to occupying this dynamic habitat.

Bone Traces

Dozens of examples of insect-modified dinosaur bones have been recovered from the Kaiparowits Formation. Borings observed in bones represent a range of morphologies from simple tunnels to previously unnamed complex surficial feeding traces. The latter morphology of insect traces consist of shallow, meandering surface trails, composed of successive arcuate grooves bored into outer cortical bone (Fig. 8.8C). Roberts, Rogers, and Foreman (2007) recognized these as new traces, which they named *Osteocallis mandibulus*. These traces have been interpreted as the feeding activity of necrophagous or osteophagous carrion insects of unknown affinity, but perhaps a group of coleopterans. These insects were likely feeding on the outer nutrient-rich periostracum after the carcass was mostly decayed and dried out. The range of other insect-generated borings, such as small pits and tunnels (Fig. 8.8D), may be incidental excavations or the result of feeding or nesting activities.

Other Trace Fossils

Sediment-filled invertebrate burrows of several varieties are common throughout the Kaiparowits (Fig. 8.5CC) in both mudstone and sandstone facies. Small (<1 cm in diameter) coprolite material often occurs in FA8 and FA9 (of Roberts, 2007) localities (Fig. 8.8B), and these likely were produced by crocodiles or other small reptiles whose bones also occur in these facies. Several well-preserved casts of wood borings also are recorded from the Kaiparowits Formation. Borings in these specimens (Fig. 8.5BB) form aggregates of curved cylindrical cavities oriented subparallel to the wood surface, and they are consistent with descriptions of the ichnogenus *Teredolites* (Kelly and Bromley, 1984; Kelly, 1988). *Teredolites* is the product of brackish- to marine-water wood-boring

bivalves, and its occurrence here supports sedimentological evidence in the middle Kaiparowits for the proximity of the fluvial system to the tidal influence of the marine seaway to the east. Other wood-boring trace fossils are absent from the Kaiparowits, including the smaller pouch-shaped wood boring, *Asthenopodichnium xylobiotum*, which was described from the underlying Wahweap Formation (Moran et al., 2009).

DEPOSITIONAL SETTINGS OF INVERTEBRATES

Nine facies associations (FA) define various depositional environments within the Kaiparowits Formation (Roberts, 2007): intraformational conglomerate (FA1); mollusk-shell conglomerate (FA2); major tabular sandstone (FA3); major lenticular sandstone (FA4); minor tabular and lenticular sandstone (FA5); finely laminated, calcareous siltstone (FA6); inclined heterolithic sandstone and mudstone (FA7); sandy mudstone (FA8); and carbonaceous mudstone (FA9). Body fossils of mollusks and other invertebrates occur in all of these facies associations, although they are most abundantly preserved in FA1, FA2, FA3, FA4, FA8, and FA9. In general, FA1–4 represent swift-water channel environments, whereas FA8 and FA9 represent overbank pond, lake, and wetland environments. A brief summary of the sedimentology and depositional environments associated with each of the major invertebrate-bearing facies associations is presented below.

Intraformational Conglomerate (FA1) and Mollusk-Shell Conglomerate (FA2)

Intraformational conglomerates are abundant throughout the Kaiparowits Formation and are typically associated with both FA3 and FA4. They are dominated by intraformational pebble- to cobble-sized clasts of siltstone and mudstone, with microsite accumulations of fragmentary gastropod and bivalve shells, along with carbonized logs and isolated, worn vertebrate bones. FA1 is interpreted to represent fluvial channel lag deposits.

Despite their rarity, mollusk-shell conglomerates (FA2) represent the single most voluminous invertebrate-preserving localities in the formation, including the clam calamity bed (~300 m level), which preserves over 1 million individual bivalves (Roberts, Tapanila, and Mijal, 2008). These units are diagnosed as clast-supported conglomerates composed almost exclusively of mollusk (mostly bivalve) shells. They are at least 10 cm thick and extend laterally for at least 3 m, but in several instances are much bigger. These distinctive units are interpreted as storm-generated mass-mortality-event beds (Roberts, Tapanila, and Mijal, 2008).

Molluscan diversity in the clam calamity bed is the highest for any locality in the Kaiparowits Formation; it includes nine taxa of medium- to large-sized bivalve, and 11 gastropod taxa including mostly physids and viviparids. This is the only locality in the Kaiparowits Formation where encrusting *Conopeum* bryozoans are preserved. These observations support the taphonomic interpretation that the clam calamity bed was the product of event deposition, where allochthonous and parautochthonous faunal elements were combined to form this exceptional taphocoenosis.

Major Tabular (FA3) and Lenticular (FA4) Sandstones

Both FA3 and FA4 are diagnosed by thick sandstone-dominated packages with trough and tabular cross-stratification. Major tabular sandstones (FA3) are differentiated from major lenticular sandstones (FA4) by width/depth ratios, which exceed 15 for FA3 and are less than 15 for FA4. FA3 is most common in the lower and upper members, whereas FA4 dominates the middle member. Both units preserve invertebrate and vertebrate fossils along with abundant carbonized wood. A wealth of bivalve concentrations, including FA2 units, is preserved with these facies associations. Moreover, the majority of articulated vertebrate skeletons found in the Kaiparowits Formation come from these two facies associations.

Major tabular and lenticular sandstones are both interpreted to represent fluvial channel deposits. FA3 units tend to be much more laterally extensive and are commonly multistory; significantly, they commonly preserve lateral accretion macroforms. Considered together, these data suggest that FA3 bodies are representative of meandering-style river systems. In contrast, the lenticular nature of FA4 bodies within thick overbank mudstone successions and the common occurrence of multiple FA4 bodies at the same stratigraphic level are characteristic of anastomosing-style river systems. Although difficult to assess, the upper and lower units of the Kaiparowits Formation fit models for meandering-style channels, whereas the middle unit is suggestive of anastomosing channels. Regardless of precise channel morphology, paleochannel reconstruction indicates that the formation was dominated by large, deep channels with stable banks (Roberts, 2007).

Large, robust bivalves (e.g., Bivalve indet. sp. 2 and Bivalve indet. sp. 7) typical of higher-discharge rivers (Ortmann, 1920; Watters, 1994) are most common in FA3, but are scarce to absent in FA4 deposits. Both facies share similar viviparids (*Lioplacodes nebrascensis*) as a common and recurrent aquatic gastropod taxon. Overall diversity in these two facies was generally low (five taxa maximum).

Sandy Mudstone (FA8) and Carbonaceous Mudstone (FA9)

Both FA8 and FA9 units are the finest-grained units in the formation, dominated by silt and clay fractions, with FA8 containing a modest sand-sized sediment fraction as well (Fig. 8.3). They are dominated by drab gray and green colors, with evidence of poorly developed soil features including slicken sides, minor caliche, blocky ped structures, and weak horizon development. An important lithology within both FA8 and FA9 is bentonite. Bentonite layers are interpreted as volcanic ash horizons, which range from pristine to heavily reworked. Five of the 12 pristine bentonite layers discovered so far in the formation have been dated (Roberts, Deino, and Chan, 2005; Roberts et al., this volume, Chapter 6). Bentonite layers are also useful marker horizons for correlation in the formation.

Invertebrate and vertebrate fossils are ubiquitous in FA8 and common in FA9. FA9, and to a lesser extent FA8, preserve abundant carbonized plant debris, well-preserved leaves, and other plant macrofossils (Miller et al., this volume, Chapter 7). Considered together, both FA8 and FA9 are interpreted as overbank environments, characterized by perennial pond, lake, and marsh environments. Weak paleosol development, indicative of hydromorphic or waterlogged conditions, is also interpreted for some units (Roberts, 2007).

Diversity of mollusks is moderate in both FA8 and FA9, reaching a maximum of nine taxa. The overall quality of preservation, including nacreous carbonate, is exceptional in these two facies and suggests that fossils are autochthonous. Development of pedogenic structures, particularly gleying of clay-rich units, fractures most shell material, but they remain associated and identifiable. Viviparids (*Lioplacodes nebrascensis*, *L. subtortuosa*, *Tulotomops thomsponi*, and *Viviparus* cf. *V. paludinaeformis*), a large physid (*Physa felixi*), and small bivalves (*Sphaerium*) are the most typical invertebrates for these facies. These mollusks are consistent with pond and small lake environments suggested by the sedimentology. The only terrestrial gastropods (stylommatophorans) of the Kaiparowits Formation occur in FA8 and FA9 localities; they may represent residents of the floodplain that were washed into ponds at some time after death.

DIVERSITY PATTERNS IN MOLLUSK ASSEMBLAGES

The occurrences of 52 mollusk taxa throughout the 860-m section of the Kaiparowits Formation are displayed in Table 8.1. The alpha diversity at each sampled locality ranges up to nine taxa, with the exception of the clam calamity outcrop,

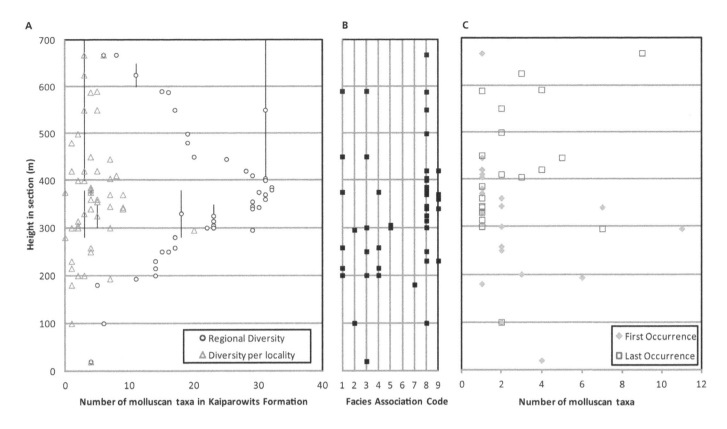

8.9. Summary of diversity trends and facies occurrence through the sampled interval of the Kaiparowits Formation. (A) Both diversity by locality and regional diversity of mollusks show maximum values during the 300–400-m level of the formation, which coincides with (B) frequent facies associations 8 and 9; (C) first and last occurrence of each molluscan taxon overlap in the 300–400-m interval.

where diversity reaches 20 taxa. A preliminary basin-wide (regional) stratigraphic range chart (Table 8.1, gray bars) for the mollusks is based on the lowest and highest occurrence of each molluscan taxon in the Kaiparowits Formation. Adding together all possible concurrent mollusk ranges through the formation (Fig. 8.9A) gives a minimum regional diversity of mollusks for the basin and shows the highest number of concurrent molluscan ranges for the middle part of the section (~300–400 m), with an upper value of 33 taxa at 380 m in section. The accuracy of this regional diversity curve is susceptible to biases in sampling evenness through the section and boundary effects that would concentrate possible concurrent ranges to the middle of any section. Nevertheless, three observations suggest that the middle part of the section was particularly enriched in molluscan taxa.

First, the plot of alpha diversity observed at each locality (Fig. 8.9A) demonstrates the highest values in the 300–400-m stratigraphic interval. Boundary effects and sampling biases that might underestimate concurrent diversity at the regional scale are not applicable to the alpha diversity values because of the broad stratigraphic coverage of samples.

Second, the pattern of first and last occurrences of each molluscan taxon (Fig. 8.9C) suggests that boundary effects

do not account for all the increased diversity in the 300–400-m interval of the section. The segregation of the first and last occurrences into the lower and upper halves of the section, respectively, demonstrates why the approach of generating regional diversity in this manner can underestimate diversity above and below the middle part of the section. However, closer scrutiny of the first and last occurrences shows a fair amount of overlap of these data between the 300- and 425-m interval, suggesting that regional diversity is affected here by the dynamics of incoming and outgoing taxa, and cannot be explained by the simple addition of throughgoing ranges.

Third, the general pattern of alpha diversity increase in the 300–400-m interval corresponds to an increase in FA8 (Fig. 8.9B). From field observation, the FA8 depositional environment typically preserves mollusks very well, often including the primary nacreous shell material, which greatly improves the chances of identification. The diversity observed at FA8 localities tends to be consistently higher among other FA (except the clam calamity bed), which shows that the environment supported a highly diverse molluscan community, and with the taphonomic advantage of preserving nacre, this facies is particularly good at recording the biocoenosis.

COMPARISON TO CAMPANIAN CONTINENTAL FAUNAS OF THE WESTERN INTERIOR

The Kaiparowits invertebrate fauna is dominated by aquatic gastropods and bivalves. This fauna is consistent with sedimentary (Roberts, 2007), paleobotanical (Miller et al., this volume, Chapter 7), and vertebrate (Wiersma, Hutchison, and Gates, 2004) evidence for a wet, warm, and humid continental environment. With the exception of *Teredolites*-bored driftwood and *Conopeum* bryozoans in the lower part of the middle unit, characteristically brackish or marine mollusks such as ostreids are absent throughout the formation. Even without brackish-marine environments to add to the diversity, the 52 different mollusk taxa supported by the Kaiparowits environment rivals (and may exceed) that of any comparable Western Interior Basin for the Campanian, from Canada to Mexico.

The Belly River Group of Alberta preserves 48 continental bivalve and gastropod taxa, of which only 21 taxa are described from Dinosaur Provincial Park (reviewed by Johnston and Hendy, 2005). The Belly River Group fauna resembles the Kaiparowits assemblage in its diversity, with overlapping taxa including *Lioplacodes subtortuosa*, *L. nebrascensis*, and *Physa copei*. The Judith River Formation in Montana records more bivalve taxa than are found in the Kaiparowits, but over half of its 34 species represent brackish-water fauna (Stanton, 1905). Thirty species of gastropods were identified from the Judith River and include *Physa felixi* and *Lioplacodes subtortuosa* described here from Utah. Mollusks of the Fruitland Formation of the San Juan Basin bear some resemblance to the Kaiparowits fauna, as first recognized by Gregory and Moore (1931). The Fruitland fauna is less diverse, however, with 18 bivalve species (at least 10 are freshwater taxa) and only nine gastropods (Stanton, 1917). The bivalve "*Unio*" *reesidei* and four gastropods (*Neritina baueri*, *Tulotomops thompsoni*, *Lioplacodes subtortuosa*, and *Planorbis* [B.] *chacoensis*) appear to be common among Fruitland and Kaiparowits faunas. In Mexico, the Cerro del Pueblo Formation records a small, unspecified number of Campanian aquatic bivalve taxa and 17 gastropods (Perrilliat et al., 2008). This assemblage of gastropods appears distinct in composition by comparison to the Kaiparowits taxa. Other basins contemporary with the Kaiparowits (e.g., Dinosaur Park Formation of Alberta; Aguja Formation of Texas) are either poorly studied or poorly preserve mollusks for comparison.

CONCLUSIONS

The Kaiparowits Formation preserves at least 52 gastropod and bivalve taxa in the Campanian wetlands facies of southern Utah. Ostracodes, rare bryozoans, and trace fossils linked to insect activity broaden the spectrum of invertebrates, which comprise an important part of the Kaiparowits biodiversity, along with better-known vertebrate and paleobotanical fossils. Aquatic mollusks dominate the assemblages, with minor occurrence of terrestrial pulmonates and very rare brackish-indicating animals.

Preliminary data reported here support a modest increase in recorded diversity for the localities in the middle part of the section, although it is difficult to isolate the roles of sampling, depositional environment, and taphonomy without further collections. As knowledge of the Kaiparowits Formation progresses with further sampling, the regional diversity curve likely will increase in the lower and upper parts of the section. In particular, sampling the underlying Wahweap Formation will provide broader range control on the various molluscan taxa and provide more information on the role of depositional environments and taphonomy in the record of continental mollusks.

ACKNOWLEDGMENTS

Fieldwork and funding for this project were supported by the BLM-Grand Staircase–Escalante National Monument grant JSA055008. D. Powell, A. Titus, and S. Foss (Bureau of Land Management) provided significant logistical support with permits and fieldwork. Fossil specimens were collected under Bureau of Land Management permit UT06–007S-GS (LT). We thank the field crews of the UMNH and Denver Museum of Nature & Science, along with students from Southern Utah University, Idaho State University, and Idaho Museum of Natural History, especially R. Myers for his help in assembling figures and J. Nichols for organizing collections.

REFERENCES CITED

Beck, H. 1838. Fasciculus Secundus; Mollusca Gastraepoda Pulmonata. Index Molluscorum Præsentis ævi Musei Principis Augustissimi Christiani Frederici. Hafniae (Copenhagen), Denmark.

Brouwers, E., and J. G. Eaton. 2000. Ostracode faunas from Upper Cretaceous units of the Kaiparowits Plateau, southern Utah. Geological Society of America Abstracts with Programs 32(7):A449.

Cox, L. R. 1959. Thoughts on the classification of the Gastropoda. Proceedings of the Malacological Society of London 33:239–261.

Cuvier, G. L. C. 1797. Tableau Élémentaire de l'Histoire Naturelle des Animaux [des Mollusques]. Baudonin, Paris.

Cuvier, G. L. C. 1817. Mémoire pour Servir à l'Histoire et à l'Anatomie des Mollusques. Deterville, Paris.

Draparnaud, J. P. R. 1801. Tableau des mollusques terrestres et fluviatiles de la France. Montpellier, Paris.

Fassett, J. E., and M. B. Steiner. 1997. Precise age of C33N-C32R magnetic polarity reversal,

San Juan Basin, New Mexico and Colorado. New Mexico Geological Society Guidebook 48:239–247.

Fitzinger, L. I. 1833. Systemaatisches Verzeichniß der im Erzherzogthume Oesterreich vorkommenden Weichthiere, als Prodrom einer Fauna desselben. Beiträge Landesk Oesterreich Enns 3:88–122.

Gates, T. A., and S. D. Sampson. 2007. A new species of *Gryposaurus* (Dinosauria: Hadrosauridae) from the late Campanian Kaiparowits Formation, southern Utah, U.S.A. Zoological Journal of the Linnaean Society 151:351–376.

Gray, J. E. 1824. Shells; pp. 240–246 in A Supplement to the Appendix of Captain Parry's Voyage for the Discovery of a North-west Passage. John Murray, London.

Gray, J. E. 1847. A list of the genera of Recent Mollusca, their synonyma and types. Proceedings of the Zoological Society of London 15:129–242.

Gregory, H. E. 1950. Geology and Geography of the Zion Park Region Utah and Arizona. U.S. Geological Survey Professional Paper 220.

Gregory, H. E. 1951. The Geology and Geography of the Paunsaugunt Region, Utah. U.S. Geological Survey Professional Paper 226.

Gregory, H. E., and R. C. Moore. 1931. The Kaiparowits Region: A Geographic and Geologic Reconnaissance of Parts of Utah and Arizona. U.S. Geological Survey Professional Paper 164.

Hall, J. 1845. Descriptions of organic remains collected by Capt. J. C. Frémont, in the geographical survey of Oregon and North California, appendix B; pp. 304–310 in Brevet Capt. J. C. Frémont's Report of the Exploring Expedition to the Rocky Mountains in the Year 1842, and to Oregon and North California in the Years 1843–1844. U.S. 28th Congress, 2nd Session House Executive Document 166.

Jablonski, D., J. A. Finarelli, and K. Roy. 2006. What, if anything, is a genus? Testing the analytical units of paleobiology against molecular data. Geological Society of America Abstracts with Programs 38(7):146.

Johnston, P. A., and A. J. W. Hendy. 2005. Paleoecology of mollusks from the Upper Cretaceous Belly River Group; pp. 139–166 in P. J. Currie and E. B. Koppelhus (eds.), Dinosaur Provincial Park: A Spectacular Ancient Ecosystem Revealed. Indiana University Press, Bloomington, Indiana.

Kelly, S. R. A. 1988. Cretaceous wood-boring bivalves from western Antarctica with a review of the Mesozoic Pholadidae. Palaeontology 31:341–372.

Kelly, S. R. A., and R. G. Bromley. 1984. Ichnological nomenclature of clavate borings. Palaeontology 27:793–807.

Lamarck, J. B. 1799. Prodrome d'une nouvelle classification des coquilles. Memoires de la Société d'Histoire Naturelle de Paris 1:63–91.

Linnaeus, C. 1758. Systemae Naturae. A Photographic Facsimile of the First Volume of the Tenth Edition (1758). Regnum Animale. Trustees of the British Museum, London.

Meek, F. B. 1864. Checklist of the invertebrate fossils of North America: Cretaceous and Jurassic. Smithsonian Miscellaneous Collections 7:1–40.

Meek, F. B. 1876. A Report on the Invertebrate Cretaceous and Tertiary Fossils of the Upper Missouri Country. United States Geological Survey of the Territories. Government Printing House, Washington, D.C.

Meek, F. B., and F. V. Hayden. 1856. Descriptions of new species of Gastropoda from the Cretaceous formations of Nebraska Territory. Proceedings of the Academy of Natural Sciences of Philadelphia 8:63–69.

Montfort, D. de 1810. Conchyliologie systématique, et classification méthodique des coquilles; offrant leurs figures, leur arrangement générique, leurs descriptions caractéristiques, leurs noms; ainsi que leur synonymie en plusieurs langues. Ouvrage destiné à faciliter l'étude des coquilles, ainsi que leur disposition dans les cabinets d'histoire naturelle. Coquilles univalves, non cloisonnées. Schoelle, Paris, France, Volume 2, 676 pp.

Moran, K. L., E. L. Simpson, H. L. Hilbert-Wolf, M. C. Wizevich, K. B. Golder, and S. E. Tindall. 2009. The recognition and implications of the wood-boring trace fossil *Asthenopodichnium xylobiotum* in Upper Cretaceous strata of Grand Staircase Escalante National Monument, Utah; p. 262 in Geological Society of America Abstracts with Programs 41.

Müller, O. F. 1774. Vermium terrestrium et fluviatilum, seu animalium infusorium, helminthicorum et testaceorum, non marinorum, succincta historia 2. Heineck & Faber, Havinae.

Müller, O. F. 1781. Geschichte der Perlen-Blasen. Naturforscher 15:1–20.

Ortmann, E. A. 1920. Correlation of shape and station in freshwater mussels (naiads). Proceedings of the American Philosophical Society 59:269–312.

Perrilliat, M. C., F. J. Vega, B. Espinosa, and E. Naranjo-Garcia. 2008. Late Cretaceous and Paleogene freshwater gastropods from northeastern Mexico. Journal of Paleontology 82:255–266.

Pilsbry, H. A. 1895. Catalogue of the Marine Mollusks of Japan, with Descriptions of New Species, and Notes on Others Collected by Frederick Stearns. F. Stearns, Detroit, Michigan.

Pilsbry, H. A. 1921. Mollusks; pp. 30–36 in H. E. Wanner (ed.), Some Faunal Remains from the Trias of York County, Pennsylvania. Proceedings of the Academy of Natural Sciences of Philadelphia 73.

Rafinesque, C. S. 1815. Analyse de la Nature, ou Tableau de l'Universe et des Corps Organisés. J. Barravecchia, Palermo, Italy.

Rafinesque, C. S. 1819. Prodrome de 70 nouveaux genres d'animaux découverts dans l'intérieur des Etas-Unis d'Amérique, durant l'année 1818. Journal de Physique, de Chimie et d'Histoire Naturelle 88:417–429.

Retzius, A. J. 1788. Dissertatio historico-naturalis sistens nova testaceorum genera. Lund.

Roberts, E. M. 2007. Facies architecture and depositional environments of the Upper Cretaceous Kaiparowits Formation, southern Utah. Sedimentary Geology 197:207–233.

Roberts, E. M., and L. Tapanila. 2006. A new social insect nest from the Upper Cretaceous Kaiparowits Formation of southern Utah. Journal of Paleontology 80:768–774.

Roberts, E. M., A. L. Deino, and M. A. Chan. 2005. $^{40}Ar/^{39}Ar$ age of the Kaiparowits Formation, southern Utah, and correlation of contemporaneous Campanian strata and vertebrate faunas along the margin of the Western Interior Basin. Cretaceous Research 26:307–318.

Roberts, E. M., R. R. Rogers, and B. Z. Foreman. 2007. Continental insect borings in dinosaur bone: examples from the Late Cretaceous of Madagascar and Utah. Journal of Paleontology 81:201–208.

Roberts, E. M., L. Tapanila, and B. Mijal. 2008. Taphonomy and sedimentology of storm-generated continental shell beds; a case example from the Cretaceous Western Interior Basin. Journal of Geology 116:462–479.

Russell, L. S. 1934. Reclassification of the fossil Unionidae (freshwater mussels) of western Canada. Canadian Field Naturalist 48:1–4

Sampson, S., M. Loewen, E. Roberts, J. Smith, L. Zanno, and T. Gates. 2004. Provincialism in Late Cretaceous terrestrial faunas: new evidence from the Campanian Kaiparowits Formation of Utah. Journal of Vertebrate Paleontology 24:108A.

Say, T. 1818. Account of two new genera, and several new species, of fresh water and land shells. Journal of the Academy of Natural Sciences of Philadelphia 1:276–284.

Schmidt, A. 1855. Der Geschlechtsapparat der Stylommatophoren in taxonomischer Hinsich. Abhandlungen des Naturwissenschaftlichen Vereines fuer Sachsen und Thueringen in Halle 1:1–52.

Scopoli, J. A. 1777. Introductio ad historiam naturalem, sistens genera lapidum, plantarum et animalium hactenus detecta, caracteribus essentialibus donata, in tribus divisa, subinde ad leges naturae. Prague.

Stanton, T. W. 1905. Invertebrate fauna; pp. 104–123 in T. W. Stanton and J. B. Hatcher (eds.), Geology and Paleontology of the Judith River Beds, with a Chapter on the Fossil Plants by F. H. Knowlton. U.S. Geological Survey Bulletin 257.

Stanton, T. W. 1917. Contributions to the Geology and Paleontology of San Juan County, New Mexico 3. Nonmarine Cretaceous Invertebrates of the San Juan Basin. U.S. Geological Survey Professional Paper 98-R.

Thiele, J. 1931. Handbuch der systematischen Weichtierkunde, 1. Verlag von Gustof Fischer, Jena.

Titus, A. L., J. D. Powell, E. M. Roberts, S. D. Sampson, S. L. Pollock, J. I. Kirkland, and L. B. Albright. 2005. Late Cretaceous stratigraphy, depositional environments, and macrovertebrate paleontology of the Kaiparowits Plateau, Grand Staircase–Escalante National Monument, Utah; pp. 101–128 in J. Pederson and C. M. Dehler (eds.), Interior Western United States. Geological Society of America Field Guide 6.

Watters, G. T. 1994. Function and form of unionoidean shell shape and sculpture. American Malacological Bulletin 11:1–20.

Wenz, W. 1939. Gastropoda, Allgemeiner Teil und Prosobranchia; in O. H. Schindewolf (ed.), Handbuch der Paläozoologie. Volume 6, Part 1. Gebrüder Borntraeger, Berlin.

White, C. A. 1876. Invertebrate paleontology of the Plateau Province; pp. 74–135 in Report on the Geology of the Eastern Portion of the Uinta Mountains (Powell Survey). U.S. Geological and Geographical Survey of the Territories. U.S. Government Printing Office, Washington, D.C.

White, C. A. 1877. Descriptions of Unionidae and Physidae collected by Professor E. D. Cope from the Judith River group of Montana Territory during the summer of 1876; pp. 599–602 in United States Geological and Geographical Survey of Territories (Hayden Survey) Bulletin 3.

White, C. A. 1880. On the antiquity of certain subordinate types of freshwater and land Mollusca. American Journal of Science, 3rd ser., 20:44–49.

White, C. A. 1883. Fossils of the Laramie group; pp. 49–103 in United States Geological and Geographical Survey of the Territories (Hayden Survey), 12th Annual Report (for 1878), Part 1. U.S. Government Printing Office, Washington, D.C.

Wiersma, J., J. H. Hutchison, and T. A. Gates. 2004. Crocodilian diversity in the Upper Cretaceous Kaiparowits Formation (Upper Campanian), Utah. Journal of Vertebrate Paleontology 24:129A.

Yen, T.-C. 1954. Nonmarine Mollusks of Late Cretaceous Age from Wyoming, Utah and Colorado. U.S. Geological Survey Professional Paper 254-B.

Zanno, L. E., and S. D. Sampson. 2006. A new oviraptosaur (Theropoda, Maniraptora) from the Late Cretaceous (Campanian) of Utah. Journal of Vertebrate Paleontology 25:897–904.

Tapanila and Roberts

Elasmobranchs from Upper Cretaceous Freshwater Facies in Southern Utah

9

James I. Kirkland, Jeffrey G. Eaton, and Donald B. Brinkman

ELASMOBRANCH TEETH FROM FRESHWATER FACIES ARE common in many microvertebrate assemblages in the Upper Cretaceous of western North America. Research on the essentially complete Upper Cretaceous terrestrial microvertebrate record of southern Utah has resulted in the collection of specimens from many stratigraphic horizons not sampled elsewhere. Two genera of hybodont shark and two clades of primitive (rhinobatoid and sclerorhynchoid) ray are present throughout the sequence. Orectoloboid sharks first appear in freshwater facies in the Coniacian, and two orectoloboid shark genera are present in the Campanian. Two new genera, *Cristomylus* and *Pseudomyledaphus*, are identified as part of a rapidly evolving line of myledaphine rhinobatoid rays; they include the new species *Cristomylus nelsoni*, *Cristomylus bulldogensis*, *Cristomylus cifellii*, and *Pseudomyledaphus madseni*. Additionally, the sclerorhynchoid sawfish *Texatrygon brycensis* and *Columbusia deblieuxi* and the orectoloboid shark *Cantioscyllium markaguntensis* are described.

INTRODUCTION

Wet screen washing for microvertebrates in the Cretaceous of the Western Interior of North America following the techniques outlined by Cifelli, Madsen, and Larson (1996) has resulted in vastly improved knowledge of the diversity of terrestrial faunas (Estes, 1964; Eaton et al., 1999). These efforts have mostly been driven by workers looking for fossil mammals; thus, the largest collections have been secured from the floodplain facies that yield the highest numbers of mammal teeth. Because 30 mesh (0.525 mm) screens are used when processing for fossil mammal teeth, an abundance of shark and ray material is recovered that is not typically obtained when using coarser standard window screens of 18 mesh (1.0 mm). From these new data, it is clear that sharks and rays were important and consistent members of Upper Cretaceous freshwater and estuarine communities (Eaton and Kirkland, 2001, 2003). Modern freshwater sharks, sawfish, and stingrays are widespread in lowland environments globally (Compagno and Cook, 1995a, 1995b; Lovejoy,

1996; Compagno, 2002), with some river systems supporting representatives of all three groups (Thorburn et al., 2004). Among living elasmobranch species, 29 are obligate freshwater species, 18 species are euryhaline ranging from inshore marine up into fully freshwater habitats, only one species is restricted to brackish water, and at least 25 species of marine shark and ray penetrate into freshwater but generally are not found far from the sea (Compagno and Cook, 1995a, 1995b). Most Cretaceous elasmobranch taxa reported on herein are assumed to be euryhaline species, and although they do not represent the same suborders as today, they fill the same spectrum of behaviors as do the modern groups of freshwater elasmobranchs.

The first of these taxa to be described was *Myledaphus bipartitus* from the upper Campanian Judith River Formation of Montana, not the Paleocene Fort Union Beds, as initially reported by Cope (1876). Large ray teeth representing this genus are abundant in these strata and are the most likely elasmobranch to be recovered by visually searching the surface of the outcrop. In his groundbreaking research on the microvertebrate fauna from the late Maastrichtian Lance Formation of Wyoming, Richard Estes distinguished four species of freshwater elasmobranch, of which three were new. However, we now recognize that all of Estes's (1964) taxa are in need of systematic revision as a result of advances in elasmobranch taxonomy and uncertainties created by Estes's well-intentioned and forward-thinking attempt to incorporate perceived heterodonty into his taxonomic definitions. Unfortunately, some of the misconceptions created by Estes's work have become well established in the literature; nearly all succeeding researchers investigating freshwater Cretaceous elasmobranchs have used Estes's (1964, 1969) problematic concepts. For this reason, we reproduce herein the figures of elasmobranch teeth published by Estes, rearranged so that they reflect the current state of taxonomy. Additionally, although most of Estes's (1964) species are not known from Utah yet, they represent the end members of several freshwater elasmobranch lineages described herein, just prior to their extinction. Here, we have the advantage of examining many of these lineages from their first appearance

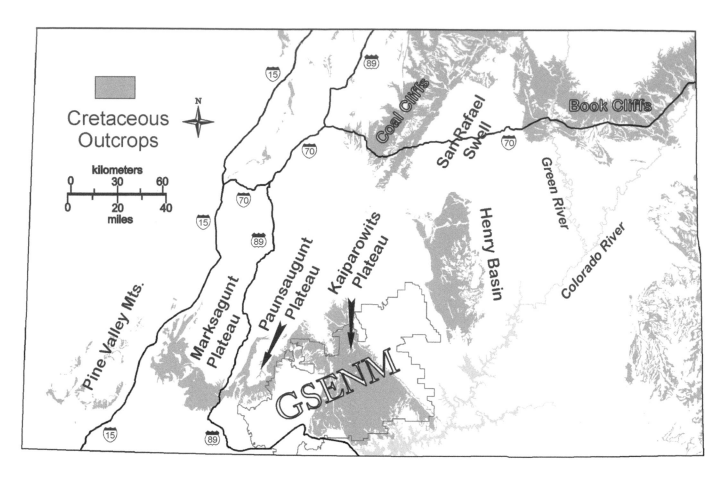

9.1. Cretaceous outcrops in southwestern Utah.

within freshwater habitats, permitting us to propose hypotheses concerning their origins.

None of the North American freshwater elasmobranch lineages survived into the Tertiary, although many actinopterygians survive the terminal Cretaceous extinction event relatively unscathed (Bryant, 1989; Archibald and Bryant, 1990; Archibald, 1996; Wroblewski, 2004; Cook et al., in press). The low reproduction (K selection) rate characteristic of elasmobranchs may have been a critical factor in this selective extinction. Likewise, marine elasmobranchs in general suffered a major extinction event at the end of the Cretaceous. (Cappetta, 1987a; Kriwet and Benton, 2004). The ease with which shark and ray teeth are reworked into younger deposits has limited the degree to which the terminal Cretaceous extinction among the elasmobranchs has been appreciated (Cappetta, 1987a; Eaton, Kirkland, and Doi, 1989; Archibald and Bryant, 1990). Cook et al. (in press) noted a marked decrease in diversity with increasing latitude from six to two freshwater taxa at the end of the Cretaceous. Additionally, studies of stable oxygen isotopes would permit an independent test as to what degree these elasmobranch species lived in fresh, brackish, or marine habitats (Kocsis, Vennemann, and Fontignie, 2007).

This chapter must be viewed as a preliminary attempt to analyze hundreds of isolated teeth spanning close to 35 million years of Earth history. No two specimens are identical, and there is an enormous amount of variation present in each sample, representing variations due to ontogeny and tooth position. Therefore, we focus on large samples of teeth in order to avoid the noise of variability and the erection of more species than actually existed. We realize that in taking this more conservative approach we may not recognize some valid species and may not identify the possible overlap of species ranges within some lineages or clades. We recognize that these specimens will provide a continuing source of research in systematics, paleobiogeography, and paleoecology for years to come and hope that this report facilitates this future research.

GEOLOGICAL SETTING

Utah preserves a remarkably complete sequence of Cretaceous terrestrial strata extending from at least the middle of the Lower Cretaceous up to the top of the Cretaceous (Kirkland, Lucas, and Estep, 1998; Eaton et al., 1999, 2001; Titus et al., 2005, Chapter 2, this volume; Kirkland and Madsen,

Kirkland, Eaton, and Brinkman

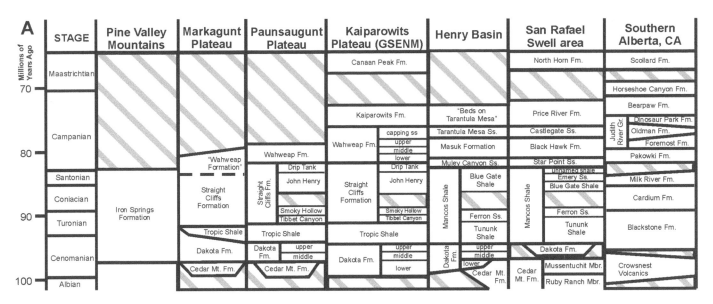

9.2. Stratigraphic framework for Upper Cretaceous in southern Utah. (A) Generalized correlation chart of Upper Cretaceous sections in Utah with southern Alberta. (B) Stratigraphic distribution of microvertebrate localities in southern Utah from which specimens discussed herein were collected. Kb, "beds on Tarantula Mesa"; Kc, Cedar Mountain Formation; Kk, Kaiparowits Formation; Km, Mule Canyon Sandstone; Kmu, Masuk Formation; Ksd, Drip Tank Member of Straight Cliffs Formation; Ksj, John Henry Member of Straight Cliffs Formation; Kss, Smoky Hollow Member of Straight Cliffs Formation; Kst, Tibbet Canyon Member of Straight Cliffs Formation; Kt, Tarantula Mesa Sandstone; Kw, Wahweap Formation; T.S., Tropic Shale. Cross-hatching indicates strata of that age not preserved.

2007). Our understanding of the stratigraphic resolution of these rocks at Grand Staircase–Escalante National Monument and elsewhere across southern Utah is still improving, as may be witnessed by reading many of the chapters in this volume. We present only a general overview of the stratigraphy of these strata so that the reader may be apprised of the relative age and geographic location of the localities described herein (Figs. 9.1, 9.2). Qualified researchers seeking more detailed locality information than we provide here are directed to make inquires with the specific repositories listed below or with the Grand Staircase–Escalante National Monument.

Cedar Mountain Formation in the San Rafael Swell Region – Basal Cenomanian

The rays described here were recovered from microvertebrate localities in the Mussentuchit Member of the Cedar Mountain Formation. These smectitic strata, at the top of the Cedar Mountain Formation, preserve one of North America's most diverse terrestrial faunas based on collections made in the process of wet screen washing for mammals (Eaton and Cifelli, 2001; Cifelli, 2004). A radiometric date of 98.39 ± 0.07 Ma (Cifelli et al., 1997) and dates ranging from about 97 to 99 Ma (Garrison et al., 2007) place the member close to the 99.6 Ma boundary between the Early and Late Cretaceous in the basal Cenomanian, as denoted by both the international (Gradstein, Ogg, and Smith, 2004; Ogg, Agterberg, and Gradstein, 2004) and Geological Society of America (Walker and Geissman, 2009) timescales. Subsequent dating of smectitic sediments near the base of the Cretaceous section in southern Utah west of the Grand Staircase–Escalante National Monument on the south side of the Markagunt Plateau and southwest of the Pine Valley mountains provides comparable dates (Biek et al., 2009).

Dakota Formation – Upper Cenomanian

A large number of elasmobranch teeth were recovered from the middle member of the Dakota Formation within Grand Staircase–Escalante National Monument near the town of Cannonville and Bureau of Land Management lands near the town of Alton. It is likely that the middle member is middle to lower upper Cenomanian, on the basis of the age of the overlying marine invertebrate faunas in the upper member and a radiometric age of 95.97 ± 0.22 Ma from within this member, below the Dakota microvertebrate sites on Bulldog Bench (Dyman et al., 2002; Titus et al., 2005).

Smoky Hollow Member, Straight Cliffs Formation – Turonian

Peterson (1969) named the Smoky Hollow Member and divided it into three parts: (1) a lower coal zone that has considerable brackish-water influence; (2) a barren middle zone representing floodplain deposits; and (3) an upper conglomeratic bed he referred to as the Calico bed. The fossils in this report come from the middle barren unit. Peterson (1969) dated the underlying Tibbet Canyon Member on the basis of the late middle Turonian index fossil *Inoceramus howelli* White, which indicates the upper part of the late middle Turonian *Prionocyclis hyatti* zone (Obradovich and Cobban, 1975; Cobban et al., 2006). A detrital zircon date of 91.9 ± 0.3 Ma (Titus et al., Chapter 2, this volume) from the

middle zone just below the Calico bed – a Middle Turonian date – suggests upper middle Turonian ages for all the elasmobranch material discussed herein.

JOHN HENRY MEMBER OF THE STRAIGHT CLIFFS FORMATION – CONIACIAN–SANTONIAN

The John Henry Member of the Straight Cliffs is mostly marine to marginally marine in its type area along the west side of the Kaiparowits Plateau. The strand line for the member oscillates about the middle of the Kaiparowits Plateau so that most of the member is nonmarine to the west on the Paunsaugunt Plateau. The ray specimens described here come from the western margin of the Kaiparowits Plateau or the eastern margin of the Paunsaugunt Plateau. Peterson (1969) reported a late Coniacian age for the marine molluscan fauna at the base of the member on the basis of the presence of *Inoceramus stantoni* (=*Inoceramus undabundas*, Walaszczyk and Cobban, 2006) and *Protexanites shoshonensis* (=*Protexanites bourgeoisinus*, Kennedy and Cobban, 1991), reflecting the uppermost Coniacian Zone of *Magdiceramus crenulatus*. *Volviceramus involutus* was found low in the John Henry Member (sandstone bed A) by Peterson (1969), although the name bearer of the upper middle Coniacian age Zone of *Volviceramus involutus* (Walaszczyk and Cobban, 2007) ranges upward to the top of the Coniacian (Walaszczyk and Cobban, 2006).

Underlying the John Henry Member is the Calico bed (Peterson, 1969) of the Smoky Hollow Member of the Straight Cliffs Formation. The middle Turonian zircon date from just below the Calico suggests there is an unconformity (or condensed interval represented by the conglomeratic Calico bed) of at least the upper Turonian through middle Coniacian at the base of the John Henry Member. A similar unconformity is recognized at the base of the Smoky Hill Member of the Mancos Shale in the Four Corners region (Leckie, Kirkland, and Elder, 1997) and extends throughout the region (Merewether, Cobban, and Obradovich, 2007).

The age of the upper part of the John Henry Member is less well constrained. The ammonite *Desmoscaphites* from near the top of the John Henry Member limits the age of the member to late Santonian (Landman and Cobban, 2007). An overall Santonian age for the John Henry Member on the Paunsaugunt and Kaiparowits Plateaus is also supported by comparison of its preserved mammalian fauna with that preserved in the Milk River Formation of Alberta, Canada (Eaton, 2006a).

The John Henry Member of the Straight Cliffs Formation is not formally recognized in the Markagunt Plateau to the west and has recently been referred to as the "upper" Straight

Cliffs Formation (Eaton, 2006b). Fossiliferous brackish-water facies preserving a Turonian molluscan fauna (Eaton et al., 2001; Eaton, 2006b) define the underlying Smoky Hollow Member in this area, which is difficult to map because the capping Calico conglomerate is thin and has only been identified sporadically in the southern part of the plateau (Bob Biek, pers. comm., 2010). At 286 m above the top of a prominent sandstone ledge marking the top of the Tibbet Canyon Member, a volcanic ash has yielded a late Coniacian radiometric age of 86.72 ± 0.58 Ma (Eaton et al., 2001; Eaton, 2006b), suggesting that the strata near the base of the "upper" Straight Cliffs Formation may be older than the base of the John Henry Member in outcrops to the east. Overlying the Straight Cliffs Formation on the western Markagunt Plateau, strata that have been questionably referred to as the Wahweap Formation or as the "formation of Cedar Canyon" yield a mammalian fauna intermediate between the Straight Cliffs and Wahweap formations of the Paunsaugunt and Kaiparowits plateaus (Eaton, 2006b).

Wahweap Formation – Middle? Campanian

In the Wahweap Formation sample, large ray teeth are rare, but overall, the diversity of tooth forms is greater than in any other formation or member sampled. This formation is about 400 m thick on the Kaiparowits Plateau. Radiometrically datable ashes are rare, but Jinnah et al. (2009) reported an $^{40}Ar/^{39}Ar$ date from about 60 m above the base of the formation of 80.1 ± 0.3 Ma (middle Campanian age). A lower Kaiparowits date of 75.96 ± 0.14 Ma and estimates of depositional rates suggest the upper age limit on the Wahweap Formation is 76.1 Ma (Roberts, Deino, and Chan, 2005), an earliest late Campanian age. This would indicate the age of most of the Wahweap Formation is middle Campanian; however, some of the basal localities may be lower Campanian. Eaton and Cifelli (1988) originally correlated the mammalian fauna from the Wahweap Formation with the fauna from the Milk River Formation in Alberta, Canada (Aquilan Land Mammal Age); however, later studies of mammals from the upper part of the John Henry Member of the Straight Cliffs Formation (Santonian) revealed a stronger correlation (Eaton, 2006a) to the Milk River fauna than to the Wahweap fauna. Ongoing study by Eaton of the mammalian faunas from the John Henry Member and the Wahweap Formation on the Paunsaugunt Plateau shows that the Wahweap fauna is much more similar to that of the John Henry Member than it is to the faunas recovered from the Kaiparowits Formation. The discrepancy between radiometric ages and faunal comparisons is unresolved at this time.

The nomenclature of the highest Cretaceous units in the area of the western Markagunt Plateau has been controversial. Traditionally they were identified as Kaiparowits Formation and more recently as the "formation of Cedar Canyon" (Moore et al., 2004). The most recent geological mapping of the region has these strata denoted as Wahweap Formation (Biek et al., 2011), and there is still dispute among the authors of this article on which term should be applied to them (Fig. 9.2B).

Kaiparowits Formation – Upper Campanian

Extensive research in the Kaiparowits Formation, as reported in the various chapters within this volume and references therein, firmly documents this thick sequence of floodplain strata as having been rapidly deposited during the early late Campanian. Radiometric dates have provided the strongest age controls on the deposition of this sequence (Roberts, Deino, and Chan, 2005; Jinnah et al., 2009; Titus et al., Chapter 2, this volume).

Institutional Abbreviations AMNH, American Museum of Natural History, New York, New York; FHSM, Sternberg Memorial Museum, Fort Hays State University, Hays, Kansas; MNA, Museum of Northern Arizona, Flagstaff, Arizona; OMNH, Oklahoma Museum of Natural History, Norman, Oklahoma; UCM, University of Colorado Museum, Boulder, Colorado; UCMP, University of California, Paleontological Museum, Berkeley, California; UMNH, Utah Museum of Natural History, University of Utah, Salt Lake City, Utah; UWM, University of Wyoming Geological Museum, Laramie, Wyoming.

SYSTEMATIC PALEONTOLOGY

ELASMOBRANCHII Bonaparte, 1838
EUSELACHI Hay, 1902
HYBODONTOIDEA Owen, 1846
HYBODONTIDAE Owen, 1846
HYBODUS Agassiz, 1837

Type Species *Hybodus reticulatus* Agassiz (1837).

Discussion Numerous species of Mesozoic sharks have been assigned to *Hybodus*, mostly on the basis of isolated teeth and dorsal fin spines. Of the 20 original *Hybodus* species proposed by Agassiz (1833–1844), on the basis of teeth and dorsal fin spines, none was assigned as the type species. Woodward (1916) designated *Hybodus reticulatus* from the Lower Jurassic as the type species, which has been accepted by most researchers (Cappetta, 1987b). Following a detailed analysis of the anatomy of *Hybodus reticulatus*, Maisey (1987) showed that teeth and dorsal fin spines by themselves were not sufficient to refer a hybodont shark species to the genus *Hybodus*, and assigned the Upper Jurassic taxon *Hybodus*

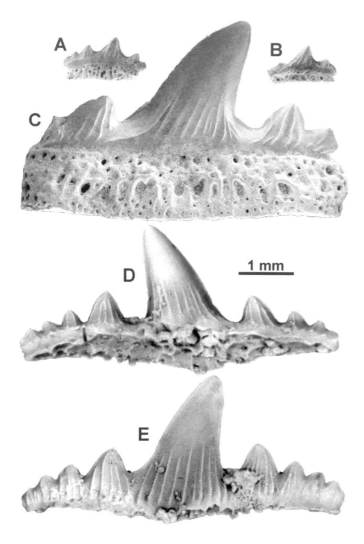

9.3. Hybodus from Dakota Formation at Bulldog Bench, westernmost Kaiparowits Basin (MNA locality 1067). Small teeth (MNA V10303) in (A) labial view and MNA V10304 in (B) labial view. Large tooth with complete root (MNA V10302) in (C) labial view. Typical large tooth missing root (MNA V10388) in (D) labial and (E) lingual views.

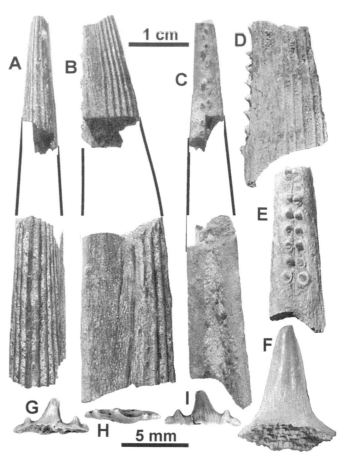

9.4. *Hybodus* sp. from Straight Cliffs and Wahweap formations. Dorsal fin spines from the basal (UMNH Loc. 1064, upper Coniacian) John Henry Member, Straight Cliffs Formation (UMNH VP 17797), in (A) anterior, (B) lateral, and (C) posterior views; (UMNH VP 18027) in (D) lateral, and (E) posterior views. Tooth from upper John Henry Member Straight Cliffs Formation (Santonian) (UMNH VP 18941, Loc. 799) in (F) labial view. Tooth from Wahweap Formation (middle Campanian) (MNA V10387, Loc. 456–2) in (G) labial (H) occlusal, and (I) lingual views.

fraasi (Maisey, 1986) and the Lower Cretaceous taxon *Hybodus basanus* (Maisey, 1983) to the new genus *Egertonodus* on the basis of their being known from nearly complete specimens that differ significantly from *Hybodus reticulatus* in skull structure and body form. Maisey (1987) recommended that other species described on less complete material should be retained in *Hybodus*. However, Underwood and Cumbaa (2010) described the genus *Meristodonoides* from a population of isolated, well-preserved teeth from Upper Cenomanian strata on the eastern side of the Western Interior Seaway in central Canada. *Meristodonoides* teeth superficially resemble some of the teeth examined during the course of this project.

Another point of confusion arises with differentiating teeth of the superficially similar *Polyacrodus* from *Hybodus*.

Histologically, *Hybodus* teeth are strikingly different, with *Polyacrodus* possessing tooth crowns composed of orthodentine and the more derived *Hybodus* possessing tooth crowns of osteodentine (Maisey, 1982, 1986, 1987; Cappetta, 1987b; Welton and Farish, 1993).

HYBODUS sp.

Discussion *Hybodus* and related genera (Figs. 9.3, 9.4) are widely distributed in marginal marine environments throughout the Cretaceous of the Western Interior of North America (Underwood, pers. comm., 2007). No attempt is made to distinguish species among the specimens discussed here. However, because hybodont teeth found in shallow open marine facies are much more abundant than those found in what are interpreted as freshwater habitats, we postulate that these marine sharks were frequent visitors to

Kirkland, Eaton, and Brinkman

riverine environments along the southwestern margin of the Western Interior Seaway. These are by far the largest sharks found in these environments and would have commonly reached 2 m or more in length. Ranging throughout the Mesozoic, they are considered to be a successful group of generalized fish-eating sharks.

Within proposed freshwater habitats, *Hybodus* teeth have been most commonly encountered in the upper Cenomanian middle carbonaceous member of the Dakota Formation, where they are commonly well preserved with complete roots (Fig. 9.3A–C). At other sites, the fragile roots are not preserved (Fig. 9.4G, H). Within the large population of *Hybodus* teeth from MNA 1067, Maisey (pers. comm., 1998) recognized two tooth morphologies, with one form bearing more than six striae on the labial surface of the principal cusp and labial surface of the root, and with small pores and a concave face, while the second form bore five to six striae, and a root with relatively large pores and a flat to rounded face. *Hybodus* teeth are less common from stratigraphically higher microvertebrate sites; they invariably have the roots broken off and, in the majority of examples, the lateral margins of the teeth as well (Fig. 9.4G, H).

Dorsal fin spines are only known from one site just to the west of Grand Staircase–Escalante National Monument in the basal John Henry Member (?upper Coniacian) at UMNH Loc. 1064. At this site, three fragments representing two large dorsal fin spines were recovered. These may represent associated spines from one large hybodont shark. In overall form, they compare well to other hybodont fin spines (Maisey, 1978; Cappetta, 1987b; Welton and Farish, 1993). As is diagnostic of hybodont fin spines, alternating pairs of enameloid, recurved denticles run up the posterior side of the fin spines above the basal groove. However, the apices of all of the denticles are broken off. A primary longitudinal ridge runs the midline of the anterior edge, with five additional equispaced longitudinal ridges on each side becoming more weakly expressed and not extending past the midflank of the spine. Unlike many hybodont shark species, the longitudinal ridges are continuous and without any enameloid on them. Between the longitudinal ribs and on the flank of the spines posterior to them, a fine reticulate ornamentation is oriented longitudinally on the teeth as well (Fig. 9.4A–E), representing Maisey's (1978) intercostal grooves. The fin spines of the Jurassic *Hybodus reticulatus* have many more longitudinal costae covering most of the lateral surfaces of the spine (Maisey, 1987), whereas the Jurassic hybodont *Egertonodus fraasi* has six longitudinal costae that extend back a bit more posteriorly than the Utah spines (Maisey, 1986). Fin spines associated with *Meristodonoides* are probably tuberculate and lack strong longitudinal ridges (Underwood and Cumbaa,

2010). Case (1987) figured two dorsal fin spine fragments that differ in possessing smaller more closely spaced rows of denticles and in having more posteriorly situated longitudinal ridges as pertaining to *Hybodus wyomingensis*.

LONCHIDIIDAE Herman, 1977b

Discussion Cappetta (1987b) placed both *Lissodus* and *Polyacrodus* together in the family Polyacrodontidae. Milner and Kirkland (2006) accepted this in describing a Lower Jurassic species of *Lissodus* from southern Utah. However, Maisey (1990) provided strong arguments for retaining the family Lonchidiidae for these taxa, which has been followed by most subsequent authors (Duffin, 2001; Rees, 2002; Rees and Underwood, 2002) and is followed herein.

LONCHIDION Estes, 1964

Type Species *Lonchidion selachos* Estes (1964).

Emended Diagnosis Teeth gracile, narrow labiolingually; central cusp low, distinct; dorsal margin of crown forms crest that may bear lows cusplets, labial protuberance narrow, well developed; lateral margins of crown pointed, often forming distinct lateral cusps, root generally more extensive than base of crown; labial face of dorsoventrally concave; small foramina irregularly spaced along the crown–root margin; cephalic spine base plate "convict arrow–shaped"; lateral sides of dorsal fin spines smooth.

Discussion *Lonchidion selachos* was initially described from the upper Maastrichtian Lance Formation of Wyoming on the basis of numerous oral teeth, cephalic spines, and dorsal fin spines (Estes, 1964). Also included were tiny orectoloboid teeth described below, which have been interpreted as symphysial teeth (Herman, 1977b). Subsequently, most small polyacrodontid teeth from the late Paleozoic through Mesozoic were all synonomized within the genus *Lissodus* (Duffin, 1985), with the type species being the Lower Triassic South African species *Lissodus africanus*. This has been followed by many subsequent researchers (Rowe et al., 1992). In describing the skeleton of *Lissodus cassangensis* from the Lower Triassic of Angola, a close relative of *Lissodus africanus*, Maisey (1990) noted that both Duffin (1985) and Cappetta (1987b) in diagnosing *Lissodus*, relied heavily on teeth previously assigned to *Lonchidion* rather than the type species. Maisey (1990) first noted differences between the cephalic spines of *Lissodus* and *Lonchidion selachos* and commented that Duffin's (1985) revised criteria for *Lissodus* were extremely broad and that greater generic subdivision was preferable. Rees and Underwood (2002), in their reappraisal of the shark taxa assigned to *Lissodus*, noted

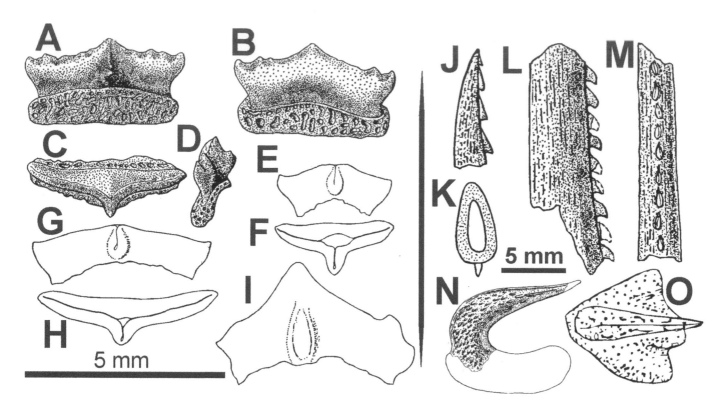

9.5. *Lonchidion selachos* type material from Estes (1964). Holotype tooth (UCMP 53897) from Estes (1964:fig. 1), in (A) labial, (B) lingual, (C) occlusal, and (D) lateral views. Tooth (UCMP 56273) from Estes (1964:fig. 3b) in (E) labial and (F) occlusal views. Tooth (UCMP 56274) from Estes (1964:fig. 3a) in (G) labial and (H) occlusal views. Large tooth (UCMP 56275) from Estes (1964:fig. 4c) in (I) labial view. Unnumbered tip of dorsal fin spine from (UCMP V-5620) from Estes (1964:fig. 4b) in (J) lateral view. Midsection of dorsal fin spine UCMP 56275 from Estes (1964:fig, 4a) in (K) cross-sectional, (L) lateral, and (M) posterior views. Cephalic spine with base restored (UCMP 53906) from Estes (1964:fig. 3d) in (N) lateral and (O) dorsal views. Scale bars = 5 mm.

distinctive features in the tooth crowns of *Lonchidion* that facilitated diagnosis as a distinct taxon.

We also note that the dorsal fin spines of *Lonchidion selachos* are smooth laterally (Fig. 9.5J, K) and lack the lateral ridges characteristic of *Lissodus* (Estes, 1964; Carpenter, 1979; Bryant, 1989; Maisey, 1990; Milner and Kirkland, 2006). All the dorsal fin spines associated with *Lonchidion* from the Upper Cretaceous of the Western Interior are of this distinct morphology, whereas no other known fin spines assigned to *Lonchidion* are known to have this morphology. The ridged dorsal fin spines from the Lower Cretaceous of Great Britain (Patterson, 1966; Cappetta, 1987b) and the Upper Triassic of southwestern North America ("*Lonchidion*" *humblei*, Murry, 1981) suggest these species should be assigned to another genus because the ornamentation of the dorsal fin spines appears to be a conservative character.

A large number of species have been assigned to *Lonchidion* (Cappetta, 1987b; Rees and Underwood, 2002) on the basis of the morphology of oral teeth. We distinguish these teeth from *Lonchidion selachos* but made no attempt to separate these teeth into species.

LONCHIDION SELACHOS Estes, 1964
Lonchidion selachos Estes, 1964:7–12
in part, figs. 1, 3a, b, d, 4
Lonchidion selachos Estes, 1964,
Carpenter, 1979:38, figs. 2–3

Type Specimen UCMP 53917.

Diagnosis Large *Lonchidion* taxon with more trenchant teeth than other species.

Distribution Only known from the upper Maastrichtian Laramie, Ferris, and Lance formations of Wyoming and the Hell Creek Formation of Montana and North and South Dakota (Estes, 1964, 1969; Carpenter, 1979; Wroblewski, 2004; Cook et al., in press).

Discussion This taxon was figured extensively by Estes (1964), who included a diversity of oral teeth, dorsal fin spines, and part of a cephalic spine (Fig. 9.5). He also figured a large partial tooth, demonstrating that this, the terminal species of *Lonchidion*, is also the largest species. He later noted (Estes, 1969) that the symphysial teeth he figured did not belong to *Lonchidion*. These teeth have been reassigned to the orectoloboid species *Cantioscyllium estesi* (Herman, 1977b).

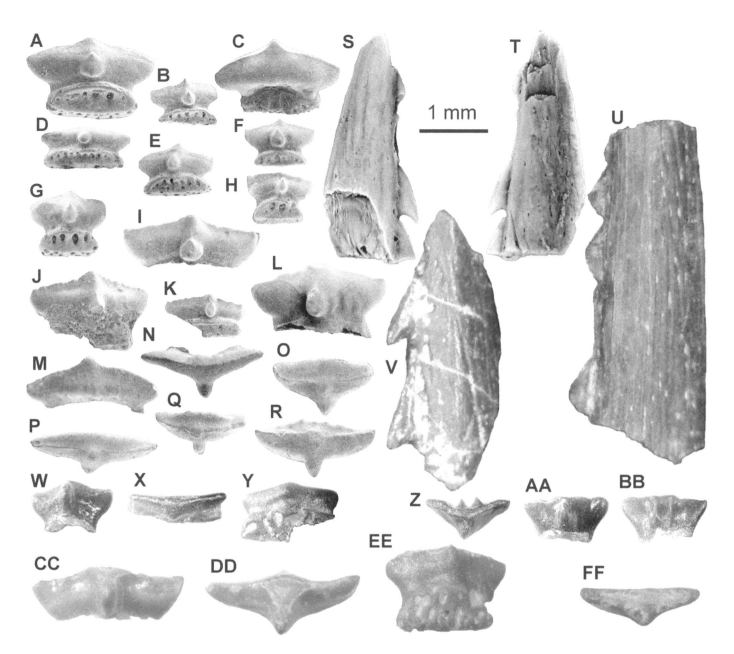

9.6. *Lonchidion* from southern Utah. Teeth from the (middle–upper Cenomanian) Dakota Formation at Bulldog Bench, easternmost Kaiparowits Basin (MNA locality 1067). Teeth (MNA V10284) in (A) labial view; (MNA V10285) in (B) labial view; (MNA V10286) in (C) labial view; (MNA V10287) in (D) labial view; (MNA V10288) in (E) labial view; (MNA V10289) in (F) labial view: (MNA V10290) in (G) labial view; (MNA V10291) in (H) labial view; (MNA V10294) in (I) labial view; (MNA V10296) in (J) labial view; (MNA V10297) in (K) labial view; (MNA V10293) in (L) labial view; (MNA V10292) in (M) labial view; (MNA V10295) in (N) labial view; (MNA V10300) in (O) labial view; (MNA V10299) in (P) labial view; (MNA V10298) in (Q) labial view; and (MNA V10301) in (R) labial view. Dorsal fin spine fragments (MNA V10381) in (S) left lateral and (T) right lateral views; (MNA V10382) in (U) right lateral view; and (MNA V10383) in (V) right lateral view. Teeth from lower John Henry Member, Straight Cliffs Formation (UMNH Loc. 8, upper Coniacian), in Cedar Canyon, western Markagunt Plateau (UMNH VP 18911), in (W) labial view; (UMNH VP 18912) in (X) occlusal and (Y) labial views; and (UMNH VP 18913) in (Z) occlusal, (AA) labial, and (BB) lingual views. Tooth from upper John Henry Member, Straight Cliffs Formation (UMNH Loc. 417, Santonian), on eastern Paunsaugunt Plateau (UMNH VP 18922 (Loc. 417) in (CC) occlusal and (DD) labial views. Tooth from Wahweap Formation (Loc. 82, middle Campanian) (UMNH VP 18917) in (EE) occlusal and (FF) labial views.

Examples of *Lonchidion selachos* reported from the Aguja Formation (Campanian) of Texas (Rowe et al., 1992; Welton and Farish, 1993; Sankey, 2008) and the Fruitland Formation (Campanian) of New Mexico (Armstrong-Ziegler, 1978; Hutchinson and Kues, 1985) need to be reexamined in order to assess the presence of this species in older strata. It should be noted that none of the figured specimens in these articles

are as elongate or trenchant as are the lateral teeth of *L. selachos* from late Maastrichtian strata.

LONCHIDION sp.

Discussion No attempt is made here to differentiate Utah's species of *Lonchidion* (Fig. 9.6). Numerous species have been described from the Cretaceous globally, but there have been no attempts to differentially diagnose these species relative to each other. Within Utah, teeth that can be assigned to this genus are recognized in nearly all fresh to estuarine microvertebrate localities in the Upper Cretaceous, ranging from the basal Cenomanian Mussentuchit Member of the Cedar Mountain Formation (Cifelli et al., 1997, 1999b) up through the upper Campanian Kaiparowits Formation (Eaton et al., 1997, 1999). Specimens from the upper Cenomanian Dakota Formation at Grand Staircase–Escalante National Monument are particularly abundant and well preserved with complete roots (Fig. 9.6A–R). These sites also preserve fragmentary dorsal fin spines (Figs. 9.6S–U) and cephalic spines.

Although mostly recorded from freshwater microvertebrate sites, Kirkland (1989) has identified specimens of *Lonchidion* in a marine phosphate pebble bed in the middle Turonian Carlile Shale in northeastern Nebraska. Other marine occurrences from Texas (Welton and Farish, 1993), suggest these species range from marine to freshwater habitats. C. Suarez (pers. comm., 2010) found that stable oxygen isotopes from *Lonchidion* teeth from the basal Cenomanian Mussentuchit Member of the Cedar Mountain Formation were consistent with these sharks living in freshwater, although she reports that additional analyses are needed.

All the *Lonchidion* teeth examined from Utah are considerably smaller than and not as trenchant as teeth of *Lonchidion selachos* from the Upper Maastrichtian. The dorsal fin spines from the Dakota Formation are the oldest examples we have found of the smooth hybodont fin spines of the morphology documented for *Lonchidion selachos*.

NEOSELACHII Compagno, 1977
GALEOMORPHII Compagno, 1973
ORECTOLOBIFORMES Applegate, 1972

Discussion Cappetta (1987b) noted many weaknesses in the systematics of orectoloboid sharks, particularly in regard to the diversity of species within recent genera and in "middle" Mesozoic lagerstätten and the diversity of intervening tooth taxa, as described below. This has been made clear to us during the course of this research, and the small orectolobiform sharks from the Western Interior should continue to be a fruitful area for further research as the taxonomic relationships within this group become clarified.

No orectoloboid sharks are known to inhabit freshwater environments today (Cappetta, pers. comm., 2011). Thus, the sharks species reported herein from these facies are possibly unique. However, it is clear that these occurrences include fully freshwater habitats, as indicated by the distance of these sites (in some cases more than 100 km) from the paleocoastline; the co-occurrence of frogs, salamanders, lizards, and a variety of mammals; and an absence of clearly marine and brackish-water vertebrate and invertebrate fossils at most of these sites, as noted in numerous chapters in this volume.

Given the classification of orectolobiform taxa applied here, two endemic orectoloboid lineages inhabited freshwater habitats in the Late Cretaceous in the Western Interior of North America. These sharks represent ginglymostomatids (nurse sharks), and hemiscyllids (bamboo sharks). These are all bottom-dwelling sharks that utilize suction feeding and their small teeth to secure a variety of food resources.

GINGLYMOSTOMATIDAE Gill, 1862
CANTIOSCYLLIUM Woodward, 1889

Type Species *Cantioscyllium decipiens* Woodward (1889), Cenomanian, Great Britain.

Emended Diagnosis Tiny teeth with broad central cusp inclined 45–60° lingually, bearing well-developed lateral cusplets, and broad, somewhat bifid lingual apron; root with broad, subtriangular central foramina opening to labial side, giving root a heart-shaped appearance in ventral view in the holaulcorhize condition. Labial surface of crown is typically ornamented by irregular plicae. In addition to a pair of marginolingual foramina, there is a central lingual duct, which in a few examples from large populations is open basally, forming a medial canal in the holaulcorhize condition.

Discussion Cappetta (2006) reported that both *Brachaelurus estesi* (Herman, 1977b) and *Brachaelurus bighornensis* (Case, 1987) belonged to the genus *Protoginglymostoma* (Herman, 1977a). He has more recently supported assigning these taxa to *Cantioscyllium* (Cappetta, in press, pers. comm., 2011), as has been done by Kirkland, Eaton, and Brinkman (2009).

The ginglymostomatid *Cantioscyllium* is known from marine strata of Cenomanian through Campanian age in North America and Europe (Cappetta, 1987b, in press; Cappetta and Case, 1999). Within the Western Interior, *Cantioscyllium decipiens* is particularly abundant in shallow marine rocks of Turonian age (Cappetta, 1973; Kirkland, 1989; Williamson, Kirkland, and Lucas, 1993), with a similar form ranging up into the Coniacian of Texas (*Cantioscyllium* aff. *C. decipiens* Cappetta and Case, 1999). The similarities in morphology of Coniacian and Santonian examples of *Cantioscyllium* from freshwater deposits in Utah, together with their first occurrence in freshwater facies of Coniacian age, suggest to us that this group of fresh and brackish-water elasmobranchs may be derived from marine species of *Cantioscyllium* at the beginning of the Coniacian (Kirkland, Eaton, and Brinkman, 2009).

Kirkland, Eaton, and Brinkman

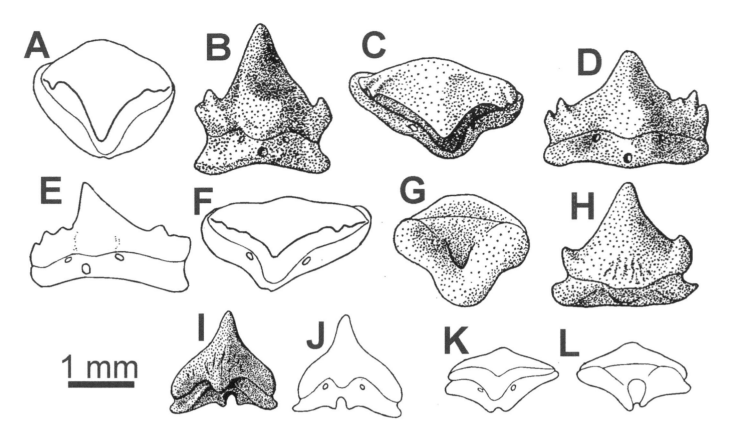

9.7. Orectoloboid shark teeth from Estes (1964). *Cantioscyllium estesi* (Herman, 1977b) originally figured as symphysial teeth of *Lonchidion* (UCMP 56272) from Estes (1964:fig. 2a) in (A) occlusal and (B) lingual views; (UCMP 53902) from Estes (1964:fig. 2c) in (C) occlusal and (D) lingual view; (UCMP 53903) from Estes (1964:fig. 3c) in (E) lingual and (F) occlusal view; (UCMP 53901) Estes (1964:fig. 2b) in (G) labial, and (H) basal view. *Restesia americana* (Estes, 1964) originally figured as *Squatirhina americana,* holotype (UCMP 55820) from Estes (1964:fig. 5) in (I) labial, (J) lingual, (K) occlusal, and (L) basal views.

The variation in root vascularization reported here for *Cantioscyllium* was also documented by Case (1987:pl. 6, fig. 5) in *Cantioscyllium bighornensis* in that AMNH 12085 had a root with an open canal in the holaulcorhize condition.

CANTIOSCYLLIUM ESTESI (Herman, 1977b)

Lonchidion selachos Estes, 1964:7–
12 in part, figs. 2, 3c
?*Squatina* sp. Langston, 1975:1598–
1599 in part, fig. 16B
Mesiteia? *estesi* nov. sp. Herman, 1977b:148–150
?*Squatirhina* sp. Langston, 1975:1597–
1598 in part, fig. 16D
?*Brachaelurus hornerstownensis* n. sp.
Case, 1979:117, pl. 2, figs. 23–26
Protoginglymostoma estesi (Herman,
1977b) Cook et al., in press

Type Specimens Holotype UCMP 56272. Paratypes UCMP 53901, 53902, and 53903 UCMP Loc. v-5620 (Herman, 1977b).

Diagnosis Species of *Cantioscyllium* (Figs. 9.7A–H, 9.8) with one to two lateral cusplets and smooth, or with few weakly developed short plications on central labial crown surface.

Distribution Upper Maastrichtian Lance Formation of Wyoming and Hell Creek Formation of Montana, and possibly upper Maastrichtian of New Jersey and uppermost Campanian St. Mary's River Formation of Alberta.

Discussion Estes (1964) initially described these teeth as symphysial teeth of *Lonchidion selachos* but later noted that they did not belong to *Lonchidion* (Estes, 1969). Herman (1977b) subsequently recognized these teeth as being from an orectoloboid shark, questionably assigned them to the genus *Mesiteia*, and described them as the new species *Mesiteia? estesi,* assigning the teeth figured by Estes (1964, Fig. 2; U.C. 56.272, 53.901, and 53.903) as the type specimens. Subsequent authors (Case, 1979; Bryant, 1989) have placed this species into the orectoloboid genus *Brachaelurus,* from whose species it differs in having an ornate crown and lateral cusplets not so well separated from the crown. Cappetta (1987b) has noted that most fossil species assigned to *Brachaelurus* belong to other genera.

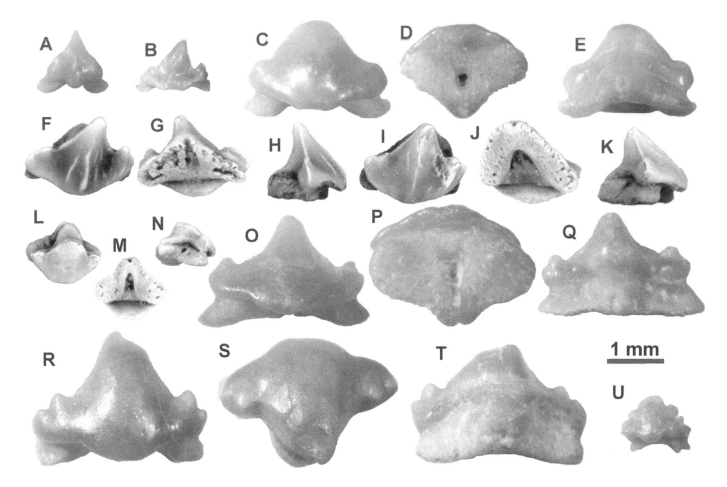

9.8. *Cantioscyllium estesi* (Herman, 1977b) teeth from the Wahweap Formation. Teeth from MNA Loc. 705 (MNA V10232) in (A) labial view; (MNA V10230) in (B) labial view; and (MNA V10231) in (C) labial, (D) basal, and (E) lingual views. Teeth from MNA Loc. 455–1 (MNA V10384) in (F) occlusal, (G) labial, and (H) lateral views; (MNA V10385) in (I) occlusal, (J) labial, and (K) lateral views; (MNA V10390) in (L) occlusal, (M) labial, and (N) lateral views. Teeth from UMNH Loc. 82 (UMNH VP 18915) in (O) labial, (P) basal, and (Q) lingual views; (UMNH VP 18916) in (R) labial, (S) basal, and (T) lingual view; and (UMNH VP 18919) in (U) labial view.

CANTIOSCYLLIUM BIGHORNENSIS (Case, 1987)

Brachaelurus bighornensis Case,
1987:20, pl. 6, figs. 1a–6e
?*Chiloscyllium missouriense* Case, 1987, Peng,
Russell, and Brinkman, 2001:7–8, figs. 18–2

Diagnosis Species of *Cantioscyllium* (Fig. 9.9K–DD) with one to two low lateral cusplets and weakly developed short labial plications on central labial crown surface.

Distribution Present throughout the John Henry Member of the Straight Cliffs Formation of the Paunsaugunt and Kaiparowits Plateaus in southern Utah, the upper Campanian of Wyoming, and possibly the Oldman and Foremost formations of southern Alberta.

Discussion Case (1987) described this species from the upper Campanian Teapot Sandstone of the Mesaverde Formation. Teeth figured from the middle Campanian Foremost and Oldman formations of Alberta by Peng, Russell,

and Brinkman (2001) are similar, but appear to be narrower with taller lateral cusps.

In a few examples, the central lingual duct is open basally, forming a medial canal in the holaulcorhize condition.

As with *P. estesi*, some specimens assigned to *P. bighornensis* have a smooth labial surface. It is tempting to synonymize *P. bighornensis* with *P. estesi* because there is really nothing different in these taxa except that *P. estesi* teeth tend to be more weakly ornamented on the labial surface of crown, have fewer plications, and are more commonly smooth. If future research supports synonymizing these species, it would extend the range of *Cantioscyllium estesi* from the upper Coniacian up to the top of the Maastrichtian.

CANTIOSCYLLIUM MARKAGUNTENSIS n. sp.

Type Specimen UMNH VP 18909 from UMNH Loc. 8, Straight Cliffs Formation (possibly equivalent to lower John Henry Member or older), Cedar Canyon, western Markagunt Plateau.

Kirkland, Eaton, and Brinkman

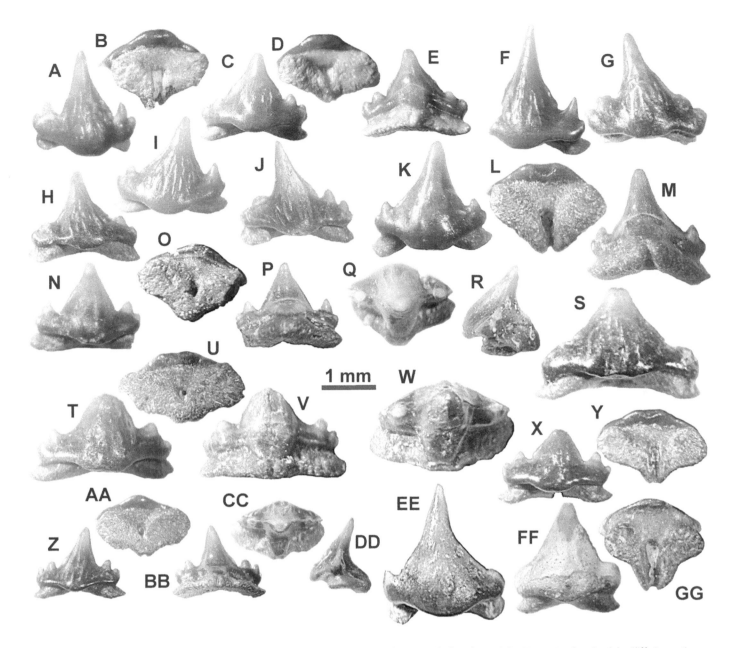

9.9. Orectoloboid shark teeth from the Straight Cliffs Formation. *Cantioscyllium markaguntensis,* from lower John Henry Member, Straight Cliffs Formation (Coniacian, UMNH Loc. 8) in Cedar Canyon, western Markagunt Plateau. (UMNH VP 18905) in (A) labial and (B) basal views; (UMNH VP 18910) in (C) labial, (D) basal, and (E) lingual views; (UMNH VP 18904) in (F) labial view; (UMNH VP 18908) in (G) labial view; (UMNH VP 18907) (H) in labial view; holotype (UMNH VP 18909) in (I) labial view; (UMNH VP 18906) (J) in labial view. *Cantioscyllium bighornensis,* John Henry Member, Straight Cliffs Formation (Santonian), from UMNH Loc. 567 (UMNH VP 17061) in (K) labial, (L) basal, and (M) lingual view; (UMNH VP 17059) in (N) labial, (O) basal, (P) lingual, (Q) occlusal, and (R) lateral views; and (UMNH VP 17055) in (T) labial, (U) basal, (V) lingual, and (W) occlusal views; and from UMNH Loc. 98 (UMNH VP 18903) in (S) labial view; (UMNH VP 18901) in (X) labial and (Y) basal views; (UMNH VP 18899) in (Z) labial, (AA) basal, (BB) lingual, (CC) occlusal, and (DD) lateral views. A possible specimen of *Cantioscyllium estesi?* (Herman, 1977b) from UMNH Loc. 98 (UMNH VP 18902) in (EE) labial, (FF) labial, and (GG) basal views.

Paratype Specimens UMNH VP 18904, 18905, 18906, 18907, 18908, all from UMNH Loc. 8, lower Straight Cliffs Formation, Cedar Canyon, western Markagunt Plateau.

Etymology For the Markagunt Plateau, where the only known examples of this species were recovered.

Diagnosis Tiny species (less than 2 mm wide) characterized by usually having at least two lateral cusplets and strong placations on the labial surface of crown (Fig. 9.9A–J).

Description The holotype UMNH VP 18909 (Fig. 9.9 I) is a typical anterior lateral tooth of the species. It is 1.98 mm across and is 1.77 mm tall from the apex of the central cusp to the tip of the uvula with a root 2.07 mm across. There is a slight cingulum arising from the lateral cusps and extending across the base of the lingual face of the tooth crown. From this cingulum, 10 discontinuous and bifurcating placations arise and extend toward the apex of the tooth. There

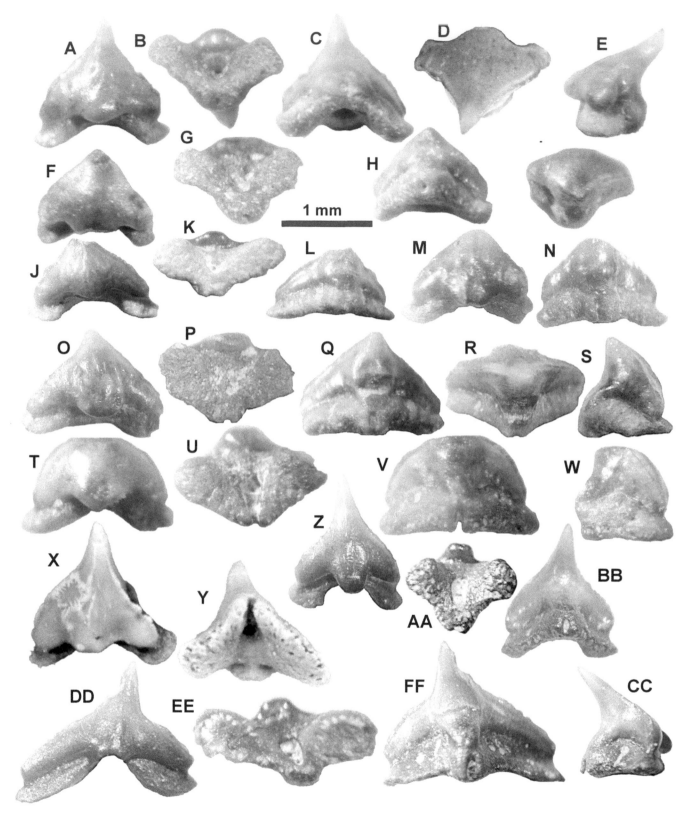

9.10. *Chiloscyllium* from Utah. *Chiloscyllium missouriense* Case (1987) from the Wahweap Formation (middle Campanian) at UMNH Loc. 77 on the eastern side of the Paunsaugunt Plateau (UMNH VP 18880) in (A) labial, (B) basal, (C) lingual, (D) occlusal, and (E) lateral views; (UMNH VP 18882) in (F) labial, (G) basal, (H) lingual, and (I) occlusal views; (UMNH VP 18886) in (J) labial, (K) basal, and (L) lingual views; (UMNH VP 18887) in (M) labial and (N) lingual views; (UMNH VP 18888) in (O) labial, (P) basal, (Q) lingual, (R) occlusal, and (S) lateral views; (UMNH VP 18881) in (T) labial, (U) basal, (V) lingual, and (W) lateral views. *Chiloscyllium missouriense Case (1987)* tooth from MNA Loc. 455–1 with divided root similar in morphology to *Restesia americana* (MNA V 10386) in (X) labial and (Y) basal views. *Chiloscyllium missouriense* Case (1987) from UMNH Loc. 51 in the upper Campanian Kaiparowits Formation (UMNH VP 18896) in (Z) labial, (AA) basal, (BB) lingual, and (CC) lateral views; and (UMNH VP 18897) in (DD) labial, (EE) basal, and (FF) lingual views.

Kirkland, Eaton, and Brinkman

are two lateral cusps on each side of the crown. The root is heart shaped with a subtriangular central foramina extending within the root to a tiny foramen on labial side of the root. In some teeth, such as UMNH VP 18905 (Fig. 9.9 A, B), this canal is open, bifurcating the base of the tooth.

Distribution Known only from Straight Cliffs Formation of the Markagunt Plateau in Cedar Canyon (UMNH Loc. 8, west of the Grand Staircase–Escalante National Monument) well above the highest brackish-water mollusk fauna (Eaton et al., 2001).

Discussion The strongly ornamented labial surface of the crown in this taxon sets it apart from any other orectoloboids collected from freshwater facies in southern Utah. As this distinctive taxon is not known from the base of the John Henry Member to the east on the Paunsaugunt and Kaiparowits plateaus, it is suggested that perhaps older Coniacian strata are present in the western Markagunt Plateau area. Thus, *Cantioscyllium markaguntensis* would be the oldest species of orectolobiform shark from freshwater facies in southern Utah. Alternatively, this ornate species may be restricted to the west in waters more proximal to the upland Sevier Orogenic Belt.

HEMISCYLLIDAE Gill, 1862
CHILOSCYLLIUM Muller and Henle, 1838–1841

Type Species *Chiloscyllium plagiosum* (Bennett, 1830), recent.

Diagnosis Tooth crowns in modern taxa essentially smooth with broad central cusps and well-developed lateral cusplets and broad somewhat bifid lingual apron or with narrow, tall central cusps and weakly developed or no lateral cusplets. Root with broad, subtriangular central foramina opening to labial side giving root a heart-shaped appearance in ventral view in the holaulcorhize condition. In addition to a pair of marginolingual foramina, there is a central lingual duct.

Discussion Cappetta (1987b) noted that within the Hemiscyllidae, many of the Cretaceous taxa are strongly ornamented by folds in their enamel, while the teeth of modern taxa are practically smooth except for the short ridge on central cusp of modern *Chiloscyllium*. There are few differences in the teeth among the many modern species.

CHILOSCYLLIUM MISSOURIENSE Case, 1987
?*Squatirhina americana* Estes, 1964.
Langston, 1975:1594, fig. 16C
Chiloscyllium missouriense Case,
1987:19, text-fig. 7a–e
Non–*Chiloscyllium missouriense* Case, 1987, Peng,
Russell, and Brinkman, 2001:7–8, figs. 18–20
?*Chiloscyllium missouriense* Case,
1987, Cook et al., in press

Diagnosis Tiny teeth (1–2 mm), squat, triangular with cusp leaning lingually; no definitive lateral cusplets, carinae arise from central cusp extend along broad, sloping shoulders may appear serrate from faint wrinkling; the labial surface of crown usually bears traces of weak, flexuous enamel folds or plications oriented with apex crown. Some teeth are smooth.

Distribution Middle Campanian Wahweap Formation and upper Campanian Kaiparowits Formation of Utah, upper Campanian Mesaverde Formation of Wyoming (Case, 1987), uppermost Campanian St. Mary's River Formation of Alberta, and possibly the upper Maastrichtian Scollard Formation of Alberta (Cook et al., in press).

Discussion Case (1979) erected *Chiloscyllium missouriensis* (Fig. 9.10) on the basis of teeth recovered from estuarine deposits of the Judith River Formation (Campanian) of Montana. This taxon is also recognized in the Mesaverde Formation of Wyoming (Case, 1987). Langston (1975) describes several teeth as *Squatirhina americana* from the latest Campanian St. Mary's River Formation of Alberta that compare well with *Chiloscyllium missouriensis*. The report of Peng, Russell, and Brinkman (2001) of this taxon from the Foremost and Oldman formations of Southern Alberta, Canada, appears to actually represent a species of *Cantioscyllium* as these teeth have well-developed lateral cusplets. The report of this taxon from the upper Maastrichtian by Cook et al. (in press) appears to actually represent an undescribed species close to *C. missouriense* because the former's teeth have a transverse ridge across the labial face of the crown that is not present in the type specimens of *C. missouriense*.

In a few examples, UMNH VP 18888 (Fig. 9.10T–W) and MNA V10386 (Fig. 9.10X, Y), the central lingual duct is open basally, forming a medial canal in the holaulcorhize condition. That this variation in root morphology is also represented in *Cantioscyllium* leads us to believe that *C. missouriense* is derived from *Cantioscyllium bighornensis* by reduction and loss of the lateral cusplets. Examples of *C. missouriense* from the upper Campanian Kaiparowits Formation of Grand Staircase–Escalante National Monument also occur with both root morphologies (Fig. 9.10X–CC) and compare well with the crown morphology of *Restesia americana* and share the narrower, pointed lingual apron. Therefore, *C. missouriense* may not belong in this genus or even belong to the Hemiscyllidae. If so, then *C. missouriense* may represent an orectoloboid shark intermediate between *P. bighornensis* and *R. americana*, which creates taxonomic problems beyond the scope of this chapter.

ORECTOLOBIDAE Gill, 1895
RESTESIA Cook et al., in press

Type Species *Squatirhina americana* Estes, 1964, upper Maastrichtian, Wyoming.

Diagnosis Minute teeth with sharp central cusp, taller on anterior teeth, extending to narrow, pointed lingual apron extending beyond root; well-developed convex shoulders may be slightly wrinkled, extending down to level of lingual apron with no lateral cusplets; heart-shaped root in ventral view with completely open root canal dividing root into two lobes in the holaulcorhize condition; a pair of marginolingual foramina may be aligned with one or more smaller foramina.

Discussion *Squatirhina americana* has long been a problematic taxon. Casier (1947b) considered the original genotype (*S. lonzeensis*) from the Santonian of Lonzee, Belgium, to be intermediate between those of squatinids and orectolobids on the basis of crown morphology, and intermediate between those of scyliorhinids and rhinobatids on the basis of root morphology (Estes, 1964; Cappetta, 1987b). Arambourg (1952), Herman (1977b), and Estes (1964) suggested *Squatirhina* had Orectolobiformes (nurse and carpet shark) affinity, with the latter author noting that the general tooth morphology of *S. americana* was that of an orectolobiform.

In true *Squatirhina*, two different tooth forms have been recognized: one with a short cusp, the other with a long cusp. According to Herman (1977b) these two types might reflect a sexual dimorphism within one species (i.e., different tooth shapes in male and female individuals). Such sexual dental dimorphism is widespread in the batoidea (Cappetta, 1987b), lending support to his assignment to the rhinobatoids on the basis of root morphology. There is no evidence of sexual dimorphism in the teeth that were assigned to *Squatirhina americana* by Estes (1964).

Cappetta (2006) agreed with Estes (1964) that *Squatirhina americana* belonged in Orectolobiformes and proposed that it represented a new genus. Cook et al. (in press) erected the genus *Restesia* for this species based on examining a population of teeth from the Hell Creek Formation.

The origins of *Restesia americana* are problematic. One scenario would be to derive the species from a possible squatinid tooth figured by Rowe et al. (1992) from the Aguja Formation, by decreasing the number of accessory nutritive foramina and extending the height of the shoulders onto the margin of the central cusp. However, a second and simpler scenario would have the species originating from *Chiloscyllium missouriensis* by lengthening the lingual apron and opening the root canal in all individuals, the view that is favored here.

RESTESIA AMERICANA (Estes, 1964)
Squatirhina americana Estes, 1964:12–13, fig. 5
Squatirhina americana Estes, 1964,
Carpenter, 1979:38, 40, fig. 4
Non–*Squatirhina americana* Estes,
1964, Rowe et al., 1992:476, fig. 3E
?*Squatirhina americana* Estes,
1964, Cifelli et al., 19991:379
Orectolobid, gen. et sp. indet., Neuman
and Brinkman, 2005:168, fig. 9.1A

Diagnosis As for genus.

Discussion *Squatirhina americana* (Figs. 9.7I–L, 9.10X, Y) was named by Estes (1964) with its type specimen being UCMP 55820 from the upper Maastrichtian Lance Formation. This taxon has not been recognized in Utah, but it may represent a species derived from one of the orectoloboid lineages described herein from Utah.

A single tooth from the Wahweap Formation, MNA V10386 (Fig. 9.10X, Y), assigned to *Chiloscyllium missouriense* appears to be nearly identical to teeth assigned to *Restesia americana* (Estes, 1964; Cook et al., in press). However, its only distinguishing character is the divided root, which occurs sporadically in every population of orectolobiforms examined from southern Utah.

Rowe et al. (1992) figured a tooth from the upper Campanian Aguja Formation of west Texas that is superficially similar to *Restesia americana* but differs in having a broad, slightly bifid lingual apron and multiple pairs of well-developed lingual foramina. The tooth is also superficially similar to the larger *Cretorectolobus* (Case, 1978) and teeth figured as *Squatina haasi* from the St. Mary's River Formation (Langston, 1975), and Oldman and Foremost formations (Peng, Russell, and Brinkman, 2001), but differing in its broader lingual apron. Examination of a population of these Aguja teeth is required to determine whether these teeth represent an older species of *Restesia* or a species of *Squatina*, *Cretorectolobus*, or other carpet shark (Siverson, 1995).

Restesia americana may be present in the Maastrichtian North Horn Formation of central Utah, listed as *Squatirhina americana* in Cifelli et al. (1999b). However, these specimens were not reexamined in light of observations reported herein.

Kirkland, Eaton, and Brinkman

BATOMORPHII Cappetta, 1980
RAJIFORMES Berg, 1940
RHINOBATOIDEA incert. fam.

Discussion Isolated teeth of durophagous rays are generally common in Cretaceous nonmarine microvertebrate localities throughout the U.S. Western Interior. We have now recovered ray teeth from throughout the Upper Cretaceous fluvial sequence of Utah (Figs. 9.1, 9.2). Although there is considerable variation in the teeth of these rays, they are united by possessing a transverse crest extending laterally across the oral face of the flattened tooth crown. As a group, these crested teeth are referred to the *Myledaphus* lineage herein.

Durophagous teeth of the *Myledaphus bipartitus* have been assigned to the Batomorphii (Batoidea) since they were first described by Cope (1876) from the upper Campanian Judith River Formation of Montana. Cappetta (1987b, 1992) assigned the genus to the Rhinobatoidea incertae sedis, proposing that their morphological similarities to the durophagous teeth of myliobatiform stingrays were a matter of convergence and not shared characteristics. We have been able to trace the lineage back to tiny, smooth, crested teeth at the base of the Upper Cretaceous, suggesting that this group of durophagous rays was derived from a rhinobatoid with low-crowned teeth similar to *Rhompterygia* (Cappetta, 1987b). Although a possible *Myledaphus bipartitus* chondrocranium has been described (Langston, 1970), the rhinobatoid affinities of *Myledaphus bipartitus* waited to be proven until the discovery of a large *Myledaphus bipartitus* mummy from the upper Campanian Dinosaur Park Formation at Dinosaur Provincial Park in southern Alberta (Neuman and Brinkman, 2005). Thus, *Myledaphus* and its close relatives represent a line of guitarfish and not stingrays.

These crested-toothed rays have been recovered from both nonmarine and brackish environments and may well have ranged into the marine environments. Certain observations suggest these freshwater rays have links to the marine environments. These rays have not been recovered from lacustrine or paludal environments, suggesting that they required water bodies with open connections to the sea. During marked phases of regression, rays appear to be relatively rare in freshwater environments of Utah as marine waters shift eastward into the basin. However, Albian through Campanian shallow marine strata commonly contain the smooth-crowned crushing teeth of the rhinobatoid *Pseudohypolophus* but not teeth of the *Myledaphus* lineage (McNulty, 1964; Kirkland, 1989, 1990, 1996; Welton and Farish, 1993; Case et al., 2001).

Our attempt to quantify the range of tooth morphology in these rays has been greatly facilitated by Frampton's (2006) M.S. thesis, which included a study of the articulated jaws of *Myledaphus bipartitus* from the Dinosaur Provincial Park mummy, which permits assessment of tooth positions. The process used here was to identify typical teeth of a sample, focusing as much as possible on selecting typical teeth of each of the main morphotypes (hexagonal, rhomboidal, and symphysial) to represent each taxon. This does not reflect the intensive long-term study that needs to be done, including a statistical tooth-by-tooth analysis of the entire sample to determine whether some of the size distinctions observed here have taxonomic significance. We view such an intensive analysis, which is beyond the scope of this chapter, as the next necessary step toward advancing the systematics of this biostratigraphically useful group.

The large size range of teeth is remarkable, and tooth form may change throughout ontogeny, making it difficult to be certain whether very small teeth are simply from juveniles. As can be discerned in the descriptions of the other elasmobranchs herein, with taxa only represented by isolated elements, it is important to examine the morphology of many elements from a population of material from single localities. Thus, although it is nearly impossible to differentiate ontogenetic and positional differences with these isolated teeth, by attempting to describe the main tooth forms of symphysial, rhombic, and hexangular teeth for each species, based on Frampton (2006), a simplistic view of the main tooth morphologies of each species becomes clearer. Thus, variation based on extreme lateral position of teeth and the development of pathologic teeth resulting from injuries to the jaw is not considered in this taxonomy. Additionally, we recognize that small teeth from succeeding stratigraphic intervals tend to be more similar to each other than are large teeth, suggesting that identifying ontogenetic variation is also important in the taxonomy of these rays. Thus, instead of developing taxa with strongly overlapping ranges due to ontogeny and variability, we utilize a conservative taxonomy that requires a population of teeth from various tooth positions to identify species, whereas isolated teeth may be readily identified to genus.

We focus primarily on ray teeth recovered from the Upper Cretaceous sequence of southern Utah (Figs. 9.1, 9.2), but in order to extend the duration of the history presented, ray teeth were also examined from the Mussentuchit Member of the Cedar Mountain Formation in the San Rafael Swell (earliest Cenomanian?) and from the Lance Formation of Wyoming (Maastrichtian). Additionally, specimens were examined from a number of stratigraphic horizons in southern Alberta, Canada.

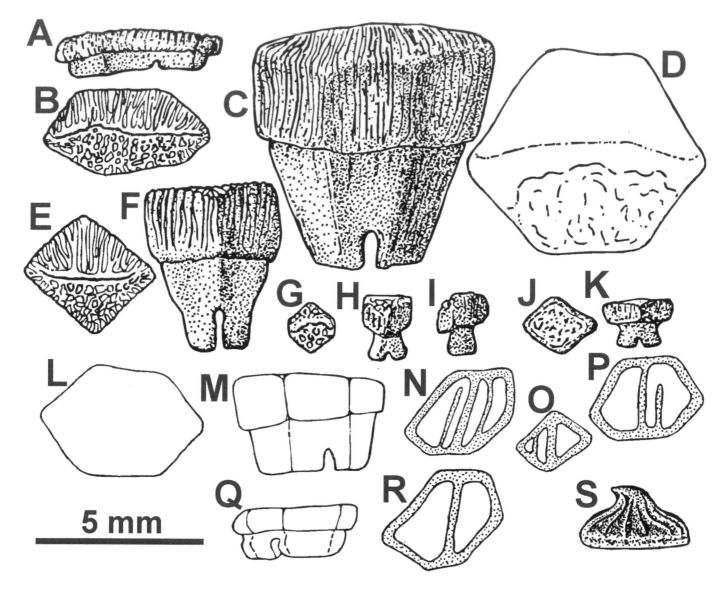

9.11. *Myledaphus pustulatus* Cook et al. (in press) from Estes (1964). Lower lateral tooth (AMNH 9302) from Estes (1964:fig. 7a) in (A) labial and (B) occlusal views. Upper lateral tooth (AMNH 9305) from Estes (1964:fig. 7d) in (C) labial and (D) occlusal views. Lower medial tooth (AMNH 9306) from Estes (1964:fig. 7e) in (E) occlusal and (F) labial views. Upper symphysial tooth (AMNH 9303) from Estes (1964:fig. 7b) in (G) occlusal, (H) lingual, and (I) lateral views. Possible symphysial tooth (AMNH 9304) from Estes (1964:fig. 7c) in (J) occlusal and (K) lingual views. Upper lateral tooth (AMNH 9309) from Estes (1964:fig. 8c), (L) shape of crown, and (M) labial profile. Root patterns of specimens with multiple roots from Estes (1964:fig. 8f), UWM 1777 in (N) basal view. AMNH 9312 (O) in basal view. AMNH 9313 in (P) basal view. Lower lateral tooth (AMNH 9311) from Estes (1964:fig. 8e) in (Q) labial profile and (R) basal view. Dermal denticle AMNH 9310 (Estes, 1964:fig. 8d) in (S) lateral view. All originally figured as *Myledaphus bipartitus* Cope (1876). Scale bar = 5 mm.

MYLEDAPHUS Cope, 1876

Type Species *Myledaphus bipartitus* Cope, 1876.

Emended Diagnosis A moderately large guitarfish with a durophagous dentition. Teeth characterized by moderate to high crowns with oral surface divided by a low transverse ridge that is slightly arched labially; oral crown also ornamented by labiolingually directed enamel folds; vertical faces of crown ornamented by vertical ridges and grooves that are continuations of folds on oral surface. Root often tapering away from crown and equal to, or more often taller than, crown, with a pair of lateral lingual foramina on smaller teeth, replaced with a series of slit-shaped foramina just below the crown on larger rhombic and hexangular teeth.

Discussion The first modern analysis of *Myledaphus* was by Estes (1964), where he figured a number of teeth from the upper Maastrichtian Lance Formation of Wyoming (Fig. 9.11). We have reinterpreted teeth that were referred to as symphysial teeth of *Myledaphus* as being oral teeth of *Texatrygon* (described below). Over the past 30 years we have observed that the crown ornamentation of upper Campanian and upper Maastrichtian teeth differ consistently. The ridges on the oral surface of the upper Campanian teeth are somewhat finer, and on the upper Maastrichtian teeth, the coarser ridges become reticulate on the labial half and break up into pustules. Thus, the upper Maastrichtian teeth represent a distinct unrecognized species that includes all of the teeth figured by Estes (1964) from the Lance Formation of Wyoming.

The genus *Myledaphus* is readily identified because even the smallest teeth (~1 mm across) preserve some labiolingual folding of the enamel on the oral surface of the crown (Fig. 9.11A–C).

Myledaphus has been described from the Turonian of south-central Asia in the Bessekty Formation, Kizylkum Desert, Uzbekistan, as *Myledaphus tritus* (Nessov, 1981; Nessov, Sigogneau-Russell, and Russell, 1994). Although the teeth are smaller than the teeth assigned to *Myledaphus* in North America, the overall morphology and ornamentation of the crown are nearly identical. The lateral margins of the crown are not as straight as the North American species.

The sudden appearance of *Myledaphus* at the base of the upper Campanian is noted everywhere from Mexico northward into Alberta, suggesting the possibility that the appearance of this genus represents a migration event wherein *Myledaphus* entered western North America, rapidly dispersed along the western margin of the Western Interior Seaway, and replaced *Cristomylus* and *Pseudomyledaphus* (both new genera described below) as a result of the increased efficiency of its dentition for feeding on coarse, abrasive food.

However, it is possible that *Myledaphus* originated within the Campanian of North America. We have observed a couple of large specimens of *Myledaphus* from the basal Campanian Tombigbee Member of the Eutaw Formation of Mississippi in the collections of the University of Colorado (e.g., UCM 99724, UCM 50229) that combine the heavily worn crown typically observed in *Pseudomyledaphus* with striations on the sides of the crown as in species of *Myledaphus*. Thus, it is possible to consider the origins of *Myledaphus* as being derived from *Pseudomyledaphus*. Only a careful analysis of the central Asiatic "myledaphid" teeth relative to the North American species will clear up this issue.

MYLEDAPHUS BIPARTITUS Cope, 1876

Non–*Myledaphus bipartitus* Cope,
1876, Estes, 1964:15–20, figs. 7, 8
Myledaphus bipartitus Cope, 1876,
Hutchinson and Kues, 1985:45, figs. 13, 1–3
Non–*Myledaphus bipartitus* Cope,
1876, Case, 1987:11, 12, fig. 11
?*Myledaphus bipartus* Cope, 1876,
Cifelli et al., 1999a:379
Myledaphus bipartitus Cope, 1876, Neuman
and Brinkman, 2005:168–169, fig. 9.2

Diagnosis Species of *Myledaphus* (Fig. 9.12A–U) ornamented on oral surface of crown, covered by fine ridges of enamel arising from transverse ridge and extending to margins of tooth both lingually and labially.

Description Symphysial teeth with equidimensional rhombic crown with root that is longer than crown is high (Fig. 9.12C, D). Even the smallest tooth crowns exhibit ridges extending to the lingual side of the tooth (Fig. 9.12A, B).

MNA V10211 (Loc. 458–1) is a well-worn hexagonal (upper) tooth (Fig. 9.12K–N). The crown is relatively shallow compared to the root. There is only a hint of the remaining rugosity of the crown surface, which bears a faint, slightly arched transverse crest. The sides of the teeth bear deep grooves (visible even in occlusal view). The root is robust and nearly as wide in all dimensions, as is the crown, and tapers only slightly downward. The lateral foramina are small and slit-like on both sides of the root. The nutritive groove is keyhole shaped or S shaped in dorsal view. MNA V10213 (Loc. 458–2) is a large hexagonal tooth with ornamentation as in typical Judithian *M. bipartitus* in having ridges and grooves developed essentially perpendicular to the transverse crest that continue across the crown and down the sides of the crown (Fig. 9.12O–Q). The crown is somewhat arched and the blocky roots taper downward somewhat from the base of the crown. The lateral foramina are small and slit-like. The nutritive groove is keyhole shaped and sinuous in dorsal view. This tooth is the largest myledaphine recorded from any stratigraphic unit below the middle unit of the Kaiparowits Formation.

At nearly 2 cm across, MNA V10215 (Loc. 1310), 187 m above the base of the Kaiparowits Formation, is an even larger hexagonal tooth than MNA V10213. The crown is complexly crenulated. The vertical striae on the sides of the crown continue down the root, forming shallow lateral foramina (Fig. 9.12R, S). The nutritive groove is keyhole shaped, very narrow at its base, and relatively straight. UCM 99703 (Loc. 83238) was recovered from about 300 m above the base of the Kaiparowits Formation (Fig. 9.12T, U). The six sides of the

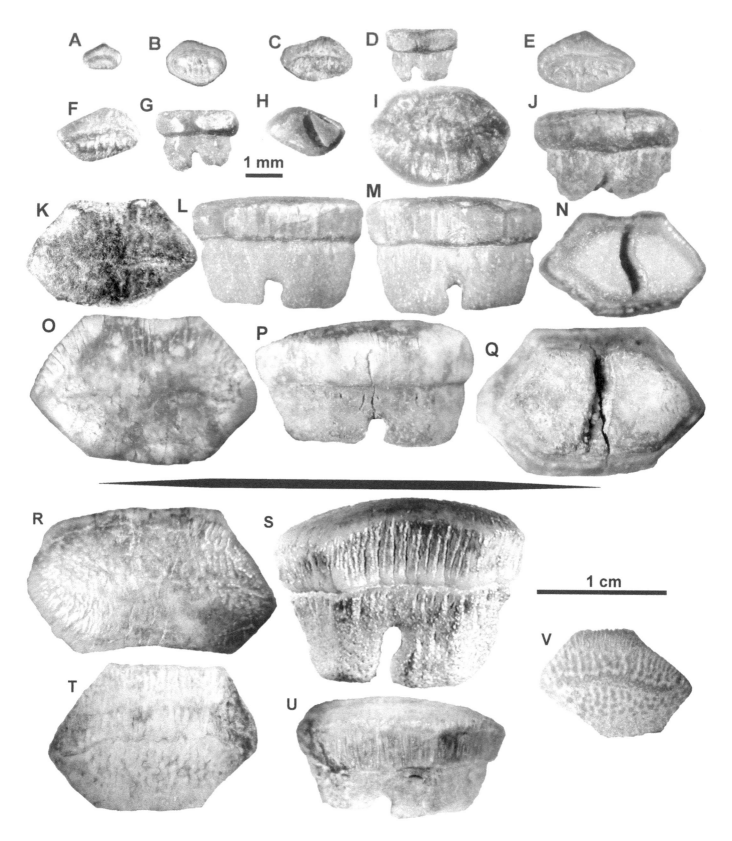

9.12. Myledaphine rays from the Kaiparowits Formation (upper Campanian) and Lance Formation of Wyoming (upper Maastrichtian). Small specimens of *Myledaphus bipartitus* Cope (1876) from MNA Loc. 458–1 (MNA V10206) in (A) occlusal view; (MNA V10207) in (B) occlusal view; (MNA V10197) in (C) occlusal and (D) lingual views; (MNA V10198) in (E) occlusal view; (MNA V10201) in (F) occlusal, (G) lingual, and (H) basal views. Medium-sized specimens of *Myledaphus bipartitus* (UCM 100173 Loc. 83268) in (I) occlusal and (J) lingual views; (MNA V10211, Loc. 458-1) in (K) occlusal, (L) lingual, (M) labial, and (N) basal views; (MNA V10213, Loc. 458-2) in (O) occlusal, (P) lingual, and (Q) basal views. Large specimens of *Myledaphus bipartitus* (MNA V10215, Loc. 1310) in (R) occlusal and (S) lingual views; (UCM 99703, Loc. 83238) in (T) occlusal and (U) lingual views. *Myledaphus "pustulatus"* Cook et al. (in press), UCM Loc. 77067, upper Maastrichtian, Lance Formation, Wyoming (UCM 99475) in (V) occlusal view.

Kirkland, Eaton, and Brinkman

crown are well flattened, and striae from the crown continue down the faces.

Distribution Upper Campanian Kaiparowits Formation, southern Utah. Commonly found in contemporaneous strata throughout the U.S. Western Interior.

Discussion The Kaiparowits Formation has yielded fewer ray specimens than strata lower in the section. Most rays have been recovered from the base of the formation (MNA localities 458–1, 458–2), but a few have been recovered up to 300 m above the base. The decreasing abundance probably reflects the gradual withdrawal of the Cretaceous epeiric seaway from the region, precluding rays from reaching that far upstream. The requirement for contact with the epicontinental sea is consistent with the absence of any ray specimens from pond and lake environments throughout the Cretaceous section on the Kaiparowits Plateau. Rays were diverse in most of the Upper Cretaceous section, but in the Kaiparowits Formation, except for the base of the formation, ray diversity was essentially reduced to *Myledaphus bipartitus*.

MYLEDAPHUS "PUSTULATUS"
Cook et al., in press
Myledaphus bipartitus Cope, 1876, Estes,
1964, pp.15–19, figs. 7, 8c–f, non 8a, b
Myledaphus bipartitus Cope, 1876,
Carpenter, 1979:41, figs. 7, 8
Myledaphus sp., Larson, Brinkman,
and Bell, 2010:1162, figs. D–F

Type Specimen UCMP 191578, lower Hell Creek Formation of Montana.

Referred Specimen UCM 99475 (Loc. 77067).

Distribution Maastrichtian.

Diagnosis Large *Myledaphus* teeth with coarse ornamentation on the occlusal surface of crown, which is commonly broken up in small tubercles, giving the occlusal surface a pustulate appearance.

Discussion Several workers have recognized that *Myledaphus* teeth from the Maastrichtian of the Western Interior were consistently more ornate with a reticulate ornament on the top of the tooth crown (Cook et al., in press). As such, UCMP 191578 from the lower Hell Creek Formation of Montana was designated as the holotype of the new species *Myledaphus "pustulatus."* We illustrate a typical specimen from the Lance Formation of southeastern Wyoming (Fig. 9.12V).

The crown of UCM 99475 is pustulate lingual of the crest and bears deep linguolabially oriented grooves labial of the crest. This is the typical morphology for *Myledaphus "pustulatus"* from the Lance and Hell Creek formations (Estes, 1964; Cook et al., in press). A range of smaller ray teeth are included under this catalog number, including four-sided

and marginal teeth. All of these teeth are pustulate on the occlusal face of the crown. By failing to differentiate specimens we ascribe to *M. pustulatus* from *Myledaphus bipartitus*, Estes (1964) inadvertently obscured the species definition to that of the genus as defined herein.

Larson, Brinkman, and Bell (2010) recognized that *Myledaphus* teeth from the *Albertosaurus* bonebed in unit 4 of the lower Maastrichtian were more coarsely ornamented than *Myledaphus bipartitus* and illustrated three teeth as *Myledaphus* sp. that compare well with *Myledaphus "pustulatus"* from the upper Maastrichtian. *Myledaphus "pustulatus"* may be present in the Maastrichtian North Horn Formation of central Utah but was listed as *Myledaphus bipartus* (Cifelli et al., 1999b). However, these specimens were not reexamined.

CRISTOMYLUS n. gen.

Type Species *Cristomylus nelsoni* n. sp.

Etymology From the Latin, *cristos*, "crest," *mylus*, "millstone."

Diagnosis Smaller than *Pseudomyledaphus* n. gen. and *Myledaphus*. Flattened sides of crown curved in dorsal view such that most upper and lower teeth have an oval appearance. Vertically, the sides of the crown are well rounded except where small wear facets are developed. The crowns bear a distinct crest on hexagonal, rhomboidal, and transitional teeth that is usually arched lingually, or less often nearly straight or weakly sinuous; crest may divide crown of tooth relatively evenly or, more commonly, closer to labial margin of the crown; crown often slightly depressed lingual to transverse crest. Symphysial teeth often appear to be flexed into a saddle shape in posterior view of the crest, with the lateral margins of the crown the tallest points. A distinct pair of lateral foramina is present on the labial side of root.

Distribution Southern Utah, earliest Cenomanian (Mussentuchit Member, Cedar Mountain Formation); late (?) Cenomanian (Dakota Formation); late Turonian (Smoky Hollow Member of the Straight Cliffs Formation); Coniacian–Santonian (John Henry Member of the Straight Cliffs Formation); middle Campanian (Wahweap Formation). Dakota Formation, northern Arizona (Kirkland, 1990, 1996). Widespread across North America in correlative strata. Examples of the genus from the Ellisdale, New Jersey, microvertebrate site may represent the survival of this genus into the late Campanian in eastern North America, where *Myledaphus* is not known.

Discussion *Cristomylus* is similar in overall morphology to the shallow marine ray *Pseudohypolophus*, which lacks the transverse crest characteristic of all the durophagous, freshwater ray teeth discussed here (McNulty, 1964; Thurmond,

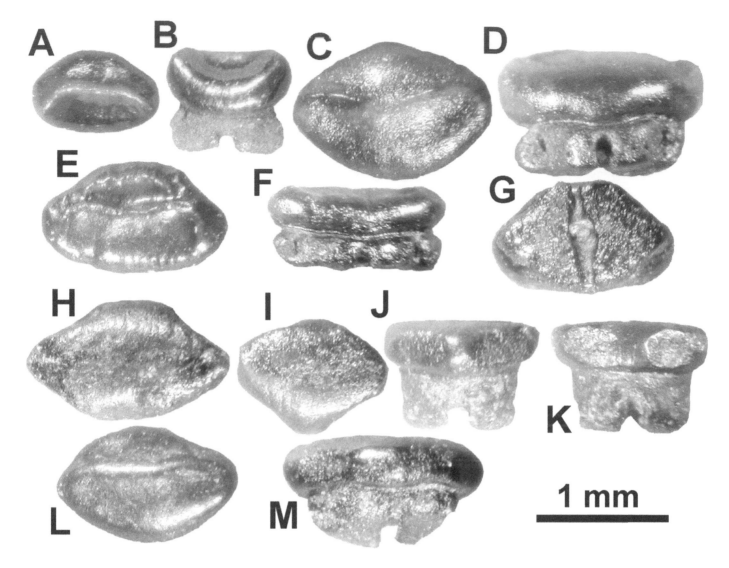

9.13. *Cristomylus nelsoni* n. sp. from the Mussentuchit Member of the Cedar Mountain Formation (basal Cenomanian) on the western side of the San Rafael Swell, (FHSM VP 16486, Loc. REQ 40) possible symphysial tooth in (A) occlusal and (B) lingual views; holotype (FHSM VP 16482, Loc. Travis KDGL REQ) in (C) occlusal and (D) lingual views; (FHSM VP 16485, Loc. REQ 40) in (E), occlusal, (F) lingual, and (G) basal views; (OMNH 66339, Loc. V694) in (H) occlusal view; (OMNH 66338, Loc. V694) in (I) occlusal, (J) lingual, and (K) basal views; (FHSM VP 16481, Loc. Travis KDGL REQ) in (L) occlusal and (M) lingual views.

1971; Meyer, 1974; Cappetta, 1987b; Welton and Farish, 1993). This may reflect a close phylogenetic relationship between these taxa as *Cristomylus* and *Pseudohypolophus* first appear in the Albian of Texas.

Case (1987) figures a single tooth from the Teapot Sandstone Member, Mesaverde Formation, Bighorn Basin, Wyoming, as *Myledaphus bipartitus* that lacks all the diagnostic features of *Myledaphus*. He reports the occurrence of numerous teeth (AMNH 12120–12123), which need to be examined as a population of teeth. Almost certainly the figured tooth represents a species of *Cristomylus* and indicates that these strata are older than late Campanian (Judithian).

CRISTOMYLUS NELSONI n. sp.

Type Specimens Holotype FHSM VP 16482, paratype FHSM VP 16481, both from Sternberg Museum locality Travis KDGL REQ, basal Cenomanian Mussentuchit Member of the Cedar Mountain Formation, western San Rafael Swell, Emery County, Utah.

Etymology Named for Michael Nelson, who pioneered microvertebrate research in the Cedar Mountain Formation during the 1980s and collected the type specimen.

Diagnosis Teeth of this taxon (Fig. 9.13) are the smallest and shallowest crowned of this genus. The teeth are the most poorly packed such that the teeth tend to be oval rather than well-developed hexagons or rhomboids. The top of the crown is rounded where it meets the sides of the crown and usually

Kirkland, Eaton, and Brinkman

bears well-developed facets only on one side of the tooth. The transverse crests on unworn lateral teeth medially arch labially and flex lingually toward the margin of the tooth to form a distinctive handlebar moustache–shaped transverse crest.

Description FHSM VP 16842 (Loc. Travis KDGL REQ, holotype, Fig. 9.13C, D) is an indistinct rhomboidal tooth similar to FHSM 16181 (Fig. 9.13L, M) with a handlebar moustache transverse crest that forms the highest part of the crown. There is no depression lateral of the transverse crest, and the crown slopes evenly away from the crest. There are two partially flattened areas on the walls of the crown at opposite sides of the tooth. There are well-developed marginal foramina on only one side of the tooth. At nearly 1.6 mm long, it is one of the largest teeth in the sample. OMNH 66339 (Loc. v694; Fig. 9.13H) is a quasi-hexagonal (upper?) tooth and is unusual in this sample in having four flattened facets near the marginal angles. The lingual and labial sides of the crown are rounded. The crest is deeply worn but nonetheless clearly has handlebar morphology. The roots are damaged, and this specimen is figured only in occlusal view. FHSM VP 16481 (Loc. Travis KDGL REQ; Fig. 9.13B) is a rhomboidal tooth with a distinct handlebar moustache–shaped transverse crest. The roots are damaged but are clearly smaller than the crown. There is a slight flattening of the sides of the crown on both sides at the margins of the crown. The crown is high relative to the shallow roots. FHSM VP 16485 (Loc. REQ 40; Fig. 9.13E–G) is an unworn oval tooth. One side bears two flattened surfaces laterally and a round side mesially. The other side is completely rounded. The transverse crest has a well-developed handlebar moustache shape. On one side of the crest there is a deep pit such that another minor ridge is developed. On the other side of the crest the basin is crenulated and slopes down to the rounded side. The marginal foramina are equally developed on both sides of the tooth. The roots are almost as broad and long as the tooth. The fact that one side is unfaceted and the tooth is proportionally shallow for its size suggests it may be a lower lateral tooth. OMNH 66338 (Loc. v694; Fig. 9.13I–K) is the only clearly rhombic tooth in the sample. The roots are deeper than the crown. Marginal foramina are variable in size even on the same side of the tooth. FHSM VP 16486 (Loc. REQ 40, Fig. 9.13A, B) is a symphysial tooth with a very sharp, ridge-like transverse crest which is arched strongly at both ends. The concave side of the transverse crest descends steeply to the crown wall and the convex side of the crown descends more gradually and bears broad crenulations.

Distribution The taxon is only known from microvertebrate localities in the basal Cenomanian Mussentuchit Member of the Cedar Mountain Formation on the west side of the San Rafael Swell in central Utah.

Discussion In this taxon, the teeth are not well packed, as shown by crowns that tend to have rounded sides and only sometimes bear flattened facets. The oval teeth tend to be bulbous, with wear often restricted to the transverse crest. There are no true hexagonal teeth. OMNH 66339 is a quasi-hexagonal tooth with four flattened facets near the marginal angles (Fig. 9.13H). This morphology is rare in the sample and most oval teeth have two flattened facets. Equally rare are teeth that could be considered to be truly rhombic. OMNH 66338 is the only truly rhombic tooth found in the sample; almost all other teeth have rounded sides, so it is difficult in this sample to differentiate upper from lower teeth.

An examination of a small sample of ray teeth from the Paluxy Formation (Albian) of Texas suggests that a species *Cristomylus* cf. *C. nelsoni* is present there as well, but is far less abundant than are the unornamented teeth of *Pseudohypolophus mcnultyi*. A study of a larger sample would be required to establish the presence of *Cristomylus nelsoni* in the Paluxy Formation with certainty.

CRISTOMYLUS BULLDOGENSIS n. sp.

Type Specimen Holotype MNA V10139, and paratypes MNA V10147, MNA V10148, MNA V10186, MNA V10188, MNA V10189, MNA V10190, MNA V10191, MNA V10192, MNA V10193, MNA V10194, MNA V10195, all from basal upper Cenomanian, upper part of the middle carbonaceous member of the Dakota Formation at MNA Loc. 1067, Bulldog Bench, Garfield County, Utah.

Etymology Name refers to Bulldog Bench on the west side of Grand Staircase–Escalante National Monument, where many specimens, including the type population, were recovered.

Diagnosis Immature teeth nearly identical to those of *Cristomylus nelsoni*, but reaching considerably larger size of as much as 4 mm across, often with faint crenulations on labial face of crown. These larger teeth are higher crowned and can be distinguished from *Cristomylus cifellii* in having a stronger transverse crest, and flatter lateral and oral surfaces of the crown (Fig. 9.14).

Description There is considerably more variation, both in terms of size and morphology, in the examples of *Cristomylus* from the Dakota Formation than observed in *Cristomylus nelsoni* from the Cedar Mountain Formation. There are many small ray teeth in the Dakota, similar to those from the Cedar Mountain Formation, which are low crowned and probably represent the continuation of *Cristomylus nelsoni*.

The holotype tooth MNA V10139 (Fig. 9.14V–X) is an essentially oval (quasi-hexagonal) upper lateral tooth and is 3.8 mm long. It has a tall, slightly arched transverse crest that is positioned lingually. The lingual side of the tooth is

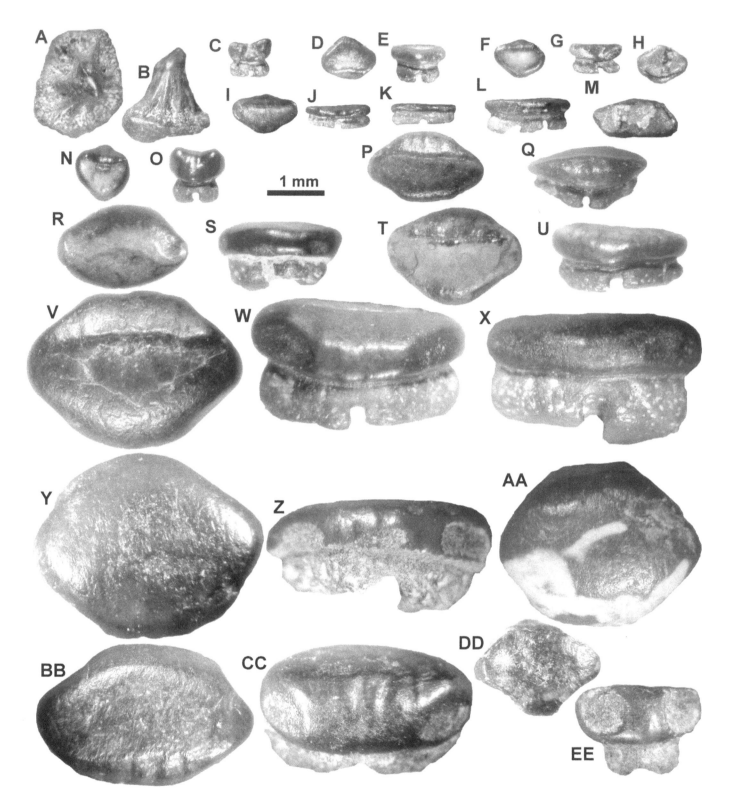

9.14. Myledaphine rays from the Dakota Formation (upper Cenomanian). Batoid dermal denticle from Bulldog Bench, westernmost Kaiparowits Basin, MNA Loc. 1067 (MNA V10187), in (A) dorsal and (B) lateral views. Tiny *Cristomylus bulldogensis* teeth of *C. nelsoni* morphology from Bulldog Bench, westernmost Kaiparowits Basin, MNA Loc. 1067, symphysial tooth (MNA V10190) in (C) lingual view; (MNA V10189) in (D) occlusal and (E) lingual views; (MNA V10188) in (F) occlusal, (G) lingual, and (H) basal views; (MNA V10191) in (I) occlusal, (J) lingual, and (K) labial views. Tooth with two nutritive grooves (MNA V10195) in (L) lingual and (M) basal views. Typical *Cristomylus bulldogensis* n. sp. tooth from MNA Loc. 939 on the southwest side of the Paunsaugunt Plateau (MNA V10128) in (N) occlusal and (O) lingual views. Tooth from UCM Loc. 85275 (UCM 99701) in (P) occlusal and (Q) lingual views. Typical *Cristomylus bulldogensis* n. sp. tooth from Bulldog Bench, westernmost Kaiparowits Basin, at MNA Loc. 1067 (MNA V10147) symphysial tooth in (R) occlusal and (S) lingual views; (MNA V10148) in (T) occlusal and (U) lingual views; holotype (MNA V10139) in (V) occlusal, (W) lingual, and (X) labial views; (MNA V10186) in (Y) occlusal and (Z) lingual view; (MNA V10193) in (AA) occlusal view; (MNA V10194) in (BB) occlusal and (CC) lingual views. Tooth with morphology of *Pseudomyledaphus* (MNA V10192) in (DD) occlusal and (EE) lingual views.

Kirkland, Eaton, and Brinkman

flatter than the labial side. Labial side of the crest the crown is lower, lingual of the crest the crown is higher and slopes lingually with lingual–labial-oriented crenulations that continue down the side of the crown. There are moderately developed lingual and labial marginal foramina. MNA V10194 (Fig. 9.14BB, CC) is somewhat smaller than MNA V10186 but shows flattened facets on the sides of the crown only near the marginal angles. Otherwise, the top of the crown curves to the sides rather than there being a sharp angle. Lingual crenulation is again apparent on this specimen.

The most hexagonal tooth of the sample is MNA V10193 (Fig. 9.14AA), which has a large bulbous crown with flattened sides near the marginal angles; however, the crown is well rounded where it joins the sides of the crown. If there was a transverse crest, it has been obliterated by wear. Vertical crenulations are present on the lingual side of the crown. Roots are damaged on this specimen. Another large bulbous tooth, MNA V10186 (Fig. 9.14Y, Z), is essentially oval but is quasi-hexagonal in appearance. This large tooth is well worn, providing only a slight hint of the transverse crest and lingual crenulations. The tooth has four flattened facets on the side of the crown near the marginal angles. The middle of the lingual side is rounded and crenulated, and there is a small, flattened facet low at the middle of the labial side of the tooth. Both of these large teeth may be uppers of a taxon with prehexagonal-form teeth.

MNA V10192 (Loc. 1067) is a moderate-sized tooth that approaches the true rhomboidal form by having four well-developed facets on the side of the tooth such that the crown surface is at right angles to the sides of the tooth (Fig. 9.14DD, EE). It has similarities to many of the smaller teeth of *Pseudomyledaphus* from the John Henry Member of the Straight Cliffs Formation. However, there are still rounded segments in the mesial position both labially and lingually. If packing were somewhat tighter, this would form a fully rhomboidal tooth. The lingual marginal foramina are well developed.

Tiny *Cristomylus nelsoni*–like teeth are exemplified by MNA V10189 (Fig. 9.14D, E), which is a quasi-rhomboidal to somewhat hexangular tooth 3.8 mm long. The transverse crest is positioned on the lingual margin of the crown and is essentially straight with a slight mesial–labial flexure. There is a deep pit labial of the transverse crest. The sides of the crown are flatted on both the lingual and labial sides of the teeth, toward the marginal angles, but are more extensively developed on the labial side. The lingual placement of the transverse crest and the asymmetry of the crown suggest a marginal or transitional position for this tooth. There are distinct lingual marginal foramina, but no labial foramina are evident.

MNA V10188 (Fig. 9.14F–H) is a tiny subrhombic tooth similar to MNA V10189 with a straight, lingually placed transverse crest, and the foramina only well developed marginally. The root is asymmetric. MNA V10191 (Fig. 9.14I–K) is rhomboidal in form. This tooth is very shallow crowned and has a lingually placed transverse crest, and hints of crenulation are present along the lingual margin of the tooth. Foramina are abundant on both the lingual and labial sides of the root (both lingual and labial sides shown).

MNA V10190 (Fig. 9.14C) is a somewhat asymmetrical saddle-shaped, symphysial tooth and has one well-developed lingual foramina. MNA V10195 (Fig. 9.14L, M) is an elongate quasi-hexagonal tooth. This tooth is very long relative to its width. The lingual side of the crown has two flattened and one rounded face, as does the labial side. The root is unusual in having two nutritive grooves (as has been observed in *Myledaphus bipartitus* [Estes, 1964]). This condition is uncommon but is found consistently in any large population in this lineage of *Pseudohypolophus* and myledaphine rays (Fig. 9.12). Although this is the closest tooth in the sample to being truly hexagonal, it is quite elongated, and the strange root configuration indicates a marginal position.

UCM 99701 (Loc. 83275; Fig. 9.14P, S) is an oval (or weakly hexagonal) transitional (?) upper (?) tooth. The specimen has a crown only slightly broader than the root, and the specimen lacks the complete root. The crest has a distinctive handlebar crest. There is a hint of crenulated enamel lingual of the crest. The tooth is oval and semihexagonal in occlusal view, and the faces are rounded, except for two areas of slight flattening. MNA V10128 (Loc. 939; Fig. 9.14N, O) is a hexagonal upper tooth and is morphologically a smaller version of holotype MNA 10139 described above. The tooth has a strong lingually positioned transverse crest. The portion of the crown labial of the crest is lower, while the position lingual is higher and bears weak crenulations. The transverse crest curves labially to meet both marginal angles. MNA V10148 (Loc. 1067; Fig. 9.14T, U) is a rhomboidal lower tooth that is larger than both UCM 99701 and MNA V10128. It has the distinct handlebar-style transverse crest, but it is too worn laterally to see the labial curling of the transverse crest and this is common on more worn specimens. Similarly, the lingual grooves become more subtle with wear. MNA V10147 (Loc. 1067; Fig. 9.14R, S) is a small symphysial tooth with a prominent transverse crest that forms two high cusps at each end of the crest, giving it a strong saddle shape. The lingual side of the crown is much higher than the labial and has two distinct creases that continue down the side of the crown.

Distribution The new species is widespread in the Dakota Formation of southern Utah with nearly identical specimens ranging up into the middle Turonian Smoky Hollow Member of the Straight Cliffs Formation.

Discussion A large number of ray teeth have been recovered from the middle member of the Dakota Formation

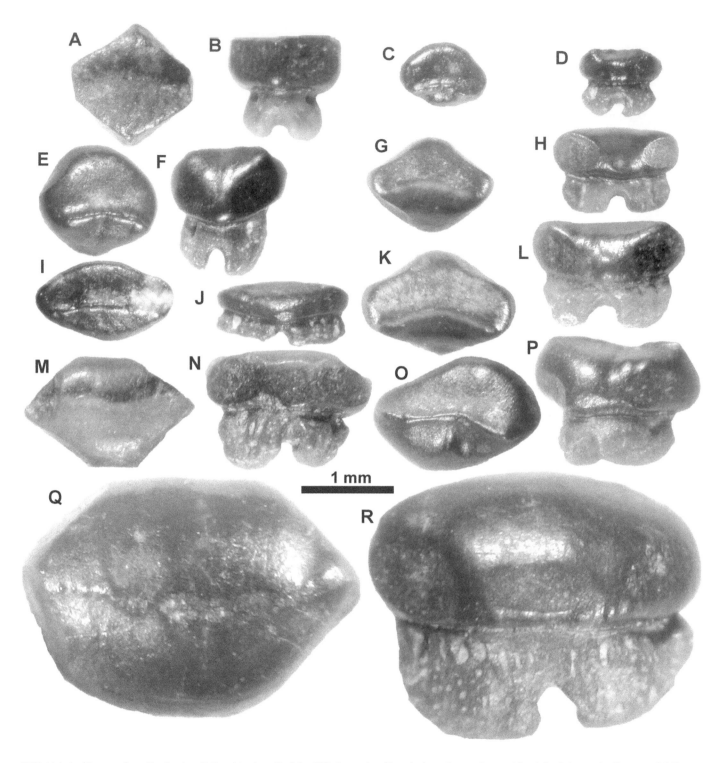

9.15. Myledaphine rays from the Smokey Hollow Member, Straight Cliffs Formation (Turonian), on the northwest side of the Kaiparowits Plateau at MNA Loc. 995. Tooth similar to a typical symphysial tooth of *Pseudomyledaphus madseni* (MNA V10176) in (A) occlusal and (B) lingual views. Teeth assigned to *Cristomylus* sp. cf. *C. bulldogensis* (MNA V10159) symphysial tooth in (C) occlusal and (D) lingual views; (MNA V10165) possible symphysial tooth in (E) occlusal and (F) lingual views; (MNA V10157) in (G) occlusal and (H) lingual views; (MNA V10163) in (I) occlusal and (J) lingual views; (MNA V10164) in (K) occlusal and (L) lingual views; (MNA V10172) in (M) occlusal and (N) lingual views; (MNA V10155) in (O) occlusal and (P) lingual views; (MNA V10178) in (Q) occlusal and (R) lingual views.

within the Grand Staircase–Escalante National Monument near the town of Cannonville and on Bureau of Land Management lands near the town of Alton. It is likely that middle member is basal upper Cenomanian. The teeth are considerably larger than *Cristomylus nelsoni* from the Cedar Mountain sample. However, small *Cristomylus* teeth from these same sites compare well with *Cristomylus nelsoni* teeth in size and morphology, suggesting juveniles are

Kirkland, Eaton, and Brinkman

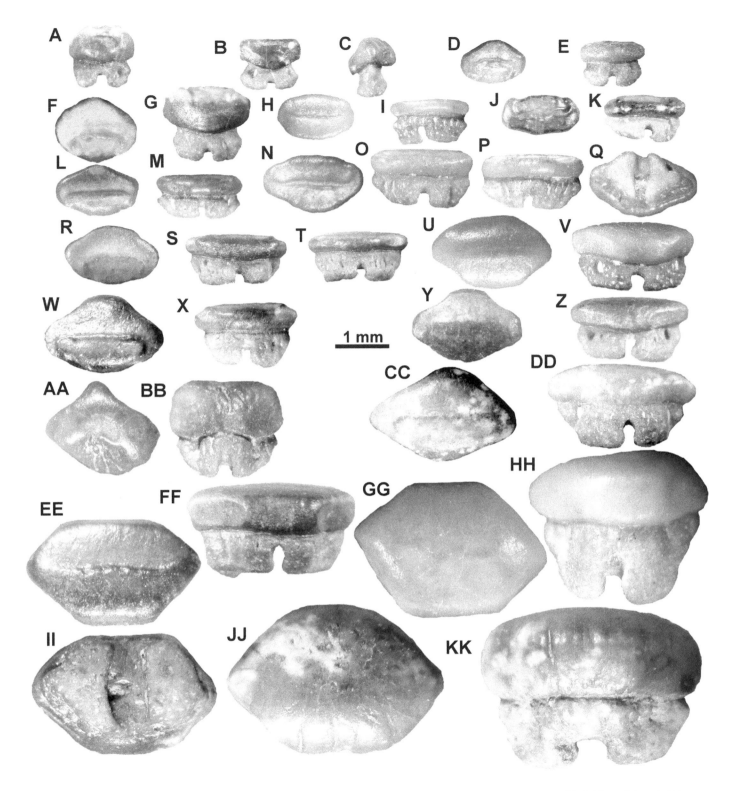

9.16. *Cristomylus cifellii* n. sp. from the Wahweap Formation (middle Campanian). Symphysial teeth (MNA V9600, Loc. 456–1) in (A) lingual view; (MNA V9568, Loc. 456–1) in (B) lingual and (C) lateral views; (UMNH VP 17390, Loc. 82) in (D) occlusal and (E) lingual views; (MNA V9531, Loc. 456–1) in (F) occlusal and (G) lingual views. Typical small teeth (UMNH VP 17393, Loc. 82) in (H) occlusal and (I) lingual views; (UMNH VP 17404, Loc. 77) in (J) occlusal and (K) lingual views; (MNA V9502, Loc. 156–1) in (L) occlusal and (M) lingual views; (UMNH VP 17395, Loc. 82) in (N) occlusal, (O) lingual, (P) labial, and (Q) basal views; (MNA V9569, Loc. 456–1) in (R) occlusal, (S) lingual, and (T) labial views; (UMNH VP 17399, Loc. 82) in (U) occlusal and (V) lingual views; (MNA V9635, Loc. 456–1) in (W) occlusal and (X) lingual views; (MNA V9525, Loc. 456–2) in (Y) occlusal and (Z) lingual views; (MNA V9652, Loc. 456–1) in (AA) occlusal and (BB) lingual views; (MNA V9633, Loc. 456–1) in (CC) occlusal and (DD) lingual views. Typical large teeth (UMNH VP 17401, Loc. 77) in (EE) occlusal and (FF) lingual views; holotype (MNA V9683, Loc. 456–1) in (GG) occlusal and (HH) lingual views; (MNA V9593, Loc. 465–1) in (II) basal view; (MNA V9686, Loc. 1294) in (JJ) occlusal and (KK) lingual views.

Elasmobranchs from Upper Cretaceous Freshwater Facies

nearly indistinguishable or that *Cristomylus nelsoni* ranges up into the upper Cenomanian. Therefore, only larger teeth from sizable populations of specimens should be utilized in specific identifications of these taxa.

CRISTOMYLUS cf. C. BULLDOGENSIS

Discussion *Cristomylus* teeth from the Turonian Smoky Hollow Member, Straight Cliffs Formation, are similar in size and in range of morphology with those from the upper Cenomanian Dakota Formation (Fig. 9.14). Until a more rigorous analysis is made, these teeth are referred to *Cristomylus cf. C. bulldogensis* (Fig. 9.15).

MNA V1076 (Loc. 995, Fig. 9.15A, B) is similar to MNA V10192 (Fig. 9.14DD, EE) from the Dakota Formation in being most similar in morphology to symphysial teeth assigned to *Pseudomyledaphus*. Other teeth in the large collection from MNA Loc. 995 are of typical *Cristomylus bulldogensis* morphology (Fig. 9.15).

CRISTOMYLUS CIFELLII n. sp.

Type Specimens Holotype MNA V9683, Paratypes MNA V9502, MNA V9568, MNA V9569, MNA V9600, MNA V9631, MNA V9633, MNA V9635, MNA V9652, all from MNA Loc. 456–1, Long Shot locality, middle member Wahweap Formation, northern Kaiparowits Plateau, Garfield County, Utah.

Etymology Named for Sam Noble Oklahoma Museum of Natural History paleontologist Richard Cifelli in recognition of his pioneering paleontological explorations of the Wahweap Formation during the 1980s, while at the Museum of Northern Arizona.

Diagnosis *Cristomylus* species (Fig. 9.16) in which crown and root are consistently of near equivalent height; symphysial teeth crowns are not saddle shaped, and oral surface of crown tends to be more domed with the transverse crest invariably more weakly expressed than in other species of *Cristomylus*; roots flare slightly outward at their midpoints toward the lateral margins of the roots, often with broad nutritive grooves.

Description The holotype, MNA V9683 (Loc. 456–1), is a *Cristomylus* hexagonal tooth with a bulbous crown (Fig. 9.16GG–II). The crown is 3.5 mm long by 2.4 mm wide, with a crown height of 1.4 mm versus a total tooth height of 3.4 mm. This tooth has a faint transverse crest, and the sides of the crown are inflated. The root is laterally inflated just below crown before tapering to base. There is only a single well-developed lateral foramen on the lingual side of the root, and no distinct foramen labially. The root is well perforated with a few slit-like lateral foramina. The nutritive groove is arched but not keyhole in form.

MNA V9686 (Loc. 1294) is a large hexagonal upper tooth (Fig. 9.16JJ, KK). The transverse crest has been obliterated by wear, but weak crenulations of the enamel are still evident lingual of the crest. The crown is bulbous, and all six sides of the crown have flattened facets, with only the mesial lingual facet bearing crenulated enamel. The roots taper strongly under the marginal angles of the crown. The nutritive groove is broad, and there is no evidence of the keyhole form in lateral view. There are two large lingual marginal foramina. MNA 9593 (Loc. 465–1) is another large hexagonal tooth with the root better preserved (Fig. 9.16 II) than on MNA V9686. The nutritive groove is shallow and broad, and the sides are straight (no keyhole form in side view). The root appears highly perforated by small holes over the entire root. This form has not been observed lower in the section and may represent a new taxon.

MNA V9633 (Fig. 9.16CC, DD) is a rhomboidal lower tooth with a straight transverse crest that curves lingually at both terminae (moustache form). The crown is bulbous (not flat-topped). Two side facets are moderately well developed labially; only one is present lingually. The root extends to the lateral margins of the crown and tapers strongly downward. The nutritive groove is arched.

Symphysial teeth are somewhat oval in dorsal view and have a weak (MNA V9600, Fig. 9.16A, and MNA V9531, Fig. 9.16F, G) or relatively strong uvula (MNA V9568, Fig. 9.16B, C) lingually. The roots are constricted beneath the crown and flare out laterally. No true saddle-formed bicuspid symphysial teeth were found in this sample, suggesting they are not present in this species of *Cristomylus*. There are two distinct lateral foramina on the lingual side of the root.

UMNH VP 17399 (Fig. 9.16U, V) is a juvenile hexagonal upper tooth. The crown is rounded, and only a hint of the transverse crest is preserved. There is no evidence of crenulated enamel on the top or sides of the crown. The sides of the crown adjacent to the marginal angles are flattened, but the mesial sides of the crown are rounded. The nutritive groove is keyhole shaped in side view. There are two deep lingual marginal foramina and four slit-like foramina labially.

MNA V9635 (Fig. 9.16W, X) is a small rhomboidal tooth. The crown is shallow. The transverse crest is straight but curves sharply lingually at both ends of the crest. Two facets are well developed on the side of the crown labially, but there are no facets on the lingual side. There are two distinct lingual lateral foramina, but no distinct labial foramina. The root tapers strongly downward, and the nutritive groove is flat-walled labially and slightly keyhole shaped lingually.

MNA V9525 (Fig. 9.16Y, Z) is a rhomboidal tooth with well-developed facets on four sides of the crown. The facets are not laterally continuous, such that there are rounded areas between the facets (i.e., marginal angles). There is a hint of a transverse crest very much like that on MNA V9635 (Fig. 9.16W, X). Labially, there are two well-developed lateral foramina and the nutritive groove is keyhole in shape. On the labial side of the root, the smaller foramina are randomly placed lateral foramina, and the nutritive groove is keyhole shaped in lateral view.

Many of the tiniest teeth from the Wahweap Formation are similar in morphology to *Cristomylus nelsoni* except they have much higher crowns and longer roots. MNA V9531 (Fig. 9.16F, G) is a tooth that is on the margin of the symphysial tooth battery. This form has the terminus of each side of the transverse crest cusp-like, and the root is pillar-like and not developed under the outer parts of the crown. There is only a single well-developed lateral foramen on the lingual side of the root and no distinct foramen labially. MNA V9652 (Loc. 1015–10) is a lower tooth from the margin of the symphysial tooth region (Fig. 9.16AA, BB). It has two flattened facets on both sides of the crown, and the lingual side of the crown is crenulated and slopes ventrally. The root is almost as wide as the crown and tapers strongly downward. There are two prominent lateral foramina on the lingual side of the root and a series of slits labially. The nutritive groove is arched in side view, not keyhole shaped. UMNH VP 17395 (Loc. 82) is an oval-shaped upper tooth (Fig. 9.16N–Q). The transverse crest is relatively straight, except at the ends. The crown is depressed labial of the transverse crest. The root flares laterally. The lateral foramina are slit-like, and the larger ones are on the lingual side. The nutritive groove is arched (not keyhole shaped) in lateral view. MNA V9569 (Loc. 456–1) is a shallow crowned oval–hexagonal upper tooth (Fig. 9.16R–T). The transverse crest is worn, but the terminae of the crest turn strongly labially (moustache form). The crown is not depressed on either side of the transverse crest. The crown is shallow, and weakly developed facets are present on the four sides nearest the marginal angles. The medial sides of the crown are rounded. The lateral foramina on the labial side of the root are a series of randomly place vertical grooves. The nutritive groove is weakly keyhole shaped. The two main lateral foramina on the lingual side of the root are widely separated and are present near the margins of the root.

UMNH VP 17393 (Loc. 82) is an upper tooth that is oval to somewhat hexagonal (Fig. 9.16H, I). The transverse crest is moustache in form, and there is a slight depression labial of the crest. The roots have a series of slit-like lateral foramina on both sides of the root, and the root flares laterally. The root tapers strongly downward below the flared area. The root is arched in side view. UMNH VP 17404 (Loc. 77) is a small, oval, possibly upper tooth (Fig. 9.16J, K). The transverse crest forms a distinct handlebar moustache form, and the top of the crest is worn flat. The elevation of the crown is not different on either side of the crest. The root is reduced at the base of the crown. The root is highly asymmetric, suggesting this tooth is from a lateral tooth position, and there is only one distinct slit-like lateral foramen lingually. UMNH VP 17390 (Loc. 82) is a small rhomboidal lower tooth (Fig. 9.16D, E). The moustache-like transverse crest is well developed. There is a slight hint of enamel crenulation lingual of the transverse crest. There are four moderately developed facets on the side of the crown. The only distinct lateral foramina are on the labial (?) side of the roots. The roots flare outward slightly at their midpoints toward their lateral margins. MNA V9502 (Loc. 456–1) is a somewhat more elongate rhomboidal tooth (Fig. 9.16L, M) than UMNH VP 17390. The transverse crest is straight and curls sharply at its terminae. The transverse crest is broad, and there is no distinct depression or pit on either side of the crest. Four facets are weakly developed on the sides of the crown. The roots flare at their midpoints anteriorly and posteriorly. There are two distinct lingual lateral foramina.

Distribution Wahweap Formation (middle Campanian) of Utah.

Discussion Very small specimens of *Cristomylus* are the most common teeth in the Wahweap sample. Clearly hexagonal teeth are rare, suggesting the teeth were not as tightly packed in the dental battery as in other species of *Cristomylus*. Significant in this group are the inflated crown and flaring of the root at its lateral midpoints. The lateral foramina are slit-like, giving a strong vertical component to the roots, and the nutritive canal is mostly arched to slightly keyhole shaped in lateral view.

Significantly, these teeth appear to differ from the species described from the slightly younger (Brinkman et al., Chapter 10, this volume) Foremost Formation of Alberta (Fig. 9.2) by Frampton (2006). The Foremost species has distinctively flatter crowns with straight, smooth sides and a generally more strongly recessed root. In these features, they appear to be closer to species we assign to *Pseudomyledaphus*, described below.

PSEUDOMYLEDAPHUS n. gen.

Type Species *P. madseni* n. sp.

Diagnosis Relatively large myledaphine teeth that compare well with those of *Myledaphus*, except that labial–lingually directed folds in the enamel are not present on the oral surface of teeth or running down the vertical sides of crown, except for slight vertical crenulations on the mesial lingual face.

Distribution Coniacian through middle Campanian strata.

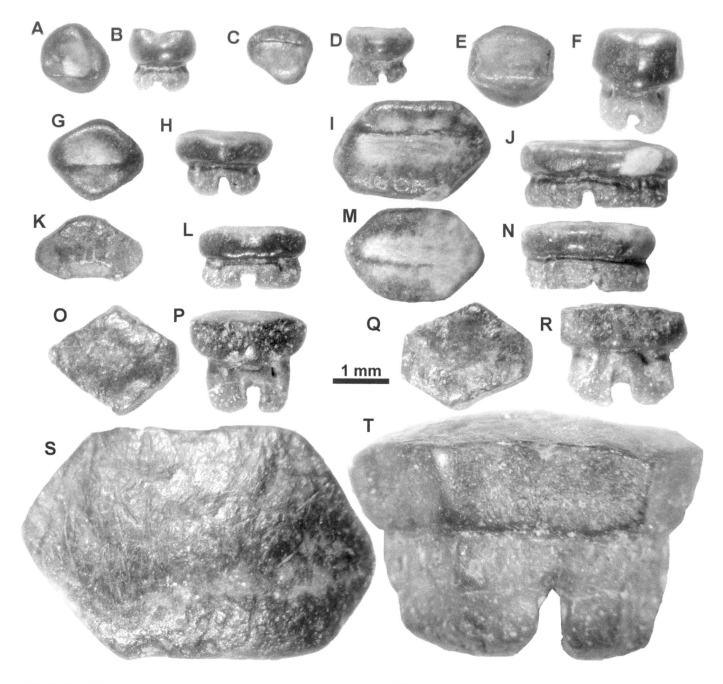

9.17. *Pseudomyledaphus madseni* n. sp. from the lower John Henry Member, Straight Cliffs Formation (upper Coniacian). Symphysial teeth (UMNH VP 14171, Loc. 567) in (A) occlusal and (B) lingual views; (UMNH VP 14451, Loc. 663) in (C) occlusal and (D) lingual views; (UMNH VP 14175, Loc. 567) in (E) occlusal and (F) lingual views. Atypical, more *Cristomylus*-like teeth (UMNH VP 14450, Loc. 663) in (G) occlusal and (H) lingual views; (UMNH VP 14448, Loc. 663) in (I) occlusal and (J) lingual views; (UMNH VP 17066, Loc. 567) in (K) occlusal and (L) lingual views; (UMNH VP 14181, Loc. 567) in (M) occlusal and (N) lingual views. Typical specimens of *Pseudomyledaphus madseni* n. sp., possible symphysial tooth (UMNH VP 17060, Loc. 567), in (O) occlusal and (P) lingual views; (UMNH VP 14168, Loc. 567) in (Q) occlusal and (R) lingual views; (UMNH VP 14166, Loc. 567) in (S) occlusal and (T) lingual views.

Discussion Reaching larger size than the teeth of *Cristomylus*, the teeth are deep crowned with large roots. The crowns are occlusally flattened to arched, with straight sides. The crowns of many symphysial teeth are strongly rhombic and overhang the root. These large, deep-crowned, and deep-rooted teeth that commonly have the root height greater than the crown height have often been identified as *Myledaphus*. The occlusal face of the crown bears little ornamentation and meets the sides of the crown at close to a right angle. The depth of the roots usually exceeds the crown height. The lingual marginal foramina are well developed. The diagnosis of *Myledaphus* Cope, 1876, as expressed in Cappetta (1987b:140), is that "the marginal, labial and lingual faces have clear vertical wrinkles in continuation of the oral face's folds." These teeth show the first phase of that in

Kirkland, Eaton, and Brinkman

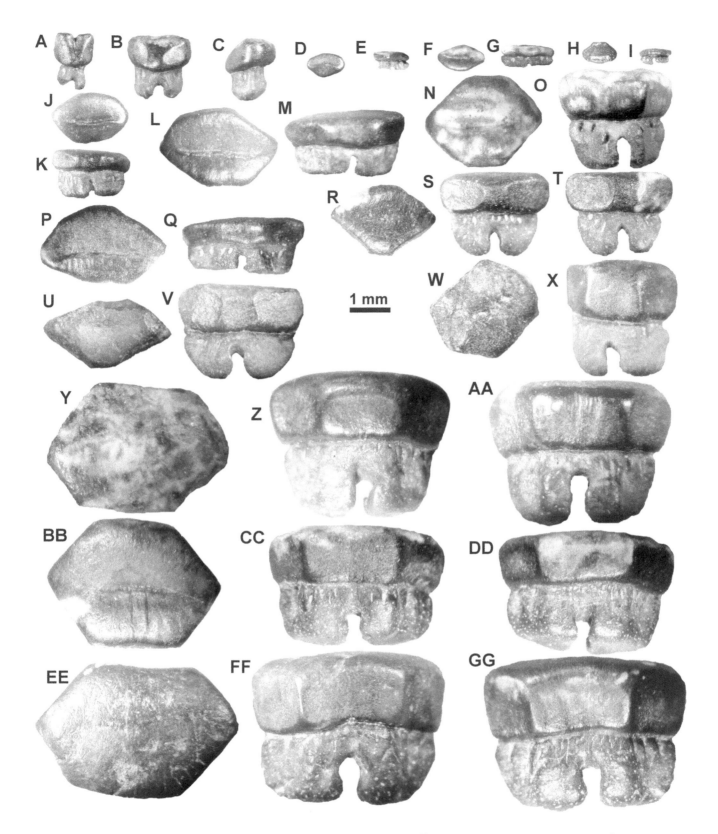

9.18. *Pseudomyledaphus madseni* n. sp. from the upper John Henry Member, Straight Cliffs Formation (Santonian). Symphysial teeth from MNA Loc. 706-2 (MNA V9802) in (A) lingual view; (MNA V9911) in (B) lingual and (C) lateral views. Tiny teeth of *Cristomylus*-like morphology from MNA Loc. 706-2 (MNA V9795) in (D) occlusal and (E) lingual views; (MNA V9816) in (F) occlusal and (G) lingual views; (MNA V9805) in (H) occlusal and (I) lingual views. Teeth of typical *Pseudomyledaphus madseni* n. sp. morphology (MNA V9925, Loc. 706-2) in (J) occlusal and (K) lingual views; (MNA V9833, Loc. 706-2) in (L) occlusal and (M) lingual views; (UMNH VP 13511, Loc. 98) in (N) occlusal and (O) lingual views; (MNA V9903, Loc. 706-2) in (P) occlusal and (Q) lingual views; (MNA V9890, Loc. 706-2) in (R) occlusal, (S) lingual, and (T) labial views; (UMNH VP 13587, Loc. 98) in (U) occlusal and (V) lingual views; (MNA V9893, Loc. 706-2) in (W) occlusal and (X) lingual views; holotype (MNA V9840, Loc. 706-2) in (Y) occlusal, (Z) lingual, and (AA) labial views; holotype (MNA V9820, Loc. 706-2) in (BB) occlusal, (CC) lingual, and (DD) labial views; (UMNH VP 12866, Loc. 98) in (EE) occlusal, (FF) lingual, and (GG) labial views.

Elasmobranchs from Upper Cretaceous Freshwater Facies

having slight vertical crenulations on the mesial lingual face, but clearly do not fit the definition of *Myledaphus*.

Many smaller to medium ray teeth from the John Henry Member have characteristics that are similar to those of *Cristomylus bulldogensis* in that the smallest teeth have low crowns similar to *C. nelsoni*, symphysial teeth are saddle shaped, and larger teeth have more curved vertical faces. These observations reflect the ontogenic development of the teeth in *Pseudomyledaphus*.

PSEUDOMYLEDAPHUS MADSENI n. sp.
Myledaphus bipartitus Cope, 1876; Fox, 1972:1482

Type Specimens Holotype MNA v9840, paratypes MNA v9795, MNA 9802, MNA v9805, MNA v9816, MNA v9820, MNA v9825, MNA v9833, MNA v9890, MNA v9893, MNA v9903, and MNA v9911, all from the upper John Henry Member, Straight Cliffs Formation, Santonian, MNA Loc. 706–2, northern Kaiparowits Plateau, Garfield County, Utah.

Etymology Named for Utah Geological Survey paleontologist Scott Madsen in recognition of his pioneering paleontological exploration of the Straight Cliffs Formation during the 1980s, while at the Museum of Northern Arizona.

Diagnosis Species of *Pseudomyledaphus* in which vertical sides of crown form sharp angle with both top and base of crown (Figs. 9.17, 9.18); oral surface of crown including transverse crest weakly developed. The oral surfaces of crowns are invariably deeply worn. Symphysial teeth highly distinctive with subcubic crowns overlying significantly more slender roots.

Description At 4.5 mm long, the holotype, MNA v9840 (Fig. 9.18Y–AA), is one of the largest hexagonal teeth from the upper John Henry Member. The lateral facets are flat and meet the occlusal surface of the crown at almost right angles. The root is almost the same size as the crown. On the labial side of the crown there are no large marginal foramina, but at the base of the crown there is a row of small foramina. On the lingual size of the crown there are two large marginal foramina. The nutritive groove is deep and narrow. MNA v9820 (Fig. 9.18BB–DD) is similar to MNA v9840, but some of the transverse crest is preserved as well as crenulation of the enamel lingual of the transverse crest. UMNH VP 12866 (Loc. 98) is larger yet, at 5.0 mm long. The marginal foramina are well developed on the labial side of the root, and there is a slight hint of crenulation on the lingual mesial facet. UMNH VP 13511 (Loc. 98) is a smaller hexagonal upper tooth with a well-developed transverse crest (Fig. 9.18N, O). The crown is lower labial of the crest and the enamel is crenulated lingual of the crest. Facets are well developed on all six sides of the crown. The nutritive groove is keyhole in form, and the root is deeper than the crown. The labial marginal foramina are

small and form a row just below the base of the crown. The lingual marginal foraminifera are slit-like, and some are present lower on the root.

Teeth from the basal John Henry Member tend to be smaller than those from the upper part, although large teeth do occur, such as UMNH VP 14166 (Fig. 9.17S, T). UMNH VP 14166 (Loc. 567) is a large, blocky, and bulbous crowned hexagonal tooth with occlusal surface features deeply worn. The only other surface features are the suggestions of some lingual crenulation. There is no crenulation of enamel on the sides of the crown. The highest part of the crown (the area that bore the transverse crest prior to wear) is on the labial side of the tooth. All six sides of the tooth are flattened, and the occlusal part of the crown meets the sides at a right angle. UMNH VP 14168 (Loc. 567) is a blocky six-sided tooth (Fig. 9.17Q, R), but it is probably a lower tooth, and the marginal angles have been flattened and the roots are deeper than the crown. UMNH VP 17060 (Loc. 567) is a five-sided lower tooth with only one of the marginal angles flattened (Fig. 9.17O, P). UMNH VP 14175 (Loc. 567) is a symphysial tooth with a lingual apron (Fig. 9.17E, F).

There are many much smaller hexagonal teeth in the John Henry collections, and many compare well with *Cristomylus* (Fig. 9.18D–O), but because there are few teeth in this size range that are morphologically like large *Pseudomyledaphus madseni* teeth, these are thought by the authors to represent ontogenetically young individuals of similar morphology to *Cristomylus*.

UMNH VP 13587 is a small elongate hexagonal tooth, probably from a marginal position (Fig. 9.19U, V). It exhibits the broadening of the nutritive canal above the base of the root such that it forms a tunnel (like a keyhole) that is characteristic of the roots of *Pseudomyledaphus*, but not *Cristomylus*. The root has abundant slit-like foramina on both the lingual and labial sides of the root just below the base of the crown. There are no indications of crenulated enamel on the mesial side facet.

MNA v9890 (Loc. 706–2) is a rhomboidal tooth with four well-developed facets on the sides of the crown (Fig. 9.18R–T). The labial side of the root has a row of lateral foramina along the top, and on the lingual side, there are two large foramina near the base of the crown with some slit-like foramina lower on the root. MNA v9893 (Loc. 706–2) is a six-sided lower tooth (Fig. 9.18W, X). The nutritive foramen does not come out under the longest lateral facets of the crown but rather under the intermediate length facets.

MNA v9833 (Loc. 706–2) is a hexagonal tooth with a distinct crest that is positioned somewhat lingual and curves labially as the ends of the crest approach the marginal angles (Fig. 9.18L, M). The crown is lower labial of the transverse crest, and there is a slight hint of broad crenulation lingually of the crest. The mesial side facets are not flattened. There are only a few tiny lateral foramina labially and two distinct slit-like foramina lingually. The sides of the nutritive groove are flat. MNA V9903 (Loc. 706–2) is about the same size as MNA V9833 but has a distinct series of small grooves perpendicular to and lingual of the transverse crest (Fig. 9.18P, Q). The crown is lower labial of the crest. The sides of the tooth are rounded and facets poorly developed. The lateral foramina, both on the lingual and labial sides, are slit like and the walls of the nutritive foramina are flat.

MNA V9925 (Loc. 706–2) is a small rhomboidal tooth (Fig. 9.18J, K). The transverse crest is sinuous. The crown is lower labial of the crest. There is only one well-developed flattened lateral face; the rest of the sides of the crown are rounded. Slit-like lateral foramina are present on both sides of the root.

Symphysial teeth of this species are relatively common in the sample. The distinct saddle form of the symphysial tooth is rare. MNA V9802 (Loc. 706–2) represents the saddle form of the symphysial tooth (Fig. 9.18A). There is a large cusp at each marginal angle with a deep groove between them. The root is distinctly narrower than the crown. The more typical tooth from the symphysial area is represented by MNA V9911 (Loc. 706–2), which has a flattened crown with the lingual part of the crown much deeper than the labial (Fig. 9.18B, C).

Distribution Widespread throughout the Coniacian and Santonian John Henry Member of the Straight Cliffs Formation of southern Utah and in the Milk River Formation of Alberta, Canada (Fox, 1972; Larson, 2010).

Discussion There are many *Pseudomyledaphus madseni* teeth in the upper John Henry Member, based on their relatively large size, deep crowns and roots, and roots narrowing from the base of the crown to the root base (Cappetta, 1987b:140). Some smaller examples of *Pseudomyledaphus* from the John Henry Member can be distinguished from *Cristomylus* in possessing roots that tend to have slit-like foramina, even though mesial side facets on hexagonal teeth are absent or poorly developed, and in some specimens present on only on one side. The upper John Henry Member (Santonian) specimens are similar but are on average larger than the Coniacian specimens. However, currently there is no objective way to distinguish Coniacian specimens from Santonian.

SCLERORHYNCHIFORMES Kriwet, 2004
SCLERORHYNCHIIDAE Cappetta, 1974
COLUMBUSIA Case et al., 2001

Type Species *Columbusia fragilis* Case et al., 2001.
Diagnosis Tiny teeth with a long, slender central cusp extending from long, slender labial apron extending well

beyond margin of root, giving the teeth a cruciform shape, with the long, low lateral shoulders extending from base of central cusp; elongate root with completely open root canal dividing root into two lobes in the holaulcorhize condition.

Discussion Two species of *Columbusia* have been described (Case et al., 2001). *Columbusia roessingi* was initially described as *Squatirhina roessingi* by Case (1987) from the upper Campanian Teapot Member of the Mesaverde Formation in the Bighorn Basin, Wyoming. Cappetta (2006) followed Case in assigning this genus to the orectoloboids but has recently thought it might better be assigned to the sclerorhynchids (Cappetta, in press, pers. comm., 2011).

The nominate species *Columbusia fragilis* is from the lower to middle Santonian Eutaw Formation of Georgia; it is diagnosed by its carinate, noncuspate lateral shoulders (Case et al., 2001:pl. 2, figs. 32–36). *Columbusia roessingi* is most readily diagnosed on possessing one to several lateral cusps along the lateral shoulders (Case, 1987:7, figs.1–4). It is noteworthy that Case (1987) considered that this taxon was close to Estes's (1964) *Squatirhina americana*. The teeth described from the Wahweap Formation are intermediate in morphology between the *C. fragilis* and *C. roessingi*.

It is tempting to assign all sclerorhynchoid rostral teeth from freshwater facies in southern Utah to *Columbusia*, and some may in fact belong here. However, *Columbusia* is, as is so far known, restricted to the Campanian, whereas tiny sclerorhynchoid rostral teeth from freshwater facies in the Western Interior of North America are found throughout the Upper Cretaceous section, as are the oral teeth assigned to *Texatrygon* described below.

COLUMBUSIA DEBLIEUXI n. sp.
Onchopristis dunklei Stromer, 1917,
Rowe et al., 1992:476. fig. 3D

Type Specimens Holotype, UMNH VP 18878, Paratypes UMNH VP 18836, 18876, 18877, 18879 all from UMNH Loc. 77, Wahweap Formation, east side of Paunsaugunt Plateau, Bryce Canyon National Park, Utah.

Etymology Named for Utah Geological Survey paleontologist Donald DeBlieux in recognition of the many paleontological sites he has discovered in the Wahweap Formation.

Diagnosis *Columbusia deblieuxi* (Fig. 9.19) differs from the type species *C. fragilis* in bearing a single distinct blade-like lateral cusp on either side of the main cusp that tends to overlap the root laterally, and in the distinct crest formed on the lingual side of the central cusp. *C. roessingi* differs in commonly possessing multiple lateral cusps on the lateral carinae.

Description The holotype (UMNH VP 18878) is 1.35 mm wide across the lateral cusp, with the root extending out

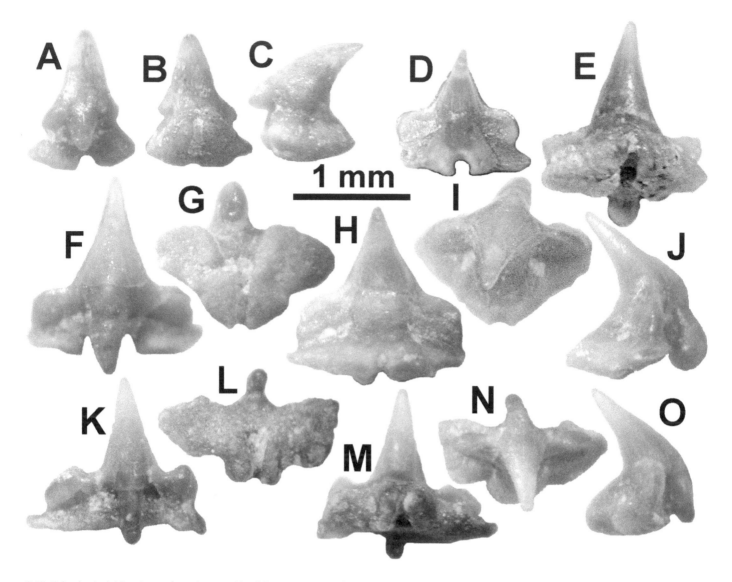

9.19. *Columbusia deblieuxi* n. sp. from the east side of the Paunsaugunt Plateau at UMNH Loc. 77, in the middle Campanian Wahweap Formation; (UMNH VP 18877) possible symphysial tooth in (A) labial, (B) lingual, and (C) lateral views; (UMNH VP 18879) in (D) lingual view; (UMNH VP 18836) in (E) lingual view; (UMNH VP 18876) in (F) labial, (G) basal, (H) lingual, (I) occlusal, and (J) lateral views; and holotype (UMNH VP 18878) in (K) labial, (L) basal, (M) lingual, (N) occlusal, and (O) lateral views.

laterally more than 0.1 mm to each side of the overhanging cusp (Fig. 9.19 F–J). The tooth measures 1.45 mm from the tip of the crown to the end of the narrow uvula. It has a fully divided root and, as with most of the teeth from the type locality, a pair of well-developed lateral cusps. UMNH VP 188876 differs from most of these teeth in lacking a lateral cusp.

UMNH VP 188877 (Fig. 9.19A–C) is an unusual tooth in being narrow and having a short uvula and short, sloping lateral shoulders lacking cusps. It may represent a symphysial tooth.

Distribution *Columbusia deblieuxi* teeth are common in the middle Campanian Wahweap Formation and are present, but less common, in the upper Campanian Kaiparowits Formation.

Discussion *Columbusia deblieuxi* teeth from southern Utah were for a long time considered to be teeth of *Squatirhina* because of their high central cusp and fully divided root (Eaton et al., 1999; Eaton and Kirkland, 2003). More recently, it was thought they might pertain to the sclerorhynchoid sawfish *Onchopristis* in their superficially similar cruciform shape and divided root (Kirkland, Eaton, and Brinkman, 2009).

These teeth are never more than 1.5 mm tall, and the slender cusps are rarely taller. One narrow tooth, UMNH VP 18877 (Fig. 9.11A–C), is unusual in nearly lacking lateral carinae and may represent a symphysial tooth.

Kirkland, Eaton, and Brinkman

Discussion Kriwet, Nunn, and Klug (2009, p. 333) diagnosed the Ptychotrygonidae among the sclerorhynchoid sawfish on the basis of oral teeth with "transversal crest differentiating the labial crown face and the presence of a very well-developed labial visor." This diagnosis is more specific in describing the oral teeth as small with broad tooth crowns that are triangular in outline in occlusal view and a variable number of transverse crests on the labial face. We amend this diagnosis by adding that the tooth crown is triangular to diamond shaped in occlusal view.

Kriwet, Nunn, and Klug (2009) noted that the placement of the Ptychotrygonidae within Batoidea and Sclerorhynchiformes is dubious and can be validated only by the discovery of skeletal material to document the presence or absence of rostral spines. More than 20 species have been assigned to *Ptychotrygon* on the basis of oral teeth. Known from complete specimens, both *Libanopristis* and *Micropristis* have oral teeth that are similar to *Ptychotrygon* and small rostral spines (Cappetta, 1980, 1987b).

The systematic position of *Ptychotrygon* has been the source of much debate, although the teeth compare best with some species of sclerorhynchoid sawfish. Numerous species of *Ptychotrygon* have been described for which there are no associated rostral teeth (Cappetta and Case, 1999; Kriwet, Nunn, and Klug, 2009). Case (1978, 1987) and Kriwet (1999) tentatively assigned enlarged placoid scales as rostral spines to *Ptychotrygon*, but similar placoid scales are present on the ligamentous band between the rostrum and skull of sclerorhynchiforms (Kriwet, 1999). However, perfectly reasonable associations of teeth assigned to *Ischyrhiza avonicola* with teeth of *Ptychotrygon* morphology have been reported (Wolberg, 1985; Welton and Farish, 1993). The association of ptychotrygonid teeth and *Ischyrhiza avonicola* rostral teeth is recorded in the Lance Formation by Estes (1964), and because no other sclerorhynchoid oral teeth are present in his samples, the reassignment of *Ischyrhiza avonicola* to the ptychotrygonids is a logical hypothesis.

Cappetta and Case (1999) erected the genus *Texatrygon* for ptychotrygonid teeth with fewer transverse ridges than *Ptychotrygon* and bearing ornamentation on the labial face of the crown, as in the teeth illustrated by Estes (1964:fig. 8a, b). Therefore, the sclerorhynchoid sawfish described herein are tentatively assigned to *Texatrygon*.

TEXATRYGON Cappetta and Case, 1999

Type Species *Texatrygon hooveri* (McNulty and Slaughter, 1972).

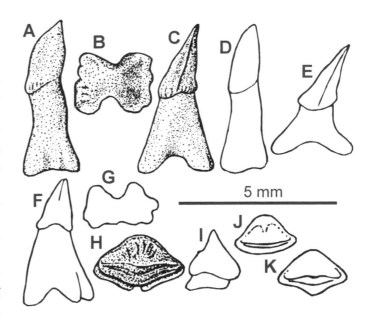

9.20. *Texatrygon avonicola* from Estes (1964). Holotype, right rostral tooth (UCMP 53917) from Estes (1964:fig. 6a), in (A) dorsal, (B) basal, and (C) posterior views. Right rostral tooth, from Estes (1964:fig. 6b as unnumbered), in (D) dorsal view. Right rostral tooth (UCMP 53915) from Estes (1964:fig. 6c) in (E) posterior view. Right rostral tooth (UCMP 53913) from Estes (1964:fig. 6d) in (F) posterior and (G) basal views. Oral tooth (AMNH 9308) from Estes (1964:fig. 8b) in (H) occlusal and (I) lateral views. Two oral teeth assigned to AMNH 9307 from Estes (1964:fig. 8a) in (J, K) occlusal view. Rostral teeth identified as *Onchosaurus avonicolus* n. sp. = *Ischyrhiza avonicola* of subsequent authors and oral teeth identified as symphysial teeth of *Myledaphus bipartitus*. Scale bar = 5 mm.

Emended Diagnosis Oral teeth of ptychotrygonid morphology bearing a reduced transverse ornamentation and stronger plicae on labial face of tooth. Rostral teeth tiny with expanded collar at base of crown and weak plicae.

Discussion All the teeth discussed herein were initially assigned to the genus *Ischyrhiza* on the basis of rostral teeth. The type of *Ischyrhiza*, *I. mira* Leidy (1856), is a large species of sclerorhynchoid sawfish with proportionally much larger rostral teeth than oral teeth, and with rostral tooth crowns that more smoothly merge into the flanks of the base. Oral teeth assigned to *I. mira* are different from those of ptychotrygonids in having smooth crowns bearing a much taller central cusp that bears a strong lingual crest, overlaps the lingual face of crown, and is supported laterally by strong, sloping shoulders (Case, 1978; Cappetta, 1987b; Welton and Farish, 1993), giving it a robust cruciform shape similar to that of many other sclerorhynchoid sawfish.

In overall morphology of the oral and rostral teeth, *Texatrygon* is most directly comparable to *Micropristis*, which is known from complete specimens from the Cenomanian of Lebanon (Cappetta, 1980). The type species of *Micropristis* was actually referred to *Ischyrhiza* sp. cf. *I. avonicola* by

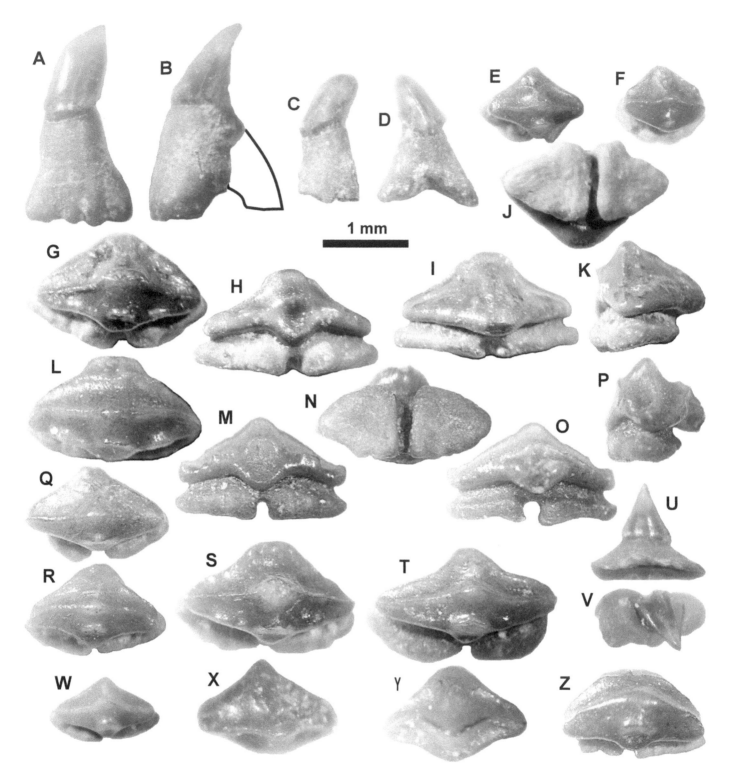

9.21. Sclerorhynchoid sawfish from southern Utah. *Texatrygon brycensis* n. sp., type population from UMNH Loc. 77 on the east side of the Paunsaugunt Plateau in the middle Campanian Wahweap Formation. Right rostral teeth (UMNH VP 18883) in (A) dorsal and (B) posterior views. Left rostral tooth (UMNH VP 18882) in (C) dorsal and (D) anterior views. Oral teeth (UMNH VP 18891) in (E) occlusal view; (UMNH VP 18892) in (F) occlusal view; holotype (UMNH VP 18885) in (G) occlusal, (H) lingual, (I) labial, (J) basal, and (K) lateral views; (UMNH VP 18889) in (L) occlusal, (M) lingual, (N) basal, (O) labial, and (P) lateral views; (UMNH VP 18890) in (Q) occlusal view; (UMNH VP 18894) in (R) occlusal view; (UMNH VP 18893) in (S) occlusal view; (UMNH VP 18836) in (T) occlusal view. Other Wahweap specimens from UMNH Loc. 82, denticle (UMNH VP 18918) in (U) anterior and (V) occlusal views; oral tooth UMNH VP 18920 in (W) occlusal view. Oral tooth of *Texatrygon* sp. from upper Campanian beds on Tarantula Mesa in Henry Basin at UMNH Loc. 414 (UMNH VP 18921) in (X) occlusal view. Oral teeth of *Texatrygon* sp. from UMNH Loc. 799 in the Santonian upper John Henry Member, Straight Cliffs Formation (UMNH VP 18940), in (Y) occlusal view; (UMNH 18901) in (Z) occlusal view.

Kirkland, Eaton, and Brinkman

Casier (1947a). *Micropristis* is distinctive in that its tiny rostral teeth are barely visible when the entire fish is figured.

TEXATRYGON AVONICOLA (Estes, 1964)

Ischyrhiza avonicola n. sp., Estes, 1964:13–15, 16, fig. 6
Myledaphus bipartitus Cope, 1876,
Estes, 1964:16–17, fig. 8a, b
Non–*Ischyrhiza avonicola* Estes, 1964, Slaughter
and Steiner, 1968:237, in part, fig. 3K
?*Ischyrhiza avonicola* Estes, 1964, Slaughter
and Steiner, 1968:237–238, in part, fig. 3L, M
Non-*Ischyrhiza* cf. *I. avonicola* Estes, 1964,
Cappetta and Case, 1975b:29–31, text-figs. A–G
Ischyrhiza cf. *I. avonicola* Estes, 1964,
Carpenter, 1979:41, figs. 5–6
Non–*Ischyrhiza avonicola* Estes, 1964,
Rowe et al., 1992:476, fig. 3G
?*Ischyrhiza* sp., Cifelli et al., 1999a:379

Type Specimen UCMP 53917, Fig. 9.20A–C, UCMP Loc. v-5620, upper Maastrichtian Lance Formation, Wyoming.

Emended Diagnosis Small sclerorhynchoid sawfish, rostral teeth lancelate crown slightly shorter than root, dorsoventrally flattened, inclined slightly posteriorly and ventrally; carinae extend from apex of tooth to just above base of crown; base of crown slightly expanded into collar above roots ornamented by weak folds; root expands slowly to subrectangular base bearing an anteroposteriorly oriented depression flanked dorsally and ventrally by two to four plicate extensions of the root, forming the attachment area. Smaller rostral teeth are more strongly ornate with a proportionally more extensive root base. Oral teeth marked by strong medial carinae drawn up into pyramidal central cusp; steep lingual face bearing distinct medial crest paralleling main carina; broad labial face ornamented by transverse crest formed of arching segments extending from near midline of tooth or ornamented by weak radial ridges.

Discussion *Texatrygon avonicola* (Fig. 9.20) was initially diagnosed on the basis of the small size of the rostral teeth with weak folds in the enamel around the base of its crown. We note that the teeth figured by Estes (1964), which we have identified as oral teeth of *T. avonicola*, are poorly figured (Fig. 9.20H–K) and need to be restudied. Welton and Farish (1993) figure oral teeth that they assign to *T. avonicola* from the Maastrichtian Kemp Clay of central Texas, which, although being of general *Texatrygon* morphology, differ from all the oral teeth figured herein by being ornamented by multiple arched ridges on the labial side of the crown.

Texatrygon avonicola may be present in the Maastrichtian North Horn Formation of central Utah, listed as *Ischyrhiza sp.* (Cifelli et al., 1999b). However, these specimens were not figured and were not reexamined in light of the new information reported herein.

TEXATRYGON BRYCENSIS n. sp.

Type Specimens Holotype, UMNH VP 18885, oral tooth, UMNH Loc. 77, Fig. 9.21G–K. Paratypes UMNH VP 18882, 18883, rostral teeth, UMNH VP 18891, 18892, 18889, 18890, 18893, 18894, 18896, oral teeth, all from UMNH Loc. 77, Wahweap Formation, east side of Paunsaugunt Plateau, Bryce Canyon National Park, Utah.

Diagnosis Small species of *Texatrygon* similar to *T. avonicola* in nearly all characters except that oral teeth have a marked depression at center of uvula expansion and weak cusplets extend up from around margin of tooth (Fig. 9.21A–W).

Description The holotype (UMNH VP 18885) is a typical oral tooth with a crown 1.81 mm across, 1.10 mm wide, and 0.85 mm tall between central cusp and base of uvula. Its more expansive root is 2.02 mm across (Fig. 9.21G–K). The tooth is divided by a transverse crest that crossed at the central cusp. As is typical for this genus, the lingual face of crown is shorter than is the labial face of crown. The uvula is indented by a circular depression approximately 0.3 mm in diameter. The dorsal margin of this depression is formed by the transverse lingual crest. There are four cusplets on either side of the labial face of crown and three cusplets on either side of the lingual face of crown. There are weak radial ridges extending part way down the labial face of crown from the central cusp.

The labial crest is variably developed in this taxon and, as in the holotype, is often represented by the line of labial cusplets along the margin of the tooth. Additionally, the marginal cusplets are not developed in the smallest teeth such as UMNH VP 18891, UMNH VP 18892, and UMNH VP 18920 (Fig. 9.21E, F, W).

Distribution Middle Campanian Wahweap Formation, southern Utah.

Discussion The population of teeth from UMNH Loc. 77 have the diagnostic features on the oral teeth consistently developed on each tooth; however, the rostral teeth are indistinguishable from rostral teeth of *I. avonicola*. The widespread distribution of rostral teeth assigned to *T. avonicola* suggests that the rostral teeth are morphologically more conservative than are the oral teeth (Wolberg, 1985).

The teeth of *Ischyrhiza texana* from the Coniacian of central Texas (Welton and Farish, 1993) represent a different

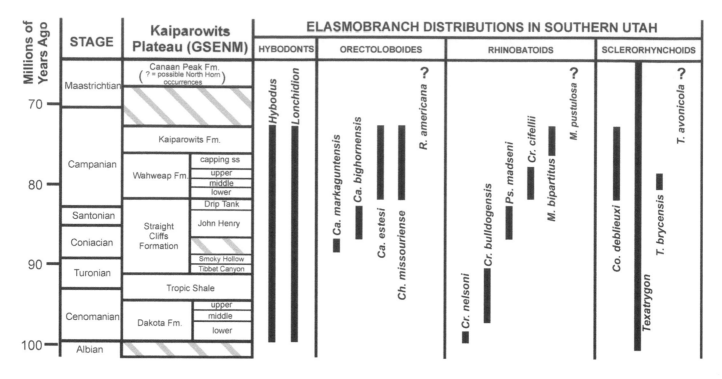

9.22. Stratigraphic ranges of freshwater elasmobranchs in the Upper Cretaceous of southern Utah plotted against section at Grand Staircase–Escalante National Monument. Cross-hatching indicates strata of that age not preserved.

taxon from those originally described as the sclerorhynchid *Kiestus texana* by Cappetta and Case (1975a; Case, 1978), which were similar to those of *I. mira*. The materials figured by Welton and Farish (1993) are similar to those *T. brycensis* but differ in being more strongly ornamented and in possessing ridges extending toward the apex of the cusp on the labial surface of the crown. Likewise, the rostral teeth they figure differ from those of *T. brycensis* and *T. avonicola* in possessing stronger folds that extend farther up the sides of the crown. An oral tooth and rostral tooth figured by Rowe et al. (1992) from the upper Campanian Aguja Formation as *Ptychotrygon* and *Ischyrhiza avonicola*, respectively, appear to pertain to the same species as that figured by Welton and Farish (1993). This species should be assigned to *Texatrygon* and may represent a new species.

CONCLUSION

Elasmobranchs were an important component of riverine systems during the Late Cretaceous in central and southern Utah. Basal Cenomanian microvertebrate sites preserve the hybodont sharks *Lonchidion* and *Hybodus*, in addition to tiny durophagous teeth of *Cristomylus nelsoni*, marking the first appearance of the myledaphine lineage of rhinobatid rays. Tiny rostral and oral teeth of the sclerorhynchoid sawfish *Texatrygon* first appear in the Cenomanian Dakota Formation. All four of these elasmobranch lineages are characteristic

elements of microvertebrate assemblages up through the Upper Cretaceous section in southern Utah (Fig. 9.22).

Oral teeth of the possible sclerorhynchoid sawfish *Columbusia deblieuxi* occur in both the Wahweap and Kaiparowits formations and are thus restricted to the Campanian.

Myledaphine rays evolved rapidly through the Late Cretaceous. However, variation in tooth morphology relative to tooth position and ontogeny makes taxonomy difficult and probably leads to lumping of taxa. Evolutionary trends of increasing tooth size, increasing relative crown and root height, increasing efficiency of tooth packing in the dental battery, and increased complexity are all observed as features that improve the efficiency of processing abrasive foods apparently favored by these bottom feeders. *Pseudomyledaphus* appears to be intermediate in morphology between *Cristomylus* and *Myledaphus*, so it is surprising to see *Pseudomyledaphus madseni* in the Upper Straight Cliffs Formation replaced by *Cristomylus cifellii* in the Wahweap Formation. To the north, in Alberta, *Pseudomyledaphus madseni* occurs in the Milk River Formation. Following a short marine incursion, it was replaced by another as yet undescribed species of *Pseudomyledaphus* in the Foremost Formation before *Myledaphus bipartitus* first appears in the Oldman and Dinosaur Park formations (Peng, Russell, and Brinkman, 2001; Frampton, 2006). The widespread occurrence, abundance, and apparent restriction of *Myledaphus bipartitus* to the late Campanian Judithian faunas demonstrate its utility as an index

Kirkland, Eaton, and Brinkman

fossil and indicate that other myledaphine rays may be of biostratigraphic utility as well (Fig. 9.22).

Small orectolobiform sharks are represented by teeth assigned to *Cantioscyllium*, and first appear in the microvertebrate assemblages of the John Henry Member. There appears to be an evolutionary trend toward decreasing ornamentation from the Coniacian through the Campanian. However, whereas *Cantioscyllium bighornensis* occurs only a short distance above the Calico Bed near the base of the John Henry Member in strata interpreted as representing the upper Coniacian in the eastern Paunsaugunt Plateau region, the more coarsely ornamented *Cantioscyllium markaguntensis* occurs in the "upper" Straight Cliffs Formation more than 200 m above the highest occurrence of Turonian brackish-water faunas (Eaton et al., 2001; Eaton, 2006b) on the west side of Markagunt Plateau in Cedar Canyon, where the unconformity represented by the Calico Bed has not yet been recognized. Moreover, they occur 29 m below a volcanic ash dated as 86.72 ± 0.58 Ma (Eaton et al., 2001; Eaton, 2006b), late Coniacian. Thus, this fossil occurrence suggests that older Coniacian or possible upper Turonian strata may be preserved to the west of the Paunsaugunt Plateau. The weakly ornamented to smooth *Cantioscyllium estesi* occurs in both the Campanian Wahweap and Kaiparowits formations. Likewise, the orectolobiforms *Chiloscyllium missouriense* occurs in both the Wahweap and Kaiparowits formations, documenting an increase in the diversity of orectolobiform sharks in riverine habitats in southern Utah during the Campanian (Fig. 9.22).

Freshwater elasmobranchs have been well documented as having undergone a complete extinction at the end of the Cretaceous in the northern Western Interior of North America (Bryant, 1989; Archibald and Bryant, 1990; Archibald, 1996; Cook et al., in press). At present, our knowledge of freshwater elasmobranchs across the Cretaceous–Tertiary boundary in southern Utah is limited, so the nature of this extinction locally cannot be addressed. Needless to say, each of the hybodont shark, ray, and sawfish lineages documented herein for Utah are not known to survive into the Tertiary anywhere, and any taxonomic relationships of the orectolobiform sharks described here and those documented in the Tertiary are suspect at best (Cappetta, 1987a, 1987b).

ACKNOWLEDGMENTS

J. Maisey and H. Cappetta are thanked for answering many of the senior author's questions over the years and are completely blameless for what is presented herein. Most of the collections were made under permit from the Bureau of Land Management over the past 25 years. Specimen(UMNH) collection from Bryce Canyon National Park and vicinity by J.G.E. was funded by the Colorado Plateau Cooperative Ecosystem Studies Unit of the National Park Service. Thanks are due the many collectors who lugged sacks of matrix, screen washed it, and picked the teeth out of the resulting concentrate. Important specimens were made available to the authors for study by D. Fox at the University of Alberta, P. Robinson at the University of Colorado, R. Cifelli at the Oklahoma Museum of Natural History, and M. Nelson while at Fort Hays State College. The chapter benefited from reviews by H. Cappetta, C. Underwood, D. DeBlieux, M. Lowe, and R. Ressetar.

REFERENCES CITED

Agassiz, L. 1833–1844. Recherches sur les poisson fossiles. Neuchatel. 5 volumes, 1420 pp.

Applegate, S. P. 1972. Revision of the higher taxa of orectolobids. Journal of the Marine Biology Association of India 14:743–751.

Arambourg, C. 1952. Les Vertébrés fossiles des Gisements de Phosphates (Maroc-Algérie-Tunisie). Service Géologique Maroc, Notes et Mémoires 92:1–372.

Archibald, J. D. 1996. Dinosaur Extinction and the End of an Era. Columbia University Press, New York.

Archibald, J. D., and L. Bryant. 1990. Differential Cretaceous–Tertiary extinctions of nonmarine vertebrates: evidence from northeastern Montana; pp. 549–562 in L. Sharpton, and P. Ward (eds.), Global Catastrophes in Earth History: An Interdisciplinary Conference on Impacts, Volcanism and Mass Mortality. Geological Society of America Special Paper 247.

Armstrong-Ziegler, J. G. 1978. An aniliid snake and associated vertebrates from the Campanian of New Mexico. Journal of Paleontology 52:480–483.

Bennett, E. T. 1830. Fishes; pp. 686–694 in S. Raffels (ed.), Memoir of the Life and Public Services of Sir Thomas Stamford Raffels. F.R.S. etc., London.

Biek, R. F., P. D. Rowley, J. M. Hayden, D. B. Hacker, G. C. Willis, L. F. Hintze, R. E. Anderson, and K. D. Brown. 2009. Geological map of the St. George and east part of the Clover Mountains 30'×60' quadrangles, Washington and Iron Counties, Utah. Utah Geological Survey Map 242.

Biek, R. F., F. Maldonado, D. W. Moore, J. J. Andreson, P. D. Rowley, V. S. Williams, L. D. Nealey, and E. G. Sable. 2011. Interim Geological Map of the West Part of the Panguitch 30'×60' Quadrangle, Garfield, Iron, and Kane Counties – Year 2 Progress Report. Utah Geological Survey Open-File Report 577.

Bonaparte, C. L. J. L. 1838. Selachorum tabula analytica. Nuovi Annali Scienze Naturli 2:195–214.

Bryant, L. J. 1989. Non-dinosaurian Lower Vertebrates across the Cretaceous–Tertiary Boundary in Northeastern Montana. University of California Publications in Geological Sciences 134.

Cappetta, H. 1973. Selachians of the Carlile Shale (Turonian) of South Dakota. Journal of Paleontology 37:504–514.

Cappetta, H. 1974. Sclerorhynchidae nov. fam., Pristidae et Pristiophoridae: un exemple de parallélisme chez les sélaciens. Academie des Sciences, Compte Rendu Hebdomadaires des Seances, Paris, ser. D, 278:225–228.

Cappetta, H. 1980. Les sélachiens du Crétacé superior du Liban. 2: Batoidea. Palaeontographica Abteilung A 168:149–229.

Cappetta, H. 1987a. Extinctions et renouvellements faunique chez les sélaciens post-jurassique. Memoire Societe Géologique de France 150:113–131.

Cappetta, H. 1987b. Chondrichthyes II: Mesozoic and Cenozoic Elasmobranchs. Handbook of Paleoichthyology 3B. Gustav Fisher Verlag, Stuttgart.

Cappetta, H. 1992. Nouveaux Rhinobatoidei (Neoselachii, Rajiformes) a denture spécialisée du Maastrichtian du Maroc. Remarques sur l'évolution dentaire des Rajiformes et des Myliobatiformes. Neues Jahrbuch für Geologie und Paläontologie, Abhandlungen 187:31–52.

Cappetta, H. 2006. Elasmobranchii Post Triadici (Index specierum et generum); in W. Riegraf (ed.), Fossilium Catalogus, Pars 142. Backhuys Publishers, Leiden.

Cappetta, H. In press. Chondrichthyes (Mesozoic and Cenozoic Elasmobranchii: teeth); in H.-P. Schultze (ed.), Handbook of Paleoichthyology 3E. Verlag Dr. Friedrich Pfeil, Munich.

Cappetta, H., and G. R. Case. 1975a. Sélaciens nouveaux de Crétacé du Texas. Geobios 8:303–307.

Cappetta, H., and G. R. Case. 1975b. Contribution a l'étude des sélaciens du Groupe Mon Mouth (Campanian–Maestrichtien) du New Jersey. Palaeontographica Abteilung A 151:1–46.

Cappetta, H., and G. R. Case. 1999. Additions aux faunes de sélaciens du Crétacé du Texas. Palaeo Ichthyologica 9:1–112.

Carpenter, K. 1979. Vertebrate fauna of the Laramie Formation (Maestrichtian), Weld County, Colorado. University of Wyoming Contributions to Geology 17:37–49.

Case, G. R. 1978. A new selachian fauna from the Judith River Formation (Campanian) of Montana. Palaeontographica Abteilung A 160:176–205.

Case, G. R. 1979. Additional fish records from the Judith River Formation of Montana. Geobios 12:223–233.

Case, G. R. 1987. A new selachian fauna from the late Campanian of Wyoming (Teapot Sandstone Member, Mesaverde Formation, Big Horn Basin). Palaeontographica Abteilung A 197:1–37.

Case, G. R., and D. R. Schwimmer, P. D. Borodon., and J. J. Leggett. 2001. A new selachian fauna from the Eutaw Formation (Upper Cretaceous/Early to middle Santonian) of Chattahoochee County, Georgia. Palaeontographica Abteilung A 261:83–102.

Casier, E. 1947a. Constitution and évolution de la racine dentaire des Euselachii. Bulletin de Musée Royal d'Histoire Naturelle de Belgique 23(13):1–15, 23(14):1–32, 23(15):1–45.

Casier, E. 1947b. Contribution a. l'étude des poissons fossiles de la Belgique. XIII. Présence de ganopristinés dans la glauconie de Lonzée et le Tuffeau de Mastrichtian. Bulletin de Musée Royal d'Histoire Naturelle de Belgique 40:1–24.

Cifelli, R. L. 2004. Marsupial mammals from the Albian–Cenomanian (Early–Late Cretaceous boundary), Utah; pp. 62–79 in G. C. Gould and S. K. Bell (eds.), Tributes to Malcolm C. McKenna: His Students, His Legacy. American Museum of Natural History Bulletin 285.

Cifelli, R. L., S. K. Madsen, and E. M. Larson. 1996. Screenwashing and associated techniques for the recovery of microvertebrate fossils. Oklahoma Geological Survey Special Publication 96:1–24.

Cifelli, R. L., J. I. Kirkland, A. Weil, A. L. Deino, and B. J. Kowallis. 1997. High precision ^{40}Ar/^{39}Ar geochronology and the advent of North America's Late Cretaceous fauna. Proceedings of the National Academy of Sciences 94:11163–11167.

Cifelli, R. L., R. L. Nydam, J. D. Gardner, A. Weil, J. G. Eaton, J. I. Kirkland, and S. K. Madsen. 1999a. Medial Cretaceous vertebrates from the Cedar Mountain Formation, Emery County, the Mussentuchit Local Fauna; pp. 219–242 in D. D. Gillette (ed.), Vertebrate Paleontology in Utah. Utah Geological Survey Miscellaneous Publication 99-1.

Cifelli, R. L., R. L. Nydam, J. D. Gardner, J. G. Eaton, and J. I. Kirkland. 1999b. Vertebrate faunas of the North Horn Formation (Upper Cretaceous–Lower Paleocene), Emery and Sanpete Counties, Utah; pp. 377–388 in D. D. Gillette (ed.), Vertebrate Paleontology in Utah. Utah Geological Survey Miscellaneous Publication 99-1.

Cobban, W. A., I. Walaszczyk, J. D. Obradovich, and K. C. McKinney. 2006. A USGS Zonal Table for the Middle Cenomanian–Maastrichtian of the Western Interior of the United States Based on Ammonites, Inoceramids, and Radiometric Ages. U.S. Geological Survey Open-File Report 2006-1250.

Compagno, L. J. V. 1973. Interrelationships of living elasmobranches; pp. 15–61 in P. H. Greenwood, R. S. Miles, and C. Patterson (eds.), Interrelationships of Fishes. Zoological Journal of Linnean Society 53, Supplement 1.

Compagno, L. J. V. 1977. Phyletic relationships of living sharks and rays. American Zoologist 17:303–322.

Compagno, L. J. V. 2002. Freshwater and estuarine elasmobranch surveys in the Indo-Pacific region: threats, distribution and speciation; pp. 168–193 in S. L. Fowler, T. M. Reed, and F. A. Dipper (eds.), Elasmobranch Biodiversity, Conservation and Management. Proceedings of the International Seminar and Workshop, Sabah, Malaysia, July 1997, Occasional Paper of the International Union for Conservation of Nature and Natural Resources Species Survival Commission 25.

Compagno, L. J. V., and S. F. Cook. 1995a. The exploitation and conservation of freshwater elasmobranchs: status of taxa and prospects for the future. Journal of Aquaculture and Aquatic Sciences 7:62–90.

Compagno, L. J. V., and S. F. Cook. 1995b. Freshwater elasmobranchs: a questionable future. Shark News 3:4–6.

Cook, T. D., M. G. Newberry, D. B. Brinkman, and J. I. Kirkland. In press. Freshwater Euselachians from the Hell Creek Formation of Montana with a Comparison to Other Maastrichtian North American Localities. Geological Society of America Special Paper.

Cope, E. D. 1876. Description of some vertebrate remains from the Fort Union beds of Montana. Proceedings of the Academy of Natural Sciences 28:248–261.

Duffin, C. J. 1985. Revision of the hybodont selachian genus *Lissodus* Brough (1935). Palaeontographica Abteilung A 188:105–152.

Duffin, C. J. 2001. Synopsis of the selachian genus *Lissodus* Brough, 1935. Neues Jahrbuch für Geologie und Paläontologie Abhandlungen 221:145–218.

Dyman, T. S., W. A. Cobban, A. L. Titus, J. D. Obradovich, L. E. Davis, R. L. Eves, G. L. Pollock, K. I. Takahashi, and T. C. Hester. 2002. New biostratigraphic and radiometric ages for Albian–Turonian Dakota Formation and Tropic Shale at Grand Staircase Escalante National Monument and Iron Springs Formation near Cedar City, Parowan, and Gunlock in SW Utah. Geological Society of America Abstracts with Programs 34(4):13.

Eaton, J. G. 2006a. Santonian (Late Cretaceous) Mammals from the John Henry Member of the Straight Cliffs Formation, Grand Staircase–Escalante National Monument, Utah. Journal of Vertebrate Paleontology 26:446–460.

Eaton, J. G. 2006b. Late Cretaceous mammals from Cedar Canyon, southwestern Utah; pp. 373–402 in S. G. Lucas and R. M. Sullivan (eds.), Late Cretaceous Vertebrates from the Western Interior. New Mexico Museum of Natural History and Science Bulletin 35.

Eaton, J. G., and R. L. Cifelli. 1988. Preliminary report on the Late Cretaceous mammalian faunas of the Kaiparowits Plateau, Utah. Contributions to Geology of the University of Wyoming 26:45–51.

Eaton, J. G., and R. L. Cifelli. 2001. Multituberculates from near the Early–Late Cretaceous boundary, Cedar Mountain Formation, Utah. Acta Palaeontologica Polonica 46:453–518.

Eaton, J. G., and J. I. Kirkland. 2001. Diversity patterns of nonmarine vertebrates in the Cretaceous western interior of North America; pp. 85–89 in H. A. Leanza (ed.), Proceedings of the VII International Symposium on Mesozoic Terrestrial Ecosystems. Buenos Aires, Argentina. Asociacion Paleontlogica Argentina Publication Especial 7.

Eaton, J. G., and J. I. Kirkland. 2003. Nonmarine Cretaceous vertebrates of the Western Interior Basin; pp. 263–313 in P. J. Harries (ed.), High-Resolution Approaches in Stratigraphic Paleontology. Kluwer Academic Publishers, Boston, Massachusetts.

Eaton, J. G., J. I. Kirkland, and K. Doi. 1989. Evidence of reworked Cretaceous fossils and their bearing on the existence of Tertiary dinosaurs. Palaios 4:281–286.

Eaton, J. G., R. L. Cifelli, J. H. Hutchinson, J. I. Kirkland, and J. M. Parrish. 1999. Cretaceous vertebrate faunas of the Kaiparowits Basin (Cenomanian–Campanian), southern Utah; pp. 345–353 in D. D. Gillette (ed.), Vertebrate Paleontology in Utah. Utah Geological Survey Miscellaneous Publication 99-1.

Eaton, J. G., J. I. Kirkland, J. H. Hutchinson, R. Denton, R. C. O'Neill, and M. J. Parrish. 1997. Nonmarine extinction across the Cenomanian–Turonian (C-T) boundary, southwestern Utah, with a comparison to the Cretaceous–Tertiary (K-T) extinction event. Geological Society of America Bulletin 59:129–151.

Eaton, J. G., J. Laurin, J. I. Kirkland, N. E. Tibert, R. M. Leckie, B. B. Sageman, P. M. Goldstrand, D. W. Moore, A. W. Straub, W. A. Cobban, and J. D. Dalebout. 2001. Cretaceous and Early Tertiary geology of Cedar and Parowan canyons, western Markagunt Plateau, Utah; pp. 337–363 in M. C. Erskine, J. E. Faulds, J. M. Bartley, and P. D. Rowley (eds.), The Geologic Transition: High Plateaus to Great Basin. Utah Geological Association Publication 30.

Estes, R. 1964. Fossil Vertebrates from the Late Cretaceous Lance Formation, Eastern Wyoming. University of California Publications in Geological Sciences 49.

Estes, R. 1969. Lower vertebrates from the Late Cretaceous Hell Creek Formation, McCone County, Montana. Breviora 337:1–33.

Fox, R. C. 1972. A primitive therian mammal from the Upper Cretaceous of Alberta. Canadian Journal of Earth Sciences 9:1479–1494.

Frampton, E. K. 2006. Taphonomy and paleoecology of mixed invertebrate–vertebrate fossil assemblage in the Foremost Formation (Cretaceous, Campanian), Milk River Valley, Alberta. M.S. thesis, University of Calgary, Calgary, Alberta.

Garrison, J. R., Jr., D. B. Brinkman, D. J. Nichols, P. Layer, D. Burge, and D. Thayn. 2007. A multidisciplinary study of the Lower Cretaceous Cedar Mountain Formation, Mussentuchit Wash, Utah: a determination of the paleoenvironment and paleoecology of the dinosaur quarry. Cretaceous Research 28:461–494.

Gill, T. C. 1862. Notes on some genera of fishes in western North America. Proceedings of the National Academy of Natural Sciences, Philadelphia 14:329–332.

Gill, T. C. 1895. Notes on the genus *Cephaleutherus* of Rafinesque, and other rays with aberrant pectoral fins (Propterygia and Hieroptera). Proceedings of the United States National Museum 18:195–198.

Gradstein, F. M., J. G. Ogg, and A. G. Smith. 2004. A Geologic Time Scale, 2004. Cambridge University Press, Cambridge.

Hay, O. P. 1902. Bibliography and Catalogue of the Fossil Vertebrata of North America. U.S. Geological Survey Bulletin 179.

Herman, J. 1977a. Additions to the Eocene fish fauna of Belgium. 3. Revision of the Orectolobiformes. Tertiary Research 1:127–138.

Herman, J. 1977b. Les sélaciens des terrains néocrétacés & paléocènes de Belgique & des contrées limitrophes. Eléments d'une biostratigraphie intercontinentale. Mémoires pour servir à l'explication de Cartes Géologiques et Minières de la Belgique 15.

Hutchinson, P. J., and B. S. Kues. 1985. Depositional environments and paleontology of Lewis Shale to lower Kirtland Shale sequence (Upper Cretaceous), Bisti area, northwestern New Mexico; pp. 25–54 in D. L. Woolberg (ed.), Contributions to Late Cretaceous Paleontology and Stratigraphy of New Mexico, Part 1. New Mexico Bureau of Mines and Mineral Resources Circular 195.

Jinnah, Z. A., E. M. Roberts, A. L. Deino, J. S. Larsen, P. K. Link, and C. M Fanning. 2009. New 40Ar-39Ar and detrital zircon U-Pb ages from the Upper Cretaceous Wahweap and Kaiparowits formations on the Kaiparowits Plateau, Utah: implications for regional correlation, provenance, and biostratigraphy. Cretaceous Research 30:287–299.

Kennedy, W. J., and W. A. Cobban. 1991. Coniacian Ammonite Faunas from the United States Western Interior. Special Papers in Palaeontology 45.

Kirkland, J. I. 1989. Fossil elasmobranchs from the mid-Cretaceous (Middle Cenomanian–Middle Turonian) Greenhorn Cyclothem of eastern Nebraska. Proceedings of the Nebraska Academy of Science 99:52.

Kirkland, J. I. 1990. Paleontology and paleoenvironments of the Middle Cretaceous (Late Cenomanian–Middle Turonian) Greenhorn Cyclothem at Black Mesa, northeastern Arizona. Ph.D. dissertation, University of Colorado, Boulder, Colorado.

Kirkland, J. I. 1996. Paleontology of the Greenhorn Cyclothem (Cretaceous: Late Cenomanian to Middle Turonian) at Black Mesa, Northeastern Arizona. New Mexico Museum of Natural History and Science Bulletin 9.

Kirkland, J. I., and S. K. Madsen. 2007. The Lower Cretaceous Cedar Mountain Formation, eastern Utah: the view up an always interesting learning curve; in W. R. Lund (ed.), Field Guide to Geological Excursions in Southern Utah. Geological Society of America Rocky Mountain Section 2007 Annual Meeting, Grand Junction Geological Society, Utah Geological Association Publication 35. CD-ROM.

Kirkland, J. I., J. G. Eaton, and D. B. Brinkman. 2009. Freshwater elasmobranchs from the Cretaceous of Utah with emphasis on fossils from Grand Staircase–Escalante National Monument (GSENM); p. 31 in A. Titus (ed.), Advances in Western Interior Paleontology and Geology: Abstracts with Program. Bureau of Land Management, Kanab, Utah.

Kirkland, J. I., S. G. Lucas, and J. W. Estep. 1998. Cretaceous dinosaurs of the Colorado Plateau; pp. 67–89 in S. G. Lucas, J. I. Kirkland, and and J. W. Estep (eds.), Lower to Middle Cretaceous Non-marine Cretaceous Faunas. New Mexico Museum of Natural History and Science Bulletin 14.

Kocsis, L., T. W. Vennemann, and D. Fontignie. 2007. Migration of sharks into freshwater systems during the Miocene and implications for Alpine paleoelevation. Geology 35:451–454.

Kriwet, J. 1999. *Ptychotrygon geyeri* n. sp. (Chondrichthyes, Rajiformes) from the Utrillas Formation (Upper Albian) of the central Iberian Ranges (east Spain). Profil 16:337–346.

Kriwet J. 2004. The systematic position of the Cretaceous sclerorhynchids sawfishes (Elasmobranchii, Pristiorajea); pp. 57–74 in G. Arratia and A Tintori (eds.), Mesozoic Fishes 3 – Systematics, Palaeoenvironment, and Biodiversity. Dr. Friedrich Pfiel, Munich.

Kriwet, J., and M. J. Benton. 2004. Neoselachian (Chondrichthyes, Elasmobranchii) diversity across the Cretaceous–Tertiary boundary. Palaeogeography, Palaeoclimatology, Palaeoecology 214:181–194.

Kriwet, J., E. V. Nunn, and S. Klug. 2009. Neoselachians (Chondrichthyes, Elasmobranchii) from the Lower and lower Upper Cretaceous of north-eastern Spain. Zoological Journal of the Linnean Society 155:316–345.

Landman, N. H., and W. A. Cobban. 2007. Redescription of the Late Cretaceous (late Santonian) ammonite *Desmoscaphites bassleri* Reeside, 1927, from the Western Interior of North America. Rocky Mountain Geology 42:67–94.

Langston, W., Jr. 1970. A fossil ray braincase, possibly *Myledaphus* (Elasmbranchii: Batoidea) from the Late Cretaceous Oldman Formation of Western Canada. National Museum of Canada Publications in Paleontology 6.

Langston, W., Jr. 1975. The ceratopsian dinosaurs and associated lower vertebrates from the St. Mary River Formation (Maestrichtian) at Scabby Butte, southern Alberta. Canadian Journal of Earth Science 12:1576–1608.

Larson, D. W. 2010. The occurrences of vertebrate fossils in the Deadhorse Coulee Member of the Milk River Formation and their implications for provincialism and evolution in the Santonian (Late Cretaceous) of North America. M.S. thesis, University of Alberta, Edmonton, Alberta.

Larson, D. W., D. B. Brinkman, and P. R. Bell. 2010. Faunal assemblages from the Horseshoe Canyon Formation, an early Maastrichtian cool-climate assemblage from Alberta, with special reference to the *Albertosaurus sarcophagus* bonebed. Canadian Journal of Earth Science 47:1159–1181.

Leckie, R. M., J. I. Kirkland, and W. P. Elder. 1997. Stratigraphic framework and correlation of a principle reference section of the Mancos Shale (Upper Cretaceous) Mesa Verde, Colorado; pp. 163–216 in New Mexico Geological Society Guidebook, 48th Field Conference, Mesozoic Geology and Paleontology of the Four Corners Region.

Leidy, J. 1856. Notices of remains of extinct vertebrate animals of New Jersey collected by Prof. Cook. Proceedings of the Academy of Natural Sciences, Philadelphia 3:220–221.

Lovejoy, N. R. 1996. Systematics of myliobatoid elasmobranchs: with emphasis on the phylogeny and historical biogeography of neotropical freshwater stingrays (Potamotrygonidae: Rajiformes). Zoological Journal of the Linnean Society 117:207–257.

Maisey, J. G. 1978. Growth and form of fin spines in hybodont sharks. Palaeontology 21:657–666.

Maisey, J. G. 1982. The anatomy and interrelationships of Mesozoic hybodont sharks. American Museum Novitates 2724:1–48.

Maisey, J. G. 1983. Cranial anatomy of *Hybodus basanus* Egerton from the Lower Cretaceous of England. American Museum Novitates 2758:1–64.

Maisey, J. G. 1986. Anatomical revision of the fossil shark *Hybodus fraasi* (Chondrichthyes: Elasmobranchii). American Museum Novitates 2857:1–16.

Maisey, J. G. 1987. Cranial anatomy of the Lower Jurassic shark *Hybodus reticulatus* (Chondrichthyes: Elasmobranchii), with comments on hybodontid systematics. American Museum Novitates 2878:1–39.

Maisey, J. G. 1990. Selachii; pp. 16–19 in R. P. de Angola and M. T. Antunes (eds.), Triassic Fishes from the Cassange Depression. Ciências da Tierra, Número Especial.

McNulty, C. L., Jr. 1964. Hypolophid teeth from the Woodbine Formation, Tarrant County, Texas. Eclogae Geologicae Helvetiae 57:537–539.

McNulty, C. L., Jr., and B. H. Slaughter. 1972. The Cretaceous selachian genus *Ptychotrygon* Jaekel, 1894. Eclogae Geologicae Helvetiae 65:647–656.

Merewether, E. A., W. A. Cobban, and J. D. Obradovich. 2007. Regional disconformities in Turonian and Coniacian (Upper Cretaceous) strata in Colorado, Wyoming, and adjoining states – biochronological evidence. Rocky Mountain Geology 42:95–122.

Meyer, R. L. 1974. Late Cretaceous elasmobranchs from the Mississippi and East Texas embayments of the Gulf Coastal plain. Ph.D. dissertation, Southern Methodist University, Dallas, Texas.

Milner, A. R. C., and J. I. Kirkland. 2006. Preliminary review of the Early Jurassic (Hettangian) freshwater Lake Dixie fish fauna in the Whitmore Point Member, Moenave Formation in southwest, Utah; pp. 510–521 in J. D. Harris, S. G. Lucas, J. A. Spielmann, M. G. Lockley, A. R. C. Milner, and J. I. Kirkland (eds.), The Triassic–Jurassic Terrestrial Transition. New Mexico Museum of Natural History and Science Bulletin 37.

Moore, D. W., L. D. Nealey, P. D. Rowley, S. C. Hatfield, D. J. Maxwell, and E. Mitchell. 2004. Geologic map of the Navajo Lake quadrangle, Kane and Iron Counties, Utah. Utah Geological Survey Map 199.

Muller, J., and J. Henle. 1838–1841. Systematische Beschreibung der Plagiostomen. Veit and Co., Berlin.

Murry, P. A. 1981. A new species of freshwater hybodont from the Dockum Group (Triassic) of Texas. Journal of Paleontology 55:603–607.

Nessov, L. A. 1981. Cretaceous salamanders and frogs of Kizylkum Desert. Akademiya Nauk S.S.S.R. 101:57–88.

Nessov, L. A., D. Sigogneau-Russell, and D. E. Russell. 1994. A survey of Cretaceous tribosphenic mammals from middle Asia (Uzbekistan, Kazakhstan and Tajikistan), of their geological setting, age, and faunal environment. Palaeovertebrata 23:51–92.

Neuman, A. G., and D. B. Brinkman. 2005. Fishes of the fluvial beds; pp. 167–185 in P. J. Currie and E. B. Koppelhus (eds.), Dinosaur Provincial Park: A Spectacular Ancient Ecosystem Revealed. University of Indiana Press, Bloomington, Indiana.

Obradovich, J. D., and W. A. Cobban. 1975. A time scale for the Late Cretaceous of the Western Interior of North America; pp. 31–54 in W. G. E. Caldwell (ed.), The Cretaceous System in the Western Interior of North America. Geological Association of Canada Special Paper 13.

Ogg, J. G., F. P. Agterberg, and F. M. Gradstein. 2004. Cretaceous; pp. 344–383 in F. M. Gradstein, J. G. Ogg, J. G., and A. G. Smith (eds.), A Geologic Time Scale. Cambridge University Press, Cambridge.

Owen, R. 1846. Lectures on the comparative anatomy and physiology of the vertebrate animals, delivered at the Royal College of Surgeons of England in 1844 and 1846. Part 1. Fishes. Longman, London.

Patterson, C. 1966. The British Wealden sharks. Bulletin of the British Museum (Natural History), Geology 11:281–350.

Peng, J.-H., A. P. Russell, and D. B. Brinkman. 2001. Vertebrate microsite Assemblages (Exclusive of Mammals) from the Foremost and Oldman Formations of the Judith River Group (Campanian) of Southeastern Alberta: An Illustrated Guide. Provincial Museum of Alberta Natural History Occasional Papers 25.

Peterson, F. 1969. Four New Members of the Upper Cretaceous Straight Cliffs Formation in Southeastern Kaiparowits Region, Kane County, Utah. U.S. Geological Survey Bulletin 1274-J.

Rees, J. 2002. Shark fauna and depositional environment of the earliest Cretaceous Vitabäck Clays at Eriksdal, southern Sweden. Transactions of the Royal Society of Edinburgh, Earth Sciences 93:59–71.

Rees, J., and C. J. Underwood. 2002. The status of the shark genus Lissodus Brough, 1935, and the position of nominal Lissodus species within the Hybodontoidea (Selachii). Journal of Vertebrate Paleontology 22:471–479.

Roberts, E. M., A. L. Deino, and M. A. Chan. 2005. ^{40}Ar-^{39}Ar age of the Kaiparowits Formation, southern Utah, and correlation of contemporaneous Campanian strata and vertebrate faunas along the margin of the Western Interior Basin. Cretaceous Research 26:307–318.

Rowe, T., R. L. Cifelli, T. M. Lehman, and A. Weil. 1992. The Campanian Terlingua Local Fauna, with a summary of other vertebrates from the Aguja Formation, Trans-Pecos Texas. Journal of Vertebrate Paleontology 12:472–493.

Sankey, J. T. 2008. Vertebrate paleoecology from microsites, Talley Mountain, upper Aguja Formation (Late Cretaceous), Big Bend National Park, Texas; pp. 61–77 in J. T. Sankey and S. Baszio (eds.), The Unique Role of Vertebrate Microfossil Assemblages in Paleoecology and Paleobiogeography. Indiana University Press, Bloomington, Indiana.

Siverson. M. 1995. Revision of Cretorectolobus (Neoselachii) and description of Cederstroemia n. gen., a Cretaceous carpet shark (Orectolobiformes) with a cutting dentition. Journal of Paleontology 69:974–979.

Slaughter, B. H., and M. Steiner. 1968. Notes on the rostral of ganopristine sawfishes, with special reference to Texas material. Journal of Paleontology 42:233–239.

Stromer, E. 1917. Ergebnisse der Forschungsreisen Prof. Stromers in den Wusten Agyptens; die Sage des pristiden Onchopristis. Abhandlungen Bayerische Akademie der Wissenschaften. Mathematisch-Naturwissenschaftliche Klasse 27:1–21.

Thorburn, D. C., D. L. Morgan, A. J. Rowland, and H. S. Gill. 2004. Elasmobranchs in the Fitzroy River, Western Australia. Report to the National Heritage Trust, Centre for Fish and Fisheries Research, Murdock University, Perth. Australia, 30 p. http://155.187.2.69/coasts/publications/pubs/elasmo-wa.pdf.

Thurmond, J. T. 1971. Cartilaginous fishes of the Trinity Group and related rocks (Lower Cretaceous) of north-central Texas. Southeastern Geology 13:207–227.

Titus, A. L., J. D. Powell, E. M. Roberts, S. D. Sampson, S. L. Pollack, J. I. Kirkland, and L. B. Albright. 2005. Late Cretaceous stratigraphy, depositional environments, and macrovertebrate paleontology of the Kaiparowits Plateau, Grand Staircase–Escalante National Monument, Utah; pp. 101–128 in J. Pederson and C. M. Dehler (eds.), Interior Western United States. Geological Society of America Field Guide 6.

Underwood, C. J., and S. L. Cumbaa. 2010. Chondrichthyans from a Cenomanian (Late Cretaceous) bonebed, Saskatchewan, Canada. Palaeontology 53:903–944.

Walaszczyk, I., and W. A. Cobban. 2006. Paleontology and biostratigraphy of the upper Middle–Upper Coniacian–Santonian inoceramids of the U.S. Western Interior. Acta Geologica Polonica 56:241–348.

Walaszczyk, I., and W. A. Cobban. 2007. Inoceramid fauna and biostratigraphy of the upper Middle Coniacian–lower Middle Santonian of the Pueblo section (SE Colorado, U.S. Western Interior). Cretaceous Research 28:132–142.

Walker, J. D., and J. W. Geissman, compilers. 2009. Geologic Time Scale. Geological Society of America, Boulder, Colorado.

Welton, B. J., and R. F. Farish. 1993. The Collector's Guide to Fossil Sharks and Rays from the Cretaceous of Texas. Before Time, Lewisville, Texas.

Williamson, T. E., J. I. Kirkland, and S. G. Lucas. 1993. Selachians from the Greenhorn Cyclothem ("Middle" Cretaceous, Cenomanian–Turonian), Black Mesa, Arizona, and the paleogeographic distribution of Late Cretaceous selachians. Journal of Paleontology 67:447–474.

Wolberg, D. L. 1985. Selachians from the Atarque Sandstone Member of the Tres Hermanos Formation (Upper Cretaceous: Turonian), Sevilleta Grant near La Joya, Socorro County, New Mexico; pp. 7–19 in D. L. Woolberg (ed.), Contributions to Late Cretaceous Paleontology and Stratigraphy of New Mexico, Part 1. New Mexico Bureau of Mines and Mineral Resources Circular 195.

Woodward, A. S. 1889. Catalogue of the Fossil Fishes in the British Museum (Natural History): Part 1–Containing the Elasmobranchii. British Museum (Natural History), London.

Woodward, A. S. 1916. The Fossil Fishes of the English Wealden and Purbeck Formations: Part 1. Palaeontogical Society Monograph 69.

Wroblewski, A. F.-J. 2004. New Selachian paleofaunas from "fluvial" deposits of the Ferris and Lower Hanna formations (Maastrichtian–Selandian 66–58 Ma), southern Wyoming. Palaios 19:249–258.

Freshwater Osteichthyes from the Cenomanian to Late Campanian of Grand Staircase–Escalante National Monument, Utah

10

Donald B. Brinkman, Michael G. Newbrey, Andrew G. Neuman, and Jeffrey G. Eaton

FOSSIL ASSEMBLAGES FROM THE GRAND STAIRCASE RE-gion of Utah provide data on patterns of diversity of bony fish from the Cenomanian through the late Campanian from the southern parts of the Western Interior of North America. Basal actinopterygians are prominent members of the assemblages and generally can be identified to at least family level. A high diversity of teleosts is also present, including hiodontiforms, osteoglossiforms, elopomorphs, clupeomorphs, ostariophysans, esocoids, and acanthomorphs, as well as taxa that cannot presently be placed in any lower-level taxonomic group. The small size of teleosts from the Late Cretaceous of the Grand Staircase region indicates that they would have occupied basal positions in the vertebrate food webs. A major faunal change occurred between the Cenomanian and Turonian assemblages, with a 47% change in taxa across the boundary. Amiids decrease in diversity, pycnodonts and *Lepidotes* become rare, and the Lepisosteidae, Vidalamiinae, and ostariophysan type U-3/BvD all first appear in the Turonian. A significant faunal turnover also occurred in the mid- to late Campanian, when esocoids and a characiform first appear. Turonian to middle Campanian assemblages show a successive appearance of five new taxa, including acanthomorphs, and a loss of two taxa. Latitudinal patterns are identified by comparing assemblages from the Kaiparowits Formation with contemporaneous assemblages from more northern localities. Lepisosteidae gen. et sp. indet. type 1 (the gar represented by lanceolate teeth), *Lepidotes*, *Micropycnodon*, *Melvius*, and ostariophysan type U-3/BvD have a more southern distribution, while sturgeon, holostean A, holostean B, *Belonostomus* and the teleosts *Paratarpon*, teleost indet. type H, the clupeomorph *Horseshoeichthyes*, and *Coriops* have a more northern distribution. Several of the biostratigraphic changes that are observed can be interpreted as shifts in the geographic distributions of these taxa that are associated with changes in mean annual temperatures during the Late Cretaceous.

INTRODUCTION

Vertebrate microfossil localities have provided an enormous amount of data on Late Cretaceous paleofaunas. This is particularly the case for the Grand Staircase–Escalante region of southern Utah, where an intensive study of such localities has provided a detailed record of vertebrate faunal change from the Cenomanian to the late Campanian of this geographically restricted area. The vertebrate microfossil localities at the Kaiparowits Plateau were initially screened for mammalian remains, and although the resulting samples have also provided important data on the entire vertebrate assemblage (Eaton, 1991; Eaton et al., 1997; Hutchison et al., 1998), fish, particularly teleosts, remain a poorly understood aspect of the fauna. In this chapter, we describe osteichthyan (bony fish) microfossils from a series of localities ranging from the Cenomanian Dakota Formation through to the late Campanian Kaiparowits Formation and assess their importance both biostratigraphically and biogeographically.

Study of the remains of bony fish from vertebrate microfossil sites presents significant challenges because they are represented exclusively by isolated elements, while phylogenetic studies of most groups of fish relies on data from articulated skeletons. In some cases, autapomorphic features allow isolated elements to be referred to established taxa (Estes, 1964; Wilson, Brinkman, and Neuman, 1992; Brinkman and Neuman, 2002), but, especially in the case of teleost fish, many of their distinctive elements can be identified only to high taxonomic levels and thus are of limited use in resolving phylogenetic questions. Despite these limitations, Brinkman and Neuman (2002) and Neuman and Brinkman (2005) demonstrated that study of fish remains from vertebrate microfossil localities can yield useful data for paleoecological studies, especially those involving diversity and distribution patterns. As well, for some of the high-level taxa, the fossils can be of significance in providing minimum age-of-origin estimates.

METHODS

The fish material described here was collected by quarrying unweathered rock at productive horizons and subsequent screen washing of the quarried rocks. The wet screen washing was accomplished by using nested screens as described in Cifelli, Madsen, and Larson (1996). Most of the localities in this report showed bone at the surface, but a few were located using blind-washing techniques described by Eaton (2004).

The goals of this study are to identify, describe, and estimate the diversity of bony fish in the Upper Cretaceous sediments of the Grand Staircase–Escalante National Monument region of Utah and to assess geographic and biostratigraphic patterns of distribution of nonmarine fish through the Late Cretaceous in the Western Interior of North America. To achieve these goals, it was necessary to use an open nomenclature because many of the taxa present could only be identified at higher taxonomic levels. To incorporate these taxa into a meaningful faunal list, an informal alphanumeric system was used when referring to them. Where elements were considered to be diagnostic for a low-level taxon, the name used is applied to that fish. Where an element is distinctive but it is uncertain whether or not it is from a taxon already represented by other elements, the name is applied to that kind of element. A special challenge is presented by teleosts. To interpret the relationships of the remains of teleost fish as fully as possible, extensive comparison with recent osteological material was undertaken. The osteological collections of the Royal Ontario Museum (Toronto, Ontario, Canada), Canadian Museum of Nature (Ottawa, Ontario, Canada), and University of Michigan (Ann Arbor, Michigan), as well as the Royal Tyrrell Museum of Palaeontology (Drumheller, Alberta, Canada) were examined. Examination of fish remains from the Eocene Bridger Basin Formation in the collections of the University of California Museum of Paleontology (Berkeley, California) was useful for identifying isolated elements of clupeomorphs and ostariophysans. To ensure that future researchers will be able to build on this system, an extensive series of photographs of the elements reported here is included. Most specimens were photographed through a binocular microscope with a Canon Powershot A95 camera. Most specimens were whitened with ammonium chloride before photography.

Both presence/absence and abundance data were considered in identifying patterns of distribution. Presence/absence data were based on all available evidence, including surface-collected elements as well as material obtained from screen-washed samples. Following the protocols suggested by Brinkman (2008), only taphonomically and hydraulically equivalent elements were considered when estimating the relative abundance of taxa. For example, estimates of the relative abundance of taxa from different sets of localities were based on comparisons of the abundance of teeth, centra, or jaws, but not combinations of these elements. Abundance of tooth-based taxa was estimated using four categories based on the number of elements observed in samples of more than 300 identifiable elements: very rare (fewer than five elements), rare (between five and 10 elements), common (between 10 and 20 elements), and very abundant (more than 20 elements). Because generally only a small sample of centra is available, the number of each morphotype relative to the total number of centra in representative samples was used to estimate abundance of centra.

Institutional Abbreviations FOBU, Fossil Butte National Monument, Kemmerer, Wyoming; MNA, Museum of Northern Arizona, Flagstaff, Arizona; OMNH, Oklahoma Museum of Natural History, Norman, Oklahoma; ROM, Royal Ontario Museum, Toronto, Ontario, Canada; UAMZ, University of Alberta Museum of Zoology, Edmonton, Alberta, Canada; UMMZ, Museum of Zoology, University of Michigan, Ann Arbor, Michigan; UCMP, University of California, Museum of Paleontology, Berkeley, California; UMNH, Utah Museum of Natural History, Salt Lake City, Utah.

GEOLOGY

Seven distinct stratigraphic intervals are represented: Cenomanian (Dakota Formation); Turonian (Smoky Hollow Member, Straight Cliffs Formation), middle Coniacian, lower Santonian, and upper Santonian (John Henry Member, Straight Cliffs Formation), mid-Campanian (Wahweap Formation), and upper Campanian (Fig. 10.1A).

Cenomanian fish assemblages are represented by material from the Dakota Formation. The Dakota Formation contains sediments deposited in fluvial, paludal, lacustrine, brackish-water, and marine environments. Dakota fish assemblages are represented exclusively by localities in the middle member of the Dakota Formation. Material studied here was recovered from MNA Locs. 939 (=UMNH.VP.LOC.123), 1008, 1064 (=UMNH.VP.LOC. 804) and 1067 (=UMNH.VP.LOC. 27). These localities are preserved in floodplain deposits and yielded large sample sizes of both teeth and centra.

Turonian fish assemblages are documented by material from the Smoky Hollow Member of the Straight Cliffs Formation. This unit was deposited during a regression of the Western Interior Sea. The basal beds of this member contain coal and associated sediments deposited in brackish paludal paleoenvironments, and these are overlain by sediments deposited in lacustrine and floodplain paleoenvironments. Fish remains studied here are from MNA Locs. 994 and 995 (=UMNH.VP.LOC. 129), both of which are in the upper beds of the Smoky Hollow Member, which is upper middle

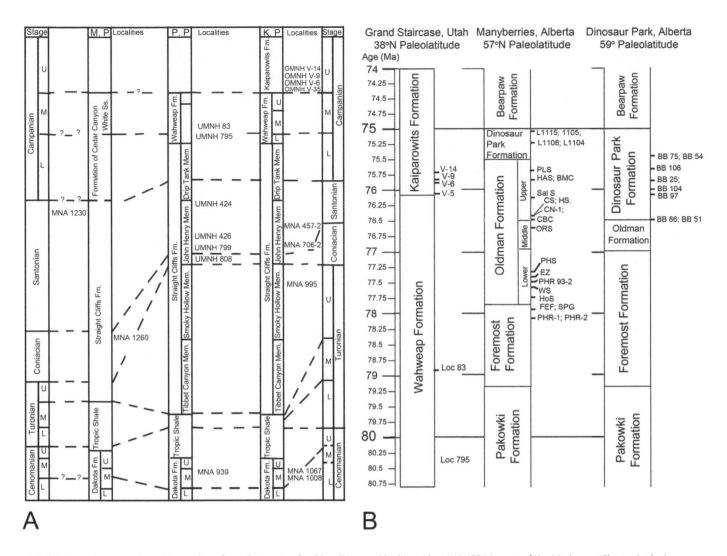

10.1. (A) Approximate stratigraphic position of vertebrate microfossil localities used in this study. MNA 1230 is west of the Markagunt Plateau in the Iron Springs Formation; (B) chronostratigraphic correlation of Wahweap and Kaiparowits formations of Utah with the Belly River Group in the Manyberries and Dinosaur Park regions of Alberta. Age estimates are from Jinnah et al. (2008), Roberts, Deino, and Chan (2005), and Eberth (2005). Localities in the Manyberries area are from Peng, Russell, and Brinkman (2001) and Eberth and Brinkman (1997). Localities in the Dinosaur Provincial Park section are from Brinkman (1990). M.P., Markagunt Plateau; P.P., Paunsaugunt Plateau; K.P., Kaiparowits Plateau.

Turonian in age. Large sample sizes of both teeth and centra were available.

Coniacian fish assemblages are documented by material from MNA Loc. 1260 (=UMNH.VP.LOC. 8) in the Straight Cliffs Formation (or its equivalent) in Cedar Canyon on the Markagunt Plateau, and UMNH.VP.LOC. 808 on the Paunsaugunt Plateau. The age of MNA Loc. 1260 is constrained by a biotite ash above the locality that has been dated at 86.72 ± 0.58 Ma (Eaton, Maldonado, and McIntosh, 1999; Eaton et al., 2001), placing this locality in the middle Coniacian. The samples available for study are large but consist mostly of teeth, scales, and compact bones.

Early Santonian fish assemblages are represented by UMNH VP Locs. 799, 781, and 426, all in the lower part of the John Henry Member of the Straight Cliffs Formation. An early Santonian age has been suggested for these localities,

principally on the basis of their stratigraphic position within the unit (Eaton, 2006). The John Henry Member of the Straight Cliffs Formation was deposited during a significant transgression of the seaway that began in the mid-Coniacian and reached the middle of the Kaiparowits Plateau by the Santonian (Eaton, 1991). The unit is primarily marine along the eastern margin of the plateau but is brackish water and nonmarine along the western margin. The three localities from the John Henry Formation are dominated by fish, with teeth, centra, and jaw elements being well represented.

Late Santonian microvertebrate assemblages are represented by MNA Loc. 1230 (=UMNH.VP.LOC. 12) in the Iron Springs Formation and MNA Locs. 457–2 (=field number PC2) and 706–2 (=field number HC2; UMNH.VP.LOC. 98) in the upper John Henry Member of the Straight Cliffs Formation. MNA Loc. 1230 is at the very top of the Iron Springs

Formation at the north end of Pine Valley Mountains, east of the town of Pinto. Eaton (1999, 2006) concluded that this locality is likely no younger than Santonian, although there is little in the way of age constraints, and an early Campanian age also is possible. MNA Locs. 457–2 (=field number PC2) and 706–2 (=field number HC2; UMNH.VP.Loc. 98) are both on the west side of the Kaiparowits Plateau. Various vertebrate and invertebrate taxa from the middle and upper part of the unit, including the ammonite *Desmoscaphites*, suggest that the upper part of the John Henry Member does not extend beyond the Santonian (Eaton, 1991). The late Santonian age assigned to fish material from this area is based on their stratigraphic position within the unit (Eaton, 2006). The relatively small samples from these localities contain several brackish-water taxa and are dominated by teeth.

Middle Campanian assemblages are documented by localities in the Wahweap Formation. The Wahweap Formation is separated from the John Henry Member by the Drip Tank Member of the Straight Cliffs Formation, indicating a considerable span of time between the upper part of the John Henry Member and the Wahweap Formation. Although an early Campanian age was suggested for the Wahweap Formation, in part because of similarities of the mammals to those from the Milk River Formation southern Alberta (Eaton and Cifelli, 1988), ^{40}Ar/^{39}Ar dating of sanidine phenocrysts from a bed near the base of the formation gave an age of 80.1 ± 03 Ma, younger than early Campanian. These results were recently combined with radiometric dates from the overlying Kaiparowits Formation, and the Wahweap is now estimated to be between 80.8 and 76.1 Ma (Jinnah et al., 2008) and thus mid-Campanian in age. Sediments of the Wahweap Formation were deposited predominantly by meandering streams. However, a few crab claws recovered from fluvial sandstones are interpreted as representing brackish or marine taxa, indicating that a marine influence was present (Eaton, 1987, 1991). Data on the fish from the Wahweap were obtained from three localities: UMNH.VP.Loc. 83, UMNH.VP.Loc. 795, and MNA 455–1. UMNH.VP.Loc. 83 is from low in the middle member on the top of the Paunsaugunt Plateau, and UMNH.VP.Loc. 795 from the lower member on the east side of the Paunsaugunt Plateau. MNA 455–1 is on the Kaiparowits Plateau at the base of upper member of the formation. Large sample sizes are available but are dominated by teeth. Moderate numbers of centra and jaws are present.

Late Campanian assemblages are represented by fossils from the Kaiparowits Formation. This formation was deposited in a relatively wet, low-relief, inland alluvial plain setting with periodic aridity (Roberts, 2007). On the basis of dates obtained from volcanic ashes, the Kaiparowits Formation was deposited between 76.1 and 74 Ma (Roberts, Deino, and

Chan, 2005). Data on fish assemblages from the Kaiparowits Formation were available from localities OMNH V14 (=MNA 453–2, field number TB-2), OMNH V35 (=MNA 458–1, field number FB-1), OMNH V-6 (=MNA 454–6, field number HM-6), and OMNH V-9, all of which are in the lower 200 m of the formation. The largest samples available were from OMNH V-9 and OMNH V-6. Teeth, centra, and tooth-bearing elements are all well represented in these samples.

Because this study compares Late Cretaceous fish assemblages from Utah and Alberta, it was necessary to correlate as precisely as possible the Wahweap and Kaiparowits formations in Utah and the Belly River Group in Alberta. Radiometric ages are available for each of these units, which, together with estimates of sedimentation rates, provide a basis for exceptionally precise correlation between these units (Fig. 10.1B). The 450-m-thick Wahweap is dated as having been deposited between 80.8 and 76.1 Ma, giving an average sedimentation rate of 9.6 cm/Ka (Jinnah et al., 2008). The material examined is from three localities. The oldest of these, UMNH.VP.Loc. 795, is from very low in the Wahweap Formation and thus would be older than any of the localities in the Belly River Group, which at its base is 79.1 Ma (Eberth, 2005). The youngest, UMNH.VP.Loc. 83, is 200 m above the base of the formation. Given an average sedimentation rate, locality Loc. 83 should be approximately 78.9 Ma, temporally equivalent with the Foremost Formation, the basal unit in the Belly River Group. Data on fish faunas from the Foremost Formation were provided by Peng, Russell, and Brinkman (2001) and Brinkman et al. (2004), although these are from localities that are younger than Loc. 83 in the Wahweap Formation (Fig. 10.1B).

The Kaiparowits Formation, which at its maximum thickness is 860 m, has been dated as having been deposited between 76.1 and 74 Ma, with an average sedimentation rate of approximately 41 cm/Ka (Roberts, Deino, and Chan, 2005). The samples examined from the Kaiparowits Formation are from localities within the lower 200 m of the formation. Given an average sedimentation rate of 41 cm/Ka, these localities should occur between 76.1 and 75.6 Ma. In Dinosaur Provincial Park, the contact between the Oldman and Dinosaur Park formations is 76.5 Ma (Eberth, 2005). A volcanic ash in the Dinosaur Park Formation 40 m above the base of the formation has been dated at 75.3 Ma (Eberth, 2005). Thus, the Kaiparowits localities are bracketed by localities in the lower 40 m of the Dinosaur Park Formation of Dinosaur Provincial Park (Fig. 10.1B). Data on the fish faunas from this interval were provided by Brinkman (1990), Brinkman and Neuman (2002), and Neuman and Brinkman (2005). The Oldman–Dinosaur Park formational contact is diachronous, becoming younger in the southern part of the province. Thus, in the

Brinkman et al.

Table 10.1. Faunal list of fish from Cenomanian to late Campanian of the Grand Staircase–Escalante National Monument region, Utah[a]

Class Actinopterygii
  Subclass Chondrostei
    Order Acipenseriformes
      Family Polyodontidae
        Genus and species indet.
  Subclass Neopterygii
    Order Lepisosteiformes
      Family Lepisosteidae
        Lepisosteus sp.
        Genus and species indet. type 1
    Order Semionotiformes
      Family Semionotidae
        Lepidotes spp.
    Order Pycnodontiformes
      Family Pycnodontidae
        Coelodus sp.
        Micropycnodon sp.
    Order Amiiformes
      Family Amiidae
        Subfamily Amiinae
         Genus and species indet.
        Subfamily Vidalamiinae
         ?*Melvius* sp.
        Subfamily indet.
         Genus and species indet. type AWSS
  Division Teleostei
    Subdivision indet.
      Genus and species indet. type O

Subdivision Osteoglossomorpha
  Order and family indet.
    Coriops sp.
  Order Hiodontiformes
    Family Hiodontidae
      Genus and species indet.
Subdivision Elopomorpha
  Order Albuliformes
    Suborder Albuloidei
      Family Albulidae
        Subfamily Phylodontinae
         Paralbula sp.
  Order Elopiformes
    Suborder indet.
      Genus and species indet.
Subdivision Otocephala
SuperorderClupeomorpha
  Order Ellimmichthyiformes
    Family Sorbinichthyidae
      Diplomystus sp.
      Genus and species indet. type LvD
      Genus and species indet. type U-7
  Order Clupeiformes
    Genus and species indet. type G
Superorder Ostariophysi
  Series Otophysi
    Order Indet
      Family indet.

        Genus and species indet. type U-3/BvD
    Order Characiformes
      Family indet.
        Genus and species indet.
Subdivision Euteleostei
  Order indet.
    Family indet.
      Genus and species indet. type U-4
  Order Salmoniformes
    Suborder Esocoidea
      Family indet.
        Estesesox foxi
        Estesesox sp.
        Genus and species indet.
  Order and Family indet.
      Genus and species indet. type H
      Genus and species indet. type BvE
Subdivision Neoteleostei
  Superorder Acanthomorpha
    Order and Family indet.
      Genus and species indet.
  Division indet.
      Genus and species indet. (fish with
saber-like teeth)
Class Sarcopterygia
  Subclass Dipnoi
    Order Ceratodontiforems
      Ceratodus gustasoni

[a] To avoid redundancy of taxa, two morphotypes of lepisosteid scales and five morphotypes of tooth-bearing elements are not included because these elements may be from fish taxa recognized on the basis of other elements.

Manyberries region of southern Alberta, localities that are temporally equivalent to the Kaiparowits localities are in the upper unit of the Oldman Formation (Fig. 10.1B). Data on the fish faunas from the upper Oldman Formation in the Manyberries region were provided by Peng, Russell, and Brinkman (2001) and Brinkman et al. (2004).

DIVERSITY OF THE OSTEICHTHYIAN ASSEMBLAGES

A faunal list of actinopterygians from the Late Cretaceous of the Grand Staircase–Escalante National Monument region is listed in Table 10.1. Osteichthyian classification follows Nelson (2006), with the exception of the Esocoidea, which we include in the Salmoniformes following Wilson

10.2. Paddlefish denticle in dorsal, ventral, and lateral views. Specimen MNA 10356, from the Wahweap Formation, MNA Loc. 456–2. Scale bar = 2 mm.

and Williams (2010). This list includes taxa of uncertain taxonomic position based on distinctive jaws and centra in addition to formally named taxa. Brief descriptions of the newly recognized taxa, as well as reviews of those previously recognized, follow.

ACTINOPTERYGII Klein, 1885
CHONDROSTEI Müller, 1844 (sensu
Grande and Bemis, 1996)
ACIPENSERIFORMES Berg, 1940
(sensu Grande et al., 2002)
POLYODONTIDAE Bonaparte, 1838
Genus et sp. indet.

Chondrostean-grade fish are notably rare in of all the Utah samples. Paddlefish are present but extremely rare in the mid-Campanian Wahweap Formation, where they are documented by a single denticle (Fig. 10.2). As described by McAlpine (1947), paddlefish denticles have a thin dorsal plate ending in posteriorly projecting spines and a pair of ventral projections that typically extend at right angles to the dorsal plate.

NEOPTERYGII Regan, 1923
(sensu Rosen et al., 1981)
LEPISOSTEIFORMES Hay, 1929
LEPISOSTEIDAE Cuvier, 1825
LEPISOSTEUS Lacépède, 1803

Lepisosteus is represented by isolated teeth, centra, and scales (Fig. 10.3). Among the most distinctive of these elements are the opisthocoelous centra, which have well-developed concave and convex articular surfaces on the ends of the centrum, a feature seen in no other fish. Teeth are conical and striated, except for a clear tip, and often have a constriction at the base of the tip (Fig. 10.3A, B). The teeth vary in stoutness, presumably reflecting variation in the length of teeth along the jaw.

Lepisosteus is not known in the Cenomanian Dakota Formation. Reports of *Lepisosteus* from localities of Cenomanian age from the Cedar Mountain Formation by Garrison et al. (2007) are based on scales; this is problematic because the scales of *Lepidotes* Agassiz, 1832, also present in the assemblage, cannot be distinguished from scales of *Lepisosteus*. Isolated teeth from the Cedar Mountain Formation have been tentatively referred to the Lepisosteidae by Cifelli et al. (1999), but this identification could not be

confirmed. In the Grand Staircase region, lepisosteids first occur in Turonian localities of the Smoky Hollow Member of the Straight Cliffs Formation, where they are represented by both centra and teeth. Lepisosteids are consistently present in younger localities.

Genus et sp. indet. type 1

Distinctive lepisosteid teeth from Coniacian through late Campanian localities document the presence of a second lepisosteid (Fig. 10.3C–E). These teeth differ from those of known species of *Lepisosteus* in being taller and more slender, relatively more elongate, and having tips with well-developed but short lateral carinae, giving the crown a lance-like morphology. The lance-like teeth from the late Campanian Kaiparowits Formation (Fig. 10.3D), differ from those found in earlier beds in being relatively shorter and in having more strongly asymmetrical tips but are not treated as taxonomically distinct.

Lepisosteid Scales Two distinct scale morphologies are present in the samples examined. One morphotype (lepisosteid scale type 1) includes scales similar to those of *Lepisosteus occidentalis* (Leidy, 1856) Estes, 1964, from the late Campanian Belly River Group in being thick, having a distinct tab on the ventrolateral corner, and lacking a well-developed peg-and-socket joint (Fig. 10.3H–K). Variation is present in these scales reflecting different positions along the body. A second scale morphotype (lepisosteid scale type 2) is thinner and has a distinct ridge on the ventral surface (Fig. 10.3L, M). In these features, lepisosteid scale type 2 resembles those of the probable semionotiform from the Belly River Group referred to as holostean A by Brinkman (1990). However, they differ in not having the well-developed deep peg and socket joint between adjacent scales seen in holostean A. Furthermore, lepisosteid scale type 2 has a distinct tab developed on the ventrolateral corner, a feature that is not seen in scales of holostean A. Lepisosteid scale types 1 and 2 are assumed to be from distinct taxa because in addition to the morphological differences, they have a different distribution. Although both scale types occur in the Grand Staircase region of Utah, only type 1 occurs in the Dinosaur Park Formation of Alberta. It is assumed that if the morphological differences were owing to ontogeny or variation along the body of a single taxon, they would also have the same stratigraphic and geographic distribution. At present, there is no way to determine which scale morphology is associated with which tooth morphology.

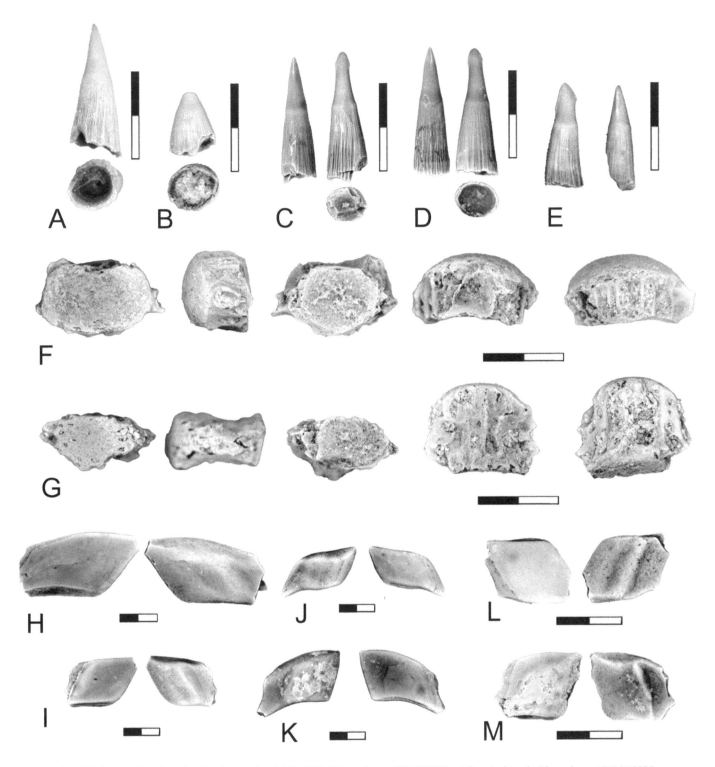

10.3. Lepisosteid elements. (A, B) Teeth of *Lepisosteus* Lacépède, 1803; (A) specimen MNA V10338, a tall conical tooth; (B) specimen MNA V10339, a low blunt tooth, both from Smoky Hollow Member of Straight Cliffs Formation, MNA Loc. 995; (C–E) teeth of Lepisosteidae gen. et sp. indet. showing lanceolate tooth morphology; (C) specimen MNA V10334; (D) specimen MNA V10335, both from John Henry Member of the Straight Cliffs Formation, MNA Loc. 706–2; (E) specimen OMNH 23500, from Kaiparowits Formation, OMNH Loc. V-6; (F, G) centra of *Lepisosteus;* (F) specimen MNA V10343, anterior precaudal centrum; (G) specimen MNA V10344, midprecaudal centrum, both from Smoky Hollow Member of Straight Cliffs Formation, MNA Loc. 995; (H–K) Lepisosteidae type 1 scales; (H) specimen MNA V10345; (I) specimen MNA V10340; (J) specimen MNA V10346; (K) specimen MNA V10347. All from Smoky Hollow Member of Straight Cliffs Formation, MNA Loc. 995; (L–M) Lepisosteidae type 2 scales; (L) specimen MNA V10341, from Smoky Hollow Member of Straight Cliffs Formation, MNA Loc. 995; (M) specimen MNA V10342, from Kaiparowits Formation, OMNH Loc. V-6. Views arranged as follows: teeth in lateral, occlusal, and underside of the crown perspectives; centra from left to right in anterior, lateral, posterior, dorsal, and ventral views; and scales from left to right in lateral and medial views. Scale bars = 2 mm.

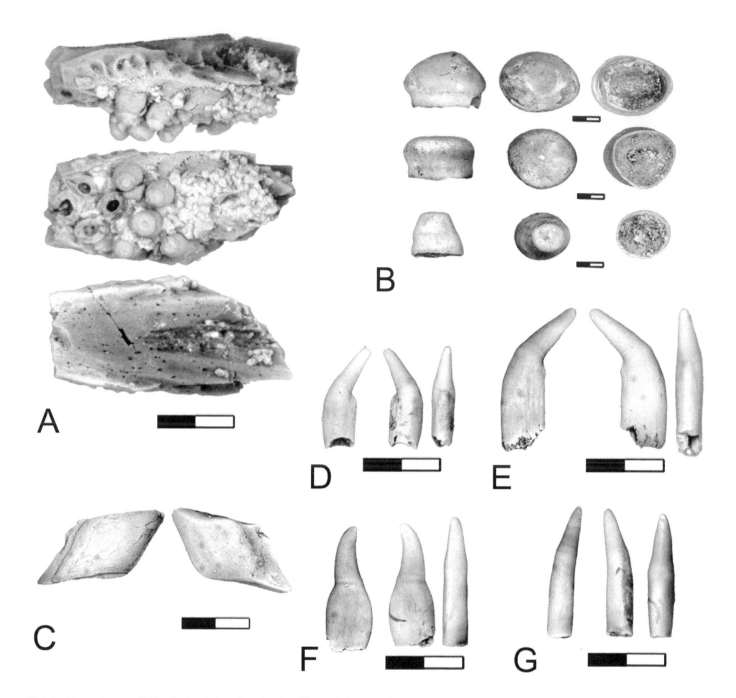

10.4. *Lepidotes* elements. (A) Tooth plate in lateral, occlusal, and internal views, specimen MNA V10308; (B) isolated oral teeth of *Lepidotes,* specimen MNA V10309, arranged in lateral, occlusal, and basal perspectives; (C) scale, specimen MNA V10310, in lateral and medial views; (D–G) isolated pharyngeal teeth; (D) specimen MNA V10311; (E) specimen MNA V10312; (F) specimen MNA V10313; and (G) specimen MNA V10314. All specimens from MNA Loc. 1067, Dakota Formation. Scale bars = 2 mm.

SEMIONOTIFORMES
Arambourg and Bertin, 1958
SEMIONOTIDAE Woodward, 1890
(sensu Olsen and McCune, 1991)
LEPIDOTES Agassiz, 1832

The semionotid *Lepidotes* is represented in the Cenomanian Dakota Formation by tooth plates, isolated oral and pharyngeal teeth, and ganoid scales (Fig. 10.4). Oral teeth are low crowned, circular in occlusal view, and have a distinct enamel cap. They vary in lateral profile from very low and squat to bluntly conical. Pharyngeal teeth are rod-like, have a swollen, laterally compressed base, and have a straight, conical crown that is angled relative to the base. The crowns are low and blunt but vary in proportions from low stout teeth to long and slender. Ganoid scales from the Dakota Formation are assumed to be those of *Lepidotes* because their teeth are abundant in this unit and *Lepisosteus* absent.

Brinkman et al.

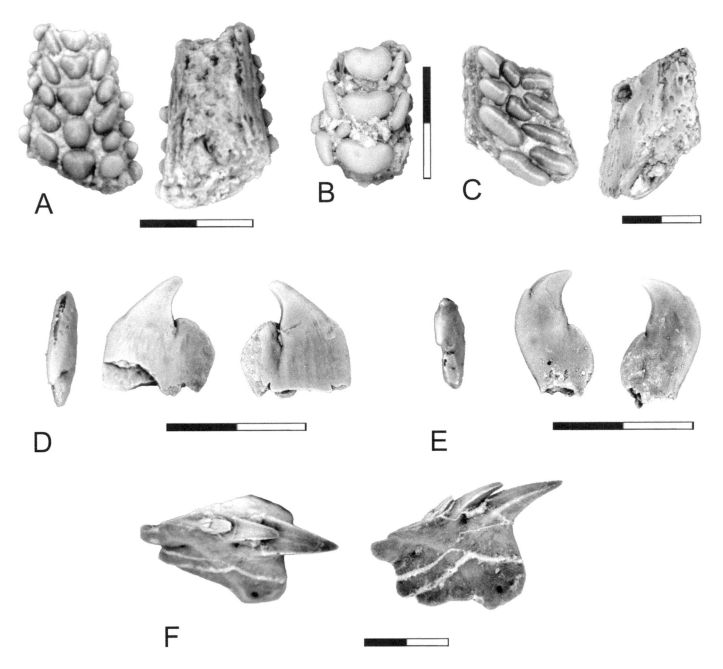

10.5. *Coelodus* sp. elements. (A) Posterior end of vomerine tooth plate in occlusal and dorsal views MNA V10315; (B) anterior end of vomerine tooth plate in occlusal view, specimen MNA V10316; (C) prearticular tooth plates in occlusal and dorsal views, specimen MNA V10317; (D–E) isolated pharyngeal teeth shows anterior edge, medial, and lateral views; (D) specimen MNA V10318; (E) specimen MNA V10319; (F) midline scale in lateral and edge views, specimen MNA V10320; all specimens from MNA Loc. 1067, Dakota Formation. Scale bars = 2 mm.

These scales are similar to those of *Lepisosteus* in the following features: (1) a tab is present on the anteroventral corner of the scale; (2) a ridge on the inner surface running between the dorsal and ventral edges is absent or poorly developed; and (3) distinct peg-and-socket joints between the scales are absent. Because these scales cannot be readily differentiated from those of some species of *Lepisosteus* on morphological grounds, identification of these genera on the basis of scales alone is problematic.

Lepidotes is most abundant in the Dakota Formation, where two distinct sizes are present: a smaller size with teeth about 1 mm in diameter, and a larger size with teeth up to 1 cm in diameter. *Lepidotes* is known from isolated teeth from the Turonian Smoky Hollow Member and the Campanian Wahweap and Kaiparowits Formations. Specimens from Turonian and Campanian localities are all of small size.

PYCNODONTIFORMES Berg, 1937
PYCNODONTIDAE sensu Nursall, 1996
COELODUS Heckel, 1856
COELODUS sp.

The pycnodont *Coelodus* is well documented by vomerine and prearticular tooth plates, isolated oral and branchial teeth, and midline scales (Fig. 10.5). Vomerine tooth plates are anteroposteriorly elongate, and they have a central row of larger teeth with two rows of smaller lateral teeth (Fig. 10.5A, B). The teeth in the central row are kidney shaped to triangular, broader than long, and concave anteriorly. The first of the lateral rows has teeth that are oval, whereas the outer of the lateral rows has small round teeth. The teeth of the central row and the first of the lateral rows have dorsal crest. In the midline row, this crest is concave anteriorly. This dental morphology is similar to that in species of *Coelodus* from Spain described by Kriwet (1999) but differs in that the disparity in size of the teeth is greater in the Utah specimens and the concave portions portion of the teeth in the median row faces anteriorly rather than posteriorly. Prearticular tooth plates have crushing teeth (Fig. 10.5C). A principal series of transversely elongated obliquely placed teeth with smaller, more circular teeth medially and laterally loosely arranged in rows is present. The hooked pharyngeal teeth (Fig. 10.5D, E) vary in width and development of the ventral blade similar to those described by Kriwet (1999). Dorsal and ventral ridge scales have posteriorly inclined spines (Fig. 10.5F).

Coelodus sp. is very abundant in the Cenomanian Dakota Formation and is also present in the Turonian and Coniacian. The genus is represented by partial vomers, isolated vomerine teeth, pharyngeal teeth, and midline scales in the Turonian Smoky Hollow Member, whereas it is represented only by rare isolated vomerine and branchial teeth in the Coniacian.

MICROPYCNODON Hibbard
and Graffham (1941), 1945
MICROPYCNODON sp.

Micropycnodon is represented by isolated teeth that bear blunt, rounded crowns with crenulated occlusal surfaces (Fig. 10.6A–C), a morphology that is considered characteristic of the genus by Cumbaa et al. (1999). *Micropycnodon* teeth were recovered from the late Santonian John Henry Member of the Straight Cliffs Formation and the mid-Campanian Wahweap Formation. Pycnodont pharyngeal teeth from the Wahweap Formation (Fig. 10.6D) are probably referable to *Micropycnodon*.

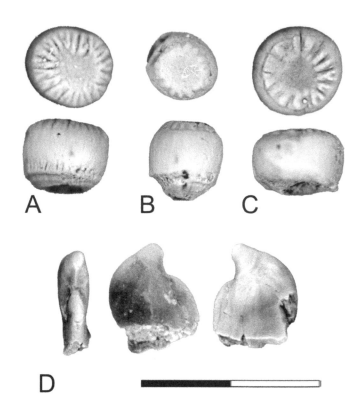

10.6. *Micropycnodon* sp. teeth. (A–C) Three isolated teeth arranged in occlusal (top) and lateral views (bottom), UMNH VP 17511, from the John Henry Member of the Straight Cliffs Formation, UMNH.VP.LOC. 799; (D) pharyngeal tooth from left to right shows anterior edge, medial, and lateral views, specimen MNA V10336, from the John Henry Member of the Straight Cliffs Formation, MNA Loc. 706-2. Scale bar = 2 mm.

AMIIFORMES Hay, 1929 (sensu
Grande and Bemis, 1998)
AMIIDAE Bonaparte, 1838
AMIINAE Bonaparte, 1838
Genus et sp. indet.

Amiinae are documented in the Grand Staircase region by centra (Fig. 10.7A–D) that are similar to those of the extant genus *Amia* Linnaeus, 1766, in being short and wide, having neural arch articular pits that are hourglass in shape, and having a pair of closely spaced, elongate narrow pits ventrally. Differences in the morphology of the centra indicate that more than one species is present. The parapophyses are not fused to the centrum in samples from the Dakota Formation and the lower part of the John Henry Member (Fig. 10.7A–C), in contrast to those from the Kaiparowits Formation (Fig. 10.7D). Also, amiine centra from the Dakota Formation have a series of large pits separated by a few large buttresses extending between the ends of the centrum (Fig. 10.7A, B), whereas the lateral surfaces of the centra from the lower John Henry Member are formed by fibers of

Brinkman et al.

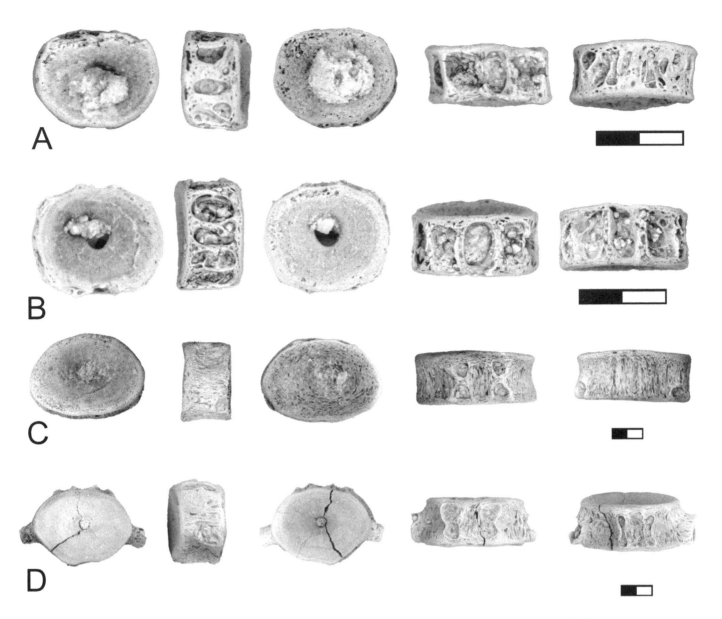

10.7. Amiinae gen. et sp. indet., centra arranged from left to right in anterior, lateral, posterior, dorsal, and ventral views. (A, B) Dakota Formation, MNA Loc. 1067; (A) specimen MNA V10322; (B) specimen MNA V10323; (C) specimen UMNH VP 17574, lower part of John Henry Member of the Straight Cliffs Formation, UMNH.VP.LOC. 781; (D) centrum from Kaiparowits Formation, OMNH Loc. V-9; OMNH 25938. Scale bars = 2 mm.

bone extending between the ends of the centra (Fig. 10.7C). In both of these centrum morphotypes, the parapophyseal articular pit is a small, circular opening located near the posterior end of the centrum. Amiine centra from the Kaiparowits Formation bear fine pits distributed across the lateral surface of the centrum, and the parapophyses have a broad base that extends across the lateral surface of the centrum (Fig. 10.7D). Although these morphological features indicate that several species are present, it was not possible to resolve the level of diversity.

VIDALAMIINAE Grande and Bemis, 1998
?*MELVIUS* Bryant, 1987
?*MELVIUS* sp.

Vidalamiinae are represented by tooth crowns and centra. Tooth crowns (Fig. 10.8A, B) match those of *Melvius* from the late Maastrichtian in being elongate, laterally compressed symmetrical blades. Typically these teeth occur in two distinct size classes, and although this may be a reflection of two closely related taxa differing in size, the consistency with which these two size classes occurs suggests that it is more likely a result of variation in the tooth size in a single

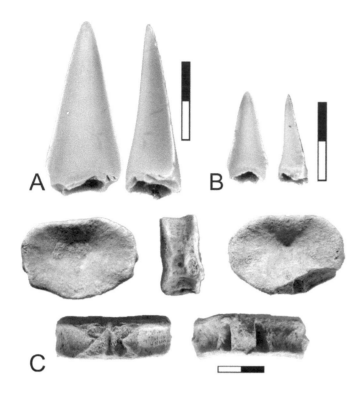

individual. Such variation is seen in the extant genus *Amia*, where the teeth on the dentary and premaxilla are larger than those on the maxilla. A large centrum from the John Henry Member of the Straight Cliffs Formation (Fig. 10.8C) measuring 5.5 cm in width is comparable in size to centra of the *Melvius thomasi* Bryant, 1987, from the Hell Creek Formation of Montana (Bryant, 1987). It differs from centra of *Melvius thomasi* in that the ventrolateral edges are less strongly excavated.

Vidalamiinae teeth are unknown in the Dakota Formation but are consistently abundant in microvertebrate assemblages from the Turonian and younger units in the study area.

Subfamily indet.
Genus et sp. indet. type AWSS

Centra characterized by having the area between the neural arch and parapophyseal pits covered by solid, smooth bone document the presence of a previously unrecognized amiid in the Dakota Formation, and are referred to here as amiid type AWSS (Fig. 10.9). These specimens are best referred to the Amiidae because the neural arch articular pits have a medial constriction and are formed by unformed bone. Amiid type AWSS is restricted to the Cenomanian Dakota Formation.

10.8. ?*Melvius* teeth and centra. (A, B) Tooth crowns, in two views; (A) specimen MNA V10353; (B) specimen MNA V10354. Both specimens from Smoky Hollow Member of the Straight Cliffs Formation, MNA Loc. 995; (C) centrum in anterior, lateral, posterior, dorsal, and ventral views, specimen UMNH VP 17634, John Henry Member of the Straight Cliffs Formation, UMNH.VP.LOC. 129. Scale bars = 2 mm.

10.9. Amiidae type AWSS, centra arranged from left to right in anterior, lateral, posterior, dorsal, and ventral views. (A) Midprecaudal centrum, specimen MNA V10332; (B) posterior precaudal centrum, specimen MNA V10333. Both specimens from Dakota Formation, MNA Loc. 1067 (Cenomanian). Scale bars = 2 mm.

Brinkman et al.

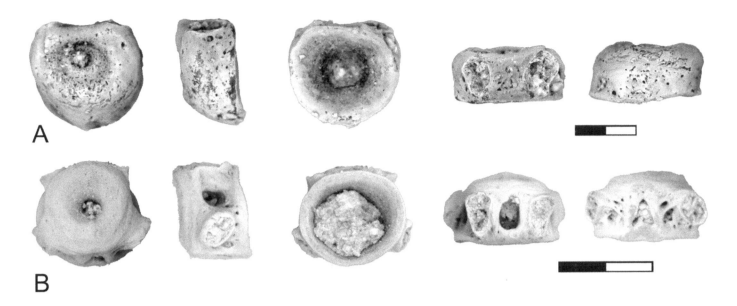

10.10. Teleost centrum type O arranged from left to right in anterior, lateral, posterior, dorsal, and ventral views. (A) Anterior precaudal centrum, specimen MNA V10324, Dakota Formation, MNA Loc. 1067; (B) midprecaudal centrum, specimen OMNH 31322, Smoky Hollow Member, Straight Cliffs Formation, MNA Loc. 995. Scale bars = 2 mm.

TELEOSTEI Müller, 1845 (sensu Patterson and Rosen, 1977)

As with most Late Cretaceous vertebrate assemblages, the teleosts of the Grand Staircase region present significant taxonomic challenges. In interpreting the diversity and relationships of teleosts at the Grand Staircase, particular attention was paid to the centra and tooth-bearing elements. Some of these could be referred to established taxa, whereas others could only be indeterminately referred to the Teleostei. A challenge in using centrum morphology to estimate diversity of fish in an assemblage is presented by variation along the vertebral column. In addition to comparisons with extant taxa, distribution patterns were considered in evaluating whether morphologically distinct centra were from distinct kinds of fish because different elements from a single kind of fish should have the same distribution pattern. Distribution patterns were also used to test hypotheses of association of centra and tooth-bearing elements.

Subdivision indet.
Genus et sp. indet. type O

Perhaps one of the most distinctive centrum morphotypes has the unusual feature of a convex anterior surface and a deeply concave posterior surface (Fig. 10.10). Variation in these centra suggests that this morphology extends along the length of the precaudal series. Centra that are assumed to be from an anterior position lack distinct pits in the wall of the centrum and are without parapophyseal articular pits (Fig. 10.10A). The convex surface is well developed, with only a small dimple representing the notochord pit. A second centrum morphology, assumed to be from a more posterior position, has pits on the side of the centrum and fused parapophyses (Fig. 10.10B). Although we consider genus and species indet. type O to be a teleost because of the general shape of the centra and the presence of deep pits and buttresses extending between the ends of the centra, the bone texture and the shape of the neural arch articular pits resembles that of amiids, so it is possible that these centra are amiid rather than teleost.

In the Grand Staircase region, teleost type O ranges from the Cenomanian to the late Campanian.

OSTEOGLOSSOMORPHA Greenwood et al., 1966
Order and Family indet.
CORIOPS Estes, 1969
CORIOPS sp.

Coriops was erected on the basis of basibranchial tooth plates that are oval in outline and have low, blunt teeth. In ventral view, these elements are formed by a network of woven bone fibers (Fig. 10.11A). Although *Coriops* was originally included in the Albulidae, Neuman and Brinkman (2005), after associating dentaries and centra with basibranchial tooth plates on the basis of size–frequency distributions and through comparison with extant osteoglossomorphs, referred the genus to the Osteoglossomorpha. *Coriops* dentaries are short and deep

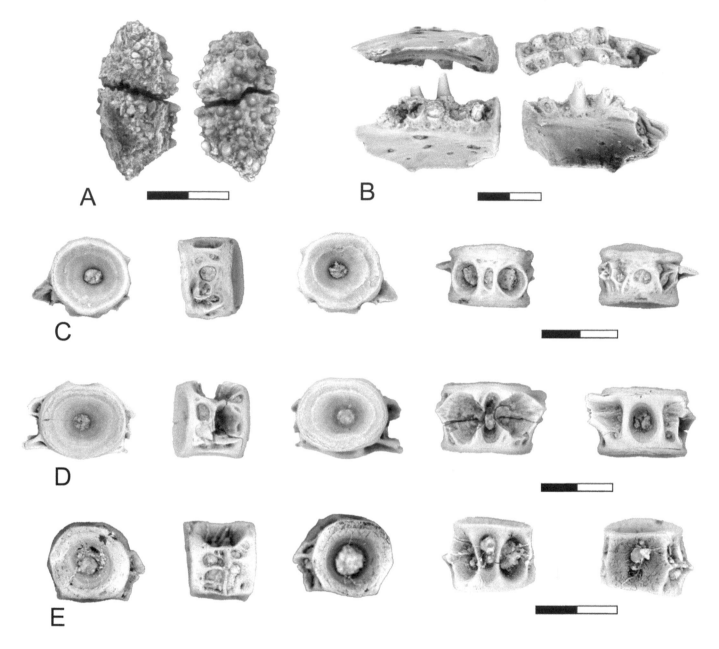

10.11. *Coriops* Estes, 1969, basibranchial tooth plate, dentary, and centra. (A) Basibranchial tooth plate, specimen MNA 4581, Kaiparowits Formation, MNA Loc. 458-1; (B) right dentary in ventral, occlusal, lateral, and medial views, specimen MNA V10337, Kaiparowits Formation, OMNH Loc. V-9; (C) anterior precaudal centrum arranged from left to right in anterior, lateral, posterior, dorsal, and ventral views, specimen OMNH 67123, Kaiparowits Formation, OMNH Loc. V-9; (D, E) midprecaudal centra, arranged from left to right in anterior, lateral, posterior, dorsal, and ventral views; (D) specimen MNA V10306; (E) specimen MNA V10307. Both from Dakota Formation, Loc. 1067. Scale bars = 2 mm.

and have multiple rows of teeth, with a row of large teeth on the lateral surface of the dentary (Fig. 10.11B). Midprecaudal centra have autogenous neural arches and long parapophyses fused to the centrum (Fig. 10.11C–E). The neural arch articular pits are large and oval and are separated by an oval middorsal pit. The parapophyses are fused to the centrum. Those on anterior precaudal centra are short and extend ventrolaterally (Fig. 10.11C), while those on midprecaudal centra form long, laterally directed transverse processes (Fig. 10.11D,

E). A circular articular pit posterior to the transverse process demonstrates that the ribs articulate with the centrum rather than the parapophyses. A midventral pit subequal to or larger in size than the middorsal pit is present.

In the Grand Staircase region of Utah, *Coriops* has a disjunct distribution, only occurring in the Cenomanian Dakota Formation and the late Campanian Kaiparowits Formation. It is of low abundance in the Dakota Formation and is represented only by centra. The size of the sample from the

Brinkman et al.

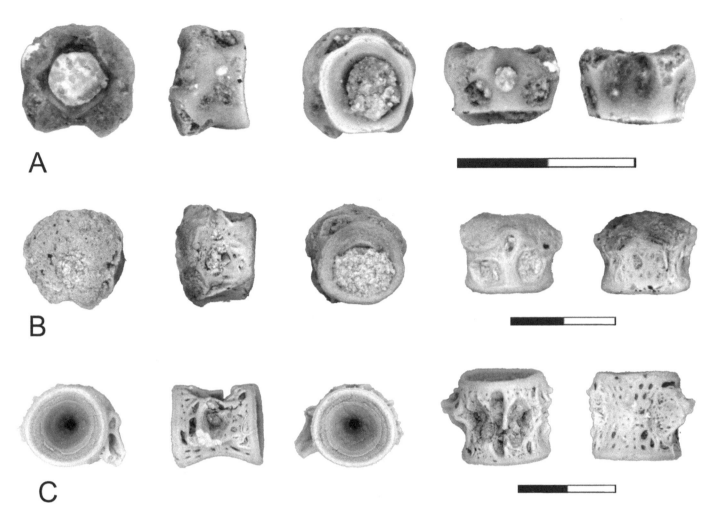

10.12. Centra attributed to Hiodontiformes and arranged from left to right in anterior, lateral, posterior, dorsal, and ventral views. (A) Atlas centrum, Dakota Formation, MNA Loc. 1067 (Cenomanian); (B) atlas centrum, specimen OMNH 31248, Smoky Hollow Member, Straight Cliffs Formation, OMNH V-9; (C) midprecaudal centrum, specimen OMNH 67122, Kaiparowits Formation, OMNH Loc. V-9. Scale bars = 2 mm.

Kaiparowits Formation is larger and includes tooth-bearing elements as well as centra. *Coriops* tooth-bearing elements and centra from the Grand Staircase region do not differ appreciably from specimens from the Dinosaur Park Formation described by Neuman and Brinkman (2005:fig. 9.8D).

HIODONTIFORMES Taverne, 1979
HIODONTIDAE Cuvier and Valenciennes, 1846
Genus et sp. indet.

The presence of a hiodontid in Grand Staircase region of Utah is documented by centra that closely resemble those of the hiodontid from the Campanian Dinosaur Park Formation (see Brinkman and Neuman, 2002, morphoseries IIB-1), particularly in the shape of the neural arch articular pits, transverse process, and shape of the rib articular pits on the side of the centrum posterior to the transverse processes (Fig.

10.12). The extant *Hiodon alosoides* Rafinesque, 1819, also has relatively deep rib articular pits on the side of the centrum posterodorsally to the transverse process (Hilton, 2002). Also, as is the case with the hiodontid centra from the Dinosaur Park Formation, the ventral surface of the centra from Utah are covered by fine pits. The atlas centrum has the distinctive hiodontid feature of a quadratic anterior articular surface.

In the Grand Staircase region, hiodontids extend from the Cenomanian Dakota Formation to the upper Campanian Kaiparowits Formation. Variation is present in the shape of the atlas centrum. Those in the Dakota Formation have a large notochordal canal (Fig. 10.12A). Turonian specimens (Fig. 10.12B) have a notochordal canal that is intermediate in size when compared with specimens from the Dakota Formation and specimens from the upper Campanian Kaiparowits Formation (Fig. 10.12C) and the Dinosaur Park Formation.

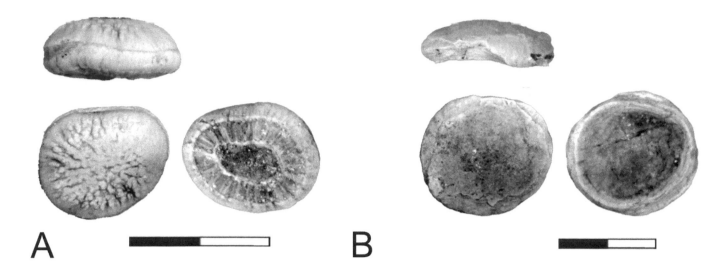

10.13. *Paralbula* Blake, 1940, teeth compared with a fossil crayfish gastrolith. (A) *Paralbula* teeth, specimen MNA V8123, Kaiparowits Formation, MNA Loc. 1073; (B) crayfish gastrolith, specimen MNA N9692, Kaiparowits Formation, OMNH Loc. V-9. Views of the teeth are arranged in lateral, occlusal, and underside of the crown perspectives. Scale bars = 2 mm.

ELOPOMORPHA Greenwood et al., 1966
ALBULIFORMES sensu Forey et al., 1996
ALBULOIDEI sensu Forey et al., 1996
ALBULIDAE Bleeker, 1859
PHYLODONTINAE Estes and Hiatt, 1978
PARALBULA Blake, 1940
PARALBULA sp.

The albulid Paralbula is represented by flat, button-like teeth (Fig. 10.13A). These are absent in samples from the Cenomanian through the late Santonian but are present in the samples from Wahweap and Kaiparowits formations. Tooth-like elements that are generally similar to *Paralbula* teeth but are flatter and have an unconstricted base are also present (Fig. 10.13B). Such elements have been referred to as Phylodontinae gen. et sp. indet. by Peng, Russell, and Brinkman (2001) and Albulidae indet. by Case and Schwimmer (1988) However, these are now recognized to be crayfish gastroliths, which are disc-like structures that serve as calcium reservoirs (Johnson, 1984).

ELOPIFORMES Greenwood et al., 1966
Suborder indet.
Genus et sp. indet.

A second elopomorph is documented by centra that resemble those of the extant genus *Elops* Forsskål, 1775, in being anteroposteriorly short, in having widely spaced neural arch and paraphophyseal articular pits, and in that the lateral wall of the centrum has numerous fibers of bone extending between the ends of the centrum (Fig. 10.14). The neural arch articular pits are small and circular, covered by finished bone, and are bordered posterolaterally by a raised edge. The parapophyseal pits are larger and more nearly rectangular and are more widely separated from one another than are the neural arch pits. Precaudal centra are subrectangular in end view and are typically of small size (Fig. 10.14A, B). Caudal centra are more nearly oval in end view, and are taller than wide (Fig. 10.14C). Centra from the Dakota Formation are nearly 1 cm in diameter, whereas specimens from the Smoky Hollow Member do not exceed 2 mm in diameter. These centra are included in the Elopiformes, rather than the Albuliformes, on the basis of the arrangement of the bony fibers that extend between the ends of the centrum. The extant genus *Albula*, and the *Albula*-like centra that Neuman and Brinkman (2005) referred to *Paralbula*, differ from those of elopiforms in that the bony fibers are grouped together into bundles forming ridges. The centra of the extant elopiforms *Megalops* (Fig. 10.14 D) and *Elops*, (Fig. 10.14 E) are evenly spaced, resembling those of the small *Elops*-like taxon here included in the Elopiformes. Further evidence that these centra are not those of *Paralbula* is provided by differences in distribution patterns: the elopiform centra are present in the Cenomanian, Turonian, and early Santonian localities while *Paralbula* teeth were recovered only in the Campanian Wahweap and Kaiparowits formations.

Brinkman et al.

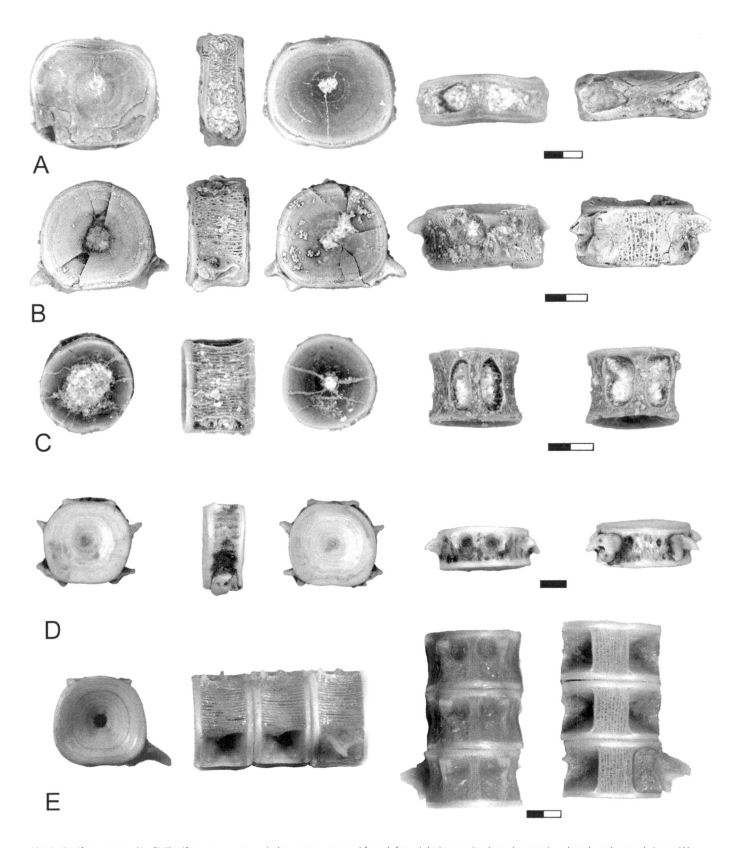

10.14. Elopiform centra. (A–C) Elopiformes gen. et sp. indet. centra arranged from left to right in anterior, lateral, posterior, dorsal, and ventral views; (A) anterior precaudal centrum, specimen MNA V10325; (B) midprecaudal centrum, specimen MNA V10321; (C) caudal centrum, specimen MNA V10326. (D) *Megalops atlanticus,* anterior precaudal centrum, specimen ROM R6903; (E) *Elops saurus,* midprecaudal centra, specimen ROM R2175. A–C are from the Dakota Formation, MNA Loc. 1067. Scale bars for A–C, E = 2 mm, D = 5 mm.

Freshwater Osteichthyes from Cenomanian to Late Campanian

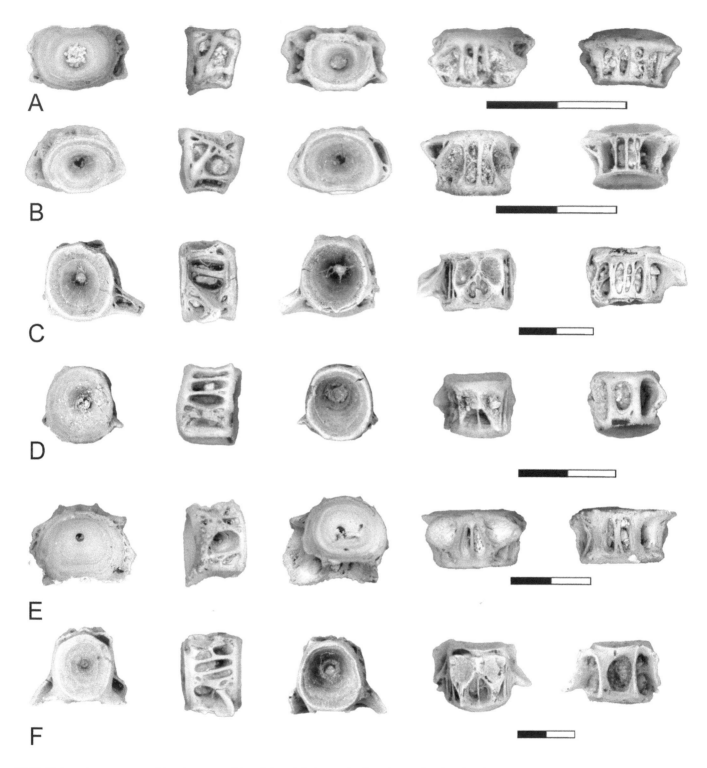

10.15. *Diplomystus* spp., precaudal centra arranged from left to right in anterior, lateral, posterior, dorsal, and ventral views. (A, B) Anterior precaudal centra; (A) specimen UMNH VP17586, John Henry Member of the Straight Cliffs Formation, UMNH.VP.LOC. 781; (B) specimen UMNH VP 17509, John Henry Member of the Straight Cliffs Formation, UMNH.VP.LOC. 799; (C, D) mid- to posterior precaudal centra; (C) specimen MNA V10327, Dakota Formation, MNA Loc. 1067; (D) specimen UMNH VP 17617, John Henry Member of the Straight Cliffs Formation, UMNH.VP.LOC. 799. (E, F) *Diplomystus* sp., Wasatch Formation, Eocene, Bitter Creek Locality; (E) anterior precaudal centrum, UCMP 198890; (F) posterior precaudal centrum, UCMP 198888. Scale bars = 2 mm.

Brinkman et al.

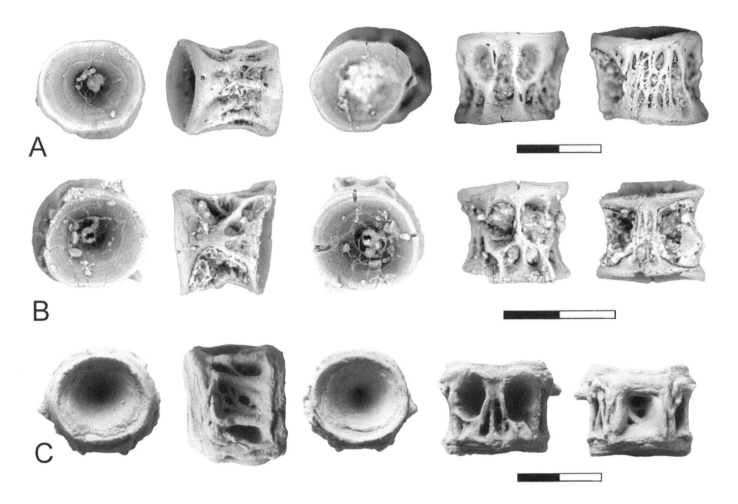

10.16. (A, B) Ellimmichthyiformes gen. et sp. indet. type LvD arranged from left to right in anterior, lateral, posterior, dorsal, and ventral views; (A) anterior precaudal centrum, specimen MNA V10328; (B) posterior precaudal centrum, specimen MNA V10328. Both specimens from Dakota Formation, MNA Loc. 1067 (Cenomanian). (C) *Horseshoeichthyes* precaudal centrum, specimen TMP 2001.045.0093. Scale bars = 2 mm.

OTOCEPHALA Johnson and Patterson, 1996
(=OSTARIOCLUPEOMORPHA Arratia, 1997
CLUPEOMORPHA Greenwood et al., 1966
ELLIMMICHTHYIFORMES Grande, 1985
SORBINICHTHYIDAE Bannikov
and Bacchia, 2000
DIPLOMYSTUS Cope, 1877
DIPLOMYSTUS sp.

Diplomystus is documented by centra like those from the Dinosaur Park Formation referred to by Brinkman and Neuman (2002) as morphoseries IIB-2. These centra (Fig. 10.15A–D) are here referred to the clupeomorph *Diplomystus* because they compare favorably with those of *Diplomystus dentatus* Cope, 1877, from the Eocene Green River Formation of Wyoming (Fig. 10.15 E–F).

The neural arches are autogenous and the shallow neural arch articular pits are about half the length of the centrum. On the more anterior centra, the neural arch articular pits are separated from each other by a middorsal pit as in *D. dentatus* (Newbrey et al., 2010, TMP 1986.224.0135) while on the more posterior centra these articular surfaces contact one another. The Utah centra resemble those of *D. dentatus* (represented by TMP 1986.224.0135) in having a middorsal ridge that extends the length of the centrum, posterodorsal processes posterior to the neural arch articular pits, a single ridge between the parapophysis and the neural arch articular pits, and three to four ridges extending between the ends of the centrum ventrally. Anterior centra are circular in end view and have long laterally directed transverse processes as in *D. dentatus* (Newbrey et al., 2010:fig. 4, FOBU 11661). More posterior precaudal centra are higher than wide and have shorter posterolaterally directed transverse processes as in *D. dentatus* (Newbrey et al., 2010:fig. 4, FOBU 11661).

Diplomystus has a patchy distribution in the Grand Staircase region of Utah, being present only in the Cenomanian Dakota Formation and the early Santonian portion of the John Henry Member of the Straight Cliffs Formation.

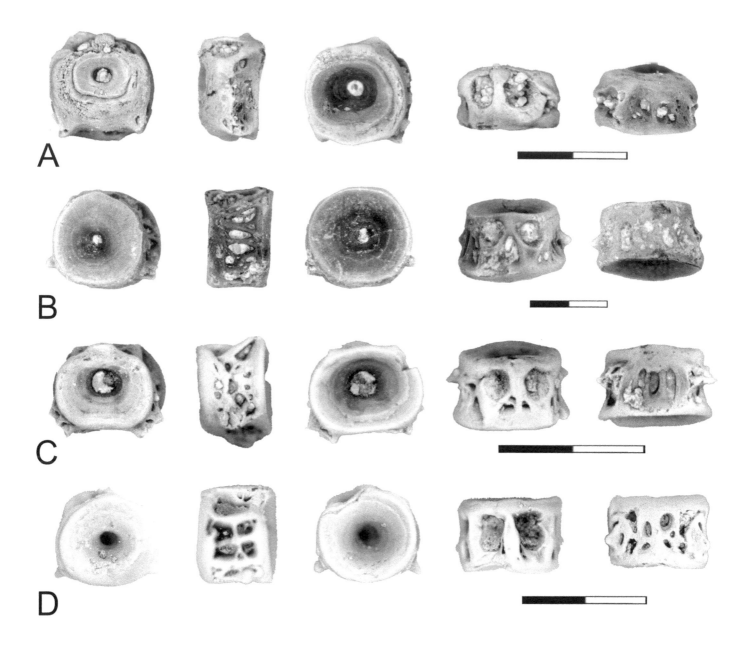

10.17. Ellimmichthyiformes gen. et sp. indet. type U-7 arranged from left to right in anterior, lateral, posterior, dorsal, and ventral views. (A–C) Dakota Formation centra, MNA Loc. 1067; (A) specimen MNA V10329; (B) specimen MNA V10330; (C) specimen MNA V10331; (D) centrum specimen TMP 2007.027.0022, Dinosaur Park Formation of Alberta. Scale bars = 2 mm.

Genus et sp. indet. type LvD

Ellimmichthyiformes gen. et sp. indet. type LvD is recognized on the basis of centra from the Cenomanian Dakota Formation (Fig. 10.16) that resembles those of both *Horseshoeichthyes* Newbrey et al., 2010, an ellimmichthyiform from the early Maastrichtian Horseshoe Canyon Formation (TMP 2001.045.0093) and *Diplomystus*. As in the precaudal centra of *Hoseshoeichthyes* (Fig. 10.16C), the neural arch articular pits in the centra of teleost type LvD are relatively small circular pits located near the anterior end of the centrum, and a middorsal pit is bordered by ridges that extend from the posterior end of the neural arch articular pit to the posterior end of the centrum. The middorsal pit is occasionally subdivided by a middorsal ridge. In some specimens, the parapophyses are autogenous, while in others they are fused to the centrum. The centra of teleost LvD are similar to those of *Diplomystus* (Fig. 10.15E) in that short posterodorsal processes are present on the posterior end of the centra and in that the ventral surface has of a series of bony rods extending between the ends of the centrum rather than a midventral pit. Differences in the development of the parapophyseal pits and the posterodorsal processes are present. These differences are interpreted as a result of variation along the vertebral column. For example, the parapophyseal pits on centra from a more anterior position (Fig. 10.16A) have more

Brinkman et al.

weakly developed paraphophyseal pits and the posterodorsal processes are short compared with centra from a more posterior position.

The centra of Ellimmichthyiformes gen. et sp. indet. type LvD are only present in the Dakota Formation.

Genus et sp. indet. type U-7

Ellimmichthyiformes gen. et sp. indet. type U-7 is represented by small centra (Fig. 10.17) with an arrangement of pits and ridges on the dorsal surface of the centrum similar to that of *Horseshoeichthyes*: the neural arch articular pits are round and are restricted to the anterior end of the centrum, and ridges extend from these to the posterior end of the centrum. As in *Horseshoeichthyes* the length of the centrum varies with some being short and others equidimensional. Short centra typically have a flat anterior end.

Centra of Ellimmichthyiformes gen. et sp. indet. type U-7 are present in samples from the Cenomanian Dakota Formation (Fig. 10.17A–C) and the Turonian Smoky Hollow Member but were not recovered from younger beds in Utah. The centra from these units differ in that in the Cenomanian the neural arch pits are separated by a single bar, while in the Turonian the middorsal pit extends between the centra. Also, in the Cenomanian short dorsal processes are absent or weakly developed on the posterior end of the centrum, while in the Turonian dorsal processes are well developed. Centra of Ellimmichthyiformes gen. et sp. indet. type U-7 are also present in the Dinosaur Park Formation of Alberta (Fig. 10.17D). These are like those of the Cenomanian in not having dorsal processes at the posterior end of the centrum. They differ in that the neural arch articular pits are relatively larger.

CLUPEIFORMES Bleeker, 1859
Genus et sp. indet. type G

A clupeiform is documented at Staircase Grand region by centra (Fig. 10.18A) similar to those form the Dinosaur Park Formation described as morphoseries IC-1 (Brinkman and Neuman, 2002) (Fig. 10.18B). Centra of this type are included in the Clupeiformes on the basis of their small neural arch articular pits located close to the anterior end of the centrum, parasagittal processes extending dorsally from the centrum posterior to the neural arch articular pits, and the cone-like shape of the posterior end of the centrum when seen in lateral view (Fig. 10.18C).

Centra of clupeiform genus and species indet. type G were only encountered in OMNH Loc. v-6 of the Kaiparowits Formation.

OSTARIOPHYSI Sagemehl, 1885
(sensu Fink and Fink, 1996)
OTOPHYSI Garstang, 1931 (sensu
Rosen and Greenwood, 1970)
Order indet.
Genus et sp. indet. type U-3/BvD

An otophysan possibly related to catfish is documented by both dentaries (Fig. 10.19) and centra (Figs. 10.20, 10.21A–C). Its relationship with catfish is suggested from comparisons with the extant catfish *Ariopsis felis* (Linnaeus, 1766) (UMMZ 186995–1 of 2) and *Ictalurus furcatus* (Valenciennes, 1840, in Cuvier and Valenciennes, 1840) (Fig. 10.19E, F) and with fossil catfish material from the Eocene Bridger Formation in the collections of the UCMP (Fig. 10.19D). The dentaries from Utah (Fig. 10.19A–C) are relatively short and deep, and have multiple rows of small teeth – features that are shared with catfish (Fig. 10.19D–F). Esocoids also have multiple rows of teeth, but these differ from catfish in that the number of rows of teeth is not consistent through most of the length of the tooth row. Furthermore, the size of the tooth bases in catfish shows little variation mediolaterally or along the length of the jaw but not in the esocoids. Additional features in which the Utah dentaries are similar to those of extant catfish include the presence of a tooth-bearing surface that is strongly convex in cross section; a symphysis that is reduced and dorsoventrally flattened (Fig. 10.19A–C); and a flattened and smooth ventral surface of the dentary in the region of the symphysis (Fig. 10.19E, F). Additionally, one specimen has a flat region with a dorsally excavated, longitudinal groove (Fig. 10.19B), as those in a sample from the Eocene Bridger Formation (Fig. 10.19D) and the extant *Ictalurus furcatus* (Fig. 10.19F). A large, enclosed (in life) sensory canal extends along the ventral edge of the dentary below the Meckelian fossa. A sensory canal of similar position is present in extant and Eocene catfish examined but is smaller in diameter. In the Utah specimens, the bony canal, although enclosed in life by delicate bone, is usually broken ventrally (Fig. 10.19B, C) so the ventral edge of the dentary is rarely preserved. One specimen from the Kaiparowits Formation that does preserve the ventral edge of the dentary shows that this element is triangular in lateral view (Fig. 10.19A). Variation in the shape of the dentary when seen in lateral view is present in the extant and fossil catfish that were examined.

Ostariophysan centra with possible affinities to catfish are referred to here as centrum type U-3 (Figs. 10.20A–C, 10.21A) and centrum type BvD (Fig. 10.21B, C). Centrum type U-3 are from the anterior portion of the precaudal column and centrum type BvD are from a more posterior position. Centrum type U-3 includes a centrum morphotype that

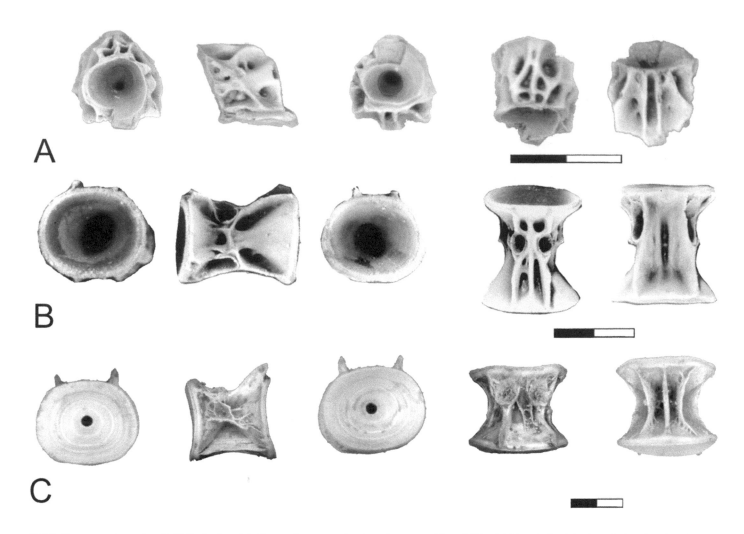

10.18. Clupeomorpha centra. (A, B) Centra from late Campanian nonmarine deposits arranged from left to right in anterior, lateral, posterior, dorsal, and ventral views; (A) specimen OMNH 23546, Kaiparowits Formation, OMNH Loc. V-6; (B) specimen TMP 1993.093.0013, Dinosaur Park Formation; (C) centrum from the extant clupeid *Clupia harengus*, specimen ROM R 5454. Scale bars = 2 mm.

is identified as the anteriormost centrum on the Weberian apparatus from a comparison with recent and Eocene ostariophysans. As in the recent and fossil catfish and catostomids examined, this centrum is greatly foreshortened and higher than wide in end view (Fig. 10.20A). The dorsal surface of the centrum bears small circular pits that are separated by a rounded bar of bone about equal in width to the diameter of the neural arch articular pits. In otophysans, these pits receive the second pair of Weberian ossicles–the scaphium. The remainder of the centra included in morphotype U-3 are interpreted as from a more posterior position. Variation in the morphology of these centra is interpreted as variation along the vertebral column, with those that are foreshortened being from immediately posterior to the Weberian apparatus and those more elongate being from a more posterior position. As in catfish, including *Ictalurus furcatus* (UMMZ 220057-M608), *Ameiurus melas* Rafinesque, 1820 (Fig. 10.20E), and *Ariopsis felis* (UMMZ 186995–1 of 2), but in contrast to catostomids, a

large, circular midventral pit is present on centra immediately posterior to the Weberian apparatus. In many of the centra of the Utah taxon, this pit is greatly enlarged and bordered by a raised edge. Parapophyses are variably developed on centra type U-3, ranging in size from long, laterally directed processes to low ridges. Reflecting the narrowness of the centra, the paraphyses are blade-like. The more elongate centra have short posterolaterally directed processes extending from the posterior edge of the centrum (Fig. 10.21A). They also have a narrow but long midventral pit as in extant *I. furcatus* (UMMZ 220057-M607) and *A. felis* (UMMZ 186995–1 of 2).

Centra type BvD (Fig. 10.21B) are elongate, low and wide, the neural arch is fused to the centrum, and a middorsal pit occurs between the bases of the neural arch, similar to extant catfish (e.g., UMMZ 220057-M607 and Fig. 10.21D). The parapophyses are autogenous, and large parapophyseal articular pits are present on the sides of the centrum. Ventrally the centrum has a flat surface pierced by small pores arranged

Brinkman et al.

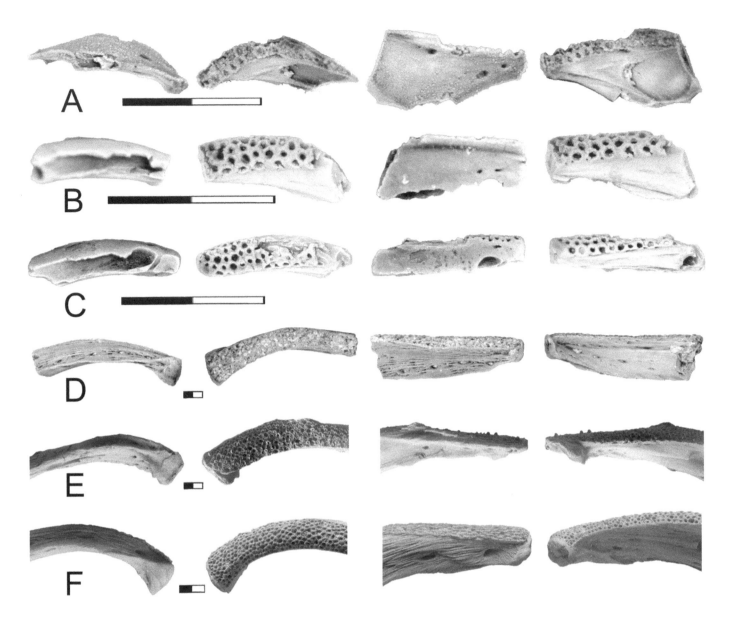

10.19. Dentaries attributed to ostariophysan type U-3/BvD compared with catfish dentaries, all arranged from left to right in ventral, occlusal, lateral, and medial views. (A) Specimen OMNH 67120, Kaiparowits Formation, OMNH Loc. V-9; (B) specimen MNA V10350, Smoky Hollow Member of the Straight Cliffs Formation, MNA Loc. 995; (C) specimen MNA V10351, Smoky Hollow Member of the Straight Cliffs Formation, MNA Loc. 995; (D) catfish dentary, Eocene Bridger Formation of Wyoming, specimen UCMP198899/V96246; (E) left dentary from extant *Ariopsis felis* (Linnaeus, 1766), specimen UMMZ 186995; (F) left dentary from extant *Ictalurus furcatus* (Valenciennes, 1840, in Cuvier and Valenciennes, 1840), specimen UMMZ 220057-M607. A–E, Scale bars = 2 mm; F, scale bar equals 5 mm.

in loose rows. Centrum type BvD from the Kaiparowits tend to have a midventral ridge (Fig. 10.21C), whereas those from older formations tend to be smooth ventrally (Fig. 10.21B).

Although these centrum morphotypes have dramatically different morphologies, the range of morphological variation present is consistent with variation observed along the vertebral column in catfish with centrum type U-3 being from a more anterior position and centrum type BvD being from a more posterior position. Centrum type BvD differs from the posterior precaudal centra of catfish in having autogenous parapophyses, but in other respects is similar.

Further, centrum types U-3 and BvD have similar patterns of abundance in the sites sampled, as would be expected if they were from a single kind of fish: both centrum morphotypes are absent in the Cenomanian Dakota Formation, both first occur in the Turonian, and both are the most abundant centra in Turonian and younger localities (Table 10.2). Thus, on the basis of these morphological and distribution patterns, it is concluded that these centrum morphologies are from a single kind of fish.

The hypothesis that the ostariophysan dentaries and centra described here are from a single kind of fish is also

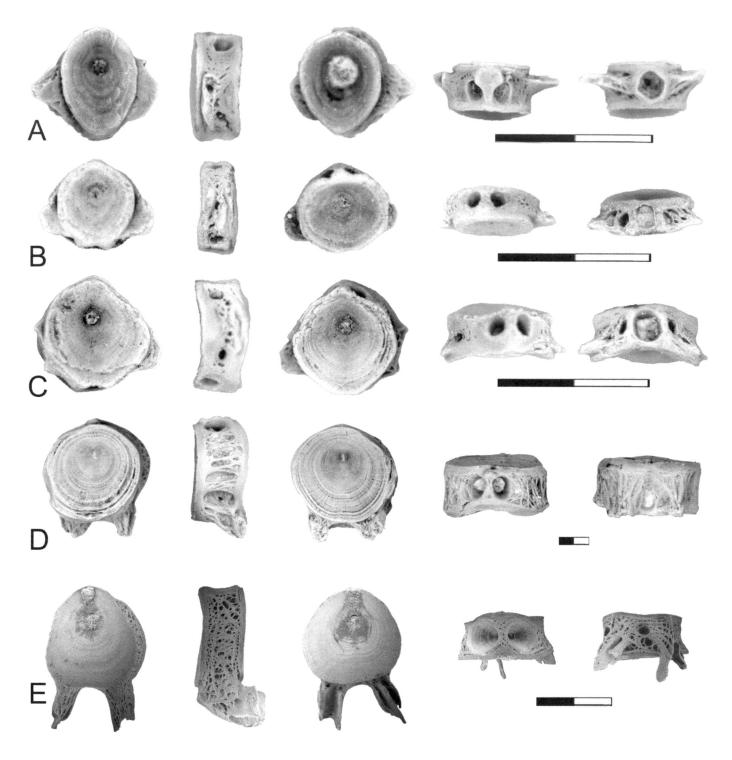

10.20. Ostariophysan type U-3 centra arranged from left to right in anterior, lateral, posterior, dorsal, and ventral views. (A–C) Centra type U-3 interpreted as anteriormost centrum of Weberian apparatus; (A) specimen UMNH VP 19137, from John Henry Member of the Straight Cliffs Formation, UMNH.VP.LOC. 424; (B) specimen MNA V10348; (C) specimen MNA V10349 both from Smoky Hollow member of the Straight Cliffs Formation, MNA Loc. 995; (D) catfish anteriormost centrum of Weberian apparatus from the Eocene Bridger Formation of Wyoming, UCMP 198906/V96246; (E) anteriormost centrum from the Weberian apparatus of extant *Ameiurus melas* Rafinesque, 1820, TMP 2010.004.0003. Scale bar = 2 mm.

supported by their pattern of distribution. The ostariophysan dentaries first occur in the Turonian along with the centra. In all Turonian to late Campanian assemblages from the Grand Staircase region, centra type U-3 and BvD were overwhelmingly dominant, and in samples of relatively small size, they were often the only kinds of centra present (Table 10.2). The ostariophysan dentaries have a similar pattern of relative abundance. Thus, in the faunal list presented in

Brinkman et al.

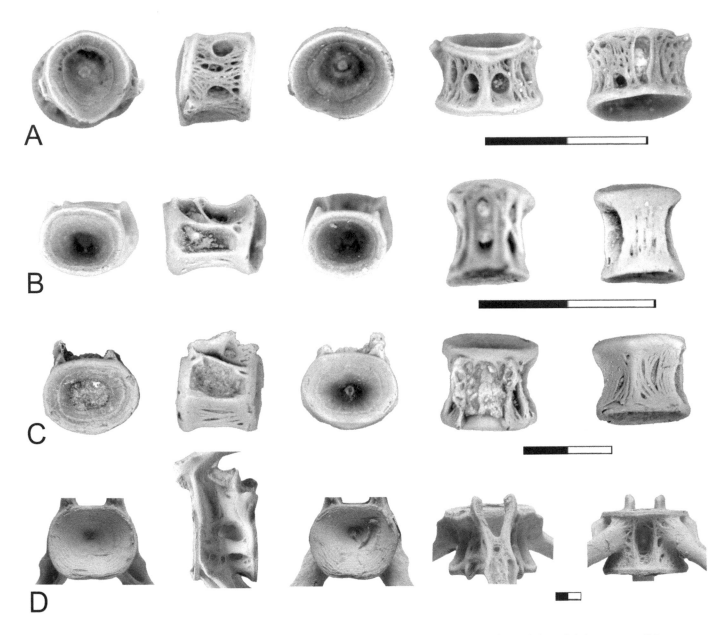

10.21. Ostariophysan type U-3/BvD centra arranged from left to right in anterior, lateral, posterior, dorsal, and ventral views. (A) Elongate type U-3 centrum interpreted as midprecaudal centrum, specimen UMNH VP 19138, John Henry Member of the Straight Cliffs Formation, UMNH.VP.LOC.. 424; (B) centra type BvD, specimen MNA V10352; Smoky Hollow Member of the Straight Cliffs Formation, MNA Loc. 995; (C) centrum, Kaiparowits Formation, Loc. V-9, OMNH 67124; (D) posterior precaudal centrum from extant *Ariopsis felis* (Linnaeus, 1766), UMMZ 186995. Scale bars = 2 mm.

Table 10.1, a single ostariophysan is included, referred to as ostariophysan type U-3/BvD.

Although similarities of the dentaries and centra strongly suggest affinities with catfish, additional skeletal elements, particularly spines, which would be expected, have not yet been recovered. The absence of spines may in part be a result of a bias against the preservation of long, slender elements (Brinkman, pers. obs.). Also, given the small size of the fish, these would be narrow in diameter and so could be lost through even very fine screens.

CHARACIFORMES Goodrich, 1909
(sensu Fink and Fink, 1996)
Family, genus, and species indet.

The presence of a characiform in the Grand Staircase region is documented by a centrum that is interpreted as the anteriormost centrum of the Weberian apparatus (Fig. 10.22A, B). This centrum is distinctive in being foreshortened and wider than tall, having a saddle-shaped anterior articular surface, lateroposterior flaring on the anterior surface, and a dorsomedian pair of small round fossae for the second pair

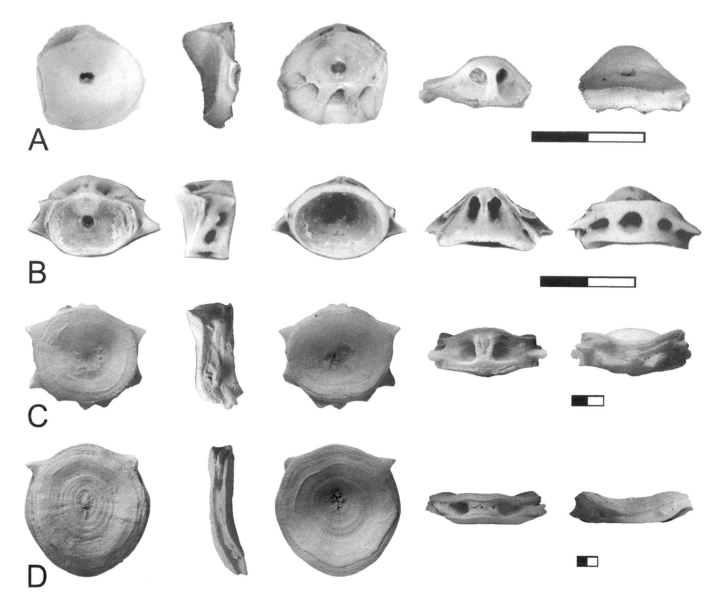

10.22. Characiform centra from the late Campanian of North America compared to an extant characiform and catostomid; arranged from left to right in anterior, lateral, posterior, dorsal, and ventral views. (A) Specimen OMNH 23171, Kaiparowits Formation, OMNH Loc. V-6; (B) specimen TMP 1993.023.0023, Dinosaur Park Formation; (C) first centrum from the Weberian apparatus of the extant characiform *Hydrolycus scomberoides* (Cuvier, 1819) UMMZ 204957; (D) first centrum from the Weberian apparatus of the extant catostomid *Moxostoma anisurum* (Rafinesque, 1820), UAMZ 6492. Scale bars = 2 mm.

of Weberian ossicles. These features are generally similar to those of extant otophysans and are particularly similar to those of characiforms. For example, the first centrum in the Weberian apparatus of the characiform *Hydrolycus scomberoides* (Cuvier, 1819) (UMMZ 204957–1 of 3) shows all the features listed above (Fig. 10.22C). Catfish and catostomids also have foreshortened centra and similar dorsal fossae, but they differ by being taller than wide and lacking a well-developed saddle-shaped anterior articular surface and lateroposterior flaring on the anterior surface (UMMZ 220057-M608, UAMZ 6492) (Fig 10.22D).

Centra of this morphology were described as centrum morphoseries IIA-3 by Brinkman and Neuman (2002). They

were associated with jaws that were subsequently identified as characiform by Newbrey et al. (2009) because they have the same pattern of occurrence in the Dinosaur Park Formation: the centra of morphoseries IIA-3 are found in only nine of 58 vertebrate microfossil sites that were examined (Brinkman, 1990; Eberth and Brinkman, 1997). The localities that contained these elements were all in a distinctive quiet-water environment of deposition (Eberth and Brinkman, 1997), but not all the localities preserved in that environment of deposition included these centrum morphotypes. Characiform dentaries were only found in sites in which centra of morphoseries IIA-3 are abundant. No other centrum morphology has this pattern of distribution supporting the conclusion that

Brinkman et al.

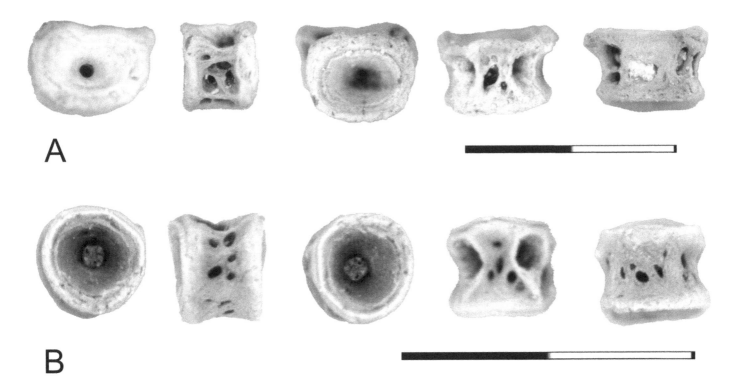

A

B

10.23. Precaudal centra of teleost type U-4 arranged from left to right in anterior, lateral, posterior, dorsal, and ventral views. (A) Specimen UMNH VP 19135, Dakota Formation, MNA Loc. 1067; (B) specimen UMNH VP 19136, John Henry Member of the Straight Cliffs Formation, UMNH.VP.LOC. 424. Scale bars = 2 mm.

the dentaries and centra that are identified as characiform on the basis of morphological evidence are from a single kind of fish.

Characiform centra were only encountered in OMNH Loc. v-6 of the Kaiparowits Formation.

EUTELEOSTEI sensu Arratia, 1999
Order indet.
Genus et sp. indet. type U-4

Teleost indet. type U-4 is represented by centra that are small, typically less than 1 mm in diameter, and have autogenous neural arches and parapophyses (Fig. 10.23). The anterior end of the centrum is relatively flat, whereas the posterior end is deeply excavated. Neural arch articular pits are about half the length of the centrum, located near the anterior end of the centrum, and are widely separated from one another. A short dorsally directed process is present posterior to the neural arch articular pits. Parapophyseal articular pits are located low on the sides of the centrum. The surfaces of the centrum between the articular pits are pierced by foramina of moderate size. Teleost indet. type U-4 is tentatively included in the Euteleostei because these centra are generally similar to those of salmoniforms in being simple cylinders with autogenous neural arches and at most weakly developed parapophyses.

Teleost indet. type U-4 first occurs in the Cenomanian and extends to the upper Campanian localities. Cenomanian specimens (Fig. 10.23A) differ from those of younger strata (Fig. 10.23B) in that a middorsal pit is present and the pits on the side of the centrum are typically larger.

SALMONIFORMES Bleeker, 1859
(sensu Greenwood et al., 1966)
ESOCOIDEA Berg, 1936

Late Cretaceous esocoideans were recognized by Wilson, Brinkman, and Neuman (1992) on the basis of isolated dentaries and palatal elements that had the apomorphic feature of C-shaped tooth bases, a feature associated with a type of tooth implantation that allows the teeth to be folded posteriorly. Subsequently centra of salmoniform morphology were described by Brinkman and Neuman (2002) and Neuman and Brinkman (2005).

Family indet.
ESTESESOX Wilson, Brinkman, and Neuman, 1992

In *Estesesox* (Fig. 10.24A, B) the anterior part of the dentary supports between two to four rows of teeth with at least two rows extending well posterior to the symphyseal region. The lateral opening of a canal for the mandibular branch of the

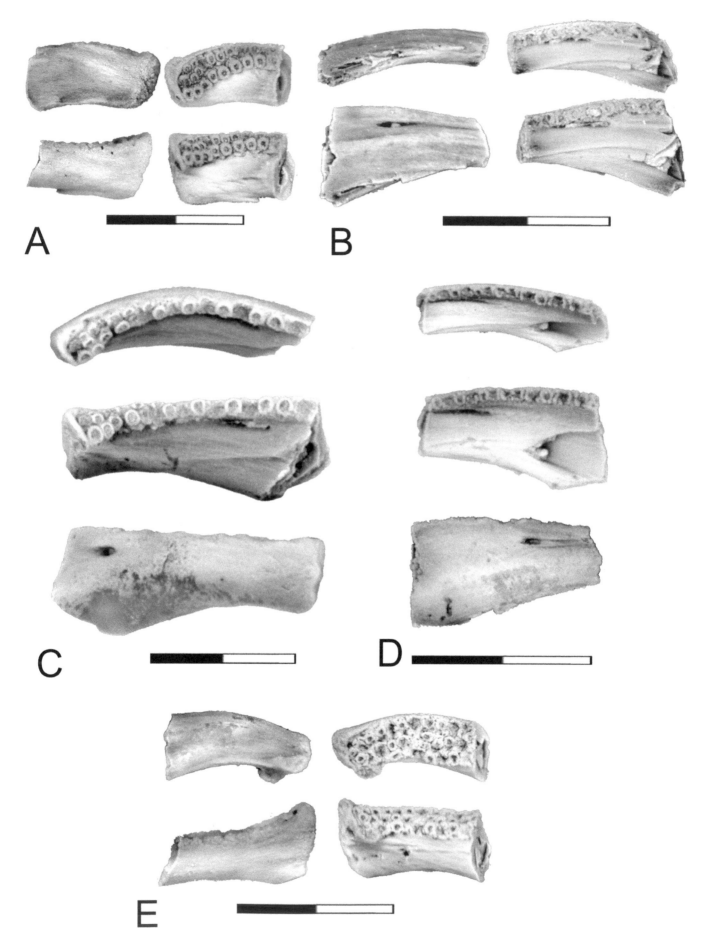

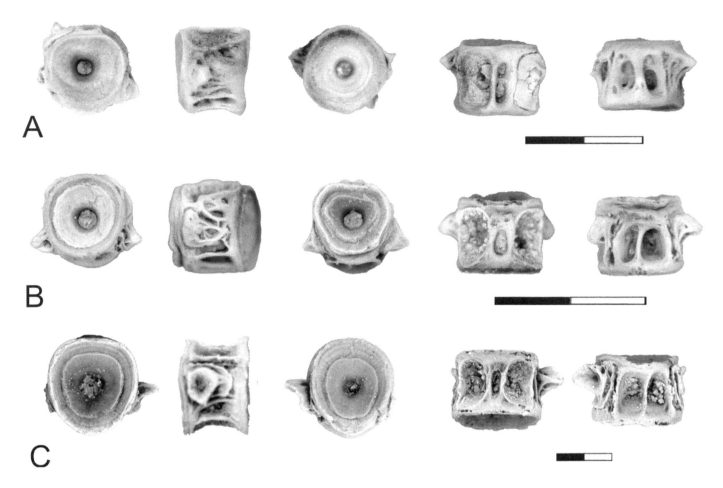

10.25. Salmoniform centra with parapophyses fused to centrum arranged from left to right in anterior, lateral, posterior, dorsal, and ventral views. (A) Specimen OMNH 67125; (B) specimen OMNH 67126, both Kaiparowits Formation, OMNH Loc. V-9; (C) specimen TMP 1986.07.0066, Dinosaur Park Formation, Alberta, Loc. L1115. Scale bars = 2 mm.

trigeminal nerve, termed the lateral trigeminal foramen by Wilson, Brinkman, and Neuman (1992), is a large opening located just below the tooth bases.

ESTESESOX FOXI Wilson, Brinkman, and Neuman, 1992

Estesesox foxi is represented by dentaries that are similar to those of the type specimen in having two to three rows of teeth anteriorly and in having relatively small teeth (Fig. 10.24A, B). The teeth of the inner row are slightly larger than the teeth of the outer row.

In southern Utah *Estesesox foxi* is only present in the Kaiparowits Formation.

ESETESESOX sp.

The presence of a second species of *Estesesox* is documented by dentaries that differ from those of *Estesesox foxi* in having larger teeth (Fig. 10.24C, D). Two rows of teeth are present at the anterior end of the dentary, but the outer row is short, extending only a short distance from the symphysis. Teeth of the outer row are smaller than those of the inner row.

Estesesox sp. is only present in the Kaiparowits Formation.

Genus et sp. indet.

The presence of a third esocoid in the Kaiparowits Formation is documented by dentaries having a very broad, flat tooth-covered area extending well posterior along the jaw

10.24. (facing) Esocoid dentaries. (A, B) *Estesesox foxi* Wilson, Brinkman, and Neuman, 1992; (A) symphyseal region, specimen OMNH 67131, Kaiparowits Formation, OMNH Loc. V-9; (B) middle of dentary, specimen OMNH 67119, Kaiparowits Formation, OMNH Loc. V-9; (C, D) *Estesesox* sp.; (C) symphyseal region of dentary, specimen MNA V 10359, Kaiparowits Formation, OMNH Loc. V-9; (D) middle of dentary, specimen OMNH 67132, Kaiparowits Formation, OMNH Loc. V-9; (E) dentary of esocoid genus and species indet., specimen OMNH 67116 (part), Kaiparowits Formation, OMNH Loc. V-9. A, B, E, views are arranged from left to right in ventral, occlusal, lateral, and medial views. C, D, views are arranged from top to bottom in occlusal, medial, and lateral views. Scale bars = 2 mm.

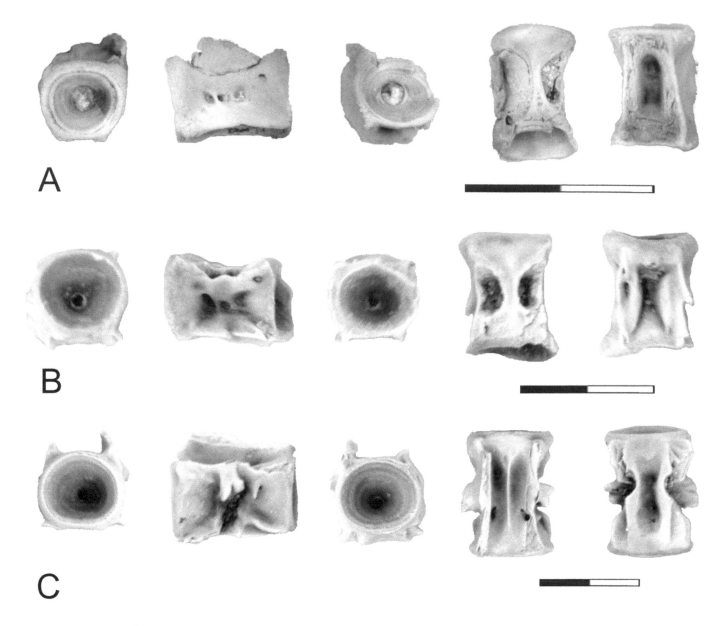

10.26. Precaudal centra of teleost type HvB arranged from left to right in anterior, lateral, posterior, dorsal, and ventral views. (A, B) Centra type H; (A) specimen OMNH 31322, Smoky Hollow Member of the Straight Cliffs Formation, MNA Loc. 995; (B) specimen TMP 2000.022.0011 from the Dinosaur Park Formation of Alberta; (C) centrum of morphoseries IIIA-1 specimen TMP 1997.018.0007, Dinosaur Park Formation, Alberta. Scale bars = 2 mm.

(Fig. 10.24E). In this morphology, esocoid genus and species indet. is similar to *Estesesox foxi*, but it differs in that the width of the tooth-bearing surface narrows only slightly posterior to the symphysis. Also, a distinctive flange bearing a single, relatively large tooth is present on the internal surface of the jaw at the symphysis. This taxon is included in the Esocoidea because the general shape of the dentary resembles that of *Estesesox*, although it was not possible to confirm the presence of C-shaped tooth bases.

Precaudal centra provide independent evidence regarding the abundance and distribution of salmoniforms in the upper Cretaceous beds of the Grand Staircase region (Fig. 10.25). The centra of extant salmoniforms are simple spools

with unfused transverse processes or neural arches, seen, for example, in *Salvelinus namaycush* (Walbaum, 1792) (ROM R2636), *Coregonus clupeaformis* (Mitchill, 1818) (ROM R1608), and *Esox lucius* Linnaeus, 1758 (ROM R7376). Neural arch and parapophyseal articular pits are located close together (in contrast to those in elopomorphs where they are widely separated), and the centra tend to be as long as they are wide (in contrast to the short, plate-like centra of elopomorphs).

Centra conforming to this generalized salmoniform morphology were only encountered in the Kaiparowits Formation. These are likely associated with one of the dentaries described above and so do not provide evidence of an additional salmoniform in the assemblage. A single esocoid centrum

Brinkman et al.

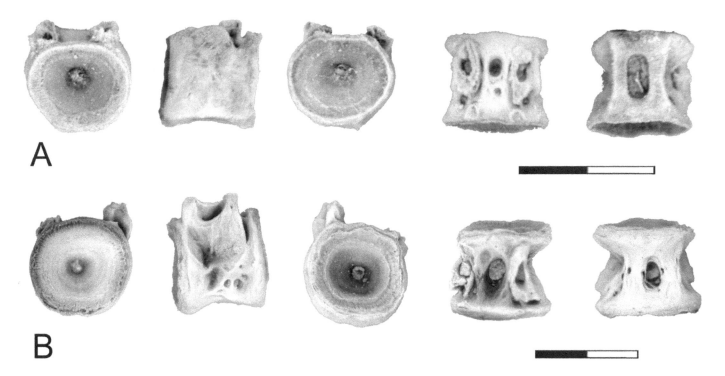

10.27. Precaudal centra of teleost BvE arranged from left to right in anterior, lateral, posterior, dorsal, and ventral views. (A) Centrum showing weakly developed parapophyseal pit and reduced middorsal pit, specimen OMNH 67127; (B) specimen showing distinct parapophyseal articular surface and deep midventral pit, specimen OMNH 67128. Both specimens from the Kaiparowits Formation, OMNH Loc. V-9. Scale bars = 2 mm.

morphotype was observed (Fig. 10.25A, B). This centrum morphology is similar to the salmoniform centrum from the Hell Creek Formation described by Brinkman, Newbrey, and Neuman (in press) as morphotype NvC. In both these centra the neural arches extend the full length of the centrum, the neural arch and parapophyseal articular pits are well separated and the ventral edges of the parapophyseal pits converge when seen in ventral view. The Utah centra differ from centrum type NvC of the Hell Creek in that parapophyses are preserved in place, partially or completely fused to the centrum and the parapophyseal articular pit contacts the neural arch articular pit. Salmoniform centra like those of the Kaiparowits morphotype are also present in the Belly River Group (Fig. 10.25C).

Order and Family indet.
Genus et sp. indet. type HvB

Euteleostei gen. et sp. indet. type HvB is represented by centra that are elongate and have autogenous neural arches and parapophyses (Fig. 10.26A). The neural arch articular pits are elongate, oval in outline, and separated from each other by a narrow bar of bone. Parapophyseal articular pits are indistinct. The ventral surface is flat, giving the centrum a subrectangular cross section, and a large midventral pit is present. Centra of this morphology were not reported from

the Dinosaur Park Formation by Brinkman and Neuman (2002) although here we recognize them in that unit (Fig. 10.26B). The centra of teleost HvB are generally similar to those described by Brinkman and Neuman (2002) as morphoseries IIIA-1 (Fig. 10.26C) in the presence of an elongate centrum with a flat base and large midventral pit. They differ, however, in that the centra of morphoseries IIIA-1 have fused neural arches and distinct parapophyseal pits on the lateral surface of the centrum while the centra of HvB have autogenous neural arches and are without distinct parapophyseal pits. These differences are thought to be most likely a result of variation along the vertebral column with the anteriormost centra having autogenous neural arches. If correct, we predict that centra similar to those of morphoseries IIIA-1 should be encountered in future samples from the Smoky Hollow Member localities.

In the Grand Staircase region, teleost HvB was only rarely encountered in the Turonian Smoky Hollow Member of the Straight Cliffs Formation.

Genus et sp. indet. BvE

Euteleostei indet. BvE is represented by centra in which the neural arch is fused to the centrum, the parapophyses are autogenous, a large shallow excavated area is present on the side of the centrum for the parapophysis, and a deep midventral

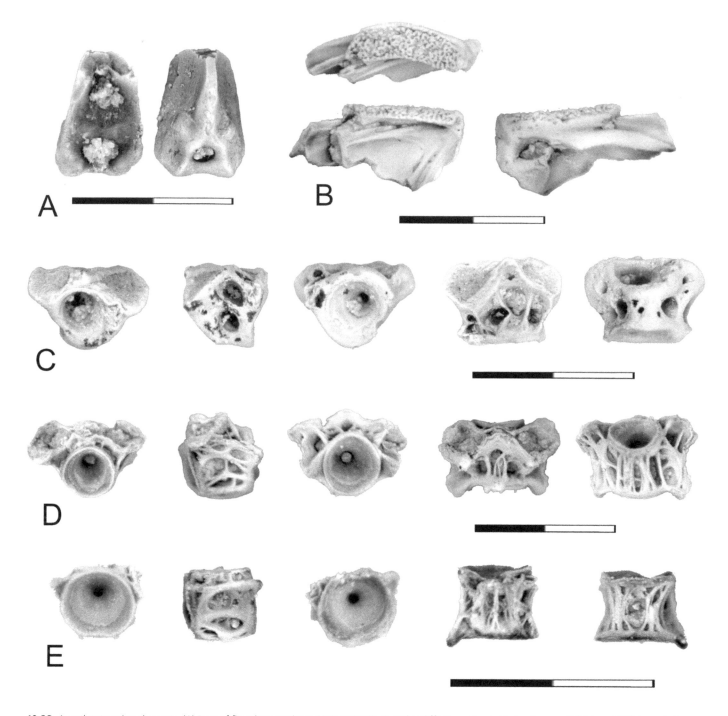

10.28. Acanthopterygian elements. (A) Base of fin spine, specimen MNA V10358, Straight Cliffs Formation in Cedar Canyon, MNA Loc. 1260 in posterior and anterior views; (B) Acanthomorph dentary sp. 2 in occlusal, medial, and lateral views, specimen OMNH 67118, Kaiparowits Formation, OMNH Loc. V-9; (C) atlas centrum, specimen UMNH VP 19139, upper John Henry Member, UMNH.VP.LOC. 424; (D) atlas centrum, specimen OMNH 67130, Kaiparowits Formation, OMNH Loc. V-9; (E) midprecaudal centrum, specimen OMNH 67129, Kaiparowits Formation, OMNH Loc. V-9. Centra are arranged from left to right in anterior, lateral, posterior, dorsal, and ventral views. Scale bars = 2 mm.

pit is present (Fig. 10.27). Middorsal pits are also present, although they are variably developed. No processes are present on the posterior end of the centrum. In general proportions, the centra of euteleost BvE are similar to anterior precaudal centra of acanthomorphs in being wedge shaped in anterior

view but they lack the zygapophyseal articulations that are present in that group and the parapophyses articulate low on the centrum, rather than on the neural arch.

In the Grand Staircase region teleost indet. BvE was present only in the Kaiparowits Formation. It is also present in

Brinkman et al.

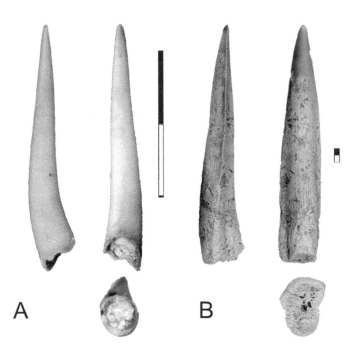

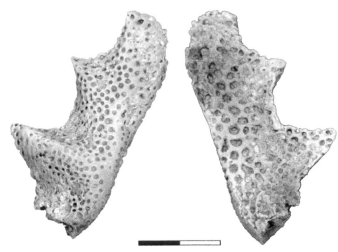

10.30. *Ceratodus gustasoni* Kirkland, 1987, tooth plate in occlusal and ventral views, specimen MNA V10380, Dakota Formation, MNA Loc. 1067. Scale bar = 2 mm.

10.29. Saber-like tooth of an unknown teleost from the Kaiparowits Formation compared with tooth of *Enchodus* in two different lateral views and one view from the underside of the crown. (A) Specimen OMNH V23506, from OMNH Loc. V-6; (B) Tooth of *Enchodus* specimen TMP 2004.013.0002 from the late Campanian Lethbridge Coal Zone of the Dinosaur Park Formation. Scale bars = 2 mm.

the Dinosaur Park Formation of Alberta and the Hell Creek Formation of Montana (Brinkman, Newbrey, and Neuman, in press).

NEOTELEOSTEI sensu Arratia, 1999
ACANTHOMORPHA sensu Stiassny, 1986
Order and Family indet.
Genus et sp. indet.

Acanthomorphs are represented by centra, fin spines, and dentaries (Fig. 10.28). Atlas centra have a tripartite anterior articular surface and lack neural arches (Fig. 10.28C, D). Zygapophyseal articulations are present between the precaudal centra, and the parapophyseal articular surfaces are located high on the neural arch. Acanthomorph precaudal centra from the Kaiparowits Formation have a middorsal ridge (Fig. 10.28E). This is a feature present in one morphotype from the Hell Creek Formation (Brinkman, Newbrey, and Neuman, in press), but not in acanthomorph centra from the Belly River Group.

An acanthomorph dentary from the Kaiparowits Formation (Fig. 10.28B) matches that described by Neuman and

Brinkman (2005) as acanthomorph dentary 2. As in that dentary morphotype, the tooth row consists of a battery of tiny teeth, the sensory canal is open laterally except for a buttress crossing the canal near its anterior end, and a broad area is present between the sensory canal posterior to this buttress and the tooth row. On the basis of the morphology of the sensory canal, Brinkman, Newbrey, and Neuman (in press) referred this dentary morphotype to the Percopsiformes.

In the Grand Staircase region, acanthomorphs first appear in the Coniacian(?) part of the Straight Cliffs Formation in Cedar Canyon (MNA Loc. 1260), where they are represented by a fin spine (Fig. 10.28A).

Division indet.
Saber-like fish tooth

An exceptionally long and slender tooth (Fig. 10.29A) from the Kaiparowits Formation documents the presence of a fish of uncertain relationships in the late Campanian of southern Utah. The length of the tooth is six times its diameter at the base. These proportions are similar to those of the saber-like teeth of *Enchodus* Agassiz, 1835 (Fig 10.29B), and approximately twice those of the lance-like gar teeth from the upper part of the John Henry Member described above. In contrast with the teeth of *Enchodus* the saber-like fish teeth of the Kaiparowits Formation are round in cross section, are hollow, have a more strongly curved base, and lack longitudinal striations. Teeth of this morphology have not been observed from any other Upper Cretaceous microvertebrate locality.

SARCOPTERYGII Romer, 1955
DIPNOI Müller, 1844
CERATODONTIFORMES Berg, 1940
CERATODUS Agassiz, 1837*CERATODUS*
GUSTASONI Kirkland, 1987

The lungfish *Ceratodus gustasoni* was first described by Kirkland (1987) on the basis of two tooth plates from the Dakota Formation of southern Utah. This species was characterized by having elongate tooth plates with a long, straight lingual margin and thin strongly projecting crests. Specimens were encountered during this survey include tooth plates from small individuals (Fig. 10.30). These have a distinctive honeycomb structure. Fragmentary specimens showing this honeycomb structure indicate that *Ceratodus gustasoni* is common in the Dakota Formation but is not present in younger beds.

DISCUSSION

The fish assemblages from the Grand Staircase region of Utah provide data on patterns of diversity in nonmarine osteichthyans assemblages from the Cenomanian through the late Campanian, a time during which nonmarine osteichthyan assemblages are poorly known. As with other Late Cretaceous nonmarine fish assemblages, basal actinopterygians are prominent and generally can be identified to at least family by isolated elements. The level of diversity of basal actinopterygians recognized in this study is generally similar to that recognized by Eaton et al. (1999), although several taxonomic identifications differ between the two. The results of this study differ significantly from those of previous studies of the Utah assemblages (Eaton et al., 1999) in recognizing a high diversity of teleosts, and in this regard they are more congruent with recent studies of teleosts from other Late Cretaceous vertebrate microfossil assemblages in the Western Interior of North America (Wilson, Brinkman, and Neuman, 1992; Brinkman and Neuman, 2002; Neuman and Brinkman, 2005; Brinkman, Newbrey, and Neuman, in press). These studies have suggested that teleosts are diverse in nonmarine environments from the Late Cretaceous, and many of the fossils can be referred to extant orders.

A striking aspect of the teleost assemblages from the Western Interior of North America is that they are dominated by taxa of small size. In a guild analysis of the freshwater communities of the Campanian of Alberta (Brinkman, 2008), it was concluded that basal acanthopterygians would have occupied positions high in the vertebrate food web, and with a few exceptions, teleosts would have occupied more basal positions. The small sizes of teleosts from the Late Cretaceous of the Grand Staircase region indicate that this was also the case in paleocommunities from this area.

Although taxonomic challenges remain, fossils from the Cenomanian to mid-Campanian microvertebrate sites of Kaiparowits Plateau, together with near-contemporaneous assemblages from other geographic areas in North America, provide a basis for considering the biostratigraphic and biogeographic distributions of osteichthyans in the Late Cretaceous of the Western Interior of North America.

BIOSTRATIGRAPHIC PATTERNS

Cenomanian–Turonian Faunal Change

One of the striking stratigraphic patterns recognized by Eaton et al. (1997), partly on the basis of osteichthyan assemblages, was a major faunal change between the Cenomanian and Turonian. Our study found a 47% change in osteichthyan taxa across the boundary, the highest between any stage or substages examined (Fig. 10.31). Three types of changes can be recognized: (1) loss of taxa; (2) appearance of new taxa; and (3) changes in relative abundance of a particular taxon.

Loss of Taxa

The stratigraphic ranges of osteichthyan taxa through the Late Cretaceous of the Grand Staircase region of Utah are shown in Fig. 10.31. Three osteichthyan taxa are present in the Cenomanian but are absent in younger localities, including the lungfish *Ceratodus gustasoni*, the ellimmichthyiform type LvD, and the amiid type ASWW. The loss of these taxa is a significant indicator of a reorganization of the Western Interior freshwater paleocommunities during the Cenomanian–Turonian because they are relatively abundant in the Dakota Formation.

Appearance of New Taxa

A major restructuring of the Western Interior freshwater paleocommunities during the Cenomanian–Turonian faunal event is also supported by the first appearances of four taxa in the Turonian: *Lepisosteus, Melvius,* ostariophysan type U-3/BvD, and teleost HvB (Fig. 10.31). The appearance of the first three of these taxa is of particular paleoecological significance because they are abundant in Turonian and younger localities from the Grand Staircase.

Changes in Abundance

A restructuring of the freshwater paleocommunities during the Cenomanian–Turonian faunal event is also supported by a decrease in the relative dominance (i.e., a decrease in abundance of fossil elements in vertebrate microfossil

Brinkman et al.

10.31. Stratigraphic distributions of fish taxa during the Late Cretaceous of the Grand Staircase–Escalante National Monument area of Utah. A thick solid line indicates intervals within the Utah section for which the taxon is known. A thick dashed line indicates intervals within the Utah section for which the taxon is not known. A thin, black, solid line indicates that the taxon is not known from younger localities in Utah but it is known from elsewhere in the Western Interior of North America. Taxa are arranged top to bottom according to their first and last occurrences. Percentage change refers to the absolute change in the number of taxa across time boundaries and assumes ghost lineages were present in North America.

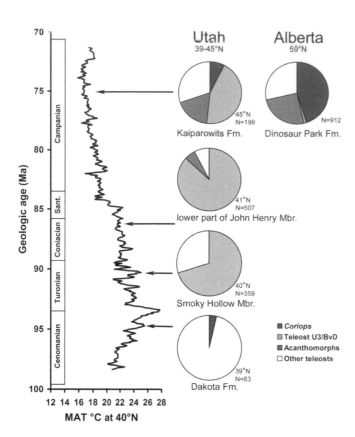

10.32. Relative abundance of teleosts in a series of assemblages with regard to climate change from the Cenomanian to late Campanian of Utah (paleolatitude 39–45°N) and the late Campanian of Alberta (paleolatitude 59°N), showing abundance patterns of *Coriops* Estes, 1969, ostariophysan type U-3/BvD, acanthomorphs, and other teleosts. Abundance is based on counts of precaudal centra. Counts of precaudal centra for Utah assemblages are listed in Table 10.2. Counts of precaudal centra from the Dinosaur Park Formation of Alberta are from table 4 in Brinkman, Newbrey, and Neuman (in press). Thermal data (sea-surface temperature) were provided by Jenkyns et al. (2004) from a paleolatitude of about 40°N and indicates a proxy for global climate change. Paleolatitudes were generated by the PLATES program (2009). MAT, mean annual temperature; Mbr., member; Sant., Santonian.

assemblages) of two taxa: the pycnodont *Coelodus* and the semionotiform *Lepidotes*. These taxa are common members of freshwater aquatic assemblages of the Early Cretaceous and Cenomanian, being present in microvertebrate assemblages from formations of these ages even in samples of small size (Garrison et al., 2007). Whereas *Coelodus* and *Lepidotes* extend across the Cenomanian–Turonian boundary, these taxa are rare in Turonian and younger localities.

Turonian to Middle Campanian Osteichthyan Faunas

Turonian to middle Campanian assemblages show successive appearances of five new taxa and a loss of three taxa (Fig. 10.31). Of the five first occurrences, two occur in the Coniacian (the gar with lanceolate teeth and an acanthomorph teleost), one in the late Santonian (the pycnodont *Micropycnodon*), and two in the middle Campanian (a paddlefish,

and *Paralbula*). The pycnodont, paddlefish, and *Paralbula* are all rare, each represented by fewer than five elements in total. However, the gar with lanceolate teeth is abundant throughout the sequence after its first appearance. Acanthomorph remains are rare in the Coniacian but increase in abundance in younger assemblages. In Santonian assemblages, about 6% of the teleost centra are from acanthomorphs, and in Campanian assemblages, 16% of the teleost centra are acanthomorph (Fig. 10.32; Table 10.2).

Three taxa present in the Turonian Smoky Hollow Member do not extend into the middle Campanian. These are the pycnodont *Coelodus*, the elmmichthyiform type U-7, and teleost HvB. *Coelodus* is rare in Turonian localities and rare in Coniacian localities. Elmmichthyiform type U-7 and teleost HvB are present only in the Turonian Smoky Hollow Member and are rare in that interval, although both have been recovered from the late Campanian Dinosaur Park Formation of Alberta, so their absence in Utah may be a result of extirpation from the Grand Staircase area rather than extinction of the taxa.

Middle–Late Campanian Faunal Change

Differences in faunal assemblages from the Kaiparowits and Wahweap formations suggest that a significant faunal turnover occurred during the middle to late Campanian. Seven taxa first appear in the Kaiparowits Formation: teleost indet. type BvE, the esocoids *Estesesox foxi*, *Estesesox* sp. Esocoidea gen. et sp. indet., a characiform, a clupeiform, and the osteichthyan of uncertain relationships represented by the saber-like tooth. The presence of a characiform is of particular interest because this group of ostariophysans is primarily a southern hemisphere group, and its occurrence in the late Campanian of North America has been taken as evidence for a nonmarine connection between North and South America at that time (Newbrey et al., 2009). However, because these taxa are represented primarily by centra and tooth-bearing elements, and because the number of these elements from the Wahweap localities is relatively small (Table 10.2), the timing and dynamics of this faunal change are poorly constrained. Thus, it is uncertain whether the appearance of these taxa in the late Campanian is a result of a single episode of reorganization or is the accumulation of a series of changes that occurred between the early Santonian, for which very large sample sizes are available (Table 10.2), and the late Campanian. However, because other aspects of the Wahweap Formation, particularly the mammals, appear to be more similar to those of the late Santonian than to the late Campanian (Cifelli and Madson, 1986), a major faunal change between the Wahweap and the Kaiparowits localities is likely. Further, a faunal change has also been suggested in

Brinkman et al.

Table 10.2. Specimen counts of precaudal centra of teleost fish from the Grand Staircase–Escalante National Monument area of Utah[a]

Specimen	Cenomanian	Turonian	Coniacian	Santonian		Campanian	
	Dakota	Smoky Hollow	Straight Cliffs	Lower	Upper	Middle	Upper
				John Henry (lower)	John Henry (upper)	Wahweap	Kaiparowits
	Loc. 1067	Loc. 995	Loc. 1260	Loc. 426, 799	Loc. 424	Loc. 83, 795	Loc. L-9
Amiddae AWSS	25						
Amiidae	71	7					2
Lepisosteus		2			5	1	5
Teleost type O	23	9		5	1		4
Coriops	9						15
Hiodontid	6	13		10	1		4
Elopomorph	12	21		3			
Diplomystus	2			4			
Ellimmichthyiformes 1	47						
Ellimmichthyiformes 2	135	8					
Clupeiformes type G							1
Centra type U-3		219	10	305	11	32	73
Centra type BvD		33		134		6	15
Characiform							1
Salmoniform							4
Teleost type U-4	3	56		18	2		20
Teleost type BvE	1					1	26
Teleost type HvB		2					
Acanthopterygian				28	1	7	36

[a] Fossils are from the Dakota Formation, Smoky Hollow member of the Straight Cliffs Formation, Straight Cliffs Formation, lower and upper parts of the John Henry Member of the Straight Cliffs Formation, Wahweap Formation, and the Kaiparowits Formation.

the basal Belly River Group in Alberta on the basis of turtles (Brinkman, 2003), suggesting that a faunal change at this time was regional rather than local.

LATITUDINAL PATTERNS

Latitudinal patterns can be inferred by comparing the assemblages from Southern Utah with contemporaneous assemblages from more northern regions of North America (Fig. 10.32). Of particular significance are comparisons of the Wahweap and Kaiparowits formations of Utah with the Belly River Group of Alberta because absolute ages and estimates of sedimentation rates provide a basis for exceptionally precise correlation between these units (Fig. 10.1B). The Wahweap localities overlap with localities in the basal Belly River Group (the Foremost Formation and lower muddy unit of the Oldman Formation), and the localities in the Kaiparowits Formation are bracketed by localities in the lower 40 m of the Dinosaur Park Formation in Dinosaur Provincial Park area and with localities in the upper Oldman Formation in the Manyberries area. The osteichthyan assemblages of these stratigraphic intervals are compared in Table 10.3. These can be divided into three groups: group 1 includes taxa that are present in both Alberta and Utah for at least one of

the time intervals; group 2 includes taxa that are present only in Utah; and group 3 includes taxa that are present only in Alberta. Taxa that are present in both the Wahweap and Kaiparowits formations, but not the Belly River Group, include: Lepisosteidae gen. et sp. indet. type 1 (the gar represented by lanceolate teeth), Lepidotes, Micropycnodon, and ?Melvius. In addition, the fauna of the Kaiparowits Formation differs from that of temporally equivalent beds in Dinosaur Park in the presence of the teleosts teleost O, a clupeomorph, a characiform, and the saber-toothed teleost. Basal actinopterygian taxa that are present throughout the Belly River localities but not the Wahweap or Kaiparowits formations include a sturgeon, holostean A, holostean B, and Belonostomus. Teleosts that are present in the Belly River Group but not the equivalent beds in Utah include teleost type HvB (the indeterminate teleost represented by centrum type IIIA-1 of Brinkman and Neuman, 2002) and the clupeomorph Horseshoeichthyes, which is represented in the Belly River Group by centra described by Brinkman and Neuman (2002) as centrum morphoseries IB-2 and IIA-2 (Newbrey et al., 2010). In addition, the large elopomorph Paratarpon is present only in the Dinosaur Park Formation.

For latitudinal patterns to be the preferred explanation for these differences, the possibility that their occurrences are

Table 10.3. Comparison of osteichthyian assemblages of similar age but different paleolatitudes[a]

	Utah (43–45°N)		Alberta (59°N)	
	Wahweap 43°N	Kaiparowits 45°N	Foremost lower Oldman	Dinosaur Park (lower 40 m)
Group 1				
Paddlefish	———			———
Paralbula	———	———	———	———
Lepisosteus	———	———	———	———
Amiinae gen. indet.	———	———	———	———
Acanthormorpha	———	———	———	———
Hiodontidae gen. indet.	- - -	———	———	———
Coriops	- - -	———	———	———
Teleost type U-4	- - -	- - -	———	———
Estesesox foxi		———		
Estesesox sp.		———		———
Esocoid gen. indet.		———		———
Teleost BvE		———		———
Ostariophysan type U-3/BvD	———	———		———
Group 2				
Lepisosteidae gen. indet.	———	———		
Lepidotes	———	———		
Micropycnodon	———	———		
?Melvius	———	———		
Teleost O	- - -	———		
Clupeoforme gen. indet.		———		
Saber tooth teleost		———		
Group 3				
Sturgeon			———	———
Holostean A			———	———
Holostean B			———	———
Belonostomus			———	———
Horseshoeichthyes			———	———
Teleost type HvB			———	———
Paratarpon				———

[a] Wahweap and Kaiparowits formations in Utah (paleolatitude 43–45°N) versus the basal Belly River Beds (Foremost Formation and the lower muddy unit of the Oldman Formation) and the lower 40 m of the Dinosaur Park Formation of Alberta (paleolatitude 59°N). A solid line indicates the presence of the taxon. A dashed line indicates intervals within the Utah section for which the taxon is not known but is present in older units. Group 1 includes taxa that are present in both Utah and Alberta during at least one of the time intervals. Group 2 includes taxa that are only present in Utah. Group 3 includes taxa that are only present in Alberta.

a result of local environmental differences needs to be considered. Aspects of the osteichthyan distribution pattern that argue in favor of regional latitudinal patterns, rather than local environmental differences, include high abundance and/or consistent presence across a broad suite of environmental settings in the area in which a taxon occurs. Also, consistency in the distribution patterns over the full temporal range of the taxon favors the hypothesis that its presence or absence is a reflection of regional latitudinal patterns.

Two of the taxa that occur in Utah and not Alberta, *Melvius* and Lepisosteidae gen. et sp. indet. type 1 (the gar represented by lanceolate teeth), are consistently present in high abundance across a broad range of localities throughout their stratigraphic range. *Lepidotes* occurs in both the Wahweap and Kaiparowits formations but is rare; however, it is present in earlier localities in Utah, supporting the hypothesis that its distribution is latitudinally restricted. Teleost O is not present in the Wahweap Formation, but it ranges from the Cenomanian Dakota Formation, and its absence in some stratigraphic intervals appears more closely related to sample

size than any other factor (Fig. 10.31, Table 10.2). Thus, for these taxa, the presence of a restricted latitudinal pattern, with these taxa not extending as far north as central Alberta, is the preferred hypothesis explaining their presence in Utah but not Alberta. The remaining taxa (the characiform, the clupeomorph, and the saber-toothed teleost) are all rare in the Utah section, both only being found at one locality, OMNH 6, and all occur in an unusual environment of deposition in the uppermost Dinosaur Park Formation. Thus, the presence of these three taxa in Utah and their absence in the lower 40 m of the Dinosaur Park Formation is more likely a result of restrictive ecological requirements of these taxa rather than regional latitudinal patterns.

All of the taxa that occur in strata of the Belly River Group in Alberta, but not in Utah, can be interpreted as representing taxa restricted to more northern regions. Holostean A is consistently present in high abundance in nonmarine localities in the Belly River Group (Brinkman, 1990). Sturgeon, *Belonostomus*, and teleost type HvB generally are not abundant but are consistently present across a broad range of localities. Holostean B is more ecologically restricted, being typically found in localities that are near the shoreline or quiet-water settings (Brinkman, 1990), but this setting is well represented in the Grand Staircase region of Utah, so a latitudinally restricted distribution pattern can be accepted as the best explanation for its absence in Utah. The large elopomorph *Paratarpon* has not been found outside the Belly River Group, but it is widespread in the Dinosaur Park Formation.

As well as differences in the presence/absence of various taxa, differences in the relative abundances of the dominant teleosts in Utah and Alberta may vary latitudinally. In the Utah assemblage, ostariophysan type U-3/BvD consistently is the dominant teleost after its first appearance in the Turonian (Table 10.2), but it is rare in the Belly River Group (Fig. 10.32). Conversely, *Coriops* is the dominant teleost in the Belly River Group but is of low abundance in the Kaiparowits Formation (Fig. 10.32). These patterns are interpreted as owing to latitude because they are consistent across a broad range of environmental settings in these geographic areas.

Shifts in Distribution Patterns in Response to Climate Change

Latitudinal differences in the osteichthyan assemblages are not surprising: during the late Campanian, the Belly River Group in the Dinosaur Provincial Park region of Alberta was located at a paleolatitude of 59°N, whereas near-contemporaneous strata in Utah were at a paleolatitude of approximately 45°N (Fig. 10.33) (PLATES, 2009). Although the climate was generally warm at both of these paleolatitudes, these assemblages would have been subject to extreme differences in

Brinkman et al.

daylight over the course of the year, which would have resulted in seasonal differences in productivity of the primary producers. The effects of seasonal differences in productivity would be especially significant on ectothermic vertebrates. Because of the seasonal differences in productivity that would be associated with changes in amount of light per day, changes in temperature that would be minor at low latitudes may have had a greater impact at higher latitudes, resulting in changes in the distributional patterns of nonmarine ectotherms. Thus, the possibility that changes in distribution patterns resulting from changes in temperature are correlated with stratigraphic patterns of distribution was considered.

The climate of the Late Cretaceous has been inferred by studies of oxygen isotopes and of faunal and floral evidence. Oxygen isotope curves from deep-sea cores provide a basic pattern of climate change (i.e., cool in the Cenomanian, a period of extreme global temperature in the Turonian, warm in the Cenomanian to Santonian, then a gradual cooling through the Campanian to the early Maastrichtian) (Huber, Norris, and MacLeod, 2002; Jenkyns et al., 2004). Faunal evidence in the form of high-latitude occurrences of ectothermic mesoreptiles in the late Turonian–early Coniacian provides additional support for extreme global temperatures at this time (Tarduno et al., 1998; Vandermark et al., 2009). Floral evidence also provides support for a cooling from the late Santonian to the early Maastrichtian (Upchurch and Wolfe, 1993).

The Cenomanian–Turonian faunal event may in part be related to the period of high global temperature in the late Turonian because three of the taxa that first appear at that time, *Melvius*, *Lepisosteus*, and the ostariophysan represented by vertebral morphotypes U-3/BvD, all are members of a southern assemblage. The higher diversity of lepisosteids in the Campanian of Utah (minimum of two taxa) compared to the Alberta localities, where no more than one lepisosteid is thought to be present, suggests that members of this family were also sensitive to environmental conditions correlated with latitude. The species of *Lepisosteus* present in the Campanian of Alberta, *Lespisosteus occidentalis*, also has a limited northern range, as it is absent at high-latitude localities (paleolatitude 63°N) of late Campanian age (Fanti and Miyashita, 2009) and is absent in central Alberta (paleolatitude 58°N) in the early Maastrichtian during a time of climatic cooling (PLATES, 2009; Larson, Brinkman, and Bell, 2010).

Some aspects of late Santonian–late Campanian faunal change may be related to cooling during this time, most notably the reappearance of *Coriops* in the Kaiparowits Formation. *Coriops* has a temporally disjunct distribution in Utah, being present the Cenomanian and Campanian but not in any intervening assemblages. Because this taxon appears to be member of a northern, cool-temperature assemblage,

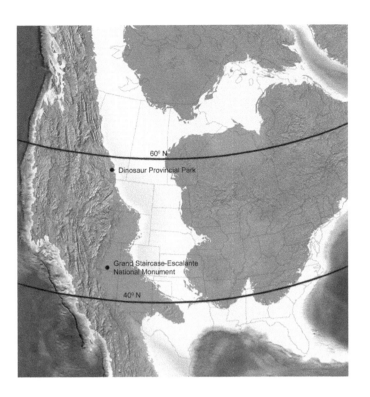

10.33. Late Cretaceous paleocoastline map showing locations of Dinosaur Provincial Park, Alberta, Canada, and Grand Staircase–Escalante National Monument, Utah, at 75 Ma. Localities in Grand Staircase–Escalante National Monument were deposited during the Cenomanian to Campanian and at times were located close to the shoreline of the Western Interior Seaway. Paleocoastline map (85 Ma) is based on Blakey (2009).

the presence of *Coriops* in Utah in the Cenomanian before Turonian thermal maxima and during the relatively cooler late Campanian can be interpreted to be the result of a southern shift in distribution of a northern taxon in response to cooling temperatures.

ACKNOWLEDGMENTS

We are grateful for the Late Cretaceous paleotemperature data provided by H. Jenkyns, University of Oxford, U.K. We thank I. Dalziel and L. Gahagan, PLATES Program, University of Texas Institute for Geophysics, for the paleolatitude data. We appreciate the assistance with collections, equipment, and lab space provided by H.-P. Schultze, D. Miao, and A. Falk, University of Kansas Natural History Museum, G. R. Smith and D. Nelson, University of Michigan, and K. Seymour, Royal Ontario Museum. We thank Arvid Aase, Fossil Butte National Monument, for sending us comparative material of *Diplomystus dentatus*. Thanks also to R. Cifelli for making material from the collections of the Oklahoma Museum of Natural History available for study. Special thanks to G. R. Smith, Univeristy of Michigan, for sending us additional comparative material and providing information on ostariophysans that aided in the identification of material;

E. Smith for the identification of crayfish gastroliths; and J. Kirkland, Utah Geological Survey, for his extensive experience on the fauna of the Late Cretaceous of Utah. We thank R. Blakey, Northern Arizona University, who provided the paleocoastline map used in Fig. 10.33. C. Redman provided access to material that was part of his Ph.D. research project, and support for his travel to Alberta was provided by the Royal Tyrrell Museum Cooperating Society through the Heaton Fund. J. Newbrey, University of Alberta, provided assistance with images in Fig. 10.21C, D. Specimens in the collections of the Utah Museum of Natural History were photographed courtesy of that museum. We thank F. Therrien, Royal Tyrrell Museum of Palaeontology, for assistance with the use of an environmental scanning electron microscope. Earlier versions of this chapter were read by J. Gardner and C. Scott of the Royal Tyrrell Museum. Reviews by J. Smith and J. Lundberg improved the chapter. Three travel grants were provided to M.G.N. by the Royal Tyrrell Museum Cooperating Society for the study of comparative material. Postdoctoral funding for M.G.N. was provided by the Royal Tyrrell Museum Cooperating Society and NSERC Discovery Grants A9180 (M. V. H. Wilson, University of Alberta) and 327448 (A. M. Murray, University of Alberta). Collection of UMNH specimens from Bryce Canyon National Park and vicinity by J.G.E. was funded by the Colorado Plateau Cooperative Ecosystem Studies Unit of the National Park Service.

REFERENCES CITED

Agassiz, L. 1832. Untersuchungen uber die fossilen Fische der Lias-Formation, Jahrbuch fur Mineralogie, Geognoise, Geologie und Petrefaktenkunde 1832, part 3:139–149.

Agassiz, L. 1835. Description des quelques espèces de cyprins du lac de Neuchâtel, qui sont encore inconnues aux naturalistes. Mémoirs de la Société des sciences naturelles de Neuchâtel 1:33–48.

Agassiz, L. 1837. Rescherches sur les poisons fossils, 3. Imprimerie de Petitpierre, Neuchâtel, 390 + 32 pp.

Arambourg, C., and L. Bertin. 1958. Super-ordre des Holostéens et des Halecostomi (Holostei et Halecostomi); pp. 2173–2203 in P.-P. Grassé (ed.), Traité de Zoologie: Anatomie, systématique, biologie 13. Masson et Cie, Paris.

Arratia, G. 1997. Basal teleosts and teleostean phylogeny. Palaeo Ichthyologica 7:1–168.

Arratia, G. 1999. The monophyly of Teleostei and stem-group teleosts. Consensus and disagreements; pp. 265–334 in G. Arratia and H.-P. Schultze (eds.), Mesozoic Fishes 2 – Systematics and Fossil Record. Verlag Dr. Friedrich Pfeil, Munich.

Bannikov, A. F., and F. Bacchia. 2000. A remarkable clupeomorph fish (Pisces, Teleostei) from a new Upper Cretaceous marine locality in Lebanon. Senckenbergianna Lethaea 80:3–11.

Berg, L. S. 1936. The suborder Esocoidei (Pisces). Izvestia Biologicheskogo Nauchno-Issledovatel'skogo institute pri Permskom 10:385–391.

Berg, L. S. 1937. A classification of fish-like vertebrates. Bulletin de l'Academie des Sciences de l'U.R.S.S., Classe des Sciences Mathematiques et Naturelles 4:1277–1280.

Berg, L. S. 1940. [Classification of fishes, both recent and fossil]]. Travaux de l'Institut Zoologique de l'Académie des Sciences de l'URSS 5:1–45 [Russian; reprint with English translation of text published by J. W. Edwards, Ann Arbor, Michigan, 1947].

Blake, S. 1940. Paralbula, a new fossil fish based on dental plates from the Eocene and Miocene of Maryland. Journal of the Washington Academy of Science 30:205–209.

Blakey, R. 2009. North American paleogeographic maps, 85 Ma. Available at jan.ucc.nau .edu/~rcb7/namK85.jpg. Accessed August 24, 2012.

Bleeker, P. 1859. Enumeratio speciorum piscium hucusque in Archipelago Indico observatarum. 4 sheets. Batavia.

Bonaparte, C. L. 1838. Selachorum tabula analytica. Nouvelles Annales des Sciences Naturelles 2:195–214.

Brinkman, D. B. 1990. Paleoecology of the Judith River Formation (Campanian) of Dinosaur Provincial Park, Alberta, Canada: evidence from vertebrate microfossil localities. Palaeogeography, Palaeoclimatology, Palaeoecology 78:37–54.

Brinkman, D. B. 2003. A review of nonmarine turtles from the Late Cretaceous of Alberta. Canadian Journal of Earth Sciences 40:557–571.

Brinkman, D. B. 2008. The structure of Late Cretaceous (Late Campanian) non-marine aquatic communities: a guild analysis of two vertebrate microfossil localities in Dinosaur Provincial Park, Alberta, Canada; pp. 33–60 in J. T. Sanke and S. Baszio (eds.), Vertebrate Microfossil Assemblages: Their Role in Paleoecology and Paleobiogeography. Indiana University Press, Bloomington, Indiana.

Brinkman, D. B., and A. G. Neuman. 2002. Teleost centra from uppermost Judith River Group (Dinosaur Park Formation, Campanian) of Alberta, Canada. Journal of Paleontology 76:138–155.

Brinkman, D. B., M. G. Newbrey, and A. G. Neuman. In press. Diversity and paleoecology of actinopterygian fish from vertebrate microfossil localities in the Maastrichtian Hell Creek Formation of Montana; in G. P. Wilson, W. A. Clemens, J. H. Hartman, J. R. Horner (eds.), Through the End of the Cretaceous in the Type Locality of the Hell Creek Formation in Montana and Adjacent Areas. Geological Society of America Special Paper.

Brinkman, D. B., A. P. Russell, D. A. Eberth, and J. H. Peng. 2004. Vertebrate Microsite Assemblages (Exclusive of Mammals) from the Foremost and Oldman Formations of the Judith River Group (Campanian) of Southeastern Alberta: An Illustrated Guide. Provincial Museum of Alberta Natural History Occasional Paper 25.

Bryant, L. 1987. A new genus and species of Amiidae (Holostei; Osteichthyes) from the Late Cretaceous of North America, with comments on the phylogeny of the Amiidae. Journal of Vertebrate Paleontology 7:349–361.

Case, G. R., and D. R. Schwimmer. 1988. Late Cretaceous fish from the Blufftown Formation (Campanian) in western Georgia. Journal of Paleontology 62:290–301.

Cifelli, R. L., and S. K. Madson. 1986. An Upper Cretaceous symmetrodont (Mammalia) form southern Utah. Journal of Vertebrate Paleontology 6:258–263.

Cifelli, R. L., S. K. Madsen, and E. M. Larson. 1996. Screen washing and concentration techniques; pp. 1–24 in R. L. Cifelli (ed.), Techniques for Recovery and Preparation of Microvertebrate Fossils. Oklahoma Geological Survey Special Publication 96.

Cifelli, R. L., R. L. Nydam, A. Weil, J. D. Gardner, J. G. Eaton, J. I. Kirkland, and S. K. Madsen. 1999. Medial Cretaceous vertebrates from the Cedar Mountain Formation, Emery County–the Mussentuchit local fauna; pp. 219–242 in D. D. Gillette (ed.), Vertebrate Paleontology in Utah. Utah Geological Survey Miscellaneous Publication 99–1.

Cope, E. D. 1877. A contribution to the knowledge of the ichthyological fauna of the Green River shales. Bulletin of the United States Geological Survey of the Territories 3:807–819.

Cumbaa, S. L., C. Schröder-Adams, R. G. Day, and A. J. Phillips. 1999. Cenomanian bonebed faunas from the northeastern margin, Western Interior Seaway, Canada. New Mexico Museum of Natural History and Science Bulletin 35:139–155.

Cuvier, G. 1819. Sur les poissons du sous-genre Hydrocyon, sur deux nouvelles especes de Chalceus, sur trois nouvelles especes de Serrasalmes. et sur l' Argentine glossodonta de Forskahl, qui est l'Albula gonorhynchus de Bloch. Memoires du Muséum d"Histoire Naturelle, Paris 5:351–379.

Cuvier, G. 1825. Recherches sur les ossemens fossiles, où l'on rétablit les caractères de plusieurs animaux dont les révolutions du globe ont détruit les espèces. 3rd ème éd. Paris.

Cuvier, G., and A. Valenciennes. 1840. Histoire naturelle des poissons. Tome quinzième. Suite du livre dix-septième. Siluroïdes. Histoire naturelle des poissons. v. 15: i–xxxi + 1–540, pls. 421–455. [Valenciennes authored volume; i–xxiv + 1–397 in Strasbourg edition]

Cuvier, G., and A. Valenciennes. 1846. Histoire Naturelle Des Poissons, Volume 19: Société Géologique de France, Strasbourg. [1969 facsimile reprint; A. Asher, Amsterdam]

Eaton, J. G. 1987. Stratigraphy, depositional environments, and age of Cretaceous mammal-bearing rocks in Utah, and systematics of the Multituberculata (Mammalia). Ph.D. dissertation, University of Colorado, Boulder, Colorado.

Eaton, J. G. 1991. Biostratigraphic framework for Upper Cretaceous rocks of the Kaiparowits Plateau, southern Utah; pp. 47–63 in J. D. Nations, and J. G. Eaton (eds.), Stratigraphy, Depositional Environments, and Sedimentary Tectonics of the Western Margin, Cretaceous Western Interior Seaway. Geological Society of America Special Paper 260.

Eaton, J. G. 1999. Vertebrate paleontology of the Iron Springs Formation, Upper Cretaceous, southwestern Utah; pp. 339–343 in D. D. Gillette (ed.), Vertebrate Paleontology in Utah. Utah Geological Survey Miscellaneous Publication 99–1.

Eaton, J. G. 2004. New screen-washing approaches to biostratigraphy and paleoecology, Cretaceous of southwestern Utah; pp. 21–30 in M. R. Dawson, and J. A. Lillegraven (eds.), Fanfare for an Uncommon Paleontologist: Papers on Vertebrate Evolution in Honor of Malcolm C. McKenna. Carnegie Museum of Natural History Bulletin 36.

Eaton, J. G. 2006. Santonian (Late Cretaceous) Mammals from the John Henry Member of the Straight Cliffs Formation, Grand Staircase–Escalante National Monument, Utah. Journal of Vertebrate Paleontology 26:446–460.

Eaton, J. G., and R. L. Cifelli. 1988. Preliminary report on Late Cretaceous mammals of the Kaiparowits Plateau, southern Utah. Rocky Mountain Geologist 26:45–55.

Eaton, J. G., F. Maldonado, and W. C. McIntosh. 1999. New radiometric dates from Upper Cretaceous rocks of the Markagunt Plateau, southwestern Utah, and their bearing on subsidence histories. Geological Society of America Abstracts with Programs 31:A11.

Eaton, J. G., R. L. Cifelli, J. H. Hutchison, J. I. Kirkland, and J. M. Parrish. 1999. Cretaceous vertebrate faunas from the Kaiparowits Plateau, South-Central Utah; pp. 345–353 in D. D. Gillette (ed.), Vertebrate Paleontology in Utah. Utah Geological Survey Miscellaneous Publication 99–1.

Eaton, J. G., J. I. Kirkland, J. H. Hutchison, R. Denton, R. C. O'Neill, and J. M. Parrish. 1997. Nonmarine extinction across the Cenomanian–Turonian boundary, southwestern Utah, with a comparison to the Cretaceous–Tertiary extinction event. Geological Society of America Bulletin 109:560–567.

Eaton, J. G., J. Laurin, J. I. Kirkland, N. E. Tibert, R. M. Leckie, B. B. Sageman, P. M. Goldstrand, D. W. Moore, A. W. Straub, W. A. Cobban, and J. D. Dalebout. 2001. Cretaceous and Early Tertiary geology of Cedar and Parowan canyons, western Markagunt Plateau, Utah; pp. 337–363 in M. C. Erskine, J. E. Faulds, J. M. Bartley, and P. D. Rowley (eds.), The Geologic Transition, High Plateaus to Great Basin–A Symposium and Field Guide. Utah Geological Association Publication 30.

Eberth, D. A. 2005. The geology; pp. 54–82 in P. J. Currie and E. B. Koppelhus (eds.), Dinosaur Provincial Park: A Spectacular Ancient Ecosystem Revealed. Indiana University Press, Bloomington, Indiana.

Eberth, D. A., and D. B. Brinkman. 1997. Paleoecology of an estuarine, incised-valley fill in the Dinosaur Park Formation (Judith River Group, Upper Cretaceous) of Southern Alberta, Canada. Palios 12:43–58.

Estes, R. 1964. Fossil Vertebrates from the Late Cretaceous Lance Formation, Eastern Wyoming. University of California Publications in Geological Sciences 49.

Estes, R. 1969. Two new Late Cretaceous fishes from Montana and Wyoming. Breviora 335.

Estes, R., and R. Hiatt. 1978. Studies on fossil phyllodont fishes: a new species of *Phyllodus* (Elopiformes, Albuloidea) from the Late Cretaceous of Montana. Paleobios 28:1–10.

Fanti, F., and T. Miyashita. 2009. A high latitude vertebrate fossil assemblage from the Late Cretaceous of west-central Alberta, Canada: evidence for dinosaur nesting and vertebrate latitudinal gradient. Palaeogeography, Palaeoclimatology, Palaeoecology 275:37–53.

Fink, S. V., and W. L. Fink. 1996. Interrelationships of Ostariophysan fishes (Teleostei); pp. 209–250 in M. L. J. Stiassny, L. R. Parenti, and G. D. Johnson (eds.), Interrelationships of Fishes. Academic Press, San Diego, California.

Forey, P. L., D. T. J. Littlewood, P. Ritchie, and A. Meyer. 1996. Interrelationships of elopomorph fishes; pp. 175–191 in M. L. J. Stiassny, L. R. Parenti, and G. D. Johnson (eds.), Interrelationships of Fishes. Academic Press, San Diego, California.

Forsskål, P. 1775. Descriptiones animalium avium, amphibiorum, piscium, insectorum, vermium; quae in itinere orientali observavit; Post mortem auctoris edidit Carsten Niebuhr. Hauniae. Descriptiones animalium quae in itinere ad Maris Australis terras per annos, 1–20 + i–xxxiv + 1–164, map.

Garrison, J. R., Jr., D. B. Brinkman, D. J. Nichols, P. Layer, D. Burge, and D. Thayn. 2007. A multidisciplinary study of the Lower Cretaceous Cedar Mountain Formation, Mussentuchit Wash, Utah: a determination of the paleoenvironment and paleoecology of the *Eolambia caroljonesa* dinosaur quarry. Cretaceous Research 28:461–494.

Garstang, W. 1931. The phyletic classification of Teleostei. Proceedings of the Leeds Philosophical and Literary Society, Scientific Section 2:240–260.

Goodrich, E. S. 1909. Vertebrata Craniata, first fascicle: cyclostomes and fishes; in R. Lankester (ed.), A Treatise on Zoology. Volume 9. A. & C. Black, London.

Grande, L. 1985. Recent and fossil clupeomorph fishes with materials for revision of the subgroups of clupeoids. Bulletin of the American Museum of Natural History 181:231–372.

Grande, L., and W. E. Bemis. 1996. Interrelationships of Acipenseriformes, with comments on "Chondrostei"; pp. 85–115 in M. L. J. Stiassny, L. R. Parenti, and G. D. Johnson (eds.), Interrelationships of Fishes. Academic Press, San Diego, California.

Grande, L., and W. E. Bemis. 1998. A Comprehensive Phylogenetic Study of Amiid Fishes (Amiidae) Based on Comparative Skeletal Anatomy. An Empirical Search for Interconnected Patterns of Natural History. Society of Vertebrate Paleontology Memoir 4.

Grande, L., F. Jin, Y. Yabumoto, and W. E. Bemis. 2002. *Protopsephurus liui,* a well-preserved primitive paddlefish (Acipenseriformes: Polyodontidae) from the Early Cretaceous of China. Journal of Vertebrate Paleontology 22:209–237.

Greenwood, P. H., D. E. Rosen, S. H. Weitzman, and G. S. Myers. 1966. Phyletic studies of teleostean fishes, with a provisional classification of living forms. Bulletin of the American Museum of Natural History 131:339–455.

Hay, O. P. 1929. Second Bibliography and Catalogue of the Fossil Vertebrata of North America. Publications of the Carnegie Institute of Washington 390.

Heckel, J. J. 1856. Beitrage zur Kenntniss der fossilen Fische osterreichs. Denkschriften der Kaiserlichen Akademie der Wissenschaften, Mathematisch-Naturwissenschaftliche Classe, Wien 11:187–274.

Hibbard, C. W., and A. Graffham. 1941. A new pycnodont fish from the Upper Cretaceous of Rooks County, Kansas. *Quarterly Bulletin University of Kansas* 27:71–77.

Hibbard, C. W., and A. Graffham. 1945. *Micropycnodon,* new name for Pycnomicrodon Hibbard and Graffham not Hay. Transactions of the Kansas Academy of Science 47:1–404.

Hilton, E. J. 2002. Osteology of the extant North American fishes of the genus *Hiodon* Lesueur, 1818 (Teleostei: Osteoglossomorpha: Hiodontiformes). Fieldiana Zoology (n.s. 100) 1520.

Huber, B. T., R. D. Norris, and K. G. MacLeod. 2002. Deep-sea paleotemperature record of extreme warmth during the Late Cretaceous. Geology 30:123–126.

Hutchison, J. H., J. E. Eaton, P. A. Holroyd, and M. A. Goodwin. 1998. Larger vertebrates of the Kaiparowits Formation (Campanian) in the Grand Staircase–Escalante National Monument and adjacent areas; pp. 391–398 in L. M. Hill (ed.), Learning from the Land: Grand Staircase–Escalante National Monument Science Symposium Proceedings. U.S. Department of the Interior, Bureau of Land Management.

Jenkyns, H. C., A. Forester, S. Schouten, and J. S. S. Damste. 2004. High temperatures in the Late Cretaceous Arctic Ocean. Nature 432:888–892.

Jinnah, Z. A., E. M. Roberts, A. L. Deino, J. S. Larsen, P. K. Link, and C. M. Fanning. 2008. New ^{40}Ar/^{39}Ar and detrital zircon U-Pb ages for the Upper Cretaceous Wahweap and Kaiparowits formations on the Kaiparowits Plateau, Utah: implications for regional correlation, provenance, and biostratigraphy. Cretaceous Research 30:287–299.

Johnson, G. D., and C. Patterson. 1996. Relationships of Lower Euteleostean fishes; pp. 251–332 in M. L. J. Stiassny, L. R. Parenti, and G. D. Johnson (eds.), Interrelationships of Fishes. Academic Press, San Diego, California.

Johnson, H. T. 1984. A method of using gastroliths to calculate length and weight of the freshwater crayfish, *Cherax destructor*, for use in predatory-prey studies. Australian Zoologist 21:435–444.

Kirkland, J. I. 1987. Upper Jurassic and Cretaceous lungfish tooth plates from the Western Interior, the last Dipnoan faunas of North America. Hunteria 2:1–15.

Klein, E. E. 1885. Beiträge zur Bildung des Schadels der Knochenfische II. *Jahreshefte des Vereins für vaterländische Naturkunde in Württemberg* 41:107–261.

Kriwet, J. 1999. Pycnodont fishes (Neopterygii, Pycnodontiformes) from the Lower Cretaceous of Una (E-Spain) with comments on branchial teeth in pycnodontid fishes; pp. 215–238 in G. Arratia and H.-P. Schultze (eds.), Mesozoic Fishes 2 – Systematics and Fossil Record. Verlag Dr. Friedrich Pfiel, Munich.

Lacépède, B. G. E. 1803. Histoire naturelle des poisons, v. 5. Plassan, Paris.

Larson, D. W., D. B. Brinkman, and P. R. Bell. 2010. Microvertebrate assemblages from the upper Horseshoe Canyon Formation, an early Maastrichtian cool-climate assemblage from Alberta, with special reference to the *Albertosaurus sarcophagus* (Theropoda: Tyrannosauridae) bonebed. Canadian Journal of Earth Sciences 47:1159–1181.

Leidy, J. 1856. Notices on remains of extinct vertebrate animals of New Jersey, collected by Professor Cook, of the state geological survey, under the direction of Dr. W. Kitchell. Academy of Natural Sciences of Philadelphia Proceedings 8:220–221.

Linnaeus, C. 1758. Systema Naturae (10th edition). Laurentii Salvii, Holmiae [Stockholm].

Linnaeus, C. 1766. Systema Naturae (revised 2nd edition). Laurentii Salvii, Holmiae [Stockholm].

McAlpin, A. 1947. *Palaeopsephurus wilsoni*, a new polyodontid fish from the Upper Cretaceous of Montana, with a discussion of allied fishes, living and fossil. Contributions from the Museum of Paleontology University of Michigan 6:167–234.

Mitchill, S. L. 1818. Memoir on ichthyology. The fishes of New York described and arranged. American Monthly Magazine and Critical Review 2:241–248; 321–328.

Müller, J. 1845. Ueber den Bau und die Grenzen der Ganoiden und über das natürliche System der Fische. Bericht über die zur Bekanntmachung geeigneten Verhandlungen der Akademie der Wissenschaften, Berlin 1845 (for 1844):117–216. [English translation by J. W. Griffith (1846), Scientific Memoirs 4:499–558]

Nelson, J. S. 2006. Fishes of the World (4th edition). John Wiley and Sons, Hoboken, New Jersey.

Neuman, A. G., and D. B. Brinkman. 2005. Fishes of the fluvial beds; pp. 167–185 in P. J. Currie and E. B. Koppelhus (eds.), Dinosaur Provincial Park: A Spectacular Ancient Ecosystem Revealed. Indiana University Press, Bloomington, Indiana.

Newbrey, M, G., A. M. Murray, D. B. Brinkman, M. V. H. Wilson, and A. G. Neuman. 2010. A new articulated freshwater fish (Clupeomorpha, Ellimmichthyiformes) from the Late Cretaceous Horseshoe Canyon Formation, Alberta, Canada. Canadian Journal of Earth Sciences 47:1183–1196.

Newbrey, M. G., A. M. Murray, M. V. H. Wilson, D. B. Brinkman, and A. G. Neuman. 2009. Seventy-five-million-year-old tropical tetra-like fish from Canada tracks Cretaceous global warming. Proceedings of the Royal Society B 276:3829–3833.

Nursall, J. R. 1996. The phylogeny of pycnodont fishes; pp. 125–152 in G. Arratia and G. Viohl (eds.), Mesozoic Fishes: Systematics, Palaeoecology. Verlag Dr. Friedrich Pfeil, Munich.

Olsen, P. E., and A. R. McCune. 1991. Morphology of the *Semionotus elegans* group from the Early Jurassic part of the Newark Supergroup of eastern North America with comments on the Family Semionotidae (Neopterygii). Journal of Vertebrate Paleontology 11:269–292.

Patterson, C., and D. E. Rosen. 1977. A review of the ichthyodectiform and other Mesozoic teleost fishes, and the theory and practice of classifying fossils. Bulletin of the American Museum of Natural History 158:81–172.

Peng, J. H., A. P. Russell, and D. B. Brinkman. 2001. Vertebrate Microsite Assemblages (Exclusive of Mammals) from the Foremost and Oldman Formations of the Judith River Group (Campanian) of Southeastern Alberta: An Illustrated Guide. Provincial Museum of Alberta Natural History Occasional Paper 25.

PLATES. 2009. The PLATES Project; L. Lawver and I. Dalziel (principal investigators), L. Gahagan (database and software manager). University of Texas Institute for Geophysics, Austin, Texas.

Rafinesque, C. S. 1819. Prodrome de 70 nouveaux genres d'animaux découverts dans l'intérieur des États-Unis d'Amérique, durant l'année 1818. Journal de Physique, de Chimie et d'Histoire Naturelle, Paris 88:417–429.

Rafinesque. C. S. 1820. Ichthyologia Ohioensis, or Natural History of Fishes Inhabiting the Ohio and Its Tributary Streams Preceded by a Physical Description of the Ohio and Its Branches. W. O. Hunt, Lexington, Kentucky.

Regan, C. T. 1923. The skeleton of *Lepisosteus*, with remarks on the origin and evolution of the lower neopterygian fishes. Proceedings of the Zoological Society of London 1923:445–461.

Roberts, E. M. 2007. Facies architecture and depositional environments of the Upper Cretaceous Kaiparowits Formation, southern Utah. Sedimentary Geology 197:207–233.

Roberts, E. M., A. L. Deino, and M. A. Chan. 2005. ^{40}Ar/^{39}Ar age of the Kaiparowits Formation, southern Utah, and correlation of contemporaneous Campanian strata and vertebrate faunas along the margin of the Western Interior Basin. Cretaceous Research 26:307–318.

Romer, A. S. 1955. Herpetichthyes, Amphibioidei, Choanichthyes or Sarcopterygii? Nature 176:126.

Rosen, D. E., and P. H. Greenwood. 1970. Origin of the Weberian apparatus and the relationships of the ostariophysan and gonorhynchiform fishes. American Museum Novitates 2428:1–5.

Rosen, D. E., P. L. Forey, B. G. Gardiner, and C. Patterson. 1981. Lungfishes, tetrapods, paleontology and plesiomorphy. Bulletin of the American Museum of Natural History 167:159–276.

Sagemehl, M. 1885. Beiträge zur vergleichenden Anotomie der Fische. III. Das Cranium der Characiniden nebst allgemeinen Bemerkungen über die mit einen Weber'schen Apparat versehenen Physostomenfamilien. Gegenbauers Morphologisches Jahrbuch 10:1–119.

Stiassny, M. L. J. 1986. The limits and relationships of the acanthomorph teleosts. Journal of Zoology, London, B 1:411–460.

Tarduno, J. A., D. B. Brinkman, P. R. Renne, R. D. Cottrell, H. Scher, and P. Castillo. 1998. Evidence for extreme climatic warmth from Late Cretaceous Arctic vertebrates. Science 282:2241–2244.

Taverne, L. 1979. Ostéologie, phylogénèse et systématique des Téléostéens fossiles et actuels du super-ordre des osteoglossomorphes. Troisième partie. Évolution des structures ostéologiques et conclusions générales relatives à la phylogénèse et à la systématique du super-order: Mémoires de la Classe des Sciences, Académie Royale de Belgique 43:1–168.

Upchurch, G. R., and J. A. Wolfe. 1993. Vegetation and warm climates during the Late Cretaceous; pp. 243–281 in W. G. E. Caldwell and E. G. Kauffman (eds.), Evolution of the Western Interior Basin. Geological Association of Canada Special Paper 39.

Vandermark, D., J. A. Tarduno, D. B. Brinkman, R. D. Cottrell, and S. Mason. 2009. New Late Cretaceous macrobaenid turtle with Asian affinities from the High Canadian Arctic: dispersal via ice-free polar routes. Geology 37:183–186.

Walbaum, J. J. 1792. Petri Artedi sueci genera piscium. In quibus systema totum ichthyologiae proponitur cum classibus, ordinibus, generum characteribus, specierum differentiis, observationibus plurimis. Redactis speciebus 242 ad genera 52. Ichthyologiae Pars III. Ant. Ferdin. Rose, Grypeswaldiae (Greifswald). Artedi Piscium Part 3. [Reprint, J. Cramer (1966), Wheldon and Wesley, New York].

Wilson, M. V. H., and R. R. G. Williams. 2010. Salmoniform fishes: key fossils, supertree, and possible morphological synapomorphies; pp. 379–409 in J. S. Nelson, H.-P. Schultze, and M. V. H. Wilson (eds.), Origin and Phylogenetic Interrelationships of Teleosts: Honoring Gloria Arratia. Verlag Dr. Friedrich Pfeil, Munich.

Wilson, M. V. H., D. B. Brinkman, and A. G. Neuman. 1992. Cretaceous Esocoidea (Teleostei): early radiation of the pikes in North American fresh waters. Journal of Paleontology 66:839–846.

Woodward, A. S. 1890. The fossil fishes of the Hawkesbury series at Gosford. Memoirs of the Geological Survey of New South Wales, Palaeontology 4:1–56.

Brinkman et al.

Preliminary Report on Salamanders (Lissamphibia; Caudata) from the Late Cretaceous (Late Cenomanian– Late Campanian) of Southern Utah, U.S.A.

11

James D. Gardner, Jeffrey G. Eaton, and Richard L. Cifelli

HERE WE REPORT ON SALAMANDER FOSSILS (VERTEBRAE and jaws) and taxa identified from 19 microvertebrate localities of late Cenomanian–late Campanian age (an interval of about 25 million years) from the Dakota, Straight Cliffs, Iron Springs, Wahweap, and Kaiparowits formations in southwestern Utah, U.S.A. All three salamander families known from better-sampled upper Campanian–terminal Maastrichtian units elsewhere in the North American Western Interior are present in the Utah sequence: Scapherpetontidae and Batrachosauroididae occur throughout the late Cenomanian–late Campanian interval, whereas Sirenidae are limited to the Santonian–late Campanian. The scapherpetontid record consists of *Scapherpeton* (?Coniacian–late Campanian), *Lisserpeton* (late Campanian), and indeterminate older occurrences, including *Lisserpeton*-like vertebrae from the ?Coniacian and late Cenomanian, *Piceoerpeton*-like vertebrae from the ?Coniacian, and vertebrae of a probable new genus from the late Cenomanian. Batrachosauroidids are represented by *Opisthotriton* (Santonian–late Campanian), *Prodesmodon* (late Campanian), and a pair of indeterminate genera, one each from the late Turonian and late Cenomanian; the last occurrence is the oldest unequivocal record for batrachosauroidids in North America. The presence in pre-Santonian localities of scapherpetontid and batrachosauroidid specimens that cannot be assigned to known later Cretaceous and Paleogene genera indicates that both families were already present and diversifying by the early Late Cretaceous. The sirenid record is founded on Santonian–late Campanian atlantes of *Habrosaurus*; the Santonian occurrences are the oldest North American records for both the family and the genus. A previously unrecognized salamander of uncertain familial affinities (but showing some similarities to sirenids) is documented by distinctive trunk vertebrae and an atlantal centrum from the late Turonian–early or middle Campanian. Other enigmatic vertebrae likely pertaining to additional salamander taxa are reported from the late Cenomanian, late Turonian, and ?Coniacian. Most of the sampled localities contain multiple salamander genera and families; these diversities compare favorably with better sampled latest Cretaceous salamander assemblages elsewhere in the Western Interior, even though the compositions of those assemblages differ.

INTRODUCTION

Modern amphibians (Lissamphibia) consist of three clades with extant and fossil members—salamanders (Caudata), frogs (Salientia), and caecilians (Gymnophiona)—and one fossil clade, the Albanerpetontidae. Salamanders are readily identified by their small to moderate body size (typically under 30 cm in total length) and generalized tetrapod body plan that primitively consists of two pairs of similarly sized limbs and a moderately elongate trunk and tail. Salamanders include fully aquatic, semiaquatic, and terrestrial forms, and all require moist, generally temperate conditions to survive. Extant salamanders number about 550 species distributed among 10 families and have a primarily Holarctic and Neotropical distribution (Duellman and Trueb, 1986; Frost et al., 2006). Salamanders have a patchy fossil record that extends back to the Middle Jurassic (Bathonian) of England (Evans, Milner, and Mussett, 1988; Evans and Milner, 1994), Kyrgyzstan (Nessov, 1988; Averianov et al., 2008; Skutschas and Martin, 2011), Siberia (Skutschas and Krasnolutskii, 2011), and potentially China (Gao and Shubin, 2003; but see Wang, 2004) and is biased toward isolated bones and rare articulated skeletons of paedomorphic taxa (Estes, 1981; Milner, 2000).

North America is an important continent for salamanders. Twelve families (nine extant and three fossil) are known from North America, and about a third of those are endemic or largely restricted to the continent (Estes, 1981; Milner, 1983; Duellman and Trueb, 1986; Petranka, 1998; Holman, 2006). The Western Interior is an important source for salamander fossils and consequently has played a critical role in interpreting the evolutionary history of North American salamanders. The oldest unequivocal salamander fossils in North America are isolated bones and several articulated skeletons from the Upper Jurassic (Kimmeridgian and Tithonian) Morrison Formation of Wyoming and Utah (Hecht and Estes, 1960; Evans and Milner, 1993; Evans et al., 2005). North American Cretaceous salamanders are known only by isolated bones, mostly vertebrae and jaws recovered from microvertebrate localities, and are best documented from the Albian of Texas (Estes, 1969a; Winkler, Murry, and Jacobs, 1990) and from the Campanian and Maastrichtian of Texas north into southern Alberta and Saskatchewan (Estes, 1964, 1965, 1969b, 1981; Sahni, 1972; Carpenter, 1979; Armstrong-Ziegler, 1980; Naylor and Krause, 1981; Breithaupt, 1985; Rowe et al., 1992; Gardner, 2000a, 2005; DeMar and Breithaupt, 2006, 2008). Although salamander fossils and taxa have been reported in preliminary faunal articles, faunal lists, and conference abstracts from the Cenomanian of Texas (Winkler and Jacobs, 2002) and from the latest Albian/ early Cenomanian to late Campanian of Utah (Gardner, 1994; Eaton et al., 1997; Eaton, Munk, and Hardman, 1998; Eaton, Cifelli et al., 1999; Munk, 1998; Cifelli et al., 1999; Gardner et al., 2009), our knowledge of salamanders from that interval in North America remains extremely poor, especially in comparison to the latest Cretaceous record elsewhere in the Western Interior, where six named (and several as yet unnamed) species in three families are known (Estes, 1981; Gardner, 2000a, 2005; Holman, 2006). Fieldwork over the past quarter century in southern Utah has resulted in discoveries of numerous nonmarine microvertebrate localities of late Cenomanian to late Campanian age (a span of about 25 million years according to the timescale of Ogg, Agterberg, and Gradstein, 2004) that have produced modest numbers of fossil salamander bones.

Our objectives here are to identify and describe representative salamander fossils and taxa from select Upper Cretaceous localities in Utah and to discuss what these findings contribute to our understanding of Late Cretaceous salamander diversity and evolutionary history. To facilitate future research, in the descriptive accounts, we have made a conscious decision to list voucher specimens from each formation and to present detailed descriptions and remarks, supplemented by numerous (191 images for 57 specimens) photographs and scanning electron micrographs.

Institutional Abbreviations MNA, Museum of Northern Arizona, Flagstaff, Arizona; OMNH, Oklahoma Museum of Natural History, University of Oklahoma, Norman, Oklahoma; UMNH, Utah Museum of Natural History, Salt Lake City, Utah.

Notes on Locality Numbers Many of the localities listed in this chapter have been worked by field parties associated with different institutions, most notably the MNA, OMNH, and UMNH. The University of Colorado Museum in Boulder also holds material from some of the Utah localities; however, no salamander specimens from that institution were included

in our study. Each institution uses its own unique numbering wsystem for localities and, in many cases, field numbers also have been applied to localities. Consequently, a given locality may be known by one or more formal institutional numbers and by an informal field number. This plethora of numbers can be confusing, so in Table 11.1 we list the equivalent numbers (where known) for each of the 19 localities included in this study; for a similar table, see Eaton and Cifelli (1988:table 1). Our convention is to use MNA locality numbers for MNA specimens, OMNH locality numbers for OMNH specimens, and UMNH locality numbers for UMNH specimens, even in cases where specimens from the same locality are housed in different institutional collections. For example, even though the vertebrae MNA V10255 and UMNH VP 13431 are from the same locality, we identify those specimens as being from MNA Loc. 995 and UMNH Loc. VP 129, respectively.

Methods Specimens used in this study were recovered from microvertebrate localities in southern Utah. Some of the localities were identified by surface prospecting for accumulations of fossil bone, teeth, and scales, but others were blind-washed localities where few, if any, vertebrate fossils were seen on the surface (for more details, see Eaton, 2004). The amount of matrix collected and processed from each locality varied from a few bags to many dozens of bags. Matrix was wet screen washed following the procedure described by Cifelli, Madsen, and Larson (1996; see also Eaton, 2004), with the finest screen size being 30 mesh. Some of the recovered specimens are tiny and delicate, so we are reasonably confident that most of the preserved salamander bones were recovered from the samples.

GEOLOGICAL SETTING

An extensive sequence of richly fossiliferous, Upper Cretaceous sedimentary rocks is exposed in southern Utah. Overviews of the regional geology and vertebrate paleontology of those deposits that are relevant to our chapter have been presented by Eaton (1991, 1999a, 1999b), Eaton, Cifelli et al. (1999), Eaton, Diem et al. (1999), Titus et al. (2005), Roček et al. (2010), and Chapter 2 of this volume. The salamander fossils reported in this chapter come from 19 nonmarine localities, mostly representing fluvial and overbank settings, located near or on, from west to east, the Markagunt, Paunsaugunt, and Kaiparowits plateaus (Eaton, Diem et al., 1999:fig. 1). These localities range in age from late Cenomanian–late Campanian and are distributed among the following five units (Table 11.1): Dakota, Straight Cliffs, Iron Springs, Wahweap, and Kaiparowits formations. Below we briefly introduce each of these formations and provide

Gardner, Eaton, and Cifelli

Table 11.1. Upper Cretaceous salamander localities (n = 19) included in this study[a]

Age	Formation	Plateau	Equivalent locality numbers			
			MNA number	OMNH number	UMNH number	Field number
Late Campanian	Kaiparowits Fm., upper half	Kaiparowits	MNA 1004-1	OMNH V5[b]	–	TB8
Late Campanian	Kaiparowits Fm., upper half	Kaiparowits	–	OMNH V61[b]	–	8642
Late Campanian	Kaiparowits Fm., lower half	Kaiparowits	MNA 454-6	OMNH V9[b]	–	HM6
Late Campanian	Kaiparowits Fm., lower half	Kaiparowits	–	–	UMNH VP 51[b]	9517
Late Campanian	Kaiparowits Fm., lower half	Kaiparowits	–	–	UMNH VP 108[b]	90-H-7-22-2
Early–middle Campanian	Wahweap Fm., upper member	Kaiparowits	–	–	UMNH VP 130[b]	JGE-0005
Early–middle Campanian	Wahweap Fm., middle member	Paunsaugunt	–	–	UMNH VP 77[b]	9703
?Santonian	Iron Springs Fm.	West of Markagunt	MNA 1230[b]	–	UMNH VP 12	JGE 9025
Late Santonian	Straight Cliffs Fm., upper part of John Henry Mbr.	Paunsaugunt	–	–	UMNH VP 424[b]	JGE 03-008
Late Santonian	Straight Cliffs Fm., upper part of John Henry Mbr.	Paunsaugunt	–	–	UMNH VP 569[b]	JGE 04-019
Early Santonian	Straight Cliffs Fm., lower part of John Henry Mbr.	Kaiparowits	MNA 706-2	OMNH V27[b]	UMNH VP 98	HC2
Early Santonian	Straight Cliffs Fm., lower part of John Henry Mbr.	Paunsaugunt	–	–	UMNH VP 799[b]	Casey8-07-06-01
?Coniacian	Straight Cliffs Fm., basal John Henry Mbr.	Kaiparowits	–	OMNH V856[b]	–	–
?Coniacian	Undifferentiated Straight Cliffs Fm.	Markagunt	MNA 1260[b]	–	UMNH VP 8	JGE 9104
Late Turonian	Straight Cliffs Fm., Smoky Hollow Mbr.	Kaiparowits	MNA 995[b]	OMNH V843	UMNH VP 129[b]	JGE 8708
Late Turonian	Straight Cliffs Fm., Smoky Hollow Mbr.	Kaiparowits	MNA 1003-1	OMNH V4[b]	–	PC3
Late Cenomanian	Dakota Fm., middle member	Kaiparowits	MNA 1067[b]	OMNH V808[b]	UMNH VP 27[b]	JGE 8803
Late Cenomanian	Dakota Fm., middle member	Kaiparowits	MNA 1064[b]	–	UMNH VP 804	JGE 8510
Late Cenomanian	Dakota Fm., middle member	Paunsaugunt	MNA 939[b]	–	UMNH VP 123	ACF-87-1

Fm, Formation; Mbr, Member; MNA, Museum of Northern Arizona; OMNH, Oklahoma Museum of Natural History; UMNH, Utah Museum of Natural History.

[a] All localities are microvertebrate sites located on or adjacent to (from west to east) the Markagunt, Paunsaugunt, and Kaiparowits plateaus in southern Utah. For a map depicting those plateaus, see Eaton et al. (1999a:fig. 1). Formations (or portions thereof) are arranged from youngest to oldest; localities within each grouping are listed in ascending numerical order (see text for comments about relative positions of localities within each unit). Some localities are known by several institutional numbers and a field number, as indicated.

[b] Institutional locality number used in this chapter.

information on the relative stratigraphic positions and ages of the 19 salamander-bearing localities. In the following accounts, we identify localities by the institutional locality numbers used in the text, figure captions, and Table 11.2.

Dakota Formation The stratigraphically lowest unit considered in our study is the Dakota Formation. This formation consists of fine-grained sediments, coal, and conglomerates deposited in a mix of marine, brackish, and nonmarine environments (Eaton, 1991; Eaton, Cifelli et al., 1999; Titus et al., 2005). The age of the formation is widely considered to be Cenomanian on the basis of biostratigraphic correlations using ammonites and inoceramid clams and on radiometric dates (see summaries by Eaton and Cifelli, 1988; Eaton, 1993; Roček et al., 2010). The Dakota Formation is informally divided into lower, middle, and upper members (Peterson, 1969a; Eaton, 1991). The three localities considered here are

in the middle member and are late Cenomanian in age (Eaton, 1993; Roček et al., 2010). The most productive locality for salamander fossils is MNA 1067 (=OMNH V808/UMNH VP 27) and occurs within an overbank deposit. The other two localities (MNA 939 and 1064) are in channel lags.

Straight Cliffs Formation The Straight Cliffs Formation is a sandstone-dominated unit that was deposited under a mix of marine, brackish, and nonmarine environments during the Turonian through Santonian (Peterson, 1969b; Eaton, 1991; Titus et al., 2005). In exposures on the Paunsaugunt and Kaiparowits plateaus, the Straight Cliffs Formation can be subdivided into four formal members (Peterson, 1969b). In ascending order, these are the Tibbet Canyon, Smoky Hollow, John Henry, and Drip Tank members. Salamander fossils reported here come from two localities in the Smoky Hollow Member and five localities in the John

Henry Member. The Smoky Hollow Member is largely non-marine and is considered to be late Turonian in age (Eaton and Cifelli, 1988; Eaton, 1991; Titus et al., 2005). The two salamander-bearing localities (MNA 995/UMNH VP 129 and OMNH V4) occur in the lower part of the member (Eaton and Cifelli, 1988; Nydam and Fitzpatrick, 2009), in floodplain deposits of the middle "barren zone" (sensu Peterson, 1969b:7). The overlying John Henry Member is a thicker unit (approximately 200 m), with a strong brackish influence and currently is considered to be middle(?) Coniacian to late Santonian in age (Eaton and Cifelli, 1988; Eaton, 1991; Roček et al., 2010). The Coniacian portion is limited to the basalmost several tens of meters. One salamander-bearing locality (OMNH V856) may occur low enough in the John Henry Member to be Coniacian in age (Gardner, 1999), whereas the other four localities (OMNH V27 and UMNH VP 424, VP 569, and VP 799) occur stratigraphically higher in the member and thus are all Santonian in age (Eaton, 2006; Roček et al., 2010). In this chapter, we use the term *postbasal* for the thicker, Santonian part of the John Henry Member. Another locality (MNA 1260) occurs farther to the west, in Cedar Canyon on the Markagunt Plateau, in rocks that were mapped as the Straight Cliffs Formation by Moore et al. (2004). Unfortunately, Peterson's (1969b) members cannot be recognized in exposures of the Straight Cliffs Formation in that area; in this chapter, we refer to those exposures as *undifferentiated Straight Cliffs Formation*. Radiometric dates (Eaton et al., 2001) from an ash layer a few meters above MNA 1260 indicate that this locality is likely Coniacian in age.

Iron Springs Formation The Iron Springs Formation is a thick (1000 m) sandstone-dominated unit of Albian(?) to Santonian(?) age that crops out on and to the west of the Markagunt Plateau (Fillmore, 1991; Eaton, 1999b; Roček et al., 2010). Salamander fossils have been recovered from one locality (MNA 1230), located near the top of the formation to the west of the Markagunt Plateau. The absolute stratigraphic position and age of MNA 1230 is uncertain, but it is most likely Santonian in age (Eaton, unpubl. data).

Wahweap Formation The Wahweap Formation consists largely of interbedded sandstones and mudstones that were deposited on an alluvial plain, with some brackish influence in the rivers (Peterson, 1969a; Eaton, 1991; Titus et al., 2005; Jinnah et al., 2009). Radiometric dates from an ash layer 40 m above the base of the Wahweap Formation (Jinnah et al., 2009), coupled with radiometric dates from the overlying lower unit of the Kaiparowits Formation and estimated depositional rates (Roberts, Deino, and Chan, 2005), suggest the Wahweap Formation is early to middle Campanian in age. Eaton (1991) informally subdivided the Wahweap Formation into four units; in ascending sequence, those are the lower, middle, upper, and capping sandstone members. Two

localities have yielded significant salamander fossils: UMNH VP 77 is in a sandstone lag in the middle member on the Paunsaugunt Plateau, and UMNH VP 130 is in a floodplain mudstone in the upper member on the Kaiparowits Plateau.

Kaiparowits Formation The Kaiparowits Formation is composed of sandstones, mudstones, and siltstones that generally represent freshwater and terrestrial deposition across an alluvial plain (Eaton, 1991; Roberts, Deino, and Chan, 2005; Titus et al., 2005), but according to Roberts (2007), there is some indication of tidal influence or brackish conditions in rivers and estuaries in the lower part of the informal middle unit (Roberts, 2007). The age of the Kaiparowits Formation is reliably dated as late Campanian on the basis of mammals and radiometric dates (Eaton and Cifelli, 1988; Eaton, 1991; Roberts, Deino, and Chan, 2005). Roberts, Deino, and Chan (2005; see also Roberts, 2007) recently proposed an informal tripartite subdivision of the Kaiparowits Formation into lower, middle, and upper units. The Kaiparowits Formation is about 855 to 860 m thick on the Kaiparowits Plateau (Eaton, 1991; Roberts, Deino, and Chan, 2005). Most of the known microvertebrate localities are in the lower 400 m of the formation ("lower half" in Table 11.1), which would be equivalent to the lower unit and about the lower two-thirds of the middle unit of the scheme of Roberts, Deino, and Chan (2005). Of the five salamander-bearing localities considered here, three are in the lower part of the formation (Cifelli, 1990; Eaton, 2002): OMNH V9 and UMNH VP 51 and VP 108. The other two localities are located in the upper half of the formation, about 520 m (OMNH V61) and 640 m (OMNH V5) above its base (Eaton, 2002), or within the upper part of the middle unit and the lower part of the upper unit, respectively, of the scheme of Roberts, Deino, and Chan (2005).

SYSTEMATIC PALEONTOLOGY

LISSAMPHIBIA Haeckel, 1866
CAUDATA Scopoli, 1777
URODELA Duméril, 1806

Remarks Here we follow the convention first proposed by Milner (1988) and subsequently adopted by most fossil salamander workers (for a historical review and detailed justification, see Evans and Milner, 1996), in using the name Urodela for crown-clade salamanders (i.e., all living salamanders and fossil taxa descended from the same common ancestor) and the name Caudata for the more inclusive group of crown-clade salamanders plus Jurassic stem salamanders such as *Karaurus, Kokartus, Marmorerpeton,* and *Urupia*.

Studies of fossil salamanders rely heavily in vertebrae because those are the most commonly recovered bones of the salamander skeleton and they have proven useful for

Table 11.2. Faunal list of salamanders from the Late Cretaceous (late Cenomanian–late Campanian) of southern Utah[a]

Kaiparowits Formation (Late Campanian)
    Scapherpetontidae
        Scapherpeton sp. [OMNH V5, V9; UMNH VP 108]
        Lisserpeton sp. [OMNH V9]
    Batrachosauroididae
        Opisthotriton sp. [OMNH V5, V9, V61; UMNH VP 51]
        Prodesmodon sp. [OMNH V61]
    Sirenidae
        Habrosaurus sp. [OMNH V9]

Wahweap Formation (Early–Middle Campanian)
    Scapherpetontidae
        Scapherpeton sp. [UMNH VP 77]
    Batrachosauroididae
        Opisthotriton sp. [UMNH VP 130]
    Family Indet.
        New genus and species [UMNH VP 77]

Iron Springs Formation (?Santonian)
    Scapherpetontidae
        Genus and species indet. [MNA 1230]
    Batrachosauroididae
        Genus and species indet. [MNA 1230]
    Sirenidae
        Habrosaurus sp. [MNA 1230]

Postbasal (Santonian) Part of John Henry Member, Straight Cliffs Formation
    Scapherpetontidae
        Scapherpeton sp. [UMNH VP 424, VP 569, VP 799]
    Batrachosauroididae
        Opisthotriton sp. [UMNH VP 424, VP 569]
        Genus and specires indet. [OMNH V27]

Sirenidae
    Habrosaurus sp. [UMNH VP 424]
Family Indet.
    New genus and species [UMNH VP 424]

Basal (?Coniacian) Part of John Henry Member, Straight Cliffs Formation
    Scapherpetontidae
        Scapherpeton sp. [OMNH V856]
        Genus and species indet. [OMNH V856]

Undifferentiated Straight Cliffs Formation (?Coniacian)
    Scapherpetontidae
        Genus and species indet. [MNA 1260]
    Family Indet.
        Genus and species indet. [MNA 1260]

Smoky Hollow Member (Late Turonian), Straight Cliffs Formation
    Batrachosauroididae
        Genus and species B [OMNH V4]
    Family Indet.
        New genus and species [MNA 995/UMNH VP 129]
    Family Indet.
        Genus and species indet. [MNA 995]

Dakota Formation (Late Cenomanian)
    Scapherpetontidae
        Genera and species indet. [MNA 1067/UMNH VP 27; MNA 939 and/or MNA 1064]
    Batrachosauroididae
        Genus and species A [MNA 1067]
    Family(ies) Indet.
        Genera and species indet.
            Trunk vertebrae [MNA 1067/UMNH VP 27]
            Dentary morphs A–C [MNA 1067/OMNH V808]

[a] Lists are arranged by geological rock unit and from youngest to oldest. Taxonomic identifications appear in the text. Localities that have produced voucher specimens are listed in brackets for each taxon; locality numbers are listed in alphanumerical order; and numbers separated by a slash denote the same locality. See Table 11.1 for a list of institutional locality numbers.

differentiating taxa (Naylor, 1978; Estes, 1981; Holman, 2006). The following combination of features are useful for identifying salamander vertebrae (see also Estes, 1981; Milner, 2000; Holman, 2006; Skutschas, 2009; osteological terms generally follow Gardner, 2003a, 2003b): no sutures or divisions between centrum and neural arch; small bony processes (the nerve cord supports of Wake and Lawson, 1973; the spinal cord supports of Skutschas, 2009) along inner walls of neural arch that project into neural canal; neural spine may be finished in bone or cartilage; posterior cotyle in atlas and anterior and posterior cotyles in postatlantal vertebrae may have an open notochordal pit; atlantal centrum is variably flattened dorsoventrally, with a concave (i.e., excavated) posterior cotyle, a pair of anteriorly directed anterior cotyles for articulation with the paired occipital condyles on the skull and, between the anterior cotyles, an anteriorly directed odontoid process (known as the intercotylar process or interglenoid tubercle of some authors) that articulates with the floor of the foramen magnum and may bear one or more articular surfaces for this contact; atlantal neural arch bears postzygapophyseal processes and its neural crest and spine are variably developed;

postatlantal vertebrae have centra that typically are longer than wide and constricted midway along their length (i.e., hourglass shaped), may be amphicoelous, opisthocoelous or, rarely, procoelous, and spinal foramina variably present in posterior part of neural arch wall; trunk vertebrae may bear anterior or posterior basapophyses or entirely lack those processes, rib bearers (the transverse processes of some authors) consistently present and bicipital or unicipital (i.e., paired or single), subcentral keel variably developed, and depending on the taxon, trunk vertebrae typically have additional foramina, fossa, crests, flanges, or processes. Atlantes of crown salamanders have a foramen or, less commonly, an anteriorly open notch for passage of the first spinal nerve in the base of the neural arch wall, whereas atlantes of stem salamanders lack such a foramen or notch (Evans, Milner, and Mussett, 1988; Evans and Milner, 1996; Averianov and Voronkevich, 2002; Averianov et al., 2008; Evans and Borsuk-Białynicka, 2009). Vertebral features that are useful for differentiating salamander families, genera, and species include: form and proportions of centrum and neural arch; presence and form of foramina, fossa, rib bearers, neural crest and spine, subcentral keel, and other processes, ridges, and crests; outlines of cotyles; presence of notochordal pits; and form of odontoid process on atlas.

Vertebrae included in our study fall into three categories. The first consists of specimens that are diagnostic for one of three families (Scapherpetontidae, Batrachosauroididae, and Sirenidae). A subset of those specimens can further be assigned to five named genera within those families. We have conservatively refrained from trying to assign any of the latter specimens to named species because the available specimens from Utah are too incomplete to be meaningfully compared with specimens previously described for those named species. The second category consists of a dozen late Turonian–early or middle Campanian vertebrae that exhibit a distinctive and unique combination of features that indicate they belong to a previously unrecognized salamander genus and species of uncertain familial affinities. The third category consists of six problematic vertebrae from three late Cenomanian–?Coniacian localities that cannot be associated with any other specimens included in this report or assigned to any currently recognized salamander families. Our report also includes six broken dentaries, one from the Kaiparowits Formation referable to *Prodesmodon* sp. (Batrachosauroididae) and five from the Dakota Formation that cannot be identified to the family level but are worth documenting because they provide additional insights into the diversity of salamanders in that formation. Salamander faunal lists for each of the sampled rock units are presented in Table 11.2.

SCAPHERPETONTIDAE (Auffenberg and Goin, 1959) spelling emended by Kuhn, 1967:37
Scapherpetonidae Auffenberg and Goin, 1959:5

Remarks The Scapherpetontidae are an endemic North American family that are known from the latest Albian/earliest Cenomanian to late Paleocene of the Western Interior and from the early Eocene of the Canadian High Arctic (Estes, 1981; Naylor and Krause, 1981; Cifelli et al., 1999; Milner, 2000; this study). Four genera and five species currently are recognized: *Scapherpeton* (one species; Santonian–late Paleocene); *Lisserpeton* (one species; late Campanian–middle Paleocene); *Piceoerpeton* (two species; late Maastrichtian–early Eocene); and an unnamed monotypic genus (late Campanian–?early Paleocene) (Estes, 1964, 1965, 1969b, 1981; Meszoely, 1967; Naylor and Krause, 1981; Naylor, 1983; Gardner, 2000a, 2005, in press). Two Late Cretaceous genera, *Eoscapherpeton* and *Horezmia*, from Middle Asia that originally were regarded as scapherpetontids (Nessov, 1981, 1988, 1997) have recently been reinterpreted as cryptobranchoids

(Skutschas, 2009). Scapherpetontids are known only by isolated bones, mostly trunk vertebrae and atlantes; skull bones have also been described for *Scapherpeton*, *Lisserpeton*, and *Piceoerpeton* (Auffenberg and Goin, 1959; Estes, 1964, 1965, 1981; Naylor and Krause, 1981; Gardner, 2000a; Holman, 2006). Although no autapomorphies have been recognized for the Scapherpetontidae, their vertebrae are distinctive and can easily be differentiated from those of other salamanders found in the same localities. For example: the posterior cotyle in the atlas and the anterior and posterior cotyles in the trunk vertebrae retain an open notochordal pit and the inner walls are variably lined with calcified cartilage; atlantes have anterior cotyles that are subcircular to dorsoventrally compressed in outline and typically have flat to shallowly concave articular surfaces (more deeply concave in *Piceoerpeton*), the odontoid process typically is elongate (reduced in *Piceoerpeton*), and the neural arch is tall, lacks any elaboration along its anterior end, and the neural spine projects dorsally and slightly posteriorly and is variably finished in cartilage; and trunk vertebrae have amphicoelous centra, consistently lack spinal foramina and basapophyses, the subcentral keel is variably present, the neural spine is undivided, projects posterodorsally, and is finished in cartilage, and the rib bearers are divergently bicipital along most, if not all, of the trunk series.

Scapherpetontid vertebrae are present in all of the formations in southern Utah included in this study. *Scapherpeton* sp. is identified from the ?Coniacian to late Campanian, *Lisserpeton* is identified only from the late Campanian, and indeterminate scapherpetontids are identified from the late Cenomanian to Santonian.

SCAPHERPETON Cope, 1876
SCAPHERPETON sp.
Voucher Specimens (Fig. 11.1A–I)

Basal (?Coniacian) Part of John Henry Member, Straight Cliffs Formation OMNH Loc. v856: OMNH 67041–67045, trunk vertebrae.

Postbasal (Santonian) Part of John Henry Member, Straight Cliffs Formation UMNH Loc. VP 424: UMNH VP 19179, 19180, atlantes; UMNH VP 19181, 19182, trunk vertebrae; UMNH Loc. VP 569: UMNH VP 19183, atlas; UMNH Loc. VP 799: UMNH VP 19184, 19185, atlantes.

Wahweap Formation UMNH Loc. VP 77: UMNH VP 19186, atlas.

Kaiparowits Formation OMNH Loc. v5: OMNH 67046, atlas. OMNH Loc. v9: OMNH 67047–67052, atlantes; OMNH 67053–67055, trunk vertebra. UMNH Loc. VP 108: UMNH VP 19187, atlas; UMNH VP 19188, trunk vertebra.

Gardner, Eaton, and Cifelli

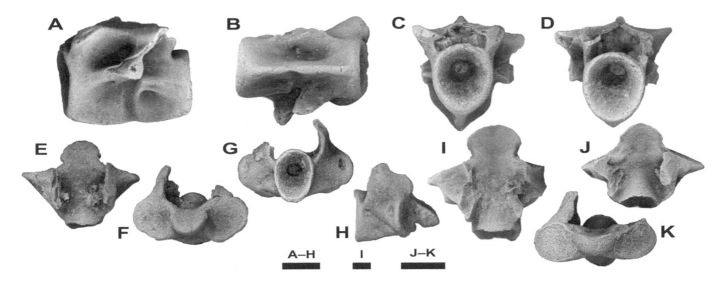

11.1. Trunk vertebra and atlantes of *Scapherpeton* sp. and atlas of *Lisserpeton* sp. (Scapherpetontidae) from southern Utah. (A–D) *Scapherpeton* sp., trunk vertebra, OMNH 67042, from OMNH Loc. V856, basal (?Coniacian) part of John Henry Member, Straight Cliffs Formation: (A) left lateral view; (B) ventral view with anterior toward right of figure; (C) anterior view; (D) posterior view. (E–H) *Scapherpeton* sp., atlantal centrum, UMNH VP 19185, from UMNH Loc. VP 799, postbasal (Santonian) part of John Henry Member, Straight Cliffs Formation: (E) dorsal view with anterior toward top of figure; (F) anterior view; (G) posterior view; (H) right lateral view. (I) *Scapherpeton* sp., atlantal centrum, OMNH 67047, from OMNH Loc. V9, Kaiparowits Formation, dorsal view with anterior toward top of figure. (J, K) *Lisserpeton* sp., atlantal centrum, OMNH 67056, from OMNH Loc. V9, Kaiparowits Formation: (J) dorsal view with anterior toward top of figure; (K) anterior view. Specimens at different magnifications; scale bars = 1 mm.

Descriptions None of the trunk vertebral specimens is intact, but insofar as they are known, they resemble specimens previously reported for *Scapherpeton tectum* (Auffenberg and Goin, 1959:fig. 1F–H, 2A–D; Estes, 1969b:fig. 6; Gardner, 2000a:fig. 11–8JJ–TT; Peng, Russell, and Brinkman, 2001:pl. 5, figs.1–3). One of the best preserved specimens, OMNH 67042 (Fig. 11.1A–D) is about 3.0 mm long and consists of the centrum and incomplete neural arch. Several other trunk centra are slightly more than 5 mm long. The centra are moderately elongate and amphicoelous. The anterior and posterior cotyles are laterally compressed and either elliptical or oval (in the latter case, the ventral end is narrower) in outline; none of the specimens has the markedly narrow cotyles seen in some *S. tectum* trunk vertebrae (Auffenberg and Goin, 1959:fig. 1F, 2D; Estes, 1969b:fig. 6A, D). Internally, both cotyles are moderately concave, with a thick lining of calcified cartilage and an open notochordal pit in the upper half. The subcentral keel is a laterally thin structure that typically projects below the ventral rims of the cotyles and, in some specimens, has a vertical groove on either side. No basapophyses or spinal nerve foramina are present. The rib bearers are bicipital, moderately divergent, and located high on the side of the centrum and neural arch wall. Where the neural arch is preserved, it is relatively narrow and flat, with a low neural crest and a spike-like, posterodorsally directed neural spine.

Numerous atlantal centra are available that are similar in size and structure to atlantes from the Dinosaur Park,

Lance, and Hell Creek formations that were referred to *Scapherpeton tectum* by Gardner (2000a:fig. 11–8A–II; see also Peng, Russell, and Brinkman, 2001:pl. 5, figs. 4, 5). The two specimens figured here (Fig. 11.1E–I) are the smallest and largest of the Utah specimens, and these adequately document the structure of the atlantal centrum. UMNH VP 19185 (Fig. 11.1E–H) is 3.0 mm wide across the anterior cotyles and 2.2 mm long; OMNH 67047 (Fig. 11.1I) is 5.8 mm long and has a preserved maximum width of 5.9 mm, but when intact was probably about 8.0 mm wide across the anterior cotyles. The anterior cotyles are subcircular to slightly depressed dorsoventrally in anterior outline, have a shallowly convex to nearly flat articular surface, and face anteriorly and slightly laterally. The prominent odontoid process is somewhat scoop shaped. In dorsal view the process has a slightly constricted base, expands laterally, and has a broadly convex anterior edge. In lateral view the odontoid process is deep at its base and becomes shallower toward its anterior end. The dorsal surface of the odontoid process is concave from side to side and typically is tilted ventrally at a shallow angle; the ventral surface is broadly convex and more nearly horizontal. On the anterior portion of the process, an articular surface that is separate from those on the anterior cotyles wraps around the anteroventral and ventrolateral surfaces of the process. The posterior cotyle is slightly laterally compressed and oval in posterior outline, with the narrower end directed ventrally. The interior of the cotyle is moderately deeply concave, the inner walls are lined with a thick layer of calcified cartilage, and the notochordal pit remains open and relatively large.

Remarks *Scapherpeton* is a monotypic genus. The type species *S. tectum* is of historical interest because it was the first Mesozoic salamander to be named (Cope, 1876) from North America. Trunk vertebrae of *Scapherpeton* are distinctive (e.g., centrum amphicoelous; cotyles laterally compressed and oval or elliptical in outline, moderately deeply concave, thick interior lining of calcified cartilage and open notochordal pit; subcentral keel narrow, typically deep, may have vertical groove along lateral surfaces and ventral surface may be notched; no basapophyses or spinal nerve foramina), and most of these features are evident in the Utah specimens. Atlantes of *Scapherpeton* historically have proven more challenging to identify, and several workers (Estes, 1965, 1981; Naylor, 1983) noted that there was considerable overlap between referred atlantes of *Scapherpeton* and *Lisserpeton*. On the basis of a pair of well-preserved scapherpetontid atlantes with nearly complete neural arches and size series of atlantal centra, Gardner (2000a) identified a suite of features that could be used to differentiate atlantes of *Scapherpeton* and *Lisserpeton*. Although the most pronounced differences involve the neural arch, atlantal centra of *Scapherpeton* can be differentiated from those of *Lisserpeton* using the following combination of features: anterior cotyles less dorsoventrally compressed; odontoid process scoop shaped and dorsal surface concave from side to side (versus more arrowhead shaped and dorsal surface flatter); and posterior cotyle more laterally compressed and shallower as a result of thicker interior lining of calcified cartilage.

Scapherpeton has been reported from over a dozen formations of late Santonian to late Paleocene age in the Western Interior (Estes, 1981; Naylor and Krause, 1981; Gardner, 2000a), so its presence in several Upper Cretaceous formations in Utah is not unexpected. Specimens documented here confirm a preliminary report (Eaton, Munk, and Hardman, 1998:table 1) of *Scapherpeton* in the Wahweap Formation and also document its presence in the Kaiparowits Formation and in the John Henry Member of the Straight Cliffs Formation.

LISSERPETON Estes, 1965
LISSERPETON sp.Voucher
Specimens (Fig. 11.1J–K)

Kaiparowits Formation OMNH Loc. V9: OMNH 67056 and 67057, atlantes.

Description The two atlantal centra (Fig. 11.1J, K) resemble specimens of late Campanian–middle Paleocene age that have been referred to *Lisserpeton bairdi* from the Tongue River Formation (Estes, 1976:text-fig. 3A, B) and the Dinosaur Park, Lance, and Hell Creek formations (Gardner, 2000a:fig. 11.9A–CC) in their structure, although both of the Utah specimens are at the small end of the size range for *Lisserpeton*. The better preserved specimen, OMNH 67056 (Fig. 11.1J, K) is 2.7 mm wide across the anterior cotyles and 2.0 mm long; OMNH 67057 (not figured) is slightly smaller. The specimens are similar to those described above for *Scapherpeton* sp., but differ in having the anterior cotyles more dorsoventrally compressed, the odontoid process more arrowhead shaped in dorsal outline and with a flatter dorsal surface, and the posterior cotyle deeper because the inner walls are coated with a thinner layer of calcified cartilage.

Remarks Like *Scapherpeton*, *Lisserpeton* is a monotypic genus. The type species *L. bairdi* is best known by its distinctive trunk vertebrae, which differ from *Scapherpeton* in a number of features (Estes, 1965; Gardner, 2000a), including having cotyles that are more subcircular in outline (versus laterally compressed) and relatively deeper with a thinner internal lining of calcified cartilage and bearing a shallower, broader subcentral keel. No examples of that kind of trunk vertebrae have been identified from any of the Utah localities in our study. As mentioned in the preceding account for *Scapherpeton*, atlantal centra of the two genera are more similar, but they can be separated by differences in the odontoid process and in the anterior and posterior cotyles.

Compared to *Scapherpeton*, *Lisserpeton* is known from fewer formations and a shorter temporal range (late Campanian to middle Paleocene) in the Western Interior. The two atlantal centra reported here from the same locality in the Kaiparowits Formation are within the previously documented temporal range for the genus. Munk (1998) reported a *Lisserpeton* atlantal centrum from a locality (UMNH VP 83) that may be in the Wahweap Formation and thus potentially older than the record from the Kaiparowits Formation, but published photographs of the specimen (Munk, 1998:fig. 3A–C) suggest it more likely pertains to *Scapherpeton*. Several geologically older vertebrae reported in the next account – two trunk vertebrae from the Dakota Formation (late Cenomanian) and an atlantal centrum from the basal (?Coniacian) part of the John Henry Member of the Straight Cliffs Formation – show some resemblances to *Lisserpeton*, but cannot confidently be assigned to that genus.

Scapherpetontidae Genera and
Species Indeterminate
Voucher Specimens (Figs. 11.2–11.4)

Dakota Formation MNA Loc. 1067: MNA V10360–V10364, atlantes; MNA V10233–10236, V10365, trunk vertebrae; MNA

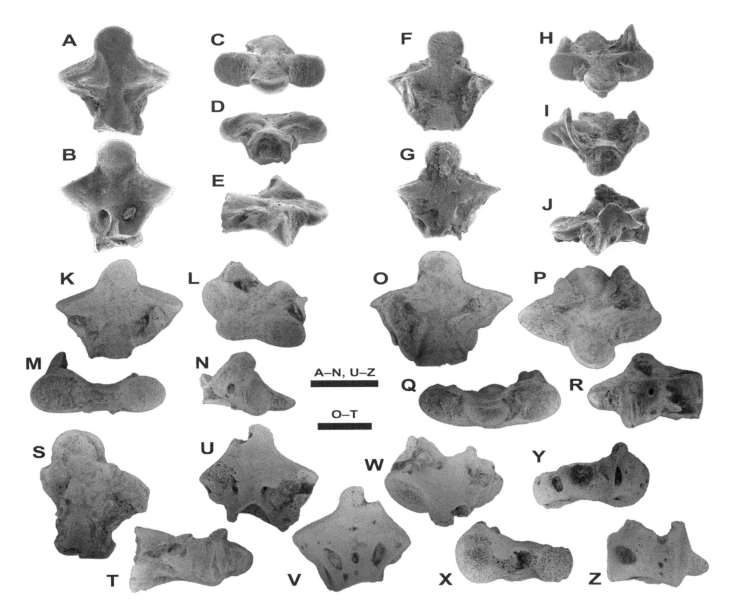

11.2. Atlantes of Scapherpetontidae Genera and Species Indeterminate from the Dakota Formation, southern Utah. (A–E) atlantal centrum, UMNH VP 13080, from UMNH Loc. VP 27: (A) dorsal view; (B) ventral view; (C) anterior and slightly dorsal view; (D) posterior and slightly dorsal view; (E) right lateral and slightly dorsal view. (F–J) atlantal centrum preserving bases of neural arch walls and with matrix adhering to medial portions of anterior cotyles and underside of odontoid process, UMNH VP 13105, from UMNH Loc. VP 27: (F) dorsal view; (G) ventral view; (H) anterior and slightly dorsal view; (I) posterior and slightly dorsal and left lateral view; (J) left lateral and slightly dorsal view. (K–N) atlantal centrum, MNA V10360, from MNA Loc. 1067: (K) dorsal view; (L) oblique dorsoanterior and left lateral view; (M) anterior view; (N) right lateral view. (O–R) atlantal centrum, MNA V10363, from MNA Loc. 1067: (O) dorsal view; (P) oblique dorsoanterior view; (Q) anterior view; (R) left lateral and slightly dorsal view. (S, T) atlantal centrum missing most of left anterior cotyle and adjacent portion of centrum, MNA V10362, from MNA Loc. 1067: (S) dorsal view; (T) right lateral view. (U–Z) atlantal centrum missing left side of odontoid process and margins of left anterior cotyle, MNA V10364, from MNA Loc. 1067: (U) dorsal view; (V) ventral view; (W) oblique dorsoanterior view; (X) anterior view; (Y) posterior view; (Z) right lateral view. All dorsal and ventral views oriented with anterior end toward top of figure. Specimens at different magnifications; scale bars = 1 mm.

Loc. 939 and/or 1064: V10244–10246, trunk vertebrae; UMNH Loc. VP 27: UMNH VP 13080 and 13105, atlantes; UMNH VP 13093 and VP 13112, trunk vertebrae.

Descriptions Vertebrae from the Dakota Formation – All of the atlantal centra and most of the trunk vertebrae are from one locality (MNA 1067/UMNH VP 27). The provenance of the other three trunk vertebrae is uncertain; they are from MNA Loc. 939 and/or 1064. The seven atlantal centra (Fig. 11.2) are all small (midline length 1.5–2.4 mm; width across anterior cotyles 1.6–2.7 mm). Neural arches are broken off in all specimens, but enough of the base remains on one or both sides to show that the spinal foramen was fully enclosed in bone. The following description is based on six specimens (MNA V10360–V10363 and UMNH VP 13080 and VP 13105; Fig. 11.2A–T); the anomalous seventh specimen, MNA V10364, is described separately. The odontoid process is broken off in

MNA V10361 (not figured) but is preserved intact on the other five specimens as a moderately elongate (length subequal to width), prong- or scoop-like projection that is relatively thick dorsoventrally through its base. The dorsal surface of the process is nearly flat to shallowly concave, whereas the ventral surface is broadly convex and covered with an articular surface that is separate from those on the anterior cotyles. Variation in the odontoid process is seen in its dorsal or ventral outline (base slightly constricted, process broadens anteriorly, and anterior margin shallowly indented at midline in UMNH VP 13105 versus base unconstricted, width of process relatively uniform along most of its length, and anterior margin broadly rounded in the other specimens) and in lateral view (process becomes notably shallower toward its anterior end in MNA V10360 versus thickness relatively uniform along much of its length in the other specimens). The anterior cotyles are dorsoventrally flattened to varying amounts in anterior outline and have nearly flat articular surfaces that, in dorsal view, are angled slightly to more strongly posteriorly (e.g., MNA V10362 versus V10363, respectively). The posterior cotyle is subcircular in posterior outline. Although the posterior cotyle is at least partially infilled with matrix in each specimen, it is evident that this cotyle is deeply concave and that its inner walls are lined with a thin coating of calcified cartilage; however, the condition of the notochordal pit is unknown. When seen in lateral view, the ventral rim of the posterior cotyle lies well below the level of the ventral rims of the anterior cotyles in MNA V10362, whereas in the other specimens the ventral rims of the anterior and posterior cotyles are in approximately the same horizontal plane.

MNA V10364 is a small atlantal centrum (midline length 1.6 mm; width across anterior cotyles 1.8 mm; Fig. 11.2U–Z). In contrast to the usual atlantal pattern, this specimen is bilaterally asymmetrical: in dorsal view the left side is about 80% as wide as the right side and in anterior view the left anterior cotyle is about 60% as tall as the right anterior cotyle. The ventral margins of both anterior cotyles are in essentially the same horizontal plane, so the height discrepancy is limited to the dorsal part of the cotyles. In anterior outline, the left anterior cotyle is dorsoventrally flattened and resembles those in the above-described atlantes, whereas the right anterior cotyle is relatively taller and subcircular; there is no way to know which outline is normal and which is anomalous for this specimen. Both anterior cotyles are angled strongly posteriorly, but the profiles of their articular surfaces differ – flat on the left cotyle versus shallowly concave on the right cotyle. Although missing a chunk from its left side, the odontoid process clearly is similar to those described above. The rim of the posterior cotyle is broken, which means that its original outline cannot be determined (as preserved, the

outline is squarish), and the measured length of 1.8 mm for the specimen slightly underestimates the actual length of the bone. However, the interior of what remains of the posterior cotyle is free of matrix – unlike in the above-described atlantes – and a moderate-sized notochordal pit can be seen. The ventral and lateral surfaces are perforated by abundant pits of various sizes, more so than in any of the atlantes described above.

Ten small- to moderate-sized trunk vertebrae (centrum length 2.4–4.2 mm) are available from the Dakota Formation; the seven most informative specimens are shown in Fig. 11.3. Specimens UMNH VP 13112 and MNA V10365 (Fig. 11.3A–D and S–X, respectively) both preserve a relatively complete centrum and neural arch; the remaining specimens preserve lesser amounts of the arch or just the centrum. MNA V10365 is relatively blocky (robustly built, anteroposteriorly foreshortened, and centrum and neural arch not medially constricted), the midventral portion of the centrum is broad and flattened, the rib bearers are stout and only moderately divergent, and the anterior end of the neural crest is swollen; this suite of features indicates that MNA V10365 is probably an anteriormost trunk vertebra. UMNH VP 13112 is interpreted as being from farther back in the trunk series on the basis of the observation that it is relatively more gracile, elongate, and the centrum and neural arch are weakly constricted medially, the subcentral keel is narrow, the rib bearers are less heavily built and more widely divergent, and the anterior end of the neural crest is less swollen. The largest specimen, MNA V10233 (Fig. 11.3E–H), is a robustly built and simple centrum. Collectively, the 10 specimens indicate that the centrum is amphicoelous, is moderately elongate (ratio of centrum length versus maximum width across anterior cotyle 1.4–2.3), is weakly constricted medially midway along its length (but not in MNA V10365), lacks basapophyses and spinal nerve foramina, and bears divergently bicipital rib bearers. The anterior and posterior cotyles are deeply concave and subcircular in outline, and in lateral view, their rims are oriented dorsoventrally. Internally, the cotylar walls are lined with a thin layer of calcified cartilage, and a prominent notochordal pit is located in the approximate center or upper half of each cotyle. MNA V10235 is an incomplete centrum split sagittally and slightly to the left of the midline; when viewed in medial aspect (Fig. 11.3L), the preserved left side of the centrum clearly shows that the anterior and posterior cotyles are separated by a disc-shaped bony plug and that the posterior cotyle is deeper anteroposteriorly than the anterior cotyle. The subcentral keel is absent or weakly developed (Fig. 11.3F, J, K, U versus B, N, P). Where present, the keel extends along about the median two thirds of the centrum and does not project ventrally past the level

Gardner, Eaton, and Cifelli

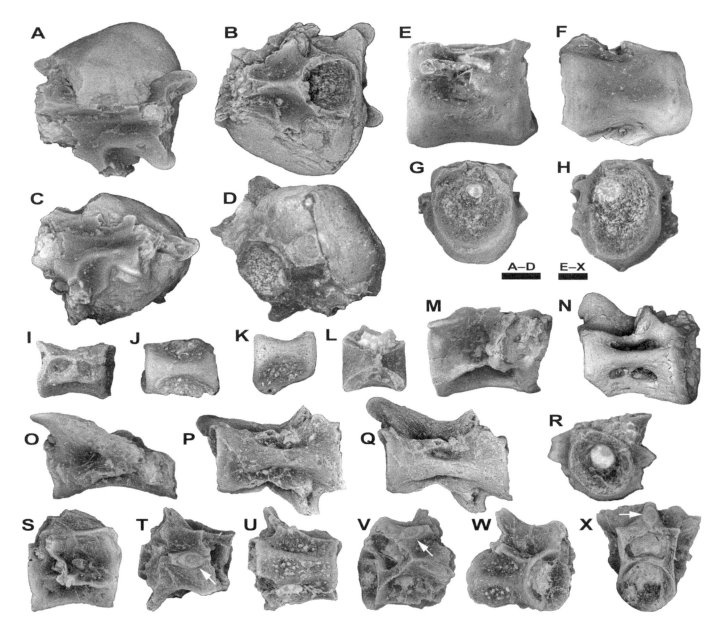

11.3. Trunk vertebrae of Scapherpetontidae Genera and Species Indeterminate from the Dakota Formation, southern Utah. (A–D) moderately complete trunk vertebra, partially encased in nodule, UMNH VP 13112, from UMNH Loc. VP 27: (A) dorsal view with anterior end toward left of figure; (B) ventral view with anterior end toward left of figure; (C) oblique left lateral and dorsoposterior view; (D) oblique right lateral and ventroposterior view. (E–H) trunk centrum, MNA V10233, from MNA Loc. 1067: (E) right lateral view; (F) ventral view with anterior end toward right of figure; (G) anterior view; (H) posterior view. (I, J) trunk centrum, UMNH VP 13093, from UMNH Loc. VP 27: (I) left lateral view; (J) ventral view with anterior end toward left of figure. (K, L) MNA V10235, from MNA Loc. 1067, left half of trunk centrum split longitudinally slightly to left of sagittal plane: (K) ventral view; (L) medial view; both views with anterior end toward right of figure. (M, N) trunk centrum with broken neural arch, MNA V10234, from MNA Loc. 1067: (M) left lateral view; (N) ventral view with anterior end toward left of figure. (O–R) trunk centrum, with broken neural arch, MNA V10236, from MNA Loc. 1067: (O) left lateral view; (P) ventral view with anterior end toward left of figure; (Q) ventral and slightly left lateral view with anterior end toward left of figure; (R) anterior view. (S–X) moderately complete trunk vertebra, probably from anteriormost part of trunk series, MNA V10365, from MNA Loc. 1067: (S) right lateral view; (T) dorsal view with anterior end toward right of figure; (U) ventral view with anterior end toward right of figure; (V) oblique left lateral and anterodorsal view; (W) oblique right lateral and anteroventral view; (X) anterior and slightly dorsal view; arrows point to expanded anterior end of neural crest. Specimens at different magnifications; scale bars = 1 mm.

of the ventral margins of the cotyles. The subcentral keel may be thin and knife-like (Fig. 11.3B) or moderately broad, with the ventral surface either shallowly convex or flattened from side to side (Fig. 11.3N versus P, respectively). Where no keel is developed, the midventral surface of the centrum in transverse profile may be broad and shallowly convex (Fig. 11.3F), narrower and shallowly convex (Fig. 11.3J, K), or moderately broad and shallowly concave (Fig. 11.3U). The ventral and ventrolateral surfaces of the centrum are perforated by small, shallow pits; these pits are abundant in most of the specimens, but are smaller and fewer in UMNH VP 13112. Four specimens (MNA V10234, V10236, V10365, and UMNH VP 13112) preserve portions of the neural arch, and these collectively demonstrate that the roof is low and broad and that the pre- and postzygapophyses are moderate in size, project slightly laterally, and lie nearly flat. In MNA V10365 the anterior end of the neural crest is elaborated into a drumlin-shaped swelling that tapers posteriorly and has a dorsal surface that is flattened and shallowly inclined posteriorly (Fig. 11.3T, X). UMNH VP 13112 is missing the anteriormost end of the neural crest, but the preserved portion just behind the broken surface widens anteriorly; this suggests that the anterior end of the crest also was broad, although probably not to the same extent as in MNA V10365. On both of these specimens, the rest of the neural crest extends posteriorly as a low, knife-like ridge. The preserved base of the neural spine in UMNH VP 13112 indicates that when complete, this process was spike-like, strongly inclined posteriorly, and at least moderately elongate.

Undifferentiated Straight Cliffs Formation MNA Loc. 1260: MNA V10277 and V10366, trunk vertebrae.

Description MNA V10277 and V10366 (Fig. 11.4A, B and C, D, respectively) are the ends of trunk vertebral centra that are similar in size, build, and general details of the cotyle. The shallowly concave cotyle of MNA V10277 suggests that this is an anterior cotyle, whereas the more deeply concave cotyle of MNA V10366 suggests that this is a posterior cotyle. The more nearly complete specimen, MNA V10277, is a robustly built cotyle with an intact rim that measures 1.9 mm wide and 1.8 mm high across the cotylar opening and is broadly orbicular or subtriangular in outline, with the ventral end narrower. Internally the cotyle is shallowly concave, with a thick layer of calcified cartilage lining the inner walls and a moderate-sized notochordal pit just above the center. The ventral and lateral surfaces of the cotyle are badly eroded, but closer to the rim the surface is undamaged, and there is no indication in that region of basapophyses or a subcentral keel. MNA V10366 is about the ventral half of a robustly built cotyle with a maximum width of 1.8 mm. Even though one side of the cotylar rim is broken, it is clear that the cotyle was broad and narrowed ventrally, as in MNA V10277. Compared to MNA V10277, MNA V10366 differs in that the interior of the cotyle is relatively deeper, is less infilled with calcified cartilage, and has a relatively larger notochordal pit. MNA V10366 also preserves the anterior end of a shallow, ridge-like subcentral keel that extends onto the ventral rim of the cotyle.

Basal (?Coniacian) Part of John Henry Member, Straight Cliffs Formation OMNH Loc. v856: OMNH 31143, atlas; OMNH 67058 and 67059, trunk vertebrae.

Description The atlas OMNH 31143 (Fig. 11.4E–H) is a centrum missing the lateral half of the left anterior cotyle and the dorsolateral edge of the right anterior cotyle. The centrum is large, with a maximum length of 3.4 mm, and, when intact, would have been about 4.2 mm wide across the anterior cotyles. The odontoid process is a moderately elongate prong (length 0.8 mm; width across base 1.2 mm) that, in dorsal view, has an unconstricted base and tapers slightly toward its bluntly rounded anterior end; the outline of the process is marred by an anomalous indentation along the anterior half of its right margin. A broad articular surface that is not continuous with the anterior cotyles wraps around the ventrolateral and ventroanterior surfaces of the odontoid process. The anterior cotyles are shallowly concave, are tilted slightly posteriorly and dorsally, and, in anterior outline, are only shallowly depressed dorsoventrally. In posterior outline, the posterior cotyle is weakly compressed laterally and has a broadly convex or nearly flat ventral rim. Internally the posterior cotyle is deeply concave, the inner walls are lined with a thin layer of calcified cartilage, and a large notochordal pit opens in the upper third of the cotyle. The ventral surface of the centrum is shallowly concave and perforated by deep pits and grooves.

Both trunk vertebral specimens are centra that do not preserve rib bearers. OMNH 67058 (Fig. 11.4I–L) has a midventral length of 3.0 mm. It is broken obliquely through the posterior cotyle, but the ventral part of the cotyle is intact. The centrum is weakly constricted medially and is amphicoelous. Both cotyles are deeply concave, are lined with a thin coating of calcified cartilage, and have a large notochordal pit. The broken posterior cotyle exposes the disc-shaped, bony plug that separates the posterior and anterior cotyles. The anterior cotyle is weakly compressed laterally and broadly oval in anterior outline, with the ventral end slightly narrower. The outline of the posterior cotyle is unknown because only its ventral rim is preserved. A prominent subcentral keel extends between the cotylar rims and is flanked on either side by a subcentral foramen. In lateral view, the ventral margin of the keel is ventrally concave and does not extend below the cotyles. In transverse view, the keel is moderately broad and convex. No basapophyses are present. The smaller trunk centrum, OMNH 67059 (Fig. 11.4M–P) is 1.6 mm long and has a more nearly complete posterior cotyle. This specimen resembles OMNH 67058, but differs in that the centrum is relatively shorter (centrum length versus width across anterior cotyle 1.5 versus 2.1), the anterior cotyle is more elliptical (ventral and dorsal portions of equivalent widths), and the ventral surface of the subcentral keel is straighter in lateral profile and relatively broader in ventral view.

Iron Springs Formation MNA Loc. 1230: MNA V10273, trunk vertebra.

Description MNA V10273 (Fig. 11.4Q, R) is the incomplete end of a centrum measuring 1.9 mm wide across the cotyle. MNA V10273 preserves more of the dorsal portion than

Gardner, Eaton, and Cifelli

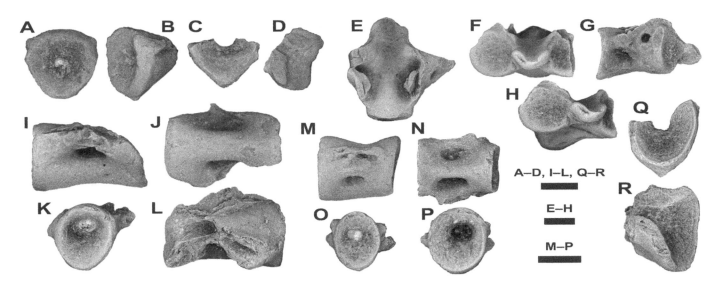

11.4. Vertebrae of Scapherpetontidae Genera and Species Indeterminate from the Straight Cliffs and Iron Springs formations, southern Utah. (A–D) Trunk vertebrae from MNA Loc. 1260, undifferentiated Straight Cliffs Formation: (A, B) anterior cotyle, MNA 10277: (A) anterior view; (B) oblique left lateral and anterior view; (C, D) ventral part of ?posterior cotyle, MNA V10366: (C) posterior view; (D) oblique left lateral and dorsoposterior view. (E–P) Atlas and trunk vertebrae from OMNH Loc. V856, basal (?Coniacian) part of John Henry Member, Straight Cliffs Formation: (E–H) atlantal centrum, OMNH 31143: (E) dorsal view with anterior end toward top of figure; (F) anterior view; (G) right lateral view; (H) oblique anterior, ventral, and right lateral view; (I–L) trunk centrum, OMNH 67058: (I) left lateral view; (J) ventral view with anterior end toward to left of figure; (K) anterior view; (L) oblique right lateral, dorsal, and slightly posterior view; (M–P) trunk centrum, OMNH 67059: (M) right lateral view; (N) ventral view with anterior end toward right side of figure; (O) anterior view; (P) posterior view. (Q, R) ventral part of ?posterior cotyle, MNA V10273, from MNA Loc. 1230, Iron Springs Formation: (Q) posterior view; (R) oblique left lateral and dorsoposterior view. Specimens at different magnifications; scale bars = 1 mm.

does MNA V10366 from the undifferentiated Straight Cliffs Formation, but otherwise is comparable in size and structure to that Coniacian specimen and, probably, also is the posterior end of a centrum.

Remarks The 17 vertebrae from the Dakota Formation can be assigned to the Scapherpetontidae on the basis of the following suite of features: posterior cotyles in atlantes and anterior and posterior cotyles in trunk vertebrae deeply concave, all having inner walls lined with a thin layer of calcified cartilage; notochordal pit present and large in trunk vertebrae and in at least one atlas (condition unknown in remaining atlantes); atlantes have dorsoventrally compressed anterior cotyles with relatively flat articular surfaces and a prominent, thick odontoid process with articular surface on ventral face; and trunk vertebrae lack basapophyses and spinal nerve foramina and bear divergently bicipital rib bearers and, evidently, an elongate neural spine. Differences among the atlantes (e.g., forms of odontoid processes and anterior cotyles) and the trunk vertebrae (e.g., overall build and form of midventral surface) suggest that these specimens probably do not all belong to the same species. None of the atlantes can reliably be associated with any of the trunk vertebrae; in fact, judging by cotylar widths, the trunk vertebral specimens are from individuals at least twice as large as the atlantal specimens. Considering that all the atlantes and most of the trunk vertebrae are from the same locality (MNA 1067/

UMNH VP 27), it is curious that all the atlantes are considerably smaller than the trunk vertebrae; presumably this is an artifact of small sample sizes. Among named scapherpetontids, the atlantes differ from *Piceoerpeton* and more closely resemble *Scapherpeton* and *Lisserpeton* in having a prominent odontoid process and anterior cotyles with nearly flat articular surfaces, whereas the trunk vertebrae differ from *Scapherpeton* and more closely resemble *Lisserpeton* and *Piceoerpeton* in having subcircular cotyles (versus teardrop shaped in *Scapherpeton*). None of those resemblances are particularly compelling, so pending the recovery of more diagnostically informative specimens, we conservatively identify the seven atlantes and 10 trunk vertebrae as Scapherpetontidae indeterminate, with the caveat that several species likely are represented among the specimens.

MNA V10277 and V10366 from the undifferentiated Straight Cliffs Formation can be assigned to the Scapherpetontidae on the basis of their moderately deep, concave cotyles that internally have a thick coating of calcified cartilage and retain open notochordal pits and the lack of basapophyses. Aside from minor differences in the interiors of their cotyles and the presence in MNA V10366 of a low subcentral keel, both specimens are similar enough that they probably pertain to the same taxon.

Most of the features listed above for the vertebrae from the Dakota Formation also can be used to assign the atlas and two trunk vertebral centra from the John Henry Member to the Scapherpetontidae. The two trunk vertebrae are similar enough that they probably belong to the same species, but it is not certain that those specimens can be associated with the atlas. The atlas OMNH 31143 resembles *Scapherpeton* in having less compressed anterior cotyles and an odontoid process with an unconstricted base, but the more tapered (rather than scoop shaped) outline of the odontoid process is more typical of *Lisserpeton*; the latter resemblance is not certain because the odontoid process is slightly deformed. By contrast, the two trunk centra are most similar to *Piceoerpeton*, and differ from trunk vertebrae from the same locality assigned above to *Scapherpeton* sp., in having a moderately wide subcentral keel with a broadly convex ventral surface. OMNH 67059 also has more rounded, *Piceoerpeton*-like cotyles.

Assignment of MNA V10273 from Iron Springs Formation to the Scapherpetontidae is supported by the structure of its anterior cotyle (moderately deeply concave, with thick coating on inner walls of calcified cartilage and open notochordal pits) and no basapophyses. The ventrally tapered outline of the cotylar rim is more suggestive of *Scapherpeton* or *Lisserpeton*, than *Piceoerpeton*.

BATRACHOSAUROIDIDAE Auffenberg, 1958

Remarks The Batrachosauroididae are reliably known by isolated bones and rare skeletons from the Cenomanian–Pliocene of North America and the Campanian–middle Eocene of Europe (Estes, 1981; Naylor, 1981:table 1; Duffaud, 1995). Depending on the affinities of *Prosiren* (Albian, Texas) and an assortment of named and unnamed genera from the Bathonian–Berriasian of Western Europe (Milner, 2000; Evans and McGowan, 2002; Holman, 2006), the record of batrachosauroidids on both continents may extend back even farther. Earlier reports (Nessov, 1981, 1988, 1997) of the family in the Late Cretaceous of Middle Asia can be disregarded because Skutschas (2009) has convincingly argued that the supposed batrachosauroidid *Mynbulakia* was a chimera originally described from isolated bones belonging to two cryptobranchoid genera. Six named genera can confidently be included within the Batrachosauroididae: *Batrachosauroides* (two species; early Eocene–middle Miocene, southern and western United States); *Opisthotriton* (one species; Santonian–late Paleocene) and *Prodesmodon* (one species; late Campanian–?early Paleocene), both in the Western Interior; *Peratosauroides* (one species, Pliocene, California); *Parrisia* (Campanian, New Jersey); and *Palaeoproteus* (two species; late Paleocene–middle Eocene, France and Germany) (Estes, 1981; Naylor, 1981; Denton and O'Neill, 1998; Gardner, 2000a; Milner, 2000). Batrachosauroidid vertebrae

can be recognized by the following combination of features: posterior cotyle in atlantes and anterior and posterior cotyles in trunk vertebrae are subcircular in outline; atlantes have an odontoid process that is greatly reduced in depth and length compared to most other salamanders, the anterior cotyles are subcircular to laterally compressed in anterior outline and deeply concave, the anterior end of the posterior cotyle may be infilled with calcified cartilage that closes the notochordal pit, and the neural arch is robust and moderately tall, with the anterior end often swollen or elaborated with bony processes and bears a low but thick neural crest and a neural spine that projects posteriorly and only slightly dorsally; trunk vertebrae are amphicoelous or opisthocoelous, the notochordal pit is variably closed, spinal foramina are absent, the subcentral keel and posterior basapophyses are variably present, the neural spine projects posterodorsally, is finished in bone, and may be forked, and the rib bearers may be unicipital along the more posterior part of the trunk series. In all North American Late Cretaceous batrachosauroidids for which trunk vertebrae are known, the centra are opisthocoelous and well-developed posterior basapophyses are consistently present.

Batrachosauroidids are known from most of the formations in southern Utah included in this study. *Opisthotriton* is identified from the Santonian to late Campanian, *Prodesmodon* is known only from the late Campanian, and a variety of indeterminate batrachosauroidids (including at least two genera) are documented from the late Cenomanian to Santonian.

OPISTHOTRITON Auffenberg, 1961
OPISTHOTRITON sp.
Voucher Specimens (Fig. 11.5A–R)

Postbasal (Santonian) part of John Henry Member, Straight Cliffs Formation UMNH Loc. VP 424: UMNH VP 19189–19191, atlantes; UMNH VP 19192, trunk vertebra; UMNH Loc. VP 569: UMNH VP 19193, atlas.

Wahweap Formation UMNH Loc. VP 130: UMNH VP 19194–19198, atlantes; UMNH VP 19199, 19200, trunk vertebrae.

Kaiparowits Formation OMNH Loc. V5: OMNH 67060, atlas; OMNH Loc. V9: OMNH 67061–67066, atlantes; OMNH 67067, trunk vertebra. OMNH Loc. V61: OMNH 67068–67070, trunk vertebrae; UMNH Loc. VP 51: UMNH VP 13275 trunk vertebra.

Descriptions All of the available specimens are centra that, at best, preserve no more than the base of the neural arch walls. Although incomplete, these specimens are comparable in size and structure to *Opisthotriton* atlantes of Campanian–Paleocene age reported from elsewhere in the Western

Interior (Estes, 1964:fig. 38f, g, 1975:fig. 2C, D, 1976:text-fig. 4A, B; Naylor, 1979:fig. 5D–F; Gardner, 2000a:figs.11–2, 11–3; Peng, Russell, and Brinkman, 2001:pl. 5, figs. 7, 8). The three specimens figured here (Fig. 11.5A–H, M–O) were selected to show the range of variation in the Utah samples. The atlantal centra are moderate in size, ranging from about 2.2–3.5 mm across the anterior cotyles and 1.3–1.8 mm in total length. In dorsal or ventral outline, the centrum is relatively short and broad. The anterior cotyles are deeply concave, subcircular in anterior outline, and face anteriorly and slightly laterally. Variation is seen in the outline of the anterior cotyles, which range from slightly dorsoventrally depressed to slightly mediolaterally compressed, and in the degree to which the anterior cotyles are angled laterally. The odontoid process is level with about the dorsoventral midpoint of the anterior cotyles and, as in all batrachosauroidids, it is anteriorly short and dorsoventrally shallow. The odontoid process varies from a prong-like structure to a horizontal bony ridge or shelf (cf., Fig. 11.5A, E, M). Typically the ventral surface and margins are smooth, but in some specimens having a better developed odontoid process, a small and slightly projecting or flattened articular surface may occur on the anteromedian face of the process. The posterior cotyle is consistently subcircular in outline. The interior of the posterior cotyle may be deeply concave, with little or no calcified cartilage lining the walls and retains an open notochordal pit (Fig. 11.5C, G) or it may be shallowly concave, with the anterior infilled with calcified cartilage and the notochordal pit is closed (Fig. 11.5O).

One specimen (UMNH VP 19192; Fig. 11.5I–L) consists of a relatively complete centrum and neural arch, but the rest of the specimens (Fig. 11.5P–R) are centra with varying amounts of the neural arch attached. Insofar as can be determined, these specimens compare favorably with more nearly complete trunk vertebrae of *Opisthotriton* reported elsewhere (Estes, 1964:fig. a–e, 1975:fig. 2C, D, 1976:text-fig. 4C–F; Gardner, 2000a:fig. 11–4) in size and in having opisthocoelous centra that are moderately elongate and hourglass shaped, with cotyles that are subcircular in outline, and bear moderately well-developed posterior basapophyses. As in previously described *Opisthotriton* trunk vertebrae, variation is seen in the amount of calcified cartilage infilling the anterior cotyle and closure of the anterior notochordal pit (varies from a thick layer of calcified cartilage lining the interior walls and an open notochordal pit to cotyle completely infilled with calcified cartilage and notochordal pit closed; cf. Fig. 11.5R versus P, Q, respectively) and, although not obvious in specimens figured here, also in the relative depth of the subcentral keel.

Remarks Two species of *Opisthotriton* have been proposed: the type *O. kayi* Auffenberg, 1961, which is known

by the holotype trunk vertebra, plus dozens of isolated vertebrae and skull bones and several skeletons (Auffenberg, 1961; Estes, 1964, 1969b, 1975, 1976, 1981; Naylor, 1979; Gardner, 2000a), and *O. gidleyi* Sullivan, 1991, which is known only from a poorly preserved skull and was argued by Gardner (2000a) to be a nomen dubium. *Opisthotriton* has been reported from over 20 formations of late Santonian to Paleocene age in the Western Interior (Estes, 1981; Naylor, 1981; Gardner, 2000a). Specimens reported here confirm a preliminary report (Eaton, Cifelli et al., 1999:table 4) of *Opisthotriton* in the Wahweap Formation and also document the presence of the genus in the Kaiparowits Formation and in the postbasal (Santonian) part of the John Henry Member of the Straight Cliffs Formation.

Variation reported above in the atlantal centra and trunk vertebrae from Utah is consistent with that seen for *Opisthotriton* elsewhere in the Western Interior. Based largely on specimens from the Milk River, Dinosaur Park, Lance, and Hell Creek formations, Gardner (2000a) identified two atlantal morphs for *O. kayi* that differed in absolute size, structure of the neural arch, and whether the notochordal pit was open or closed. Although the Utah specimens exhibit a fairly narrow size range and none preserves an intact neural arch, they too vary in having the notochordal pit either open or closed (Fig. 11.5C, G versus O). In the samples that Gardner (2000a) examined, the odontoid process was similarly variable in both atlantal morphs; this also appears to be the case for the Utah specimens.

PRODESMODON Estes, 1964
PRODESMODON sp.
Voucher Specimens (Fig. 11.5S–AA)

Kaiparowits Formation OMNH Loc. v61: OMNH 23951 and 67071, trunk vertebrae; OMNH 67072, dentary.

Descriptions Two incomplete trunk vertebrae are available. OMNH 23951 (Fig. 11.5S–V) consists of a nearly complete centrum, whereas OMNH 67071 (Fig. 11.5W–Z) consists of about the anterior three quarters of the centrum and adjacent part of the neural arch wall and roof on the right side. OMNH 23951 has a total centrum length (including the anterior cotyle) of 2.9 mm. OMNH 67071 has a preserved length of 3 mm and, when complete, probably was about 3.5 mm long. In both specimens, the centrum is relatively narrow and shallow. The most obvious feature in both is that the anterior cotyle is completely infilled with a ball-shaped mass of calcified cartilage that projects anteriorly well beyond the cotylar rim to form a bony condyle. The anterior face of the condyle is convex and smooth in OMNH 23951, but slightly flattened and indented by a shallow, median concavity in

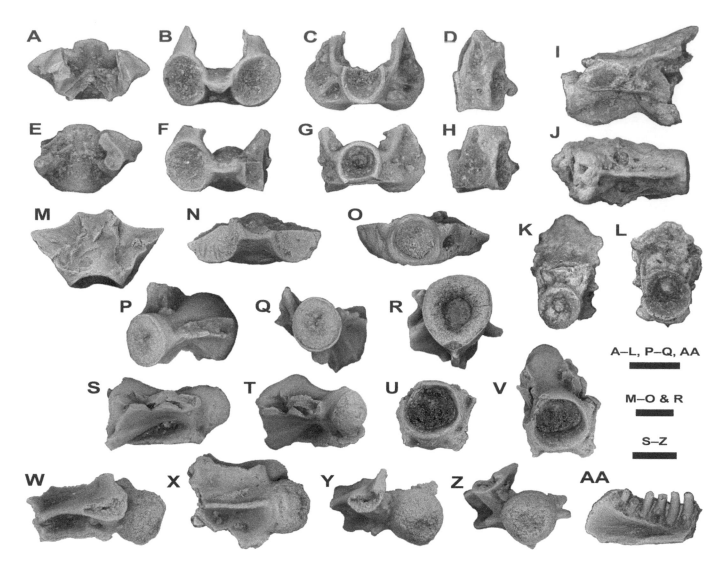

11.5. Atlantes and trunk vertebrae of *Opisthotriton* sp. and trunk vertebrae and dentary of *Prodesmodon* sp. (Batrachosauroididae) from southern Utah. (A–L) *Opisthotriton* sp., from UMNH Loc. VP 424, postbasal (Santonian) part of John Henry Member, Straight Cliffs Formation: (A–D) atlantal centrum preserving bases of neural arch wall (note relatively well-developed odontoid process), UMNH VP 19189: (A) dorsal view with anterior end toward top of figure; (B) anterior view; (C) posterior view; (D) right lateral view; (E–H) atlantal centrum missing lateral part of left anterior cotyle (note reduced odontoid process), UMNH VP 19190: (E) dorsal view with anterior end toward top of figure; (F) anterior view; (G) posterior view; (H) right lateral view; (I–L) trunk vertebra, UMNH VP 19192: (I) left lateral view; (J) ventral view with anterior end toward right of figure; (K) anterior view (note thick ring of calcified cartilage around and inside anterior cotylar rim and small notochordal pit in center); (L) posterior view. (M–R) *Opisthotriton* sp., from Kaiparowits Formation: (M–O) atlantal centrum missing upper part of both anterior cotyles (note extremely reduced odontoid process is little more than a horizontal ridge), OMNH 67066, from OMNH Loc. V9: (M) dorsal view with anterior end toward top of figure; (N) anterior view; (O) posterior view; (P, Q) trunk vertebral centrum (note calcified cartilage infills entire anterior cotyle with only a concave depression in center), UMNH VP 13275, from UMNH Loc. VP 51: (P) oblique anterodorsal and left lateral view; (Q) anterior view; (R) trunk vertebral centrum (note moderately thick ring of calcified cartilage around and inside anterior cotylar rim and large notochordal pit in center), OMNH 67067, from OMNH Loc. V9, anterior view. (S–AA) *Prodesmodon* sp., from OMNH Loc. V61, Kaiparowits Formation: (S–V) trunk centrum, OMNH 23951: (S) right lateral view; (T) oblique right lateral, anterior, and slightly dorsal view; (U) posterior view; (V) oblique dorsoposterior and slightly left lateral view, showing large plug of calcified cartilage infilling anterior end of posterior cotyle and smooth inner walls adjacent to rim; (W–Z) trunk vertebrae preserving anterior three quarters of centrum and adjacent part of neural arch wall and roof on right side, OMNH 67071: (W) right lateral view; (X) ventral view with anterior end toward right of figure; (Y) oblique anterior and right lateral view; (Z) anterior view; (AA) anteriorly and posteriorly incomplete left dentary preserving posterior six tooth positions, but no intact tooth crowns, OMNH 67072, lingual view. Specimens at different magnifications; scale bars = 1 mm.

OMNH 67071. OMNH 23951 provides details about the posterior part of the centrum. In ventral aspect, the centrum widens posteriorly and, in posterior view, the posterior cotyle is subcircular in outline. In contrast to other salamander postatlantal vertebrae reported herein, the interior of the posterior cotyle is shallowly excavated because all but about the posterior quarter of the cotyle is infilled with a large plug of calcified cartilage. The posterior surface of this infill is broad, smooth, and shallowly concave, with no trace of a notochordal pit. The interior walls of the posterior cotyle

Gardner, Eaton, and Cifelli

are exposed only as an anteroposteriorly thin band and their surfaces are smooth. In lateral view, OMNH 23951 lacks spinal foramina, but bears a pair of relatively prominent posterior basapophyses that extend ventroposteriorly slightly past the posterior cotylar rim. Both specimens bear a shallow, subcentral keel that projects ventrally slightly below the level of the cotylar rims. The preserved, anterior portion of the subcentral keel in OMNH 67071 is a narrow flange that, in lateral view, is shallowly concave dorsally. OMNH 23951 preserves an intact subcentral keel that differs from OMNH 67071 in being slightly wider along the posterior third of the centrum and more anteriorly the keel expands laterally into more of a broad, flattened plate with a bluntly rounded anterior end. The rib bearers are broken on both specimens. On its right side, OMNH 67071 preserves most of the anterior alar process, which is developed as a laterally narrow and horizontally projecting shelf that extends posteriorly from the base of the broken prezygapophyseal process along the neural arch wall. In anterior view, the preserved portion of the neural arch on OMNH 67071 shows that, when intact, the arch would have been low and broad.

OMNH 67072 (Fig. 11.5AA) is an incomplete left dentary that preserves the posteriormost six tooth positions and adjacent part of the area for attachment of the postdentary bones. Comparisons with an intact dentary figured by Estes (1964:fig. 68) indicates that OMNH 67072 is missing about the anterior half of the tooth-bearing portion of the bone and the posterior half of the area for attachment of the postdentary bones. The specimen is small (preserved length ~2.4 mm), but relatively stout. In labial or lingual outline the specimen is triangular; it is deepest immediately behind the tooth row and rapidly becomes shallower anteriorly, the occlusal edge above the tooth row is broadly concave, and the ventral edge is relatively straight. The labial surface is smooth, with no indication of external nutritive foramina. In lingual view, the prominent Meckelian trough resembles a posteriorly open V that extends anteriorly below the tooth row to the level of the fifth locus from the posterior end. A low ridge extends along the upper margin of the Meckelian trough, below the tooth row. The ventral edge of the dentary is intact below the Meckelian trough, and in that region, it projects lingually as a bony ridge or flange (the ventromedial ridge of Estes, 1964:139). Anterior to the Meckelian trough, the Meckelian canal is closed and a low, medially narrow subdental shelf extends forward for a short distance to the broken anterior end. The six teeth are relatively closely spaced, moderately pleurodont in their attachment, and nonpedicellate. The shafts are conical, taper slightly toward the apex, and are weakly curved both distally and lingually. No tooth has an intact crown, but each has a lingual replacement pit in its base.

Remarks The type and only species *Prodesmodon copei* is known by isolated, but distinctive, atlantes, trunk vertebrae, dentaries (originally assigned to *Cuttysarkus mcnallyi*), and vomers (Estes, 1964, 1981; Naylor, 1979; Gardner, 2000a). The two broken trunk vertebrae reported above compare favorably with *P. copei* trunk vertebrae from the Lance Formation and Bug Creek Anthills (Estes, 1964:fig. 42e–g; Gardner, 2000a:fig. 11–6) in having elongate and narrow centra that are strongly opisthocoelous, with the notochordal pit closed in the posterior cotyle and prominent posterior basapophyses, and the neural arch low and broad. The anteriorly broadening, plate-like subcentral keel on OMNH 23951 has not been reported in any other *Prodesmodon* trunk vertebrae. This specimen may be from the sacral region because in salamanders a broad subcentral keel is typical of vertebrae from that region (Naylor, 1978). The preserved portion of the broken dentary is similar to referred dentaries of *P. copei* from the Lance Formation (Estes, 1964:fig. 68; Naylor, 1979:fig. 1) in such features as its triangular outline, lingually directed ventral edge of the bone below the Meckelian canal, and structure of the teeth. Considering how incomplete the trunk vertebrae and dentary are from the Kaiparowits Formation and that *P. copei* is best known from younger (Lancian and ?Puercan) specimens, we conservatively identify the specimens from Utah as *Prodesmodon* sp. Regardless of their specific identity, these specimens are the first reported occurrence of *Prodesmodon* in the Kaiparowits Formation and only the second report of the genus from the Judithian; the other report is of undescribed specimens from the Mesaverde Formation of Wyoming (DeMar and Breithaupt, 2006, 2008).

Genus and Species A
Voucher Specimens (Fig. 11.6A–M)

Dakota Formation MNA Loc. 1067: MNA V10367, atlas; MNA V10237, trunk vertebra.

Descriptions MNA V10367 (Fig. 11.6A–G) is a small atlantal centrum (midline length 1.3 mm; width across anterior cotyles 2.0 mm) that is missing the dorsolateral portion of the right anterior cotyle and smaller amounts of the rim around the left anterior cotyle. The odontoid process is dorsoventrally shallow and relatively short anteroposteriorly (midline length less than width across base). The base of the odontoid process is broad and approximately in line with the dorsoventral midpoint of the anterior cotyles. In dorsal or ventral outline, the odontoid process is broadly acute. The dorsal surface of the process is shallowly concave, whereas the ventral surface is shallowly convex and smooth. The anterior cotyles are moderately deeply concave, extend slightly posteriorly, and, when intact, would have been subcircular and slightly compressed mediolaterally in anterior outline. The posterior cotyle is circular in posterior outline, deeply concave, retains an open notochordal pit in the upper half of the centrum, and the inner walls along the anterior half are lined with thick coating of calcified cartilage.

MNA V10237 (Fig. 11.6H–M) is a small trunk centrum (centrum length 1.4 mm) that is slightly crushed, preserves only the bases of the neural arch wall, and has some matrix adhering to the ventrolateral surface on the left side. The centrum is relatively elongate (ratio of centrum length versus maximum width across anterior cotyle 2.0), strongly constricted medially and distinctly hourglass shaped in ventral view, has smooth ventral and lateral surfaces, and lacks basapophyses. The centrum is weakly opisthocoelous: the anterior cotyle is partially infilled with a plug-like mass of calcified cartilage that is perforated by a dorsoventrally compressed notochordal pit, whereas the posterior cotyle is deeply concave, has a circular notochordal pit in its center, and its inner walls are lined with only a thin coating of calcified cartilage. The anterior cotyle is subcircular in outline. The rim of the posterior cotyle is fractured and slightly distorted, but in life it probably also was subcircular. The subcentral keel is moderately narrow (i.e., not knife-like) and has a convex ventral surface, is limited to the median third of the centrum, and is only moderately deep (i.e., ventral edge in line with ventral rim of anterior and posterior cotyles). The base of a posterolaterally projecting rib bearer is preserved on the left side. Because the more dorsal portion of the vertebra is missing, it is unknown whether a dorsal rib bearer also was present in life. It is possible that only one rib bearer was present because in batrachosauroidids known from articulated skeletons, such as *Opisthotriton*, vertebrae from the posterior portion of the trunk series have only one rib bearer. Portions of the anterior and posterior alar processes are preserved on both sides, where they extend in a horizontal plane between the basal portion of the preserved rib bearer and the lateral surfaces of the centrum; when viewed in ventral aspect the former process is little more than a ridge, whereas the latter is a triangular flange.

Remarks The combination of a reduced odontoid process, anterior cotyles that are subcircular and concave, and a thick lining of calcified cartilage limited to the anterior portion of the posterior cotyle in the atlas and the weakly opisthocoelous trunk centrum are all diagnostic for the Batrachosauroididae. Compared to trunk vertebrae of named batrachosauroidids, MNA V10237 has a unique combination of three features that suggest it represents a new taxon: centrum weakly opisthocoelous; posterior basapophyses absent; and subcentral keel weakly developed. The North American batrachosauroidids *Opisthotriton* and *Parrisia* also have weakly

opisthocoelous trunk centra (versus strongly opisthocoelous in *Prodesmodon* or amphicoelous in other genera), but differ from MNA V10237 in having prominent posterior basapophyses and a deeper subcentral keel (Denton and O'Neill, 1998:fig. 4; Gardner, 2000a:fig. 11–4). The atlantal centrum MNA V10367 is less distinctive; in fact, this specimen resembles the centrum portion of atlantes of *Parrisia* (Denton and O'Neill, 1998:fig. 2) and morph I atlantes of *Opisthotriton* (Gardner, 2000a:fig. 11–2) in having subcircular anterior cotyles that are only weakly compressed mediolaterally and in having an open notochordal pit in the posterior cotyle. MNA V10237 and V10367 are from similarly small individuals and were recovered from the same locality (MNA 1067) and thus may be from the same species. Although we suspect that these two vertebrae belong to a previously unrecognized batrachosauroidid taxon, additional vertebral specimens – ideally ones preserving the neural arch – are needed to confirm this proposed association and to establish the distinctiveness of the Dakota Formation batrachosauroidid.

Genus and Species B
Voucher Specimen (Fig. 11.6N–R)

Smoky Hollow Member, Straight Cliffs Formation OMNH Loc. V4: OMNH 67073, trunk vertebra.

Description OMNH 67073 (Fig. 11.6N–R) is a small (centrum length 1.7 mm) trunk vertebral centrum that is broken anterodorsally–posteroventrally through the posterior cotyle, but preserves the ventral margin of that cotyle intact. The centrum is relatively robust in build, slightly medially compressed, and moderately elongate (ratio of centrum length versus maximum width across anterior cotyle 2.1). The anterior half of the ventral surface is flattened and tapers posteriorly into a moderately narrow and extremely shallow subcentral keel that extends along the posterior half of the centrum, almost to the rim of the posterior cotyle. The left and right sides of the centrum are each perforated by four prominent, sediment-infilled pits or, perhaps, foramina. The rib bearers are broken off at their bases; although it is evident that these processes were located high up on the lateral sides of the centrum, it is uncertain whether they were single or paired. There is no indication of posterior basapophyses, although if small ones were present they might have been lost along with the missing portions of the lateral walls of the posterior cotyle. The centrum is weakly opisthocoelous. The anterior cotyle is largely infilled with calcified cartilage and the notochordal pit is closed. The central part of the calcified cartilage infill is shallowly concave; toward the periphery, the calcified cartilage extends anteriorly almost to the level of the cotylar rim and its anterior surface is flattened. The

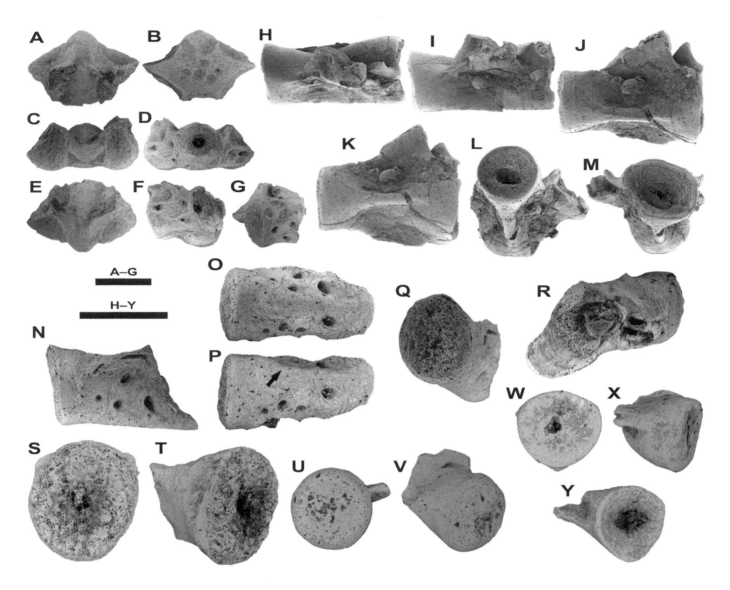

11.6. Vertebrae of Batrachosauroididae from the Dakota, Straight Cliffs, and Iron Springs formations, southern Utah. (A–M) Genus and Species A, from MNA Loc. 1067, Dakota Formation: (A–G) atlantal centrum, missing dorsolateral edge of right anterior cotyle, MNA V10367: (A) dorsal view with anterior end toward top of figure; (B) ventral view with anterior end toward top of figure; (C) anterior view; (D) posterior view; (E) oblique dorsoanterior view; (F) oblique left lateroposterior and slightly ventral view; (G) left lateral and slightly dorsal view; (H–M) trunk centrum, MNA V10237: (H) left lateral view; (I) left lateral and slightly ventral view; (J) oblique left lateral and ventral view; (K) ventral view with anterior end toward left of figure; (L) anterior and slightly ventral view; (M). posterior and slightly ventral view. (N–R) Genus and Species B, trunk centrum missing lateral and dorsal portions of posterior cotyle, OMNH 67073, from OMNH Loc. V4, Smoky Hollow Member, Straight Cliffs Formation: (N) left lateral view with anterior end toward left of figure; (O) ventral view with anterior end toward left of figure; (P) ventral and slightly right lateral view with anterior end toward left of figure, showing shallow subcentral keel marked by an arrow; (Q) oblique anterior and slightly ventral and left lateral view; (R) oblique right lateral and dorsoposterior view, showing interior of posterior cotyle. (S–X) Genera and species indeterminate, anterior ends of trunk vertebrae, from OMNH Loc. V27, postbasal (Santonian) part of John Henry Member, Straight Cliffs Formation: (S, T) OMNH 67074: (S) anterior view; (T) oblique right lateral and anterior view; (U, V) OMNH 67075: (U) anterior view; (V) oblique right lateral and dorsoanterior view; (W, X) OMNH 67076: (W) anterior view; (X) oblique right lateral and slightly anterior and dorsal view. (Y) Genus and Species Indeterminate, anterior end of trunk vertebra, MNA V10272, oblique right lateral and anterior view, from MNA Loc. 1230, Iron Springs Formation. Specimens at different magnifications; scale bars = 1 mm.

anterior cotyle is subcircular and slightly compressed laterally in anterior outline. The outline of the posterior cotyle is unknown because its lateral and dorsal edges are missing. The preserved ventral portion of the posterior cotyle extends ventrally below the level of the ventral rim of the anterior cotyle. Consequently, in lateral view the ventral surface of the cotyle descends posteroventrally at a shallow angle. Inside the posterior cotyle, the preserved, ventral portion of the inner walls in the posterior half are smooth; more anteriorly, the inner walls are covered with a thick layer of calcified cartilage and a notochordal pit opens in the dorsal half.

Remarks The weakly opisthocoelous nature of its centrum supports assigning OMNH 67073 to the Batrachosauroididae. OMNH 67073 resembles the above-described trunk vertebra (MNA V10237) and differs from other batrachosauroidids in having a weakly opisthocoelous centrum, a weakly developed subcentral keel, and posterior basapophyses either absent or weakly developed. However, OMNH 67073 differs from MNA V10237 in at least three notable features: (1) anterior cotyle almost completely infilled with calcified cartilage, to the extent that the notochordal pit is no longer present (versus anterior cotyle partially infilled with thick layer of calcified cartilage lining inner walls and notochordal pit present); (2) inner walls of posterior cotyle lined with thick layer of calcified cartilage along anterior half and smooth along posterior half (versus entire walls covered with thin layer of calcified cartilage); and (3) centrum weakly medially compressed and subcentral keel limited to posterior half of centrum (versus centrum strongly medially compressed and subcentral keel extends along median third of centrum). Differences in medial compression and extent of the subcentral keel could be due to differences in the relative positions each vertebra occupied along the vertebral column, but differences in the first two features are sufficiently great that they are better interpreted as indicating that OMNH 67073 pertains to a distinct batrachosauroidid taxon.

Genera and Species Indeterminate
Voucher Specimens (Fig. 11.6S–Y)

Postbasal (Santonian) Part of John Henry Member, Straight Cliffs Formation OMNH Loc. v27: OMNH 67074–67076, trunk vertebrae.

Iron Springs Formation MNA Loc. 1230: MNA V10272, trunk vertebra.

Descriptions The three John Henry Member specimens are the anterior ends of trunk vertebral centra that resemble one another in small size, in lacking anterior basapophyses, and in having the anterior cotyle completely infilled with calcified cartilage. OMNH 67074 (Fig. 11.6S, T) is the largest specimen, with a maximum width of 1.1 mm across the anterior cotyle. The cotyle is slightly oval in outline, with the ventral end narrower. The interior of the cotyle is filled with calcified cartilage, to the extent that toward its periphery the infilling extends anteriorly beyond the cotylar rim; in that region, the anterior surface of the infilling is flat, whereas the central part of the infilling is moderately concave and no notochordal pit is present. The ventral surface of the centrum is too damaged to determine whether a subcentral keel was present. OMNH 67075 and 67076 (Fig. 11.6U, V and W, X, respectively) are slightly smaller, with each being 0.9 mm wide across their anterior cotyles. As in OMNH 67074, the calcified cartilage infilling of the anterior cotyle in OMNH 67075 also extends anteriorly beyond the cotylar rim and no notochordal pit is present, but the anterior surface is more convex with

only a shallow central concavity. The anterior cotyle is more nearly subcircular in outline and no subcentral keel is present. In OMNH 67076 the anterior limit of the calcified cartilage infilling is approximately level with the cotylar rim and its anterior surface is nearly flat, with a small pit in the upper half of the cotyle. It is not clear whether this pit is natural or the result of a break in the surface of the calcified cartilage. The cotyle is oval in anterior outline, with the ventral end narrower. A shallow, thin subcentral keel arises just behind the ventral rim of the cotyle and extends back to the broken end of the centrum.

MNA V10272 (Fig. 11.6Y) from Iron Springs Formation is the anterior end of a small (maximum width across anterior cotyle 0.9 mm) trunk vertebral centrum. Although broken posteriorly at about the midpoint of the centrum, it is evident that the centrum was medially constricted and no anterior basapophyses were present. The anterior cotyle is broadly orbicular in anterior outline, with the ventral rim subpointed. The interior of the cotyle is thickly infilled with calcified cartilage; the anterior surface of the calcified cartilage is convex and lies well behind the cotylar rim. A relatively large, dorsoventrally compressed notochordal pit opens in the center of the cotyle. Although not visible in Fig. 11.6Y, an extremely shallow, narrow, and broadly V-shaped subcentral keel extends posteriorly from behind the ventral rim of the cotyle.

Remarks The three John Henry Member specimens can be assigned to the Batrachosauroididae on the basis of their anterior cotyles that are completely infilled with calcified cartilage; however, they are too fragmentary to be identified more precisely. Considering that the form of the calcified cartilage infilling differs among the three specimens, it is not certain that they all belong to the same species. Among other batrachosauroidids, OMNH 67074 and 67075 resemble some *Opisthotriton* and *Parrisia* trunk vertebrae in that the calcified cartilage infill extends anteriorly slightly beyond the cotylar rim and that the anterior surface of the infill is convex around the periphery and concave in the center. OMNH 67076 resembles OMNH 67073 (Batrachosauroididae Genus and Species B) from the Smoky Hollow Member in having the anterior face of the calcified cartilage infill being level with the cotylar rim, but differs in that the anterior face is flat and relatively smooth (versus shallowly concave and cancellous) and in having a subcentral keel that extends forward to and projects below the level of the anterior cotyle. All three specimens differ from MNA V10237 (Batrachosauroididae Genus and Species A) from the Dakota Formation and from MNA V10272 from the Iron Springs Formation in having the anterior cotyle completely infilled with calcified cartilage and in lacking an open notochordal pit.

The thick layer of calcified cartilage partially infilling the anterior cotyle of MNA V10272 from Iron Springs Formation is

Gardner, Eaton, and Cifelli

diagnostic for the Batrachosauroididae. Although this specimen is too fragmentary to be identified more precisely, it is worth noting that it differs from all of the other indeterminate batrachosauroidid trunk centra reported above in that the anterior surface of the calcified cartilage infill is convex and lies behind the cotylar rim, an open notochordal pit is retained, and a weakly developed subcentral keel is present.

SIRENIDAE Gray, 1825

Remarks The Sirenidae are Late Cretaceous to Recent salamanders that include three North American genera: *Habrosaurus* (two species; Santonian–middle Paleocene, Western Interior); *Pseudobranchus* (four species; Pliocene–Recent); and *Siren* (six species; Eocene–Recent) (Estes, 1981; Gardner, 2003a; Holman, 2006). The monotypic *Paleoamphiuma* from the Eocene of Wyoming was described as an amphiumid (Rieppel and Grande, 1998), but it also is a sirenid (Gardner, unpubl. obs., 2008). Several Gondwanan taxa–*Noterpeton* (one species; Maastrichtian, Bolivia), *Kababisha* (two species; Cenomanian, Sudan), and cf. *Kababisha* (Cenomanian, Morocco)–known by unusual vertebrae, jaws, and braincases (Rage, Marshall, and Gayet, 1993; Evans, Milner, and Werner, 1996; Rage and Dutheil, 2008) have been interpreted as sirenids (Evans, Milner, and Werner, 1996), but that assignment has been questioned (Gardner, 2003a). Unequivocal sirenids have distinctive trunk vertebrae that, for example, have amphicoelous centra with open notochordal pits, bear a Y-shaped configuration of crests on the dorsal surface of the neural arch (comprising a median neural crest anteriorly and, more posteriorly, a pair of aliform processes that diverge lateroposteriorly and extend onto the postzygapophyseal processes) and prominent anterior basapophyses, have unicipital rib bearers on all but the anteriormost trunk vertebrae, and have spinal foramina. Those features are not especially useful for the Utah samples because no sirenid trunk vertebrae were identified. We did, however, identify five atlantal centra in three formations of Santonian–late Campanian age that are diagnostic for *Habrosaurus* sp.

HABROSAURUS Gilmore, 1928
HABROSAURUS sp.
Voucher Specimens (Fig. 11.7)

Postbasal (Santonian) Part of John Henry Member, Straight Cliffs Formation UMNH Loc. VP 424: UMNH VP 19201–19203, atlantes.

Iron Springs Formation MNA Loc. 1230: MNA V10275, atlas.

Kaiparowits Formation OMNH Loc. V9: OMNH 67077, atlas.

Descriptions The five specimens are all atlantal centra that are essentially identical to *Habrosaurus* atlantes described previously from the Lance, Hell Creek, and Dinosaur Park formations (Estes, 1964; Gardner, 2003a). Consequently, the following description is brief and is based on the best-preserved and largest of the Utah specimens: UMNH VP 19201 (Fig. 11.7A–F) from the Santonian part of the John Henry Member. UMNH VP 19201 is broken through the bases of the neural arch walls and is missing the lower right portion of the posterior cotylar rim. The specimen is 3.3 mm long and 4.5 mm wide across the anterior cotyles. The centrum is robust and moderately flat. In anterior view, the anterior cotyles are dorsoventrally flattened and broad, with the dorsal edge nearly flat and horizontal and the ventral edge shallowly convex and deepest about two thirds of the distance laterally across the cotyle. The surface of the anterior cotyles is flat and tilted slightly dorsally. In dorsal view, the anterior cotyles project laterally. The odontoid process is relatively deep (i.e., subequal to the height of the anterior cotyles) and, in dorsal view, it is moderately broad and elongate, tapers slightly and has a broadly rounded anterior end. The anterior articular surface on the atlas for contact with the skull is continuous across the anterior cotyles and the lateral and anterior surfaces of the odontoid process. The posterior cotyle is subcircular and slightly dorsoventrally depressed in posterior outline. The interior of the posterior cotyle is deeply concave, the inner walls are lined with a thin coating of calcified cartilage, and a large notochordal pit opens in the center of the cotyle. On either side, a pair of horizontal flanges extends along the ventrolateral and dorsolateral margins of the centrum, between the rims of the anterior and posterior cotyles. The posterior half of the ventral surface of the centrum bears a shallow, flattened, subcentral keel. More anteriorly and to either side of the keel, the ventral surface of the centrum is shallowly concave and perforated by scattered pits. The spinal foramen extends through the base of the neural arch and is enclosed by bone. The other two atlantal centra from the same locality (UMNH VP 19202, 19203; not figured) are similar in size to UMNH VP 19201, but are less complete and have relatively shorter odontoid processes.

MNA V10275 (Fig. 11.7G, H) from the Iron Springs Formation preserves the left anterior quadrant of an atlantal centrum and, when intact, was probably about three quarters the size of UMNH VP 19201. The preserved portions of MNA V10275 resemble those of UMNH VP 19201, except the odontoid process is relatively longer and distinctly shallower than the anterior cotyle. The smallest of the five specimens is OMNH 67077 (Fig. 11.7I, J) from the Kaiparowits Formation, which is 2.0 mm wide across its anterior cotyles and has a preserved length of 1.3 mm; when complete, the length was probably closer to 1.5 mm. Compared to the other four specimens, the odontoid process in OMNH 67077 is more tapered or subtriangular in dorsal or ventral outline.

Remarks The five atlantal centra are diagnostic for *Habrosaurus* in the form of their anterior cotyles and odontoid process and in having a continuous articular surface extending across the anterior cotyles and wrapping around the odontoid process (Estes, 1964; Gardner, 2003a). Two species of *Habrosaurus* are recognized in the North American Western Interior on the basis of dental characters: the type species *H. dilatus* from the late Maastrichtian–middle Paleocene and *H. prodilatus* from the late Campanian (Gardner, 2003a). We do not know whether atlantes differ between the two congeners because although numerous atlantal centra are known for *H. dilatus* (Estes, 1964:fig. 37a; Gardner, 2003a:text-fig. 10), currently only one poorly preserved specimen is known for *H. prodilatus* (Gardner, 2003a:text-fig. 12L, M). Consequently, the specimens from Utah can only be identified to genus. Nevertheless, these specimens are significant because they (1) confirm the presence of *Habrosaurus* in the Kaiparowits Formation (*H. dilatus* previously was listed by Eaton, Cifelli et al., 1999, in a preliminary faunal list for the unit), (2) are the first records for the genus in both the Iron Springs formations and the John Henry Member of the Straight Cliffs Formation, and (3) the three atlantes from the John Henry Member and possibly the one atlas from the Iron Springs Formation extend the temporal range of *Habrosaurus* and of unequivocal sirenids back to the Santonian from their previously oldest reported occurrences (middle or late Campanian) in the Dinosaur Park Formation of Alberta, Judith River Formation of Montana, and Mesaverde Formation of Wyoming (Gardner, 2003a; DeMar and Breithaupt, 2006, 2008; Rogers and Brady, 2010). Munk (1998) reported *Habrosaurus* from UMNH Loc. VP 61, which may be in the Wahweap Formation (early or middle Campanian), but that identification cannot be corroborated because the tooth plate (UMNH VP 7365) upon which that identification was made has not been figured or examined by us.

Family Indeterminate
New Genus and Species
Voucher Specimens (Fig. 11.8)

Smoky Hollow Member, Straight Cliffs Formation MNA Loc. 995: MNA V10251–V10253, V10255, trunk vertebrae, MNA V10254, atlas; UMNH Loc. VP 129 (=MNA Loc. 995): UMNH VP 13431, trunk vertebra.

Postbasal (Santonian) Part of John Henry Member, Straight Cliffs Formation UMNH Loc. VP 424: UMNH 19204–19208, trunk vertebrae.

Wahweap Formation UMNH Loc. VP 77: UMNH 19209, trunk vertebra.

Descriptions None of the trunk vertebrae is complete, but the 11 specimens collectively document the entire structure of the bone. Figure 11.8 depicts examples from each of the three units from which vertebrae have been identified. UMNH VP 13431 (Fig. 8A–E) and MNA V10251 (Fig. 11.8F, G) are from the Smoky Hollow Member. The former specimen preserves most of the centrum and neural arch, but lacks the rib bearers, both postzygapophyses, the anterior portion of the right prezygapophysis, and the leading edge of the neural arch roof. Although the latter specimen consists only of the anterior half, it preserves the leading edge of the neural arch roof intact and a complete rib bearer on the left side. UMNH VP 19204 (Fig. 11.8H, I), from the John Henry Member, consists of about the anterior two thirds of a smaller trunk vertebra, whereas UMNH VP 19209 (Fig. 11.8J), from the Wahweap Formation, consists of only about the anterior half but retains an intact right rib bearer. Much of the following description is based on UMNH 13431. Trunk vertebrae are small and relatively elongate (e.g., for UMNH VP 13431: centrum length 1.8 mm; ratio of centrum length versus maximum width across anterior cotyle 3.0). The centrum is weakly constricted medially and is amphicoelous, with the anterior and posterior cotyles circular in outline, deeply concave, and having smooth inner walls and a moderate sized notochordal pit. A low and narrow subcentral keel extends along about the posterior two thirds of the centrum. Toward its anterior end, the keel begins to widen and grades into the base of a pair of broader, flat, anterior basapophyses (Fig. 11.8B, arrows) that project lateroanteriorly and slightly ventrally to extent past the rim of the anterior cotyle. Each of these prominent anterior basapophyses is connected by a vertical bony strut to the overlying prezygapophyseal process. The pre- and postzygapophyses extend laterally at moderate angles. The neural arch is low and its surface is dimpled with shallow, small, and irregular pockmarks. In dorsal view, the neural arch roof is subrectangular and its anterior end is narrow, squared off or shallowly convex anteriorly, and approximately in line with the anterior cotylar rim. A low, ridge-like neural crest extends along most of the neural arch roof. A pair of low, forked neural spines extends posterodorsally slightly past the level of the posterior cotylar rim. MNA V10251 and UMNH VP 19209 preserve a unicipital rib bearer high up on the side of the centrum; this process is relatively short, projects strongly posteriorly, and has a large, triangular posterior alar process and less prominent, but still relatively well developed, anterior and dorsal alar processes.

The sole atlantal specimen, MNA V10254 (Fig. 11.8K–O) consists of a centrum that is missing about the right quarter. The centrum is 1.1 mm long and, when intact, would have been about 1.2 mm wide across its anterior cotyles. The odontoid process is shallow in anterior view and is moderately

Gardner, Eaton, and Cifelli

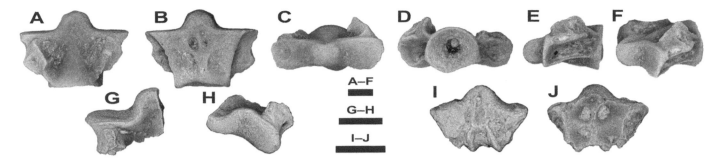

11.7. Atlantal centra of *Habrosaurus* sp. (Sirenidae) from southern Utah. (A–F) atlantal centrum, UMNH VP 19201, from UMNH Loc. VP 424, postbasal (Santonian) part of John Henry Member, Straight Cliffs Formation: (A) dorsal view; (B) ventral view; (C) anterior view; (D) posterior view; (E) left lateral view; (F) oblique left lateral and slightly anterodorsal view. (G, H) atlantal centrum missing posterior cotyle and most of right side, MNA V10275, from MNA Loc. 1230, Irons Springs Formation: (G) dorsal and slightly anterior view; (H) oblique anterodorsal and left lateral view. (I, J) atlantal centrum, OMNH 67077, from OMNH Loc. V9, Kaiparowits Formation: (I) dorsal view; (J) ventral view. All dorsal and ventral views oriented with anterior end toward top of figure. Specimens at different magnifications; scale bars = 1 mm.

elongate (i.e., length subequal to width) and tapers to a blunt point in dorsal view. The articular surfaces of the left and right anterior cotyles extend along the lateral surfaces of the odontoid process, but do not wrap around the anterior end of the process. In anterior view, the anterior cotyles are dorsoventrally depressed, with the dorsal margin being essentially straight and the ventral margin asymmetrically convex. The articular surface of the anterior cotyles is flat and faces anteriorly. The posterior cotyle is circular in posterior outline. The interior of the posterior cotyle is partially plugged by several grains of sediment, but even so it is clear that the cotyle is deeply concave and at least the ventral portions of its inner walls are smooth. Along its ventral midline, the centrum bears a ventrally low and anteroposteriorly short, irregular-shaped ridge. The base of the neural arch wall is preserved on the left side and, along its leading edge, is indented by an anteriorly open notch for passage of the first spinal nerve.

Remarks Even though the trunk vertebrae reported here come from three different units, they can be associated with one another on the basis of their distinctive structure, especially their elongate and amphicoelous centra, cotyles with smooth inner walls, prominent Y-shaped anterior basapophyses, unicipital rib bearers, and neural arch roofs that are low, dimpled, have a narrow anterior end, a low neural crest, and short, paired neural spines. These trunk vertebrae are easily differentiated from all other salamander trunk vertebrae that we have seen in collections from the Straight Cliffs and Wahweap formations. We associate the atlantal centrum MNA V10254 with the trunk vertebrae based on its correspondingly small size, the fact that it was recovered from one of the most productive localities (MNA 995) for the trunk vertebrae, and because it too differs from other salamander atlantes we have seen from the Straight Cliffs and Wahweap formations. The combination of a relatively well developed odontoid process, forward facing and low anterior cotyles, and articular

surfaces that extend from the anterior cotyles along the lateral surfaces of the odontoid process, but do not wrap around the anterior end of the process, are atypical of other atlantes from those formations. The structure of the trunk vertebrae and atlas is unique among salamanders known to us and, for that reason, we provisionally interpret these specimens as belonging to a new genus and species. The higher level affinities of this putative new taxon are unclear. Several features of the atlas (e.g., forms of odontoid process and anterior cotyles; articular surfaces that extend onto lateral surfaces, but do not wrap around anterior end of odontoid process) and of the trunk vertebrae (e.g., amphicoelous centrum; prominent anterior basapophyses; unicipital rib bearer with relatively well developed alar processes; dimpled neural arch roof) are reminiscent of sirenids. However, the Utah trunk vertebrae lack the Y-shaped arrangement of dorsal crests (a median neural crest and a pair of lateroposteriorly projecting aliform processes) on the neural arch roof that are diagnostic for unequivocal sirenids (Estes, 1964, 1981; Gardner, 2003a; Holman, 2006).

Family(ies) Indeterminate
Genera and Species Indeterminate
Voucher Specimens (Figs. 11.9 and 11.10)

Trunk Vertebrae from the Dakota Formation MNA Loc. 1067: MNA V10238, V10239, and MNA V10368; UMNH Loc. VP 27 (=MNA Loc. 1067): UMNH VP 13103.

Dentaries from the Dakota Formation MNA Loc. 1067: MNA V9095, V10369, and V10370; OMNH Loc. v808 (=MNA Loc. 1067): OMNH 67078 and 67079.

Trunk Vertebra from the Smoky Hollow Member, Straight Cliffs Formation MNA Loc. 995: MNA V10371.

Trunk Vertebra from the Undifferentiated Straight Cliffs Formation MNA Loc. 1260: MNA V10372.

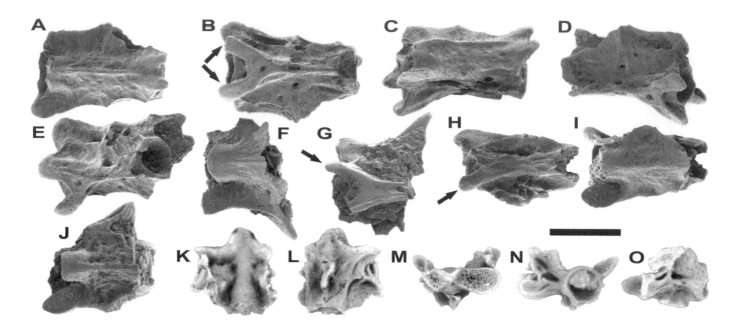

11.8. Vertebrae of Urodela Family Indeterminate, New Genus and Species, from the Straight Cliffs and Wahweap formations, southern Utah. (A–J) trunk vertebrae: (A–E) most nearly complete trunk vertebra, UMNH VP 13431, from UMNH Loc. VP 129, Smoky Hollow Member, Straight Cliffs Formation: (A) dorsal view with anterior end toward left of figure; (B) ventral view with anterior end toward left of figure and arrows pointing to intact anterior basapophyses; (C) oblique left lateral, dorsal, and slightly anterior view; (D) oblique right lateral, ventral, and slightly anterior view; (E) oblique ventral, left lateral, and posterior view; (F, G) anterior half of trunk vertebra preserving intact, uniciptal rib bearer on left side, MNA V10251, from MNA Loc. 995 (=UMNH Loc. VP 129) Smoky Hollow Member, Straight Cliffs Formation: (F) dorsal view with anterior end toward left of figure; (G) ventral view with anterior end toward left of figure and arrow pointing to intact anterior basapophysis on left side; (H, I) anterior two thirds of trunk vertebra, UMNH VP 19204, from UMNH Loc. VP 424, postbasal (Santonian) part of John Henry Member, Straight Cliffs Formation: (H) ventral view with anterior end toward left of figure and arrow pointing to intact anterior basapophysis on right side; (I) dorsal view with anterior end toward left of figure; (J) anterior half of trunk vertebra preserving intact, uniciptal rib bearer on right side, UMNH VP 19209, from UMNH Loc. VP 77, Wahweap Formation, dorsal view with anterior end toward left of figure. (K–O) atlantal centrum missing part of right side, MNA V10254, from MNA Loc. 995 (=UMNH Loc. VP 129) Smoky Hollow Member, Straight Cliffs Formation: (K) dorsal view with anterior end toward top of figure; (L) ventral view with anterior end toward top of figure; (M) anterior view; (N) posterior view; (O) left lateral view; photographs of atlantal centrum courtesy of Donald Brinkman (Royal Tyrrell Museum of Palaeontology). Specimens at same magnification; scale bar = 1 mm.

Descriptions The four trunk vertebrae from the Dakota Formation are all incomplete: MNA V10238 (Fig. 11.9A–F) consists of a fractured centrum with the neural arch walls; UMNH VP 13103 (Fig. 11.9G–J) and MNA V10239 (Fig. 11.9K–M) each consist of about the anterior three quarters of the centrum; and MNA V10368 (Fig. 11.9N–R) is a nearly complete centrum. MNA V10238 and MNA V10368 have centrum lengths of 2.0 and 2.8 mm, respectively; when intact, the other two specimens each would have been similar in size to MNA V10238 or slightly longer. Besides being small, the four specimens resemble one another as follows: centra relatively elongate (ratio of centrum length versus maximum width across anterior cotyle 2.9 in MNA V10238 and 3.5 in MNA V10368), moderately constricted medially, ventral and ventrolateral surfaces smooth, and ventral surface shallowly concave dorsally in lateral outline; amphicoelous, with calcified cartilage lining inner walls of cotyles and large notochordal pit in center; and basapophyses absent. Other features are more variable among the specimens. The anterior and posterior cotyles in MNA V10238 are slightly compressed laterally

and oval, with the narrower end directed ventrally. The anterior cotyle in UMNH 13103 and MNA V10368 also are oval in outline; the outlines of the posterior cotyles are unknown, however, because that cotyle is missing from UMNH 13103 and the ventral margin of the cotyle is too broken in MNA V10368 to determine its outline. MNA V10239 also is missing its posterior cotyle, but its anterior cotyle is intact and differs from the previous three specimens in being more nearly subcircular in outline, with only minimal lateral compression. MNA V10238 and UMNH 13103 further resemble one another in having the inner walls of their cotyles covered with a thick layer of calcified cartilage and in having a subcentral keel that is shallow (i.e., does not extend ventrally below the lower edges of the cotyles), narrow, and extends almost the entire length of the centrum. By contrast, in the other two specimens the inner walls of the cotyles have a thinner coating of calcified cartilage and the subcentral keel is either reduced to an indistinct ridge (MNA V10239) or is absent (MNA V10368). When seen in lateral view, the rims of the anterior and posterior cotyles in MNA V10238 are vertical, whereas in MNA V10368 the rims are

Gardner, Eaton, and Cifelli

tilted slightly ventrally. Although their posterior cotyles are not preserved, the anterior cotyles in MNA V10239 and UMNH VP 13103 also are tilted slightly ventrally. On one or both sides, MNA V10238, UMNH VP 13103, and MNA V10239 each preserve the base of a single rib bearer low on the lateral side of the centrum; the upper part of the centrum is sufficiently intact in the first two specimens that it appears unlikely a second (i.e., dorsal) rib bearer was present in life. By contrast, MNA V10368 clearly had bicipital rib bearers. On its right side, that vertebra preserves the proximal portions of a pair of rib bearers, which are notable for being close together and parallel (i.e., not divergent), but without an interconnecting web of bone between them as typically occurs in salamanders (e.g., cryptobranchids and hynobiids) with closely appressed rib bearers. MNA V10238 preserves enough of the neural arch wall on the left side to establish that a spinal foramen is present at the junction between the wall and the centrum, slightly in front of the posterior edge of the wall. When the specimen was first recovered, the wall on the left side was more nearly complete and the spinal foramen was fully enclosed by bone; during photography the posterior part of the wall broke, fortuitously through the foramen and exposed the ventral floor of the canal extending anteromedially–posterolaterally between the internal and external openings of the spinal foramen (Fig. 11.9A versus B, C).

Four of the dentary specimens from the Dakota Formation each preserve a complete symphyseal end and the anterior portion of the tooth row, but none retains any intact teeth. Three distinct morphs can be recognized among these four specimens on the basis of differences in absolute size, form of the bone, and inferred tooth sizes and attachment patterns. Morph A, represented by OMNH 67079 (Fig. 11.9S, T), is the most generalized of the three dentary morphs and is characterized by the following suite of features: small size (preserved length 2.4 mm; maximum depth of symphyseal end 0.9 mm); moderately robust build; symphyseal end squared off in lingual view and not significantly expanded lingually or ventrally, subrectangular in symphyseal outline, and symphyseal surface flat; labial surface smooth; ventral surface labiolingually broad and nearly flat; subdental shelf shallow, moderately broad lingually, dorsal surface shallowly gutter-like, and lingual surface convex; dental parapet moderately high; teeth moderate in size, with highly pleurodont attachments. Morph B is known from two specimens of different sizes: MNA V10369 (Fig. 11.9U) is 3.8 mm long and MNA V10370 (Fig. 11.9V, W) is 1.2 mm long. Judging by the maximum depths of their symphyseal ends (3.0 and 1.3 mm, respectively), MNA V10369 is from an individual at least twice as large as MNA V10370. Characteristic features of Morph B dentaries are: large size; robust build; symphyseal

end expanded ventrally and lingually (especially in larger individuals), resulting in symphyseal end being subtriangular in symphyseal outline, and symphyseal surface shallowly sigmoidal; labial surface smooth; ventral surface moderately broad labiolingually, surface shallowly concave and horizontal well behind symphysis, toward symphysis ventral surface flattens, expands medially, and curves ventrolabially; subdental shelf moderately deep and lingually broad, lingual surface shallowly convex or nearly flat, and dorsal surface shallowly convex; dental parapet high; teeth relatively large, with highly pleurodont attachments. Morph C, represented by OMNH 67078 (Fig. 11.9X–Z), is characterized as follows: small size (preserved length 2.0 mm; maximum depth of symphyseal end 0.9 mm); moderately robust build; symphyseal end expanded ventrally but not lingually, labiolingually compressed and oval in symphyseal outline with dorsal end slightly narrower, and symphyseal surface flat but with labial two thirds indented by a deep and narrow groove (arrow in Fig. 11.9Z); labial surface smooth, with row of small external nutritive foramina below and paralleling dorsal edge and with shallow, narrow groove (mentioned above) along ventrolabial surface that extends mesially and deepens across symphyseal surface; ventral surface labiolingually narrow and steeply convex; subdental shelf deep, narrow lingually, lingual surface broadly convex, and dorsal surface narrow and deeply concave, (i.e., groove-like) adjacent to tooth bases; dental parapet low; teeth tiny, with weakly pleurodont attachments. When each of the above-described specimens are viewed in dorsal aspect, the ramus extends nearly perpendicular from the symphyseal end.

A fifth Dakota specimen, MNA V9095 (Fig. 11.9AA, BB), is an incomplete left dentary that measures 3.9 mm in preserved length and preserves nine empty tooth slots, the anterior portion of the opening for the Meckelian canal, and the posterior portion of the subdental shelf. In extant salamanders the opening for the Meckelian canal typically extends far forward, to at least the midpoint of the tooth-bearing portion of the dentary. On the basis of that observation, MNA V9095 likely is from the middle or posterior portion of the dentary. It is not clear, however, whether the broken anterior end of this specimen overlaps with the broken posterior ends of any of the above-described dentaries. In labial or lingual outline, the dorsal and ventral margins of MNA V9095 taper anteriorly at a pronounced angle. In lingual view, the opening for the Meckelian canal extends forward to about the level of the fifth locus from the broken posterior end; more anteriorly, the subdental shelf is slightly expanded lingually, is moderately deep, and its dorsal surface is flat and tilted ventrolingually; and more dorsally the dental parapet is moderately high and the tooth slots are moderately large.

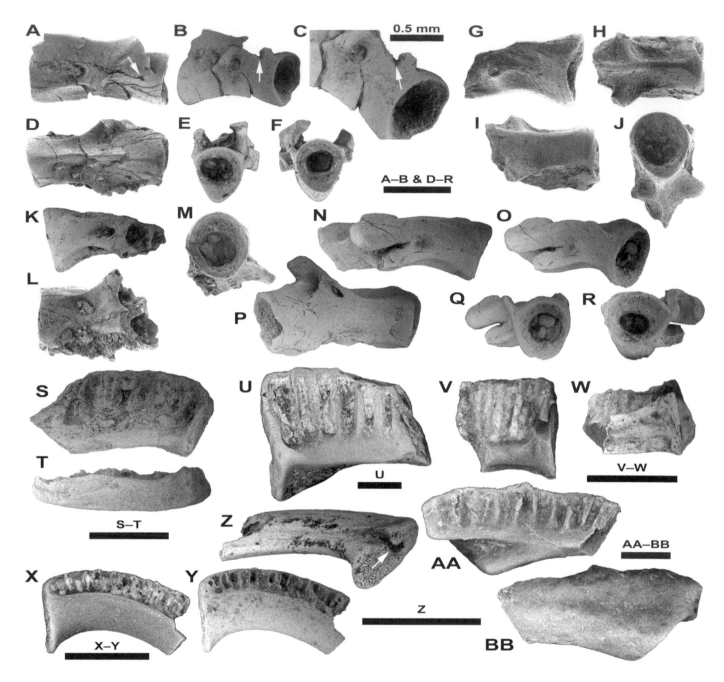

11.9. Vertebrae and dentaries of Urodela Indeterminate from the Dakota Formation, southern Utah. (A–R) Trunk vertebrae: (A–F) centrum preserving broken neural arch walls and bases of a unicipital rib bearer on both sides, MNA V10238, from MNA Loc. 1067: (A) left lateral view, image taken before posterior part of left neural arch wall was broken and with arrow pointing at spinal foramen fully enclosed in bone at base of arch wall; (B) left lateral and slightly ventroposterior view, image taken after posterior part of left neural arch wall was broken and with arrow pointing at lower rim of broken spinal foramen; (C) magnified and slightly rotated view of previous image with arrow pointing at floor of canal for spinal foramen exposed in broken base of neural arch wall; (D) ventral view with anterior end toward left of figure; (E) anterior view, image taken after posterior part of left neural arch wall was broken; (F) posterior view, image taken before posterior part of left neural arch wall was broken; (G–J) centrum missing posterior end, but preserving base of a unicipital rib bearer on both sides, UMNH VP 13103, from UMNH Loc. VP 27 (=MNA 1067): (G) right lateral and slightly anterior view; (H) ventral view with anterior end toward right of figure; (I) oblique ventral and left lateral view with anterior end toward right of figure; (J) ventroanterior view; (K–M) centrum missing posterior end, but preserving much of a unicipital rib bearer on left side, MNA V10239, from MNA Loc. 1067: (K) left lateral view; (L) ventral view with anterior end toward left of figure; (M) anterior and slightly ventral view; (N–R) centrum preserving nearly parallel, bicipital rib bearers on right side, MNA V10368, from MNA Loc. 1067: (N) right lateral view; (O) oblique right lateral and anterior view; (P) ventral view with anterior end toward right of figure; (Q) anterior view; (R) posterior view. (S–BB) Dentaries, none with intact teeth: (S, T) anterior end of left dentary, morph A, OMNH 67079, from OMNH Loc. V808 (=MNA 1067): (S) lingual view; (T) ventral view; (U) anterior end of right dentary, morph B, MNA V10369, from MNA Loc. 1067, lingual view; (V, W) anterior end of left dentary, morph B, MNA V10370, from MNA Loc. 1067: (V) lingual view; (W) ventral view; (X–Z) anterior end of left dentary, morph C, OMNH 67078, from OMNH Loc. V808: (X) photograph in lingual view to highlight shallow subdental groove; (Y) scanning electron micrograph in lingual view to highlight tooth loci; (Z) scanning electron micrograph, slightly magnified from previous and in oblique ventrolabial and symphyseal view to show groove (arrow) extending from ventrolabial surface onto symphyseal surface; (AA, BB) anteriorly and posteriorly incomplete left dentary, MNA V9095, from MNA Loc. 1067: (AA) lingual view; (BB) labial view. Specimens at different magnifications; except where labeled otherwise, all scale bars = 1 mm.

Gardner, Eaton, and Cifelli

The labial surface is smooth and indented posteriorly by a shallow, V-shaped trough whose apex is directed anteriorly and ends slightly anterior to the level of the anterior end of the opening for the Meckelian canal in the lingual surface.

MNA V10371 from the Smoky Hollow Member (Fig. 11.10A–D) is the anterior half, or slightly less, of a delicate and tiny trunk centrum (preserved length 0.6 mm) that is broken through the bases of its rib bearers and posteriormost end of its anterior cotyle, but preserves the base of the neural arch walls. Although the anterior cotyle is mostly infilled with matrix that cannot safely be removed, it obviously is deeply concave and the anteriormost portion of its inner walls is smooth. The anterior cotyle is broadly orbicular in anterior outline, with its ventral end narrower. The ventral surface of the centrum bears a ridge-like subcentral keel; in transverse profile, the keel is shallow, narrow, and has relatively straight sides and a broadly convex ventral surface. The anterior end of the keel lies slightly behind the anterior rim of the anterior cotyle and is broadly rounded in lateral view; the keel becomes shallower posteriorly along the preserved portion of the centrum. The base of a rib bearer is preserved high up on both sides, at the junction between the centrum and neural arch wall; there is no indication of a second, more dorsal rib bearer. The preserved bases of the rib bearers extend posterolaterally at a pronounced angle and a thin, posterior alar process extends between the rib bearer and the centrum wall. The lateral surfaces of the centrum are smooth, except for a sediment-infilled opening on either side in the junction between the centrum wall and the underside of the rib bearer. No anterior basapophyses are present.

MNA V10372 from the undifferentiated Straight Cliffs Formation (Fig. 11.10E–G) is half of a delicately built centrum; it is not clear whether this is the anterior or posterior end of a centrum. The cotyle is intact, measures 1.1 mm in maximum width, and is broadly orbicular in outline, with its ventral end narrower. The interior of the cotyle is deeply concave, its inner walls are coated with a thin layer of calcified cartilage, and a small notochordal pit lies in the approximate center. A subcentral keel extends from the ventral rim of the cotyle to the broken end of the centrum. In transverse profile, the subcentral keel is shallow and broadly triangular. The rest of the exterior surface is smooth, without any trace of basapophyses.

Remarks Although the small sizes of the four trunk specimens from the Dakota Formation and the apparent presence of unicipital rib bearers in three of them are reminiscent of albanerpetontid trunk vertebrae, they can be assigned to the Urodela because all four specimens lack the thickened rims around the anterior and posterior cotyles that are characteristic for albanerpetontids (McGowan, 1996:figs. 12, 13). The bicipital rib bearers in MNA V10368, the subcentral keels in

MNA V10238 and UMNH VP 13103, and the spinal foramen in MNA V10238 further exclude these specimens from the Albanerpetontidae and support assigning them to the Urodela. The four trunk vertebrae resemble one another in their small size and general form of the centrum. Although there are numerous differences, most of those are relatively minor and there is enough overlap among the specimens that they could conceivably belong to the same species, but this is not certain. None of the four trunk vertebrae can be assigned to a known salamander family. The laterally compressed cotyles and the thick layer of calcified cartilage lining the inner walls of the cotyles in MNA V10238 and UMNH 13103 are reminiscent of the Scapherpetontidae; however, the presence of a spinal foramen in MNA V10238 and what appears to be unicipital rib bearers in three of the specimens argues against that familial assignment because all known scapherpetontids have divergently bicipital rib bearers and lack spinal foramina in their trunk vertebrae. The closely spaced and parallel rib bearers on MNA V10368 recall the condition in anterior trunk vertebrae of hynobiids and cryptobranchids, except that in those families there is an interconnecting web of bone between the rib bearers. Although it is possible that a web of bone extended between the dorsal and ventral rib bearers on MNA V10368, but has since been broken away, there is no clear indication of broken surfaces to suggest that such a structure originally was present. The four trunk vertebrae are similar in size and certain aspects of their morphology (e.g., centra elongate and amphicoelous) to some described trunk vertebrae of *Prosiren* and are not too far removed geographically or temporally from the known occurrences (early–middle Albian, Antlers Formation, north-central Texas) of that taxon. However, the Utah trunk vertebrae differ in lacking the anterior basapophyses that are present on examples of the elongate trunk vertebral morph of *Prosiren* (these are the only *Prosiren* trunk vertebrae that are comparable with the Utah specimens; cf. Fig. 11.9A, G, K, N versus Estes, 1969a:fig. 1n) and thus cannot be assigned to *Prosiren*.

The four dentary anterior ends from the Dakota Formation can be grouped into three distinct morphs. The sole examples known for both morph A and morph C, and the smaller of the two specimens known for morph B are similar in size, judging by their symphyseal ends, so the pronounced differences among those morphs cannot be attributed to ontogenetic variation. Instead, we interpret those differences as being taxonomic. At present, we cannot confidently assign any of the dentary morphs to a known salamander family or associate them with any of the vertebrae, other than to note that the large dentary MNA V10369 probably did not come from the same kind of salamander represented by the small, indeterminate trunk vertebrae from the Dakota Formation.

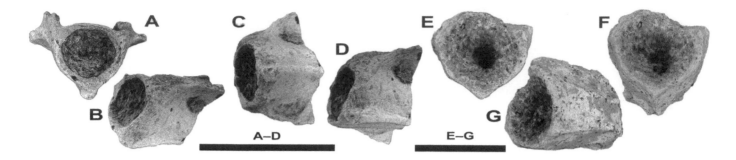

11.10. Vertebrae of Urodela Indeterminate from the Straight Cliffs Formation, southern Utah. (A–D) Anterior half of trunk centrum, MNA V10371, from MNA Loc. 995, Smoky Hollow Member, Straight Cliffs Formation: (A) anterior view; (B) oblique left lateral, anterior, and slightly ventral view; (C) oblique ventral and slightly anterior view; (D) oblique ventral, left lateral, and slightly anterior view; last three views all with anterior end toward left of figure. (E–G) Anterior half of trunk centrum, MNA V10372, from MNA Loc. 1260, undifferentiated Straight Cliffs Formation: (E) anterior view; (F) oblique anterior and slightly ventral view; (G) oblique ventral, left lateral, and slightly anterior view with anterior end toward left of figure. Specimens at different magnifications; scale bars = 1 mm.

The remaining dentary specimen, MNA V9095, clearly is not a morph C dentary because its dental parapet is too high and its subdental shelf is too shallow. Both of those structures are a better match for morph A and B, but MNA V9095 cannot be assigned to either morph because its subdental shelf does not match those in either morph. MNA V9095 is of historical interest because it was listed in the MNA records as "cf. *Batrachosauroides*" and presumably was the basis for "cf. *Batrachosauroides* sp." being recorded in two preliminary faunal lists (Eaton et al., 1997:table 1; Eaton, Cifelli, et al. 1999:table 1) for the Dakota Formation. *Batrachosauroides* is a Tertiary batrachosauroidid known by two species, one from the early Eocene of North Dakota and one from the early–middle Miocene of Florida and Texas (Estes, 1981). The anteriorly directed, V-shaped trough on the labial surface of MNA V9095 resembles a similar trough described for both Tertiary species of *Batrachosauroides* (Estes, 1969b, 1981) and may have been the basis for that identification. An indentation along the posterior portion of the labial surface of the dentary is fairly widespread among salamanders, so its presence in MNA V9095 is not particularly diagnostic and, for the time being, the specimen is better regarded as being from an indeterminate salamander.

MNA V10371 from the Smoky Hollow Member exhibits no features that ally it with any known salamander taxa. Compared to other salamander trunk vertebrae reported herein from the Smoky Hollow Member, MNA V10371 differs from the unnamed new genus and species in lacking anterior basapophyses and from the batrachosauroidid Genus and Species B in having a smooth centrum and more prominent subcentral keel and in not having its anterior cotyle infilled with cartilage. The distinct subcentral keel and apparent presence of unicipital rib bearers high on the centrum differentiates MNA V10371 from the similarly small, although geologically older, indeterminate salamander trunk vertebrae from the Dakota Formation.

MNA V10372 from the undifferentiated Straight Cliffs Formation most closely resembles MNA V10371 and, like that older Turonian specimen, it exhibits no features that can be used to assign it with confidence to any salamander family. Although MNA V10372 and V10371 resemble one another in their gracile build, in having a cotyle that is deeply concave and has an orbicular outline, and in having a low subcentral keel that extends to the cotylar rim, they differ in size (based on cotylar widths, MNA V10372 is about half the size of MNA V10371) and transverse profiles of their subcentral keels. Those differences are sufficiently minor that they could reasonably be attributed to differences in body sizes of individuals or positional variation along the vertebral column. The presence of a thin coating of calcified cartilage inside the cotyle of MNA V10372 and the apparent absence, at least closer to the rim, of such a coating in MNA V10371 is a more substantial difference that argues against these vertebrae being from the same taxon. In having only a thin coating of calcified cartilage lining the inner cotylar wall, MNA V10372 also differs from the two indeterminate scapherpetontid trunk vertebral cotyles (MNA V10277 and V10366) described above from the same locality.

DISCUSSION

Vertebrae and dentaries reported herein demonstrate that salamanders were moderately abundant and diverse during the late Cenomanian through late Campanian in southern Utah. Previous reports of salamanders in the Dakota, Wahweap, and Kaiparowits formations and in the Smoky Hollow Member of the Straight Cliffs Formation (Eaton, Munk, and Hardman, 1998; Munk, 1998; Eaton, Cifelli et al., 1999) are confirmed and the presence of salamanders in the Iron Springs Formation and the John Henry Member of the Straight Cliffs Formation is established for the first time. Despite the modest sample sizes—both in terms of specimens

Gardner, Eaton, and Cifelli

and localities–available for this preliminary report, a surprising diversity of salamander taxa and some interesting patterns are revealed (Tables 11.2 and 11.3).

All three of the salamander families (Scapherpetontidae, Batrachosauroididae, and Sirenidae) known from better-sampled uppermost Cretaceous (upper Campanian–upper Maastrichtian) units elsewhere in the Western Interior (Estes, 1981; Gardner, 2000a, 2003a, 2005; Holman, 2006) are present in the Utah localities. Scapherpetontids and batrachosauroidids occur throughout the sampled interval (late Cenomanian–late Campanian), whereas sirenids are limited to the Santonian–late Campanian.

Scapherpetontids are represented in most of the sampled units in Utah. The absence of scapherpetontid specimens in the Smoky Hollow Member is puzzling because although only two localities (MNA 995/UMNH VP 129 and OMNH V4) were included in our study, those yielded moderate numbers of salamander vertebrae from at least two other families (Table 11.2). Two named scapherpetontid genera are identified in Utah: *Scapherpeton* in the John Henry Member of the Straight Cliffs Formation and in the Wahweap and Kaiparowits formations and *Lisserpeton* in the Kaiparowits Formation. Both of those genera are well represented through the latest Cretaceous and into the Paleocene in the Western Interior, so their presence in Utah is not unexpected. Providing that OMNH Loc. v856 is Coniacian in age, the presence of *Scapherpeton* vertebrae at that locality extends the record for the genus back one stage, from its previous earliest known occurrence (Fox, 1972; Naylor and Krause, 1981; Gardner, 2000a) in the Deadhorse Coulee Member (Santonian) of the Milk River Formation, Alberta. Outside of Utah, the oldest occurrences of *Lisserpeton* are in two late Campanian units: the Dinosaur Park Formation of Alberta (Gardner, 2000a) and the Mesaverde Formation of Wyoming (DeMar and Breithaupt, 2006, 2008). The pair of *Lisserpeton* atlantes identified here in the contemporaneous Kaiparowits Formation extends the geographic range of the genus southward into Utah, but does not alter the lower limit of its temporal range. However, several geologically older specimens–a slightly deformed atlas from the basal (?Coniacian) part of the John Henry Member and several trunk vertebrae from the middle member (late Cenomanian) of the Dakota Formation–identified above as Scapherpetontidae indeterminate have some *Lisserpeton*-like features and thus may indicate that *Lisserpeton* or a similar scapherpetontid genus extended back farther in time. Additional indeterminate scapherpetontid vertebrae from those same two units likely pertain to other scapherpetontid taxa: several trunk vertebrae from the basal part of the John Henry Member are reminiscent of *Piceoerpeton* (currently known from the late Maastrichtian–early Eocene; Estes, 1981; Naylor and Krause, 1981; Naylor, 1983; Gardner,

in press) and the atlantes and remaining trunk vertebrae from the Dakota Formation probably belong to at least one other scapherpetontid genus. Regardless of their generic identities, scapherpetontid vertebrae in the middle member (late Cenomanian) of the Dakota Formation are slightly younger (~3 million years according to the timescale of Ogg, Agterberg, and Gradstein, 2004) than the oldest known occurrence of scapherpetontids, which is in the Mussentuchit Member (latest Albian/earliest Cenomanian; Cifelli et al., 1997) of the Cedar Mountain Formation, farther to the north in central Utah (Cifelli et al., 1999).

Batrachosauroidids also are known from most of the sampled units in Utah. Two previously named genera are recognized from the younger units: *Opisthotriton* in the postbasal part of the John Henry Member and in the Wahweap and Kaiparowits formations (Santonian–late Campanian) and *Prodesmodon* in the Kaiparowits Formation (late Campanian). Outside of Utah, the earliest occurrence for *Opisthotriton* is in the Deadhorse Coulee Member (Santonian) of the Milk River Formation, Alberta (Fox, 1972; Naylor, 1981; Gardner, 2000a) and for *Prodesmodon* is in the Mesaverde Formation (late Campanian), Wyoming (DeMar and Breithaupt, 2006, 2008); consequently, the Utah occurrences do not extend the temporal ranges of either genus. Elsewhere in the Western Interior, *Opisthotriton* is a common salamander whose range extends into the late Paleocene, whereas *Prodesmodon* is less common and extends only into the early Paleocene (Estes, 1964, 1981; Naylor, 1981; Gardner, 2000a). The presence of at least two other, geologically older batrachosauroidid genera in Utah is indicated by a distinctive atlas and trunk vertebra from the Dakota Formation ('Genus and Species A') and by a distinctive trunk centrum from the Smoky Hollow Member of the Straight Cliffs Formation ('Genus and Species B'). The late Cenomanian age specimens from the Dakota Formation are the oldest unequivocal occurrence for the Batrachosauroididae in North America and extend the record for the family on the continent back about 10 million years, according to the timescale of Ogg, Agterberg, and Gradstein (2004), from the previous oldest known occurrence (Santonian) of *Opisthotriton* in the Milk River Formation. Reports of geologically older batrachosauroidids in North American are unsubstantiated. Several trunk vertebrae from the Mussentuchit Member (latest Albian/earliest Cenomanian) of the Cedar Mountain Formation, Utah, that initially were identified as batrachosauroidid (Gardner, 1994) have since been reinterpreted as being from scapherpetontids (Cifelli et al., 1999). *Prosiren* (Albian, Texas) originally was assigned to its own family, the Prosirenidae (Estes, 1969a, 1981; Fox and Naylor, 1982; Duellman and Trueb, 1986), but more recently it has been suggested that the genus may be a batrachosauroidid (Milner, 2000; Holman, 2006). However, pending a

Table 11.3. Comparison of late Cenomanian–late Campanian salamander assemblages from Utah versus elsewhere in the North American Western Interior[a]

Age	Utah	Elsewhere in Western Interior

Late Campanian	Scapherpetontidae
	  Scapherpeton sp.
	  Lisserpeton sp.
	Batrachosauroididae
	  Opisthotriton sp.
	  Prodesmodon sp.
	Sirenidae
	  Habrosaurus sp.
		Scapherpetontidae
		  Scapherpeton tectum
		  Lisserpeton bairdi
		  New genus and species[b]
		Batrachosauroididae
		  Opisthotriton kayi
		  Prodesmodon copei
		Sirenidae
		  Habrosaurus prodilatus

Early–middle Campanian	Scapherpetontidae
	  Scapherpeton sp.
	Batrachosauroididae
	  Opisthotriton sp.
	Family Indet.
	  New genus and species
		NA

Santonian	Scapherpetontidae
	  Scapherpeton sp.
	Batrachosauroididae
	  Opisthotriton sp.
	  Genus and species indet.
	Sirenidae
	  Habrosaurus sp.
	Family Indet.
	  New genus and species
		Scapherpetontidae
		  Scapherpeton tectum
		Batrachosauroididae
		  Opisthotriton kayi
		  New genus and species[b]

?Coniacian	Scapherpetontidae
	  Scapherpeton sp.
	  Genus and species indet.
	Family Indet.
	  Genus and species indet.
		NA

Late Turonian	Batrachosauroididae
	  Genus and species B
	Family Indet.
	  New genus and species
	Family Indet.
	  Genus and species indet.
		NA

Late Cenomanian	Scapherpetontidae
	  Genera and species indet.
	Batrachosauroididae
	  Genus and species A
	Family(ies) Indet.
	  Genus and species indet.
		NA

[a] Taxonomic lists for Utah are from this study, as follows: late Campanian based on Kaiparowits Formation; early–middle Campanian based on Wahweap Formation; Santonian based on postbasal part of John Henry Member, Straight Cliffs Formation; ?Coniacian based on basal part of John Henry Member, Straight Cliffs Formation, and on undifferentiated Straight Cliffs Formation; late Turonian based on Smoky Hollow Member, Straight Cliffs Formation; and late Cenomanian based on middle member of Dakota Formation. Taxonomic lists for outside of Utah follow Gardner (2005:table 10.2); NA entry indicates unknown.

[b] Unnamed new taxa reported by Gardner (2000a, 2005).

detailed restudy of the available *Prosiren* specimens—these include dozens of vertebrae not included in Estes's (1969a) original description (Gardner, pers. obs., 1995)—the familial affinities of *Prosiren* are best considered uncertain. Although vertebrae from the Dakota Formation are the oldest record for the Batrachosauroididae in North America, a variety of European fossils (Milner, 2000) indicate that the family is considerably older, extending back to at least the earliest Cretaceous (Berriasian; Evans and McGowan, 2002) and, perhaps, even into the Middle Jurassic (late Bathonian; Evans and Milner, 1994).

Distinctive atlantal centra of *Habrosaurus* from the post-basal part of the John Henry Member and from the Iron Springs and Kaiparowits formations establish the presence of sirenids in Utah during the Santonian to late Campanian. The first two occurrences extend the North American record for both *Habrosaurus* and the Sirenidae back about 10 million years, according to the timescale of Ogg, Agterberg, and Gradstein (2004), from the previous oldest known occurrences in the late Campanian of Wyoming (DeMar and Breithaupt, 2006, 2008), Montana (Rogers and Brady, 2010), and Alberta (Gardner, 2003a). If the enigmatic Gondwanan salamanders *Noterpeton* (Maastrichtian, Bolivia) and *Kababisha* (Cenomanian, Morocco and Sudan) are not sirenids (for differing opinions, cf. Evans, Milner, and Werner, 1996; Gardner, 2003a), then the Utah specimens are the oldest worldwide record for the Sirenidae.

In summary, fossils reported herein include the second oldest occurrence for the Scapherpetontidae, the oldest North American occurrence of the Batrachosauroididae, and the oldest North American occurrence and, possibly, the oldest worldwide occurrence for the Sirenidae. Although the occurrences in Utah of scapherpetontids during the latest Albian/earliest Cenomanian and late Cenomanian and of sirenids during the Santonian are consistent with those families being endemic to North America, two lines of evidence indicate that the earliest occurrences of those two families and of batrachosauroidids (late Cenomanian) in Utah underestimate their probable times of origins. First, as noted above, batrachosauroidid-like vertebrae from Europe indicate that the family had arisen by at least the Early Cretaceous and, possibly, earlier by the Middle Jurassic. Second, two recent molecular phylogenetic analyses of higher-level salamander relationships (Roelants et al., 2007; Zhang and Wake, 2009) that used fossil occurrences, biogeographic distributions, and tectonic events to constrain their phylogenetic trees found that extant salamander families had diverged from each other by the Late Cretaceous, with several families (e.g., the Sirenidae, which occupied a relatively basal position in both analyses) having originated even earlier. Although fossil

Gardner, Eaton, and Cifelli

families were not included in either analysis, if we accept that batrachosauroidids are most closely related to proteids and that scapherpetontids are most closely related to ambystomatids and dicamptodontids (Naylor, 1978, 1981; Estes, 1981; Duellman and Trueb, 1986), we can use the divergence times provided by Roelants et al. (2007:fig. 1) and Zhang and Wake (2009:fig. 4) for those extant families to estimate that batrachosauroidids arose between the basal Jurassic–middle Early Cretaceous and that scapherpetontids arose between the basal Jurassic–late Early Cretaceous. Although those estimates predate the first appearances of batrachosauroidids and scapherpetontids in Utah, it is interesting to note that the late Early Cretaceous upper limit on the estimated divergence time for the Scapherpetontidae is not much older than the first fossil evidence for that family at the Albian/Cenomanian boundary in the Cedar Mountain Formation of central Utah.

Other distinctive vertebrae indicate the existence of additional salamander taxa in Utah during the Late Cretaceous. A new late Turonian–early or middle Campanian genus of uncertain familial affinities is recognized by an atlantal centrum and 11 trunk vertebrae from the Straight Cliffs (Smoky Hollow and postbasal part of John Henry members) and Wahweap formations. Certain features of the atlas and trunk vertebrae are reminiscent of sirenids, which raises the intriguing possibility that these fossils may belong to a primitive sirenid (one lacking the characteristic Y-shaped configuration of dorsal crests on the trunk vertebrae), to a sirenid-like taxon, or to a taxon that convergently resembles sirenids. Further study of the available specimens and, hopefully, the recovery of additional specimens may help resolve the affinities of this new taxon. The second group of specimens consists of four trunk vertebrae from the middle member (late Cenomanian) of the Dakota Formation that resemble each other in their small size and some general features, but differ in others (most notably rib bearers unicipital in three and bicipital in one). These four vertebrae may pertain to the same taxon, but that is not certain. One of the trunk vertebrae preserves a spinal nerve foramen. Among extant salamander families, only sirenids, ambystomatids, plethodontids, and salamandrids have spinal nerve foramina in their trunk vertebrae (Edwards, 1976); however, the Cenomanian specimen is not otherwise similar enough to be assigned to any of those families. The third group of specimens consists of two small, broken trunk vertebral centra – one from the Smoky Hollow Member (late Turonian) and one from a probable Coniacian locality in the undifferentiated Straight Cliffs Formation – that show some resemblances to each other and differ from the other salamander vertebrae included in this study. Although intriguing, both specimens

are too fragmentary to say anything more definitive about their higher level affinities.

On a regional scale, it has long been evident that salamander assemblages in the North American Western Interior during the latest Cretaceous and well into the Paleocene were dominated by three families – scapherpetontids, batrachosauroidids, and sirenids (Estes, 1964, 1981; Gardner, 2005; Holman, 2006). The Utah record suggests that this basic framework may have been established in two phases: the early appearances of the Scapherpetontidae by the Albian/Cenomanian boundary and the Batrachosauroididae by the late Cenomanian, followed by the later Santonian appearance of the Sirenidae. Within those families, genera typical of later Cretaceous assemblages began appearing sequentially: *Scapherpeton* in the Coniacian, *Opisthotriton* and *Habrosaurus* in the Santonian, and *Lisserpeton* and *Prodesmodon* in the late Campanian. Whereas the first appearance of the Sirenidae is coincident with the first appearance of *Habrosaurus*, the presence in pre-Santonian localities of distinctive scapherpetontid and batrachosauroidid specimens that cannot be assigned to known latest Cretaceous genera indicates that both families were already diversifying in the Western Interior during the first part of the Cretaceous. Three lots of distinctive vertebrae (one from the late Cenomanian, the second from the late Turonian–early or middle Campanian, and the third from the late Turonian and ?Coniacian) that cannot be assigned to any recognized family(ies) indicate that other salamander families might also have been present in the Western Interior during the early and middle Late Cretaceous.

The youngest and oldest of the sampled units in the Utah sequence are approximately comparable in age to units that have produced salamander fossils elsewhere in the Western Interior. Winkler and Jacobs (2002) reported unidentified Cenomanian salamanders in the Woodbine Formation of Texas, but because those fossils have not yet been described they cannot be compared with the scapherpetontid, batrachosauroidid, and indeterminate specimens reported herein from the Dakota Formation. There are no Turonian or Coniacian rocks outside of Utah in North America that have produced salamander fossils, so localities from that interval in the Smoky Hollow and basal part of the John Henry members are the only direct source of information about the evolution of Turonian–Coniacian salamanders on the continent. Higher up in the Utah sequence, the Santonian part of the John Henry Member of the Straight Cliffs Formation can be correlated with the Deadhorse Coulee Member of the Milk River Formation, Alberta, the Wahweap Formation (early–middle Campanian) may be at least partly correlated with the Foremost and Oldman formations, Alberta, and the Kaiparowits Formation (late Campanian) can be correlated with the type Judith River Formation, Montana, the Mesaverde Formation, Wyoming, and the Dinosaur Park Formation, Alberta (Roberts, Deino, and Chan, 2005; Eaton, 2006; Jinnah et al., 2009; Brinkman et al., this volume, Chapter 10). The overall compositions of the Santonian–late Campanian salamander assemblages in Utah and farther to the north in the Western Interior are broadly similar (Table 11.3) in that both have *Scapherpeton* and *Opisthotriton* extending from the Santonian through late Campanian, *Lisserpeton* and *Prodesmodon* in the late Campanian, and *Habrosaurus* in the late Campanian. Differences include: *Habrosaurus* is also known from the Santonian of Utah, but not until the late Campanian outside of Utah; the sirenid-like new genus and species makes its last appearance during the early or middle Campanian in the Wahweap Formation, but is unknown from outside of Utah; and there is no indication in contemporaneous Utah localities of either of the undescribed new genera recognized by Gardner (2000a, 2005) in Alberta, namely a batrachosauroidid in the Milk River Formation (late Santonian) and a scapherpetontid in the Dinosaur Park Formation (late Campanian).

Outside of North America, the only comparable sequence of salamander-bearing rocks occurs in a series of Albian–Campanian localities in Uzbekistan, Kazakhstan, and Tajikistan. Initially scapherpetontids (*Eoscapherpeton* and *Horezmia*) and batrachosauroidids (*Mynbulakia*) were identified from the region (Nessov, 1981, 1988, 1997). More recent and ongoing work (Skutschas, 2009) indicates that all Cretaceous salamanders (*Eoscapherpeton*, *Horezmia*, and *Nesovtriton*) currently recognized from Middle Asia are cryptobranchoids and that there is no direct fossil evidence for salamander families being shared between Asia and North America during the Cretaceous. In Europe, the only records for salamanders of comparable age are: indeterminate vertebrae from La Neuve (middle or late Campanian), France (Garcia et al., 2000); batrachosauroidid and indeterminate vertebrae and dentaries from Champ-Garimond (late Campanian or early Maastrichtian), France (Estes, 1981; Duffaud, 1995; Sigé et al., 1997); and salamandrid- or plethodontid-like vertebrae from Laño (late Campanian or early Maastrichtian), Spain (Duffaud and Rage, 1999). The only similarity between the North American and European Late Cretaceous salamander assemblages is the shared presence of batrachosauroidids.

Most of the sampled localities in Utah contain multiple salamander taxa, representing several genera and families; these diversities compare favorably with those seen in latest Cretaceous salamander assemblages elsewhere in the Western Interior (Estes, 1964; Sahni, 1972; Gardner, 2000a, 2005;

DeMar and Breithaupt, 2006, 2008). The most productive of the oldest localities (late Cenomanian: MNA 1067/OMNH V808/UMNH VP 27) contains at least three salamander taxa, based on the presence of three distinctive dentary morphs. The diversity seen in the Dakota Formation contrasts with that reported for older Cretaceous units in the Western Interior, where one or possibly two indeterminate scapherpetontids were identified in the Mussentuchit Member (latest Albian/earliest Cenomanian) of the Cedar Mountain Formation, Utah (Cifelli et al., 1999), and the monotypic *Prosiren* is the only salamander recognized in the Antlers Formation (Albian), Texas (Estes, 1969a). Some of the vertebrae recovered from the Utah localities are tiny (< 2 mm) and delicate, which indicates that even small-bodied individuals are being sampled, and thus we can be reasonably confident that the samples are not biased against smaller specimens and taxa. In most of the Utah localities, fossils of anurans and albanerpetontids also are present (Gardner, 1999, 2000b; Gardner et al., 2009; Roček et al., 2010). The co-occurrence of all three lissamphibian groups is typical of uppermost Cretaceous nonmarine microvertebrate localities throughout the Western Interior (Estes, 1964; Rowe et al., 1992; Peng, Russell, and Brinkman, 2001; Gardner, 2005; Szentesi et al., in press). The Utah localities indicate that this characteristic lissamphibian community structure, which was established by at least the Albian in North America on the basis of the presence of all three groups in the Antlers Formation (Winkler, Murry, and Jacobs, 1990), was maintained throughout the Late Cretaceous.

The results of our preliminary survey of salamander fossils and taxa from 19 localities ranging in age from late Cenomanian to late Campanian in southern Utah demonstrate that salamanders were relatively abundant and diverse during that interval. Specimens reported here provide new information about the ranges of known taxa and document the existence of new taxa. Considering that many of the sampled localities continue to produce fossiliferous matrix, that dozens of other fossiliferous microsites in Utah have yet to be examined for salamanders, and that ongoing fieldwork continues to identify new microsites, we predict that new salamander specimens and taxa will continue to be recognized, and almost certainly that some of the preliminary identifications and interpretations in this chapter will need to be modified. The prospect of discovering additional localities, specimens, and taxa from the Cenomanian–Coniacian is especially exciting because (1) on a global scale, that interval is poorly sampled for salamanders and (2) except for the Woodbine Formation (Cenomanian) in Texas, the Utah sequence is the only source of information about salamanders from that interval in North America. The Utah record thus holds considerable

Gardner, Eaton, and Cifelli

promise for providing important insights into the evolutionary history of Cretaceous salamanders.

ACKNOWLEDGMENTS

We thank all the persons who have helped with J.G.E. and R.L.C.'s microvertebrate fieldwork in Utah over the past quarter century. Specimens (UMNH) collected from Bryce Canyon National Park and vicinity by J.G.E. were funded by the Colorado Plateau Cooperative Ecosystem Studies Unit of the National Park Service. For assistance with obtaining and curating specimens, arranging loans, tracking down locality information, and numerous other favors (often cheerfully provided at short notice), we thank J. Whitmore-Gillette (MNA), N. Czaplewski and J. Larsen (OMNH), J. Krishna and N. Kubricky (UMNH), R. Nydam (Midwestern University, Glendale, Arizona), D. Larson (formerly University of Alberta, Edmonton), and D. Brinkman (Royal Tyrrell Museum of Palaeontology, RTMP). Some of the specimens reported herein were originally part of C. Redman's (Texas A&M University, College Station, Texas) PhD samples and were hand carried by him to the RTMP in December 2009; that trip was supported by a grant from the RTMP Cooperating Society's Heaton Research Fund. JDG thanks his wife, J. Marklund, for her support and tolerance during the course of this project, G. Braybrook (University of Alberta) for some of the scanning electron micrographs, and the following RTMP colleagues: M. Newbrey, C. Scott, and F. Therrien for assistance with photography and access to equipment, E. Davis for obtaining an interlibrary loan of the Kuhn (1967) publication, and T. Williams for sharing the secret of caps lock in Photoshop. We appreciate the invitation from A. Titus to contribute to this volume and the constructive and supportive comments by D. DeMar, Jr. (University of Washington, Seattle), A. Milner (Natural History Museum, London), and J.-C. Rage (Muséum National d'Histoire Naturelle, Paris) on the submitted version of this chapter.

REFERENCES CITED

Armstrong-Ziegler, J. G. 1980. Amphibia and Reptilia from the Campanian of New Mexico. Fieldiana Geology, n.s., 4:1–39.

Auffenberg, W. 1958. A new family of Miocene salamanders from the Texas Coastal Plain. Quarterly Journal of the Florida Academy of Sciences 21:169–176.

Auffenberg, W. 1961. A new genus of fossil salamander from North America. American Midland Naturalist 66:456–465.

Auffenberg, W., and C. J. Goin. 1959. The status of the salamander genera *Scapherpeton* and *Hemitrypus* of Cope. American Museum Novitates 1979:1–12.

Averianov, A. O., and A. V. Voronkevich. 2002. A new crown-group salamander from the Early Cretaceous of Western Siberia. Russian Journal of Herpetology 9:209–214.

Averianov, A. O., T. Martin, P. P. Skutschas, A. S. Rezvyi, and A. A. Bakirov. 2008. Amphibians from the Middle Jurassic Balabansai Svita in the Fergana Depression, Kyrgyzstan (Central Asia). Palaeontology 51:471–485.

Breithaupt, B. H. 1985. Nonmammalian vertebrate faunas from the Late Cretaceous of Wyoming; pp. 159–175 in G. E. Nelson (ed.), The Cretaceous Geology of Wyoming. Wyoming Geological Association, 36th Annual Field Conference Guidebook.

Carpenter, K. 1979. Vertebrate fauna of the Laramie Formation (Maestrichtian), Weld County, Colorado. University of Wyoming Contributions to Geology 17:37–49.

Cifelli, R. L. 1990. Cretaceous mammals of southern Utah. IV. Eutherian mammals from the Wahweap Formation (Aquilan) and Kaiparowits (Judithian) formations. Journal of Vertebrate Paleontology 10:346–360.

Cifelli, R. L., S. K. Madsen, and E. M. Larson. 1996. Screen-washing and associated techniques for the recovery of microvertebrate fossils; pp. 1–24 in R. L. Cifelli (ed.), Techniques for Recovery and Preparation of Microvertebrate Fossils. Oklahoma Geological Survey Special Publication 96–4.

Cifelli, R. L., J. I. Kirkland, A. Weil, A. R. Deino, and B. J. Kowallis. 1997. High precision $^{40}Ar/^{39}Ar$ geochronology and the advent of North America's Late Cretaceous terrestrial fauna. Proceedings of the National Academy of Science of the United States of America 94:11163–11167.

Cifelli, R. L., R. L. Nydam, J. D. Gardner, A. Weil, J. G. Eaton, J. I. Kirkland, and S. K. Madsen. 1999. Medial Cretaceous vertebrates from the Cedar Mountain Formation, Emery County, Utah: the Mussentuchit local fauna; pp. 219–242 in D. D. Gillette (ed.), Vertebrate Paleontology in Utah. Utah Geological Survey Miscellaneous Publication 99–1.

Cope, E. D. 1876. On some extinct reptiles and Batrachia from the Judith River and Fox Hills beds of Montana. Proceedings of the Academy of Natural Sciences of Philadelphia 28:340–359.

DeMar, D. G., Jr., and B. H. Breithaupt. 2006. The nonmammalian vertebrate microfossil assemblages of the Mesaverde Formation (Upper Cretaceous, Campanian) of the Wind River and Bighorn basins, Wyoming; pp. 33–53 in S. G. Lucas and R. M. Sullivan (eds.), Late Cretaceous vertebrates from the Western Interior. New Mexico Museum of Natural History and Science Bulletin 35.

DeMar, D. G., Jr., and B. H. Breithaupt. 2008. Terrestrial and aquatic vertebrate paleocommunities of the Mesaverde Formation (Upper Cretaceous, Campanian) of the Wind River and Bighorn basins, Wyoming; pp. 78–103 in J. T. Sankey and S. Baszio (eds.), Vertebrate Microfossil Assemblages: Their Role in Paleoecology and Paleobiogeography. Indiana University Press, Bloomington, Indiana.

Denton, R. K., Jr., and R. C. O'Neill. 1998. *Parrisia neocesariensis*, a new batrachosauroidid salamander and other amphibians from the Campanian of eastern North America. Journal of Vertebrate Paleontology 18:484–494.

Duellman, W. E., and L. Trueb. 1986. Biology of Amphibians. McGraw-Hill, New York.

Duffaud, S. 1995. A Batrachosauroididae (Amphibia, Caudata) from the late Cretaceous of Champ-Garimond (Southern France); in First European Workshop on Vertebrate Paleontology. Geological Society of Denmark, DGF On Line Series 1. Available at 2dgf.dk/publikationer/dgf_on_line/vol_1/duffaud.html. Accessed August 6, 2010.

Duffaud, S., and J.-C. Rage. 1999. Amphibians from the Upper Cretaceous of Laño (Basque Country, Spain). Estudios del Museo de Ciencias Naturales de Alava 14(Número Especial 1):111–120.

Duméril, A. M. C. 1806. Zoologie analytique ou methode naturelle de classification des animaux. Allais Libraire, Paris.

Eaton, J. G. 1991. Biostratigraphic framework for the Upper Cretaceous rocks of the Kaiparowits Plateau, southern Utah; pp. 47–63 in J. D. Nations and J. G. Eaton (eds.), Stratigraphy, Depositional Environments, and Sedimentary Tectonics of the Western Margin, Cretaceous Western Interior Seaway. Geological Society of America Special Paper 260.

Eaton, J. G. 1993. Therian mammals from the Cenomanian (Upper Cretaceous) Dakota Formation, southwestern Utah. Journal of Vertebrate Paleontology 13:105–124.

Eaton, J. G. 1999a. Vertebrate paleontology of the Paunsaugunt Plateau, Upper Cretaceous, southwestern Utah; pp. 335–338 in D. D.

Gillette (ed.), Vertebrate Paleontology in Utah. Utah Geological Survey Miscellaneous Publication 99-1.

Eaton, J. G. 1999b. Vertebrate paleontology of the Iron Springs Formation, Upper Cretaceous, southwestern Utah; pp. 339–343 in D. D. Gillette (ed.), Vertebrate Paleontology in Utah. Utah Geological Survey Miscellaneous Publication 99-1.

Eaton, J. G. 2002. Multituberculate Mammals from the Wahweap (Campanian, Aquilan) and Kaiparowits (Campanian, Judithian) Formations, Within and Near the Grand Staircase–Escalante National Monument, Southern Utah. Utah Geological Survey Miscellaneous Publication 02-4.

Eaton, J. G. 2004. New screen-washing approaches to biostratigraphy and paleoecology of nonmarine rocks, Cretaceous of Utah. Bulletin Carnegie Museum of Natural History 36:21–30.

Eaton, J. G. 2006. Santonian (Late Cretaceous) mammals from the John Henry Member of the Straight Cliffs Formation, Grand Staircase–Escalante National Monument, Utah. Journal of Vertebrate Paleontology 26:446–460.

Eaton, J. G., and R. L. Cifelli. 1988. Preliminary report on Late Cretaceous mammals of the Kaiparowits Plateau, southern Utah. University of Wyoming Contributions to Geology 26:45–55.

Eaton, J. G., H. Munk, and M. A. Hardman. 1998. A new vertebrate fossil locality within the Wahweap Formation (Upper Cretaceous) of Bryce Canyon National Park and its bearing on the presence of the Kaiparowits Formation on the Paunsaugunt Plateau; pp. 36–40 in V. Santucci and L. McClelland (eds.), National Park Service Paleontological Research. Technical Report NPS/NRGRD/GRDTR-98/01.

Eaton, J. G., R. L. Cifelli, J. H. Hutchison, J. I. Kirkland, and J. M. Parrish. 1999. Cretaceous vertebrate faunas from the Kaiparowits Plateau, south-central Utah; pp. 345–353 in D. D. Gillette (ed.), Vertebrate Paleontology in Utah. Utah Geological Survey Miscellaneous Publication 99-1.

Eaton, J. G., S. Diem, J. D. Archibald, C. Schierup, and H. Munk. 1999. Vertebrate paleontology of the Upper Cretaceous rocks of the Markagunt Plateau, southwestern Utah; pp. 323–333 in D. D. Gillette (ed.), Vertebrate Paleontology in Utah. Utah Geological Survey Miscellaneous Publication 99-1.

Eaton, J. G., J. I. Kirkland, J. H. Hutchison, R. Denton, R. C. O'Neill, and J. M. Parrish. 1997. Nonmarine extinction across the Cenomanian–Turonian boundary, southwestern Utah, with a comparison to the Cretaceous–Tertiary extinction event. Geological Society of America Bulletin 109:560–567.

Eaton, J. G., J. Laurin, J. I. Kirkland, N. E. Tibert, R. M. Leckie, B. B. Sageman, P. M. Goldstrand, D. W. Moore, A. W. Straub, W. A. Cobban, and J. D. Dalebout. 2001. Cretaceous and Early Tertiary Geology of Cedar and Parowan canyons, western Markagunt Plateau, Utah; pp. 337–363 in M. C. Erskine, J. E. Faulds, J. M. Bartley, and P. D. Rowley (eds.), The Geologic Transition, High Plateaus to Great Basin–A Symposium and Field Guide. Utah Geological Association Publication 30.

Edwards, I. L. 1976. Spinal nerves and their bearing on salamander phylogeny. Journal of Morphology 148:305–328.

Estes, R. 1964. Fossil vertebrates from the Late Cretaceous Lance Formation, eastern Wyoming. University of California Publications in Geological Sciences 49:1–180.

Estes, R. 1965. A new fossil salamander from Montana and Wyoming. Copeia 1965:90–95.

Estes, R. 1969a. Prosirenidae, a new family of fossil salamanders. Nature 224:87–88.

Estes, R. 1969b. The Batrachosauroididae and Scapherpetontidae, Late Cretaceous and Early Cenozoic salamanders. Copeia 1969:225–234.

Estes, R. 1975. Lower vertebrates from the Fort Union Formation, late Paleocene, Big Horn Basin, Wyoming. Herpetologica 31:365–385.

Estes, R. 1976. Middle Paleocene lower vertebrates from the Tongue River Formation, southeastern Montana. Journal of Paleontology 50:500–520.

Estes, R. 1981. Gymnophiona, Caudata; in P. Wellnhofer (ed.), Encyclopedia of Paleoherpetology, Part 2. Gustav Fischer Verlag, Stuttgart.

Evans, S. E., and M. Borsuk-Białynicka. 2009. The Early Triassic stem-frog Czatkobatrachus from Poland. Palaeontologica Polonica 65:79–105.

Evans, S. E., and G. J. McGowan. 2002. Lissamphibian remains from the Purbeck Limestone Group, southern England. Special Papers in Palaeontology 68:103–119.

Evans, S. E., and A. R. Milner. 1993. Frogs and salamanders from the Upper Jurassic Morrison Formation (Quarry Nine, Como Bluff) of North America. Journal of Vertebrate Paleontology 13:24–30.

Evans, S. E., and A. R. Milner. 1994. Middle Jurassic microvertebrate assemblages from the British Isles; pp. 303–321 in N. C. Fraser and H.-D. Sues (eds.), In the Shadow of the Dinosaurs: Early Mesozoic Tetrapods. Cambridge University Press, New York.

Evans, S. E., and A. R. Milner. 1996. A metamorphosed salamander from the early Cretaceous of Las Hoyas, Spain. Philosophical Transactions of the Royal Society of London B 351:627–646.

Evans, S. E., A. R. Milner, and F. Mussett. 1988. The earliest known salamanders (Amphibia, Caudata): a record from the Middle Jurassic of England. Geobios 21:539–552.

Evans, S. E., A. R. Milner, and C. Werner. 1996. Sirenid salamanders and a gymnophionan amphibian from the Cretaceous of the Sudan. Palaeontology 39:77–95.

Evans, S. E., C. Lally, D. C. Chure, A. Elder, and J. A. Maisano. 2005. A Late Jurassic salamander (Amphibia: Caudata) from the Morrison Formation of North America. Zoological Journal of the Linnean Society 143:599–616.

Fillmore, R. P. 1991. Tectonic influence on sedimentation in the southern Sevier foreland, Iron Springs Formation (Upper Cretaceous), southwestern Utah; pp. 9–25 in J. D. Nations and J. G. Eaton (eds.), Stratigraphy, Depositional Environments, and Sedimentary Tectonics of the Western Margin, Cretaceous Western Interior Seaway. Geological Society of America Special Paper 260.

Fox, R. C. 1972. A primitive therian mammal from the Upper Cretaceous of Alberta. Canadian Journal of Earth Sciences 9:1479–1494.

Fox, R. C., and B. G. Naylor. 1982. A reconsideration of the relationships of the fossil amphibian Albanerpeton. Canadian Journal of Earth Sciences 19:118–128.

Frost, D. R., T. Grant, J. Faivovich, R. H. Bain, A. Haas, C. F. B. Haddad, R. O. De Sá, A. Channing, M. Wilkinson, S. C. Donnellan, C. J. Raxworthy, J. A. Campbell, B. L. Blotto, P. Moler, R. C. Drewes, R. A. Nussbaum, J. D. Lynch, D. M. Green, and W. C. Wheeler. 2006. The amphibian tree of life. American Museum of Natural History Bulletin 297:1–370.

Gao, K.-Q., and N. H. Shubin. 2003. Earliest known crown-group salamanders. Nature 422:424–428.

Garcia, G., S. Duffaud, M. Feist, B. Marandat, Y. Tambareau, J. Villatte, and B. Sigé. 2000. La Neuve, gisement à plantes, invertébrés et vertébrés du Bégudien (Sénonien supérieur continental) du bassin d'Aix-en-Provence. Geodiversitas 22:325–348.

Gardner, J. D. 1994. Amphibians from the Lower Cretaceous (Albian) Cedar Mountain Formation, Emery County, Utah. Abstracts of Papers, Fifty-Fourth Annual Meeting of the Society of Vertebrate Paleontology. Journal of Vertebrate Paleontology 14(3, Supplement):26A.

Gardner, J. D. 1999. New albanerpetontid amphibians from the Albian to Coniacian of Utah, U.S.A.–bridging the gap. Journal of Vertebrate Paleontology 19:632–638.

Gardner, J. D. 2000a. Systematics of albanerpetontids and other lissamphibians from the Late Cretaceous of western North America. Ph.D. dissertation, University of Alberta, Edmonton, Alberta.

Gardner, J. D. 2000b. Albanerpetontid amphibians from the Upper Cretaceous (Campanian and Maastrichtian) of North America. Geodiversitas 22:349–388.

Gardner, J. D. 2003a. Revision of Habrosaurus Gilmore (Caudata; Sirenidae) and relationships among sirenid salamanders. Palaeontology 46:1089–1122.

Gardner, J. D. 2003b. The fossil salamander Proamphiuma cretacea Estes (Caudata; Amphiumidae) and relationships within the Amphiumidae. Journal of Vertebrate Paleontology 23:769–782.

Gardner, J. D. 2005. Lissamphibians; pp. 186–201 in P. J. Currie and E. B. Koppelhus (eds.), Dinosaur Provincial Park: A Spectacular Ancient Ecosystem Revealed. Indiana University Press, Bloomington, Indiana.

Gardner, J. D. In press. Revision of Piceoerpeton Meszoely (Caudata: Scapherpetontidae) and description of a new species from the late Maastrichtian and ?early Paleocene of western North America. Bulletin de la Société Géologique de France.

Gardner, J. D., Z. Roček, J. G. Eaton, and R. L. Cifelli. 2009. Amphibians from the Late Cretaceous of southwestern Utah; p. 20 in Advances in Western Interior Late Cretaceous Paleontology and Geology, Grand Staircase–Escalante National Monument Cretaceous Symposium, Abstracts with Program.

Gilmore, C. W. 1928. Fossil lizards of North America. Memoir of the National Academy of Sciences 22:1–201.

Gray, J. E. 1825. A synopsis of the genera of reptiles and Amphibia, with a description of some new species. Annals of Philosophy, n.s., 10:193–217.

Haeckel, E. 1866. Generelle Morphologie der Organismen. 2 volumes. Reimer, Berlin.

Hecht, M. K., and R. Estes. 1960. Fossil amphibians from Quarry Nine. Postilla 46:1–19.

Holman, J. A. 2006. Fossil Salamanders of North America. Indiana University Press, Bloomington, Indiana.

Jinnah, Z. A., E. M. Roberts, A. L. Deino, J. S. Larsen, P. K. Link, and C. M. Fanning. 2009. New ^{40}Ar-^{39}Ar and detrital zircon U-Pb ages for the Upper Cretaceous Wahweap and Kaiparowits formations on the Kaiparowits Plateau, Utah: implications for regional correlation, provenance, and biostratigraphy. Cretaceous Research 30:287–299.

Kuhn, O. 1967. Amphibien und Reptilien: Katalog der Subfamilien und höheren Taxa mit Nachweis des ersten Auftretens. Gustav Fischer Verlag, Stuttgart.

McGowan, G. J. 1996. Albanerpetontid amphibians from the Jurassic (Bathonian) of southern England; pp. 227–234 in M. Morales (ed.), The Continental Jurassic. Museum of Northern Arizona Bulletin 60.

Meszoely, C. A. M. 1967. A new cryptobranchid salamander from the Early Eocene of Wyoming. Copeia 1967:346–349.

Milner, A. R. 1983. The biogeography of salamanders in the Mesozoic and Early Caenozoic: a cladistic–vicariance model; pp. 431–468 in R. W. Sims, J. H. Price, and P. E. S. Whalley (eds.), Evolution, Time and Space: The Emergence of the Biosphere. Academic Press, London.

Milner, A. R. 1988. The relationships and origin of living amphibians; pp. 59–102 in M. J. Benton (ed.), The Phylogeny and Classification of the Tetrapods, Volume 1: Amphibians, Reptiles, Birds. Clarendon Press, Oxford.

Milner, A. R. 2000. Mesozoic and Tertiary Caudata and Albanerpetontidae; pp. 1412–1444 in H. Heatwole and R. L. Carroll (eds.), Amphibian Biology, Volume 4: Paleontology: The Evolutionary History of Amphibians. Surrey Beatty and Sons, Chipping Norton.

Moore, D. W., L. D. Nealey, P. W. Rowley, S. C. Hatfield, D. J. Maxwell, and E. Mitchell. 2004. Geologic Map of the Navajo Lake Quadrangle, Kane and Iron counties, Utah. Utah Geological Survey Map 199.

Munk, H. 1998. A preliminary report on Late Cretaceous herptiles within or near Bryce Canyon National Park, Utah; pp. 41–44 in V. Santucci and L. McClelland (eds.), National Park Service Paleontological Research. Technical Report NPS/NRGRD/GRDTR-98/01.

Naylor, B. G. 1978. The systematics of fossil and recent salamanders (Amphibia: Caudata), with special reference to the vertebral column and trunk musculature. Ph.D. dissertation, University of Alberta, Edmonton, Alberta.

Naylor, B. G. 1979. The Cretaceous salamander Prodesmodon (Amphibia: Caudata). Herpetologica 35:11–20.

Naylor, B. G. 1981. A new salamander of the family Batrachosauroididae from the late Miocene of North America, with notes on other batrachosauroidids. PaleoBios 39:1–14.

Naylor, B. G. 1983. New salamander (Amphibia: Caudata) atlantes from the Upper Cretaceous of North America. Journal of Paleontology 57:48–52.

Naylor, B. G., and D. W. Krause. 1981. Piceoerpeton, a giant Early Tertiary salamander from western North America. Journal of Paleontology 55:507–523.

Nessov, L. A. 1981. [Cretaceous salamanders and frogs of Kyzylkum Desert]. Trudy Zoologicheskogo Instituta, Akademiya Nauk SSSR 101:57–88. [Russian]

Nessov, L. A. 1988. Late Mesozoic amphibians and lizards of Soviet Middle Asia. Acta Zoologica Cracoviensia 31:475–486.

Nessov, L. A. 1997. Cretaceous Nonmarine Vertebrates of Northern Eurasia. Institute of Earth Crust, University of Saint Petersburg, Saint Petersburg, Russia. [In Russian with English abstract]

Nydam, R. L., and B. M. Fitzpatrick. 2009. The occurrence of Contogenys-like lizards in the Late Cretaceous and early Tertiary of the Western Interior of the U.S.A. Journal of Vertebrate Paleontology 29:677–701.

Ogg, J. G., F. P. Agterberg, and F. M. Gradstein. 2004. The Cretaceous period; pp. 344–383 in F. M. Gradstein, J. G. Ogg, and A. G. Smith (eds.), A Geologic Time Scale 2004. Cambridge University Press, Cambridge.

Peng, J., A. P. Russell, and D. B. Brinkman. 2001. Vertebrate Microsite Assemblages (Exclusive of Mammals) from the Foremost and Oldman Formations of the Judith River Group (Campanian) of Southeastern Alberta: An Illustrated Guide. Provincial Museum of Alberta Natural History Occasional Paper 25.

Peterson, F. 1969a. Cretaceous Sedimentation and Tectonism in the Southeastern Kaiparowits Region, Utah. U.S. Geological Survey Open-File Report 69–202.

Peterson, F. 1969b. Four New Members of the Upper Cretaceous Straight Cliffs Formation in the Southeastern Kaiparowits Region, Kane County, Utah. U.S. Geological Survey Bulletin 1274-J.

Petranka, J. W. 1998. Salamanders of the United States and Canada. Smithsonian Institution Press, Washington, D.C.

Rage, J.-C., and D. B. Dutheil. 2008. Amphibians and squamates from the Cretaceous (Cenomanian) of Morocco. Palaeontographica Abteilung A 285:1–22.

Rage, J.-C., L. G. Marshall, and M. Gayet. 1993. Enigmatic Caudata (Amphibia) from the Upper Cretaceous of Gondwana. Geobios 26:515–519

Rieppel, O., and L. Grande. 1998. A well-preserved fossil amphiumid (Lissamphibia: Caudata) from the Eocene Green River Formation of Wyoming. Journal of Vertebrate Paleontology 18:700–708.

Roberts, E. M. 2007. Facies architecture and depositional environments of the Upper Cretaceous Kaiparowits Formation, southern Utah. Sedimentary Geology 197:207–233.

Roberts, E. M., A. L. Deino, and M. A. Chan. 2005. ^{40}Ar/^{39}Ar age of the Kaiparowits Formation, southern Utah, and correlation of contemporaneous Campanian strata and vertebrate faunas along the margin of the Western Interior Basin. Cretaceous Research 26:307–318.

Roček, Z., J. G. Eaton, J. D. Gardner, and T. Přikryl. 2010. Evolution of anuran assemblages in the Late Cretaceous of Utah, USA. Palaeobiodiversity and Palaeoenvironments 90:341–393.

Roelants, K., D. J. Gower, M. Wilkinson, S. P. Loader, S. D. Biju, K. Guillaume, L. Moriau, and F. Bossuyt. 2007. Global patterns of diversification in the history of modern amphibians. Proceedings of the National Academy of Sciences of the United States of America 104:887–892.

Rogers, R. R., and M. E. Brady. 2010. Origins of microfossil bonebeds: insights from the Upper Cretaceous Judith River Formation of north-central Montana. Paleobiology 36:80–112.

Rowe, T., R. L. Cifelli, T. M. Lehman, and A. Weil. 1992. The Campanian Terlingua local fauna, with a summary of other vertebrates from the Aguja Formation, Trans-Pecos Texas. Journal of Vertebrate Paleontology 12:472–493.

Sahni, A. 1972. The vertebrate fauna of the Judith River Formation, Montana. Bulletin of the American Museum of Natural History 147:321–412.

Scopoli, G. A. 1777. Introductio ad historiam naturalem, sistens genera lapidium, planatarum, et animalium: hactenus detecta, caracteribus essentialibus donata, in tribus divisia, subinde ad leges naturae. Apud Wolfgangum Gerle, Prague.

Sigé, B., A. D. Buscalioni, S. Duffaud, M. Gayet, B. Orth, J.-C. Rage, and J. L. Sanz. 1997. Etat des données sur le gisement Crétacé Supérieur continental de Champ-Garimond (Gard, Sud de la France). Münchner Geowissenschaftliche Abhandlungen 34:111–130.

Skutschas, P. P. 2009. Re-evaluation of Mynbulakia Nesov, 1981 (Lissamphibia: Caudata) and description of a new salamander genus from the Late Cretaceous of Uzbekistan. Journal of Vertebrate Paleontology 29:659–664.

Skutschas, P. P., and S. A. Krasnolutskii. 2011. A new genus and species of basal salamanders from the Middle Jurassic of Western Siberia, Russia. Proceedings of the Zoological Institute of the Russian Academy of Sciences 315:167–175.

Skutschas, P. P., and T. Martin. 2011. Cranial anatomy of the stem salamander Kokartus honorarius (Amphibia: Caudata) from the Middle Jurassic of Kyrgyzstan. Zoological Journal of the Linnean Society 161:816–838.

Sullivan, R. M. 1991. Paleocene Caudata and Squamata from Gidley and Silberling Quarries, Montana. Journal of Vertebrate Paleontology 11:293–301.

Szentesi, Z., J. D. Gardner, and M. Venczel. In press. Albanerpetontid amphibians from the Late Cretaceous (Santonian) of Iharkút, Hungary, with remarks on regional differences in Late Cretaceous Laurasian amphibian assemblages. Canadian Journal of Earth Sciences.

Titus, A. L., J. D. Powell, E. M. Roberts, S. D. Sampson, S. L. Pollock, J. I. Kirkland, and L. B. Albright. 2005. Late Cretaceous stratigraphy, depositional environments, and macrovertebrate paleontology of the Kaiparowits Plateau, Grand Staircase–Escalante National Monument, Utah; pp. 101–128 in J. Pederson and C. M. Dehler (eds.), Interior Western United States. Geological Society of America Field Guide 6.

Wake, D. B., and R. Lawson. 1973. Developmental and adult morphology of the vertebral column in the plethodontid salamander *Eurycea bislineata,* with comments on vertebral

evolution in the Amphibia. Journal of Morphology 139:251–300.

Wang, Y. 2004. Taxonomy and stratigraphy of Late Mesozoic anurans and urodeles from China. Acta Geologica Sinica 78:1169–1178.

Winkler, D. A., and L. L. Jacobs. 2002. Cenomanian vertebrate faunas of the Woodbine Formation, Texas. Abstracts of Papers, Sixty-Second Annual Meeting of the Society of Vertebrate Paleontology.

Journal of Vertebrate Paleontology 22(3 Supplement):120A.

Winkler, D. A., P. A. Murry, and L. L. Jacobs. 1990. Early Cretaceous (Comanchean) vertebrates of central Texas. Journal of Vertebrate Paleontology 10:95–116.

Zhang, P., and D. B. Wake. 2009. Higher-level salamander relationships and divergence dates inferred from complete mitochondrial genomes Molecular Phylogenetics and Evolution 53:492–508.

Gardner, Eaton, and Cifelli

12

Anuran Ilia from the Upper Cretaceous of Utah – Diversity and Stratigraphic Patterns

Zbyněk Roček, James D. Gardner, Jeffrey G. Eaton, and Tomáš Přikryl

THANKS TO THEIR RELATIVELY ROBUST BUILD AND DIS-tinctive structure, isolated ilia are among the most commonly recovered anuran bones from fossil micovertebrate sites. Across the spectrum of known anurans, there is considerable variation in features of the ilium. With some caveats, these features may be useful for assigning anuran ilia to biological taxa or, more conservatively, for estimating taxonomic diversities in fossil assemblages. A stratigraphically extensive sequence of 37 microvertebrate sites, ranging in age from the middle? Cenomanian–late Campanian (i.e., an interval of about 25 million years), in southwestern Utah, U.S.A., has yielded a relatively large sample of about 180 anuran ilia. Three major groups of ilia can be identified in the Utah sequence: those with an oblique groove on the medial surface, those with a dorsal tubercle, and those with neither structure. Within each group, specimens further can be subdivided into morphotypes based on other features (e.g., outline and relative size of acetabulum; extent and path of oblique groove; shape and position of dorsal tubercle). Some of the iliac morphotypes are discrete and easily recognizable, whereas others are less distinct. Certain of the iliac morphotypes (especially the more distinct ones) undoubtedly represent biological species, and the occurrence in many of the sampled localities and horizons of multiple morphotypes implies the presence in those areas of at least moderately diverse anuran assemblages. Only one iliac morphotype with an oblique groove can be assigned to a named genus, *Scotiophryne*, and this extends the temporal range for the genus back from the late Maastrichtian into the late Campanian. Although anuran ilia are not useful for stratigraphic correlations within the Utah sequence, several interesting patterns are evident; for example, the rarity both of specimens with a dorsal tubercle in the early–middle Santonian and middle Campanian and of specimens with an oblique groove in the late Santonian or early Campanian. The Utah ilia are typical for Mesozoic anurans in that none has a dorsal crest and only a minority have a dorsal tubercle; this contrasts with the situation in the Cenozoic, when most anurans have one or both of those iliac structures.

INTRODUCTION

Anurans (frogs) are the most diverse and widespread group of living amphibians, consisting of about 5400 extant species and occurring on every continent except Antarctica (Bossuyt and Roelants, 2009). Anurans also were a characteristic component of Late Cretaceous ecosystems in the North American Western Interior, judging by how common their remains are in nonmarine deposits in that region (Estes, 1964; Sahni, 1972; Estes and Sanchíz, 1982; Brinkman, 1990; Gardner, 2005; Roček et al., 2010). In contrast to the situation elsewhere in the Western Interior, where Late Cretaceous anurans are known from more limited stratigraphic intervals (e.g., Campanian and Maastrichtian of Wyoming and Montana; Santonian and Campanian of Alberta, Canada), in Utah anurans are known from every stage of the Late Cretaceous (Cifelli et al., 1999a, 1999b; Roček et al., 2010). Long-term quarrying by one of us (J.G.E.) of Upper Cretaceous strata in southwestern Utah, and subsequent processing of the rock by wet screen washing, has recovered more than 700 isolated anuran bones from 37 microvertebrate localities of middle? Cenomanian–late Campanian age. This interval is about 25 million years, according to the timescale of Ogg, Agterberg, and Gradstein (2004), and the Utah sequence represents one of the most stratigraphically extensive records of Late Cretaceous anurans from anywhere in the world. Recently we presented a preliminary report on this collection (Roček et al., 2010). Ilia are the most commonly recovered bones and account for about one quarter of the total collection of anuran bones from southwestern Utah. Historically, ilia have been important for identifying fossil anuran species. To cite two examples: first, in the most recent compendium of fossil anurans, about 40% of the listed species had the ilium as their holotype (Sanchiz, 1998:6); and second, four of the five anuran species recognized by Holman (2003) from the North American Late Cretaceous had the ilium as their holotype. In the past decade, justifiable concerns have been raised about the uncritical use of iliac features to differentiate among anuran species (Jones, Evans, and Ruth, 2002; Bever, 2005; Gardner, 2008). Nevertheless, it is evident that across the spectrum of extant and fossil anurans that

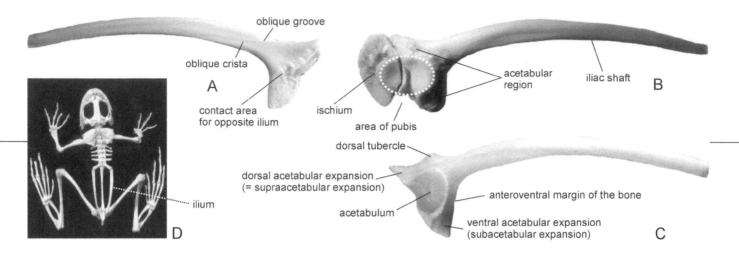

12.1. Principal morphological features of the anuran ilium. (A) Right ilium with an oblique groove (group 1 ilia) in medial view, *Pelobates fuscus* (Recent). (B) Pelvis, in right lateral view and with outlines of acetabulum marked by white dotted line, *Pelobates fuscus* (Recent). (C) Right ilium with a dorsal tubercle (group 2 ilia) in lateral view, *Bufo calamita* (Recent). (D) Articulated anuran skeleton showing location of ilium, in dorsal view, *Rana ridibunda* (Recent).

there is considerable variation in the structure of this bone (Lynch, 1971:figs. 36–40; Trueb, 1973:fig. 2–11; Tyler, 1976:fig. 3) and that some of those features are phylogenetically and taxonomically significant. Although it is no longer accepted practice to erect fossil anuran species solely on the basis of subtle differences in iliac structure, especially when known from only one isolated bone, anuran ilia remain useful for estimating or gauging taxonomic diversities in fossil assemblages, especially ones in which isolated bones are the only or a major source of information (Henrici, 1998; Prasad and Rage, 2004; Gardner, 2008). As a follow-up to our 2010 article, here we present more detailed identifications, descriptions, and comparisons of anuran iliac morphotypes from the Upper Cretaceous (middle? Cenomanian–upper Campanian) of southwestern Utah, and we evaluate whether these morphotypes can be used for stratigraphic correlations.

All the ilia collected from southwestern Utah and used for this study are curated in the vertebrate paleontological collections of the Utah Museum of Natural History (UMNH) in Salt Lake City, Utah. The complete list of those specimens and their localities may be found in Roček et al. (2010). Examples of extant anuran ilia and important osteological features are shown in Fig. 12.1.

GEOLOGICAL SETTING

A thick succession of Upper Cretaceous sediments was deposited in the Sevier foreland basin of Utah. The anuran specimens reported here are from three adjacent plateaus in southwestern Utah (Figs. 12.2, 12.3). On these plateaus, the Upper Cretaceous sequence unconformably overlies Jurassic rocks and in turn is overlain unconformably by lower

Cenozoic rocks. The youngest unit in the sequence, the Kaiparowits Formation, is found only on the Kaiparowits Plateau. With the exception of the Kaiparowits Formation, the same units found on the Kaiparowits Plateau are found on the next plateau to the west, the Paunsaugunt Plateau. Units that are time equivalent to those on the Paunsaugunt Plateau are present on the westernmost Markagunt Plateau, but on the latter, the lithology is different enough that it has been difficult to use the Paunsaugunt Plateau terminology on the Markagunt sequence. Moore et al. (2004) suggested using the "Formation of Cedar Canyon" for rocks possibly equivalent to the upper part of the Straight Cliffs Formation and the Wahweap Formation until a more secure basis for lithologic correlation has been established. The correlation used for the Markagunt Plateau has been based on mammalian faunas (Eaton, 2006) and radiometric dates (Eaton, Maldonado, and McIntosh, 1999; Eaton et al., 2001). Further details about the anuran-bearing localities and formations can be found in Roček et al. (2010).

GENERAL STRUCTURE OF ANURAN ILIA

The anuran ilium consists of two portions: an expanded acetabular portion and an anteriorly elongate shaft (Fig. 12.1). Compared to the more primitive tetrapod condition, anuran ilia are distinctive in the following ways: (1) the left and right ilia articulate with one another along the medial surface of the acetabular region (primitively the ilia are separated by the ischia and pubes); (2) the ilia and other bones of the pelvic girdle are rotated posteriorward about 90°, so the iliac shaft lies in the horizontal plane (primitively in a more vertical plane); and (3) the iliac shaft is greatly elongate (primitively

Roček et al.

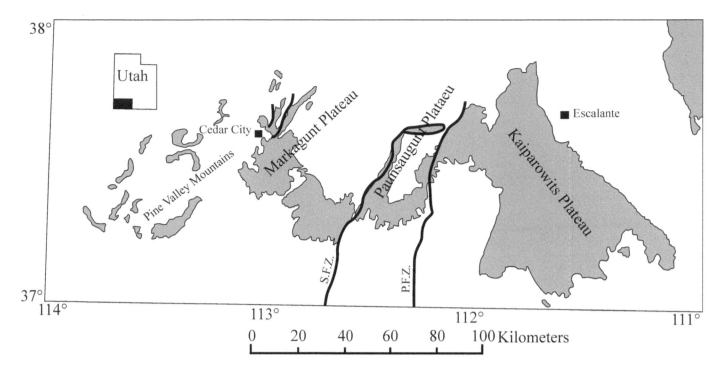

12.2. Location map showing Cretaceous outcrops in southwestern Utah from which anuran fossils were recovered. Most of the anuran specimens were recovered from Cedar Canyon on the west side of the Markagunt Plateau, from Bryce Canyon National Park area on the east side of the Paunsaugunt Plateau, and from the Kaiparowits Plateau, which is within the Grand Staircase–Escalante National Monument. P.F.Z., Paunsaugunt fault zone; S.F.Z., Sevier fault zone.

it is relatively shorter). Thanks to their distinctive structure, anuran ilia can be easily recognized in fossil deposits, even when they are isolated and broken. This is fortunate, because all of the 183 Upper Cretaceous ilia collected to date from southwestern Utah are isolated and incomplete, with each missing part of the shaft and some missing portions of the acetabular region. For a more detailed description of anuran ilia and features that are useful for identifying that bone, see Gardner et al. (2010, and references therein).

The acetabular portion of the anuran ilium articulates by immovable sutures with two other pelvic bones–ventrally with the pubis, which typically is cartilaginous and thus not capable of fossilization (the pubis is ossified in some anurans; e.g., Pipidae) and posteriorly with the ischium. On its lateral surface, the acetabular portion bears the anterior part of a bowl-like cavity, termed the acetabulum, for articulation with the femur; posteriorly, the acetabulum extends onto the ischium. The shape, size, prominence, and position of the acetabulum may differ considerably among species. The posterior part of the acetabular region of the ilium may be strongly extended both dorsally and ventrally, resulting in a fan-like or triangular shape. As a result, when viewed in lateral aspect, a considerable portion of the bone may be visible below the acetabulum (termed the ventral acetabular or sub-acetabular expansion) and above the acetabulum (termed the dorsal acetabular or supraacetabular expansion). The medial surface of the acetabular portion usually bears a rugosity that is the contact area for attachment with the ilium on the opposite side (Fig. 12.1A). The iliac shaft is anteriorly elongate, and moderately convex dorsally and, to a lesser extent, also laterally. Unlike in more advanced Cenozoic anurans, most Mesozoic taxa have an iliac shaft that is suboval in cross section and lacks a longitudinal dorsal crest.

The transition between the acetabular region and the iliac shaft may be marked by a depression between the dorsal margins of the supraacetabular expansion and the iliac shaft. The dorsal margin of the supra-acetabular expansion slants anteroventrally and continues onto the medial surface of the shaft, thus producing a more or less prominent crista (Fig. 12.1A). As a result, an oblique groove (called the spiral groove by some authors) that is delimited posteroventrally by that crista crosses the dorsal margin of the ilium and extends onto the medial surface, sometimes extending as far as the lower margin of the shaft. On the lateral surface of the ilium, the oblique groove may begin as a horizontal depression above the acetabulum. The oblique groove is a typical feature of many Mesozoic anurans.

In many extant anurans, the dorsal surface of the acetabular portion bears a raised outgrowth called the dorsal tubercle. Although its base may be anteroposteriorly elongate, the top of this tubercle is uniformly located at the level of the anterior margin of the acetabulum. The dorsal tubercle

Stage		Markagunt Plateau	Paunsaugunt Plateau	Kaiparowits Plateau
Campanian	U			
	M			Kaiparowits Fm.
	L	"Formation of Cedar Canyon"	Wahweap Fm.	Wahweap Fm.
Santonian	U		Drip Tank Member	Drip Tank Member
	M			
	L	Straight Cliffs Formation	John Henry Member	John Henry Member
Coniacian	U			
	M		?	?
	L			
Turonian	U		Smoky Hollow Member	Smoky Hollow Member
	M		Tibet Canyon Member	Tibet Canyon Member
	L	Tropic Shale	Tropic Shale	Tropic Shale
Cenomanian	U	Dakota Fm.	Dakota Fm.	Dakota Fm.
	M			
	L			

(Straight Cliffs Fm. labeled vertically in the Paunsaugunt and Kaiparowits columns.)

12.3. Generalized Upper Cretaceous stratigraphic units on the Markagunt, Paunsaugunt, and Kaiparowits Plateaus in southwestern Utah. Diagonal lines indicate missing strata.

is the area of origin for three muscles that are functionally important for the preparatory (crouching) and initial phases of jumping. However, the same muscles also occur in those taxa that lack a dorsal tubercle. In such cases, these muscles originate on the lateral surface of the ilium (Přikryl et al., 2009). It is not known whether the shift in the origin of these muscles is associated with differences in jumping capabilities. It is, however, important to emphasize that ilia with a dorsal tubercle never have an oblique groove crossing their dorsal margin onto the medial surface, so their medial surface is mostly smooth. Some anurans, including some Mesozoic taxa, lack both an oblique groove and a dorsal tubercle on their ilia.

MORPHOTYPIC DIVERSITY OF ILIA IN THE UPPER CRETACEOUS OF SOUTHWESTERN UTAH

To make comparisons easier and try to make some sense of the diverse array of ilia, we created an informal classification based on the morphotypes listed below. Such morphotypes can be used to make limited inferences about the taxonomic diversity of fossil assemblages, without erecting potentially redundant biological taxa on the basis of isolated bones from different parts of the skeleton that may be difficult to associate with each other. (For a similar approach using isolated Upper Cretaceous fish elements, see Brinkman and Neuman, 2002; Brinkman et al., this volume, Chapter 10.) Three

principal groups of anuran ilia can be recognized easily in the Utah samples: those with an oblique groove (group 1), those with a dorsal tubercle (group 2), and those with neither structure (group 3). Within each group, many morphotypes could be recognized, mostly on the basis of subtle differences. The main challenge we encountered in identifying and defining those morphotypes was how to handle variation within each of the three principal groups. Especially in the first group, morphotypes that initially seemed to be clearly defined often were bridged, as sample sizes increased, by specimens with intermediate features. For instance, the oblique groove in some ilia is delimited by sharp, prominent cristae and in others by faint, rounded ridges; however, there is a series of specimens that are intermediate between those two extremes. That pattern could suggest extensive or broad individual variation. However, other specimens exhibit minimal or no obvious variation. For example, two group 2 specimens (UMNH VP 13482 and 13494; Fig. 12.9, upper part of middle column) from the Santonian or Campanian Paul's locality in Cedar Canyon (Roček et al., 2010) are right ilia (and therefore from different individuals) that are identical even in minor details.

The first group consists of ilia with an oblique groove but no dorsal tubercle. The oblique groove originates on the lateral surface of the bone, above the dorsal margin of the acetabulum, then it crosses over the dorsal edge of the bone along a shallow depression between the acetabular region and iliac shaft, and continues onto the medial surface of the posterior portion of the shaft where it may extend as far as the ventral margin of the shaft. In *Pelobates* (Fig. 12.1A), *Scaphiopus*, and to a certain degree also *Megophrys*, which are the only extant frogs in which the oblique groove is known, the portion of the groove on the medial surface of the shaft is the area of origin of muscles important for locomotion (Přikryl at al., 2009). The second group consists of ilia that have a dorsal tubercle but no oblique groove. The dorsal tubercle is located on the upper margin of the ilium at the level of the anterior margin of the acetabulum. In extant anurans, the dorsal tubercle serves as the area of origin for muscles responsible for extension and flexion of the hind limb (Přikryl at al., 2009). It should be noted that the earliest anurans (*Prosalirus*; Early Jurassic) lacked the dorsal tubercle, and that in the Early Triassic proanurans (*Triadobatrachus* and *Czatkobatrachus*) there is a similar, prominent tubercle, but it is located more anteriorly relative to the acetabulum. The third group consists of ilia that have neither an oblique groove nor a dorsal tubercle.

In the following accounts, each morphotype is briefly diagnosed, followed by information on stratigraphic occurrences and a list of voucher specimens. To allow for better comparisons, key characters of all morphotypes are

Roček et al.

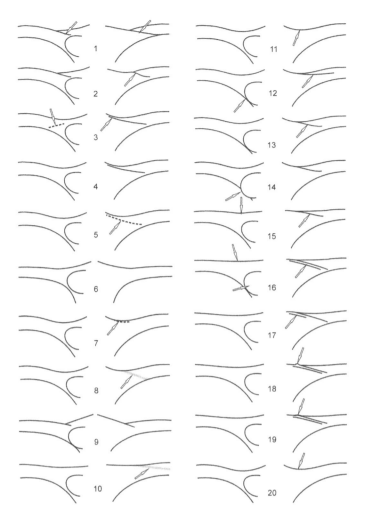

Figure 12.4. Semidiagrammatic line drawings of the first 20 of 26 iliac morphotypes with an oblique groove (group 1 ilia) left ilium in lateral (left) and medial (right) views (continues onto Fig. 12.5. Morphotype numbers are the same as in the text. Key characters are marked by arrows. Full lines within the outlines of the bone denote crests, broken lines denote rounded ridges, gray strips denote a broad shallow groove. See the text for diagnoses and voucher specimens.

illustrated in Figs. 12.4–12.7, and photographs of most of the voucher specimens are presented in approximate stratigraphic sequence in Figs. 12.8–12.11.

GROUP 1: ILIA WITH AN OBLIQUE GROOVE (FIGS. 12.4–12.5; FIGS. 12.8–12.11, LEFT COLUMN)

Morphotype 1

Description The dorsal margin of the pars ascendens extends anteriorly as a distinct, oblique crista onto the medial surface of the ilium and reaches the ventral margin of the shaft; the iliac shaft is markedly convex dorsally, and consequently there is a distinct concavity on the dorsal margin of the bone in lateral view; an oblique groove passes the dorsal

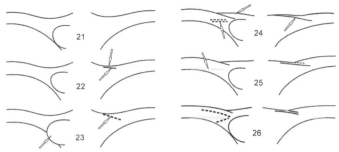

12.5. Semidiagrammatic line drawings of the last six of 26 iliac morphotypes with an oblique groove (group 1 ilia) left ilium in lateral (left) and medial (right) views (continued from Fig. 12.4).

margin of the bone from its lateral surface onto the medial surface on the bottom of this concavity; the groove begins in the horizontal depression above the acetabulum and terminates on the lower margin of the iliac shaft; acetabulum not extending beyond the anteroventral outline of the bone.

Distribution Dakota Formation, middle? Cenomanian: UMNH VP 12936.

Morphotype 2

Description Similar to 1, except the dorsal margin of the pars ascendens continues anteroventrally onto the medial surface of the bone as a prominent but gradually lowering osseous lamina that disappears before reaching the lower margin of the ilia shaft; the lamina may extend in an obtuse point and may be paralleled by a shallow groove posteriorly; the medial surface of the acetabular portion is depressed; acetabulum does not reach the level of the anteroventral outline of the bone. These features correspond to those given by Estes (1969:fig. 1c, d) in his illustration of the holotype ilium of *Scotiophryne pustulosa*.

Distribution Wahweap Formation, lower–middle Campanian: UMNH VP 18182, 18263, 18283, 18338, perhaps also UMNH VP 18287, 18288, 18140 (the last may be from a juvenile).

Morphotype 3

Description Similar to 1, except the pars ascendens is broad and widely rounded on its dorsal margin, consequently the crista delimiting the oblique groove is shifted onto the medial surface of the bone; the oblique groove is wide and continues onto the medial surface of the shaft, consequently the medial surface of the shaft is horizontally concave or flat; the oblique groove originates on the dorsal margin of the bone, not on its lateral surface; on the lateral surface, a faint oblique mound or groove runs anteroventrally from the supraacetabular region.

Distribution Straight Cliffs Formation, John Henry Member, Coniacian: UMNH VP 19371. Kaiparowits Formation, upper Campanian: UMNH VP 19311.

Morphotype 4

Description Similar to 3, except the lateral surface of the bone is uniformly convex and smooth, because the oblique groove is confined to the medial surface.

Distribution Straight Cliffs Formation, John Henry Member, upper Santonian: UMNH VP 18480.

Morphotype 5

Description Similar to 4, except instead of the crista delimiting the oblique groove posteriorly there is only a faint rounded ridge.

Distribution Straight Cliffs Formation, Smoky Hollow Member, Turonian: UMNH VP 18367. Straight Cliffs Formation, John Henry Member, upper Santonian: UMNH VP 18476. Wahweap Formation, lower–middle Campanian: UMNH VP 18106.

Morphotype 6

Description Similar to 5, except the margins of the acetabular region are markedly divergent from the longitudinal axis of the bone; the shaft is slender; the anteroventral margin of the acetabulum is markedly prominent; the medial surface of the acetabular region is concave, consequently this part of the bone is very thin.

Distribution Wahweap Formation, lower–middle Campanian: UMNH VP 18104.

Morphotype 7

Description Similar to 5, except the dorsal margin of the pars ascendens joins the dorsal margin of the shaft by a short crista shifted medially, which, however, does not continue onto the medial surface; although this gives an impression that there is a broad oblique groove crossing the dorsal surface of the bone, it does not continue onto either the lateral or medial surfaces.

Distribution Straight Cliffs Formation, John Henry Member, lower Santonian: UMNH VP 18299. Straight Cliffs Formation, John Henry Member, upper Santonian: UMNH VP 18484, 18506.

Morphotype 8

Description Similar to 1, except the oblique groove originates on the dorsal surface of the bone and continues as a broad shallow depression toward the lower margin of the bone.

Distribution Straight Cliffs Formation, John Henry Member, lower Santonian: UMNH VP 18544 (note: spine projecting ventrally from anteroventral margin of the acetabulum is considered a malformation). Kaiparowits Formation, upper Campanian: UMNH VP 19314, 19330.

Morphotype 9

Description Similar to 8, except the tip of the pars ascendens extends dorsally beyond the level of the shaft; the acetabulum is prominent but shallow; the oblique groove deep, with rounded margins; the dorsal part of the shaft is markedly swollen medially.

Distribution Wahweap Formation, lower–middle Campanian: UMNH VP 18102.

Morphotype 10

Description Similar to 1, except the oblique groove is only a faint depression on the medial surface of the shaft; the lateral surface of the shaft is smooth; the dorsal margins of the shaft and the pars ascendens meet in a shallow concavity.

Distribution Formation uncertain, lower–middle Campanian: UMNH VP 18304.

Morphotype 11

Description Similar to 1, except the oblique groove is developed only in the concavity on the rounded dorsal margin of the bone; the medial surface of the acetabular portion is moderately depressed; the acetabulum is in the middle of the acetabular region.

Distribution Straight Cliffs Formation, John Henry Member, middle Santonian: UMNH VP 18213, perhaps also UMNH VP 18220, 19356. Straight Cliffs Formation, John Henry Member, upper Santonian: UMNH VP 18344. Wahweap Formation, lower–middle Campanian: UMNH VP 18154, 18282.

Morphotype 12

Description Similar to 11, except the oblique groove carries onto the medial surface of the shaft, extends anteriorly and slightly ventrally, and is delimited ventrally by a short, faint ridge; the acetabulum is extremely prominent ventrally.

Roček et al.

Distribution Straight Cliffs Formation, John Henry Member, middle Santonian: UMNH VP 18227.

Morphotype 13

Description Similar to 12, except the faint ridge deliminating the ventral margin of the oblique groove extends anteroventrally at a steeper angle and for a shorter distance along the medial surface of the shaft.

Distribution Straight Cliffs Formation, John Henry Member, middle Santonian: UMNH VP 18539.

Morphotype 14

Description Similar to 13, except the acetabulum is extremely large, exceeding the anteroventral margin of the bone.

Distribution Wahweap Formation, lower–middle Campanian: UMNH VP 18286.

Morphotype 15

Description Similar to 1, except the shaft is only slightly convex and the pars ascendens only moderately declined from the longitudinal axis of the bone; consequently, the dorsal margin of the bone has only a very shallow concavity; the acetabulum is located in the middle of the acetabular region and is not too extensive; lateral surface of the bone is smooth; only a faint ridge runs down from the pars ascendens onto the medial surface of the bone, but is not accompanied by an obvious oblique groove. Although this morphotype lacks an obvious oblique groove, we have retained it in the group 1 ilia because the faint ridge that typically accompanies that groove is present and thus implies that the same muscle inserted in that region.

Distribution Straight Cliffs Formation, John Henry Member, lower Santonian: UMNH VP 18509.

Morphotype 16

Description Similar to 10 in that a shallow oblique groove begins on the dorsal margin of the bone, but the dorsal margin of the bone is straight; the oblique groove on the medial surface of the bone is paralleled by another groove originating on the medial surface of the pars ascendens; the acetabulum is shifted ventrally.

Distribution Straight Cliffs Formation, John Henry Member, lower Santonian: UMNH VP 18293. Straight Cliffs Formation, John Henry Member, upper Santonian: UMNH VP 18395, 18471. Wahweap Formation, lower–middle Campanian: UMNH VP 18180, perhaps also UMNH VP 18183. Formation uncertain, lower–middle Campanian: UMNH VP 18307.

Morphotype 17

Description Similar to 16, except the groove that runs parallel to the main oblique groove is faint and short; the dorsal margin of the bone is shallowly concave; the acetabulum is not extensive, located in the middle of the acetabular region.

Distribution Wahweap Formation, lower–middle Campanian: UMNH VP 18264.

Morphotype 18

Description Similar to 16, except the posterior margin of the oblique groove is delimited by a prominent crista that may appear, in lateral view, as a dorsal tubercle-like structure; also the ridge coming from the medial surface of the pars ascendens and delimiting the parallel groove on the medial surface of the bone is better pronounced than in 16.

Distribution Straight Cliffs Formation, Smoky Hollow Member, Turonian: UMNH VP 13461.

Morphotype 19

Description Similar to 18, except the cristae are rounded and the dorsal tubercle-like structure is less prominent.

Distribution Wahweap Formation, lower–middle Campanian: UMNH VP 19345.

Morphotype 20

Description Similar to 11, except the medial surface of the bone is entirely smooth, because the oblique groove does not extend onto that surface.

Distribution Straight Cliffs Formation, John Henry Member, lower Santonian: UMNH VP 18556. Straight Cliffs Formation, John Henry Member, middle Santonian: UMNH VP 19303. Straight Cliffs Formation, John Henry Member, upper Santonian: UMNH VP 18473.

Morphotype 21

Description Similar to 11, except the medial surface of the bone is entirely smooth, because the oblique groove does not extend onto that surface, and the acetabulum is extensive and shifted to the lower part of the acetabular region.

Distribution Straight Cliffs Formation, John Henry Member, lower Santonian: UMNH VP 19270, 19273.

Morphotype 22

Description Similar to 11, except the oblique groove (which is restricted to the dorsal margin of the bone) is accompanied by a similar, short groove posteriorly.

Distribution Straight Cliffs Formation, John Henry Member, upper Santonian: UMNH VP 18396.

Morphotype 23

Description Similar to 20, except the acetabulum extends beyond the anteroventral margin of the bone; a rounded ridge runs down from the medial surface of the dorsal margin of the acetabular region onto the medial surface of the iliac shaft, where it soon disappears.

Distribution Formation uncertain, upper Santonian or lower Campanian: UMNH VP 13480. Wahweap Formation, lower–middle Campanian: UMNH VP 18103.

Morphotype 24

Description Similar to 1 in that the dorsal margin of the shaft arises on the lateral margin of the bone above the anterior margin of the acetabulum, and that a shallow concavity between the shaft and the pars ascendens continues onto the medial surface of the bone, delimited posteriorly by a faint ridge; the shaft is widely convex dorsally; however, the pars ascendens is low, producing a moderate convexity and not terminating in a posterodorsally directed point; the lateral surface of the proximal section of the shaft bears faint, anteroventrally directed grooves and longitudinal elevations; a posteroventrally directed crista splits off from the anteroventral margin of the acetabulum and the ventral margin of the acetabulum continues posteriorly and even posterodorsally; consequently, a triangular field occurs between the crista and margin of the acetabulum. (Note that although this crista is a distinctive structure, it is too small to be depicted in the corresponding drawing in Fig. 12.5.)

Distribution Straight Cliffs Formation, Smoky Hollow Member, Turonian: UMNH VP 18366.

Morphotype 25

Description Similar to 24, except the oblique groove is only on the dorsal margin of the bone, delimited posteriorly by a rounded ridge coming from the margin of the pars ascendens; posteriorly, the oblique groove is paralleled by another groove, delimited by a crista; on the lateral surface of the bone is a shallow horizontal depression.

Distribution Wahweap Formation, lower–middle Campanian: UMNH VP 18247.

Morphotype 26

Description Similar to 25, except the acetabulum is large and shifted ventrally, with its anteroventral margin markedly prominent; a horizontal mound derives from the dorsal portion of the acetabulum, and another, less prominent mound is directed anteroventrally; a distinct depression is developed between them; the crista delimiting posteroventrally the oblique groove is rather prominent, accompanied by a parallel groove that is delimited by only a faint ridge.

Distribution Kaiparowits Formation, upper Campanian: UMNH VP 18435.

GROUP 2: ILIA WITH A DORSAL TUBERCLE (FIG. 12.6; FIGS. 12.8– 12.11, MIDDLE COLUMN)

Morphotype 1

Description The iliac shaft is moderately convex and its dorsal margin widely rounded; the dorsal margin of the pars ascendens is continuous with the dorsal margin of the shaft; the dorsal tubercle is mediolaterally compressed, triangular in lateral view, with its upper part widely rounded and slightly declined anteriorly; the dorsal margin of the acetabulum is markedly prominent; consequently, the dorsal surface of the pars ascendens is horizontal or even concave; a broad mound runs from the anterior margin of the acetabulum anteriorly; a broad shallow groove extends from the base of the tubercle onto the posterior section of the shaft where it disappears.

Distribution Straight Cliffs Formation, Smoky Hollow Member, Turonian: UMNH VP 13459. Formation uncertain, upper Santonian or lower Campanian: UMNH VP 13488, 13496.

Morphotype 2

Description Similar to 1, except the dorsal tubercle is knoblike, not compressed mediolaterally, and is slightly squarish in lateral outline.

Distribution Straight Cliffs Formation, John Henry Member, upper Santonian: UMNH VP 18483. Formation uncertain, upper Santonian or lower Campanian: UMNH VP 13500, 13504.

Morphotype 3

Description Similar to 1, except the dorsal tubercle is knoblike, not compressed mediolaterally, and the medial surface of the bone is smooth.

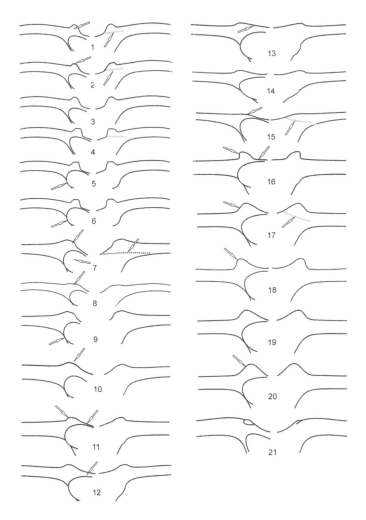

12.6. Semidiagrammatic line drawings of the 21 iliac morphotypes with a dorsal tubercle (group 2 ilia) left ilium in lateral (left) and medial (right) views. Symbols are the same as in Fig. 12.4.

Distribution Dakota Formation, upper Cenomanian: UMNH VP 13159. Straight Cliffs Formation, Smoky Hollow Member, Turonian: UMNH VP 18355.

Morphotype 4

Description Similar to 1, except the dorsal tubercle is rather quadrangular in lateral view and the acetabulum is prominent but shallow; because of the prominent dorsal margin of the acetabulum, there is a groove between the acetabulum and the dorsal tubercle.

Distribution Straight Cliffs Formation, Smoky Hollow Member, Turonian: UMNH VP 13460.

Morphotype 5

Description Similar to 4, except the acetabulum is deeply concave with a raised edge along the entire margin and it extends from the level of the dorsal margin of the pars

ascendens to beyond the anteroventral margin of the bone; the medial surface of the bone is smooth.

Distribution Straight Cliffs Formation, Smoky Hollow Member, Turonian: UMNH VP 13458.

Morphotype 6

Description Similar to 4, except the dorsal tubercle is not compressed laterally and the dorsal margin of the acetabulum is not prominent.

Distribution Formation uncertain, upper Santonian or lower Campanian: UMNH VP 13483.

Morphotype 7

Description The posterior section of the iliac shaft is straight;; the dorsal tubercle has an anteroposteriorly extended base but is low, with its upper part rather swollen; in cross section, the dorsal surface of the pars ascendens is a broadly rounded, nearly horizontal plane extending onto the dorsal margin of the acetabulum; the acetabulum extends from nearly the level of the dorsal surface of the pars ascendens almost to the anteroventral margin of the bone; the medial surface of the bone is flat and almost smooth, except for a faint crista running from the medial surface of the pars ascendens to the lower margin of the iliac shaft.

Distribution Dakota Formation, upper Cenomanian: UMNH VP 13158.

Morphotype 8

Description Similar to 7, except the dorsal tubercle is a low, anteroposteriorly elongate crista; the iliac shaft is moderately convex dorsally; the medial surface of the bone is smooth.

Distribution Dakota Formation, upper Cenomanian: UMNH VP 13156. Wahweap Formation, lower–middle Campanian: UMNH VP 18112.

Morphotype 9

Description Similar to 7 in size and lateral shape of the tubercle; however, the upper part of the tubercle is sharp (not swollen) and its lateral surface slants down toward the prominent dorsal margin of the acetabulum; the medial surface of the tubercle is vertical, continuous without any distinct border with the flat and smooth medial surface of the bone; the anteroventral margin of the acetabulum reaches the level of the bone; the iliac shaft is straight.

Distribution Straight Cliffs Formation, John Henry Member, middle Santonian: UMNH VP 19355.

Morphotype 10

Description Similar to 7, except the dorsal tubercle is more extensive anteroposteriorly, and slightly declined anteriorly; the medial surface of the bone is smooth.

Distribution Straight Cliffs Formation, John Henry Member, upper Santonian: UMNH VP 18474.

Morphotype 11

Description Similar to 7, except the dorsal tubercle is small and declined anteriorly; the acetabulum extends to the dorsal margin of the bone.

Distribution Straight Cliffs Formation, John Henry Member, upper Santonian: UMNH VP 18394.

Morphotype 12

Description Similar to 7, except the acetabulum extends beyond the anteroventral margin of the bone.

Distribution Wahweap Formation, lower–middle Campanian: UMNH VP 18114, 18122.

Morphotype 13

Description Similar to 12, except the dorsal margin of the pars ascendens is above the level of the dorsal margin of the iliac shaft.

Distribution Wahweap Formation, lower–middle Campanian: UMNH VP 18099.

Morphotype 14

Description Similar to 13, except the dorsal margin of the pars ascendens is slightly concave and the acetabular region of the bone is low.

Distribution Wahweap Formation, lower–middle Campanian: UMNH VP 18098.

Morphotype 15

Description Similar to 11, except the dorsal tubercle is only a faint though extensive elevation; a shallow groove extends on the medial surface of the bone as in morphotype 1.

Distribution Formation uncertain, upper Santonian or lower Campanian: UMNH VP 13497.

Morphotype 16

Description Similar to 11, except the dorsal tubercle is comparatively large and prominent; the dorsal margin of the pars

ascendens is horizontal, continuous with that of the iliac shaft.

Distribution Formation uncertain, upper Santonian or lower Campanian: UMNH VP 13481.

Morphotype 17

Description Similar to 1, except the dorsal tubercle is a triangular, comparatively thin lamella with its rounded tip slightly declined anteriorly; the tubercle is well delimited, both laterally and medially, from the surrounding horizontal surface of the bone.

Distribution Formation uncertain, upper Santonian or lower Campanian: UMNH VP 13482, 13494, 13499.

Morphotype 18

Description Similar to 17, except the dorsal tubercle is more prominent; the medial surface of the bone is smooth.

Distribution Wahweap Formation, lower–middle Campanian: UMNH VP 18095, 18100, 18113.

Morphotype 19

Description Similar to 17, except the medial surface of the dorsal tubercle is continuous with the medial surface of the bone, the latter of which is smooth.

Distribution Wahweap Formation, lower–middle Campanian: UMNH VP 18101.

Morphotype 20

Description Similar to 10, except the dorsal tubercle is comparatively large, triangular in lateral view, compressed mediolaterally either only in its posterior part or in the whole of its extent, with a rugosity in its lateral side; the medial surface of the tubercle is continuous with the medial surface of the bone, which is flat and smooth; the acetabulum extends up to the dorsal margin of the bone.

Distribution Formation uncertain, upper Santonian or lower Campanian: UMNH VP 13501. Wahweap Formation, lower–middle Campanian: UMNH VP 18108, 18121.

Morphotype 21

Description Large ilium; the dorsal margin of the pars ascendens is declined posteroventrally from the longitudinal axis of the bone; the acetabulum is comparatively small and shifted ventrally; the iliac shaft is strongly compressed mediolaterally but not bearing a dorsal crest; the dorsal tubercle is small, declined dorsolaterally, separated from the dorsal

Roček et al.

edge of the shaft by a shallow depression; both the lateral and medial surfaces of the bone are smooth.

Distribution Formation uncertain, upper Santonian or lower Campanian: UMNH VP 13551.

GROUP 3: ILIA LACKING AN OBLIQUE GROOVE AND A DORSAL TUBERCLE (FIG. 12.7; FIGS. 12.8–12.11, RIGHT COLUMN)

Morphotype 1

Description Between the pars ascendens and shaft is only a very faint concavity on the dorsal surface of the bone; the acetabulum is cup-like, located in the middle of the acetabular region; on the lateral surface of the posterior section of the shaft are longitudinal, anteroventrally directed elevations, well delimited posteroventrally; the medial surface of the acetabular portion is depressed.

Distribution Straight Cliffs Formation, John Henry Member, Coniacian: UMNH VP 19366.

Morphotype 2

Description Similar to 1, except the lateral surface bears a moderately well-developed (i.e., not too prominent) crista.

Distribution Wahweap Formation, lower–middle Campanian: UMNH VP 18139.

Morphotype 3

Description Similar to 1, except the lateral surface of the shaft bears no elevations.

Distribution Straight Cliffs Formation, John Henry Member, middle Santonian: UMNH VP 18211, 18218. Straight Cliffs Formation, John Henry Member, upper Santonian: UMNH VP 18475. Formation uncertain, ?Campanian: UMNH VP 18554, possibly also UMNH VP 18555. Wahweap Formation, lower–middle Campanian: UMNH VP 18134.

Morphotype 4

Description Similar to 1, except there are no longitudinal elevations on the lateral surface of the shaft; the pars ascendens is extensive and produces a robust posterodorsal point; the posterior section of the shaft is comparatively slender; the acetabulum is shifted ventrally.

Distribution Straight Cliffs Formation, John Henry Member, middle Santonian: UMNH VP 18228, 18233; Formation uncertain, upper Santonian or lower Campanian: UMNH

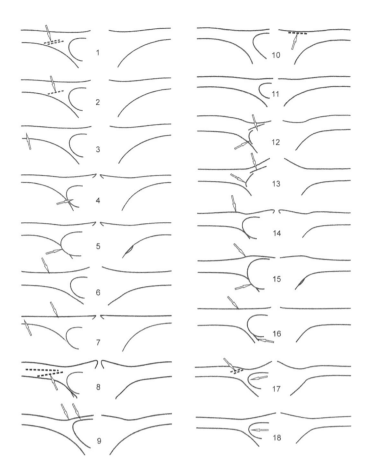

12.7. Semidiagrammatic line drawings of the 18 iliac morphotypes lacking both an oblique groove and a dorsal tubercle (group 3 ilia) left ilium in lateral (left) and medial (right) views. Morphotypes 17 and 18 are cf. *Nezpercius* sp. (Caudata) each depicted with long axis rotated about 90° from life orientation (i.e., iliac shaft would have extended vertically) to emphasize similarities between isolated ilia of anurans and caudates. Symbols are the same as in Fig.12.4.

VP 13484, 13495, possibly also 13498 and 19293. Formation uncertain, ?Campanian: UMNH VP 18552.

Morphotype 5

Description Similar to 4, except the posterior section of the shaft is not slender; the acetabulum is large, exceeding the anteroventral margin of the bone.

Distribution Straight Cliffs Formation, John Henry Member, upper Santonian: UMNH VP 18479. Perhaps also Wahweap Formation, lower–middle Campanian: UMNH VP 18088, 18320.

Morphotype 6

Description Similar to 4, except the margins of the acetabular portion are moderately divergent; the dorsal margin of the shaft is straight and meets the dorsal margin of the pars

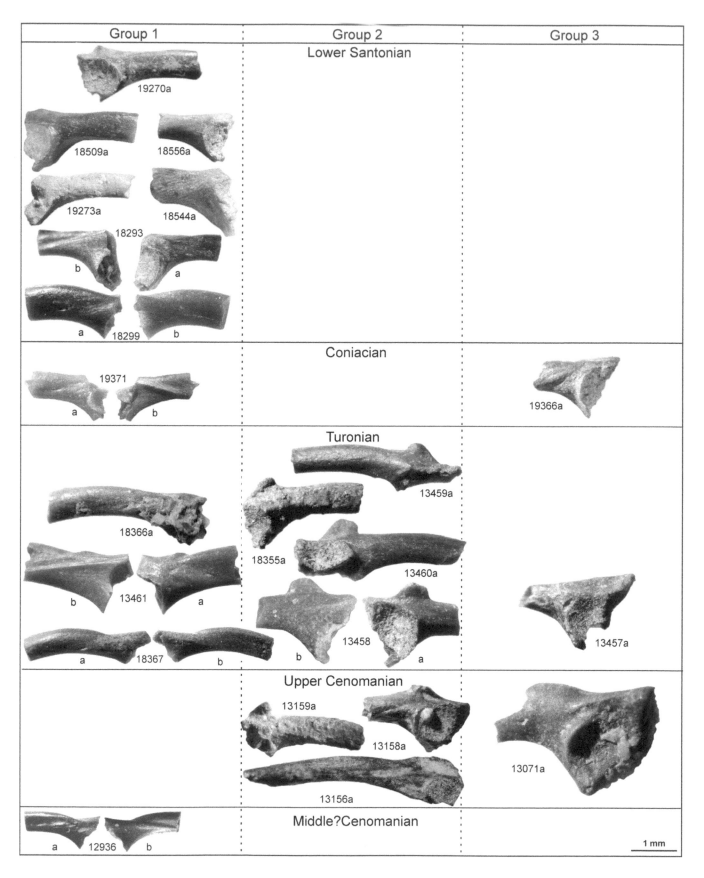

12.8. Stratigraphic distribution of representative anuran ilia from middle? Cenomanian through lower Santonian localities in south–western Utah (continues onto Figs. 12.9–12.11). Left column is group 1 ilia (with an oblique groove); center column is group 2 ilia (with a dorsal tubercle); and right column is group 3 ilia (lacking both an oblique groove and a dorsal tubercle). Most of the voucher specimens listed in the text are depicted here. For each specimen, the accompanying five-digit number is its formal UMNH catalog number; a, b, and c indicate lateral, medial, and dorsal views, respectively. Specimens shown at same magnification (see scale bar at lower right).

Roček et al.

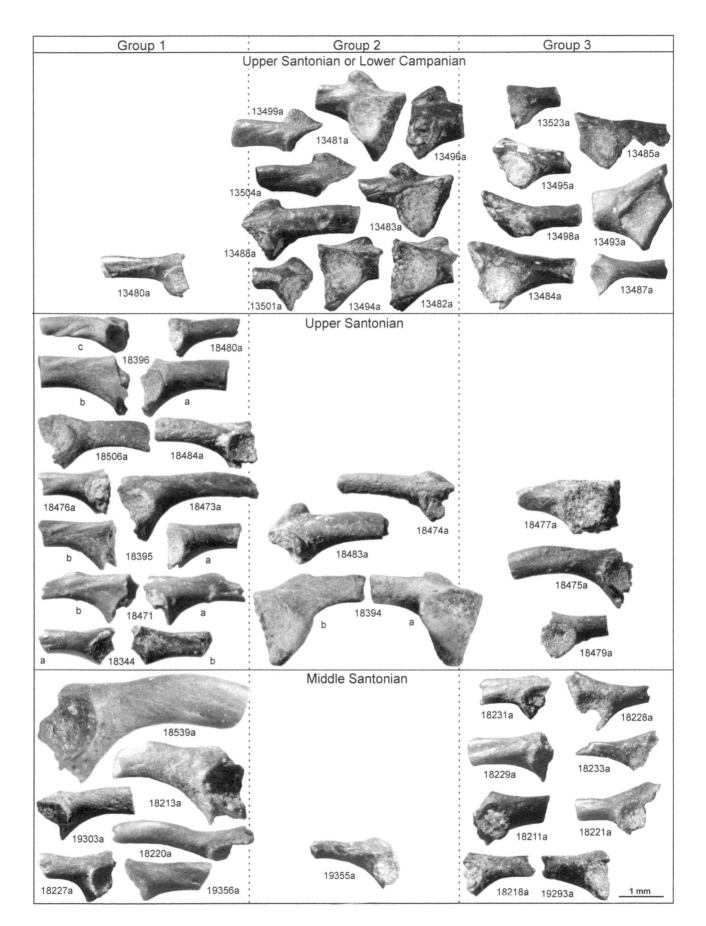

12.9. Stratigraphic distribution of representative anuran ilia from middle Santonian through upper Santonian or lower Campanian localities in southwestern Utah (continued from Fig. 12.8; see Fig. 12.8 caption for details).

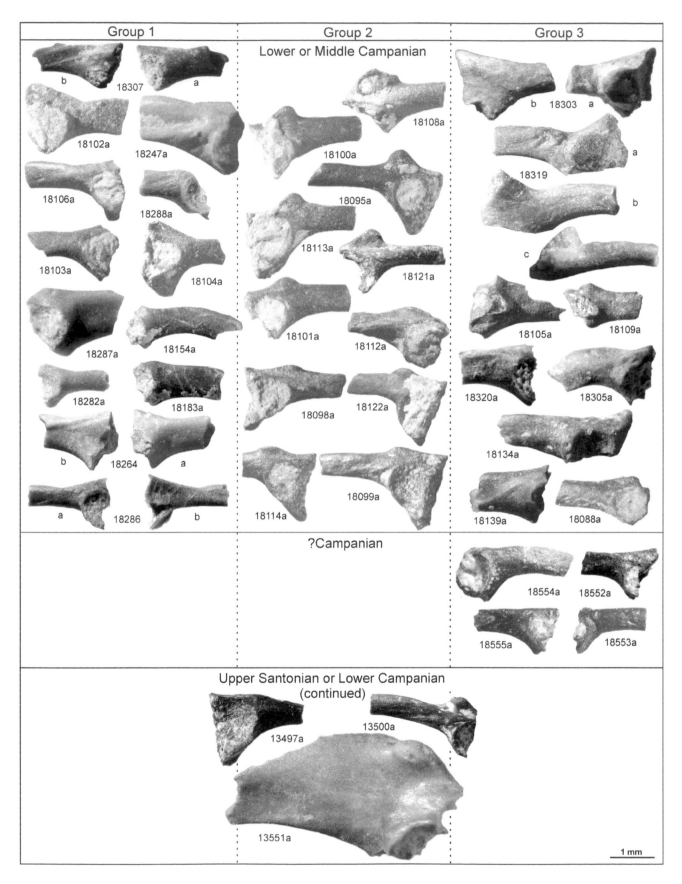

Group 1

b
18307
a

18102a
18247a

18106a
18288a

18103a
18104a

18287a
18154a

18282a
18183a

b
18264
a

a
18286
b

Group 2

Lower or Middle Campanian

18108a

18100a

18095a

18113a
18121a

18101a
18112a

18098a
18122a

18114a
18099a

Group 3

b
18303
a

a
18319

b

c

18105a
18109a

18320a
18305a

18134a

18139a
18088a

?Campanian

18554a
18552a

18555a
18553a

Upper Santonian or Lower Campanian
(continued)

13497a
13500a

13551a

1 mm

12.10. Stratigraphic distribution of representative anuran and caudate ilia from upper Santonian or lower Campanian through lower or middle Campanian localities in southwestern Utah (continued from Figs. 12.8 and 12.9; see Fig. 12.8 caption for details). Four examples of group 3 ilia (UMNH VP 18303, 18319, 18105, and 18109; all in upper right) are cf. *Nezpercius* sp. (Caudata) each depicted with long axis rotated about 90° from life orientation (i.e., iliac shaft would have extended vertically) to emphasize similarities between isolated ilia of anurans and caudates.

Roček et al.

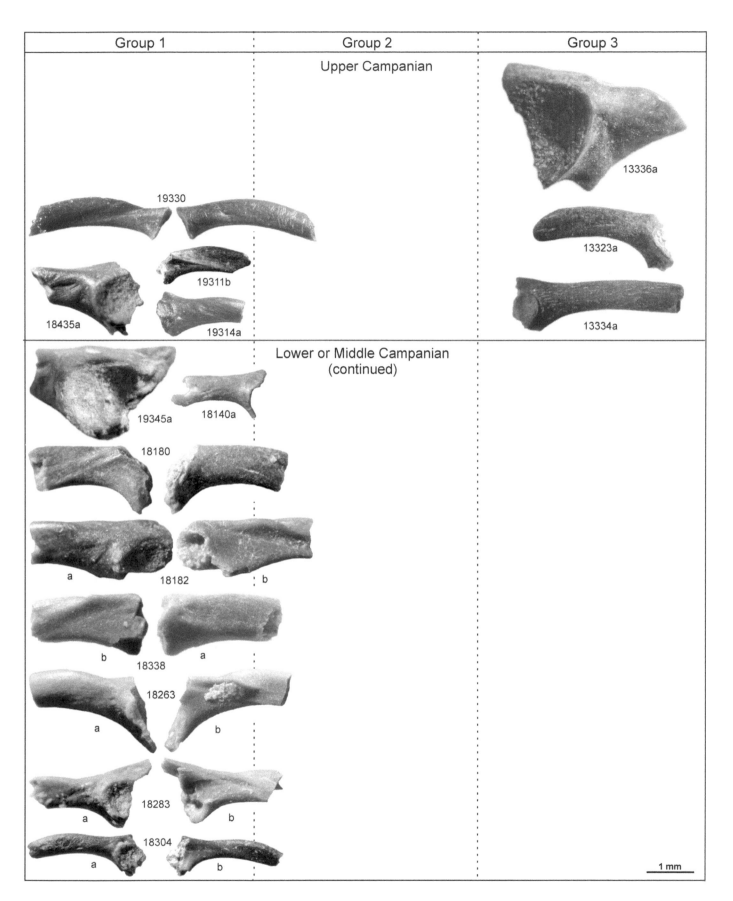

Group 1	Group 2	Group 3

Upper Campanian

12.11. Stratigraphic distribution of representative anuran ilia from lower or middle Campanian through upper Campanian localities in southwestern Utah (continued from Figs. 12.8–12.10; see Fig. 12.8 caption for details).

ascendens in a shallow depression; the acetabulum is prominent but shallow; the medial surface of the bone is concave and smooth.

Distribution Formation uncertain, upper Santonian or lower Campanian: UMNH VP 13487, 13523.

Morphotype 7

Description Similar to 4, except the dorsal margin of the bone is straight.

Distribution Straight Cliffs Formation, Smoky Hollow Member, Turonian: UMNH VP 13457.

Morphotype 8

Description Similar to 4, except there are two elongate and anteriorly divergent mounds on the lateral surface of the posterior section of the shaft; the medial surface of the pars ascendens is concave.

Distribution Straight Cliffs Formation, John Henry Member, middle Santonian: UMNH VP 18221, 18229, possibly also UMNH VP 18231.

Morphotype 9

Description Acetabulum large, extending closer to the dorsal margin of the acetabular region than to the ventral one; on the dorsal margin of the bone, the transition between the pars ascendens and the iliac shaft is marked by a moderate indentation; the shaft seems to be comparatively slender; the medial surface of the bone is flat or even slightly concave posteriorly.

Distribution Kaiparowits Formation, upper Campanian: UMNH VP 13336.

Morphotype 10

Description Acetabulum small, its anterodorsal part is a mere depression in the bone; the shaft is comparatively robust; a prominent though rounded ridge runs horizontally on the dorsal part of the medial surface.

Distribution Kaiparowits Formation, upper Campanian: UMNH VP 13334.

Morphotype 11

Description Similar to 10, except the acetabulum is well delimited and located in the middle of the acetabular region; the acetabular region is comparatively small; the shaft is only moderately convex dorsally; the medial surface of the acetabular region is smooth and flat.

Distribution Kaiparowits Formation, upper Campanian: UMNH VP 13323.

Morphotype 12

Description The dorsal margin of the pars ascendens is declined posterodorsally and its posterior tip is swollen and slightly declined laterally above the acetabulum; the iliac shaft is laterally compressed; the posterior section of the dorsal margin of the iliac shaft is widely convex; consequently, it meets the dorsal margin of the pars ascendens in a shallow concavity; an oblique rounded ridge splits from the dorsal margin of the pars ascendens and runs down anteroventrally onto the lateral surface of the shaft where it terminates abruptly at the level of the anterior margin of the acetabulum; the acetabulum is a mere depression in the bone, except for a well-delimited anterior part that is prominent; the medial surface of the bone is moderately convex and smooth.

Distribution Dakota Formation, upper Cenomanian: UMNH VP 13071. Formation uncertain, lower or middle Campanian: UMNH VP 18305.

Morphotype 13

Description Similar to 12, except the margins of the acetabular region are strongly divergent from the longitudinal axis of the bone; the shaft is relatively stout; the acetabulum is a mere depression in the bone, except for the anterior margin, which is slightly prominent; a rounded ridge coming from the dorsal margin of the pars ascendens terminates abruptly above the anterior acetabular margin; the medial surface of the bone is moderately convex and smooth.

Distribution Formation uncertain, upper Santonian or lower Campanian: UMNH VP 13493.

Morphotype 14

Description The iliac shaft is convex in its posterior section, whereas its anterior part is straight; the acetabulum is a mere depression with rounded margins; the medial surface of the bone is moderately convex and smooth.

Distribution Formation uncertain, upper Santonian or lower Campanian: UMNH VP 13485.

Morphotype 15

Description The pars ascendens is not developed, consequently the dorsal margin of the acetabulum extends to the dorsal margin of the bone; the acetabulum is large, exceeding beyond the anteroventral margin of the bone; the shaft is comparatively slender; a very shallow convexity is present

Roček et al.

on the dorsal surface of the acetabular region, above the anterior margin of the acetabulum (although this convexity occurs in the same position as the dorsal tubercle, we hesitate to identify it as such because it is very shallow and it is not prominently developed as the knob- or ridge-like structure seen in most group 2 ilia); the medial surface of the acetabular portion is concave.

Distribution Straight Cliffs Formation, John Henry Member, upper Santonian: UMNH VP 18477.

Morphotype 16

Description The dorsal margin of the bone is straight; the acetabulum is small but markedly prominent; a posteroventrally directed crista splits from the anteroventral margin of the acetabulum.

Distribution Formation uncertain, ?Campanian: UMNH VP 18553.

Morphotype 17

Description The posterior part of the acetabular portion is strongly extended both dorsally and ventrally so it has a fan-like, triangular shape; the posterior, articular surface of the bone is thick but concave (it was probably filled with cartilage in the living animal); the anterior part of the acetabulum is convex and extremely prominent, so the acetabulum is strongly declined posteriorly and very shallow; the acetabulum does not reach the level of the anteroventral outline of the bone; the medial surface of the acetabular portion is smooth; there is a distinct rugosity on the lateral surface of the iliac shaft. Those features are similar (but not all identical) to those of *Nezpercius dodsoni* (Blob et al., 2001), so referral to *Nezpercius* is justified.

Distribution Wahweap Formation, lower–middle Campanian: UMNH VP 18319.

Morphotype 18

Description Same as in 17 (and also referable to *Nezpercius*), except there is no rugosity on the lateral surface of the iliac shaft. The surface of the anterior, convex margin of the acetabulum may be abraded.

Distribution Wahweap Formation, lower–middle Campanian: UMNH VP 18105, 18109, 18303.

STRATIGRAPHIC OCCURRENCES OF ILIA IN THE UPPER CRETACEOUS OF SOUTHWESTERN UTAH

Group 1 ilia

For the first group of ilia, namely those with an oblique groove, we defined our morphotype 1 on the basis of a typical ilium with an oblique groove (UMNH VP 12936) from the lowest locality in our stratigraphic sequence (middle? Cenomanian). On the basis of its stratigraphic occurrence, this morphotype can be postulated as the basic or most primitive iliac morphotype, from which others in the same group could have been derived. Interestingly, morphotype 1 or its slight modifications persisted throughout the entire stratigraphic sequence, as indicated by morphotype 8 (represented by UMNH VP 19330, which differs from morphotype 1 only in that there is a shallow depression instead of a groove) in the upper Campanian portion of the sequence.

Morphotype 2 could have been derived from morphotype 1. It also is one of the few morphotypes in our samples that corresponds with ilia from elsewhere in the Western Interior that have been assigned to a named anuran species. Estes (1969) chose an ilium having an oblique groove delimited posteriorly by an expanded crista from the Bug Creek Anthills (mixed upper Maastrichtian and lower Paleocene), Hell Creek Formation of Montana, as the holotype of *Scotiophryne pustulosa*. Estes (1969) also referred to the species some distinctive skull bones (maxillae and squamosals), humeri, and several ilia in which the oblique groove was less strongly marked than in the holotype. Gardner (2008) described additional, isolated bones and presented a revised diagnosis for *S. pustulosa*. Because the holotype ilium (Estes, 1969:fig. 1c, d) of *S. pustulosa* appears virtually identical to our morphotype 2 ilium, we assign the Utah specimens to *Scotiophryne*. In our samples, this morphotype occurs in the lower–middle Campanian Wahweap Formation, at both the Campbell Canyon and White Flats Road localities. These Utah occurrences extend the range for *Scotiophryne* from the late Maastrichtian/early Paleocene (Estes, 1969; Gardner, 2008) back to the early–middle Campanian.

Morphotype 3 mainly differs from the previous two in that its crista delimits a broad depression in the medial surface of the iliac shaft and bears an oblique depression on its lateral surface. It was recorded by only two specimens, one (UMNH VP 19371) from the Coniacian and one (UMNH VP 19311) from the upper Campanian. Morphotype 4, represented by UMNH VP 18480 from the upper Santonian, is similar and differs only in that its lateral surface is smooth. If these differences express individual variation and morphotypes 3 and 4 belong to a single taxon, then its stratigraphic range is from the Coniacian to late Campanian.

Morphotypes 5, 7, and 8 are also similar to each other and vary only in different expressions of the oblique groove. Considering that the oblique groove is an area for attachment of the medial head of the iliacus externus muscle (Přikryl et al., 2009), variation in the groove may indicate differences in how well developed this muscle was in individuals of the same or closely related taxa. These three morphotypes were recorded from the Turonian to the uppermost level (upper Campanian) in our stratigraphic sequence. This lengthy stratigraphic range supports the view that morphotypes 1–5, 7, and 8 are generalized ilia with an oblique groove, from which only morphotype 2 is markedly derived by its peculiar morphology of having a prominent lamina delimiting the groove posteriorly, which is a feature that substantiates the taxonomic distinctiveness of the genus *Scotiophryne* (Estes, 1969). Morphotypes 10, 11, and 15 may also be derived from the generalized scheme, as are morphotypes 1–5, 7, and 8, in that their oblique groove is poorly developed (i.e., very shallow or entirely absent). Morphotypes 10, 11, and 15 have a more restricted range and are known from the middle Santonian through middle Campanian.

In contrast to the previously discussed ilia, morphotype 6 (UMNH VP 18104) differs in that the margins of its pars ascendens and pars descendens are strongly divergent from the longitudinal axis of the bone (they form an angle of more than 90°) and that its shaft is straight and comparatively slender. The only example of this morphotype was recovered from the lower–middle Campanian Barker Reservoir Road locality.

Morphotype 9 (UMNH VP 18102) also is markedly different in having a massive pars ascendens and an acetabulum that is shifted to the anteroventral margin of the acetabular region and is markedly prominent from the surface of the bone, but with a flat acetabular surface. It also was recovered only from the lower–middle Campanian Barker Reservoir Road locality.

Nine morphotypes (12–14, 16–19, 20, and 21) still maintain the basic morphology of the oblique groove, especially on the medial surface of the bone; however, there is some variation, especially in the position and size of their acetabula.

Morphotypes 12 and 13 are characterized by their large and ventrally shifted acetabula, which resembles the condition in morphotype 9. However, morphoptypes 12 and 13 are older (middle Santonian) and thus provide the earliest evidence for expansion and ventral shift of the acetabulum in ilia with an oblique groove from Utah. Otherwise, the two morphotypes are typical for ilia with an oblique groove and differ from each other only in minor details, most notably the course of the oblique groove on the medial surface of the bone. Morphotype 14 could also be included with the previous two morphotypes, but it differs from them in having a larger acetabulum and a stratigraphically higher (lower Campanian) occurrence.

Morphotypes 16 through 19 share two features: (1) the oblique groove is paralleled posteriorly by another groove arising on the medial surface of the pars ascendens, and (2) the nearly straight dorsal margin of the bone. Morphotype 16 is characterized by a large and ventrally shifted acetabulum; its stratigraphic occurrence is lower Santonian through lower–middle Campanian. Morphotypes 17 through 19 differ from morphotype 16 and resemble one another in having the acetabulum smaller and located in the middle of the acetabular region, but they differ from each other in how prominently developed is the crista that posteriorly delimits the oblique groove. In morphotype 18, represented by UMNH VP 13461, this crista is so prominent that where it crosses the dorsal margin of the bone it resembles a small tubercle. However, because morphotype 18 occurs in the Turonian and the similar morphotype 19 (UMNH VP 19345) occurs much later in the lower–middle Campanian, it appears that these two morphotypes (and probably also other morphotypes with two parallel grooves) are just variations of the basic morphotype with a single oblique groove and, judging by their stratigraphic occurrence, they have no stratigraphic value.

Morphotypes 20 and 21 are characterized by having the oblique groove restricted to the dorsal margin of the bone and by the smooth medial surface of the bone. These morphotypes differ from one another only in the position of their acetabula. Both occur only in the Santonian.

The remaining two morphotypes, 23 and 26, are notable for their large and ventrally located acetabula. Besides UMNH VP 13480 from the upper Santonian or lower Campanian Pinto Flats locality; this is the only example of an ilium with an oblique groove recovered from this locality, even though the other two principal groups of ilia are well represented at that locality.

Although ilia with an oblique groove have a continuous distribution throughout the stratigraphic sequence in southern Utah and we have recognized 26 different morphotypes, we were not able to recognize correlations between any particular morphotype and stratigraphic intervals. Even if there were morphotypes restricted to certain stratigraphic units, those would be difficult to recognize on the basis of our small sample sizes, which, in some cases, constituted only a single specimen for a particular morphotype. Moreover, because the morphotypes in this group of ilia generally differ only subtly from one another, it is difficult to decide whether those differences are taxonomically significant or simply reflect individual variation; regardless, most of the morphotypes appear to have no stratigraphic value. The only exception seems to be morphotype 2 (*Scotiophryne*), which is clearly defined by a prominent lamina that posteriorly delimits the oblique groove and, in the Utah samples, is stratigraphically restricted to the lower–middle Campanian. Elsewhere in the Western Interior, *Scotiophryne* is known from the late Maastrichtian and early Paleocene of Montana and Wyoming (Estes, 1969; Gardner, 2008).

Group 2 ilia

The second principal group consists of ilia with a dorsal tubercle. The most generalized morphotype (morphotype 1) is characterized by having a slightly convex shaft, by a dorsal tubercle that is triangular, laterally compressed, has a rounded apex and is slightly declined anteriorly, and by the prominent dorsal margin of the acetabulum. This morphotype occurs in the Turonian and upper Santonian or, possibly, lower Campanian.

Morphotype 3 is similar, except its dorsal tubercle is not laterally compressed. It occurs in the upper Cenomanian and Turonian. The relatively early (Turonian) co-occurrence and the overall similarity of morphotypes 1 and 3 might suggest that the generalized morphotype for this group of ilia was one in which the dorsal tubercle could vary in its mediolateral compression. In other words, it could include ilia with both compressed and knob-like dorsal tubercles. If this is correct, then morphotypes 1 and 3 could both represent the generalized morphotype for this group of ilia.

Morphotype 2 from the upper Santonian or lower Campanian is similar to morphotypes 1 and 3, except the outline of its dorsal tubercle tends to be a more squarish and it is not laterally compressed. A squarish outline of the dorsal tubercle is also characteristic of morphotypes 4, 5, and 6; these vary only in the shape and location of their acetabula. However, they occur in the Turonian (UMNH VP 13458 and 13460 in Fig. 12.8) and upper Santonian or lower Campanian (UMNH VP 13483 in Fig. 12.9). It is unknown why there is a gap between these stratigraphic occurrences and whether the squarish outline of the dorsal tubercle indicates the ilia are from closely related taxa. If the squarish outline of the dorsal tubercle arose independently more than once, then at least some of the ilia with that feature might be from unrelated taxa.

Morphotype 7 differs from the previous six in that its acetabular region is declined ventrally (i.e., the dorsal margin behind the tubercle is declined posteroventrally, whereas the dorsal margin of the shaft is located horizontally; consequently, the dorsal tubercle is a protuberance located at the point where both meet together). It is represented by a single specimen (UMNH VP 13158) from the upper Cenomanian.

Morphotype 8 is similar, except its dorsal tubercle is compressed laterally and elongated anteroposteriorly, so it is more a crista than a tubercle. This morphotype is very characteristic, and therefore it is rather surprising that examples occur in the upper Cenomanian (UMNH VP 13156 in Fig. 12.8) and Campanian (UMNH VP 18112 in Fig. 12.10).

Morphotype 9 possibly could be related to morphotype 7 (there are only minor differences between them); however, they are separated by a large stratigraphic gap (upper Cenomanian versus middle Santonian). On the other hand, morphotype 9 likely is related to morphotypes 10, 11, and 12, represented by UMNH VP 18474, 18394, and 18114 and 18122, respectively. All (including morphotype 9) are from the middle Santonian to lower or middle Campanian and all are characterized by a massive acetabular region that is declined posteroventrally from the longitudinal axis of the shaft. Morphotype 13, which also is from the lower or middle Campanian, differs from morphotype 12 only by having the posterior margin of its dorsal tubercle being confluent with the dorsal margin of the pars ascendens. Consequently, the dorsal tubercle is only slightly prominent.

Morphotypes 14 and 15 (represented by UMNH VP 18099 and 18098, respectively) are similar to each other in their acetabular region being shallow and slender, and their dorsal tubercle is not prominent. These two morphotypes are also restricted to the upper Santonian or lower Campanian (morphotype 15) and lower–middle Campanian (morphotype 14).

Morphotype 16 is characterized by an extensive acetabular region (the dorsal margin of which is continuous with the dorsal margin of the shaft) and by a prominent dorsal tubercle. It occurs in the upper Santonian or lower Campanian.

Morphotypes 17–20 are all characterized by a dorsal tubercle that is large and triangular in lateral aspect and by an extremely large acetabulum. All are restricted to the upper Santonian–middle Campanian.

Morphotype 21 strongly deviates from all other ilia in that the dorsal margin of the shaft is a comparatively sharp edge. This morphotype is similar to ilia of *Enneabatrachus hechti* Evans and Milner, 1993, from the Upper Jurassic Morrison Formation, in that the dorsal tubercle is declined laterally and that its shaft has a medially declined dorsal crest. However, in UMNH VP 13551 from the upper Santonian or lower Campanian, the dorsal tubercle is separated from the dorsal crest by a deep depression, which argues against assigning this specimen to *Enneabatrachus*.

To sum up the situation in the group of the ilia with a dorsal tubercle, one can conclude that although there are intriguing irregularities in the occurrences of this type of ilia (e.g., their complete absence or relative scarcity in the lower and middle Santonian and in the upper Campanian), it is not possible to recognize any morphotypes that can be correlated with a particular stratigraphic interval. Even those morphotypes that seemingly are well defined (i.e., those with squarish tubercle, such as morphotypes 2, 4, 5, and 6) occur in the upper Santonian or lower Campanian, but also in the Turonian. If differences in the size and shape of the dorsal tubercle are ignored, one might argue that ilia with an enlarged acetabular region (morphotypes 11–20) are typical of the upper Santonian–middle Campanian.

Group 3 ilia

Within the third group – ilia lacking both the oblique groove and dorsal tubercle – morphotypes 1 to 3 differ from each other only in presence or absence of rounded ridges on the lateral surface of the iliac shaft. Their shafts are rather stout. However, their stratigraphic range spans the Coniacian to lower–middle Campanian. Morphotype 8 from the middle Santonian is similar in general proportions, but differs in having two rounded and slightly divergent mounds on the lateral surface of the iliac shaft.

Morphotype 5 is similar to the previous four in general shape of the bone; however, it differs in the relatively larger size of its acetabulum. It was recorded from the upper Santonian and lower–middle Campanian.

Morphotype 4 is clearly different from morphotypes 1–3 in its slender shaft and in having the acetabulum shifted toward the anteroventral margin of the bone. It is restricted to the middle Santonian through lower Campanian.

Morphotype 6 is similar to morphotypes 1–3, but differs in that its acetabular region is symmetrical, with the acetabulum in the middle. Its stratigraphic record is from the upper Santonian or lower Campanian.

Morphotype 7 is well distinguished from all others in this group by a thin shaft, a straight dorsal margin terminated by a point, and by its ventrally shifted acetabulum. It was recorded from the Turonian.

Morphotype 9 clearly differs from all others (irregardless of the size of the bone) by its large acetabulum extending almost to the dorsal margin of the bone and by its slender iliac shaft. Unfortunately, it is represented by only a single specimen (UMNH VP 13336), from the upper Campanian. This specimen is also notable because it documents the presence of large anurans in the upper Campanian.

Morphotypes 10 and 11 resemble one another, and differ only in the size and position of the acetabulum and in the presence or absence of a horizontal rounded ridge. Both are from the upper Campanian.

Morphotypes 12 and 13 are similar to one another in the shape of the anterodorsal margin of the acetabulum, which is unusual in being discontinuous. Morphotype 12 is known from the upper Cenomanian (UMNH VP 13071) and also the lower or middle Campanian (UMNH VP 18305). UMNH VP 13493 (morphotype 13) is from the upper Santonian or lower Campanian and it differs in the shape of the acetabular region and the iliac shaft.

Morphotype 14 is based on one fragmentary specimen, UMNH VP 13485, but it is distinct in having a dorsal margin that is moderately depressed at the level of the anterior margin of the iliac shaft. It was recovered from the upper Santonian or lower Campanian Paul's locality.

Morphotype 15, although also based on a fragmentary specimen (UMNH VP 18477), is profoundly different from all others in having an extremely large acetabulum and by the unusual shape of its acetabular region. It is known only from the upper Santonian.

Morphotype 16 is dubious, but it seems to be characterized by its straight and horizontal dorsal margin. It is from the upper Santonian or lower Campanian Pinto Flats locality.

The remaining two morphotypes (17 and 18) generally resemble the other morphotypes, but differ in that the acetabular surface closest to the iliac shaft is broadly convex, but the rest of the surface is steeply declined in the opposite direction. Such an acetabular morphology is characteristic of *Nezpercius*, as described by Blob et al. (2001) from the Campanian age Judith River Formation in Montana. One of the Utah morphotypes (17) bears an elevated rugosity on the dorsolateral side of its shaft, which is a key diagnostic character of *Nezpercius*. All our specimens referable to these two morphotypes (and thus to the genus *Nezpercius*) are from the lower or middle Campanian, which is in agreement with the stratigraphic occurrence of the three specimens described by Blob et al. (2001). Although *Nezpercius* was described as an anuran, its distinctive ilia (which are the only bones known for the taxon) have recently been reinterpreted as belonging to a caudate amphibian (Gardner et al., 2010). We have retained the *Nezpercius*-like ilia from Utah here to highlight their presence in the Utah sequence and to emphasize that these problematic ilia are similar in many respects to, and thus may easily be mistaken for, ilia of unequivocal anurans. For a more detailed list of similarities and differences between ilia of anurans and caudates, see Gardner et al. (2010).

DISCUSSION

The collection used in this study from the Upper Cretaceous of southwestern Utah is the largest sample of anuran ilia yet reported from a comparable area and stratigraphic interval in North America. Even a cursory glance at Figs. 12.4–12.7 and 12.8–12.11 indicates a diversity of ilia were recovered. The question naturally arose whether these ilia could be used for stratigraphic correlations. Unfortunately, this is not feasible at present. Although we have identified 65 iliac morphotypes among the nearly 200 specimens used in this study, there are several problems with using those morphotypes for biostratigraphic correlation. Many of the morphotypes differ only subtly from one another. From a practical standpoint, these morphs may be challenging for nonspecialists to identify in their screen-wash samples. Equally importantly, some of these morphotypes may simply be variants within a broader range of biologically distinct taxa. What we have recognized here as different morphotypes could conceivably be some combination of ontogenetic, sexually dimorphic, temporal, or individual variants of a lesser number of temporally longer-ranging species. The very real possibility that some of those iliac morphotypes are from conspecific individuals is difficult to evaluate, because many of the morphotypes are known by just one or only a few specimens and from only one or a few localities. It is possible that certain iliac morphotypes had a broader stratigraphic range than what is suggested by our current, relatively small samples. Some morphotypes admittedly are more distinctive (e.g., group 3, morphotypes 14 and 15), yet their utility remains limited because they are known by a single specimen each from just one locality. Fortunately, collecting and processing of previously sampled and new microvertebrate localities in the study area continues. As sample sizes of anuran ilia increase, it should be possible to refine identifications of anuran iliac morphotypes (perhaps even assign some of those to biological taxa) and better resolve their stratigraphic ranges.

At a general morphological level, however, we were able to record some potentially interesting patterns in the stratigraphic occurrences of the three groups of ilia. The most remarkable is the relatively poor record of the ilia with a dorsal tubercle in the lower and middle Santonian and in the upper Campanian. Another peculiarity is the relative absence of ilia with an oblique groove in the upper Santonian or lower Campanian. Although these patterns may accurately reflect the presence or absence (or relative abundances) of taxa at different times, they may simply be an artifact of sample sizes (which generally were small; i.e., less than a dozen ilia per locality) or particular depositional conditions in the localities.

The predominance of ilia without a dorsal tubercle in Mesozoic anurans is surprising, and it contrasts with the situation in the Cenozoic when most anurans have a dorsal tubercle. It is interesting that almost all known Mesozoic frogs also lack a dorsal crest on the iliac shaft (in life, this crest separates the origins of the iliacus externus and coccygeoiliacus muscles from each other; Přikryl et al., 2009). A notable exception from North America is *Paradiscoglossus americanus* Estes and Sanchíz, 1982, which is known by two ilia from the upper Maastrichtian Lance Formation of Wyoming that each bear a relatively tall and anteriorly elongate dorsal crest (Estes and Sanchíz, 1982; Gardner, 2008); no examples of this kind of ilia were identified from any of our older Utah localities. It is not clear whether the dorsal tubercle and the dorsal crest appeared at the same or different times during the evolution of anurans as a consequence of a shift in locomotor behavior or whether the widespread occurrence of these structures in Cenozoic anurans is associated with the appearance and diversification during the Paleogene and Neogene of more derived groups (e.g., Bufonidae, Ranidae, and Hylidae) that now dominate most contemporary anuran faunas (Rage and Roček, 2003). We do not even know whether the dorsal tubercle in the Early Triassic proanurans *Triadobatrachus* and *Czatkobatrachus* (Rage and Roček, 1989; Evans and Borsuk-Białynicka, 2009) is homologous with those in true anurans. Additional material is needed to resolve these kinds of problems. Further collection and study of specimens from stratigraphically extensive sequences of localities–such as the sequence in southwestern Utah used for this study–should help establish when key iliac features appeared during the evolution of anurans and may allow us to better understand how significant those novelties might have been to the success of particular anuran groups.

ACKNOWLEDGMENTS

This research was funded by the Ministry of Education, Youth and Sports of the Czech Republic (grant ME08066) through the American Science Information Center, Prague. Funds for collecting in Utah were provided by grants to Eaton from the National Park Service and the Bryce Canyon Natural History Association, the Petroleum Research Fund (ACS-PRF# 30989-GB8; ACS-PRF# 34595-B8), the National Science Foundation (EAR-9004560), and the National Geographic Society. We thank J.-C. Rage for his constructive review and A. L. Titus for his helpful suggestions and editorial help.

Bever, G. S. 2005. Variation in the ilium of North American *Bufo* (Lissamphibia; Anura) and its implications for species-level identification of fragmentary anuran fossils. Journal of Vertebrate Paleontology 25:548–560.

Blob, R. W., M. T. Carrano, R. R. Rogers, C. A. Forster, and N. R. Espinoza. 2001. A new fossil frog from the Upper Cretaceous Judith River Formation of Montana. Journal of Vertebrate Paleontology 21:190–194.

Bossuyt, F., and K. Roelants. 2009. Frogs and toads (Anura); pp. 357–364 in S. B. Hedges and S. Kumar (eds.), The Timetree of Life. Oxford University Press, New York.

Brinkman, D. B. 1990. Paleoecology of the Judith River Formation (Campanian) of Dinosaur Provincial Park, Alberta, Canada: evidence from vertebrate microfossil localities. Palaeogeography, Palaeoclimatology, Palaeoecology 78:37–54.

Brinkman, D. B., and A. G. Neuman. 2002. Teleost centra from uppermost Judith River group (Dinosaur Park Formation, Campanian) of Alberta, Canada. Journal of Paleontology 76:138–155.

Cifelli, R. L., R. L. Nydam, J. G. Eaton, J. D. Gardner, and J. I. Kirkland. 1999a. Vertebrate faunas of the North Horn Formation (Upper Cretaceous–lower Paleocene), Emery and Sanpete counties, Utah; pp. 377–388 in D. D. Gillette (ed.), Vertebrate Paleontology in Utah. Utah Geological Survey Miscellaneous Publication 99–1.

Cifelli, R. L., R. L. Nydam, J. D. Gardner, A. Weil, J. G. Eaton, J. I. Kirkland, and S. K. Madsen. 1999b. Medial Cretaceous vertebrates from the Cedar Mountain Formation, Emery County: The Mussentuchit local fauna; pp. 219–242 in G. D. Gillette (ed.), Vertebrate Paleontology in Utah. Utah Geological Survey Miscellaneous Publication 99–1.

Eaton, J. G. 2006. Late Cretaceous mammals from Cedar Canyon, southwestern Utah; pp. 373–402 in S. G. Lucas and R. M. Sullivan (eds.), Late Cretaceous Vertebrates from the Western Interior. New Mexico Museum of Natural History and Science Bulletin 35.

Eaton, J. G., F. Maldonado, and W. C. McIntosh. 1999. New radiometric dates from Upper Cretaceous rocks of the Markagunt Plateau, southwestern Utah, and their bearing on subsidence histories. Geological Society of America, Abstracts with Programs 31:A-11.

Eaton, J. G., J. Laurin, J. I. Kirkland, N. E. Tibert, R. M. Leckie, B. B. Sageman, P. M. Goldstrand, D. W. Moore, A. W. Straub, W. A. Cobban, and J. D. Dalebout. 2001. Cretaceous and Early Tertiary Geology of Cedar and Parowan canyons, western Markagunt Plateau, Utah; pp. 337–363 in M. C. Erskine, J. E. Faulds, J. M. Bartley, and P. D. Rowley (eds.), The Geologic Transition, High Plateaus to Great Basin–A Symposium and Field Guide. Utah Geological Association Publication 30.

Estes, R. 1964. Fossil Vertebrates from the Late Cretaceous Lance Formation, Eastern Wyoming. University of California Publications in Geological Sciences 49.

Estes, R. 1969. A new fossil discoglossid frog from Montana and Wyoming. Breviora 328:1–7.

Estes, R., and B. Sanchíz. 1982. New discoglossid and palaeobatrachid frogs from the Late Cretaceous of Wyoming and Montana, and a review of other frogs from the Lance and Hell Creek formations. Journal of Vertebrate Paleontology 2:9–20.

Evans, S. E., and M. Borsuk-Białynicka. 2009. The Early Triassic stem-frog *Czatkobatrachus* from Poland. Palaeontologia Polonica 65:79–105.

Evans, S. E., and A. R. Milner. 1993. Frogs and salamanders from the Upper Jurassic Morrison Formation (Quarry Nine, Como Bluff) of North America. Journal of Vertebrate Paleontology 13:24–30.

Gardner, J. D. 2005. Lissamphibians; pp. 186–201 in P. J. Currie and E. B. Koppelhus (eds.), Dinosaur Provincial Park: A Spectacular Ancient Ecosystem Revealed. Indiana University Press, Bloomington, Indiana.

Gardner, J. D. 2008. New information on frogs (Lissamphibia: Anura) from the Lance Formation (late Maastrichtian) and Bug Creek anthills (late Maastrichtian and early Paleocene), Hell Creek Formation, USA; pp. 219–249 in J. T. Sankey and S. Baszio (eds.), Vertebrate Microfossil Assemblages: Their Role in Paleoecology and Paleobiogeography. Indiana University Press, Bloomington, Indiana.

Gardner, J. D., Z. Roček, T. Přikryl, J. G. Eaton, R. W. Blob, and J. Sankey. 2010. Comparative morphology of the ilium of anurans and urodeles (Lissamphibia) and a re-assessment of the anuran affinities of *Nezpercius dodsoni* Blob et al., 2001. Journal of Vertebrate Paleontology 30:1684–1696.

Henrici, A. C. 1998. New anurans from the Rainbow Park Microsite, Dinosaur National Monument, Utah. Modern Geology 23:1–16.

Holman, J. A. 2003. Fossil Frogs and Toads of North America. Indiana University Press, Bloomington, Indiana.

Jones, M. E. H., S. E. Evans, and B. Ruth. 2002. Ontogenetic variation in the frog ilium and its impact on classification. Abstracts of the Palaeontological Association 46th Annual Meeting. Palaeontological Association Newsletter 51:132.

Lynch, J. D. 1971. Evolutionary Relationships, Osteology, and Zoogeography of Leptodactyloid Frogs. University of Kansas Museum of Natural History Miscellaneous Publication 53.

Moore, D. W., L. D. Nealey, P. D. Rowley, S. C. Hatfield, D. G. Maxwell, and E. Mitchell. 2004. Geological map of the Navajo Lake Quadrangle, Kane and Iron counties, Utah. Utah Geological Survey Map 199.

Ogg, J. G., F. P. Agterberg, and F. M. Gradstein. 2004. Cretaceous time scale; pp. 344–383 in F. M. Gradstein, J. G. Ogg, and A. G. Smith (eds.), A Geologic Time Scale 2004. Cambridge University Press, Cambridge.

Prasad, G. V. R., and J.-C. Rage. 2004. Fossil frogs (Amphibia: Anura) from the Upper Cretaceous intertrappean beds of Naskal, Andrhra Pradesh, India. Revue de Paléobiologie, Genève 23:99–116.

Přikryl, T., P. Aerts, P. Havelková, A. Herrel, and Z. Roček. 2009. Pelvic and thigh musculature in frogs (Anura) and origin of anuran jumping locomotion. Journal of Anatomy 214:100–139.

Rage, J.-C., and Z. Roček. 1989. Redescription of *Triadobatrachus massinoti* (Piveteau, 1936), an anuran amphibian from the Early Triassic. Palaeontographica Abteilung A 206:1–16.

Rage J.-C., and Z. Roček. 2003. Evolution of anuran assemblages in the Tertiary and Quaternary of Europe, in the context of palaeoclimate and palaeogeography. Amphibia-Reptilia 24:133–167.

Roček, Z., J. G. Eaton, J. D. Gardner, and T. Přikryl. 2010. Evolution of anuran assemblages in the Late Cretaceous of Utah. Palaeobiodiversity and Palaeoenvironments 90:341–393.

Sahni, A. 1972. The vertebrate fauna of the Judith River Formation, Montana. American Museum of Natural History Bulletin 147:321–412.

Sanchiz, B. 1998. Salientia; in P. Wellnhofer (ed.), Encyclopedia of Paleoherpetology, Part 4. Verlag Dr. Friedrich Pfeil, Munich.

Trueb, L. 1973. Bones, frogs, and evolution; pp. 65–132 in J. L. Vial (ed.), Evolutionary Biology of the Anurans: Contemporary Research on Major Problems. University of Missouri Press, Columbia, Missouri.

Tyler, M. J. 1976. Comparative osteology of the pelvic girdle of Australian frogs and description of a new fossil genus. Transactions of the Royal Society of South Australia 100:3–14.

Turtles from the Kaiparowits Formation, Utah

13

J. Howard Hutchison, Michael J. Knell, and Donald B. Brinkman

FOSSIL TURTLE REMAINS ARE ABUNDANT IN THE KAIP-arowits Formation and add to our understanding of turtle diversity, taxonomic relationships, and biogeography during the Late Cretaceous. A minimum of 14 taxa are present. Latitudinal patterns are identified by comparing the Kaiparowits assemblage with the contemporaneous assemblage from the Dinosaur Park Formation of Alberta. Non-trionychid taxa present in the Kaiparowits Formation but not in the Dinosaur Park Formation include *Compsemys*, *Denazinemys nodosa*, *Plesiobaena* sp. nov., a small smooth-shelled kinosternid, and the trionychids *Helopanoplia* and *Derrisemys*. These are considered members of a southern vertebrate assemblage. Taxa that are absent in the Kaiparowits Formation but present in more southerly localities are *Hoplochelys* and pleurodires. The absence of these taxa in the Kaiparowits Formation suggests that even further latitudinal subdivision of the turtle assemblages was present, and that the change from southern to northern assemblages is gradational. Several taxa that are restricted to southern localities in the late Campanian are present in the Late Maastrichtian Hell Creek of Montana. This change in distribution patterns is correlated with changes in climate.

INTRODUCTION

Turtles are conspicuous members of vertebrate assemblages within the Kaiparowits Formation, and a diverse assemblage is documented by specimens from numerous localities. The collection includes disarticulated fragments, articulated shells, and in one case, a skeleton containing eggs. Because the Kaiparowits Formation is, in part, contemporaneous with the Dinosaur Park Formation of Alberta, which also includes a well-documented turtle assemblage (Brinkman, 2005), it provides a basis for interpreting latitudinal patterns of distribution of turtles during the late Campanian. In this chapter, we review the turtles from the Kaiparowits Formation and discuss the paleobiogeographical implications of turtle distributions.

Institutional Abbreviations BYU, Brigham Young University, Provo, Utah; UCMP, University of California, Museum of Paleontology, Berkeley, California; UMNH, Utah Museum of Natural History, Salt Lake City, Utah.

GEOLOGY

The Kaiparowits Formation was deposited in a relatively wet, low-relief, inland alluvial plain setting with periodic aridity (Roberts, 2007; Roberts et al., this volume, Chapter 6). The primary depositional facies comprise fluvial channel and overbank deposits with occasional paludal lenses (Roberts, 2007). On the basis of dates obtained from volcanic ashes, the Kaiparowits Formation is thought to have been deposited between 76.1 and 74 Ma (Roberts, Deino, and Chan, 2005), overlapping in age with the Dinosaur Park Formation, which extends from 76.5 Ma to 74.8 Ma (Eberth, 2005). The majority of fossil turtle material collected is from the lower and middle units, and thus it is contemporaneous with the Dinosaur Park Formation material.

DIVERSITY OF THE TURTLE ASSEMBLAGES

A faunal list of the turtles from the Late Cretaceous of the Kaiparowits Formation is given in Table 13.1. This classification follows Gaffney and Meylan (1988), and uses stem-based definitions of the taxa. Stem-based rather than crown-based definitions of taxa are used because, as argued by Martin and Benton (2008), they are more consistent with historical usage of the taxa. Thus, paracryptodires, such as *Compsemys* and *Denazinemys*, are included in the cryptodira as has traditionally been done (Gaffney, 1972; Brinkman, 2003a) even though they are outside the clade that includes all extant cryptodires.

A special problem is presented by elements that document the presence of unnamed genera or species but are not adequate for erecting new species. Such taxa are recognized by a letter designation, as for example, trionychid type A.

295

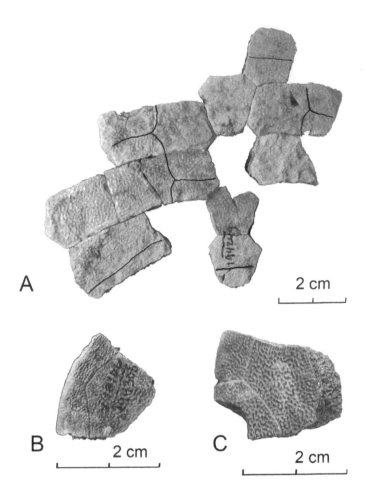

13.1. Shell elements of *Compsemys victa*. (A) UCMP 194249, partial carapace in dorsal view; (B) UCMP 194305, complete peripheral from anterior edge of carapace; (C) UMNH VP 16068, complete peripheral from posterior edge of carapace.

SYSTEMATIC PALEONTOLOGY

TESTUDINATA Klein, 1760
CRYPTODIRA Cope, 1868
PARACRYPTODIRA Gaffney, 1975
PLEUROSTERNIDAE Cope, 1868
COMPSEMYS Leidy, 1856
COMPSEMYS VICTA Leidy, 1856

Reference Specimens UCMP 194249, V93095: weathered carapace including posterior costals and neurals preserved in articulation and fragments of plastron (Fig. 13.1). UCMP 194305, V93106: shell fragments including a complete

peripheral of moderate size and a hyoplastron preserving inguinal notch and the sulci in this area. UCMP 194334, V94026: about 20 shell fragments including a complete peripheral. UCMP 194336, V94029: four fragments including a hyoplastron that preserves the notch region and a complete first neural. UCMP 194109, V94064: seven fragments including 2 complete neurals and a complete posterior peripheral. UCMP 194135, V95086: five fragments including a lateral end of a costal showing suture surface for the axillary buttress located on middle of its ventral surface. UCMP 194195, V98054: a complete hypoplastron. UCMP 194369, V99832: about a dozen fragments including a complete peripheral. UCMP 194287, V93103: about 20 fragments including a complete peripheral. UMNH VP 16068, fragments, including a complete peripheral and a costal.

Comments *Compsemys* is easily recognizable on the basis of sculpture pattern, which consists of compactly arranged, low flat-topped tubercles (Hay, 1908). Thus, fragmentary elements can be referred to this taxon even though they may not be identified to element. Elements that have

Hutchison, Knell, and Brinkman

been recovered include a portion of the posterior of a carapace and complete peripheral elements (Fig. 13.1) as well as fragmentary costals and plastron elements. These are generally well preserved with distinct sulci. The pattern of sulci on these elements does not differ from those of *Compsemys victa*, the only currently recognized species. The material from the Kaiparowits Formation of Utah is thus included in this species.

BAENIDAE Cope, 1882

Comments The Baenidae is an extinct clade of primitive cryptodire turtles found only in North America (Gaffney, 1972). The group appears in the Early Cretaceous and persists to the middle Eocene. Although baenids did not develop the ability to retract their head fully within the shell, the cryptodiran affinities of the family are indicated by the structure of the braincase: the otic capsule forms a pulley over which a tendon from the adductor muscles pass (Gaffney, 1972). Two morphologies are present, a primitive group in which a pygal is present and the marginals meet at the posterior end of the shell, excluding the fifth vertebral from the posterior edge of the shell (the primitive condition), and a derived group in which the fifth vertebral enters the posterior edge of the shell. Both groups are present in the Kaiparowits Formation. The primitive group is represented by a single genus, *Neurankylus*. The derived group is more diverse, with three genera present: *Denazinemys*, *Boremys*, and *Plesiobaena*.

NEURANKYLUS Lambe, 1902

Comments *Neurankylus* is one of the larger freshwater aquatic turtles of the Late Cretaceous. The shell is oval to subrectangular in shape, relatively low domed, and has an upturned edge along the lateral margin of the carapace. The shape of the shell of *Neurankylus* is similar to that of large riverine turtles found today in tropical rivers such as *Dermatemys*.

Among Late Cretaceous baenids, *Neurankylus* is primitive in that the middle pair of scales on the anterior portion of the plastron (intragulars in the terminology of Zangerl, 1969) separates the lateral pair (gulars). In other Late Cretaceous baenids, including *Thescelus*, which Holroyd and Hutchison (2002) considered to be closely related to *Neurankylus*, the lateral pair meets at the midline behind the median pair. Two species of Neurankylus are currently recognized in the Late Cretaceous of North America, *N. eximius*, the type of which is from Dinosaur Park Formation, and *N. baueri*, the type of which is from the Fruitland/Kirtland Formations. These were synonymized by Gaffney (1972), but treated as different species by Brinkman (2005) because *Neurankylus*

eximius has a strong middorsal ridge along the posterior half of the shell, while in *N. baueri* this region is smooth or has at most a subdued middorsal ridge. Two species of *Neurankylus* are here recognized in the Kaiparowits Formation. These differ from both *N. baueri* and *N. eximius* in features of the plastron and carapace and are designated *Neurankylus* type A and *Neurankylus* type B.

NEURANKYLUS sp. indet. type A

Reference Specimens BYU 12001: complete carapace and plastron. BYU 9411, shell visible in ventral view. UCMP 194201, Loc. v98054: partial shell including nuchal and the anterolateral quarter of the shell and anterior half of plastron. UMNH VP 20443: partial skeleton including pelvis and hind limb elements. UMNH VP 13979: anterior lobe of plastron. UMNH VP 20450: posterior end of plastron visible in ventral view. UMNH VP 16159: epiplastron. UCMP 194216, Loc. v93073: fragmentary anterior lobe of plastron and fragmentary shell pieces. UCMP 194122, Loc. v95083: first peripheral of very large size. UCMP 194136, Loc. v95087: pygal in articulation with suprapygal and 11th peripheral, showing the sulci of the 12th marginal. UCMP 194136, Loc. v95090: weathered remains of a shell including part of the posterior lobe of plastron. UCMP 158884, Loc. v93079: partial plastron including potion of anterior and posterior lobes (Fig. 13.2).

Description *Neurankylus* type A is well represented in the Kaiparowits Formation by both complete shells and isolated elements. The carapace, best documented in BYU 1200 (Fig. 13.2A, C) is similar to that of *N. baueri* from the Fruitland/Kirtland formations and *N. eximius* from the Dinosaur Park Formation in general shape and in the development of sulci on the carapace. In all of these, the shell is an elongate oval in shape. The size of the shell of *Neurankylus* type A, which in BYU 1200 is approximately 75 cm long, exceeds that of both *N. baueri*, which reaches a length of 60.5 cm (Gilmore, 1935) and *N. eximius*, which reaches a length of 52 cm (TMP 2003.13.171). The pattern of scales on the carapace of these three taxa is similar. The vertebral scales are wide and have relatively straight lateral margins. The anteriormost edge of the carapace is missing in specimen BYU 1200 but is present in UCMP 194201 (Fig. 13.2B). The latter specimen shows that, as in *N. eximius* and *N. baueri*, the nuchal scale is rectangular, slightly longer than wide. As in *N. baueri*, a low middorsal ridge is poorly developed on the posterior portion of the shell.

The plastron, completely preserved in BYU 1200 (Fig. 13.2C) and BYU 9411 (Fig. 13.2D), differs from that of both *N. baueri* and *N eximius* in that the anterior lobe is longer than the posterior lobe and is rounded. In *N. baueri* and *N eximius* the posterior lobe is more elongate and the anterior

A

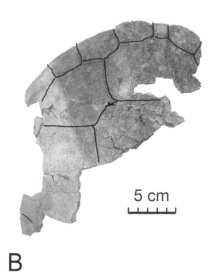

B

13.2. *Neurankylus* type A. (A) Specimen BYU 1200, carapace in dorsal view; (B) specimen UCMP 194201, anterior end of carapace in dorsal view; (C) specimen BYU1200, plastron in ventral view; (D) specimen BYU 9411, plastron in ventral view.

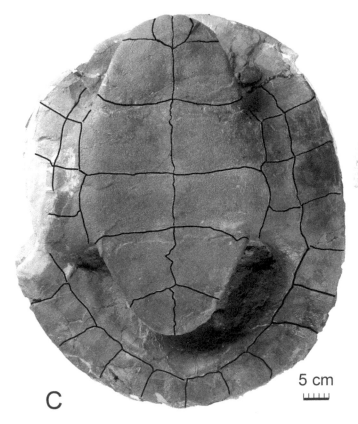

C

D

Hutchison, Knell, and Brinkman

lobe has a truncated anterior edge so is more rectangular in shape. *Neurankylus* sp. A differs little from *N. baueri* and *N eximius* in the development of scales on the plastron. In all of these, two pairs of scales are present anteriorly and the lateral pair is separated by the median pair.

Comments Material from the Kaiparowits Formation included in *Neurankylus* type A is similar to *N. baueri* in the reduced middorsal ridge (Fig. 13.2). However, the proportions of the plastron differ in that the anterior lobe is rounded and longer than the posterior lobe. In both *N. eximius* and *N. baueri*, the anterior lobe is shorter than the posterior lobe and has a truncated anterior end, giving it a subrectangular shape.

NEURANKYLUS sp. indet. type B

Reference Specimen UCMP 154450, Loc. V93118: partial skeleton, including left half of carapace, complete pelvis, both scapulae, a humerus, two cervical vertebrae, two sacral vertebrae, and two caudal vertebrae (Figs. 13.3–13.4).

Description *Neurankylus* type B is represented by a partial skeleton of a very large individual with fused sutures. The shell, represented by the left half of a carapace (Fig. 13.3A), measures approximately 100 cm long. As preserved, the shell is oval in shape with the widest area located at the inguinal buttress. In dorsal view, the shell differs from that of *Neurankylus* type A in having a more strongly curved lateral margin. Also, the shell is highly domed compared to *Neurankylus* type A. Sulci are lightly impressed but can be traced through most of the dorsal surface of the carapace. The pattern of scales on the carapace is distinct from that that of *Neurankylus* type A in that the nuchal scale is relatively longer, the third marginal scale has a long tapering point extending between the first and second neurals, and the fifth to tenth marginals are wider (Fig. 13.3B).

Postcranial elements preserved in UCMP 154450 include cervical and caudal vertebrae, parts of both girdles, and a humerus. Cervical vertebrae (Fig. 13.4A) conform to the baenid pattern as seen in *Boremys* (Brinkman and Nicholls, 1991) in being relatively short, having poorly formed articular surfaces, and having transverse processes located midway along the centrum. The posterior articular surfaces are tall and narrow, suggesting that lateral movements of the head were emphasized over dorsoventral movements.

Centra of the dorsal vertebrae are very narrow, forming a blade-like structure along the middle of the carapace, and the heads of the rib are well separated from the shell. This region of the skeleton was not preserved in any available specimens of *N. eximius* or *N. baueri*.

The shoulder girdle is represented by a complete right and partial left scapula (Fig. 13.4B). The dorsal and acromion

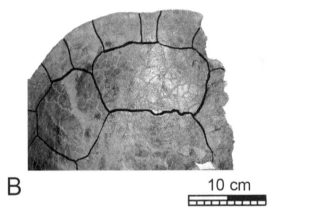

13.3. *Neurankylus* type B. (A) Carapace of UCMP 154450 in left lateral and (B) anterior dorsal views.

processes are both rod-like and are at about 90° to each other. The dorsal process is about 1.5 times the length of the acromion process.

The pelvis is complete (Fig. 13.4C). The anterior apron of the pubis is elongate and narrow. The ilium slopes posteriorly from the acetabulum, ending in a slightly expanded blade. The pubic tubercles are robust and extend forward from the shaft of the ilium. The ischium is reduced to a narrow bar of bone extending between the acetabular fossa, with a relatively short posterior ischial process. The obturator fenestrae are separated by a midventral bar formed by the ischium and pubis.

The humerus (Fig. 13.4D) is notable in being elongate and relatively straight. These are interpreted as primitive baenid features.

Comments *Neurankylus* type B differs from *Neurankylus* type A, *N. baueri*, and *N. eximius* most obviously in its size and more rounded shape. At a meter in length, the carapace of *Neurankylus* type B matches or exceeds that of *Basilemys*, and thus may be the largest nonmarine turtle

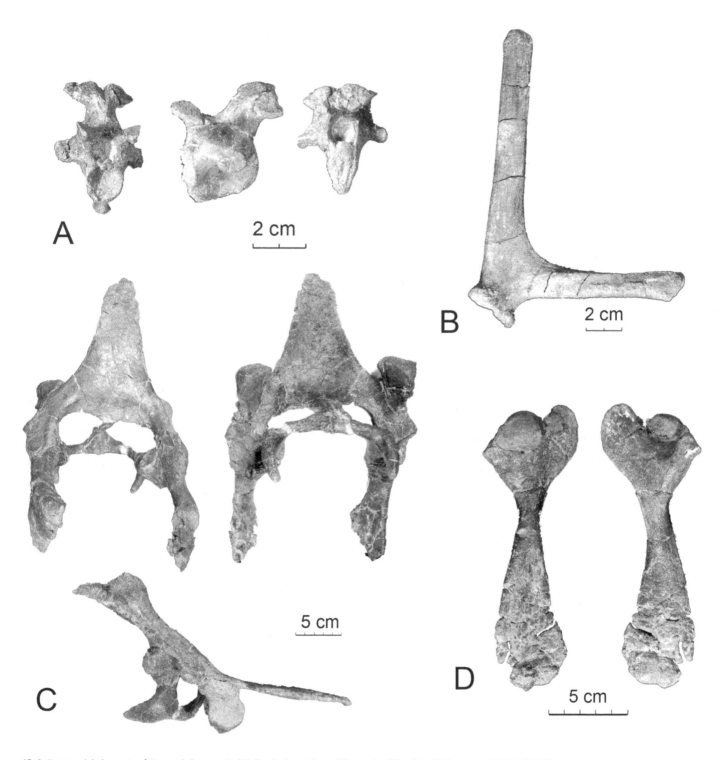

13.4. Postcranial elements of *Neurankylus* type B. (A) Cervical vertebrae; (B) scapula; (C) pelvis; (D) humerus. UCMP 154450.

in the Cretaceous of North America. As well, *Neurankylus* type B is distinct in the relative size of the marginal scales. The posteromedial corner of the third marginal scale, which extends between the first and second pleural scales, is more greatly developed, and marginal scales four to 10 are relatively wider. The strongly laterally compressed dorsal centra differs from the typical condition in baenids and may be an autapomorphy of *Neurankylus* type B as well.

DENAZINEMYS Lucas and Sullivan, 2006
DENAZINEMYS NODOSA (Gilmore, 1916)

Reference Specimens UMNH VP 20447, complete carapace visible in dorsal view. UMNH VP 2046, nearly complete shell visible in dorsal and ventral view, carapace missing posterior end, plastron complete. UMNH VP 18628, anterolateral peripherals. UMNH VP 18872, partial shell visible in dorsal and

Hutchison, Knell, and Brinkman

13.5. Shell of *Denazinemys nodosa*. (A) Specimen UMNH VP 20447, in dorsal view; (B) specimen UMNH VP 2046, plastron in ventral view; (C) specimen UCMP 159703, front part of carapace showing sulci of the nuchal region; (D) specimen UCMP 194335, anterior lobe of plastron showing sulci of the gular region.

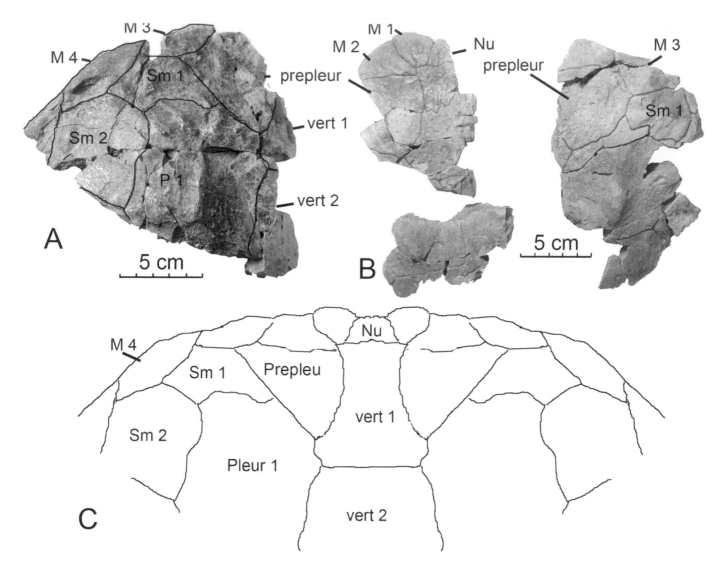

13.6. Anterior end of carapace of *Boremys grandis*. (A) Specimen UCMP 156997, front left quarter of carapace; (B) specimen UCMP 151773, front of carapace showing presence of supramarginal scales. M1 to M4, first to fourth marginals; Nu, nuchal scale; P1, first pleural scale; prepleur, prepleural scale; Sm1, first supramarginal; Sm2, second supramarginal; vert 1, first vertebral scale; vert 2, second vertebral scale;

ventral views. BYU 1441: skeleton including skull, fore limb, and nearly complete shell, carapace missing posterolateral edge, plastron complete. UCMP V159703, Loc. v95087: front part of shell, shows sulci of prepleural but supramarginals absent. Nuchal present and cervical scale preserved. UCMP V194125, Loc. v95083: a nearly complete first costal and a partial nuchal including the nuchal scale. UCMP 159399, Loc. v99441: carapace preserved as a largely steinkern with some adhering bone. The anterior and posterior ends of carapace are missing. Plastron includes complete posterior lobe. UCMP 194342, Loc. v94039: pieces of a once complete shell including the anterior end of the carapace and anterior lobe of plastron. UCMP 194207, Loc. v93070: front left quarter of carapace including costals 1–3, half of the nuchal, and peripherals one to four. UCMP 194248, Loc. v93084: shell

fragment of a juvenile individual including first and second costal in articulation. UCMP 19271, Loc. v93096: partial carapace. UCMP 194335, Loc. v94028: fragments of an articulated carapace, including anterior and posterior lobe of plastron and anterior right quarter of carapace. UCMP 194332, Loc. v94010: shell fragments of eroded individual (Fig 13.5).

Comments *Denazinemys* is a shell-based taxon distinguished by the presence of rounded node-like elevations of irregular shape and size on the carapace (Lucas and Sullivan, 2006). It is the most abundant baenid in the Kaiparowits Formation, and both articulated shells and isolated elements are present. The shape of the carapace is distinctive in that the anterior end projects distinctly forward and the posterior edge is strongly serrated (Fig. 13.5A, C). Although frequently distorted, articulated shells indicate that the carapace was

Hutchison, Knell, and Brinkman

highly domed. The plastron has a relatively long narrow anterior lobe compared with that of *Neurankylus* (Fig. 13.5B) Two pairs of scales are present anteriorly (gulars and intragulars in the terminology of Zangerl, 1969). The lateral pair (gulars) meet behind the median pair (intragulars) (Fig. 13.5C, D). Notches may be present on the edge of the anterior lobe where the sulci between these scales meet the edge of the plastron (Fig. 13.5D).

Lucas and Sullivan (2006) recognize two species of *Denazinemys*; *D. nodosa* and *D. ornata*. They differ in that a prepleural scale is present in *D. nodosa* but not *D. ornata*, the posterior peripherals are wider in *D. nodosa*, and two pairs of scales are present on the anterior lobe of the plastron of *D. nodosa* but a single pair of scales is present in *D. ornata*. On the basis of this combination of features all the specimens of *Denazinemys* from the Kaiparowits Formation are members of the species *D. nodosa*.

BOREMYS Lambe, 1906
BOREMYS GRANDIS Gilmore, 1935

Reference Specimen UCMP 151773, Loc. v94009: back half of carapace visible in ventral view (Fig 13.6). Associated fragments include a portion of the front left quarter of the carapace. UCMP 156997, Loc. v97098: Front quarter of carapace.

Comments The carapace of *Boremys* is similar to that of *Denazinemys nodosa* (Fig. 13.5) in having prepleural scales and a serrated posterior margin of the carapace but differs in the presence of supramarginal scales. Both UCMP 151773 and UCMP 156997 are referred to *Boremys* because supramarginal scales are present (Fig. 13.6A, B). Surface texture on the carapace of these specimens is similar to that of *Denazinemys* in the presence of nodes, some of which are round and others are elongate, but differs in that the ornamentation is subdued with the nodes and bulges low and widely scattered, rather that distinct and tightly packed.

The plastron, preserved in UCMP 151773, is similar to that of *Denazinemys* in having a relatively long narrow anterior lobe. The gular scales meet behind the intragular and slight notches are present where the sulci separating these scales meet the edge of the plastron.

In their review of *Boremys*, Brinkman and Nicholls (1993) recognized two species, *Boremys pulchra* from the Judith River Group and *Boremys grandis* from the Fruitland/Kirtland Formations. In addition to size, these species are distinct in that supramarginal scales are present lateral to the first pleural scale in *B. grandis* but not *B. pulchra*. UCMP 151773 and 156997 are referred to *B. grandis* because of the presence of supramarginal scales lateral to the first pleural scale.

PLESIOBAENA Gaffney, 1972
PLESIOBAENA sp. indet.

Reference Specimens UMNH VP 20451; complete carapace. UMNH VP 20183; partial carapace, missing posterior end (Fig 13.7).

Description *Plesiobaena* sp. indet. is represented by two shells. UMNH VP 20451 (Fig. 13.7A) is a nearly complete and largely uncrushed. The surface appears weathered, but sulci could be identified. The second, UMNH VP 20451 (Fig. 13.7B) is also largely uncrushed, but is missing the posterior end and part of the left lateral side. The shell is oval, with a tapered posterior end, giving it a distinct torpedo shape. The surface is smooth. A midline ridge is present along the full length of the shell, although this is most strongly developed posteriorly where distinct nodes are present. The anterior end of the carapace is rounded, lacking both a nuchal emargination and an anterior projection of the nuchal like that of *Denazinemys nodosa*. The posterior edge of the carapace is strongly serrated.

In both of the available shells of *Plesiobaena* sp. the individual elements are fused so sutures cannot be identified. However, the sulci are distinct. A distinctive feature is that the nuchal scale is wide. In other respects the pattern of scales on the carapace is similar to that of previously described species of *Plesiobaena*. Prepleural scales and supramarginal scales are absent. The vertebral scales are wider than in specimens of *Plesiobaena antiqua* from the Dinosaur Park Formation (Brinkman, 2003a). However, the proportions of these scales match those of some specimens of *Plesiobaena* from the Late Maastrichtian.

The plastron of UMNH VP 20451 is complete (Fig. 13.7A). It is similar to that of other species of *Plesiobaena* in the shape and development of scales on the anterior lobe of the plastron. The shape is distinctive in being triangular and relative short compared to that in *Boremys* and *Denazinemys*. A single pair of scales is present anteriorly. The humeral/pectoral sulcus is located midway along the anterior lobe of the plastron. The remainder of the plastron differs little from that of other baenids.

Comments *Plesiobaena* differs from *Denazinemys nodosa* and *Boremys* in having a smooth shell, lacking prepleural scales, having a triangular anterior lobe of the plastron and in that the intergular scales are greatly reduced or absent. Specimens UMNH VP 20451 and UMNH VP 20183 are included in *Plesiobaena* because of the absence of a prepleural scale, undivided nuchal scale, triangular anterior lobe of the plastron, and presence of a single pair of scales on the anterior end of the plastron. The Kaiparowits specimens of *Pleisobaena* differ from *Plesiobaena antiqua* in the presence of a strongly

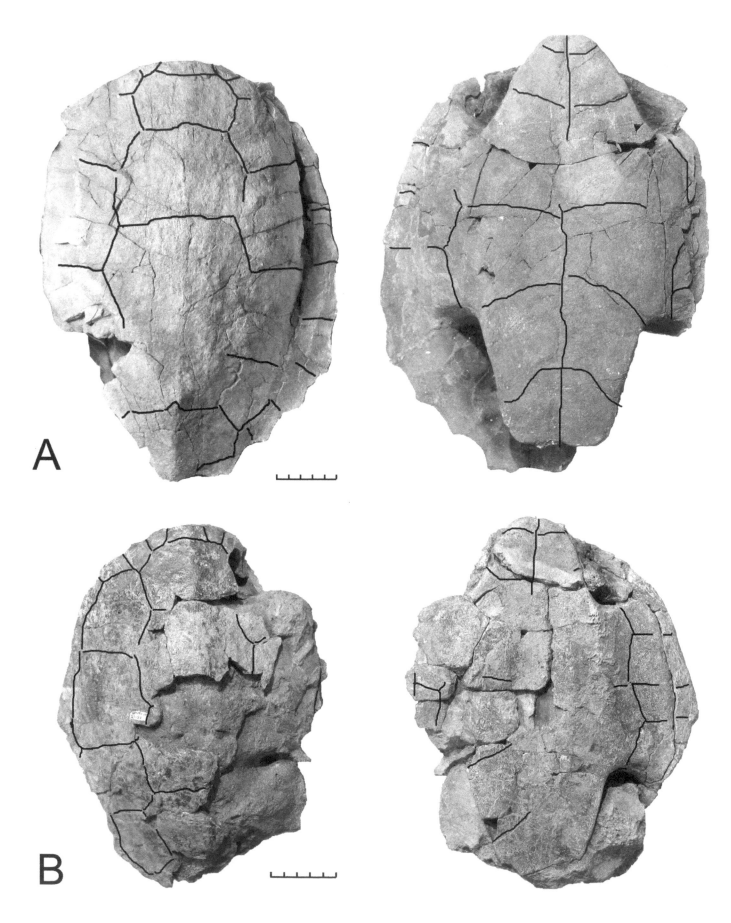

13.7. Shell of *Plesiobaena* sp. (A) UMNH VP 20451, complete shell in dorsal and ventral views; (B) UMNH VP 20183, partial shell in dorsal and ventral views.

Hutchison, Knell, and Brinkman

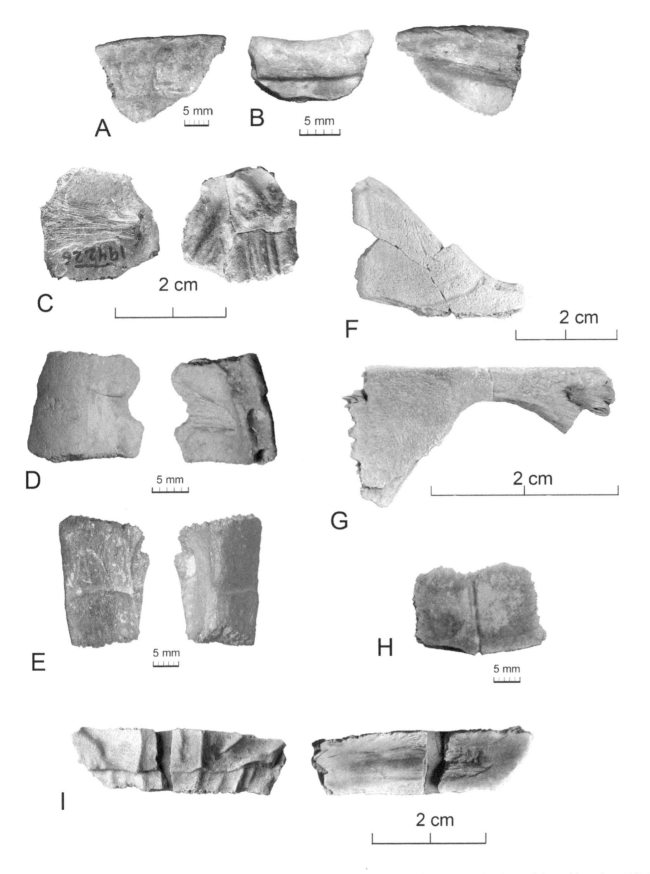

13.8. Chelydridae gen. et sp. indet. shell element. (A) Specimen UCMP V 193655, first peripheral, in dorsal and ventral views; (B) specimen UCMP V 193654 in dorsal view; (C) UCMP V 194226, proximal end of costal; (D) specimen UCMP V 193658, right fourth peripheral in dorsal and ventral views; (E) Specimen UCMP V 193662, right eighth peripheral in dorsal and ventral views; (F) specimen UCMP V 194314, medial portion of hyoplastron in ventral view; (G) specimen UCMP V 193672, left hyoplastron; (H) specimen UCMP V 193675, pygal, in dorsal view; (I) specimen UMNH VP 16046, small costal showing hypertrophied plications (RA).

serrated posterior edge of the carapace, a middorsal ridge, wide vertebral scales, and absence of intergular scales. In the presence of wide vertebral scales and a serrated posterior edge, it is similar to some specimens of *Plesiobaena* from the Maastrichtian but differs from *Plesiobaena antiqua* from the Dinosaur Park Formation.

In some features, the pattern of scales on the carapace and plastron of *Pleisobaena* sp. is similar to *Denazinemys ornata*. In both, prepleural scales are absent and the plastron has a single pair of scales anteriorly. However, *Plesiobaena* sp. lacks the nodular surface texture typical of *Denazinemys*. Also, the anterior lobe of the plastron is more triangular in shape, a feature typical of *Plesiobaena*.

EUCRYPTODIRA Gaffney, 1975
CHELYDRIDAE Gray, 1870
Genus et sp. indet.

Reference Specimens UMNH VP 12541; proximal end of costal. UMNH VP 16046; costal of juvenile individual (Fig 13.8). UCMP 16068; costal of juvenile individual. UCMP Loc. v93072:194213, fragment of costal. UCMP V93076:194226, three costal fragments, one of which includes the proximal end. UCMP V93111:194314, medial end of hyoplastron, showing very narrow bridge and retracted abdominal scales. UCMP V94063: 194104, fragment of the proximal end of a costal. UCMP V97093:194167, fragment of a small costal showing change in elevation across the pleural sulcus and plications extending from sulcus. UCMP V98053:194193, medial end of the first costal. 194362, costal fragments. UCMP V99831: 193654, nuchal fragment; 193655, peripheral 1; 193656, peripheral 2? fragment; 193657, two peripheral 3? fragments; 193658, peripheral 4; 193659, peripheral 4 fragment; 193660, partial peripheral 5?; 193661, peripheral 6; 193662, peripheral 8; 193663, 193663, peripheral 9?; 193665, three peripheral fragments; 193666, 193667, neural; 193668, costal fragments; 193669, five hyoplastron fragments; 193670, two hypoplastron fragments. UCMP V99832: 194372, costal fragment. UCMP V99833: 993650, five neural fragments; 193651, five peripheral fragments; 193652, 36 costal fragments; 194494, five hyoplastron fragments; 194496, three hypoplastron fragments; 194496, partial peripheral 1; 194497, 194498, peripheral 11; 194499, two first neural fragment.

Comments The Chelydridae (snapping turtles) first occurs in North America in the Turonian and is restricted to that region until the Paleogene. Cretaceous remains are largely fragmentary but are easily identified on the basis of surface texture of the carapace elements and other details. The texture is distinctive in having a strong change in elevation across the sulci with plications extending approximately ninety degrees posterior from the sulci. Specimens from the Kaiparowits Formation are referred to the Chelydridae on the basis of this feature (Fig. 13.8A–B). Among the Kaiparowits specimens referred to the Chelydridae are a series of small costals with extreme development of plications (Fig. 13.8I). In these costals, the plications extending posteriorly from the sulci form sharp edged crests. Such extreme development of the plications is not present in the Dinosaur Park Formation. Pending the discovery of more complete specimens these are assumed to be juvenile individuals.

The plastron is reduced and contacts the peripherals via a series of peg and socket joints. A distinctive feature is the presence of abdominal scutes that are retracted from the midline. The presence of this feature in the chelydrid from the Kaiparowits Formation is documented by a partial hyoplastron that shows the sulcus separating the pectoral and abdominal scutes meeting the suture for the hypoplastron well lateral to the midline (Fig. 13.8C).

KINOSTERNIDAE Agassiz, 1857
Stem Kinosternidae gen. et sp. nov.

Reference Specimens UCMP Loc. v93013: 194297, distal end of a costal (Fig 13.9). UCMP Loc. v93072: 194212, peripheral from just adjacent to bridge. UCMP Loc. v93088: 194256, medial portion of hyoplastron. UCMP Loc. v93094: 194264, medial portion of hypoplastron. UCMP Loc. v93095: 194269, well-preserved posterior peripheral, probably the ninth. UCMP Loc. v93098: 194278, eight shell fragments including a bridge peripheral and a neural; 194284, eight shell fragments including the lateral end of a hyoplastron and the proximal end of a costal. UCMP Loc. v93108: 194308, lateral end of hyoplastron UCMP Loc. v93111: 194313, fragments including a nearly complete costal and a bridge peripheral. UCMP Loc. v93117: 194320, ninth peripheral. UCMP V93120: 194324, proximal end of costal. UCMP Loc. v94038: 194341, four pieces including the medial end of the hyoplastron and a second peripheral. UCMP Loc. v94041: 194069, two hypoplastra, on missing the lateral end, and one missing the medial portion, and a complete bridge peripheral; 194351, complete hypoplastron. UCMP Loc. v94050: 194080, complete third peripheral. UCMP V94056: 194088, two neurals. UCMP Loc. v94061: 194009, three pieces including the medial end of a costal. UCMP Loc. v95083: 194123, a complete but eroded hypoplastron. UCMP Loc. v95085: 194436, 194437, epiplastron; 194440, hyoplastron fragments; 194442, seven hypoplastron fragments; 194443, peripheral 2; 194448, peripheral 5 fragments; 194449, peripheral 6; 194450, peripheral 7; 194451 peripheral 9; 194452 peripheral 10; 194454 peripheral 10 fragment; 194455–194457, neurals; 194458, two costal 1 proximal ends; 194459, shell fragments; UCMP Loc. v95090: 194145, lateral end of hypoplastron. UCMP Loc.

Hutchison, Knell, and Brinkman

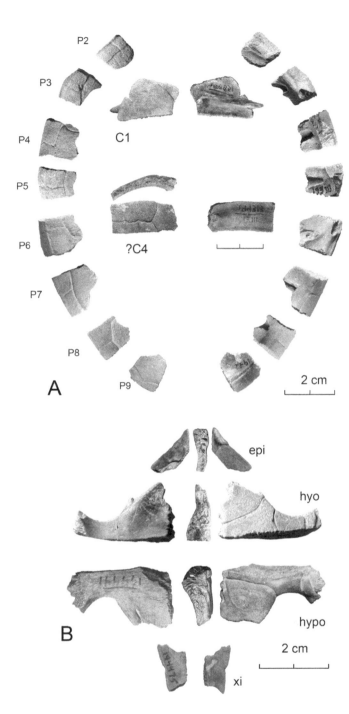

13.9. Kinosternidae gen. indet. (A) Composite reconstruction of carapace based on isolated costal and peripheral elements. Costal 1 UMNH VP 58851; ?costal 4: UCMP 194313; peripheral 2: UMNH VP 16074; peripheral 3: UMNH VP 18663; peripheral 4: UMNH VP 9534; peripheral 5: UCMP 194313; peripheral 6: UMNH VP 17013; peripheral 7: UMNH VP 16902; peripheral 8: UMNH VP 12544 (figure reversed for comparison); peripheral 9: UCMP 194269 (figure reversed for comparison). (B) Composite reconstruction of plastron: epiplastron: UCMP V 194437, hyoplastron: UMNH VP 16074; hypoplastron, UCMP 194441, xiphiplastron, UCMP 194475. C1, first costal; ?C4, fourth costal; epi, epiplastron; hyo, hyoplastron; hypo, hypoplastron; P1 to P9, first to ninth peripheral; xi, xiphiplastron.

v95092: 194157, three small shell fragments, including the proximal end of a costal. UCMP Loc. v98054: 194197, shell fragments, including partial costals and a peripheral. UCMP Loc. v99831: 194364, hypoplastron; 194460, six hypoplastron fragments; 194461, peripheral 4; 194462–194465, peripheral 5; 194466, 194467, peripheral 9; 194468 peripheral fragment; 194469, neural; 194470, costal 1 fragment; 194471, costal 8; 194472, 19 coastal fragments; 194493, peripheral 7. UCMP Loc. v99833: 194473, three hyoplastron fragments; 194474, four hypoplastron fragments; 194475, anterior half of xiphiplastron; 194476, nuchal fragment; 194477, three peripheral 4 fragments; 194478, peripheral 5 fragment; 194479, peripheral 7; 194480, peripheral 7 fragment; 194481–194485, peripheral 9; 194486, peripheral fragments; 194488, three costal fragments; 194489, 194490, costal 7 fragments; 194491, proximal end of costal 8; 194492, 17 costal 2, 4 or 6 fragments; 194493, 13 costal 3, 5 or 7 fragments.

Comments The presence of a small, smooth-shelled kinosternid in the Late Cretaceous (Maastrichtian) was first recognized by Hutchison and Archibald (1986). Subsequently a similar taxon was reported in the Cerro del Pueblo Formation of Coahuila, Mexico (Rodriguez-de la Rosa and Cevallos-Ferriz, 1998). Brinkman and Rodriguez-de la Rosa (2006) illustrated additional elements from the Cerro del Pueblo Formation and included the taxon in the family Kinosternidae. These specimens show marked similarities to Eocene to extant kinosternids in the general shape, sculpture and small size of the shell, probable absence of the peripheral 11, probable loss of some neurals, reduction of the inframarginals to two, short marginal elevation, and elevated marginal 10. However, at least the Campanian species from the Kaiparowits and Cerro del Pueblo Formations fall outside the crown group Kinosternidae in lacking musk duct grooves and inguinal scales crossing the hyo-hypoplastron suture and presence of well-developed abdominal scales. Taken together, these suggest that the Campanian species are sister taxa to the crown group and that the retracted abdominal scales of *Hoplochelys*, the next postulated outgroup to all these (Hutchison and Bramble, 1981), is of independent derivation.

Isolated elements of this small smooth-shelled kinosternid are abundant in the Kaiparowits Formation. From comparison with extant kinosternids, the shell is less than 10 cm. long. An isolated costal, tentatively identified as the fourth, shows that the shell was highly domed (Fig. 13.9A). The sulci marking the lateral edges of the vertebral scutes are strongly angled, indicating that the vertebral scutes were generally diamond shaped as in extant kinosternids.

Peripherals are frequently preserved, and, from comparison with extant kinosternids and fossil kinosternoid material from the Cerro del Pueblo Formation, all but the first peripheral could be identified, allowing for a composite reconstruction of the peripheral series (Fig. 13.9A). The marginal scales are restricted to a narrow band around the edge of the shell (possibly a kinosternoid feature). The presence of costiform processes is indicated by grooves on the ventral surface of the anterior peripherals. These processes extend across the first peripheral to end in the second peripheral. Sutural surfaces for the plastron extend from the posterior edge of the third peripheral to the anterior edge of the sixth peripheral. The fifth peripheral contacts both the hyo and hypoplastron and the articular surfaces for these is separated by a sharp ridge.

The plastron is represented by the epiplastron, hyoplastron, hypoplastron, and anterior part of the xiphiplastron (Fig. 13.9B). The epiplastron and xiphiplastron are narrow as in the Eocene kinosternid *Baltemys* (Hutchison, 1991). The bridge region is narrow also as in *Baltemys*. In cross section, the bridge is oval in shape, with a single thickened area extending from the center of the plastron to the peripherals. This buttress is thickest in the hypoplastron just posterior to the hyo-hypoplastron suture. The pattern of scales on the plastron is primitive relative to extant kinosternids in that distinct abdominal scales are present and these contact one another at the midline, although the area of contact is reduced almost to a point.

ADOCIDAE Cope, 1870
ADOCUS Cope, 1868
ADOCUS sp.

Reference Specimens UMNH VP 20452, complete carapace in dorsal view. BYU 1071: anterior end of plastron (Figs. 13.10, 13.11A). UMNH VP 16925: nuchal. UCMP 194272, Loc. v93096: anterior portion of hyoplastron. UCMP 159702, Loc. v95084: complete plastron. UCMP 194286, Loc. v93100: fragments of a very large shell, including the anterior end of the nuchal and three costals in articulation. UCMP 194328, Loc. v94004: fragmentary carapace with associated fragments of a scapula and humerus. UCMP 194066, Loc. v94046: complete costal. UCMP 194119, Loc. v95083; four shell fragments including part of the anterior lobe of the plastron. UCMP 194142, Loc. v95088: shell fragments of a single individual including a nuchal. UCMP 194170, Loc. v97096: weathered remains of a carapace, including a hyoplastron that that preserves the axillary notch region. UCMP 194165, Loc. v97093: proximal end of costal. UCMP 194360, Loc. v99831: about a dozen fragments, including a nearly complete peripheral.

Comments *Adocus* is a relatively large turtle with an elongate subrectangular shell. The shell is ornamented by small pits that are in the shape of elongate diamonds and arranged in rows, allowing even small fragments to be easily identified. In addition to numerous isolated elements, many

preserving distinct sulci and sutural surfaces, several articulated specimens were collected. These include a specimen (UMNH VP 16868) in which the skull and eggs are preserved inside the carapace (Knell et al., 2011), a complete shell visible in dorsal view (Fig. 13.10A), a complete plastron (Fig. 13.10D), and the anterior lobe of a plastron (Fig. 13.10C). The shell conforms to that of *Adocus* specimens from the Judith River Group and the Fruitland/Kirtland Formations in shape and arrangement of sulci on the shell. A characteristic feature of the genus that was noted by Meylan and Gaffney (1989) is that the sulcus separating the fifth marginal and pleural scales abruptly crosses from the peripherals onto the second costal, and posterior to this the marginals extend onto the costals. This feature is present in specimens of *Adocus* from the Kaiparowits Formation (Fig. 13.10A). Numerous species of *Adocus* have been named (Hay, 1908; Gilmore, 1919) and a full revision of the species level diversity of the genus is needed. Variation in the morphology of the anterior lobe of the plastron suggests that two distinct species are present in the Kaiparowits Formation. One plastron morphotype, represented by BYU 1071 (Fig. 13.10C), has a short, rounded anterior lobe, and the second, represented by UCMP 159702 (Fig. 13.10D) has a larger, more rectangular lobe. However, the number of specimens available at present is too small to show that these differences are taxonomic significance, rather than a result of variation in the shape of the plastron in a single species.

NANHSIUNGCHELYIDAE Yeh, 1966
BASILEMYS Hay, 1902
BASILEMYS NOBILIS Hay, 1911

Reference Specimens UMNH VP 6138: complete plastron. UCMP 194086, Loc. v94055: fragment plastron with sulci of inguinal scale region. UCMP 150818, Loc. v94049: complete peripheral. UCMP 194090, Loc. v94058: a large plastron fragment including the area adjacent to the inguinal notch. UCMP 194114, Loc. v95083: a fragment from the inguinal notch of a very large specimen. UCMP 159500; Loc. v98156: 11th peripheral and pygal in articulation. UCMP 194164: fragment from posterior edge of shell (Figs. 13.11B, 13.12).

Comments *Basilemys*, with a shell reaching just under a meter in length, is one of the largest nonmarine turtles of the Late Cretaceous of North America. The surface of the carapace is covered by a sculpture consisting of shallow pits generally arranged in rows and with small pyramidal elevations between the pits, allowing even small fragments to be identified (Fig. 13.11B).

In addition to isolated shell fragments, a complete plastron was recovered (Fig. 13.12). Four species of *Basilemys* are currently recognized, and these are distinguished on the

Hutchison, Knell, and Brinkman

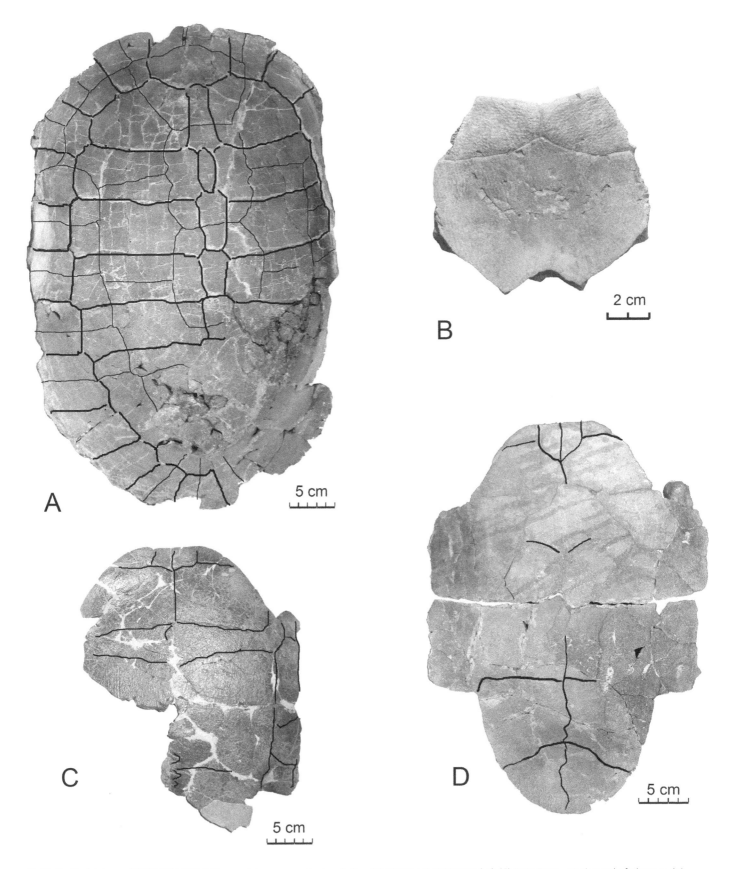

13.10. Shell of *Adocus*. (A) UMNH VP 20452: complete carapace in dorsal view; (B) UMNH VP 16925: nuchal; (C) BYU 1071: anterior end of plastron; (D) UCMP 159702: complete plastron.

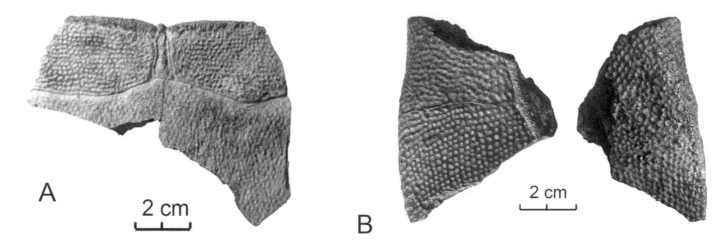

13.11. Comparison of the sculpture of *Adocus* and *Basilemys*. (A) *Adocus* nuchal, specimen UCMP 194286; (B) *Basilemys,* shell fragment from edge of carapace, specimen UCMP 194164.

basis of features of the plastron, including the arrangement of the gular scales and the degree of sinuosity of the midventral sulcus (Brinkman and Nicholls, 1993). Two of the species are Campanian, *B. variolosa* from the Judith River Group of Alberta and specimens from the Fruitland/Kirtland of New Mexico attributed to *B. nobilis* by Gilmore (1935) and Wiman (1933), and two are Late Maastrichtian, *B. praeclara* from the Lance and Frenchman Formations and *B. sinuosus* from the Hell Creek Formation. Sullivan et al. (in review) regard *B. nobilis*, the type of which is from the Ojo Alamo, to be a nomen dubium. The Kaiparowits specimen is similar to the Campanian species in that the gulars are separated by the intragular. The two Campanian species differ in that specimens referred to *B. nobilis* have a distinct beak on the anterior end of the plastron and the posterior end of the midplastral sulcus is more strongly sinuous. In these features, the Kaiparowits Formation specimens of *Basilemys* are similar to those ascribed to *B. nobilis*. Thus, it is included in that species.

TRIONYCHIDAE (Fitzinger, 1826)

Discussion The Trionychidae are found today in Africa, southern Asia and North America. As fossils they were even more widespread, occurring in the Tertiary of South America, Australia, and Europe. Trionychids first occur late in the Early Cretaceous of Asia. The oldest records in North America are in the early Late Cretaceous Dunvegan Formation of northern Alberta and the upper portion of the Mussentuchit Member of the Cedar Mountain Formation of Utah (Fiorillo, 1999; Hirayama et al., 2000), both of which are Cenomanian in age. Currently, no more than three species of trionychid occur sympatrically, and these differ in size (Pritchard, 2001). Trionychid assemblages of the Late Cretaceous are sometimes much more diverse, with possibly six species being present in the Dinosaur Park Formation (Brinkman, 2005), and a minimum of four taxa in the Hell Creek Formation (Holroyd and Hutchison, 2002). Resolving

13.12. Plastron of *Basilemys nobilis,* specimen UMNH VP 6138.

13.13. Carapace and plastron of *Helopanoplia* sp. Specimen UMNH VP 17025.

the level of diversity present in the Kaiparowits Formation in the absence of large series of well-articulated shells is difficult because of variation in the shell (Gardner and Russell, 1994). However, a minimum of four species-level taxa are present in the Kaiparowits Formation. Three of these are referred to established genera, *Helopanoplia*, *Aspideretoides*, and *Derrisemys*. The third is generically indeterminate and is referred to as Trionychid type B.

HELOPANOPLIA Hay, 1908
HELOPANOPLIA sp.

Reference Specimens UMNH VP 17025: partial skeleton including carapace, complete right and partial left hyoplastron, hypoplastron, both xiphiplastra, and associated skeletal fragments. UMNH VP 11799: partial hyoplastron and hypoplastron. UMNH VP 16284: lateral end of two costals preserved in articulation. UCMP 194260, Loc. V93092: lateral half of costal. UCMP 194261, Loc. V93092: partial costal of moderate size. UCMP 194254, Loc. V93088: plastron fragments. UCMP 194095, Loc. V94059: lateral edge of the costal. UCMP 194365: lateral edge of carapace. UCMP 194149: two plastron fragments. UCMP 194149, Loc. V93103, complete costal (Figs. 13.13, 13.14A–B).

Comments *Helopanoplia* is a large trionychid previously known only from the Hell Creek Formation. *Helopanoplia*

is represented in the Kaiparowits Formation by a partial skeleton including carapace and most of the plastron (Fig. 13.13), as well as associated skeletal elements. The carapace is distinctive in that the sculpture extends to the lateral edge of the carapace and a groove is present along the edge of the shell. On the basis of this feature, a series of isolated costal elements are also referred to *Helopanoplia*. The specimens with this feature all have a distinctive sculpture in which pits are subequal in size across the carapace and the border between the pits is formed by a sharp ridge varying in height. The available shell elements shows that this sculpture pattern does not grade into other sculpture patterns (Fig. 13.14), so for the estimates of relative abundance, costal fragments with this sculpture pattern are identified as *Helopanoplia*.

The plastron of UMNH VP 17025 (Fig. 13.13) provides the first nearly complete plastron of this genus. Epiplastra are missing. The entoplastron has a callosity near the center of the element. As in plastomenines, the hyoplastron and hypoplastron are broad, having an extensive web of bone extending between the anteromedial and anterolateral ends of the hyoplastron. However, the medial edges of the hyo-hypoplastron are convex, so although the medial plastron fontanelle is reduced, these elements do not have a strong midline contact as is the case in plastomenines. However, a straight midline edge of the xiphiplastra indicates that these elements contacted one another. The sculpture on the

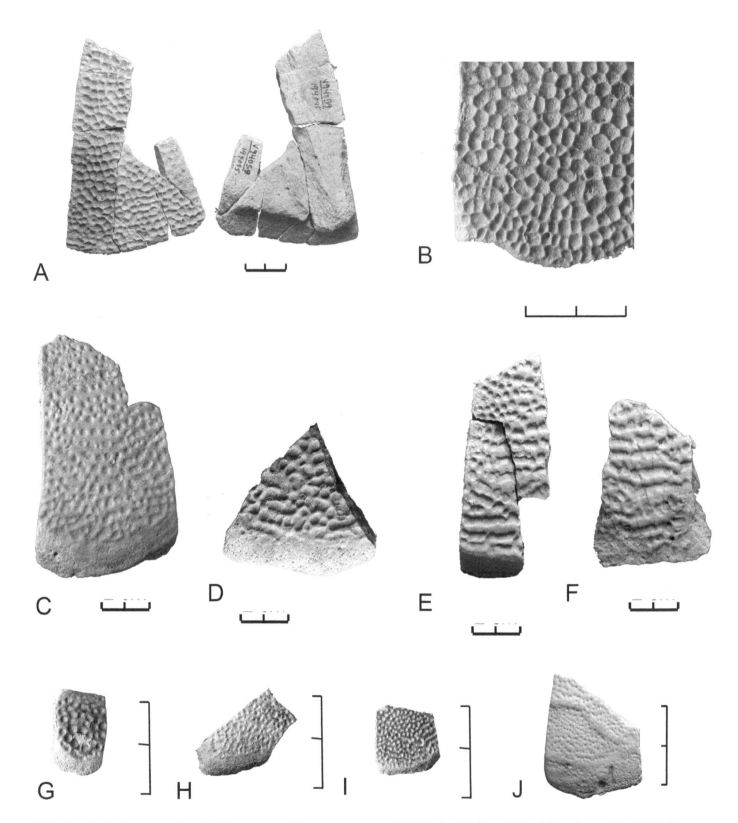

13.14. Comparison of sculpture trionychid sculpture patterns from the Kaiparowits Formation. (A–B) *Helopanoplia* sp. (A) Specimen UCMP 194095; (B) specimen UCMP 194260. (C, D) *Aspideretoides* sp. costals in which the sculpture consists of round pits; (C) specimen UCMP 194124; (D) specimen UCMP 194238. (E–F) *Aspideretoides* sp. costals in which the sculpture consists of elongate grooves; (E) specimen UCMP 194082; (F) specimen UCMP 194129. (G–I) small trionychid costals showing fine sculpture pattern; (G) specimen UCMP 194159; (H) specimen UCMP 194148; (I) specimen UCMP 194159. (J) costal with plications, specimen UCMP 194120. Scale bar equals two cm.

Hutchison, Knell, and Brinkman

plastron elements of *Helopanoplia* is distinctive in that the surface texture consists of isolated, closely spaced tubercles.

ASPIDERETOIDES Gardner,
Russell, and Brinkman, 1995
ASPIDERETOIDES sp. indet.

Reference Specimens UMNH VP 17026, complete carapace; UMNH VP 20453, partial carapace with associated hyo-hypoplastron and xiphiplastron; UMNH VP 20449, carapace missing anterior right quarter; UMNH 16729: anterior left portion of carapace, including complete nuchal; UCMP 555842, Loc. V95082: anterior end of carapace (Figs. 13.14C–F, 13.15).

Description A large trionychine with a broad hyo-hypoplastron is documented by a series of shells, one of which is associated with plastron elements (Fig. 13.15). These are included in the genus *Aspideretoides* because of the presence of a subdivided first neural (Fig. 13.15B). The carapace is wider than long, although the disparity in width and length is not as great as in *A. splendidus* (Hay, 1908). The carapace also differs from *A. splendidus* in the shape of the nuchal and first costal (Fig. 13.15A–B). The nuchal is relatively short and anteroposteriorly elongate relative to *A. foveatus* (Leidy, 1856) and *A. splendidus*. The posterior border of the first costal curves forward, rather than extending directly laterally as in *A. splendidus*, so the exposure or the lateral end of this costal on the edge of the shell is reduced (Fig. 13.15A–B). The posterior end of the carapace is distinctive in having a truncated posterior end, the flat portion marked laterally by distinct bulges on the seventh costal (Fig. 13.15A). This feature was not observed on other species of *Aspideretoides*. The sculpture on the carapace consists of pits evenly developed across the shell, becoming smaller toward the edge of the shell. The borders between the pits are rounded or flat topped. This pattern of sculpture is frequently encountered in isolated shell fragments (Fig. 13.14C). Isolated elements show that in larger individuals the pits tend to align (Fig. 13.14D), and ultimately coalesce to form grooves separated by broad, flat-topped ridges (Fig. 13.14E–F).

The plastron is documented by a hyo-hypoplastron and xiphiplastron preserved in association with a partial carapace (Fig. 13.15C–E). The carapace is distinct in having larger lateral projections of the ribs despite being larger than any of the more complete shells that are preserved. This difference is interpreted as a result of individual variation. The hyo-hypoplastron and xiphiplastron are similar in shape to those of *Helopanoplia*, differing only in being larger and having a sculpture formed by ridges and pits rather than fine tubercles.

Comments Alpha-level taxonomy of Late Cretaceous trionychids is largely unresolved, preventing resolution of the specific relationships of *Aspideretoides* sp. It is distinct from species of *Aspideretoides* in the Dinosaur Park Formation in the proportions of the shell shape of the nuchal, and pattern of sculpture. The shape of the hyo-hypoplastron and xiphiplastron is similar to that of *Helopanoplia*, suggesting a relationship with that taxon. Also, in the shape of the nuchal and first costal, *Aspideretoides* sp. is similar to both *Helopanoplia* sp. and the Late Maastrichtian species *Aspideretoides beecheri* as illustrated by Hay (1908:plate 92). Thus, it is possible that *Aspideretoides* sp. is more closely related to Late Maastrichtian trionychids than to contemporaneous species from the Dinosaur Park Formation.

DERRISEMYS Hutchison, 2009
DERRISEMYS sp.

Reference Specimens UMNH VP 16040, hypoplastron and lateral end of costal (Fig 13.16).

Description *Derrisemys* is recognized in the Kaiparowits Formation on the basis of an isolated hypoplastron (Fig. 13.16A). This element is broad and has a straight midline border, indicating complete reduction of the median fontanelles as in *Derrisemys* and *Hutchemys* (Joyce et al., 2009) (=*Plastomenoides* of Hutchison, 2009). The bridge is narrow and the ventral surface posterior to the bridge is broad and extends well posterior to the bridge. The sculpture on this element is distinctive in being very subdued, so that the ventral surface is nearly smooth. A costal fragment that is similar to Late Cretaceous plastomenines in having a downturned lateral edge with the rib not extending beyond the costal margin is assumed to also be from *Derrisemys*.

UMNH V16040 differs from *Hutchemys tetanetran* (Hutchison, 2009) and resembles *Derrisemys sterea* Hutchison, 2009, and *D. acupictus* (Hay, 1907) in having both the hyo-hypoplastron suture and the hypoxiphiplastron suture curving anteromedially, although the curvature of the hypoxiphiplastron suture curves to a lesser extent.

Comments The Plastomeninae is a group of trionychids with a well-ossified plastron showing a reduced midline fontanelle and a tight contact between the hypoplastron and xiphiplastra. Traditionally all plastomenines were included in the genus *Plastomenus*, but recently, a series of small plastomenines from the Late Campanian and Early Maastrichtian were placed in their own genera (Hutchison, 2009; Joyce et al., 2009). The Kaiparowits plastomenine is tentatively included in the genus *Derrisemys* on the basis of similarity in shape with that of UCMP 197898 (Hutchison, 2009; Fig. 20), particularly in the presence of a narrow plastral bridge and that the ventral surface posterior to the bridge is broad and extends well posterior to the bridge. In other small plastomenines, such as *Huchemys*,

13.15. Carapace and plastron of *Aspideretoides* sp. (A) Specimen UMNH VP 17026. (B) BYU specimen; (C–E) carapace and plastron elements of specimen UMNH VP 20453; (C) partial carapace; (D) right hyo-hypoplastron; (E) left xiphiplastron. Scale bar equals 10 cm.

Hutchison, Knell, and Brinkman

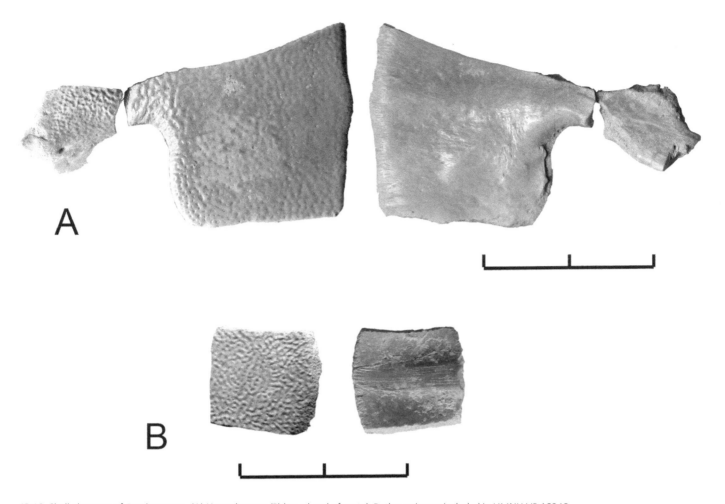

13.16. Shell elements of *Derrisemys* sp. (A) Hypoplastron; (B) lateral end of costal. Both specimens included in UMNH VP 16040.

the hypoplastron does not extend as far posteriorly (Joyce et al., 2009).

Genus indet. type B

Reference Specimens UMNH VP 16049: lateral end of costal; UMNH VP 16248: lateral end of costal. UMNH VP 18596: partial costal. UCMP 194085, Loc. v94053: shell fragments with plications extending anteroposteriorly along carapace. Probably remains of once complete carapace. UCMP 194252, Loc. v93088: costal. UCMP 194160, Loc. v95095: costal. UCMP 194120, Loc. v95083: eighth costal. UCMP 194200, Loc. v98054: eighth costal, complete. UCMP 194323, Loc. v93120: costal. UCMP 194310, Loc. v93110: lateral end of costal (Figs. 13.14J, 13.17).

Comments A small trionychine with very fine sculpture and with plications extending lengthwise along the shell is recognized here as taxonomically distinct and is designated trionychid type B (Figs. 14J, 17A–C). The plications are evenly spaced across the costals and are typically curved, indicating that in the complete carapace the plications are concave toward the lateral edge of the carapace. Support

13.17. Carapace elements of Trionychidae type B. (A) Specimen UCMP 194252; (B) specimen UCMP 194310; (C) specimen UCMP 194120.

for the interpretation that the elements with this feature are taxonomically distinct rather than juvenile is provided by the eighth costal, which is unusually elongate anteroposteriorly (Fig. 13.17C).

This feature of plications on the carapace is present in some trionychids from the Paleocene of New Mexico, such as *Plastomenus torrrejonensis* and *"Aspideretes" vegetus* (Gilmore, 1919), suggesting a relationship with those species. However, Trionychid B is distinct in that the eighth costal is much longer than wide. Thus, additional material will be necessary to resolve the relationships of this trionychid.

Trionychid indet.

Comments A series of very small costals with fine sculpture suggest the possibility of a small-bodied species of a trionychid in the Kaiparowits Formation (Fig. 13.14G–I). These do not show evidence for an elongate rib, which would be expected if they were from juveniles of one of the large-bodied species. However, pending discovery of complete specimens that would confirm that these are from adult individuals and allow the species to be diagnosed, they are considered taxonomically problematic and are not included in the faunal list.

DISCUSSION

Taphonomy

A few taphonomic patterns are discernible with Kaiparowits fossil turtles based upon both field observations and examinations of museum specimens. As mentioned previously, the majority of turtle specimens are recovered from the lower and middle units. The quality of preservation is greatest for those specimens recovered from the base of fluvial channels within these units. Fluvial channel specimens frequently consist of whole or partial shells with associated skeletal elements and lack evidence of bone modification, thus indicating rapid burial. Adocids and baenids are the primary taxa recovered from these channel deposits. Occasionally, these turtle specimens are recovered in association with fossil wood, possibly indicating the presence of a log jam or woody debris in which the turtle carcasses became entangled during transport. Those specimens recovered in overbank deposits are often incomplete and exhibit signs of abrasion and weathering, indicators of prolonged transport and exposure prior to burial.

In terms of taxa abundance, the number of trionychid, baenid, and adocid specimens far exceeds specimens from other families. The rarest taxa are also the smallest taxa, which include the kinosternids and chelydrids, and they are frequently found incomplete as fragmentary material. This pattern is possibly caused by a preservation bias in which smaller turtles are more susceptible to breakage prior to burial, or perhaps a depositional bias in which fluvial deposits outnumber those of ponds, the probable habitat of the smaller turtle taxa. Those turtle families that are thought to have inhabited fluvial environments, which are the trionychids, baenids, and adocids (Hutchison and Archibald, 1986), are well represented and well preserved within the fluvial and overbank deposits of the Kaiparowits Formation. In contrast, *Basilemys*, which is thought to have been more terrestrial in habit (Hutchison and Archibald, 1986; Brinkman, 2003b), is poorly represented and poorly preserved.

Latitudinal Differentiation of Turtle Assemblages

The abundant and well-preserved record of turtles in the Kaiparowits Formation adds to the evidence that turtles were one of the dominate mesovertebrates in Late Cretaceous paleocommunities. With a minimum of 14 taxa present, the Kaiparowits assemblage is the second most diverse assemblage known, exceeding that of the Dinosaur Park Formation, which has 11 taxa present (Brinkman, 2005). Because the Kaiparowits Formation was located at a paleolatitude of approximately 37.5°N, while the Dinosaur Park Formation was at a paleolatitude of 59°N (PLATES, 2009) the presence of a higher diversity of turtles in the Kaiparowits Formation compared to the Dinosaur Park Formation is consistent with Brinkman's (2003b) assumptions that diversity decreases with increased paleolatitude. As well, where closely related species occur in both the Kaiparowits and the Dinosaur Park Formations (*Neurankylus*, *Boremys*, and *Adocus*), those from the Kaiparowits Formation tend to be larger. This pattern is also consistent with the difference in paleolatitude: the ectothermic body size of higher-latitude taxa is smaller than comparable taxa at lower latitudes.

Because the turtle remains are primarily from the lower portion of the Kaiparowits Formation, this assemblage is contemporaneous with that of the Dinosaur Park Formation of Alberta. Thus, differences in the taxonomic composition of the Kaiparowits and Dinosaur Park formations provide further insight into the latitudinal differentiation of vertebrate assemblages in North America west of the interior seaway during the late Campanian. Latitudinal differentiation of vertebrate assemblages in North America in the late Campanian were documented by Lehman (1997), partly on the distribution of turtles, particularly the presence of a pleurodire in the Aguja Formation of southern Texas, but not more northerly localities. Subsequent descriptions of southern turtle assemblages (Tomlinson, 1997; Hutchison et al., 1998; Rodriguez-de la Rosa and Cevallos-Ferriz, 1998; Brinkman

Hutchison, Knell, and Brinkman

Table 13.2. Comparison of nonmarine turtle assemblages from late Campanian the Dinosaur Park Formation and Kaiparowits Formation[a]

Taxa	Dinosaur Park Formation	Kaiparowits Formation
Southern taxa		
Compsemys		X
Denzinemys nodosa		X
Helopanoplia		X
Plastomenoides		X
Kinosternid indet.		X
Northern taxa		
Plesiobaena antiqua	X	
Judithemys	X	
"*Plastomenus*"	X	
Cosmopolitan taxa		
Neurankylus	X	X
Boremys	X	X
Chelydrid indet.	X	X
Adocus	X	X
Basilemys	X	X
Aspideretoides	X	X
Taxa with restricted distributions		
Plesiobaena sp.		X
"*Apalone*" *latus*	X	
Trionychid type B		X

[a] Data on the turtles from the Dinosaur Park Formation are from Brinkman (2005).

and Rodriguez-de la Rosa, 2006; Sankey, 2006) allowed additional latitudinal patterns to be identified (Brinkman, 2003b). Taxa that were considered to be members of a northern assemblage include the Macrobaenidae, represented in the Dinosaur Park Formation by *Judithemys*, and the baenid *Plesiobaena antiqua*. Taxa that were considered members of a southern assemblage on the basis of presence/absence data include *Denazinemys*, kinosternoids (represented by *Hoplochelys* and a small smooth-shelled kinosternid), pleurodires, and *Compsemys*.

The turtle assemblages of the Kaiparowits and Dinosaur Park Formations are compared in Table 13.2. Nontrionychid taxa present in the Kaiparowits Formation but not the Dinosaur Park Formation include *Compsemys*, *Denazinemys*, *Plesiobaena* sp. nov., and the small smooth-shelled kinosternid. The presence of these taxa is consistent with the hypothesis that these taxa are latitudinally restricted.

The trionychid assemblage of the Kaiparowits Formation and the Dinosaur Park Formation are both diverse but have no overlap at the species level. Trionychid genera that are present in the Kaiparowits Formation but not in contemporaneous formations further north include *Helopanoplia* and *Derrisemys*. Both of these taxa also occur in the Late Maastrichtian Hell Creek Formation. This is consistent with the suggestion proposed by Brinkman and Eberth (2006) that changes in climate in the Maastrichtian resulted in changes in turtle distributions, with southern taxa moving north during the warming of the late Maastrichtian.

Taxa that were previously considered to be members of a southern turtle assemblage do not occur in the Kaiparowits Formation are *Hoplochelys* and pleurodires. Both of these taxa are present in the Aguja and Cerro del Pueblo formations of Texas and Coahulia, Mexico. The absence of these in the Kaiparowits Formation, suggests that even further latitudinal subdivision of the turtle assemblages was present, and that the change from southern to northern assemblages is gradational.

ACKNOWLEDGMENTS

The authors thank P. Holroyd, M. Getty, R. Irmis, R. Scheetz, and B. Britt for access to specimens in their care. Thanks to B. Britt for help in photographing large *Neurankylus* specimens in the collections of the Museum of Paleontology of Brigham Young University. Thanks to A. Titus for his encouragement of this project and for help in gaining access to specimens. Reviews by W. Joyce led to improvement of the content.

REFERENCES CITED

Agassiz, L. 1857. Contributions to the Natural History of the United States of America. Volumes 1 and 2. Little, Brown, Boston, Massachusetts, i–li + 1–643 pp.

Brinkman, D. B. 2003a. Anatomy and systematics of *Plesiobaena antiqua* (Testudines; Baenidae) from the mid-Campanian Judith River Group of Alberta, Canada. Journal of Vertebrate Paleontology 23:146–155.

Brinkman, D. B. 2003b. A review of nonmarine turtles from the Late Cretaceous of Alberta. Canadian Journal of Earth Sciences 40:557–571.

Brinkman, D. B. 2005. Turtles: diversity, paleoecology and distribution; pp. 202–220

in P. J. Currie and E. B. Kopplelhus (eds.), Dinosaur Provincial Park: A Spectacular Ancient Ecosystem Revealed. Indiana University Press, Bloomington, Indiana.

Brinkman, D. B., and D. A. Eberth. 2006. Turtles of the Horseshoe Canyon and Scollard Formations–further evidence for a biotic response to Late Cretaceous climate change. Fossil Turtle Research 1:11–18.

Brinkman, D. B., and E. L. Nicholls. 1991. Anatomy and relationships of the turtle *Boremys pulchra* (Testudines: Baenidae). Journal of Vertebrate Paleontology 11:302–315.

Brinkman, D. B., and E. L. Nicholls. 1993. A new specimen of *Basilemys praeclara* Hay and its bearing on the relationships of the Nanhsiungchelyidae (Reptilia: Testudines). Journal of Paleontology 67:1027–1031.

Brinkman, D. B., and R. Rodriguez-de la Rosa. 2006. Nonmarine turtles from the Cerro del Pueblo Formation (Campanian), Coahila State, Mexico. New Mexico Museum of Natural History and Science Bulletin 35:229–233.

Cope, E. D. 1868. On some Cretaceous Reptilia. Proceedings of the Academy of Natural Sciences, Philadelphia 1868:233–242.

Cope, E. D. 1870. On the Adocidae. Proceedings of the American Philosophical Society 11:547–553.

Cope, E. D. 1882. Contributions to the history of the Vertebrata of the Lower Eocene of Wyoming and New Mexico, made during 1881. Proceedings of the American Philosophical Society 20:139–197.

Eberth, D. A. 2005. The geology; pp. 54–82 in P. J. Currie and E. B. Koppelhus (eds.), Dinosaur Provincial Park: A Spectacular Ancient Ecosystem Revealed. Indiana University Press, Bloomington, Indiana.

Fiorillo, A. R. 1999. Non-mammalian microvertebrate remains from the Robinson Eggshell Site, Cedar Mountain Formation (Lower Cretaceous), Emery County, Utah; pp. 259–268 in D. D. Gillette (ed.), Vertebrate Paleontology in Utah. Utah Geological Survey Miscellaneous Publication 99–1.

Fitzinger, L. F. F. J. de. 1826. Neue Classification der Reptilien nach ihren natürlichen Verwandtschaften. Vienna.

Gaffney, E. S. 1972. The systematics of the North American family Baenidae (Reptilia, Cryptodira). Bulletin of the American Museum of Natural History 147:241–320.

Gaffney, E. S. 1975. A phylogeny and classification of the higher categories of turtles. Bulletin of the American Museum of Natural History 155:387–436.

Gaffney, E. W., and P. Meylan. 1988. A phylogeny of turtles; pp. 157–291 in M. J. Benton (ed.), The Phylogeny and Classification of the Tetrapods. Volume 1: Amphibians, Reptiles, Birds. Clarendon Press, Oxford.

Gardner, J. D., and A. P. Russell. 1994. Carapacial variation among soft-shelled turtles (Testudines: Trionychidae), and its relevance to taxonomic and systematic studies of fossil taxa. Neues Jahrbuch für Geologie und Paläontologie, Abhandlungen 193:209–244.

Gardner, J. D., A. P. Russell, and D. B. Brinkman. 1995. Systematics and taxonomy of soft-shelled turtles (Family Trionychidae) from the Judith River Group (mid-Campanian) of North America. Canadian Journal of Earth Sciences 32:631–643.

Gilmore, C. W. 1916. Vertebrate Faunas of the Ojo Alamo, Kirtland and Fruitland Formations. U.S. Geological Survey Professional Paper 98Q.

Gilmore, C. W. 1919. Reptilian Faunas of the Torrejon, Puerco, and Underlying Upper Cretaceous Formations of San Juan County, New Mexico. U.S. Geological Survey Professional Paper 119.

Gilmore, C. W. 1935. On the Reptilia of the Kirtland Formation of New Mexico, with descriptions of new species of fossil turtles. Proceedings of the U.S. National Museum 83:159–188.

Gray, J. E. 1870. Supplement to the Catalogue of Shield Reptiles in the Collection of the British Museum. Part 1. Testudinata (Tortoises). Taylor and Francis, London.

Hay, O. P. 1902. Bibliography and Catalogue of Fossil Vertebrates of North America. U.S. Geological Survey Bulletin 179.

Hay, O. P. 1907. Descriptions of seven new species of turtles from the Tertiary of the United States. Bulletin of the American Museum of Natural History 23:847–863.

Hay, O. P. 1908. The fossil turtles of North America. Publications of the Carnegie Institute Washington 75:1–555.

Hay, O. P. 1911. Descriptions of eight new species of fossil turtles from west of the one hundredth meridian. Proceedings of the United States National Museum 38:307–326.

Hirayama, R., D. B. Brinkman, and I. G. Danilov. 2000. Distribution and biogeography of non-marine Cretaceous turtles. Russian Journal of Herpetology 7:181–198.

Holroyd, P. A., and J. H. Hutchison. 2002. Patterns of geographic variation in latest Cretaceous vertebrates; evidence from the turtle component; 177–190 in J. H. Hartman, K. R. Johnson, and D. J. Nichols (eds.), The Hell Creek Formation and the Cretaceous–Tertiary Boundary in the Northern Great Plains. Geological Society of America Special Paper 361.

Hutchison, J. H. 1991. Early Kinosternidae (Reptilia: Testudines) and their phylogenetic significance. Journal of Vertebrate Paleontology 11:145–167.

Hutchison, J. H. 2009. New soft-shelled turtles (Plastomeninae, Trionychidae, Testudines) from the Late Cretaceous and Paleocene of North America. Paleobios 29:36–47.

Hutchison, J. H., and J. D. Archibald. 1986. Diversity of turtles across the Cretaceous/Tertiary boundary in northeastern Montana. Palaeogeography, Palaeoclimatology, Palaeoecology 55:1–22.

Hutchison, J. H., and D. M. Bramble. 1981. Homology of the plastral scales of the Kinosternidae and related turtles. Herpetologica 37:73–85.

Hutchison, J. H., J. E Eaton., P. A. Holroyd, and M. A. Goodwin. 1998. Larger vertebrates of the Kaiparowits Formation (Campanian) in the Grand Staircase–Escalante National Monument and adjacent areas; pp. 391–398 in L. M. Hill (ed.), Learning from the Land: Grand Staircase–Escalante National Monument Science Symposium Proceedings. U.S. Department of the Interior, Bureau of Land Management.

Joyce, W., G., A. Revan, T. R. Lyson, and I. G. Danilov. 2009. Two new plastomenine softshell turtles from the Paleocene of Montana and Wyoming. Bulletin of the Peabody Museum of Natural History 50:307–325.

Klein, I. T. 1760. Klassification und kurze Geschichte der Vierfübigen Thiere. Jonas Schmidt, Lübeck.

Knell, M. J., F. D. Jackson, A. L. Titus, and L. B. Albright. 2011. A gravid fossil turtle from the Upper Cretaceous (Campanian) Kaiparowits Formation, southern Utah. Historical Biology 23:57–62.

Lambe, L. 1902. On the Vertebrata of the Mid-Cretaceous of the Northwestern Territory 2: New genera and species from the Belly River series (Mid-Cretaceous). Contributions to Canadian Palaeontology 3:25–81.

Lambe, L. 1906. Boremys, a new chelonian genus from the Cretaceous of Alberta. Ottawa Naturalist 12:232–234.

Lehman, T. M. 1997. Late Campanian dinosaur biogeography in the western interior of North America; pp. 223–240 in D. L. Wolberg, E. Stump, and G. D. Rosenberg (eds.), Dinofest International: Proceedings of a Symposium Held at Arizona State University. Academy of Natural Sciences, Philadelphia.

Leidy, J. 1856. Notice of remains of extinct reptiles and fishes, discovered by D. F. V. Hayden in the Bad Lands of the Judith River, Nebraska Territory. Proceedings of the Academy of Natural Sciences of Philadelphia 8:72–73.

Lucas, S. G., and R. M. Sullivan. 2006. Denazinemys, a new name for some Late Cretaceous turtles from The Upper Cretaceous of the San Juan Basin, New Mexico; pp. 223–227 in S. G. Lucas and R. M. Sullivan (eds.), Late Cretaceous Vertebrates from the Western Interior. New Mexico Museum of Natural History and Science Bulletin 35.

Martin, J. E., and J. Benton. 2008. Crown clades in vertebrate nomenclature: correcting the definition of crocodylia. Systematic Biology 57:173–181.

Meylan, P. A., and E. S. Gaffney. 1989. The skeletal morphology of the Cretaceous cryptodiran turtle, Adocus, and the relationships of the Trionychoidea. American Museum Novitates 2941:1–60.

PLATES program. 2009. The PLATES Project; L. Lawver, and I. Dalziel, principal investigators, L. Gahagan, database and software manager. University of Texas Institute for Geophysics, Austin, Texas.

Pritchard, P. C. H. 2001. Observations on body size, sympatry and niche divergence in softshell turtles (Trionychidae). Chelonian Conservation and Biology 4:5–27.

Roberts, E. M. 2007. Facies architecture and depositional environments of the Upper Cretaceous Kaiparowits Formation, southern Utah. Sedimentary Geology 197:207–233.

Roberts, E. M., A. L. Deino, and M. A. Chan. 2005. $^{40}Ar/^{39}Ar$ age of the Kaiparowits Formation, southern Utah, and correlation of contemporaneous Campanian strata and vertebrate faunas along the margin of the Western Interior Basin. Cretaceous Research 26:307–318.

Rodriguez-de la Rosa, R. A., and R. S. Cevallos-Ferriz. 1998. Vertebrates of the El Pelillal locality (Campanian, Cerro del Pueblo Formation), Southeastern Coahuila, Mexico. Journal of Vertebrate Paleontology 18:751–764.

Sankey, J. 2006. Turtles of the upper Aguja Formation (Late Campanian), Big Bend National Park, Texas; pp. 235–243 in S. G. Lucas and R. M. Sullivan (eds.), Late Cretaceous Vertebrates from the Western Interior. New Mexico Museum of Natural History and Science Bulletin 35.

Tomlinson, S. L. 1997. Late Cretaceous and Early Tertiary turtles from the Big Bend Region, Brewster County, Texas. Ph.D. dissertation, Texas Tech University, Lubbock, Texas.

Wiman, C. 1933. Über Schildkröten aus der Oberen Kreide in New Mexico. Nova Acta Regiae Societatis Scientiarum Upsaliensis 9:1–34.

Yeh, H. K. 1966. A new Cretaceous turtle of Nanhsiung, northern Kwanlung. Vertebrata Palasiatica 10:191–200.

Zangerl, R. 1969. The turtle shell; pp. 311–339 in C. Gans, A. Bellairs, and T. S. Parsons (eds.), Biology of the Reptilia. Volume 1. Academic Press, New York.

Hutchison, Knell, and Brinkman

Review of Late Cretaceous Mammalian Faunas of the Kaiparowits and Paunsaugunt Plateaus, Southwestern Utah

14

Jeffrey G. Eaton and Richard L. Cifelli

MAMMALS HAVE BEEN RECOVERED BY WET SCREEN washing of microvertebrate localities in the Dakota Formation (Cenomanian); the Smoky Hollow (Turonian) and John Henry (Coniacian–Santonian) members of the Straight Cliffs Formation; the Wahweap Formation (lower–middle Campanian); and the Kaiparowits Formation (upper Campanian). The faunas of the Dakota Formation and the Smoky Hollow Member are unique to the North American record and are dominated by *Paracimexomys*-group multituberculates, archaic therian taxa, and metatherians. The fauna from the John Henry Member of the Straight Cliffs Formation is correlative with the fauna from the Milk River Formation (Aquilan), and its multituberculate fauna includes genera that are common later in the Cretaceous. The fauna from the Wahweap Formation includes taxa from the Paunsaugunt Plateau assigned previously to the Kaiparowits Formation. The fauna from the Wahweap Formation is not completely correlative with the Aquilan because it is younger.

The first undoubted eutherians appear in this interval. The fauna from the Kaiparowits Formation is correlative with the Judithian but also exhibits noticeable endemism. Archaic taxa have disappeared, and there is a greater diversity of eutherians. Although there are differences between these faunas and more northerly assemblages, there are also marked faunal differences between the mammalian faunas of the Markagunt, Paunsaugunt, and Kaiparowits plateaus.

INTRODUCTION

Eaton and Cifelli (1988) presented a preliminary report on the Late Cretaceous mammalian faunas of the Kaiparowits Plateau. Vertebrate faunal lists were presented, with updated compendia of the mammals, by Eaton et al. (1999). Eaton (1993b) published on the mammalian fauna from the uppermost Cretaceous strata of the adjacent Paunsaugunt Plateau (Figs. 14.1, 14.2). More than a decade has passed since

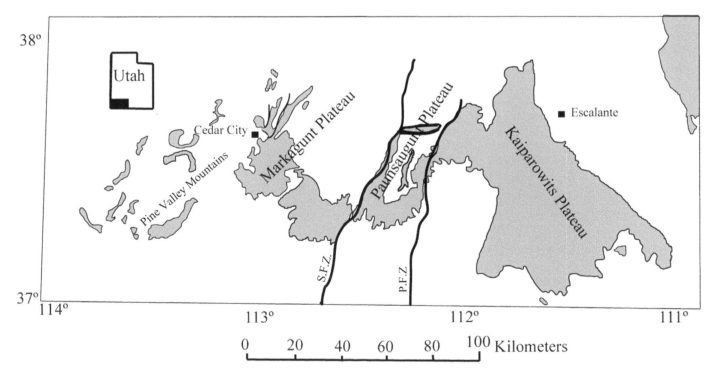

14.1. Shaded areas indicated exposures of Cretaceous rocks in southwestern Utah. P.F.Z., Paunsaugunt Fault Zone; S.F.Z., Sevier Fault Zone.

these original summaries, and considerable work has been undertaken since then (and more will continue in the future). The goal of this chapter is to bring the mammalian faunal lists of Eaton (1993b) and Eaton et al. (1999) up to date and to reconsider their ages in light of new radiometric and biostratigraphic data. Unlike many of the taxonomic groups discussed in this volume, the mammalian fauna has been well illustrated in publications, so we felt no need to duplicate that here. Alongside the name of each taxon in the faunal list is the citation of the literature in which that taxon has been documented from the Kaiparowits region.

Institutional Abbreviations MNA, Museum of Northern Arizona, Flagstaff, Arizona; OMNH, Oklahoma Museum of Natural History, Norman, Oklahoma; UMNH, Utah Museum of Natural History, University of Utah, Salt Lake City, Utah.

DAKOTA FORMATION

All nonmarine vertebrate fossils were recovered from the middle member of the Dakota Formation. The upper member is well dated as late Cenomanian on the basis of ammonites (Cobban, 1984; Tibert et al., 2003). Because the middle member becomes increasing brackish up section, it seems likely that the middle member is also late Cenomanian, or at oldest, middle Cenomanian. Kowallis, Christiansen, and Deino (1989) provided $^{40}Ar/^{39}Ar$ dates of 94.7 ± 0.2 Ma and 94.5 ± 0.1 Ma from horizons low in the overlying Tropic Shale Formation, which also constrain the upper member of the Dakota Formation as late Cenomanian in age. Bohor et al. (1991) had reported $^{40}Ar/^{39}Ar$ dates of 92.9 ± 0.2 Ma and 90.5 ± 0.1 from the base and near the top of the Dakota Formation, respectively, but these ages would place the Dakota in the Turonian. Dyman et al. (2002) also considered the upper member to be of late Cenomanian age and published a $^{40}Ar/^{39}Ar$ date of 96.06 ± 0.30 Ma date for the middle member, which would place the middle member in the early middle Cenomanian (timescale of Ogg, Agterberg, and Gradstein, 2004). Eaton examined the horizon from which the sample that yielded this date was taken and found that the unit is not air-fall ash but bentonitic mudstone, and therefore the date may reflect contamination.

The multituberculates consist almost entirely of "Paracimexomys group" taxa. There is only one taxon that represents a family that becomes common later, the Cimolodontidae. *Dakotadens* is tentatively placed here in the Subclass Boreosphenida because of its similarity to *Potamotelses*, a younger boreosphenidan. *Arcantiodelphys*, represented by scant, hard-won specimens from the Cenomanian of France, is similar to *Dakotadens* (Vullo et al., 2009). Taxa referred to *Alphadon* in Eaton (1993a) no longer fit in that genus based

Stage	Kaiparowits Plateau	Paunsaugunt Plateau
Campanian	Kaiparowits Fm.	
Campanian	Wahweap Fm.	Wahweap Fm.
Santonian	Straight Cliffs Fm. — Drip Tank	Straight Cliffs Fm. — Drip Tank
Santonian	Straight Cliffs Fm. — John Henry	Straight Cliffs Fm. — John Henry
Con.	Straight Cliffs Fm. — Smoky Hollow	Straight Cliffs Fm. — Smoky Hollow
Turonian	Straight Cliffs Fm. — Tibbet Canyon	Straight Cliffs Fm. — Tibbet Canyon
Turonian	Tropic Shale	Tropic Shale
Cenomanian	Dakota Fm.	Dakota Fm.

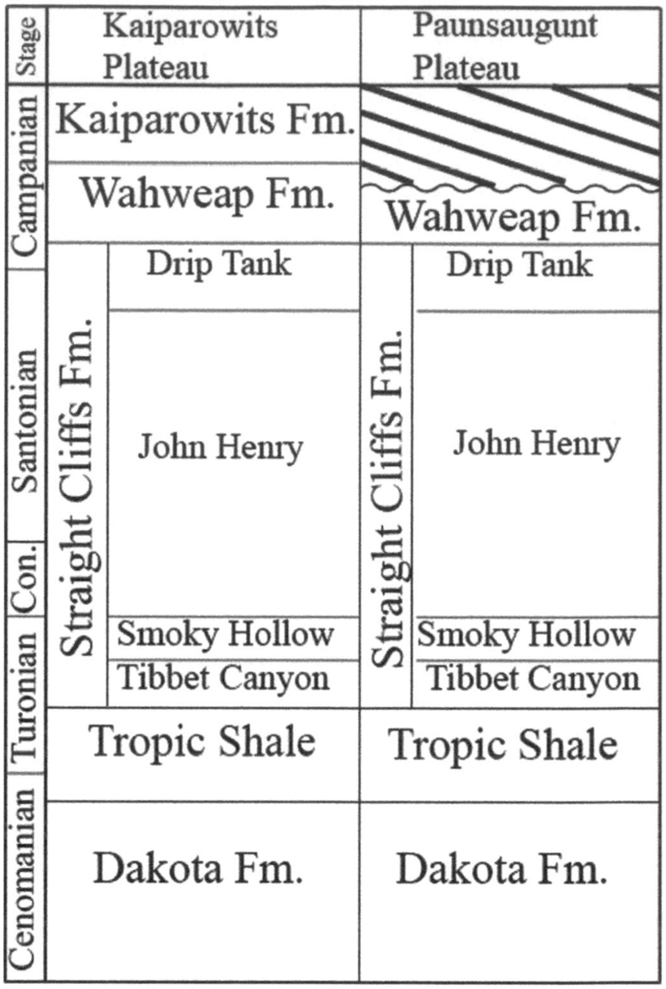

14.2. Cretaceous stratigraphy of the Kaiparowits and Paunsaugunt plateaus. Con., Coniacian.

on Johanson's (1996) revised definition and were referred to *Eoalphadon* by Eaton (2009).

Order Multituberculata Cope, 1884
    Suborder Cimolodonta McKenna, 1975
        Family incertae sedis
            Paracimexomys group
                Paracimexomys sp. cf. *P. robisoni* Eaton and Nelson, 1991 (Eaton, 1995:fig. 6A–F)
                Paracimexomys sp. (Eaton, 1995:fig. 6G–N)
                Cf. *Paracimexomys* sp. (Eaton, 1995:fig. 6O, P)
                Dakotamys malcolmi Eaton, 1995:fig. 7A–M
                ?*Dakotamys* sp. (Eaton, 1995:fig. 7N)
                Genus and species undetermined A (Eaton, 1995:fig. 7O)
                Genus and species undetermined B (Eaton, 1995:fig. 7P)
        Family Cimolodontidae Marsh, 1889a
            Cimolodon sp. cf. *similis* Fox, 1971a (Eaton, 1995:fig. 4A–J)
            Genus and species undetermined (Eaton, 1995:fig. 5A–M)
Subclass ?Boreosphenida Luo, Cifelli, and Kielan-Jaworowska, 2001
    Order, family, genus, and species undetermined (Eaton, 1993a:fig. 3A–G)
    Order and Family incertae sedis
        Dakotadens morrowi Eaton, 1993a:fig. 4A–M
        Dakotadens sp. (Eaton, 1993a:fig. 4N, O)
    Infraclass Metatheria Huxley, 1880

Eaton and Cifelli

Cohort Marsupialia Illiger, 1811
  Order "Didelphomorpha" Gill, 1872
  Family "Alphadontidae" Marshall, Case, and Woodburne, 1990
    Eoalphadon clemensi Eaton, 1993a:fig. 5A–J
    Eoalphadon lillegraveni Eaton, 1993a:fig. 6A–H
    Eoalphadon sp. (Eaton, 1993a; fig. 6I–L)
    Protalphadon sp. (Eaton, 1993a:fig. 7A–C)
    Genus and species undetermined (Eaton, 1993a:fig. 7D)
  Family ?Stagodontidae Marsh, 1889b
    Pariadens kirklandi Cifelli and Eaton, 1987:fig. 1 (also see Eaton, 1993a:fig. 8A–I)

STRAIGHT CLIFFS FORMATION

Smoky Hollow Member

Peterson (1969) named the Smoky Hollow member and divided it into three parts: (1) a lower coal zone that has considerable brackish-water influence; (2) a barren middle zone representing floodplain deposits; and (3) an upper conglomeratic bed he named the Calico bed. The fossils in this report all come from the middle barren unit. Recent U-Pb analysis of a bentonite bed located a few meters below the Calico bed yielded a middle Turonian age of 91.86 ± 0.34 Ma (Titus et al., this volume, Chapter 2), which is close to the age of the underlying Tibbet Canyon Member, which contains the late middle Turonian index fossils *Inoceramus howelli* White and *Prionocyclus hyatti* (Stanton). Both of these are typical of the middle Turonian *Prionocyclis hyatti* ammonoid biozone of Cobban et al. (2006).

As is the case with the Dakota Formation, the multituberculate assemblage consists almost entirely of "*Paracimexomys* group" taxa and a possible cimolomyid. In Eaton et al. (1999), several taxa were listed that have not yet been described and documented (*Picopsis, Dakotadens, Alphadon*, and a possible eutherian); they are omitted from this list of figured taxa. Additional taxa of boreosphenidan mammals from the Smoky Hollow Member as exposed on the Kaiparowits Plateau are almost certainly represented in existing collections, but the material is too fragmentary for diagnosis and circumscription.

Order Multituberculata Cope, 1884
  Suborder ?Taeniolabiboidea Granger and Simpson, 1929
    Family, genus, and species undetermined (Eaton, 1995:fig. 8A, B)
        Suborder Cimolodonta McKenna, 1975
    Family incertae sedis
      Paracimexomys group
        Paracimexomys sp. cf. *P. robisoni* (Eaton, 1995:fig. 8C, D)
        Bryceomys fumosus Eaton, 1995:figs. 8E–T–9A–H
        Bryceomys sp. cf. *B. fumosus* (Eaton, 1995:fig. 9I)
        Bryceomys hadrosus Eaton, 1995:fig. 9J–M
        Bryceomys spp. (Eaton, 1995:fig. 9N–P)
        Genus and species undetermined (Eaton, 1995:fig. 10A–H)
Order Symmetrodonta Simpson, 1925
  Family, gen., et sp. indet. (Cifelli and Gordon, 1999:fig. 7)
  Family Spalacotheriidae Marsh, 1887

Symmetrodontoides oligodontos Cifelli, 1990a:fig. 1A–D (Cifelli and Gordon, 1999, figs. 4, 5)
Spalacotheridium mckennai Cifelli, 1990a:fig. 1E, F (Cifelli and Gordon, 1999:fig. 6)
Order Aegialodontia Butler, 1978
  Family Deltatheridiidae Gregory and Simpson, 1926
    Unnamed genus and species (Cifelli, 1990a:fig. 2A–C)
  Family uncertain
    Genus and species undetermined (Cifelli, 1990a:fig. 2D–J)
Cohort Marsupialia Illiger, 1811
  Family uncertain
    ?*Varalphadon delicatus* Cifelli, 1990a:fig. 3
    Unnamed genus and species (Cifelli, 1990a:fig. 4A–G)
  Family ?Stagodontidae Marsh, 1889b
    Genus undetermined (Cifelli, 1990a:fig. 4H–J)

John Henry Member

Peterson (1969) reported *Volviceramus involutus* from low in the John Henry Member, a taxon considered to represent the middle Coniacian (Merewether, Cobban, and Obradovich, 2007:fig. 5). Taxa from the middle and upper parts of the member, including the ammonite *Desmoscaphites*, suggest that the upper part of the John Henry Member does not extend beyond the Santonian (Eaton, 1991). As such, the member ranges from middle Coniacian to possibly late Santonian. Much of the faunal list below has been generated from UMNH VP Loc. 99, which is high in the member, but not in the uppermost part. As such, the age of this fauna is likely to be middle or late Santonian. Recently, several new localities have been discovered high in the John Henry Member on the Paunsaugunt Plateau, and these specimens are described by Eaton in Chapter 15 of this volume.

Interestingly, the multituberculate fauna is much different than that from the Smoky Hollow Member or the Dakota Formation. There are relatively few taxa from the "*Paracimexomys* group" and many taxa representing families that continue through the rest of the Late Cretaceous. ?*Anchistodelphys* sp. was originally listed in Eaton (2006a:fig. 9A–C), but the genus *Anchistodelphys* is probably not a valid genus according to Johanson (1996). However, because she provided no diagnostic criteria for lower teeth, it is not known to what taxon this lower molar should be assigned.

Order Multituberculata Cope, 1884
  Suborder Cimolodonta McKenna, 1975
      Genus and species undetermined (Eaton, 2006a:fig. 3A)
    Family incertae sedis
      Paracimexomys group
        Cedaromys sp. cf. *C. hutchisoni* Eaton, 2002 (Eaton, 2006a:fig. 5E, F) (Eaton, this volume, Chapter 15:fig. 15.6M, N)
        Cedaromys sp. (Eaton, 2006a:fig. 6A–D)
        Dakotamys shakespeari Eaton, this volume, Chapter 15:fig. 15.6K, L
      Family Neoplagiaulacidae Ameghino, 1890
        Mesodma sp. Jepsen, 1940 (Eaton, this volume, Chapter 15:fig. 15.4I, J)

Mesodma sp. cf. *M. minor* Eaton, 2002 (Eaton, 2006a:fig.
   3B–F) (Eaton, this volume, Chapter 15:fig. 15.4C–H)
?*Mesodma* sp. Jepsen, 1940 (Eaton, this volume, Chapter
   15:fig. 15.4K–N)
Family Cimolodontidae Marsh, 1889a
   Cimolodon foxi Eaton, 2002 (Eaton, 2006a:fig. 4D)
   Cimolodon sp. cf. *C. foxi* Eaton, 2002 (Eaton, this volume,
      Chapter 15:fig. 15.5F)
   Cimolodon similis Fox, 1971a (Eaton, this volume, Chapter
      15:fig. 15.5G–K)
   Cimolodon sp. cf. *C. similis* Fox, 1971a (Eaton, this volume,
      Chapter 15:fig. 15.6B)
   Cimolodon spp. Marsh, 1889b (Eaton, 2006a:fig. 4E)
   ?*Cimolodon* sp. Marsh, 1889b (Eaton, 2006a:fig. 4F, G; this
      volume, Chapter 15:fig. 15.6C)
Family Cimolomyidae Marsh, 1889a
   Cimolomys sp. A Marsh, 1889a (Eaton, this volume, Chapter
      15:fig. 15.6D–F)
   Cimolomys sp. B Marsh, 1889b (Eaton, this volume, Chapter
      15:fig. 15.6G)
   ?*Cimolomys* sp. A Marsh, 1889b (Eaton, 2006a:fig. 5A–D; this
      volume, Chapter 15:fig. 15.6H, I)
   ?*Cimolomys* sp. B (Eaton, this volume, Chapter 15:fig. 15.6J)
Order Triconodonta Osborn, 1888
   Family Triconodontidae Marsh, 1887
      Genus et sp. indet. (Eaton, this volume:fig. 15.4A)
      Cf. *Alticonodon* sp. Fox, 1969 (Eaton, this volume:fig. 15.4B)
Order Symmetrodonta Simpson, 1925
   Family Spalacotheriidae Marsh, 1887
      Spalacotheridium sp. Cifelli, 1990a (Eaton, 2006a:fig. 7A–C)
      ?*Spalacotheridium* sp. Cifelli, 1990a (Eaton, this volume,
         Chapter 15:fig. 15.7A–D)
      Symmetrodontoides sp. cf. *S. oligodontos* (Cifelli and Gordon,
         1999:fig. 5F, G)
      Symmetrodontoides sp. Fox, 1976 (Eaton, this volume,
         Chapter 15:fig. 15.7E–G)
Subclass Boreosphenida Luo, Cifelli, and Kielan-Jaworowska, 2001
   Order incertae sedis
      Family Picopsidae Fox, 1980
         Picopsis sp. Fox, 1980 (Eaton, 2006a:fig. 7G–I)
      Family incertae sedis
         Potamotelses sp. Fox, 1972 (Eaton, 2006a:fig. 7D–F)
Infraclass Metatheria Huxley, 1880
   Cohort Marsupialia Illiger, 1811
      Order and Family incertae sedis
         Genus and species undetermined (Eaton, 2006a:fig. 9D–F)
      Order "Didelphomorpha" Gill, 1872
         Family incertae sedis
            Genus et sp. indet. (Eaton, 2006a:fig., this volume, Chapter 15:figs.
               15.8A, B, 15.10A–N)
            Apistodon sp. cf. *A. exiguus* (Fox, 1971b) (Eaton, this volume,
               Chapter 15:fig. 15.8C–M)
            Cf. "*Anchistodelphys*" sp. Cifelli, 1990b (Eaton, this volume,
               Chapter 15:fig. 15.9A)
         Family "Alphadontidae" Marshall, Case, and Woodburne, 1990
            Alphadon sp. cf. *A. halleyi* Sahni, 1972 (Eaton, 2006a:fig.
               8A–C)
            Varalphadon sp. Johanson, 1996 (Eaton, 2006a:fig. 8D–G)
            ?*Varalphadon* sp. Johanson, 1996 (Eaton, this volume,
               Chapter 15:fig. 15.9B–E)
         Family ?Stagodontidae Marsh, 1889b
            Eodelphis sp. Matthew, 1916 (Eaton, this volume, Chapter
               15:fig. 15.11A)
            Genus and species undetermined (Eaton, 2006a:fig. 8H–J)
         Family Pediomyidae Simpson, 1927
            Genus et sp. indet. (Eaton, this volume, Chapter 15:fig. 15.11B)
            ?*Leptalestes* sp. Davis, 2007 (Eaton, this volume, Chapter
               15:fig. 15.11C)

Wahweap Formation

This formation is over 400 m thick on the Kaiparowits Plateau. Radiometrically datable ashes are rare. Jinnah et al. (2009) have reported a $^{40}Ar/^{39}Ar$ date of 80.1 ± 0.3 Ma (early middle Campanian age based on Ogg, Agterberg, and Gradstein, 2004) from about 60 m above the base of the formation. On the basis of a date of 75.96 ± 0.14 Ma for the lower Kaiparowits Formation and on estimates of depositional rates, an upper age limit on the Wahweap Formation has been suggested to be 76.1 Ma (Roberts, Deino, and Chan, 2005), or earliest late Campanian age. This would suggest that the age of most of the Wahweap Formation is middle Campanian; however, some of the basal localities are probably uppermost lower Campanian. Eaton and Cifelli (1988) originally correlated the mammalian fauna from the Wahweap Formation with the fauna from the Milk River Fauna in Alberta, Canada (Aquilan Land Mammal Age, now correlated to the upper Santonian; Leahy and Lerbekmo, 1995); however, later studies of mammals from the upper part of the John Henry Member of the Straight Cliffs Formation (Santonian) revealed a stronger correlation (Eaton, 2006a) to the Milk River fauna than it did to the Wahweap fauna. A Santonian age for the Milk River fauna has also been well documented on the basis of palynomorphs (Braman, 2002) and U-Pb geochronology (Payenberg et al., 2002).

Included in the faunal list below are taxa from the Paunsaugunt Plateau that were tenuously considered to be from the Kaiparowits Formation by Eaton (1993b). Lateral walking of strata and additional measured sections (by Eaton) now indicate that these taxa should be considered to come from the middle member of the Wahweap Formation (Fig. 14.2). The Kaiparowits Plateau Wahweap localities and those on the Paunsaugunt Plateau are only separated by tens of kilometers, yet it is striking how different the faunas and lithologies are. Most of the Wahweap material on the Kaiparowits comes from stream lag deposits or dark, organic-rich mudstones representing relatively low-lying coastal floodplains. The Paunsaugunt localities (MNA Loc. 1073, 1074) that have a distinctive mammalian fauna are about 100 m above the base of the formation in a series of brightly colored variegated mudstones representing a relatively well-drained floodplain with well-developed paleosols, a lithology not found in the Wahweap Formation on the Kaiparowits Plateau. This suggests that some subtle paleoecologic differences may have profound controls on the nature of mammalian faunas.

It has been pointed out (Eaton, 1993b, 2002) that the fauna from the Wahweap Formation includes taxa similar to those from the Aquilan North American LandMammal Age (NA-LMA), taxa similar to those from the Judithian NALMA, and also some unique taxa. It is likely that at least the upper

Table 14.1. Occurrences of mammals from the Wahweap or Kaiparowits formations recorded from other geographic areas

Taxon	W	K	MR	JR+	K-F	"Ed"	LA
Cimexomys antiguus	cf.		X		cf.		
Cimexomys judithae	cf.			X	X		
Dakotamys magnus		X		X			
Mesodma formosa	cf.						X
Mesodma hensleighi	cf.						X
Cimolodon electus	X		X		X		
Cimolodon similis	X	cf.	X				
Cimolodon nitidus	cf.	cf.				X	X
Cimolomys trochuus	cf.						X
Cimolomys clarki		cf.		X			
Meniscoessus intermedius	cf.	cf.		X	X		
Meniscoessus major		cf.		X			
Symmetrodontoides	X		X				
Varalphadon	X	X	X				
Turgidodon russelli	cf			X			
Alphadon halleyi		X		X	X		
Alphadon marshi	cf.				cf.	X	X
Paranyctoides	X	X	X	X	X		
Gypsonictops		X		X	X	X	X

"Ed," Edmontonian suggested by Lillegraven and McKenna (1986); JR +, Judith River Formation and other Judithian faunas such as the Oldman and Dinosaur Park formations; KF, Kaiparowits Formation; K-F, Kirtland–Fruitland formations; LA, Lancian; MR, Milk River Formation; WF, Wahweap Formation.

portion of the Wahweap is equivalent to the Judithian NALMA (Jinnah et al., 2009:fig. 7), but it also now clear that most of the lower half of the formation correlates with the Clagget Shale and thus predates the type Judith River Formation. As such, the Wahweap would be expected to contain taxa that predate the Judithian NALMA.

Two multituberculates taxa identified with certainty would be considered index fossils of the Aquilan NALMA (*Cimolodon electus* and *C. similis*). One taxon suggests affinities to the Aquilan, *Cimexomys* sp. cf. *C. antiguus*, but is only conferred to that species, and two genera, *Symmetrodontoides* and *Varalphadon*, are also suggestive of the Aquilan. Only one conferred taxon (*Meniscoessus* sp. cf. *M. intermedius*) suggests a Judithian NALMA; however, some conferred taxa suggest an age younger than Judithian (Table 14.1). As such, the mammalian fauna of the Wahweap Formation only weakly suggests equivalence to that recovered from the older Milk River Formation (Santonian). It may be necessary at some point to establish a biostratigraphic zone that represents the early to middle Campanian.

Order Multituberculata Cope, 1884
  Suborder Cimolodonta McKenna, 1975
    Family incertae sedis
      Paracimexomys group
        Genus and species undetermined (Eaton, 2002)
        ?*Paracimexomys* sp. Archibald, 1982 (Eaton, this volume, Chapter 15:fig. 15.12A, B)
        Cf. *Paracimexomys* sp. A (Eaton, 2002:fig. 24C, D)
        Cf. *Paracimexomys* sp. B (Eaton, 2002:fig. 24E)
        Bryceomys sp. cf. *B. fumosus* Eaton, 1995 (Eaton, 2002:fig. 24F)

        Cedaromys sp. cf. *C. hutchisoni* Eaton, 2002 (Eaton, this volume, Chapter 15:fig. 15.12C, D)
        Cedaromys sp. Eaton, 2002:fig. 24G
        Cf. *Cedaromys* sp. Eaton, 2002:fig. 24H
      ?*Paracimexomys* group
        Cimexomys sp. cf. *C. antiguus* Fox, 1971a (Eaton, 2002:fig. 24A, B)
        ?*Cimexomys gregoryi* Eaton, 1993b:fig. 4i
        Genus and species undetermined (Eaton, 2002:fig. 23H, I)
    Family Neoplagiaulacidae Ameghino, 1890
      Mesodma sp. cf. *M. formosa* (Marsh, 1889b) (Eaton, 1993b:fig. 4a; Eaton, 2002:fig. 18A–G) (Eaton, this volume, Chapter 15:fig. 15.12G, H)
      Mesodma sp. cf. *M. hensleighi* Lillegraven, 1969 (Eaton, 1993b:fig. 4b)
      Mesodma sp. cf. *M. minor* Eaton, 2002 (Eaton, this volume, Chapter 15:fig. 15.12E)
      Mesodma sp. cf. *M. archibaldi* Eaton, 2002 (Eaton, this volume, Chapter 15:fig. 15.12F)
      Mesodma sp. (Eaton, 1993b:fig. 4c)
    Family Cimolodontidae Marsh, 1889a
      Cimolodon electus Fox, 1971a (Eaton, 2002:fig. 19A–G)
      Cimolodon similis Fox, 1971a (Eaton, 2002:fig. 20A–H)
      Cimolodon sp. cf. *C. nitidus* Marsh, 1889a (Eaton, 1993b:fig. 4d–f; Eaton, 2002:fig. 21G–I)
      Cimolodon sp. cf. *C. foxi* Eaton, 2002 (Eaton, this volume, Chapter 15:fig. 15.12I)
      Cimolodon sp. (small) (Eaton, 2002:fig. 22A–C)
      ?*Cimolodon* sp. (Eaton, 1993:fig. 4g)
    Family Cimolomyidae Marsh, 1889a
      Cimolomys milliensis Eaton, 1993:fig. 4h
      Cimolomys sp. cf. *C. trochuus* Lillegraven, 1969 (Eaton, 2002:fig. 23A–C)
      Cimolomys sp. Marsh, 1889b (Eaton, this volume, Chapter 15:fig. 15.12J)
      ?*Cimolomys* sp. (Eaton, this volume, Chapter 15:fig. 15.12K)
      ?*Cimolomys* sp. A (Eaton, 2002:fig. 23D)
      ?*Cimolomys* sp. B (Eaton, 2002:fig. 23E, F)

?*Cimolomys* sp. C (large) (Eaton, 2002:fig. 23G)
    Meniscoessus sp. cf. *M. intermedius* Fox, 1976 (Eaton, 2002:fig. 22D–G)
    Meniscoessus sp. Cope, 1882 (Eaton, this volume, Chapter 15:fig. 15.12L)
Order Symmetrodonta Simpson, 1925
  Family, gen., et sp. indet. (Cifelli and Gordon, 1999:fig. 7)
  Family Spalacotheriidae Marsh, 1887
    Symmetrodontoides foxi Cifelli and Madsen, 1986:fig. 1 (Cifelli and Gordon, 1999:fig. 3) (Eaton, 1993b:fig. 5a, b)
  Order uncertain
    Zygiocuspis goldingi Cifelli, 1990c:fig. 1
Infraclass Metatheria Huxley, 1880
  Cohort Marsupialia Illiger, 1811
    Family incertae sedis
      Genus et sp. indet. A (Eaton, this volume, Chapter 15:fig. 15.13C)
      Genus et sp. indet. B (Eaton, this volume, Chapter 15:fig. 15.13D)
      Cf. *Apistodon* sp. Davis, 2007 (Eaton, this volume, Chapter 15:fig. 15.13E, F)
  Order "Didelphomorpha" Gill, 1872
    Family "Alphadontidae" Marshall, Case, and Woodburne, 1990
      Varalphadon crebreforme (Cifelli, 1990b:fig. 2A)
      Varalphadon wahweapensis (Cifelli, 1990b:figs. 4, 5; Johanson, 1996:pl. 2:fig. C, L)
      Varalphadon sp. cf. *V. creber* (Fox, 1971b) (Eaton, this volume, Chapter 15:fig. 15.13G–I)
      Cf. *Varalphadon* sp. Johanson, 1996 (Eaton, this volume, Chapter 15:fig. 15.13J–N)
      Genus and species undetermined (Cifelli, 1990b:fig. 2B–D)
      Turgidodon sp. cf. *T. russelli* (Fox, 1979a) (Eaton, 1993b:fig. 6e–h)
      Turgidodon sp. Cifelli, 1990d (Eaton, 1993b:fig. 7f, g)
    Family Pediomyidae Simpson, 1927
      Genus and sp. unknown (Eaton, this volume, Chapter 15:fig. 15.14A).
  Cohort ?Marsupialia Illiger, 1811
    Family uncertain
      Iugomortiferum thoringtoni Cifelli, 1990b:fig. 3A–J
      Cf. *Iugomortiferum* sp. (Cifelli, 1990b:fig. 3K–M; Eaton, this volume, Chapter 15:fig. 15.13A, B)
  Order Insectivora Cuvier, 1817
    Family ?Nyctitheriidae Simpson, 1928
      Paranyctoides sp. Cifelli, 1990e:fig. 3

KAIPAROWITS FORMATION

The Kaiparowits Formation is about 855 m thick (Eaton, 1991; 860 m in Roberts, Deino, and Chan, 2005) and was mainly deposited in a fluvial regime. Radiometric dates (Roberts, Deino, and Chan, 2005) and mammalian faunas (Eaton and Cifelli, 1988; Cifelli, 1990d; Eaton, 2002) have suggested a late (but not latest) Campanian age for the Kaiparowits Formation. The youngest ^{40}Ar/^{39}Ar determination reported from the Kaiparowits Formation is 74.21 ± 0.18 Ma (Roberts, Deino, and Chan, 2005), which places the Kaiparowits Formation in the lower part of the upper Campanian. Most of the fauna has been recovered from the lower 400 m of the formation, with the exception of locality TB8 (Eaton, 1991:fig. 16); UMNH VP Loc. 1082 at 747 m above the base of the formation was discovered in 2008, but the mammals from this locality have not yet been studied. Depositional rates and poor soil development suggest rapid deposition,

and a rate as high as ~41 cm/ka has been suggested by Jinnah et al. (2009).

Roberts, Deino, and Chan (2005) considered the Kaiparowits fauna to be at least in part correlative to the Judith River Formation and the Judithian NALMA. Jinnah et al. (2009:fig. 7) have correlated the Wahweap Formation with the Judith River Formation (and Judithian NALMA) and suggested that the Kaiparowits fauna is younger than the Judith River Formation (and the Judithian NALMA) on the basis of a comparison of their dates with a ^{40}Ar/^{39}Ar date of 79.52 ± 0.61 Ma from Hicks, Obradovich, and Tauxe (1995) from near the middle of the Judith River Formation (Jinnah et al., 2009, provide a date of 79.3 Ma, but that is the estimated age of Hicks, Obradovich, and Tauxe, 1995, of the C33n–C33r geomagnetic reversal), which makes the Judith River Formation older than the Kaiparowits Formation. The date by Hicks, Obradovich, and Tauxe (1995) places the lower part of the Judith River Formation in Elk Basin, Wyoming, within the lower Campanian (based on Ogg, Agterberg, and Gradstein, 2004) and they correlate the top of the formation to the lower part of the *B. perplexus* ammonite zone, which is earliest middle Campanian. Jinnah et al. (2009) have made an error in assuming the Judithian NALMA is equivalent to only the Judith River and Oldman formations as currently defined. When Lillegraven and McKenna (1986) redefined the Judithian NALMA, the Dinosaur Park Formation had not yet been recognized and at that time was the upper part of the Oldman Formation. Part of the fauna definitive of the Judithian NALMA is based on fossil mammals described by Fox (1979a, 1979b, 1979c, 1980, 1981) from what is now called the Dinosaur Park Formation, but was then referred to as the Oldman Formation. Eberth and Hamblin (1993), based on a regional unconformity and a distinct change in lithology, separated the Dinosaur Park Formation from the Oldman Formation along a diachronous discontinuity. The base of the Dinosaur Park Formation is time transgressive but probably slightly younger than a bentonite 4 m below the Oldman-Dinosaur Park contact with a date of 76.5 ± 0.5 m (Eberth and Hamblin, 1993:185); a date from the basal Bearpaw Formation of 74.98 ± 0.24 Ma (Eberth and Deino, 2005) suggests an upper age of ~75 Ma for the Dinosaur Park Formation, which is coeval with the lower part of the Kaiparowits Formation. As some of the mammalian taxa used to define the Judithian NALMA in Lillegraven and McKenna (1986) are from what is now considered to be the Dinosaur Park Formation, it is still appropriate to consider at least the lower part of the Kaiparowits Formation to be correlative to the Dinosaur Park Formation and the Judithian NALMA. The Kaiparowits Formation may extend ~0.8 Ma beyond Dinosaur Park deposition, but there is as of yet no known distinctive faunal change in the one described locality from the upper part of the Kaiparowits Formation (TB8, OMNH V5; see Cifelli 1990d for contrasting view).

An examination of the multituberculate faunal list below indicates only one species definitively described from the Judithian NALMA (*Dakotamys magnus*), while the remaining taxa are either conferred to species or are new. The conferred taxa *Cimolomys* sp. cf. *C. clarki*, *Meniscoessus* sp. cf. *M. intermedius*, and *M.* sp. cf. *M. major* are suggestive of a Judithian age. *Alphadon halleyi* is a Judithian taxon, and on that basis, the mammalian fauna from the Kaiparowits Formation correlates most favorably with the Judithian NALMA (Table 14.1). The mammals from the Fruitland-Kirtland (Sullivan and Lucas, 2006) include many taxa diagnostic of the Judithian NALMA (Table 14.1), and on the basis of mammals there is no reason for establishing a Kirtlandian land vertebrate age.

Alphadon attaragos Lillegraven and McKenna (1986) is included in quotation marks because the species is based on a single lower molar and Johanson (1996) has questioned the validity of the species, but it has not been formally synonymized. Study of eutherian mammals, including a number of specimens collected since the preliminary work of Cifelli (1990e), is in progress, and shows that a number of endemic species are represented in the assemblage. Interestingly, fragmentary yet diagnostic specimens demonstrate the presence of a new form of Cimolestidae, which are common in Lancian assemblages but are rare or lacking from earlier faunas.

Order Multituberculata Cope, 1884
  Suborder and family incertae sedis
    Paracimexomys group
      Cedaromys hutchisoni Eaton, 2002:fig. 16A–H
      Cf. *Cedaromys* sp. (Eaton, 2002:fig. 17A–F)
      Dakotamys magnus (Sahni, 1972) (Eaton, 2002:fig. 17G, H)
    ?*Paracimexomys* group
      Cimexomys sp. cf. *C. judithae* Sahni, 1972 (Eaton, 2002:fig. 14A–J)
      Cimexomys or *Mesodma* sp. (Eaton, 2002:fig. 15A–D)
  Suborder Cimolodonta McKenna, 1975
    Family Neoplagiaulacidae Ameghino, 1890
      Mesodma archibaldi Eaton, 2002:figs. 3H, 4B
      Mesodma sp. cf. *M. archibaldi* (Eaton, 2002:fig. 4C)
      Mesodma minor Eaton, 2002:fig. 4D–G
      Mesodma sp. (large) (Eaton, 2002:figs. 7A–H, 8A–D)
    Family Cimolodontidae Marsh, 1889a
      Cimolodon foxi Eaton, 2002:figs. 8F, G, 9A–I, 10A, B
      Cimolodon sp. cf. *C. nitidus* Clemens, 1963 (Eaton, 2002:fig. 10C–F)
      Cimolodon sp. cf. *C. similis* Fox, 1971a (Eaton, 2002:fig. 10G, H)
    Family ?Cimolodontidae Marsh, 1889a
      Kaiparomys cifellii Eaton, 2002:fig. 11A–D
    Family Cimolomyidae Marsh, 1889a
      Cimolomys sp. A cf. *C. clarki* Sahni, 1972 (Eaton, 2002:fig. 12A, B)
      Cimolomys sp. B cf. *C. clarki* Sahni, 1972 (Eaton, 2002:fig. 12C–F)
      Meniscoessus sp. cf. *M. intermedius* Fox, 1976 (Eaton, 2002:fig. 11E, F)
      Meniscoessus sp. cf. *M. major* Sahni, 1972 (Eaton, 2002, unfigured)
    Family ?Cimolomyidae Marsh, 1889a
      ?*Cimolomys butleria* Eaton, 2002:fig. 13A–G
  Cohort Marsupialia Illiger, 1811
    Family uncertain

      Aenigmadelphys archeri Cifelli and Johanson, 1994:fig. 1E (Cifelli, 1990d:fig. 2)
    Family "Alphadontidae" Marshall, Case, and Woodburne, 1990
      Varalphadon wahweapensis (Cifelli, 199da:fig. 3)
      Turgidodon lillegraveni Cifelli, 1990d:fig. 4
      Turgidodon sp. cf. *T. lillegraveni* (Cifelli, 1990d:fig. 6)
      Turgidodon madseni Cifelli, 1990d:fig. 7
      Turgidodon sp. (Cifelli, 1990d, unfigured)
      Alphadon halleyi Sahni, 1972 (Cifelli, 1990d:fig. 8)
      "*Alphadon attaragos*" Lillegraven and McKenna, 1986 (Cifelli, 1990d:fig. 10)
Order Insectivora Cuvier, 1817
  Family Leptictidae Gill, 1872
    Gypsonictops spp. (Cifelli, 1990e:fig. 2)
  Family ?Nyctitheriidae Simpson, 1928
    Paranyctoides spp. (Cifelli, 1990e:figs. 4, 5)
Order Uncertain
  Family uncertain
    Avitotherium utahensis Cifelli, 1990e:fig. 6

Paleobiogeographic Implications

There has been considerable discussion of latitudinal differences in faunas, including mammalian assemblages (Gates et al., 2010). Certainly there is not a great overlap in mammalian species or even genera between Utah and Canada (Table 14.1); however, there are several taxa that are conspecific or conferred. This may indicate that strong latitudinal taxonomic gradients were present affecting the distribution of mammalian taxa.

Although this may well have played a role, small mammals are good at partitioning niches and having relatively small ranges that may reflect controls other than latitude. Late Cretaceous faunas have been described from the Paunsaugunt (Eaton, 1993b) and Markagunt (Eaton, 2006b, 2009) plateaus west of Kaiparowits region. The stratigraphy from the Kaiparowits Plateau can be carried into the Paunsaugunt Plateau, but the stratigraphy of the Markagunt Plateau is far less certain (Eaton et al., 2001). Although the stratigraphic relations of the Kaiparowits and Paunsaugunt plateaus are known (Fig. 14.2), the mammalian faunas of these two adjacent plateaus have no species in common. There are most certainly Santonian and possibly Campanian fossil-bearing horizons on the Markagunt Plateau. Eaton (this volume, Chapter 15) has recently compared large new collections of Santonian material from Bryce Canyon National Park and all previously described Paunsaugunt and Kaiparowits plateaus mammals recovered from the Wahweap Formation, and there is not a single certainly conspecific taxon. He has also compared this new Santonian fauna to potentially correlative faunas on the Markagunt Plateau described in Eaton (2006b), and again, there are no conspecific taxa and the faunas are remarkably different. This represents either the sampling of different time intervals (to some degree likely between all faunal comparisons) or strong lateral changes in mammalian faunal composition. The floodplain sediments are distinctly different from east to west, with dark, organic-rich coastal floodplains to the east and better-drained, more upland variegated floodplains to the west. These subtle lateral changes in facies may reflect ecologic differences that may have a strong impact on the distribution of mammalian taxa, perhaps as marked as latitudinal changes.

ACKNOWLEDGMENTS

The specimens recovered from Bryce Canyon National Park were collected with funding from the National Park Service and the Bryce Canyon Natural History Association (to J.G.E., 2006–2009). Other specimens were collected on grants from the Petroleum Research Fund (ACS-PRF# 30989-GB8; ACS-PRF# 34595-B8 to J.G.E.; ACS-PFR# 20311-G8 to R.L.C.), National Science Foundation (EAR-9004560, to J.G.E. and J. I. Kirkland; DEB-8020265 and BSR-8416540, 8507598, 8796225, 8906992 to R.L.C.) and the National Geographic Society (to 3965–88 to J.G.E.; 2881–84 to R.L.C.). Reviews by A. Titus and M. Woodburne were very helpful.

REFERENCES CITED

Ameghino, F. 1890. Los plagiaulácidos Argentinos y sus relaciones zoológicas, geológicas y geográficas. Boletin del Instituto Geográfico Argentino 11:143–208.

Archibald, J. D. 1982. A study of Mammalia and geology across the Cretaceous–Tertiary boundary in Garfield County, Montana. University of California Publications in Geological Sciences 122:1–286.

Bohor, B. F. D., B. Dalrymple, D. Triplehorn, and M. Kirschbaum. 1991. Argon/argon dating of tonsteins from the Dakota Formation, Utah. Geological Society of America Abstracts with Programs 23:A85.

Braman, D. R. 2002. Terrestrial palynomorphs of the upper Santonian–?lowest Campanian Milk River Formation, southern Alberta. Palynology 25:57–107.

Butler, P. M. 1978. A new interpretation of the mammalian teeth of tribosphenic pattern from the Albian of Texas. Breviora 446:1–27.

Cifelli, R. L. 1990a. Cretaceous mammals from southern Utah. III. Therian mammals from the Turonian (early Late Cretaceous). Journal of Vertebrate Paleontology 10:332–345.

Cifelli, R. L. 1990b. Cretaceous mammals from southern Utah. II. Marsupials and marsupial-like mammals from the Wahweap Formation (early Campanian). Journal of Vertebrate Paleontology 10:320–331.

Cifelli, R. L. 1990c. A primitive higher mammal from the Late Cretaceous of southern Utah. Journal of Mammalogy 71:343–350.

Cifelli, R. L. 1990d. Cretaceous mammals from southern Utah. I. Marsupial mammals from the Kaiparowits Formation (Judithian). Journal of Vertebrate Paleontology 10:295–319.

Cifelli, R. L. 1990e. Cretaceous mammals from southern Utah. IV. Eutherian mammals from the Wahweap (Aquilan) and Kaiparowits (Judithian) formations. Journal of Vertebrate Paleontology 10:343–350.

Cifelli, R. L., and J. G. Eaton. 1987. Marsupial from the earliest Late Cretaceous of Western U.S. Nature 325:520–522.

Cifelli, R. L., and C. L. Gordon. 1999. Symmetrodonts from the Late Cretaceous of southern Utah and distribution of archaic mammals in the Cretaceous of Utah. Brigham Young University Geology Studies 44:1–15.

Cifelli, R. L., and Z. Johanson. 1994. New marsupial from the Upper Cretaceous of Utah. Journal of Vertebrate Paleontology 14:292–295.

Cifelli, R. L., and S. K. Madsen. 1986. An Upper Cretaceous symmetrodont (Mammalia) from southern Utah. Journal of Vertebrate Paleontology 6:258–263.

Clemens, W. A. 1963. Fossil mammals of the type Lance Formation, Wyoming. Part 1. Introduction and Multituberculata. University of California Publications in Geological Sciences 48:1–105.

Cobban, W. A. 1984. Mid-Cretaceous ammonite zones, Western Interior, United States. Geological Society of Denmark Bulletin 33:71–89.

Cobban, W. A., I. Walaszczyk, J. D. Obradovich, and K. C. McKinney. 2006. A USGS zonal table for the Upper Cretaceous middle Cenomanian–Maastrichtian of the Western Interior of the United States based on ammonites, inoceramids, and radiometric ages. U.S. Geological Survey Open-File Report 2006–1250.

Cuvier, Baron G. L. C. F. D. 1817. Le Règne animal distribué d'après son Organization. Volume 1, Introduction, les mammifères et les oiseaux. Deterville, Paris.

Cope, E. D. 1882. Mammalia in the Laramie Formation. American Naturalist 16:830–831.

Cope, E. D. 1884. The Tertiary Marsupialia. American Naturalist 18:686–697.

Davis, B. M. 2007. A revision of "pediomyid" marsupials from the Late Cretaceous of North America. Acta Palaeontologica Polonica 52:217–256.

Dyman, T. S., W. A. Cobban, A. L. Titus, J. D. Obradovich, L. E. Davis, R. L. Eves, G. L. Pollock, K. I. Takahashi, and T. C. Hester. 2002. New biostratigraphic and radiometric ages for Albian–Turonian Dakota Formation and Tropic Shale at Grand Staircase–Escalante National Monument and Iron Springs Formation near Cedar City, Parowan, and Gunlock in SW Utah. Geological Society of America, Abstracts with Programs 34:13.

Eaton, J. G. 1991. Biostratigraphic framework for the Upper Cretaceous rocks of the Kaiparowits Plateau, southern Utah; pp. 47–63 in J. D. Nations and J. G. Eaton (eds.), Stratigraphy, Depositional Environments, and Sedimentary Tectonics of the Western Margin, Cretaceous Western Interior Seaway. Geological Society of America Special Paper 260.

Eaton, J. G. 1993a. Therian mammals of the Cenomanian (Late Cretaceous) Dakota Formation, southwestern Utah. Journal of Vertebrate Paleontology 13:105–124.

Eaton, J. G. 1993b. Mammalian paleontology and correlation of uppermost Cretaceous rocks of the Paunsaugunt Plateau, Utah; pp. 163–180 in M. Morales (ed.), Aspects of Mesozoic Geology and Paleontology of the Colorado Plateau. Museum of Northern Arizona Bulletin 59.

Eaton, J. G. 1995. Cenomanian and Turonian (early Late Cretaceous) multituberculate mammals from southwestern Utah. Journal of Vertebrate Paleontology 15:707–882.

Eaton, J. G. 2002. Multituberculate mammals from the Wahweap (Campanian, Aquilan) and Kaiparowits (Campanian, Judithian) formations, within and near the Grand Staircase–Escalante National Monument, southern Utah. Utah Geological Survey Miscellaneous Publication 02–4:1–66.

Eaton, J. G. 2006a. Santonian (Late Cretaceous) mammals from the John Henry Member of the Straight Cliffs Formation, Grand Staircase–Escalante National Monument, Utah. Journal of Vertebrate Paleontology 26:446–460.

Eaton, J. G. 2006b. Late Cretaceous mammals from Cedar Canyon, southwestern Utah; pp. 373–402 in S. G. Lucas and R. M. Sullivan (eds.), Late Cretaceous Vertebrates from the Western Interior. New Mexico Museum of Natural History and Science Bulletin 35.

Eaton, J. G. 2009. Cenomanian (Late Cretaceous) mammals from Cedar Canyon, southwestern Utah, and a revision of Cenomanian Alphadon-like marsupials; pp. 97–110 in L. B. Albright III (ed.), Papers on Geology, Vertebrate Paleontology and Biostratigraphy in Honor of Michael O. Woodburne. Museum of Northern Arizona Bulletin 65.

Eaton, J. G., and R. L. Cifelli. 1988. Preliminary report on the Late Cretaceous mammalian faunas of the Kaiparowits Plateau, Utah. Contributions to Geology, University of Wyoming 26:45–51.

Eaton, J. G., and M. Nelson. 1991. Multituberculate mammals from the Lower Cretaceous Cedar Mountain Formation, San Rafael Swell, Utah. Contributions to Geology, University of Wyoming 29:1–12.

Eaton, J. G., R. L. Cifelli, J. H. Hutchison, J. I. Kirkland, and J. M. Parrish. 1999. Cretaceous vertebrate faunas from the Kaiparowits Plateau, south central Utah; pp. 345–353 in D. D. Gillette (ed.), Vertebrate Paleontology in Utah. Utah Geological Survey Miscellaneous Publication 99-1.

Eaton, J. G., J. Laurin, J. I. Kirkland, N. E. Tibert, R. M. Leckie, B. B. Sageman, P. M. Goldstrand, D. W. Moore, A. W. Straub, W. A. Cobban, and J. D. Dalebout. 2001. Cretaceous and Early Tertiary Geology of Cedar and Parowan canyons, western Markagunt Plateau, Utah; pp. 337–363 in M. C. Erskine, J. E. Faulds, J. M. Bartley, and P. D. Rowley (eds.), The Geologic Transition, High Plateaus to Great Basin–A Symposium and Field Guide. Utah Geological Association Publication 30.

Eberth, D. A., and A. Deino. 2005. New ^{40}Ar/^{39}Ar ages from three bentonites in the Bearpaw, Horseshoe Canyon, and Scollard formations (Upper Cretaceous–Paleocene) of southern Alberta, Canada; pp. 23–24 in D. R. Braman, E. B. Therrien, and W. Taylor (eds.), Dinosaur Park Symposium. Special Publication of the Royal Tyrrell Museum.

Eberth, D. A., and A. P. Hamblin. 1993. Tectonic, stratigraphic, and sedimentologic significance of a regional disconformity in the upper Judith River Group (Belly River wedge) of southern Alberta, Saskatchewan, and northern Montana. Canadian Journal of Earth Sciences 30:174–200.

Fox, R. C. 1969. Studies of Late Cretaceous vertebrates. III. A triconodont mammal from Alberta. Canadian Journal of Zoology 47:1253–1256.

Fox, R. C. 1971a. Early Campanian multituberculates (Mammalia: Allotheria) from the upper Milk River Formation, Alberta. Canadian Journal of Earth Sciences 8:916–938.

Fox, R. C. 1971b. Marsupial mammals from the early Campanian Milk River Formation, Alberta, Canada. Zoological Journal Linnean Society 50(Supplement 1):145–164.

Fox, R. C. 1972. A primitive therian mammal from the upper Cretaceous of Alberta. Canadian Journal of Earth Sciences 9:1479–1494.

Fox, R. C. 1976. Cretaceous mammals (*Meniscoessus intermedius,* new species, and *Alphadon* sp.) from the lowermost Oldman Formation, Alberta. Canadian Journal of Earth Sciences 13:1216–1222.

Fox, R. C. 1979a. Mammals from the Upper Cretaceous Oldman Formation, Alberta I. *Alphadon* Simpson (Marsupialia). Canadian Journal of Earth Sciences 16:91–102.

Fox, R. C. 1979b. Mammals from the Upper Cretaceous Oldman Formation, Alberta II. *Pediomys* Marsh (Marsupialia). Canadian Journal of Earth Sciences 16:103–113.

Fox, R. C. 1979c. Mammals from the Upper Cretaceous Oldman Formation, Alberta III. Eutheria. Canadian Journal of Earth Sciences 16:114–125.

Fox, R. C. 1980. Mammals from the Upper Cretaceous Oldman Formation, Alberta IV. *Meniscoessus* Cope (Multituberculata). Canadian Journal of Earth Sciences 17:1480–1488.

Fox, R. C. 1981. Mammals from the Upper Cretaceous Oldman Formation, Alberta V. *Eodelphis* Matthew, and the evolution of the Stagodontidae (Marsupialia). Canadian Journal of Earth Sciences 18:350–365.

Gates, T. A., S. D. Sampson, L. E. Zanno, E. M. Roberts, J. G. Eaton, R. L. Nydam, J. H. Hutchison, J. A. Smith, M. A. Loewen, and M. A. Getty. 2010. Biogeography of terrestrial and freshwater vertebrates from the Late Cretaceous (Campanian) Western Interior of North America. Palaeogeography, Palaeoclimatology, Palaeoecology 291:371–387.

Gill, T. N. 1872. Arrangement of the families of mammals. With analytical tables. Smithsonian Miscellaneous Collections 11:1–98.

Granger, W., and G. G. Simpson. 1929. A revision of the Tertiary Multituberculata. Bulletin of the American Museum of Natural History 56:601–676.

Gregory, W. K., and G. G. Simpson. 1926. Cretaceous mammal skulls from Mongolia. American Museum Novitates 225:1–20.

Hicks, J. F., J. D. Obradovich, and L. Tauxe. 1995. A new calibration point for the Late Cretaceous Time Scale: The ^{40}Ar/^{39}Ar isotopic age of the C33r/C33n geomagnetic reversal from the Judith River Formation (Upper Cretaceous), Elk Basin, Wyoming, U.S.A. Journal of Geology 103:243–256.

Huxley, T. H. 1880. On the application of the laws of evolution to the arrangement of the Vertebrata and more particularly of the Mammalia. Proceedings of the Zoological Society of London 43:649–662.

Illiger, C. 1811. Prodrodomus systematis mammalium et avium additis terminis zoographicis ultriusque classis. C. Salfeld, Berlin.

Jepsen, G.L. 1940. Paleocene faunas of the Polecat Bench Formation, Park County, Montana. Proceedings of the American Philosophical Society 83:217–341.

Jinnah, Z. A., E. M. Roberts, A. L. Deino, J. S. Larsen, P. K. Link, and C. M. Fanning. 2009. New ^{40}Ar-^{39}Ar and detrital zircon U-Pb ages from the Upper Cretaceous Wahweap and Kaiparowits formations on the Kaiparowits Plateau, Utah: implications for regional correlation, provenance, and biostratigraphy. Cretaceous Research 30:287–299.

Johanson, Z. 1996. Revision of the Late Cretaceous North American marsupial genus *Alphadon.* Palaeontographica Abt. A 242:127–184.

Kowallis, B. J., E. H. Christiansen, and A. Deino. 1989. Multi-characteristic correlation of Upper Cretaceous volcanic ash beds from southwestern Utah to central Colorado. Utah Geological and Mineral Survey Miscellaneous Publication 89–5:1–22.

Leahy, G. D., and J. F. Lerbekmo. 1995. Macrofossil magnetostratigraphy for the upper Santonian–lower Campanian interval in the Western Interior of North America: comparisons with European stage boundaries and planktonic foraminiferal zonal boundaries. Canadian Journal of Earth Sciences 32:247–260.

Lillegraven, J. A. 1969. Latest Cretaceous mammals of upper part of Edmonton Formation of Alberta, Canada, and review of marsupial-placental dichotomy in mammalian evolution. Paleontological Contributions, University of Kansas 50(Vertebrata 12):1–122.

Lillegraven, J. A., and M. C. McKenna. 1986. Fossil mammals from the "Mesaverde" Formation (Late Cretaceous, Judithian) of the Bighorn and Wind River basins, Wyoming, with definitions of Late Cretaceous North American land-mammal "ages." American Museum Novitates 2840:1–68.

Luo, Z. -X., R. L. Cifelli, and Z. Kielan-Jaworowska. 2001. Dual origin of tribosphenic mammals. Nature 409:53–57.

Marsh, O. C. 1887. American Jurassic mammals. American Journal of Science 33:326–348.

Marsh, O. C. 1889a. Discovery of Cretaceous Mammalia. American Journal of Science 38:81–92.

Marsh, O. C. 1889b. Discovery of Cretaceous Mammalia. Part II. American Journal of Science 38:177–180.

Marshall, L. G., J. A. Case, and M. O. Woodburne. 1990. Phylogenetic relationships of the families of marsupials. Current Mammalogy 2:433–502.

Matthew, W. D. 1916. A marsupial from the Belly River Cretaceous. Bulletin of the American Museum of Natural History 35:477–500.

McKenna, M. C. 1975. Towards a phylogenetic classification of the Mammalia; pp. 21–46 in W. P. Luckett and F. S. Szalay (eds.), Phylogeny of Primates. Plenum Press, New York.

Merewether, E. A., W. A. Cobban, and J. D. Obradovich. 2007. Regional disconformities in Turonian and Coniacian (Upper Cretaceous) strata in Colorado, Wyoming, and adjoining states; biochronological evidence. Rocky Mountain Geology 42:95–122.

Ogg, J. G., F. P. Agterberg, and F. M. Gradstein. 2004. The Cretaceous Period; pp. 344–383 in F. M. Gradstein, J. G. Ogg, and A. G. Smith (eds.), A Geologic Time Scale 2004. Cambridge University Press, Cambridge.

Osborn, H. F. 1888. On the structure and classification of the Mesozoic mammals. Journal of the Academy of Natural Sciences, Philadelphia 59:186–265.

Payenberg, T. H. D., D. R. Braman, D. W. Davis, and A. D. Miall. 2002. Litho- and chronostratigraphic relationships of the Santonian–Campanian Milk River Formation in southern Alberta and Eagle Formation in Montana utilizing stratigraphy, U-Pb geochronology, and palynology. Canadian Journal of Earth Sciences 39:1553–1577.

Peterson, F. 1969. Four new members of the Upper Cretaceous Straight Cliffs Formation in southeastern Kaiparowits region, Kane County, Utah. U.S. Geological Survey Bulletin 1274-J.

Roberts, E. M., A. L. Deino, and M. A. Chan. 2005. ^{40}Ar-^{39}Ar age of the Kaiparowits Formation, southern Utah, and correlation of contemporaneous Campanian strata and vertebrate faunas along the margin of the Western Interior Basin. Cretaceous Research 26:307–318.

Sahni, A. 1972. The vertebrate fauna of the Judith River Formation, Montana. Bulletin of the American Museum of Natural History 147:321–412.

Simpson, G. G. 1925. Mesozoic Mammalia. II. *Tinodon* and its allies. American Journal of Science 10:451–470.

Simpson, G. G. 1927. Mesozoic Mammalia. VIII. Genera of Lance mammals other than multituberculates. American Journal of Science 14:121–130.

Simpson, G. G. 1928. Affinities of the Mongolian Cretaceous insectivores. American Museum Novitates 330:1–11.

Sullivan, R. M., and S. G. Lucas. 2006. The Kirtlandian land-vertebrate "age"–faunal composition, temporal position and biostratigraphic correlation in the nonmarine Upper Cretaceous of Western North America; pp. 7–29 in S. G. Lucas and R. M. Sullivan (eds.), Late Cretaceous Vertebrates from the Western Interior. New

Mexico Museum of Natural History and Science Bulletin 35.

Tibert, N. E., R. M. Leckie, J. G. Eaton, J. I. Kirkland, J.-P. Colin, E. L. Leithold, and M. McCormick. 2003. Recognition of relative sea-level change in Upper Cretaceous coal-bearing strata: a paleoecological approach using agglutinated foraminifera and ostracodes to detect key stratigraphic surfaces; pp. 263–299 in H. Olson and R. M. Leckie (eds.), Microfossils as

Proxies for Sea Level Change and Stratigraphic Discontinuities. Society of Sedimentary Geologists (SEPM) Special Publication 75.

Vullo, R., E. Gheerbrant, C. D. Muizon, and D. Néraudeau. 2009. The oldest modern therian mammal from Europe and its bearing on stem marsupial paleobiogeography. Proceedings of the National Academy of Sciences of the United States of America 106:19910–19915.

Eaton and Cifelli

Late Cretaceous Mammals from Bryce Canyon National Park and Vicinity, Paunsaugunt Plateau, Southwestern Utah

15

Jeffrey G. Eaton

THE MAMMALIAN FAUNA FROM THE JOHN HENRY MEM-ber of the Straight Cliffs Formation (Santonian) of the Paunsaugunt Plateau described here includes an unidentified triconodont, cf. *Alticonodon, Mesodma* sp. cf. *M. minor, Mesodma* sp., *?Mesodma* sp., *Cimolodon* sp. cf. *C. foxi, Cimolodon similis, Cimolodon* sp. cf. *C. similis, ?Cimolodon* sp., *Cimolomys* sp. A, *Cimolomys* sp. B, *?Cimolomys* sp. A, *?Cimolomys* sp. B, *Dakotamys* sp., *D. shakespeari* sp. nov., *Cedaromys* sp. cf. *C. hutchisoni, ?Spalacotheridium* sp., *Symmetrodontoides* sp., an unidentified didelphid, *Apistodon* sp. cf. *A. exiguus,* cf. *"Anchistodelphys"* sp., *?Varalphadon* sp., *Eodelphis* sp., an unidentified pediomyid, and *?Leptalestes* sp. The fauna from the Wahweap Formation (Campanian) of the Paunsaugunt Plateau described here includes *?Paracimexomys* sp., *Cedaromys* sp. cf. *C. hutchisoni, Mesodma* sp. cf. *M. minor, M.* sp. cf. *M. archibaldi, M.* sp. cf. *M. formosa, Cimolodon* sp. cf. *C. foxi, Cimolomys* sp., *?Cimolomys* sp.,

Meniscoessus sp., cf. *Iugomortiferum* sp., unidentified didelphids sp. A and B., cf. *Apistodon* sp., *Varalphadon* sp. cf. *V. creber,* cf. *Varalphadon* sp., and an unidentified pediomyid. The fauna of the Wahweap Formation from the Paunsaugunt Plateau compares well to the fauna from the Wahweap Formation on the Kaiparowits Plateau, but also compares well to the fauna recovered from the underlying John Henry Member. However, the fauna does not compare closely to that recovered from the Kaiparowits Formation. This is unexpected because the Wahweap fauna should be essentially middle Campanian and the Kaiparowits faunas early Late Campanian, as based on radiometric dates, and a substantial unconformity (the Early Campanian) should separate the Wahweap Formation from the Santonian John Henry Member if these dates are correct. The fauna from the Wahweap Formation also correlates well with that of a locality 200 m below the top of the Cretaceous on the Markagunt

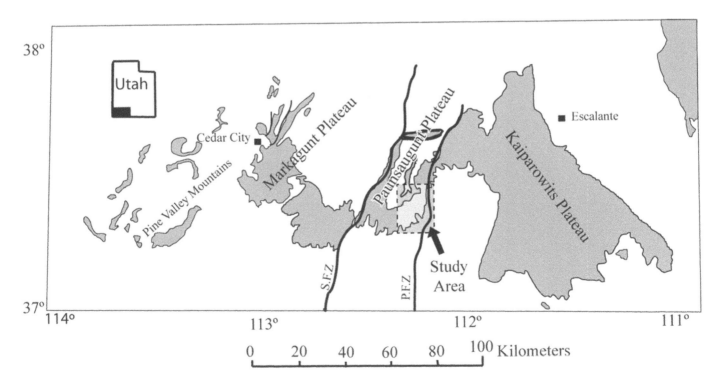

15.1. Study area location map, central and eastern Paunsaugunt Plateau encompassing Bryce Canyon National Park. S.F.Z., Sevier Fault Zone; P.F.Z., Paunsaugunt Fault Zone.

329

Plateau suggesting that locality may also be Campanian in age. Neither fauna compares well to that recovered from the highest known fauna from the Cretaceous sequence on the Markagunt Plateau.

INTRODUCTION

The first Cretaceous microvertebrate localities found near Bryce Canyon National Park were discovered in the Wahweap Formation on USDA Forest Service lands west of the park in the center of the Paunsaugunt Plateau (Fig. 15.1) in 1988 (MNA Loc. 1073, 1074). The faunas recovered from these localities were described by Eaton (1993). More specimens have been subsequently collected from these and nearby other localities (UMNH VP Loc. 63, 83, 84) and are included in this chapter.

The first Cretaceous microvertebrate sites (MNA Loc. 1186, 1187, 1188) actually in Bryce Canyon National Park were found in 1989 in the John Henry Member of the Straight Cliffs Formation near an area referred to as the Hat Shop. Because other important vertebrate localities were discovered in Bryce Canyon National Park in the Wahweap Formation in 1997 (UMNH VP Loc. 77, 79) and in the John Henry Member in 2003 (UMNH VP 424), it became apparent that the Cretaceous sequence in Bryce Canyon National Park had significant vertebrate fossil resources. As a result of these initial discoveries, a paleontologic inventory of the Cretaceous rocks of the park and its margins was undertaken during the summers of 2006 through 2010. During this time, we discovered 93 vertebrate localities, 17 invertebrate localities, and four paleobotany localities. Only a few of these localities have been extensively screen washed and the mammals recovered from these localities to date are described herein.

The systematic approach employed here is conservative and only one new taxon is named. There has been a great increase in the number of Cretaceous mammalian taxa in the past few decades and it is difficult in a large review paper to review all material pertaining to these taxa. It is likely that careful review of individual taxonomic groups, such as Davis's (2007) review of taxa attributed to the Pediomyidae, are more likely to produce justifiable new taxa as well as identify synonyms.

Initially I considered the uppermost Cretaceous beds on the Paunsaugunt Plateau to possibly belong to the Kaiparowits Formation (Eaton, 1993; Eaton, Goldstrand, and Morrow, 1993) because they were very fine grained and variegated, unlike the drab and often sandy Wahweap Formation on the Kaiparowits Plateau to the east (Eaton, 1991), lacked sharks (although one shark tooth was subsequently recovered) and rays, and contained abundant fossils in fine-grained floodplain facies. These characteristics are much more like those

Age		Formation	Member	Thickness	UMNH Localities *MNA Localities*
Paleogene		Claron	Red		
Campanian	Middle	Wahweap		145-180 m	61, 83, 110 *1073, 1074* (77) (792) 807
	Lower	unconformity?			
			Drip Tank	25-30 m	
Santonian	U	Straight Cliffs	John Henry Member	250 m	420, 424
	M				419, 1144
	L				(781, 799, 826)
Coniacian	U				823
	M				
	L	unconformity?			

15.2. Age and stratigraphic position of localities. The position of localities in parentheses are estimated; all others have been placed in section on the basis of measured sections. The placement of Lower, Middle, and Upper Santonian is estimated. MNA locality numbers are indicated in italics.

of the Kaiparowits Formation than the Wahweap Formation on the Kaiparowits Plateau. However, subsequent walking out of key marker horizons and contacts convinced me that the uppermost unit is equivalent to the Wahweap Formation, and that name is used here (Fig. 15.2). The only units on the Paunsaugunt Plateau from which vertebrates have been recovered are the John Henry Member of the Straight Cliffs Formation and the Wahweap Formation (see Fig. 15.2 for the stratigraphic distribution of localities).

The stratigraphic positions of localities in the John Henry Member are mostly determined by measuring stratigraphic sections; however, those localities in parentheses have not yet been placed into section, and their positions are estimated. Also, the placement of Santonian subdivision boundaries (lower, middle, upper) in Fig. 15.2 are only estimated as there is no formal way to subdivide these substages in terrestrial sequences. The relative positions of localities in the Wahweap Formation are more difficult. The top of the Wahweap is usually missing in this area as there is an angular unconformity between the top of the Wahweap and the base of the overlying Claron Formation. Also, in many areas the base of the Wahweap is not exposed so there is no base to measure a section up from. Several localities have been tied to measured sections (not in parentheses in Fig. 15.2) and the stratigraphic position of other localities has been estimated (in parentheses).

The age of the John Henry Member of the Straight Cliffs Formation is thought to be upper Coniacian through Santonian, as based on marine invertebrates (Peterson, 1969; see also Eaton, 2006a). Peterson (1969) reported a Middle Coniacian age for the marine molluscan fauna at the base of the member on the basis of the presence of *Inoceramus*

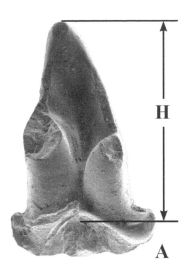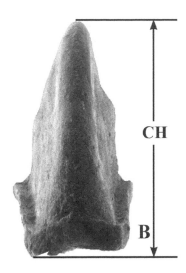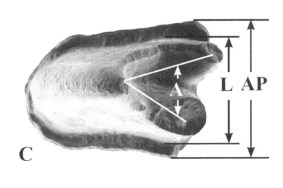

15.3. Measurements of symmetrodont lower molars. (A) Lingual view; (B) labial view; (C) occlusal view. H, height; CH, crown height; L, length; AP, anteroposterior; A, angle.

undabundus (=*I. stantoni*) and *Protexanites bourgeoisianus* (=*P. shoshonensis*), taxa now used, along with *Scaphites depressus*, to define the Late Coniacian (Kennedy and Cobban, 1991; Walaszczyk and Cobban, 2006) in North America. The age of the upper part of the John Henry Member is less well constrained. The ammonite *Desmoscaphites* from near the top of the John Henry Member limits the age of the member to Late Santonian (Landman and Cobban, 2007).

The Wahweap is considered to be mostly Middle Campanian on the basis of radiometric dates of Jinnah et al. (2009).

METHODS

All measurements are in millimeters. Multituberculate tooth measurements follow Eaton, 2002. Symmetrodont lower tooth measurements are a bit more complicated. Cifelli and Madsen (1986:table 1) provide a measurement H (=height), but this measurement was not defined. Cifelli (pers. comm., 2010) measured height on the lingual side of the tooth as is measured from the tip of the protoconid to the base of the cingulum where it flexes upward between roots (this orientation can be seen in Cifelli and Madsen, 1986:fig. 1c). This measurement does not represent the complete crown height as the crown is higher labially than lingually. The measurements provided here include H as measured by Cifelli and Madsen (1986), but also include CH (=crown height) which is measured labially from the base of the lowest part of the cingulum to the apex of the protoconid. Cifelli and Madsen (1986) measure of length (L) of lower molars is the total AP dimension of the tooth including the cingulum. In Cifelli and Madsen (1999:fig. 2A) and Cifelli and Gordon (1999:fig. 1c), length (L) is the AP dimension of only the paraconid and metaconid. As such, the measurement L is not comparable

between Cifelli and Madsen (1986), and Cifelli and Madsen (1999) and Cifelli and Gordon (1999). I measured both the total anteroposterior dimension (AP) and the length (L) as defined in fig. 2A of Cifelli and Madsen (1999). I made the length (L) measurement at the base of the paraconid to the metaconid in lingual view as these cusps flair anteroposteriorly and the apices of these cusps are usually worn. The trigonid angle (A) was measured in photographs following Cifelli and Madsen (1999:fig. 2B). Upper teeth are measured as in Cifelli and Madsen (1999:fig. 2c) except that here AP = L, LBa = ANW, and LBp = POW. The angle (A) was measured by drawing lines on occlusal photographs of specimens following Cifelli and Madsen (1999:fig. 2B). The angles of specimens described in Cifelli (1990c:fig. 1) provided here were measured as described above. This measurement is approximate in all cases as invariably the apices of the trigonid cusps are worn making exact measurements impossible. The measurements used for symmetrodonts lower molars are shown in Fig. 15. 3.

Many localities in the area of Bryce Canyon National Park occur on relatively steep slopes so if a few small pieces of either invertebrate or vertebrate material were located on the surface we attempted to find the producing horizon. This was often accomplished by blind washing (Eaton, 2004). A small test sack (25–30 kg) was recovered from likely horizons, usually dark colored mudstones. These were then soaked until the sample broke down and then wet screen washed in double-nested screens which included a standard window screen and a nested finer mesh (0.3 mm) screen. The amount sampled for specimens in this report depended both on productivity and ease of access. Some localities have had as little 500 kg removed for processing, while other localities more than 3000 kg. The specimens from both the coarse and fine concentrates were sorted with the use of stereomicroscopes.

All specimens were photographed on a scanning electron microscope. All specimens in figures are oriented with the anterior of the tooth toward the bottom of the page, except for side views of symmetrodonts or lingual views of marsupial uppers molars, where the occlusal surface faces the top of the page. All images were treated in Adobe Photoshop and pasted up with Abode InDesign.

Institutional Abbreviations BRCA, Bryce Canyon National Park, southwestern Utah, National Park Service catalog number; MNA, Museum of Northern Arizona, Flagstaff, Arizona; OMNH, Oklahoma Museum of Natural History, Norman, Oklahoma; UALVP, University of Alberta Laboratory for Vertebrate Paleontology, Edmonton, Alberta, Canada; UMNH, Utah Museum of Natural History, University of Utah, Salt Lake City, Utah.

Abbreviations A, angle (for symmetrodont lower molars, see Fig. 15.3); AP, greatest anteroposterior dimension; CH, crown height (for symmetrodont lower molars only, see Fig. 15.3); CL, climb length all used in measurements of multituberculate P4s (see Eaton, 2002:fig. 2); CR, climb rate; CRH, climb rate height; Dp, deciduous lower premolar; DP, deciduous upper premolar; ER, number of external ridges; H, height (for symmetrodont lower molars, see Fig. 15.3; for *Alticonodon*, see discussion in text and for multituberculate p4s, see Eaton, 2002:fig. 2); IR, number of internal ridges; all used in descriptions of multituberculate p4s, see Eaton, 2002; L, length (for symmetrodont lower molars only, see Fig. 15.3); LB, greatest lingual–buccal dimension; LBa, anterior lingual–buccal dimension; LBp, posterior lingual–buccal dimension; LB:AP, ratio between the linguobuccal dimension and the anteroposterior dimension; mx, unspecified lower molar; Mx, unspecified upper molar; osp, occlusal stereo pair (in figures); PL, posterior length; Ri, ridge in cusp formulae where individual cusps merge to form a continuous ridge; S, number of serrations; *, uncertain value (except for Table 15.7).

SYSTEMATIC PALEONTOLOGY

John Henry Member, Straight Cliffs Formation

TRICONODONTA Osborn, 1888
TRICONODONTIDAE Marsh, 1887
Genus et sp. indet.

Referred Specimen UMNH VP 19003 (BRCA 8458), Loc. VP 424, incomplete right lower molar (Fig. 15.4A).

Description mx-UMNH VP 19003 (Fig. 15.4A) is an incomplete right lower molar (AP = 2.08*; H = 0.94) with four cusps. The anterior root has a well-developed groove that locks the posterior root of the penultimate molar. The first cusp (cusp b) is low and rounded. The second cusp (cusp a) is the tallest of the molar. The third cusp (cusp c) is lower but well separated from the second cusp. The fourth cusp (cusp d) appears to have been considerably lower and smaller than the third cusp. The lingual cingulum is weakly developed. The labial side has been heavily worn.

Discussion The specimen is too small and lightly constructed to represent *Alticonodon* Fox, 1969. The specimen is very much different from the lower molars of *Corviconodon*, *Jugalator*, or *Arundelconodon* in many aspects but certainly in that the cusp are upright. The significant difference is cusp height and small size may suggest a similarity to the upper premolar(?) described in Fox (1976) as a possible new genus and species of ?Triconodontidae. The simple structure of the tooth, the small size, and relatively weak lingual cingulid perhaps suggest a relationship to *Astroconodon*, except for the well-developed wear on the labial side of this molar.

cf. *ALTICONODON* sp. Fox, 1969

Referred Specimens UMNH VP 17292 (BRCA 7899), Loc. VP 424, Rm?

Description mx-UMNH VP 17292 (Fig. 15.4B) is a tricuspid tooth with a lingual cingulid (AP = 1.95; H = 1.21). Tooth tapers anteriorly and is probably the anteriormost molar. There is an anterior groove which interlocks with the next most anterior tooth (a premolar?). There is no significant difference between the heights of the cusps.

Discussion The penultimate molar of *Alticonodon lindoei* figured by Fox (1969:fig. 1) has the cusps separated only apically which is unique for the genus. This specimen lacks even apical cusp separation but it is likely that labial shear worn the cusp apices down below were they were separated. It is not uncommon for cusp separation to be less pronounced on more anterior molars (as in *Arundelconodon hottoni*, Cifelli et al., 1999:fig. 1A). I measured height from the base of the lingual cingulid, it is unclear how Fox (1969:table 1) measured height of the penultimate and ultimate molars, and the height (1.21) and length (1.95) of this molar suggests that it is too small for the specimen to be assigned to *A. lindoei* (height of penultimate molar = 1.7; L = 2.3; Fox, 1969:table 1), unless the more anterior molar of that species is significantly smaller. Because the more anterior molars of *Alticonodon* are unknown, this specimen is only tentatively conferred to that genus.

The presence of two probable species of triconodont stratigraphically high in the John Henry Member of the Straight Cliffs Formation, with neither species representing the late Santonian species found in the Milk River

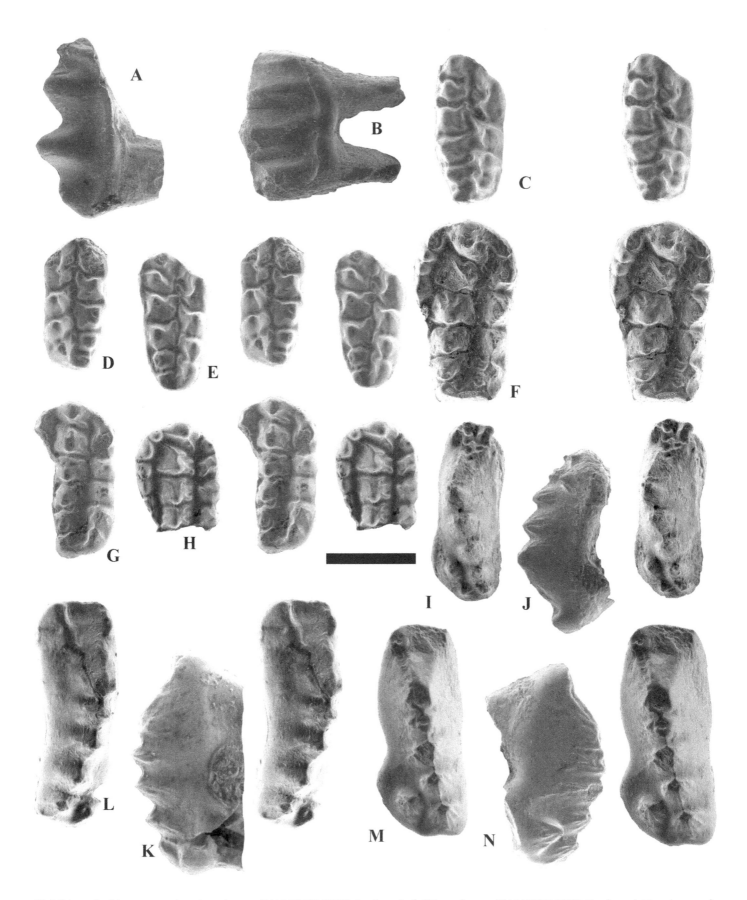

15.4. Triconodontidae genus and species unknown: (A) UMNH VP 19003, Rm, lingual; cf. *Alticonodon* sp.: (B) UMNH VP 17292, Rm, lingual; *Mesodma* sp. cf. *minor:* (C) UMNH VP 14128, Rm1, osp; (D) UMNH VP 14130, Lm1, osp; (E) UMNH VP 18019, Rm1. osp; (F) UMNH VP 14144, RM1, osp; (G) UMNH VP 14413, RM1, osp; (H) UMNH VP 14433, RM1, osp; *Mesodma* sp.: (I) UMNH VP 17332, LP4, osp; (J) labial; ?*Mesodma* sp.: (K) UMNH VP 14246, RP4, osp; (L) labial; (M) UMNH VP 14131, LP4, osp; (N) labial. Scale bar = 1 mm.

Formation, suggests that tricondonts were diverse through the Santonian. No tricondonts have been described from the Campanian.

MULTITUBERCULATA Cope, 1884
CIMOLODONTA McKenna, 1975
?PTILODONTOIDEA Sloan and Van Valen, 1965
NEOPLAGIAULACIDAE Ameghino, 1890
MESODMA Jepsen, 1940
MESODMA sp. cf. *M. MINOR* Eaton, 2006a

Referred Specimens UMNH VP 14128 (BRCA 7204), Loc. 424, Rm1; UMNH VP 14130 (BRCA 7206), Loc. 424, Lm1; UMNH VP 14253 (BRCA 7261), Loc. 424, Rm1; UMNH VP 18019, Loc. 419, Rm1; UMNH VP 18021, Loc. 419, Lm1; UMNH VP 17818, Loc. 823, Lm1; UMNH VP 14250 (BRCA 7258), Loc. 424, Rm1 (posterior half); UMNH VP 14144 (BRCA 7220), Loc. 424, RM1; UMNH VP 14413 (BRCA 7278), Loc. 424, RM1 (posterior part); UMNH VP 14133 (BRCA 7209), Loc. 424, RM1 (posterior part); UMNH VP 17273 (BRCA 7880), Loc. 424, RM1, posterior part; UMNH VP 17283 (BRCA 7890), Loc. 424, RM1 with posteriormost part missing; UMNH VP 14135 (BRCA 7211), Loc. 424, a deeply worn RM1; UMNH VP 14125 (BRCA 7201), Loc. 424, deeply worn RM2 (Table 15.1, Fig. 15.4C–H).

Description m1-UMNH VP 14128 (Fig. 15.4C) is a Rm1 (AP = 1.63; LB = 0.83, LB:AP = 0.51) with a cusp formula of 6:4. The first cusp of the external row is small and conical. The second cusp is slightly larger and conical. The third cusp is taller than all but the sixth cusp, is the broadest cusp of the row, and is conical with a slightly concave posterior face. The fourth cusp is the only truly pyramidal cusp of the row, and the fifth and sixth cusps are essentially conical. The broad third cusp expands the tooth labially such that the widest part of the tooth labially is adjacent to the third cusp. The first two cusps of the internal row are weakly divided, the third cusp is only slightly better separated, and the valley between the third and fourth cusps is the most deeply developed. The central valley is aligned with the AP axis of the tooth and bears deep pits and ribs crossing the valley. The posterior part of the molar has a waist in occlusal view. Compared to the m1s of *Mesodma minor* described in Eaton (2002), the cusps of the internal row of this taxon are less well divided, but otherwise the molars are similar.

UMNH VP 14130 (Fig. 15.4D) is a Lm1 (AP = 1.46; LB = 0.72; LB:AP = 0.49) with a cusp formula of 5:4. The first cusp of the external row is small, and the remaining cusps become progressively larger and more pyramidal in form posteriorly. The fifth cusp is damaged. The labial wall projects outward at the valley between the second and third cusps. The first two cusps of the internal row are conical and the third cusp has a concave posterior face. The valleys between the interior cusps are progressively deeper posteriorly. The labial wall of the fourth cusp is deeply grooved. The cusp formula is less than for any known species of *Mesodma*. These are tentatively referred to *Mesodma* rather than *Cimexomys* on the basis of the narrow and blade-like internal cusp row, the molar narrows anteriorly, the internal row cusps are well divided and at least the third cusp is crescentic, and the central valley is approximately aligned with the AP axis of the tooth (not transverse) (see discussion in Eaton, 2002:26). The molar is slightly smaller than m1s of *Mesodma minor* from the Kaiparowits Formation (Eaton, 2002:table 2) or the single m1 described from the John Henry Member of the Straight Cliffs Formation as *Mesodma* sp. cf. *M. minor* in Eaton, 2006a).

UMNH VP 14253 is a Rm1 (AP = 1.44; LB = 0.73; LB:AP = 0.51) with a cusp formula of 6:4. The last two cusps of the external row are weakly subdivided and not as distinct as on UMNH VP 14128. The first three cusps of the internal row are only separated at their apices.

UMNH VP 18019 (Fig. 15.4E) is a Rm1 with a cusp formula of 6:4 (AP = 1.50; LB = 0.81; LB:AP = 0.54). The molar tapers strongly anteriorly. The first cusp of the external row is small, low, and positioned more anteriorly than the first cusp of the internal row. The second is slightly larger, conical, and is worn on its labial side. There is a narrow platform at the labial base of the cusp. The third cusp is well separated from the second and is almost pyramidal in form, with a concave posterior face. The fourth cusp is essential triangular and is the broadest cusp of the row. The fifth cusp is anteroposteriorly compressed and is well separated from both the fourth and sixth cusps. The sixth cusp is distinct and formed on the posterior ridge of the molar. The central valley is crossed by ribs. The first two cusps of the internal row are closely appressed and only separated at their apices. The third cusp is more deeply separated than are the first and second cusps, and is essentially conical in form but slightly flattened labially. The fourth cusp is the largest and tallest of the row and is deeply separated from the third cusp. Both cusp rows broaden posteriorly. The cusps of the external row in *Cimexomys* tend to be narrowly divided (*C. gratis*, Archibald, 1982:fig. 37d; *C. judithae*, Montellano, Weil, and Clemens, 2000:fig. 2A), and the strong anterior taper of the molar and the broadly separated cusps of both rows are more *Mesodma*-like (*M. thompsoni*, Lillegraven, 1969:fig. 5b; *M. hensleighi*, Lillegraven, 1969:fig. 3a, b) than *Cimexomys*-like.

UMNH VP 18021 is a Lm1 with a cusp formula of 6:4 (AP = 1.52; LB = 0.84; LB:AP = 0.55). This specimen is similar to UMNH VP 18019 (Fig. 15.4E) from the same locality, but is more worn, particularly along the labial side of the molar.

Table 15.1. Comparisons of measurements, ratios, and cusp formulae of *Mesodma* sp. cf. *M. minor*, *Mesodma minor*, *M. archibaldi*, and *M.* sp. cf. *M. archibaldi* from the John Henry Member of the Straight Cliffs Formation, Wahweap and Kaiparowits formations, and UMNH VP Loc. 10, Markagunt Plateau, unit uncertain

Tooth	Taxon	Specimen	AP	LB	LB:AP	Formula	Unit	Source[a]
m1	*M.* sp. cf. *M. minor*	UMNH VP 14128	1.63	0.83	0.51	6:4	John Henry Member	1
	M. sp. cf. *M. minor*	UMNH VP 14130	1.46	0.72	0.49	5:4	John Henry Member	1
	M. sp. cf. *M. minor*	UMNH VP 14253	1.44	0.73	0.54	6:4	John Henry Member	1
	M. sp. cf. *M. minor*	UMNH VP 18019	1.50	0.81	0.54	6:4	John Henry Member	1
	M. sp. cf. *M. minor*	UMNH VP 18021	1.52	0.84	0.55	6:4	John Henry Member	1
	M. sp. cf. *M. minor*	UMNH VP 17818	1.53	0.80	0.52	6:4	John Henry Member	1
	M. sp. cf. *M. minor*	Means	1.51	0.79	0.53	5–6:4	John Henry Member	1
	M. minor	Means	1.55	0.81	0.53	6:4	Kaiparowits Formation	2
M1	*M.* sp. cf. *M. minor*	UMNH VP 14144	2.00	1.18	0.59	5:6:5?	John Henry Member	1
	M. sp. cf. *M. minor*	UMNH VP 14413	1.72	0.94[b]		5:6:[b]	John Henry Member	1
	M. sp. cf. *M. minor*	UMNH VP 14133		0.95		5?:6?:6	John Henry Member	1
	M. sp. cf. *M. minor*	UMNH VP 17273		1.03			John Henry Member	1
	M. sp. cf. *M. minor*	UMNH VP 17283				5?:6?:6	John Henry Member	1
	M. sp. cf. *M. minor*	UMNH VP 14135	1.75	0.92	0.53		John Henry Member	1
	M. sp. cf. *M. minor*	Mean	1.82	1.00	0.56		John Henry Member	1
	M. sp. cf. *M. minor*	UMNH VP 6794	2.09	1.29	0.62	4:6:4	Wahweap Formation	1
	M. sp. cf. *M. minor*	Means	1.96	1.15	0.59	4:6:5	Loc. 10, Markagunt Plateau	4
	M. minor	Mean	2.02	1.12	0.56	4–6:6:4–5?	Kaiparowits Formation	3
	M. sp. cf. *M. archibaldi*	UMNH VP 6798	2.23	1.40	0.63	5:6:5	Wahweap Formation	1
	M. sp. cf. *M. archibaldi*	MNA V5291	2.21	1.39	0.63	5:6:5	Kaiparowits Formation	3
	M. archibaldi	Means	2.49	1.26	0.51	5:6:4	Kaiparowits Formation	3
M2	*M.* sp. cf. *M. minor*	UMNH VP 14125	1.44	1.23	0.85		John Henry Member	1
	M. minor	Means	1.23	1.13	0.92	Ri:3:3	Kaiparowits Formation	3

[a] Source: 1 – this chapter; 2 – Eaton (2002:table 2); 3 – Eaton (2002:table 3); 4 – Eaton (2006b:table 7). [b] Uncertain value.

UMNH VP 17818 is a Lm1 with a cusp formula of 6:4 (AP = 1.53; LB = 0.80; LB:AP = 0.52). This worn molar is clearly the same taxon as UMNH VP 14128 (Fig. 15.4C), but this specimen is from very low in the John Henry Member, and is possibly of Coniacian age. This suggests that *M. minor*, or a similar taxon, is present throughout the John Henry Member.

UMNH VP 14250 is a partial Rm1 (posterior half; LBp = 0.86). The anteriormost preserved cusp in the external row is crescentic with a distinct posterior pit between it and the next cusp. The next cusp is attached with a ridge to the more distant posterior cusp. The central valley is deep and aligned with the AP axis of the tooth. The internal row cusps are taller and better separated than those of the external row.

M1-UMNH VP 14144 (BRCA 7220), Fig. 15.4F, is a RM1 with a cusp formula of 5:6:5?. (AP = 2.00; LB = 1.18). The first two cusps of the external row are connected over half of their height, but the third cusp is well separated from the second and fourth cusps. The fourth and fifth cusps are closely positioned and are not well separated. There is evidence of grooves and pits in the central valley. The cusps of the medial row are pyramidal, lean anteriorly, and become wider and taller posteriorly. The first cusp of the internal row is positioned adjacent to the valley separating the third and

fourth cusps of the medial row. There is an indentation posterior to the first cusp, then an elongate second cusp and three smaller equal sized cusps which together form a ridge. This molar is within the size range of *Mesodma minor* from the Kaiparowits Formation (Eaton, 2002:table 3) and is morphologically similar including the rather irregular nature of the external cusp row (Eaton, 2002:fig. 4E, F).

UMNH VP 14413 (Fig. 15.4G) is a worn RM1 with a cusp formula of 5:6:* (AP = 1.72; LB = 0.94*). The cusps of the external row are conical and are progressively better separated posteriorly. The cusps maintain their width posteriorly. The valley between the external and medial cusp rows in closed posteriorly. The cusps of the medial row are pyramidal and lean anteriorly. The internal row is mostly missing and now only bears a single cusp and is worn anteriorly making it difficult to accurately assess the length of the row, but it was probably about half the length of the molar. The row appears to join the main body of the tooth posterior of the fourth cusp of the medial row. This specimen is 14% smaller in its AP dimension than UMNH VP 14144 (Fig. 15.4F) described above. The sample of M1s is inadequate to determine if there is a bimodal size distribution in the sample possibly representing two species.

UMNH VP 14133 (Fig. 15.4H) is a partial RM1 (posterior part; LB = 0.95) with an estimated cusp formula of 5?:6?:6 (formula also based on UMNH VP 17283). Cusps of the external row are separated throughout about half of the height and lean anteriorly. The valley separating the external and medial cusp rows is continuous to the back of the tooth. The first two cusps of the medial row are not as deeply separated as are the last two cusps. The cusps are broad, deeply grooved, and lean anteriorly. The last cusp of the row is conical and shifted slightly to the center of the tooth. The cusps of the internal row are conical and diminish in height and size anteriorly. This molar appears similar in size to those of *Mesodma minor* (Eaton, 2002:table 3) but has two more cusps in the internal cusp row than known M1s of *M. minor*.

UMNH VP 17273 is a partial RM1 (posterior part; LB = 1.03) with five cusps in the internal row. The cusps are slightly more robust than those of UMNH VP 14133. The last cusp of the central row is much taller than the other cusps of the molar. The internal row bears five distinct cusps.

UMNH VP 17283 is a RM1 with the posteriormost part missing and no standard dimensions can be measured. Cusps of the external row are conical and progressively better separated posteriorly. Cusps of the medial row are pyramidal, lean anteriorly, broaden and become taller posteriorly. The internal cusp row connects at the third cusp of the medial row. This tooth has the same cusp formula as given for UMNH VP 14133 above.

M2-UMNH VP 14125 is a deeply worn RM2 (AP = 1.44; LB = 1.23). This worn M2 is included here tentatively because of its fit with the posterior facet of the M1s.

Discussion All of the M1s clearly represent the same species of *Mesodma* as described previously (Eaton, 2006a) from the John Henry Member as *M.* sp. cf. *M. minor*. Although this taxon is close to *Mesodma minor* Eaton, 2002, from the younger Kaiparowits Formation, only one M1 (UMNH VP 14144) really appears conspecific with the type material from the Kaiparowits Formation. Most of the M1s are smaller and have a shorter internal cusp row. It is uncertain if these specimens should be considered to be conspecific with *M. minor*, although they are certainly close to that species, or if they represent a new smaller species.

MESODMA sp.

Referred Specimen UMNH VP 17332 (BRCA 794), Loc. 424, damaged LP4 (Fig. 15.4I–J).

Description P4-UMNH VP 17332 (BRCA 794), Fig. 15.4I–J, is a damaged LP4 with a cusp formula of 2:5 (AP = 1.05; LBa = 0.83; LBp = 0.71; H = 1.06). The anteroexternal platform is not greatly expanded. The anteriormost cusp on the platform is tiny, the second is much larger. The first cusp of the central

row is tiny, and the cusps climb at a high rate (CR = 0.61). The posterior heel is bordered on both sides by distinct cusps and the basin between them is deeply crenulated.

Discussion Lack of a well-expanded anterior platform bearing large cusps indicates that this specimen is closer to *Mesodma* than *Cimolodon*. The high CR (=0.61) is also more like previously described P4s of *Mesodma* than P4s of *Cimolodon*. The tooth is well ribbed and has crenulated enamel, but less like that of the P4s of *Cimolodon* and more like those of *M. primaeva*. This specimen is longer than the M1s above attributed to *M. minor*, and P4s of *Mesodma* are generally shorter than the associated M1s (except possibly for *M. primaeva*, Sahni, 1972). Although this P4 is *Mesodma*-like, it is morphologically different from the P4 attributed to *M. minor* in Eaton (2006a:fig. 3B–D).

?MESODMA sp.

Referred Specimens UMNH VP 14246 (BRCA 7254), Loc. 424, RP4; UMNH VP 14131 (BRCA 7207), Loc. 424, LP4 (Fig. 15.4K–N).

Description P4-UMNH VP 14246 (Fig. 15.4K–N), is a RP4 with a cusp formula of 1:5 (AP = 2.57; PL = 0.81*; PL:AP = 0.32; LBa = 0.99; LBp = 1.03; H = 1.09; CRH = 0.34*; CL = 1.12*; CR = 0.30*). This P4 has an only slight expanded anteroexternal platform bearing a single cusp such that the LBp exceeds the LBa. The central row has a low rate of climb (CR = 0.30*), but it is difficult to calculate accurately as the last cusp is worn and lower than the fourth cusp. It appears that the fifth cusp may actually have been lower than the fourth, but because of wear, this is not certain. UMNH VP 14131 (BRCA 7207), Fig. 15.4M–N, is a LP4 with a cusp formula of 1:5. (AP = 2.33; PL = 0.87*; PL:AP = 0.37*; LBa = 0.99; LBp = 0.96; H = 1.12*; CRH = 0.33*; CL = 0.87*; CR = 0.38*). This specimen is similar to UMNH VP 14246 (described above) but is not identical. The anteroexternal platform bears a single cusp but it is more distinctly expanded than on UMNH VP 14246. The central cusp row climbs at a low rate, but as with UMNH VP 14246, the last cusp of the row is worn lower than the fourth cusp and it is unclear if it was taller than the fourth cusp prior to wear. The posterior basin is small and was probably bordered by a single cusp on each side.

Discussion The posterior heel of UMNH VP 14246 is short (PL:AP = 0.32) and bears two small cuspules labially, and a single posterior cusp lingually. The low climb rate of the cusps of the central row and the moderately expanded anteroexternal basin most likely suggest *Mesodma*; however, this specimen is too large to represent a P4 of *M. minor*. It is possible that these two P4s are variants of the same taxon, a genus similar to *Mesodma*, but they also may represent two different taxa.

CIMOLODONTIDAE Marsh, 1889b
CIMOLODON Marsh, 1889b
CIMOLODON sp. cf. *C. FOXI* Eaton, 2002

Referred Specimens UMNH VP 17582, Loc. 781, Lp4; UMNH VP 14142 (BRCA 7218), Loc. 424, Lm1 (some damage to labial wall of the tooth); UMNH VP 17336 (BRCA 7974), Loc. 424, Rm1 (posterior part); UMNH VP 17278 (BRCA 7885), Loc. 424, RM1 posterior part; UMNH VP 18016, Loc. 419, incomplete RM2 (Fig. 15.5A–F).

Description p4-UMNH VP 17582 (Fig. 15.5A, B) is a Lp4 (S:ER:IR = 10:9:8; AP = 2.89; H = 1.79; AL = 1.43; H:AP = 0.62; AL:AP = 0.50). The first and second serrations are widely spaced, the remaining serrations are approximately equally spaced. The crest reaches its apogee at the fourth serration, almost exactly at the middle of the blade. The posterior serrations are larger than the anterior serrations.

m1-UMNH VP 14142 (Fig. 15.5C) is a Lm1 (some damage to labial wall of the tooth) with a cusp formula of 6?:3 (AP = 2.22*; LB = 1.26*). The tooth is worn and some missing enamel probably results in the measurements being slightly less than they would be if the specimen was better preserved. I interpret the first low bump on the external row as a small cusp that was not deeply separated from the second cusp. The second cusp is separated from the pyramidal third cusp by a deep and narrow valley. The third and fourth cusps are separated by a wider valley. The valley separating the fourth and fifth cusps is arcuate and narrow. The fifth and sixth cusps are only weakly separated and could also be considered a subdivided fifth cusp. The central valley is straight and deeply pitted. There is a cuspule on the anterior face of the first cusp of the internal row. The first and second cusps are divided over about half of their height. The second and third cusps are more deeply divided. The interior row cusps are worn flat labially, but deep grooves are still apparent on the labial wall of the third and fourth cusps.

UMNH VP 17336 (Fig. 15.5D) is a fragmentary Rm1 (posterior part; LB = 1.40). The last two cusps of the external row are separated mostly on the lingual wall of the cusps. The third to last cusp is well separated from the second to last cusp and is pyramidal with a concave posterior wall. The central valley is crossed by sharp crested ridges and deep pits are present.

M1–UMNH VP 17278 (Fig. 15.5E) is a fragmentary RM1 (posterior part, LB = 1.83). The molar is somewhat broader than those of *C. foxi* from the Kaiparowits Formation (Eaton, 2002:table 4). The five cusps of the internal row are all well connected (unlike the holotype of *C. foxi*).

M2-UMNH VP 18016 (Fig. 15.5F) is an incomplete RM2 with a cusp formula of ?2:4 (AP = 1.95, LB = 1.48). The external ridge is missing on this specimen, but the breadth of

the base of the tooth indicates it was not greatly broadened. There is a high anterior ridge (not counted as a cusp) and a large central cusp posterior of the ridge. The central valley is sinuous. The internal row forms a crest and the cusps are only divided apically.

Discussion When *C. foxi* was originally described (Eaton, 2002) no m1s where known, but a p4 was referred to that taxon (OMNH 20483). The apogee of OMNH 20483 (Eaton, 2002:fig. 8E, F) is reached anteriorly and the rest of the crown is relatively flat, more like the condition seen in p4s of *Mesodma* (Clemens, 1964:fig. 5, 7, 10). Lower first molars of *C. foxi* were first described in Eaton (2006a) and relative to those m1s OMNH 20483 is probably too large to be a p4 of *C. foxi*. The AP ratio m1:p4 in species of *Cimolodon* appears to be highly variable (0.93 in *C. similis*, Fox, 1971a; 0.57 in *C. electus*, Fox, 1971a; 0.76 in *C. nitidus*, Lillegraven, 1969) and the same ratio in these specimens, relative to the only relatively complete m1 described here (UMNH VP 14142), is 0.76 (similar to that of *C. nitidus*). Calculation of the m1:p4 ratio of this p4 with the m1s of *C. foxi* described from UMNH VP Loc. 10 on the Markagunt Plateau (unknown horizon, possibly equivalent to basal Wahweap Formation) in Eaton (2006b:table 1) yields an m1:p4 AP ratio of 0.92 (similar to *C. electus*). The specimen described here is also similar to the specimen (MNA V4525) described as *Cimolodon* sp. (small) from the Wahweap Formation in Eaton (2002). MNA V4525 is almost the same size (AP = 2.99), has the same number of serrations (although fewer internal and external ridges), and a similar configuration. It is likely that *C.* sp. (small) in Eaton (2002) actually represents the p4 of *C. foxi*. UMNH VP 14142 is close in size to the m1s of *Cimolodon foxi* described previously from the John Henry Member in Eaton (2006a:table 1).

There is no doubt that UMNH VP 17336 (partial m1) represents *Cimolodon*. The LB dimension of the specimen is less than that provided by Fox (1971a) for *C. similis*, and just slightly larger than the AP range of m1s of *C. foxi* (Eaton, 2006b:table 1) from UMNH VP locality 10 possibly equivalent to basal Wahweap Formation. This specimen appears to be significantly larger than UMNH VP 14142, and may indicate the presence of two small taxa of *Cimolodon* in the sample.

UMNH VP 17278 (M1) appears to be slightly larger than those of *C. foxi* and the internal row is different than the M1 of this species described from the Kaiparowits Formation (Eaton, 2002:fig. 10A); however, the internal row of UMNH VP 17278 is similar to the M1s previously described from the John Henry Member (Eaton, 2006a) as *C. foxi*. For this reason, these specimens are only conferred to *C. foxi*, and it is possible that the specimens described in Eaton (2006a), perhaps should have also only been conferred to that species.

UMNH VP 18016 (M2) is included here because it fits well against the posterior wall of UMNH VP 17278, a RM1 (described above, Fig. 15.5E). This tooth is longer than any of the m2s of *C. foxi* described from the Kaiparowits Formation (Eaton, 2002), the cusps of the internal row are not as closely placed, and the tooth is not as complexly pitted and grooved as the Kaiparowits specimens. This suggests that this species, although close in size to *C. foxi*, may represent another species.

It is clear that a species (or several species) close to *Cimolodon foxi* is present in this fauna, but there are enough differences in the sample to warrant caution in assigning all of the specimens to *C. foxi*, and the sample is inadequate to recognize multiple species.

CIMOLODON SIMILIS Fox, 1971a

Referred Specimens UMNH VP 14254 (BRCA 7262), Loc. 424, partial Rp4; UMNH VP 14228, Loc. 420, the anterior part of a Lp4; UMNH VP 17527, Loc. 799, Lm1; UMNH VP 17516, Loc. 799, Rm1; UMNH VP 18022, Loc. 419, posterior part of a Lm1; UMNH VP 17524, Loc. 799, Rm2 (worn); UMNH VP18999, Loc. 419, Lm2 (worn); UMNH VP 14227 (BRCA 7234), Loc. 420, RM2 (Fig. 15.5G–K).

Description p4-UMNH VP 14254 (Fig. 15.5G–H) is a partial Rp4 (AP = 4.1*, H = 2.54*, H:AP = 0.62*). It is not possible to accurately determine the number of serrations or ridges on this specimen, but it has at least 10 serrations and eight external ridges The arch of the p4 is high and symmetrical, a characteristic of p4s of *Cimolodon* (Fox, 1971a:figs. 3d, e, 5c, d).

m1-UMNH VP 17527 (Fig. 15.5I) is a Lm1 with a cusp formula of 6:4 (AP = 2.83; LB = 1.51). The first cusp of the external row is small and low on the crown. The second cusp is larger, pyramidal, and not completely separated from the third cusp which is the largest of the row. The fourth cusp is smaller than and completely separated from the third cusp. The fifth and sixth cusps are small and only separated at their apices. The central valley is straight and deeply pitted. It is open anteriorly and closed posteriorly. The first cusp of the internal row is essentially conical, small, and low. The remaining three cusps of the row get progressively taller and larger posteriorly. These three cusps are well separated and conical with the sides facing the central valley flattened and bearing grooves that ascend from pits in the central valley. The molar has a slight waist in occlusal view and is broader posteriorly than anteriorly.

UMNII VP 17516 (Fig. 15.5J) is a Rm1 with a cusp formula of 7:4 (AP = 3.24; LB = 1.63). The first cusp of the external row is low and small, but demonstrates occlusal wear. The remaining cusps of the row are pyramidal. The third cusp is the largest and tallest of the row (although the posterior part of the cusp is damaged). The medial valley is deep and pitted. The first three cusps of the internal row are essentially conical with a flattened face along the medial valley. There is a small accessory cusp anterior of the first cusp of the row. The first three cusps are separated only apically, but the fourth and largest cusp is separated from the third by a deep valley. All of the internal row cusps have wear on their internal sides. This m1 is within the size range and has the same cusp formula as the m1s described by Fox (1971a) as *Cimolodon similis* and therefore confirms the presence of *C. similis* in the John Henry Member.

m2-UMNH VP 17524 is a worn Rm2 with a cusp formula of 4:2 (AP = 2.30; LB = 1.91). This worn specimen fits well with the larger m1 (UMNH VP 17516) described above and is close in size to the m2 of *C. similis* described from the Wahweap Formation (Eaton, 2002:table 14) and the m2 described by Fox (1971a). UMNH VP 18999 is a worn Lm2 with a cusp formula of 4:2 (AP = 2.10; LB = 1.85). This worn tooth is slightly narrower and significantly shorter than UMNH VP 17524 but fits well against the posterior face of the Lm1 UMNH VP 17527.

M2-UMNH VP 14227 (Fig. 15.5K) is a RM2 with a cusp formula of Ri:3:? (AP = 2.46; LB = 2.15). This molar is strongly ribbed and pitted. The only distinct cusps are the three of the medial row. The internal row is essentially a long sharp crest with deep ribbing on the labial side of the row.

Discussion UMNH VP 14254 (incomplete p4) was probably in the size range of Fox's specimens of *C. similis* from the Milk River Formation. UMNH VP 14228, the anterior part of a Lp4, probably also represents *C. similis*. The size of UMNH VP 17527 (m1) is below the range of *C. similis* of Fox, 1971a, but is in the range of the Wahweap Formation measured specimens in Eaton (2002:table 14). UMNH VP 17527 is 13% shorter and 8% narrower than UMNH VP 17516, but falls within the range of variation of m1s from the Wahweap Formation (Eaton, 2002; table 14). UMNH VP 17527 is at the low end of the size range, and UMNH VP 17516 is at the high end of the size range as compared to the specimens from the Wahweap Formation. This bimodal size distribution is also present in the Wahweap sample (Eaton, 2009:table 14) and it might be possible to distinguish two species on the basis of size; however, Fox's (1971a) sample of m1s of *C. similis* is intermediate in size relative to these samples. UMNH VP 14227 (M2) is just slightly below the size range provided in Fox (1971a) for M2s of *C. similis*, and no M2s of this taxon have been described previously from the Wahweap Formation.

CIMOLODON sp. cf. C. SIMILIS Fox, 1971a

Referred Specimens UMNH VP 17619, Loc. 799, LM1; UMNH VP 17325 (BRCA 7936), Loc. 424, anterior part LM1 (Fig. 15.6A–B).

Description M1-UMNH VP 17619 (Fig. 15.6A) is a LM1 with a cusp formula of 5?:6?:2 (AP = 3.37; LB = 1.99). The

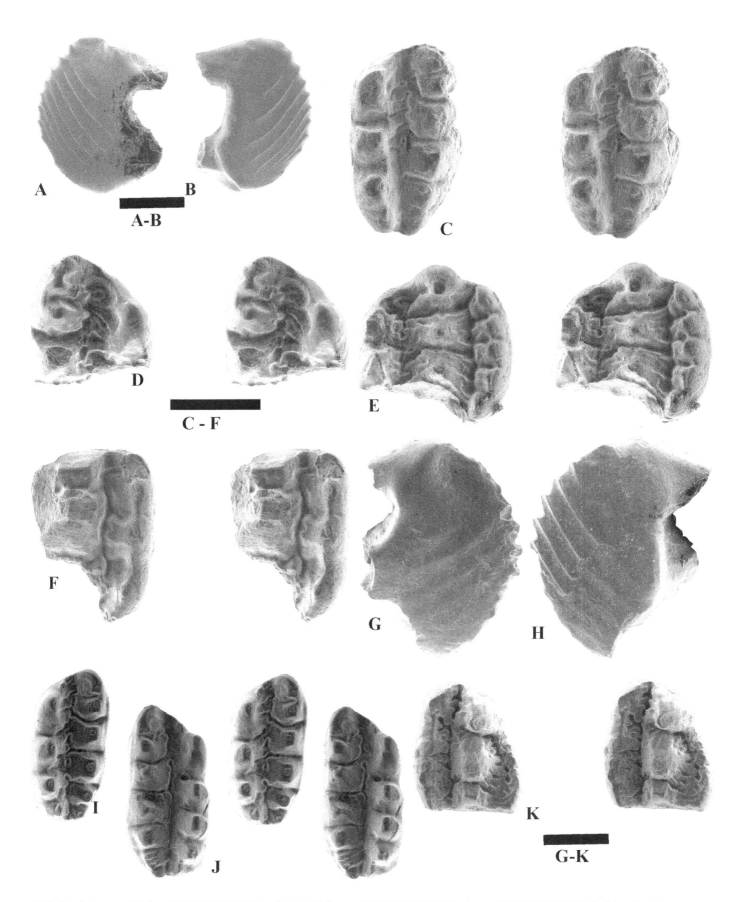

15.5. *Cimolodon* sp. cf. *C. foxi*: (A) UMNH VP 17582, Lp4, labial; (B) lingual; (C) UMNH VP 14142, Lm1, osp; (D) UMNH VP 17336, Rm1 (posterior), osp; (E) UMNH VP 17278, RM1 (posterior), osp; (F) UMNH VP 18016, RM2 (partial), osp; *Cimolodon similis*: (G) UMNH VP 14254, Pp4 (partial), labial; (H) lingual; (I) UMNH VP 17527, Lm1, osp; (J) UMNH VP 17516, Rm1, osp; (K) UMNH VP 14227, RM2 (partial), osp. Scale bars = 1 mm.

Late Cretaceous Mammals from Bryce Canyon National Park

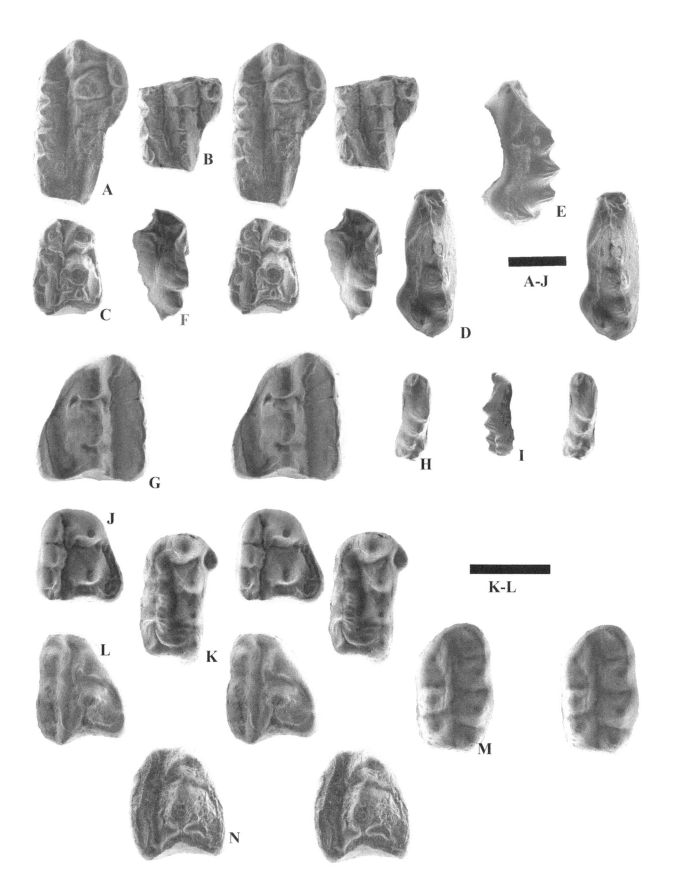

15.6. *Cimolodon* sp. cf. *C. similis:* (A) UMNH VP 17619, LM1, osp; (B) UMNH VP 17325, LM1 (partial), osp; ?*Cimolodon* sp.: (C) UMNH VP 12862, RM2, osp; *Cimolomys* sp. A: (D) UMNH VP 17534, LP4, osp; (E) labial; (F) UMNH VP 19002, LP4 (partial), osp; *Cimolomys* sp. B: (G) UMNH VP 12859, LM2, osp; ?*Cimolomys* sp. A: (H) UMNH VP 17208, RP4, osp; (I) labial; ?*Cimolomys* sp. B: (J) UMNH VP 19001, RM2, osp; *Dakotamys shakespeari* sp. nov.; (K) UMNH VP 19160, LM1, osp; (L) UMNH VP 14231, RM2, osp; *Cedaromys* sp. cf. *C. hutchisoni:* (M) UMNH VP 14149, Lm1, osp; (N) UMNH VP 14125, RM2, osp. Scale bars = 1 mm.

anterior part is deeply worn, and it is not possible to determine if there was another anterior cusp prior to wear on the external cusp row. The first two cusps of the external row are not deeply divided. The second and third cusps are deeply divided and there is also an indentation between the cusps on the labial wall of the molar. The third and fourth cusps are less well divided, and the fourth and fifth cusps are weakly divided by a narrow lingual valley. The cusps of the external row narrow slightly posteriorly. The central valley is deep and marked by ribs and pits. The medial cusp row consists of five or six cusps. The anterior part of the cusp row is deeply worn, and the posterior part increases progressively in height. The interior row has only two distinct cusps and joins the main body of the tooth posterior of the third (or possibly second) cusp of the medial cusp row. UMNH VP 17325 (Fig. 15.6B) is the anterior part of a LM1. The internal cusp row joins the main body of the tooth at the posterior part of the second cusp of the medial cusp row.

Discussion UMNH VP 17619 is too large to represent *C. foxi* and is within the AP range of the original hypodigm of *C. similis* Fox, 1971a, but it has fewer cusps on its internal (2 compared to 4) and medial (probably six rather than seven) cusp rows. This suggests a different, probably new species is present in the lower Santonian, but a less worn M1 would be required to erect a new species with certainty.

The two preserved internal row cusps of UMNH VP 17325 are not deeply separated, unlike in the holotype of *C. foxi* (Eaton, 2002:fig. 10A). The specimen is clearly too large to represent *C. foxi* and approaches the size of *C. similis*, but because of its incomplete nature, comparison of measurements is not possible. However, the internal cusp row meets the main body of the tooth well anterior of that of UMNH VP 17619 (described above, Fig. 15.6A) presumably indicating a cusp formula similar to that of M1s of *C. similis* (5:7:4; Fox, 1971a), but as this is not certain the molar is only conferred to *Cimolodon similis*.

?*CIMOLODON* sp.

Referred Specimen UMNH VP 12862, Loc. 426, RM2 (Fig. 15.6C).

Description M2-UMNH VP 12862 (Fig. 15.6C) is a RM2 with a cusp formula of 2?:3? (AP = 1.97*; LB = 1.55*) with some enamel missing on parts of the molar. The external ridge on this M2 is quite narrow, as is the entire molar relative to its length (LB:AP = 0.79). It is not clear if the anterior ridge on the medial cusp row functions as a cusp and it is not counted as one here. The first cusp of the medial row is far larger than any other and bears deep pits and striae. The second and much smaller cusp of the medial row is separated from the first by two deep pits. The central valley is slightly

sinuous, deep, and pitted. It is not clear if the internal cusp row had more than three cusps. The cusps of the internal row are not well separated and are lower than the cusps of the medial row.

Discussion This tooth is thought to represent *Cimolodon* because of the strong pitting and ribbing on the molar; however, this is the lowest LB:AP ratio reported for an M2 attributed to *Cimolodon* which usually has relatively equidimensional teeth. This molar, although much smaller, approaches the LB:AP ratio seen in the much younger *Cimolodon nitidus* (mean LB:AP = 0.82: Eaton, 2002:table 6 [based on Archibald, 1982]; see also Lillegraven, 1969:fig. 11a–c). M2s of *Paracimexomys* are similarly long relative to width (Archibald, 1982:fig. 39; Eaton, 2002:fig. 24D), but have a more sinuous central valley, a ridge forming internal row, and a slight waist in occlusal view (Archibald, 1982:fig. 39b). The specimen is also much smaller and proportionally narrower than the M2 described above as *C. similis*. Collectively this specimen is different enough that its assignment to *Cimolodon* is questionable despite its characteristic deep pitting and ribbing.

CIMOLOMYIDAE Marsh, 1889a
CIMOLOMYS sp. A Marsh 1889a

Referred Specimens UMNH VP 17534, Loc. 799, LP4; UMNH VP 19002, Loc. 1144, partial LP4 (Fig. 15.6D–F).

Description P4-UMNH VP 17534 (Fig. 15.6D, E) is a LP4 with a cusp formula of 1:4 (AP = 3.22; LB = 1.48; AL = 1.55, PL = 1.14). There is one cusp on the moderately expanded anteroexternal platform. There was a small, and low first cusp on the medial cusp row but the enamel for the cusp is missing and is only hinted at by the shape of the interior of the tooth. The second and third cusps increase height noticeably but the fourth cusp is only slightly taller than the third. Overall, this is a low crowned P4. The posterior basin is grooved and bears one cusp lingually.

UMNH VP 19002 (Fig. 15.6F) is an unworn fragmentary LP4 that is about the same size as UMNH VP 17534, and is identifiable as a cimolomyid because the cusps are not deeply striated (unlike *Cimolodon*), the posterior basin is complex (unlike *Mesodma* or *Cimexomys*), and the posterior basin is relatively short relative to overall tooth length (unlike *Paracimexomys*).

Discussion The low cusp formula of UMNH VP 17534, the low climb rate of the cusps of the medial row, and the relatively large size of the posterior basin relative to the AP dimension of the tooth indicates *Cimolomys*. This specimen has the lowest cusp formula for a P4 of *Cimolomys* recorded, except for the specimen of ?*Cimolomys* sp. reported in Eaton (2006a), and ?*Cimolomys* sp. B in Fox, 1971a. This specimen is close in size and morphology to Fox's specimens and most certainly represent the same species. Archibald (1982, p. 122) suggested Fox's specimens may represent *Paracimexomys magister,* but this is doubtful because of the low cusp formula, striated cusps, and complex posterior basin.

CIMOLOMYS sp. B

Referred Specimen UMNH VP 12859 (BRCA 5894), Loc. 420, LM2 (Fig. 15.6G).

Description and Discussion M2-UMNH VP 12859 (Fig. 15.6G) is a LM2 with a cusp formula of R:3:4 (AP = 2.69; LB = 2.55). The external ridge is worn to a low cuspless crest. The three large pyramidal cusps of the medial row become taller posterior. The four cusps of the internal row form an almost continuous crest. In overall form, the tooth is close in size and morphology to *?Cimolomys* sp. B from the Wahweap Formation (Eaton, 2002:fig. 23E:table 16). This specimen appears to be too large to represent the same species as the P4s described above as *Cimolomys* sp. A.

?CIMOLOMYS sp. A

Referred Specimen UMNH VP 17208 (BRCA 7446), Loc. 424, RP4 (Fig. 15.6H–I).

Description P4-UMNH VP 17208 (Fig. 15.6H, I) is a RP4 with a cusp formula of 2:4 (AP = 1.78; LB = 0.68*, AL = 1.03; PL = 0.75, CRL = 0.81, CRH = 0.77, CR = 0.95). The specimen is well preserved but only includes the enamel cap, so the width and length dimensions may be incomplete. The anteroexternal platform is distinct and bears two striated cusps. The first cusp is smaller than the second. There is a distinct recess in the anterior tip of the premolar for the posterior part of P3. The cusps of the central row increase in size posteriorly, are weakly striated, and have a low climb rate (0.95). A distinct and sharp anteroposteriorly oriented ridge connects the last cusp of the central row to a high cusp on the labial margin of the posterior basin. The posterior basin is deep and closed posteriorly by a ridge and a large posterolingual cusp.

Discussion This P4 is considered to be *Cimolomys*-like because of the distinct anterior platform bearing few cusps, the low number of cusps in the central cusp row (same as for *C. gracilis*), the anteroposteriorly directed posterior ridge, and the low rate of climb for the cusps of the central row. The posterior basin is well developed and deep, unlike P4s of *Mesodma.* The only similar P4s previously described (*?Cimolomys* sp., UMNH VP 7574:fig. 15.5B–D, and UMNH VP 12797; incorrectly listed as UMNH VP 12772 in Eaton, 2006a), both from the John Henry Member, are somewhat larger and broader than this specimen.

?CIMOLOMYS sp. B

Referred Specimen UMNH VP 19001, Loc. 1144, RM2 (Fig 15.6J).

Description M2-UMNH VP 19001 (Fig. 15.6J) is a RM2 with a cusp formula of Ri:2:4 (AP = 2.03, LB = 2.00). The external ridge has a series of tiny cuspules, but no distinct cusps. The platform lingual of the ridge is crenulated and has one distinct pit. The platform is broad such that the AP and LB dimensions of the tooth are almost equal. The anterior ridge in front of the central cusp row is low and shows no wear and is not counted as a cusp. The first cusp of the central row is an anterior leaning pyramid. The second cusp is slightly lower and smaller. The central valley is sinuous and deep. The first cusp of the internal row is missing, but clearly was closely appressed to the conical second cusp. The second and third cusps are more completely separated, and the third cusp is the tallest. The fourth cusp is smaller than the third and well separated from it.

Discussion The squared nature of this molar, resulting from the expansion of the anteroexternal platform, suggests *Cimolomys* (e.g., *C. gracilis* in Archibald, 1982:fig. 23d; *C. clarki* in Lillegraven and McKenna, 1986:fig. 7D); however, many illustrated specimens show deep incised grooves not present on this specimen. The cusps of the internal row in *Cimolomys* are typically well separated (Sahni, 1972:373) and the first cusp of the row is well developed, unlike the condition seen in this specimen. This M2 is smaller than for any described species of *Cimolomys,* but the *Cimolomys*-like M2s described in Eaton (2006a:figs. 5C, 11B) are smaller yet. It may be that there are smaller species of *Cimolomys,* or a new genus of small *Cimolomys*-like multituberculates in this sample.

Family incertae sedis
DAKOTAMYS sp. Eaton, 1995

Referred Taxa *Paracimexomys robisoni* Eaton and Nelson, 1991; cf. *P. robisoni* Eaton and Cifelli, 2001.

Discussion Eaton and Cifelli (2001) restricted *Paracimexomys* to taxa with smooth enamel. Hunter, Heinrich, and Weishample (2010:888) have pointed out that this left several taxa that were not transferred to other genera (*Paracimexomys robisoni,* Eaton and Nelson, 1991; *?P. crossi* Cifelli et al., 1997; cf. *P. perplexus,* Eaton and Cifelli, 2001; *P.* sp. A, Eaton and Nelson, 1991; *P.* sp. B, Eaton and Nelson, 1991; and *P.* sp. A, Eaton, 2002). Of these taxa, only *Paracimexomys robisoni* can be transferred with certainty to *Dakotamys.* The holotype of *P. robisoni* is an M1 with parallel medial and external cusp rows, a medial row that projects strongly anteriorly, and cusps

Table 15.2. Comparisons of measurements, ratios, and cusp formulae of *Dakotamys shakespeari*, *D. malcolmi*, and cf. *Paracimexomys perplexus* from the John Henry Member of the Straight Cliffs Formation

Tooth	Taxon	Specimen	AP	LB	LB:AP	Formula	APM2:M1	Source[a]
M1	*Dakotamys shakespeari*	UMNH VP 19160	1.61	1.04	0.65	3–4:4:1	0.89	1
	Dakotamys malcolmi	Mean	2.00	1.15	0.58	4:4:1–w2	0.80	2
	Dakotamys robisoni	Mean	2.20	1.30	0.6	4:4:1		3
	cf. *Paracimexomys perplexus*	Mean	1.70	1.06	0.62	5:4:1		4
M2	*Dakotamys shakespeari*	UMNH VP 14231	1.43	1.21	0.85	R:3:3?		1
	Dakotamys malcolmi	Mean	1.60	1.35	0.81	R:3:3?		2

[a] Source: 1–this chapter; 2–Eaton (1995:table 3); 3–Eaton and Nelson (1991:table 1); 4–Eaton and Cifelli (2001:table 1B).

that are ribbed and the valleys deeply pitted (Eaton and Nelson, 1991:fig. 2F).

?P. crossi is only known from a p4 and cannot be further evaluated systematically from the current knowledge of p4s. Another taxon, cf. *Paracimexomys perplexus* Eaton and Cifelli (2001) probably should be transferred to *Dakotamys* but variability in the specimens assigned to this taxon raises questions. M1s assigned to cf. *P. perplexus* are highly variable. The holotype (OMNH 27615, Eaton and Cifelli, 2001:fig. 7E) has parallel medial and external cusp rows as does OMNH 26400 (Eaton and Cifelli, 2001:fig. 7B), a character of *Dakotamys*. OMNH 26624 (Eaton and Cifelli, 2001:fig. 7A) and 29729 (Eaton and Cifelli, 2001:fig. 7D) the cusp rows appear to diverge as in *Paracimexomys*, and in OMNH 27580 (Eaton and Cifelli, 2001:fig. 7C) the cusp rows actually appear to converge. Most of the figured M1s in Eaton and Cifelli (2001) have an anteriorly projecting medial cusp row, but the holotype (OMNH 27615, Eaton and Cifelli, 2001:fig. 7E) has a squared anterior margin as in *Paracimexomys*. If these characters are variable in cf. *P. perplexus*, but if so this precludes inclusion in *Dakotamys* without a significant revision of the diagnosis.

Several other taxa that have been referred to *Paracimexomys* (P. sp. A and P. sp. B in Eaton and Nelson, 1991; and P. sp. A Eaton, 2002) require larger samples to evaluate their assignment to *Paracimexomys* with greater certainty.

DAKOTAMYS SHAKESPEARI sp. nov.

Type Specimen UMNH VP 19160 (BRCA 3020), Loc. 1156, LM1 (Table 15.2, Fig. 15.6K–L).

Hypodigm UMNH VP 14231 (BRCA 7239), Loc. 424, RM2.

Distribution John Henry Member of the Straight Cliffs Formation, Late Cretaceous (Santonian), Utah.

Diagnosis Smaller than *Dakotamys robisoni* (Eaton and Nelson, 1991), *D. malcolmi* Eaton, 1995, or *D. magnus* (Sahni, 1972), the M1 has fewer cusps in the external row (3:4:1) than other species of *Dakotamys*, and the valley posterior of the third cusp of the medial row is not deep and does not extend to the internal row as in other species of *Dakotamys*.

Etymology In recognition of the discovery of this specimen near Shakespear Point in Bryce Canyon National Park.

Description M1-UMNH VP 19160 (Fig. 15.6K) is a LM1 with a cusp formula of 3:4:1 (AP = 1.61, LB = 1.02). The cusps of the external row are striated internally and externally. The cusps are joined together such that the unjoined apices are less than half the height of the external cusp row. The central valley is deep, sinuous, and closed anteriorly and posteriorly. The medial cusp row extends anteriorly much more so than the external row. The cusps have a steep anterior face and a more gently sloping posterior. The third cusp of the medial row is connected by the gently sloping ridge to the fourth cusp such that there is not a continuous valley leading to the single internal cusp. The medial and external cusp rows are essential parallel. The single internal cusp is low on the side of the tooth but nonetheless prominent.

M2-UMNH VP 14231 (Fig. 15.6L) is a RM2 with a cusp formula of R:3:3? (AP = 1.43; LB = 1.21). The external ridge has an anterolabial swelling that may have functioned as a cusp and then a low ridge that extends to the last cusp of the central row. The first cusp of the central row forms the anterior wall of the molar and is well connected to the second cusp. The second and third cusps of the row are deeply separated. The third cusp of the tooth is the largest and it leans anteriorly. The deep central valley runs diagonal to the AP axis of the molar, and is slightly sinuous. The cusps of the internal row are weakly separated.

Discussion This M1 is distinctly smaller than *Paracimexomys magister* (AP = 3.2, LB = 1.8–1.9, Fox, 1971a), *P. propriscus* (mean AP = 2.26; mean LB = 1.47, Hunter et al., 2010:table 1), and *P. priscus* (mean AP = 2.14, mean LB = 1.45, Archibald, 1982:table 17). The M1s of *Paracimexomys* are defined by having smooth enamel and diverging medial and external cusp rows (Eaton and Cifelli, 2001) whereas the M1 of *D. shakespeari* has the anteriorly projecting medial cusp row (Eaton, 1995:fig. 7K), striated cusps, and parallel medial and external cusp rows characteristic of *Dakotamys*.

CEDAROMYS Eaton and Cifelli, 2001
CEDAROMYS sp. cf. *C.*
HUTCHISONI Eaton, 2002

Referred Specimens UMNH VP 14149 (BRCA 7225), Loc. 424, Lm1; UMNH VP 19000, Loc. 1144, LM1; UMNH VP 14125 (BRCA 7201), Loc. 424, RM2 (Table 15.3, Fig. 15.6M–N).

Description m1-UMNH VP 14149 (Fig. 15.6M) is a Lm1 with a cusp formula of 4:3 (AP = 1.66; LB = 1.02). The first cusp of the external row is low and bulbous. The second cusp is the tallest of the row and well separated from the first and third cusps. The third and fourth cusps are not deeply separated. The central valley is sinuous and has occasional pits. The first two cusps of the internal row are not deeply separated. The posterior wall of the second cusp is concave and descends into a deep pit that separates the second and third cusps. The third cusp is the tallest and largest of the tooth. The cusp rows converge anteriorly. UMNH VP 19000 is a LM1 with a cusp formula of 4:3 (AP = 1.50; LB = 0.93) and is morphologically similar to the m1 described above, but is smaller and more worn.

M2-UMNH VP 14125 (Fig. 15.6N) is a RM2 with a cusp formula of R:3:? (AP = 1.47; LB = 1.22). This worn molar has a deeply worn external ridge. The first cusp of the central row is narrow and formed on the anterior wall of the tooth. It is connected by a low ridge to the large second cusp, the dominant cusp of the tooth. The third cusp is much smaller but the cusps of the medial row are consistently higher than those of the internal row. The internal row is deeply worn so the number of cusps is uncertain. Even though the molar is deeply worn, it is likely it bore heavy ribs and pits along the central valley.

Discussion The low cusp formula of UMNH VP 17534, the low climb rate of the cusps of the medial row, and the relatively large size of the posterior basin relative to the AP dimension of the tooth indicates *Cimolomys*. This specimen has the lowest cusp formula for a P4 of *Cimolomys* recorded, except for the specimen of ?*Cimolomys* sp. reported in Eaton (2006a), and ?*Cimolomys* sp. B in Fox, 1971a. This specimen is close in size and morphology to Fox's specimens and most certainly represent the same species. Archibald (1982:122) suggested Fox's specimens may represent *Paracimexomys magister*, but this is doubtful because of the low cusp formula, striated cusps, and complex posterior basin.

The rugosity of the UMNH VP 14125 (M2) and the large dominant cusp of the central row are similar to that seen on M2s of *Cimolodon* (Fox, 1971a:fig. 5a; Eaton, 2002:fig. 10B), but this specimen is well below the size range of M2s of the smallest species of *Cimolodon*, *C. foxi* (Eaton, 2002:table 4). The overall shape of the tooth, and the dominant central

cusp is also similar to the M2s described as *Cedaromys* from the Cedar Mountain Formation (Eaton and Cifelli, 2001:figs. 13F, 16F), but the specimen described here is much smaller. The specimen is closest in size and morphology to MNA v6760 (Eaton, 1995:fig. 10D) described as genus and species undetermined (*Cedaromys* had not yet been described) from the Smoky Hollow Member of the Straight Cliffs Formation (Turonian).

These specimens suggest the presence of a new small species of *Cedaromys* in the Turonian and Santonian of southern Utah or that these, along with the sample of C. sp. cf. *C. hutchisoni* reported from lower in the John Henry Member (Eaton, 2006a) may represent a single species of *Cedaromys* that increases slightly in size over time.

THERIA Parker and Haswell, 1897
TRECHNOTHERIA McKenna, 1975
SPALACOTHERIIDAE Marsh, 1887
?*SPALACOTHERIDIUM* Cifelli, 1990c
?*SPALACOTHERIDIUM* sp.

Referred Specimens UMNH VP 14151 (BRCA 7227), Loc. 424, Rm6 or 7; UMNH VP 17294 (BRCA 7901), Loc. 424, Lm4?; UMNH VP 14129 (BRCA 7205), Loc. 424, posterior mx; UMNH VP 14132 (BRCA 7208), Loc. 424, Rm3 or 4?; UMNH VP 14138 (BRCA 7214), Loc. 424, Lmx (Table 15.4, Fig. 15.7A–D)

Description mx-UMNH VP 14151 (Fig. 15.7A, B) is a Rm6 or 7 (AP = 0.96, L = 0.62. LB = 0.96, CH = 1.47, H = 0.86, A = 35°). The specimen has a well-developed basal cingulid that encircles the base of the crown and is most likely an m6 or m7 because of the acute angle formed by the proto- and paracristids. The mesial cingular cusp is not distinct. The distal cingular cusp projects slightly posteriorly. The paraconid and metaconid are both somewhat worn but appear to not have had greatly differential heights prior to wear, whereas in *Symmetrodontoides* the metaconid and protocristid are markedly taller than the paraconid and the paracristid (Fox, 1976; Cifelli and Madsen, 1986). This specimen, if it does represent an m7, is smaller than *Symmetrodontoides canandensis* and *S. foxi* (Cifelli and Madsen (1986:table 1) but larger than *Spalacotheridium noblei* (Cifelli and Madsen, 1999:table 5). UMNH VP 17294 (Fig. 15.7C, D) is a Lm4? (AP = 1.00, L = 0.69, LB = 1.06, CH = 1.60; H = 1.09, A = 41°). It is considered to represent an m4 because of the angle formed by the trigonid cusps, the trigonid is linguobuccally broader than in UMNH VP 14151 and the tooth is higher crowned. The mesial conule (cusp e) is formed by a slight elevation on the cingulid. The distal conule (cusp d) projects slightly posteriorly. The paraconulid is slightly lower than the metaconid.

Table 15.3. Comparisons of measurements, ratios, and cusp formulae of *Cedaromys hutchisoni*, *C.* sp. cf. *C. hutchisoni*, and *Cedaromys* sp.

Taxon	Unit–plateau	Specimen	AP	LB	LB:AP	Formula	Source[a]
C. hutchisoni	Kk–Kaiparowits	Means	1.89	1.21	0.64	4–5:3	1
C. cf. *C. hutchisoni*	Kw–Kaiparowits	UMNH VP 12771	1.73	1.10	0.64	4:3	2
Cedaromys sp.	Kw–Kaiparowits	UMNH VP 7714	1.58	0.95	0.6	4:3	2
C. cf. *C. hutchisoni*	Kw–Paunsaugunt	UMNH VP 17491	1.6[b]	0.93	0.58	4:3	3
C. cf. *C. hutchisoni*	Kscjh–Paunsaugunt	UMNH VP 14149	1.66	1.02	0.61	4:3	3
C. cf. *C. hutchisoni*	?–Markagunt (Loc. 10)	Means	1.66	1.06	0.64	4:3	4

Kk, Kaiparowits Formation of the Kaiparowits Plateau; Kw, Wahweap Formation of the Paunsaugunt and Kaiparowits plateaus; Kscjh, John Henry Member of the Straight Cliffs Formation of the Paunsaugunt Plateau; ?, uncertain unit on the Markagunt Plateau (UMNH VP Loc. 10).

[a] Source: 1–Eaton (2002:table 11); 2–Eaton (2006a); 3–this chapter; 4–Eaton (2006b:table 2).

[b] Uncertain value.

Similar, but less complete or deeply worn teeth represent the lower dentition of this species including UMNH VP 14129 a posterior mx; UMNH VP 14132 an m3 or 4? (AP = 1.15; LB = 1.27); and UMNH VP 14138, a small posterior lower molar with deeply worn cusps (AP = 0.85, L = 0.62, LB = 0.90).

Discussion These teeth are similar to UMNH VP 12810 described as *Spalacotheridium* sp. (Eaton, 2006a:fig. 7a–c) from the John Henry Member in being small (AP = 0.92, L = 0.59, LB = 0.91, CH not measurable as a result of damaged protoconid) and having relatively less height differential between the metaconid and paraconid relative to described species of *Symmetrodontoides*. UMNH VP 12810 was described as *Spalacotheridium* sp. by Eaton (2006a) because of the relative equal development and heights of the metaconid–protocristid and paraconid–paracristid (Cifelli, 1990c:334).

Fox (1976:1110) diagnosed lower molars of *Symmetrodontoides* as "having a well developed basal cingulum encircling the lower molar crowns, a broadly obtuse trigonid angle formed by the major cusps of the lower first molar, and a progressively more acutely angled trigonid angle on each of at least four molars posterior to the lower first molar." The lower molars described here have a well-developed basal cingulid surrounding the entire tooth, but in isolated teeth, it is difficult to determine whether these are posterior teeth or simply teeth of a more acutely angled taxon. Cifelli (1990c:334) diagnosed lower molars of *Spalacotheridium* "in equal development of the paraconid and metaconid on lower molars, with paracristid and protocristid of equivalent height." In his description of *Spalacotheridium mckennai*, he indicated that "the lingual cingulum is faint; the labial side of the crown base has been damaged, and it cannot be determined whether the labial cingulum was complete, as in *Symmetrodontoides* and *Spalacotherium* (in contrast to the condition in *Spalacotheroides*; Fox, 1976)" (Cifelli, 1990c:335). The teeth described here have the relatively equally elevated metaconid and paraconid, but have a complete cingulid. Cifelli and Madsen (1999:195) revised the diagnosis of lower

molars of *Spalacotheridium* as "differing from other members of the family in their small size (maximum length and width of molars generally less than 0.75 mm) . . . differ from those of *Spalacotheroides* in having a complete labial cingulum; and from *Symmetrodontoides* and *Spalacolestes* in being lower crowned, with paraconid and metaconid subequal in development and of approximately equivalent height, and only slightly lower than the protoconid, and in having posterior molars that are proportionally narrower and have more obtuse trigonid angles." *Spalacotheridium* has a complete labial cingulid, based on the description of *S. noblei* (Cifelli and Madsen, 1999:197) as do the specimens described here, but the latter are not as small as any previously described as *Spalacotheridium*. Because Fox's (1976) diagnosis of *Symmetrodontoides* was primarily based on the presence of a complete basal cingulid on lower molars, it is unclear why the new genus *Spalacotheridium* was established. Assuming its validity, these lower molars are tentatively assigned to *Spalacotheridium* despite their relatively large size because they possess a complete basal cingulid and paraconids and metaconids of equal height.

SYMMETRODONTOIDES Fox, 1976
SYMMETRODONTOIDES sp. Fox, 1976

Referred Specimens UMNH VP 12860 (BRCA 5885), Loc. 424, Lm6? UMNH VP 12861 (BRCA 5886), Loc. 420; UMNH VP 14147 (BRCA 7223), Loc. 424, RM4? (Table 15.4, Fig. 15.7E–G).

Description mx–UMNH VP 12860 (Fig. 15.7E, F) is a Lm6? (AP = 1.01, L = 0.81, LB = 1.31, CH = 1.78, H = 1.24; A = 38°). This specimen has a basal cingulid surrounding the tooth. The trigonid angle is quite narrow on this specimen, so it seems to represent a posterior molar, but is significantly larger than specimens described as ?*Spalacotheridium* sp. above (UMNH VP 17294, 14151). The metaconid, although broken, appears to be significantly larger and taller than the paraconid and similarly the protocristid is

Table 15.4. Comparisons of measurements of species of *Spalacotheridium* and *Symmetrodontoides*

Taxon	Tooth	Specimen no.	AP	LB	Source[a]
?*Spalacotheridium* sp.	m6 or 7	UMNH VP 14151	0.96	0.96	1
	m4?	UMNH VP 17294	1.00	1.06	1
Spalacotheridium sp.	m4?	UMNH VP 12810	0.92	0.91	2
Spalacotheridium mckennai	m4	MNA V5792	0.65	0.63	3
Spalacotheridium noblei	m4	Mean	0.49	0.68	4
	m6	Mean	0.35	0.54	4
Symmetrodontoides sp.	m6?	UMNH VP 12860	1.01	0.81	1
	M4	UMNH VP 14147	1.04[b]	1.52	1
Symmetrodontoides canadensis	m4	Mean	1.30	1.10	5
	m7	Mean	1.10	1.30	5
Symmetrodontoides foxi	m4	Mean	1.10	1.10	5
	m7	Mean	1.10	1.70	5
	M4	Mean	0.80	1.24	6
Symmetrodontoides oligodontos	m4	Mean	0.58	0.92	7

[a] Source: 1 – this chapter; 2 – Eaton (2006a); 3 – Cifelli (1990c); 4 – Cifelli and Madsen (1999:table 5); 5 – Cifelli and Madsen (1986:table 1); 6 – Cifelli and Gordon (1999:table 2); 7 – Cifelli and Gordon (1999:table 1).

[b] Uncertain value.

much more elevated than the paracristid. These cristids are much more equal in ?*Spalacotheridium* sp. described above. UMNH VP 12861 is a poorly preserved m4 or m5 with nearly all of the cingulid and the metaconid missing. Although not measureable, the molar corresponds well in size to UMNH VP 12860.

Mx-UMNH VP 14147 (Fig. 15.7G) is considered to possibly represent a RM4 because the tooth is broad relative to its narrow AP dimension and the posterior width (LBp), as interpreted here, is greater than the anterior width (LBa; Cifelli and Madsen, 1999:fig. 11) (AP = 1.04*, LBa = 1.34*, LBp = 1.52).

Discussion These lower molars are considered to represent *Symmetrodontoides* because they have a complete basal cingulid and a substantial height difference between the paraconid and metaconid. These specimens appear to be closer in size to *S. canadensis* than *S. oligodontos* (Table 15.4). The cusps on the upper molar is too worn to be diagnostic and it is placed here on the basis of the LB dimension which is somewhat greater than the LB dimension of lower molars assigned here.

MARSUPIALIA Illiger, 1811
"DIDELPHIMORPHIA" Gill, 1872
Family incertae sedis, gen. et sp. indet.

Referred Specimens UMNH VP 18769 (BRCA 7679), Loc. 424, labial half of LM2; UMNH VP 14244 (BRCA7252), Loc. 424, labial half LMX; UMNH 18770 (BRCA 7680), Loc. 424, labial half LM2? (Fig. 15.8A–B)

Description Mx-UMNH VP 18769 (Fig. 15.8A) is the labial half of a LM2 (AP = 1.76). The AP dimension and paracone being slightly taller than the metacone is very much in line with the molars described below as *Apistodon*, but this molar has some significant differences. There is a well-developed stylar cusp C positioned directly in the ectoflexus which is positioned opposite the base of the centrocrista. The preparacrista connects to stylar shelf between stylar cusp A and B. The entire stylar shelf is well elevated above the centrocrista. The stylar cusps and paracone–metacone are more inflated than the specimens described below as *Apistodon*. There is a slight postmetaconular crista which extends labially past the apex of the metacone, but certainly no extensive postmetaconular crista is present (a character of *Varalphadon* according to Johanson, 1996).

UMNH VP 14244 is the labial half of a LMX (AP = 1.80). This molar has a weakly developed ectoflexus, with a distinct stylar cusp C that is only slightly smaller than stylar cusp D. Stylar cusp B is badly worn, but was probably the largest stylar cusp.

UMNH 18770 (Fig. 15.8B) is the labial half of a LM2? (AP = 1.74). There is a distinct stylar cusp A, and cusp B is the largest stylar cusp. There is a tiny cuspule in the C position, the ectoflexus is aligned with the bottom of the centrocrista, and stylar cusp D is distinct. The stylar shelf slopes downward to meet the bottom of the centrocrista. The paracone is taller than the metacone. The preparacrista connects to stylar cusp B. A postmetaconular crista terminates well anterior of the apex of the metacone.

Discussion These partial molars all have a distinct cusp in the C position, at the ectoflexus, unlike *Apistodon* and the paracone is wider than the metacone in labial view (unlike *Varalphadon*) on the two specimens in which that character can be evaluated (UMNH VP 18770, 18769). This suggests the possibility of the presence of another taxon of primitive marsupial in the fauna.

Eaton

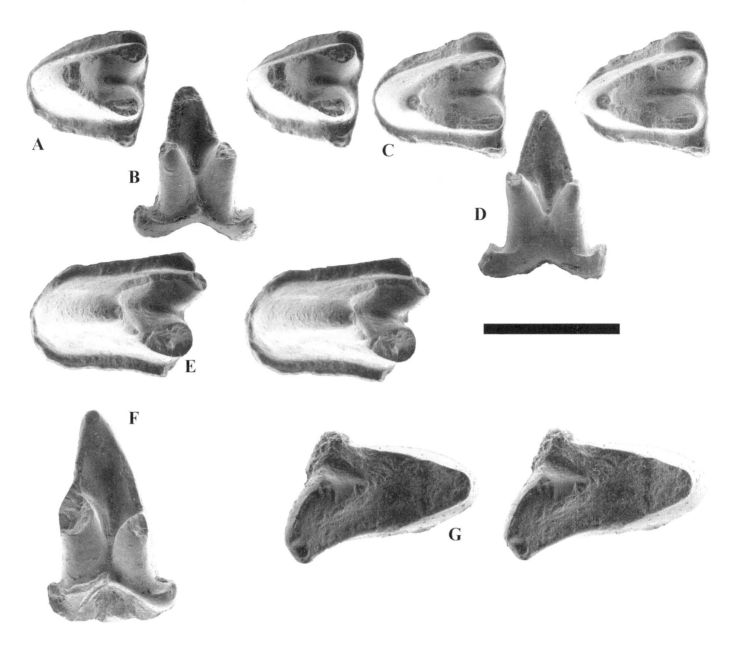

15.7. *?Spalacotheridium* sp.: (A) UMNH VP 14151, Rm6 or 7, osp; (B) lingual; (C) UMNH VP 17294, Lm4?, osp; (D) lingual; *Symmetrodontoides* sp.: (E) UMNH VP 12860, Lm6?, osp; (F) lingual; (G) UMNH VP 14147, RM4, osp. Scale bar = 1 mm.

APISTODON Davis, 2007
APISTODON sp. cf. *A. EXIGUUS* (Fox, 1971b)

Referred Specimens UMNH VP 14136 (BRCA 7212), Loc. 424, RM2; UMNH VP 14232 (BRCA 7240), Loc. 424, RM2; UMNH VP 17270 (BRCA 7877), Loc. 424, labial half RM3; UMNH VP 17205 (BRCA 7443), Loc. 424, LM2; UMNH VP 14150 (BRCA 7226), Loc. 424, RM2; UMNH VP 18017, Loc. 419, LM2; UMNH VP 17331 (BRCA 7942), Loc. 424, RM2; UMNH VP 17324 (BRCA 7935), Loc. 424, RM2; UMNH VP 17380 (BRCA 7992), Loc. 424, RM3; UMNH VP 17583, Loc. 781, labial part LM3?; UMNH VP 17274 (BRCA 7881), Loc. 424, RM4; UMNH VP 17385, Loc. 424, lingual part LM4 (Table 15.5, Fig. 15.8C–M).

Description Mx-UMNH VP 14136 (Fig. 15.8C) is a RM2 (AP = 1.45, LBa = 1.43, LBp = 1.53). The stylar shelf is narrowed, but not so much as in pediomyids. Stylar cusps B and C (following the designation of Davis, 2007; the same cusp was referred to a D by Fox, 1971) are subequal in size, but cusp C has a sharp posterolabial crest that connects to stylar cusp D. No cusp is present in the ectoflexus. The paracone is slightly taller than the metacone. The paraconule and metaconules are worn but do not appear to be strongly developed. The trigon basin is deep and the protocone region is anteroposteriorly compressed. The specimen is similar to *Eoalphadon* (Eaton, 2009) from the Dakota Formation except for the lack of a stylar cusp in the ectoflexus. An almost identical tooth is

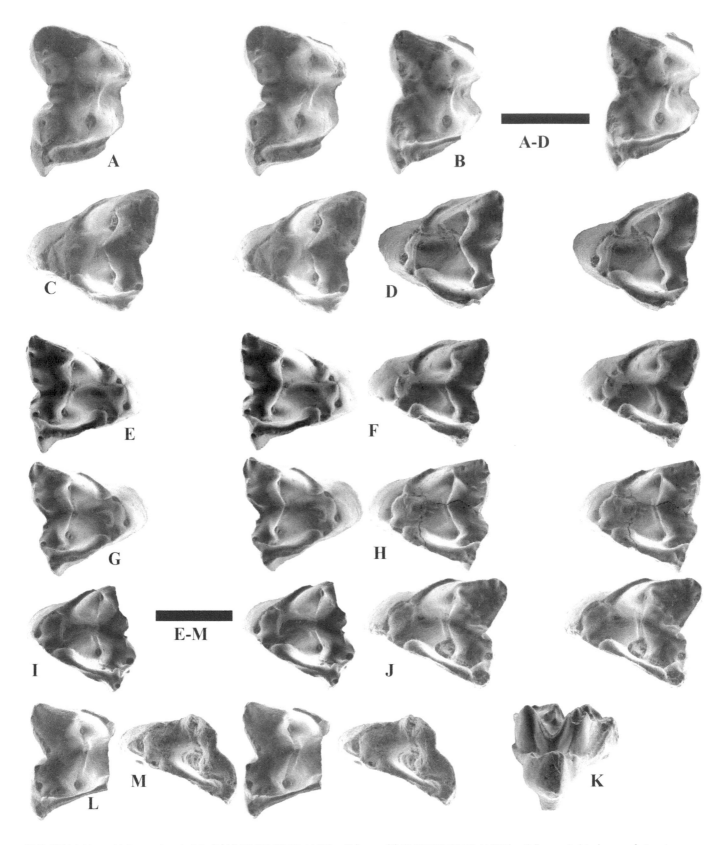

15.8. "Didelphimorphia" gen. et sp. indet.: (A) UMNH VP 18769, LM2 (partial), osp; (B) UMNH VP 18770, LM2? (partial), osp; *Apistodon* sp. cf. *A. exiguus:* (C) UMNH VP 14136, RM1, osp; (D) UMNH VP 14232, RM1, osp; (E) UMNH VP 17205, LM2, osp; (F) UMNH VP 14150, RM2, osp; (G) UMNH VP 18017, LM2, osp; (H) UMNH VP 17331, RM2, osp; (I) UMNH VP 17324, RM2?, osp; (J) UMNH VP 17380, RM3, osp; (K) lingual; (L) UMNH VP 17583, LM3?, osp; (M) UMNH VP 17274, RM4, osp. Scale bars = 1 mm.

UMNH VP 14232, Fig. 15.8D, a slightly smaller RM2 (AP = 1.40, LBa = 1.39, LBp = 1.42). The metacone and the anterolabial corner of the tooth are damaged. On this molar, it is apparent that the stylar shelf narrows anteriorly and that the angle formed by the apices of the paracone and metacone and the labial margin of the tooth opens broadly posteriorly. The conules are weakly developed but are distinctly winged. These molars are similar to the M1 figured by Davis (2007:fig. 6b).

UMNH VP 17205 (Fig. 15.8E) is a LM2 (AP = 1.79; LBa = 1.73; LBp = 1.86). Stylar cusp A is low on the anterior face of the molar and is part of an anterior directed crest. The tallest stylar cusp is B. Stylar cusp C is positioned posterior of the ectoflexus, as large as cusp B, but is slightly lower. There is a small additional cuspule (D?) on the labial margin of the molar, and a tiny stylar cusp E(?) at the posterolabial corner of the tooth. The paracone is slightly taller than the metacone. The paraconule is winged and the preparaconular crista attaches to stylar cusp A. The metaconule is also winged, but the postmetaconular crista fades below the apex of the metacone. The protocone is tall and the conules are much closer to the protocone than the metacone or paracone. The metaconule originates higher on the postprotocrista than the paraconule does on the preprotocrista. The postmetaconular crista does not extend far labially.

UMNH VP 14150 (Fig. 15.8F) is a RM2 (AP = 1.73, LBa = 1.73, LBp = 1.82) which lacks a stylar cusp in the ectoflexus, has a large cusp in the C position, and an additional cusp (D?) anterior of the posterolabial corner of the tooth. The paracone and metacone are subequal in height but are also approximately equal in width. The conules are strongly developed. UMNH VP 18017, a LM2, (Fig. 8G; AP = 1.78; LBa = 1.69, LBp = 1.77) is a similar upper molar but has an additional cuspule on the postmetacrista and is from stratigraphically much lower, about the middle of the John Henry Member (Loc. 419).

UMNH VP 17331 (Fig. 15.8H) is a RM2 (AP = 1.72, LBa = 1.83. LBp = 1.86) similar to UMNH VP 17205 described above (Fig. 15.8E). Stylar cusp B is only slightly taller than stylar cusp C and the posterolabial ridge from that extends from cusp C bears a weakly developed accessory cusp. The paracone is distinctly taller than the metacone, but the metacone appears to be slightly larger at the base than the paracone.

UMNH VP 17324 (Fig. 15.8I) is a RM2 lacking the posterolabial corner (LBa = 1.75). Stylar cusp B is tall and conical. It appears that no stylar cusp was present in the ectoflexus. The paracone is taller than the metacone, but they are similar in overall size. The trigon basin is deep.

UMNH VP 17380 (Fig. 15.8J, K) is a RM3 (AP = 1.71, LBa = 2.00, LBp = 2.12 AP:LB = 0.81). In overall size and form this molar is similar to UMNH VP 14126 (Fig. 15.8J) described below as ?Varalphadon, but there are significant differences. There is a distinct notch at the ectoflexus and the posterially

directed crest of stylar cusp B extends posterolabial of the notch. The posterior part of stylar cusp C forms a ridge that descends posteriorly. The area between stylar cusp C and the metacone is complexly crenulated and is there are small crests originating on the lingual side of cusp C and the labial side of the metacone. The postmetacrista is cuspate. The tip of the paracone is missing but it was clearly taller than the metacone. The internal cristae run up the lingual face of the metacone and paracone forming sharp lingual crests as on other upper molars of this taxon (Fig. 15.8K). A postmetaconular crista extends as far labially as the apex of the metacone, the paraconule is positioned more lingually than the metaconule, and the protocone is anteroposteriorly compressed such that the lingual margin of the molar is sharp.

UMNH VP 17583 (Fig. 15.8L) is the labial part of a LM3 (AP = 1.84). It is from stratigraphically much lower in the John Henry Member (Loc. 781; Fig. 2) than most of the specimens described here and is probably early rather than middle or late Santonian. Stylar cusp A is distinct, stylar cusp B is the largest stylar cusp and the preparacrista attaches to the anterior part of the cusp. There is a small cuspule on the posterior descending crest of stylar cusp B. The ectoflexus is deep and incised and there is another stylar cusp (C) on the posterior side of the ectoflexus and a sharp crest runs from this cusp to the posterolabial margin of the tooth. There are two distinct wear facets on the posterolabial side of stylar shelf. This unusual feature of the ectoflexus is similar to that found on UMNH VP 17380 described above (Fig. 15.8J, K). The stylar shelf slopes downward making a groove as it descends through the bottom of the centrocrista. The paracone and metacone are subequal in height, but the paracone is a larger cusp. The postmetaconular crista is relatively broad and extends posterior of the metacone.

UMNH VP 17270 is the labial half of a RM3 (AP = 1.59). There is no stylar cusp in the ectoflexus and stylar cusp C is blade like, positioned posterolabial of the centrocrista, and extends almost to the posterolabial corner of the molar. The paracone is taller than the metacone and shifted labially relative to the metacone.

UMNH VP 17274 (Fig. 8M) is a RM4 (AP = 1.39, LB = 1.87) is tentatively assigned here based its size and proportions relative to the M3s described above. No M4s of Apistodon have been previously described. The tooth is deeply eroded, possibly by digestion. The anterolabial cingulum is well developed and continuous. The paraconule is much more developed than the metaconule. There appears to have been small stylar cusps in the position of B and C. UMNH VP 17385 is the lingual part of a LM4 that clearly represents that same taxon as UMNH VP 17274 and is better preserved. The paraconule is well developed but the metaconule is only hinted at as a slight swelling on the postprotocrista. Stylar cusp C(?)

is well developed and positioned labial of the base of the centrocrista.

Discussion These specimens do not appear to represent *Alphadon* or *Turgidodon* based on the criteria in Johanson (1996), as the paracone and metacone are not inflated and are closely positioned on the molar. They appear to be similar to *Protalphadon* because the paracone is usually slightly taller than the metacone, and the metacone does not appear to be wider than the paracone in labial view. This last character (metacone wider than paracone) was considered by Johanson (1996) as a distinguishing character of *Varalphadon* relative to *Protalphadon*. In the specimens described here, stylar cusp C is positioned posterior of the ectoflexus. The presence of stylar cusp C is variable for both *Varalphadon* and *Protalphadon*; however, in *Protalphadon* the cusp is either absent or small on the specimens described in Cifelli (1990a) but can be relatively prominent in *Protalphadon foxi* described in Johanson (1996:pl. 1B). The lack of a postmetacingulum is a characteristic of *Protalphadon*, and a postmetacingulum is present on these upper molars. The specimens illustrated as *Protalphadon foxi* (Johanson, 1996:pl. 1) have relatively flattened labial walls on the metacone and paracone, and both cusps appear to lean lingually such that the centrocrista is weakly V shaped in occlusal view, unlike the condition in *P. lulli* (Clemens, 1966:fig. 7B). The specimens described here are closer to those of *P. lulli* than *P. foxi*, but have a postmetaconular crista that extends to about the apex of the metacone. These teeth are not greatly different than those of *Varalphadon*, but the metacone is not wider than the paracone in labial view and the paracone is generally taller than the metacone. These molars are also somewhat like *Eoalphadon* from the Dakota Formation of Cedar Canyon (Eaton, 2009:fig. 5D), except for the lack of a stylar cusp in the ectoflexus, which is variably developed but present in all known species of *Eoalphadon*; the stylar cusps are linguobuccally compressed rather than being essentially conical; and these specimens have much better developed conular cristae (also present on both *Protalphadon* and *Varalphadon*). These specimens are close in size and morphology to those of *Apistodon exiguus* as stylar cusp B is consistently present, a well-developed stylar cusp C is placed posterior of the ectoflexus, and the ectoflexus is anteriorly placed relative to the deepest point of the centrocrista. However, on some of the specimens in this sample, the paracone is not distinctly taller than the metacone and the conules, particularly the metaconule, often appear to be better developed (particularly on M2) than on figured specimens of *Apistodon exiguus*. UMNH VP 17380 (M3; Fig. 15.8J, K) and 17583 (M3; Fig. 15.8L) have a somewhat different ectoflexus geometry and may not belong to the same taxon. The sample described by Davis (2007) appears to be slightly smaller than the sample described here on the

basis of AP measurements provided by Davis (pers. comm., 2010, as no measurements were provided in Davis, 2007); however, measurements provided by Fox (1971b) indicate the specimens he described were as large as those described here. For these reasons the specimens are only conferred to *A. exiguus* until the sample at University of Alberta can be examined.

cf. "ANCHISTODELPHYS" sp. Cifelli, 1990b

Referred Specimens UMNH VP 17530, Loc. 799, RMX (labial part) (Fig. 15.9A)

Description UMNH VP 17530 (Fig. 15.9A) is the labial part of a RMX (AP = 1.52). The specimen has a distinct stylar cusp A and a larger and taller stylar cusp B, no other stylar cusps are developed on the stylar shelf but a slightly cuspate ridge extends posteriorly from cusp B. The paracone is attached to stylar cusp B by a preparacrista and the paracone appears to have been taller than the metacone prior to wear (both cusp apices are worn). The paracone is shifted labially relative to the metacone.

Discussion This partial upper molar is most like *Anchistodelphys* (Cifelli, 1990b) and particularly ?*Anchistodelphys delicatus* Cifelli, 1990c:fig. 3A, which was recovered from the Turonian Smoky Hollow Member of the Straight Cliffs Formation. This specimen was recovered from low in the John Henry Member and is either of Coniacian or, most likely, Early Santonian age. Cifelli (1990b) diagnosed upper molars of the type species, *A. archibaldi*, in their variable absence of both stylar cusps C and D. Johanson, 1996, synonymized *A. archibaldi* (the type and only species) with *Varalphadon wahweapensis*. All of her figured molars have well-developed stylar cusps D (Johanson, 1996:pl. 2) and seem quite distinct compared to specimens referred to *Anchistodelphys* by Cifelli (1990b:fig. 4c, *A. archibaldi*; 1990c:fig. 3A, ?*A. delicatus*). The specimen described here clearly lacks stylar cusps C or D and could not be referred to *Varalphadon*. It would take a more extensive collection to verify that *Anchistodelphys* is a valid taxon distinct from *Varalphadon*.

"ALPHADONTIDAE" Marshall, Case, and Woodburne, 1990
VARALPHADON Johanson, 1996
?VARALPHADON sp.

Referred Specimens UMNH VP 17518, Loc. 799, LMX; UMNH VP 14249 (BRCA 7257), Loc. 424, RM3; UMNH VP 14126 (BRCA 7202), Loc. 424, RM3; UMNH VP 17215 (BRCA 7453), Loc. 424, labial half LM4; UMNH VP 17216 (BRCA 7454), Loc. 424, LM4 (Fig. 15.9B–E).

Table 15.5. Comparisons of measurements of upper molars of *Apistodon* sp. cf. *A. exiguus* from the John Henry Member of the Straight Cliffs Formation (Kscjh) to *Apistodon exiguus* from the Milk River Formation

Upper Ms	Specimen no.	AP	LBa	LBp	Unit	Source[a]
M2	UMNH VP 14136	1.45	1.43	1.53	Kscjh	1
	UMNH VP 14232	1.40	1.39	1.42	Kscjh	1
	UMNH VP 17205	1.79	1.73	1.86	Kscjh	1
	UMNH VP 14150	1.73	1.73	1.82	Kscjh	1
	UMNH VP 18017	1.78	1.69	1.77	Kscjh	1
	UMNH VP 17331	1.72	1.83	1.86	Kscjh	1
	UMNH VP 17324		1.75		Kscjh	1
Means		1.65	1.65	1.71		
M2	UALVP 29691	1.6			–	2
M3	UALVP 5537			1.8	–	3
	UALVP 5548	1.6		1.9	–	3
	UMNH VP 17380	1.71	2.00	2.12	Kscjh	1
	UMNH VP 17583	1.84			Kscjh	1
M4	UMNH VP 17270	1.59			Kscjh	1
	UMNH VP 17274	1.39	1.87		Kscjh	1
	UALVP 29695	1.6			–	2

[a] Source: 1–this chapter; 2–Davis (pers. comm., 2010); 3–Fox (1971b:table 4).

Description Mx-UMNH VP 17518 is an upper molar missing the anterior part of the stylar shelf (LBp = 2.01). A distinct stylar cusp D is preserved. The paracone and metacone are subequal in height and width. The preparaconular crista extends up the lingual face of the paracone, but the premetaconular crista extends to the valley between the paracone and metacone. The tooth is compressed anteroposteriorly.

UMNH VP 14249 (Fig. 15.9B) is a RM3 (AP = 2.29, LBa = 2.18, LBp = 2.24). This badly damaged molar is important as it is the largest upper molar in the sample. Stylar cusp A is part of a large anterior directed flange, much more prominent than on any other specimen in the sample. The preparacrista attaches to a damaged stylar cusp B, which appears to be the largest stylar cusp. Stylar cusp C is small, conical and positioned slightly posterior on the ectoflexus. Stylar cusp D is larger than C, but smaller than B. The apices of the paracone and metacone are broken, so their relative heights cannot be determined, but the metacone has a broader base than the paracone in lingual view. The stylar shelf is not elevated relative to the base of the centrocrista. The internal cristae continue up the lingual face of the paracone and metacone. The postmetaconular crista is broad on the posterior side of the metacone. The expanded flange of stylar cusp C is similar to that seen on M3s of *Varalphadon creber* (Johanson, 1996:pl. 3D). This specimen is also similar to UMNH VP 11621, from stratigraphically lower in the John Henry Member described in Eaton (2006a:fig. 8G). The specimen described here has less anteroposterior compression of the protocone, the protocone is large, and has a small stylar cusp C.

UMNH VP 14126 (Fig. 15.9C) is a RM3 (AP = 1.98, LBa = 2.04, LBp = 2.15). Stylar cusp A is a well-developed prominent

cusp on an anteriorly projecting flange. Stylar cusp B is the tallest stylar cusp. There is a slight swelling posterolingual of the ectoflexus in the position of stylar cusp C. Stylar cusp D is broken, but on the basis of the size of its base, it was probably slightly smaller than stylar cusp B. The paracone is tall, but the metacone is broken so their relative heights cannot be determined. It does appear that the metacone is broader (AP dimension) than the paracone in labial view. The postparaconular crista forms a distinct ridge on the lingual face of the paracone, and the metaconule originates higher on the postprotocrista than the paraconule does on the preprotocrista (i.e., the metaconule is positioned more lingually).

UMNH VP 17215 (Fig. 15.9D) is the labial half of a LM4 (AP = 1.58). The stylar cusp A projects strongly anteriorly. Stylar cusp B is conical and the tallest of the stylar cusps. There is a small stylar cusp C anterior of the deeply developed ectoflexus. Stylar cusp D is distinct and labiolingually compressed. The paracone is much taller than the much smaller metacone. The preparacrista is tall, blade-like, and connects to the base of stylar cusp B. The anterolabial cingulum is broad. The stylar shelf is lower than the bottom of the centrocrista. Despite the presence of stylar cusp C, I include this specimen within ?*Varalphadon* because of the deeply incised ectoflexus. It is unlikely an M4 would have any posterior cingulum, so the use of the postmetacingulum as a character for identifying *Varalphadon* is not meaningful on M4s. This molar also has similarities to the M4 figured in Fox (1987:fig. 3) as *Albertatherium primum* in terms of the indentation of the ectoflexus, the tall preparacrista, and the diminutive metacone. *Albertatherium primum* differs in being larger, having a prominent stylar cusp C posterior of the

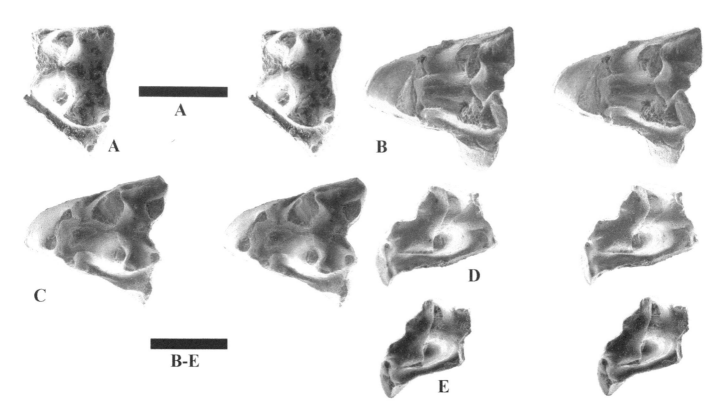

15.9. cf. *"Anchistodelphys"* sp.: (A) UMNH VP 17530, RMX (partial), osp; *?Varalphadon* sp.: (B) UMNH VP 14249, RM3, osp; (C) UMNH VP 14126, RM3, osp; (D) UMNH VP 17215, LM4 (partial), osp; (E) UMNH VP 17216, LM4 (partial). Scale bars = 1 mm.

ectoflexus (according to Fox, 1971b:151) and the preparacrista does not connect to stylar cusp B. Another incomplete specimen is UMNH VP 17216 (Fig. 15.9E; AP = 1.58). Stylar cusp A is well developed on the anterolabial cingulum. Stylar cusp B is tall and distinct and is connected to the paracone by a sharp preparacrista. There is a small stylar cusp just posterior of cusp B. There is no cusp in the slight ectoflexus, and a distinct stylar cusp C(?) posterior of the ectoflexus. The paracone is much taller than the metacone.

Discussion Several small distinctions have led me to questionably assign these molars to *Varalphadon* based on the diagnosis of Johanson (1996). The protocone is anteroposteriorly compressed as in *Varalphadon* and the metacone is broader than the paracone in labial view. The postmetaconular crista is broader and extends labially more than on specimens of *Apistodon* described above. Unlike *Varalphadon*, the paracone is usually slightly taller than the metacone. However, it should be noted that *Apistodon* and *Varalphadon* are similar taxa and differ primarily in *Apistodon* having a taller paracone, a character that appears to be variable in the sample described above, and in *Varalphadon* having a wider metacone than paracone in labial view. These are relatively slight differences and rather than adding to the plethora of generic names for primitive marsupials, I have attempted to retain existing terminology and noting the variation in these samples. UMNH VP 11621, also from the John Henry Member,

was assigned to *Varalphadon* in Eaton (2006a); however, the metacone does not appear to be wider than the paracone in labial view (see discussion in Eaton, 2006a:457) and the ectoflexus is more anteriorly placed as in *Apistodon*. If UMNH VP 11621 represents *Apistodon*, it is clearly a different species (much more AP compressed) than that described here.

Indet. didelphid lower molars

Discussion These specimens (Table 15.6, Fig. 15.10A–N) were originally assigned to *Apistodon* sp. cf. *A. exiguus* or cf. *Varalphadon* sp. by the author but it became apparent in review that I could not defend the assignments. Distinguishing lower molars is difficult as no lower molars were referred to *A. exiguus* by either Fox (1971b) or Davis (2007), and Johanson (1996) did not refer any lower molars to or diagnose lower molars of *Protalphadon*, *Varalphadon*, or *Alphadon*. Cifelli (1990a, 1990b) did assign taxa to *Varalphadon* (*Protalphadon wahweapensis*, *Anchistodelphys archibaldi*, both transferred to *Varalphadon wahweapensis* by Johanson, 1996), but the systematic status of these lowers has not been addressed in or after Johanson's 1996 revision. Lowers assigned to *Protalphadon wahweapensis*, *Anchistodelphys archibaldi* and particularly *?A. delicatus* by Cifelli (1990b, 1990c) have some distinct characteristics that suggests these lower molars do not all belong to the same taxon (e.g., the thumb-like hypoconulid of *A. archibaldi*). It is unlikely that only the upper molars of metatherians evolve while lower molars remain static. It may be that the changes in lower molar morphologies are more subtle and require a detailed morphometric analysis to reveal taxonomic associations. In any event, these lower molars are described here in the hope of stimulating study of this neglected part of the dentition. There are distinct morphological differences in these lower molars such as the trigonid to talonid relative heights and degree of separation of the entoconid and hypoconulid.

Referred Specimens UMNH VP 17269 (BRCA 7876), Loc. 424, Lm1; UMNH VP 14241 (BRCA 7249), Loc. 424, RDp3; UMNH VP 17373 (BRCA 7985), Loc. 424, Lm1; UMNH VP 17225 (BRCA 7463), Loc. 424, Rm1; UMNH VP 17269 (BRCA 7876), Loc. 424, Lm1; UMNH VP 14331 (BRCA 6024), Loc. 424, Lm1; UMNH VP 14146 (BRCA 7281), Loc. 424, partial Lm1; UMNH VP 19159 (BRCA 3019), Loc. 1156, Rm1 or 2; UMNH VP 14416 (BRCA 7222), Loc. 424, Lm2; UMNH VP 17235 (BRCA 7473), Lm2; UMNH VP 17275 (BRCA 7882), Loc. 424, Rm2; UMNH 14247 (BRCA 7255), Loc. 424, Lm2; UMNH VP 14127 (BRCA 14127), Loc. 424, Rm2; UMNH VP 17419 (BRCA 8011), Loc. 424, Rm2; UMNH 17386, Loc. 426, Lm2 (too damaged to measure); UMNH VP 14140, Loc. 424, Lm4; UMNH VP 19166 (BRCA 3026), Loc. 1156, Lm4.

Description Dp3-UMNH VP14241 (Fig. 15.10A, B) lacks roots and has an entoconid and hypoconid that are subequal in height and only slightly higher than the hypoconulid (AP = 1.40, LBa = 0.67, LBp = 0.72). The paraconid is low and anteriorly directed and the precingulid forms the labial margin of the cusp. The trigonid is low relative to the talonid. The cristid obliqua originates medially on the posterior face of the trigonid. The entoconid and hypoconulid are well separated.

m1-UMNH VP 17269 (Fig. 15.10C, D) is a Lm1 (AP = 1.46, LBa = 0.76, LBp = 0.84) that is well rooted, and its cusps have considerable apical wear. There is no distinct basin in the trigonid. The paraconid is damaged and appears smaller than it originally was; however, it is a relatively low forwardly directed cusp. The valley between the paraconid and protoconid is lower than the valley between the metaconid and protoconid. The cristid obliqua originates medially on the posterior face of the trigonid. The entoconid is the tallest cusp of the talonid, and the hypoconid is taller than the hypoconulid. The entoconid and hypoconulid are well separated. The talonid basin is deep. The cusps are delicate and uninflated. UMNH VP 19159 (Fig. 15.10E, F) is a Rm1 or 2 (AP = 1.77, LBa = 0.96, LBp = 0.98) with a small, low paraconid. The metaconid is larger and taller, and not markedly lower than the protoconid. The cristid obliqua originates at the protocristid notch. The entoconid and hypoconulid are closely twinned and form a tall wall that is significantly higher than the deeply worn hypoconid. The talonid basin is deep.

m2-UMNH VP 14146 (Fig. 15.10G, H) is a Lm2 (AP = 1.78; LBa = 1.06, LBp = 1.13) on which the paraconid leans anteriorly and is distinctly lower than the larger and more erect metaconid. There is a distinct notch between the protoconid and paraconid and there is an additional small cuspule on the paracristid. The protoconid is about one-third taller (measured from the trigonid basin) than the metaconid. The cristid obliqua originates slightly labial of the middle of the posterior face of the trigonid and does not connect to the protocristid notch. The entoconid is the tallest cusp of the talonid. There is a small cuspule on the entocristid. The hypoconulid and hypoconid are about the same size and height. The talonid basin is deep. A similar but more deeply worn m2 is UMNH VP 17235 (AP = 1.67; LBa = 0.89, LBp = 0.95). UMNH VP 17275 is similar morphologically to UMNH VP 14146 but is smaller (AP = 1.74, LBa = 0.90, LBp = 0.97). UMNH VP 17419 (Fig. 15.10I, J; AP = 1.83, LBa = 1.16; LBp = 1.15), is a large Rm2 with a wider talonid than trigonid. The trigonid is different from all other lower molars in this sample in having a very high trigonid relative to the talonid. The talonid possesses a tall sharp entoconid this is deeply separated from the hypoconulid. The hypoconulid and hypoconid are deeply worn.

m4-UMNH VP 14140 (Fig. 15.10K, L) is a Lm4? (AP = 1.66, LBa = 0.97, LBp = 0.90). This is tentatively considered to be an m4 as the talonid is narrower than the trigonid and the talonid is elongated. The trigonid is low relative to the talonid. In Fig. 10K the paraconid looks like it is placed medially, but much of the lingual part of that cusp is either broken or worn. The cristid obliqua originates medially on the posterior the trigonid. The talonid cusps are badly worn but the entoconid appears to have been the tallest cusp prior to wear. The hypoconid is deeply worn. UMNH VP 19166 (Fig. 15.10M, N) is a larger well-preserved Lm4 (AP = 2.01, LBa = 1.26, LBp = 0.96). The paraconid is slightly lower than the metaconid and has a sharp anterior keel. The precingulid is broad. The protoconid is distinctly taller than the metaconid. The cristid oblique originates at the protocristid notch. The hypoconulid is recumbent (typical of m4s), and is closely twinned with the entoconid and the two cusps are not deeply separated; both cusps are taller than the hypoconid.

STAGODONTIDAE Marsh, 1889b
EODELPHIS Matthew, 1916
EODELPHIS sp.

Referred Specimens UMNH VP 18767 (BRCA 7677), Loc. 424, Rmx; UMNH VP 18768 (BRCA 7678), Loc. 424, Rmx (trigonid) (Fig. 15.11A).

Description and Discussion mx-UMNH VP 18767 (Fig. 15.11A, LBp = 2.24) is the talonid of a large lower molar. The

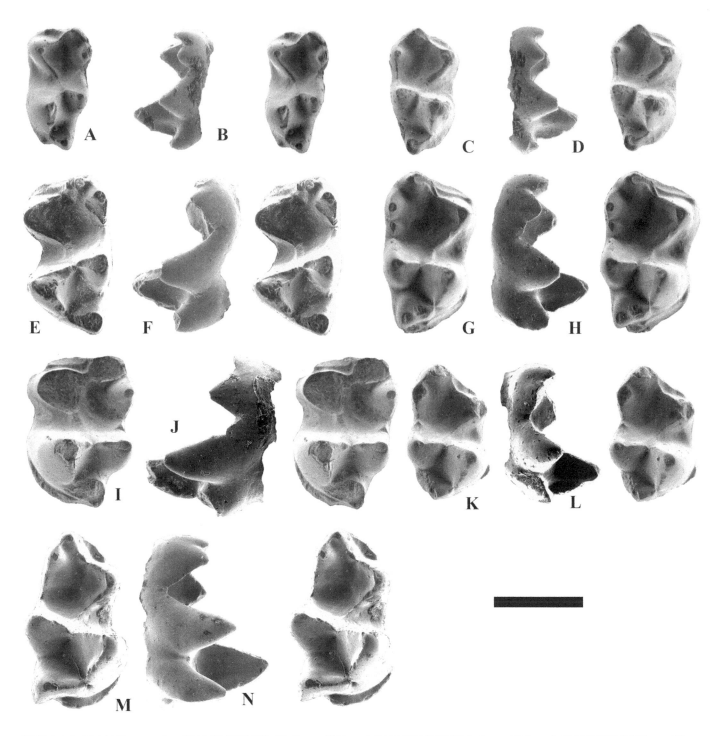

15.10. Indet. didelphid lower molars: (A) UMNH VP 14241, RDp3, osp; (B) lingual; (C) UMNH VP 17269, Lm1, osp; (D) lingual; (E) UMNH VP 19159, osp; (F) lingual; (G) UMNH VP 14146, Lm2, osp; (H) lingual; (I) UMNH VP 17419, Lm3, osp; (J) lingual; (K) UMNH VP 14140, Lm4, osp; (L) lingual; (M) UMNH VP 19166, osp; (N) lingual. Scale bar = 1 mm.

entoconid is the tallest cusp of the talonid and it is deeply separated from the distinctly lower and smaller hypoconulid. There is a distinct cusp anterior of the entoconid. The hypoconid is about the same height as the hypoconulid. The talonid basin does not form a deep central pit but rather bears a distinct anteroposterior crease along the lingual part of the basin. UMNH VP 18768 is a damaged trigonid of a right lower molar (LBa = 2.10) with only the protoconid complete and is included here on the basis of its size. This talonid is somewhat smaller than the mx (UA 5330; W = 2.5) reported by Fox (1971b) as *Eodelphis* sp. from the Milk River Formation and is close in size to *Eodelphis* sp. from the "Oldman" Formation (Fox, 1981:table 4).

Eaton

Table 15.6. Measurements of indet. didelphid lower molars from the John Henry Member of the Straight Cliffs Formation

Lower ms	Specimen no.	AP	Lba	LBp
Dp3	UMNH VP 14241	1.40	0.67	0.72
m1	UMNH VP 17269	1.46	0.76	0.84
	UMNH VP 17373	1.43	0.76	0.77
	UMNH VP 17225	1.47	0.76	0.78
	UMNH VP 17269	1.48	0.74	0.84
	UMNH VP 14331	1.62	0.85	0.91
	UMNH VP 14146			0.85
	UMNH VP 19159	1.77	0.96	0.98
m2	UMNH VP 14416	1.78	1.06	1.13
	UMNH VP 17235	1.67	0.89	0.95
	UMNH VP 17275	1.74	0.90	0.97
	UMNH VP 14247	1.69	1.00	0.98
	UMNH VP 14127	1.47	0.98	1.02
	UMNH VP 17419	1.83	1.12	1.16
m4	UMNH VP 14140	1.66	0.97	0.90
	UMNH VP 19166	2.01	1.26	0.96

PEDIOMYIDAE Simpson, 1927
Genus et sp. indet.

Referred Specimens UMNH VP 17356 (BRCA 7967), Loc. 424, LDP3 (Fig. 15.11B).

Description DP3-UMNH VP 17356 (Fig. 15.11B) is a LDP3 (AP = 2.48; LB = 2.01). This tooth is considered to be a DP3 based on DP3s figured in Clemens (1966:figs. 28, 30), although this premolar appears to be well rooted. The AP dimension far exceeds the LB dimension. There is a distinct and sharp parastyle. The paracone is almost at the labial margin of the stylar shelf. A small cuspule (a remnant of a stylocone?) is positioned posterolabial of the paracone. A larger stylar cusp (stylar cusp C) is positioned between the paracone and metacone. There is a smaller stylar cusp (D?) labial of the metacone. The paracone is taller than the metacone. The paraconule is tall and pointed and no lateral cristae are evident. The metaconule is more weakly developed than the paraconule and does possess a premetaconular crista. The trigon basin is deep and the protocone is anteroposteriorly broad.

Discussion DP3s are not figured or discussed in Davis (2007), but the extreme narrowing of the stylar shelf anteriorly and the anteroposteriorly broadened protocone with small conules suggests inclusion within the Pediomyidae. It is too large to belong to the small ?*Leptalestes* described below and must represent another pediomyid in the fauna.

?*LEPTALESTES* sp. Davis, 2007

Referred Specimens UMNH VP 14139 (BRCA 7215), Loc. 424, RM1?; UMNH VP 14145, Loc. 424, RM1? (Fig. 15.11C).

Description Mx-UMNH VP 14139 (Fig. 15.11C) is a RM1? (AP = 1.50, LBa = 1.25, LBp = 1.33). This is one of the rare pediomyid upper molars in the sample, but unfortunately it is not well preserved. There is a distinct parastyle. The paracone is erect and extends almost to the labial margin of the molar. There is no stylocone. The ectoflexus is weakly developed and there is a small stylar cusp D. The metacone leans lingually and although worn appears to have been lower than the paracone prior to wear. The para- and metaconule are present, but the cristae adjacent to the trigon basin were either weakly developed or obliterated by wear. UMNH VP 14145, a RM1?, is too incomplete to measure and is similar to UMNH 14139, but is also in poor condition.

Discussion Fox (1971b) named three pediomyids from the Milk River Formation, *Pediomys exiguus*, *Aquiladelphis incus*, and *A. minor*. In 1987 Fox indicated that *Pediomys exiquus* was probably not a pediomyid and Davis (2007) erected a new genus, *Apistodon*, outside of the Pediomyidae for this taxon. Davis (2007) also erected a new family (Aquiladelphidae) within the Pediomyoidea and the specimens described here clearly are not part of that family. The specimens described here fall within the family Pediomyidae as defined by Davis (2007) except for not having a strong postmetacrista, although a postmetacrista is present. Davis (2007) included *Pediomys elegans*, *Leptalestes* ssp., and *Protolambda* ssp. in the Pediomyidae. The specimens described here are clearly unlike *Pediomys elegans*, and are also unlike those referred to as *Protolambda* in being much smaller, lacking the extreme lingual shift of the parastyle, and a V-shaped centrocrista. These molars are similar to those of *Protolambda* in lacking stylar cusp C and having a small stylar cusp D. In overall characteristics, the molars described here are closest to those of *Leptalestes* in their small size, having the protocone and metacone well connected at their bases, having small conules, and having no stylar cusp C and a reduced stylar cusp D. These specimens lack a tall postmetacrista and the protocone is not broadened posteriorly as in figured specimens of the genus. Of described species of *Leptalestes*, these specimens are closest to *L. prokrejcii* in having weak conules and no protoconule cingula, although these specimens are smaller. These molars represent the smallest taxon ever described as a belonging to the Pediomyidae, but are not well enough preserved to establish a new taxon.

Wahweap Formation Multituberculates

Some multituberculates from the area of Bryce Canyon National Park were described previously in Eaton (1993, 2002). These taxa include *Cimolodon similis*, ?*Cimolomys* sp. B., and family, genus, and species unknown all recovered from

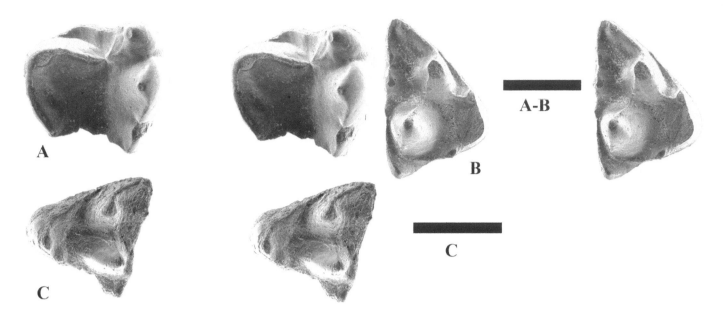

15.11. *Eodelphis* sp.: (A) UMNH VP 18767, Rmx (partial), osp; Pediomyidae gen. et sp. indet.: (B) UMNH VP 17356, LDP3, osp; *?Leptalestes* sp.: (C) UMNH VP 14139, RM1?, osp. Scale bars = 1 mm.

UMNH VP Loc. 77, Bryce Canyon National Park. Other taxa were described from localities in the Wahweap Formation on top of the Paunsaugunt Plateau in Eaton (1993): *Mesodma* sp. cf. *M. formosa*, *Mesodma* sp. cf. *M. hensleighi*, *Mesodma* sp., *Cimolodon* sp. cf. *C. nitidus*, *Cimolodon* sp. cf. *C. foxi* (listed in Eaton, 1993, as *?Cimolodon* sp. and revised in Eaton, 2002), *Cimolomys milliensis*, *Cimexomys gregoryi*, and *Paracimexomys* sp.

CIMOLODONTA McKenna, 1975
Family incertae sedis
?PARACIMEXOMYS sp. Archibald, 1982

Referred Specimens UMNH VP 17972, Loc. 807, Rm1; VP 17309 (BRCA 7913), Loc. 792, LM2 (Fig. 15.12A–B).

Description m1-UMNH VP 17972 (Fig. 15.12A) is a Rm1 with a cusp formula of 5:3 (AP = 1.85*; LB = 1.03*, LB:AP = 0.56). This tooth broke during preparation and when glued backed together the pieces did not fit perfectly resulting in inaccurate measurements. The first cusp of the external row is low and worn. The second cusp is taller and more squared. The third cusp is somewhat narrower and the fourth and fifth cusps are formed by a lingually subdivided posteroexternal wall. There is an indentation on the labial wall of the tooth adjacent to the third cusp. The internal cusps are taller than the external. The first internal cusp is elongate, the second cusp distinctly short, and the third cusp is again elongated. The cusps of the external row are divided labially and lingually.

M2-UMNH VP 17309 (Fig. 15.10B) is a LM2 with a cusp formula of Ri:2:3 (AP = 1.35; LB=1.29).

Discussion UMNH VP 19972 is morphologically closer to the m1 of *Paracimexomys magister* described by Fox (1971a) in that the cusp formula is the same, the middle cusp of the internal row is the shortest, and the LB:AP is similar (0.6, Fox, 1971a:table 1). The m1 of *P. magister* is more ovate (broadens medially) and is much larger than this taxon. There appears to be no pitting or ribbing on the molar, and the poor condition of this molar precludes certain assignment to *Paracimexomys*. *Cedaromys* has a higher LB:AP (Eaton, 2002:table 12, mean LB:AP = 0.64) than does this specimen. UMNH VP 17309 is in within the range of widths for M2s of cf. *Paracimexomys* sp. A from the Wahweap Formation described by Eaton (2002:table 18), this specimen is proportionally broader (LB:AP = 0.96; mean for Wahweap M2s in Eaton, 2002:table 18 = 0.86). The specimen hints at having some ribs and enamel crenulations on the platform and may not be referable to *Paracimexomys*.

CEDAROMYS Eaton and Cifelli, 2001
CEDAROMYS sp. cf. *C.*
HUTCHISONI Eaton, 2002

Referred Specimens UMNH VP 17491 (BRCA 8337), Loc. 807, Lm1 (incomplete); UMNH VP 7664 (BRCA 5930), Loc. 77, Rm2; UMNH VP 17687, Loc. 83, Lm2 (Table 15.3, Fig. 15.12C–D).

Description m1-UMNH VP 17491 (Fig. 15.12C) is an incomplete Lm1 with a cusp formula of 4:3 (AP estimated = 1.6; LB = 0.93). The molar has a waist in occlusal view. The first cusp of the external row is low, small, and essentially conical. It is deeply separated from the much larger second cusp of the row which is weakly pyramidal anteriorly with a concave posterior face. The second cusp is well separated from the third cusp which is weakly pyramidal anteriorly and connected directly to the posterior wall of the tooth which bears the missing fourth cusp. The central valley is slightly sinuous. The first cusp of the internal row is large with three faces and is connected for about half its height with the more pyramidal and subequal in size second cusp. The third cusp is broadly and deeply separated from the second and is pyramidal anteriorly and flattened along the central valley.

m2-UMNH VP 7664 (Fig. 15.12D) is a Rm2 with a cusp formula of ?:2 (AP = 1.19; LB = 1.13, LB:AP = 0.95). This squared m2 has a relatively flat anterior face. The first cusp of the external row is anteroposteriorly compressed. The remaining part of the row is formed by a somewhat weakly cuspate ridge that expands posteriorly creating a broad posterior basin that is continuous with the posterior side of the second cusp on the internal row. The first cusp of the internal row is the largest of the tooth and bears a weak posterolabial crest. Lingual of this crest the posterior of the cusp is slightly concave. The second cusp is low and linguobucally compressed adding to the broadness of the posterior basin. Another m2, UMNH VP 17687 has a cusp formula of ?:2 (AP = 1.28, LB = 1.16, LB:AP = 0.91) and is a somewhat larger tooth than UMNH VP 7664 but appears to be similar. The external row is worn to a ridge and only the suggestion of a distinct first cusp is present. The worn central valley appears straight, but there are hints of some pitting or ribbing posteriorly. The first cusp of the internal row is short, tall, and prominent. The second cusp is lower, anteroposteriorly elongated, and has a slight notch posterior of the cusp.

Discussion UMNH VP 19491 (m1) compares closely in size and morphology with UMNH VP 14149 (Fig. 15.6M) from the uppermost part of the John Henry Member and is therefore conferred to the same species (Table 15.3). UMNH VP 7664 (m2) morphologically compares most closely to the larger m2 described as *Cedaromys* sp. cf. *C. hutchisoni* in Eaton (2006a:fig. 5F), but that specimen is more elongate relative to width (LB:AP = 0.82). This specimen does not compare well with another m2 described by Eaton (2006a, UMNH VP 13653:fig. 4D) as *Cedaromys* sp. cf. *C. hutchisoni*. UMNH VP 7664 has a crest-like labial cusp row, whereas UMNH VP 13653 has distinct and bulbous cusp; the first cusp of the internal row in UMNH VP 7664 is larger than the second but in UMNH VP 13653 both cusps are subequal in size; and, the central basin of UMNH VP 7664 is open and deep but on UMNH VP 13653 ribs cross a relatively narrow basin. UMNH VP 7664 is closer to the morphology of *C. bestia* (Eaton and Cifelli, 2001:fig. 12F). It is unlikely that both of these molars could represent *Cedaromys*, and one of them must be inappropriately referred

to the genus. Part of the problem arises from the lack of m2s described from the original sample of *Cedaromys hutchisoni* from the overlying Kaiparowits Formation (Eaton, 2002). The original description of the genus *Cedaromys* is from the much older Cedar Mountain Formation (Eaton and Cifelli, 2001) and the m2s from the Cedar Mountain Formation tend to have broad posterior basins (Eaton and Cifelli, 2001:fig. 12 F, 13C, 13D). For this reason, this m2 is tentatively assigned to *Cedaromys* sp. cf. *C. hutchisoni*.

?PTILODONTOIDEA Sloan and Van Valen, 1965
NEOPLAGIAULACIDAE Ameghino, 1890
MESODMA Jepsen, 1940
MESODMA sp. cf. *M. MINOR* Eaton, 2002

Referred Specimen UMNH VP 6794, Loc. 61, RM1 (Table 15.1, Fig. 15.12E).

Description M1-UMNH VP 6794 (Fig. 15.12E) is a RM1 with a cusp formula of 4:6:4 (AP = 2.09, LB = 1.29, LB:AP = 0.62). The internal cusp row is broad. It bears two closely positioned blade-like cusps anteriorly, then, posterior of a break in the crest, two closely adjoined smaller cusps. The internal row joins the main body of the tooth posterior of the third cusp of the medial row. The cusps of the medial row become taller, larger, and better separated posteriorly. The first two cusps of the medial are small and weakly separated. The next three cusps lean anteriorly, and the sixth cusp is upright and anteroposteriorly compressed. There is a small cuspule attached to the anterior ridge of the tooth just before the external row, and is not counted as an external cusp in the formula. The cusps of the external row are well separated and formed by essentially conical cusps sheared along their lingual side.

Discussion This M1 has fewer cusps in the external row than do most specimens of *Mesodma minor* from the Kaiparowits Formation, which it is close to in size (Table 15.1) but is somewhat broader relative to length as the internal cusp row is distinctly more expanded than on previously described M1s of *M. minor* (and much more so than the similar sized *M. hensleighi*). Of described specimens, this M1 is closest to UMNH VP 5110, Loc. 10, from Cedar Canyon that was identified as *Mesodma* sp. cf. *M. minor* (Eaton, 2006b). It is also similar to UMNH VP 14144 (Fig. 15.4F) from the John Henry Member described earlier in the chapter as *Mesodma* sp. cf. *M. minor*, but UMNH VP 6794 does not compare as well to other M1s from the John Henry Member conferred to that species. This specimen, as well as the specimens from UMNH VP Loc. 10 in Cedar Canyon, may represent a new species of *Mesodma* that has not been observed from the Wahweap Formation on the Kaiparowits Plateau.

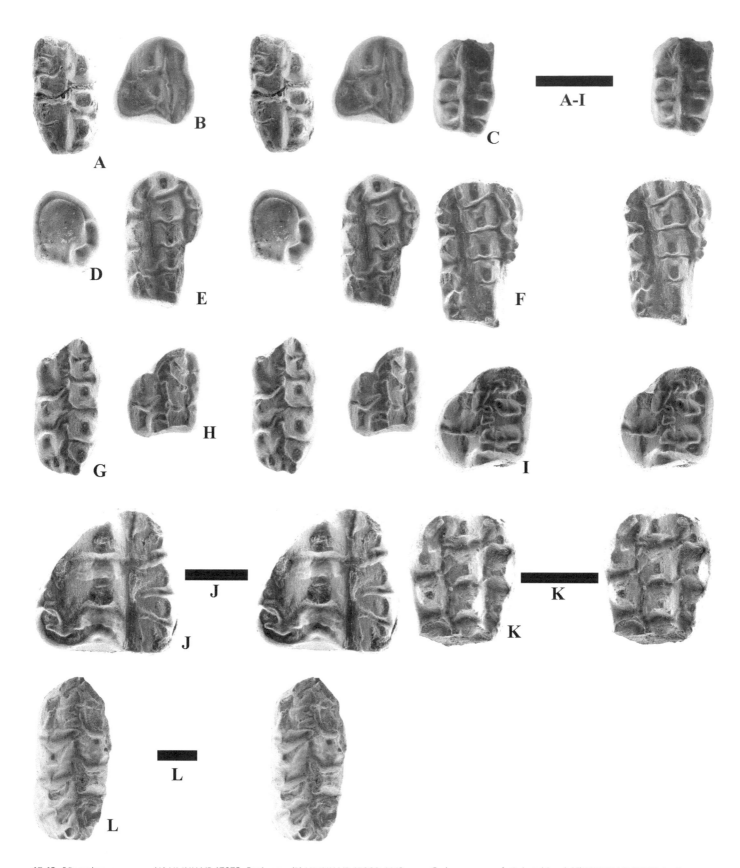

15.12. *?Paracimexomys* sp.: (A) UMNH VP 17972, Rm1, osp; (B) UMNH VP 17903, LM2, osp; *Cedaromys* sp. cf. *C. hutchisoni*: (C) UMNH VP 17491, Lm1 (partial), osp; (D) UMNH VP 7664, Rm2, osp; *Mesodma* sp. cf. *M. minor*: (E) UMNH VP 6794, RM1, osp; *Mesodma* sp. cf. *M. archibaldi*: (F) UMNH VP 6798, LM1, osp; *Mesodma* sp. cf. *M. formosa*: (G) UMNH VP 7744, Lm1, osp; (H) UMNH VP 14273, Lm2, osp. *Cimolodon* sp. cf. *C. foxi*: (I) UMNH VP 18995, Lm2, osp; *Cimolomys* sp.: (J) UMNH VP 17975, LM2, osp; *?Cimolomys* sp.: (K) UMNH VP 17690, LM1 (partial), osp; *Meniscoessus* sp.: (L) UMNH VP 17492, Lm1, osp. Scale bars = 1 mm.

MESODMA sp. cf. *M. ARCHIBALDI* Eaton, 2002

Referred Specimen UMNH VP 6798, Loc. 61, LM1 (Table 15.1, Fig. 15.12F).

Description M1-UMNH VP 6798 (Fig. 15.12F) is a LM1 with a cusp formula of 5:6:5 (AP = 2.23, LB = 1.40, LB:AP = 0.63). The first two cusps of the external row are closely appressed, there are deep valleys between the second and third and third and fourth cusps. The fifth and sixth cusps are small and only slightly separated. The cusp row maintains an approximately even width from anterior to posterior. The medial cusp row broadens posteriorly. The first two cusps are closely adjoined, the third cusp is not deeply separated from the second. The fourth and fifth cusps are well-separated pyramidal cusps. The fifth cusp extends posteriorly into the valley between the medial and internal cusp rows. The sixth cusp of the medial row is anteroposteriorly compressed and is the tallest cusp of the molar. The internal row terminates at the third cusp of the medial row. The first two cusps of the row are small, low, and closely adjoined. The third, fourth and fifth cusps form a broad ridge on the widely expanded interior row.

Discussion Of described specimens, this specimen is closest to MNA V5291, *Mesodma* sp. cf. *M. archibaldi*, from the Kaiparowits Formation (Eaton, 2002). They have the same cusp formula and are almost identical in size and their LB:AP proportions are identical (0.63, which is high for an M1 of *Mesodma*). The two teeth differ from the other Kaiparowits Formation specimens assigned to *M. archibaldi* in that the interior of the cusp walls are deeply grooved and the internal cusp row extends further anteriorly. The consistent differences in these two specimens, relative to those recovered from the Kaiparowits Formation, may suggest they represent a species other than *M. archibaldi*.

MESODMA cf. *M. FORMOSA* Clemens, 1964

Referred Specimens UMNH VP 7744, Loc. 61, Lm1; UMNH VP 14273 (BRCA 5964), Loc. 77, Lm2; MNA V5802, Loc. 1074, Rm1 (Fig. 15.12G–H).

Description mx-UMNH VP 7744 (Fig. 15.12G) is a Lm1 with a cusp formula of 6:3 (AP = 2.05, LB = 1.02, LB:AP = 0.50). The molar is slightly broader anteriorly than posteriorly and has a slight waist in occlusal view. The first cusp of the external row is small and is counted as a cusp as it shows signs of wear. It is closely positioned with the larger second cusp. The third cusp is much larger, crescentic, and well separated from the second cusp. The fourth cusp is well separated from the third, is subequal to it in height, but is slightly smaller. The fifth cusp is separated for about half its height from the fourth

cusp and is only apically separated from the sixth cusp. The central valley is somewhat sinuous and has a few shallow pits. There is a small labially placed cuspule attached to the much larger first cusp of the internal row. It is not counted as a cusp as it is much smaller than the other cusps of the row and is attached to the base of the first cusp. Despite its diminutive size it has wear on its apex. The first cusp is tall, conical, and recurved posteriorly. The second cusp is similar and deeply separated from the first cusp. The third cusp is well separated from the second, and it is larger and slightly taller than the first and second cusps.

UMNH VP 14273 (Fig. 15.12H) is a Lm2 with a cusp formula of 4:2 (AP = 1.31; LB = 1.15; LB:AP = 0.88). The cusps of the external row were only separated slightly at their apices, and with slight wear, as in this specimen, form a continuous ridge which becomes progressively taller posteriorly. The cusps are separated lingually by deep grooves. The central valley is essentially straight but intersected by grooves and ridges. The first cusp of the internal row is taller than the anterior part of the external row. The cusp on its lingual side and bears a deep vertical groove labially. The second cusp of the row is slightly larger and taller than the first cusp. The two cusps are well separated.

Discussion UMNH VP 7744 is too large to represent an m1 of *M. archibaldi*, *M. hensleighi*, or *M. minor*. It is at the very bottom size range of *M. formosa* (Eaton, 2002:table 2), and the tiny anterior cusp on the internal row is similar to that seen on *M. formosa* (Lillegraven, 1969:21) indicating that this is a taxon close *M. formosa*, but having one fewer cusp in the internal row. This specimen does not represent the same taxon as the M1 (UMNH VP 6794) of *M.* sp. cf. *M. minor* described above, as the M1s of all species of *Mesodma* are significantly longer than their m1s. This specimen is smaller, has fewer cusps, and is narrower in proportion to its length than the m1 (UMNH VP 7989:fig. 18A, Eaton, 2002) described as *M.* sp. cf. *M. formosa* from the Wahweap Formation on the Kaiparowits Plateau. A smaller m1 (MNA V5802), from a locality in the same area that UMNH VP 7744 was recovered, was described as *Mesodma* sp. in Eaton (1993) and it is in the upper size range of *M. hensleighi* but the specimen morphologically closer to *M. formosa* and is considered here to represent the same species as UMNH VP 7744.

The m2:m1 AP ratio of UMNH VP 14273 and UMNH VP 7744 is 0.64, slightly higher than the for *M. formosa* (Eaton, 2002:table 2). UMNH VP 14273 compares well in size and morphology with the m2s from the Wahweap Formation assigned to *Mesodma* sp. cf. *M. formosa* in Eaton, 2002 (e.g., MNA V4571:fig. 18E:table 13). It is unlikely that this represents the later "Edmontonian"–Lancian taxon *M. formosa*, but material is inadequate to diagnose a new species.

CIMOLODONTIDAE Marsh, 1889b
CIMOLODON Marsh, 1889b
CIMOLODON sp. cf. *C. FOXI* Eaton, 2002

Referred Specimens UMNH VP 18995, Loc. 83, Lm2; UMNH VP 7649, Loc. 61, Rm2 (Fig. 15.12I).

Description m2-UMNH VP 18995 (Fig. 15.12I) is a Lm2 with a cusp formula of 4:2 (AP = 1.62; LB = 1.49; LB:AP = 0.92). The external cusp row is deeply divided lingually, but not labially. The first three cusps are approximately of equal size, the fourth has been deeply worn by wear or transport. The central valley is deeply pitted. The first cusp of the internal row is the tallest cusp of the tooth and the second cusp of the row is about the same size and only slightly lower. UMNH VP 7649 (cusp formula 4:4, AP = 1.57; LB = 1.46; LB:AP = 0.93) is an m2 very similar to UMNH VP 18995 but is not as well preserved.

Discussion UMNH VP 18995 is close both in size and proportion to OMNH 20492, *Cimolodon foxi* (Eaton, 2002:table 5), from the Kaiparowits Formation, but is also morphologically similar to MNA V5267 (Eaton, 2002:fig. 7E), *Mesodma* sp. (large) also from the Kaiparowits Formation, but MNA V5267 is proportionally more elongate relative to width (Eaton, 2002:table 2) than UMNH VP 18995. The complex pitting present on MNA V5267 suggests it should also have been assigned to a small species of *Cimolodon* rather than to a large species of *Mesodma*. Two slightly larger m2s (MNA V6413, V6453 from the same locality as UMNH VP 18995) described in Eaton (1993) as *?Cimolodon* sp. are deeply pitted and probably represent the same species. Another small m2 (MNA V4537) of *Cimolodon* from the Wahweap Formation was described in Eaton (2002:fig. 22C:table 14).

These specimens suggest a species of *Cimolodon* close to *C. foxi* is present in the Wahweap Formation, but it is essential to have other elements of the dentition to be assured these represent that species.

CIMOLOMYIDAE Marsh, 1889b
CIMOLOMYS Marsh, 1889b
CIMOLOMYS sp.

Referred Specimen UMNH VP 17975 (BRCA 7640), Loc. 807, LM2 (Fig. 15.12J).

Description M2-UMNH VP 17975 (Fig. 15.12J) is a LM2 with a cusp formula of 2?:3:4 (AP = 2.75, LB = 2.82). Two distinct small cusps can be identified on the external ridge. The anterior ridge is broad and worn and is counted as the first cusp of the central row. The second cusp is a large anterior leaning pyramidal cusp. The third cusp is much smaller but is taller than the second. The central valley is open and straight. The first cusp of the interior row is lower than

(perhaps as a result of wear) and closely positioned with the second cusp. The separations between the second and third and third and fourth cusps are deeper, but less that half the height of the row.

Discussion UMNH VP 17975 is larger than M2s of *?Cimolomys* sp. B and smaller than *?Cimolomys* sp. C reported from the Wahweap Formation (Eaton, 2002:table 16). The tooth is about at the lower end of the size range of *Cimolomys clarki* (Lillegraven and McKenna, 1986:table 3) but this tooth is broader than long and is not deeply pitted or grooved as the specimens from the "Mesaverde" Formation (Lillegraven and McKenna, 1986:fig. 7D). Fox (1971a) assigned some specimens from the Milk River Formation to *?Cimolomys*, but unfortunately there were no M2s in his sample. An M2 of *Cimolomys* from the Cedar Canyon was described in Eaton (2006b) but it is significantly smaller than this specimen. The described M2 closest to this specimen in size and morphology is UMNH VP 12859 (Fig. 6G) from the John Henry Member described earlier is this chapter as *Cimolomys* sp. B. That specimen is slightly smaller and the amount of wear on it restricts detailed comparison of the morphology of the two specimens.

?CIMOLOMYS sp.

Referred Specimen UMNH VP 17690, Loc. 83, LM1 (posterior part).

Description M1-UMNH VP 17690 (Fig. 15.12K) is the posterior part of a LM1 (LB = 1.74). The second to last cusp of the external row is weakly pyramidal, erect and worn lingually. The second and third to last cusps of the medial row are pyramidal, lean anterior, and have concave anterior faces. The cusps of the medial row are broader then the cusps of the external row. The cusps of the internal row are low and with apices well spread out. The last cusp is low and small, weakly triangular in form, and connected to the posterolingual corner of the second to last cusp of the medial row. The second to last cusp is much larger and broader than the last cusp, but it retains the triangular form. The next anterior cusp is smaller, lower, and is probably the last cusp of the internal row.

Discussion The spacing of the cusps of the internal row, the small posteriormost cusp of the row connected to the second to the last cusp of the medial row, and the shapes of the cusps of the medial and external rows are most similar to the M1s of *?Cimolomys butleria* from the Kaiparowits Formation (MNA V5226, Eaton, 2002:fig. 13H), which is slightly broader than this specimen (LB = 1.87, Eaton, 2002:table 8); however, because that molar is incomplete, it cannot be referred to that species with certainty.

MENISCOESSUS Cope, 1882
MENISCOESSUS sp.

Referred Specimens UMNH VP 17492 (BRCA 8338), Loc. 807, Lm1; UMNH VP 7743, Loc. 6, RM1 (partial).

Description m1-UMNH VP 17492 (Fig. 15.12L) is a Lm1 with a cusp formula of 5?:4 (AP = 4.97). The labial side of this molar is damaged, with only the third cusp well preserved. The cusp is identical in form to the opposite cusp on the lingual cusp row. The central valley is crossed by ribs and is deeply pitted. There is a small conical cuspule at the labial base of the first cusp of the lingual row. The first cusp is the smallest of the row, and it is not deeply separated from the second cusp. The cusp is almost conical in form with a worn posterolabially directed crest. The second cusp is an anteroposteriorly compressed pyramidal cusp with a well-worn posterolabial crest. The cusp has vertical crenulations in the enamel on the lingual side. The third cusp is a squared pyramidal cusp with much less wear on the posterolabial corner than the first two cusps of the row. The last cusp of the row is anteroposteriorly elongated and slightly compressed linguobuccally. It has distinct ridges on its labial face that cross the central valley.

M1-UMNH VP 7743 is a badly worn posterior half of a RM1 (LB = 2.52). The external cusp row is narrower than the medial, and the internal row bows lingually and had at least six cusps.

Discussion UMNH VP 17492 is smaller than m1s of *Meniscoessus major*, and might possibly be in the lower part of the size range of *M. intermedius*, but this m1 has a much lower cusp formula than that of *M. intermedius* (7:5; Fox, 1976). It is possible that this m1 might represent *M. ferox* Fox, 1971a, but the m1 of *M. ferox* is unknown and the dental proportions for the m1 of *M. ferox* would have to be much different than for m1s of *M. robustus* in which the m1 is almost twice as long as the P4 (Clemens, 1964:table 10). The AP dimension for the P4 of *M. ferox* is 4.5 mm, only slightly smaller than UMNH VP 17492, while the m1 of *M. ferox* would be expected to be much larger. In terms of the size and erect pyramidal cusps this specimen compares most closely with OMNH 20507 (cusp formula 6:4, AP = 4.67), *M. sp. cf. M. intermedius*, from the Kaiparowits Formation described in Eaton (2002:fig. 11E). This suggests the presence of a small species of *Meniscoessus* in the Campanian of southern Utah, as does the M1 UMNH VP 7743.

UMNH VP 7743 is narrower than all described M1s of *Meniscoessus*, but it is closest in size to MNA V5281 (LB = 2.63), *M. sp. cf. M. intermedius* from the Wahweap Formation (Eaton, 2002:fig. 22G) and UMNH VP 5101 (LB = 2.79), *M. sp. cf. M. intermedius*, from locality 11 in Cedar Canyon (Eaton, 2006b:fig. 11A). The spacing of the valleys on this M1

is appropriate for the occlusion of the cusp rows of the m1 described above.

MARSUPIALIA Illiger, 1811

Some marsupial taxa have been previously described by Eaton (1993) from exposures of the Wahweap Formation exposed on top of the Paunsaugunt Plateau, just west of Bryce Canyon National Park. These were described prior to Johanson's (1996) revision of *Alphadon* and Davis's (2007) revision of pediomyid-like marsupials. The MNA specimens are redescribed and figured here to reflect more recent taxonomy.

Order and Family incertae sedis
IUGOMORTIFERUM Cifelli, 1990b
cf. IUGOMORTIFERUM sp.

Referred Specimens UMNH VP 18994, Loc. 83, LMX; UMNH VP 6793, Loc. 83, Rm1? (Fig. 15.13A, B).

Description UMNH VP 18994 (Fig. 15.13A) (AP = 2.31, LBa = 2.39, LBp = 2.37) is similar to the M1? described as *Iugomortiferum thoringtoni* in Cifelli (1990b) in size, triangular form, inflated cusps, weakly developed conules (although wear in this area makes this character difficult to assess), and no extension of the postmetaconular crista labial of the base of the metacone. The paracone on this specimen was destroyed in preparation, but like *I. thoringtoni* it was significantly more robust than the metacone. The ectoflexus is shallow as in *Iugomortiferum*. Stylar cusp B is the tallest cusp of the stylar cusps but is not as inflated as on the specimen described by Cifelli (1990b).

m1-UMNH VP 6793 (Fig. 15.13B) is probably an m1 (AP = 1.54, LBa = 0.73, LBp = 0.79). The apices of all of the trigonid cusps are deeply worn, but the tooth seems to belong to a taxon with low trigonid cusps. The base of the paraconid is larger than the base of the metaconid and is forward projecting. The anterior of the paraconid is deeply worn but there is no evidence of a precingulid. The cristid obliqua originates labial of the notch in the metacristid. The entoconid and the hypoconulid are closely twinned with the entoconid being the tallest cusp of the worn talonid. The hypoconid is deeply worn and its original height relative to the entoconid cannot be estimated. The postcristid is deeply worn but not well developed.

Discussion UMNH VP 18994 differs significantly from *I. thoringtoni* in not having a stylar cusp C (also unlike *Turgidodon* in this characteristic) but having a stylar cusp D. *Iugomortiferum* lacks a stylar cusp A, but the presence or absence of that cusp is difficult to determine as a result of damage on this specimen. If a stylar cusp A was present, then this tooth is certainly incorrectly conferred to *Iugomortiferum*.

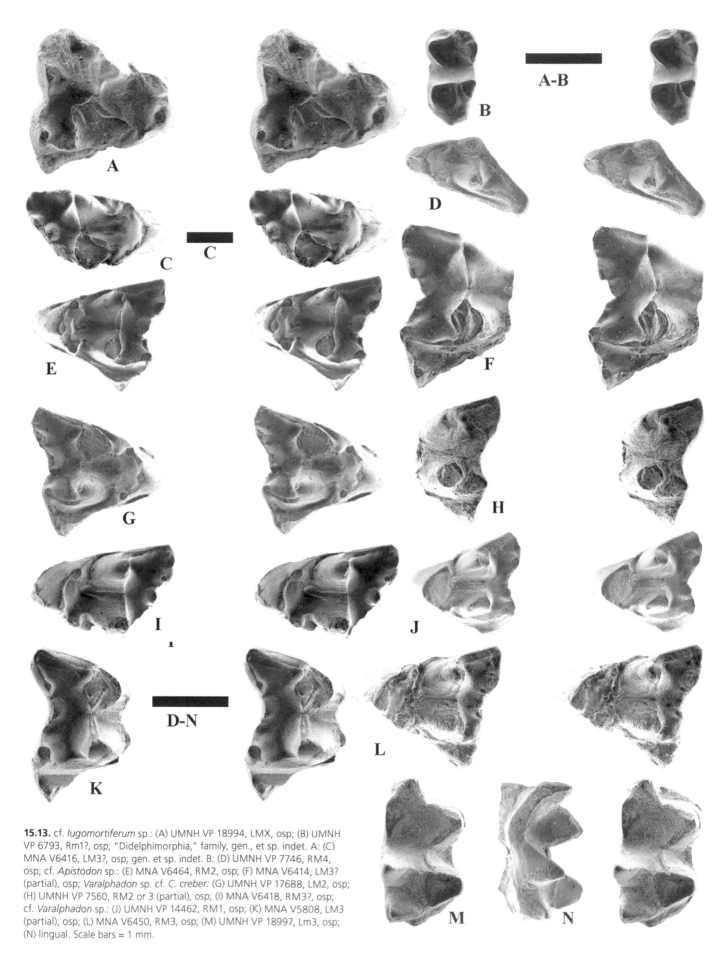

15.13. cf. *Iugomortiferum* sp.: (A) UMNH VP 18994, LMX, osp; (B) UMNH VP 6793, Rm1?, osp; "Didelphimorphia," family, gen., et sp. indet. A: (C) MNA V6416, LM3?, osp; gen. et sp. indet. B: (D) UMNH VP 7746, RM4, osp; cf. *Apistodon* sp.: (E) MNA V6464, RM2, osp; (F) MNA V6414, LM3? (partial), osp; *Varalphadon* sp. cf. *C. creber:* (G) UMNH VP 17688, LM2, osp; (H) UMNH VP 7560, RM2 or 3 (partial), osp; (I) MNA V6418, RM3?, osp; cf. *Varalphadon* sp.: (J) UMNH VP 14462, RM1, osp; (K) MNA V5808, LM3 (partial), osp; (L) MNA V6450, RM3, osp; (M) UMNH VP 18997, Lm3, osp; (N) lingual. Scale bars = 1 mm.

Eaton

UMNH VP 6793 is similar to, but smaller than, OMNH 20928 described as *Iugomortiferum thoringtoni* in Cifelli (1990b:fig. 3b–d). It has the low cusps, the labial origination of the cristid obliqua, and weak cristids of that specimen. The specimen is tentative considered to represent *Iugomortiferum* but it certainly represents a smaller species than UMNH VP 18994.

"DIDELPHOMORPHIA" Gill, 1872
Family incertae sedis, gen. et sp. indet. A

Referred Specimen MNA V6416, Loc. 1073, LM3?

Description MNA V6416 (Fig. 15.13C) is a LM3? with the anterolabial part of this molar missing and as such the relative height of the paracone and metacone cannot be assessed (LBP = 3.12). The conular cristae are moderately developed. Stylar cusps C and D are both small with C being slightly larger than D. Stylar cusp C is positioned posterior of the ectoflexus.

Discussion This specimen was originally assigned to *Alphadon* sp. cf. *A. russelli* in Eaton (1993) based largely on its size. The rediagnosis of *Alphadon* by Johanson (1996) clearly indicates this specimen lacks the inflated cusps of that genus. This molar does not appear to represent *Varalphadon* as a result of the posterior position of stylar cusp C, and the conular cristae are not as well developed as on *Apistodon*. The specimen is larger than described species of *Varalphadon* or *Apistodon* but lacks the inflated cusps of either *Alphadon* or *Turgidodon*. The specimen may represent a new taxon, but is too incomplete to provide the basis for a diagnosis.

Genus et sp. indet. B

Referred Specimen UMNH VP 7746, Loc. 110, RM4.

Description M4-UMNH VP 7746 (Fig. 15.13D) has a dominant stylar cusp A that is connected to a strong preparacrista (AP = 1.19, LBa = 1.92). There is no ectoflexus. A small cuspule is present at the middle of the stylar shelf (stylar cusp C?). The paracone is tall and well formed. The metacone is a small low cusp formed at the posterolabial corner of the tooth. There is a weak paraconule with cristae, and the metaconule is weakly developed, with no cristae evident. The protocone is low and worn lingually. The margins of the molar are straight, giving it a triangular appearance.

Discussion M4s have been rarely associated with more anterior molars or figured (e.g., Johanson, 1996, figures no M4s). This M4 is similar to the M4 (UMNH VP 17274, Fig. 15.7M) attributed above to *Apistodon* from the John Henry Member, but differs significantly in having a narrower stylar shelf, a stylar cusp in the C rather than D position, and in the complete absence of an ectoflexus. The tooth is too small and delicately constructed to represent either *Varalphadon*,

Turgidodon, *Alphadon*, or *Iugomortiferum*. This M4 is closest in size to the small M4, OMNH 20034 (AP = 1.16, measured from cast), described by Cifelli (1990c:fig. 3C) as ?*Anchistodelphys delicatus* (the genus was synonymized with *Varalphadon* by Johanson, 1996). OMNH 20374, another M4 Cifelli (1990c:fig. 3B) assigned to ?*A. delicatus*, has a relatively straight labial margin with no distinct ectoflexus similar to the UMNH specimen. However, both specimens assigned to ?*A. delicatus* have a distinct anterior projection of the parastylar area lacking on the UMNH specimen.

cf. *APISTODON* sp. Davis, 2007

Referred Specimens MNA V6464, Loc. 1073, RM2; MNA V6414, Loc. 1073, incomplete RM3? (Fig. 15.13E–F)

Description Mx-MNA V6464 (Fig. 15.13E) is a RM2 (AP = 1.78; LBa = 1.94; LBp = 2.00) with a damaged stylar cusp A region and a broken paracone. Stylar cusp B is the tallest and largest stylar cusp. There is a posteriorly oriented ridge on stylar cusp B that bears a small cuspule at its base. Stylar cusp C is positioned slightly posterior of the ectoflexus and is subequal in size to stylar cusp D. Stylar cusp E, if there was one, has become part of the worn metacrista. The broken paracone appears to have been larger than the metacone, but it is not clear if it was taller. The conules have well-developed cristae, and the protoconule is positioned lower than the metaconule. There as short narrow cristae developed on both sides of the AP compressed protocone.

MNA V6414 (Fig. 15.13F; AP = 2.54) is an incomplete LM3? missing the lingual part of the molar and having a damaged paracone. The metacone is about the same width as the paracone in lingual view. There is a well-developed stylar cusp A, stylar cusp B is the largest stylar cusp, there is a cuspule present anterior of the ectoflexus and another small cusp (stylar cusp C?) posterior of the ectoflexus, and stylar cusp D is larger than the cusp interpreted to be C.

Discussion MNA V6464 was originally described as *Alphadon* sp. cf. *A. attaragos* in Eaton (1993) based largely on size but appears to represent *Apistodon* because of its uninflated cusps, well-developed conular cristae, narrow protocone, and the presence of stylar cusp C posterior of the ectoflexus and opposite the centrocrista notch. Stylar cusp D on this specimen appears to be larger than on specimens of *Apistodon* illustrated by Davis (2007:fig. 6A–C). As in all *Apistodon* upper molars (based on figures in Davis, 2007), the ectoflexus is positioned behind the paracone and not centrally along the labial margin of the tooth. However, on this specimen, it cannot be determined whether the paracone is taller than the metacone, a definitive character of *Apistodon*, and this specimen appears to be larger than described species of *Apistodon*, but this is based on only a few measured specimens (data from B. Davis, pers. comm., 2010; Table 15.5), so the specimen is only conferred to *Apistodon* here.

MNA V6414 was originally referred by Eaton (1993) to *Alphadon* sp. cf. *A. russelli* based largely on its size. This molar appears to be a larger version of MNA V6464 described above including an additional cuspule posterior of stylar cusp B, but the tooth is too incomplete to be assigned to *Apistodon* with certainty. If these two specimens do represent *Apistodon*, then there is a species present in the Wahweap Formation that is larger than *A. exiguus* and has differences in the size and arrangement of stylar cusps. The inability to determine the relative heights of the paracone and metacone on these two specimens limits assignment of these specimens to *Apistodon* with certainty.

VARALPHADON Johanson, 1996
VARALPHADON sp. cf. *V. CREBER* (Fox, 1971b)

Referred Specimens UMNH VP 17688, Loc. 83, LM2; UMNH VP 7560 (BRCA 5667), Loc. 77, RM2 or 3; MNA V6418, Loc. 1073, incomplete RM3 (Fig. 15.13G–I).

Description Mx-UMNH VP 17688 (Fig. 15.13G) has a distinct stylar cusp A which rests on top of an anterior projecting flange (AP = 1.99, LBa = 1.93, LBp = 2.09). Stylar cusp B is the tallest stylar cusp. Stylar cusp C is lower and linguobucally compressed such that it forms a blade. Stylar cusp D is small and sharply pointed. The paracone is tall and slender and narrower than the broken metacone. The conules are well developed, close to the apex of the protocone, and have strong cristae. UMNH VP 7560 (Fig. 15.13H; AP = 2.08) is the labial half of an M2 or M3. Stylar cusp B is the largest stylar cusp. There is a small cuspule posterior of the ectoflexus (stylar cusp C?) and another subequal cusp positioned more posteriorly (stylar cusp D?). The paracone (the tip of which was broken during preparation) is delicate (narrower than the metacone), taller than the metacone, and leans labially.

M3?-MNA V6418 (Fig. 15.13I, LBp = 2.25) is a RM3? described as *Alphadon* sp. cf. *A. wilsoni* in Eaton (1993:fig. 7C), but lacks characters of the genus as defined by Johanson (1996). The paracone and metacone are subequal in height, but the metacone is not noticeably broader than the paracone in labial view. Stylar cusp C is positioned in the ectoflexus and is subequal in size to stylar cusp D.

Discussion UMNH VP 17688 and 7560 are considered to represent *Varalphadon* because of having a broader metacone broader than paracone in labial view. UMNH VP 17688 is unusual in having small cingula low on the protocone both anteriorly and posteriorly. MNA V6418 is referred to *Varalphadon* because of the equal height of the uninflated and closely positioned paracone and metacone, and the placement of

stylar cusp C in the ectoflexus, but the equal width of the paracone and metacone is problematic. The ectoflexus is relatively deep as is diagnostic for *Varalphadon creber* relative to the similar sized *V. wahweapensis* (Johanson, 1996) and this specimen is closest to UALVP 29525 (Johanson, 1996:fig. 3F), *V. creber*, of figured specimens but is slightly smaller (Johanson, 1996:152).

cf. *VARALPHADON* sp.

Referred Specimens UMNH VP 14462, Loc. 61, RM1; MNA V5808, Loc. 1073, incomplete LM3; MNA 6450, Loc. 1074, damaged RM3; UMNH VP 18997, Loc. 83, Lm3 (Fig. 15.13J–N).

Description Mx-UMNH VP 14462 (Fig. 15.13J) is a deeply worn RM1 in the region of stylar cusp A (AP = 1.50, LBa = 1.58, LBp = 1.69). Stylar cusp B is worn, but was the largest and tallest stylar cusp. There is a small cuspule anterior of the ectoflexus. The ectoflexus is distinct and there is no stylar cusp present at the ectoflexus. Stylar cusp D is smaller than cusp B and also shows apical wear. The paracone is larger than the metacone but apical wear precludes assessment of relative cusp height. The conules are weakly developed, far from the protocone, and have weak internal cristae. The protoconal region is anteroposteriorly compressed.

MNA V5808 (Fig. 15.13K, AP = 2.44) is an M3 missing the labial portion and the metacone. There is a well-developed stylar cusp A. Stylar cusp B is the largest and tallest stylar cusp, cusp C is low and positioned at the ectoflexus, and cusp D is subequal in size to C. There is no distinct stylar cusp E. The paracone is only slightly taller than stylar cusp B and is probably the same height and not taller than the now missing metacone. The base of the metacone is broader than the paracone in labial view. MNA V6450 (Fig. 15.13L; AP = 2.55, LBa = 2.64*; LBp = 2.82*) is an M3 that was damaged in preparation and did not repair well. The molar is similar to MNA V5808 but larger and with a lower stylar cusp B (possibly as a result of wear). Stylar cusp C is small and positioned in the ectoflexus and cusp D is the larger of the stylar cusp C. The paracone and metacone are subequal in height and in width in lingual view.

mx-All lower molars recovered from the Wahweap Formation of the Bryce Canyon National Park area are either poorly preserved or fragmentary. The most complete lower molar is UMNH VP 18997 (Fig. 15.13M–N) an m3 (AP = 2.25, LBa = 1.30, LBp = 1.30). The metaconid is only slightly larger than the paraconid. The protoconid has been broken. The anterior face of the molar is relatively straight with a broad precingulid. The cristid obliqua originates medially on the posterior wall of the trigonid. The entoconid is broken. The hypoconulid is tall and sharp, and is almost as tall as the hypoconid. There is a broad postcingulid.

Discussion UMNH VP 14462 does have the AP compressed protocone found on M1s of *Varalphadon* (Johanson, 1996:pl. 2A–H). The specimen is slightly smaller than M1s of *V. wahweapensis* or *V. creber* (Johanson, 1996; Cifelli, 1990a:table 2, 1990b:table 2). MNA V5808 and V6450 were referred to *Alphadon* sp. *A. wilsoni* in Eaton (1993), but on the basis of the diagnosis of Johanson (1996), these specimens can no longer be considered to be *Alphadon*. MNA V5808 is likely to represent a large species of *Varalphadon* on the basis of the position of stylar cusp C and the broader metacone than paracone. MNA V6450 is somewhat more problematic. The metacone is not distinctly wider than the paracone as is diagnostic of *Varalphadon* and the stylar shelf appear to be narrower than on the M3s of *Varalphadon* figured in Johanson (1996). On the basis of the incomplete nature of MNA V5808 and the few characteristics of MNA V6450 it seems most appropriate to only confer these two specimens to *Varalphadon*. If these do represent *Varalphadon*, then there is a taxon in the Wahweap Formation that is larger than either *V. wahweapensis* or *V. creber*.

UMNH VP 18997 is close in length and morphology to MNA V5818, a lower molar described in Eaton (1993:fig. 6a) from the Wahweap Formation on the top of the Paunsaugunt Plateau as *Alphadon* sp. cf. *A. wilsoni* (a species considered to be conspecific with *A. marshi* in Johanson, 1996); however, the specimen described here is significantly broader than that specimen. The trigonid is less compressed and lower than *Varalphadon* (=*crebreforme* described by Cifelli, 1990b) from the Wahweap Formation. This specimen is larger than any molar described by Cifelli (1990b:table 3) as *Varalphadon wahweapensis* from the Wahweap Formation but is close in size to OMNH 20597 (AP = 2.26; LBa = 1.26; LBp = 1.34; Cifelli, 1990a:table 2) from the Kaiparowits Formation. However, this specimen has the thumb-like projecting hypoconulid that characterized "*Anchistodelphys*" (Cifelli, 1990b:fig. 5I). Because the synonymy of all specimens included with "*Anchistodelphys*" with *Varalphadon* by Johanson (1996) is uncertain (particularly lower molars), this lower molar is only conferred to *Varalphadon*.

PEDIOMYIDAE Simpson, 1927
Genus et sp. indet.

Referred Specimen UMNH VP 7600 (BRCA 7661), LDP3?, Loc. 77.

Description UMNH VP 7600 (Fig. 15.14A) is a rootless LDP3? with extensive wear (AP = 2.81; LBa = 2.02; LBp = 2.06). The molar is triangular in outline. The anterolabial cingulum is broad, extends anteriorly, and supports a small stylar cusp A. The paracone is much large than the metacone and extends almost to the labial margin of the tooth. Stylar cusp C is positioned between the paracone and metacone and bears a small crista which attaches to the base of the premetacrista. There is a cuspule (D?) posterior of stylar cusp C. The postmetacrista bears a large cusp. A paraconule and metaconule are present and have short and weak internal cristae.

Discussion The tooth is similar to UMNH VP 7600 (UMNH VP 6894, identified as an M1) described in Eaton (2006b, fig.13A) as pediomyid gen. et sp. indet. from UMNH VP Loc. 11 on the Markagunt Plateau. No DP3s or M1s of *Aquiladelphis* have been described, but this specimen (and UMNH VP 6894) are probably smaller than would be expected for DP3s or M1s of known species of that genus.

COMPARISONS TO OTHER
SOUTHERN UTAH FAUNAS

The correlation of units based on mammals across southern Utah is still a work in progress. Possibly correlative faunas (UMNH VP Loc. 10, 11) were described from uncertain stratigraphic units (possible Wahweap equivalent) on the Markagunt Plateau in Eaton (2006b), from the John Henry Member of the Straight Cliffs Formation from the Kaiparowits Plateau (Eaton, 2006a), from the Wahweap Formation of the Kaiparowits Plateau (Cifelli, 1990b; Eaton, 2002; the latter also included a few specimens from UMNH VP Loc. 77, Paunsaugunt Plateau) and from the Kaiparowits Formation of the Kaiparowits Plateau (Cifelli, 1990a; Eaton, 2002). The results are shown in Table 15.7.

Few taxa are shared across these three plateaus and those taxa have been omitted from Table 15.7. Lower first molars of *Cedaromys* close to *C. hutchisoni* were found in almost all areas; however, it appears that m1s of *Cedaromys* have a broad but subtle range of variation all of which are close to the m1s of *C. hutchisoni*. In any event, the analysis yielded nothing of biostratigraphic significance other than the taxon from the John Henry Member and locality 10 on the Markagunt Plateau appears to be the same.

UMNH VP LOC. 10, *Markagunt Plateau*

The fauna from UMNH VP Loc. 10 shares more taxa (5) with the Wahweap fauna from the Paunsaugunt Plateau than with the fauna from the John Henry Member (3). Locality 10 and the Wahweap fauna share *Cedaromys* sp. cf. *hutchisoni*, *Cimolodon similis*, *Cimolodon* sp. cf. *C. nitidus*, *Mesodma* sp. cf. *M. minor*, *Symmetrodontoides foxi* and the genus *Turgidodon*. Although locality 10 has more taxa in common with the fauna from the Wahweap Formation than it does with the fauna from the John Henry Member, the presence of *Picopsis* at locality 10 and several taxa similar to John Henry

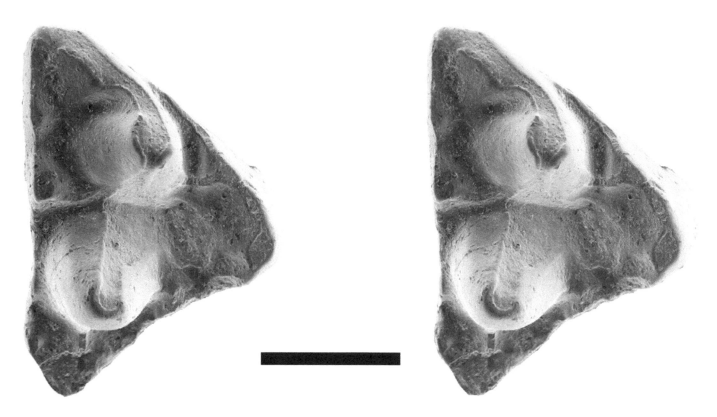

15.14. Pediomyidae, genus and species unknown: UMNH VP 7600, LDP3?, osp. Scale bar = 1 mm.

forms may indicate the locality represents a fauna transitional between John Henry and Wahweap faunas to the east. Locality 10 is a few meters above a possible Drip Tank Member of the Straight Cliffs correlative (Eaton et al., 2001:fig. 5) and may represent an interval of time (Early Campanian) not sampled on the Paunsaugunt or Kaiparowits plateaus.

UMNH VP LOC. 11, *Markagunt Plateau*

Two specimens from locality 11 (UMNH VP 5635, Eaton, 2006b:fig. 12E–F; UMNH VP 7490, Eaton, 2006b:fig. 15.12G) were referred to cf. *Protalphadon* and are not anteroposteriorly compressed, have a straight centrocrista, have a distinct stylar cusp C, the paracone is wider and taller than the metacone and extends further lingually than the metacone, and stylar cusp C is centrally placed in the ectoflexus and is taller than cusp D on both specimens. These specimens do not represent *Protalphadon*, *Varalphadon* or *Apistodon* and probably represent a new genus. Nothing like these molars were found in the Paunsaugunt faunas. "Pediomyidae" gen. et sp. indet. (UMNH VP 6894), described in Eaton (2006b, Fig. 13A) is almost identical to UMNH VP 7600 (described above, Fig. 15.14A) from the Wahweap Formation. This is the only taxon from locality 11 found on another southern Utah plateau. As such, the fauna from locality 11, which is near the top of the Cretaceous section on the Markagunt Plateau (Eaton et

al., 2001:fig. 5), does not compare well with the faunas from either the John Henry Member or the Wahweap Formation of the Paunsaugunt Plateau.

Kaiparowits Plateau Faunas

The fauna from the John Henry Member of the Paunsaugunt Plateau does not compare well to the John Henry Member fauna of the Kaiparowits Plateau (Table 15.7) and only shares two taxa with certainty (there are four other similar taxa). It more strongly resembles the Wahweap fauna from the Paunsaugunt Plateau with which it shares three taxa and possibly three others (Table 15.7). This is unlikely to reflect sampling of significantly different time intervals as most of the described samples on both plateaus come from high in the John Henry Member, and instead may reflect paleoecologic differences. The John Henry Member on the Kaiparowits Plateau consists of drab coastal floodplain deposits and associated coals whereas the upper part of the member on the Paunsaugunt Plateau is well variegated and represents better-drained floodplain deposits.

The fauna from the Wahweap Formation on the Paunsaugunt Plateau was thought to possibly correlate to the Kaiparowits Formation in Eaton (1993) and Eaton, Goldstrand, and Morrow (1993). This was largely due to trying to force marsupial taxa into the existing taxonomy and the lack of

Eaton

Table 15.7. Taxa from the Paunsaugunt Plateau that are also recorded from the Kaiparowits and Markagunt Plateaus

Unit	Paunsaugunt Plateau taxa	Paunsaugunt Plateau		Kaiparowits Plateau		Markagunt Plateau[a]	
		Kw	Kscjh	Kw	Kk	Loc. 10	Loc. 11
Kscjh	*Mesodma* sp. cf. *M. minor*	X	X			X	
Kscjh	*Cimolodon* sp. cf. *foxi*	X	?		?		
Kscjh	*Cimolodon similis*	X		X		X	
Kscjh	*Cimolodon* sp. cf. *C. similis*				?		?
Kscjh	*Cimolomys* sp. A		?				
Kscjh	*Cimolomys* sp. B	?		?			
Kscjh	*Cedaromys* sp. cf. *C. hutchisoni*		X		?	X	
Kscjh	*Symmetrodontoides* sp.	?	?	?			
Kscjh	*Apistodon* sp. cf. *A. exiguus*	?					
Kscjh	?*Varalphadon* sp.		?			?	?
Kscjh	?*"Anchistodelphys"* sp.			?			
Kscjh	Certain co-occurences	3	2	1	0	3	0
Kw	?*Paracimexomys/Parcimexomys* sp.[b]			?			
Kw	*Cedaromys* sp. cf. *hutchisoni*		X			X	
Kw	*Cimolodon similis*			X	?	X	?
Kw	*Cimolodon* sp. cf. *C. nitidus*[b]			?	?	X	
Kw	*Cimolodon* sp. cf. *C. foxi*			?			
Kw	*Mesodma* sp. cf. *M. formosa*[b]			X			
Kw	*Mesodma* sp. cf. *M. archibaldi*				?		
Kw	*Mesodoma* sp. cf. *M. minor*		X			X	
Kw	?*Cimolomys* sp.			X			
Kw	*Meniscoessus* sp.			?			?
Kw	*Symmetrodontoides foxi*[b]			X		X	
Kw	cf. *Iugomortiferum* sp.			?			
Kw	cf. *Apistodon* sp.						?
Kw	*Varalphadon* sp. cf. *V. creber*		?				
Kw	*Turgidodon* sp.[b]				?	?	
Kw	Pediomyidae genus and species unknown						X
Kw	Certain co-occurences	2	4		0	5	1

Kscjh, John Henry Member of the Straight Cliffs Formation; Kw, Wahweap Formation; Kk, Kaiparowits Formation; X, certain shared taxon; ?, possible shared taxon.

[a] UMNH VP Locs. 10 and 11 are from an uncertain unit (possible Wahweap equivalent) on the Markagunt Plateau.

[b] Taxa from the Paunsaugunt Plateau described in Eaton (1993).

shark or ray specimens in the Paunsaugunt samples (shark and ray teeth are common in the Wahweap Formation of the Kaiparowits Plateau, but rare in the overlying Kaiparowits Formation). However, revision of the marsupial fauna based on Johanson (1996) and Davis (2007) and a more complete description of the multituberculates from the Wahweap Formation of the Kaiparowits Plateau (Eaton, 2002) indicates instead that the Wahweap Formation on both plateaus correlate well to each other (Table 15.7) and not to the Kaiparowits Formation.

The fauna from the Wahweap Formation on the Paunsaugunt Plateau shares no certain taxa with the fauna from the Kaiparowits Formation (Table 15.7). Roberts, Deino, and Chan (2005) consider the age range of the Kaiparowits Formation to be 76.1–74 Ma (early Late Campanian based on $^{40}Ar/^{39}Ar$ dates on bentonites. Jinnah et al. (2009) have reported a $^{40}Ar/^{39}Ar$ date from a bentonite about 40 m above the base of the Wahweap Formation of 80.1 ± 0.3 Ma (an early–middle Campanian age based on Ogg, Agterberg, and Gradstein, 2004). This would suggest at least a 3 million year gap between the fauna from the John Henry Member and that of the Wahweap Formation. On the basis of a lower Kaiparowits date of 75.96 ± 0.14 Ma and estimates of depositional rates it is likely that the upper limit on the Wahweap Formation is 76.1 Ma (Roberts, Deino, and Chan, 2005), earliest Late Campanian. This would indicate the age of most of the Wahweap Formation is mostly Middle Campanian. Assuming the Wahweap Formation represents several million years of deposition there would be little time missing between the Wahweap and Kaiparowits formations. Jinnah et al. (2009) suggest the Wahweap Formation is at least in part coeval with the type-Judithian Judith River Formation.

However, the fauna from the Wahweap Formation is much more similar to that of the John Henry Member than it is to the Kaiparowits Formation and is more similar to the fauna of the Aquilan than to the Judithian (based on the original definitions of Lillegraven and McKenna, 1986; see discussion in Cifelli et al., 2004; it is recognized that these definitions are in serious need of revision). This suggests one of these possibilities: (1) the Wahweap Formation is older than the radiometric date of Jinnah et al. (2009) indicates; (2) the Wahweap Formation was deposited rapidly and there was a substantial hiatus between Wahweap and Kaiparowits deposition; or (3) some marked change in environmental conditions resulted in a major faunal shift. Data are currently not available to resolve this problem.

ACKNOWLEDGMENTS

Bryce Canyon National Park, and in particular L. Schrage, has been extremely helpful with permitting, funding, and logistics within the park. Most of the fossils in this report were recovered as part of a 5-year inventory of paleontological resources funded by the Colorado Plateau Ecosystems Study Unit (CESU) of the National Park Service, and the considerable help from and patience of Dr. J. Bischoff is appreciated. The USDA Forest Service has been very helpful throughout this project. I am indebted to the patience and hard work of my reviewers, A. Titus and B. Davis, who toiled to help me improve this chapter. I must also acknowledge the hard work of many Weber State University students, both in the field and in the lab, who have contributed greatly to this project.

REFERENCES CITED

Ameghino, F. 1890. Los plagiaulácidos Argentinos y sus relaciones zoológicas, geológicas y geográficas. Boletin del Instituto Geográfico Argentino 11:143–208.

Archibald, J. D. 1982. A study of Mammalia and geology across the Cretaceous–Tertiary boundary in Garfield County, Montana. University of California Publications in Geological Sciences 122:1–286.

Cifelli, R. L. 1990a. Cretaceous mammals of southern Utah. I. Marsupials from the Kaiparowits Formation. Journal of Vertebrate Paleontology 10:295–319.

Cifelli, R. L. 1990b. Cretaceous mammals of southern Utah. II. Marsupials and marsupial-like mammals from the Wahweap Formation. Journal of Vertebrate Paleontology 10:320–331.

Cifelli, R. L. 1990c. Cretaceous mammals of southern Utah. III. Therian mammals from the Turonian (early Late Cretaceous). Journal of Vertebrate Paleontology 10:332–345.

Cifelli, R. L., and C. L. Gordon. 1999. Symmetrodonts from the Late Cretaceous of southern Utah and the distribution of archaic mammals in the Cretaceous of North America. Brigham Young University Geology Studies 44:1–15.

Cifelli, R. L., and S. K. Madsen. 1986. An Upper Cretaceous symmetrodont (Mammalia) from southern Utah. Journal of Vertebrate Paleontology 6:258–263.

Cifelli, R. L., and S. K. Madsen. 1999. Spalacotheriid symmetrodonts (Mammalia) from the medial Cretaceous (upper Albian or lower Cenomanian) Mussentuchit local fauna, Cedar Mountain Formation, Utah, U.S.A. Geodiversitas 21:167–213.

Cifelli, R. L., J. D. Gardner, R. L. Nydam, and D. L. Brinkman. 1997. Additions to the vertebrate fauna of the Antlers Formation (Lower Cretaceous), southeastern Oklahoma. Oklahoma Geology Notes, Oklahoma Geological Survey 75:124–131.

Cifelli, R. L., T. R. Lipka, C. R. Schaff, and T. B. Rowe. 1999. First Early Cretaceous mammal from the eastern seaboard of the United States. Journal of Vertebrate Paleontology 19:199–203.

Cifelli, R. L., J. J. Eberle, D. L. Lofgren, J. A. Lillegraven, and W. A. Clemens. 2004. Mammalian biochronology of the latest Cretaceous; pp. 21–42 in M. O. Woodburne (ed.), Late Cretaceous and Cenozoic Mammals of North America: Biostratigraphy and Geochronology. Columbia University Press, New York.

Clemens, W. A. 1964. Fossil mammals of the type Lance Formation, Wyoming. Part I. Introduction and Multituberculata. University of California Publications in Geological Sciences 48:1–105.

Clemens, W. A. 1966. Fossil mammals of the type Lance Formation, Wyoming. Part II. Marsupialia. University of California Publications in Geological Sciences 62:1–122.

Cope, E. D. 1882. Mammalia in the Laramie Formation. American Naturalist 16:830–831.

Cope, E. D. 1884. The Tertiary Marsupialia. American Naturalist 18:686–697.

Davis, B. M. 2007. A revision of "pediomyid" marsupials from the Late Cretaceous of North America. Acta Palaeontologica Polonica 52:217–256.

Eaton, J. G. 1991. Biostratigraphic framework for the Upper Cretaceous rocks of the Kaiparowits Plateau, southern Utah; pp. 47–63 in J. D. Nations and J. G. Eaton (eds.), Stratigraphy, Depositional Environments, and Sedimentary Tectonics of the Western Margin, Cretaceous Western Interior Seaway. Geological Society of America Special Paper 260.

Eaton, J. G. 1993. Mammalian paleontology and correlation of uppermost Cretaceous rocks of the Paunsaugunt Plateau, Utah; pp. 163–180 in M. Morales (ed.), Aspects of Mesozoic Geology and Paleontology of the Colorado Plateau. Museum of Northern Arizona Bulletin 59.

Eaton, J. G. 1995. Cenomanian and Turonian (early Late Cretaceous) multituberculate mammals from southwestern Utah. Journal of Vertebrate Paleontology 15:761–784.

Eaton, J. G. 2002. Multituberculate mammals from the Wahweap (Campanian, Aquilan) and Kaiparowits (Campanian, Judithian) formations, within and near the Grand Staircase–Escalante National Monument, southern Utah. Utah Geological Survey Miscellaneous Publication 02–4.

Eaton, J. G. 2004. New screen-washing approaches to biostratigraphy and paleoecology, Cretaceous of southwestern Utah; pp. 21–30 in M. R. Dawson and J. A. Lillegraven (eds.), Fanfare for an Uncommon Paleontologist: Papers on Vertebrate Evolution in honor of Malcolm C. McKenna. Carnegie Museum of Natural History Bulletin 36.

Eaton, J. G. 2006a. Santonian (Late Cretaceous) Mammals from the John Henry Member of the Straight Cliffs Formation, Grand Staircase–Escalante National Monument, Utah. Journal of Vertebrate Paleontology 26:446–460.

Eaton, J. G. 2006b. Late Cretaceous mammals from Cedar Canyon, southwestern Utah; pp. 373–402 in S. G. Lucas and R. M. Sullivan (eds.), Late Cretaceous vertebrates from the Western Interior. New Mexico Museum of Natural History and Science Bulletin 35.

Eaton, J. G. 2009. Cenomanian (Late Cretaceous) Mammals from Cedar Canyon, southwestern Utah, and a Revision of Cenomanian Alphadon-like Marsupials; pp. 97–110 in L. B. Albright III (ed.), Papers on Geology, Paleontology, and Biostratigraphy in Honor of Michael O. Woodburne. Museum of Northern Arizona Bulletin 65.

Eaton, J. G., and R. L. Cifelli. 2001. Additional multituberculate mammals from near the Early–Late Cretaceous boundary, Cedar Mountain Formation, San Rafael Swell, Utah. Acta Palaeontologica Polonica 46:453–518.

Eaton, J. G., and M. E. Nelson. 1991. Multituberculate mammals from the Lower Cretaceous Cedar Mountain Formation, San Rafael Swell, Utah. Contributions to Geology, University of Wyoming 29:1–12.

Eaton, J. G., P. M. Goldstrand, and J. Morrow. 1993. Composition and stratigraphic interpretation of Cretaceous strata on the Paunsaugunt Plateau, Utah; pp. 153–162 in M.

Morales (ed.), Aspects of Mesozoic Geology and Paleontology of the Colorado Plateau. Museum of Northern Arizona Bulletin 59.

Eaton, J. G., J. Laurin, J. I. Kirkland, N. E. Tibert, R. M. Leckie, B. B. Sageman, P. M. Goldstrand, D. W. Moore, A. W. Straub, W. A. Cobban, and J. D. Dalebout. 2001. Cretaceous and Early Tertiary Geology of Cedar and Parowan canyons, western Markagunt Plateau, Utah; pp. 337–363 in M. C. Erskine, J. E. Faulds, J. M. Bartley, and P. D. Rowley (eds.), The Geologic Transition, High Plateaus to Great Basin – A Symposium and Field Guide. Utah Geological Association Publication 30.

Fox, R. C. 1969. Studies of Late Cretaceous vertebrates. III. A triconodont mammal from Alberta. Canadian Journal of Zoology 47:1253–1256.

Fox, R. C. 1971a. Early Campanian multituberculates (Mammalia: Allotheria) from the upper Milk River Formation, Alberta. Canadian Journal of Earth Sciences 8:916–938.

Fox, R. C. 1971b. Marsupial mammals from the early Campanian Milk River Formation, Alberta, Canada. Zoological Journal Linnean Society 50 (Supplement 1):145–164.

Fox, R. C. 1976. Additions to the mammalian local fauna from the Upper Milk River Formation (Upper Cretaceous), Alberta. Canadian Journal of Earth Sciences 13:1105–1118.

Fox, R. C. 1981. Mammals from the Upper Cretaceous Oldman Formation, Alberta. V. *Eodelphis* Matthew, and the evolution of the Stagodontidae (Marsupialia). Canadian Journal of Earth Sciences 18:350–365.

Fox, R. C. 1987. An ancestral marsupial and its implications for early marsupial evolution; pp. 101–105 in P. J. Currie and E. H. Koster (eds.), Fourth Symposium on Mesozoic Terrestrial Evolution, Short Papers. Occasional Paper of the Tyrrell Museum of Paleontology 3.

Gill, T. N. 1872. Arrangement of the families of mammals. With analytical tables. Smithsonian Miscellaneous Collections 11:1–98.

Hunter, J. P., R. E. Heinrich, and D. B. Weishample. 2010. Mammals from the St. Mary River Formation (Upper Cretaceous), Montana. Journal of Vertebrate Paleontology 30:885–898.

Illiger, C. 1811. Prodrodomus systematic mammalium et avium additis terminis zoographicis ultriusque classis. C. Salfeld, Berlin.

Jepsen, G. L. 1940. Paleocene fauna from the Polecat Bench Formation, Park County, Montana. Proceedings of the American Philosophical Society 83:217–341.

Jinnah, Z. A., E. M. Roberts, A. L. Deino, J. S. Larsen, P. K. Link, and C. M. Fanning. 2009. New ^{40}Ar-^{39}Ar and detrital zircon U-Pb ages from the Upper Cretaceous Wahweap and Kaiparowits formations on the Kaiparowits Plateau, Utah: implications for regional correlation, provenance, and biostratigraphy. Cretaceous Research 30:287–299.

Johanson, Z. 1996. Revision of the Late Cretaceous North American marsupial genus *Alphadon*. Palaeontographica Abt. A 242:127–184.

Kennedy, W. J., and W. A. Cobban. 1991. Coniacian Ammonite Faunas from the United States Western Interior. Special Papers in Palaeontology 45.

Landman, N. H., and W. A. Cobban. 2007. Redescription of the Late Cretaceous (late Santonian) ammonite *Desmoscaphites bassleri* Reeside, 1927, from the Western Interior of North America. Rocky Mountain Geology 42:67–94.

Lillegraven, J. A. 1969. Latest Cretaceous mammals of the upper part of Edmonton Formation of Alberta, Canada, and review of marsupial–placental dichotomy in mammalian evolution. Paleontological Contributions, University of Kansas 50(Vertebrata 12):1–122.

Lillegraven, J. A., and M. C. McKenna. 1986. Fossil mammals from the "Mesaverde" Formation (Late Cretaceous, Judithian) of the Bighorn and Wind River basins, Wyoming, with definitions of Late Cretaceous North American Land-Mammal "Ages." American Museum Novitates 2840:1–68.

Marsh, O. C. 1887. American Jurassic mammals. American Journal of Science 33:326–348.

Marsh, O. C. 1889a. Discovery of Cretaceous Mammalia. Part II. American Journal of Science 38:81–92.

Marsh, O. C. 1889b. Discovery of Cretaceous Mammalia. Part III. American Journal of Science 38:177–180.

Marshall, L. G., J. A. Case, and M. O. Woodburne. 1990. Phylogenetic relationships of the families of marsupials. Current Mammalogy 2:433–505.

Matthew, W. D. 1916. A marsupial from the Belly River Cretaceous. Bulletin of the American Museum of Natural History 35:477–500.

McKenna, M. C. 1975. Towards a phylogenetic classification of the Mammalia; pp. 21–46 in W. P. Luckett and F. S. Szalay (eds.), Phylogeny of Primates. Plenum Press, New York.

Montellano, M., A. Weil, and W. A. Clemens. 2000. An exceptional specimen of *Cimexomys judithae* (Mammalia: Multituberculata) from the Campanian Two Medicine Formation of Montana, and phylogenetic status of *Cimexomys*. Journal of Vertebrate Paleontology 20:333–340.

Ogg, J. G., F. P. Agterberg, and F. M. Gradstein. 2004. The Cretaceous period; pp. 344–383 in F. M. Gradstein, J. G. Ogg, and A. G. Smith (eds.), A Geologic Time Scale 2004. Cambridge University Press, Cambridge.

Osborn, H. F. 1888. On the structure and classification of the Mesozoic mammals. Journal of the Academy of Natural Sciences, Philadelphia 59:186–265.

Parker, T. J., and W. A. Haswell. 1897. A Text-book of Zoology. Volume 2. Macmillan, London.

Peterson, F. 1969. Four New Members of the Upper Cretaceous Straight Cliffs Formation in Southeastern Kaiparowits Region, Kane County, Utah. U.S. Geological Survey Bulletin 1274-J.

Roberts, E. M., A. L. Deino, and M. A. Chan. 2005. ^{40}Ar-^{39}Ar age of the Kaiparowits Formation, southern Utah, and correlation of contemporaneous Campanian strata and vertebrate faunas along the margin of the Western Interior Basin. Cretaceous Research 26:307–318.

Sahni, A. 1972. The vertebrate fauna of the Judith River Formation, Montana. American Museum of Natural History Bulletin 147:321–412.

Simpson, G. G. 1927. Mesozoic Mammalia: VIII. Genera of Lance mammals other than multituberculates. American Journal of Science 14:121–130.

Sloan, R. E., and L. Van Valen. 1965. Cretaceous mammals from Montana. Science 148:220–227.

Walaszczyk, I., and W. A. Cobban. 2006. Paleontology and biostratigraphy of the upper Middle–Upper Coniacian–Santonian inoceramids of the U.S. Western Interior. Acta Geologica Polonica 56:241–348.

Lizards and Snakes from the Cenomanian through Campanian of Southern Utah: Filling the Gap in the Fossil Record of Squamata from the Late Cretaceous of the Western Interior of North America

16

Randall L. Nydam

A RICH AND DIVERSE FAUNA OF LIZARDS AND SNAKES has been recovered from the Cenomanian–Campanian of southern Utah. The specimens reported herein represent eight new taxa, 19 named taxa, and 34 distinct, unnamed morphotypes from the Dakota Formation (Cenomanian), Smoky Hollow Member of the Straight Cliffs Formation (Turonian), Iron Springs Formation (Cenomanian–Santonian), John Henry Member of the Straight Cliffs Formation (Coniacian), Wahweap Formation (Santonian–early Campanian), and Kaiparowits Formation (mid-Campanian). Although each fauna is taxonomically and morphologically distinct (although only a very few taxa are currently known from the Iron Springs and Wahweap formations), most of them share the presence of Chamopsiidae, Polyglyphanodontinae, Contogeniidae, cf. Xenosauridae, Platynota, and Serpentes. Additionally, many taxa and/or unnamed morphotypes from the reported faunas are referable to Paramacellodidae, a taxon with its origins in the Jurassic. The presence of jaws referable to two distinct species of *Odaxosaurus* in the Kaiparowits Formation represents the first record of two co-occurring species of *Odaxosaurus* in the Western Interior and the earliest conclusive evidence of anguids in southern Utah, although anguid-grade osteoderms are known from as early as the Cenomanian. The overall composition of the described faunas represents a very consistent taxonomic/morphologic diversity with the greatest change being a substantial increase in chamopsiid and platynotan diversity by the mid-Campanian. Only a small sample of squamate specimens has been recovered from the Santonian–early Campanian, and additional investigation of these horizons is still necessary.

INTRODUCTION

The Cretaceous-age sedimentary rocks of southern Utah preserve a nearly continuous sequence of fossil-rich, terrestrially deposited mudstones and sandstones that span the mid-Cenomanian through the mid-Campanian (Eaton, 1991). Much of this temporal range is not represented by vertebrate fossil–bearing strata elsewhere in the Western Interior, making the fossil-bearing exposures in Utah's Iron, Garfield, Kane, and Washington counties uniquely important both geologically and paleontologically. Unlike most other Cretaceous-age terrestrial fossil localities of the Western Interior, the initial work on vertebrate fossils from southern Utah were studies of mammals and associated microvertebrates instead of macrovertebrates (e.g., dinosaurs). These early studies were conducted by J. Eaton and R. Cifelli and resulted in a series of reports on the mammalian faunas (Cifelli, 1990a, 1990b, 1990c, 1990d, 1990e; Eaton, 1993a, 1993b, 1995). These reports were then followed by series of papers listing the identified mammalian and nonmammalian vertebrates collected during the microvertebrate sampling (Eaton, 1999a, 1999b; Eaton et al., 1999a, 1999b), which are valuable as the initial documentation of the overall faunal richness of Cretaceous of southern Utah. More recently, there have been a growing number of comprehensive analyses (e.g., systematic, taxonomic, paleobiogeographical) of specific nonmammalian microvertebrate taxa including chondrichthyeans (Fults, 2009), Lissamphibia (Gardner, 2000), and Squamata (McCord, 1998; Nydam, 1999; Nydam and Voci, 2007; Nydam, Eaton, and Sankey, 2007; Nydam and Fitzpatrick, 2009).

My own work on the squamates from the Cretaceous of southern Utah has been an ever-evolving project, a result of the continually increasing number of specimens recovered by myself and others (e.g., J. Eaton and R. Cifelli) from numerous microvertebrate-producing localities in the region (Fig. 16.1). My work on squamate fossils from this region has focused on the systematics, distribution, and evaluation/reevaluation of the evolutionary patterns of specific groups of lizards such as Polyglyphanodontinae (Nydam, Eaton, and Sankey, 2007), Contogeniidae (Nydam and Fitzpatrick, 2009), and Chamopsiidae (Nydam and Voci, 2007; taxonomy here follows Nydam, Caldwell, and Fanti, 2010). However, the taxa from the Cretaceous of southern Utah that are the basis of these studies represent only a fraction of the recognizable squamates from the region. It is the goal of this chapter to describe, illustrate, and discuss the remaining currently known fossil squamate material from the Dakota Formation (Cenomanian), the Smoky Hollow Member of the Straight Cliffs Formation (Turonian), the John Henry Member of the Straight Cliffs Formation (Coniacian–Santonian)–these units will be referred to henceforth as SHMbr SCFm and JHMbr SCFm, respectively–the Iron Springs Formation (Cenomanian–Santonian), the Wahweap Formation (early Campanian), and the Kaiparowits Formation (middle Campanian).

I have two primary goals here: first, to provide a comparison-based description of the recognizable and definable squamate taxa from the Cretaceous of southern Utah, and second, to demonstrate the overall diversity of the squamates by also describing the distinct morphotypes of squamate fossils that are recognizably unique, but represented by insufficiently complete material to warrant formal taxonomic treatment. Some of this work was alluded to in a preliminary report of my research given at the 2006 Learning from the Land symposium celebrating the 10-year anniversary of the Grand Staircase–Escalante National Monument. That symposium was organized by Marietta Eaton, and through her persistent endeavors, those contributions, including an earlier outline of my work on the lizards (Nydam, 2010), were recently published. Additional preliminary information regarding lacertilian from the Kaiparowits Formation was also included in a large-scale biogeographical analysis of paracontemporaneous, freshwater/terrestrial faunas from the Campanian of the Western Interior (Gates et al., 2010). It is my goal herein to provide a more comprehensive account of my research on the lizards from southern Utah. This opportunity is in large part due to the kind invitation by Dr. Alan Titus to participate in the 2009 Learning from the Land special symposium on the Advances in Western Interior Late Cretaceous Paleontology and Geology. Dr. Titus is a true champion of geological and paleontological research in the Grand Staircase–Escalante National Monument and southern Utah in general.

Institutional Abbreviations BMNH, British Museum of Natural History, London; FUB, Lerstuhl für Paläontologi, Frie Universität, Berlin; MG-LNEG, Museu Geológico, Laboratório Nacional de Energia e Geologia, Lisbon; MNA, Museum of Northern Arizona, Flagstaff, Arizona; OMNH, Sam Noble Oklahoma Museum of Natural History, Norman, Oklahoma; UMMZ, University of Michigan Museum of Zoology, Ann Harbor, Michigan; UMNH, Utah Museum of Natural History, Salt Lake City, Utah; USNM, United States

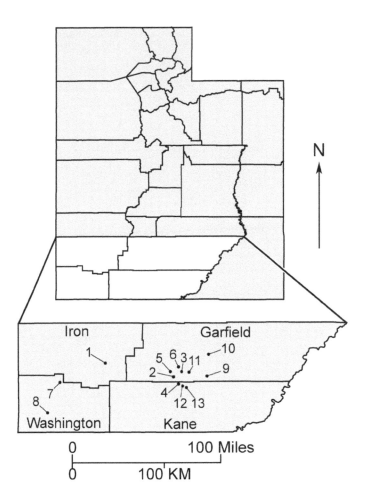

16.1. Squamate producing microvertebrate localities of southern Utah. (1, 2) Dakota Formation: (1) UMNH Loc. 162E; (2) MNA 1067. (3, 4) Smoky Hollow Member of the Straight Cliffs Formation: (3) MNA 995 (=OMNH V843, UMNH Loc. 129); (4) OMNH V1404. (5, 6) John Henry Member of the Straight Cliffs Formation: (5) UMNH Loc. 424 and UMNH Loc. 569 (too close to discriminate at scale of map); (6) OMNH V60 and UMNH Loc. 99 (to close to discriminate at scale of map). (7, 8) Iron Springs Formation: (7) UMNH Loc. 12; (8) UMNH LOC. 416. (9, 10) Wahweap Formation: (9) OMNH V2 and UMNH Loc. 103 (to close to discriminate at scale of map); (10) UMNH Loc. 130. (11–13) Kaiparowits Formation: (11) OMNH V5, V61, and UMNH Loc. 108 (to close to discriminate at scale of map); (12) OMNH V6; (13) OMNH V9; (14) UMNH Loc. 1268.

National Museum, Washington, D.C.; UALVP, University of Alberta Laboratory of Vertebrate Paleontology, Edmonton.

Anatomical Terminology Unless otherwise noted, I follow Oelrich (1956) and Gao and Fox (1996) for most osteological terminology with regard to lizard skull and jaw morphology. Exceptions include following Smith and Dodson (2003) with regard to general orientation of dental structures, Richter (1994; see also Evans and Searle, 2002), and Kosma (2004) for scincomorphan-specific tooth crown terminology (Fig. 16.2), and Klembara (2008) for the skull table. Numerical reference to tooth position is based on counting in situ teeth from anterior to posterior. For descriptions of the vertebral features of snakes, I follow Rage (1984).

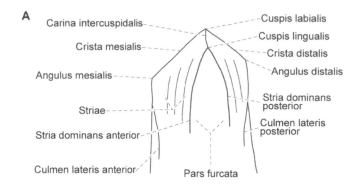

A

Carina intercuspidalis
Crista mesialis
Angulus mesialis
Striae
Stria dominans anterior
Culmen lateris anterior

Cuspis labialis
Cuspis lingualis
Crista distalis
Angulus distalis
Stria dominans posterior
Culmen lateris posterior
Pars furcata

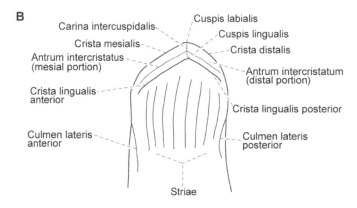

B

Carina intercuspidalis
Crista mesialis
Antrum intercristatus (mesial portion)
Crista lingualis anterior
Culmen lateris anterior

Cuspis labialis
Cuspis lingualis
Crista distalis
Antrum intercristatum (distal portion)
Crista lingualis posterior
Culmen lateris posterior
Striae

16.2. Terminology used in descriptions of scincomorphan teeth. (A) Schematic tooth of a paramacellodid/cordylid-grade lizard (redrawn from Richter, 1994). (B) Schematic tooth of a scincid-grade lizard (redrawn from Kosma, 2004).

GEOLOGICAL SETTING

The Cretaceous-age sediments of southern Utah have been very well described and discussed in numerous other works, some of which appear in this volume. In the interest of avoiding unnecessary repetition, I refer the reader to those works (and their included references) that describe and discuss the Dakota Formation (Eaton, 1991; am Ende, 1991; Dyman et al., 2002; Heller, Dueker, and McMillan, 2003), the Straight Cliffs Formation (Peterson and Waldrop, 1965; Peterson, 1969; Eaton, 1991; Titus, Roberts, and Albright, this volume, Chapter 2), the Wahweap Formation (Peterson and Waldrop, 1965; Eaton, 1991; Jinnah, this volume, Chapter 4), the Iron Springs Formation (Eaton, 1999a; Dyman et al., 2002), and the Kaiparowits Formation (Eaton, 1990, 1991; Roberts, Deino, and Chan, 2005). I will discuss, as relevant, some aspects of the geology as part of the systematic description of the squamates.

All microvertebrate-producing localities referred to in the systematic descriptions are indicated in Fig. 16.1.

Dakota Formation (Cenomanian)

REPTILIA Linnaeus, 1758
SQUAMATA Oppel, 1811
SCINCOMORPHA Camp, 1923
BORIOTEIIOIDEA Nydam,
Eaton, and Sankey, 2007
POLYGLYPHANODONTINAE Estes, 1983
BICUSPIDON Nydam and Cifelli, 2002a
BICUSPIDON SMIKROS sp. nov.

Etymology From the Greek *smikros*, "small," in recognition of the diminutive size of this species in comparison to other species of *Bicuspidon* (Fig. 16.3A–G).

Holotype UMNH VP 11627 (Fig. 16.3A–B).

Paratype Specimens MNA V9086, MNA V10020, MNA V10029, UMNH VP 12900, UMNH VP 12923, UMNH VP 12949, UMNH VP 13001, UMNH VP 13004–13007.

Type Locality UMNH 162E (type locality), Iron County, Utah, approximately 25 km southwest of the town of Parowan; MNA Loc. 1067, Kane County, Utah, approximately 1.5 km west of Cannonville.

Diagnosis Differs from *Bicuspidon numerosus* Nydam and Cifelli, 2002a, in smaller size and in bicuspid teeth with more closely spaced labial and lingual cusps. Differs from *Bicuspidon hatzegiensis* Folie and Codrea, 2005, in lacking apical striae on the tooth crowns and lacking a shelf around the base of the lingual cusp.

Description UMNH VP 11627 (Fig. 16.3A, B) is a partial right dentary with three in situ teeth. The teeth are subpleurodont, conical, become more robust distally, and have heavy deposits of cementum at the bases. The crowns are transversely bicuspid with a tall, prominent labial cusp and a shorter lingual cusp. The cusps are closely spaced and joined by an asymmetrical V-shaped ridge with a longer labial than lingual portion. As is also seen in *B. numerosus*, the smaller lingual cusp is weakly trifid. The subdental shelf is increasingly narrow posteriorly to accommodate the splenial. The subdental gutter is narrow and shallow where preserved. UMNH VP 11627 (Fig. 16.3C) is a fragment of a left dentary with two in situ teeth. UMNH VP 19225 (Fig. 16.3D–E) and UMNH VP 19226 (Fig. 16.3F–G) are jaw fragments (element indeterminate) with a single in situ tooth.

Discussion The presence of *Bicuspidon smikros* in the Dakota Formation extends the range of the genus in Utah from the Albian–Cenomanian boundary (*B. numerosus*) to the mid-Cenomanian. There are two records of *Bicuspidon*

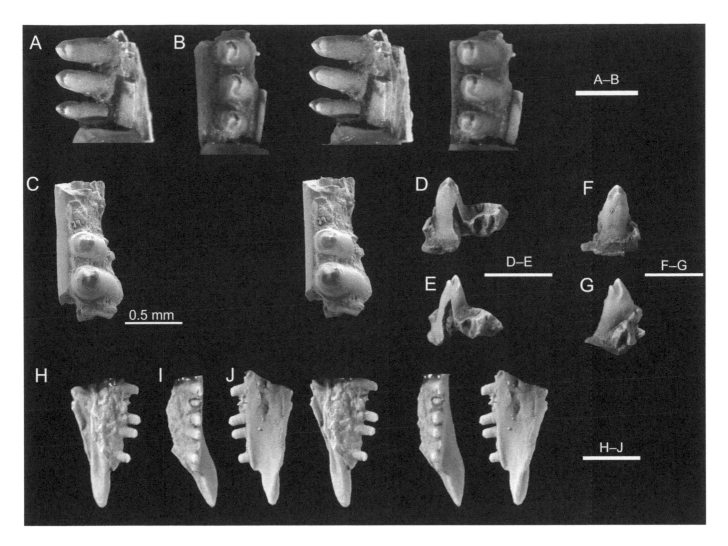

16.3. Borioteiioidea and Contogeniidae from the Dakota Formation (Cenomanian). (A–G) *Bicuspidon smikros* sp. nov.: (A, B) stereopairs of UMNH VP 11627, broken right mandible in (A) lingual, and (B) occlusal views; (C) scanning electron micrograph stereopair of MNA V10029, jaw fragment in occlusal view; (D, E) UMNH VP 19225 in (D) lingual and (E) distal views; (F, G) UMNH VP 19226 in (F) lingual and (G) mesial views. (H–J) *Utahgenys antongai* sp. nov.: stereopairs of UMNH VP 12912, broken left maxilla in (H) lingual; (I) occlusal; and (J) labial views. All scale bars = 1 mm unless otherwise noted.

from Europe: *Bicuspidon* cf. *B. hatzegiensis* from the Santonian of Hungary (Makádi, 2006) and *Bicuspidon hatzegiensis* from the Maastrichtian of Romania (Folie and Codrea, 2005), the latter being the youngest record for the genus. This dramatic temporal and geographic separation raises the question as to the center of origin and subsequent distribution of *Bicuspidon*. Nydam (2002) identified taxonomic evidence supporting the existence of an Asian–North American exchange of squamate taxa (primarily Asia to North America; see also references in Nydam, 2002, discussing previously recognized important evidence for other taxonomic groups). It is possible that *Bicuspidon* originated in Asia and distributed east into North America during the Early Cretaceous as well as west into Europe. Alternatively, the presence of the oldest representative in central Utah suggests a North American origin for *Bicuspidon* with subsequent dispersal

to Europe either (1) east across North America prior to the completion of the midcontinental seaway of the Late Cretaceous and opening of the North Atlantic, or (2) west across Asia and into Europe. Dispersal option 1 would have to have happened no later than the pre-Albian Early Cretaceous, and dispersal option 2 requires the presence of the paleo-Berigean land bridge, which was very likely established during the Aptian–Albian and intermittently through at least the early part of the Late Cretaceous (see discussion in Nydam, 2002). Both scenarios have their difficulties (e.g., migration across a likely European archipelago in both options and the lack of this distinct taxon from the Cretaceous of Asia for option 2) and must await the discovery of additional specimens from either Asia or the Eastern Seaboard of North America that can help support either of the options or indicate another possibility.

SCINCOMORPHA Camp, 1923
CONTOGENIIDAE Nydam and Fitzpatrick, 2009
UTAHGENYS Nydam and Fitzpatrick, 2009
UTAHGENYS ANTONGAI sp. nov.

Etymology For Antonga (also known as Chief Black Hawk), the leader of the Ute Indians in the Utah territory during the 1860s.

Holotype UMNH VP 12912 (Fig. 16.3H–J).

Type Locality UMNH 162E, Iron County, Utah, approximately 25 km southwest of the town of Parowan.

Diagnosis Differs from *Utahgenys evansi* Nydam and Fitzpatrick, 2009, in having tooth crowns not trifid and with less well-developed apical striae.

Description UMNH VP 12912 (Fig. 16.3H, I) is the posterior portion of a left maxilla preserving six tooth spaces, four in situ teeth, and a complete posterior (or jugal) process. The teeth are pleurodont and delicate, and have chisel-like crowns with a mesiodistally oriented apical groove. The tooth bases are obscured by sediment. The labial surface of the element is smooth and pierced by two foramina. The posterior process projects posterolaterally and has well-developed dorsal groove for articulation with the jugal.

Discussion Nydam and Fitzpatrick (2009) recognized the presence of *Utahgenys* in the Dakota Formation, but the material available at the time was insufficient for diagnosis. *Utahgenys antongai* is the oldest representative of a lineage that extends across the K-T boundary and into the middle Paleocene.

cf. SCINCOIDEA Oppel, 1811
PARAMACELLODID/CORDYLID GRADE

This is an informal designation for lizards of indeterminate family-level affinities that share specific dental characteristics identified by Richter (1994; see also Fig. 16.2), Evans and Searle (2002), and extensively illustrated by Kosma (2004). Referral to Scincoidea follows Estes (1983) and Estes, de Quieroz, and Gauthier (1988) in defining Scincoidea as Scincidae + Cordylidae and inferring a close relationship between Cordylidae and Paramacellodidae, but Evans and Chure (1998a, 1998b) recovered a sister taxon relationship between *Paramacellodus* and Scincoidea. The dental characteristics used to identify these morphotypes represent a grade that is common to known paramacellodid lizards, cordylids lizards, and some scincids. Although useful in illustrating the diversity of the lizard fauna based on variations in tooth morphology, I do not think these characteristics are sufficiently diagnostic for the erection of formal taxonomic designations for those specimens that lack supporting nondental morphological data.

DAKOTASAURUS gen. nov.

Etymology For Dakota Formation and the Greek *sauros* (lizard); in recognition of the rock unit from which the type and only know species has been recovered.

Type and Only Known Species *Dakotasaurus gillettorum* sp. nov.

Distribution Dakota Formation (Cenomanian), Garfield County, Utah.

Diagnosis As for type and only known species.

Remarks Referral to a family is not warranted, but recognition of this taxon as being of a paramacellodid/cordylid grade is supported by the shared presence of tooth crown characteristics distinct to paramacellodids as noted by Richter (1994) and Evans and Searle (2002) as well as cordylids (Kosma, 2004). These include the presence of well-developed striae dominans (anterior and posterior), anguli mesialis and distalis, cristae mesialis and distalis, and a pars furcata. Characteristics of dental crown morphology are not recognized as part of the more taxonomically inclusive diagnosis of Paramacellodidae of Estes (1983) or in the more taxonomically exclusive diagnosis of Evans and Chure (1998a). However, the tooth morphology of this taxon (as well as all other "paramacellodids" noted by Evans and Searle, 2002) closely resembles the tooth morphology illustrated by Kosma (2004) of many living gerrhosaurine and cordyline lizards (following the taxonomy of Estes, 1983, and Estes, de Quieroz, and Gauthier, 1988). However, referral to Gerrhosaurinae or Cordylinae is unlikely because the size of the splenial of *Dakotasaurus* is more like that in other paramacellodids in that it extends much further rostrally. More complete material and a comprehensive reevaluation of "paramacellodid" lizards are necessary for a more confident family-level referral.

I do not follow Conrad (2008) in assigning the "paramacellodids" to the Anguimorpha. There is significant evidence for a scincomorphan origin of these lizards (e.g., dental morphology, tooth replacement mode, morphology of the dentary and maxilla). Although paramacellodid monophyly requires additional testing, the placement of these taxa in Anguimorpha is in conflict with significant evidence to the contrary.

DAKOTASAURUS GILLETTORUM
gen. et sp. nov.

Etymology Named in honor of David Gillette and Janet Whitmore-Gillette of the MNA who have supported, encouraged, and contributed much to paleontological research of the Colorado Plateau.

Holotype MNA V9110 (Fig. 16.4A–C).

Hypodigm Type specimen, MNA V10374.

Type Locality MNA Loc. 1067, Kane County, Utah, approximately 1.5 km west of Cannonville.

Diagnosis Differs from species of *Paramacellodus* and *Becklesius* in having a fused dentary and splenial, mesial teeth less robust and with less widely separated anterior and posterior striae dominans than distal teeth. Differs further from *P. keebleri* in tooth crowns lacking a hypertrophied and offset angulus distalis. Differs further from *Becklesius hoffstetteri* in tooth crowns not mesiodistally expanded or as strongly concave lingually. Among Cretaceous-age lizards of the Western Interior it is most similar to *Aocnodromeus corrugatus* Gao and Fox, 1996, but it differs from this taxon in having fused splenial, shorter tooth row, and less prominent lingual striae on tooth crowns.

Description MNA V9110 (Fig. 16.4A–C) is a partial right mandible preserving the posterior half (approximately) of a dentary with eight preserved tooth spaces, seven in situ teeth, and a fused splenial. The specimen is missing the mandibular symphysis but preserves a portion of the postdentary articulations (Fig. 16.4B). The teeth are pleurodont and closely spaced, and they have heavy deposits of cementum at the bases. There are replacement pits of varying levels of development at the bases of the second and sixth preserved teeth, and there is an empty tooth space between the third and fourth preserved teeth. The tooth crowns are labiolingually compressed (Fig. 16.4C), and they have a slightly longer crista mesialis than crista distalis, giving the crowns a weak, false recurved appearance (see Nydam and Cifelli, 2002b, for definition). The stria dominans anterior and stria dominans posterior are best preserved on the first, second, fourth, and sixth preserved teeth and border an increasingly larger pars furcata in the more distally positioned teeth. Well-developed striae are present on the surface of the pars furcata and best preserved on the sixth tooth. The subdental shelf is short, and the narrow sulcus dentalis (subdental gutter) is mostly filled with cementum. Overall, the preserved portion of the mandible is labiolingually compressed with a weakly concave–vertically flat lingual surface and a convex labial surface. Although there are some compression fractures on the labial surface, the overall specimen appears to be only minimally distorted. There is a single preserved external inferior alveolar foramen on the labial surface (Fig. 16.4B) near the anterior end of the specimen. The rest of the labial surface is smooth. The anterior portion of the surangular notch is preserved, but only the proximalmost portions of the coronoid and posteroventral (surangular) processes are preserved. Although incomplete, the coronoid process formed an obviously robust lappet onto the labial surface of the coronoid. Lingually there is a large, subcircular notch between the splenial and the coronoid process for a blunt, tabular anteromedial process of the coronoid. The preserved portion

of the splenial is weakly concave posteriorly and grades to a vertically flat surface rostrally. The anterior mylohyoid foramen is smaller and posteroventral to the anterior inferior alveolar foramen.

MNA V10374 (Fig. 16.4D, E) is a partial left mandible preserving much of the dentary, 11 tooth spaces, seven in situ teeth, and a fused splenial. The tooth implantation, tooth spacing, tooth crowns, and degree of basal cementum (Fig. 16.4D) is nearly identical to that of MNA V9110. The tooth crowns are better preserved than in MNA V9110 and show a very distinct carina intercuspidalis between the tall cuspis labialis and the shorter cuspis lingualis (Fig. 16.4E). The labial surface of the element is convex and pierced by three external inferior alveolar foramina. The anteriormost portion of the subdental shelf that is preserved is expanded lingually, which is consistent with it giving rise to the mandibular symphysis. Posteriorly, the lateral parapet is inflected dorsally, which is consistent with it giving rise to the coronoid process. As such, the 11 preserved tooth spaces represent nearly the entire tooth row.

Discussion In comparison to known "paramacellodids," the presence of a fused splenial is unique to *Dakotasaurus gillettorum*. Even though it is fused, it is possible to compare at least the general size (e.g., length) of the splenial to other relevant taxa. With only a dozen or so teeth in the tooth row, *Dakotasaurus gillettorum* is a relatively short-jawed lizard. As noted in the diagnosis, among other lizards from the Cretaceous of the Western Interior, *D. gillettorum* is most similar *Aocnodromeus corrugatus* from the early Campanian of Alberta (Gao and Fox, 1996). Gao and Fox (1996) tentatively referred this taxon to Scincidae pending a better understanding of the dental and associated jaw characteristics. When compared to the results of a comprehensive survey of fossils and extant scincomorphan dentitions by Kosma (2004), *A. corrugatus* should more appropriately be referred to a paramacellodid/cordylid-grade taxon. If *A. corrugatus* should prove to be a paramacellodid, then it would be one of the youngest representatives of a signature taxon (though monophyly remains in question) of the Laurasian lizard faunas of the Late Jurassic and Early Cretaceous. If a cordylid, then *A. corrugatus* represents one of the earliest examples of a taxon currently restricted to the old world.

Dakota Formation Morphotype A

Referred Specimen UMNH VP 19224 (Fig. 16.4F).

Locality UMNH 162E, Iron County, Utah, approximately 25 km southwest of the town of Parowan.

Description UMNH VP 19224 (Fig. 16.4F) is a jaw fragment (dentary?) with one in situ tooth and one broken tooth base. The tooth is pleurodont, short, mesiodistally broad and

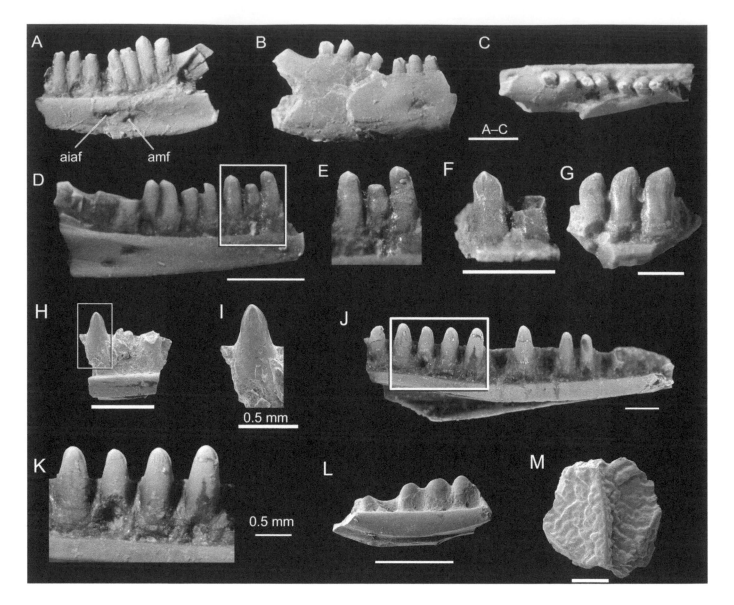

16.4. Scincoidea and Scincomorpha indeterminate from the Dakota Formation (Cenomanian). (A–E) *Dakotasaurus gillettorum* gen. et sp. nov.: (A–C) MNA V9110, broken right mandible in (A) lingual; (B) labial; and (C) occlusal views; (D, E) MNA V10374, broken left mandible in (D) lingual view with box indicating area of detail in (E). (F) Dakota Formation morphotype A: UMNH VP 19224, jaw fragment in lingual view. (G) Dakota Formation morphotype B: UMNH VP 12918, jaw fragment in lingual view. (H, I) Dakota Formation morphotype C: scanning electron micrograph of MNA V10033, jaw fragment in (H) lingual view with box indicating area of detail in (I). (J, K) *Webbsaurus lofgreni* gen. et sp. nov.: MNA V9083, broken left dentary in (J) lingual view with box indicating area of detail in (K). (L) Dakota Formation morphotype D: scanning electron micrograph (SEM) of MNA V10031, jaw fragment in lingual view. (M) cf. Scincomorpha: SEM of MNA V10037, osteoderm, in dorsal view. aiaf, anterior inferior alveolar foramen; amf, anterior mylohyoid foramen. All scale bars = 1 mm unless otherwise noted.

labiolingually compressed, and has moderate deposits of cementum at the base. The tooth crown has well-developed cristae mesialis and distalis that extend down the tooth and turn lingually as a prominent culmen lateris anterior and a culmen lateris posterior. There are very closely spaced striae dominans anterior and posterior descending from a small cuspis lingualis. The pars furcata is restricted and very narrow.

Discussion The dentition of the specimen is of a paramacellodid/cordylid grade, lacks a well-defined angulus mesialis and angulus distalis on the tooth crown. This tooth

morphology is unique among the scincomorphans known from the Dakota Formation, but the specimen is otherwise too incomplete for meaningful comparison and is left as an unnamed morphotype.

Dakota Formation Morphotype B

Referred Specimen UMNH VP 12918 (Fig. 16.4G).

Locality UMNH 162E, Iron County, Utah, approximately 25 km southwest of the town of Parowan; Dakota Formation.

Nydam

Description UMNH VP 12918 (Fig. 16.4G) is a jaw fragment (element uncertain) with three moderately spaced, pleurodont teeth that have distinct, but not heavy, amounts of basal cementum. There are replacement pits at the base of the anterior two teeth (to the left in Fig. 16.4G). The tooth crowns have a much longer crista mesialis than crista distalis resulting in the posterior displacement of the cuspis labialis and a dramatic false recurve appearance to the teeth. This false recurve is exaggerated by a moderate real recurvature to the tooth shafts. The tooth crowns are labiolingually compressed and the angulus mesialis and angulus distalis are rounded and blunt. The cuspis labialis and cuspis lingualis are nearly conjoined and there is no crista intercuspidalis. The striae dominans anterior and posterior descend from the cuspis lingualis and enclose a moderate pars furcata with weakly developed lingual striae on its surface.

Discussion As with Dakota morphotype A, this specimen has a unique dental morphology among scincomorphans of the Dakota Formation, but also lacks sufficient comparable morphology for a more formal diagnosis. As such, UMNH VP 12918 is recognized as second unnamed morphotype from the fauna.

Dakota Formation Morphotype C

Referred Specimen MNA V10033 (Fig. 16.4H, I).

Locality MNA Loc. 1067, Kane County, Utah, approximately 1.5 km west of Cannonville.

Description MNA V10033 (Fig. 16.4H) is a dentary fragment (side uncertain) preserving three tooth positions, but only a single complete tooth. The tooth (Fig. 16.4I) is conical with symmetrical crista mesialis and crista distalis. The angulus mesialis and angulus distalis are absent, but the striae dominans anterior and posterior are well developed and descend from a blunt cuspis lingualis and enclose a smooth pars furcata. Between the crista mesialis and the stria dominans anterior there is a single, weakly developed lingual stria. A similar weakly developed stria is located between the crista distalis and the stria dominans posterior. The preserved portion of the subdental shelf is tall and straight. The sulcus dentalis is very shallow and moderately wide.

Discussion As with Dakota Formation morphotype A, this specimen lacks a well-defined angulus mesialis and angulus distalis on the tooth crown, but is otherwise more similar in tooth structure to Paramacellodidae (or Cordylidae) than to Scincidae.

WEBBSAURUS gen. nov.

Etymology In honor of the Webb Schools and from the Greek *sauros* (lizard); in recognition of all of the paleontology students of the Webb Schools of Claremont, California whose participation in various aspects of fieldwork was instrumental to the success of this study.

Type and Only Known Species *Webbsaurus lofgreni* gen. et sp. nov.

Distribution Dakota Formation (Cenomanian), Garfield County, Utah.

Diagnosis As for type and only known species.

Remarks Referral to of this taxon to either Paramacellodidae or Cordylidae is unlikely but cannot be confidently eliminated as a possibility as a result of the lack of characteristic tooth crown features described by Richter (1994) and Kosma (2004). Most diagnoses of Scincidae (Estes, 1983; Estes, de Quieroz, and Gauthier, 1988) lack characters from the lower jaw and additionally lack all but a few cranial characters. In his attempt at a comprehensive revision of squamate phylogenetics, Conrad (2008) appropriately identified the possible paraphyletic status of Scincidae as a concern and proposed the more inclusive taxon, Scinciformes (=Scincidae + Dibamidae + Cordylidae + Serpentes). Of the diagnostic characters of Scinciformes the only applicable jaw characters are a (1) partly closed, but unfused Meckel's canal, (2) coronoid process short and broad, (3) retroarticular process posteriorly expanded, and (4) replacement teeth occur posterolingually in small replacement pit. My personal observation of many taxa of Scincidae and Cordylidae (sensu lato) is that the rostral portion of the Meckel's canal is open as a narrow groove from the tip of the splenial to the mandibular symphysis and is rarely closed. Likewise, I have observed that in scincid and cordylid tooth rows the replacement teeth are basal and any posterior displacement of the replacement crown is minor and random from position to position, jaw to jaw, and taxon to taxon (anguimorphans and snakes do have consistent posterolingual replacement crowns). With regard to *Webbsaurus*, the coronoid process of the dentary and the retroarticular process of the mandible are both missing from the holotype and only known specimen, but the shape of Meckel's canal is consistent with a referral to Scinciformes.

WEBBSAURUS LOFGRENI gen. et sp. nov.

Etymology Named in recognition of Dr. Don Lofgren of the Raymond Alf Museum and the Webb Schools for all of his many contributions to the advancement of our understanding of the Cretaceous and Paleogene of the Western Interior.

Holotype MNA v9083 (Fig. 16.4J, K).

Type Locality MNA Loc. 1067, Kane County, Utah, approximately 1.5 km west of Cannonville.

Diagnosis Differs from *Paramacellodus*, *Becklesius*, *Dakotasaurus*, and most modern cordylid taxa in having tooth crowns lacking well-defined anguli mesialis and distalis and lacking a sharply pointed cuspis labialis. Among scincids *Webbsaurus lofgreni* is most similar in dental morphology to *Eumeces obsoletus* and *Mabuya multifaciata* in having bluntly pointed teeth and well-developed lingual striae, but *W. lofgreni* differs from both taxa in having more robust, widely spaced, less numerous teeth in the dentary. It further differs from *M. multifaciata* in lacking concave lingual surfaces of the tooth crowns, and from *E. obsoletus* in having a taller subdental shelf with a vertically flat, not rounded, lingual surface. Closest comparison of the tooth crown morphology is to that of the morphotype 1 (Evans and Searle, 2002:pl. 1, fig. 7, text-fig. 3F) from the Sunnydown Farm Quarry in the Purbeck Formation, but *Webbsaurus lofgreni* differs in having more pronounced striae on the surface of the pars furcata and more bluntly pointed apices of the teeth.

Description MNA V9083 (Fig. 16.4J) is a left dentary missing part of the mandibular symphysis and all of the postdentary articulations. There are 14 tooth spaces and eight in situ teeth. The teeth are pleurodont, conical, widely spaced, and with heavy deposits of basal cementum. The teeth become progressively more robust from the mesial to the distal portions of the tooth row. There are replacement pits at the bases of the first, third, fourth, and sixth preserved teeth (counting from mesial to distal). The tooth crowns (Fig. 16.4K) are bluntly tipped, have well-defined cristae mesialis and distalis that join apically to form a blunt cuspis labialis. The crista mesialis is longer than the crista distalis giving the tooth crown a false recurve. On the lingual surface of the tooth crowns there are prominent striae that parallel the cristae mesialis and distalis similar to the cristae lingualis anterior and posterior of scincid teeth (sensu Kosma, 2004), but also extend toward the base of the tooth similar to the striae dominans anterior and posterior of paramacellodid and cordylid teeth (sensu Richter, 1994, and Kosma, 2004). These striae form the borders of a prominent lingual pars furcata with numerous well-developed striae on its surface. The subdental shelf is tall for all of its preserved length with a horizontal anterior half and a posterior half that slopes dorsally. The ventral surface of the posterior portion of the subdental shelf is grooved from the articulation with the splenial. This groove extends anteriorly and ends as a shallow notch where the subdental shelf becomes horizontal, indicating the anterior extent of the splenial. The anteriormost portion of the subdental shelf is expanded for the mandibular symphysis, but the symphysis is not completely preserved. The Meckelian fossa (Meckel's groove) is widely open posteriorly, but is a shallow, parallel-sided groove along the anterior portion of the element. The sulcus dentalis is shallow, narrow, and mostly filled with the cementum surrounding the tooth bases. The labial surface of the dentary is convex, smooth, and pierced by six external inferior alveolar foramina. The anteriormost portion of the lateral parapet is displaced laterally forming a narrow boss along the lateral margin of the anteriormost three tooth positions. It is possible that this feature is an injury or a postmortem distortion, but the retained symmetry of the tooth row suggests that it is a normal feature of this element.

Discussion The overall morphology of the dentition suggests that this is a scincid, but crown morphology seems to be transitional between the paramacellodid/cordylid grade and the scincid grade. Additionally, the relatively large and widely spaced teeth differs from most scincids that tend to have numerous, small, closely spaced teeth.

SCINCOMORPHA indeterminate
Dakota Formation Morphotype D

Referred Specimen MNA V10031 (Fig. 16.4L).

Locality MNA Loc. 1067, Kane County, Utah, approximately 1.5 km west of Cannonville.

Description MNA V10031 (Fig. 16.4L) is a jaw fragment of a right (?) dentary preserving five tooth positions and four teeth. The teeth are low crowned, blunt, and closely spaced with moderate deposits of basal cementum. There are posteriorly displaced resorption pits at the base of the second and fourth preserved teeth. The crowns are unremarkable and possibly eroded. The preserved subdental shelf is tall, the subdental gutter is narrow and shallow. The Meckelian fossa is open, but much of the element is not preserved and it is not possible to determine the fossas' exact nature.

Discussion Although it is possible that the tooth crowns are eroded I consider this unlikely because similar dentition is known from lizard specimens from the Smoky Hollow Member of the Straight Cliffs Formation (Fig. 16.6R, S) and the Kaiparowits Formation (Fig. 16.13U, V). Referral to Scincomorpha is tentative, but supported by the tall subdental shelf and degree of basal cementum along the preserved portion of the tooth row.

?SCINCOMORPHA

Referred Specimen MNA V10037 (Fig. 16.4M).

Locality MNA Loc. 1067, Kane County, Utah, approximately 1.5 km west of Cannonville.

Description MNA V10037 (Fig. 16.4M) is a lizard osteoderm missing much of the periphery. There is a well-developed keel along the midline of the specimen. The ornamentation is weakly pustulate.

Discussion Tentative referral to Scincomorpha is based on the presence of the well-defined central keel. However, the presence of a pustulate surface (uncommon, but not unknown in scincomorphan osteoderms) and the inability to confirm that the osteoderm was rectangular preclude confident referral of this element to Scincomorpha. As such, referral of any of the jaw material is similarly unclear.

ANGUIMORPHA Fürbringer, 1900
ANGUIFORMES Conrad, 2008
aff. XENOSAURIDAE Cope, 1886
CNODONTOSAURUS gen. nov.

Etymology From the Greek *knodon* (sword), *odous* (tooth), and *saurus* (lizard); in recognition of the tall, trenchant teeth of this taxon.

Type and Only Known Species *Cnodontosaurus suchockii*

Distribution Dakota Formation (Cenomanian), Garfield County, Utah.

Diagnosis As for type and only known species.

CNODONTOSAURUS
SUCHOCKII gen. et sp. nov.

Etymology Named for Larry Suchocki, senior laboratory technician, Midwestern University, in recognition of his many hours of patient expertise in recovering microvertebrate specimens from fossil concentrate.

Holotype MNA v9084 (Fig. 16.5A–D).

Hypodigm MNA v10373 (Fig. 16.5E, F).

Type Locality MNA Loc. 1067, Kane County, Utah, approximately 1.5 km west of Cannonville.

Diagnosis Differs from *Xenosaurus* in subdental shelf does not participate in the anterior inferior alveolar foramen; teeth are taller; tooth crowns labiolingually compressed and blade-like, dentary teeth not recurved; tooth replacement is alternating. Differs further from *Xenosaurus* and from *Restes* in lacking an incipient mesial cusp on posterior teeth. Differs further from *Xenosaurus*, *Restes* and *Exostinus* in maxilla lacking fused osteoderms. Differs further from *Exostinus* in teeth lacking replacement pits, intramandibular septum extends further posteriorly, tooth crowns of dentary teeth not recurved, more distally positioned dentary teeth more gracile than mesial teeth. Among anguimorphans dentition most similar to *Dorsetisaurus* in being trenchant and with alternate, interdental tooth replacement, but differs from *Dorsetisaurus* in teeth not conical, tooth crowns more blade-like, tooth crowns not sharply pointed, presence of a sulcus dentalis, well-developed intramandibular septum, and dentary not dorsally concave.

Description MNA v9084 (Fig. 16.5A–D) is a right dentary with 17 tooth spaces and seven in situ teeth. The teeth are pleurodont with tall, labiolingually compressed, blade-like crowns with blunt tips. The teeth near the distal portion of the tooth row are narrower than more mesially positioned teeth. The lingual and labial surfaces of the crowns bear well-developed striae. The base of each tooth is heavily cemented with a well-developed nutrient foramen opening posterolingually just posterior to the midpoint of the base of the tooth. The subdental shelf is present, rounded, and dorsoventrally short. There is a small piece of subdental shelf missing that was lost during preparation, but was similarly smooth and continuous with the preserved portions of the shelf on either side of the gap (Fig. 16.5D). There is no indication of participation of the subdental shelf in the formation of the border of the anterior inferior alveolar foramen as is seen in Anguidae. The sulcus dentalis is present, but very shallow. The articulation facet for the splenial is on the ventral surface of the subdental shelf in the form a long narrow, shallow groove. The mandibular symphysis is not preserved and it is not possible to determine the exact anterior extent of the splenial, but it certainly approached the symphysis. The Meckelian fossa is widely open posteriorly and gradually narrows anteriorly to a narrow, deep groove. The intramandibular septum is fused to the floor of the Meckelian fossa and extends posteriorly to near the end of the preserved portion of the element where it is notched for the articulation with the surangular.

MNA v10373 (Fig. 16.5E, F) is a broken left maxilla with six tooth spaces and four in situ teeth. The teeth are slightly recurved and tips broken, but otherwise are identical to those in MNA v9084. The supradental shelf is very narrow and the labial surface of the maxilla is smooth. A small posterolingual resorption pit is located at the base of the second preserved tooth (from anterior).

Discussion Referral of *Cnodontosaurus suchockii* to Anguiformes (sensu Conrad, 2008, contrary to Conrad, 2006) is supported by the presence of the (1) reduced subdental shelf, (2) posterolingual replacement pit, and (3) basal nutrient foramina of teeth. Although none of these characteristics are currently recognized as diagnostic of Anguiformes, they are common to many of the included taxa to the exclusion of representatives of nonanguiform taxa. Referral to Xenosauridae is less clear (hence the use of aff. instead of cf.) and other family-level alternatives are possible. However, the tooth bases of *Cnodontosaurus suchockii* are not expanded (contra the platynotan characteristic) and are more similar to those of *Dorsetisaurus*, anguids, xenosaurids, and shinisaurids. The presence of a subdental shelf that borders a sulcus dentalis is shared with *Exostinus lancensis*, some, but not all, extant xenosaurids, and possibly the anguid *Barisia imbricata* (e.g., UMMZ 149593). However, *Cnodontosaurus suchockii* lacks the accessory cusp ("shoulder") that Conrad (2008) identifies as one of the diagnostic characters of Xenosauridae. Another fossil xenosaurid, *Exostinus lancensis* has been described as having this dental feature (Gao and Fox, 1996), but my observation of this specimen (UALVP 29838) indicates that the teeth are chipped and/or worn and were in life more like those described herein for *Cnodontosaurus suchockii*.

As noted in the above diagnosis, the dentition of *Cnodontosaurus suchockii* shares some similarities to that of *Dorsetisaurus*, an anguimorphan from the Late Jurassic–Early Cretaceous with a pan-Laurasian distribution. This could be easily attributed to parallel feeding specializations selecting for similar dentitions, but a phylogenetic basis for this similarity is also possible. *Dorsetisaurus*, as with *Paramacellodus*, is known from the Late Jurassic of the Western Interior (Prothero and Estes, 1980; Evans and Chure, 1999) and it is possible that *Cnodontosaurus suchockii* represents a continuation of that lineage that persisted to the early Late Cretaceous. On the basis of the presence of a modestly developed subdental shelf I have proposed a closer affinity with xenosaurs. Additional and more complete specimens are needed for a more decisive resolution of taxonomic status and interrelationships of this taxon.

PLATYNOTA Duméril and Bibron, 1836
Family indeterminate
Dakota Formation Morphotype E

Referred Specimens MNA V9086, UMNH VP 12920 (Fig. 16.5G–I).

Localities MNA Loc. 1067 (specimen MNA V9086), Kane County, Utah, approximately 1.5 km west of Cannonville; UMNH 162E (specimen UMNH VP 12920), Iron County, Utah, approximately 25 km southwest of the town of Parowan.

Description MNA V9086 (Fig. 16.5G, H) is the anterior portion of the left dentary with the base of a single tooth. The tooth base is relatively broad, lacks infolding of the enamel, is surrounded by a small ring of cementum, and has a distinct nutrient foramen that opens posterolingually (Fig. 16.5G). The tooth crown is missing and the hollow pulp chamber is visible in occlusal view (Fig. 16.5H). Plicidentine is absent. There is no subdental shelf and the preserved portion of the Meckelian fossa is narrow and shallow.

UMNH VP 12920 is a jaw fragment (element indeterminate) with a single, nearly complete tooth. The tooth is conical, recurved, has a shallow groove along the mesial margin, and lacks basal infolding of the enamel.

Discussion Both MNA V9086 and UMNH VP 12920 are referable to Platynota on the basis of the morphological features associated with the tooth bases, tooth crown (i.e., UMNH VP 12920), and the lack of a subdental shelf (i.e., MNA V9086). The absence of basal infolding of the enamel as well as plicidentine often confused, but the former is an external dental feature and the latter an internal dental feature indicates

that these specimens are from taxa with a closer affinity to *Colpodontosaurus*, but both of these specimens have thicker-walled teeth and *Colpodontosaurus* has been described by Gao and Fox (1996) as retaining a weakly developed subdental shelf, although lacking a sulcus dentalis.

ANGUIMORPHA indeterminate

Referred Specimens MNA V10038, UMNH VP 13004, UMNH VP 13005, UMNH VP 13007 (Fig. 16.5J–M).

Localities MNA Loc. 1067 (MNA V10038), Kane County, Utah, approximately 1.5 km west of Cannonville; UMNH 162E (UMNH VP 13004–13005, 13007), Iron County, Utah, approximately 25 km southwest of the town of Parowan.

Description MNA V10038 (Fig. 16.5J) is a polygonal osteoderm with a pit-and-ridge dorsal ornamentation. UMNH VP 13004 (Fig. 16.5K) is a rectangular osteoderm with a pit-and-ridge dorsal ornamentation in which the pits are more interconnected almost to the point of being vermiculate than in MNA V10038. There are both anterior and lateral imbrication surfaces. UMNH VP 13005 (Fig. 5L) is a rectangular osteoderm with a dorsal ornamentation of vermiculate grooves separated by rugose ridges. It has a lateral imbrication surface, but the anterior imbrication surface is missing (broken). UMNH VP 13007 (Fig. 16.5M) is a broken osteoderm that is rectangular to suboval in shape, has a vermiculate pattern of grooves separated by a series of pustulate ridges. This osteoderm lacks imbrication surfaces.

Discussion Referral of these osteoderms to Anguimorpha is based on the patterns of ornamentation and the lack of central keels. Indeed, the more rugose pattern of UMNH VP 13004–13005, 13007 is very similar to the cephalic osteoderms of the xenosaurid *Exostinus lancensis*. However, because these osteoderms were recovered from a different locality than the jaws of *Cnodontosaurus suchockii*, I am reluctant to refer them to that taxon.

SERPENTES Linnaeus, 1758
CONIOPHIS Marsh, 1892
CONIOPHIS sp.

Referred Specimen MNA V9102 (Fig. 16.5N–Q).

Localities MNA Loc. 1067, Kane County, Utah, approximately 1.5 km west of Cannonville.

Description MNA V9102 is a thoracic vertebra missing the condyle. In dorsal view (Fig. 16.5N) the zygosphene is broad, has a nearly mediolaterally straight anterior margin with only slight projections of the anterolateral corners, and is displaced to the left as a result of postmortem compression. The prezygapophyses project anteriorly and laterally. The neural arch is low, nearly flat and lacks any indication

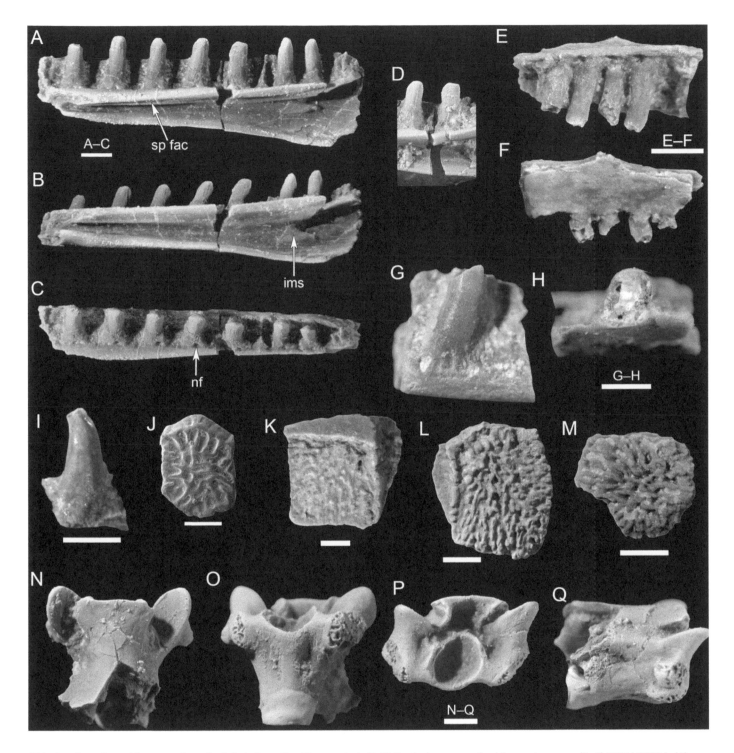

16.5. Anguimorpha and Serpentes from the Dakota Formation (Cenomanian). (A–F) *Cnodontosaurus suchockii* gen. et sp. nov.: (A–C) MNA V9084 in (A) lingual; (B) oblique inferior; (C) occlusal views; and (D) detail of prepreparation specimen with now lost segment of the subdental shelf. (E, F) MNA V10373 in (E) lingual and (F) labial views. (G–I) Dakota Formation morphotype E: (G, H) MNA V9086, broken left dentary in (G) lingual and (H) occlusal views; (I) UMNH VP 12920, jaw fragment in lingual view. (J, K) indeterminate anguimorphan osteoderms in dorsal view: (J) scanning electron micrograph of MNA V10038; (K) UMNH VP 13004; (L) UMNH VP 13005; (M) UMNH VP 13007. (N–Q) MNA V9102, trunk vertebra of *Coniophis* sp. in (N) dorsal; (O) ventral; (P) anterior; and (Q) right lateral views. ims, intramandibular septum; nf, nutrient foramen; sp fac, splenial articulation facet. All scale bars = 1 mm.

of dorsally projecting neural spine, though the posterior portion of the neural arch is missing and the exact condition of the neural spine is not known. The centrum is triangular in ventral view (Fig. 16.5O) with shallow grooves bordering a very weakly developed ventral keel. There is a singly small

subcentral foramen on either side of the keel (somewhat obscured by matrix). In anterior view (Fig. 16.5P), the cotyle is clearly circular and presumably would accommodate a similarly circular condyle. The synapophyses are large with the right synapophysis showing a distinct diapophysis and parapophysis, and is kidney shaped in lateral view (Fig. 16.5Q).

Discussion The laterally expanded anterior portion of the centrum, low and broad neural arch, round cotyle, and weakly developed hemal keel are all characteristics of *Coniophis* (sensu Rage, 1984). The missing posterior margin of the neural arch prevents meaningful comparison with *C. precedens* Marsh, 1892, though it is notable that the neural arch of *C. precedens* (Marsh, 1892) is more arched than in MNA V10375. The lack of a preserved neural spine makes comparison with *C. carinatus* Hecht, 1959, problematic, but the zygapophyses of MNA V9102 are not sharply tilted upward so referral to either *C. carinatus* or *C. platycarinatus* Hecht, 1959, is unlikely (see also discussion in Gardner and Cifelli, 1999, regarding these and other *Coniophis* species). On the basis of the lack of preservation of important aspects of the morphology, a species-level evaluation is not warranted.

The presence of snake, particularly *Coniophis*, in the Dakota Formation is not surprising because the genus has been reported from the slightly older (Albian–Cenomanian boundary) Mussentuchit Local Fauna, Cedar Mountain Formation, central Utah (Gardner and Cifelli, 1999). The specimen from the Dakota Formation differs from the specimens from the Cedar Mountain Formation in having a broader zygosphene and broader prezygapophyses such that there is no discernible gap between these features in dorsal view. However, the Dakota Formation specimen is weakly dorsoventrally compressed and it is possible, though unlikely, that these differences are an artifact of preservation. Additional specimens from both horizons and representatives of multiple regions of the vertebral column are necessary for formal taxonomic treatment of these and of other vertebrae referred to *Coniophis* sp. described below.

Smoky Hollow Member of the Straight Cliffs Formation (Turonian)

SCINCOMORPHA Camp, 1923
BORIOTEIIOIDEA Nydam,
Eaton, and Sankey, 2007
POLYGLYPHANODONTINAE Estes, 1983
POLYGLYPHANODONTINI Nydam,
Eaton, and Sankey, 2007
DICOTHODON Nydam, 1999
DICOTHODON CIFELLII Nydam,
Eaton, and Sankey, 2007

Dicothodon cifellii Nydam, Eaton, and Sankey, 2007:fig. 2.

Dicothodon cifellii was described by Nydam, Eaton, and Sankey (2007) as a second known species of the genus. Two addition species are known; the type species *D. moorensis* Nydam, 1999, from the Cedar Mountain Formation of central Utah (Albian–Cenomanian boundary) and *D. bajaensis* Nydam, 1999 (revised by Nydam, Eaton, and Sankey, 2007) from the "El Gallo Formation" (Campanian) of Baja California del Norte. The teeth of the species of *Dicothodon* are intermediate between those of *Bicuspidon* (conical, cusps united only by central ridge) and *Polyglyphanodon* (teeth greatly expanded transversely, cusps connected by a central ridge and mesial and distal ridges. The mammal-like teeth of polyglyphanodontine *Peneteius* (*P. aquilonius*, *P. saueri*) also share several similar features of this transverse oriented tooth specialization (see Nydam, Eaton, and Sankey, 2007, for further discussion).

POLYGLYPHANODONTINAE Estes, 1983
POLYGLYPHANODONTINI Nydam,
Eaton, and Sankey, 2007
CHAMOPSIIDAE Nydam,
Caldwell, and Fanti, 2010
CHAMOPS Marsh, 1892
CHAMOPS cf. *C. SEGNIS* Marsh, 1892

Referred Specimen OMNH 63666 (Fig. 16.6A, B).

Locality OMNH V1404, Kane County, Utah, approximately 13 km southeast of the town of Henrieville.

Description OMNH 63666 is a fragment from the posterior portion of a left maxilla. It preserves two tooth spaces and a single tooth. The tooth is subpleurodont, has a barrel-shaped shaft, and has a trifid crown with a tall central cusp and smaller, subequal mesial and distal accessory cusps. There is a small remnant of the facial process preserved dorsal to the in situ tooth. Posterior to this facial process there is a shallow depression on the dorsal surface of the supradental shelf presumably for the squamous articulation with the jugal.

Discussion *Chamops segnis* was the first Cretaceous-age fossil lizard described (Marsh, 1892) and is a common taxon from the Campanian–Maastrichtian of the northern portions of the Western Interior (Marsh, 1892; Estes, 1964; Sahni, 1972). As pointed out by Gao and Fox (1996), the report of *Chamops* from New Mexico is actually a premaxilla of *Albanerpeton* and an indeterminate lizard(?) jaw. Nydam, Caldwell, and Fanti (2010) erected the family-level taxon Chamopsiidae in recognition of several closely related taxa that include *Chamops* that have formally been referred to the Teiidae. This is the oldest known and southernmost confirmed occurrence of *Chamops*.

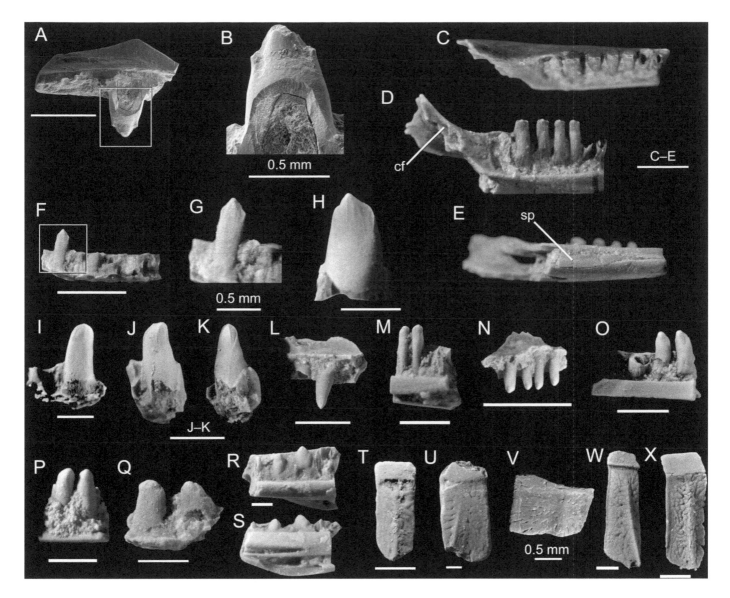

16.6. Scincomorphan taxa and morphotype from the Smoky Hollow Member of the Straight Cliffs Formation (Turonian). (A, B) *Chamops* cf. *C. segnis:* scanning electron micrograph of OMNH 63666, broken left maxilla in (A) lingual view with box indicating area of detail in (B). (C–E) *Utahgenys evansi:* OMNH 31223, broken left mandible in (C) occlusal; (D) lingual; and (E) ventral view. (F, G) SHMbr SCFm morphotype A: MNA V10011, jaw fragment in (F) lingual view with box indicating area of detail in (G). (H, I) SHMbr SCFm morphotype B: (H) UMNH VP 18041, jaw fragment with tooth in lingual view; (I) OMNH 23679, jaw fragment in lingual view. (J, K) SHMbr SCFm morphotype C: UMNH VP 18043, jaw fragment with tooth in (J) linguomesial view and (K) linguodistal view. (L) SHMbr SCFm morphotype D: MNA V10016, fragment of left maxilla in lingual view. (M, N) SHMbr SCFm morphotype E: (M) MNA V10019 fragment of right dentary in lingual view, (N) MNA V10013, broken left maxilla in lingual view. (O, P) SHMbr SCFm morphotype F: (O) MNA V10010, fragment of left dentary in lingual view, (P) MNA V10020, fragment of right dentary in lingual view. (Q) SHMbr SCFm morphotype G: MNA V10012, jaw fragment in lingual view. (R, S) SHMbr SCFm morphotype H: OMNH 64717, broken left dentary in (R) lingual and (S) ventral views. (T–X) Scincomorpha indeterminate osteoderms in dorsal view: (T) UMNH VP 19231; (U) OMNH 34322; (V) OMNH 67037; (W) OMNH 64342; (X) OMNH 67040. cf, coronoid facet; sp, splenial. All scale bars = 1 mm unless otherwise noted.

SCINCOMORPHA Camp, 1923
CONTOGENIIDAE Nydam and Fitzpatrick, 2009
UTAHGENYS Nydam and Fitzpatrick, 2009
UTAHGENYS EVANSI Nydam
and Fitzpatrick, 2009

Utahgenys evansi Nydam and Fitzpatrick, 2009:fig. 7A–P.
Referred Specimen OMNH 31223 (Fig. 16.6C–E).

Locality OMNH V843 (=MNA 995), Garfield County, Utah, approximately 7.5 km northeast of the town of Henrieville.

Description OMNH 31223 (Fig. 16.6C–D) is a broken left mandible preserving part of the dentary and the splenial (Fig. 16.6C–E). The dentary portion includes the five posteriormost tooth positions and four in situ teeth (Fig. 16.6C, D) as well as much of the coronoid process with a distinct coronoid facet on the medial surface (Fig. 16.6D). The preserved portion of the lateral surface of the dentary inferior to the coronoid process is expanded laterally and shelf-like (Fig. 16.6C). The preserved portion of the splenial is the small, wedge-shaped portion from the anteriormost portion of the element and it is firmly articulated within the Meckelian fossa (Fig. 16.6E). The Meckelian fossa anterior of the splenial is open only as a narrow slit.

Discussion I was unaware of this specimen until recently and it is included here because it provides a more complete representation of the posterior portion of the dentary than does the original paratypes MNA V10061 (Nydam and Fitzpatrick, 2009:fig. 7E–G) and MNA V10025 (Nydam and Fitzpatrick, 2009:fig. 7H–J). In particular, OMNH 31223 preserves a more complete coronoid process and lateral expansion of the posterior portion of the dentary. Additionally, OMNH 31223 preserves more of the dentary anterior to the splenial than does MNA V10025 and shows that the Meckelian fossa is a narrower slit than evident in MNA V10025. *Utahgenys evansi* is one of the most commonly recovered lizards from the Smoky Hollow Member of the Straight Cliffs Formation.

cf. SCINCOIDEA Oppel, 1811
PARAMACELLODID/CORDYLID GRADE
SHMbr SCFm Morphotype A

Referred Specimen MNA V10011 (Fig. 16.6F, G).

Locality MNA 995, Garfield County, Utah, approximately 7.5 km northeast of the town of Henrieville.

Description MNA V10011 (Fig. 16.6F, G) preserves only the tooth bearing portion of either a dentary or maxilla from a very small lizard. There are seven preserved tooth spaces, but only one complete tooth. The tooth is pleurodont, narrow, tall, and sharply pointed. The crown has well-developed cristae mesialis and distalis that extend from the tip to well-defined anguli mesialis and distalis (Fig. 16.6G). There is a small cuspis lingualis from which two well-developed striae dominans anterior and posterior descend and border a pars furcata. There are relatively heavy deposits of cementum at the tooth bases and there is a replacement pit at the base of the broken tooth adjacent to (i.e., to the right in Fig. 16.6G) complete tooth.

Discussion My grade-level referral of this specimen is based on the paramacellodid/cordylid-like tooth crown morphology (sensu Richter, 1994; Kosma, 2004). This diminutive taxon is not represented by sufficiently complete material for formal diagnosis and designated here an unnamed morphotype from the Smoky Hollow Member of the Straight Cliffs Formation.

SHMbr SCFm Morphotype B

Referred Specimens UMNH VP 18041, OMNH 23679 (Fig. 16.6H, I).

Localities UMNH Loc. 129 (=MNA 995), Garfield County, Utah, approximately 7.5 km northeast of the town of Henrieville; OMNH V60, Garfield County, Utah, approximately 10 km northwest of the town of Henrieville.

Description UMNH VP 18041 (Fig. 16.6H) is a small fragment of jaw with a single tooth. The tooth is short and stoutly built. The crista mesialis is longer than the crista distalis resulting in a posterior displacement of the cuspis labialis. The anguli mesialis and distalis are rounded, not angular, and do not form a distinct transition between the cristae mesialis and distalis and the associated culmen lateris anterior and posterior, respectively. The stria dominans anterior and stria dominans posterior are separated by a narrow pars furcata and there is a distinct striation on the labial surface of the crown between the stria dominans anterior and the culmen lateris anterior.

OMNH 23679 (Fig. 16.6I) is a jaw fragment (element indeterminate) that preserves two tooth spaces and one tooth. The tooth is subpleurodont, tall, robust, and has a rounded apex. The cristae mesialis and distalis are continuous with the culmen lateri anterior and posterior, respectively. The anguli mesialis and distalis are absent. There is a distinct, narrow pars furcata bordered by striae dominans anterior and posterior. The tip of the crown is worn and the condition and relation of the cuspis labialis and cuspis lingualis cannot be determined.

Discussion This unnamed morphotype of paramacellodid/cordylid-grade lizard is similar in tooth crown morphology with Dakota Formation morphotype A. The only significant difference between the Cenomanian specimen (UMNH VP 19224) and this Turonian-age specimen is the unequal lengths of the cristae mesialis and distalis in the latter.

SHMbr SCFm Morphotype C

Referred Specimen UMNH VP 18043 (Fig. 16.6J, K).

Locality UMNH Loc. 129 (=MNA 995), Garfield County, Utah, approximately 7.5 km northeast of the town of Henrieville.

Description UMNH VP 18043 is a small fragment of jaw with a single in situ tooth. The tooth is conical with a well-developed cuspis labialis and cuspis lingualis connected by a carina intercuspidalis. The crista mesialis (Fig. 16.6J) and crista distalis (Fig. 16.6K) are well developed, each border a shallow channel, and are continuous with an associated culmen lateri without an intervening anguli mesialis or distalis. The culmen lateris anterior nearly comes into contact with the stria dominans anterior as both structures course toward the tooth base. The pars furcata is bordered by well-developed striae dominans anterior and posterior that descends from the cuspis lingualis.

Discussion The lack of an angulus mesialis and angulus distalis to demarcate the transition from crista mesialis to culmen lateris anterior and crista distalis to culmen lateris posterior is more typical of scincids (sensu Kosma, 2004), but the very distinct pars furcata is typical of paramacellodid/cordylids-grade tooth morphology (Fig. 16.2). This unnamed SHMbr SCFm morphotype C is similar to Dakota Formation morphotype C (MNA V10033), but the culmen lateris anterior in the later does not approach the stria dominans anterior.

SHMbr SCFm Morphotype D

Referred Specimen MNA V10016 (Fig. 16.6L).

Locality MNA 995, Garfield County, Utah, approximately 7.5 km northeast of the town of Henrieville.

Description MNA V10016 (Fig. 16.6L) is a broken right (?) maxilla preserving three tooth positions and a single tooth. The tooth is tall, conical, and has a basal resorption pit. The crista mesialis and distalis are well developed but are not continuous with culmen lateri anterior or posterior, respectively (i.e., they do not turn lingually as they course toward the tooth base). The stria dominans anterior and stria dominans posterior descend from a cuspis lingualis and border a distinct pars furcata. The broken posterior portion of the specimen (right side in Fig. 16.6L) shows a distinct channel for the superior alveolar neurovascular bundle. The facial process is mostly broken away.

Discussion This unnamed SHMbr SCFm morphotype D is also very similar in tooth morphology to Dakota Formation morphotype C, but the latter specimen has a more robust tooth with a wider base. This difference in overall tooth shape may be attributable to positional and/or ontogenetic differences and is not necessarily a potential taxonomic difference.

SCINCOIDEA indeterminate
SHMbr SCFm Morphotype E
(Fig. 16.6M, N)

Referred Specimens MNA V10019, MNA V10013 (Fig. 16.6M, N).

Locality MNA 995, Garfield County, Utah, approximately 7.5 km northeast of the town of Henrieville.

Description MNA V10019 (Fig. 16.6N) is a fragment of a right dentary preserving three tooth positions and two teeth. The teeth are pleurodont, narrow, gracile, closely spaced, and columnar. There are moderately heavy deposits of cementum at the tooth bases and the mesialmost tooth has a small resorption pit at the base. The crowns are slightly eroded, but the crista mesialis and shorter crista distalis are clearly visible. The subdental shelf is tall, vertical, and narrows posteriorly, presumably as part of the articulation with the splenial. The Meckelian fossa is wide posteriorly, but narrows anteriorly indicating that the specimen likely preserves the portion of the dentary associated with the anterior portion of the splenial.

MNA V10013 (Fig. 16.6N) is a poorly preserved fragment of the left maxilla preserving five tooth positions and four teeth. The teeth are nearly identical to those in MNA V10019, but are not so closely spaced. The fourth tooth (farthest to left in Fig. 16.6N) is markedly larger than the other preserved teeth and has a well-developed replacement pit at its base. The supradental shelf turns ventrally toward the base of the fourth tooth.

Discussion Although the teeth of this unnamed SHMbr SCFm morphotype E are slightly worn, these specimens are not referable to another known taxon or morphotype from the Turonian of southern Utah. Closest similarity, based on dental morphology, is to ?Cordylidae genus and species undetermined (A) from the Campanian of Alberta and ?Cordylidae genus and species undetermined (B) from the "Lancian" (=Maastrichtian; my quotes) of Montana (Gao and Fox, 1996:fig. 25). However, the teeth of the specimens from Alberta and Montana are much taller and have cristae mesialis and distalis developed such that some of the tooth crowns are weakly trifid. Gao and Fox (1996) attributed specimens recovered from the Bug Creek Anthills (BCA) locality to be from the Hell Creek Formation and therefore Maastrichtian in age. However, the BCA locality is a Paleocene deposit of the Tullock Formation with both Paleocene-age fossils and reworked Cretaceous-age fossils (Lofgren, 1995). It is possible that the "?Cordylidae genus and species undetermined (B)" specimen described by Gao and Fox (1996:60) is from the Paleocene.

SHMbr SCFm Morphotype F

Referred Specimens MNA V10010, MNA V10020 (Fig. 16.6O, P).

Locality MNA 995, Garfield County, Utah, approximately 7.5 km northeast of the town of Henrieville.

Description MNA V10010 (Fig. 16.6O) is a fragment of a left dentary preserving four tooth positions and two complete teeth. The teeth are pleurodont, slightly recurved, and have only a small amount of cementum around the bases. The distalmost of the preserved teeth has a well-developed resorption pit at the base. The crowns are pointed, labiolingually compressed, and have mesial and distal expansions. The subdental shelf is tall and borders a narrow and shallow sulcus dentalis.

MNA V10020 (Fig. 16.6P) is a fragment of a right dentary preserving two teeth. The teeth are similar in morphology to those of MNA V10010, but are more robust. The subdental shelf is very short. Both the stoutness of the dentition and the shorter subdental shelf are indicative of MNA V10020 preserving a more posterior portion of the dentary than that of MNA V10010.

Discussion The tooth morphology of this unnamed morphotype is very similar to that of *Ptilotodon wilsoni* Nydam and Cifelli, 2002b, from the Aptian–Albian of Oklahoma. Nydam and Cifelli (2002a) described the taxon *Ptilotodon wilsoni* from a fragmentary right dentary and assigned it to Teiidae on the basis of the presence of heavy deposits of basal cementum, widely open and circular resorption pits (both of which are not characteristics of SHMbr SCFm morphotype F), and mesial and distal expansions of the crown. Since then, the evidence for the presence of Teiidae (sensu stricto) in the Cretaceous of North America has been questioned and most of the referred taxa have been reassigned to Borioteiioidea, as sister taxon to Teiioidea (Nydam and Voci, 2007; Nydam, Eaton, and Sankey, 2007; Nydam, Caldwell, and Fanti, 2010). The dental and jaw characteristics noted by Nydam and Cifelli (2002a) as indicative of the teiid affinities of *Ptilotodon* are now recognized as common to both teiids and borioteiioids (see Nydam, Eaton, and Sankey, 2007, for further discussion). Although diagnostic at the genus and species level, the specimen of *Ptilotodon* is not sufficiently complete for confident referral to Borioteiioidea but should instead be considered as Scincomorpha indeterminate.

SHMbr SCFm Morphotype G

Referred Specimen MNA V10012 (Fig. 16.6Q).

Locality MNA 995, Garfield County, Utah, approximately 7.5 km northeast of the town of Henrieville.

Description MNA V10012 (Fig. 16.6Q) is a jaw fragment (element indeterminate) that preserves two eroded teeth. The teeth are pleurodont, robust, have a heavy deposit of cementum at the bases, and the tooth to the right in Fig. 16.6Q has a large, circular resorption pit at the base. The tooth crowns are eroded and detailed features of the enamel are lost.

Discussion The robust structure of the teeth, pleurodont attachment, and large resorption pit are all consistent with a referral to *Chamops*. However, the structure of the tooth base is also consistent with SHMbr SCFm morphotype B. The lack of detailed crown morphology precludes definitive referral of this specimen and it is conservatively regarded as an unnamed morphotype.

SHMbr SCFm Morphotype H

Referred Specimen OMNH 64717 (Fig. 16.6R, S).

Locality OMNH V1404, Kane County, Utah, approximately 13 km southeast of the town of Henrieville.

Description OMNH 64717 is a portion of a left dentary preserving four tooth positions and two teeth. The teeth are pleurodont, very low crowned (distalmost preserved tooth does not extend beyond the lateral parapet), and triangular in lingual view (Fig. 16.6R). There are heavy deposits of cementum at the tooth bases and the mesial preserved tooth has a small, circular resorption pit at the base. On the basis of the greater size of the broken base, the missing tooth from the mesialmost tooth position was likely much larger than the preserved teeth. The subdental shelf is tall and rounded and the sulcus dentalis is nearly flat and not gutter-like. The ventral surface of the subdental shelf (Fig. 16.6S) is deeply notched for what appears to be the articulation facet for the dorsal process of an anteriorly forked splenial.

Discussion Scincoidean lizards with very low-crowned teeth have also been recovered from the Dakota Formation (Fig. 16.4L) and the Kaiparowits Formation (Fig. 16.13V, W). Because of the fragmentary nature of these specimens it is not possible to determine if they represent one or several taxa (at least at the genus level), but closest similarity is between OMNH 64717 and OMNH 30751 (Fig. 16.13V) from the Kaiparowits Formation. The very low crowns of OMNH 64717 likely did not extend beyond the gingiva and were effectively nonfunctional. Myrmecophagous lizards, such as *Phrynosoma*, also have very low-crowned marginal teeth that contribute little, if anything, to prey capture and/or prey handling (Montanucci, 1989; pers. obs.). However, pending the recovery of more complete specimens the possible dietary habits of SHMbr SCFm morphotype G are speculative at best.

SCINCOMORPHA indeterminate
Osteoderms

Referred Specimens UMNH VP 19231, OMNH 64322, OMNH 64323, OMNH 67037, OMNH 67040 (Fig. 16.6T–X).

Localities UMNH Loc. 129 (specimen UMNH VP 19231), Garfield County, Utah, approximately 7.5 km northeast of the town of Henrieville; OMNH V60 (specimens OMNH 64322, 67037, 67040), Garfield County, Utah, approximately 10 km northwest of the town of Henrieville; OMNH V1404 (specimen OMNH 64342), Kane County, Utah, approximately 13 km southeast of the town of Henrieville.

Description UMNH VP 19231 (Fig. 16.6T) is a rectangular scute that is robust (thick) has an anterior imbrication facet, and a central dorsal keel. The dorsal surface is worn, but some of the ornamentation is preserved as a series of weakly developed vermiculate channels that generally course laterally from the central keel. OMNH 64322 (Fig. 16.6U) is a robust, rectangular osteoderm missing the anterior right and posterior left corners. There is an anterior imbrication facet and a dorsal keel that deviates to the left as it courses posteriorly. The ornamentation is better preserved than in UMNH VP 19231; a series of shallow vermiculate channels coursing laterally/posteriorly and connected to a series of small foramina on either side of the dorsal keel. OMNH 67037 (Fig. 16.6V) is a portion of a large, robust, rectangular(?) osteoderm. There is a portion of the anterior imbrication facet, but its full extent cannot be determined. There is a dorsal keel offset to the right side and the ornamentation is a series of short, unconnected channels in a pattern that radiates from near the anterior end of the dorsal keel. OMNH 64342 (Fig. 16.6W) is an anteroposteriorly elongate osteoderm with an anterior imbrication facet that is offset from the ornamented portion of the osteoderm by a well-developed step. The dorsal keel of OMNH 64342 is offset to the right and does not extend to the imbrication facet. The ornamentation consists of a network of numerous, small vermiculate channels and a series of foramina on either side of the dorsal keel. OMNH 67040 (Fig. 16.6X) is also elongate, has an anterior imbrication facet (not as dramatically offset as the facet in OMNH 64342), and a dorsal keel offset to the right. The ornamentation is similar to that of OMNH 64342.

Discussion These five specimens are representative of numerous, thick, rectangular, keeled osteoderms recovered from the microvertebrate localities of the Smoky Hollow Member of the Straight Cliffs Formation. The general morphology of these osteoderms is consistent with those from known scincomorphan taxa with body armor. The differences in ornamentation between UMNH VP 19231, OMNH 64322, and OMNH 67037, and between OMNH 64342 and OMNH 67040 may be taxonomically significant but could also be due to differential degrees of wear. Pending the recovery of osteoderms in unquestionable association with identifiable jaw elements, referral to any one taxon or morphotype from the horizon is not possible.

ANGUIMORPHA Fürbringer, 1900
ANGUIFORMES Conrad, 2006
ANGUIDAE Gray, 1825
ODAXOSAURUS Gilmore, 1928

Revised Diagnosis (from Estes, 1983, and Gao and Fox, 1996) Anguid lizard that differs from all other known anguids in having the following combination of character traits: tooth shafts expanded lingually; tooth crowns labiolingually compressed, squared off, striated labially and lingually, slightly posteriorly rotated, with distinct lingual cuspule.

Further differs from *Elgaria* (e.g., *E. multicarinatus*) and *Gerrhonotus* (e.g., *G. infernalis*) in teeth lacking an offset, or step, of the mesial portion of the apex.

Type Species *Odaxosaurus piger* Gilmore, 1928.

Discussion When Gilmore (1928) originally erected *Peltosaurus piger* (later to be recognized as *Odaxosaurus piger*; see Gao and Fox, 1996, for a thorough discussion of the taxonomic history) it was not yet a common and necessary practice to provide a differential diagnosis. However, Gilmore did identify the type specimen as a nearly complete right dentary (USNM 10687) and a paratype partial right maxilla with similar teeth (USNM 10688). Estes (1983) provided the first differential diagnosis of the species *O. piger*, which as the only known species was also the diagnosis for the genus. Estes included in this diagnosis characteristics of the frontals and osteoderms based on his earlier (Estes, 1964) referral of these elements to the species, but he did not establish these specimens as a hypodigm. In their review of *Odaxosaurus*, Gao and Fox (1996:67–68) retained Estes's (1983) diagnosis of *Odaxosaurus piger* as the diagnosis of the genus *Odaxosaurus*. However, they later discuss in excellent detail (Gao and Fox, 1996:69–70) the problematic referral of nonjaw specimens (i.e., frontal and osteoderms) to *O. piger* and the resulting inherent weaknesses in making meaningful comparisons based on those specimens and associated characteristics. I agree with Gao and Fox and propose the above revised diagnosis as a logical extension of their argument to reflect only those characteristics demonstrated by the type material for *Odaxosaurus* sensu *O. piger*, against which meaningful taxonomic comparisons should be made.

aff. *ODAXOSAURUS*

Referred Specimens MNA V9129, MNA V10017 (Fig. 16.7A, B).

Locality MNA 995, Garfield County, Utah, approximately 7.5 km northeast of the town of Henrieville.

Description MNA V9129 (Fig. 16.7A) is a broken left dentary preserving six tooth spaces and three partial teeth. The teeth are pleurodont and have a basal ring of cementum pierced by a posterolingually directed nutrient foramen. The tooth shafts have a nearly vertical lingual surface (not expanded) and are slightly recurved. The distalmost tooth reserves tooth crown that is chisel-like crown that is expanded mesially and distally, slightly rotated posteriorly, has a lingually displaced apical cuspule, and is only weakly striated (but is also slightly eroded and may not preserve all details). Tooth replacement was apparently alternate and interdental. The medial margin of the element is broken and it is not possible to evaluate the condition of the subdental shelf or sulcus dentalis. The labial surface of the element is weakly convex and pierced by three external inferior alveolar foramina.

MNA V10017 (Fig. 16.7B) is a fragment of a right dentary that preserves three tooth spaces and one complete tooth. This specimen is from a smaller individual than MNA V9129. The tooth attachment is subpleurodont (approximately one half of the tooth height extends beyond the lateral parapet; sensu Gao and Fox, 1996) and the tooth is recurved, convex lingually, and the crown tapers to a chisel-like apex. There is a ring of cementum around the base of the tooth that is pierced by a distinct nutrient foramen which opens postero-lingually. There is no subdental shelf or sulcus dentalis.

Discussion The single complete tooth of both MNA V9129 and V10017 are characteristic of an *Odaxosaurus*-grade taxon, but are too poorly preserved to confirm referral to that genus, but as this is closest comparison, additional discussion of characteristics is warranted. The lack of well-developed apical striae on the teeth of both specimens and the less expanded tooth crown of MNA V10017 are features characteristic of *O. priscus* Gao and Fox, 1996. However, the more expanded crown and apparently closely spaced teeth (based on spacing of empty tooth positions in addition to teeth in MNA V9129) are equally characteristic of *O. piger.* These contradictions are likely the result of the fragmentary nature of the specimens and confirmation of the generic referral and a formal taxonomic treatment will have to await the recovery of more complete material from the Smoky Hollow Member of the Straight Cliffs Formation. If correctly referred, these specimens represent the oldest known occurrences of *Odaxosaurus* extending the record from the Campanian to the Turonian.

aff. XENOSAURIDAE Cope, 1886
CNODONTOSAURUS sp.

Referred Specimens OMNH 24453, MNA V9125 (Fig. 16.7C, D).

Localities OMNH V60 (specimen OMNH 24453), Garfield County, Utah, approximately 10 km northwest of the town of Henrieville; MNA 995 (specimen MNA V9125), Garfield County, Utah, approximately 7.5 km northeast of the town of Henrieville.

Description OMNH 24453 is a fragmentary jaw, possibly a left dentary, that preserves six tooth spaces, one broken tooth base, and one nearly complete tooth. The tooth is pleuro-dont, has a shaft that is subcircular in cross section, and a labiolingually compressed crown. The tip of the crown is missing, but it is clear that the mesial and distal margins of the crown are more compressed than the central portion. The specimen is broken such that the subdental shelf (if it was present) and associated lingual edges of the tooth bases are missing.

MNA V9125 is a fragment of a possible left maxilla preserving three tooth spaces, the base of a broken tooth, and one nearly complete tooth. The nearly complete tooth is sub-pleurodont, slightly recurved, has a shaft that is subcircular in cross section, a labiolingually compressed and trenchant crown, and is missing the lingual margin of its base. The mesial and distal margins of the crown are more compressed than the central portion. The lingual and labial surfaces of the crown are smooth, except for a small wear facet on the lingual face of the apex. The tooth base of the broken tooth has a nearly complete lingual margin surrounded by a small, thin band of cementum. A small nutrient foramen opens anterolingually (assuming identification as a left maxilla is correct). The supradental shelf is not preserved.

Discussion The morphology of the preserved portion of the tooth crowns of both OMNH 24453 and MNA V9125 are nearly identical to that of *Cnodontosaurus suchockii* (Fig. 16.5A–F) from the Dakota Formation. On the basis of its diminutive size, MNA V9125 most likely represents an onto-genetically younger individual.

Eaton et al. (1999b) reported the presence of *Dorsetisaurus* (an anguimorphan taxon known from the Late Jurassic–earliest Cretaceous of Laurasia; (Prothero and Estes, 1980; Evans, 1993; Evans and Chure, 1998b, 1999; Broschinski, 2000) in the fauna of the Smoky Hollow Member of the Straight Cliffs Formation based on MNA V9125. Although both have trenchant teeth, the posterior teeth of *Dorsetisaurus* have more mesiodistally expanded crowns (nearly to the state of being leaf shaped in the distalmost portions of the tooth row) and the bases are more swollen where they attach to the jaw (pers. obs. of specimens BMNH R8058, R8065, R8067, R8076, R8079, R8108; MG-LNEG Gui Squ 15, and Gui Squ 21–the Gui Squ specimens were originally curated as FUB specimens, but have been recently transferred to the geology museum in Lisbon and I use the original "Gui Squ" designations with the new acronym pending assignment of new numbers from MG-LNEG). At this time the more parsimonious solution is to place this and OMNH 24453 in *Cnodontosaurus*.

PLATYNOTA Duméril and Bibron, 1836
SHMbr SCFm Morphotype I

Referred Specimen OMNH 64530 (Fig. 16.7E).

Locality OMNH V1404, Kane County, Utah, approximately 13 km southeast of the town of Henrieville.

Description OMNH 64530 (Fig. 16.7E) is fragment of jaw (element indeterminate) of a large platynotan that preserves the broken base of a single tooth. The tooth base is broad with external enamel infolding and evidence of internal

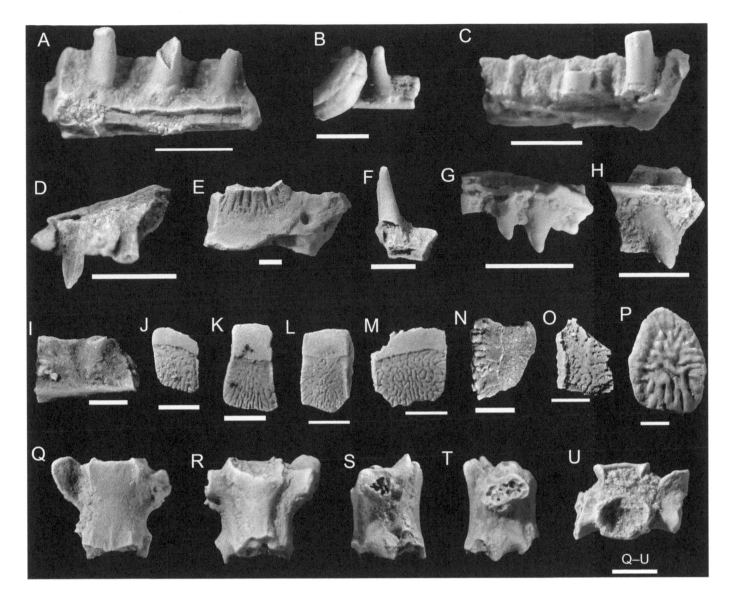

16.7. Anguimorpha and Serpentes from the Smoky Hollow Member of the Straight Cliffs Formation (Turonian). (A, B) aff. *Odaxosaurus* sp.: (A) MNA V9129, broken left dentary in lingual view; (B) MNA V10017, broken right dentary in lingual view. (C, D) *Cnodontosaurus* sp.: (C) OMNH 24453, broken left dentary (?) in lingual view; (D) MNA V2125, broken left maxilla in lingual view. (E) SHMbr SCFm morphotype I: OMNH 64530, jaw fragment of in lingual view. (F–I) SHMbr SCFm morphotype J: (F) UMNH VP 19235, jaw fragment in lingual view; (G) UMNH VP 18045, fragment of left maxilla (?) in lingual view; (H) UMNH VP 18044, broken left maxilla in lingual view; (I) UMNH VP 19236, broken left dentary in lingual view. (J–P) indeterminate anguimorphan osteoderms in dorsal view; (J) UMNH VP 19232; (K) UMNH VP 19234; (L) OMNH 67038; (M) OMNH 67039; (N) UMNH VP 19230; (O) UMNH VP 19233; (P) OMNH 64308. (Q, R) *Coniophis* sp.: MNA V10375, trunk vertebra in (Q) dorsal; (R) ventral; (S) left lateral, (T) right lateral, and (U) anterior views. All scale bars = 1 mm.

plicidentine, but the specimen is too fragmentary to unquestionably substantiate the presence of plicidentine.

Discussion The presence of a large platynotan in the lizard fauna from the Smoky Hollow Member of the Straight Cliffs Formation is not unexpected considering the presence of *Primaderma nessovi* Nydam, 2000, at the Albian–Cenomanian boundary in the Cedar Mountain Formation of central Utah (Nydam, 2000) and the presence of platynotans in the overlying John Henry Member of the Straight Cliffs Formation and the Kaiparowits Formation.

SHMbr SCFm Morphotype J

Referred Specimens UMNH VP 19235, UMNH VP 18045, UMNH VP 18044, UMNH VP 18045, UMNH VP 19236 (Fig. 16.7F–I).

Locality MNA 995, Garfield County, Utah, approximately 7.5 km northeast of the town of Henrieville.

Description UMNH VP 19235 (Fig. 16.7F) is small fragment of a jaw (element indeterminate) that preserves two tooth spaces and a single tooth. The tooth is conical, tall, and has an offset mesial(?) carina that borders a distinct groove.

The base of the tooth is poorly preserved, but the intact portion of the tooth–jaw contact preserves weak basal infolding of the enamel. There is no indication of plicidentine. UMNH VP 18045 (Fig. 16.7G) is a fragment of a possible left maxilla that preserves four tooth spaces and two teeth. The teeth are conical, low crowned, and slightly recurved, and each has an offset mesial carina. The bases of the teeth are slightly eroded, and there is no indication of infolded enamel.

UMNH VP 18044 (Fig. 16.7H) is a broken left maxilla preserving the anterior portion of the element, but is missing the premaxillary process. UMNH VP 18044 preserves the base of a tooth. This base is wider than the preserved portion of the tooth shaft, lacks basal infolding of the enamel, and has a nutrient foramen at its base. The shaft of the tooth is broken too close to the crown to determine whether or not plicidentine is present. UMNH VP 19236 (Fig. 16.7I) is small fragment of a left dentary preserving two tooth spaces and the slightly eroded base of one tooth. The base is wider than the cross section of the tooth shaft at the lateral parapet. There is a nutrient foramen at the base, but no evidence of basal infolding of the enamel or plicidentine. There is no subdental shelf and the Meckelian fossa is narrow indicating that the specimen is from a more rostral portion of the element.

Discussion These specimens are united in the unnamed SHMbr SCFm morphotype J based on the platynotan grade of tooth attachment (broad tooth bases with nutrient foramina) and weakly developed/lack of basal infolding of the enamel. Additionally, UMNH VP 19235 and UMNH VP 18045 share the presence of conical teeth with offset mesial carina. An offset carina bordering a shallow groove is also known to occur in other fossil platynotans such as *Primaderma nessovi* (Nydam, 2000:figs. 2, 3) and *Paraderma bogerti* Estes, 1964 (Gao and Fox, 1996:fig. 35B). However, unlike these two taxa, SHMbr SCFm morphotype J lacks extensive and well-developed basal infoldings of enamel. The gross size difference between SHMbr SCFm morphotype J and SHMbr SCFm morphotype I indicates that there are at least two platynotan taxa in the lizard fauna from the Smoky Hollow Member of the Straight Cliffs Formation. However, the lack of tooth crowns in UMNH VP 18044 and UMNH VP 19236 make referral of these specimens more tentative and they may represent at least one additional taxon.

ANGUIMORPHA indeterminate
Osteoderms

Referred Specimens UMNH VP 19232, UMNH VP 19234, UMNH VP 19230, UMNH VP 19233, OMNH 64308, OMNH 67038, OMNH 67039 (Fig. 16.7J–P).

Localities UMNH Loc. 129 (specimens UMNH VP 19230, 19232–19234), Garfield County, Utah, approximately 7.5 km northeast of the town of Henrieville; OMNH V60 (specimens OMNII 64308, 67038, 67039) Garfield County, Utah, approximately 10 km northwest of the town of Henrieville.

Description UMNH VP 19232 (Fig. 16.7J) is a small, rhomboidal osteoderm with a diagonal posterior margin, large anterior imbrication facet, narrow lateral imbrication facet on the right side, and vermiculate ornamentation. UMNH VP 19234 (Fig. 16.6N) is a trapezoidal osteoderm with a wider posterior margin than anterior margin, large imbrication facet anteriorly (approximately ⅓ of overall length), narrow lateral imbrication facets on both the right and left sides, and a vermiculate ornamentation. OMNH 67038 (Fig. 16.7L) is a rectangular osteoderm with a large anterior imbrication facet, a narrow lateral imbrication facet on the right side, and a vermiculate ornamentation pattern. OMNH 67039 (Fig. 16.7M) is a broken osteoderm with nearly parallel right and left sides, a large anterior imbrication facet with a weakly concave posterior margin, and a vermiculate ornamentation pattern. UMNH VP 19230 (Fig. 16.7N) is a broken osteoderm with no evidence of an anterior articulation facet and an ornamentation pattern that grades from weakly vermiculate (right side on Fig. 16.7N) to heavily rugose (left side of Fig. 16.7N). UMNH VP 19233 (Fig. 16.7O) is a broken rectangular osteoderm lacking an anterior imbrication facet, with narrow and raised lateral imbrication facet on the left(?) side, and rugose ornamentation. OMNH 64308 (Fig. 16.7P) is an ovoid osteoderm lacking imbrication facets and ornamented with pustulate ridges separated by deep grooves.

Discussion UMNH VP 19232, UMNH VP 19234, OMNH 67038, and OMNH 67039 are identical to osteoderms from the Campanian and Maastrichtian that have been referred to *Odaxosaurus* (Estes, 1983), but Gao and Fox (1996) argue that referral to *Odaxosaurus* is tentative at best. Jaw specimens referable to aff. *Odaxosaurus* in the Smoky Hollow Member lizard fauna (e.g., MNA V9129) may be associated with these osteoderms (following the interpretation of Estes, 1983), but on the basis of the lack of unquestionable association, I follow Gao and Fox (1996) in not assigning these osteoderms to a taxon more exclusive than Anguimorpha.

UMNH VP 19230 and UMNH VP 19233 are distinct in their apparent lack of anterior imbrication facets and rugose ornamentation pattern. Rugose patterns on osteoderms are widespread in Anguimorpha, but with respect to fossils from the Cretaceous of the Western Interior, the pattern on these two specimens is most similar to that of the cranial osteoderms of the xenosaurid *Exostinus lancensis* from the Maastrichtian of Wyoming and Montana (Gao and Fox, 1996:fig. 27). As such, these osteoderms may be referable to cf. *Cnodontosaurus*.

Among the osteoderms recovered from the Smoky Hollow Member of the Straight Cliffs Formation, OMNH 64308 is unique in being ovoid and with a pustulate ridge + deep groove ornamentation. Ovoid osteoderms are known to occur in modern diploglossine anguids (Strahm and Swartz, 1977), helodermatids (Bogert and del Campo, 1956), *Lanthanotus borneensis* (see reconstructed computed tomographic scans at digimorph.org/specimens/Lanthanotus_borneensis/), and the fossil platynotans *Necrosaurus eucarinatus* Kuhn, 1940 (see also Estes, 1983) from the Eocene of Germany, and possibly *Necrosaurus cayluxi* Filhol, 1876 (see also Estes, 1983). Of these, the diplogossine osteoderms have vermiculate ornamentation and well-developed imbrication facets, while the platynotans (excluding possibly *Necrosaurus eucarniantus* and *N. cayluxi*) typically have pustulate + pit ornamentation without clear grooves. The osteoderms referred to both species of *Necrosaurus* are keeled. Clearly, referral of OMNH 64308 to any one taxon is not possible at this time.

SERPENTES Linnaeus, 1758
CONIOPHIS Marsh, 1892
CONIOPHIS sp. indet.

Referred Specimen MNA V10375 (Fig. 16.7Q–U).

Locality MNA 995, Garfield County, Utah, approximately 7.5 km northeast of the town of Henrieville.

Description MNA V10375 is a thoracic vertebra missing the condyle, postzygapophyses, posterior margin of the neural arch (and associated zygantra), and the right prezygapophyses. The neural arch (Fig. 16.7Q, S, T) is broad and flat with wide zygosphenes. It narrows at its midpoint and widens posteriorly, but the final posterior width is unknown. The neural spine is a very low, subtle, keel-like ridge at the posterior end of the neural arch, but its full development is unknown. The ventral surface of the centrum (Fig. 16.7R) has a relatively broad hemal keel that expands anteriorly toward the cotyle, but its posterior extent is unknown. The left synapophysis (Fig. 16.7S) is mostly preserved and is continuous with the left prezygapophysis. The left parapophysis is complete and has a weakly convex articular surface, but the left diapophysis is missing its articular surface. The articular surfaces of the right synapophysis (Fig. 16.7T) are missing. The cotyle (Fig. 16.7R, U) appears to be deep and round, but the ventral margin is eroded.

Discussion Unfortunately, lack of the posterior margin of the neural arch precludes meaningful comparison with many of the known species of *Coniophis*. However, some comparisons are warranted. The zygapophyses of MNA V10375 are not oriented sharply dorsally so it is unlikely that this specimen is referable to *C. carinatus*. Although both are missing the posterior margins of the neural arch, the posterior expansion of the arch in MNA V10375 does not appear to be as dramatic as in MNA V9102 from the Dakota Formation.

Additionally, the rim of the cotyle in MNA V10375 projects more anteriorly than that of MNA V9102 and the hemal keel of MNA V10375 is better developed than in MNA V9102.

It is likely that the Dakota Formation and Smoky Hollow Member of the Straight Cliffs Formation specimens represent different, possibly unique, species of *Coniophis*. However, I agree with Gardner and Cifelli (1999) that identification of new species of *Coniophis* should not precede a much needed review of (1) positional variation in vertebral morphology and (2) diagnostic characters of the genus and referred species.

Iron Springs Formation (Cenomanian–Santonian)

SCINCOMORPHA Camp, 1923
BORIOTEIIOIDEA Nydam,
Eaton, and Sankey, 2007
POLYGLYPHANODONTINAE Estes, 1983
POLYGLYPHANODONTINI Nydam,
Eaton, and Sankey, 2007
DICOTHODON Nydam, 1999
DICOTHODON sp.

Referred Specimen UMNH VP 14155 (Fig. 16.8A–C).

Locality and Horizon UMNH Loc. 416, southwestern Washington County, Utah, near base of Iron Springs Formation; Cenomanian.

Description UMNH VP 14155 is an isolated posterior tooth that is oval in occlusal view with well-developed mesial and distal accessory ridges enclosing shallow basins on either side of the central ridge (Fig. 16.8A). There is a tall labial cusp, a lower lingual cusp, and the central ridge connecting the cusps is U shaped in mesiodistal view and lacks any secondary development of a central swelling (Fig. 16.8B). Both cusps have small apical wear facets.

Discussion This specimen is similar to *D. moorensis* and differs from *D. cifellii* in lacking any development of a central swelling on the central ridge. However, it is more similar to *D. cifellii* in having a U-shaped central ridge instead the V-shaped central ridge specific to *D. moorensis*. Referral to either species, or erection of a new species, must await recovery of additional teeth for more meaningful comparison.

The Iron Springs Formation was deposited well west of the fluctuating margin of Cretaceous Western Interior Seaway and represents a thick accumulation of terrestrial deposits that extend from the Cenomanian to at least the Santonian (Eaton, 1999a). UMNH Loc. 416 is near the bottom of the unit and is paraequivalent to the Dakota Formation (Eaton, pers. comm.). As such, this record adds *Dicothodon* to the lizard fauna from the Cenomanian of southern Utah.

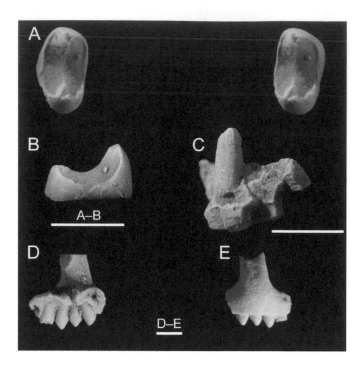

16.8. Scincomorpha from the Iron Springs Formation (Cenomanian, Santonian). (A, B) *Dicothodon* sp.: UMNH VP 14155, isolated tooth in (A) stereopair of occlusal view, and (B) mesial view. (C) Iron Springs Formation morphotype A: UMNH VP 19253, jaw fragment in lingual view. (D, E) Iron Springs Formation morphotype B: UMNH VP 19254, broken premaxilla in (D) lingual and (E) rostral views. Scale bars = 1 mm.

cf. SCINCOIDEA Oppel, 1811
PARAMACELLODID/CORDYLID GRADE
Iron Springs Morphotype A

Referred Specimen UMNH VP 19253 (Fig. 16.8C).

Locality UMNH Loc. 12, north-central Washington County, Utah; near top of Iron Springs Formation; ?Santonian.

Description UMNH VP 19253 (Fig. 16.8C) preserves two tooth spaces and single tooth. The tooth is tall, conical, and has a posteriorly rotated tooth crown. The crown has weakly developed (albeit slightly eroded) cristae mesialis and distalis, but the cuspis lablialis is eroded. There is a well-developed cuspis lingualis from which descend prominent striae dominans anterior and posterior. The pars furcata is narrow. The lingual surface of the tooth crown lacks any additional striae.

Discussion This specimen is from a locality near the top of the Iron Springs Formation (J. Eaton, pers. comm.) and is most likely Santonian in age. The tooth crown morphology is classically paramacellodid/cordylid grade, but the specimen is too incomplete to be referred to any known taxa. However, it is relevant to make some comparison to similar morphotypes from southern Utah discussed herein. Of the morphotypes from the Dakota Formation and the Smoky Hollow Member of the Straight Cliffs Formation, Iron Springs Formation morphotype A is most similar to

Dakota Formation morphotype C (Fig. 16.4H, I) and SHMbr SCFm morphotype D (Fig. 16.6L), but differs from both in having a posteriorly rotated tooth crown, distinguishable anguli mesialis and distalis, and more bluntly tipped apex. Iron Springs Formation morphotype A is not immediately comparable to any scincomorphan morphotypes from the John Henry Member of the Straight Cliffs Formation, Wahweap Formation, or Kaiparowits Formation.

SCINCOMORPHA indeterminate
Iron Springs Morphotype B

Referred Specimen UMNH VP 19254 (Fig. 16.8D, E).

Locality UMNH Loc. 12, north-central Washington County, Utah; near top of Iron Springs Formation; ?Santonian.

Description UMNH VP 19254 (Fig. 16.8D, E) is a premaxilla missing part of the left tooth-bearing portion and the nasal process. It preserves five tooth spaces, one incomplete tooth, and three complete teeth. The tooth bases are heavily cemented to the tooth-bearing surface and the complete teeth have conical, low crowns (with slight wear). There is a shallow, round resorption pit in the cementum at the base of the broken tooth at the leftmost tooth space. The incisive process (Fig. 16.8D) is short, bilobed, and incised by a deep central notch. The nasal process is wide at its base, but the full length is not known. There is a shallow median furrow on the external surface of the nasal process (Fig. 16.8E) superior to the tooth-bearing portion of the element. The maxillary process is preserved on the left side and is pierced by a small foramen (Fig. 16.8E), likely for an ethmoidal neurovascular bundle.

Discussion Very few premaxillae of fossil lizards have been described and even more rarely are they recognized to provide taxonomically diagnostic features. This specimen can be referred tentatively to Scincomorpha on the basis of the heavy deposits of basal cementum and the basal resorption pit. Because of the paucity of comparable materials, I assign this specimen to the unnamed Iron Springs morphotype B.

John Henry Member of the Straight Cliffs Formation (Santonian)

cf. SCINCOIDEA Oppel, 1811
PARAMACELLODID/CORDYLID GRADE
MONOCNEMODON gen. nov.

Etymology From the Greek *mono* (single, one), *knemos* (shoulder), and *odous* (tooth); in recognition of the expanded and shoulder-like crista mesialis of the tooth crown.

Type and Only Known Species *Monocnemodon syphakos* sp. nov.

Distribution Upper portion of the John Henry Member of the Straight Cliffs Formation (Santonian), Washington County, Utah.

Diagnosis As for type and only known species.

MONOCNEMODON SYPHAKOS gen. et sp. nov.

Etymology From the Greek *syphax* (sweet new wine) in recognition of the productive evenings with J. Eaton discussing the Cretaceous of southern Utah.

Holotype Specimen UMNH VP 17384 (Fig. 16.9A).

Referred Specimen UMNH VP 7699 (Fig. 16.9B).

Localities and Horizon UMNH Loc. 424 (type locality; uppermost John Henry Member [J. Eaton, pers. comm.] within Bryce Canyon National Park), Garfield County, Utah, approximately 7 km west of the town of Tropic; UMNH Loc. 99, Garfield County, Utah, approximately 11 km north of the town of Henrieville (see also Eaton, 2006).

Diagnosis A paramacellodid/cordylid-grade scincomorphan most similar to *Becklesius* in tooth morphology, but differing from *Becklesius* in having only weakly developed lingual striae on the tooth crown; hypertrophied crista mesialis well below level of tooth apex does not form an anterior cuspule; distal teeth more posteriorly rotated, not recurved. Differs from other scincomorphans with similar tooth crown morphology in that the hypertrophied crista mesialis lacks an anterior cusp or cuspule.

Description UMNH VP 17384 (Fig. 16.9A) is a partial right dentary lacking the anteriormost and posteriormost portions of the element. It preserves nine, closely spaced tooth positions, two complete/nearly complete teeth, and six teeth missing crowns. The bases of the teeth are heavily cemented. The best preserved tooth is pleurodont, tall, and posteriorly rotated. The crown has a very long crista mesialis that turns ventrally at a rounded angulus mesialis, and descends nearly vertically until it ends near the level of the lateral parapet. This results in a shoulder-like mesial expansion of the tooth crown such that there is a shallow mesially displaced fossa on the lingual surface of the crown. The shorter crista distalis joins the crista mesialis at distinct cuspis labialis. There is a short carina intercuspidalis joining the cuspis labialis to a shorter cuspis lingualis. From the cuspis lingualis descends poorly developed striae dominans anterior and posterior, which border a relatively broad, but poorly defined, pars furcata. The tooth just distal to this one has a similar morphology, but is less posteriorly rotated and has a broken tip. The next successively distal tooth position preserves only the tooth shaft, which is larger in diameter than either of the former teeth. Beginning with the mesialmost preserved

tooth position there are resorption pits at the first, third, fifth, and eighth tooth positions and the second tooth position is empty. The subdental shelf is tall anteriorly and narrows posteriorly commensurate with the posterior expansion of the splenial. The sulcus dentalis is narrow and deep. The labial surface of the dentary is convex, smooth, and pierced by three widely spaced external inferior alveolar foramina. The splenial is weakly concave, tapers gradually from posterior to anterior and then narrows abruptly near the anterior end of the specimen. The splenial completely encloses the anterior inferior alveolar foramen and the mylohyoid foramen.

UMNH VP 7699 (Fig. 16.B) is a fragment of a possible right maxilla preserving three teeth. The teeth are similar to those of UMNH VP 17384 in being pleurodont, closely spaced, but with a reduced amount of basal cementum (consistent with a maxilla). The crowns of the teeth all exhibit the hypertrophied crista mesialis seen in the teeth of UMNH VP 17384, but the crowns are only weakly posteriorly rotated (similar to the tooth in the dentary specimen missing only its apex). The lingual surfaces of the crowns have very subtle pars furcata.

Discussion The unique dental morphology of the type and referred specimens of *Monocnemodon syphakos* supports erection of a new genus and species.

SCINCOMORPHA Camp, 1923
JHMbr SCFm Morphotype A

Referred Specimen UMNH VP 7701 (Fig. 16.9C).

Locality UMNH Loc. 99, Garfield County, Utah, approximately 11 km north of the town of Henrieville.

Description UMNH VP 7701 is a jaw fragment (element indeterminate) preserving four tooth positions and two teeth. The teeth are pleurodont, tall, columnar, closely spaced, have a small amount of basal cementum, and have trifid crowns with a tall central cusp and symmetrical mesial and distal accessory cusps. A well-developed culmen lateris anterior and posterior descend from the mesial and distal accessory cusps, respectively, and terminate near the level of the lateral parapet. The lingual surfaces of the teeth are vertical and nearly flat. The labial surfaces of the tooth crowns are strongly convex.

Discussion A trifid tooth crown is a common feature of chamopsiid lizards (e.g., *Chamops*, *Leptochamops*, *Meniscognathus*) common to the Late Cretaceous of the Western Interior. However, the columnar shape of the teeth and the small amount of basal cementum do not support referral of UMNH VP 7701 to Chamopsiidae. The closest comparison is to OMNH 33867 (Fig. 16.13F, G) from the Kaiparowits Formation, but JHMbr SCFm morphotype A has a less prominent central cusp. The dental morphology appears to be unique, but as a result of the fragmentary nature of the specimen, it

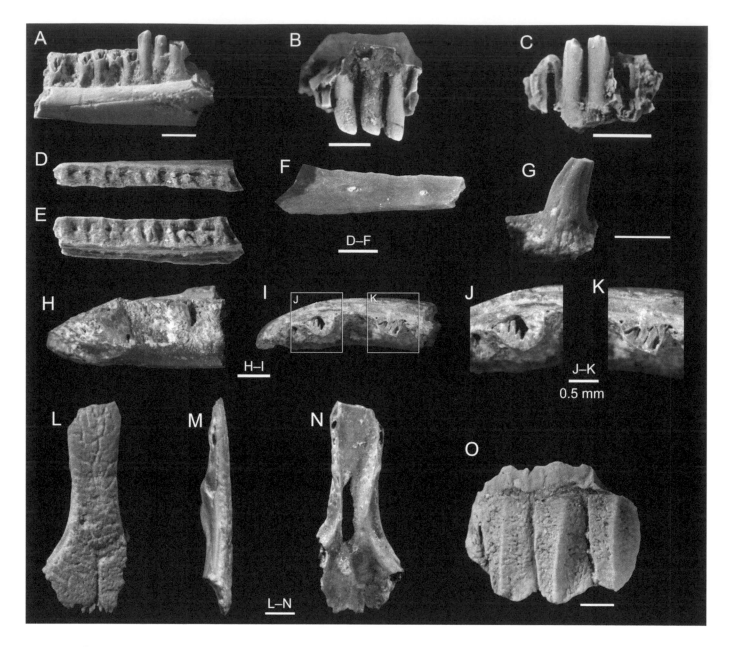

16.9. Lizards from the John Henry Member of the Straight Cliffs Formation (Santonian). (A, B) *Moncnemodon syphakos* gen. et sp. nov.: (A) UMNH VP 17384, partial right dentary in lingual view, (B) UMNH VP 7699, fragment of left maxilla(?) in lingual view. (C) JHMbr SCFm morphotype A: UMNH VP 7701, broken jaw in lingual view. (D–F) cf. *Colpodontosaurus,* UMNH VP 19246, broken left dentary of in (D) occlusal, (E) lingual, and (F) labial views. (G) JHMbr SCFm morphotype B, UMNH VP 19245, jaw fragment of in lingual view. (H–K) JHMbr SCFm morphotype C: UMNH VP 19247, broken right dentary of in (H) lingual and (I) occlusal views with boxes on (I) indicating areas of detail in (J) and (K). (L–N) JHMbr SCFm morphotype D: UMNH VP 17298, broken frontal in (L) dorsal, (M) left lateral, and (N) ventral views. (O) Scincomorpha indeterminate: UMNH VP 17342, three articulated osteoderms in dorsal view. Scale bars = 1 mm unless otherwise indicated.

is placed in an unnamed morphotype pending the recovery of more diagnostic material.

ANGUIMORPHA indeterminate
cf. *COLPODONTOSAURUS*

Referred Specimens UMNH VP 19246 (Fig. 16.9D–F).

Locality UMNH Loc. 424, Garfield County, Utah, approximately 7 km west of the town of Tropic.

Description UMNH VP 19246, is a partial right dentary missing the anteriormost and posteriormost portions as well as much of the lateral and inferior wall of the Meckelian fossa (Fig. 16.9D, E). The specimen preserves nine tooth positions as represented by remnants of tooth bases (Fig. 16.9D, E). The preserved tooth-bearing portion of the dentary is straight. The labial surface is smooth, weakly convex, and pierced by three small external inferior alveolar foramina (Fig. 16.9F).

Discussion Tentative referral to *Colpodontosaurus* is based on the similarity in the build of the dentary and tooth attachment mode to the specimens described by Estes (1964) and Gao and Fox (1996) for *C. cracens* from the Maastrichtian of Montana, Wyoming, and Alberta. As noted by Estes (1983) and Gao and Fox (1996), *Colpodontosaurus cracens* is a problematic taxon with regard to higher taxonomic referral, but both each cautiously retained it within Necrosauridae. "Necrosauridae" is widely considered to be para- or polyphyletic grouping of enigmatic platynotans (Pregill et al., 1986; Gao and Fox, 1996; Lee, 2002; Rieppel et al., 2007) and is currently under revision by B. Barr (University of Alberta). When compared with the known "necrosaurs" *Necrosaurus. eucarinatus* and *Eosaniwa koehni* from Germany, *N. cayluxi* from France, and *Parasaniwa wyomingensis* from North America (pers. obs. *Provaranosaurus* was reassigned to Xenosauridae by Smith, 2011), I find no morphological basis for the association of *Colpodontosaurus* with Necrosauridae or Platynota. I am tentatively retaining referral to Anguimorpha on the basis of the tooth shape (as illustrated for the type specimen of *Colpodontosaurus cracens*) and absence of a subdental gutter. This referral is pending a revision of *Colpodontosaurus* (Nydam and Caldwell, pers. comm.).

PLATYNOTA Duméril and Bibron, 1836
JHMbr SCFm Morphotype B

Referred Specimen UMNH VP 19245 (Fig. 16.9G).

Locality UMNH Loc. 424, Garfield County, Utah, approximately 7 km west of the town of Tropic.

Description UMNH VP 19245 (Fig. 16.9G) is a fragment of a jaw (element indeterminate) with a single tooth. The tooth is trenchant, recurved, has an expanded base with conspicuous, but weakly developed basal infolding of the enamel, and a basal nutrient foramen that opens lingually. The tip of the tooth is missing.

Discussion This specimen is classically platynotan in tooth morphology, but lacks sufficient comparable characteristics for taxonomic assessment. The basal infolding of the enamel is more pronounced than Dakota Formation morphotype E (Fig. 16.5G, I), but not nearly as well developed as in SHMbr SCFm morphotype I (Fig. 16.7E). The basal infolding is similar in degree of development to SHMbr SCFm morphotype J (Fig. 16.7F), but the tooth is more trenchant than conical. There is a possibility that SHMbr SCFm morphotype J and JHMbr SCFm morphotype B do represent the same taxon with morphological differences possibly attributable to ontogenetic variation at the genus level, if not species level. The teeth of the maxilla of *Primaderma nessovi* are conical anteriorly and trenchant posteriorly (Nydam,

2000:fig. 2) and taken together SHMbr SCFm morphotype J and JHMbr SCFm morphotype B represent a similar range of positional variation. However, both morphotypes are represented by specimens that are too fragmentary for more conclusive comparison.

JHMbr SCFm Morphotype C

Referred Specimen UMNH VP 19247 (Fig. 16.9H–K).

Locality UMNH Loc. 569, Garfield County, Utah, approximately 7 km west of the town of Tropic.

Description UMNH VP 19247 (Fig. 16.9H–K) is the anterior portion of right dentary preserving four tooth spaces and the bases of two teeth. The tooth bases are expanded and have infolded enamel and plicidentine (Fig. 16.9H, J, K). The mesialmost tooth base also has a basal nutrient foramen that opens posterolingually. There is no subdental shelf and the preserved portion of the Meckelian fossa is narrow, shallow, and terminates prior to the mandibular symphysis.

Discussion This specimen is classically platynotan in tooth base morphology and is placed in its own unnamed morphotype because of its much greater size than JHMbr SCFm morphotype B.

AUTARCHOGLOSSA Wagler, 1830
JHMbr SCFm Morphotype D

Referred Specimen UMNH VP 17298 (Fig. 16.9L–N).

Locality UMNH Loc. 424, Garfield County, Utah, approximately 7 km west of the town of Tropic.

Description UMNH VP 17298 is a fused frontal missing the posterolateral portions and the nasal articulation. The dorsal surface (Fig. 16.9L) is covered in fused osteoderms. There is a large frontal shield extending along the interorbital segment and this is separated from the paired frontoparietal shields by a V-shaped groove. There is also a narrow V-shaped groove between the posteromedian margins of the frontoparietal shields that represents the contact with the interfrontal or interparietal shields (indeterminate as a result of the lack of parietal). The frontal shield is ornamented by a series of vermiculate channels and the frontoparietal shields are ornamented anteriorly by similar channels that grade posteriorly to an unconnected series of small pits. The interorbital portion of the frontal is the narrowest part of the element, but the constriction is not dramatic and as the anteriormost portion of the element is not preserved it is not clear whether this qualifies as an hourglass-shaped frontal (sensu Conrad, 2008, and references therein). The crista cranii frontalis (=descending processes, Fig. 16.9M, N) are distinct and short, and they do not approach each other. There is a

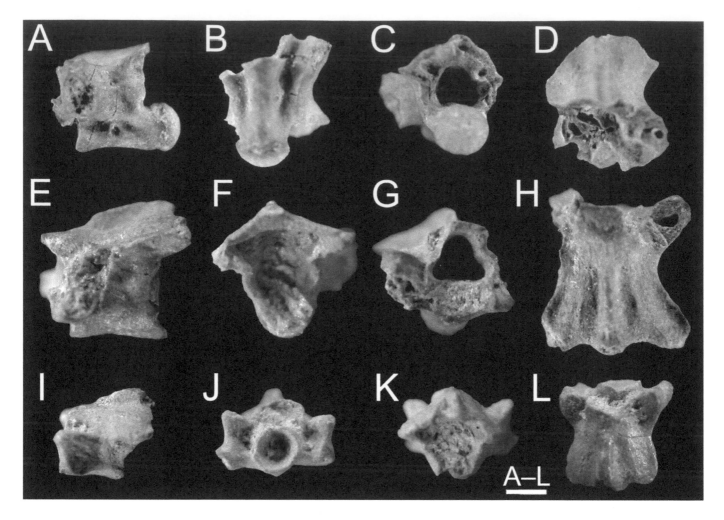

16.10. Serpentes from the John Henry Member of the Straight Cliffs Formation (Santonian). (A–L) *Coniophis* sp.: (A–D) UMNH VP 19242, vertebra in (A) left lateral, (B) ventral, (C) posterior, and (D) dorsal views; (E–H) UMNH VP 19243, vertebra in (E) right lateral, (F) posterior, (G) anterior, and (H) dorsal views; (I–L) UMNH VP 19243, vertebra in (I) right lateral, (J) anterior, (K) posterior, (L) dorsal views. Scale bars = 1 mm.

narrow groove along the anterolateral margins (Fig. 16.9M) for the articulation with the prefrontal. This groove has a raised central ridge bordered by a narrow channel superiorly, posteriorly, and inferiorly.

Discussion This specimen lacks sufficient morphological features to confidently place it within Anguimorpha, but the presences of cristae cranii frontales as simple downgrowths is cited by Conrad (2008:character 61) numerous times as one of the diagnostic features of numerous anguimorphan groups. The character state is similarly scored in his data set for nearly all lepidosauromorphans exclusive of Platynota. However, the osteodermal pattern on the specimen is currently known for some scincomorphans (e.g., Cordylidae) and anguids (e.g., Anguinae). On the basis of this distribution of such limited characters, I can only refer this frontal to an unnamed morphotype of an indeterminate autarchoglossan.

SCINCOMORPHA indeterminate
Osteoderm

Referred Specimen UMNH VP 17342 (Fig. 16.9O).

Locality UMNH Loc. 424, Garfield County, Utah, approximately 7 km west of the town of Tropic.

Description UMNH VP 17342 is a series of three osteoderms articulated along their lateral margins. The left margin of each osteoderm overlaps the right margin of the adjacent osteoderm indicating that these are from the right flank of the body (sensu Richter, 1994). The osteoderms are elongate, parallel sided, have convex posterior margins, anterior imbrication facets, and narrow lateral imbrication facets along the right margin. There is a well-defined dorsal keel and a rugose dorsal ornamentation. The rightmost osteoderm lacks a lateral imbrication facet and is most likely in a position of transition (e.g., lateral to ventral scuttelation).

Nydam

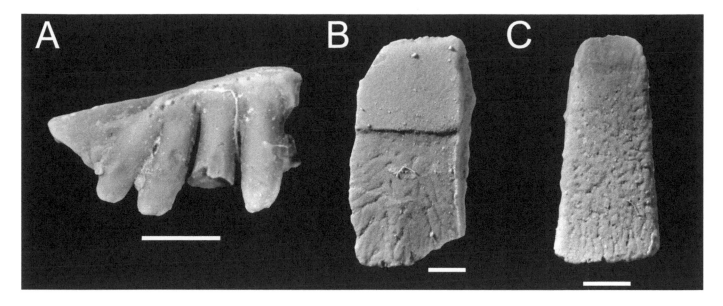

16.11. Lizards of the Wahweap Formation (early–middle Campanian). (A) Wahweap Formation morphotype A: OMNH 21995, broken left maxilla in lingual view. (B) cf. Scincomorpha: UMNH VP 19252, osteoderm in dorsal view. (C) cf. Anguimorpha: UMNH VP 19251, osteoderm in dorsal view. Scale bars = 1 mm.

Discussion Referral to Scincomorpha is based on the elongate shape and presence of dorsal keel on the osteoderms.

SERPENTES Linnaeus, 1758
CONIOPHIS Marsh, 1892
CONIOPHIS sp. indet.

Referred Specimens UMNH VP 19242, UMNH VP 19243, UMNH VP 19244 (Fig. 16.10A–L).

Locality UMNH Loc. 424, Garfield County, Utah, approximately 7 km west of the town of Tropic.

Description UMNH VP 19242 (Fig. 16.10A–D) is a trunk vertebra missing the anterior right portion. The neural arch is weakly convex dorsally and has a narrow, low spinous process (Fig. 16.10A, C, D). The zygosphenes are missing, but the deepest parts of the notch-like zygantra are still present (Fig. 16.10C) indicating a well-developed complementary zygosphene. The condyle is round and is slightly constricted at the neck. The hemal keel is blunt, low, and narrow at the condylar neck and widens anteriorly. There are deep lateral channels on either side of the hemal keel (Fig. 16.10B), and a large subcentral foramen in the medial wall of each channel near the midpoint of the centrum (Fig. 16.10A). The left synapophysis is partially preserved (Fig. 16.9A) and is as tall as the neural arch.

UMNH VP 19243 (Fig. 16.10E–H) is a trunk vertebra missing the condyle and the cotyle, as well as the prezygapophysis. The neural arch is nearly flat (Fig. 16.10E, F) with a narrow, low, posteriorly displaced neural spine (Fig. 16.10H). The postzygapophyses are nearly horizontal and there is a transverse zygantral slot in the superior, posterior portion of the

neural canal, but the zygantral facets are not preserved (Fig. 16.10F). The neural canal is triangular, particularly anteriorly (Fig. 16.10G). The ventral surface of the centrum is weakly convex and lacks a distinct hemal keel (Fig. 16.10E–G). The tall synapophysis is partially preserved on the left size (Fig. 16.10E).

UMNH VP 19244 (Fig. 16.10I–L) is a small caudal vertebra. The neural arch is low and has a short, posteriorly displaced neural spine (Fig. 16.10I, L). The prezygapophyses are tilted upward gently and are directly superior to the inferolaterally directed pleurapophysis (Fig. 16.10J). The cotyle is small and circular (Fig. 16.10J) and the condyle is missing. The postzygapophyses are poorly preserved, but do not appear to be as tilted as steeply as the prezygapophyses (Fig. 16.10K) possibly indicating a change in orientation along the vertebral column. The ventral surface of the centrum is weakly convex and has a weakly developed hemal keel (Fig. 16.10I, J).

Discussion The presence of a more distinct, albeit very low, neural spine and better developed hemal keel differentiate these vertebrae from those referred to *Coniophis* sp. from the Dakota Formation or the Smoky Hollow Member of the Straight Cliffs Formation.

Wahweap Formation (early–middle Campanian)

cf. SCINCOMORPHA indeterminate
Wahweap Formation Morphotype A

Referred Specimen OMNH 21995 (Fig. 16.11A).

Locality OMNH V2, Garfield County, Utah, approximately 23 km East of the town of Henrieville.

Description OMNH 21995 (Fig. 16.11A) is the posterior portion of a left maxilla preserving the posteriormost four tooth positions, one broken tooth, and three complete teeth. The teeth are columnar, have worn, blunt tips, and become more posteriorly directed in successively more posterior positions. There is a moderate amount of cementum at the tooth bases. The jugal process is posterolaterally oriented and deeply grooved for the maxillary process of the jugal.

Discussion The specimen is tentatively referred to Scincomorpha on the basis of the relatively heavy cementum (for a maxilla). The orientation of the jugal process of the maxilla is similar to that of contogeniids (sensu Nydam and Fitzpatrick, 2009), but the tooth crowns do not exhibit this group's distinctive apical groove.

cf. SCINCOMORPHA indeterminate
Osteoderm

Referred Specimen UMNH VP 19252 (Fig. 16.11B).

Locality UMNH Loc. 103, Garfield County, Utah, approximately 23 km East of the town of Henrieville.

Description UMNH VP 19252 (Fig. 16.11B) is a rectangular osteoderm with a large anterior imbrication facet and narrow lateral imbrication facet on the right side. There is very weakly developed median dorsal keel and the dorsal ornamentation consists of a series of small channels coursing posterolaterally from the median keel.

Discussion Tentative referral to Scincomorpha is based on the elongate rectangular shape and weak dorsal keel.

cf. ANGUIMORPHA indeterminate
Osteoderm

Referred Specimen UMNH VP 19251 (Fig. 16.11C).

Locality UMNH Loc. 103, Garfield County, Utah, approximately 23 km east of the town of Henrieville.

Description UMNH VP 19251 (Fig. 16.11C) is an elongate, trapezoidal osteoderm that is narrower anteriorly. There is a small anterior imbrication facet and the ornamentation is a subtle series of vermiculate channels.

Discussion Referral to Anguimorpha is tentative and is based on the lack of a keel, the subtle vermiculate ornamentation, and the thin build of the osteoderm.

SERPENTES Linnaeus, 1758
CONIOPHIS Marsh, 1892
CONIOPHIS sp.

Referred Specimens UMNH VP 19248, UMNH VP 19249, UMNH VP 19250 (Fig. 16.12A–J).

Locality UMNH Loc. 103, Garfield County, Utah, approximately 23 km east of the town of Henrieville.

Description UMNH VP 19248 (Fig. 16.12A–C) is a trunk vertebra missing the right prezygapophyses. The specimen is weakly dorsoventrally compressed. The cotyle is oval with a longer mediolateral axis than dorsoventral axis (Fig. 16.12A). The left synapophyses is the best preserved (Fig. 16.12A) and is bilobed with a larger diapophysis than parapophysis. The prezygapophyses is gently tilted superiorly (Fig. 16.12A) and the postzygapophyseal facets are more flat (Fig. 16.12B). The condyle is round and is unconstructed at the neck (Fig. 16.12B). The neural arch is broad and flat with a low, posteriorly displaced neural spine that extends posteriorly as a small, midline process (Fig. 16.12A–C). The zygsophene is flat and narrow and constricted near the level of the anterior extension of the neural spine (Fig. 16.12C).

UMNH VP 19249 (Fig. 16.12D–F) is a trunk vertebra with a weakly oval cotyle, smaller synapophysis than UMNH VP 19251 and nearly flat prezygapophyseal articulation facets (Fig. 16.12D). The condyle is round (Fig. 16.12E), posterodorsally oriented, and has a weal constriction of the neck. The postzygapophyseal facets are nearly horizontal (Fig. 16.12E). The neural arch is broad and nearly flat (Fig. 16.12D–F) with a weakly developed neural arch that becomes a small posterior process superior to the zygantral facets (Fig. 16.12F). The zygosphene is broad and flat with a weakly constricted neck (Fig. 16.12F).

UMNH VP 19250 (Fig. 16.12G–J) is a trunk vertebra missing portions of the right side. The anterior portion (Fig. 16.12G) is heavily damaged, but the left prezygapophyses is preserved and oriented dorsoventrally. The condyle is oval and slightly upturned and the neural arch and zygantrum are obscured by matrix (Fig. 16.12H, J). The postzygapophyseal facet on the left side is slightly rotated anteriorly (Fig. 16.12H, J). The neural arch is broad and flat with a narrow, low neural spine that forms a small posterior process (Fig. 16.12I, J). There is a constriction posterior to the zygosphene on the left side; the right side is missing (Fig. 16.12I).

Discussion With respect to the presence of the midline posterior process of the posterior margin of the neural arch, this specimen cannot be compared to the *Coniophis* sp. vertebrae from the Dakota Formation or the Smoky Hollow Member of the Straight Cliffs Formation as the known specimens from those formations are missing this portion of the neural arch. However, this feature is also seen in the vertebrae of *Coniophis* sp. from the John Henry Member of the Straight Cliffs Formation. A smaller process is also present on the vertebrae of *C. cosgriffi* from the Fruitland Formation of New Mexico (Armstrong-Ziegler, 1978) and *Coniophis* sp. from the Cedar Mountain Formation (Gardner

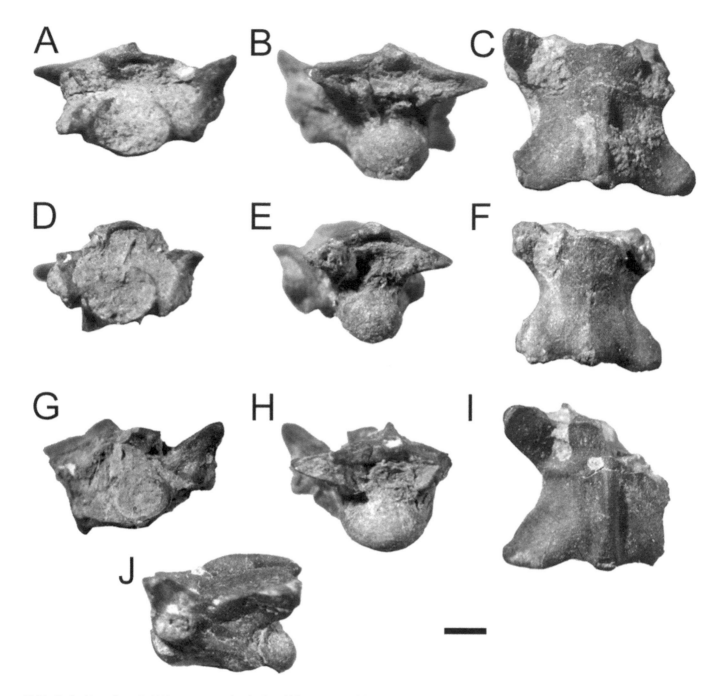

16.12. *Coniophis* sp. from the Wahweap Formation (early–middle Campanian). (A–C) UMNH VP 19248, vertebra in (A) anterior, (B) posterior, and (C) dorsal views. (D–F) UMNH VP 19249, vertebra in (D) anterior, (E) posterior, and (F) dorsal views. (G–J) UMNH VP 19250, vertebra in (G) anterior, (H) posterior, (I) dorsal, and (J) left lateral views. Scale bar = 1 mm.

and Cifelli, 1999:fig. 2), but is lacking in vertebrae from the Kaiparowits Formation (Fig. 16.17A, C). The constriction of the neck of the zygosphene in the Wahweap Formation specimens is more pronounced than the subtle constriction seen in the *Coniophis* sp. from the Smoky Hollow Member of the Straight Cliffs Formation (Fig. 16.7Q) and the

Cedar Mountain Formation or in *C. precedens*. However, a more pronounced, but still less fully developed, constriction is seen in vertebrae of *Coniophis* sp. from the Dakota Formation (Fig. 16.5N) and the Kaiparowits Formation (Fig. 16.17C).

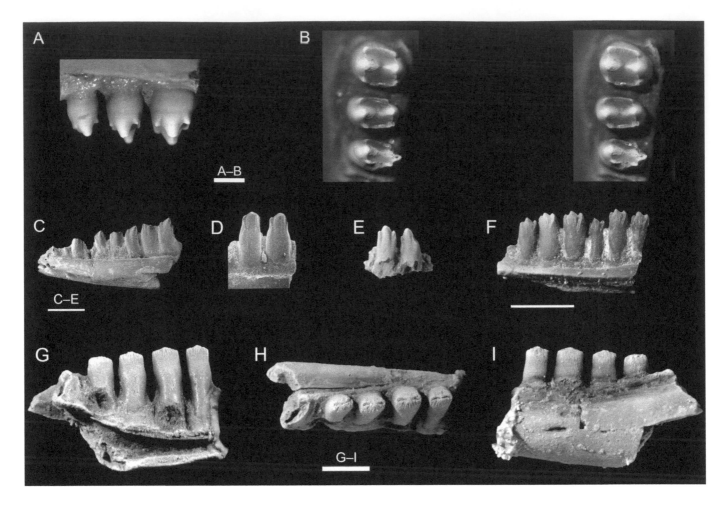

16.13. Previously described lizards from the Kaiparowits Formation (upper Campanian). (A–B) *Peneteius saueri:* OMNH 20904, partial maxilla of in (A) lingual and (B) stereopair occlusal views. (C) *Meniscognathus molybrochoros:* OMNH 23743, partial right dentary in lingual view. (D) *Chamops* sp.: scanning electron micrograph (SEM) of OMNH 33852, jaw fragment in lingual view. (E) cf. *Leptochamops:* OMNH 65142, jaw fragment in lingual view. (F) *Trippenaculus eatoni:* SEM of OMNH 23146, partial right dentary in lingual view. (G–I) *Paleoscincosaurus pharkidodon:* OMNH 33852, partial left dentary in (G) lingual; (H) occlusal; and (I) labial views. All scale bars = 1 mm. (A) and (B) modified from Nydam, Eaton, and Sankey (2007) (C–F) modified from Nydam and Voci (2007) (G–I) modified from Nydam and Fitzpatrick (2009).

Kaiparowits Formation (upper Campanian)

SCINCOMORPHA Camp, 1923
BORIOTEIIOIDEA Nydam,
Eaton, and Sankey, 2007
POLYGLYPHANODONTINAE Estes, 1983
POLYGLYPHANODONTINI Nydam,
Eaton, and Sankey, 2007
PENETEIUS Estes, 1969
PENETEIUS SAUERI McCord, 1998

Manangyasaurus saueri McCord, 1998:285, fig. 2 (original description)
Peneteius saueri Nydam, Eaton, and Sankey, 2007:544, figs. 4.1–4.21 (new combination)
Holotype OMNH 20904 (Fig. 16.13A–B).
Paratypes OMNH 21832, OMNH 23524, OMNH 23862, UMNH VP 11651, UMNH VP 11652, UMNH VP 11653, UMNH VP11654 (includes UMNH VP 11651), UMNH VP 11655, UMNH VP11658.

Discussion *Peneteius* was erected by Estes (1969) for the species *Peneteius aquilonius* from the Hell Creek Formation (Maastrichtian) of Montana. Nydam, Gauthier, and Chiment (2000) redescribed *P. aquilonius* on the basis of additional materials from the Hell Creek Formation. McCord (1998) described *Manangyasaurus saueri*, a broken maxilla with three teeth that was recovered from the Kaiparowits Formation. The type specimen and additional previously undescribed tooth and jaw specimens of *Manangyasaurus saueri* were transferred to *Peneteius* (*P. saueri*) by Nydam, Eaton, and Sankey (2007) on the basis of similarities in tooth morphology to *P. aquilonius*. Additional records of isolated teeth of *Peneteius* are known from the Campanian of Texas (Nydam, Eaton, and Sankey, 2007) and the Maastrichtian of New Mexico (Williamson and Weil, 2004).

MENISCOGNATHUS Estes, 1964
MENISCOGNATHUS MOLYBROCHOROS
Nydam and Voci, 2007

Meniscognathus molybrochoros Nydam and Voci, 2007:fig. 3
Holotype OMNH 23743 (Fig. 16.13C).
Discussion The presence of a new species of *Menis-cognathus* in the Kaiparowits Formation represents both a stratigraphic and geographical range extension of the genus (Nydam and Voci, 2007).

CHAMOPS Marsh, 1892
CHAMOPS sp. cf. C. SEGNIS Marsh, 1892

Chamops sp. cf. *C. segnis* Marsh, 1892; Nydam and Voci, 2007, fig. 3
Referred Specimen OMNH 33852 (Fig. 16.13D).
Discussion The presence of *Chamops* in the Kaiparowits Formation is based on fragmentary jaw remains with teeth indistinguishable from those of *Chamops segnis* (Nydam and Voci, 2007). The known specimens are, however, too fragmentary to evaluate other potentially diagnostic features of jaw morphology (i.e., structure of Meckelian fossa, degree of heterodonty, structure of mandibular symphysis). The presence of *Chamops* in the Kaiparowits Formation provides a geographically occurrence of the taxon between the Judith River Formation of Montana and the Fruitland Formation of New Mexico.

Comment For over 100 years, since Marsh's 1892 original description, the presence of *Chamops* (either *C. segnis* or *Chamops* sp.) has been ubiquitous in nearly all microvertebrate faunas of the Late Cretaceous of the Western Interior of North America, and it has been identified from nearly all Campanian–Maastrichtian terrestrial faunas ranging geographically from New Mexico to Canada (Marsh, 1892; Sternberg, 1951; Estes, 1964; Waldman, 1970; Sahni, 1972; Armstrong-Ziegler, 1978, 1980; Sullivan, 1981; Bryant, 1989; Gao and Fox, 1996; Eaton et al., 1999b; Peng, Russell, and Brinkman, 2001; Nydam and Voci, 2007). Gao and Fox (1996) described a series of "teiid" lizards from the Late Cretaceous of Canada that share with general jaw and tooth characteristics with *Chamops*. It has been suggested that this diversity represents no more than variation within *Chamops* (Denton and O'Neill, 1995), but on the basis of personal observation of the specimens, I support the taxonomic conclusions of Gao and Fox (1996) including their concerns regarding the accuracy of some of the previous identifications of *Chamops* (Sternberg, 1951; Waldman, 1970; Armstrong-Ziegler, 1978, 1980; Sullivan, 1981). The occurrence of Chamops in the Kaiparowits Formation continues the trend of the presence of this taxon in the Late Cretaceous of the Western Interior.

Additionally, the presence of *Chamops* in the Turonian of Kaiparowits Plateau indicates that this taxon, at least at the generic level, had its greatest known duration in southern Utah.

LEPTOCHAMOPS Estes, 1964
cf. LEPTOCHAMOPS

cf. *Leptochamops*, Nydam and Voci, 2007, fig. 3
Referred Specimens OMNH 23863, OMNH 64151, OMNH 64152 (Fig. 16.13E).
Discussion As with *Chamops*, the specimens from the Kaiparowits Formation that are referable to *Leptochamops* are fragmentary jaws that are too incomplete for more precise taxonomic treatment (Nydam and Voci, 2007). The presence of Leptochamops in the Kaiparowits Formation if confirmed is the southernmost record for the genus in the Western Interior.

TRIPENNACULUS Nydam and Voci, 2007
TRIPENNACULUS EATONI
Nydam and Voci, 2007

Tripennaculus eatoni Nydam and Voci, 2007, fig. 2
Holotype OMNH 23146 (Fig. 16.13F).
Discussion *Tripennaculus* is closely related to *Chamops*, *Leptochamops*, and *Meniscognathus*, but unlike these taxa, the teeth of *Tripennaculus* have well-developed mesial and distal cusps flanking a tall central cusp (Nydam and Voci, 2007).

CONTOGENIIDAE Nydam and Fitzpartrick, 2009
PALAEOSCINCOSAURUS PHARKIDODON
Nydam and Fitzpatrick, 2009
(Fig. 16.13G–I; Nydam and Fitzpatrick, 2009:fig. 6)

Holotype OMNH 33855.
Discussion *Palaeoscincosaurus* is the largest of the contongeniid lizards from the Cretaceous. In southern Utah the contogeniids are also represented by *Utahgenys atongai* and *U. evansi* (above). Nydam and Fitzpatrick (2009) review in detail the fossil record of contongeniids in North America.

cf. SCINCOIDEA Oppel, 1811
PARAMACELLODID/CORDYLID GRADE
Kaiparowits Formation Morphotype A

Referred Specimens OMNH 33863, OMNH 33883 (Fig. 16.14A–E).
Locality OMNH V9, Kane County, Utah, approximately 26 km southeast of the town of Henrieville.

Description OMNH 33883 (Fig. 16.14A, B) is a partial right dentary missing the anteriormost and posteriormost portions of the element as well as much of the lateral wall of the Meckelian fossa. The three in situ teeth are subpleurodont with heavy deposits of cementum at the bases. Resorption pits are present at the bases of the two posteriormost preserved teeth and at the anteriormost preserved tooth space (tooth crown missing). The tooth crown has long, well-developed cristae mesialis and distalis (Fig. 16.14B). The cuspis labialis is offset posteriorly such that the crista mesialis is longer than the crista distalis. The effect of this inequity in crista length is a false recurve appearance of the crown. Well-defined and widely separated anterior and posterior striae dominans converge apically at a cuspis lingualis. Additional striae present on the crown's lingual surface between the cristae and striae dominans as well as between both striae dominans. OMNH 33863 (Fig. 16.14C–E) is a premaxilla missing the nasal process and preserves six tooth spaces with three in situ teeth. The rostral surface is smooth. The lateral aspects of the element preserve well-defined fossa for articulation with the maxilla (Fig. 16.14C). The teeth are pleurodont (less than one half of tooth height extends beyond the lateral parapet; sensu Gao and Fox, 1996) and closely spaced. The leftmost preserved tooth has a resorption pit at its base and the other preserved teeth have small nutrient foramina at their bases (Fig. 16.14D). The tooth crown morphology of the best preserved tooth of OMNH 33863 (Fig. 16.14E) is very similar to that of OMNH 33883. The main difference between them is the reduced number of lingual striae on the former. This difference is likely attributable to positional differences and both OMNH 33863 and 33883 are referred to the same morphotype (i.e., unnamed taxon).

Discussion The closest comparison of dental characteristics of morphotype A is to *Bothriagenys mysterion* Nydam, 2002, from the basal Upper Cretaceous of Utah (Cedar Mountain Formation). *Bothriagenys mysterion* was originally referred to family incertae sedis, but the very close similarity of the tooth crown morphology to *Paramacellodus oweni* Hoffstetter, 1967, and *Paramacellodus* sp. (Seiffert, 1973; Evans and Chure, 1999) supports referral to a paramacellodid/cordylids grade taxon. Similar tooth morphology is also described for the putative skink *Aocnodromeus corrugatus* from the Upper Cretaceous of Canada (Gao and Fox, 1996) equally supportive of the opposing interpretations of (1) close relationship of *Aocnodromeus* and *Paramacellodus*, or, (2) the presence of this morphology in more than a single clade. Both *Paramacellodus* and *Aocnodromeus* have anterior and posterior striae dominans that are more closely spaced than is seen in morphotype A or *Bothriagenys*.

Kaiparowits Formation Morphotype B

Referred Specimens OMNH 33867, OMNH 33879 (Fig. 16.14F–H).

Locality OMNH V9, Kane County, Utah, approximately 26 km southeast of the town of Henrieville (OMNH specimens).

Description The teeth of OMNH 33867 (Fig. 16.14F) are subpleurodont (more than half the tooth height extends beyond the lateral parapet; sensu Gao and Fox, 1996), have swollen bases, and crowns that are mesiodistally constricted and laterally compressed. The largest tooth has a large resorption pit at its base. The apex of the crown is the cuspis labialis, which is continuous with broad, low angle cristae mesialis and distalis (Fig. 16.14G). The cristae mesialis and distalis extend mesially and distally to large, rounded anguli mesialis and distalis, respectively. Prominent anterior and posterior culmen lateri extend basally from the anguli nearly approximately half the remaining distance to the lateral parapet of the jaw. The lingual surface of the crown has a prominent cuspis lingualis subequal in height to the cuspis labialis. The carina intercuspidalis is not well developed. Descending basally from the cuspis lingualis are prominent anterior and posterior striae dominans that extend nearly as far as the culmen lateris. Weakly developed striae are present near the apex of the tooth crowns between the cristae and the striae prominens.

OMNH 33879 (Fig. 16.14H) is a fragment of a left dentary preserving two tooth spaces that are each occupied by a complete tooth. The teeth are subpleurodont, tall, weakly conical, have truncated crowns, and a ring of basal cementum. The crowns have well-developed cristae mesialis and distalis, but lack anguli mesialis or distalis. There is a well-developed cuspis lingualis from which descends widely spaced, weakly developed striae dominans anterior and posterior which form the borders of a wide pars furcata. The subdental shelf and sulcus dentalis are absent.

Discussion The tooth crown morphology of OMNH 30748 is most similar to that of the specimens of the paramacellodid *Paramacellodus oweni* Hoffstetter, 1967 (i.e., BMNH R8131) from the Purbeck of England as well as to the specimens referred to *Becklesius* by Seiffert (1973; but see Evans and Searle, 2002) from the Guimarota locality in Portugal. In particular, the expanded, rounded, shoulder-like anguli mesialis and distalis in combination with the unstriated lingual surface is very similar to *Becklesius* sp. as figured in Kosma (2004:pl. 16, figs. 6, 7). OMNH 30748 differs from both *Becklesius* sp. and *B. hoffsetteri* Seiffert, 1973, in having cristae mesialis and distalis that are equivalent in length giving the crown a symmetrical shape. In the specimens of *Becklesius* from the Upper Jurassic of Portugal, the crista mesialis is

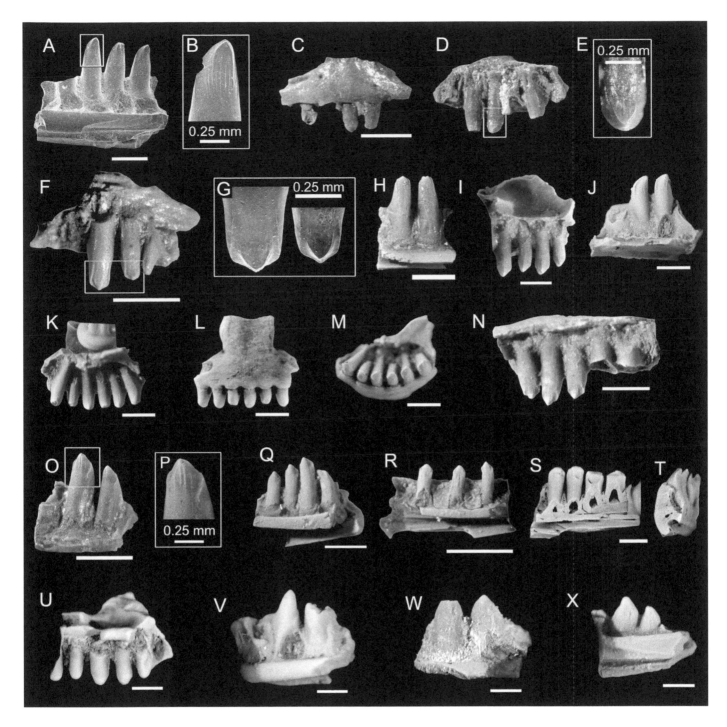

16.14. Scincomorphan jaws from the Kaiparowits Formation (upper Campanian). (A–E) Kaiparowits Formation morphotype A: (A, B) scanning electron micrograph of OMNH 33883, broken right dentary in (A) lingual view with box indicating area of detail in (B); (C, D) OMNH 33863, broken premaxilla in (C) rostral and (D) lingual views with box indicating area of detail in (E). (F–H) Kaiparowits Formation morphotype B: (F, G) OMNH 33867, broken maxilla in (F) lingual view with box indicating area of detail in (G); (H) OMNH 33879, broken left dentary in lingual view. (I–M) Kaiparowits Formation morphotype C: (I) OMNH 20264, broken left maxilla in lingual view; (J) OMNH 30750, broken right dentary in lingual view; (K–M) OMNH 20113, premaxilla in (K) lingual, (L) rostral, and (M) occlusal views. (N–R) Kaiparowits Formation morphotype D: (N) OMNH 30754, broken left maxilla in lingual view; (O, P) OMNH 30757, jaw fragment in (O) lingual view with box indicating area of detail in (P); (Q) OMNH 23590, broken left dentary in lingual view; and (R) OMNH 30772, broken left dentary in lingual view. (S, T) Kaiparowits Formation morphotype E: OMNH 63529, broken right(?) dentary of in (S) lingual and (T) mesial views. (U, V) Kaiparowits Formation morphotype F: (U) OMNH 33865, broken left maxilla in lingual view; (V) OMNH 20267, jaw fragment in lingual view. (W, X) Kaiparowits Formation morphotype G: (W) OMNH 30751, jaw fragment in lingual view; (X) OMNH 21167, broken left dentary in lingual view. Scale bars = 1 mm unless otherwise noted.

longer than the crista distalis (Seiffert, 1973:fig. 20; Kosma, 2004:pl. 16, figs. 1–8). This gives the crown an asymmetrical shape or what has previously been referred to as false recurvature (Nydam and Cifelli, 2002b). OMNH 33879 is referred to this morphotype on the basis of the structure of the tooth crowns. However, the lack of a subdental shelf and sulcus dentalis is more typically found in anguimorphans and is unusual for a scincomorphan.

With the exception of a taller central cuspule, the teeth of this morphotype are very similar to those of JHMbr SCFm morphotype A (Fig. 16.9C) and likely represent the same taxon, at least at the genus level.

Kaiparowits Formation Morphotype C

Referred Specimens OMNH 20113, OMNH 20264, OMNH 30750 (Fig. 16.14I–M).

Localities OMNH V6 (specimens OMNH 20113 and 20264), Kane County, Utah, approximately 22 km southeast of the town of Henrieville, and, OMNH V9 (specimen OMNH 30750), Kane County, Utah, approximately 26 km southeast of the town of Henrieville.

Description OMNH 20264 (Fig. 16.14I) is the anterior portion of a left maxilla that partially preserves the premaxillary and dorsal processes. The preserved portions of these processes enclose a deep fossa for the Jacobson's organ. The four preserved teeth are pleurodont and have moderate amounts of cementum at their bases. There are resorption pits in various stages of development at the bases of the three anterior teeth. The tooth shafts are straight and near the tip of the tooth the mesial and distal cristae taper quickly from shoulder-like mesial and distal anguli that converge at a cuspis labialis. The central portion of the crown just basal to the apex is swollen lingually and bears one, or two closely spaced, prominent striae dominans. The striae extend basally from a weakly developed cuspis lingualis. Additional more weakly developed lingual striae are inconsistently present mesially or distally of the prominent central striae. The tooth crowns of OMNH 30750 (Fig. 16.14J) are very similar except that they are rotated posteriorly, have a slightly more pronounced mesial crista, and slightly more prominent anterior and posterior striae dominans.

The premaxilla (OMNH 20113; Fig. 16.14K–M) has a broad nasal process with a shallow central depression on the rostral surface. The process is broken and its length cannot be determined. The lateral portions of the premaxilla are incised with short facets for articulation with the maxilla. The external surface of the parapet is scalloped at each tooth position. The basal process (sensu Conrad, 2004) is partially preserved on the right side and indicates that there was likely a broad articulation with the vomers (Fig. 16.14M). The teeth of the premaxilla are pleurodont and have similar crown morphology to OMNH 30750.

Discussion As with morphotypes A and B, the tooth crown morphology of morphotype C suggests that these specimens represent another possible paramacellodid taxon in the Kaiparowits Formation. Kaiparowits Formation morphotype C differs from Kaiparowits Formation morphotypes A and B in having tooth crowns with more closely spaced anterior and posterior striae dominans and less prominent cuspis lingualis.

Kaiparowits Formation Morphotype D

Referred Specimens OMNH 23590, OMNH 30754, OMNH 30757, OMNH 30772 (Fig. 16.14N–R).

Localities OMNH V6 (specimen OMNH 23590), Kane County, Utah, approximately 22 km southeast of the town of Henrieville, and, V9 (specimens OMNH 30754, 30757, 30772), Kane County, Utah, approximately 26 km southeast of the town of Henrieville.

Description OMNH 30754 (Fig. 16.14N) is an incomplete left maxilla preserving the middle (approximately) portion of the tooth row. The teeth are closely spaced, subpleurodont, and have modest amounts of cementum at the tooth bases. A small resorption pit is at the base of the posteriormost preserved tooth. OMNH 30757 (Fig. 16.14O, P) preserves two teeth with a similar mode of attachment. The four preserved teeth on OMNH 23590 (Fig. 16.14Q) have thicker bands of basal cementum. OMNH, 23590, 30754, and 30757 all have teeth with identical crown morphology. The mesial and distal cristae are well developed and extend between the apical cuspis labialis and the pronounced anguli mesialis and distalis, respectively. The anterior culmen lateralis and posterior culmen lateralis turn lingually as they extend basally from the anguli mesialis and distalis (Fig. 16.14P). The lingual surface of the crown has two prominent striae dominans that converge at a weakly developed cuspis lingualis. The carina intercuspidalis is very short.

OMNH 30772 is a broken left dentary preserving six tooth spaces and three complete teeth. The teeth are similar to those of OMNH 30757 and 23590 (Fig. 16.14N, O, and P, respectively). The diminutive size of this specimen suggests that it is from a juvenile.

Discussion Kaiparowits Formation morphotype D differs from Kaiparowits Formation morphotype C in having teeth with broader bases, more strongly developed anguli mesialis and distalis, and in having an anterior culmen lateralis and a posterior culmen lateralis that courses basolingually as opposed to basally. It is possible that these differences in tooth morphology reflect positional differences along the tooth row of a single taxon. Complete or nearly complete dentaries of

Becklesius hoffstetteri from the Isle of Wight, England (BMNH R8120, left dentary) and Guimarota, Portugal (e.g., MG-LNEG Gui Squ 715, a left dentary) have teeth that gradually change their crown morphology along the tooth row. Tooth crowns in the anterior portion of the tooth row are similar to Kaiparowits Formation morphotype C and those in the posterior portion of the tooth row are more similar to Kaiparowits Formation morphotype D (pers. obs.). However, none of the specimens of either Kaiparowits Formation morphotype C or D from the Kaiparowits Formation preserves both types of teeth or teeth with crowns of an acceptable intermediate grade of morphology. As such I retain separate morphotypes for the exhibited crown morphologies of these specimens. However, as shown by Evans and Searle (2002) the specimens from Guimarota referred to *Becklesius hoffstetteri* do not have the same dramatic expansion of the tooth crowns that are diagnostic of type specimens of *Becklesius hoffstetteri* from the Purbeck Limestone. As such the referral of the Portuguese material to this taxon should be treated as suspect. This uncertainty and the relatively incomplete nature of the specimens from the Kaiparowits Formation preclude a more formal taxonomic designation.

Scincomorpha Family indeterminate
Kaiparowits Formation Morphotype E

Referred Specimen OMNH 63529 (Fig. 16.14S, T).

Locality OMNH V5, Garfield County, Utah, approximately 14 km northeast of the town of Henrieville.

Description OMNH 63529 is most likely a broken right dentary missing much of the ventral portion of the element and both the anterior and posterior ends. In addition, the subdental shelf and lingualmost portions of the tooth bases have been sheared away (Fig. 16.14S). The teeth are pleurodont and lingually concave (Fig. 16.14T). The tooth crowns are broad and spatulate.

Comments Kaiparowits Formation morphotype E clearly differs from the other scincomorphans in the fauna in having distinctly lingually concave teeth.

Kaiparowits Formation Morphotype F

Referred Specimens OMNH 3386, OMNH 20267 (Fig. 16.14U, V).

Localities OMNH V6 (specimen OMNH 20267), Kane County, Utah, approximately 22 km southeast of the town of Henrieville, and, V9 (specimen OMNH 33865), Kane County, Utah, approximately 26 km southeast of the town of Henrieville.

Description OMNH 33865 (Fig. 16.14U) is the anterior portion of a left maxilla that is missing the premaxillary process and much of the dorsal process. The preserved teeth are pleurodont, widely spaced, and have a moderate amount of cementum at the bases. There are resorption pits at the bases of two teeth (first and third from anterior). The tooth shafts are columnar and the tooth crowns are bluntly conical, lacking any additional distinguishing features. OMNH 20267 (Fig. 16.14V) is a jaw fragment of uncertain position. The complete tooth is simple, conical, with a large resorption pit at its base. The incomplete tooth has a larger base surrounded by a moderate amount of cementum.

Discussion These specimens are grouped into this morphotype on the basis of a shared simplicity of their teeth. It is possible that they belong to separate taxa, but the lack of any distinguishing features makes it unlikely they are referable to any of the potential taxa represented by the Kaiparowits Formation morphotypes A–E.

Kaiparowits Formation Morphotype G

Referred Specimens OMNH 21167, OMNH 30751 (Fig. 16.14W, X).

Locality OMNH V5 (specimen OMNH 21167), Garfield County, Utah, approximately 14 km northeast of the town of Henrieville and OMNH V9 (specimen OMNH 30751), Kane County, Utah, approximately 26 km southeast of the town of Henrieville.

Description OMNH 30751 (Fig. 16.14W) is an indeterminate jaw fragment with two teeth that are subpleurodont and laterally compressed. In lateral view these teeth are triangular. The tooth crowns are smooth. OMNH 21167 (Fig. 16.14X) is a posterior portion of a left dentary. The subdental shelf is tall anteriorly with a posteriorly broadening facet for the splenial. The teeth are subpleurodont, weakly laterally compressed, and weakly recurved. There are 2–3 widely spaced shallow grooves that extend from the base to the apex of each tooth.

Discussion As with the specimens of Kaiparowits Formation morphotype F, these specimens are grouped into Kaiparowits Formation morphotype G on the basis of shared triangular tooth shape. It is also possible they belong to separate taxa, but the lack of distinguishing features makes it unlikely that they are referable to any of the potential taxa represented by Kaiparowits Formation morphotypes A–E. It may turn out that Kaiparowits Formation morphotypes F and G belong to the same taxon. However, on the basis of the observable differences (e.g., lateral compression of tooth crowns and greater mesiodistal width of tooth bases in morphotype G) in all of these specimens, I retain two distinct morphotypes to account for the apparent diversity.

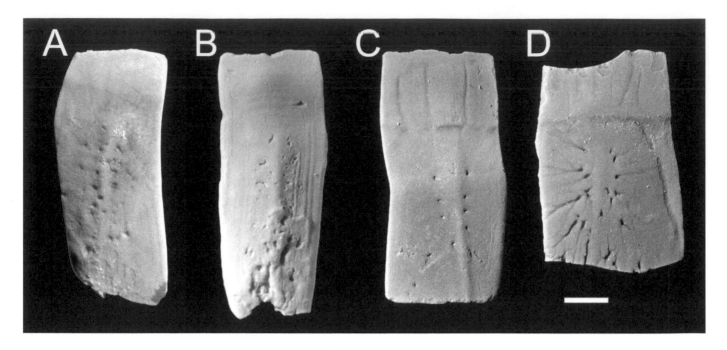

16.15. Scincomorphan osteoderms from the Kaiparowits Formation (upper Campanian) all in dorsal view. (A) UMNH VP 18056; (B) OMNH 66080; (C) OMNH 66073; (D) OMNH 66089. Scale bar = 1 mm.

Osteoderms

Referred Specimens OMNH 66073, OMNH 66080, OMNH 66089, UMNH VP 18056 (Fig. 16.15A–D).

Localities OMNH V5 (specimens OMNH 66073, 66089), Garfield County, Utah, approximately 14 km northeast of the town of Henrieville; OMNH V61 (specimen OMNH 66080), Garfield County, Utah, approximately 17 km northeast of the town of Henrieville; and UMNH Loc. 108 (specimen UMNH VP 18056), Garfield County, Utah, approximately 14 km northeast of the town of Henrieville (equivalent to OMNH V5).

Description UMNH VP 18056 (Fig. 16.15A) is a rectangular osteoderm with a weak central dorsal keel. There are numerous small pits on the dorsal surface concentrated near the central keel. The anterior imbrication facet is smooth and accounts for approximately ¼ of the total length of the specimen. OMNH 66080 (Fig. 16.15B) is similar in size and shape to UMNH VP 18056 except that the posterior end is abraded. OMNH 66073 (Fig. 16.15C) is rectangular, but slightly wider than the former two specimens. It also has a weak dorsal keel with a row of four small pits on either side of the keel. The imbrication facet is separated from the body of the osteoderm by an incomplete horizontal ridge and there are three shallow, longitudinal grooves on the dorsal surface of the facet. OMNH 66089 (Fig. 16.15D) is broken posteriorly and is missing a small portion of the imbrication shelf. The weak dorsal keel is flanked by a series of small pits. The pits are continuous with narrow grooves that radiate toward the

margins of the osteoderm. The imbrication shelf is offset by a horizontal ridge and has a series of shallow longitudinal grooves. All four of the osteoderms are relatively thick as compared to anguimorphan osteoderms.

Discussion The presence of a central keel on the dorsal surface and the large size of these osteoderms indicate that they are most likely from a scincomorphan lizard (sensu Broschinski and Sigogneau-Russell, 1996). It is not possible to confidently associate these osteoderms with any of the known scincomorphan taxa or morphotypes, but it is possible that they belong to one of the paramacellodid/cordylid-grade morphotypes based on the relative abundance of these jaw types in the fauna and on the association of similar osteoderms with the paramacellodid *Sharovisaurus karatauensis* Hecht and Hecht, 1984, from Kazakhstan. This morphotype of osteoderm is much less common in the Kaiparowits Formation than the anguimorphan types.

ANGUIMORPHA Fürbringer, 1900
ANGUIDAE Gray, 1825
ODAXOSAURUS Gilmore, 1928
(See revised diagnosis above)
ODAXOSAURUS ROOSEVELTI sp. nov.

Etymology Name in honor of Theodore Roosevelt, accomplished amateur naturalist and the 26th president of the United States. President Roosevelt established the first national monument in 1906. This set the stage for the eventual establishment of the Grand Staircase–Escalante National Monument, from within which is located the type locality.

Holotype OMNH 30748 (Fig. 16.16A).

Paratype OMNH 30747 (Fig. 16.16B–D).

Tentative Referral OMNH 24182 (Fig. 16.16E).

Type Locality OMNH V5, Garfield County, Utah, approximately 14 km northeast of the town of Henrieville.

Diagnosis Differs from *Odaxosaurus piger* Gilmore, 1928 (see Estes, 1983, and Gao and Fox, 1996), and *O. priscus* Gao and Fox, 1996, in having teeth with recurved crowns that partially overlap the next posterior tooth. Differs further from *O. piger* in having less prominent apical striae on the tooth crowns. Differs further from *O. priscus* in having teeth are more closely spaced.

Description OMNH 30748 (Fig. 16.16A) is the posterior portion of a left dentary. It is lacking much of the lateral wall of the Meckelian fossa. On the lingual surface of the posteriormost portion of the element is a distinct facet for the anteromedial process of the coronoid. The subdental gutter and shelf are absent. The medial border of the tooth-bearing portion of the element is moderately concave just anterior to the coronoid facet and broken at the anteriormost preserved tooth position. The teeth are pleurodont with labiolingually expanded bases. There is a small amount of cementum and a distinct, medially directed nutrient foramen at the base of each tooth. The tooth shafts are lingually swollen below the crown and each subsequent more posterior tooth has a shaft that is progressively shorter. The tooth crowns are labiolingually compressed, chisel-like, and weakly striated. The tooth crowns of the two anteriormost preserved teeth are posteriorly displaced giving the tooth a recurved appearance. The third tooth posteriorly is also recurved, but to a lesser extent. The crown is not completely preserved on the posteriormost preserved tooth. OMNH 30747 (Fig. 16.16B–D) is the anterior portion of a right maxilla that preserves a portion of the dorsal process and four teeth. The mesial carina of each tooth is well developed and much longer than the distal carina. As with the teeth in OMNH 30748, the crowns of the teeth of OMNH 30747 are recurved, but the anteriormost three teeth are also more pointed and less chisel-like (Fig. 16.16B, C). The lateral surface is smooth and has a single, oval foramen dorsal to the posteriormost two in situ teeth (Fig. 16.16C). The posterior superior alveolar foramen is circular and opens dorsally and posteriorly above the posteriormost two in situ teeth (Fig. 16.16D). OMNH 24182 (Fig. 16.16C) is the posterior portion of a right maxilla preserving five tooth positions and three in situ teeth. The posteriormost aspect of the jugal (or posterior) process has a small posterolaterally directed facet for articulation with the ectopterygoid. The posteriormost portion of the dorsal process rises superior to

the fourth tooth position from posterior. The supradental shelf is broken along much of its length and the position of the articulation of the palatine cannot be determined. The teeth are pleurodont, recurved and poorly preserved.

Discussion The difference in crown morphology between OMNH 30748 (chisel-like) and OMNH 30747 (pointed) is consistent with a similar observable degree of heterodonty in the living anguid *Elgaria* as well as with many other iguanian and scleroglossan lizards. The teeth of OMNH 24182 are not well preserved, the specimen is tentatively referred to *Odaxosaurus roosevelti* sp. nov. on the basis of the apparent recurvature of the teeth.

Among anguids *Odaxosaurus roosevelti* sp. nov. is most similar in morphology to *Odaxosaurus piger* in mode of tooth attachment, swollen bases of teeth, and gentle recurvature of the teeth. *Odaxosaurus piger* is currently known only from the Maastrichtian of Montana and western Canada (Estes, 1983; Gao and Fox, 1996).

ODAXOSAURUS PRISCUS Gao and Fox, 1996

Referred Specimens OMNH 24176, OMNH 24168, OMNH 30657, OMNH 30771, UMNH VP 18065, UMNH VP 19239, UMNH VP 19240 (Fig. 16.16F–K).

Localities OMNH V5 (specimens OMNH 24176, 24168), Garfield County, Utah, approximately 14 km northeast of the town of Henrieville; OMNH V9 (specimens OMNH 30657, 30771), Kane County, Utah, approximately 26 km southeast of the town of Henrieville, UMNH Loc. 108 (specimens UMNH VP 18065, UMNH VP 19239, UMNH VP 19240), Garfield County, Utah, approximately 14 km northeast of the town of Henrieville.

Description UMNH VP 19239 (Fig. 16.16F) is a fragment of a possible left dentary that preserves four tooth positions and a single complete tooth. The tooth is pleurodont, has a striated, chisel-like crown, minimal basal cementum, and a basal nutrient foramen. There is no subdental shelf or sulcus dentalis. UMNH VP 19240 (Fig. 16.16G, H) is a posterior portion of a left dentary preserving four tooth spaces and three complete teeth. The teeth are pleurodont, widely spaced, have slightly recurved tooth shafts, but the crowns do not overlap (Fig. 16.16G). The tooth bases are medially expanded, but the shafts are not. There is a basal nutrient foramen well-preserved in the middle tooth. The crowns are chisel-like and heavily striated (Fig. 16.16H). There is a narrow, diagonally directed facet on the medial side of the coronoid process of the dentary that extends below distalmost tooth position (Fig. 16.16G) for articulation with the anterior process of the coronoid.

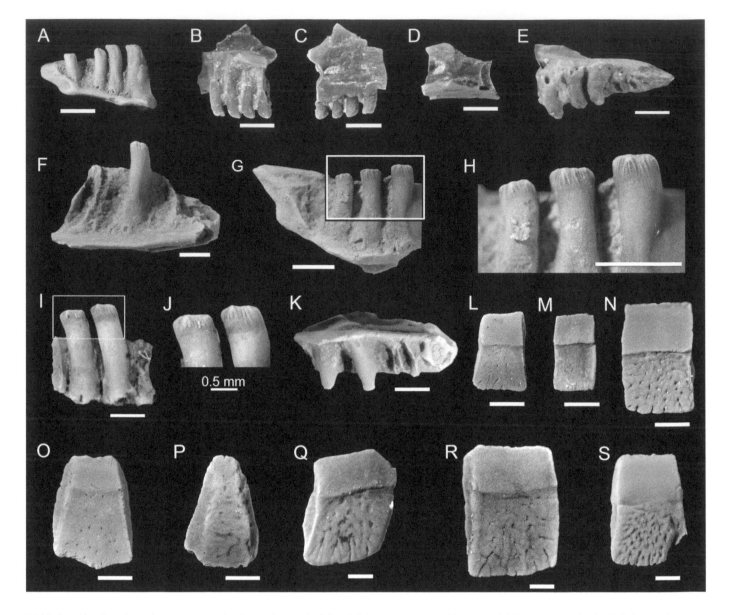

16.16. Anguidae from the Kaiparowits Formation (upper Campanian). (A–E) *Odaxosaurus roosevelti,* sp. nov.: (A) OMNH 30748, broken left dentary in lingual view; (B–D) OMNH 30747, broken right maxilla in (B) lingual; (C) labial; and (D) dorsal views; (E) OMNH 24182, broken right maxilla in lingual view. (F–K) *Odaxosaurus priscus:* (F) UMNH VP 19239, broken left dentary(?) in lingual view; (G, H) UMNH VP 19240, broken left dentary in (G) lingual view with box indicating area of detail in (H); (I) OMNH 24176, jaw fragment in (I) lingual view with box indicating area of detail in (J); (K) UMNH VP 18065, broken left maxilla in lingual view. (L–S) cf. Anguimorpha, osteoderms in dorsal view: (L) OMNH 66072; (M) UMNH VP 18057; (N) OMNH 66071; (O) OMNH 66063; (P) UMNH VP 18058; (Q) UMNH VP 18060; (R) OMNH 66065; and (S) OMNH 66059. All scale bars = 1 mm unless otherwise indicated.

OMNH 24176 (Fig. 16.16I, J) is a small fragment of a jaw (element indeterminate) with two large teeth. The teeth are pleurodont with lingually expanded bases (Fig. 16.16I) and labiolingually compressed, chisel-like tooth crowns that are heavily striated (Fig. 16.16J). The base of the best preserved tooth has a medially directed nutrient foramen. There are several unfigured, isolated, unassociated teeth (OMNH 24168, 30657, and 30771) with identical morphology to the teeth described above.

UMNH VP 18065 (Fig. 16.16K) is a right maxilla missing the anterior and posterior ends of the element as well as the dorsal process. Of the five (possibly six) preserved tooth positions there are only two in situ teeth near the anterior end of the element. The teeth are not well preserved, but the bases are clearly lingually swollen. The most completely preserved tooth has a medially directed nutrient foramen at its base and a chisel-like crown that is labiolingually compressed, but otherwise worn smooth.

Discussion Gao and Fox (1996) erected *Odaxosaurus priscus* on the basis of specimens from the Campanian (Judithian NALMA) from the Oldman Formation of southern Alberta. The specimens from the Kaiparowits Formation show a similar spacing of the teeth and reduced mesiodistal flaring of the tooth crowns (as compared to *O. piger*). In both these specimens (UMNH VP 18065; Fig. 16.16K) and the specimens described by Gao and Fox (1996:fig. 30A, D; UAL-VP 29896, 29891, respectively) the much reduced size of the tooth crowns is more a function of wear. In the unworn tooth crowns seen on UMNH VP 19240, OMNH 24176, and UALVP 33380 (Gao and Fox, 1996:fig. 30C) the mesiodistal width of the tooth crowns is equivalent to that of the tooth shaft. The presence of this taxon in the Campanian of southern Utah represents a significant extension of its paleogeographical distribution and the first recorded instance of two species of *Odaxosaurus* recovered from the same fauna.

cf. ANGUIMORPHA
Osteoderms

Referred Specimens OMNH 66059, OMNH 66063, OMNH 66065, OMNH 66071, OMNH 66072, UMNH VP 18057, UMNH VP 18058, UMNH VP 48060 (Fig. 16.16L–S).

Localities OMNH V5 (specimens OMNH 66063, 66071, 66072), Garfield County, Utah, approximately 14 km northeast of the town of Henrieville; OMNH V6 (specimen OMNH 66065), Kane County, Utah, approximately 22 km southeast of the town of Henrieville; OMNH V9 (specimen OMNH 66059), Kane County, Utah, approximately 26 km southeast of the town of Henrieville; UMNH Loc. 108 (all UMNH VP specimens), Garfield County, Utah, approximately 14 km northeast of the town of Henrieville.

Descriptions OMNH 66072 (Fig. 16.16L) is a trapezoidal osteoderm with a large anterior imbrication facet, narrow lateral imbrication facets, and a subtle ornamentation of small foramina that open into posteriorly/posterolaterally directed channels. UMNH VP 18057 (Fig. 16.16M) is a rectangular osteoderm with a large anterior imbrication facet, and a similar ornamentation to OMNH 66072, although the dorsal surface is worn. OMNH 66071 (Fig. 16.16N) is a large rectangular osteoderm with an anterior imbrication facet that is nearly one half of the dorsal surface. The ornamentation pattern is series of small foramina connected to web of vermiculate channels course through a weakly rugose pattern of ridges. OMNH 66063 (Fig. 16.16O) is a trapezoidal osteoderm with a small anterior imbrication facet and narrow lateral imbrication facets. The dorsal surface is worn and all that remains of the ornamentation pattern are the small foramina. UMNH VP 18058 (Fig. 16.16P) is a triangular osteoderm with a small anterior imbrication facet and wide lateral imbrication facets. The ornamentation is weakly rugose with small foramina. UMNH VP 18060 (Fig. 16.16Q) is a rhomboidal osteoderm with a large anterior imbrication facet and narrow lateral imbrication facet on the left side. The ornamentation is muted as a result of wear of the dorsal surface, but it is possible to discern a series of small foramina connected to posteriorly/posterolaterally directed channels. OMNH 66065 (Fig. 16.16R) is a rectangular osteoderm with a large anterior imbrication facet and narrow lateral imbrication facet on the left side. The ornamentation is a series of foramina connected to a web-like network of vermiculate channels. OMNH 66059 (Fig. 16.16S) is a trapezoidal osteoderm with a very large anterior imbrication facet and narrow lateral imbrication facet on the left side. The ornamentation pattern is nearly identical to that of OMNH 66071 (Fig. 16.16N).

Discussion Similar osteoderms have commonly been recovered from Upper Cretaceous microvertebrate localities in more northern latitudes (i.e., Montana, Wyoming, Canada) and in some cases are among the most numerous single type of fossil (pers. obs.). These osteoderms have typically been referred to the most commonly recovered anguid in these units, namely *Odaxosaurus piger* (Estes, 1964, 1983; Estes, Berberian, and Meszoely, 1969; Gao and Fox, 1996). That a similar referral for the anguimorphan osteoderms from the Kaiparowits Formation is correct is possible, if unlikely, but there is no direct evidence or unquestionable articulation or association of osteoderms with *O. piger* or *O. priscus* skeletal remains. Indeed, the specimen UALVP 33382, a maxilla of *Odaxosaurus piger* illustrated by Gao and Fox (1996:fig. 28C), has a fused osteoderm with a distinctly more rugose ornamentation pattern than seen in the rectangular osteoderms noted above. Additionally, there are other, non-*Odaxosaurus*, morphotypes of anguimorphan lizards in the Kaiparowits Formation fauna to which such osteoderms could belong. As such I am unable to refer these osteoderms with confidence to any specific taxon or morphotype of anguid lizard until such time as appropriate evidence is available.

XENOSAURIDAE Cope, 1886
?*EXOSTINUS* sp.

Referred Specimens OMNH 33885, OMNH 33860, OMNH 66100, OMNH 66104 (tentative referral), UMNH VP 18059 (tentative referral), UMNH VP 18063 (tentative referral) (Fig. 16.17A–F).

Localities OMNH V5 (specimens OMNH 66100, 66104); OMNH V9 (OMNH 33885, 33860), Garfield County, Utah, approximately 14 km northeast of the town of Henrieville; UMNH Loc. 108 (specimens UMNH 18059, 18063), Garfield County, Utah, approximately 14 km northeast of the town of Henrieville.

Description OMNH 33860 (Fig. 16.17A) is a right maxilla missing the anteriormost and posteriormost portions. All six of the preserved tooth spaces have complete teeth. The teeth are subpleurodont, narrowly built, sharply pointed, and slightly recurved. The mesial carina of each tooth is well developed and blade-like. The bases of the teeth are encircled by modest amounts of cementum. The fifth tooth (from anterior) has a shallow resorption pit at the base. The supradental shelf is weakly dorsally convex and the medial margin is recessed laterally at the anterior and posterior ends of the specimen. These recessions may be the articulation areas for the palatine and ectopterygoid, respectively, but the margin is not distinctly facetted. The posterior portion of the dorsal process rises from the lateral edge of the supradental shelf.

OMNH 33885 (Fig. 16.17B) is a left dentary missing the anteriormost and posteriormost portions of the element. Most of the lateral wall of the Meckelian fossa is also missing. There are six in situ teeth and an empty tooth position at the anterior end of the specimen. The teeth are pleurodont, conical, and weakly recurved. The mesial carina of each tooth is well developed and blade-like. The single complete tooth crown is sharply pointed. The bases of the teeth are encircled by a modest amount of cementum. There are small, medially directed nutrient foramina at the bases of the second, third, fourth, and sixth in situ teeth (from anterior). The fifth tooth appears has a small resorption pit at the base. The subdental shelf and gutter are absent. The base of a ventrally directed intramandibular septum extends from the roof of the Meckelian fossa. This ventral margin of the intramandibular septum is free, but the element is too poorly preserved to determine whether this is the actual condition or an artifact of breakage.

UMNH VP 18063 (Fig. 16.17C) is an elongate, ovoid osteoderm with a vermiculate grooves separating swollen ridges. There are no indications of imbrication surfaces. UMNH VP 18059 (Fig. 16.17D) is a rectangular osteoderm with a similar, but worn ornamentation. There is a short imbrication surface at the anterior end of the specimen and another along the left lateral margin. OMNH 66100 (Fig. 16.17E) is a rectangular osteoderm with a similar pattern of vermiculate grooves between swollen ridges. The anterior portion is broken, but a small, smooth surface is present that may be part of an imbrication surface. The right lateral margin has a narrow imbrication surface. OMNH 66104 (Fig.16.17F) is a broken osteoderm with a similar pattern of ornamentation. There is a remnant of the anterior imbrication surface still present. A faint depression along the anterior right margin may also be an imbrication surface.

Discussion Tentative referral of the two jaw specimens to *Exostinus* Cope, 1873, is based on close similarities in tooth structure and tooth attachment to that of *Exostinus lancensis* Gilmore, 1928, from the Maastrichtian of Montana (see also descriptions and figures in Estes, 1983, and Gao and Fox, 1996). The revised diagnosis of *E. lancensis* provided by Gao and Fox (1996) does not include any dental or maxillary-based characters so OMNH 33860 is of limited taxonomic

value, but some comparisons can be made. Gao and Fox (1996:fig. 27A, B) illustrate two partial maxillae of *Exostinus lancensis*. The lateral surface of UALVP 29849 (Gao and Fox, 1996:fig. 27A) is an anterior portion of a maxilla and has osteoderms fused to the lateral surface of the dorsal process. UALVP 29850 (Gao and Fox, 1996:fig. 27B) is a posterior portion of a maxilla and has a distinctive ventral step to the posterior portion of the tooth row. OMNH 33860 lacks both of these characteristics, but it preserves the portion of the maxilla not represented by the two UALVP specimens. OMNH 33885 is similar to *Exostinus serratus* Cope, 1873 (Oligocene of North America), in lacking a subdental gutter and shelf (Estes, 1983). Pending the recovery of more complete material, the taxonomic assignment of the two jaw specimens remains tentative.

Referral of the osteoderms is based on the similarity in ornamentation to those described and illustrated for *E. lancensis* (Gao and Fox, 1996; pers. obs.). The only osteoderms described for *E. lancensis*, *E. serratus*, and *Restes rugosus* Gilmore, 1942, are those of the cranium (Estes, 1983; Gao and Fox, 1996). The ovoid shape of UMNH VP 18063 is consistent with a cranial osteoderms. The rectangular shape of the other osteoderms is more consistent with cervical, trunk, or tail osteoderms of lizards. The presence of cephalic osteoderms suggests the presence of postcranial osteoderms for these fossil xenosaurids and the similarity of ornamentation of the four osteoderms described here suggest they are from the same taxon. As such I am tentatively referring them to *?Exostinus*.

PLATYNOTA indeterminate
PARASANIWA Gilmore, 1928
PARASANIWA CYNOCHOROS sp. nov.

Etymology From the Greek *kynos* (dog) and *choros* (land). Named for Dog Flat, the local area of the type locality (Fig. 16.17G–J).

Holotype UMNH VP 21180.

Type Locality UMNH VP 1268, Kane County, Utah, approximately 22 km southeast of the town of Henrieville.

Diagnosis Very similar to *Parasaniwa wyomingensis* Gilmore, 1928, in tooth shape, but differs in teeth being shorter, not trenchant. Differs from *Parasaniwa*, new species, cf. *P. wyomingensis* Gao and Fox, 1996, in teeth not trenchant, tooth crowns with conspicuous apical constriction. Further differs from differs from *Parasaniwa*, new species, cf. *P. wyomingensis*, and from *Lowesaurus matthewi* Gilmore, 1928, *Paraderma bogerti* Estes, 1964, *Primaderma nessovi* Nydam, 2000, in fused osteoderms on maxilla extending inferiorly to labial foramina. Differs further from *Primaderma nessovi* in teeth not laterally compressed and basal infolding of enamel

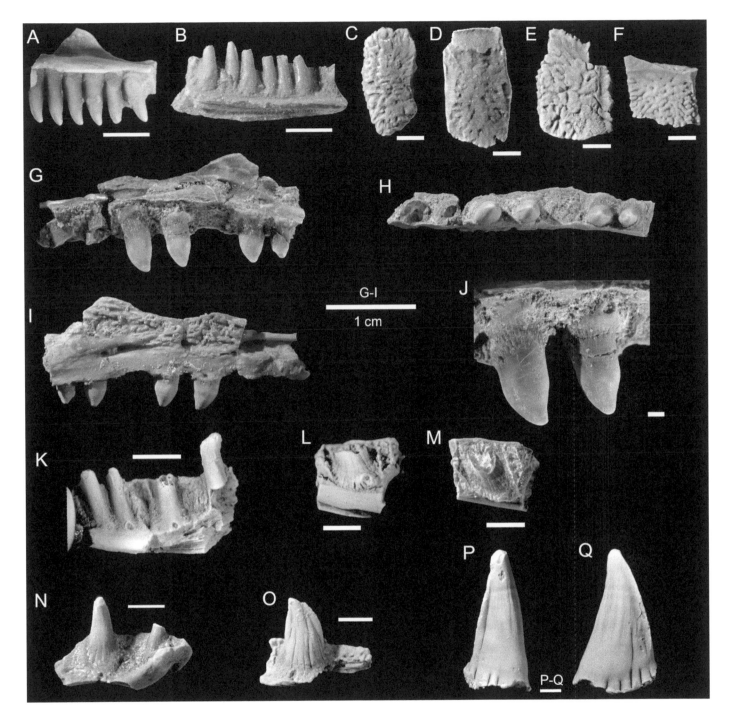

16.17. Xenosauridae and Platynota from the Kaiparowits Formation (upper Campanian). (A–F) *?Exostinus* sp.; (A) OMNH 33860, broken right maxilla in lingual view; (B) OMNH 33885, broken left dentary in lingual view. (C–F) referred osteoderms in dorsal view: (C) UMNH VP 18063; (D) UMNH VP 18059; (E) OMNH 66100; (F) OMNH 66104. (G–Q) Platynota. (G–J) *Parasaniwa cynochoros*, sp. nov.: UMNH VP 21180, broken right maxilla in (G) lingual view, (H) occlusal view, (I) labial view, and (J) detail of teeth. (K) Kaiparowits Formation morphotype H: OMNH 33858, broken right dentary in lingual view. (L–O) Kaiparowits Formation morphotype I: (L, M) OMNH 23523, broken right dentary in (L) lingual and (M) occlusal views; (N) OMNH 64153, broken left dentary in lingual view; (O) OMNH 24372, jaw fragment in lingual view. (P–Q) Kaiparowits Formation morphotype J: OMNH 21838, isolated tooth in (P) mesiolingual and (Q) lingual views. Scale bars = 1 mm unless otherwise noted.

less prominent. Differs further from *Lowesaurus matthewi* in teeth lacking venom grooves.

Description UMNH VP 21180 is a right maxilla lacking much of the facial process as well as the rostralmost and caudalmost portions of the element. As seen in lingual view

(Fig. 16.17G) the supradental shelf is weakly dorsally convex and deeply grooved at the base of the facial process. Just posterior to the groove and superior to the third in situ tooth is the small, posteriorly directed opening of the posterior inferior alveolar foramen. The tooth bearing surface is dorsally concave (complementarily to the convexity of the dorsal surface of the supradental shelf) and broad. The tooth bases are broad, heavily cemented, have large basal nutrient foramina, and have numerous, weakly developed basal infoldings of the enamel (Fig. 16.17J). The tooth shafts are parallel-sided well beyond the level of the lateral parapet and then the tooth crowns narrow quickly to a blunt tip. As seen in occlusal view (Fig. 16.17H), the tooth crowns are rotated posteriorly and have blunt carinae the meet apically to form a posteromedially directed V. This rotation of the tooth crowns gives the teeth a bulbous appearance in both lingual and labial view. There are two broken tooth bases posteriorly that show the lack of plicidentine. The preserved portion of the labial surface (Fig. 16.17I) is covered in fragmented osteoderms that extend across the base of the facial process and surround the posteriormost preserved labial foramen.

Discussion The presence of fused osteoderms on the maxilla supports the referral of UMNH VP 21180 to *Parasaniwa* (sensu Gao and Fox, 1996). In *Parasaniwa wyomingensis* the fused osteoderms are plate-like and separated by deep grooves (Gao and Fox, 1996:fig. 33A), which is also characteristic of the maxillae of *Lowesaurus matthewi*, *Paraderma bogerti*, and *Primaderma nessovi*; all three of which are considered monstersaurs (Norell and Gao, 1997) related, or possibly referable, to Helodermatidae (Nydam, 2000; Conrad, 2008). The preserved portion of the maxilla of UMNH VP 21180 does not have plate-like osteoderms, but the maxillae of both *Parasaniwa wyomingensis* and *Primaderma nessovi* are similar to UMNH VP 21180 in the plate-like osteoderms of the facial process becoming more fragmented inferiorly, near the labial foramina.

The morphology of the tooth crowns of UMNH VP 21180 are more like nonplatynotan anguimorphans (Gauthier, 1982) in that the tips describe a posteromedially directed V in occlusal view. The massive structure of the teeth and the blunt tips suggest a durophagous diet for this animal.

Kaiparowits Formation Morphotype H

Referred Specimen OMNH 33858 (Fig. 16.17K).

Locality OMNH V9, Garfield County, Utah, approximately 14 km northeast of the town of Henrieville.

Description OMNH 33858 (Fig. 16.17K) is a fragment of the anterior portion of a right dentary. It is missing the symphysis and much of the lateral wall of the Meckelian fossa. There is no subdental shelf or gutter. There are six tooth positions and four partially preserved teeth. All but the posteriormost tooth are missing their crowns. Although the lateral parapet is high, the most completely preserved tooth suggests that the tooth attachment was subpleurodont. The tooth bases are not much wider than the tooth shafts, have

numerous, closely spaced basal infoldings, but not plicidentine. The three preserved tooth bases have distinct, lingually directed nutrient foramina with the posterior two foramina surrounded by a conspicuous ridge of enamel. The one partial crown is somewhat labiolingually compressed, but not trenchant. The sharp hook of the apical part of the crown is an artifact of breakage.

Discussion The presence of basal infoldings of the enamel and the lack of a subdental shelf are clear indications that this specimen is a platynotan. The close spacing and delicate nature of the basal infoldings differs from other known platynotans of the Cretaceous and it is likely that this specimen represents a new taxon.

Kaiparowits Formation Morphotype I

Referred Specimens OMNH 23523, OMNH 24372, OMNH 64153 (Fig. 16.17L–O).

Locality OMNH V6 (specimen OMNH 23523), Kane County, Utah, approximately 22 km southeast of the town of Henrieville; OMNH V61 (specimens OMNH 24372, 64153), Garfield County, Utah, approximately 14 km northeast of the town of Henrieville.

Description OMNH 23532 (Fig. 16.17L, M) is a small fragment of the anterior portion of a right dentary. The subdental shelf and sulcus dentalis are absent and the Meckelian fossa is restricted to a narrow, ventromedially directed opening. The single tooth space has the base of a tooth in situ. The tooth base is broad and the preserved portion of the tooth shaft is circular and much smaller in diameter than the base (Fig. 16.17L). The tooth base has well-developed basal infoldings that extend to the dentine (i.e., plicidentine; Fig. 16.17M) and the cementum forms a circumferential ridge around the base. There is also a large, posteromedially directed nutrient foramen at the tooth base. OMNH 64153 (Fig. 16.17N) is the anterior portion of a left dentary. It preserves a broad, smooth symphysis and four tooth positions. There are teeth preserved in the second and fourth tooth positions. The tooth attachment is subpleurodont, with a circumferential ridge of cementum around the broad tooth bases. There is a well-defined nutrient foramen at the base of the largest preserved tooth. The tooth crowns taper quickly, but the tips are missing. OMNH 24372 (Fig. 16.17O) is a fragment of a jaw (element indeterminate) that bears a single tooth. The tooth is short, laterally compressed, blade-like, and recurved. The tooth base is broad, surrounded by a thick ring of cementum, and has a nutrient foramen at the medial margin. The infoldings of the enamel extend along the entire length of the tooth.

Discussion All three of these specimens share the presence of a thick ring of cementum around the base of the tooth. The more conical shape of the tooth of OMNH 64153 is typical of the anterior teeth of many platynotan lizards (pers. obs.). The well-developed infoldings that extend the length of the tooth in OMNH 24372 may support placing this specimen in a different morphotype/taxon, but as it is also possible that this is a positional difference so it is retained in Kaiparowits Formation morphotype I pending the recovery of more complete material.

Kaiparowits Formation Morphotype J

Referred Specimen OMNH 21838 (Fig. 16.17P, Q).

Locality OMNH V5, Garfield County, Utah, approximately 14 km northeast of the town of Henrieville.

Description OMNH 21838 is a large platynotan tooth that is weakly laterally compressed, and recurved. Most of the basal infoldings are very short, but two are much longer than the rest. The enfolding along the distal carina (Fig. 16.17P) extends nearly the length of the tooth and forms a distinct channel between the distal carina and the body of the tooth. Along the lingual surface (Fig. 16.17Q), just mesial of the center of the tooth is another long infolding of enamel that extends apically about one half the length of the tooth.

Comments The large size and possible presence of a venom groove indicate that this specimen represents a unique, possibly new platynotan taxon in the fauna. Closest comparison is to *Labrodioctes montanaensis* Gao and Fox, 1996, from the Judith River Formation (Campanian) of Montana, which also has a groove along the margin of the preserved teeth (Gao and Fox, 1996:fig. 36A). Unlike Kaiparowits Formation morphotype I, the grooves on the teeth of *L. montanaensis* are more widely open and not slit-like. The channel on OMNH 21838 is more similar to the slit-like venom groove on the teeth of *Lowesaurus matthewi* and *Heloderma* spp. (pers. obs.; see also discussion of venom grooves in Nydam, 2000)

SERPENTES Camp, 1923
CONIOPHIS sp.

Referred Specimens UMNH VP 19237, UMNH VP 19238, UMNH VP 18061 (Fig. 16.18).

Locality UMNH Loc. 108 (all UMNH VP specimens), Garfield County, Utah, approximately 14 km northeast of the town of Henrieville.

Description UMNH VP 19237 is an anterior caudal vertebra with a weakly convex neural arch that is trapezoidal in dorsal view (Fig. 16.18A). The neural spine is low, bifurcates posteriorly, but does not project posteriorly beyond the posterior margin of the neural arch. The zygosphenes project anteriorly as small processes from the anterolateral margin of the

neural arch and there is no constriction, or neck present. The left prezygapophysis is complete and has a superolaterally directed facet (Fig. 16.18B). Inferior to each of the prezygapophyses is a complex inferolaterally directed lymphapophysis. Each of these structures are missing their distal portions, but the right side is best preserved (Fig.16.18B). The superiormost persevered portion of the lymphapophysis is the broken base of a rod-like structure with a flattened oval outline. Inferior and anterior to this is a small, unbroken anterolateral process and inferoposterior to this process is the broken base of the inferior portion of the lymphapophysis. The cotyle is round (Fig. 16.18B), the hemal keel (Fig. 16.18C) is low and narrow anteriorly and broadens toward the condyle. The condyle is round (Fig. 16.18C, D), but is missing its right half.

UMNH VP 19238 is a trunk vertebra that has a weakly convex neural arch with a narrow, long neural spine (Fig. 16.18E). The left prezygapophysis is complete and has a superolaterally directed facet (Fig. 16.18E, F). The zygosphene facets are inferomedially directed (Fig. 16.18F) and the zygosphene is constricted at the level of the anteriormost portion of the neural spine. The postzygapophyses are broad (Fig. 16.18E) and have nearly horizontal facets (Fig. 16.17G). The zygantral facets are notches in the roof of the neural arch; only the right side is well preserved (Fig. 16.18G). The cotyle is round and deep and the broken synapophyses extend superolateral from its inferolateral borders (Fig. 16.18F). The condyle is round and has a weakly constricted neck (Fig. 16.18G, H). There is no hemal keel.

UMNH VP 18061 is a broken trunk vertebra (Fig. 16.18I–L) preserving only portions of the anterior end of the element. The zygosphene is missing and there is only a low ridge on the neural arch to indicate the presence of a neural spine (Fig. 16.18I). The prezygapophyses have broad, oval-shaped facets (Fig. 16.18I, K) and are positioned superior to massive synapophyses (Fig. 16.18K, L). The cotyle is deep and slightly oval shaped (Fig. 16.18K), and the ventral surface of the preserved portion of the centrum has a low, broad hemal keel pierced laterally by asymmetrical subcentral foramina (Fig. 16.18J, L). This specimen is from an individual that is much larger than that are represented by UMNH VP 19237, and UMNH VP 19238.

Discussion The size difference between these specimens may represent an ontogenetic difference or, more likely, indicates the presence of two snakes in the squamate fauna from the Kaiparowits Formation.

Taxon indet.

Referred Specimen OMNH 23539 (Fig. 16.18M).

Locality OMNH Loc. v6, Kane County, Utah, approximately 22 km southeast of the town of Henrieville.

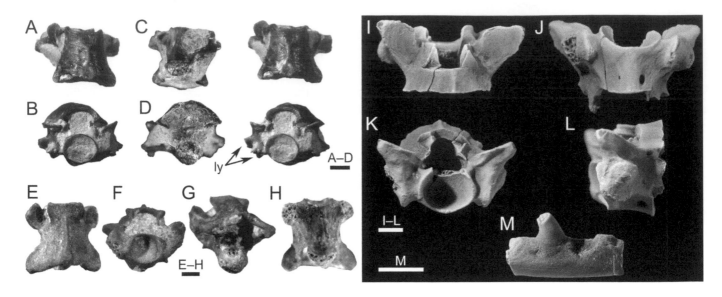

16.18. *Coniophis* sp. from the Kaiparowits Formation (upper Campanian). (A–D) UMNH VP 19237, anterior caudal vertebra in (A) stereopair of dorsal view; (B) stereopair of anterior view, (C) ventral view, and (D) posterior view. (E–H) UMNH VP 19238, trunk vertebra in (E) dorsal; (F) anterior; (G) posterior, and (H) ventral views. (I–L) UMNH VP 18061, trunk vertebra in (I) dorsal; (J) ventral; (K) anterior; and (L) left lateral views. (M) OMNH 23539, jaw fragment in lingual view. Ly, lymphapophysis. Scale bars = 1 mm.

Description The specimen is a fragment from a gracile and presumably long tooth-bearing element. The element is solid (i.e., lacks an internal neurovascular channel), and the closest comparison in many modern snakes (*Anilius scytale, Cylindrophis rufus, Lampropeltis getula*) is to the anterior ramus of a left pterygoid, but exact identification of such a limited specimen is not possible. There is a single preserved tooth with a broad base pierced by a nutrient foramen. The crown of the tooth is circular, narrows towards the tip, and shows evidence of recurvature, but the tip of the crown is missing. Anterior to the tooth is an empty alveolus that is notched on one side as is typical with many snakes (Budney, Caldwell, and Albino, 2006). Posterior to the tooth is an alveolus with the ankylosed remains of a tooth covering the opening.

Discussion Although the specimen is in the general size range for the smaller vertebrae from this horizon (Fig. 16.18A–H), I believe it is premature to refer nonvertebral elements to the vertebral form taxon *Coniophis* pending discovery of an articulated skeleton.

DISCUSSION

The systematic study of the lizards and snakes from the Cretaceous of southern Utah (summarized in Fig. 16.19) provides an unusual opportunity in the Western Interior of North America; the chance to compare and contrast sequential faunas from a single region. The following is a review that takes into account all of the known taxa (both from above and from previously published reports only summarized here)

as well as the morphotypes identified here and how they affect our understanding of lizard diversity and distribution in the Western Interior of North America. As noted earlier, the use of morphotypes is not as satisfactory as named taxa, but their inclusion permits a more thorough understanding of the diversity of each fauna and allows for more informative comparisons of the faunas to each other. Because they were only recently dealt with in some detail, I will not review the Contogeniidae here, but instead refer the reader to Nydam and Fitzpatrick (2009).

Scincomorpha: Specimens with a Paramacellodid/Cordylid-Grade Dentition

The iterative presence of fossil jaw fragments with teeth that have a paramacellodid/cordylid-grade morphology (Fig. 16.2) is one of the consistent themes in the lizard faunas of southern Utah's Cretaceous and represents the potentially greatest diversity of lizards from these horizons. These referrals are based nearly entirely on the dental morphology of the specimens following the results of the extensive review and illustration of scincomorphan tooth morphology by Kosma (2004) as well as those of Evans and Searle (2002). However, with only a few exceptions, none of the specimens are complete enough for formal taxonomic treatment and remain unnamed, but distinctly recognized, morphotypes of scincomorphan (?scincoid) lizards. Fossil taxa from the Western Interior of North America with a similar grade of dentition includes *Paramacellodus* from the Late Jurassic of the Morrison Formation (Prothero and Estes, 1980; Evans, 1996;

Evans and Chure, 1998a, 1998b, 1999), the Early Cretaceous of the Cloverly (*Atokasaurus*, Nydam and Cifelli, 2002b) and Antlers (cf. *Paramacellodus* and an unnamed morphotype) formations (Nydam and Cifelli, 2002b); *Dimekodontosaurus* Nydam, 2002, and *Bothriogenys* Nydam, 2002, from the early Late Cretaceous Cedar Mountain Formation (Nydam, 2002); a purported scincid from the Late Cretaceous Milk River Formation (Gao and Fox, 1996); and a specimen referred to "*Leptochamops*" from the Late Cretaceous Oldman Formation (Peng, Russell, and Brinkman, 2001:pl. 6, fig. 14; the figured specimen is not *Leptochamops*, but more closely resembles Kaiporowits Formation morphotype D). Of these occurrences, the greatest morphotypic diversity is in the Dakota Formation (five morphotypes/taxa), and Smoky Hollow Member of the Straight Cliffs Formation and Kaiparowits Formation (four morphotypes/taxa each). Although this is potentially significant, I caution again that the morphotypes presented herein are not intended to be equated with formal taxonomic designations. For example, *Dimekodontosaurus madseni* Nydam, 2002, from the Mussentuchit Member of the Cedar Mountain Formation has paramacellodid/cordylid-grade dentition, but there are three distinct morphotypes of teeth in the dentary based on whether they are in the anterior, middle, or posterior portions of the tooth row (Nydam, 2002:fig. 3).

An additional confounding issue is the question of the systematic status of paramacellodid lizards. Aspects of this problem have been discussed previously (Nydam and Cifelli, 2002b; Evans, 2003) and I will only briefly summarize here the current status of Paramacellodidae. Estes (1983:115) placed his newly erected Paramacellodidae within Cordyloidea as the sister family to Cordylidae. Phylogenetic analyses by Evans and Chure (1998a), Evans and Barbadillo (1998), and Reynoso and Callison (2000) recovered a sister-group relationship between *Paramacellodus* and Scincidae + Cordylidae (other paramacellodid taxa were not tested). In a later phylogenetic analysis, Conrad (2008:111, fig. 60B, C), recovered *Pseudosaurillus*, *Paramacellodus* and *Becklesius* at the base of Anguimorpha (fig. 55) or—when excluding soft tissue characters—as members of a series of polytomies at the crown of his Evansauria (fig. 60B) or the base of his Anguimorpha (fig. 60C). However, the anguimorphan affinities recovered by Conrad (2008) may be biased by the lack of a good fossil record for scincids, cordylids, and gerrhosaurids and the resulting uncertainties of plesiomorphic morphology for these taxa (J. Conrad, pers. comm.). In his expanded phylogenetic analysis of the systematics of Anguimorpha, Conrad et al. (2010) did not include any paramacellodid taxa. The formal name Paramacellodidae, as proposed by Estes (1983) is still useful for communication (see also the

discussion in Nydam and Cifelli, 2002b:288) and is used to refer to an extinct lineage of closely related scincomorphans (Evans, 2003; Evans and Wang, 2010), but paramacellodids are possibly, even likely, to be paraphyletic (summary of Gao and Fox, 1996:61) and their scincomorphan affinities are even questioned (Conrad, 2008; but see discussion of *Dakotasaurus*). As such, meaningful taxonomic treatment of the numerous morphotypes described herein of lizards with paramacellodid/cordylid-grade teeth requires not only more complete specimens, but also may require further analysis of the systematics of paramacellodid lizards.

Taxonomy and systematics aside, the consistent presence of lizards with this grade of tooth morphology is thematic of most of the known microvertebrate producing localities from the Late Jurassic through the Late Cretaceous of the Western Interior. The only notable exceptions the apparent absence of such lizards from the San Juan Basin of New Mexico and the Aguja Formation of Texas, but this may be a sampling bias related to the relatively limited amount of attention the lizards from these regions have received to date. This iterative presence of lizards with this grade of tooth morphology may be indicative of a relatively uniform diet preference of these lizards and/or the presence of a long successful clade of lizards in the Western Interior.

Scincomorpha: Scincidae

The fossil record for skinks is frustratingly sparse with most confidently referred paleotaxa not occurring prior to the Tertiary. Cretaceous records of scincids are decidedly rare and include only tentatively referred taxa from the early Campanian (*Penemabuya antecessor*, Gao and Fox, 1996) and mid-Campanian (*Orthrioscincus mixtus*, Gao and Fox, 1996). The addition of *Webbsaurus lofgreni* gen. et sp. nov. from the Cenomanian (Dakota Formation) potentially extends the record of scincids in North America by nearly 20 mya. But, the problem of scincid monophyly remains unresolved and it is becoming increasingly likely that Scincidae sensu lato is most likely paraphyletic (see review and discussion in Conrad, 2008). The solution to the problem of scincid monophyly proposed by Conrad (2008) is an evolutionary scenario in which snakes evolved from within Scincidae (i.e., the erection of Scincophidia Conrad, 2008, a clade comprising all fossorial squamate taxa except legless anguids—recovered even with the "nullification of fossorial/limbless characters"; Conrad, 2008:139). This adds yet another hypothesis to the debate of snake origins that requires additional testing and implies that resolution of scincid interrelationships will require a commensurate resolution of snake origins.

The typically large, multicuspid teeth of the borioteiioids make them conspicuous scincomorphan lizards from the Late Cretaceous of the Western Interior. On the basis of reviews of shared characteristics (Nydam, Eaton, and Sankey, 2007; Nydam, Caldwell, and Fanti, 2010) the borioteiioids include *Chamops, Leptochamops* and closely related taxa of the Chamopsiidae (sensu Nydam, Caldwell, and Fanti, 2010), the Polyglyphandontinae taxa *Dicothodon, Peneteius,* and *Polyglyphanodon* (Nydam, Eaton, and Sankey, 2007) as well as *Bicuspidon,* the sister taxon to the Polyglyphanodontinae. The temporal and geographic distribution of Polyglyphanodontinae has already been established (Nydam, 1999, 2002; Nydam and Cifelli, 2002a; Nydam, Eaton, and Sankey, 2007). What can now be added to that record are the occurrences in the Cenomanian of southern Utah of *Bicuspidon* (*B. smikros;* Dakota Formation) and *Dicothodon* (*Dicothodon* sp., base of Iron Springs Formation). Neither of these occurrences is surprising as both genera are known from the Mussentuchit Member of the Cedar Mountain Formation (Albian–Cenomanian boundary) of central Utah (Nydam, 1999, 2002; Nydam, Eaton, and Sankey, 2007). The occurrence of *Dicothodon* in the Iron Springs Formation is significant in that it fills the gap between the Albian–Cenomanian and Turonian records of the genus.

Chamopsiidae includes the earliest published record of fossil squamates from the Cretaceous of the Western Interior: Marsh's (1892) description of the lizard *Chamops segnis* and the snake *Coniophis precedens* from the Maastrichtian of Wyoming. This newly recognized family and the included taxa are described and discussed by Nydam, Caldwell, and Fanti (2010). However, the presence of *Chamops* sp. in the Turonian of southern Utah (Smoky Hollow Member of the Straight Cliffs Formation) is the first confirmed report of *Chamops* predating the known occurrences of the genus from the Campanian (see review in Gao and Fox, 1996). The earliest record of a taxon referable to Chamopsiidae is tentatively recognized as *Harmondontosaurus emeryensis* from the Cedar Mountain Formation (Nydam, 2002). The known diversity of chamopsiids in southern Utah reaches its peak in the Kaiparowits Formation (upper Campanian) with four taxa (Nydam and Voci, 2007) whereas six taxa are known from paracontemporaneous localities in Canada (Gao and Fox, 1996). Of these Campanian-age chamopsiids, only *Chamops* co-occurs in Canada and southern Utah. Similar observations of potential regional endemism have also been noted for mammals (Eaton and Cifelli, 1988; Cifelli, 1990b, 1990e; Eaton et al., 1999b). The direct evidence (presence of *Chamops* in the Turonian of southern Utah) and the indirect evidence (taxonomically distinct chamopsiids in southern versus northern Campanian-age faunas of the Western Interior) indicate that like Polyglyphanodontinae, Chamopsiidae was likely established in North America prior to the Late Cretaceous. If correctly referred, *Harmondontosaurus emeryensis* may represent an antecedent form of chamopsiid from the Early–Late Cretaceous transition. Alifanov (2000:375) placed *Gerontoseps, Socognathus,* and *Sphenosiagon* (all North American chamopsiid taxa sensu Nydam, Caldwell, and Fanti, 2010) in his "Mongolochamopsidae." Although his taxonomy is in conflict with that of current studies of the North American taxa, the proposal by Alifanov that chamopsiid (based on the taxonomy of Nydam, Caldwell, and Fanti, 2010) taxa, in addition to polyglyphanodontid taxa, are shared between Asia and North America during the Cretaceous seems to lends further support to the hypothesis of an interchange of squamates during the early part of the Cretaceous (Nydam, 2002). However, I am also cautious with regard to this conclusion. Although my own observations of the Asian polyglyphanodontid taxa *Darchansaurus, Cherminsaurus, Erdenetesaurus,* and *Gilmoreteius* (see nomenclatural revision by Langer, 1998) in the collections of the Palaeontological Institute in Warsaw, as well as the type material for *Polyglyphanodon* at the USNM, support Estes's (1983) conclusions with regard to his definition of Polyglyphanodontidae, I have not had the opportunity to study the specimens at the Paleontological Institute in Moscow.

Anguimorpha: Xenosauridae

The recognition of the presence of Xenosauridae in the Cretaceous of the Western Interior was first established by Gilmore (1928) with the erection of *Exostinus lancensis* from the Lance Formation (Maastrichtian of Wyoming) – though Gilmore originally referred the taxon to Iguanidae – as a Cretaceous-age species of a genus erected for Oligocene-age specimens (Cope, 1873). Both Estes (1964, 1983) and Gao and Fox (1996) have reviewed and discussed the fossil record of *Exostinus* in detail and I will only summarize a couple of pertinent points here and refer the reader to those works for more detailed discussions. The referral of the fossil specimens for *Exostinus lancensis* to Xenosauridae is supported by Estes (1983) and Gao and Fox (1996) on the basis of comparative study and is additionally supported by the phylogenetic analysis of Conrad (2008). Estes (1983) and Gao and Fox (1996) also give cautious support to the suggestion by Gauthier (1982) that referral to the same genus as the original Oligocene material may be problematic, but both choose to retain referral to *Exostinus.* As noted by Estes (1964) *Exostinus lancensis* is also important in that it, as well as other

Xenosauridae, represent anguimorphans that primitively retain direct tooth replacement (i.e., formation of resorption pits at tooth bases) and a sulcus dentalis. The record of Mesozoic and Tertiary fossils of xenosaurids is relatively rich and is currently under review by B.-A. Bhullar of Harvard University, who has recently questioned the monophyly of *Exostinus* (Bhullar, 2007, 2010). The fossils from southern Utah that are referable to Xenosauridae potentially extend the record of this family back to the Cenomanian (i.e., *Cnododontosaurus suchockii*, Dakota Formation). Although referral to Xenosauridae is even more tentative, specimens from the Smoky Hollow Member of the Straight Cliffs Formation and the Kaiparowits Formation suggest that xenosaurids were likely a small component of the lizard faunas of southern Utah until at least the late Campanian.

Anguimorpha: Anguidae

The Cretaceous record for Anguidae is an interesting combination of tentative and well-supported records of occurrence. The earliest evidence for Anguidae (sensu stricto) in the Western Interior is the referred osteoderms recovered from the Dakota, Straight Cliffs (Smoky Hollow and John Henry members), Wahweap, and Kaiparowits formations. A possible record of anguids from the Aptian–Albian of Texas (Winkler, Murry, and Jacobs, 1990) is doubtful as the supporting evidence given is a ventrally open Meckelian fossa (but the illustration, Winkler, Murry, and Jacobs, 1990:fig. 8E, shows a ventromedial opening of the fossa) and striae on the lingual surface of the tooth crowns (now widely understood to be as an independently nondiagnostic character present in many scincomorphan as well as anguimorphan taxa). Less common, but more reliable for systematic and taxonomic analyses, are anguid jaws and/or other nonosteoderm elements. In southern Utah, such specimens have only been recovered from the Smoky Hollow Member of the Straight Cliffs Formation and the Kaiparowits Formation. Of these, the presence of *Odaxosaurus roosevelti* and *O. priscus* in Kaiparowits Formation are the earliest incontrovertible record of anguids in southern Utah and are paracontemporaneous with other records of *Odaxosaurus* (sensu Gao and Fox, 1996). The discrepancy between the more extensive records of anguid osteoderms and diagnostic jaw material is likely due to a couple of factors. Firstly, an armored anguid has hundreds of osteoderms, but only four tooth-bearing elements that are useful for identifications (premaxillae are not well known for fossil anguids and within lizards tend to have more simplified dental features that are less useful for comparative analysis of incomplete specimens; pers. obs.), giving a strong bias to the recovery of osteoderms in comparison to jaws. Second, referral of osteoderms to any one

taxon is problematic, particularly in a mixed taxon fauna recovered through screen washing concentrations of specimens, and it is possible, although not likely, that the thin, flat, unkeeled osteoderms from the Dakota and/or Wahweap formations (units from which anguids are represented only by osteoderms) are from nonanguids. Regardless, the record of presence of Anguidae in the Western Interior by the mid-Campanian is well substantiated, and potential earlier occurrences need to be confirmed by the recovery of diagnostic, nonosteodermal elements.

Anguimorpha: Platynota

Unlike their counterparts, the fossil record of platynotans (which includes the Necrosauridae, Monstersauria/Helodermatidae, *Lanthanotus*/Lanthanotidae, and Varanidae; I refer the reader to Conrad, 2008, for a review of the possible interrelationships of these taxa) in the Western Interior can be confidently traced back to at least the Early–Late Cretaceous transition (i.e., *Primaderma nessovi* in the Mussentuchit Member of the Cedar Mountain Formation, central Utah). Although they are not numerically dominant in their respective faunas, platynotans are easily recognized by their conspicuous tooth and jaw morphology (e.g., expanded tooth bases with basal enamel infolding, lack of a subdental shelf and sulcus dentalis). In southern Utah, platynotans have been recovered from the Dakota, Straight Cliffs (Smoky Hollow and John Henry members), and Kaiparowits formations. These specimens partially fill in the gap in the record of platynotans between the Albian–Cenomanian boundary (*Primaderma nessovi*) and the upper Campanian (Estes, 1983; Gao and Fox, 1996). With the addition of the records from southern Utah, it is clear that platynotans of southern Utah, as well as those of the more northern latitudes, were a conspicuous and successful part of nearly all the known lizard faunas from the Late Cretaceous of the Western Interior.

Serpentes

With the exception of the Iron Springs Formation, all of the rock units of southern Utah that have produced lizards have also produced snake vertebrae. The presence of snakes in the Late Cretaceous of southern Utah is not unexpected. Indeed, what is notable is the apparent lack of diversity of the snakes recovered. All of these specimens are generally similar in morphology to the vertebrae of the enigmatic taxon *Coniophis precedens* in having very low neural spines and lacking a notched posterior border of the neural arch. As such, I have referred these specimens described herein to the genus *Coniophis*. Side-by-side comparison of the specimens

16.19. (above and facing) Approximate stratigraphic (not to scale) occurrences of fossil squamates recovered from the Cretaceous of Utah. Taxonomic occurrences in the Cedar Mountain Formation are based on published accounts from Nydam (1999, 2000, 2002) and Nydam and Cifelli (2002b). Taxonomic occurrences in the North Horn Formation are based on published accounts in Gilmore (1940), Cifelli et al. (1999), and Nydam (1999). Southern Utah occurrences are based on the information presented herein as well as Nydam, Eaton, and Sankey (2007); Nydam and Voci (2007); and Nydam and Fitzpatrick (2009).

Age	Formation	Member	Polyglyphanodontini	Chamopsiidae	Contogeniidae	Paramacellodid/ Cordylid grade	aff. Scincidae	Scincoidea/ Scincomorpha indet.
MAASTRICHTIAN	NORTH HORN		*Polyglyphanodon steembergi* / *Paraglyphanodon gazini* / *Pa. utahensis*	cf. *Leptochamops*				
CAMPANIAN	KAIPAROWITS		*Peneteius saueri*	*Mensicognathus molybrochoros* / *Chamops* cf. *C. segnis* / cf. *Leptochamops* / *Tripennaculus eatoni*	*Palaeoscincosaurus pharikidodon*	Kaiparowits A / Kaiparowits B / Kaiparowits C / Kaiparowits D		Kaiparowits E / Kaiparowits F / Kaiparowits G / Misc. osteoderms
	WAHWEAP							Wahweap A / Misc. osteoderm
	IRON SPRINGS	DRIP TANK				Iron Springs A		Iron Springs B
SANTONIAN	IRON SPRINGS	JOHN HENRY				*Monocnemodon syphakos*		JHMbr SCFm A
CONIACIAN								
TURONIAN	STRAIGHT CLIFFS	SMOKY HOLLOW	*Dicothodon ciffellii*	*Chamops* cf. *C. segnis*	*Utahgenys evansi*	SHMbr SCFm A, B, C, D		SHMbr SCFm E, F, G, H / Misc. osteoderms
		TIBBET CANYON						
	TROPIC SHALE							
CENOMANIAN	DAKOTA				*Utahgenys antongai*	*Dakotasaurus gilletorum* / Dakota A / Dakota B / Dakota C	*Webbsaurus lofgreni*	Dakota D / Misc. osteoderm
ALBIAN	CEDAR MOUNTAIN	MUSSENTUCHIT	*Dicothodon ciffellii*	*Harmondontosaurus emeryensis*		*Pseudosaurillus* sp.	*Bothriogenys mysterion*	*Dimekodontosaurus madseni* / Scincomorpha indet.

show that vertebrae from each horizon are slightly different from the others, but to what degree these differences provide taxonomically relevant information is not clear. It would be reasonable to assume that the differences between the specimens, in conjunction with the time intervals separating each fauna, support at least species-level diagnoses. However, I am reluctant to invoke temporal differences in a diagnosis, and there is also the more overwhelming concern of not being able to recognize positional variation of vertebral morphology versus taxonomic variation. Finally, whereas all of the specimens do share many features that are considered indicative, if not diagnostic, of *Coniophis*, *Coniophis* itself is not well understood (Gardner and Cifelli, 1999) and may well be paraphyletic. I agree with the decision of Gardner and

Cifelli (1999) to refrain from assigning a species name to their specimens, and I follow that example here. It is clear that a comprehensive review of the known specimens of *Coniophis* is needed, and adding identifications that are very likely fated for nomen dubium will only complicate an already intricate problem.

The only known nonvertebral elements of *Coniophis* are some unassociated fragmentary maxilla and dentary specimens from the Maastrichtian that were referred to *C. precedens* by Longrich, Bhullar, and Gauthier (2012). However, I consider this referral to be problematic and potentially chimeric because some of the elements appear to be more appropriately referred to a lizard instead of a snake (Caldwell, Nydam, Palci, pers. comm.). The serpentian tooth-bearing

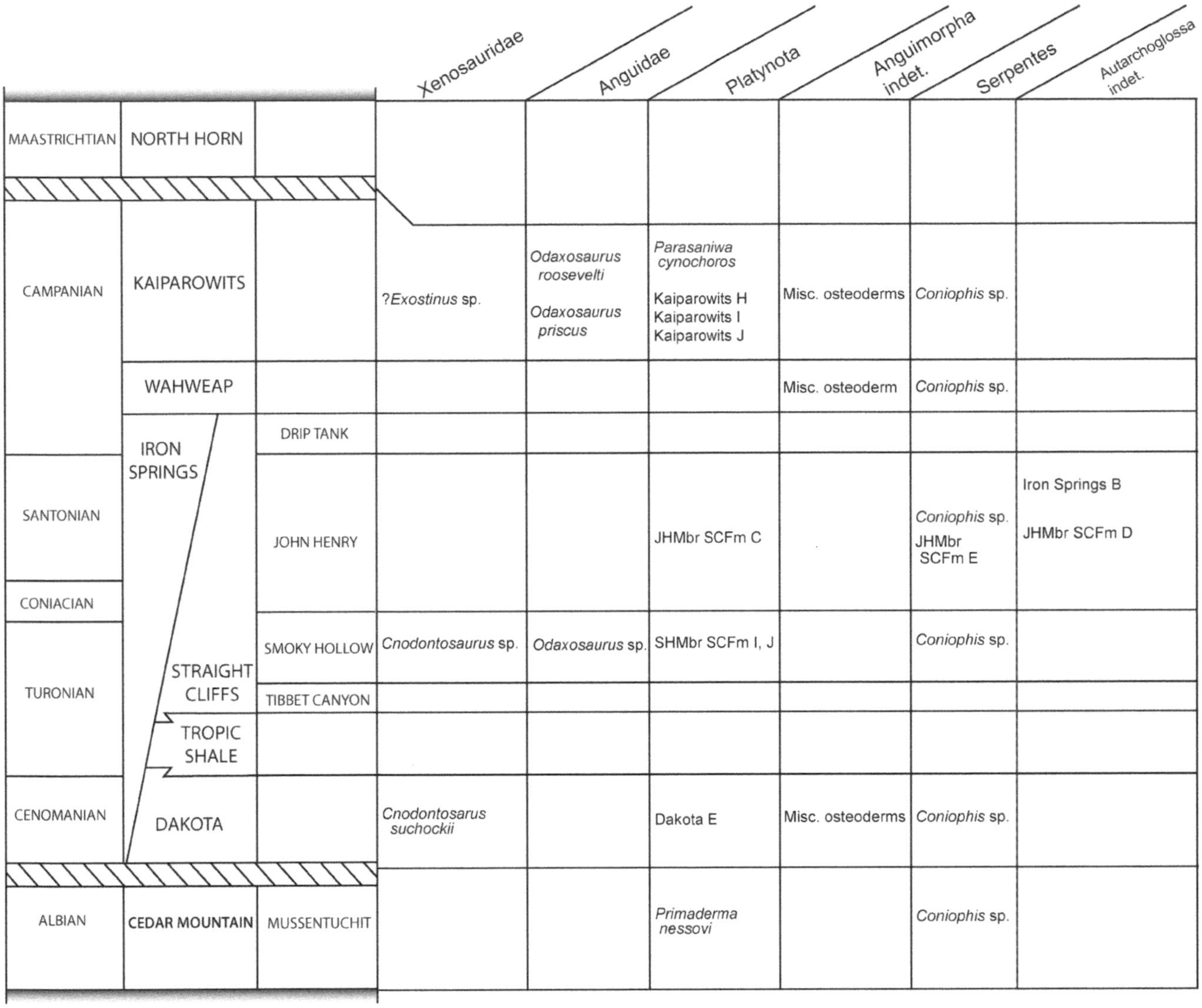

			Xenosauridae	Anguidae	Platynota	Anguimorpha indet.	Serpentes	Autarchoglossa indet.
MAASTRICHTIAN	NORTH HORN							
CAMPANIAN	KAIPAROWITS		?Exostinus sp.	Odaxosaurus roosevelti / Odaxosaurus priscus	Parasaniwa cynochoros / Kaiparowits H / Kaiparowits I / Kaiparowits J	Misc. osteoderms	Coniophis sp.	
	WAHWEAP					Misc. osteoderm	Coniophis sp.	
SANTONIAN	IRON SPRINGS	DRIP TANK						
		JOHN HENRY			JHMbr SCFm C		Coniophis sp. JHMbr SCFm E	Iron Springs B / JHMbr SCFm D
CONIACIAN								
TURONIAN	STRAIGHT CLIFFS	SMOKY HOLLOW	Cnodontosaurus sp.	Odaxosaurus sp.	SHMbr SCFm I, J		Coniophis sp.	
		TIBBET CANYON						
	TROPIC SHALE							
CENOMANIAN	DAKOTA		Cnodontosarus suchockii		Dakota E	Misc. osteoderms	Coniophis sp.	
ALBIAN	CEDAR MOUNTAIN	MUSSENTUCHIT			Primaderma nessovi		Coniophis sp.	

element recovered from the Kaiparowits Formation is the only known nonvertebral snake specimen from the Cretaceous from southern Utah. Although it is clearly a snake, I believe it is premature for a formal referral because there is potentially more than one snake taxon from this horizon and because of the need for a revision of *Coniophis*..

Although the interrelationships still need to be reassessed, there are some broad conclusions that can be drawn from the addition of these specimens to the fossil record. First, snakes are a known component of the squamate faunas of the Cretaceous of Utah from at least the Albian–Cenomanian boundary to the upper Campanian. Second, the rarity of the snake fossils in any of the horizons from which they are known indicates that they remained a relatively rare component of these squamate faunas. In contrast, snakes are the most diverse component of the modern squamate fauna, with 2700–3000 species known species (Greene, 1997; Pianka and Vitt, 2003). It seems that each of the paleofaunas from southern Utah was restricted to a single species until the upper

Campanian, when it appears that two species co-occurred in the Kaiparowits Formation.

CONCLUSIONS

The lizard and snake fossils reported herein – in addition to those already described from southern Utah – represent a significant increase to the known taxonomic and morphological record of Mesozoic squamates. These records are important in that they (1) demonstrate a consistent presence of polyglyphanodontines, contogeniids, and serpentians, possible xenosaurids, as well as diverse series of morphotypes with a paramacellodid/cordylids-grade dentition throughout the Late Cretaceous of Utah (Fig. 16.19); (2) show a marked increase in the diversity of chamopsiids and the establishment of anguids by the upper Campanian; (3) fill, or at least partially fill, long-standing stratigraphic and geographical gaps in the Cretaceous record of squamates from the Western Interior; and (4), establish a record of some of the most taxonomically and morphologically diverse squamate faunas of the Late Cretaceous of the Western Interior.

Even so, this report should not be read as a conclusion of the study of squamates from the Cretaceous of southern Utah. Instead, it should be seen as an indication that the vast potential of this region has yet to be fully appreciated. There are literally thousands of square kilometers of exposures yet to be surveyed, and there are undoubtedly numerous specimens yet to be found. The interval with the greatest need for continued effort remains the Santonian–middle Campanian (Fig. 16.19) because only a small sample of squamates has been recovered so far. I strongly encourage continued work in all aspects of the paleontology of southern Utah as it is an incredible natural resource at the southern end of the Western Interior.

ACKNOWLEDGMENTS

This work represents a long-term project that has benefitted greatly from the assistance and charity of numerous people. First, I wish to thank A. Titus of the Grand Staircase–Escalante National Monument for his support and assistance with access to the monument, bringing the specimen of *Parasaniwa cynochoros* sp. nov. to my attention, for organizing the symposium that has resulted in this book, and for his patience with my glacial writing pace. Alan's presence in southern Utah has been an irrefutable catalyst in the recent major advances in the geological and paleontological research of that region. Second, I would like to thank R. Cifelli and J. Eaton for their pioneering efforts in southern Utah in the 1980s and 1990s (and indeed up through today), generosity in letting me study the lizards they collected, and unflagging camaraderie. I would also like to thank L. Bryant and S. Foss of the Utah State Office of the Bureau of Land Management in Salt Lake City for their attention and assistance with permits. R. Cifelli (OMNH), J. Eaton, S. Sampson (UMNH), and D. Gillette (MNA) all were invaluable in providing access to specimens. Student participants in the field and/or laboratory include: MWU–B. Fitzpatrick, B. Flanagan, M. Hickman, A. Hill, C. Hills, M. Kryzek, M. Rose, G. Voci; ASU–J. Friday, S. Risso. D. Lofgren generously let me camp on his property on near the Grand Staircase–Escalante National Monument, and his staff and students from the Webb Schools provided much-appreciated assistance in the field. B. Davis (OMNH) assisted with production of scanning electron microscope figures. M. Bodeen, C. Lefubure, L. Suchocki, and C. Yacavone provided laboratory-based technical assistance at MWU. A. Titus and S. Evans provided many thoughtful and useful comments on an earlier version of this chapter (but I accept full responsibility for all errors and/or omissions). Funding for this project was made possible through a series of intramural research grants to the author from Midwestern University.

REFERENCES CITED

Alifanov, V. R. 2000. The fossil record of Cretaceous lizards from Mongolia; pp. 368–389 in M. J. Benton, M. A. Shishkin, D. M. Unwin, and E. N. Kurochkin (eds.), The Age of Dinosaurs in Russia and Mongolia. Cambridge University Press, Cambridge.

Armstrong-Ziegler, J. G. 1978. An aniliid snake and associated vertebrates from the Campanian of New Mexico. Journal of Paleontology 52:480–483.

Armstrong-Ziegler, J. G. 1980. Amphibia and Reptilia from the Campanian of New Mexico. Fieldiana Geology, n.s., 4:1–39.

Bhullar, B.-A. S. 2007. The enigmatic fossils *Exostinus* and *Restes*: resolving the stem and the crown of *Xenosaurus*, the knob-scaled lizards. Journal of Vertebrate Paleontology(3, Supplement) 27:48A.

Bhullar B.-A. S. 2010. Cranial osteology of *Exostinus serratus*, fossil sister taxon to the enigmatic clade *Xenosaurus*. Zoological Journal of the Linnean Society 159:921–953.

Bogert, C. M., and R. M. del Campo. 1956. The gila monster and its allies: the relationships, habits, and behavior of the lizards of the Family Helodermatidae. Bulletin of the American Museum of Natural History 109:1–238.

Broschinski, A. 2000. The lizards from the Guimarota mine; pp. 59–68 in T. Martin and B. Krebs (eds.), Guimarota, A Jurassic Ecosystem. Verlag Dr. Friedrich Pfeil, Munich.

Broschinski, A., and D. Sigogneau-Russell. 1996. Remarkable lizard remains from the Lower Cretaceous of Anoual (Morocco). Annales de Paléontologie 82:147–175.

Bryant, L. J. 1989. Non-dinosaurian lower vertebrates across the Cretaceous–Tertiary boundary in Northeastern Montana. Geological Sciences 134:1–107.

Budney, L. A., M. W. Caldwell, and A. Albino. 2006. Tooth socket histology in the Cretaceous snake *Dinilysia*, with a review of amniote dental attachment tissues. Journal of Vertebrate Paleontology 26:138–145.

Camp, C. L. 1923. Classification of the lizards. Bulletin of the American Museum of Natural History 48:289–481.

Cifelli, R. L. 1990a. A primitive higher mammal from the Late Cretaceous of southern Utah. Journal of Mammology 71:342–350.

Cifelli, R. L. 1990b. Cretaceous mammals of Southern Utah. I. Marsupials from the Kaiparowits Formation (Judithian). Journal of Vertebrate Paleontology 10:295–319.

Cifelli, R. L. 1990c. Cretaceous mammals of Southern Utah. II. Marsupials and marsupial-like mammals from the Wahweap Formation (Early Campanian). Journal of Vertebrate Paleontology 10:320–331.

Cifelli, R. L. 1990d. Cretaceous mammals of Southern Utah. III. Therian mammals from the Turonian (Early Late Cretaceous). Journal of Vertebrate Paleontology 10:332–345.

Cifelli, R. L. 1990e. Cretaceous mammals of Southern Utah. IV. Eutherian mammals from the Wahweap (Aquilan) and Kaiparowits (Judithian) formations. Journal of Vertebrate Paleontology 10:346–360.

Cifelli, R. L., R. L. Nydam, J. G. Eaton, J. D. Gardner, and J. I. Kirkland. 1999. Vertebrate faunas of the North Horn Formation (Upper Cretaceous–Lower Paleocene), Emery and Sanpete counties, Utah; pp. 377–388 in D. D. Gillette (ed.), Vertebrate Paleontology in Utah. Utah Geological Survey Miscellaneous Publication 99.

Conrad, J. L. 2004. Skull, mandible, and hyoid of *Shinisaurus crocodilurus* Ahl (Squamata, Anguimorpha). Zoological Journal of the Linnean Society 141:399–434.

Conrad, J. L. 2006. An Eocene shinisaurid (Reptilia, Squamata) from Wyoming, U.S.A. Journal of Vertebrate Paleontology 26:113–126.

Conrad, J. L. 2008. Phylogeny and systematics of Squamata (Reptilia) based on morphology.

Bulletin of the American Museum of Natural History 310:1–182.

Conrad, J. L., J. Ast, S. Montanari, and M. Norell. 2010. A combined evidence phylogenetic analysis of Anguimorpha (Reptilia: Squamata). Cladistics 26:1–48.

Cope, E. D. 1866. Fifth contribution to the herpetology of tropical America. Proceedings of the Academy of Natural Sciences of Philadelphia 1866:317–323.

Cope, E. D. 1873. Synopsis of New Vertebrata from the Tertiary of Colorado: Obtained during the Summer of 1873. United States Government Printing Office, 19 pp.

Denton, R. K., and R. C. O'Neill. 1995. *Prototeius stageri*, gen. et sp. nov., a new teiid lizard from the Upper Cretaceous Marshalltown Formation of New Jersey, with a preliminary phylogenetic revision of the Teiidae. Journal of Vertebrate Paleontology 15:235–253.

Duméril, A. M. C., and G. Bibron. 1836. Erpétologie générale ou histoire naturelle compléte des reptiles. Librairie Encyclopédique de Roret, Paris.

Dyman, T. S., W. A. Cobban, A. L. Titus, J. D. Obradovich, L. E. Davis, R. L. Eves, G. L. Pollock, K. I. Takahashi, and T. C. Hester. 2002. New biostratigraphic and radiometric ages for Albian–Turonian Dakota Formation and Tropic Shale at Grand Staircase–Escalante National Monument and Iron Springs Formation near Cedar City, Parowan, and Gunlock, Utah. Geological Society of America Abstracts with Programs 34(4):A-13.

Eaton, J. G. 1990. Stratigraphic revision of Campanian (Upper Cretaceous) rocks in the Henry Basin, Utah. Mountain Geologist 27:27–38.

Eaton, J. G. 1991. Biostratigraphic framework for Upper Cretaceous rocks of the Kaiparowits Plateau, southern Utah; pp. 47–63 in J. D. Nations and J. G. Eaton (eds.), Stratigraphy, Depositional Environments, and Sedimentary Tectonics of the Western Margin, Cretaceous Western Interior Seaway. Geological Society of America Special Paper 260.

Eaton, J. G. 1993a. Therian mammals from the Cenomanian (Upper Cretaceous) Dakota Formation, Southwestern Utah. Journal of Vertebrate Paleontology 13:105–124.

Eaton, J. G. 1993b. Mammalian paleontology and correlation of the uppermost Cretaceous rocks of the Pausaugunt Plateau, Utah. Museum of Northern Arizona Bulletin 59:163–180.

Eaton, J. G. 1995. Cenomanian and Turonian (early Late Cretaceous) multituberculate mammals from southwestern Utah. Journal of Vertebrate Paleontology 15:761–784.

Eaton, J. G. 1999a. Vertebrate paleontology of the Pausaugunt Plateau, Upper Cretaceous, southwestern Utah; pp. 334–338 in D. D. Gillette (ed.), Vertebrate Paleontology in Utah. Utah Geological Survey Miscellaneous Publication 99–1.

Eaton, J. G. 1999b. Vertebrate paleontology of the Iron Springs Formation, Upper Cretaceous, southwestern Utah; pp. 339–343 in D. D. Gillette (ed.), Vertebrate Paleontology in Utah. Utah Geological Survey Miscellaneous Publication 99–1.

Eaton, J. G. 2006. Santonian (Late Cretaceous) Mammals from the John Henry Member of the Straight Cliffs Formation, Grand Staircase–Escalante National Monument, Utah. Journal of Vertebrate Paleontology 26:446–460.

Eaton, J. G., and R. L. Cifelli. 1988. Preliminary report on Late Cretaceous mammals of the Kaiparowits Plateau, southern Utah. Contributions to Geology 26:45–55.

Eaton, J. G., S. Diem, J. D. Archibald, C. Schierup, and H. Munk. 1999a. Vertebrate paleontology of the Upper Cretaceous rocks of the Markagunt Plateau, southwestern Utah; pp. 323–333 in D. D. Gillette (ed.), Vertebrate Paleontology in Utah. Utah Geological Survey Miscellaneous Publication 99–1.

Eaton, J. G., R. L. Cifelli, J. H. Hutchison, J. I. Kirkland, and J. M. Parrish. 1999b. Cretaceous vertebrate faunas from the Kaiparowits Plateau, south central Utah; pp. 345–353 in D. D. Gillette (ed.), Vertebrate Paleontology in Utah. Utah Geological Survey Miscellaneous Publication 99–1.

Edmund, A. G. 1969. Dentition; 117–200 in C. Gans and T. S. Parsons eds., Biology of the Reptilia 1. Academic Press, New York.

Estes, R. 1964. Fossil vertebrates from the Late Cretaceous Lance Formation, eastern Wyoming. University of California Publications in Geological Sciences 49:1–169.

Estes, R. 1969. Relationships of two lizards (Sauria, Teiidae). Breviora 317:1–8.

Estes, R. 1983. Teil 10A, Sauria Terrestria, Amphisbaenia; pp. 179–189 in P. Wellnhofer (ed.), Handbuch der Paläoherpetologie. Gustav Fisher Verlag, Stuttgart.

Estes, R., P. Berberian, and C. A. M. Meszoely. 1969. Lower Vertebrates from the Late Cretaceous Hell Creek Formation, McCone County, Montana. Brevoria 337:1–33.

Estes, R., K. de Quieroz, and J. Gauthier. 1988. Phylogenetic relationships within Squamata; pp. 119–281 in R. Estes and G. Pregill (eds.), Phylogenetic Relationships of the Lizard Families: Essays Commemorating Charles L. Camp. Stanford University Press, Stanford, California.

Evans, S. E. 1993. Jurassic lizard assemblages. Revue de Paléobiologie Spécial 7:55–65.

Evans, S. 1996. *Parviraptor* (Squamata: Anguimorpha) and other lizards from the Morrison Formation at Fruita, Colorado; pp. 243–248 in M. Morales (ed.), The Continental Jurassic. Museum of Northern Arizona Bulletin 60.

Evans, S. E. 2003. At the feet of dinosaurs: the early history and radiation of lizards. Biological Review 28:513–551.

Evans, S. E., and L. J. Barbadillo. 1998. An unusual lizard (Reptilia: Squamata) from the Early Cretaceous of Las Hoyas, Spain. Zoological Journal of the Linnean Society 1998:235–365.

Evans, S. E., and D. J. Chure. 1998a. Paramacellodid lizard skulls from the Jurassic Morrison Formation at Dinosaur National Monument, Utah. Journal of Vertebrate Paleontology 18:99–114.

Evans, S. E., and D. J. Chure. 1998b. Morrison lizards: structure, relationships and biogeography. Modern Geology 23:35–48.

Evans, S. E., and D. J. Chure. 1999. Upper Jurassic lizards from the Morrison Formation of Dinosaur National Monument, Utah; pp. 151–159 in D. D. Gillette (ed.), Vertebrate Paleontology in Utah. Utah Geological Survey Miscellaneous Publication 99–1.

Evans, S. E., and B. Searle. 2002. Lepidosaurian reptiles from the Purbeck Limestone Group of Dorset, Southern England. Special Papers in Palaeontology 68:145–159.

Evans, S. E., and Y. Wang. 2010. A new lizard (Reptilia: Squamata) with exquisite preservation of soft tissue from the Lower Cretaceous of Inner Mongolia, China. Journal of Systematic Palaeontology 8:81–95.

Filhol, H. 1876. Sur les reptiles des phosphorites du Quercy. Bulletin des Société Philomathique de Paris 11:27–28.

Folie, A., and V. Codrea. 2005. New lissamphibians and squamates from the Maastrichtian of Haţeg Basin, Romania. Acta Palaeontologica Polonica 50:57–71.

Fults, A. A. 2009. A comparison of freshwater selachian species from before and after the Greenhorn Cyclothem, Kaiparowits Plateau, southwestern Utah. M.Sc. thesis, Midwestern University, Glendale, Arizona.

Fürbringer, M. 1900. Zur vergleichenden Anatomie des Brustschulter-apparates und der Schultermuskeln. Jenaische Zeitschrift für Naturwissenschaft 34:215–718.

Gao, K., and R. C. Fox. 1996. Taxonomy and evolution of Late Cretaceous lizards (Reptilia: Squamata) from western Canada. Bulletin of the Carnegie Museum of Natural History 33:1–107.

Gardner, J. D. 2000. Albanerpetontid amphibians from the Upper Cretaceous (Campanian and Maastrichtian) of North America. Geodiversitas 22:349–388.

Gardner, J. D., and R. L. Cifelli. 1999. A primitive snake from the Cretaceous of Utah. Special Papers in Palaeontology 60:87–100.

Gates, T. A., S. D. Sampson, L. E. Zanno, E. M. Roberts, J. G. Eaton, R. L. Nydam, J. H. Hutchison, J. A. Smith, M. A. Loewen, and M. A. Getty. 2010. Biogeography of terrestrial and freshwater vertebrates from the Late Cretaceous (Campanian) Western Interior of North America. Palaeogeography, Palaeoclimatology, Palaeoecology 291:371–387.

Gauthier, J. A. 1982. Fossil xenosaurid and anguid lizards from the early Eocene Wasatch Formation, southeast Wyoming, and a revision of the Anguiodea. Contributions to Geology University of Wyoming 21:7–54.

Gilmore, C. W. 1928. Fossil lizards of North America. Memoirs of the National Academy of Sciences 22:1–201.

Gilmore, C. W. 1940. New fossil lizards from the Upper Cretaceous of Utah. Smithsonian Miscellaneous Collections 99:1–3.

Gilmore, C. W. 1942. Paleocene faunas of the Polecat Bench Formation, Park County, Wyoming. Proceedings of the American Philosophical Society 85:159–167.

Gray, J. E. 1825. A synopsis of the genera of reptiles and amphibia, with a description of some new species. Annals of Philosophy (Thompson), n.s., 10:193–217.

Greene, H. W. 1997. Snakes: The Evolution of Mystery in Nature. University of California Press, Berkeley, California.

Hecht, M. K. 1959. Amphibians and reptiles. Bulletin of the American Museum of Natural History 117:117–176.

Hecht, M., and B. Hecht. 1984. A new lizard from Jurassic deposits of middle Asia. Paleontological Journal 18:135–138. [Russian]

Heller, P. L., K. Dueker, and M. E. McMillan. 2003. Post-Paleozoic alluvial gravel transport as evidence of continental tilting in the U.S. Cordillera. Geological Society of America Bulletin 115:1122–1132.

Hoffstetter, M. R. 1967. Coup d'oeil sur les sauriens (= lacertiliens) des couches de Purbeck (Jurassique supérieur d'Angleterre). Colloque international du Centre National de la Recherche Scientifique: Problèmes Actuels de Paléontologie (Évolution des Vertébrés) 163:349–371.

Klembara, J. 2008. A new anguimorph lizard from the lower Miocene of north-west Bohemia, Czech Republic. Palaeontology 51:81–94.

Kosma, R. 2004. The dentitions of recent and fossil scincomorphan lizards (Lacertilia, Squamata)–systematics, functional morphology, paleoecology. Ph.D. dissertation, Universität Hannover, Hannover, Germany.

Kuhn, O. 1940. Die Placosauriden und Anguiden aus dem mittleren Eozän des Geiseltales. Nova Acta Leopoldina 8:461–486.

Langer, M. C. 1998. Gilmoreteiidae new family and Gilmoreteius new genus (Squamata, Scincomorpha): replacement names for Macrocephalosauridae Sulimski, 1975 and Macrocephalosaurus Gilmore, 1943. Comunicações do Museu de Ciências e Technologia da PUCRS, Série Zoologia 11:13–18.

Lee, M. S. Y. 1997. The phylogeny of varanoid lizards and the affinities of snakes. Philosophical Transactions of the Royal Society of London Series B, Biological Sciences 352:53–91.

Linnaeus, C. 1758. Systema naturae per regna tria naturae, secundum classes, ordines, genera, species, cum characteribus, differentiis, synonymis, locis. 10th revised edition, volume 1. Salvii Nat., Holmiae [Stockholm], i–ii + 1–824 pp.

Lofgren, D. L. 1995. The Bug Creek problem and the Cretaceous–Tertiary transition at McGuire Creek, Montana. University of California Publications, Geological Sciences 140:1–185.

Longrich, N. R., B.-A. S. Bhullar, and J. A. Gauthier. 2012. A transitional snake from the Late Cretaceous period of North America. Nature 488:205–208.

Makádi, L. 2006. Bicuspidon aff. hatzegiensis (Squamata: Scincomorpha: Teiidae) from the Upper Cretaceous Csehbánya Formation (Hungary, Bakony Mts). Acta Geologica Hungarica 49:373–385.

Marsh, O. C. 1892. Notice of new reptiles from the Laramie Formation. American Journal of Science 43:449–453.

McCord, R. D. 1998. A new genus and species of Cretaceous polyglyphanodontine lizard (Squamata, Teiidae) from the Kaiparowits Plateau, Utah; pp. 281–292 in Y. Tomida, L. J. Flynn, and L. L. Jacobs (eds.), Advances in Vertebrate Paleontology and Geochronology. National Science Museum, Tokyo.

Montanucci, R. R. 1989. The relationship of morphology to diet in the horned lizard genus Phrynosoma. Herpetologica 45:208–216.

Norell, M. A., and K. Gao. 1997. Braincase and phylogenetic relationships of Estesia mongoliensis from the Late Cretaceous of the Gobi Desert and the recognition of a new clade of lizards. American Museum Novitates 3211:1–25.

Nydam, R. L. 1999. Polyglyphanodontinae (Squamata: Teiidae) from the medial and Late Cretaceous: new records from Utah, U.S.A., and Baja California del Norte, Mexico; pp. 303–317 in D. D. Gillette (ed.), Vertebrate Paleontology in Utah. Utah Geological Survey Miscellaneous Publication 99-1.

Nydam, R. L. 2000. A new taxon of helodermatid-like lizard from the Albian–Cenomanian of Utah. Journal of Vertebrate Paleontology 20:285–294.

Nydam, R. L. 2002. Lizards of the Mussentuchit Local Fauna (Albian–Cenomanian) and comments on the evolution of the Cretaceous lizard fauna of North America. Journal of Vertebrate Paleontology 22:645–660.

Nydam, R. L. 2010. An updated summary of the Cretaceous-aged lizard faunas of Grand Staircase–Escalante National Monument; pp. 133–140 in M. Eaton (ed.), Learning from the Land, Grand Staircase–Escalante National Monument Science Symposium Proceedings. Grand Staircase–Escalante Partners.United States Department of the Interior, Washington, D.C.

Nydam, R. L., and R. L. Cifelli. 2002a. A new teiid lizard from the Cedar Mountain Formation (Albian–Cenomanian boundary) of Utah. Journal of Vertebrate Paleontology 22:276–285.

Nydam, R. L., and R. L. Cifelli. 2002b. Lizards from the Lower Cretaceous (Aptian–Albian) Antlers and Cloverly formations. Journal of Vertebrate Paleontology 22:286–298.

Nydam, R., and B. Fitzpatrick. 2009. The occurrence of Contogenys-like lizards in the Late Cretaceous and Early Tertiary of the Western Interior or the U.S.A. Journal of Vertebrate Paleontology 29:677–701.

Nydam, R., and G. E. Voci. 2007. Teiid-like scincomorphan lizards from the Late Cretaceous (Campanian) of Southern Utah. Journal of Herpetology 41:215–223.

Nydam, R. L., M. W. Caldwell, and F. Fanti. 2010. Borioteiioidean lizard skulls from Kleskun Hill (Wapiti Formation; Upper Campanian), west-central Alberta, Canada. Journal of Vertebrate Paleontology 30:1090–1099.

Nydam, R. L., J. G. Eaton, and J. Sankey. 2007. New taxa of transversely-toothed lizards (Squamata: Scincomorpha) and new information on the evolutionary history of "teiids." Journal of Paleontology 81:538–549.

Nydam, R. L., J. A. Gauthier, and J. J. Chiment. 2000. The mammal-like teeth of the Late Cretaceous lizard Peneteius aquilonius Estes 1969 (Squamata, Teiidae). Journal of Vertebrate Paleontology 20:628–631.

Oelrich, T. M. 1956. The anatomy of the head of Ctenosaura pectinata (Iguanidae). Miscellaneous Publications of the Museum of Zoology, University of Michigan 94:1–122.

Oppel, M. 1811. Die Ordnung, Familien und Gattungen der Reptilien als Prodrom einer Naturgeschichte derselben. Lindauer, Munich.

Peng, J., A. P. Russell, and D. B. Brinkman. 2001. Vertebrate Microsite Assemblages (Exclusive of Mammals) from the Foremost and Oldman Formations of the Judith River Group (Campanian) of Southeastern Alberta: An Illustrated Guide. Provincial Museum of Alberta Natural History Occasional Paper 25.

Peterson, F. 1969. Four New Members of the Upper Cretaceous Straight Cliffs Formation in the Southeastern Kaiparowits Region, Kane County, Utah. U.S. Geological Survey Bulletin 1274-J.

Peterson, F., and H. A. Waldrop. 1965. Jurassic and Cretaceous stratigraphy of south-central Kaiparowits Plateau, Utah; pp. 47–69 in H. D. Goode and R. A. Robison (eds.), Geology and Resources of South-Central Utah. Utah Geological Society and Intermountain Association of Petroleum Geologists Guidebook to the Geology of Utah.

Pianka, E. R., and L. J. Vitt. 2003. Lizards: Windows to the Evolution of Diversity. University of California Press, Berkeley, California.

Pregill, G. K., J. A. Guathier, and H. W. Greene. 1986. The evolution of helodermatid squamates, with description of a new taxon and an overview of Varanoidea. Transactions of the San Diego Society of Natural History 21:167–202.

Prothero, D. R., and R. Estes. 1980. Late Jurassic lizards from Como Bluff, Wyoming, and their palaeobiogeographic significance. Nature 286:484–486.

Rage, J.-C. 1984. Teil 11, Serpentes; in P. Wellnhofer (ed.), Handbuch der Paläoherpetologie. Gustav Fisher Verlag, Stuttgart.

Reynoso, V.-H., and G. Callison. 2000. A new scincomorph lizard from the Early Cretaceous of Puebla, México. Zoological Journal of the Linnean Society 2000:183–212.

Richter, A. 1994. Lacertilia aus der Unteren Kreide von Uña und Galve (Spanien) und Anoual (Marokko). Berliner geowissenschaftliche Abhandlungen (E) 14:1–147.

Rieppel, O., J. L. Conrad, and J. A. Maisano. 2007. New morphological data for Eosaniwa koehni Haubold, 1977 and a revised phylogenetic analysis. Journal of Paleontology 81:760–769.

Roberts, E. M., A. L. Deino, and M. A. Chan. 2005. ^{40}Ar/^{39}Ar age of the Kaiparowits Formation, southern Utah, and correlation of contemporaneous Campanian strata and vertebrate faunas along the margin of the Western Interior Basin. Cretaceuos Research 26:307–318.

Sahni, A. 1972. The vertebrate fauna of the Judith River Formation, Montana. Bulletin of the American Museum of Natural History 147:321–412.

Seiffert, J. 1973. Upper Jurassic lizards from central Portugal. Separata da Memória dos Serviços Geológicos de Portugal 22:1–85.

Smith, J. B., and P. Dodson. 2003. A proposal for standard terminology for anatomical notation and orientation in fossil vertebrate dentitions. Journal of Vertebrate Paleontology 23:1–12.

Smith, K. T. 2011. The long-term history of dispersal among lizards in the early Eocene: new evidence from a microvertebrate assemblage in the Bighorn Basin of Wyoming, U.S.A. Palaeontology 54:1243–1270.

Sternberg, C. M. 1951. The lizard Chamops from the Wapiti Formation of Northern Alberta:

Polydontosaurus grandis not a lizard. Bulletin of the National Museum of Canada 123:256–258.

Strahm, M. H., and A. Swartz. 1977. Osteoderms in the anguid lizard subfamily Diploglossinae and their taxonomic importance. Biotropica 9:58–72.

Sullivan, R. M. 1981. Fossil lizards from the San Juan Basin, New Mexico; pp. 76–88 in S. G. Lucas, J. K. Rigby Jr., and B. S. Kues (eds.), Advances in San Juan Basin Paleontology.

University of New Mexico Press, Albuquerque, New Mexico.

Wagler, J. 1830. Naturliches System der Amphibien, mit vorangehender Klassification der Saügethiere und Vögel: ein Beiträg zur vergleichenden Zoologie. J. G. Cotta, Munich.

Waldman, M. 1970. A teiid lizard jaw from the Cretaceous of Alberta, Canada. Canadian Journal of Earth Sciences 7:542–547.

Williamson, T. E., and A. Weil. 2004. First occurrence of the teiid lizard *Peneteius* from the latest Cretaceous Naashoibito Member, Kirtland Formation, San Juan Basin, New Mexico. New Mexico Geology 26:65.

Winkler, D. A., P. A. Murry, and L. L. Jacobs. 1990. Early Cretaceous (Comanchean) vertebrates of central Texas. Journal of Vertebrate Paleontology 10:95–116.

17

Crocodyliforms from the Late Cretaceous of Grand Staircase–Escalante National Monument and Vicinity, Southern Utah, U.S.A.

Randall B. Irmis, J. Howard Hutchison, Joseph J. W. Sertich, and Alan L. Titus

ALTHOUGH THE KAIPAROWITS BASIN OF SOUTHERN UTAH contains an excellent Late Cretaceous stratigraphic record of nonmarine fossiliferous sediments, crocodyliforms from these deposits remain poorly known. Isolated teeth and osteoderms from the Dakota and Straight Cliffs formations document the presence of widespread large clades such as Mesoeucrocodylia and Neosuchia, but finer taxonomic resolution is not currently possible. The record from the middle Campanian Wahweap Formation is similarly fragmentary but also includes specimens assignable to Crocodylia, the crown group of crocodyliforms. By far the best-known crocodyliform assemblage from the Late Cretaceous of Utah is that of the late Campanian Kaiparowits Formation, which preserves the remains of at least three alligatoroids: *Deinosuchus hatcheri*, *Brachychampsa*, and an unnamed small alligatoroid that lacks globidont teeth. The Wahweap and Kaiparowits formations also preserve large rectangular osteoderms of an indeterminate basal neosuchian that might be related to the goniopholidid *Denazinosuchus* from the late Campanian of the San Juan Basin in northern New Mexico but could also pertain to a late surviving pholidosaurid. The Kaiparowits Formation crocodyliform assemblage is most similar to that of the Fruitland and Kirtland formations in the San Juan Basin, and differs from more northerly assemblages in Montana and Alberta. Kaiparowits crocodyliforms provide additional evidence for a distinct southern Laramidian vertebrate faunal province during the Campanian and possible climate-induced northern migration of southern taxa during the Maastrichtian.

INTRODUCTION

Crocodyliformes, a subclade of Crocodylomorpha, includes nearly all known (fossil and living) long-snouted aquatic and semiaquatic forms that are typically thought of as the prototypical crocodylian form (i.e., modern crocodylids and alligatoroids). Extinct taxa also include a host of semiterrestrial and likely fully terrestrial forms that include short-skulled omnivores/herbivores (notosuchians) and tall-skulled macropredators (peirosaurids, sebecosuchians, and pristichampsids). This clade reached its peak in both diversity and disparity during the Cretaceous and Paleogene periods (e.g., Markwick, 1998), with highly distinct Laurasian and Gondwanan assemblages (Turner, 2004).

Crocodyliforms are a common component of nonmarine vertebrate assemblages from the Late Cretaceous of the Western Interior (Markwick, 1998; Eaton, Cifelli, et al. 1999; Peng, Russell, and Brinkman, 2001; Wu, 2005; Lucas et al., 2006). Crocodyliform fossils from Laramidia have become increasingly important in understanding the origin of Crocodylia (crown-group crocodyliforms), and in particular alligatoroids (Erickson, 1972; Wu, Brinkman, and Russell, 1996a; Brochu, 1997, 1999, 2003, 2004; Wu, Russell, and Brinkman, 2001). However, nearly all of these taxa come from northern Laramidia (Montana and Canada), whereas southern Late Cretaceous assemblages are poorly known. A recent review identified only three taxa from the well-studied San Juan Basin in northern New Mexico (Lucas et al., 2006), and crocodyliforms from the Late Cretaceous of southern Utah have only been mentioned in passing or in faunal lists (Eaton, 1991; Gillette and Hayden, 1997; Eaton and Cifelli, 1998; Eaton, Cifelli, et al. 1999; Eaton, Diem, 1999; Hutchison et al., 1998; Wiersma, Hutchison, and Gates, 2004; Titus et al., 2005; Gates et al., 2010; Sampson, Loewen, et al., 2010). These southern records are critically important for understanding Late Cretaceous nonmarine vertebrate biogeography, where there is evidence for latitudinally arrayed endemic faunas that show limited mixing over time (Gates et al., 2010; Sampson, Loewen, et al., 2010; Sampson, Gates, et al., 2010; Hutchison et al., this volume, Chapter 13).

Here, we review the crocodyliform fossil record of Late Cretaceous nonmarine units in southern Utah, with a focus on the Campanian Wahweap and Kaiparowits formations of the Kaiparowits Plateau. We discuss the identity of crocodyliform fossils recovered over the past 25 years of fieldwork, and discuss their implications for Late Cretaceous Western Interior biogeography. These fossils are discussed by formation in the following stratigraphic order: Dakota Formation (upper Cenomanian), Straight Cliffs Formation (Turonian through Santonian), Wahweap Formation (middle Campanian), and Kaiparowits Formation (upper Campanian).

Institutional Abbreviations MNA, Museum of Northern Arizona, Flagstaff, Arizona; UCMP, University of California Museum of Paleontology, Berkeley, California; UMNH, Natural History Museum of Utah, Salt Lake City, Utah.

CROCODYLIFORM FOSSIL ASSEMBLAGES

Dakota Formation

CROCODYLIFORMES Hay, 1930,
sensu Benton and Clark, 1988
MESOEUCROCODYLIA
Whetstone and Whybrow, 1983

Specimens MNA V10402, 10403, and V10406 (MNA Loc. 939); MNA V10407 (MNA Loc. 1008); MNA V10408 (MNA Loc. 1064), MNA V10404 and V10405 (MNA Loc. 1067); MNA V10409 (MNA Loc. 1068–3); MNA V10410 (MNA Loc. 1071–1); and MNA V10411 (MNA Loc. 1217) (see Eaton, 1993, 1995, 2009, for stratigraphic and locality information) (Fig. 17.1A–F).

Description These specimens were recovered from the screen-wash concentrate of several microvertebrate localities within the Dakota Formation of southern Utah (Eaton, 1993, 1995, 2009). Recovered teeth are elongate and conical, with a marked labiolingual curvature, are not recurved, and have distinct longitudinal striations and well-developed carinae (Fig. 17.1A–F). All of the specimens are isolated and have not been found associated with nondental crocodylomorph remains.

Discussion These teeth may form the basis for reports of *Teleorhinus* sp. (=*Terminonaris*) and *Goniopholis* sp. from this unit (Eaton, Cifelli, et al. 1999). Unfortunately, this tooth morphology is not restricted to *Terminonaris* and/or *Goniopholis*, or even Pholidosauridae and Goniopholididae (contra Rowe et al., 1992; Sankey, 2008). Striated, conical unrecurved teeth are present in many taxa across Crocodyliformes, including notosuchians (e.g., Pol, 2003), pholidosaurids (e.g., Wu, Russell, and Cumbaa, 2001), dyrosaurids (e.g., Jouve, 2005), and thalattosuchians (e.g., Young and de Andrade, 2009), goniopholidids (e.g., Wu, Brinkman, and Russell, 1996b; Salisbury, 2002), and gavialoids (e.g., Delfino, Piras, and Smith, 2005), among others. Although elongate, conical unrecurved teeth are present in basal crocodyliforms, none of the known taxa outside Mesoeucrocodylia possess longitudinal ridges on their teeth (e.g., Colbert and Mook, 1951; Clark, 1985; Wu, Brinkman, and Lu, 1994; Wu, Sues,

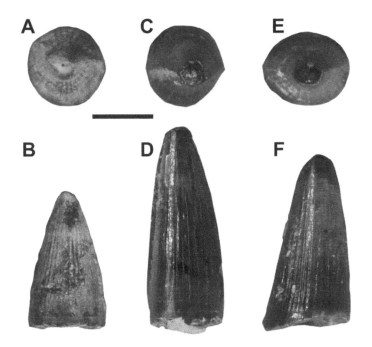

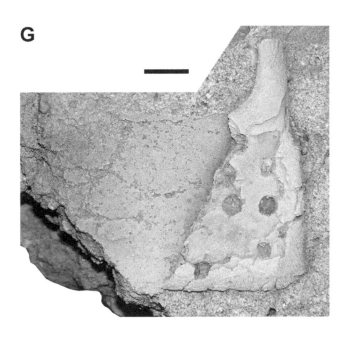

17.1. Crocodyliform specimens from the Dakota Formation of southern Utah. (A, B) MNA V10403, mesoeucrocodylian tooth in (A) occlusal and (B) lingual views. (C, D) MNA V10404, mesoeucrocodylian tooth in (C) occlusal and (D) lingual views. (E, F) MNA V10405, mesoeucrocodylian tooth in (E) occlusal and (F) lingual views. (G) UMNH VP 21148, eusuchian osteoderm in dorsal view. Scale bars = 5 mm.

and Dong, 1997; Gao, 2001; Pol and Norell, 2004a, 2004b; Peng and Shu, 2005), indicating that the Dakota teeth can be referred to the clade Mesoeucrocodylia; if thalattosuchians are outside Mesoeucrocodylia (Pol and Gasparini, 2009; Wilberg, 2010; cf. Clark, 1994), these teeth would only be referable to Crocodyliformes.

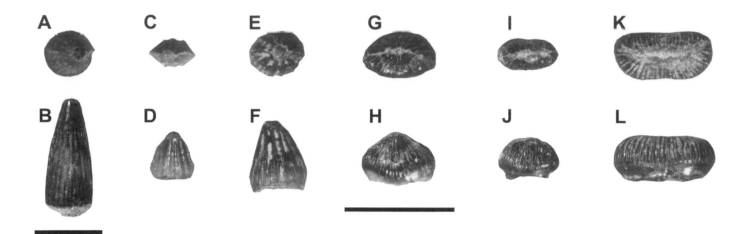

17.2. Crocodyliform teeth from the John Henry Member of the Straight Cliffs Formation. (A, B) UMNH VP 14403, mesoeucrocodylian tooth in (A) occlusal and (B) lingual views. (C, D) UMNH VP 14403, neosuchian tooth in (C) occlusal and (D) lingual views. (E, F) UMNH VP 17258, eusuchian tooth in (E) occlusal and (F) lingual views. (G, H) UMNH VP 17258, neosuchian tooth in (G) occlusal and (H) lingual views. (I, J) UMNH VP 17258, eusuchian tooth in (I) occlusal and (J) lingual views. (K, L) UMNH VP 17258, neosuchian tooth in (K) occlusal and (L) lingual views. Scale bars = 5 mm.

NEOSUCHIA Benton and Clark, 1988

Specimens UMNH VP 21148, partial paramedian osteoderm from the base of the middle member of the Dakota Formation. (Fig. 17.1G)

Description This specimen preserves only the lateral margin of the osteoderm exposed in dorsal view, with an impression of the ventral surface of the central portion of the plate. The dorsal surface of the osteoderm is covered with very widely spaced circular pitting (Fig. 17.1G). The anterolateral margin of the plate possessed a well-developed elongate anteriorly projecting process (Fig. 17.1G).

Discussion Like similar osteoderms from the Wahweap and Kaiparowits formations (see below), crocodyliform osteoderms of this morphology from the mid- to Late Cretaceous are typically referred to the Goniopholididae. The lack of unambiguous synapomorphies shared only with goniopholidids means that this specimen can only be referred to the clade Neosuchia because identical osteoderms are known from both goniopholidids and pholidosaurids among crocodyliforms. No known crocodylian or notosuchian possesses similar osteoderms.

Straight Cliffs Formation

Like the Dakota Formation material, most of our understanding of crocodyliforms from the Straight Cliffs Formation is from isolated dental specimens recovered through screen washing for microvertebrates. The recovery of these specimens primarily results from the long-term efforts of Jeff Eaton focused on the Santonian John Henry Member

(e.g., Eaton, 2006). Although figured specimens have been individually cataloged, many hundreds of teeth have been recovered, and these are generally batch cataloged. Thus, specimen numbers may be listed under more than one taxonomic assignment.

MESOEUCROCODYLIA
Whetstone and Whybrow, 1983

Specimens UMNH VP 14403, 17258, 17366, 17471, 17543, 17577, 17579, 17649, 17866, 17884, and 19092, all collections of multiple isolated teeth (Fig. 17.2A, B).

Description These specimens consist of elongate conical teeth with well-developed carinae that are not recurved but have varying degrees of labiolingual curvature and longitudinal striations (Fig. 17.2A, B). They are identical in morphology to teeth from the underlying Dakota Formation and vary in size from 4 to 20 mm.

Discussion As with similar teeth from the Dakota Formation, these can only be referred to the clade Mesoeucrocodylia. All of these teeth are from the Santonian portion of the John Henry Member of the Straight Cliffs Formation, both from the Kaiparowits Plateau in Grand Staircase–Escalante National Monument (Eaton, 2006), and further west along the Paunsaugunt Plateau in Bryce Canyon National Park (Eaton, 1999). These teeth are probably the basis for previous reports of "Goniopholidae indet." from the John Henry Member (Eaton, Cifelli, et al. 1999).

NEOSUCHIA Benton and Clark, 1988

Specimens UMNH VP 14403, 17258, 17366, 17467 17501, 17513, 17541, 17543, 17557, 17579, 17604, 17620, 17649, and 19090, all collections of multiple isolated teeth (Fig. 17.2C–F).

Description These teeth are triangular to slightly leaf shaped in outline, and constricted near the base of the crown (Fig. 17.2C–F). The teeth are not recurved but vary from slightly bulbous to laterally compressed, and there is little to no labiolingual curvature. The crown is decorated with prominent coarse to fine longitudinal ridges (Fig. 17.2C–F). This tooth morphology grades into those of the teeth described below.

Discussion When discovered in Cretaceous sediments, these teeth are typically referred to *Theriosuchus* sp. or Atoposauridae (e.g., Brinkmann, 1992; Schwarz-Wings, Rees, and Lindgren, 2009) and are probably the basis for "Atoposauridae indet." in the faunal list of Eaton, Cifelli, et al. (1999) for the John Henry Member of the Straight Cliffs Formation. Brinkman (2008) referred a tooth of this morphotype from the Campanian Dinosaur Park Formation in Alberta to the basal alligatoroid *Leidyosuchus* because this taxon is most abundant in the formation, though he noted the tooth morphology was not specifically diagnostic of that genus. Although these teeth are characteristic of atoposaurid crocodyliforms (e.g., Salisbury, 2002; Schwarz and Salisbury, 2005), they are also present in the basal neosuchian *Pachycheilosuchus* from the Albian of Texas (Rogers, 2003:fig. 4), the midportion of the maxilla and lower jaw of the basal neosuchian *Bernissartia* (Dollo, 1883; Buffetaut, 1975:pl. 2; Buffetaut and Ford, 1979:figs. 27–29; Buscalioni and Sanz, 1990; Norell and Clark, 1990), and the anterior maxilla and dentary of the globidontan alligatoroid *Brachychampsa* (UCMP 151710 and UCMP 159000, see below). All of these taxa are within the crocodyliform clade Neosuchia (e.g., Pol et al., 2009:fig. 37; Turner and Sertich, 2010:fig. 18), and to our knowledge, this tooth morphotype has not been reported from basal crocodyliforms. These teeth share some similarities to the anterior dentition of the notosuchian *Araripesuchus* (Turner, 2006; Sereno and Larsson, 2009:figs. 16A, 19A) but differ in the presence in deep longitudinal striations. Thus, the least inclusive clade these teeth can be referred to is Neosuchia.

NEOSUCHIA Benton and Clark, 1988
Unnamed Clade of Atoposauridae + Eusuchia

Specimens UMNH VP 14403, 17258, 17366, 17467, 17471, 17501, 17513, 17541, 17543, 17557, 17577, 17579, 17604, 17620, and 17649, all collections of multiple isolated teeth (Fig. 17.2G–L).

Description These teeth are low, mesiodistally elongate, not recurved, and strongly bulbous (Fig. 17.2G–L). The occlusal surface varies from having a low and broad weakly developed ridge, to being flat or broadly convex (Fig. 17.2G–L). Teeth with an occlusal ridge sometimes have a slight labiolingual curvature, whereas those with no ridge are straight. All teeth have a fine to medium density of longitudinal striations, which can be weakly or strongly developed (Fig. 17.2G–L). The variation in morphology appears to form a continuum, both within this sample of teeth and in comparison to the teeth described above. This strongly suggests, at least in part, variation along a tooth row rather than different taxa.

Discussion These teeth are likely the basis for the report of cf. *Bernissartia* sp. and *Bernissartia* sp. from the Straight Cliffs Formation of the Kaiparowits Plateau (Eaton, Cifelli, et al. 1999). Teeth of this morphology are certainly present in the posterior maxilla and dentary of *Bernissartia* (Dollo, 1883; Buffetaut, 1975; Buffetaut and Ford, 1979; Buscalioni and Sanz, 1990; Norell and Clark, 1990); however, they are also characteristic of the posterior maxilla and dentary of the basal alligatoroid *Brachychampsa* (Norell, Clark, and Hutchison, 1994; Williamson, 1996; Sullivan and Lucas, 2003), a variety of other globidontan alligatoroids (e.g., Brochu, 1999, 2004), and some taxa within the basal neosuchian clade Atoposauridae (e.g., Martin, Rabi, and Csiki, 2010). Because *Bernissartia* is Early Cretaceous in age (Dollo, 1883; Buscalioni and Sanz, 1990; Norell and Clark, 1990), and the oldest globidontans are Campanian in age (Williamson, 1996; Brochu, 2003), isolated teeth of this morphology have typically been assigned to *Bernissartia* (or Bernissartidae) when found in pre-Campanian strata (e.g., Buffetaut and Ford, 1979; Brinkmann, 1992; Lee, 1997; Schwarz-Wings, Rees, and Lindgren, 2009), and to *Brachychampsa* when found in Campanian or Maastrichtian strata (e.g., Estes, Berberian, and Meszoely, 1969; Hutchinson and Kues, 1985; Bryant, 1989; Sankey, 2008); Eaton, Cifelli, et al. (1999) assigned teeth from the Santonian of the Straight Cliffs Formation to both taxa. Peng, Russell, and Brinkman (2001) referred similar isolated teeth from the Foremost and Oldman formations (middle Campanian) of Alberta to Alligatorinae. Geologic age and similarity to geographically proximate fossils alone are not reliable criteria for the identification of fossils; discrete apomorphies are necessary for an unambiguous referral to any clade or species (Bell, Head, and Mead, 2004; Bever, 2005; Nesbitt, Irmis, and Parker, 2007; Nesbitt and Stocker, 2008; Bell, Gauthier, and Bever, 2010). Thus, because this type of tooth has a wider distribution among crocodyliforms than just *Bernissartia* or Globidonta, it is not diagnostic alone for either of those taxa. These teeth also cannot be referred to

Alligatorinae, as *Brachychampsa* is positioned outside this subclade in a basal position within Globidonta. Instead, these teeth are referable to the unnamed clade of Atoposauridae + Globidonta, which is nested within Neosuchia (Turner and Buckley, 2008; Jouve, 2009; Pol et al., 2009; Turner and Sertich, 2010).

Wahweap Formation

CROCODYLIFORMES Hay, 1930,
sensu Benton and Clark, 1988

Specimens UCMP 150157, dentary fragment (UCMP Loc. v98173) (Fig. 17.3A, B).

Description This specimen preserves three alveoli lacking teeth, which successively increase in size as the width of the dentary increases posteriorly (Fig. 17.3A). The medial surface of the element preserves a clear midline suture for the mandibular symphysis; no splenial articular surface is apparent. In cross section, the specimen is mediolaterally wide as it is dorsoventrally tall (Fig. 17.3B).

Discussion The character combination of a dentary that is equant in cross section with a clear midline symphysis indicates that this belongs to a long-snouted taxon. Relatively long skulls (in the broad sense, not just longorostrine taxa) have evolved numerous times within crocodyliforms (Clark, 1994; Brochu, 2001). Because this is the only taxonomically useful character state present in the specimen, it can only be referred to the clade Crocodyliformes.

CROCODYLIFORMES Hay, 1930,
sensu Benton and Clark, 1988
MESOEUCROCODYLIA
Whetstone and Whybrow, 1983

Specimens UCMP 150156, partial left dentary (UCMP Loc. v98173) (Fig. 17.3C, D).

Description This small dentary preserves seven alveoli, all lacking teeth; comparison with other crocodyliform dentaries (e.g., UCMP 150283, UCMP 159000) suggests that these represent the seventh through 13th alveoli. Positions 7, 8, 9, and 12 are approximately the same size, 10 and 11 are smaller, and 13 is significantly larger than all other preserved alveoli (Fig. 17.3D). The occlusal margin of the dentary in medial and lateral views is distinctly concave, with a posterior preserved end that is dorsoventrally taller than the anterior preserved end (Fig. 17.3C). The lateral surface of the dentary is convex and rugose; the medial surface is relatively flat with a distinct Meckelian groove, though the groove's relationship with the anterior tip of the splenial is unclear because the splenial is not preserved.

Discussion Although this specimen is very similar to dentaries of basal alligatoroids, it does not preserve any apomorphies that allow assignment to this clade. The broadly concave occlusal margin is similar to the condition in basal alligatoroids such as *Brachychampsa* (UCMP 159000), but this is the plesiomorphic condition for Eusuchia and differs from the deeply concave condition that is apomorphic within Alligatoridae (Brochu, 1999:fig. 49). The specimen is not complete enough to tell if it was sinusoidal in lateral view or had a single dorsal expansion with a posteriorly adjacent concavity (Pol et al., 2009:character 159), but the presence of at least one dorsal expansion and concavity allows this specimen to be assigned to the clade Mesoeucrocodylia (Pol et al., 2009).

MESOEUCROCODYLIA
Whetstone and Whybrow, 1983
NEOSUCHIA Benton and Clark, 1988

Specimens UMNH VP 9547, lower jaw fragments, tooth, and multiple osteoderms; UMNH VP 18617, partial osteoderm (Fig. 17.3E–G).

Description Specimens of large rectangular crocodyliform osteoderms are not infrequent in both the Wahweap and Kaiparowits formations. Except for variation in relative dimensions consistent with different positions within the dorsal armor, these osteoderms share a consistent suite of features. The armor plates are generally at least twice as long mediolaterally as they are anteroposteriorly, resulting in an elongate subrectangular shape (Fig. 17.3F, G). In cross section, the osteoderms are flat to slightly arched. The medial edge is always straight for articulation with the corresponding osteoderm from the other side of the body, and the lateral side of the osteoderm varies from slightly to moderately convex. The anterior and posterior edges of the osteoderm are mostly straight; the anterior margin is sometimes slightly convex, and the edge forms a smooth lamina overlapped dorsally by the preceding plate (Fig. 17.3F, G). The anterolateral margins of the osteoderms possess a distinct anteriorly projecting process. This process is not uncommon among suchian archosaurs (Sues, 1992), but it is particularly well developed in these specimens, sometimes reaching three quarters of the remaining anteroposterior length of the osteoderm (Fig. 17.4C, D).

These osteoderms are distinctly ornamented with randomly oriented, deep subcircular to oblong coarse pitting that varies considerably in size across the plate (Fig. 17.3F, G). The largest and deepest pits are generally in the central portion of the plate, with smaller shallower pits around the edges. The pitting is not tightly packed; the width of raised areas between the pits is generally one-third to one-half the maximum diameter of adjacent pits.

Irmis et al.

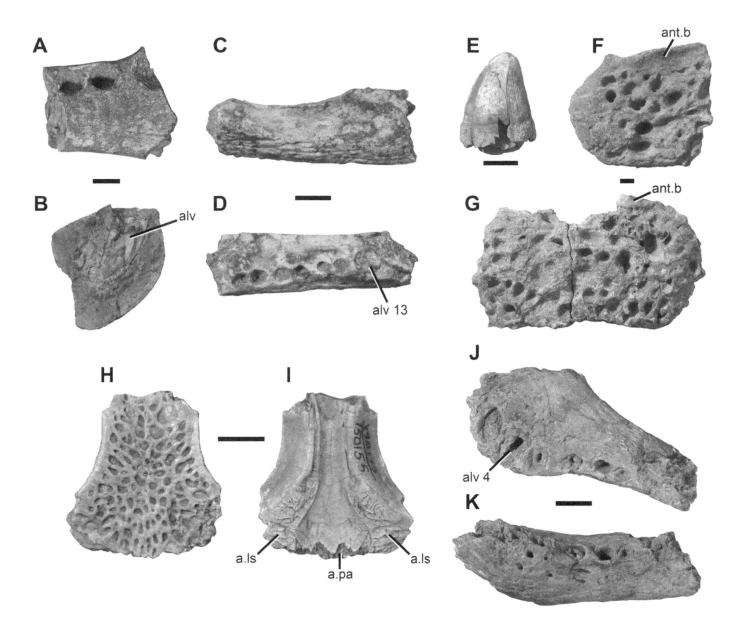

17.3. Crocodyliform specimens from the Wahweap Formation. (A, B) UCMP 150157, crocodyliform dentary fragment in (A) occlusal view and (B) internal cross section. (C, D) UCMP 150156, mesoeucrocodylian partial dentary in (C) lateral and (D) occlusal views. (E, G) Tooth (E) and osteoderms (F, G) of UMNH VP 9547, an indeterminate neosuchian. (H, I) UCMP 150155, crocodylian frontal in (H) dorsal and (I) ventral views. (J, K) UMNH VP 11948, crocodylian partial dentary in (J) occlusal and (K) lateral views. a., articulation; alv, alveolus; ant.b, anterior bar; ls, laterosphenoid; pa, parietal. Scale bars = 1 cm.

The cranial elements associated with UMNH VP 9547 are too fragmentary to be systematically informative. They include the posterior portion of the dentary behind the last alveolus, and fragments of the angular and surangular. An isolated tooth was also associated with this specimen; it is bulbous, not recurved, and possesses marked longitudinal striations (Fig. 17.3E). Teeth of this morphology are common among large crocodyliforms (e.g., Mook, 1942; Colbert and Bird, 1954; Averianov, 2000; Sereno et al., 2001; Brochu, 2007) and by themselves are not diagnostic beyond Crocodyliformes.

Discussion These osteoderms presumably form the basis of records of "Goniopholidae indet." in previous faunal lists

of the vertebrate taxa from the Late Cretaceous of southern Utah (e.g., Gillette and Hayden, 1997; Hutchison et al., 1998; Eaton, Cifelli, et al. 1999). They are nearly identical in all respects to osteoderms from goniopholidids (e.g., Mook, 1933:fig. 4; Wu, Brinkman, and Russell, 1996b:fig. 12; Salisbury, 2002:figs. 3, 4; Schwarz, 2002:fig. 6; Lauprasert et al., 2007:fig. 6). Goniopholidid osteoderms share with the Late Cretaceous Utah specimens the mediolaterally elongate subrectangular dimensions, the deep coarse subcircular pitting, and enlarged anterolateral process. This combination of character states is also characteristic of pholidosaurid crocodyliform osteoderms (e.g., Buffetaut and Taquet, 1977:pl. 28; Sereno et al., 2001:fig. 3; Wu, Russell, and Cumbaa, 2001:fig.

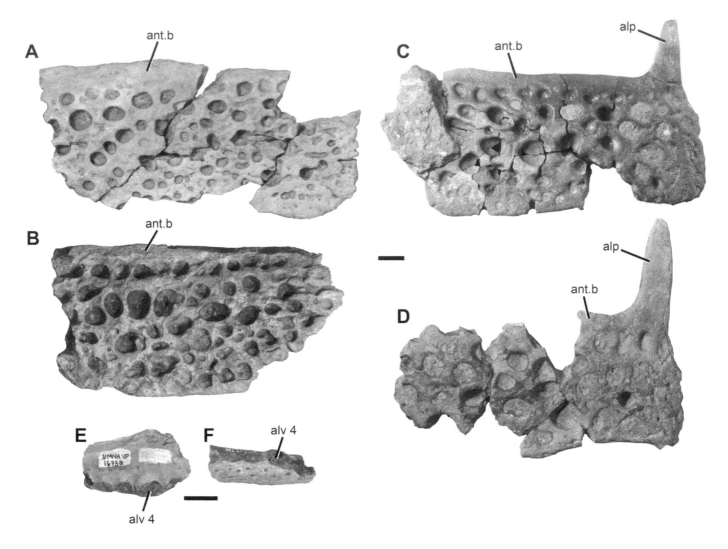

17.4. Neosuchian crocodyliform specimens from the Kaiparowits Formation. (A) UCMP 150240, osteoderm in dorsal view. (B) UMNH VP 18590, osteoderm in dorsal view. (C) UMNH VP 18589, osteoderm in dorsal view. (D) UMNH VP 13916, osteoderm in dorsal view; (E, F) UMNH VP 16732, partial dentary in (E) occlusal and (F) lateral views. alp, anterolateral process; alv, alveolus; ant.b, anterior bar. Scale bars = 1 cm.

8; Hua et al., 2007:fig. 4). Although the osteoderms of some pholidosaurids (i.e., *Sarcosuchus*; see Sereno et al., 2001:fig. 3) have a denser concentration of pitting that is even more irregularly shaped, other pholidosaurids (e.g., Hua et al., 2007:fig. 4) are indistinguishable from the condition in goniopholidids and the Utah specimens.

Because pholidosaurid and goniopholidid osteoderms are indistinguishable, they are thus not diagnostic to their respective clades. Therefore, the Utah specimens can only be referred to the least inclusive clade containing both the Pholidosauridae and Goniopholididae. Although some phylogenetic analyses have found a close relationship between these two clades (Clark, 1994; Sereno et al., 2001), nearly all recent analyses have found them to be distantly related within Mesoeucrocodylia (e.g., Wu, Sues, and Dong, 1997; Wu, Russell, and Cumbaa, 2001; Pol and Norell, 2004b;

Gasparini, Pol, and Spalletti, 2006; Turner and Buckley, 2008; Jouve, 2009; Pol et al., 2009; Turner and Sertich, 2010). Most recent analyses agree that pholidosaurids are closely related to dyrosaurids and possibly thalattosuchians, whereas goniopholidids are more closely related to atoposaurids and eusuchians (Turner and Buckley, 2008; Jouve, 2009; Pol et al., 2009; Turner and Sertich, 2010; but see Pol and Gasparini, 2009; Wilberg, 2010). Assuming these topologies are correct, the least inclusive clade containing pholidosaurids and goniopholidids is Neosuchia. Until such time as better cranial material is recovered, the Utah material can only be assigned to Neosuchia.

Both pholidosaurids and goniopholidids are typically thought of as Early to mid-Cretaceous clades, and so an assignment of these Campanian osteoderms to either group would be a late record. However, there is no geologic or geographic evidence that one clade is a more likely candidate than the other. Multiple records of the pholidosaurid *Terminonaris robusta* are found in Cenomanian–Turonian-age deposits of the Western Interior Seaway of North America (Wu, Russell, and Cumbaa, 2001; Adams et al., 2011). The youngest published goniopholidid is *Denazinosuchus kirtlandicus* from the mid-Campanian of the San Juan Basin in northern New Mexico (Wiman, 1932; Lucas, 1992; Lucas and Sullivan, 2003). Although the goniopholidid affinities of this taxon have never been questioned, it has also never been placed in a phylogenetic analysis to confirm its placement in this clade. The geologically youngest goniopholidid to have been placed in a phylogenetic analysis is probably *Siamosuchus* from the Early Cretaceous of Thailand (Lauprasert et al., 2007), though there are possible goniopholidid remains from the Cenomanian of Texas (Lee, 1997). Thus, the presence of pholidosaurids and goniopholidids in the Late Cretaceous of western North America indicate that it is equally likely that the Campanian osteoderms belong to either of these clades.

NEOSUCHIA Benton and Clark, 1988
EUSUCHIA Huxley, 1875, sensu Brochu, 1999
CROCODYLIA Gmelin, 1789,
sensu Benton and Clark, 1988

Specimens UMNH VP 16223, frontal; UCMP 150155, frontal; and UMNH VP 11948, left partial dentary (Fig. 17.3H, K).

Description Both isolated frontals are trapezoidal in shape, with concave lateral margins that form the medial edge of the orbit. The dorsal surface of the frontals are well ornamented by distinct irregular pitting with anastomosing ridges (Fig. 17.3H). The dorsolateral margin that forms the orbital margin is slightly raised relative to the rest of the dorsal surface. The posterolateral margins of the frontals preserve articular facets for the parietal and postorbital; these two facets have extensive contact (Fig. 17.3H, I). The posterior third of the ventral surface of the frontals has distinct articular facets on either side of the midline for articulation with the anterior portion of the laterosphenoid (Fig. 17.3I).

UMNH VP 11948 is a left dentary that preserves alveoli 1 through 10, most of which still contain tooth roots (Fig. 17.3J, K). The lateral margin of the dentary is rounded and expanded for alveoli 1 through 5. It then straightens and narrows posteriorly (Fig. 17.3J). Medially, the midline symphysis extends the length of the first through sixth alveolus. The end of the dentary symphysis forms an obtuse angle with the straight medial margin on the posterior portion of the dentary; no splenial is preserved and its articular contact on the dentary is indistinct. All alveoli are distinct and separate from each other; alveoli 3 and 4 are the largest, with alveolus 5 slightly smaller, and alveoli 1, 2, and 6–10 being subequal to each other and smaller than 3, 4, and 5 (Fig. 17.3J). The first four alveoli project strongly anterodorsally. The ventrolateral surface of the dentary is convex and lightly sculptured. In lateral view, the dorsal occlusal margin is gradually concave between alveolus 4 and 10 (Fig. 17.3K).

Discussion All three specimens are consistent with the morphology of the same elements in some alligatoroids, but no apomorphies are preserved that allow assignment to this clade. The extensive contact between the postorbital and parietal articular facets on the frontals is a feature present in some species of *Borealosuchus*, a number of basal alligatoroids, and several gavialoids (Brochu, 1999:character 81). This indicates these frontals are referable to Crocodylia (Brochu, 1999, 2004; Turner and Sertich, 2010) or a slightly more inclusive clade (Pol et al., 2009).

The dentary has a broadly concave occlusal margin, which is similar to several basal alligatoroids, but is plesiomorphic for Eusuchia (Brochu, 1999:fig. 49). The presence of an enlarged alveolus 4 separate from alveolus 3 is also found in the Glen Rose form and nearly all crocodylians (Brochu, 1999:character 52). This specimen may be referable to Crocodylia or a slightly more inclusive clade because the anterior teeth project anterodorsally, which is a synapomorphy for *Borealosuchus* plus all other crocodylians (Brochu, 1999:character 53).

Kaiparowits Formation

CROCODYLIFORMES Hay, 1930,
sensu Benton and Clark, 1988

Specimens UCMP 150241, skull fragments and osteoderms; UMNH VP 12668, frontal; UMNH VP 16119, frontal; UMNH VP 16280, frontal; UMNH VP 16731, frontal; UMNH VP 16744, frontal; UMNH VP 18573, frontal and osteoderms.

Description The frontals preserved in these specimens are very similar to those of alligatoroids and the crocodylian frontals from the Wahweap Formation. Unfortunately, they do not preserve the postorbital and parietal articular facets, except for UMNH VP 16280, where it is not clear if the postorbital and parietal would have contacted each other in articulation. The frontals are subtrapezoidal in outline with concave lateral margins that form the medial margin of the orbit. The dorsal surface is ornamented with distinct pits and anastomosing grooves, typical of crocodyliform cranial ornamentation. The anterolateral surfaces of UMNH VP 18573 preserve long and deep sutures for the prefrontals.

The osteoderms preserved with these specimens are typical of crocodyliforms; they are square to rectangular in shape with a distinct raised anterior bar (="anterior articulation facet" of some authors) on the dorsal surface. The ornamentation consists of irregularly sized subcircular pits separated by anastomosing ridges.

Discussion The lack of preservation or clear contact between the postorbital and parietal sutures on the frontals means this character cannot be evaluated, which limits the taxonomic resolution available for these specimens. Fusion of the frontals into a single midline element is a synapomorphy for *Edentosuchus* + all other crocodyliforms (Pol et al., 2009:character 21), indicating these specimens can be assigned to Crocodyliformes. This assignment is further supported by the osteoderms preserved for UCMP 150241 and UMNH VP 18573, as rectangular osteoderms that are wider than long are a synapomorphy for Crocodyliformes (Pol et al., 2009:character 95). UCMP 150241 was previously cataloged as *Leidyosuchus*, but does not preserve any autapomorphies of that taxon or of *Borealosuchus* (Brochu, 1997; Wu, Russell, and Brinkman, 2001).

MESOEUCROCODYLIA
Whetstone and Whybrow, 1983

Specimens UCMP 150238, fragmentary dentary, angular, and osteoderms.

Description Two dentary fragments together preserve approximately seven alveoli, but the broken nature of the specimen and missing teeth make it impossible to determine what positions they occupied. The medial surface of the dentary fragments has broken away, exposing the interior of the bone. The lateral surface is convex and ornamented. One fragment preserves the original dorsal edge, displaying at least one dorsal projection followed posteriorly by a concavity. The osteoderms are rectangular, being longer than wide. They have a distinct raised anterior bar (="anterior articulation facet" of some authors), typical crocodyliform ornamentation, and a distinct but low longitudinal ridge in the center of the dorsal surface.

Discussion The presence of at least one dorsal expansion and concavity in the dentary allows this specimen to be assigned to Mesoeucrocodylia (Pol et al., 2009:character 159). The incomplete nature of the specimen prevents any further assessment.

NEOSUCHIA Benton and Clark, 1988

Specimens UCMP 150240, skull fragments and osteoderms; UMNH VP 11927, osteoderm; UMNH VP 12495, osteoderms; UMNH VP 12635, osteoderm; UMNH VP 12695, quadrate, skull

fragments, vertebrae, ribs, and osteoderms; UMNH VP 13916, osteoderm; UMNH VP 13920, osteoderm; UMNH VP 16282, osteoderms; UMNH VP 16869, osteoderms; UMNH VP 18589, osteoderm; UMNH VP 18590, osteoderm (Fig. 17.4A–D).

Description These specimens preserve large rectangular osteoderms that are nearly identical to those found in the Wahweap and Straight Cliffs formations. The plates are flat or slightly arched and at least twice as long mediolaterally as they are anteroposteriorly, giving them a subrectangular shape. The medial edge is straight, whereas the lateral edge is slightly to moderately convex. The anterolateral margin possesses a large anterior process that is often three-quarters the anteroposterior length of the rest of the osteoderm (Fig. 17.4C, D). Both the anterior and posterior edges are straight, and the anterior edge forms a smooth lamina that is overlapped dorsally by the preceding plate (Fig. 17.4A–D). The dorsal surface is ornamented with irregularly spaced and sized subcircular pitting.

Discussion These specimens form the basis for reports of "goniopholids" from the Kaiparowits Formation (Hutchison et al., 1998; Eaton, Cifelli, et al. 1999; Wiersma, Hutchison, and Gates, 2004). As discussed for the Wahweap material, this osteoderm morphotype is indistinguishable from those found in both goniopholidids and pholidosaurids (see above). Therefore, the least inclusive clade they can be referred to is Neosuchia. Unfortunately, the fragmentary cranial material associated with UMNH VP 12695 is only diagnostic to Crocodyliformes.

NEOSUCHIA Benton and Clark, 1988
Unnamed Clade of Glen Rose Form + EUSUCHIA

Specimens UMNH VP 16732, partial right dentary (Fig. 17.4E, F).

Description This dentary fragment preserves alveoli 2 though 7. All are subequal in size except for alveolus 4, which is significantly larger and separated from alveolus 3 (Fig. 17.4E). The medial side of the dentary where the midline symphysis would be is broken away, but it is clear that the maximum possible posterior extent of the symphysis is to alveolus 4. The ventrolateral surface of the dentary is convex and distinctly ornamented.

Discussion An enlarged alveolus 4 that is separate from alveolus 3 is also found in the Glen Rose Form and nearly all crocodylians (Brochu, 1999:character 52). Because the specimen is incomplete, it is impossible to determine the procumbency of the anterior teeth and the curvature of the dorsal margin of the dentary. Thus, this specimen can be referred to the clade of Glen Rose Form + Crocodylia, a subclade within Neosuchia.

EUSUCHIA Huxley, 1875, sensu Brochu, 1999
CROCODYLIA Gmelin, 1789,
sensu Benton and Clark, 1988

Specimens UCMP 150234, osteoderms; UMNH VP 12676, osteoderm; UMNH VP 19656, right femur, osteoderms, and teeth (Fig. 17.5A, B).

Description The three ventral osteoderms of UCMP 150234 are rectangular and mediolaterally wider than long. The raised anterior bar (="anterior articulation facet" of some authors) covers three quarters of the entire dorsal surface of the osteoderm, with only a quarter of the surface being covered by irregularly sized subcircular pitting and anastomosing ridges (Fig. 17.5A, B). The medial, lateral, and posterior sides all have a well-defined suture that abuts with adjacent osteoderms (Fig. 17.5A, B). UMNH VP 12676 is broken, but was clearly rectangular. It is flat with light irregular pitting on the dorsal surface. All preserved margins have clear sutures that would abut with adjacent osteoderms. UMNH VP 19656 preserves both osteoderm morphologies, as well as tooth morphologies similar to those typically referred to "*Theriosuchus*" and "*Bernissartia*"/"*Brachychampsa*" (see discussion in Straight Cliffs Formation section about why these teeth are not diagnositic to these taxa).

Discussion The presence of an extremely large anterior bar and a suture on the posterior margin of the osteoderms of UCMP 150234 (Fig. 17.5A, B) indicates that these are bipartite ventral osteoderms (e.g., Brochu, 1997:fig. 15C). In life, the suture on the posterior edge would abut against, but not overlap osteoderms such as UMNH VP 12676, which lack any trace of an anterior bar and are lightly ornamented. This condition is present in some species of *Borealosuchus*, taxa previously referred to *Leidyosuchus* (Brochu, 1997), which probably led to these specimens being cataloged as *Leidyosuchus*; they are probably also the basis for the reports of *Leidyosuchus* from the Kaiparowits Formation (Eaton, Cifelli, et al. 1999; Wiersma, Hutchison, and Gates, 2004). However, bipartite ventral osteoderms are also present in some species of *Borealosuchus*, *Diplocynodon*, *Tsoabichi*, and Caimaninae (Brochu, 1997, 1999, 2010). Thus, the least inclusive clade this specimen can be referred to is Crocodylia, or a slightly more inclusive clade if *Borealosuchus* is outside of the crown group (e.g., Pol et al., 2009).

CROCODYLIA Gmelin, 1789,
sensu Benton and Clark, 1988

Specimens UMNH VP 11870, frontal (Fig. 17.5C, D).

Description This is the most complete isolated frontal recovered so far from the Kaiparowits Formation. The main body of the element is subtrapezoidal in outline, with

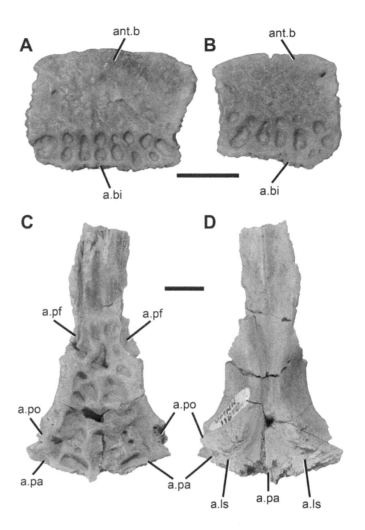

17.5. Crocodylian specimens from the Kaiparowits Formation. (A, B) UCMP 150234, bipartite osteoderms in dorsal view. (C, D) UMNH VP 11870, frontal in (C) dorsal and (D) ventral view. a., articulation; ant.b, anterior bar; bi, bipartite edge; ls, laterosphenoid; pa, parietal; pf, prefrontal; po, postorbital. Scale bars = 1 cm.

concave anterolateral margins that form the medial edge of the orbits (Fig. 17.5C). The dorsal surface of the main body is deeply sculptured with irregular pitting. The orbital rims are distinctly raised dorsally above the rest of the frontal (Fig. 17.5C). The element displays an elongate anterior process that accounts for half of the anteroposterior length of the frontal; this anterior process is depressed relative to the main body. Anteriorly, this process articulated with the nasals, and on either side of the process there are long deep sutures for articulation with the prefrontals (Fig. 17.5C, D). On the ventral side of the element, there are two curved ridges that form the medial margin of the orbit. The posterior half of these ridges is incised by a suture for articulation with the laterosphenoid (Fig. 17.5D). The posterolateral edge of the frontal has a distinct suture for the postorbital, which abuts extensively with the suture on the posterior edge for articulation with the parietal.

Discussion Raised orbital margins are found in a variety of crocodyliforms, including the Fruita form, *Theriosuchus*, *Shamosuchus*, *Bernissartia*, *Hylaeochampsa*, Glen Rose form, *Pristichampsus*, *Gavialis*, and *Alligator* (Pol et al., 2009:character 103). A significant contact between the postorbital and parietal is also present in *Borealosuchus*, some gavialoids, and a number of basal alligatoroids (Brochu, 1999:character 81). Thus, on the basis of this character state, the least inclusive clade that UMNH VP 11870 can be referred to is Crocodylia. However, this specimen is consistent with the frontal preserved in the crushed skull of UMNH VP 19268 (see below), a small alligatoroid distinct from *Brachychampsa*.

CROCODYLIA Gmelin, 1789,
sensu Benton and Clark, 1988
ALLIGATOROIDEA Gray, 1844, sensu
Norell, Clark, and Hutchison, 1994
DEINOSUCHUS HATCHERI Holland, 1909

Specimens UMNH VP 16783, right partial premaxilla, nearly complete maxilla, and partial palatine; UMNH VP 16146, osteoderm; UMNH VP 16843, osteoderm; UMNH VP 18575, osteoderms (2); UMNH VP 18591, osteoderm; UMNH VP 20193, osteoderms (4); UMNH VP 20614, osteoderm (Fig. 17.6).

Description The cranial material will be fully described elsewhere. Only small portions of the premaxilla and palatine are preserved in UMNH VP 16783, but it is clear that the premaxilla–maxilla junction formed a large embayment for occlusion with the fourth dentary tooth (Fig. 17.6B). The maxilla is subrectangular in outline; the lateral margin forms a slightly convex swelling anteriorly, and more posteriorly it is straight and oriented slightly posterolaterally (Fig. 17.6B). The dorsal and lateral faces form a broadly convex sloping surface that is rugose and ornamented with deeply incised pits and anastomosing grooves. In ventral view, the tooth row is straight. Maxillary alveoli 3 and 4 are subequal in size, but much larger than any other maxillary alveolus (Fig. 17.6B). Alveoli 9–11 are slightly larger than surrounding alveoli. A large flat rectangular palatal process extends medially from the ventral tooth row, and would have met its counterpart along the midline. Lateral to the suborbital fenestra there is a slender palatal projection that would separate the ectopterygoid from articulating adjacent to the maxillary tooth row (Fig. 17.6B).

All of the osteoderms are subcircular to elliptical in outline. They have a flat ventral surface with a broad convex mound-like dorsal surface (Fig. 17.6C–E). The dorsal surface is somewhat pustulose, and is ornamented by widely spaced irregular deep subcircular pits (Fig. 17.6C–E). These pits are sometimes separated by anastomosing ridges.

Discussion We use the binomen *Deinosuchus hatcheri* because the type tooth of "*Polyptychodon*" *rugosus* is not diagnostic and a nomen dubium, the type specimen of *D. hatcheri* (Holland, 1909) is the earliest described diagnosable material, and there is presently no evidence for more than one species of *Deinosuchus* (Schwimmer, 2002).

The subcircular to elliptical outline, mound-like dorsal surface, and irregular deep pitting of these osteoderms allow their referral to *Deinosuchus*. The characters have long been recognized as diagnostic for *Deinosuchus* osteoderms (Holland, 1909; Colbert and Bird, 1954, Brochu, 1999; Schwimmer, 2002); they are present in the type specimen of *D. hatcheri* (Holland, 1909), and differ strongly from typical alligatoroid osteoderms; *Deinosuchus* osteoderms lack typical alligatoroid characters such as a rectangular outline, distinct slightly raised anterior bar, and distinct longitudinal ridge (Brochu, 1999).

It is surprisingly difficult to refer the Kaiparowits Formation cranial material to *Deinosuchus* for several reasons: (1) the type of *D. hatcheri* does not preserve diagnostic cranial elements; (2) the skull material of the holotype of *D. riograndensis* (Colbert and Bird, 1954) from the Big Bend region in Texas only preserves the premaxillae, a single maxilla fragment, partial dentaries, and partial posterior jaw; and (3) well-preserved, nearly complete skulls from the southeastern United States and the Big Bend region in Texas (Schwimmer, 2002) have never been described or diagnosed. Schwimmer (2002:114) considered the presence of a large premaxilla–maxilla diastema that laterally accommodated the large fourth dentary tooth and the rest of the dentary tooth row occluding medial to the maxillary tooth row to be a unique character combination for *Deinosuchus*. Unfortunately, this character combination is plesiomorphic for Alligatoroidea (Brochu, 1999, 2010:characters 71 and 72), and is also present in nearly all non alligatoroids (e.g., *Borealosuchus*; see Brochu, 1999, 2010).

UMNH VP 16783 can be assigned to the Alligatoroidea because it possesses a narrow posterior palatal process that separates the maxillary tooth row from the ectopterygoid (Brochu, 1999:35). It is excluded from Globidonta because the fourth maxillary tooth is not the sole largest tooth in the maxilla, and excluded from the clade Diplocynodontinae + Globidonta because dentary tooth 4 occludes lateral to the maxilla and premaxilla rather than in a closed pit (Brochu, 1999:38). This character combination for UMNH VP 16783 is consistent with the phylogenetic position of *Deinosuchus* (Brochu, 1999). The character combination of the presence of a large premaxilla–maxilla diastema that laterally accommodated the large fourth dentary tooth, the rest of the dentary tooth row occluding medial to the maxillary tooth

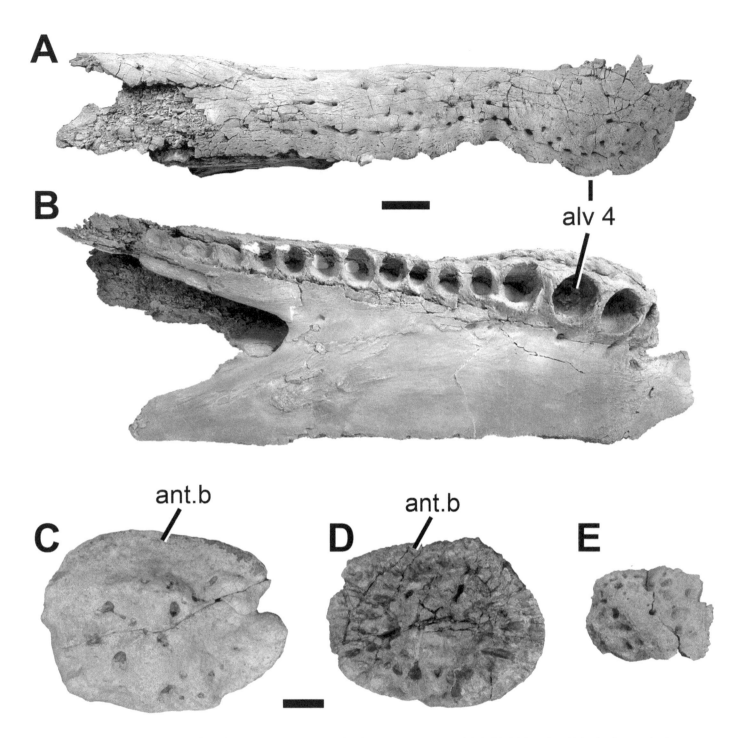

17.6. *Deinosuchus hatcheri* specimens from the Kaiparowits Formation. (A, B) UMNH VP 16783, partial skull in (A) lateral and (B) palatal views. (C–E) osteoderms in dorsal view; (C) UMNH VP 16843; (D) UMNH VP 20193; (E) UMNH VP 18591. ant.b, anterior bar; alv, alveolus. Scale bars = 10 cm (A, B) and 2 cm (D, E, F).

row to be a unique character combination for *Deinosuchus*, maxillary alveoli 4 and 5 the same size, and a linear maxillary tooth row is only present among alligatoroids in *Deinosuchus*, *Pristichampsus vorax*, and *Leidyosuchus canadensis* (Brochu, 2010:characters 71–74). However, all known specimens of *Deinosuchus* are significantly larger than *Pristichampsus* and *Leidyosuchus*. It is on the basis of this character combination and very large size that we refer UMNH VP 16783 to *Deinosuchus hatcheri*.

ALLIGATOROIDEA Gray, 1844, sensu
Norell, Clark, and Hutchison, 1994
GLOBIDONTA Brochu, 1999
BRACHYCHAMPSA Gilmore, 1911

Specimens UCMP 159000, well-preserved disarticulated partial skull; UCMP 151710, left maxilla; UCMP 149172, fragmented disarticulated partial skull (Fig. 17.7) and well-preserved, nearly complete articulated postcranial skeleton.

Description This material will be fully described elsewhere; here, we focus on describing features salient for identification. The premaxillae form a broad blunt snout terminus and each contains five alveoli. The maxilla is a broad subrectangular element with a flat but heavily sculptured dorsal surface. Laterally, this forms a sharp abrupt edge with the lateral surface of the maxilla, which curves somewhat medially in the anterior and middle portion of the element. The anterior portion of the lateral margin of the maxilla is slightly bulbous and convex, posteriorly it is straight. Within the maxillary tooth row, alveolus 5 is the largest among the first nine maxillary teeth; overall, alveoli 12–14 are the largest in the maxilla (Fig. 17.7A). Anterior maxillary teeth are blunt, unrecurved, and peg shaped, with longitudinal striations (Fig. 17.7A). Posteriorly, they become very bulbous with no carinae and a broad convex occlusal margin. The relative sizes of the posterior maxilla, postorbital, jugal, frontal, and squamosal indicate that the lateral temporal fenestra was almost as large as the orbit. The frontal–parietal–postorbital suture is broad, allowing for extensive contact between the parietal and postorbital. The dentary is broadly curved in lateral view; the dorsal margin forms two concavities separated by two peaks (Fig. 17.7B). When both lower jaws are articulated, the tooth row is U shaped and the midline symphysis extends posteriorly to a point equal to between the fourth and fifth alveolus (Fig. 17.7B). Starting at the 12th alveolus, the tooth row is inset from the lateral margin of the dentary. Alveolus 4 is the largest in the anterior dentary and is separated from the third alveolus (Fig. 17.7B). The fourth alveolus is not equaled in size until the 13th alveolus. Anterior dentary teeth are very similar to the anterior maxillary teeth, and the large bulbous posterior teeth of the dentary first appear in the 14th alveolus (Fig. 17.7C). Although the splenial is not preserved on either hemimandible of UCMP 159000, the articular facet on the medial side of the dentary is apparent. Like other *Brachychampsa* specimens (Norell, Clark, and Hutchison, 1994; Sullivan and Lucas, 2003), the splenial appears to have approached, but not entered, the dentary symphysis.

Discussion These three specimens are the basis for previous reports of *Brachychampsa* from the Kaiparowits Formation (Hutchison et al., 1998; Eaton, Cifelli, et al. 1999;

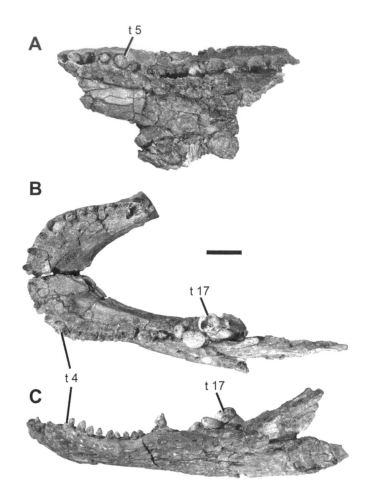

17.7. *Brachychampsa* (UCMP 159000) from the Kaiparowits Formation. (A) Maxilla in palatal view; (B) dentaries in occlusal view; (C) left dentary in lateral view. t, tooth. Scale bar = 2 cm.

Wiersma, Hutchison, and Gates, 2004; Gates et al., 2010; Sampson, Loewen, et al., 2010). UCMP 159000, the most complete cranium, can easily be assigned to *Brachychampsa* on the basis of the combination of the presence of a blunt broad snout, an enlarged narial opening more than half as wide as the rostrum, lateral temporal fenestrae that are nearly as large as the orbits, and a maxillary fifth alveolus larger than all other anterior maxillary alveoli (Norell, Clark, and Hutchison, 1994; Williamson, 1996; Brochu, 1999). On the basis of differing skull proportions and the shape of the dorsolateral edge of the maxilla, the Kaiparowits material could represent a new species of *Brachychampsa* (e.g., Wiersma, Hutchison, and Gates, 2004; Sampson, Loewen, et al., 2010), but this requires further comparison with the Hell Creek Formation *B. montana* material (Norell, Clark, and Hutchison, 1994), *B. sealeyi* from the Menefee Formation of New Mexico (Williamson, 1996), and *Brachychampsa* material from the Kirtland Formation of northern New Mexico (Sullivan and Lucas, 2003).

Irmis et al.

The isolated left maxilla, UCMP 151710, can be referred to *Brachychampsa* because it possesses the unique character combination of alveolus 5 being the largest in the anterior maxilla and mid- and posterior maxillary and dentary teeth are enlarged, bulbous, and with longitudinal striations. In all other alligatoroids with globidont teeth, another alveolus in the maxilla is largest (Brochu, 2010:character 73), and although the basal neosuchian *Bernissartia* has similar teeth, maxillary alveolus 4 is largest in that taxon (Buffetaut, 1975; Norell and Clark, 1990). Though the cranial material of UCMP 149172 is fragmentary, it can be assigned to *Brachychampsa* on the basis of the following combination of character states which are unique among alligatoroids (e.g., Brochu, 1999, 2004, 2010): alveolus 5 is the largest in the anterior maxilla; mid- and posterior maxillary and dentary teeth are enlarged, bulbous, and with longitudinal striations; and there is extensive contact between the postorbital and parietal.

ALLIGATOROIDEA Gray, 1844, sensu Norell, Clark, and Hutchison, 1994

Specimens UMNH VP 19268, crushed but complete skull and partial postcranium; UCMP 150283, articulated anterior lower jaws (Fig. 17.8).

Description We describe these two specimens together because both appear to belong to a small alligatoroid taxon that lacks globidont teeth. However, because UMNH VP 19268 is poorly preserved, and UCMP 150283 lacks a cranium, we cannot at present formally refer these specimens to a single taxon (see below).

Although complete, the extreme crushing and fragmentation of the skull of UMNH VP 19268 makes it very difficult to interpret sutures (Fig. 17.8A, B). The postcrania of this specimen are currently unprepared. The overall shape of the skull is similar to that of other alligatoroids, with a rounded, but not blunt snout (e.g., Brochu, 1999:figs. 4–8). Most of the preorbital sutures cannot be determined. The premaxillary and maxillary alveoli and teeth are visible in ventral view. These demonstrate that the fourth maxillary alveolus is the largest in the tooth row; all teeth posterior to the fifth maxillary alveolus are subequal to each other and much smaller in size than positions 1–5 (Fig. 17.8B). All preserved teeth are slender, circular, and peg-like. The postorbital and parietal have a sutural contact along the posterior edge of the frontal (Fig. 17.8A). The quadratojugal and quadrate appear to contact the postorbital at the dorsal apex of the lateral temporal fenestra, but this is difficult to evaluate because of the crushed nature of the skull. Part of the supraoccipital is exposed dorsally, but it does not fully exclude the parietal from the posterior margin of the skull table (Fig. 17.8A). The

dorsal surface of the lower jaw including the dentary teeth is not exposed because the jaws are in articulation.

The articulated dentaries of UCMP 150283 are extraordinarily well preserved and complete to the 14th alveolus where they are broken (Fig. 17.8C–E). The anterior margin of the lower jaw forms a rounded point and is not blunt, and the lateral margin is slightly convex from alveoli 1 through 4, and straight posterior to alveolus 4 (Fig. 17.8C, E). In ventrolateral view, the external surface of the dentaries is convex and lightly sculptured, and in lateral view, the dorsal margin of the dentary is broadly curved. The midline symphysis extends posteriorly to the level of the fourth alveolus, which is the largest alveolus in the anterior dentary and clearly separated from the third alveolus; alveoli 5–12 are significantly smaller (Fig. 17.8C). Alveoli 13 and 14, or possibly a more posterior alveolus in the missing portion of the dentary, are the largest in the jaw. All preserved teeth are simple straight conical structures with light longitudinal striations and no recurvature. In medial view, most of the spenials are disarticulated and preserved separately. The anteriormost portion of the splenial is still articulated to the right dentary; this portion is missing for the left dentary, but the splenial articular scar is clearly visible on the left dentary and delineates its anterior extent (Fig. 17.8E). Both dentaries demonstrate that the splenials were excluded from the midline symphysis and terminated anteriorly at the mandibular ramus for cranial nerve V. The anteriormost portion of the splenial comes to a point at the same level or just slightly above the Meckelian groove (Fig. 17.8E).

Discussion These two specimens are notable because they appear to be alligatoroids, but lack the bulbous posterior maxillary and dentary teeth characteristic of *Brachychampsa*. Thus, they provide evidence for at least one additional small alligatoroid in the Kaiparowits Formation assemblage. An articulated skull and skeleton currently under preparation (UMNH VP 21150) may also be related to these specimens.

UCMP 150283 was the basis for previous reports of a caimanine from the Kaiparowits Formation (Wiersma, Hutchison, and Gates, 2004; Sampson, Gates et al., 2010). The presence of an enlarged fourth alveolus that is separate from the third alveolus is characteristic of globidontan alligatoroids (Brochu, 1999, 2004, 2010), but has a much broader distribution among Crocodylia (Brochu, 1999:character 52). Most other characters preserved in this specimen are plesiomorphic for Alligatoroidea (e.g., tooth morphology and dentary outline). The exclusion of the splenial from the midline symphysis and its anterior termination at or just dorsal to the level of the Meckelian groove could suggest an affinity with caimanines (Brochu, 1999:fig. 31, 2010:1118). However, this character state has a wider distribution, as it is present in several *Alligator*

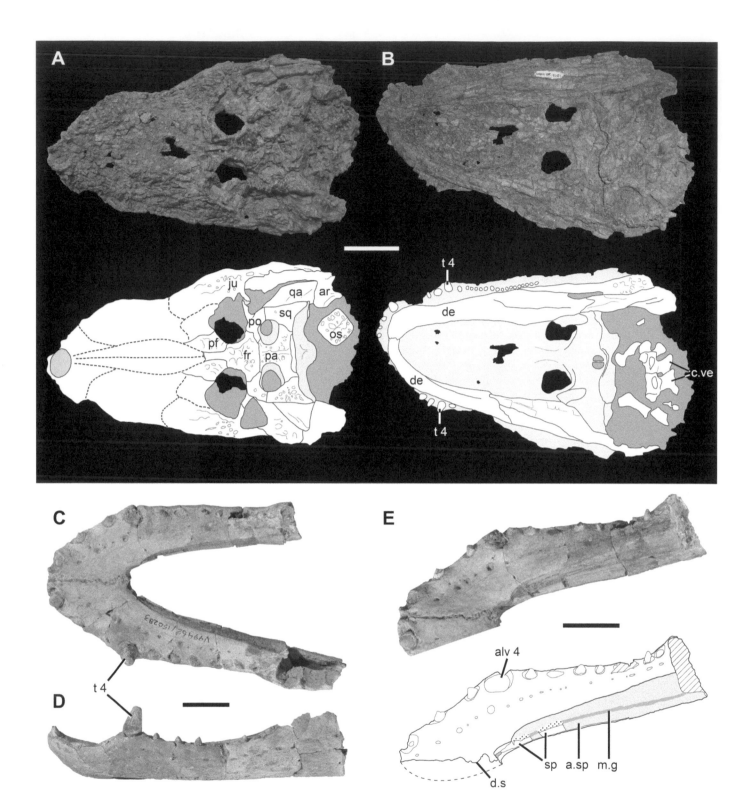

17.8. Small alligatoroid taxon from the Kaiparowits Formation. (A, B) Skull of UMNH VP 19268 in (A) dorsal and (B) ventral views. (C–E) UCMP 150283, paired dentaries in (C) occlusal, (D) lateral, and (E) oblique medial views. a., articulation; alv, alveolus; ar, articular; c.ve, cervical vertebra; de, dentary; d.s, dentary symphysis; fr, frontal; ju, jugal; m.g, Meckelian groove; os, osteoderm; pa, parietal; pf, prefrontal; po, postorbital; qa, quadrate; sp, splenial; sq, squamosal; t, tooth. Scale bars = 2 cm.

Irmis et al.

species and *Diplocynodon muelleri* (Brochu, 2010:character 40), and it is therefore only diagnostic to Alligatoroidea. Moreover, on further examination, the condition in UCMP 150283 (Fig. 17.8E) differs from that of Caimaninae. In the Kaiparowits specimen and at least some *Alligator* specimens (e.g., Brochu, 1999:fig. 31F), the splenial narrows anteriorly to a point that terminates at the level of the Meckelian groove. In contrast, the anterior process of the splenial in caimanines (e.g., Brochu, 1999:fig. 31E) narrows dorsally above an exposed Meckelian groove, and parallels the groove for some length. Thus, UCMP 150283 cannot be assigned to the Caimaninae because this character state has a wider distribution among alligatoroids, and the condition in this specimen and *Alligator* differs in detail with the condition in caimanines.

This specimen is similar in some features to the lower jaw of *Stangerochampsa* (Wu, Brinkman, and Russell, 1996a:fig. 2; Wu, 2005:fig. 15.3C) and several isolated Dinosaur Park Formation (DPF) dentaries (Wu, 2005:fig. 15.3B), though these characters are more widespread and therefore not necessarily diagnostic of this taxon. Both share a similar angle of 20–25 between the midline symphysis and the middentary tooth row and alveoli 5–12 that are much smaller than alveolus 4. In contrast, another DPF dentary morph (Wu, 2005:fig. 15.3D) has a smaller angle, and alveoli posterior to the fourth alveolus are only slightly smaller (Wu, 2005). However, *Stangerochampsa* (Wu, Brinkman, and Russell, 1996a) differs from UCMP 150283 because the splenial of *Stangerochampsa* participates in the midline symphysis. Unfortunately, Wu (2005) did not discuss the condition of the splenial for either of the DPF dentary morphs, nor did he figure them in medial view, so their condition remains unknown.

The skull of UMNH VP 19268 is possibly assignable to Globidonta because there is a significant postorbital–parietal contact along the supratemporal fenestra, the postorbital contacts the quadrate and quadratojugal at the dorsal margin of the lateral temporal fenestra, and the largest maxillary alveolus is position 4 (Brochu, 1999, 2004, 2010). It's unclear if this specimen preserves character states diagnostic of Alligatoridae; whether the postorbital–parietal suture is on the margin of the supratemporal fenestra or fully on the dorsal skull table (Brochu, 1999:52) cannot be determined because of crushing. The parietal still forms part of the posterior margin of the skull table, which is a plesiomorphic state that excludes UMNH VP 19268 from the Caimaninae; however, if the slight dorsal exposure of the supraoccipital is not a result of crushing, it is a character that appears in some caimanines (Brochu, 2010). Unfortunately, most of the informative alligatoroid characters in the lower jaw cannot be determined because the jaws are articulated closed and crushed.

These two specimens provide evidence for at least one small alligatoroid taxon distinct from *Brachychampsa*

co-occurring in the same formation. UMNH VP 19268 may be a globidontan, though it lacks globidont posterior teeth. However, there is no preserved evidence that UCMP 150283 is a member of Caimaninae (contra Wiersma, Hutchison, and Gates, 2004; Sampson Gates, et al., 2010).

PHYLOGENETIC POSITION
OF *DENAZINOSUCHUS*

The crocodyliform *Denazinosuchus kirtlandicus* (Wiman, 1932) from the late Campanian Kirtland Formation of the San Juan Basin in northern New Mexico (Lucas and Sullivan, 2003) is putatively the geologically youngest record of Goniopholididae. As such, it is an important occurrence for a clade that first appeared in the Early Jurassic (Tykoski et al., 2002). Furthermore, the inclusion in the *Denazinosuchus* holotype of a wide rectangular osteoderm with an enlarged anterior process and large subcircular pitting (Wiman, 1932:pl. V-4) has been one of the main lines of evidence for identifying similar osteoderms from Campanian strata in the Western Interior as belonging to goniopholidid crocodyliforms (e.g., Lucas, 1992; Hutchison et al., 1998; Eaton, Cifelli, et al. 1999; Lucas et al., 2006). Although these osteoderms on their own can only be referred to Neosuchia (see above), the confirmation of a Campanian goniopholidid in nonmarine strata of the Western Interior would lend support to the idea that these osteoderms (e.g., from the Wahweap and Kaiparowits formations of southern Utah) probably pertain to a goniopholidid.

Although *Denazinosuchus* has been identified as a goniopholidid since its original description (Wiman, 1932) and this placement has not been challenged (e.g., Lucas, 1992; Lucas and Sullivan, 2003; Lucas et al., 2006), it has never been placed in a published phylogenetic analysis. The characters originally cited by Wiman (1932) in support of a goniopholidid placement are now known to have a wider distribution among neosuchians (Lucas and Sullivan, 2003), and the same can be said for the features cited by Lucas and Sullivan (2003), who mainly appealed to overall similarity in support of their assignment. In fact, the only character that is both an unambiguous synapomorphy of Goniopholididae and could be evaluated for the type (and only) specimen of *Denazinosuchus* is the presence of a distinct fossa on the lateral surface of the maxilla, anterolateral to the orbits (Pol et al., 2009:character 207; Turner and Sertich, 2010:character 207). This fossa is clearly absent in *D. kirtlandicus* (e.g., Lucas and Sullivan, 2003:116; pers. obs.), so the taxon does not preserve any unambiguous synapomorphies of Goniopholididae.

Therefore, to test the phylogenetic position of *Denazinosuchus* for the first time, we included it in the most recent global phylogenetic analysis of Mesozoic crocodyliforms by

Turner and Sertich (2010). This analysis already included five goniopholidids (*Goniopholis simus*, *G. stovalli*, *Eutretauranosuchus delfsi*, *Calsoyosuchus valliceps*, and *Sunosuchus*) as well as comprehensive taxon and character sampling across Mesoeucrocodylia, and specifically within Neosuchia, so it is appropriate for a preliminary test of whether or not *Denazinosuchus* is a goniopholidid. Our codings for *Denazinosuchus* are presented in Appendix 17.1. We analyzed the phylogenetic matrix by using equally weighted parsimony in TNT version 1.1 (Goloboff, Farris, and Nixon, 2003, 2008), and applied the same search strategy and settings as used in the original analysis (Turner and Sertich, 2010:197).

Our analysis recovered 3060 equally most parsimonious trees (MPTs), each with a length of 1319 steps (Fig. 17.9). In all MPTs, *Denazinosuchus* was recovered as part of a monophyletic Goniopholididae, whose closest sister taxon is *Bernissartia*, in agreement with the original analysis (Turner and Sertich, 2010). A strict consensus of all MPTs places *Denazinosuchus* in a polytomy with *Goniopholis simus*, *G. stovalli*, *Eutretauranosuchus* and the clade of *Calsoyosuchus* + *Sunosuchus* (Fig. 17.9). The loss of resolution within Goniopholididae compared to the original analysis (Turner and Sertich, 2010:fig. 18) is a result of the large amount of missing data for *Denazinosuchus*. Nonetheless, this analysis supports a goniopholidid assignment for *Denazinosuchus*, even though it lacks a maxillary fossa (pers. obs.). Although they are not unambiguous synapomorphies, the goniopholid placement of *Denazinosuchus* is supported by the following apomorphic combination of character states: platyrostral snout (3–3), broad premaxilla anterior to the nares (5–1), nasal excluded from the narial border (13–1), antorbital fenestra absent (67–3), uncompressed maxillary teeth (140–0), posterior tip of the nasals separated by anterior sagittal projection of the frontals (165–1), first enlarged maxillary tooth is fourth or fifth alveolus (184–1), lateral margins of the frontals are flush with the skull surface (266–0), and lateral margins of the anterior half of the interfenestral bar between the supratemporal fenestrae are parallel to subparallel (278–0). The results of this phylogenetic analysis should be considered preliminary as the Turner and Sertich (2010) matrix is biased toward basal mesoeucrocodylians (as is its predecessor, Pol et al., 2009, and older versions of this matrix); however, there is currently no available matrix that has a broad sampling of both basal mesoeucrocodylians and basal eusuchians. Better resolution in the placement of goniopholidids among Neosuchia and interrelationships within Goniopholididae await a thorough review of the group, which is currently in process (Allen, 2010).

A goniopholidid phylogenetic placement for *Denazinosuchus* makes it reasonable to hypothesize that the large rectangular osteoderms from the Late Cretaceous of southern Utah

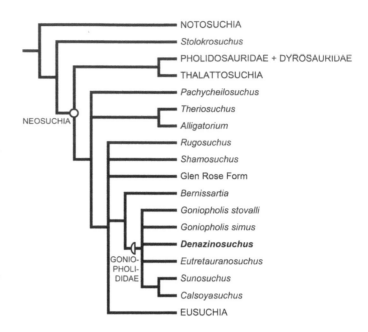

17.9. Phylogenetic position of *Denazinosuchus kirtlandicus* from the San Juan Basin of northern New Mexico based on inclusion in the phylogenetic analysis of Turner and Sertich (2010). Cladogram is a strict consensus of 3060 most parsimonious trees with a length of 1319 steps. Some larger clades have been collapsed for brevity, and taxa more basal than Notosuchia are not shown.

might belong to goniopholidid crocodyliforms. Nonetheless, this hypothesis cannot be confirmed without better cranial material associated with these osteoderms, and at present they can only be referred to Neosuchia (see above). In particular, there is still the possibility that some or all of these osteoderms actually belong to a late-surviving pholidosaurid crocodyliform.

LARAMIDIAN LATE CRETACEOUS CROCODYLIFORM BIOGEOGRAPHY

As with other vertebrate clades (e.g., Gates et al., 2010; Sampson, Loewen, et al., 2010; Sampson, Gates, et al., 2010; Hutchison et al., this volume, Chapter 13; Loewen et al., this volume, Chapter 21), the crocodyliform record from the Late Cretaceous of the Kaiparowits Basin in southern Utah occupies a critical position in understanding Late Cretaceous nonmarine biogeography of the Western Interior because it occupies a position between well-known northern assemblages in Alberta and Montana, and southern assemblages in New Mexico and Mexico. Fragmentary remains from the Dakota, Straight Cliffs, and Wahweap formations can only be assigned to large widespread clades within Crocodyliformes (e.g., Mesoeucrocodylia, Neosuchia, and Crocodylia), so they are of limited use in discerning biogeographic patterns. Thus, in the current discussion, we focus on the Kaiparowits Formation record.

Irmis et al.

The Kaiparowits crocodyliform assemblage is most similar to that of Campanian strata in the San Juan Basin of northern New Mexico. Both assemblages contain *Deinosuchus* and *Brachychampsa* (Williamson, 1996; Sullivan and Lucas, 2003; Lucas et al., 2006; Lucas, Sullivan, and Spielmann, 2006), and large rectangular osteoderms from the Kaiparowits might belong to a taxon similar to the goniopholidid *Denazinosuchus* from the San Juan Basin (see above; Lucas and Sullivan, 2003; Lucas et al., 2006). Both crocodyliform assemblages conspicuously lack any evidence of *Leidyosuchus*, an alligatoroid that is the most common crocodyliform in the penecontemporaneous Dinosaur Park Formation of Alberta, Canada (Wu, Russell, and Brinkman, 2001; Wu, 2005). The report of cf. *Leidyosuchus* sp. by Lucas et al. (2006) cannot be substantiated, as these type of teeth are found throughout many different mesoeucrocodylian taxa.

Rivera-Sylva et al. (2009, 2011) recently reported the occurrence of *Deinosuchus* from late Campanian strata of northwestern Coahuila, Mexico that are probably correlative to the Aguja Formation of southern Texas. This material consists of disarticulated teeth, vertebrae, and osteoderms. The authors surmise that much of the material belongs to a single individual (Rivera-Sylva et al., 2011) but do not provide taphonomic details. The teeth and vertebrae were assigned to *Deinosuchus* on the basis of their size, but as discussed above, size alone is not a valid taxonomic criterion. Because of their incompleteness, the osteoderms do not appear to preserve any of the diagnostic traits of *Deinosuchus* (i.e., pustulose ornamentation). Thus, this material does not preserve diagnostic character states that allow assignment to *Deinosuchus*, but there are also no character states that contradict such identification. At present this material should be considered that of an indeterminate neosuchian, but if better material confirms the presence of *Deinosuchus* in Coahuila, it would provide an important biogeographic connection with assemblages further north.

The Kaiparowits Formation of southern Utah is penecontemporaneous with a number of other nonmarine Campanian formations in the Western Interior, including the Fruitland and Kirtland formations of the San Juan Basin of New Mexico (Roberts, Deino, and Chan, 2005; Jinnah et al., 2009; Roberts et al., this volume, Chapter 6). Beyond the San Juan Basin, the best described late Campanian crocodyliform assemblage is that of the Dinosaur Park Formation in Alberta (Wu, 2005). The absence of *Leidyosuchus* and presence of *Deinosuchus* and *Brachychampsa* strongly differentiate the Kaiparowits and Fruitland/Kirtland assemblages from those of the Dinosaur Park Formation (Wu, 2005). The small alligatoroid from the Kaiparowits shares some features with isolated dentaries from the Dinosaur Park Formation (Wu,

2005) and *Stangerochampsa* from the younger Horseshoe Canyon Formation of Alberta (Wu, Brinkman, and Russell, 1996a), but these character states do not mandate a particularly close phylogenetic relationship.

The Two Medicine and Judith River formations of Montana also partially overlap in age with the Kaiparowits and Fruitland/Kirtland formations. Unfortunately, the crocodyliform assemblages from these northern formations are not well described, though the Judith River does share with the southern basins the presence of *Deinosuchus* (Holland, 1909; Schwimmer, 2002). *Deinosuchus* is also known from partially equivalent strata of the Mesaverde Formation further east in Wyoming (Schwimmer, 2002; Wahl and Hogbin, 2003; DeMar and Breithaupt, 2008), as well as southern Appalachia (i.e., southeastern United States) (e.g., Schwimmer, 2002). The younger Maastrichtian Hell Creek Formation of Montana and the Dakotas shares with the southern Campanian basins the presence of *Brachychampsa* (Gilmore, 1911; Carpenter and Lindsey, 1980; Norell, Clark, and Hutchison, 1994), but differs in the presence of the basal crocodylian *Borealosuchus* (Bryant, 1989; Brochu, 1997) and absence of *Deinosuchus*.

The close biogeographic similarity of crocodyliform assemblages from the Kaiparowits Formation and Fruitland–Kirtland formations supports the hypothesis of distinct southern and northern vertebrate biogeographic provinces in the Campanian of Laramidia (Gates et al., 2010; Sampson, Loewen, et al., 2010; Sampson, Gates, et al., 2010; Hutchison et al., this volume, Chapter 13). The presence of *Brachychampsa* in the late Campanian of southern Laramidia, and its subsequent appearance in the north during the Maastrichtian, is consistent with the hypothesis that some southern vertebrate taxa moved north as climate warmed near the Campanian–Maastrichtian boundary (Hutchison et al., this volume, Chapter 13). Nonetheless, the strength of biogeographic hypotheses derived from available Late Cretaceous Utah crocodyliform data is limited because so few specimens can be identified to less inclusive clades, genera, or species. Future discoveries of more complete crocodyliform remains from these different Late Cretaceous basins of the Western Interior will allow better taxonomic resolution and therefore more robust biogeographic hypotheses.

APPENDIX 17.1. CHARACTER CODINGS FOR *DENAZINOSUCHUS KIRTLANDICUS* (WIMAN, 1932) IN THE PHYLOGENETIC MATRIX OF TURNER AND SERTICH (2010)

203?1211??0010?????010?1????0?000??????????????????????????
?????131?????????????????????????12?0????????00??0????????1?
???1?11????????0010?????????????????????0001??????????0?1????

?1???0?????????????????1??????0???????0??0???0???????0?0????
????0????100?00???????0?????????001??????????????00????0??

ACKNOWLEDGMENTS

Our fieldwork and research in the Kaiparowits Basin, and in particular Grand Staircase–Escalante National Monument, has been generously supported by the Bureau of Land Management (to R.B.I. and A.L.T.) and the University of Utah (R.B.I.). We are deeply indebted to the UMNH and Bureau of Land Management paleontology volunteers, who contributed thousands of hours of work in the field and lab, and who discovered and prepared a number of specimens described in this chapter. In particular, J. Golden masterfully prepared the skull of UMNH VP 19268 in spite of its challenging preservation. P. Holroyd graciously provided access to the UCMP collections. We thank D. Brinkman, J. Eaton, M. Getty, Z. Jinnah, M. Loewen, S. Nesbitt, E. Roberts, and S. Sampson for helpful discussions. C. Brochu and A. Turner provided constructive reviews that greatly improved the chapter. C. Brochu kindly provided the Nexus file for the phylogenetic analysis in his 2010 paper.

REFERENCES CITED

Adams, T. L., M. J. Polcyn, O. Mateus, D. A. Winkler, and L. L. Jacobs. 2011. First occurrence of the long-snouted crocodyliform *Terminonaris* (Pholidosauridae) from the Woodbine Formation (Cenomanian) of Texas. Journal of Vertebrate Paleontology 31:712–716.

Allen, E. 2010. Phylogenetic analysis of goniopholidid crocodyliforms of the Morrison Formation. Journal of Vertebrate Paleontology 30:52A.

Averianov, A. O. 2000. *Sunosuchus* sp. (Crocodylomorpha, Goniopholidae) from the Middle Jurassic of Kirghisia. Journal of Vertebrate Paleontology 20:776–779.

Bell, C. J., J. A. Gauthier, and G. S. Bever. 2010. Covert biases, circularity, and apomorphies: a critical look at the North American Quaternary herpetofaunal stability hypothesis. Quaternary International 217:30–36.

Bell, C. J., J. J. Head, and J. I. Mead. 2004. Synopsis of the herpetofauna from Porcupine Cave; pp. 117–126 in A. D. Barnosky (ed.), Biodiversity Response to Climate Change in the Middle Pleistocene: The Porcupine Cave Fauna from Colorado. University of California Press, Berkeley, California.

Benton, M. J., and J. M. Clark. 1988. Archosaur phylogeny and the relationships of the Crocodylia; pp. 295–338 in M. J. Benton (ed.), The Phylogeny and Classification of the Tetrapods, Volume 1: Amphibians, Reptiles, Birds. Clarendon Press, Oxford.

Bever, G. S. 2005. Variation in the ilium of North American *Bufo* (Lissamphibia: Anura) and its implications for species-level identification of fragmentary anuran fossils. Journal of Vertebrate Paleontology 25:548–560.

Brinkman, D. B. 2008. The structure of Late Cretaceous (late Campanian) nonmarine aquatic communities: a guild analysis of two vertebrate microfossil localities in Dinosaur Provincial Park, Alberta, Canada; pp. 33–60 in J. T. Sankey and S. Baszio (eds.), Vertebrate Microfossil Assemblages: Their Role in Paleoecology and Paleobiogeography. Indiana University Press, Bloomington, Indiana.

Brinkmann, W. 1992. Die krokodilier-fauna aus der Unter-Kreide (Ober-Barremium) von Uña (Provinz Cuenca, Spanien). Berliner Geowissenschaftliche Abhandlungen, Reihe E 5:1–123.

Brochu, C. A. 1997. A review of *"Leidyosuchus"* (Crocodyliformes, Eusuchia) from the Cretaceous through Eocene of North America. Journal of Vertebrate Paleontology 17:679–697.

Brochu, C. A. 1999. Phylogenetics, taxonomy, and historical biogeography of Alligatoroidea. Society of Vertebrate Paleontology Memoir 6:9–100.

Brochu, C. A. 2001. Crocodylian snouts in space and time: phylogenetic approaches toward adaptive radiation. American Zoologist 41:564–585.

Brochu, C. A. 2003. Phylogenetic approaches toward crocodylian history. Annual Review of Earth and Planetary Sciences 31:357–397.

Brochu, C. A. 2004. Alligatorine phylogeny and the status of *Allognathosuchus* Mook, 1921. Journal of Vertebrate Paleontology 24:857–873.

Brochu, C. A. 2007. Morphology, relationships, and biogeographical significance of an extinct horned crocodile (Crocodylia, Crocodylidae) from the Quaternary of Madagascar. Zoological Journal of the Linnean Society 150:835–863.

Brochu, C. A. 2010. A new alligatorid from the lower Eocene Green River Formation of Wyoming and the origin of caimans. Journal of Vertebrate Paleontology 30:1109–1126.

Bryant, L. J. 1989. Non-dinosaurian lower vertebrates across the Cretaceous–Tertiary boundary in northeastern Montana. University of California Publications in Geological Sciences 134:1–107.

Buffetaut, E. 1975. Sur l'anatomie et la position systématique de *Bernissartia fagesii* Dollo, L., 1883, crocodilien du Wealdien de Bernissart, Belgique. Bulletin de l'Institut Royal des Sciences Naturelles de Belgique, Sciences de la Terre 51:1–24.

Buffetaut, E., and R. L. E. Ford. 1979. The crocodilian *Bernissartia* in the Wealden of the Isle of Wight. Palaeontology 22:905–912.

Buffetaut, E., and P. Taquet. 1977. The giant crocodilian *Sarcosuchus* in the Early Cretaceous of Brazil and Niger. Palaeontology 20:203–208.

Buscalioni, A. D., and J. L. Sanz. 1990. The small crocodile *Bernissartia fagesii* from the Lower Cretaceous of Galve (Teruel, Spain). Bulletin de l'Institut Royal des Sciences Naturelles de Belgique, Sciences de la Terre 60:129–150.

Carpenter, K., and D. Lindsey. 1980. The dentary of *Brachychampsa montana* Gilmore (Alligatorinae; Crocodylidae), a Late Cretaceous turtle-eating alligator. Journal of Paleontology 54:1213–1217.

Clark, J. M. 1985. A new crocodilian from the Upper Jurassic Morrison Formation of western Colorado, with a discussion of relationships within the "Mesosuchia." M.Sc. thesis, University of California, Berkeley, California.

Clark, J. M. 1994. Patterns of evolution in Mesozoic Crocodyliformes; pp. 84–97 in N. C. Fraser and H.-D. Sues (eds.), In the Shadow of the Dinosaurs: Early Mesozoic Tetrapods. Cambridge University Press, Cambridge.

Colbert, E. H., and R. T. Bird. 1954. A gigantic crocodile from the Upper Cretaceous beds of Texas. American Museum Novitates 1688:1–22.

Colbert, E. H., and C. C. Mook. 1951. The ancestral crocodilian *Protosuchus*. Bulletin of the American Museum of Natural History 97:147–182.

Delfino, M., P. Piras, and T. Smith. 2005. Anatomy and phylogeny of the gavialoid crocodylian *Eosuchus lerichei* from the Paleocene of Europe. Acta Palaeontologica Polonica 50:565–580.

DeMar, D. G., Jr., and B. H. Breithaupt. 2008. Terrestrial and aquatic vertebrate paleocommunities of the Mesaverde Formation (Upper Cretaceous, Campanian) of the Wind River and Bighorn basins, Wyoming, U.S.A.; pp. 78–103 in J. T. Sankey and S. Baszio (eds.), Vertebrate Microfossil Assemblages: Their Role in Paleoecology and Paleobiogeography. Indiana University Press, Bloomington, Indiana.

Dollo, M. L. 1883. Première note sur les crocodiliens de Bernissart. Bulletin du Musée Royal d'Histoire Naturelle de Belgique 2:309–338.

Eaton, J. G. 1991. Biostratigraphic framework for Upper Cretaceous rocks of the Kaiparowits Plateau, southern Utah; pp. 47–63 in J. D. Nations and J. G. Eaton (eds.), Stratigraphy, Depositional Environments, and Sedimentary Tectonics of the Western Margin, Cretaceous Western Interior Seaway. Geological Society of America Special Paper 260.

Eaton, J. G. 1993. Therian mammals from the Cenomanian (Upper Cretaceous) Dakota Formation, southwestern Utah. Journal of Vertebrate Paleontology 13:105–124.

Eaton, J. G. 1995. Cenomanian and Turonian (early Late Cretaceous) multituberculate mammals from southwestern Utah. Journal of Vertebrate Paleontology 15:761–784.

Eaton, J. G. 1999. Vertebrate paleontology of the Paunsaugunt Plateau, Upper Cretaceous, southwestern Utah. Utah Geological Survey Miscellaneous Publication 99-1:335–338.

Eaton, J. G. 2006. Santonian (Late Cretaceous) mammals from the John Henry Member of the Straight Cliffs Formation, Grand Staircase–Escalante National Monument, Utah. Journal of Vertebrate Paleontology 26:446–460.

Eaton, J. G. 2009. Cenomanian (Late Cretaceous) mammals from Cedar Canyon, southwestern Utah, and a revision of Cenomanian *Alphadon*-like marsupials. Museum of Northern Arizona Bulletin 65:97–110.

Eaton, J. G., and R. L. Cifelli. 1998. Cretaceous vertebrates of the Grand Staircase–Escalante National Monument; pp. 365–372 in L. M. Hill (ed.), Learning from the Land: Grand Staircase–Escalante National Monument Science Symposium Proceedings. U.S. Department of the Interior, Bureau of Land Management.

Eaton, J. G., R. L. Cifelli, J. H. Hutchison, J. I. Kirkland, and J. M. Parrish. 1999. Cretaceous vertebrate faunas from the Kaiparowits Plateau, south central Utah; pp. 345–353 in D. D. Gillette (ed.), Vertebrate Paleontology in Utah. Utah Geological Survey Miscellaneous Publication 99-1.

Eaton, J. G., S. Diem, J. D. Archibald, C. Schierup, and H. Munk. 1999. Vertebrate paleontology of the Upper Cretaceous rocks of the Markagunt Plateau, southwestern Utah; pp. 323–333 in D. D. Gillette (ed.), Vertebrate Paleontology in Utah. Utah Geological Survey Miscellaneous Publication 99-1.

Erickson, B. R. 1972. *Albertochampsa langstoni,* gen. et sp. nov., a new alligator from the Cretaceous of Alberta. Scientific Publications of the Science Museum of Minnesota, n.s., 2:1–13.

Estes, R., P. Berberian, and C. A. M. Meszoely. 1969. Lower vertebrates from the Late Cretaceous Hell Creek Formation, McCone County, Montana. Breviora 337:1–33.

Gao, Y.-H. 2001. A new species of *Hsisosuchus* from Dashanpu, Zigong, Sichuan. Vertebrata PalAsiatica 39:177–184.

Gasparini, Z., D. Pol, and L. A. Spalletti. 2006. An unusual marine crocodyliform from the Jurassic–Cretaceous boundary of Patagonia. Science 311:70–73.

Gates, T. A., S. D. Sampson, L. E. Zanno, E. M. Roberts, J. G. Eaton, R. L. Nydam, J. H. Hutchison, J. A. Smith, M. A. Loewen, and M. A. Getty. 2010. Biogeography of terrestrial and freshwater vertebrates from the Late Cretaceous (Campanian) Western Interior of North America. Palaeogeography, Palaeoclimatology, Palaeoecology 291:371–387.

Gillette, D. D., and M. C. Hayden. 1997. A preliminary inventory of paleontological resources within the Grand Staircase–Escalante National Monument, Utah. Utah Geological Survey Circular 96.

Gilmore, C. W. 1911. A new fossil alligator from the Hell Creek beds of Montana. Proceedings of the U.S. National Museum 41:297–302.

Gmelin, J. F. 1789. Linnei Systema Naturae. G. E. Beer, Leipzig.

Goloboff, P. A., J. S. Farris, and K. C. Nixon. 2003. TNT: tree analysis using new technology. v. 1.1. Program and documentation. Available at www.zmuc.dk/public/phylogeny/tnt.

Goloboff, P. A., J. S. Farris, and K. C. Nixon. 2008. TNT, a free program for phylogenetic analysis. Cladistics 24:774–786.

Gray, J. E. 1844. Catalogue of the Tortoises, Crocodilians, and Amphisbaenians in the Collection of the British Museum. British Museum of Natural History, London.

Hay, O. P. 1930. Second Bibliography and Catalogue of the Fossil Vertebrata of North America. Carnegie Institution, Washington, D.C.

Holland, W. J. 1909. *Deinosuchus hatcheri,* a new genus and species of crocodile from the Judith River beds of Montana. Annals of the Carnegie Museum 6:281–294.

Hua, S., E. Buffetaut, C. Legall, and P. Rogron. 2007. *Oceanosuchus boecensis* n. gen, n. sp., a marine pholidosaurid (Crocodylia, Mesosuchia) from the lower Cenomanian of Normandy (western France). Bulletin de la Société Géologique de France 178:503–513.

Hutchinson, P. J., and B. S. Kues. 1985. Depositional environments and paleontology of Lewis Shale to lower Kirtland Shale sequence (Upper Cretaceous), Bisti area, northwestern New Mexico. New Mexico Bureau of Mines & Mineral Resources Circular 195:25–54.

Hutchison, J. H., J. G. Eaton, P. A. Holroyd, and M. B. Goodwin. 1998. Larger vertebrates of the Kaiparowits Formation (Campanian) in the Grand Staircase–Escalante National Monument and adjacent areas; pp. 391–398 in L. M. Hill (ed.), Learning from the Land: Grand Staircase–Escalante National Monument Science Symposium Proceedings. U.S. Department of the Interior, Bureau of Land Management.

Huxley, T. H. 1875. On *Stagonolepis robertsoni,* and on the evolution of the Crocodilia. Quarterly Journal of the Geological Society of London 31:423–438.

Jinnah, Z. A., E. M. Roberts, A. L. Deino, J. S. Larsen, P. K. Link, and C. M. Fanning. 2009. New ^{40}Ar-^{39}Ar and detrital zircon U-Pb ages for the Upper Cretaceous Wahweap and Kaiparowits formations on the Kaiparowits Plateau, Utah: implications for regional correlation, provenance, and biostratigraphy. Cretaceous Research 30:287–299.

Jouve, S. 2005. A new description of the skull of *Dyrosaurus phosphaticus* (Thomas, 1893) (Mesoeucrocodylia: Dyrosauridae) from the lower Eocene of North Africa. Canadian Journal of Earth Sciences 42:323–337.

Jouve, S. 2009. The skull of *Teleosaurus cadomensis* (Crocodylomorpha; Thalattosuchia), and phylogenetic analysis of Thalattosuchia. Journal of Vertebrate Paleontology 29:88–102.

Lauprasert, K., G. Cuny, E. Buffetaut, V. Suteethorn, and K. Thirakhupt. 2007. *Siamosuchus phuphokensis,* a new goniopholidid from the Early Cretaceous (ante-Aptian) of northeastern Thailand. Bulletin de la Société Géologique de France 178:201–216.

Lee, Y.-N. 1997. The Archosauria from the Woodbine Formation (Cenomanian) in Texas. Journal of Paleontology 71:1147–1156.

Lucas, S. G. 1992. Cretaceous–Eocene crocodilians from the San Juan Basin, New Mexico. New Mexico Geological Society Guidebook 43:257–264.

Lucas, S. G., and R. M. Sullivan. 2003. A new crocodilian from the Upper Cretaceous of the San Juan Basin, New Mexico. Neues Jahrbuch für Geologie und Paläontologie, Monatschefte 2003:109–119.

Lucas, S. G., R. M. Sullivan, and J. A. Spielmann. 2006b. The giant crocodylian *Deinosuchus* from the Upper Cretaceous of the San Juan Basin, New Mexico. New Mexico Museum of Natural History and Science Bulletin 35:245–248.

Lucas, S. G., J. A. Spielmann, R. M. Sullivan, and C. Lewis. 2006a. Late Cretaceous crocodylians from the San Juan Basin, New Mexico. New Mexico Museum of Natural History and Science Bulletin 35:249–252.

Markwick, P. J. 1998. Crocodilian diversity in space and time: the role of climate in paleoecology and its implication for understanding K/T extinctions. Paleobiology 24:470–497.

Martin, J. E., M. Rabi, and Z. Csiki. 2010. Survival of *Theriosuchus* (Mesoeucrocodylia: Atoposauridae) in a Late Cretaceous archipelago: a new species from the Maastrichtian of Romania. Naturwissenschaften 97:845–854.

Mook, C. C. 1933. A crocodilian skeleton from the Morrison Formation at Canyon City, Colorado. American Museum Novitates 671:1–8.

Mook, C. C. 1942. Skull characters of *Amphicotylus lucasii* Cope. American Museum Novitates 1165:1–5.

Nesbitt, S. J., and M. R. Stocker. 2008. The vertebrate assemblage of the Late Triassic Canjilon Quarry (northern New Mexico, U.S.A.), and the importance of apomorphy-based assemblage comparisons. Journal of Vertebrate Paleontology 28:1063–1072.

Nesbitt, S. J., R. B. Irmis, and W. G. Parker. 2007. A critical re-evaluation of the Late Triassic dinosaur taxa of North America. Journal of Systematic Palaeontology 5:209–243.

Norell, M. A., and J. M. Clark. 1990. A reanalysis of *Bernissartia fagesii,* with comments on its phylogenetic position and its bearing on the origin and diagnosis of the Eusuchia. Bulletin de l'Institut Royal des Sciences Naturelles de Belgique, Sciences de la Terre 60:115–128.

Norell, M. A., J. M. Clark, and J. H. Hutchison. 1994. The Late Cretaceous alligatoroid *Brachychampsa montana* (Crocodylia): new material and putative relationships. American Museum Novitates 3116:1–26.

Peng, G.-Z., and C.-K. Shu. 2005. A new species of *Hsisosuchus* from the Late Jurassic of Zigong, Sichuan, China. Vertebrata PalAsiatica 43:312–324.

Peng, J., A. P. Russell, and D. B. Brinkman. 2001. Vertebrate Microsite Assemblages (Exclusive of Mammals) from the Foremost and Oldman Formations of the Judith River Group (Campanian) of Southeastern Alberta: An Illustrated Guide. Provincial Museum of Alberta Natural History Occasional Paper 25.

Pol, D. 2003. New remains of *Sphagesaurus huenei* (Crocodylomorpha: Mesoeucrocodylia) from the Late Cretaceous of Brazil. Journal of Vertebrate Paleontology 23:817–831.

Pol, D., and Z. Gasparini. 2009. Skull anatomy of *Dakosaurus andiniensis* (Thalattosuchia: Crocodylomorpha) and the phylogenetic position of Thalattosuchia. Journal of Systematic Palaeontology 7:163–197.

Pol, D., and M. A. Norell. 2004a. A new crocodyliform from Zos Canyon, Mongolia. American Museum Novitates 3445:1–36.

Pol, D., and M. A. Norell. 2004b. A new gobiosuchid crocodyliform taxon from the Cretaceous of Mongolia. American Museum Novitates 3458:1–31.

Pol, D., A. H. Turner, and M. A. Norell. 2009. Morphology of the Late Cretaceous crocodylomorph Shamosuchus djadochtaensis and a discussion of neosuchian phylogeny as related to the origin of Eusuchia. Bulletin of the American Museum of Natural History 324:1–103.

Rivera-Sylva, H. E., E. Frey, F. J. Palomino-Sánchez, J. R. Guzmán-Gutierrez, and J. A. Ortiz-Mendieta. 2009. Preliminary report on a Late Cretaceous vertebrate fossil assemblage in northwestern Coahuila, Mexico. Boletín de la Sociedad Geológica Mexicana 61:239–244.

Rivera-Sylva, H., E. Frey, J. R. Guzmán-Gutierrez, F. Palomino-Sánchez, and W. Stinnesbeck. 2011. A Deinosuchus riograndensis (Eusuchia: Alligatoroidea) from Coahuila, north Mexico. Revista Mexicana de Ciencias Geológicas 28:267–274.

Roberts, E. M., A. L. Deino, and M. A. Chan. 2005. ^{40}Ar/^{39}Ar age of the Kaiparowits Formation, southern Utah, and correlation of contemporaneous Campanian strata and vertebrate faunas along the margin of the Western Interior Basin. Cretaceous Research 26:307–318.

Rogers, J. V., II. 2003. Pachycheilosuchus trinquei, a new procoelous crocodyliform from the Lower Cretaceous (Albian) Glen Rose Formation of Texas. Journal of Vertebrate Paleontology 23:128–145.

Rowe, T., R. L. Cifelli, T. M. Lehman, and A. Weil. 1992. The Campanian Terlingua Local Fauna, with a summary of other vertebrates from the Aguja Formation, Trans-Pecos Texas. Journal of Vertebrate Paleontology 12:472–493.

Salisbury, S. W. 2002. Crocodilians from the Lower Cretaceous (Berriasian) Purbeck Limestone Group of Dorset, southern England. Special Papers in Palaeontology 68:121–144.

Sampson, S. D., M. A. Loewen, A. A. Farke, E. M. Roberts, C. A. Forster, J. A. Smith, and A. L. Titus. 2010a. New horned dinosaurs from Utah provide evidence for intracontinental dinosaur endemism. PLoS One 5:e12292.

Sampson, S. D., T. A. Gates, E. M. Roberts, M. A. Getty, L. Zanno, A. L. Titus, M. A. Loewen, J. A. Smith, E. K. Lund, and J. Sertich. 2010b. Grand Staircase–Escalante National Monument: a new and critical window into the world of dinosaurs; pp. 391–398 in M. Eaton (ed.), Learning from the Land, Grand Staircase–Escalante National Monument Science Symposium Proceedings. Grand Staircase–Escalante Partners.

Sankey, J. T. 2008. Vertebrate paleoecology from microsites, Talley Mountain, upper Aguja Formation (Late Cretaceous), Big Bend National Park, Texas, U.S.A.; pp. 61–77 in J. T. Sankey and S. Baszio (eds.), Vertebrate Microfossil Assemblages: Their Role in Paleoecology and Paleobiogeography. Indiana University Press, Bloomington, Indiana.

Schwarz, D. 2002. A new species of Goniopholis from the Upper Jurassic of Portugal. Palaeontology 45:185–208.

Schwarz, D., and S. W. Salisbury. 2005. A new species of Theriosuchus (Atoposauridae, Crocodylomorpha) from the Late Jurassic (Kimmeridgian) of Guimarota, Portugal. Geobios 38:779–802.

Schwarz-Wings, D., J. Rees, and J. Lindgren. 2009. Lower Cretaceous mesoeucrocodylians from Scandinavia (Denmark and Sweden). Cretaceous Research 30:1345–1355.

Schwimmer, D. R. 2002. King of the Crocodylians: The Paleobiology of Deinosuchus. Indiana University Press, Bloomington, Indiana.

Sereno, P. C., and H. C. E. Larsson. 2009. Cretaceous crocodyliforms from the Sahara. ZooKeys 28:1–143.

Sereno, P. C., H. C. E. Larsson, C. A. Sidor, and B. Gado. 2001. The giant crocodyliform Sarcosuchus from the Cretaceous of Africa. Science 294:1516–1519.

Sues, H.-D. 1992. A remarkable new armored archosaur from the Upper Triassic of Virginia. Journal of Vertebrate Paleontology 12:142–149.

Sullivan, R. M., and S. G. Lucas. 2003. Brachychampsa montana Gilmore (Crocodylia, Alligatoroidea) from the Kirtland Formation (upper Campanian), San Juan Basin, New Mexico. Journal of Vertebrate Paleontology 23:832–841.

Titus, A. L., J. D. Powell, E. M. Roberts, S. D. Sampson, S. L. Pollock, J. I. Kirkland, and L. B. Albright. 2005. Late Cretaceous stratigraphy, depositional environments, and macrovertebrate paleontology of the Kaiparowits Plateau, Grand Staircase–Escalante National Monument, Utah; pp. 101–128 in J. Pederson and C. M. Dehler (eds.), Interior Western United States. Geological Society of America Field Guide 6.

Turner, A. H. 2004. Crocodyliform biogeography during the Cretaceous: evidence of Gondwanan vicariance from biogeographical analysis. Proceedings of the Royal Society of London, Biological Sciences 271:2003–2009.

Turner, A. H. 2006. Osteology and phylogeny of a new species of Araripesuchus (Crocodyliformes: Mesoeucrocodylia) from the Late Cretaceous of Madagascar. Historical Biology 18:255–369.

Turner, A. H., and G. A. Buckley. 2008. Mahajangasuchus insignis (Crocodyliformes: Mesoeucrocodylia) cranial anatomy and new data on the origin of the eusuchian-style palate. Journal of Vertebrate Paleontology 28:382–408.

Turner, A. H., and J. J. W. Sertich. 2010. Phylogenetic history of Simosuchus clarki (Crocodyliformes: Notosuchia) from the Late Cretaceous of Madagascar. Society of Vertebrate Paleontology Memoir 10:177–236.

Tykoski, R. S., T. B. Rowe, R. A. Ketcham, and M. W. Colbert. 2002. Calsoyasuchus valliceps, a new crocodyliform from the Early Jurassic Kayenta Formation of Arizona. Journal of Vertebrate Paleontology 22:593–611.

Wahl, W., and J. Hogbin. 2003. Deinosuchus material from the Mesaverde Formation of Wyoming: filling in a gap. Journal of Vertebrate Paleontology 23:107A.

Whetstone, K. N., and P. J. Whybrow. 1983. A "cursorial" crocodilian from the Triassic of Lesotho (Batsutoland), southern Africa. Occasional Papers of the Museum of Natural History, University of Kansas 106:1–37.

Wiersma, J., H. Hutchison, and T. Gates. 2004. Crocodilian diversity in the Upper Cretaceous Kaiparowits Formation (Upper Campanian), Utah. Journal of Vertebrate Paleontology 24:129A.

Wilberg, E. 2010. The phylogenetic position of Thalattosuchia (Crocodylomorpha) and the importance of outgroup choice. Journal of Vertebrate Paleontology 30:187A.

Williamson, T. E. 1996. ?Brachychampsa sealeyi, sp. nov. (Crocodylia, Alligatoroidea), from the Upper Cretaceous (lower Campanian) Menefee Formation, northwestern New Mexico. Journal of Vertebrate Paleontology 16:421–431.

Wiman, C. 1932. Goniopholis kirtlandicus n. sp. aus der oberen Kreide in New Mexico. Bulletin of the Geological Institution of the University of Uppsala 23:181–190.

Wu, X.-C. 2005. Crocodylians; pp. 277–291 in P. J. Currie and E. B. Koppelhus (eds.), Dinosaur Provincial Park: A Spectacular Ecosystem Revealed. Indiana University Press, Bloomington, Indiana.

Wu, X.-C., D. B. Brinkman, and J.-C. Lu. 1994. A new species of Shantungosuchus from the Lower Cretaceous of Inner Mongolia (China), with comments on S. chuhsiensis Young, 1961, and the phylogenetic position of the genus. Journal of Vertebrate Paleontology 14:210–229.

Wu, X.-C., D. B. Brinkman, and A. P. Russell. 1996a. A new alligator from the Upper Cretaceous of Canada and the relationships of early eusuchians. Palaeontology 39:351–375.

Wu, X.-C., D. B. Brinkman, and A. P. Russell. 1996b. Sunosuchus junggarensis sp. nov. (Archosauria: Crocodyliformes) from the Upper Jurassic of Xinjiang, People's Republic of China. Canadian Journal of Earth Sciences 33:606–630.

Wu, X.-C., A. P. Russell, and D. B. Brinkman. 2001. A review of Leidyosuchus canadensis Lambe, 1907 (Archosauria: Crocodylia) and an assessment of cranial variation based upon new material. Canadian Journal of Earth Sciences 38:1665–1687.

Wu, X.-C., A. P. Russell, and S. L. Cumbaa. 2001b. Terminonaris (Archosauria: Crocodyliformes): new material from Saskatchewan, Canada, and comments on its phylogenetic relationships. Journal of Vertebrate Paleontology 21:492–514.

Wu, X.-C., H.-D. Sues, and Z.-M. Dong. 1997. Sichuanosuchus shuhanensis, a new ?Early Cretaceous protosuchian (Archosauria: Crocodyliformes) from Sichuan (China), and the monophyly of Protosuchia. Journal of Vertebrate Paleontology 17:89–103.

Young, M. T., and M. B. de Andrade. 2009. What is Geosaurus? Redescription of Geosaurus giganteus (Thalattosuchia: Metriorhynchidae) from the Upper Jurassic of Bayern, Germany. Zoological Journal of the Linnean Society 157:551–585.

Review of Late Cretaceous Ankylosaurian Dinosaurs from the Grand Staircase Region, Southern Utah

18

Mark A. Loewen, Michael E. Burns, Michael A. Getty, James I. Kirkland, and Matthew K. Vickaryous

NEW ANKYLOSAUR SPECIMENS FROM GRAND STAIRCASE–Escalante National Monument, Utah, provide data on the distribution and diversity of ankylosaurian dinosaurs of southern Laramidia. These materials are from the Dakota, Straight Cliffs, Wahweap, Kaiparowits, and laterally equivalent formations of the Grand Staircase of southern Utah. The earliest record of ankylosaurs from Grand Staircase–Escalante National Monument is based on a tooth from the Dakota Formation. Teeth and postcranial elements of both nodosaurids and ankylosaurids are present in the Straight Cliffs and Wahweap formations, and their lateral equivalents. Specimens from the Kaiparowits Formation include nodosaurid teeth and a distal cervical spine, and several ankylosaurid specimens represented by teeth, cranial fragments, postcrania, osteoderms, and tail clubs.

Analysis of the ankylosaurid material from the Kaiparowits Formation indicates that there are two distinct, essentially coeval, ankylosaurids from the lower middle unit of the Kaiparowits Formation.

INTRODUCTION

Ankylosaur dinosaurs are a characteristic component of dinosaur faunas from the Upper Cretaceous of western North America (Laramidia). Campanian strata have yielded numerous specimens of nodosaurids and ankylosaurids, mostly in the north, particularly in Alberta, Canada. Previously, our knowledge of ankylosaurs in the southern part of Laramidia has largely been limited to isolated teeth and osteoderms in Utah, New Mexico, and Texas.

In Alberta, a number of ankylosaurids including *Dyoplosaurus acutosquameus* (Parks, 1924), *Scolosaurus cutleri* (Nopsca, 1928), and *Anodontosaurus lambei* (Sternberg, 1929) were synonymized into *Euoplocephalus tutus* (Lambe, 1902, 1910) by Coombs (1978). Although this taxonomic lumping of Campanian ankylosaurids has been questioned (Penkalski, 2001), it has until recently remained unaltered. However, as part of the results of an ongoing revision of *Euoplocephalus* (Arbour, Burns, and Sissons, 2009; Arbour, 2010), *Dyoplosaurus* is now recognized as distinct from *Euoplocephalus*

(Arbour, Burns, and Sissons, 2009). Nodosaurids from northern Laramidia are best represented by three taxa from Alberta, *Edmontonia rugosidens* and *Panoplosaurus mirus* from the late Campanian Dinosaur Park Formation and *Edmontonia longiceps* from the late Campanian to early Maastrichtian Horseshoe Canyon Formation (Vickaryous, Maryańska, and Weishampel, 2004; Ryan and Evans, 2005).

Late Cretaceous ankylosaurs from the southern portion of Laramidia are less well known. *Nodocephalosaurus kirtlandensis* (Sullivan, 1999; Sullivan and Fowler, 2006) is an ankylosaurid represented by a partial skull from the Kirtland Formation of the San Juan Basin of New Mexico. This specimen is geologically younger than any of the specimens from the Grand Staircase of southern Utah (Roberts, Deino, and Chan, 2005; Sullivan and Lucas, 2006; Roberts et al., this volume, Chapter 6). Nodosaurid material has been recovered from the San Juan Basin of New Mexico but is younger than the material reviewed here. This material includes *Glyptodontopelta mimus* from the Naashoibito Member of the Maastrichtian Ojo Alamo Formation (Ford, 2000; Burns, 2008).

Finally, *Aletopelta coombsi* (Ford and Kirkland, 2001) has been described from the upper Campanian of southern California and is distinguished in having a mosaic of ankylosaurid (basally excavated osteoderms and short distal limb elements) and nodosaurid (large teeth and a mosaic of polygonal osteoderms over the pelvis) characters.

Although ankylosaur remains are considered rare in the late Cretaceous of Utah, previous workers have recovered remains of specimens from the Grand Staircase region. Most of these specimens are teeth recovered as a result of intensive screen washing of sediments by R. Cifelli and J. Eaton (Eaton, Cifelli et al., 1999, Eaton, Diem et al., 1999). Isolated ankylosaur teeth and small osteoderms (i.e., ossicles) are rarely encountered in vertebrate microfossil sites, in contrast to their relative abundance in contemporary northern formations of the Western Interior Basin. Macrofossil ankylosaur material recently collected under the Kaiparowits Basin Project, mostly from the Kaiparowits Formation, has greatly increased our knowledge of ankylosaurs from southern Utah (Getty, Vickaryous, and Loewen, 2009). Associated macrofossil ankylosaur remains represent less than 5% of over 32 recorded associated dinosaur localities in the Kaiparowits Formation (Getty, Vickaryous, and Loewen, 2009), underscoring their rarity. Although ankylosaurs represent a relatively small component of the total Kaiparowits Formation dinosaur fauna, the new material from the Kaiparowits and other Campanian formations of southern Utah is diagnostic enough to warrant a preliminary comparison with contemporaneous ankylosaurs to the north and south along Laramidia.

MATERIALS AND METHODS

Here we reevaluate all available fossil specimens referred to Ankylosauria from the Straight Cliffs, Wahweap, and Kaiparowits formations, including detailed gross anatomical and histological analyses of six osteoderms (UMNH VP 12264, UMNH VP 12562, UMNH VP 12675, UMNH VP 19472, and UMNH VP 19473). Our descriptions of osteoderm surface anatomy follow the system of Hieronymus et al. (2009). Morphological and histological descriptors for osteoderms follow the work of Scheyer and Sander (2004), Hayashi et al. (2010), and Burns (2008), and tail club terminology follows Arbour (2009).

Molds and plaster casts were made of the original samples selected for paleohistological analysis (from UMNH VP 12264, UMNH VP 12675, UMNH VP 16940, UMNH VP 19472, and UMNH VP 19473). For UMNH VP 16940, a core sample was taken for thin sectioning from the ventral surface of the osteoderm with a 0.5-inch Lapcraft diamond coring drill. All specimens were resin impregnated with Buehler EpoThin Low Viscosity Resin and Hardener. Petrographic thin sections were prepared to a thickness of 60–80 μm and polished to a high gloss with CeO_2 powder. Sections were examined and photographed on a Nikon Eclipse E600POL trinocular polarizing microscope with an attached Nikon DXM 1200F digital camera and imaged with a Nikon Super Coolscan 5000 ED with and without polarized film. Histological measurements were taken using ImageJ 1.40g.

Institutional Abbreviations AMNH, American Museum of Natural History, New York; CMN, Canadian Museum of Nature, Ottawa; MNA, Museum of Northern Arizona, Flagstaff, Arizona; OMNH, Oklahoma Museum of Natural History, Norman, Oklahoma; ROM, Royal Ontario Museum, Toronto; SMP, State Museum of Pennsylvania, Harrisburg, Pennsylvania; UCMP, University of California Museum of Paleontology, Berkeley, California; UGS, Utah Geological Survey, Salt Lake City, Utah; UMNH VP, Natural History Museum of Utah (formerly Utah Museum of Natural History), Salt Lake City, Utah.

REVIEW OF MATERIALS

A review of ankylosaur specimens presently known from southern Utah is presented here by formation.

Ankylosaurs of the Cedar Mountain and Dakota formations

The oldest Cretaceous deposits in southern Utah, the Cedar Mountain Formation, are dated radiometrically to 97.0 ± 0.5 Ma on the basis of $^{40}Ar/^{39}Ar$ sanidine date from a volcanic ash in the upper part of the formation just west of Grand Staircase–Escalante National Monument (Biek et al., 2009; Biek, Willis, et al., 2010). Although this largely Lower Cretaceous formation has yielded a large diversity of ankylosaur taxa elsewhere, to date, no ankylosaur remains have been recovered from the Cedar Mountain Formation in southern Utah. The overlying middle to upper Cenomanian Dakota Formation has produced a possible ankylosaurian tooth, although Eaton, Cifelli, et al. (1999) noted that it could instead be pachycephalosaurian. This specimen was not analyzed in this study. Additionally, two ankylosaur footprint casts in a trackway (cf. *Tetrapodosaurus*) are known from the roof of a coal seam in the Dakota Formation (Titus et al., 2005), near the Rimrocks area of Grand Staircase–Escalante National Monument (MNA Loc. DTW-2003–3). The specimen is no longer accessible because of a roof collapse during the fall of 2004 (A. Titus, pers. comm.).

Ankylosaurs of the Straight Cliffs Formation

The Straight Cliffs Formation, Turonian to early Campanian in age, has yielded a number of ankylosaur teeth (Eaton, Cifelli, et al., 1999; Parrish, 1999). The Straight Cliffs Formation is represented by four members in the Grand Staircase: the Tibbet Canyon, Smoky Hollow, John Henry, and Drip Tank members (Peterson, 1969). All four members represent at least some terrestrial sedimentation, but only the Smoky Hollow and John Henry members have produced dinosaur teeth from microsites and a single postcranial element of a nodosaurid.

The Smoky Hollow Member of the Straight Cliffs Formation has yielded teeth of both nodosaurids (OMNH 21711, OMNH 24448) and ankylosaurids (OMNH 23684) (Eaton, Cifelli, et al., 1999). Parrish (1999) also reports both nodosaurid (OMNH 21671) and ankylosaurid (OMNH 21243) teeth from the John Henry Member of the Straight Cliffs Formation.

Recent explorations of the John Henry Member have yielded the only ankylosaur skeletal material from the Straight Cliffs Formation. UMNH VP 17760 (Fig. 18.1) a distal

18.1. Nodosaurid spine (UMNH VP 17760) from the John Henry Member of the Straight Cliffs Formation in (A) dorsal; and (B) ventral views. Scale bar = 10 cm. Superficial is to the right.

cervical spine of a nodosaurid was collected as float (UMNH VP Loc. 811) on the east side of the Paunsaugunt Plateau. The asymmetrical spine is 26 cm long at its longest and has a concave base. UMNH VP 17760 was collected 6 m below (and likely derived from) an interval rich in dinosaur skeletal remains which has also produced diverse microvertebrate specimens including mammals, theropods, and hadrosauromorphs or hadrosaurids. The association of the brackish water archaeogastropod *Vellatella* at nearby sites correlative with this locality implies that UMNH VP Loc. 811 represents an estuarine environment correlative to the lower marine tongue of the John Henry Member further seaward on the eastern side of the Kaiparowits Plateau. The lower marine tongue is within the *Protexanites bourgeoisi* biozone and is late Coniacian in age (Kennedy and Cobban, 1991). This age constraint indicates that this nodosaurid from the basal John Henry Member is comparable in age to *Niobrarasaurus coleii*, a nodosaurid from the Smoky Hill Chalk of Kansas (Carpenter, Dilkes, and Weishampel, 1995; Everhart and Hamm, 2005; Carpenter and Everhart, 2007). Further excavation at this interval within the John Henry Member is ongoing as of the summer of 2011.

Scraps of a possible ankylosaurid were discovered in rocks that correlate with the John Henry Member of the Straight Cliffs Formation in the Iron Springs Formation on the west side of the Pine Valley Mountains. This specimen is further discussed in the Iron Springs Formation section.

A site on the Markagunt Plateau has also produced an ankylosaurian tooth in rocks that are chronostratigraphically equivalent to the Drip Tank Member of the Straight Cliffs Formation. This specimen is discussed in the Grand Castle Formation section.

Ankylosaurs of the Iron Springs Formation (Equivalent to the John Henry Member of the Straight Cliffs Formation)

A fragmentary possible ankylosaurid was discovered by Jim Kirkland and Jeff Eaton in the Iron Springs Formation on the west side of the Pine Valley Mountains (MNA Loc. 1229) in 1990. This specimen (MNA v8044) was encrusted by layers of iron carbonate and is poorly preserved. Identifiable materials are limited to thin osteoderms and vertebral centra. The strata are from the middle to upper parts of the Iron Springs Formation, which are correlative to the John Henry Member based on their fossil mammalian fauna (Eaton, 1999).

Ankylosaurs of the Grand Castle Formation (Equivalent to the Drip Tank Member of the Straight Cliffs Formation or possibly lower Wahweap Formation)

An ankylosaurian tooth was collected from the top of the Cretaceous section in Cedar Canyon (UMNH VP Loc. 11) at the western end of the Markagunt Plateau (Eaton, Diem et al., 1999). UMNH VP Loc. 11 is within strata that have been correlated with the Grand Castle Formation to the north (Biek, Maldonado, et al., 2010). Although the Grand Castle Formation has been previously mapped as Paleocene, it is now recognized as Cretaceous on the basis of physical correlation, the presence of dinosaur tracks (Biek, Maldonado, et al., 2010) and pollen (Nichols, 1997). The precise age of this section is subject to debate (Eaton, 2006; Biek, Maldonado, et al., 2010), but it is likely equivalent to the Drip Tank Member of the Straight Cliffs Formation or lower portion of the Wahweap Formation. We here consider that UMNH VP Loc. 11 is equivalent to the Drip Tank Member of the Straight Cliffs Formation or the unconformity separating it from the Wahweap Formation (Jinnah, this volume, Chapter 4; Titus, Roberts, and Albright, this volume, Chapter 2).

Ankylosaurs of the Wahweap Formation

The Wahweap Formation is middle to late Campanian in age (Jinnah et al., 2009, Jinnah, this volume, Chapter 4), dating from about 81 to 77 Ma assuming continuous deposition into the Kaiparowits Formation. However, most of the Wahweap Formation (over 180 m of section, from 50 m above the base to 40 m above the Middle–Upper Member contact) is constrained between 81 and 79 Ma (Titus, Roberts, and Albright, this volume, Chapter 2; Jinnah, this volume, Chapter 4).

Until recently, the only notable ankylosaur skeletal material was a partial cranium (from the nares to the orbits) tentatively identified as nodosaurid by Edwin H. Colbert (pers. comm. to J.I.K.). This specimen, surface collected from the Lower Sandstone Member of the Wahweap Formation in 1981, was last seen in 1983 at the MNA. Unfortunately, the current location of this specimen is unknown.

Excavations by UMNH and UGS crews from the Middle Member of the Wahweap Formation over the past decade have yielded numerous osteoderms (UMNH VP 13981, 15664, 16408, and 21207; Fig. 18.2) tentatively identified as nodosaurid.

Parrish (1999) reported nodosaurid (OMNH 21280 in part, OMNH 21992, OMNH and 24278) and ankylosaurid (OMNH 21280 part, OMNH 21858, and OMNH 24276) teeth from the Upper Sandstone Member of the Wahweap Formation. Recent excavations by GSENM crews have uncovered a likely nodosaurid in the Upper Sandstone Member of the Wahweap Formation that includes a skull and postcranial elements. This specimen is currently under excavation.

Ankylosaurs of the Kaiparowits Formation

The Kaiparowits Formation is represented on the Kaiparowits Plateau by three informal units: the lower, middle, and upper units (Roberts, Deino, and Chan, 2005). The Kaiparowits Formation is by far the best temporally constrained of the Late Cretaceous formations of the Grand Staircase, with over 13 recognized ash horizons, of which six have been dated, constraining the age of the formation to between 76.5 to 74.7 Ma. Additionally, macrovertebrate specimens produced by the Kaiparowits Basin Project have greatly increased the ankylosaur record.

The Kaiparowits Formation has previously yielded ankylosaur teeth (Eaton, Cifelli, et al., 1999; Parrish, 1999). Both Eaton, Cifelli, et al. (1999) and Parrish (1999) reported nodosaurid teeth (OMNH 21190, OMNH 24471, UCMP 83240 and UCMP 83263). Kaiparowits ankylosaurid teeth reported by Parrish (1999) include OMNH 21118, OMNH 24112; UCMP 8319,

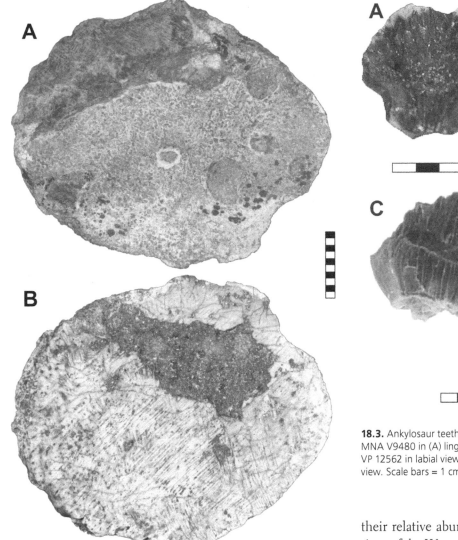

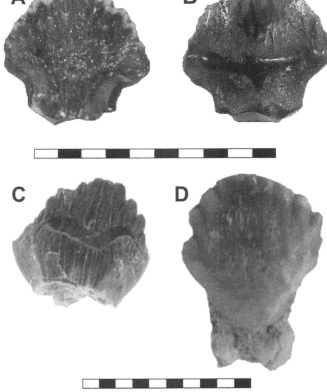

18.3. Ankylosaur teeth from the Kaiparowits Formation. Ankylosaurid tooth MNA V9480 in (A) lingual, and (B) labial views. (C) Nodosaurid tooth UMNH VP 12562 in labial view. (D) Nodosaurid tooth UMNH VP 16221 in labial view. Scale bars = 1 cm. Occlusal surface is up.

18.2. UMNH VP 21207, nodosaurid osteoderm from the Wahweap Formation in (A) superficial, and (B) deep views. Scale bar = 1 cm.

UCMP 8330, UCMP 8659 part, UCMP 8667 part and UCMP 83240 part. Eaton, Cifelli, et al. (1999) reference Parrish (1999) in their report of both nodosaurid and ankylosaurid material from the Kaiparowits Formation and tentatively assign the ankylosaurid material to *Euoplocephalus* sp. These specimens are from the middle unit of the Kaiparowits Formation based on locality information and other specimens in both collections.

The Kaiparowits Basin Project field efforts have produced the first associated ankylosaur specimens from the Kaiparowits Formation from a total of seven localities. Ankylosaur remains represent less than 5% of associated dinosaur sites (*n* = 32) in the Kaiparowits Formation and thus form a relatively small component of the Kaiparowits fauna. In contrast to

their relative abundance in contemporary northern formations of the Western Interior Basin, isolated ankylosaur teeth and ossicles are rarely encountered in microvertebrate sites. The following specimens were all recovered from the middle unit of the Kaiparowits Formation.

MNA V9480 is a small ankylosaurian tooth collected by MNA crews from the Kaiparowits Formation (Fig. 18.3A, B). Crews from the UMNH also recovered two isolated ankylosaurian teeth of nodosaurid affinity (UMNH VP 12562 and 16221; Fig. 18.3C, D).

Specimen UMNH VP 16940 from UMNH VP Loc. 904, is a single distal cervical nodosaurid spine nearly 30 cm long with a concave base (Fig. 18.4A). Another possible nodosaurid specimen is represented by UMNH VP 12264 (Fig. 18.4B–F), collected from UMNH VP Loc. 430 on Fossil Ridge. This specimen is an associated assemblage of more than 60 osteoderms ranging in size from one to 26 cm. Most of the osteoderms are keeled and some have an apex (Fig. 18.4). The largest osteoderm (Fig. 18.4C) forms a 26-cm spike with an excavated base. Another osteoderm from UMNH VP 12264 is flat, with a linear, sutural surface interpreted as being a

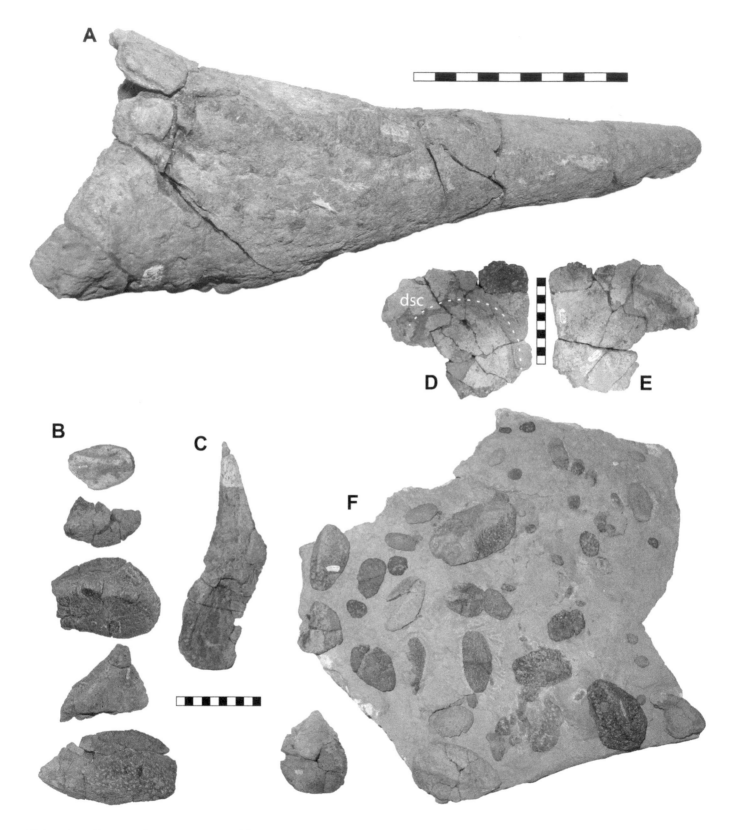

18.4. Nodosaurid material from the middle unit of the Kaiparowits Formation. (A) Nodosaurid distal cervical spine (UMNH VP 16940) superficial is to the right. (B–F) Possible nodosaurid associated specimen UMNH VP 12264 including (B) associated ossicles (in superficial view) found associated with block shown in (F); (C) lateral spine in oblique view (superficial toward top); (D, E) cervical osteoderm in (D) superficial and (E) deep views; (F) sandstone slab with associated in situ osteoderms and ossicles (in superficial view). dsc, dermal element scar. Scale bars = 10 cm.

Loewen et al.

18.5. Ankylosaurid dentary UMNH VP 12562, part of a disarticulated collection of cranial material from the middle unit of the Kaiparowits Formation in (A) medial oblique, and (B) lateral oblique views. mo, mandibular osteoderm. Scale bar = 1 cm.

sagittal midline suture, possibly from a cervical half ring. It has a circular impression of another dermal ossification on its superficial surface, possibly the attachment point for a cervical spine or some other osteoderm (Fig. 18.4D).

UMNH VP 12562 is a disarticulated but associated partial skull from UMNH VP Loc. 368 near Canaan Peak in the upper part of the middle unit of the Kaiparowits Formation. The most complete element is a partial left mandible. It has a faceted apical osteoderm on the ventrolateral surface in the position of the angular (Fig. 18.5). Some of the other elements are tentatively identified as cranial elements and osteoderms.

UMNH VP 12675, from UMNH VP Loc. 314 in The Blues section of the Kaiparowits, represents the stratigraphically highest associated specimen. This site is from the upper part of the middle unit of the Kaiparowits Formation. It consists of multiple associated small osteoderms (Fig. 18.6).

UMNH VP 19473 represents the second most complete specimen (Fig. 18.7) from the Kaiparowits Formation, and was recovered from UMNH VP Loc. 1004 on Horse Mountain. UMNH VP 19473 consists of a disarticulated but virtually complete right forelimb (fused scapulocoracoid, humerus, radius, ulna, four metacarpals, one manual phalanx and an ungual), a left scapula and coracoid, one cervical and three caudal vertebrae, a complete cervical half ring, and more than 50 associated osteoderms. This specimen comes from a bonebed, together with a partial centrosaurine ceratopsid skull and a complete *Denazinamys* turtle plastron and carapace.

The left scapula of UMNH VP 19473 has a rounded and expanded distal end. The acromion process is on the lateral surface restricted to the anterior edge. The humerus has a relatively short deltopectoral crest with a crest length that is 50% that of total humeral length. This is in contrast to UMNH VP 20202 which has a crest length of 65%. The manual ungual is rounded in dorsal view. The cervical half

ring consists of apical medial elements and two lateral elements, each with well-developed horizontal keels. Most of the osteoderms of UMNH VP 19473 are basally excavated, with larger osteoderms having greater excavations (Fig. 18.7). The morphology of the osteoderms includes circular ossicles with apices, 4–10-cm ovoid osteoderms with longitudinal keels, 10-cm circular apical osteoderms, 25-cm-long elongate osteoderms with a keel asymmetrically placed along the edge of the osteoderm, and 30-cm-long plate-like osteoderms with a low central keel (Fig. 18.7).

UMNH VP 19472 (Fig. 18.8) from UMNH VP Loc. 981 on Horse Mountain consists of ribs, vertebrae, osteoderms and a partial tail club including the handle and associated major osteoderms. Preserved vertebrae include two dorsal centra, parts of the sacrum, four proximal caudal vertebrae and six distal caudal vertebrae. There are at least eight dorsal ribs including three that are interpreted as anterior dorsal ribs as a result of a lack of overall curvature and expression of a dorsolateral shelf. There are three proximal chevrons, all of which are complete and fused to the base of the centra. The four disarticulated proximal caudal vertebrae have horizontal, anterolaterally directed transverse processes (similar to UMNH VP 19473) and a posteriorly offset neural spine.

The handle of UMNH VP 19472 consists of three articulated caudal centra surrounded by ossified tendons and a fused section of caudal vertebrae (representing parts of three caudal vertebrae) also surrounded by ossified tendons. The most proximal caudal vertebrae in the handle have elongated neural spines that extend at least half the anteroposterior length of the successive caudal vertebrae. Ventrally, there is a series of six sequential, fused chevrons that overlap each other. These chevrons form a nested pattern when viewed ventrally with the point of the chevron facing posteriorly. Whereas the entire series of fused chevrons is co-ossified,

18.6. Ankylosaurid specimen UMNH VP 12675 consisting of associated osteoderms from the middle unit of the Kaiparowits Formation, in superficial view. Scale bar = 10 cm.

the sutures between successive chevrons are visible as faint outlines both anterolaterally and medially. There is a laterally positioned triangular osteoderm associated with the three more distal fused caudals in the handle. The length of this lateral osteoderm is subequal to the cross-sectional width of the tail including caudal centra and complement of ossified tendons at the position of attachment for the lateral osteoderm. Distal to this triangular lateral osteoderm is a pair of smaller conical osteoderms positioned on either side of the chevrons lateral to the midline and ventral to the triangular lateral osteoderm. Proximally, at the position of the second most proximal caudal of the handle, is a pair of conical osteoderms corresponding to the distal osteoderms (Fig. 18.8F, G). The pattern of lateral osteoderms on the handle is interpreted as an array of two conical osteoderms positioned distally on either side of the chevrons, and two larger, flat triangular osteoderms positioned laterally. This pattern may have been expressed at the position of every other caudal centrum. UMNH VP 19472 also includes one major osteoderm

(terminology sensu Arbour, 2009) pertaining to the tail club (Fig. 18.8H). This osteoderm has a dorsoventral thickness 40% of its anteroposterior length.

By far the most complete ankylosaurid specimen from the Kaiparowits Formation, UMNH VP 20202, was excavated during the 2008, 2009, and 2010 field seasons from UMNH VP Loc. 1222 on Horse Mountain, known as the Horse Mountain Gryposaur Quarry. The Horse Mountain Gryposaur Quarry is a multitaxic bonebed that has also produced a mostly complete, partially articulated hadrosaurine (*Gryposaurus* sp.) with skin impressions, an articulated alligatoroid, possible small theropod material, and two articulated turtles. UMNH VP 20202 (Fig. 18.9) is represented by a complete skull, both dentaries, both first and second cervical half rings, a cervical vertebra, four dorsal vertebrae (two with co-ossified ribs), six disarticulated dorsal ribs, a nearly complete sacrum with nine fused vertebrae, two proximal caudal vertebrae, five midcaudal vertebrae, a handle region of the distal tail in two sections, a complete tail club knob, the right coracoid,

Loewen et al.

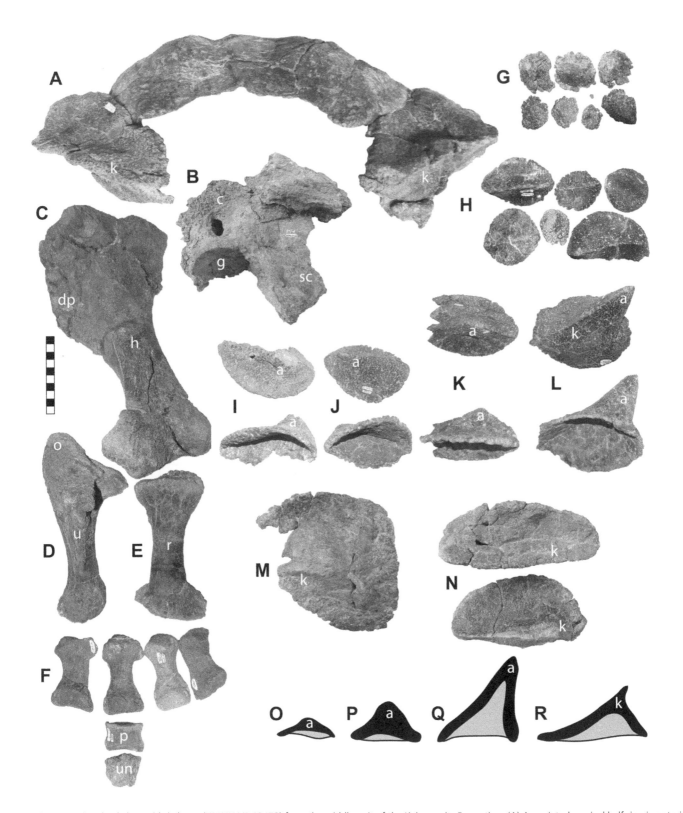

18.7. Associated ankylosaurid skeleton (UMNH VP 19473) from the middle unit of the Kaiparowits Formation. (A) Associated cervical half ring in anterior view. (B) Glenoid region of the left scapula and coracoid in oblique lateral view. (C) Right humerus in posterior view. (D) Right ulna in lateral view. (E) Right radius in anterior view. (F) Partial right manus in dorsal view. The complete left scapula is depicted in Figure 18.11. (G–N) Osteoderms of UMNH VP 19473; (G) small circular osteoderms in superficial view; (H) medium sized osteoderms with keels and apices in superficial view; (I–L) large osteoderms in superficial view (above) and deep view (below) to indicate extent of excavation along the deep surface; (M, N) Larger osteoderms (depicted in superficial view) include two distinct morphologies; (M) flat osteoderms with a laterally directed median keel; and (N) flattened, elongate osteoderms with longitudinal keels along the margin of the elements. (O–R) Cross-sectional diagrams depicting varying extent of osteoderm basal excavation; (O, P) shallower excavation in smaller osteoderms (similar to osteoderms shown in G and H); (Q) circular apical osteoderms (similar to I–L); (R) ovoid keeled osteoderms (similar to N). a, apex; c, coracoid; dp, deltopectoral crest; g, glenoid; h, humerus; k, keel; o, olecranon process; p, manual phalanx; r, radius; sc, scapula; u, ulna; un, manual ungual. Every image in this figure is scaled to the same size except for O–R, which are diagrammatic only and not to scale. Scale bars = 10 cm.

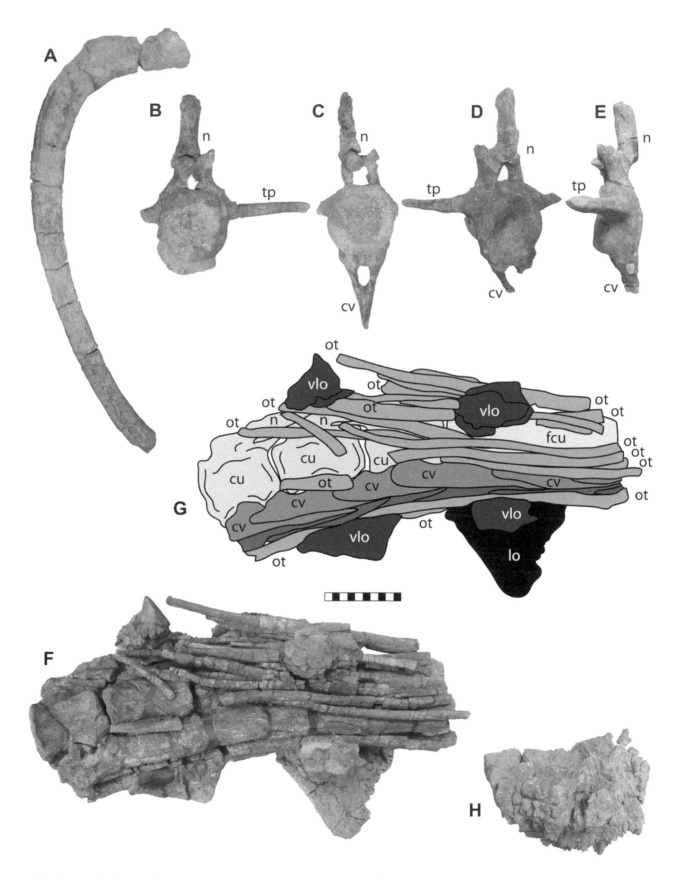

18.8. Associated ankylosaurid skeleton (UMNH VP 19472) from the middle unit of the Kaiparowits Formation. (A) Anterior dorsal rib. (B–E) Proximal caudal vertebrae in order from more proximal to more distal from left to right; (B) anterior view; (C) anterior view; (D) posterior view; (E) in left lateral view. (F) Tail club "handle" vertebrae in oblique ventral view. (G) Interpretive line drawing of the handle. (H) Major osteoderm of the tail club knob. cu, caudal vertebra; cv, chevron; fcu, fused caudal vertebrae; lo, lateral osteoderm; n, neural arch; ot, ossified tendon; tp, transverse process; vlo, ventrolateral osteoderm. Every image in this figure is scaled to the same size and scale bars = 10 cm.

both scapulae, the left humerus, the ilium, tibia and fibula, and more than 20 associated osteoderms. The transverse processes on the proximal caudals of UMNH VP 20202 project laterally from the centrum. The fused handle and knob are represented by a section of eight fused distal caudal vertebrae and a section of three distal caudal vertebrae both surrounded by ossified tendons. The posteriormost section of caudal vertebrae is articulated with a complete knob composed of two major osteoderms and a minor osteoderm. Materials of UMNH VP 20202 are currently under preparation and study.

A nearly complete ankylosaurid cranium was excavated in 2011 in the lower member of the Kaiparowits formation in Paradise Canyon. This specimen is distinct from UMNH VP 20202 and is currently under preparation and study.

Osteoderm Morphology and Histology of Kaiparowits Formation Ankylosaurs

As with most fossil material, many of the ankylosaur specimens from Grand Staircase–Escalante National Monument are fragmentary. This lack of complete specimens often renders confident taxonomic assignments difficult. Several recent studies have proposed a methodology for taxonomically distinguishing ankylosaur specimens as either ankylosaurid or nodosaurid on the basis of osteoderm histology. Scheyer and Sander (2004) were the first to systematically survey a broad sample of ankylosaur taxa and identify taxon-specific patterns in their osteoderm histology. More recently, Hayashi et al. (2010) examined the histology of modified osteoderms such as nodosaurid cervical spines and ankylosaurid tail club knob osteoderms.

To test for taxonomic affinities of fragmentary specimens several of the above ankylosaurian specimens from the Kaiparowits Formation were selected for histological analysis in combination with gross morphological analyses. Only one of the osteoderms examined histologically can be associated with a distinct position of the body: UMNH VP 16940 is most likely a distal cervical osteoderm of a nodosaurid on the basis of morphological similarities to the distal cervical spines of *Edmontonia rugosidens* (AMNH 5665; Carpenter, 1990) and *Sauropelta* (AMNH 3035; Carpenter, 1984). The rest of the osteoderms investigated histologically share a common superficial surface texture characterized by a projecting, uniformly distributed rugosity. The presence of sparse neurovascular grooves and/or foramina is variable among the osteoderms, possibly due to taxonomic differences. Their morphology is also variable. UMNH VP 12264, 12562, 12675, 19472, and 19473 each have one flat osteoderm. A keeled osteoderm was also sampled from each of UMNH VP 12264 and 12675. UMNH VP 19473 has a smaller, circular osteoderm with a posteriorly

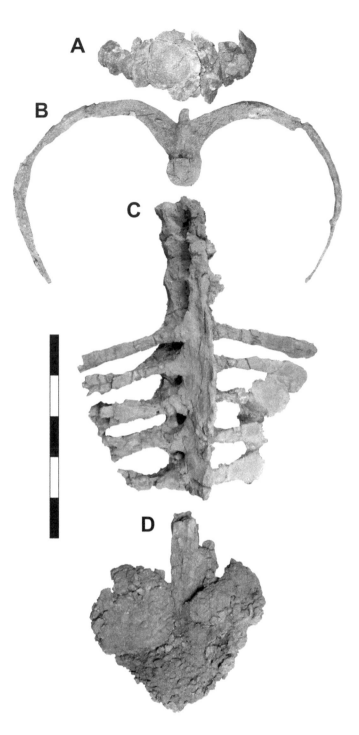

18.9. Associated ankylosaurid skeleton (UMNH VP 20202) from the lower middle unit of the Kaiparowits Formation. (A) Cervical half ring in dorsal view; (B) dorsal vertebra with co-ossified ribs in posterior view; (C) sacrum in dorsal view; (D) tail club in dorsal view. Scale bar = 50 cm.

offset apex. UMNH VP 12562 is also a circular osteoderm, although its apex is more central on the element.

A core sample from UMNH VP 16940 shows a thick layer of compact, Haversian bone grading into a central region of cancellous bone. This has also been observed in nodosaurid spine-like osteoderms (Hayashi et al., 2010). Although the superficial surface has eroded in places, a system of neurovascular grooves is evident on the dorsal surface (interpreted as the convex side of the spine). The ventral (flat) side of the spine, from which the core sample was taken, is smoother. This specimen is comparable to cervical spines described for nodosaurids on the basis of morphology (Carpenter, 1990) and histology (Hayashi et al., 2010).

A fragmented flat circular osteoderm and an oval osteoderm with an offset apex were sectioned from UMNH VP 12264. As a result of preservation, the superficial surface texture is only visible on the flat osteoderm. It is characterized by projecting, uniformly distributed rugosity and several neurovascular grooves and foramina. There is a higher concentration of neurovascular grooves on this specimen than in the other samples examined, indicating a possible nodosaurid affinity (Burns, 2008). However, the presence of superficial and deep cortices of roughly the same thickness is typical for ankylosaurid osteoderms (Scheyer and Sander, 2004; Burns, 2009).

UMNH VP 12675 includes two osteoderms, one keeled (Fig. 18.10A–C) and one flat. They share a common superficial surface texture (Fig. 18.10A), characterized by a uniform pattern of rugosity and a relatively fewer neurovascular grooves and foramina. Only the keeled osteoderm was thin sectioned (Fig. 18.10B, C). The superficial and deep cortices are relatively thin (12% and 13% total osteoderm thickness, respectively), relatively indistinct, and are composed of woven bone. These cortices grade into the secondarily remodeled core, which consists of compact bone of disorganized Haversian canals.

Two osteoderms from UMNH VP 19473 were thin sectioned: an incomplete flat osteoderm (Fig. 18.10G–I) and a small, complete circular osteoderm (Fig. 18.10J–L) with an offset apex. The surface texture (Fig. 18.10G, J) is uniform projecting rugosity, sparse neurovascular foramina, and no neurovascular grooves. The histology of the flat osteoderm (Fig. 18.10H, I) is similar to those of UMNH VP 19472 and 12675, although the cortices are either extremely thin or largely indistinguishable from the Haversian core. The histology of the circular osteoderm, however, is different (Fig. 18.10K, L). The entirety of the osteoderm is woven-fibered bone composed of a dense network of orthogonally oriented structural fibers. These fibres extend through the thin lamellar cortices (superficial 7% and deep 14 % total osteoderm

thickness) where they would have crossed the line of mineralization in vivo. A few primary and secondary osteons are scattered throughout the bone matrix. This histology is characteristic of an immature bone fabric (Vickaryous and Hall, 2008; see also Vickaryous and Sire, 2009) and as a result is not taxonomically informative (Burns, 2009). Consequently, at this time, we are unable to establish the taxonomic identity of UMNH VP 19473 on the basis of histology.

UMNH VP 19472 (Fig. 18.10D–F) is histologically similar to that of UMNH VP 12675 despite a difference in morphology (Fig. 18.10D). The superficial and deep cortices (Fig. 18.10E, F) are more distinct owing to more obvious lamellae, although they are still thin (4% and 6% respectively). The surface texture is characterized by projecting, uniformly distributed rugosity and an absence of neurovascular grooves and foramina.

DISCUSSION

Isolated Tooth Taxonomy

In general, Late Cretaceous ankylosaurid teeth are relatively smaller than those of nodosaurids, with the latter often developing prominent cingula (Coombs, 1990). Of the teeth recently collected from the Kaiparowits Formation, several relatively large teeth with well-developed cingula (Fig. 18.3) are tentatively assigned to the Nodosauridae. One tooth (MNA V9480) is tentatively assigned to the Ankylosauridae on the bases of size and lack of well-developed cingulum.

Overall Taxonomic Affinities of Postcranial Ankylosaur Materials from Southern Utah

UMNH VP 17760, from the John Henry Member of the Straight Cliffs Formation, is identified as a distal cervical spine of a nodosaurid on the basis of surface texture and overall morphology. This spine has a maximum length of 26 cm, a concave base on the deep surface and surface texturing consistent with those of nodosaurid osteoderms. Accordingly, UMNH VP 17760 is assigned to the Nodosauridae. UMNH VP 13981, 15664, 16408, and 21207 from the Middle Member of the Wahweap Formation are osteoderms with a characteristically nodosaurid pattern of surface sculpturing and are also assigned to the Nodosauridae.

18.10. Histological cross section of ankylosaurine dermal ossicles from the Kaiparowits Formation. (A–C) Keeled osteoderm from UMNH VP 12675; (A) gross morphology in superficial view; (B) thin section in plane polarized light; (C) thin section in cross-polarized light. (D–F) UMNH VP 19472; (D) gross morphology in superficial view; (E) thin section in plane polarized light; (F) thin section in cross-polarized light. (G–I) Flat osteoderm from UMNH VP 19473; (G) gross morphology in superficial view; (H) thin section in plane polarized light; (I) thin section in cross-polarized light. (J–L) Circular ossicle from UMNH VP 19473; (J) gross morphology in superficial view; (K) thin section in plane polarized light; (L) thin section in cross-polarized light. Dashed lines across whole specimens indicate the plane of thin sectioning photographed in plane (B, E, H, and K) and cross-polarized (C, F, I, and L) light (superficial is up). Scale bars for whole specimens (vertical) and thin sections (horizontal) = 1 cm.

Loewen et al.

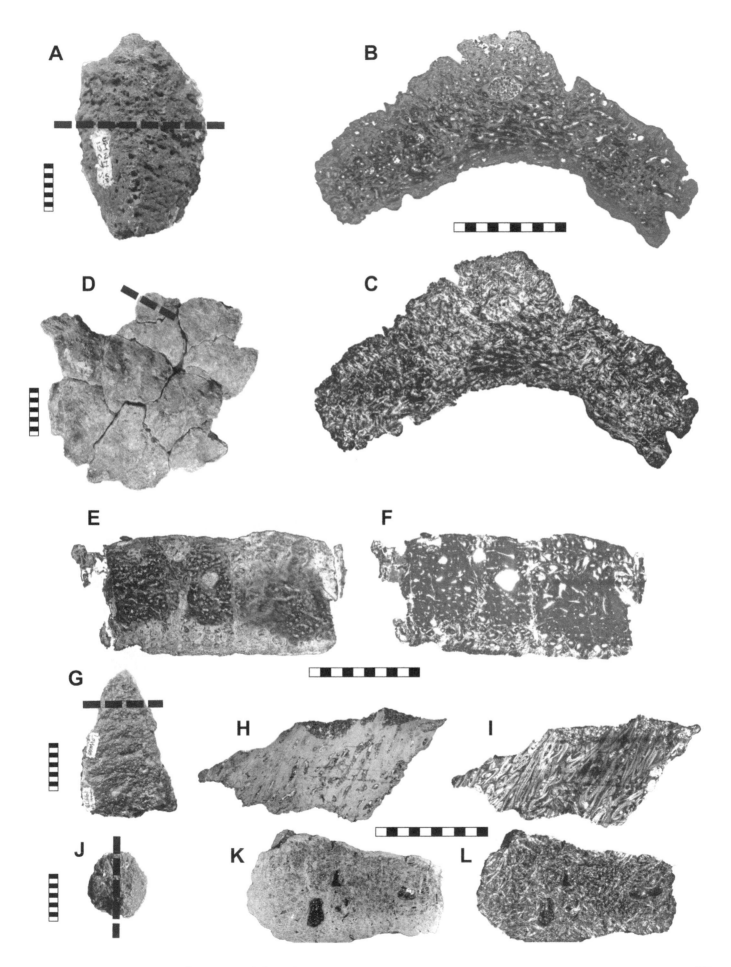

Table 18.1. Ankylosaur Specimens from Southern Utah

Formation	Member	Specimen	Specimen description	Affinity	Study[a]
Straight Cliffs	Smoky Hollow	OMNH 21711	Tooth	Nodosaurid	1
Straight Cliffs	Smoky Hollow	OMNH 24448	Tooth	Nodosaurid	1
Straight Cliffs	Smoky Hollow	OMNH 21671	Tooth	Nodosaurid	1
Straight Cliffs	Smoky Hollow	OMNH 23684	Tooth	Ankylosaurid	2
Straight Cliffs	Smoky Hollow	OMNH 21243	Tooth	Ankylosaurid	1
Straight Cliffs	Smoky Hollow	UMNH VP 17760	Distal cervical spine	Nodosaurid	3
Iron Springs[b]	Middle to upper unit	MNA V 8044	Vertebrae and osteoderms	Nodosaurid	3
Grand Castle[c]	Lower unit	UMNH VP Loc. 11	Tooth	Ankylosaurine indet.	3
Wahweap	Lower Sandstone	Lost uncataloged specimen	Skull	Nodosaurid	3
Wahweap	Middle	UMNH VP 13981	Osteoderm	Nodosaurid	3
Wahweap	Middle	UMNH VP 15664	Osteoderm	Nodosaurid	3
Wahweap	Middle	UMNH VP 16408	Osteoderm	Nodosaurid	3
Wahweap	Middle	UMNH VP 21207	Osteoderm	Nodosaurid	3
Wahweap	Upper Mudstone	OMNH 21280 part	Tooth	Nodosaurid	2
Wahweap	Upper Mudstone	OMNH 21992	Tooth	Nodosaurid	2
Wahweap	Upper Mudstone	OMNH 24278	Tooth	Nodosaurid	2
Wahweap	Upper Mudstone	OMNH 21280 part	Tooth	Ankylosaurid	2
Wahweap	Upper Mudstone	OMNH 21858	Tooth	Ankylosaurid	2
Wahweap	Upper Mudstone	OMNH 24276	Tooth	Ankylosaurid	2
Kaiparowits	Middle unit	OMNH 21190	Tooth	Nodosaurid	4
Kaiparowits	Middle unit	OMNH 24471	Tooth	Nodosaurid	4
Kaiparowits	Middle unit	UCMP 83240	Tooth	Nodosaurid	4
Kaiparowits	Middle unit	UCMP 83263	Tooth	Nodosaurid	4
Kaiparowits	Middle unit	UMNH A 16221	Tooth	Nodosaurid	3
Kaiparowits	Middle unit	UMNH VP 12562	Tooth	Nodosaurid	3
Kaiparowits	Middle unit	UMNH VP 16940	Distal cervical spine	Nodosaurid	3
Kaiparowits	Middle unit	UMNH VP 12264	Associated and articulated osteoderms	Nodosaurid	3
Kaiparowits	Middle unit	OMNH 21118	Tooth	Ankylosaurid	4
Kaiparowits	Middle unit	OMNH 24112	Tooth	Ankylosaurid	4
Kaiparowits	Middle unit	OMNH 8319	Tooth	Ankylosaurid	4
Kaiparowits	Middle unit	MNA V 9480	Tooth	Ankylosaurid	3
Kaiparowits	Middle unit	UCMP 8330	Tooth	Ankylosaurid	4
Kaiparowits	Middle unit	UCMP 8659 part	Tooth	Ankylosaurid	4
Kaiparowits	Middle unit	UCMP 8667 part	Tooth	Ankylosaurid	4
Kaiparowits	Middle unit	UCMP 83240 part	Tooth	Ankylosaurid	4
Kaiparowits	Middle unit	UMNH VP 12562	Associated skull	Ankylosaurid	3
Kaiparowits	Middle unit	UMNH VP 12657	Associated osteoderms	Ankylosaurid	3
Kaiparowits	Middle unit	UMNH VP 19472	Associated specimen	Ankylosaurid taxon A	3
Kaiparowits	Middle unit	UMNH VP 19473	Associated specimen	Ankylosaurid taxon A	3
Kaiparowits	Middle unit	UMNH VP 20202	Associated specimen	Ankylosaurid taxon B	3

[a] Study: 1–Eaton, Diem, et al. (1999); 2–Parrish (1999); 3–this chapter; 4–Easton, Parrish et al., (1999).

[b] Equivalent to the John Henry Member of the Straight Cliffs Formation.

[c] Likely equivalent to the Drip Tank Member of the Straight Cliffs Formation.

Although a genus-level taxonomic identification of Kaiparowits ankylosaur specimens is not possible at this time, most can be assigned to either Nodosauridae or Ankylosauridae (Table 18.1). UMNH VP 16940 is most likely from a nodosaurid on the basis of bone histology, surface texture and overall morphology. This type of distal cervical spine occurs only in the Nodosauridae. In addition, this confirms the histology for nodosaurid spines, of which only one has been thin sectioned before to this study by Hayashi et al. (2010). UMNH VP 16940 is here assigned to the Nodosauridae.

The osteoderms from UMNH VP 12264 show taxonomic mixed signals, with the surface texture being consistent with nodosauridids and the histology less definitive. One osteoderm sectioned shows the structural fiber pattern characteristic of nodosaurids whereas the other is more like the structural fiber pattern characteristic of ankylosaurids. The

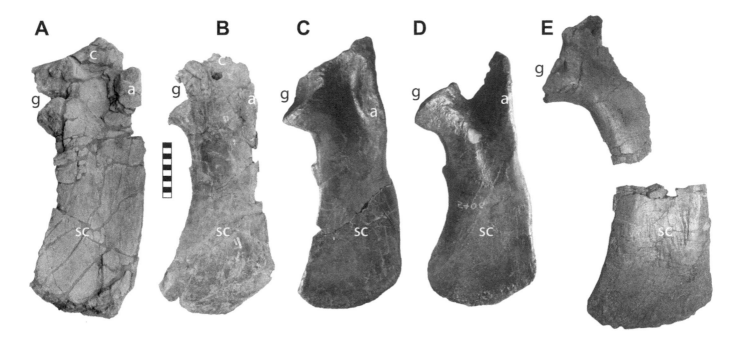

18.11. Comparison of two distinct ankylosaurid scapular morphologies from the middle unit of the Kaiparowits Formation to those of *Euoplocephalus.* (A) right scapula of UMNH VP 20202 in lateral view (photoreversed for comparison). (B) Left scapula of UMNH VP 19473 in lateral view. (C) Left scapula of *Euoplocephalus* (AMNH 5406) in lateral view. (D) Right scapula of *Euoplocephalus* (AMNH 5406) in lateral view (photoreversed for comparison). (E) Right scapula of *Euoplocephalus* (ROM 1930) in medial view. a, acromion process; c, coracoid; g, glenoid; sc, scapula. Images C–E by Victoria Arbour. Scale bar = 10 cm.

spine-like osteoderm (Fig. 18.4C) is similar in surface texture and overall morphology to nodosaurid cervical spines, and the flat osteoderm (Fig. 18.4D) is similar to the median cervical osteoderms of nodosaurids that are unfused at the midline. Because the preponderance of evidence leans toward nodosauridae, UMNH VP 12264 is here provisionally assigned to the Nodosauridae.

The partial mandible of UMNH VP 12562 is comparable with ankylosaurids on the basis of the protruding osteoderm (it is generally flatter in nodosaurids sensu Coombs, 1978); because of the fragmentary nature of the specimen, it is tentatively assigned to the Ankylosauridae. UMNH VP 12675 is assigned to the Ankylosauridae on the basis of osteoderm histology.

Materials of UMNH VP 19473 are histologically and morphologically consistent with Ankylosauridae. This specimen differs from *Dyoplosaurus* in the shape of the unguals. Those of *Dyoplosaurus* are characteristically triangular in dorsal view, not rounded as in *Euoplocephalus* and UMNH VP 19473. Interestingly, the superficial surface texture of the osteoderms, which consists of a uniform projecting rugosity, sparse neurovascular foramina, and no neurovascular grooves, is most similar to an ankylosaurid specimen (SMP VP-1930) from the Kirtland Formation of New Mexico (Burns, 2008).

UMNH VP 19472 is histologically and morphologically consistent with ankylosaurid ankylosaurs. Similar to

Euoplocephalus and *Dyoplosaurus*, the free caudal centra are circular in axial view. The transverse processes on the free caudals of UMNH VP 19472 project anterolaterally from the centrum, distinguishing it from *Nodocephalosaurus* in which the transverse processes are more laterally projecting (Arbour, Burns, and Sissons, 2009). The rugose superficial surface texture of the major osteoderm of the tail club differs from that of *Nodocephalosaurus*, which is comparatively smooth (Burns and Sullivan, in press). *Dyoplosaurus* also has relatively smooth knob osteoderms (Arbour, Burns, and Sissons, 2009). In terms of osteoderm surface texture, UMNH VP 19472 most closely resembles that of *Euoplocephalus*, with the numerous randomly distributed globular structures found on the osteoderms of UMNH VP 19472 also occurring on the osteoderms of cf. *Euoplocephalus* (i.e., CMN 2252 and ROM 788; Burns and Sullivan, in press). The pattern of lateral osteoderms on the handle of UMNH VP 19472, with lateral triangular osteoderms that are present on the fused proximal handle, is distinct from the pattern preserved in *Dyoplosaurus* (Arbour, Burns, and Sissons, 2009) but is similar to cf. *Euoplocephalus* (e.g., AMNH 5245).

UMNH VP 20202 is referred to the Ankylosauridae on the basis of diagnostic characters including a raised acromion on the dorsal edge of the scapula and the presence of a tail club. It is distinguished from *Dyoplosaurus* on the basis of the sacrum. The transverse processes on the third sacral,

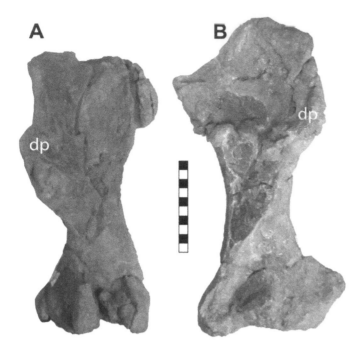

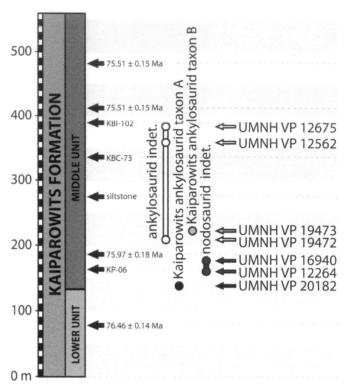

18.12. Comparison of the two distinct ankylosaurid humeral morphologies from the middle unit of the Kaiparowits Formation. (A) UMNH VP 20202, anterior view. (B) UMNH VP 19473, anterior view. dp, deltopectoral crest. Scale bar = 10 cm.

18.13. Stratigraphic position of major ankylosaur specimens from the Kaiparowits Formation.

autapomorphically, project anterolaterally in *Dyoplosaurus* (Arbour, Burns, and Sissons, 2009), not laterally as in *Euoplocephalus* and UMNH VP 20202. The sacrum is not known in *Nodocephalosaurus*. Although not yet fully prepared, observations made in the field about the cervical half ring, tail club, and skull suggest this specimen is an ankylosaurid and distinct in morphology from the contemporary genus *Euoplocephalus*.

Finally, UMNH VP 20202 can be distinguished from UMNH VP 19473 on the basis of scapular and humeral morphology (Figs. 18.11, 18.12). The scapular blade of UMNH VP 20202 has subparallel dorsal and ventral edges of the blade without the rounded distal expansion present in UMNH VP 19473. This distal expansion in UMNH VP 19473 is similar to, but more pronounced than, the morphology present in *Euoplocephalus* (AMNH 5406, ROM 1930). Both UMNH VP 19473 and 20202 have an ankylosaurid-like acromion process restricted to the dorsal edge of the element. The humerus of UMNH VP 20202 is short and robust with a relatively long deltopectoral crest compared to UMNH VP 19473. Further comparison of the tail club of UMNH VP 19472 and UMNH VP 20202 is ongoing; however, it is presently clear that there are two distinct ankylosaurid morphologies present in Kaiparowits Formation, most likely pertaining to two distinct ankylosaurid taxa. Furthermore, the two distinct ankylosaurid morphologies both occur in the lower part of the middle unit of the Kaiparowits

Formation (Fig. 18.13), geographically separated by less than 3 km.

Both UMNH VP 19743 and 20202 are recognized here as distinct from each other and from *Dyoplosaurus* and possibly *Euoplocephalus* to the north, suggesting that at 76 Ma there were distinct ankylosaurids in the north and south on Laramidia. The possible endemism of Kaiparowits ankylosaurids is consistent to that of many other groups of dinosaurs (i.e., tyrannosaurids, maniraptorans, hadrosaurines, lambeosaurines, chasmosaurines, and centrosaurines) from the Kaiparowits Formation that have been shown to be endemic rather than cosmopolitan (Gates et al., 2010). Further preparation, description and comparison of Kaiparowits ankylosaurs is ongoing and will provide further opportunities to test the hypothesis of late Campanian endemism in Laramidia.

CONCLUSION

This study illustrates for the first time definitive postcranial nodosaurid material from the Campanian Straight Cliffs, Wahweap, and Kaiparowits formations. Diagnostic ankylosaurid material is now known from multiple specimens from the middle unit of the Kaiparowits Formation. Of these, two essentially coeval ankylosaurid specimens from the middle unit of the Kaiparowits Formation are distinct from each

Loewen et al.

other. They are both distinguishable from the coeval *Nodocephalosaurus* and *Dyoplosaurus*. These specimens may also be distinguishable from *Euoplocephalus* (including cf. *Euoplocephalus* specimens). Additional anatomical data from UMNH 20202 will help to clarify this proposed distinction.

ACKNOWLEDGMENTS

The authors wish to thank Grand Staircase–Escalante National Monument and the Bureau of Land Management (BLM) for their continued support throughout the duration of this project. All specimens were collected under permits from the BLM and the U.S. Forest Service. Special thanks go to A. Titus (Grand Staircase–Escalante National Monument) for his ongoing efforts both in and out of the field, and to M. Eaton and S. Foss of the BLM for ongoing support of this project. Without dozens of dedicated students, faculty, and museum volunteers from the UMNH, the University of Utah, Utah Friends of Paleontology, the Web Schools, and collaborating institutions, this project could never have achieved the success we document here. Additional thanks go to E. Lund, J. Gentry, D. DeBlieux, J. Eaton, R. Irmis, and E. Roberts for their substantial contributions in the field and in the lab. Histological analyses were performed at the University of Alberta. P. Currie commented on an early version of this contribution. Reviews by V. Arbour and K. Carpenter greatly improved the chapter. The use of sectioning equipment was provided by R. Stockey. I. Jakab and the Earth and Atmospheric Sciences Digital Imaging Facility provided a slide scanner, and M. Caldwell provided use of a petrographic microscope. J. Gillette is thanked for tracking down information on the MNA specimens. Funding was provided by BLM assistance agreement JSA071004, Grand Staircase–Escalante National Monument, UMNH, UGS, and the National Science Foundation (NSF 0745454, Kaiparowits Basin Project).

REFERENCES CITED

Arbour, V. M. 2009. Estimating forces of tail club strikes by ankylosaurid dinosaurs. PLoS One 4(8):e6738.

Arbour, V. M. 2010. A Cretaceous armory: multiple ankylosaurid taxa in the Late Cretaceous of Alberta, Canada and Montana, U.S.A. Journal of Vertebrate Paleontology 30(3, Supplement):55A.

Arbour, V. M., M. E. Burns, and R. L. Sissons. 2009. A redescription of the ankylosaurid dinosaur *Dyoplosaurus acutosquameus* Parks, 1924 (Ornithischia: Ankylosauria) and a revision of the genus. Journal of Vertebrate Paleontology 29:1117–1135.

Biek, R. F., G. C. Willis, M. D. Hylland, and H. H. Doelling. 2010. Geology of Zion National Park; pp. 109–143 in D. A. Sprinkel, P. B. Anderson, and T. C. Chidsey, Jr. (eds.), Geology of Utah's Parks and Monuments (3rd edition). Utah Geological Association Publication 28.

Biek, R. F., P. D. Rowley, J. M. Hayden, D. B. Hacker, G. C. Willis, L. F. Hintze, R. E. Anderson, and K. D. Brown. 2009. Geologic map of the St. George and east part of the Clover Mountains 30'×60' quadrangles, Washington and Iron Counties, Utah. Utah Geological Survey Map 242.

Biek, R. F., F. Maldonado, D. W. Moore, J. J. Anderson, P. D. Rowley, V. S. Williams, D. Nealey, and E. G. Sable. 2010. Interim Geologic Map of the West Part of the Panguitch 30'×60' Quadrangle, Garfield, Iron, and Kane Counties, Utah–Year 2 Progress Report. Utah Geological Survey Open-File Report 277.

Burns, M. E. 2008. Taxonomic utility of ankylosaur (Dinosauria, Ornithischia) osteoderms: *Glyptodontopelta mimus* Ford, 2000: a test case. Journal of Vertebrate Paleontology 28:1102–1109.

Burns, M. E. 2009. Morphology, histology, and structural mechanics of ankylosaur osteoderms: implications for function and systematics. Journal of Vertebrate Paleontology 28:73A.

Burns, M. E., and R. M. Sullivan. In press. The tail club of *Nodocephalosaurus kirtlandensis* (Dinosauria: Ankylosauridae), with a review of ankylosaurid tail club morphology and homology. New Mexico Museum of Natural History and Science Bulletin.

Carpenter, K. 1984. Skeletal reconstruction and life restoration of *Sauropelta* (Ankylosauria: Nodosauridae) from the Cretaceous of North America. Canadian Journal of Earth Sciences 21:1491–1498.

Carpenter, K. 1990. Ankylosaur systematics: example using *Panoplosaurus* and *Edmontonia* (Ankylosauria: Nodosauridae); pp. 281–298 in K. Carpenter and P. J. Currie (eds.), *Dinosaur Systematics: Approaches and Perspectives.* Cambridge University Press, Cambridge.

Carpenter, K., and M. J. Everhart. 2007. Skull of the ankylosaur *Niobrarasaurus coleii* (Ankylosauria: Nodosauridae) from the Smoky Hill Chalk (Coniacian) of western Kansas. Kansas Academy of Science, Transactions 110:1–9.

Carpenter, K., D. Dilkes, and D. B. Weishampel. 1995. The dinosaurs of the Niobrara Chalk Formation (Upper Cretaceous, Kansas). Journal of Vertebrate Paleontology 15:275–297.

Coombs, W. P., Jr. 1978. The families of the Ornithischian dinosaur order Ankylosauria. Palaeontology 21:143–170.

Coombs, W. P., Jr. 1990. Teeth and taxonomy in ankylosaurs; pp. 269–279 in K. Carpenter and P. J. Currie (eds.), Dinosaur Systematics: Approaches and Perspectives. Cambridge University Press, Cambridge.

Eaton, J. G. 1999. Vertebrate Paleontology of the Iron Springs Formation, Upper Cretaceous, Southwestern Utah; pp. 339–344 in D. D. Gillette (ed.), Vertebrate Paleontology in Utah. Utah Geological Survey Miscellaneous Publication 99–1.

Eaton, J. G. 2006. Late Cretaceous mammals from Cedar Canyon, southwestern Utah; pp. 373–402 in S. G. Lucas and R. M. Sullivan (eds.), Late Cretaceous Vertebrates from the Western Interior. New Mexico Museum of Natural History and Science Bulletin 35.

Eaton, J. G., R. L. Cifelli, J. H. Hutchison, J. I. Kirkland, and J. M. Parrish. 1999. Cretaceous vertebrate faunas from the Kaiparowits Plateau, south-central Utah; pp. 345–353 in D. D. Gillette (ed.), Vertebrate Paleontology in Utah. Utah Geological Survey Miscellaneous Publication 99–1.

Eaton, J. G., S. Diem, J. D. Archibald, C. Schierup, and H. Munk. 1999. Vertebrate Paleontology of the Upper Cretaceous Rocks of the Markagunt Plateau, Southwestern Utah; pp. 323–334 in D. D. Gillette (ed.), Vertebrate Paleontology in Utah. Utah Geological Survey Miscellaneous Publication 99–1.

Everhart, M. J., and S. A. Hamm. 2005. A new nodosaur specimen (Dinosauria: Nodosauridae) from the Smoky Hill Chalk (Upper Cretaceous) of western Kansas. Kansas Academy of Science, Transactions 108:15–21.

Ford, T. L. 2000. A review of ankylosaur osteoderms from New Mexico and a preliminary review of ankylosaur armor; pp. 157–176 in S. G. Lucas and A. B. Heckert (eds.), Dinosaurs of New Mexico. New Mexico Museum of Natural History and Science Bulletin 17.

Ford, T. L., and J. I. Kirkland. 2001. Carlsbad ankylosaur (Ornithischia, Ankylosauria): an ankylosaurid and not a nodosaurid; pp. 239–260 in K. Carpenter (ed.), The Armored Dinosaurs. Indiana University Press, Bloomington, Indiana.

Gates, T. A., S. D. Sampson, L. E. Zanno, E. M. Roberts, J. G. Eaton, R. L. Nydam, J. H.

Hutchison, J. A. Smith, M. A. Loewen, and M. A. Getty. 2010. The fauna of the Late Cretaceous Kaiparowits Formation: testing previous ideas about late Campanian vertebrate biogeography in the Western Interior Basin. Paleogeography, Paleoclimatology, and Paleoecology 291:371–387.

Getty, M. A., M. Vickaryous, and M. A. Loewen. 2009. Ankylosaurid dinosaurs of the Upper Campanian Kaiparowits Formation, Grand Staircase–Escalante National Monument. Journal of Vertebrate Paleontology, SVP Program and Abstracts Book, 2009:104.

Hayashi, S., K. Carpenter, T. M. Scheyer, M. Watabe, and D. Suzuki. 2010. Function and evolution of ankylosaur dermal armor. Acta Palaeontologica Polonica 55:213–228.

Hieronymus, T. L., L. M. Witmer, D. H. Tanke, and P. J. Currie. 2009. The facial integument of centrosaurine ceratopsids: morphological and histological correlates of novel skin structures. Anatomical Record 292:1370–1396.

Jinnah, Z. A., E. M. Roberts, A. L. Deino, J. S. Larsen, P. K. Link, and C. M. Fanning. 2009. New ^{40}Ar-^{39}Ar and detrital zircon U-Pb ages for the Upper Cretaceous Wahweap and Kaiparowits formations on the Kaiparowits Plateau, Utah: implications for regional correlation, provenance, and biostratigraphy. Cretaceous Research 30:287–299.

Kennedy, W. J., and W. A. Cobban. 1991. Coniacian Ammonite Faunas from the United States Western Interior. Special Papers in Palaeontology 45.

Lambe, L. M. 1902. New genera and species from the Belly River Series (mid-Cretaceous). Contributions to Canadian Paleontology 3:25–81.

Lambe, L. M. 1910. Note on the parietal crest of *Centrosaurus apertus* and a proposed new generic name for *Stereocephalus tutus*. Ottawa Naturalist 14:149–151.

Nichols, D. J. 1997. Palynology and ages of some Upper Cretaceous formations in the Markagunt and northwestern Kaiparowits Plateaus, southwestern Utah; pp. 81–95 in F. Maldonado and L. D. Nealey (eds.), Geologic Studies in the Basin and Range–Colorado Plateau Transition Zone in Southeastern Nevada, Southwestern Utah, and Northwestern Arizona. U.S. Geological Survey Bulletin 2153-E.

Nopsca, F. 1928. Paleontological notes on reptiles. Geologica Hungarica Series Palaeontologica 1:1–84.

Parks, W. A. 1924. *Dyoplosaurus acutosquameus,* a new genus and species of armored dinosaur; and notes on a skeleton of *Prosaurolophus maximus*. University of Toronto Studies Geological Series 18:1–35.

Parrish, J. M. 1999. Dinosaur teeth from the upper Cretaceous (Turonian–Judithian) of southern Utah; pp. 323–334 in D. D. Gillette (ed.), Vertebrate Paleontology in Utah. Utah Geological Survey Miscellaneous Publication 99-1.

Penkalski, P. 2001. Variation in specimens referred to *Euoplocephalus tutus;* pp. 363–385 in K. Carpenter (ed.), The Armored Dinosaurs. Indiana University Press, Bloomington, Indiana.

Peterson, F. 1969. Four new members of the Upper Cretaceous Straight Cliffs Formation in the southeastern Kaiparowits region, Kane County, Utah; pp. J1–J28 in Contributions to Stratigraphy, 1968, U.S. Geological Survey Bulletin, 1274-J.

Roberts, E. M., A. D. Deino, and M. A. Chan. 2005. ^{40}Ar/^{39}Ar age of the Kaiparowits Formation, southern Utah, and correlation of coeval strata and faunas along the margin of the Western Interior Basin. Cretaceous Research 26:307–318.

Ryan, M. J., and D. C. Evans. 2005. Ornithischian dinosaurs; pp. 312–348 in P. J. Currie and E. B. Koppelhus (eds.), Dinosaur Provincial Park: A Spectacular Ancient Ecosystem Revealed. Indiana University Press, Bloomington, Indiana.

Scheyer, T. M., and P. M. Sander. 2004. Histology of ankylosaur osteoderms: implications for systematics and function. Journal of Vertebrate Paleontology 24:874–893.

Sternberg, C. M. 1929. A toothless armored dinosaur from the Upper Cretaceous of Alberta. Bulletin of the National Museum of Canada 54:28–33.

Sullivan, R. M. 1999. *Nodocephalosaurus kirtlandensis,* gen. et sp. nov., a new ankylosaurid dinosaur (Ornithischia: Ankylosauria) from the Upper Cretaceous Kirtland Formation (Upper Campanian), San Juan Basin, New Mexico. Journal of Vertebrate Paleontology 19:126–139.

Sullivan, R. M., and D. W. Fowler. 2006. New specimens of the rare ankylosaurid dinosaur *Nodocephalosaurus kirtlandensis* (Ornithischia: Ankylosauridae) from the Upper Cretaceous Kirtland Formation (De-na-zin Member), San Juan Basin, New Mexico; pp. 259–261 in S. G. Lucas and R. M. Sullivan (eds.), Late Cretaceous Vertebrates from the Western Interior. New Mexico Museum of Natural History and Science Bulletin 35.

Sullivan, R. M., and S. G. Lucas. 2006. The Kirtlandian land-vertebrate "age"–faunal composition, temporal position and biostratigraphic correlation in the nonmarine Upper Cretaceous of western North America. New Mexico Museum of Natural History and Science Bulletin 35:70–76.

Titus, A. L., J. D. Powell, E. M. Roberts, S. D. Sampson, S. L. Pollock, J. I. Kirkland, and L. B. Albright. 2005. Late Cretaceous stratigraphy, depositional environments, and macrovertebrate paleontology of the Kaiparowits Plateau, Grand Staircase–Escalante National Monument, Utah; pp. 101–128 in J. Pederson and C. M. Dehler (eds.), Interior Western United States. Geological Society of America Field Guide 6.

Vickaryous, M. K., and B. K. Hall. 2008. Development of the dermal skeleton in *Alligator mississippiensis* (Archosauria, Crocodylia) with comments on the homology of osteoderms. Journal of Morphology 269:398–422.

Vickaryous, M. K., and J.-Y. Sire. 2009. The integumentary skeleton of tetrapods: origin, evolution, and development. Journal of Anatomy 214:441–464.

Vickaryous, M. K., T. Maryańska, and D. B. Weishampel. 2004. Ankylosauria; pp. 363–392 in D. B. Weishampel, P. Dodson, and H. Osmólska (eds.), The Dinosauria (2nd edition). University of California Press, Berkeley, California.

Ornithopod Dinosaurs from the Grand Staircase–Escalante National Monument Region, Utah, and Their Role in Paleobiogeographic and Macroevolutionary Studies

19

Terry A. Gates, Eric K. Lund, C. A. Boyd, Donald D. DeBlieux, Alan L. Titus, David C. Evans, Michael A. Getty, James I. Kirkland, and Jeffrey G. Eaton

ORNITHOPOD DINOSAURS WERE BIPEDAL, HERBIVOROUS dinosaurs represented in the Late Cretaceous of North America by basal ornithopods ("hypsilophodontids") and a clade of derived iguanodontians containing, in part, hadrosaurids. Recent research focused on the Cretaceous macrovertebrates of Grand Staircase–Escalante National Monument and surrounding areas of southern Utah has resulted in numerous discoveries of ornithopod dinosaur skeletons, complimenting the previous record based largely on teeth. The new collections are dominated by hadrosaurid material but include multiple basal ornithopod and indeterminate iguanodontian specimens. Several new taxa of hadrosaurids have been identified from these rocks along with one new species of basal ornithopod. Isolated teeth typify the majority of basal ornithopod remains currently known from the Straight Cliffs and Wahweap formations, with several skeletal specimens from the Kaiparowits Formation representing the hypodigm of a new taxon. The Straight Cliffs Formation has recently yielded several iguanodontian specimens with at least three taxa represented from Turonian, Coniacian, and Santonian strata. Hadrosaurid diversity within the Campanian Wahweap and Kaiparowits formations now includes five taxa, at least three of which appear to be new species. Hadrosaurids from the Kaiparowits Formation include one species of *Parasaurolophus* and two stratigraphically separated species of *Gryposaurus*. The recognition of this diverse ornithopod fauna within southern Utah provides a rare opportunity in North America to examine the transition from more basal iguanodontians to those taxa within Hadrosauridae. In addition, these sediments provide a testing ground for biogeographic hypotheses of basal ornithopods.

INTRODUCTION

Ornithopods were bipedal herbivorous dinosaurs that originated in the Late Jurassic (sensu Butler, Upchurch, and Norman, 2008) and diversified through the terminal Cretaceous period. Ornithopoda was first proposed by Marsh (1881), and more recently defined in a cladistic sense by Norman et al. (2004:393) as "all cerapodans closer to *Edmontosaurus* than to *Triceratops*." These dinosaurs were extremely widespread, with fossil remains discovered from every continent, including Antarctica (Case et al., 2000).

Though sporadic work on dinosaur paleontology by Brigham Young University began in what is now Grand Staircase–Escalante National Monument in the early 1970s, it was not until after the monument's creation in 1996 that expanding knowledge of the macrovertebrate fossil record from Upper Cretaceous formations started in earnest. (See Titus, this volume, Chapter 1, for an overview of paleontology history in Grand Staircase–Escalante National Monument.) This work was built upon a rich foundation of microvertebrate work spearheaded by Jeff Eaton and Rich Cifelli (Cifelli, 1987, 1990; Eaton, 1991, 2002). To date, dinosaur paleontology of the region has focused on the Kaiparowits Plateau where three stratigraphically contiguous terrestrial formations have all yielded diagnostic macroskeletal remains of dinosaurs. These are, in ascending order, the Straight Cliffs (see Titus et al., Chapter 2, this volume), Wahweap (see Jinnah, Chapter 4, this volume), and Kaiparowits (see Roberts et al., Chapter 6, this volume). The Cedar Mountain and Dakota Formations, and the Tropic Shale have not yielded diagnostic skeletal remains of dinosaurs and are not treated here, although teeth recovered by screen washing of microvertebrate samples demonstrates the presence of both basal ornithopods ("hypsilophodontids") and more advanced hadrosaurids or hadrosauromorphs. These same two clades of ornithopods occur through the remainder of the region's Cretaceous record, and they are generally the most common dinosaur fossils found in any given formation.

Basal ornithopods were small-bodied, bipedal, cursorial ornithopods ranging from 1 to 3 m in length (Norman et al., 2004). They have relatively small heads and leaf-shaped teeth, primitive characteristics in the ornithopod clade. In contrast, virtually all hadrosaurids are large-bodied animals, and some of those found within the Kaiparowits Formation were relatively gigantic, exceeding 10 m in length. Hadrosaurids are further distinguished from basal ornithopods in possessing large, elongate heads and densely packed, multiple rooted teeth in each socket that total more than 200 teeth in each jaw quadrant—a highly derived condition among dinosaurs. The hind limbs of hadrosaurids are massively built in order to support their enormous weight, more so than the forelimbs, which were used for facultative weight-bearing only during certain types of locomotion (Horner, Weishampel, and Forster, 2004). The clade Hadrosauridae consistently resolves into two subclades, the hollow-crested Lambeosaurinae and the non-hollow-crested Hadrosaurinae (Forster, 1997). Both subclades possessed highly modified skulls with cranial ornamentations, although lambeosaurines are extreme in this regard, forming an elaborate extension of the nasal cavity within hollow, bony crests composed mostly of nasals and premaxillae.

This chapter provides a brief overview of known ornithopod diversity within the Upper Cretaceous Straight Cliffs, Wahweap, and Kaiparowits formations in the Grand Staircase region, ultimately focusing on their biostratigraphic and biogeographic significance. This chapter is an updated and expanded version of a previous survey (Gates et al., 2010).

Institutional Abbreviations BYU, Brigham Young University, Provo, Utah; FMNH, Field Museum of Natural History, Chicago, Illinois; MNA, Museum of Northern Arizona, Flagstaff, Arizona; MOR, Museum of the Rockies, Bozeman, Montana; MSMP, Mesa Southwest Museum Paleontology Collection, Mesa, Arizona; RAM, Raymond M. Alf Museum, Claremont, California; ROM, Royal Ontario Museum, Toronto; UCMP, University of California Museum of Paleontology, Berkeley, California; UMNH, Natural History Museum of Utah, Salt Lake City, Utah; USNM, United States National Museum, Washington, D.C.

STRAIGHT CLIFFS FORMATION

The Smoky Hollow and John Henry members produce the majority of well-preserved nonmarine vertebrate taxa recovered from the Straight Cliffs Formation, whereas the Drip Tank Member contains mostly logs and scrappy bone fragments (Peterson, 1969; Eaton, 1991; Eaton et al., 1999; Cobban et al., 2000; Doelling et al., 2000). Most previous research has focused on screen washing microvertebrate localities, which resulted in the collection of abundant and diverse dinosaur teeth (Eaton et al., 1999; Parrish, 1999). Consequently, our current understanding of the taxonomic diversity of dinosaurs within the Straight Cliffs Formation is limited largely to inferences based on tooth morphology—a practice that generally produces only family-grade resolution.

More specific to the present review, taxonomic resolution of ornithopod diversity from the Straight Cliffs Formation is currently constrained to basal ornithopod and Hadrosauridae incertae sedis or *confer* (Eaton et al., 1999; Parrish, 1999). Parrish (1999) identified basal ornithopod teeth from the Smoky Hollow Member, comparing them to the Maastrichtian genus *Thescelosaurus*. On the basis of recent phylogenetic and stratigraphic analysis (Roberts, Deino, and Chan, 2005; Boyd et al., 2009), it appears that the clade of basal ornithopods containing *Thescelosaurus* did not exist until the Maastrichtian, making it more appropriate to compare the basal ornithopod material from the Straight Cliffs Formation to the taxa *Zephyrosaurus* or *Oryctodromeus*. No basal ornithopods have been identified from other portions of the formation.

In recent years, the fossil record for large ornithopods within the Straight Cliffs Formation has improved, with several fragmentary skeletal elements and partial skeletons discovered. In 2006, one of us (J.G.E.) discovered the partial skeleton of a large ornithopod in the Middle Turonian Smoky Hollow Member west of the Kaiparowits Plateau and north of the town of Tropic; the specimen, subsequently excavated by the Utah Geological Survey, consists only of vertebrae and limb elements, allowing no further taxonomic assessment. In addition to the previous specimen, several others demonstrate a broadening diversity of ornithopods from the Straight Cliffs Formation. Thus far, dentary fragments provide evidence for at least two taxa of large ornithopods from the Coniacian–Santonian of southern Utah. The partial dentaries UMNH VP 17037 and UMNH VP 17187 (Fig. 19.1) from the Coniacian (less than 5 m above the base of the John Henry Member) both have near vertical tooth rows, whereas the Santonian-aged UMNH VP 17434 has distinctly more posteriorly angled tooth rows, demonstrating distinct morphotypes. Another locality discovered just to the west of Grand Staircase–Escalante National Monument in more terrestrial sediments of the Coniacian John Henry Member (less than 5 m above the base of the member; same area as above) represents a partial skeleton that has the potential to greatly increase understanding of Santonian large ornithopod diversity and hadrosaurid evolution. The only other large ornithopod known from this time period of the southwestern United States is a derived iguanodontian ornithopod from the Middle Turonian–aged lower Moreno Hill Formation in west-central New Mexico (McDonald, Wolfe, and Kirkland, 2006; McDonald, Wolf, and Kirkland, 2010). It is not possible to make detailed comparisons between the Moreno Hill taxon and the Straight Cliffs taxa, although the anterior dentary downturning of UMNH VP 17187 is similar to MSMP-4166 but the relative height of the MSMP-4166 tooth row is

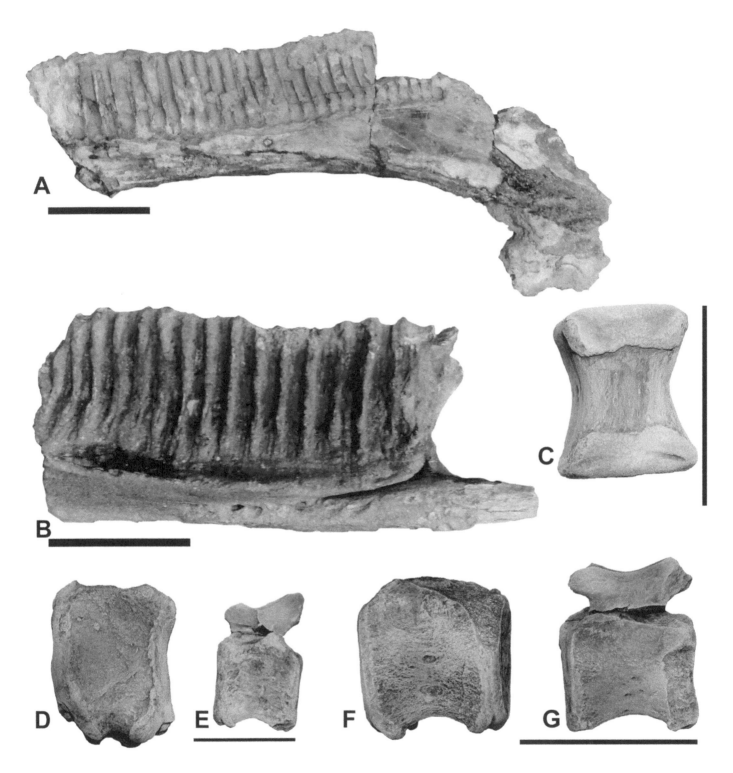

19.1. Examples of iguanodontian skeletal material discovered within the Straight Cliffs Formation. (A) UMNH VP 17187; (B) UMNH VP 17037; (C–G) vertebrae from UMNH VP 20240. Scale bars = 10 cm.

narrower, more similar to *Shuangmiaosaurus gilmorei* (You, Ji, and Li, 2003; McDonald, Wolfe, and Kirkland, 2006; McDonald, Wolf, and Kirkland, 2010).

Most recently, a multi-individual bonebed was discovered by A. Titus in the Smoky Hollow Member located along the southwest margin of the monument. Preliminary excavation and surface collection revealed a variety of skeletal elements ranging from caudal and dorsal vertebrae (UMNH VP 20240; Fig. 19.1), chevrons, phalanges, and a partial dentary, all in partial articulation or close association. The dentary is the most taxonomically useful element available for study at this time, although it is still largely unprepared and more detailed description will be forthcoming. The middentary shows what appears to be an original height of the tooth-bearing portion only around 4–5 cm tall. Given that the element is not a hatchling or yearling, this height is not uncommon for most nonhadrosaurid iguanodontian ornithopods, but it seems that a critical characteristic of the ornithopod dentary is the overall length relative to height. Within North America, this is an easily recognizable characteristic of the Zuni Basin–derived iguanodontian MSMP-4166 (McDonald, Wolf, and Kirkland, 2010). It turns out that middle Turonian MSMP-4166 is penecontemporaneous with UMNH VP 20240, allowing speculation that the two specimens might represent the same species.

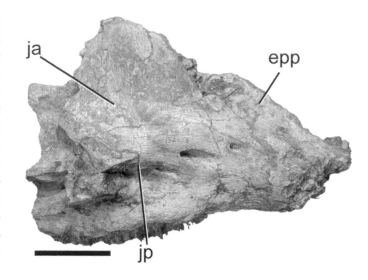

19.2. Unidentified lambeosaurine maxilla (UCMP 152028) from the Wahweap Formation shown in lateral view. epp, expansion of the premaxillary process (maxillary shelf sensu Horner, Weishampel, and Forster, 2004); ja, jugal articulation; jp, jugal process. Scale bar = 5 cm.

WAHWEAP FORMATION

Significant ornithopod localities yielding taxonomically informative specimens or an abundance of ornithopod fossils have been found in the lower three members – the lower (LM), middle (MM), and upper (UM) members – of the Wahweap Formation, although the MM is currently the most productive member. Unfortunately, to date, teeth comprise the only evidence of basal ornithopods within the formation (Eaton et al., 1999; Parrish, 1999). As with the Smoky Hollow Member teeth, Parrish (1999) compared the Wahweap ornithopod teeth with *Thescelosaurus*. Given that the thescelosaurid clade probably did not originate until the Maastrichtian, a more suitable comparison would be with the as yet unnamed basal ornithopod taxon from the Kaiparowits Formation.

Hadrosauridae

Hadrosaurids are more abundant than basal ornithopods, and numerous macroskeletal sites have been found throughout most of the Wahweap Formation. Beginning in the lowest portion of the stratigraphic section, several dozen fossils have been recovered from the hadrosaurid-dominated Tibbet Springs Quarry, located in highly indurated sandstone near the base of the LM. Multiple individuals are represented in this quarry by numerous limb and girdle elements, vertebrae, and a poorly preserved jaw. Some of this material likely represents partial associated skeletons, but more excavation at the difficult Tibbet Springs Quarry is required before further taphonomic information can be revealed. The Tibbet Springs Quarry has been the richest dinosaur locality found thus far in the lower member of the Wahweap Formation. Other hadrosaur sites occur nearby around Nipple Butte, but no diagnostic cranial material has been recovered.

All but one of the identifiable hadrosaurid specimens recovered thus far from the Wahweap Formation can be referred to the clade Hadrosaurinae. The single exception is an isolated lambeosaurine maxilla recovered from the UM (UCMP 152028; Fig. 19.2), making it the oldest lambeosaurine currently known in North America. Diagnosis of UCMP 152028 to Lambeosaurinae is based on the presence of the anterodorsal shelf that forms a flat, broad premaxillary shelf on the anterodorsal region of the maxilla (Prieto-Marquez, 2010:character 84, state 1), dorsal process of the maxilla dorsoventrally taller than it is wide, with a peaked and caudally inclined apex (Prieto-Marquez, 2010:character 91, state 1). The medial side of the preserved portion reveals similarities to an undescribed maxilla of *Parasaurolophus* sp. (UMNH VP 16666.1) from the Kaiparowits Formation, such as a dorsal expansion of the premaxillary process (maxillary shelf of Horner, Weishampel, and Forster, 2004; Fig. 2.19, epp). These same characteristics are muted in the crested hadrosaurs *Corythosaurus* and *Hypacrosaurus*. On the other hand, UCMP 152028 differs significantly from UMNH VP 16666.1 and all other known lambeosaurines, suggesting that this element pertains to a new taxon.

Published records of dinosaur faunas contemporaneous with those of the Wahweap Formation (81–76.1 Ma; Jinnah et al., 2009) do not include substantive descriptions of materials attributed to lambeosaurines. This interval corresponds in part to the Foremost and Oldman formations of Alberta, and the Judith River and lower part of the Two Medicine formations of Montana. An undescribed lambeosaurine has been reported to have been excavated from the Oldman Formation near Sandy Point, Alberta, Canada, by Horner,

Gates et al.

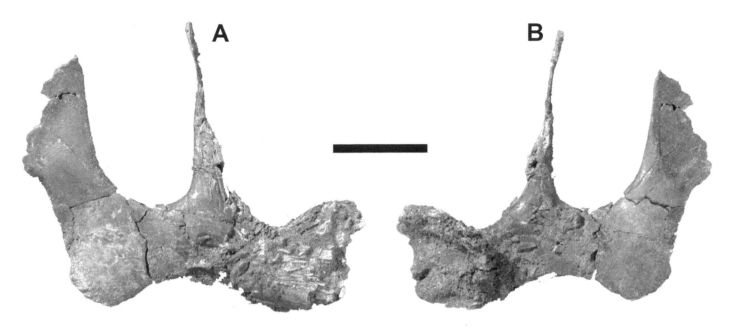

19.3. Right juvenile hadrosaurine jugal (UMNH VP 16695) found in the Upper Sandstone Member of the Wahweap Formation shown in (A) lateral and (B) medial views. Scale bar = 5 cm.

Weishampel, and Forster (2004). However, recent relocation of the quarry reveals that the specimen originated from the base of the Dinosaur Park Formation.

Among hadrosaurines, a large bonebed, an associated skeleton, and a partial skull are the three most significant discoveries to date, all found within the Middle Mudstone member. A locality known as Jim's Hadrosaur Site has yielded a number of juvenile postcranial elements, including a large portion of the fore- and hind limbs, dorsal, and cervical vertebrae. This specimen was scattered amid abundant carbonized log sections and conifer branches, as well as a disarticulated turtle, unionid clams, and freshwater crab claws. Jim's Hadrosaur Site is located near the base of the MM and appears to preserve a single juvenile hadrosaur individual.

Another locality within the MM preserved in the same stratigraphic level and taphonomic style as Jim's Hadrosaur Site is a large bonebed near Camp Flats along the Smoky Mountain Road. Deposited in a back swamp environment, two hadrosaurine individuals, an adult and a juvenile, were completely disarticulated over an area of more than 18.5 m² (Jinnah, Getty, and Gates, 2009). Additional fossils found within the site include maniraptoran theropod teeth, fish bones, a turtle pelvis, and large freshwater xanthid (mud and stone crabs) or paguroid (hermit and coconut crabs) crab claws (A. Milner, pers. comm., 2010). The site also preserves abundant plant remains consisting of mostly plant hash and unidentified conifer leaves. However, an interesting aspect of the site is that numerous coalified tree trunks crisscross both above and below the hadrosaur specimens. The excavation has thus far revealed approximately 70–80% of the adult

postcranium and numerous elements of a much smaller juvenile specimen (Jinnah, Getty, and Gates, 2009).

Unfortunately, the only skull material collected from this locality to date consists of a juvenile jugal, maxilla, and dentary, and an adult postorbital. Studies of taxonomic affinity are still underway, although the taxon does appear to be a new species but not in the same clade as the Wahweap hadrosaurine *Acristavus*. In addition to a set of possible autapomorphies, the taxon is likely diagnosable through a unique suite of characters involving at minimum the jugal, pubis, and caudal vertebrae. As such, it seems that the pubis is diagnosable enough to support the identification of the juvenile recovered from Jim's Hadrosaur Site to the same taxon as found at this locality. The site is extremely significant in that it has yielded the most complete dinosaur specimen known from the Wahweap Formation.

Other significant hadrosaur materials recovered from the MM include the following: (1) an isolated juvenile jugal diagnostic of brachylophosaurin hadrosaurines (UMNH VP 16695; Fig. 19.3; Gates et al., 2011); (2) a large partial pubis and two femora; (3) an associated hadrosaur scapula, proximal humerus, and dentary; and (4) a partial skull representing a new taxon of hadrosaurid dinosaur (UMNH VP 16607; Gates et al., 2011).

Currently, the most diagnostic hadrosaurid specimen discovered from the Wahweap Formation is the partial skull of a hadrosaurine (UMNH VP 16607) that pertains to a new genus also found in Montana (Gates et al., 2011). Dr. Riley Nelson (Brigham Young University) discovered the specimen in a massive sandstone at the top of the MM, just north of Right

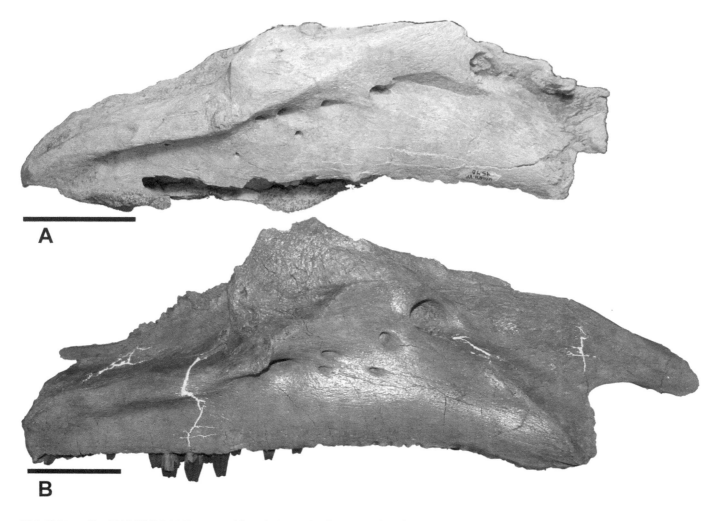

19.4. Right maxillae (A) UMNH VP 9548 recovered from the Upper Sandstone Member of the Wahweap Formation, currently attributed to cf. *Brachylophosaurus* and (B) MOR 1071-8-13-92-559, *Brachylophosaurus canadensis,* shown in lateral view. Scale bar = 5 cm.

Hand Collett Canyon. The skull consists of a complete braincase and mostly complete posterior skull roof and lacrimals. UMNH VP 16607 clearly belongs to a clade of hadrosaurines that include *Brachylophosaurus* and *Maiasaura* to the exclusion of all other known hadrosaurids (interbasipterygoid ridge descends to level of ventral basipterygoid processes of basisphenoid; alar process of basisphenoid large and highly angled; Gates et al., 2011). A marked difference between *Acristavus gagslarsoni* and the former two taxa is the overall robustness of the skull elements. For instance, the lacrimal of UMNH VP 16607 is almost twice as broad posteriorly as a *Brachylophosaurus* lacrimal (MOR 1071 7-10-98-171) of comparative external size. Another interesting feature is that it possesses orbital rugosity in similar patterns to that seen on the Moreno Hill ornithopod (MSMP-4166; McDonald, Wolf, and Kirkland, 2010).

The stratigraphically highest bonebed in the Wahweap Formation occurs near the base of the UM in the area of the monument known as The Gut. This site yielded several postcranial elements from at least two individuals entombed within a silty mudstone; representative elements include a tibia, two ilia, a partial humerus, and ribs.

Another significant specimen, discovered within a sandstone unit of the UM on Death Ridge, includes partial limb bones and an isolated partial maxilla (UMNH VP 9548; Fig. 19.4) tentatively identified as cf. *Brachylophosaurus*. More material is required to verify the generic assignment. The maxilla more closely matches the morphology of *Brachylophosaurus* than of *Maiasaura*, *Acristavus*, or the taxon *Gryposaurus* commonly found in the overlying Kaiparowits Formation. This assessment is based on the more gracile appearance of the maxilla compared to the other taxa, the presence of a well-developed shelf on the medial side of the dorsal process (smaller shelf in *Maiasaura* and no shelf in *Acristavus*; Gates et al., 2011), long low jugal articular surface differing from *Gryposaurus* which possesses a more curved articulation surface that projects dorsally, and an anterodorsal process that appears to be as broad as seen on

Gates et al.

19.5. Ornithopod footprints from the (A) Wahweap Formation and (B) Kaiparowits Formation.

Brachylophosaurus, but unfortunately is broken in UMNH VP 9548).

In addition to the body fossils discussed above, we have discovered a number of dinosaur tracksites in the Wahweap Formation (Fig. 19.5A). These tracksites consist primarily of natural casts associated with thin sandstone units most likely representing crevasse splays. The majority of these are large tridactyl tracks attributable to hadrosaurs. These tracks have been found in both the LM and MM, and are relatively abundant in the transition from the LM to the MM in the Wesses Cove and Wesses Canyon vicinity. We have noted a similar mode of track preservation in the Kaiparowits Formation. A more detailed examination and documentation of these tracksites is currently underway. Together, all of the above-mentioned specimens are providing a substantial foundation for the study of hadrosaurs within the Wahweap Formation, and the middle portion of the Campanian stage as a whole.

KAIPAROWITS FORMATION

Three informal units (lower, middle, and upper) subdivide the Kaiparowits Formation based on sandstone to mudstone ratios (Roberts, 2007). The vast majority of ornithopod specimens come from the lower and middle units, the former unit being composed mostly of sandstone and the latter being mudstone dominated. Most of the articulated ornithopod specimens collected to date have been recovered from fluvial sandstone bodies within the lower and middle units. Only a couple of basal ornithopods and a single hadrosaur have been recovered articulated from finer grained sediments, where disarticulation is much more common.

Basal Ornithopods

Over the past 10 years, at least eight basal ornithopod specimens (i.e., "hypsilophodontids") were recovered from the Kaiparowits Formation. Prior to these discoveries, basal ornithopods had not been reported from the formation. Nearly all of these specimens consist of concentrations of disarticulated elements, though two specimens from the lower unit preserve articulated portions of the skeleton (UMNH VP 12677, 16281). The basal ornithopod material from the Kaiparowits Formation displays a large range of size variation that is here interpreted as ontogenetic variation (Fig. 19.6). This interpretation is supported by the absence of fusion between the neural arches and the centra of isolated cervical, dorsal, and sacral vertebrae in all but the largest specimens and the lack of fusion between adjacent sacral vertebrae in smaller specimens (e.g., UMNH VP 12665 and 19470), features previously shown to be useful for assessing ontogenetic maturity in other archosaurian taxa (Brochu, 1996; Irmis, 2007).

Two morphologically and taxonomically informative specimens were recovered from the Lower unit of the Kaiparowits Formation. The first consists of articulated left and right feet from a single individual preserved in a sandy siltstone (UMNH VP 16281; Fig. 19.6A). The morphology of the pes is typical of most basal ornithopod taxa, except as noted. The medial distal tarsal is L shaped in proximal view, with a short process extending over the posterior portion of the dorsal surface of metatarsal II. The metatarsals are thin and elongate, with metatarsal III being more than twice as long as the combined transverse width of the proximal ends of metatarsals II through IV as in *Orodromeus*, *Oryctodromeus*,

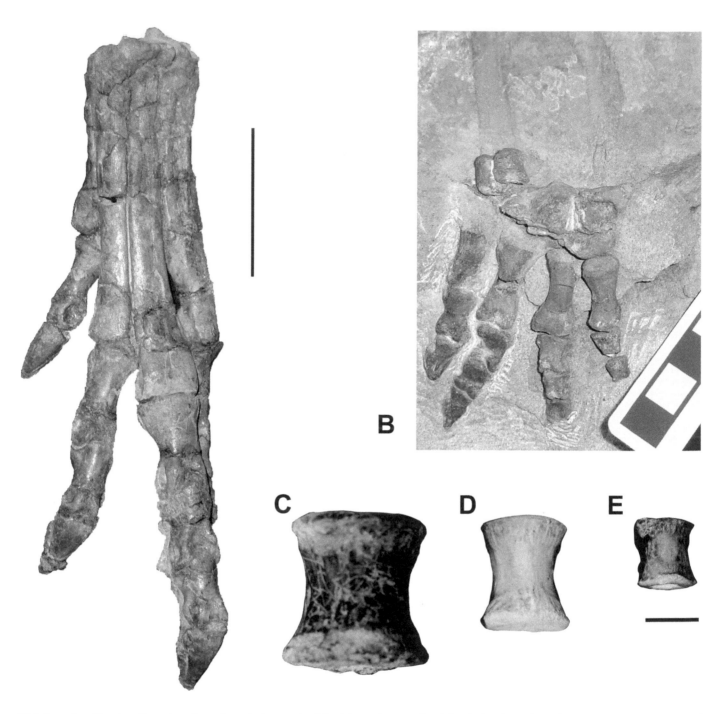

19.6. Examples of basal ornithopod specimens from the lower unit of the Kaiparowits Formation. (A) Articulated right foot of UMNH VP 16281; (B) articulated left hand of UMNH VP 12677; (C–E) vertebrae showing variation in size of basal ornithopod specimens recovered. Scale bar = 5 cm in (A) and 1 cm in (C–E).

and *Zephyrosaurus*. This morphology differs from that of *Parksosaurus* and *Thescelosaurus*, where these three metatarsals are relatively shorter and more robust. Metatarsal V is a short, thin plate of bone appressed to the posterior surface of the lateral distal tarsal proximally and then curves ventromedially against the posterior surface of metatarsal IV. In *Orodromeus* metatarsal V is also a flattened plate of bone, but in *Thescelosaurus* it is more robust, though still greatly shortened. Digit III of the right pes of the Kaiparowits basal ornithopod measures 211 mm, which is ~60% larger than that

of the contemporaneous *Orodromeus* (based on MOR 530; Scheetz, 1999), but ~25% smaller than that of the immature holotype of *Thescelosaurus* (USNM 7757; Gilmore, 1913). A second specimen recovered from the lower unit (UMNH VP 12677) consists of a mostly complete, well-preserved, articulated hand exposed in plantar view (Fig. 19.6B) and associated disarticulated vertebrae and limb fragments from a channel sandstone. The carpus is composed of five carpals as in *Thescelosaurus*, though their arrangement differs. Digits I through IV are present, but digit V is not preserved. The

Gates et al.

manual phalangeal formula is 2–3–(4?)–2–?, with the reduction from three to two phalanges in digit IV being autapomorphic among basal ornithopod taxa.

The most taxonomically informative specimen yet recovered was collected from an overbank mudstone within the middle unit (UMNH VP 19470; Fig. 19.7). This specimen consists of a partial disarticulated skull (including a partial jugal, frontal, dentary, pterygoid, and exoccipital–opisthotics) and fragmentary postcranial material. The jugal displays a prominent posterolaterally projecting boss that is present in *Orodromeus* and *Zephyrosaurus* (Varricchio, Martin, and Katsura, 2007). The morphology of the frontal more closely resembles *Orodromeus* than *Zephyrosaurus*, especially in that the articulation surface for the postorbital faces dorsolaterally rather than only laterally. The dorsal ramus of the jugal bears a dorsally oriented socket that received the ventral ramus of the postorbital. This condition is unknown in any described taxon, though in *Orodromeus* the dorsal ramus of the jugal bears a dorsoventrally oriented groove that received the postorbital. The cranial elements that are preserved in UMNH VP 19470 are slightly smaller than those preserved with the immature holotype of *Orodromeus* (MOR 294; Scheetz, 1999), and the presence of disarticulated dorsal vertebral centra that are unfused to their respective neural arches suggests that UMNH VP 19470 likely represents an immature individual (Brochu, 1996; Irmis, 2007).

The most productive basal ornithopod site within the Kaiparowits Formation is located within a crevasse splay deposit in the Middle unit. This site has produced a substantial number of disarticulated elements representing at least three ontogenetically immature individuals of various sizes (UMNH VP 12665). This material consists largely of disarticulated vertebral centra and fragmentary fore and hind limb elements, though fragments of two dentaries and a partial maxilla were also recovered. Among the more important elements recovered is an isolated sacral centrum that displays prominent articulation facets on the anterolateral margin that directly supported the medial protuberance of the pubis (sensu Scheetz, 1999). This feature is apomorphic for a clade containing *Orodromeus*, *Oryctodromeus*, and *Zephyrosaurus* (Varricchio, Martin, and Katsura, 2007). Additionally, a fragmentary left coracoid collected from this site possesses an oblong foramen in the ventral margin of the sternal hook just anterior to the glenoid fossa, a characteristic unknown in any previously described basal ornithopod taxon.

Preliminary examination of all basal ornithopod specimens from the Kaiparowits Formation indicates the presence of at least one previously undescribed taxon. This conclusion is based on the recognition of three autapomorphic traits from three different specimens. However, the fragmentary nature of all known specimens makes it difficult to accurately

resolve the taxonomic diversity these specimens represent. Thus far, only UMNH VP 12665 and 19470 from the middle unit of the Kaiparowits Formation can be confidently referred to a single taxon on the basis of shared character evidence (Boyd and Gates, unpubl. data). The combined presence of a prominent jugal boss in UMNH VP 19470 and the direct support of the pubis by the sacral centra demonstrated by UMNH VP 12665 suggest a close relationship between this new taxon and the North American basal ornithopod taxa *Orodromeus* and *Zephyrosaurus*.

Hadrosauridae

Both lambeosaurine and hadrosaurine remains have been recovered from the Kaiparowits Formation. The lambeosaurine *Parasaurolophus* is the most distinctive hadrosaur from the formation, possessing a large, curved, hollow narial tube composed almost entirely of fused premaxillae (Fig. 19.8A). *Parasaurolophus* was the first dinosaur to be identified from the Kaiparowits Formation, based upon a highly eroded, partial skull (BYU 2467; Weishampel and Jensen, 1979). Sullivan and Williamson (1999) identified these materials and another, more complete specimen (UCMP 143270; Fig. 19.8A) as pertaining to *P. cyrtocristatus*, a taxon otherwise known only from the Fruitland Formation of New Mexico.

Recent work conducted by the KBP has yielded additional *Parasaurolophus* materials from this formation. Currently, the total sample consists of seven isolated partial skulls (BYU 2467, UCMP 143270, UMNH VP 16394, 16666, 16689 and two unnumbered unprepared UMNH specimens), a partial skeleton consisting of mostly the pelvic region (UMNH VP 19471), and fragmentary associated elements within a multitaxic bonebed, all collected from or closely associated with sandstone deposits in the middle unit of the Kaiparowits Formation. One of the recently collected partial skulls (UMNH VP 16666.1; Fig. 19.8B) includes a maxilla, jugal, palatine, ectopterygoid, and quadrate–elements unknown for any *P. cyrtocristatus* specimen or for any other *Parasaurolophus* specimen collected from the Kaiparowits. The best preserved *Parasaurolophus* specimen from the Kaiparowits Formation (UCMP 143270) differs from the holotype specimen of *P. cyrtocristatus* (FMNH P27393; Fruitland Formation, New Mexico) in the curvature of the snout and the degree of descent of the posterior portion of the crest. However, FMNH P27393 is much larger than UCMP 143270 and the observed differences may represent ontogenetic variation. In fact, all of the recovered Kaiparowits *Parasaurolophus* skull material, except UMNH VP 16666.1 and one unprepared skull, is smaller than the type specimen of *P. cyrtocristatus*. Whether or not this seemingly consistent size discrepancy is taxonomically important still remains a question. The excellently preserved,

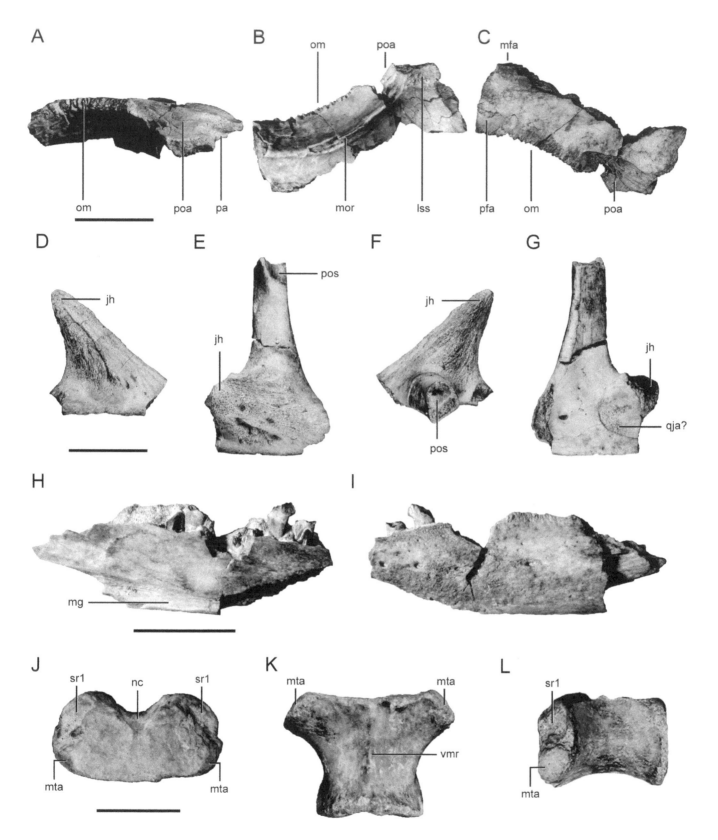

19.7. Examples of basal ornithopod material from the middle unit of the Kaiparowits Formation. (A–C) Left frontal from UMNH VP 19470 shown in lateral, ventral, and dorsal views, respectively. (D–G) Right jugal from UMNH VP 19470 shown in ventral, lateral, dorsal, and medial views, respectively. (H, I) Left dentary from UMNH VP 19470 shown in medial and lateral views, respectively. (J–L) First true sacral centrum from the same locality as UMNH VP 12665 (given its own specimen number: UMNH VP 21095) shown in anterior, ventral, and lateral views, respectively. All Scale bars = 1 cm and apply to its corresponding row of figures. jh, jugal horn; lss, laterosphenoid socket; mfa, medial frontal articulation; mg, Meckelian groove; mor, medial orbital ridge; mta, articulation for the medial tubercle of the pubis; nc, neural canal; om, orbital margin; pa, parietal articulation; pfa, prefrontal articulation; poa, postorbital articulation; pos, postorbital socket; qja, quadratojugal articulation; sr, sacral rib; vmr, ventral midline ridge.

Gates et al.

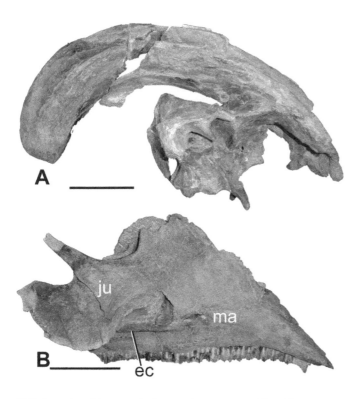

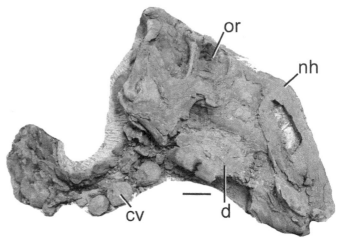

19.9. Lateral view of articulated skull of *Gryposaurus* sp. (UMNH VP 16667) recovered from of the lower unit of the Kaiparowits Formation. cv, cervical vertebrae; d, dentary; nh, nasal hump; or, orbit. Scale bar = 10 cm.

19.8. Examples of *Parasaurolophus* specimens recovered from the Kaiparowits Formation. (A) Partial articulated *Parasaurolophus* skull UCMP 143270 shown in lateral view; (B) articulated *Parasaurolophus* maxillary complex shown in lateral view. et, ectopterygoid; ju, jugal; ma, maxilla. Scale bar = 10 cm.

uncrushed nature of the UCMP 143270 specimen reveals details of crest morphology including premaxillary and nasal sutural contacts unknown or controversially interpreted in all other specimens of *Parasaurolophus*. The Kaiparowits material will provide important new information that will contribute significantly to understanding the morphology and relationships of this iconic genus. All of these materials are currently under study in order to assess whether or not the Kaiparowits taxon corresponds to *P. cyrtocristatus*.

An articulated tail, partial pelvis, and left leg (MNA v8369) from a site in the lower unit was tentatively identified as a lambeosaurine hadrosaur by Titus, Gillette, and Albright (2001). However, more recent analysis of this specimen demonstrated that it is a hadrosaurine of unknown generic affinity, currently leaving lambeosaurines undocumented in the lower portion of the Kaiparowits Formation.

The most common hadrosaur fossils discovered in the Kaiparowits Formation are of the hadrosaurine genus *Gryposaurus*, the remains of which – including several associated partial skulls (UMNH VP 12265, 13831, 13970, 16669, 18568, 20181 and RAM 6797), three of which are associated with partial postcranial skeletons (UMNH VP 12265, 18568, and 20181) – have been collected at eight localities that range from near the base of the Kaiparowits through the top of the Middle unit. Significantly, those specimens found in the

lower unit, near the base of the formation, are morphologically distinct from those found stratigraphically higher in the formation, indicating the presence of two successive species of *Gryposaurus*, and thus intergeneric faunal turnover.

The stratigraphically lowest occurring taxon, here termed *Gryposaurus* sp., is represented by several specimens, including a virtually complete articulated skull (UMNH VP 18568; Fig. 19.9) discovered by one of us (AT) within a massive sandstone near Wahweap Creek. The skull of this specimen exhibits the distinctive nasal "hump" of *Gryposaurus* but is almost 1 m long, exceeding that of any other previously described specimens attributable to this genus. Another more fragmentary skull from the Kaiparowits Formation (UMNH VP 16668), also attributable to *Gryposaurus* sp., is approximately 20% larger than UMNH VP 18568, suggesting that this taxon greatly exceeded the body size of more northern congeners. The proportions of the temporal region of the skull more closely resemble *Gryposaurus notabilis* from the Dinosaur Park Formation of Alberta, in contrast to the reduced size of the infratemporal fenestra in *G. monumentensis* (Gates and Sampson, 2007). In fact, there is no apparent morphological distinction aside from size that would infer the Kaiparowits *Gryposaurus* sp. to be classified in a different species from *Gryposaurus notabilis*. However, more detailed anatomical analysis and comparisons are required to determine if *Gryposaurus* sp. from the Kaiparowits is conspecific with the northern taxon, or if it represents a new, closely related species.

The second *Gryposaurus* species now recognized in the Kaiparowits Formation, *Gryposaurus monumentensis* (Gates and Sampson, 2007), occurs higher in section and is also represented by multiple specimens, including the mostly complete type skull (RAM 6797; Fig. 19.10) found in a muddy sand

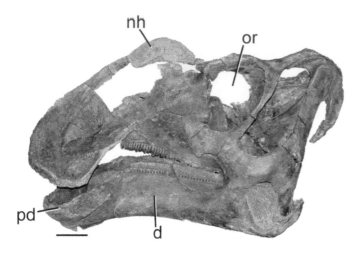

19.10. Lateral view of articulated skull of *Gryposaurus monumentensis* (RAM 6797). d, dentary; nh, nasal hump; or, orbital rim; pd, predentary. Scale bar = 10 cm.

point bar deposit. *Gryposaurus monumentensis* possesses a number of unique characteristics, mostly related to the more robust nature of the skull and lower jaws (Gates and Sampson, 2007). A partial subadult skull attributed to this taxon (UMNH VP 13970) demonstrates that the extremely robust dentary present in adults developed prior to the onset of adult size. One of the more interesting features of *G. monumentensis* is the predentary, which possesses large, clover-shaped processes along the oral margin that are much more conspicuous than similar structures seen on the predentaries of other hadrosaurs (e.g., *Gryposaurus latidens* MOR 553s). Although the precise function of these structures is uncertain, other hadrosaur taxa show evidence of a keratinous beak adhered to the snout via small processes on the predentary oral margin (Morris, 1970), and it is perhaps likely that these features had a similar role.

Modifications to chewing appear to be the driving force behind the evolution of *G. monumentensis* especially when one considers the bony correlates of a large keratinous beak, robustness of the masticating skull elements, and the relative anteroposterior shortening of the posterior skull region (Gates and Sampson, 2007). More specifically, the infratemporal fenestra does not display the gaping morphology seen in all other species of *Gryposaurus*, but instead has a much narrower fenestra. This implies that the posterior region of the skull is shorter in this taxon. An independent confirmation of posterior shortening is the observation that the fenestra present between the surangular, jugal, and quadrate is round, nearly circular, in all *Gryposaurus* taxa except *G. monumentensis*, which displays a distinctly oval fenestra (Fig. 19.10). Overall shortening of the posterior skull would shorten the moment arm of the chewing lever and allow muscles of

equivalent size as other *Gryposaurus* taxa apply more force to processing plant matter.

Recently, a crew from the Alf Museum found another important specimen higher in stratigraphic section than the *G. monumentensis* type specimen. RAM 12065 consists of a partial skull roof and partial braincase. According to A. Farke (unpubl. data, pers. comm. 2010), RAM 12065 is smaller than and possesses a slightly different morphology from the *G. monumentensis* type specimen such as a planar skull roof instead of the dorsally projecting posterior process of the postorbital seen in the latter taxon and *G. notabilis* (Herrero and Farke, 2010).

One of the most complete skeletons of *Gryposaurus monumentensis* (also one the most complete adult hadrosaur skeletons discovered to date from the Monument) is UMNH VP 12265 (Fig. 19.11), an exceptionally preserved specimen recovered from the middle unit in an expansive area of outcrop known as The Blues. Encased mostly in well-indurated sandstone, the associated and partially articulated skeleton includes a portion of the skull and lower jaws (maxillae, jugal, quadrate, dentary), most of the dorsal, sacral, and caudal vertebral series, fragmentary ribs, scapulae, coracoid, humerus, and the entire pelvis.

Gates and Sampson (2007) attributed UMNH VP 12265 to *G. monumentensis*, which is the only postcranial skeleton known for the species. The original description was based on the complete skull RAM 6797 and the authors neglected to describe the postcranial material preserved in UMNH VP 12265. Presented below is a short description of this partial skeleton.

The partial skeleton UMNH VP 12265 preserves most of the dorsal, sacral, and caudal vertebral series, a number of ribs, right scapula, right humerus, right radius and ulna, and a complete pelvis. This description focuses on the phylogenetically informative appendicular elements: humerus, scapula, ilium, and pubis (Fig. 19.11).

Humerus The UMNH VP 12265 humerus (Fig. 19.11A, B) is typical for hadrosaurine hadrosaurids with a deltopectoral crest that extends to nearly midshaft and a more gracile morphology compared to most lambeosaurines. The ulnar condyle is larger than the radial, which is an usual feature and has been speculated as being age related (Brett-Surman and Wagner, 2007). Among hadrosaurines, UMNH VP 12265 most closely resembles *Gryposaurus notabilis* (ROM 764, formerly *G. incurvimanus*) although the former is more robust. Increased robusticity is a characteristic of most skeletal elements associated with *G. monumentensis*, including the skull. As such, the humerus does not resemble the gracile elements seen in *Brachylophosaurus canadensis* (Prieto-Marquez, 2007) or even *Edmontosaurus* spp.

Gates et al.

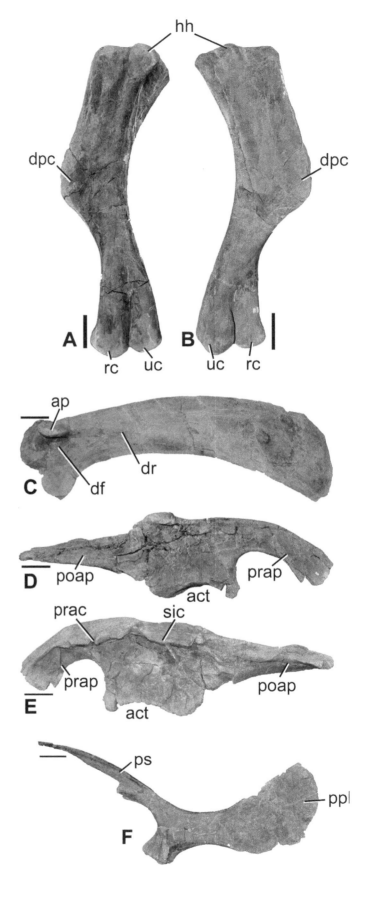

19.11. Postcranial elements of *Gryposaurus monumentensis* (UMNH VP 12265). (A, B) Humerus in posterior and anterior views; (C), scapula in lateral view; (D, E) ilium in medial and lateral views; (F) pubis. act, acetabulum; ap, acromion process; df, deltoid fossa; dpc, deltopectoral crest; dr, deltoid ridge; hh, humeral head; poap, postacetabular process; ppb, prepubic blade; prap, preacetabular process; ps, pubic shaft; rc, radial condyle; uc, ulnar condyle. Scale bar = 10 cm.

Scapula In general aspect, the shape of the scapula matches the long, narrow profile of hadrosaurines as opposed to the shorter, wider profile of lambeosaurines (Brett-Surman and Wagner, 2007). The distal end of the scapular blade is eroded so a ratio of length to width is unavailable. Proximally, as in other hadrosaurids, a convex deltoid fossa occupies the center of the proximal region (Fig. 19.11C). The acromion process protrudes laterally from the scapula. The deltoid ridge can be observed just dorsal to the acromion process extending posteroventrally to the ventral margin of the scapula, midway through its length. Angulation between the proximal scapula and the blade appears to be greater than that shown for *Brachylophosaurus* by Prieto-Marquez (2007); in other words, there is more curvature in the *Brachylophosaurus* scapula compared to *G. monumentensis*. The straighter scapula in *Gryposaurus* is confirmed by the nearly identical profile of the element preserved on *Gryposaurus notabilis* (ROM 764; formerly *G. incurvimanus*; see Prieto-Marquez, 2010).

Ilium The ilium (Fig. 19.11D, E) is the most distinctive skeletal element associated with *G. monumentensis* and contains features that make this element diagnostic of the species. The preacetabular process deflects ventrally as is seen in all hadrosaurid ilia, but the process does not taper anteriorly; instead, it increases in dorsoventral height. This feature is not known in other hadrosaurids. Additionally, the process is relatively shorter than most other taxa. The dorsal margin of the main iliac plate is flat. Other taxa have varying degrees of swelling of the pre- and postacetabular processes, yet *G. monumentensis* apparently displays none. A strongly overhanging ridge marks the dorsal boundary of the medial preacetabular process. The ridge is thick and extensive, continuing to the proximal margin of the postacetabular process. The supraacetabular crest is positioned dorsal to the iliac peduncle, and as preserved, it does not fold ventrally, as seen in ROM 764 or most other hadrosaurid specimens (Prieto-Marquez, 2010). Finally, the postacetabular process is distinctly triangular in lateral view, much more so than the squared to rounded posterior end seen in other hadrosaurids. This termination is different from that of *Gryposaurus notabilis* (ROM 764). As currently documented with known *G. monumentensis* ilia, characteristics that may be diagnostic to this species include distally expanded preacetabular process, nearly horizontal supraacetabular process, and triangular postacetabular process.

Pubis Pubic morphology varies substantially between hadrosaurid species, even between genera. This is true of *G. monumentensis* and *G. notabilis*, where the former has a ventrally deflected prepubic blade with a rounded dorsal margin (Fig. 19.11F) and the latter has a notched neck and square-like prepubic blade with rounded corners. However, it should be noted that an unknown portion of the dorsal prepubic blade of UMNH VP 12265 had eroded prior to excavation. The morphology of the prepubic blade is most similar to *Brachylophosaurus* (MOR 794).

Remarkably, UMNH VP 12265 also preserves more than 2.5 m² of fossilized skin impressions. Nonmineralized vertebrate tissues tend to be rare in the fossil record, because they are a rich source of nutrients for predators, scavengers, and microbes (Lund, 2006). Yet, more than 30 vertebrate localities preserving soft-tissue patterns have been recorded in the Kaiparowits Formation over the past six years, and all integument impressions are preserved within fine-to-coarse-grained, fluvial indurated sandstone. The majority of these localities preserve ornithopod, particularly hadrosaurid, skin impressions in both negative and positive relief. Impressions are known from nearly every portion of the body, although the best-preserved examples to date occur in association with the head, neck, and tail (Fig. 19.12). Tubercle density, shape, and size appear to vary along the body. Smaller, unornamented, circular tubercles cluster tightly around the head, whereas larger, ovoid, wider-spaced tubercles dominate along the back, tail, and limbs, many of the latter are ornamented with radiating ridges and grooves that converge at their apices. Tubercle size ranges from small (<3 mm) to large (>10 mm). The only exception is UMNH VP 12265, which preserves large, butterfly-shaped scales (~80 mm wide) and similarly sized ovoid scales found in direct association with each of the distal neural spines along the back and tail; a similar conformation has been described for a hadrosaur from the Hell Creek Formation (probably *Edmontosaurus*) (Horner, 1984). Overall, the hadrosaurid skin impressions known from Kaiparowits Formation compare favorably to others found in the Dinosaur Park Formation of Alberta, Canada, the Two Medicine and Judith River formations of Montana, and the Ringbone Formation of New Mexico (Lambe, 1914; Parks, 1920; Horner, 1984; Anderson et al., 1998; Negro and Prieto-Marquez, 2001; Evans and Reisz, 2007).

Finally, the most complete hadrosaurid specimen collected to date from Grand Staircase–Escalante National Monument (UMNH VP 16677; Fig. 19.13) consists of a juvenile specimen entombed in highly cemented sandstone. Other than the skull, lower jaws, hands, feet, and distal tail, the skeleton of this specimen appears to be complete. It is approximately

19.12. Specimens of hadrosaur skin impressions recovered from the Kaiparowits Formation. (A) Sandstone block containing large area of skin impression from near the neck region of UMNH VP 16667; (B) Close-up of skin impression, showing large tubercles with radiating ridges. Scale bar = 10 cm.

88 cm long and fully articulated except for the anteriormost cervical vertebrae that are just slightly separated from the remainder of the neck. On the basis of size-to-age comparisons by Horner, Ricqles, and Padian (2000), this individual is estimated to be approximately 3 months old, assuming that its growth rate was similar to that of *Maiasaura*. Definitive taxonomic identification is currently restricted to Hadrosaurinae, although given that only one genus of hadrosaurine has been discovered thus far from the Kaiparowits Formation, it seems

Gates et al.

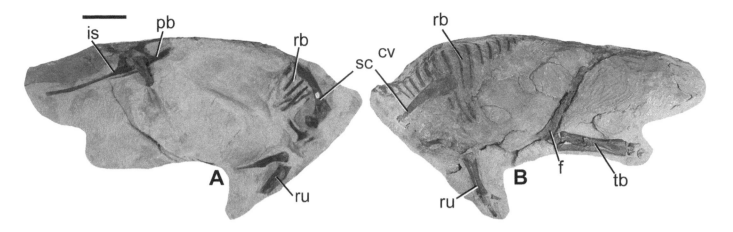

19.13. Articulated juvenile hadrosaur skeleton (UMNH VP 16677) found in the middle unit of the Kaiparowits Formation. cv, cervical vertebrae; f, femur; is, ischium; pb, pubis; rb, rib; ru, radius and ulna; sc, scapula; tb, tibia. Scale bar = 10 cm.

reasonable to suggest that UMNH VP 16677 is likely a species of *Gryposaurus*.

Ornithopod fossil foot traces are currently restricted to hadrosaurids. To date only isolated traces have been discovered, composed of sandstone and achieving sizes of 1 m in maximum width (Fig. 19.5). There has been no study of the morphology of the Kaiparowits Formation ornithopod tracks but some tracks such as that seen in Fig. 19.5 display asymmetry in pedal form.

IMPLICATIONS OF GRAND STAIRCASE–ESCALANTE NATIONAL MONUMENT FOSSIL RECORD

One of the greatest challenges within paleontology is discovering the nature of transition from one species or clade to a succeeding species or clade in terrestrial settings. More specific to the present ornithopod dinosaur survey is the transition from slightly more basal iguanodontians to those ornithopod taxa strictly within Hadrosauridae (as defined by Forster, 1997). Questions to be asked include: when did hadrosaurids arise in North America?; did hadrosaurids coexist with more basal iguanodontians in southern Utah?; if so, how long did they coexist before nonhadrosaurid iguanodontians became extinct?

The virtually continuous sequence of Upper Cretaceous sediments comprising the Straight Cliffs, Wahweap, and Kaiparowits formations provides an unparalleled opportunity to study the transitional zone between North American hadrosaurids and nonhadrosaurid iguanodontians and then the turnover of hadrosaurid taxa through the Campanian. On the basis of fossils collected thus far, it appears that nonhadrosaurid iguanodontians dominated the Smoky Hollow Member of the Straight Cliffs Formation, but forms that closely resemble hadrosaurids are preserved in the overlying

John Henry Member. This assertion is based strictly on the apparently narrow dentary of the Coniacian specimens, and therefore the evolutionary interpretation could be changed once better material has been acquired. As more attention is devoted to the Straight Cliffs Formation, the fossil record of ornithopods will undoubtedly increase and, consequently, the understanding of hadrosaurid first occurrences in southern Utah. The diversity of morphologies among dentaries collected from the John Henry Member demonstrates multiple taxa living in the region. Yet the stratigraphic relationship of these specimens is not established and further discussion of the transition toward a hadrosaurid dominated dinosaurian fauna must await more data collection.

Nevertheless, a relatively comprehensive picture of hadrosaur diversity is developing for the upper Wahweap and Kaiparowits formations. The growing picture appears to be one of relatively rapid faunal turnover and replacement. Such resolved evolutionary patterns are relatively rare, and approximately coeval examples of within-lineage turnover are otherwise documented only from geologic formations in the northern portion of the Western Interior Basin (e.g., Dinosaur Park Formation, Two Medicine Formation; Horner, Varricchio, and Goodwin, 1992; Ryan and Evans, 2005). Recognition of this pattern in the Campanian of Utah is the direct result of the large number of hadrosaur specimens collected within Grand Staircase–Escalante National Monument. To date, dozens of highly significant hadrosaur localities have been identified within the monument, with some preserving multiple specimens of one species, exquisite examples of hadrosaurids preserving integument impressions and skeletal elements in articulation, as well as sites containing taxonomically informative material from stratigraphic areas that do not contain many known fossils across the continent, making these some of the most productive Campanian strata in North America for the recovery of ornithopod dinosaurs.

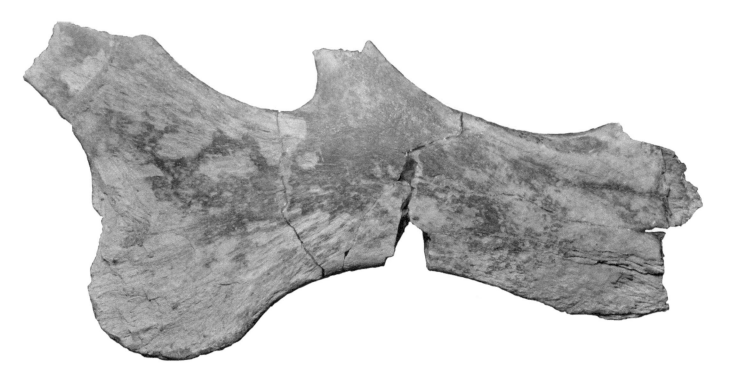

19.14. UMNH VP 16722 brachylophosaur-type hadrosaurine jugal found in the lower unit of the Kaiparowits Formation.

At present, the Wahweap and Kaiparowits formations preserve minimal evidence of faunal turnover within lambeosaurines. The new lambeosaurine taxon found in the Upper Sandstone Member of the Wahweap Formation is replaced by *Parasaurolophus* sp., which occurs in the Kaiparowits Formation. However, these two taxa are separated by a temporal gap of approximately 2–3 million years. More finely scaled faunal turnover can be documented for hadrosaurines as a result of better sampling in almost all of the subdivisions of the Wahweap and Kaiparowits. Beginning from a broader perspective, it is apparent that the "brachylophosaurs" are succeeded by the *Gryposaurus* clade. There are at least two "brachylophosaur" taxa in the Wahweap Formation and one other hadrosaurine taxon of unknown affinity. The Kaiparowits Formation has no substantial evidence of "brachylophosaur" hadrosaurs except one isolated jugal found near the base of the formation (UMNH VP 16722; Fig. 19.14). Nonetheless, this single element pushes the upper stratigraphic boundary of "brachylophosaur" hadrosaurines to younger than 76 million years, which is the latest known occurrence of this hadrosaurid clade. All other "brachylophosaurs" (*Acristavus*, *Maiasaura*, and *Brachylophosaurus*) are extinct elsewhere in the Western Interior Basin by around 76.5 million years ago (Horner et al., 2001; Gates et al., 2011).

Despite the three distinct hadrosaurines currently known from the Wahweap Formation, the exact stratigraphic distribution of these hadrosaurs is unknown and little can be said regarding turnover suffice to say that *Acristavus* may be replaced by cf. *Brachylophosaurus* over a course of 1–2 million years. More substantive evidence of faunal turnover is documented by the presence of two stratigraphically successive species of *Gryposaurus* within the Kaiparowits Formation. The large *Gryposaurus* sp. is currently only known from the lower unit of the Kaiparowits Formation whereas *G. monumentensis* is found in the middle unit. Exact timing of the interchange cannot be established at this time, as more specimens are needed to stratigraphically sandwich the turnover boundary, but the change likely took place over a course of a few hundred thousand years.

A series of recently obtained radiometric dates from Campanian-aged geologic formations within the Western Interior Basin (see Roberts, Deino, and Chan, 2005, for review) now permit time slice biogeographic comparisons. For example, the hadrosaurine *Acristavus* in the Wahweap Formation is approximately time correlative, with a specimen of the same species from the lower portion of the Two Medicine Formation of Montana and the cf. *Brachylophosaurus* sp. material found in the Upper Sandstone member of the Wahweap occurs at nearly the same time as *Brachylophosaurus canadensis* in the Judith River Formation of Montana.

Within the Kaiparowits Formation, two main time intervals are available for comparison, corresponding to the lower and middle units. The only hadrosaur genus in the lower unit is *Gryposaurus* sp. A radiometric date of 76.1 mya for the base of the formation indicates that this taxon is approximately coeval with *G. notabilis* from the lower Dinosaur Park Formation of Alberta, Canada (Ryan and Evans, 2005). Within the middle unit of the Kaiparowits Formation, *G. monumentensis* is probably contemporaneous with *Prosaurolophus maximus*, which has an extended stratigraphic range within the Dinosaur Park Formation and lower part of the marine Bearpaw Formation in Alberta, and the upper portion of the Two Medicine Formation in Montana (Horner, 1992; Gates and Evans, 2005; Ryan and Evans, 2005; Gates and Sampson, 2007; Prieto-Marquez, 2010).

The contemporaneous nature of the fossiliferous portion of the Kaiparowits and Dinosaur Park formations permits faunal comparisons over a large geographic transect in the Western Interior Basin. The Kaiparowits formation hadrosaurid fauna differs significantly from that of the contemporaneous Dinosaur Park Formation in that it is dominated by hadrosaurines. In this respect it is more typical of Late Cretaceous hadrosaur-dominated assemblages. The Dinosaur Park Formation is dominated by lambeosaurines, of which corythosaurian taxa are common and *Parasaurolophus*, although present, is rare (Ryan and Evans, 2005; Evans, Bavington, and Campione, 2009). Corythosaurian lambeosaurines, which are well documented in the series of Late Cretaceous assemblages of Alberta and Montana, are unknown from contemporaneous strata of Utah and New Mexico.

Also within the middle unit of the Kaiparowits Formation is *Parasaurolophus* sp., which is approximately contemporaneous with two northern lambeosaurines, *Hypacrosaurus stebingeri* in Montana and *Lambeosaurus lambei* in Alberta (Gates and Evans, 2005). The type and only confirmed specimen of *Parasaurolophus cyrtocristatus* occurs in the upper Fruitland Formation, in beds that are stratigraphically higher than all known occurrences of the Utah *Parasaurolophus*. Thus, if it turns out that the *Parasaurolophus* from the Kaiparowits Formation should be placed within *P. cyrtocristatus*, this finding would extend the stratigraphic distribution of the species. The other two species of *Parasaurolophus* (*P. walkeri* and *P. tubicen*) occur in the lower Dinosaur Park Formation in Alberta (~ 75.5 Ma) and in the Upper Kirtland Formation (~ 73.5 Ma) of New Mexico, respectively (Sullivan and Williamson, 1999; Gates and Evans, 2005; Ryan and Evans, 2005; Evans, Bavington, and Campione, 2009). On the basis of current evidence, none of the *Parasaurolophus* species appear to co-occur in time (Sullivan and Williamson, 1999; Evans, Bavington, and Campione, 2009). However, it is interesting to note that the geographic distribution of this genus correlates with its stratigraphic distribution, the earliest example of *Parasaurolophus*, *P. walker*, occurs in the north, whereas the youngest species occur in the southern region of the Western Interior Basin.

With regard to basal ornithopods, the only well-known basal ornithopod taxon from the late Campanian, *Orodromeus*, occurs stratigraphically higher within the Two Medicine Formation of Montana than the earliest known basal ornithopod specimen from the Grand Staircase–Escalante National Monument (UMNH VP 16281) that was recovered near the base of the Kaiparowits Formation. In fact, UMNH VP 16281 has no known stratigraphic equivalent in North America. In contrast, the basal ornithopod specimens collected from the younger middle unit of the Kaiparowits Formation (UMNH VP 12665 and 19470) appear to be approximately coeval with, though unequivocally different from, the more northerly occurring *Orodromeus*.

A single specimen of basal ornithopod collected from the lower unit of the Kaiparowits Formation (UMNH 12677) displays an autapomorphic trait of only two phalanges in the fourth digit, though the specimen is too fragmentary to accurately compare to the diagnostic specimens recovered from the middle unit (i.e., UMNH VP 12665 and 19470). This fact leaves open the possibility that the material from the lower and middle units of the Kaiparowits Formation may represent two distinct, temporally segregated basal ornithopod taxa. Alternatively, the lack of character conflict among all known specimens from the formation may indicate all of these specimens are referable to a single, previously undescribed taxon. Resolution of this taxonomic question will require the recovery of additional, diagnostic specimens from the lower unit of the Kaiparowits Formation, or a specimen with an articulated manus from the middle unit.

CONCLUSIONS

In summary, Upper Cretaceous (Campanian) sediments preserved within Grand Staircase–Escalante National Monument have yielded an abundance of ornithopod specimens that dramatically increase our knowledge of this clade, particularly for the southwestern region of the Western Interior Basin. The majority of these specimens have been recovered from the late Campanian Kaiparowits Formation, although the underlying Wahweap Formation has also produced several significant discoveries, including hadrosaur bonebeds and the partial skull of a new hadrosaurine. The Straight Cliffs Formation is producing significant specimens and will become an extremely important source of information regarding the transition between nonhadrosaurid iguanodontians and hadrosaurids.

Key specimens from the Kaiparowits Formation include several partial skulls and a partial skeleton of the lambeosaurine *Parasaurolophus* as well as multiple specimens (skulls and postcranial remains) pertaining to two species of *Gryposaurus*. Finally, several "hypsilophodontid" specimens have been recovered from the Kaiparowits Formation, the most significant of which includes partial skulls as well as manual and pedal elements representing an undescribed form (Boyd and Gates, unpubl. data). The acquisition of multiple radiometric dates from the Kaiparowits Formation permits relatively high-resolution temporal comparisons within a 2 million year window spanning approximately 76–74 Ma. Thus, for example, the *Gryposaurus* species discovered near the base of the Kaiparowits Formation (*Gryposaurus* sp.) appears to have been coeval with other species of *Gryposaurus* in the northern region of the Western Interior Basin. In contrast, *Gryposaurus monumentensis*, from the middle unit of the Kaiparowits Formation, corresponds temporally with the northern hadrosaurine genus *Prosaurolophus*. Together, these recent discoveries establish Grand Staircase–Escalante National Monument as one of the premier localities in North America for producing remains of ornithopod dinosaurs.

ACKNOWLEDGMENTS

The authors sincerely thank J. Gentry, J. Golden, S. Walkington, F. Lacey, S. Dahl, H. S. Richardson, and all of the Natural History Museum of Utah, Utah Geological Survey, and Grand Staircase–Escalante National Monument volunteers for their generous assistance in fieldwork and preparation of Grand Staircase–Escalante National Monument specimens; S. Sampson for initiating and sustaining the Kaiparowits Basin Project and for helpful comments on earlier drafts of this chapter; L. Bryant and S. Foss (Bureau of Land Management, BLM), and A. Titus and M. Eaton (Grand Staircase–Escalante National Monument) for assistance with permitting and field logistics; the staff of Grand Staircase–Escalante National Monument for ongoing support; and A. Farke, J. Horner, J. Hutchison, M. Loewen, J. Sertich, J. A. Smith, and L. Zanno for fieldwork and helpful discussions. Don Lofgren (Raymond M. Alf Museum of Paleontology) is thanked for permission to study RAM 6797. We are indebted to R. Nelson for discovering and reporting UMNH VP 16607. Many thanks to J. Hutchison for help obtaining hadrosaurid specimens on loan from the UCMP and for relocating UMNH VP 16607. P. Makovicky and L. Hertzog assisted in histological sectioning of UMNH VP 12665. For access to comparative specimens, we thank J. Horner (Museum of the Rockies), B. Simpson (Field Museum), P. Currie and J. Gardner (Royal Tyrrell Museum of Palaeontology), K. Shepherd (Canadian Museum of Nature), and K. Seymour (Royal Ontario Museum). Suggestions by D. Weishampel and P. Godefroit greatly improved the chapter. This research has been supported by funding from the BLM (Grand Staircase–Escalante National Monument), Discovery Communications Quest Grants, the University of Utah, the Utah Geological Survey, and the Jurassic Foundation.

REFERENCES CITED

Anderson, B. G., S. G. Lucas, R. E. Barrick, A. B. Heckert, and G. T. Basabilvazo. 1998. Dinosaur skin impressions and associated skeletal remains from the Upper Campanian of southwestern New Mexico: new data on the integument morphology of hadrosaurs. Journal of Vertebrate Paleontology 18:739–745.

Boyd, C. A., C. M. Brown, R. D. Scheetz, and J. A. Clarke. 2009. Taxonomic revision of the basal Neornithischian taxa *Thescelosaurus* and *Bugenasaura*. Journal of Vertebrate Paleontology 29:758–770.

Brett-Surman, M. K., and J. R. Wagner. 2007. Discussion of character analysis of the appendicular anatomy in Campanian and Maastrichtian North American hadrosaurids–variation and ontogeny; pp. 135–170 in K. Carpenter (ed.), Horns and Beaks–Ceratopsian and Ornithopod Dinosaurs. Indiana University Press, Bloomington, Indiana.

Brochu, C. A. 1996. Closure of neurocentral sutures during crocodilian ontogeny: implications for maturity assessment in fossil archosaurs. Journal of Vertebrate Paleontology 16:49–62.

Butler, R. J., P. Upchurch, and D. B. Norman. 2008. The phylogeny of the ornithischian dinosaurs. Journal of Systematic Palaeontology 6:1–40.

Case, J. A., J. E. Martin, D. S. Chaney, M. Reguero, S. A. Marenssi, S. M. Santillana, and M. O. Woodburne. 2000. The first duck-billed dinosaur (Family Hadrosauridae) from Antarctica. Journal of Vertebrate Paleontology 20:612–614.

Cifelli, R. L. 1987. Therian mammals from the Late Cretaceous of the Kaiparowits region, Utah. Journal of Vertebrate Paleontology 7:14A.

Cifelli, R. L. 1990. Cretaceous mammals of southern Utah; I, Marsupials from the Kaiparowits Formation (Judithian). Journal of Vertebrate Paleontology 10:295–319.

Cobban, W. A., T. S. Dyman, G. L. Pollock, K. I. Takahashi, L. E. Davis, and D. B. Riggin. 2000. Inventory of dominantly marine and brackish-water fossils from Late Cretaceous rocks in and near Grand Staircase–Escalante National Monument, Utah; pp. 579–589 in D. A. Sprinkel, T. C. Chidsey, and P. B. Anderson (eds.), Geology of Utah's Parks and Monuments. Utah Geological Association Publication 28.

Doelling, H. H., R. E. Blackett, A. H. Hamblin, J. D. Powell, and G. L. Pollock. 2000. Geology of Grand Staircase–Escalante National Monument, Utah; pp. 189–231 in D. A. Sprinkel, T. C. Chidsey, and P. B. Anderson (eds.), Geology of Utah's Parks and Monuments. Utah Geological Association Publication 28.

Eaton, J. G. 1991. Biostratigraphic framework for Upper Cretaceous rocks of the Kaiparowits Plateau, southern Utah; pp. 47–63 in J. D. Nations and J. G. Eaton (eds.), Stratigraphy, Depositional Environments, and Sedimentary Tectonics of the Western Margin, Cretaceous Western Interior Seaway. Geological Society of America Special Publication 260.

Eaton, J. G. 2002. Multituberculate mammals from the Wahweap (Campanian, Aquilan) and Kaiparowits (Campanian, Judithian) formations, Grand Staircase–Escalante National Monument, southern Utah, and implications for biostratigraphic methods. Geological Society of America, Abstracts with Programs 34:6.

Eaton, J. G., R. L. Cifelli, J. H. Hutchison, J. I. Kirkland, and J. M. Parrish. 1999. Cretaceous vertebrate faunas from the Kaiparowits Plateau, south-central Utah; pp. 345–353 in D. D. Gillette (ed.), Vertebrate Paleontology in Utah. Utah Geological Survey Miscellaneous Publication 99–1.

Evans, D. C., and R. R. Reisz. 2007. Anatomy and relationships of *Lambeosaurus magnicristatus*, a crested hadrosaurid dinosaur (Ornithischia) from the Dinosaur Park Formation, Alberta. Journal of Vertebrate Paleontology 27:373–393.

Evans, D. C., R. Bavington, and N. E. Campione. 2009. An unusual hadrosaurid braincase from the Dinosaur Park Formation, and the biostratigraphy of *Parasaurolophus* (Ornithischia: Lambeosaurinae) from southern Alberta. Canadian Journal of Earth Science 46:791–800.

Forster, C. A. 1997. Phylogeny of the Iguanodontia and Hadrosauridae. Journal of Vertebrate Paleontology 17:47A.

Gates, T. A., and D. C. Evans. 2005. Biogeography of Campanian hadrosaurid dinosaurs from western North America; pp. 33–39 in D. R. Braman, F. Therrien, E. B. Koppelhus, and W. Taylor (eds.), Dinosaur Park Symposium Short Papers, Abstracts, and Programs. Royal Tyrrell Museum of Paleontology, Drumheller, Alberta.

Gates, T. A., and S. D. Sampson. 2007. A new species of *Gryposaurus* (Dinosauria: Hadrosauridae) from the Late Campanian Kaiparowits Formation. Zoological Journal of the Linnean Society 151:351–376.

Gates, T. A., J. R. Horner, R. R. Hanna, and C. R. Nelson. 2011. New hadrosaurine hadrosaurid (Dinosauria, Ornithopoda) from the Campanian of North America. Journal of Vertebrate Paleontology 31:798–811.

Gates, T. A., E. K. Lund, M. A. Getty, J. I. Kirkland, A. L. Titus, D. Deblieux, C. A. Boyd, and S. D. Sampson. 2010. Late Cretaceous ornithopod dinosaurs from the Kaiparowits Plateau, Grand Staircase–Escalante National Monument, Utah; p. 159 in in M. Eaton (ed.), Learning from the Land, Grand Staircase–Escalante National Monument Science Symposium Proceedings. Grand Staircase–Escalante Partners.

Gilmore, C. W. 1913. A new dinosaur from the Lance Formation of Wyoming. Smithsonian Miscellaneous Collections 61:1–5.

Herrero, L., and A. Farke. 2010. Morphological variation in the hadrosaur dinosaur *Gryposaurus* from the Kaiparowits Formation (Late Campanian) of southern Utah. Journal of Vertebrate Paleontology Abstracts with Programs:104A.

Horner, J. R. 1984. A "segmented" epidermal tail frill in a species of hadrosaurian dinosaur. Journal of Paleontology 58:270–271.

Horner, J. R. 1992. Cranial morphology of *Prosaurolophus* (Ornithischia: Hadrosauridae) with descriptions of two new hadrosaurid species and an evaluation of hadrosaurid phylogenetic relationships. Museum of the Rockies Occasional Paper 2:1–119.

Horner, J. R., A. d. Ricqles, and K. Padian. 2000. The bone histology of the hadrosaurid dinosaur *Maiasaura peeblesorum*: growth dynamics and physiology on an ontogenetic series of skeletal elements. Journal of Vertebrate Paleontology 20:109–123.

Horner, J. R., D. J. Varricchio, and M. B. Goodwin. 1992. Marine transgressions and the evolution of Cretaceous dinosaurs. Nature 358:59–61.

Horner, J. R., D. B. Weishampel, and C. Forster. 2004. Hadrosauridae; pp. 438–463 in D. B. Weishampel, P. Dodson, and H. Osmólska

(eds.), The Dinosauria. University of California Press, Berkeley, California.

Horner, J. R., J. G. Schmitt, F. Jackson, and R. Hanna. 2001. Bones and rocks of the Upper Cretaceous Two Medicine–Judith River Clastic Wedge Complex, Montana; pp. 3–14 in C. L. Hill (ed.), Guidebook for the Field Trips, Society of Vertebrate Paleontology 61st Annual Meeting: Mesozoic and Cenozoic Paleontology in the Western Plains and Rocky Mountains. Museum of the Rockies, Bozeman, Montana.

Irmis, R. B. 2007. Axial skeleton ontogeny in the parasuchia (Archosauria: Pseudosuchia) and its implications for ontogenetic determination in archosaurs. Journal of Vertebate Paleontology 27:350–361.

Jinnah, Z. A., M. A. Getty, and T. A. Gates. 2009. Taphonomy of a hadrosaur bonebed from the Upper Cretaceous Wahweap Formation, Grand Staircase–Escalante National Monument, Utah. Advances in Western Interior Late Cretaceous Paleontology and Geology (St. George, Utah), Abstracts with Program, p. 28.

Jinnah, Z. A., E. M. Roberts, A. L. Deino, J. S. Larsen, P. K. Link, and C. M. Fanning. 2009. New ^{40}Ar-^{39}Ar and detrital zircon U-Pb ages for the Upper Cretaceous Wahweap and Kaiparowits formations on the Kaiparowits Plateau, Utah: implications for regional correlation, provenance, and biostratigraphy. Cretaceous Research 30:287–299.

Lambe, L. M. 1914. On the fore-limb of a carnivorous dinosaur from the Belly River Formation of Alberta, and a new genus of Ceratopsia from the same horizon, with remarks on the integument of some Cretaceous herbivorous dinosaurs. Ottawa Naturalist 27:129–135.

Lund, E. K. 2006. The softer side of preparation: dealing with nonmineralized vertebrate tissues. Journal of Vertebrate Paleontology 26:91A-92A.

Marsh, O. C. 1881. Principal characters of American Jurassic dinosaurs. Part IV. American Journal of Science, ser. 3, 21:167–170.

McDonald, A. T., D. G. Wolfe, and J. I. Kirkland. 2006. On a hadrosauromorph (Dinosauria: Ornithopoda) from the Moreno Hill Formation (Cretaceous, Turonian) of New Mexico; pp. 277–279 in S. G. Lucas and R. M. Sullivan (eds.), Late Cretaceous Vertebrates from the Western Interior. New Mexico Museum of Natural History and Science Bulletin 35.

McDonald, A. T., D. G. Wolf, and J. I. Kirkland. 2010. A new basal hadrosauroid (Dinosauria: Ornithopoda) from the Turonian of New Mexico. Journal of Vertebate Paleontology 30:799–812.

Morris, W. J. 1970. Hadrosaurian dinosaur bills–morphology and function. Contributions in Science of the Los Angeles County Museum of Natural History 193:1–14.

Negro, G., and A. Prieto-Marquez. 2001. Hadrosaurian skin impressions from the Judith River Formation (Lower Campanian) of Montana. North American Paleontological Conference Abstracts 2001:96.

Norman, D. B., H.-D. Sues, L. M. Witmer, and R. A. Coria. 2004. Basal Ornithopoda; pp.

438–463 in D. B. Weishampel, P. Dodson, and H. Osmólska (eds.), The Dinosauria. University of California Press, Berkeley, California.

Parks, W. A. 1920. The osteology of the trachodont dinosaur *Kritosaurus incurvimanus*. University of Toronto Studies, Geology Series 11:1–75.

Parrish, J. M. 1999. Dinosaur teeth from the Upper Cretaceous (Turonian–Judithian) of southern Utah; pp. 319–321 in D. D. Gillette (ed.), Vertebrate Paleontology in Utah. Utah Geological Survey Miscellaneous Publication 99–1.

Peterson, F. 1969. Four new members of the Upper Cretaceous Straight Cliffs Formation in southeastern Kaiparowits region, Kane County, Utah. U.S. Geological Survey Bulletin 1274-J.

Prieto-Marquez, A. 2007. Postcranial osteology of the hadrosaurid dinosaur *Brachylophosaurus canadensis* from the Late Cretaceous of Montana; pp. 91–116 in K. Carpenter (ed.), Horns and Beaks. Indiana University Press, Bloomington, Indiana.

Prieto-Marquez, A. 2010. Global phylogeny of Hadrosauridae (Dinosauria: Ornithopoda) using parsimony and Bayesian methods. Zoological Journal of the Linnean Society 159:435–502.

Roberts, E. M. 2007. Facies architecture and depositional environments of the Upper Cretaceous Kaiparowits Formation, southern Utah. Sedimentary Geology 197:207–233.

Roberts, E. M., A. L. Deino, and M. A. Chan. 2005. ^{40}Ar/^{39}Ar age of the Kaiparowits Formation, southern Utah, and correlation of contemporaneous Campanian strata and vertebrate faunas along the margin of the Western Interior Basin. Cretaceous Research 26:307–318.

Ryan, M., and D. C. Evans. 2005. Ornithischian dinosaurs; pp. 312–348 in P. J. Currie and E. B. Koppelhus (eds.), Dinosaur Provincial Park: A Spectacular Ancient Ecosystem Revealed. Indiana University Press, Bloomington, Indiana.

Scheetz, R. D. 1999. Osteology of *Orodromeus makelai* and the phylogeny of basal ornithopod dinosaurs. Ph.D. dissertation, Montana State University, Bozeman, Montana.

Sullivan, R. M., and T. E. Williamson. 1999. A New Skull of *Parasaurolophus* (Dinosauria: Hadrosauridae) from the Kirtland Formation of New Mexico and a Revision of the Genus. New Mexico Museum of Natural History and Science Bulletin 15.

Titus, A. L., D. D. Gillette, and L. B. Albright. 2001. Significance of an articulated lambeosaurine hadrosaur from the Kaiparowits Formation (Upper Formation), southern Utah. Journal of Vertebrate Paleontology 21:108A.

Varricchio, D. J., A. J. Martin, and Y. Katsura. 2007. First trace and body fossil evidence of a burrowing, denning, dinosaur. Proceedings of the Royal Society of London, Series B 274:1361–1368.

Weishampel, D. B., and J. A. Jensen. 1979. *Parasaurolophus* (Reptilia: Hadrosauridae) from Utah. Journal of Paleontology 53:1422–1427.

You, H., Q. Ji, and Y. Li. 2003. A new hadrosauroid dinosaur from the mid-Cretaceous of Liaoning, China. Acta Geologica Sinica 77.148–154.

Review of Pachycephalosaurian Dinosaurs from Grand Staircase–Escalante National Monument, Southern Utah

20

David C. Evans, Thomas Williamson, Mark A. Loewen, and James I. Kirkland

RECENT FIELDWORK IN THE UPPER CRETACEOUS (CAM-panian) Wahweap and Kaiparowits formations of southern Utah have resulted in the recovery of the first cranial remains of pachycephalosaurian dinosaurs from these units. Pachycephalosaurs from the Wahweap Formation are represented by isolated teeth and a single, incomplete frontoparietal dome, whereas the Kaiparowits Formation has yielded isolated teeth plus six fragmentary cranial specimens. The available material is incomplete and the sample remains small, but it suggests the possible presence of at least two new taxa, one in each of the two formations. These specimens significantly expand our knowledge and understanding of pachycephalosaurian dinosaur diversity from southern Laramidia.

INTRODUCTION

Pachycephalosaurs are a group of small- to medium-sized bipedal ornithischian dinosaurs characterized by a thickened cranial vault typically forming a prominent frontoparietal dome (Maryańska, Chapman, and Weishampel, 2004). Pachycephalosaur remains are consistently recovered from terrestrial Campanian–Maastrichtian dinosaur-bearing sediments of western North America and Asia, although they are typically represented by fragmentary material (Maryańska, Chapman, and Weishampel, 2004; Sullivan, 2006). In well-sampled units of the Western Interior, such as the Dinosaur Park and Hell Creek formations, at least two or three species of presumably contemporaneous pachycephalosaurs are recognized (Ryan and Evans, 2005; Sullivan, 2006). As such, pachycephalosaurs are a consistent part of the dinosaur faunas where they occur, and they should be considered in studies of Late Cretaceous paleoecology, historical biogeography, and ornithischian biodiversity dynamics.

Pachycephalosaur remains from the southern region of the Western Interior Basin, which preserves the Late Cretaceous terrestrial biotas of southern Laramidia, are rare and remain poorly understood (Lucas, Heckert, and Sullivan, 2000; Williamson and Carr, 2002; Sullivan, 2003, 2006; Lehman, 2010; Longrich, Sankey, and Tanke, 2010). Recent fieldwork focused on vertebrate macrofossils in the Upper Cretaceous Wahweap and Kaiparowits formations of southern Utah have greatly increased our knowledge of these stratigraphic units and have resulted in the recovery of the first nondental pachycephalosaurian remains from these formations (Kirkland, 2001; Kirkland and DeBlieux, 2005; Sampson, Gates, et al., 2010; Sampson et al., this volume, Chapter 28). Here we review the pachycephalosaurain fossil record from southern Utah and provide a preliminary assessment of taxonomic diversity and phylogenetic affinities.

Institutional Abbreviations ROM, Royal Ontario Museum, Toronto, Ontario; UMNH, Utah Museum of Natural History, Salt Lake City, Utah.

RESULTS

The only previous reported remains of pachycephalosaurs from Utah are based on a small number of undescribed, isolated teeth, most of which were collected via intensive screen-washing efforts by Cifelli and Eaton and colleagues (Eaton, Cifelli, et al., 1999; Eaton, Diem, et al., 1999). Recent studies have emphasized uncertainties associated with identification of isolated, foliform ornithischian teeth as pachycephalosaurian (Sullivan, 2006; Butler and Sullivan, 2009), including specimens from Utah (Eaton, Cifelli, et al., 1999). Therefore, although isolated teeth possibly attributable to pachycephalosaurs have been recovered from most of the southern Utah units discussed below, this review focuses on newly discovered cranial materials that exhibit the characteristic dense, thickened frontoparietal morphology diagnostic of the clade.

Dakota through Straight Cliffs Formations

The upper Cenomanian Dakota Formation represents the earliest record of Cretaceous sedimentation in the Grand Staircase region. In their review of Cretaceous vertebrate faunas of the Kaiparowits Plateau, Eaton, Cifelli, et al. (1999) noted the possible presence of Pachycephalosauria in the Dakota Formation on the basis of isolated teeth, but preferred a tentative ankylosaurian assignment for these specimens. The Cenomanian through early Campanian Tropic and Straight Cliffs formations have not yet produced any

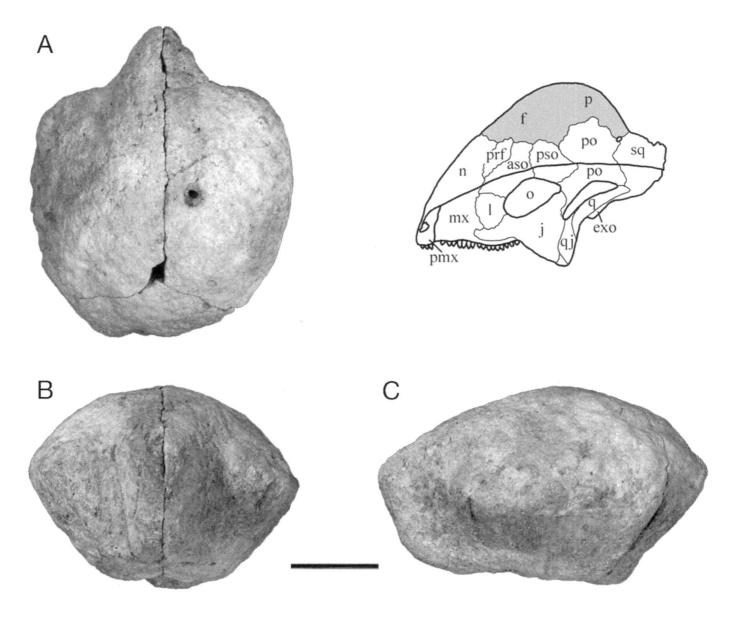

20.1. Pachycephalosaurid frontoparietal dome (UMNH VP 11939) from the Wahweap Formation in (A) dorsal, (B) anterior, and (C) left lateral views. Gray region on inset *Stegoceras* skull indicates region of the thickened frontoparietal dome (redrawn from Maryańska, Chapman, and Weishampel, 2004). Scale bar = 10 mm.

definitive remains of pachycephalosaurs (Eaton, Diem, et al., 1999; Parrish, 1999).

Wahweap Formation

A partial pachycephalosaur frontoparietal dome (UMNH VP 11939) was discovered by UMNH volunteer Jodi Vincent in the lower part of the Middle Member of the Wahweap Formation in Clints Cove (UMNH VP Loc. 261). This specimen represents the only diagnosable pachycephalosaur fossil (with the possible exception of isolated teeth) known from the Wahweap Formation (Kirkland and DeBlieux, 2005). Although this isolated dome is badly weathered, considerable anatomical information is preserved (Fig. 20.1). The

frontal portion of the dome is mostly complete and the parietal portion is largely missing. In dorsal view, the partial frontoparietal is nearly square with rounded corners, except for a triangular projection of the frontal that extends anteriorly, forming the base of the frontonasal boss. The relatively deep lateral sutural surfaces indicate that large parts of the peripheral roofing bones were incorporated into the dome, as in other mature pachycephalosaurids (Williamson and Carr, 2002). The specimen appears to represent an advanced ontogenetic growth stage, possibly of an adult.

UMNH VP 11939 resembles several other pachycephalosaurids (e.g., *Stegoceras*, *Sphaerotholus*) in having a relatively wide and broadly convex frontoparietal dome (Williamson and Carr, 2002; Sullivan, 2003). The anteromedial extension,

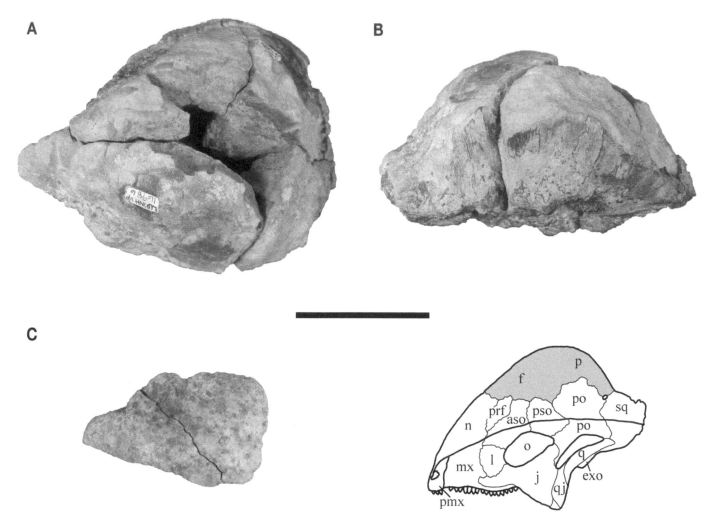

20.2. Pachycephalosaurid fossils from the Kaiparowits Formation, Utah. (A, B) UMNH VP 16986, incomplete frontal region of dome; (C) UMNH VP 18611 almost complete frontoparietal dome in lateral and dorsal views respectively. Gray region on inset *Stegoceras* skull indicates region of the thickened frontoparietal dome (redrawn from Maryańska, Chapman, and Weishampel, 2004). Scale bar = 10 mm.

or frontonasal boss, of the frontals differs from that of all other known pachycephalosaurs in being short and narrowing anteriorly, nearly converging to a point in dorsal aspect. Shallow depressions lateral to the frontonasal boss are present but weakly developed, and the specimen appears to lack the distinctive grooving present in mature specimens of *Stegoceras validum* (Williamson an Carr, 2002; Sullivan, 2003). Unfortunately, the specimen lacks the posterior portion of the parietal, so many potentially diagnostic characters—for example, relative development of the parietosquamosal shelf, presence or absence of nodal ornamentation along the posterior margin of the parietal, and morphology of the temporal fossa—cannot be determined.

Kaiparowits Formation

Eaton, Cifelli, et al. (1999) assigned isolated teeth from the Kaiparowits Formation to the pachycephalosaurid

Stegoceras. Subsequent to this study, a total of four partial cranial domes have been collected from the middle member of the formation. The most complete specimen is UMNH VP 16986, consisting of a large portion of a heavily weathered frontoparietal (Fig 20.2). The dorsal and ventral surfaces are abraded and lack much of their cortex, but portions of the marginal sutural surfaces for the peripheral cranial bones are well preserved, particularly on the left side. The frontoparietal is strongly domed, with a prominent frontonasal boss. The presence of well-developed sutural surfaces on the posterolateral region of the parietal indicates that the supratemporal fenestrae were closed in this individual. In lateral view, the prominent, rounded dome is symmetrical and the marginal sutural surfaces indicate that the posterior peripheral bones are high on the dome, as in *Stegoceras validum* and some specimens assigned to *Hanssuesia sternbergi* (Sullivan, 2003). Another nearly complete frontoparietal (UMNH VP 21112) was discovered by Mike Getty of the UMNH in the

Evans et al.

summer of 2010 (not figured). Despite its small size, doming is advanced, particularly in the region of the frontoparietal contact. The specimen is proportionately wide across this region of the dome, and the frontals appear relatively short. Although UMNH VP 21112 appears to be missing the postero-median extension of the parietal, which is broken at its base, the preserved portion of the parietal suggests the presence of large, open supratemporal fenestrae. The prominent fronto-nasal boss and the low supraorbital contact surfaces generally resemble those features in *Stegoceras*-grade taxa.

An anterior fragment of a smaller dome, UMNH VP 18611, preserves a weakly developed groove between the fronto-nasal boss and the supraorbital lobe of the frontal, and also resembles *Stegoceras*-grade taxa (Fig. 20.2; Williamson and Carr, 2002; Sullivan, 2003). UMNH VP 18594 is a badly weathered posterior fragment of a frontoparietal that preserves few anatomical features. At greater than 68 mm in thickness, it is notable for being the largest pachycephalosaur specimen recovered from the Kaiparowits Formation, within the size range of the largest specimens of *Stegoceras validum* (e.g., ROM 53555) from the Belly River Group, Alberta, yet still representative of a relatively small-bodied pachycephalosaur.

At the beginning of the 2011 field season, Alan Titus discovered the remains of a relatively complete pachycephalosaur skull, also from the middle member of the Kaiparowits Formation. This specimen is the best-preserved pachycephalosaur skull collected to date from the Kaiparowits Formation, and one of the best ever found in the Campanian of North America. Preparation of the specimen is not yet complete, but it will surely provide the most detailed data on the anatomy and relationships of the Kaiparowits pachycephalosaurs.

In addition to the frontoparietals, two thickened cranial elements that contact the dome have also been recovered from the Kaiparowits Formation. These include a reasonably complete squamosal (UMNH VP 9546) and a heavily weathered element that may be a postorbital (UMNH VP 19653). The squamosal is very similar in morphology to that of *Stegoceras validum*, with a single row of enlarged primary nodes set against a background ornamentation of diminutive, irregularly oriented nodes (Sues and Galton, 1987). The margin of a relatively large, circular supratemporal fenestra is clearly preserved in UMNH VP 9546. However, on the basis of individual and ontogenetic variation in *Stegoceras validum* (Williamson and Carr, 2002; Schott and Evans, 2012), the presence or absence of open supratemporal fenestrae is a poor indicator of phylogenetic relationships. In dorsal view, the squamosal is notably broad posterolaterally, where the otherwise flat dorsal surface is depressed. Although subtle, this condition is unusual and differs from that typically seen in the large sample of *Stegoceras* squamosals from Dinosaur Provincial Park, Alberta (Schott and Evans, 2012).

DISCUSSION

Newly discovered cranial material from the Wahweap and Kaiparowits formations, including characteristically thickened frontoparietal domes and peripheral skull roof elements, demonstrate conclusively that pachycephalosaurids occupied this region of Laramidia throughout much of the Campanian. The incomplete, worn frontoparietal (UMNH VP 11939) from the middle mudstone member of the Wahweap Formation of Utah (Fig. 20.3) represents the oldest record of pachycephalosaurs in the Grand Staircase region on the basis of unequivocal, nondental skeletal material (Kirkland and DeBlieux, 2005; Titus et al., 2005). The probable middle Campanian age of this material (~80 mya; Jinnah et al., 2009) also makes it one of the earliest incontrovertible records of this ornithischian clade (Sullivan, 2006; Butler and Sullivan, 2009). Although worn, the dome differs from all known pachycephalosaurs in possessing a narrow, triangular nasal buttress. This unique morphology, together with its geographic and temporal provenance, suggests that the specimen may pertain to a new taxon of pachycephalosaurid (Kirkland and DeBlieux, 2005). However, the poor preservation of the specimen makes any generic assignment or taxonomic distinction premature at this time.

Frontoparietals from the Kaiparowits Formation are also difficult to identify beyond the level of Pachycephalosauridae as a result of poor or incomplete preservation. Three of the domes, (UMNH VP 16986, 21112, and 9546) share a number of features with *Stegoceras*-grade taxa, including a pear-shaped dome in dorsal view, and grooved frontals that offset a relatively low supraorbital lobe (Williamson and Carr, 2002; Sullivan, 2003). Given its excellent state of preservation and the diagnostic nature of parietosquamosal morphology, the squamosal UMNH VP 9546 is the most taxonomically informative pachycephalosaurian fossil available for study from the Kaiparowits Formation. The pattern of parietosquamosal ornamentation is very similar to that of *Stegoceras validum* (Schott and Evans, in press), although it differs from the known sample of *Stegoceras* in subtle, yet potentially significant morphological details, and as such may also represent new species. However, squamosal morphology is variable intraspecifically, and any taxonomic assignment must await future discoveries of more complete remains. The available sample indicates the presence of at least one *Stegoceras*-grade taxon in the Kaiparowits Formation. Differences in the structure of the parietal in the two most complete domes (UMNH VP 16986, and 21112) may support the presence of two taxa, but this region is variable in *Stegoceras validum* (Williamson and Carr, 2002; Sullivan, 2003), and the small sample size and generally poor preservation of the material prevents a more definitive taxonomic assessment.

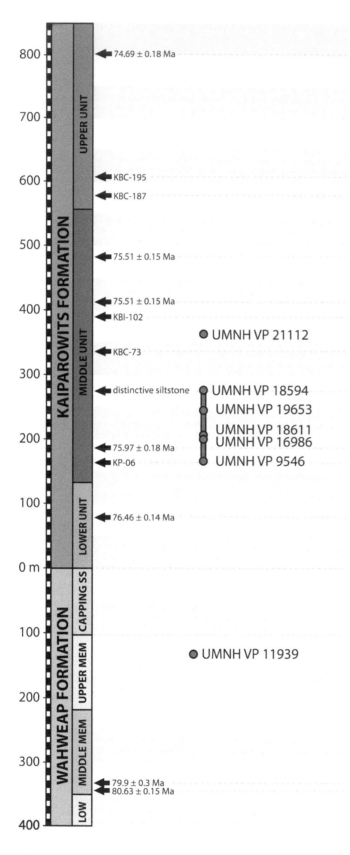

20.3. Stratigraphic distribution of Wahweap and Kaiparowits Formation pachycephalosaurid cranial material.

On the basis of the paucity of fossil remains, pachycephalosaurids appear to form a relatively minor component of Campanian dinosaur communities of southern Utah (Titus et al., 2005), both in terms of relative abundance and species richness. Taphonomic and facies factors may influence the currently known fossil record of pachycephalosaurids (Evans et al., pers. comm.). Regardless, these new records of pachycephalosaurs from Grand Staircase–Escalante National Monument are of interest in light of purported differences between the Late Cretaceous faunas of the northern and southern parts of western North America (Gates et al., 2010, this volume, Chapter 19; Sampson, Loewen, et al., 2010). The temporal spans of the Wahweap and Kaiparowits formations largely overlap with those of the Foremost and Dinosaur Park formations of southern Alberta, respectively (Eberth, 2005; Roberts, Deino, and Chan, 2005; Jinnah et al., 2009). Both of the latter are rich in pachycephalosaur remains that have been well characterized (Sullivan, 2003; Ryan and Evans, 2005; Schott et al., 2009, 2011; Schott and Evans, 2012). Although the available material is currently limited, continuing field efforts in the Kaiparowits region will surely result in more complete, diagnostic pachycephalosaur materials that will be invaluable in making definitive taxonomic assignments, as well as biogeographic comparisons between taxa inhabiting northern and southern regions of the Laramidia in the Campanian.

ACKNOWLEDGMENTS

We thank M. Getty (UMNH) for his ongoing efforts in the field and in the lab, and for his assistance with loans related to this project. We thank R. Schott, M. Goodwin, R. Sullivan, and N. Longrich for discussions about, and Caleb Brown for assistance with, the chapter. We also thank Grand Staircase–Escalante National Monument and the Bureau of Land Management for their continued support throughout the duration of this project. Additional thanks go to S. Sampson, A. Titus, E. Lund, D. DeBlieux, J. Gentry, R. Irmis, and E. Roberts for their substantial contributions in the field and in the lab. Research funding for D.C.E. was provided by an NSERC Discovery Grant and the Royal Ontario Museum. Additional funding was provided by Grand Staircase–Escalante National Monument, UMNH, and the National Science Foundation (NSF 0745454, Kaiparowits Basin Project).

REFERENCES CITED

Butler, R. J., and Sullivan, R. M. 2009. The phylogenetic position of the ornithischian dinosaur *Stenopelix valdensis* from the Lower Cretaceous of Germany and the early fossil record of Pachycephalosauria. Acta Palaeontologica Polonica 54:21–34.

Eaton, J. G., R. L. Cifelli, J. H. Hutchison, J. I. Kirkland, and J. M. Parrish. 1999. Cretaceous vertebrate faunas from the Kaiparowits Plateau, south-central Utah; pp. 345–353 in D. D. Gillette (ed.), Vertebrate Paleontology in Utah. Utah Geological Survey Miscellaneous Publication 99-1.

Eaton, J. G., S. Diem, J. D. Archibald, C. Schierup, and H. Munk. 1999. Vertebrate Paleontology of the Upper Cretaceous Rocks of the Markagunt Plateau, Southwestern Utah; pp. 323–333 in D. D. Gillette (ed.), Vertebrate Paleontology in Utah. Utah Geological Survey Miscellaneous Publication 99-1.

Eberth, D. A. 2005. The Geology; pp. 54–82 in P. J. Currie and E. B. Koppelhus (eds.), *Dinosaur Provincial Park: A Spectacular Ancient Ecosystem Revealed*. Indiana University Press, Bloomington, Indiana.

Gates, T. A., S. D. Sampson, L. E. Zanno, E. M. Roberts, J. G. Eaton, R. L. Nydam, J. H. Hutchison, J. A. Smith, M. A. Loewen, and M. A. Getty. 2010. Biogeography of terrestrial and freshwater vertebrates from the Late Cretaceous (Campanian) Western Interior of North America. Palaeogeography, Palaeoclimatology, Palaeoecology. 291:371–387.

Jinnah, Z. A., E. M. Roberts, A. L. Deino, J. S. Larsen, P. K. Link, and C. M. Fanning. 2009. New ^{40}Ar-^{39}Ar and detrital zircon U-Pb ages for the Upper Cretaceous Wahweap and Kaiparowits formations on the Kaiparowits Plateau, Utah: implications for regional correlation, provenance, and biostratigraphy. Cretaceous Research 30:287–299.

Kirkland, J. I. 2001. The Quest for New Dinosaurs at Grand Staircase–Escalante National Monument. Utah Geological Survey, Survey Notes 33.

Kirkland, J., and D. Deblieux. 2005. Dinosaur remains from the Lower to Middle Campanian Wahweap Formation at Grand Staircase–Escalante National Monument, southern Utah (abstract). Journal of Vertebrate Paleontology 25(Supplement):78A.

Lehman, T. M. 2010. Pachycephalosauridae from the San Carlos and Aguja Formations (Upper Cretaceous) of west Texas, and observations of the frontoparietal dome. Journal of Vertebrate Paleontology 30:786–798.

Longrich, N. R., J. Sankey, and D. Tanke. 2010. *Texacephale langstoni,* a new genus of pachycephalosaurid (Dinosauria: Ornithischia) from the upper Campanian Aguja Formation, southern Texas, U.S.A. Cretaceous Research 31:274–284.

Lucas, S. G., A. B. Heckert, R. M. Sullivan. 2000. Cretaceous dinosaurs of New Mexico; pp. 83–92 in S. G. Lucas, and A. B. Heckert (eds.), Dinosaurs of New Mexico. New Mexico Museum of Natural History and Science Bulletin 17.

Maryańska, T., R. E. Chapman, and D. B. Weishampel. 2004. Pachycephalosauria; pp. 464–477 in D. B. Weishampel, P. Dodson, and H. Osmólska (eds.), The Dinosauria (2nd edition). University of California Press, Berkeley, California.

Parrish, J. M. 1999. Dinosaur teeth from the Upper Cretaceous (Turonian–Judithian) of southern Utah; pp. 319–321 in D. D. Gillette (ed.), Vertebrate Paleontology in Utah. Utah Geological Survey Miscellaneous Publication 99-1.

Roberts, E. M., A. L. Deino, and M. A. Chan. 2005. ^{40}Ar/^{39}Ar age of the Kaiparowits Formation, southern Utah, and correlation of contemporaneous Campanian strata and vertebrate faunas along the margin of the Western Interior Basin. Cretaceous Research 26:307–318.

Ryan, M. J., and D. C. Evans. 2005. Ornithischian dinosaurs; pp. 313–348 in P. J. Currie and E. B. Koppelhus (eds.), *Dinosaur Provincial Park: A Spectacular Ancient Ecosystem Revealed.* Indiana University Press, Bloomington, Indiana.

Sampson, S. D., M. A. Loewen, A. A. Farke, E. M. Roberts, and C. A. Forster. 2010. New horned dinosaurs from Utah provide evidence for intracontinental dinosaur endemism. PLoS One 5(9):e12292.

Sampson, S. D., T. A. Gates, E. M. Roberts, M. A. Getty, L. E. Zanno, M. A. Loewen, J. A. Smith, E. K. Lund, J. Sertich, and A. L. Titus. 2010. Grand Staircase–Escalante National Monument: a new and critical window into the world of dinosaurs; pp. 161–179 in M. Eaton (ed.),

Learning from the Land: Grand Staircase–Escalante National Monument Science Symposium Proceedings. Grand Staircase–Escalante Partners, Escalante, Utah.

Schott, R. K., and D. C. Evans. 2012. Squamosal ontogeny and variation in the pachycephalosaurian dinosaur *Stegoceras validum* from the Dinosaur Park Formation, Alberta. Journal of Vertebrate Paleontology 32:903–913.

Schott, R. K., D. C. Evans, T. E. Williamson, T. D. Carr, and M. B. Goodwin. 2009. The anatomy and systematics of *Colepiocephale lambei* (Dinosauria: Pachycephalosauridae). Journal of Vertebrate Paleontology 29:771–786.

Schott R. K., D. C. Evans, M. B. Goodwin, J. R. Horner, C. M. Brown, and N. R. Longrich. 2011. Cranial ontogeny in *Stegoceras validum* (Dinosauria: Pachycephalosauria): a quantitative model of pachycephalosaur dome growth and variation. PLoS One 6(6):e21092.

Sues, H.-D., and P. M. Galton. 1987. Anatomy and classification of the North American Pachycephalosauria (Dinosauria: Ornithschia). Paleontographica Abteilung A 198:1–40.

Sullivan, R. M. 2003. Revision of the dinosaur *Stegoceras* Lambe (Ornithischia, Pachycephalosauridae). Journal of Vertebrate Paleontology 23:181–207.

Sullivan, R. M. 2006. A taxonomic review of the Pachycephalosauridae (Dinosauria: Ornithischia); pp. 347–365 in S. G. Lucas and R. M. Sullivan (eds.), Late Cretaceous Vertebrates from the Western Interior. New Mexico Museum of Natural History and Science Bulletin 35.

Titus, A. L., J. D. Powell, E. M. Roberts, S. D. Sampson, S. L. Pollock, J. I. Kirkland, and L. B. Albright. 2005. Late Cretaceous stratigraphy, depositional environments, and macrovertebrate paleontology of the Kaiparowits Plateau, Grand Staircase–Escalante National Monument, Utah; pp. 101–128 in J. Pederson and C. M. Dehler (eds.), Interior Western United States. Geological Society of America Field Guide 6.

Williamson, T. E., and T. D. Carr. 2002. A new genus of derived pachycephalosaurian from western North America. Journal of Vertebrate Paleontology 22:779–801.

Ceratopsid Dinosaurs from the Grand Staircase of Southern Utah

21

Mark A. Loewen, Andrew A. Farke, Scott D. Sampson,
Michael A. Getty, Eric K. Lund, and Patrick M. O'Connor

RECENT FIELDWORK AND RESEARCH IN THE UPPER CRE-
taceous Wahweap and Kaiparowits formations of southern
Utah has greatly increased our knowledge and understanding of ceratopsid dinosaurs from the Campanian of southern
Laramidia. This research, undertaken by the Kaiparowits
Basin and Horned Dinosaur projects, has documented evidence of at least six distinct taxa, three each from the Wahweap and Kaiparowits formations. The Wahweap Formation taxa consist of *Diabloceratops eatoni* and at least two
other undescribed centrosaurines. No evidence exists of
chasmosaurines from this unit. The Kaiparowits Formation
has yielded multiple individuals of two previously unknown
chasmosaurines, *Utahceratops gettyi* and *Kosmoceratops richardsoni*, as well as at least one new centrosaurine. The Kaiparowits Formation is unique among Campanian-aged units
in possessing evidence of at least three distinct, co-occurring
ceratopsid taxa.

INTRODUCTION

Ceratopsid dinosaurs are one of the most prominent Late
Cretaceous (~80.5–65.5 Ma) clades of large-bodied, herbivorous ornithischians in North America. They are characterized by double-rooted teeth, facial ornamentations such as
nasal, orbital and epijugal horns, and enlarged, ornamented
parietosquamosal frills (Dodson et al., 2004). Ceratopsidae
is divided into two subclades: Centrosaurinae, generally
characterized by enlarged, round narial regions, and short,
subcircular parietosquamosal frills; and Chasmosaurinae,
typified by rostrocaudally elongate narial regions, abbreviated nasal horns, and elongate supraorbital horn cores and
parietosquamosal frills. With the exception of a single occurrence of the centrosaurine *Sinoceratops zhuchengensis*
from China (Xu et al., 2010), ceratopsids are restricted to Laramidia, the island continent of western North America that
was isolated by incursion of the Cretaceous Interior Seaway.

In 2000, the Utah Museum of Natural History (now
the Natural History Museum of Utah), the University of
Utah, the Utah Geological Survey, and the Bureau of Land

Management initiated an comprehensive research project
to survey, collect, and document Late Cretaceous terrestrial
faunas preserved within the Kaiparowits Basin of southern
Utah. The initial focus of the project was Campanian aged
exposures of the Kaiparowits and Wahweap formations exposed within Grand Staircase–Escalante National Monument. Recent field efforts have extended the temporal scope
of the project to include the underlying Straight Cliffs Formation. Since the inception of the project, field crews from
Ohio University, the Denver Museum of Nature & Science,
Idaho State University, and the Raymond M. Alf Museum
of Paleontology have also conducted field and laboratory research contributing to these overarching efforts.

To date, this collaborative effort–known as the Kaiparowits Basin Project (KBP)–has achieved notable successes,
recovering evidence of at least 11 new dinosaur taxa, along
with many other vertebrate and invertebrate groups and documenting a unique flora. Although several of these taxa are
still under study (Sampson et al., this volume, Chapter 28),
results to date challenge a number of previous ideas regarding Late Cretaceous dinosaur evolution within the Western
Interior Basin.

In 2009, a comprehensive, corollary effort, known as the
Horned Dinosaur Project (HDP), was undertaken to document the evolution and radiation of horned dinosaurs (Ceratopsidae). This collaborative project involves several of us
(M.A.L., A.A.F., S.D.S.), together with Catherine Forster
(George Washington University). Considered together, the
results of the KBP and HDP have dramatically increased our
working knowledge of horned dinosaurs from Laramidia in
general, and the southern region of this landmass in particular (Mallon et al., 2008; Clayton et al., 2009, 2010; Farke,
Wolff, and Tanke, 2009; Farke et al., 2009, 2011; Getty et
al., 2009, 2010; Loewen et al., 2009, 2010; Sampson et al.,
2009, 2010; Brandau and Getty, 2010; Lund, Sampson, and
Loewen, 2010; Sampson and Loewen, 2010; Farke, 2011).

This chapter reviews the ceratopsid taxa recovered from
Campanian units in Grand Staircase–Escalante National
Monument as a result of the KBP and HDP projects (Fig.

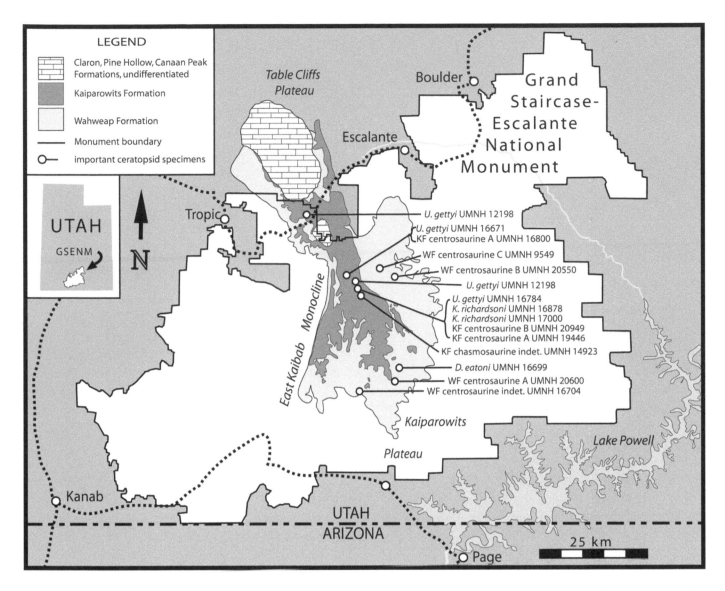

21.1. Map of Grand Staircase–Escalante National Monument, southern Utah, showing geographic distribution of important ceratopsid specimens recovered by the Kaiparowits Basin and Horned Dinosaur projects.

21.1), providing both phylogenetic and stratigraphic contexts for the developing fauna. Newly described taxa include: two chasmosaurines (*Utahceratops* and *Kosmoceratops*) from the Kaiparowits Formation (Sampson et al., 2010) and one centrosaur (*Diabloceratops*) from the Wahweap Formation (Kirkland and DeBlieux, 2010). Undescribed taxa include a total of five new centrosaurines, three from the Wahweap Formation and two from the Kaiparowits Formation.

Institutional Abbreviation NMMNH, New Mexico Museum of Natural History, Albuquerque, New Mexico; UMNH, Natural History Museum of Utah (previously Utah Museum of Natural History), Salt Lake City, Utah.

CERATOPSID DIVERSITY IN THE KAIPAROWITS BASIN

Iron Springs Formation (Equivalent to the John Henry Member of the Straight Cliffs Formation)

Milner et al. (2006) reported a ceratopsid tracksite from the Iron Springs Formation of southwestern Utah. The Iron Springs Formation is considered to be correlative with the John Henry Member of the Straight Cliffs Formation on the basis of taxa shared between respective fossil mammal faunas (Eaton, 1999; Titus, Roberts, and Albright, this volume, Chapter 2). Other than this tracksite, no other evidence exists for ceratopsids from the Straight Cliffs Formation or its lateral equivalents (Eaton, Cifelli, et al., 1999; Eaton, Diem, et al., 1999; Parrish, 1999).

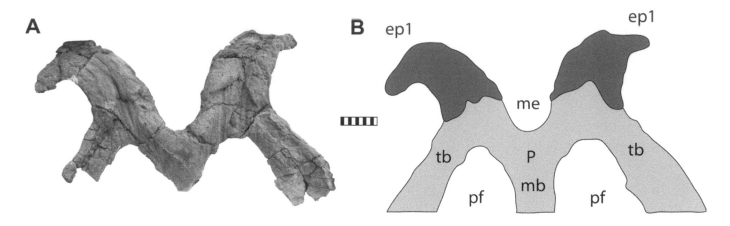

21.2. UMNH VP 20600 (Wahweap centrosaurine A), a centrosaurine parietal recovered from the lower member of the Wahweap Formation, shown in dorsal (A) and ventral (B) views. ep1, epiparietal position 1; me, midline parietal embayment; mb, midline parietal bar; P, parietal; pf, parietal fenestra; tb, transverse parietal bar. Scale bar = 10 cm.

Wahweap Formation

The Wahweap Formation exposed on the Kaiparowits Plateau consists of four subunits: the lower, middle, upper, and capping sandstone members. The formation has been estimated to span between the middle and late Campanian (Jinnah et al., 2009; Jinnah, this volume, Chapter 4; Titus, Roberts, and Albright, this volume, Chapter 2), dating from ~81–77 Ma; however, most of the Wahweap Formation (>180 m of section, from 50 m above the base to 40 m above the middle–upper member contact) is likely constrained between 81 and 79 Ma (Jinnah, this volume, Chapter 4; Titus, Roberts, and Albright, this volume, Chapter 2).

The Wahweap Formation has produced the oldest known record of diagnostic ceratopsids (all centrosaurine), with specimens recovered from the lower three of the four members. The lowest occurrence ("Wahweap centrosaurine A," UMNH VP 20600) was recovered from the lower sandstone member, approximately 45 m above the Drip Tank Member of the Straight Cliffs Formation near Pilot Knoll. It consists of a partial braincase and a nearly complete parietosquamosal frill. The rostral portion of the skull had eroded away before discovery, and the caudal portion of each squamosal was bisected by a large fissure in the sandstone matrix and subsequently eroded. Nevertheless, the specimen reveals that the parietals are relatively wide and fenestrate, with a large midline embayment, forming an M-shaped frill in dorsal view. The only epiossifications preserved on the frill are bilateral, laterally curving, dorsoventrally flattened hooks on either side of the embayment (Fig. 21.2). These parietal ossifications differ in shape, length, and orientation from those of *Diabloceratops eatoni* (Kirkland and DeBlieux, 2010), a form recovered in the middle unit, and more closely resemble those of *Albertaceratops nesmoi* (Ryan and Russell, 2005).

UMNH VP 16704 is another specimen recovered from the lower member of the Wahweap Formation near Nipple Butte about 50 m above the Drip Tank Member of the Straight Cliffs Formation. It is represented by a partial skull composed of the palate, partial maxillae, right jugal, part of the right orbit (partial lacrimal, palpebral and postorbital), pterygoids, braincase, and a partial frill consisting of a complete right squamosal and the rostral portion of the right parietal. UMNH VP 16704 was assigned to cf. *Diabloceratops* sp. by Kirkland and DeBlieux (2010). It is diagnostically centrosaurine in its short, fan-shaped squamosal with a stepped caudal margin associated with the supratemporal fenestra (Fig. 21.3). The major difference between UMNH VP 16704 and *Diabloceratops eatoni* (UMNH VP 16699) is in the shape of the squamosal (fan shaped in UMNH VP 16704 [like derived centrosaurines] versus rectangular in UMNH VP 16699 [like *Protoceratops* and neoceratopsians]) and the presence of a restricted otic notch in UMNH VP 16704 similar to other more derived centrosaurines. Further study of the recently discovered centrosaurs of the Wahweap Formation may allow refinement of the affinities of this specimen, but we currently consider it to be Centrosaurinae indet.

Diabloceratops eatoni (Kirkland and DeBlieux, 2010) from the middle member of the Wahweap represents the most complete diagnostic basal centrosaurine dinosaur known (Fig. 21.4). UMNH VP 16699 is also the oldest and most basal member of Ceratopsidae thus far diagnosed to species. Both *Zuniceratops christopheri* and *Turanoceratops tardabilis* have been interpreted as exhibiting ceratopsid synapomorphies, such as elongate postorbital horn cores and the presence of double-rooted teeth (Wolfe and Kirkland, 1998; Wolfe, 2000; Wolfe et al., 2007; Sues and Averianov, 2009). However, *Zuniceratops* lacks double-rooted teeth (Loewen and Farke, pers. obs.), and although *Turanoceratops* has been reported

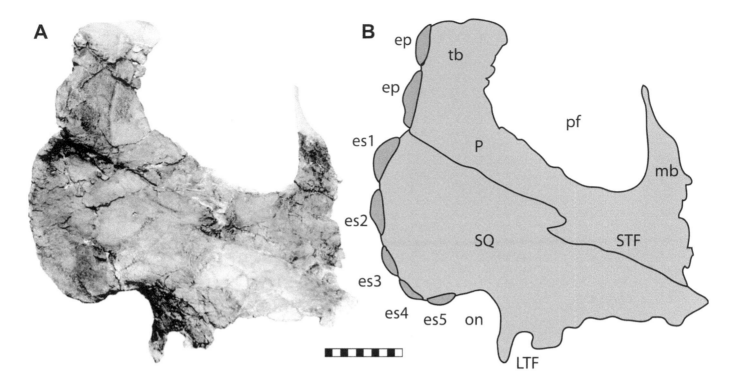

21.3. UMNH VP 16704 from the lower member of the Wahweap Formation, shown in dorsal (A) and lateral (B) views. Light area in (A) is mirrored from the other side to show the overall shape of the skull. ep, epiparietal; es, episquamosal; LTF, laterotemporal fenestra; mb, midline parietal bar; on, otic notch; P, parietal; pf, parietal fenestra; STF, supratemporal fenestra; SQ, squamosal, tb, transverse parietal bar. Scale bar = 10 cm.

to have double-rooted teeth, the single cited specimen likely belongs to a hadrosauroid dinosaur found at the same locality (P. Makovicky, pers. comm.). Elongate supraorbital horn cores appear to be plesiomorphic for Ceratopsidae (present in the basalmost members of both Centrosaurinae and Chasmosaurinae), so it is not surprising that other derived neoceratopsians should possess this character. Both *Zuniceratops* and *Turanoceratops* are considered herein to be derived neoceratopsians outside the clade Ceratopsidae (Farke et al., 2009). As such, *Diabloceratops eatoni* is currently regarded as the most basal centrosaurine on the basis of the following characters: relatively large, rounded, and expanded ectonaris with a slight ventral notch; "stepped" parietosquamosal suture; and fenestrate frill with elongate spike-shaped epiossifications on the caudal margin and crescentic epiossifications along the lateral margin of the frill. Plesiomorphic characters of *Diabloceratops* include: enlarged accessory antorbital fossa; down-dropped (ventrally displaced) tooth row with step and diastema at rostroventral margin of maxilla; elongate, rostrally curving supraorbital horn cores; and elongate, subrectangular squamosal (Kirkland and DeBlieux, 2010).

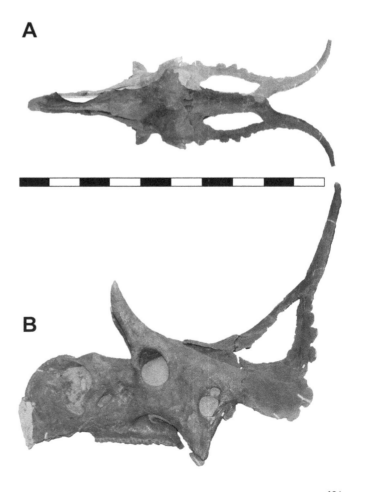

21.4. Holotype skull of *Diabloceratops eatoni* (UMNH VP 16699) from the middle member of the Wahweap Formation, shown in dorsal (A) and lateral (B) views. Light area in (A) is mirrored from the other side to show the overall shape of the skull. Scale bar = 1 m.

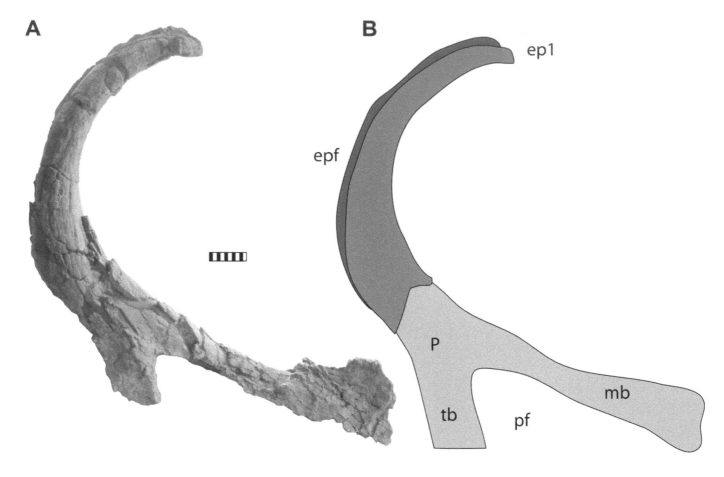

21.5. UMNH VP 20550 (Wahweap centrosaurine B), a centrosaurine parietal recovered from the upper member of the Wahweap Formation, shown in right lateral view (A). (B) Interpretative illustration of UMNH VP 20550, also in right lateral view, showing the extent of the medial epiparietal flange. ep1, epiparietal position 1; epf, epiparietal flange; mb, midline parietal bar; P, parietal; pf, parietal fenestra; tb, transverse parietal bar. Scale bar = 10 cm.

A possible third centrosaurine taxon from the Wahweap Formation is represented by a single specimen (UMNH VP 20550) recovered from near the top of the upper member near Star Seep. This animal, denoted here as "Wahweap centrosaurine B," consists of two elongate, curved supraorbital horn cores; a jugal; a partial braincase; a typical centrosaurine fan-shaped squamosal; and a nearly complete parietal adorned only by rostrally directed spikes on either side of the midline bearing a caudomedially oriented tab-like flange along their entire lengths (Fig. 21.5).

The contact between the upper and the capping sandstone members of the Wahweap has also produced a partial frill of a centrosaurine ceratopsid. This specimen (UMNH VP 9549), discovered in 2000, was the first ceratopsid found by a KBP field crew. It was recovered at the contact between the upper member and the capping sandstone, as an erosional deflation lag. It is not clear whether UMNH VP 9549 was in situ in the uppermost upper member or was derived from the lowest part of the capping sandstone. "Wahweap centrosaurine C" (UMNH VP 9549) consists of a partial frill and some postcranial elements. The incomplete parietal has distinct imbricated marginal scallops with epiparietal attachment scars in their centers. Another unique feature of UMNH VP 9549 is a raised lip on the dorsolateral edge of the frill that transitions to a corresponding lip on the ventral edge of the frill caudally (Fig. 21.6). From the preserved material, the shape of the parietal is consistent with centrosaurine ceratopsids in forming a rounded frill. There is evidence of at least six epiparietals per side. Based on the presence of thinning of the parietal in the region typically associated with parietal fenestrae, it is probable that the frill was fenestrate. The elongate, raised bases for attachment of epiparietals suggest the presence of spikes similar to those *Styracosaurus* or *Rubeosaurus*; however, lacking preserved epiparietals in UMNH VP 9549, this interpretation remains preliminary.

Kaiparowits Formation

The Kaiparowits Formation, dating to ~76.6–74.5 Ma, is divided into three informal units: lower, middle, and upper

Loewen et al.

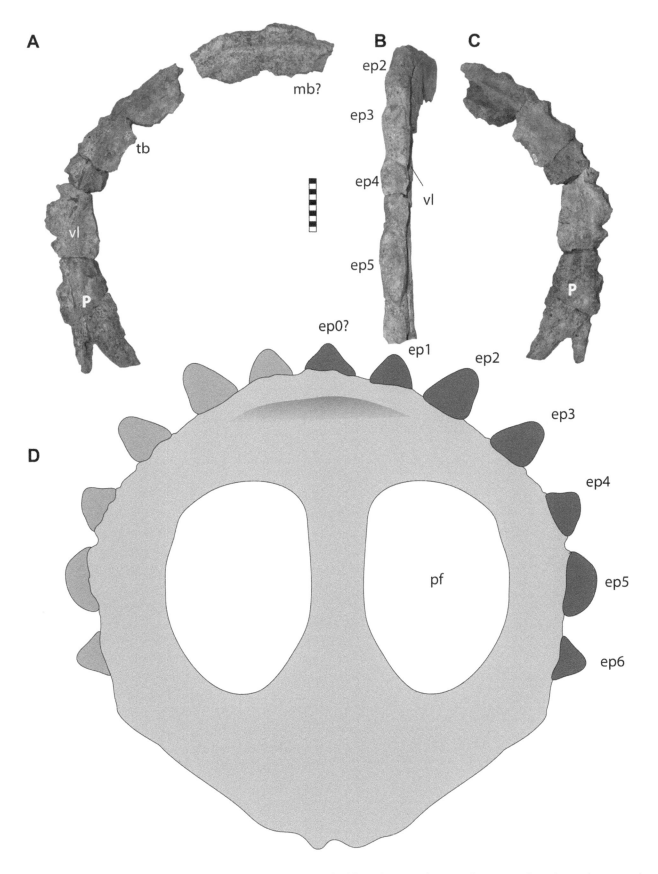

21.6. UMNH VP 9549 (Wahweap centrosaurine C), a centrosaurine parietal from the contact between the upper and capping sandstone members of the Wahweap Formation, in ventral view (A), showing two major pieces of the lateral edge of a centrosaurine parietal. The position of the smaller piece is approximate. UMNH VP 9549 in lateral oblique view (B), illustrating the ventral lip transitioning to the dorsal surface. (C) Dorsal view of UMNH VP 9549. (D) Illustration depicting reconstruction of parietosquamosal frill of UMNH VP 9849 in dorsal view. ep, epiparietal; mb, midline parietal bar; P, parietal; pf, parietal fenestra; tb, transverse parietal bar; vl, ventral lip. Scale bar = 10 cm.

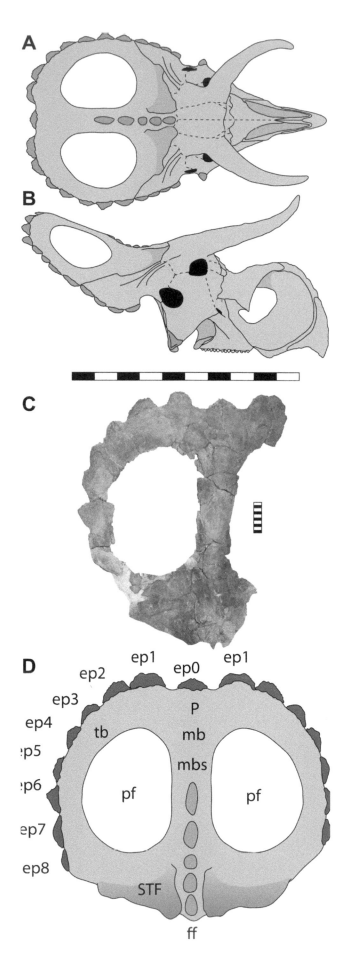

21.7. Reconstruction of Kaiparowits centrosaurine A in dorsal (A) and lateral (B) views. Parietosquamosal frill of UMNH VP 16800 from the middle unit of the Kaiparowits Formation, shown in dorsal view (C). (D) Illustration depicting reconstruction of the same specimen, also in dorsal view. ep, epiparietal; ff, frontal fontanel; ep, epiparietal; mb, midline parietal bar; mbs, midline bar swellings; P, parietal; pf, parietal fenestra; tb, transverse parietal bar. Scale bar (A, B) = 1 m; scale bar (C, D) = 10 cm.

(Roberts et al., this volume, Chapter 6). This formation has yielded a rich ceratopsid fauna that includes at least one centrosaurine taxon and two taxa of chasmosaurines.

A well-preserved centrosaurine (presently undescribed and referred to here as Kaiparowits centrosaurine A) has been recovered from the middle unit of the Kaiparowits Formation (Lund, Sampson, and Loewen, 2010; Sampson and Loewen, 2010).

Kaiparowits centrosaurine A (Fig. 21.7) is characterized by very long forward-curving supraorbital horn cores and an extremely large naris with a low pneumatized nasal ornamentation. The frill is relatively short and round with large parietal fenestrae and simple crescentic epiparietals (completely lacking large hooks or horns on the frill). This taxon is currently represented by two specimens. UMNH VP 16800 consists of a mostly complete and partially articulated skull, as well as an articulated forelimb and a small number of associated postcranial elements, including some vertebrae and ribs. A large portion of the skull was intact, including the midline of the skull from the orbits to the distal portion of the frill. The lower portion of the frill, including a large portion of both squamosals, was previously lost to erosion. The intact portion of the skull preserved both postorbital horn cores in articulation with the braincase and frill, thus retaining the overall orientation of major cranial elements. The narial region, including both premaxilla and most of the nasals is also preserved in articulation, but was found broken and displaced from the upper portion of the skull. One maxilla and a jugal horn were also found associated and displaced. The sandstone around the articulated arm also preserves some patches of skin impressions, which appear to be directly associated with this part of the carcass. Whereas skin impressions are relatively common in association with articulated hadrosaurs in the formation, this is the first and only example of ceratopsian skin impressions from the Kaiparowits. UMNH VP 19466 includes elements of a disarticulated skull including premaxilla, maxilla, and nasal, all showing nearly identical morphology as UMNH VP 16800. These elements were found in a bonebed together with a partial ankylosaur skeleton (UMNH VP 19473) and a complete *Denazinamys* turtle shell.

There is a well-preserved, isolated right squamosal (UMNH VP 19469) that may also be attributable to Kaiparowits

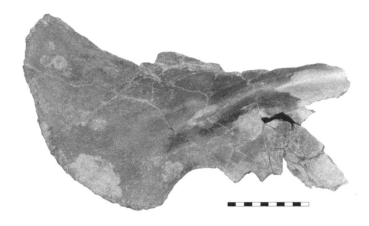

21.8. Right squamosal UMNH VP 19469 that may be attributable to Kaiparowits centrosaurine A, from the middle unit of the Kaiparowits Formation, shown in lateral view. Scale bar = 10 cm.

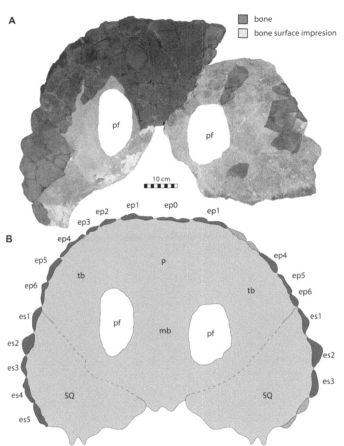

21.9. Parietosquamosal frill of UMNH VP 20949 (Kaiparowits centrosaurine B), from the middle unit of the Kaiparowits Formation, shown in dorsal view. Illustration depicting reconstruction of the same specimen, also in dorsal view. ep, epiparietal; es, episquamosal; mb, midline parietal bar; P, parietal; pf, parietal fenestra; tb, transverse parietal bar. Scale bar = 10 cm.

centrosaurine A from the lower part of the middle unit of the Kaiparowits Formation (Fig. 21.8). It shares a well-developed triangular ridge that extends from the rostrodorsal end and continues above the laterotemporal fenestra and across to the blade of the squamosal. This ridge is unusual among centrosaurines (present in a well-developed way only in centrosaurine indet. NMMNH P34906 from the Fort Crittenden Formation), but is also present in the potential holotype for Kaiparowits centrosaurine A (UMNH VP 16900).

Another centrosaurine specimen (Kaiparowits centrosaurine B) recovered from the middle unit of the Kaiparowits Formation (a partial skull; UMNH VP 20949) differs from Kaiparowits centrosaurine A in the size of the parietal fenestrae and shape of the frill (Fig. 21.9). Whereas the size of the fenestra of Kaiparowits centrosaurine B are the absolute smallest of any known centrosaurine relative to overall frill size, it is not possible at present to determine if the unique frill morphology of UMNH VP 20949 represents a distinct taxon.

The oldest chasmosaurine from the Kaiparowits Formation (and from Utah) is UMNH VP 14923, a partial skull recovered 80 m above the base of the Kaiparowits Formation, within the lower unit in the Dog Flat area (Fig 21.1). The specimen consists of the right side of an articulated partial skull, including the ventral lacrimal, jugal, epijugal, quadratojugal, postorbital (caudoventral portion only), maxilla (jugal process only), squamosal, and rostrolateral parietal in the region of the supratemporal fenestra (Fig. 21.10). This specimen is referable to Chasmosaurinae on the basis of the elongate, triangular squamosal. Much of the external surface is covered with neurovascular grooves, consistent with at least subadult bone texture morphology (Brown, Russell, and Ryan, 2009). The six rostralmost episquamosals

are preserved. The most rostral of these is larger and more strongly peaked than episquamosals at more caudal positions. Successively more distal episquamosals are all lower and less triangular. The medial margin of the squamosal is thickened, and the lateral margin has a crescentic profile.

The epijugal of UMNH VP 14923 is large and D shaped, covering approximately 20% of the lateral surface of the jugal, with a striated rostrodorsal surface. This morphology is consistent with several other chasmosaurine dinosaurs, including *Utahceratops*, but larger than that seen in *Kosmoceratops*. The laterotemporal fenestra of UMNH VP 14923 is roughly triangular, bounded dorsally and caudally by the squamosal, rostrally by the jugal, and ventrally by the squamosal (caudally), quadratojugal (in the middle), and jugal (rostrally).

Although UMNH VP 14923 lacks the diagnostic parietal and postorbitals and thus cannot be definitively identified to species, it remains significant as the stratigraphically lowest chasmosaurine known from Utah. Based on its distinctive

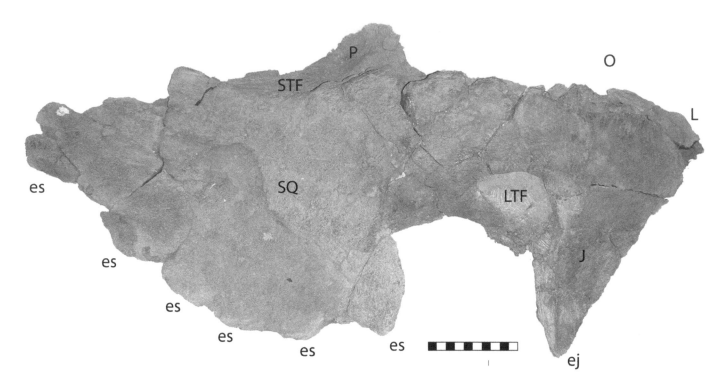

21.10. UMNH VP 14923, partial skull of a subadult chasmosaurine (Kaiparowits chasmosaurine indet.) of uncertain affinity from the lower unit of the Kaiparowits Formation, shown in lateral view. ej, epijugal; es, episquamosal; J, jugal; L, lacrimal; LTF, laterotemporal fenestra; O, orbit; P, parietal; SQ, squamosal; STF, supratemporal fenestra. Scale bar = 10 cm.

squamosal and episquamosal morphology, this specimen may pertain to a taxon distinct from *Utahceratops* and *Kosmoceratops*. However, this assessment requires testing through the recovery of additional materials.

Utahceratops gettyi is a chasmosaurine known from several articulated and disarticulated specimens recovered from the lower half of the middle unit of the Kaiparowits Formation (Fig. 21.11). This taxon is characterized by multiple autapomorphies, including abbreviated, blunt, dorsolaterally projecting supraorbital horn cores; caudally positioned nasal horn core; and medial portion of parietal transverse bar rostrally curved (Smith et al., 2004; Sampson et al., 2010).

The holotype specimen (UMNH VP 16784) consists of a partial disarticulated skull with well-preserved, nearly complete elements, including squamosal, parietal, orbit, nasal, and braincase. UMNH VP 12198 is the first specimen recovered of this taxon and preserves about 45% of the postcranial elements and 80% of cranial elements, from at least one side of the animal (Getty et al., 2010; Sampson et al., 2010). This locality (UMNH VP Loc. 145) represents the most extensive ceratopsid excavation conducted thus far in the Kaiparowits Formation, with a total of more than 280 elements and fragments from one associated individual recovered over an area of approximately 29 m². Several other localities have produced other diagnostic cranial elements of *Utahceratops*

including orbits, parietals, and squamosals of individuals of various ages that will ultimately enable us to study the ontogeny of this taxon. Two of these localities are small bonebeds preserving multiple individuals and may be indicative of some level of gregarious behavior in this taxon.

Utahceratops gettyi is the sister taxon to *Pentaceratops sternbergii*, sharing, among other features, the medially tapering transverse bar of the parietal with an upturned epiparietal on the dorsal surface (Sampson et al., 2010). The abbreviated, laterally directed supraorbital horn core is the most distinctive feature distinguishing the two taxa; multiple specimens of *Utahceratops* verify its unusual morphology. Multiple partial skulls and skeletons are now known across a range of sizes, with skull lengths varying from ~1.2 to 2.5 m. The morphologies across this size range are easily distinguishable from those seen in *Kosmoceratops richardsoni* of similar size. The smallest individual *Utahceratops* (UMNH VP 20444) is represented by a disarticulated associated skull and a complete, articulated postcranial skeleton, including the limb skeleton through the distal phalanges and the caudal vertebral series with the syncervical being the only element not recovered.

Kosmoceratops richardsoni is a chasmosaurine known from a nearly complete adult skull (the holotype, UMNH VP 17000) and a disarticulated subadult skull (UMNH VP 16878),

Loewen et al.

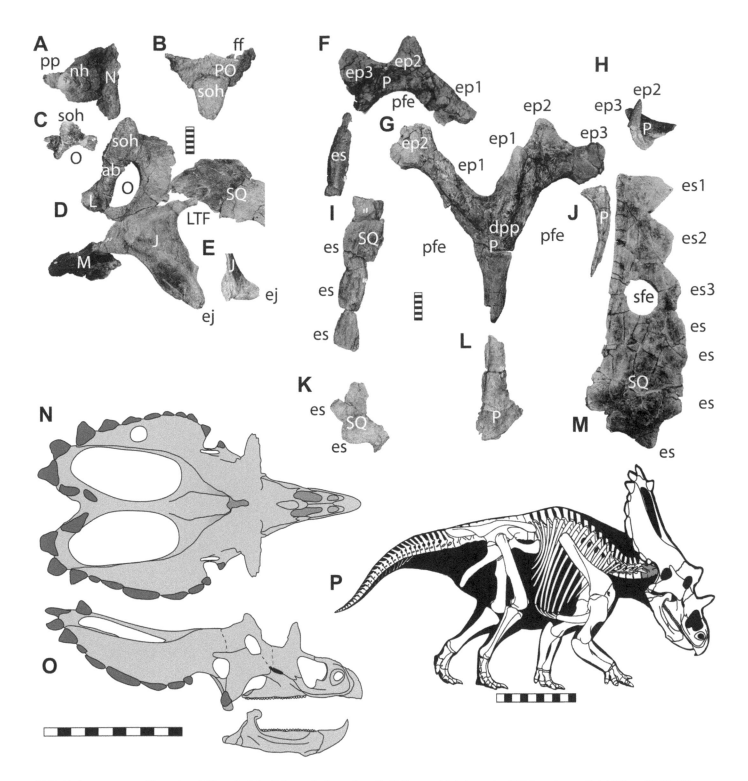

21.11. *Utahceratops gettyi* from the middle unit of the Kaiparowits Formation. (A–E) Circumorbital elements of *Utahceratops gettyi*. (A) Type (UMNH VP 16784) nasal in dorsal view. (B) Supraorbital horn core of UMNH VP 12198 in dorsal view. (C) Juvenile postorbital horn core (UMNH VP 13913) in left lateral view. (D) Circumorbital region of UMNH VP 12198 in left lateral view. (E) Jugal and epijugal of UMNH VP 12198 in rostral view. (F–M) Parietosquamosal frill elements of *Utahceratops gettyi*, all in dorsal view except (H), which is in medial view. (F) Parietal bar of type specimen (UMNH VP 16784). (G) Parietal bar of UMNH VP 16671 showing median embayment. (H) Curved cross section through parietal bar of UMNH VP 12198 at position ep1. (I) Episquamosals of UMNH VP 12198. (J) Lateral parietal bar of UMNH VP 12198. (K) Proximal right squamosal of UMNH VP 12198. (L) Proximal midline parietal bar of UMNH VP 12198. (M) Left squamosal of type specimen. Skull reconstruction in dorsal (N) and lateral (O) views. (P) Skeletal elements recovered for *Utahceratops* (in white) from the middle unit of the Kaiparowits Formation, leaving the syncervical as the only missing element. ab, antorbital buttress; dpp, dorsal parietal process; ej, epijugal horn; ep, epiparietal positions 1–3; es, episquamosal; ff, frontal fontanel; J, jugal; L, lacrimal; LTF, laterotemporal fenestra; M, maxilla; N, nasal; nh, nasal horn core; O, orbit; P, parietal; pfe, parietal fenestra; PO, postorbital; pp, premaxillary process of nasal; sfe, squamosal fenestra; SQ, squamosal, soh, supraorbital horn core. Scale bars (A–M) = 10 cm; scale bars (N–P) = 1 m.

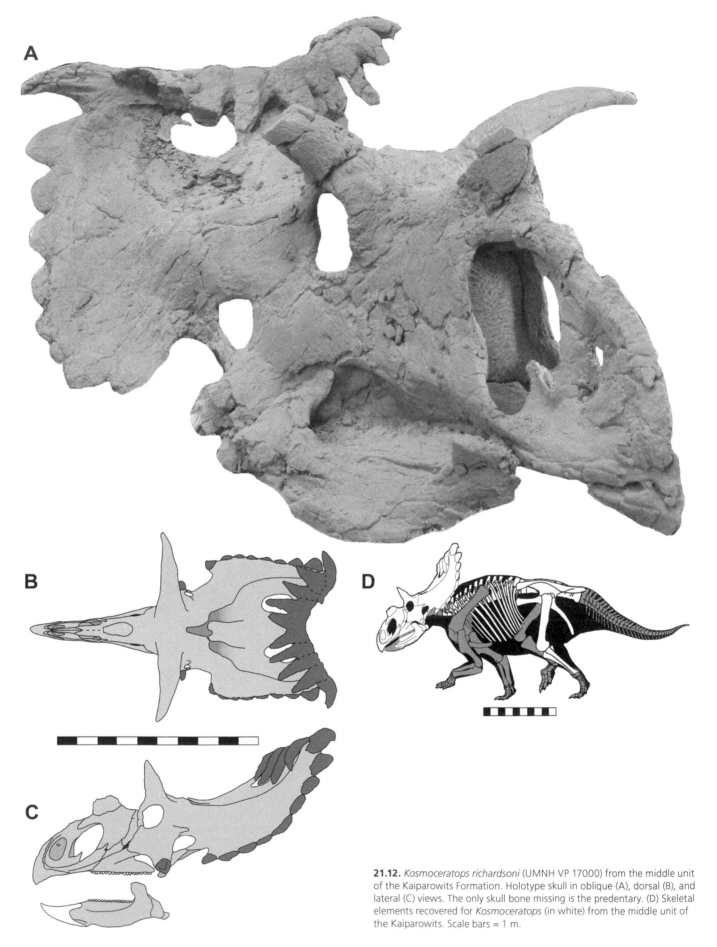

21.12. *Kosmoceratops richardsoni* (UMNH VP 17000) from the middle unit of the Kaiparowits Formation. Holotype skull in oblique (A), dorsal (B), and lateral (C) views. The only skull bone missing is the predentary. (D) Skeletal elements recovered for *Kosmoceratops* (in white) from the middle unit of the Kaiparowits. Scale bars = 1 m.

Loewen et al.

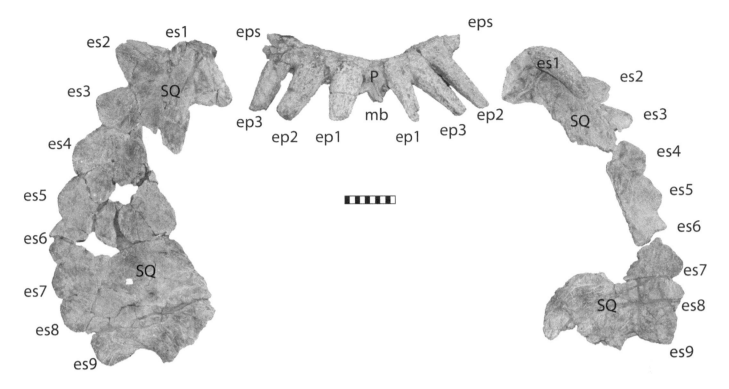

21.13. Associated parietosquamosal frill of juvenile *Kosmoceratops richardsoni* (UMNH VP 16878) from the lower middle unit of the Kaiparowits Formation, shown in dorsal view. The right spike at the es1 position and the two epiparietosquamosals are disarticulated and not figured here. This unfused associated skull conserves the epiossification count and positions of the articulated adult skull. ep, epiparietal; eps, epiparietosquamosal; es, episquamosal; mb, parietal midline bar; P, parietal; SQ, squamosal. Scale bar = 10 cm.

both recovered from upper facies of the lower half of the middle unit. UMNH VP 16878 was incorrectly listed in Sampson et al. (2010) as UMNH VP 12198 (a specimen number that had previously been given to the disarticulated *Utahceratops* skull). We here correct this typographic error and formally refer UMNH VP 16878 to *K. richardsoni*, but not UMNH VP 12198. *Kosmoceratops* displays unique narial anatomy (short and caudodorsally inclined internal naris), long supraorbital horn cores (curved dorsolaterally at their proximal end and ventrally at their distal end), a short parietosquamosal frill (with procurved, prominently elongate epiparietal hooks), and a laterally directed hook at the distal end of the squamosal (Fig. 21.12). The subadult specimen, approximately half the size of the adult, displays the identical number and pattern of epiossifications of the adult form, although they are disarticulated from the parietal and squamosals (Fig. 21.13). This distinctive morphology allows for clear separation of subadult ontogenetic stages of *Kosmoceratops* from those of *Utahceratops*. *Kosomceratops richardsoni* appears to be the sister taxon to *Vagaceratops irvinensis* from the Dinosaur Park Formation of Alberta, united by several features, including the straight caudal margin of the parietal with procurved epiparietals (Sampson et al., 2010).

DISCUSSION

The recent focus on the fossil record of ceratopsids from the Kaiparowits Basin has added much data for interpreting the biogeography, diversity, and evolutionary patterns of the group. In this discussion, we summarize implications of the new finds from the Wahweap and Kaiparowits formations (placed in section in Fig. 21.14) for the broader understanding of ceratopsid biogeography and diversity in Laramidia during the Campanian.

Centrosaurinae

Until recently, centrosaurine ceratopsids were known almost exclusively from late Campanian and early Maastrichtian deposits of the northern portion of the Western Interior Basin of North America. Possible referred centrosaurine remains from Mexico (Murray et al., 1960) and New Mexico (Gilmore, 1916) are fragmentary and nondiagnostic to subfamily. This changed with the identification of a centrosaurine squamosal in the early Campanian Allison Member of the Menefee Formation of New Mexico (Williamson, 1997), as well as the identification of a partial centrosaurine skull in the late Campanian Fort Crittenden Formation of Arizona (Heckert

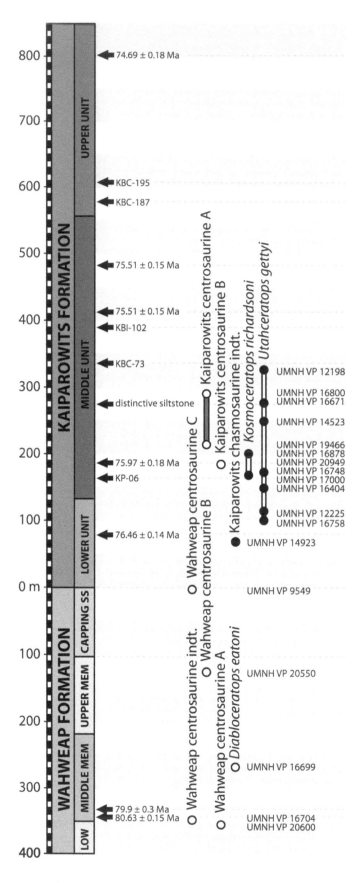

21.14. Stratigraphic distribution of major ceratopsid specimens from the Wahweap and Kaiparowits formations, with dates and member contacts after Jinnah (this volume, Chapter 4) and Roberts et al. (this volume, Chapter 6).

et al., 2003). Identification of a centrosaurine squamosal from the Late Campanian Cerro del Pueblo Formation (~74.1–72.5 Ma) confirmed the presence of centrosaurines in Mexico (Lund et al., 2007; Loewen et al., 2010). However, these materials are identifiable only as Centrosaurinae indet. and thus have not clarified the relationships between southern and northern centrosaurines. Consequently, the nearly complete skulls from the Wahweap and Kaiparowits formations are a welcome addition to the scientific record of ceratopsid dinosaurs.

The Wahweap Formation records the oldest diagnostic record of centrosaurines, with the next oldest centrosaurines dating to ~78 Ma in the Judith River Formation (*Avaceratops* and the slightly younger *Albertaceratops*). *Diabloceratops* and Wahweap centrosaurines A and B share with *Albertaceratops* a relatively deep embayment on the frill, a triangular (rather than round) frill, and elongate spikes or blades at the medialmost epiparietal loci (p1). Wahweap centrosaurine C and both of the Kaiparowits centrosaurines share with *Avaceratops* rather simple epiparietal ornamentation and the presence of a midline epiparietal (p0) is shared between *Avaceratops* and the Kaiparowits centrosaurines.

Information on the early (pre–late Campanian) biogeography and evolutionary relationships of Centrosaurinae is extremely tentative, pending full description and phylogenetic analysis of the unnamed Kaiparowits and Wahweap centrosaurines, as well as other taxa. The southern centrosaurines with the most complete material, *Diabloceratops* and Kaiparowits centrosaurine A, show many primitive features in addition to some unusual autapomorphies. Centrosaurine remains from strata equivalent to the Wahweap Formation in northern Laramidia are extremely fragmentary. Thus, it is impossible at present to determine the center of origin for Centrosaurinae, or how early centrosaurines were related to each other. The Kaiparowits centrosaurines share some characters with *Avaceratops* (including a midline epiparietal), but the fragmentary fossil record precludes a more comprehensive assessment at present.

All of the centrosaurines known from the southern part of the Western Interior Basin are relatively basal within this clade (Kirkland and DeBlieux, 2010; Lund, Sampson, and Loewen, 2010; Farke et al., 2011). There is no evidence yet of any centrosaurines with more derived features (i.e., those within the clade including the common ancestor of *Centrosaurus* and *Pachyrhinosaurus* and all of its descendents, following the cladogram in Farke et al., 2011) from southern Laramidia. This may reflect a genuine lack of that clade or sampling bias. Interestingly, no centrosaurines have yet been recovered from the relatively well-sampled Fruitland and Kirtland formations, both of which are slightly younger than the Kaiparowits Formation, or from the slightly older Aguja Formation of Texas. This may reflect extinction of centrosaurines in southern North America by the latest Campanian, or local rarity of the clade in the San Juan Basin and Texas.

Chasmosaurinae

Chasmosaurine ceratopsid dinosaurs are known exclusively from western North America and have been relatively well-known elements of southern Laramidian faunas for some time (Osborn, 1923; Lehman, 1990; Forster et al., 1993; Lucas, Heckert, and Sullivan, 2000). However, species-level identifications for chasmosaurines from the Kaiparowits Basin have only recently become possible. These finds (*Utahceratops gettyi* as well as *Kosmoceratops richardsoni*) bring important new information on the biogeography and interrelationships of Chasmosaurinae (Sampson et al., 2010).

No chasmosaurines are known from the Wahweap Formation of Utah, or from early or mid-Campanian rocks anywhere else. The oldest known chasmosaurine is *Agujaceratops mariscalensis*, from the late Campanian Aguja Formation of Texas (77.5 Ma; Roberts et al., this volume, Chapter 6). However, it is not the most basal chasmosaurine (Sampson et al., 2010); instead, *Chasmosaurus* spp. and *Mojoceratops perifania* occupy this position in the most recently published cladograms. The earliest diagnostic chasmosaurines in the Kaiparowits Basin, *Kosmoceratops richardsoni* and *Utahceratops gettyi*, occur in the Kaiparowits Formation between 76.5 and 75.5 Ma, at roughly the same time as *Chasmosaurus* spp. and *Mojoceratops perifania*, known from southern Canada.

Importantly, however, no chasmosaurine species are known from both northern and southern Laramidia. This suggests regional endemism during the late Campanian (Sampson et al., 2010). As described above, *Utahceratops* is united with *Pentaceratops sternbergii* on the basis of several cranial characters, suggesting that at least one clade of chasmosaurines was confined exclusively to southern Laramidia. *Kosmoceratops*, in contrast, is most closely related to *Vagaceratops irvinensis* from southern Alberta. This suggests some faunal interchange between northern and southern Laramidia during the latest Campanian (Sampson et al., 2010), a finding congruent with the record of hadrosaurs (Gates and Sampson, 2007; Gates et al., 2010) and other clades (Ryan and Evans, 2005). Chasmosaurines from the Maastrichtian (e.g., *Triceratops*, *Arrhinoceratops*) are most closely related to the forms from southern Laramidia, suggesting a southern origin for the latest chasmosaurines (Sampson et al., 2010).

CONCLUSIONS

Recent finds from the Kaiparowits Basin of southern Utah have greatly increased the known diversity of ceratopsid dinosaurs. *Diabloceratops eatoni* from the Wahweap Formation is a centrosaurine possessing a suite of basal features, and it represents the oldest diagnostic centrosaurine now known. Other more fragmentary centrosaurines are also known from the Wahweap Formation. Notably, no chasmosaurines are currently recognized from this formation. A nearly complete skull from the Kaiparowits Formation represents another new centrosaurine taxon, but its exact affinities and detailed morphology remain to be described. At present it is unclear if other, more fragmentary remains of centrosaurines from the Kaiparowits Formation represent the same taxon or distinct taxa.

Two co-occurring chasmosaurines in the Kaiparowits Formation, *Kosmoceratops richardsoni* and *Utahceratops gettyi*, are apparently endemic to southern Laramidia. One of these forms, *U. gettyi*, appears to be a member of an exclusively southern clade, whereas *K. richardsoni* is sister taxon to *Vagaceratops irvinensis* from Alberta.

The newly recovered ceratopsid materials from the Kaiparowits Basin are an important part of a ceratopsid renaissance that has occurred over the past decade, filling in many previously unknown aspects of the group's evolutionary and biogeographic history (Sampson and Loewen, 2010). Centrosaurines are now definitively known from southern North America, where only scrappy material was previously known. The chasmosaurines record reveals unusual patterns of endemism and faunal exchange in southern North America during the late Campanian. Additional work, particularly the description and interpretation of new or poorly known taxa, will undoubtedly clarify both the details and broader aspects of the ceratopsid radiation.

ACKNOWLEDGMENTS

The authors wish to thank Grand Staircase–Escalante National Monument and the Bureau of Land Management (BLM) for their continued support throughout the duration of this project. All specimens were collected under permits from the BLM, and we thank S. Foss and L. Bryant (BLM), and A. Titus and M. Eaton (BLM, Grand Staircase–Escalante National Monument) for assistance with permitting and field logistics; and all of the staff of Grand Staircase–Escalante National Monument for ongoing support. Special thanks go to Grand Staircase–Escalante National Monument paleontologist A. Titus for his ongoing efforts both in and out of the field. Without dozens of dedicated students, faculty, and museum volunteers from the UMNH, the University of Utah, Utah Friends of Paleontology, the Web Schools and collaborating institutions, this project could never have achieved the success we document here. Additional thanks go to J. Gentry, S. Walkington, S. Richardson, D. DeBlieux, J. Kirkland, M. Hayden, J. Eaton, R. Irmis, T. Hieronymus, L. Tapanila, Z. Jinnah, and E. Roberts for their substantial contributions in the field and in the lab. We thank M. Carrano (Smithsonian Institution National Museum of Natural History), L. Chiappe and P. Johnson (Los Angeles County Museum), P. Currie (University of Alberta Laboratory of Vertebrate Paleontology), T. Daeschler (Academy of Natural Sciences of Philadelphia), J. O. R. Ebbestad (Museum of Evolution, Uppsala University), D. Evans and K. Seymour (Royal Ontario Museum), J. Gardener and B. Strilisky (Royal Tyrrell Museum of Palaeontology), J. Gauthier (Yale Peabody Museum, Yale University), R. Gomez (Colección Paleontológica de Coahuila, at the Museo del Desierto), D. Gillette (Museum of Northern Arizona), J. Horner (Museum of the Rockies), R. McCord (Mesa Southwest Museum), M. Norell and C. Mehling (American Museum of Natural History), P. Larson (Black Hills Institute), T. Rowe (Texas Memorial Museum, University of Texas at Austin), K. Shepherd (Canadian Museum of Nature), R. Sullivan (State Museum of Pennsylvania), K. Tsogtbataar (Mongolian Institute of Geology), and T. Williamson and S. Lucas (New Mexico Museum of Natural History), for generous access to comparative specimens. Skeletal drawings were modified from artwork by L. Panzarin and S. Hartman. We thank D. Eberth, R. Rogers, R. Irmis, T. Lehman, M. Ryan, and R. Sullivan for helpful discussions and access to unpublished data. The chapter was greatly improved by comments from anonymous reviewers. Funding was provided by Grand Staircase–Escalante National Monument, the UMNH (now NHMU), the Utah Geological Survey, Ohio University, the University of Utah, the National Science Foundation (EAR 0745454, Kaiparowits Basin Project; and EAR 0819953, Horned Dinosaur Radiation Project).

REFERENCES CITED

Brandau, D., and M. A. Getty. 2010. Discovery of a new chasmosaurine bonebed from the Kaiparowits Formation (Campanian) of southern Utah. Journal of Vertebrate Paleontology 30(3, Supplement):64A.

Brown, C. M., A. P. Russell, and M. J. Ryan. 2009. Pattern and transition of surficial bone texture of the centrosaurine frill and their ontogenetic and taxonomic implications. Journal of Vertebrate Paleontology 29:132–141.

Clayton, K. E., M. A. Loewen, A. A. Farke, and S. D. Sampson. 2010. A reevaluation of epiparietal homology within chasmosaurine ceratopsids (Ornithischia) based on newly discovered taxa. Journal of Vertebrate Paleontology 30(3, Supplement):73A.

Clayton, K. E., M. A. Loewen, S. D. Sampson, A. A. Farke, and R. B. Irmis. 2009. Epiparietal homology within centrosaurine (Ornithischia, Ceratopsidae): a re-evaluation based on newly discovered basal taxa. Journal of Vertebrate Paleontology 29(3):80A.

Dodson, P., C. A. Forster, and S. D. Sampson. 2004. Ceratopsidae; pp.494–513 in D. B. Weishampel, P. Dodson, and H. Osmólska (eds.), The Dinosauria (2nd edition). University of California Press, Berkeley, California.

Eaton, J. G. 1999. Vertebrate Paleontology of the Iron Springs Formation, Upper Cretaceous, Southwestern Utah; pp. 339–344 in D. D. Gillette (ed.), Vertebrate Paleontology in Utah. Utah Geological Survey Miscellaneous Publication 99-1.

Eaton, J. G., R. L. Cifelli, J. H. Hutchison, J. I. Kirkland, and J. M. Parrish. 1999. Cretaceous vertebrate faunas from the Kaiparowits Plateau, south-central Utah; pp. 345–353 in D. D. Gillette (ed.), Vertebrate Paleontology in Utah. Utah Geological Survey Miscellaneous Publication 99-1.

Eaton, J. G., S. Diem, J. D. Archibald, C. Schierup, and H. Munk. 1999. Vertebrate Paleontology of the Upper Cretaceous rocks of the Markagunt Plateau, Southwestern Utah; pp. 323–334 in D. D. Gillette (ed.), Vertebrate Paleontology in Utah. Utah Geological Survey Miscellaneous Publication 99-1.

Farke, A. A. 2011. Anatomy and taxonomic status of the chasmosaurine ceratopsid Nedoceratops hatcheri from the Upper Cretaceous Lance Formation of Wyoming, U.S.A. PLoS One 6(1):e16196.

Farke, A. A., E. D. S. Wolff, and D. H. Tanke. 2009. Evidence of combat in Triceratops. PLoS One 4(1):e4252.

Farke, A. A., S. D. Sampson, C. A. Forster, and M. A. Loewen. 2009. Turanoceratops tardabilis–sister taxon, but not a ceratopsid. Naturwissenchaften 96:869–870.

Farke, A. A., M. J. Ryan, P. M. Barrett, D. H. Tanke, D. R. Braman, M. A. Loewen, and M. R. Graham. 2011. A new centrosaurine from the Late Cretaceous of Alberta, Canada, and the evolution of parietal ornamentation in horned dinosaurs. Acta Palaeontologica Polonica 56:691–702.

Forster, C. A., P. C. Sereno, T. W. Evans, and T. Rowe. 1993. A complete skull of Chasmosaurus mariscalensis (Dinosauria; Ceratopsidae) from the Aguja Formation (Late Campanian) of west Texas. Journal of Vertebrate Paleontology 13:161–170.

Gates, T. A., and S. D. Sampson. 2007. A new species of Gryposaurus (Dinosauria: Hadrosauridae) from the late Campanian Kaiparowits Formation, southern Utah, U.S.A. Zoological Journal of the Linnean Society 151:351–376.

Gates, T. A., S. D. Sampson, L. E. Zanno, E. M. Roberts, J. G. Eaton, R. L. Nydam, J. H. Hutchison, J. A. Smith, M. A. Loewen, and M. A. Getty. 2010. The fauna of the Late Cretaceous Kaiparowits Formation: testing previous ideas about late Campanian vertebrate biogeography in the Western Interior Basin. Paleogeography, Paleoclimatology, and Paleoecology. 291:371–387.

Getty, M. A., M. A. Loewen, E. M. Roberts, and A. L. Titus 2009. Taphonomy of associated dinosaur remains from the Kaiparowits Formation, Grand Staircase–Escalante National Monument, Utah. Advances in Western Interior Late Cretaceous Paleontology and Geology Conference Abstracts, p. 22.

Getty, M. A., M. A. Loewen, E. M. Roberts, A. L. Titus and S. D. Sampson. 2010. Taphonomy of horned dinosaurs (Ornithischia: Ceratopsidae) from the Late Campanian Kaiparowits Formation, Grand Staircase-Escalante National Monument, Utah, pp. 478–494 in M. J. Ryan, B. J. Chinnery-Allgeier, and D. A. Eberth (eds.) New Perspectives on Horned Dinosaurs: The Royal Tyrrell Museum Ceratopsian Symposium. Indiana University Press, Bloomington, Indiana.

Gilmore. C. W. 1916. Contributions to the geology and paleontology of San Juan County, New Mexico. 2. Vertebrate faunas of the Ojo Alamo, Kirtland and Fruitland Formations. United States Geological Survey Professional Paper 98-Q:279–302

Heckert, A. B., S. G. Lucas, and S. E. Krzyzanowski. 2003. Vertebrate fauna of the late Campanian (Judithian) Fort Crittenden Formation, and the age of Cretaceous vertebrate faunas of southeastern Arizona (U.S.A.). Neues Jahrbuch für Geologie und Paläontologie, Abhandlungen 227:343–364

Jinnah, Z. A., E. M. Roberts, A. L. Deino, J. S. Larsen, P. K. Link, and C. M. Fanning. 2009. New 40Ar/39Ar and detrital zircon U-Pb ages for the Upper Cretaceous Wahweap and Kaiparowits formations on the Kaiparowits

Plateau, Utah: implications for regional correlation, provenance, and biostratigraphy. Cretaceous Research 30:287–299.

Kirkland, J. I., and D. D. DeBlieux. 2010. New basal centrosaurine ceratopsian skulls from the Wahweap Formation (Middle Campanian), Grand Staircase–Escalante National Monument, southern Utah; pp. 117–140 in M. J. Ryan, B. J. Chinnery-Allgeier, and D. A. Eberth (eds.), New Perspectives on Horned Dinosaurs. Indiana University Press, Bloomington, Indiana.

Lehman, T. M. 1990. The ceratopsian subfamily Chasmosaurinae: sexual dimorphism and systematic; pp. 211–229 in P. J. Currie and K. Carpenter (eds.), Dinosaur Systematics: Perspectives and Approaches. Cambridge University Press, New York.

Loewen, M. A., S. D. Sampson, M. A. Getty, E. K. Lund, E. M. Roberts, and A. L. Titus. 2009. Horned dinosaurs from the Campanian Wahweap and Kaiparowits formations, Grand Staircase–Escalante National Monument, Utah. Advances in Western Interior Late Cretaceous Paleontology and Geology Conference Abstracts, p. 34.

Loewen, M. A., S. D. Sampson, E. K. Lund, A. A. Farke, C. de Leon, M. C. Aguillón Martínez, R. A. Rodríguez de la Rosa, M. A. Getty, and D. A. Eberth. 2010. Horned dinosaurs (Ornithischia: Ceratopsidae) from the Upper Cretaceous (Campanian) Cerro del Pueblo Formation, Coahuila, Mexico; pp. 99–116 in M. J. Ryan, B. J. Chinnery-Allgeier, and D. A. Eberth (eds.), New Perspectives on Horned Dinosaurs. Indiana University Press, Bloomington, Indiana.

Lucas, S. G., A. B. Heckert, and R. M. Sullivan. 2000. Cretaceous dinosaurs of New Mexico; pp. 83–92 in S. G. Lucas and A. B. Heckert (eds.), Dinosaurs of New Mexico. New Mexico Museum of Natural History and Science Bulletin 17.

Lund, E. K., M. A. Loewen, S. D. Sampson, M. A. Getty, A. Aguillon Martinez, R. A. Rodriguez de la Rosa, and D. A. Eberth. 2007. Ceratopsian remains from the Late Cretaceous Cerro del Pueblo Formation, Coahuila, Mexico; pp. 108–113 in D. R. Braman (ed.), Ceratopsian Symposium: Short Papers, Abstracts, and Programs. Royal Tyrrell Museum, Drumheller, Alberta, Canada.

Lund, E. K., Sampson, S. D., and M. A. Loewen. 2010. A new basal centrosaurine dinosaur (Ornithischia: Ceratopsidae) from the Upper Cretaceous of Utah: evidence of a previously unknown clade of southern centrosaurines from Laramidia. Journal of Vertebrate Paleontology 30:125A.

Mallon, J., R. Holmes, J. Anderson, A. Farke, and M. Ryan. 2008. New information on two chasmosaurine ceratopsids from the Horseshoe Canyon Formation (Late Cretaceous) of Alberta, Canada. Journal of Vertebrate Paleontology 28:111A.

Milner, A. R. C., S. V. Garrett, J. D. Harris, and M. G. Lockley. 2006. Dinosaur tracks from the Upper Cretaceous Iron Springs Formation, Iron County, Utah; pp. 105–113 in S. G. Lucas and R. M. Sullivan (eds.), Late Cretaceous Vertebrates of the Western Interior. New Mexico Museum of Natural History and Science Bulletin 35.

Murray, G. E., D. R. Boyd, J. A. Wolleben, and J. A. Wilson. 1960. Late Cretaceous fossil locality, eastern Parras Basin, Coahuila, Mexico. Journal of Paleontology 34:368–373.

Osborn, H. F. 1923. A new genus and species of Ceratopsia from New Mexico, *Pentaceratops sternbergii*. American Museum Novitates 93:1–3.

Parrish, J. M. 1999. Dinosaur teeth from the Upper Cretaceous (Turonian–Judithian) of southern Utah; pp. 319–321 in D. D. Gillette (ed.), Vertebrate Paleontology in Utah. Utah Geological Survey Miscellaneous Publication 99–1.

Roberts, E. M., A. L. Deino, and M. A. Chan. 2005. ^{40}Ar/^{39}Ar age of the Kaiparowits Formation, southern Utah, and correlation of contemporaneous Campanian strata and vertebrate faunas along the margin of the Western Interior Basin. Cretaceous Research 26:307–318.

Ryan, M. J., and D. C. Evans. 2005. Ornithischian dinosaurs; pp. 312–348 in P. J. Currie and E. Koppelhus (eds.), Dinosaur Provincial Park: A Spectacular Ancient Ecosystem Revealed. Indiana University Press, Bloomington, Indiana.

Ryan, M. J., and A. P. Russell. 2005. A new centrosaurine ceratopsid from the Oldman Formation of Alberta and its implications for centrosaurine taxonomy and systematics. Canadian Journal of Earth Science 42:1369–1387.

Sampson, S. D., and M. A. Loewen. 2010. Unraveling a radiation: a review of the diversity, stratigraphic distribution, biogeography, and evolution of horned dinosaurs. (Ornithischia: Ceratopsidae); pp. 405–427 in M. J. Ryan, B. J. Chinnery-Allgeier, and D. A. Eberth (eds.), New Perspectives on Horned Dinosaurs. Indiana University Press, Bloomington, Indiana.

Sampson, S. D., M. A. Loewen, A. A. Farke, J. A. Smith, and E. M. Roberts. 2009. Two new chasmosaurine ceratopsids from the Late Cretaceous of Utah. Vertebrate Paleontology 29:175A.

Sampson, S. D., M. A. Loewen, A. A. Farke, E. M. Roberts, C. A. Forster, J. A. Smith, and A. L. Titus. 2010. New horned dinosaurs from Utah provide evidence for intracontinental dinosaur endemism. PLoS One 5(9):e12292.

Smith, J. A., S. D. Sampson, E. Roberts, M. Getty, and M. Loewen. 2004. A new chasmosaurine ceratopsian from the Upper Cretaceous Kaiparowits Formation, Grand Staircase–Escalante National Monument, Utah. Journal of Vertebrate Paleontology 24:114A.

Sues, H. D., and A. Averianov. 2009. *Turanoceratops tardabilis*–the first ceratopsid dinosaur from Asia. Naturwissenschaften 96:645–652.

Williamson, T. E. 1997. Late Cretaceous (Early Campanian) vertebrate fauna from Allison Butte Member, Menefee Formation, San Juan Basin, New Mexico. New Mexico Museum of Natural History and Science Bulletin 11:51–59.

Wolfe, D. G. 2000. New information on the skull of *Zuniceratops christopheri,* a neoceratopsian dinosaur from the Cretaceous Moreno Hill Formation, New Mexico; 93–94 in S. G. Lucas and A. B. Heckert (eds.), Dinosaurs of New Mexico. New Mexico Museum of Natural History and Science Bulletin 17.

Wolfe, D. G., and J. I. Kirkland. 1998. *Zuniceratops christopheri* n. gen. & n. sp., a ceratopsian dinosaur from the Moreno Hill Formation (Cretaceous, Turonian) of west-central New Mexico; pp. 307–317 in S. G. Lucas, J. I. Kirkland, and J. W. Estep (eds.), Lower and Middle Cretaceous Terrestrial Ecosystems. New Mexico Museum of Natural History and Science Bulletin 24.

Wolfe, D. G., J. I. Kirkland, D. Smith, K. Poole, B. Chinnery-Allgeier, and A. McDonald. 2007. *Zuniceratops christopheri:* an update on the North American ceratopsid sister taxon, Zuni Basin, west-central New Mexico; pp. 159–163 in D. R. Braman (ed.), Ceratopsian Symposium: Short Papers, Abstracts, and Programs. Royal Tyrrell Museum, Drumheller, Alberta, Canada.

Xu, X, K. Wang, X. Zhao, and D. Li. 2010. First ceratopsid dinosaur from China and its biogeographical implications. Chinese Science Bulletin 55:1631–1635.

Late Cretaceous Theropod Dinosaurs of Southern Utah

22

Lindsay E. Zanno, Mark A. Loewen, Andrew A. Farke, Gy-Su Kim, Leon P. A. M. Claessens, and Christopher T. McGarrity

RECENT INTEREST IN UPPER CRETACEOUS FORMATIONS of southern Utah including intense collection efforts by the Kaiparowits Basin Project—a joint collaboration between the Utah Museum of Natural History, the University of Utah, and the Bureau of Land Management—has added considerably to our understanding of dinosaur diversity in the Western Interior Basin. These taxonomically unique and historically underrepresented ecosystems document a relatively high diversity of theropods, including a minimum of seven taxa known from the Kaiparowits Formation alone. Recent discoveries include at least five new taxa: *Hagryphus giganteus*, the first diagnostic North American oviraptorosaurian south of Montana; a new species of troodontid paravian; *Nothronychus graffami*, the most complete therizinosaurid skeleton yet discovered; and two new tyrannosaurid taxa, including *Teratophoneus curriei* and an undescribed taxon that represents the oldest North American tyrannosaurid recovered to date. Presently, data-rich paleobiogeographical comparison of latitudinally arrayed, coeval Western Interior Basin formations can only be made for a short temporal window that includes the upper Campanian Kaiparowits Formation. These investigations reveal that theropod diversity is relatively homogenous at higher taxonomic levels. Yet new discoveries also demonstrate a high degree of interformational, species-level endemism, indicating that the southern Utah theropod fauna is surprisingly unique and that theropod ranges in the upper Campanian Western Interior Basin were more restricted than previously understood. On the basis of these data, we argue against the referral of fragmentary dinosaur remains and teeth recovered from upper Campanian strata of the Western Interior Basin to taxa from other Western Interior Basin formations without substantial morphological evidence.

INTRODUCTION

In 2000, field crews of the Utah Museum of Natural History, the University of Utah, and the Bureau of Land Management embarked on an exhaustive research project to survey and document the Late Cretaceous dinosaur fauna of the Kaiparowits Basin, southern Utah, with a focus on upper Campanian exposures of the Wahweap and Kaiparowits formations of Grand Staircase–Escalante National Monument. This collaborative effort—known as the Kaiparowits Basin Project (KBP)—has documented at least 11 definitively new dinosaur taxa (with many more currently under study) that are challenging previous ideas regarding Late Cretaceous dinosaur evolution and diversity within Western Interior Basin.

Although the recent work undertaken by the KBP along with several other institutions, including the Raymond M. Alf Museum of Paleontology (RAM) and the Utah Geological Survey, represents the first concerted effort to collect and research the dinosaurian fauna of the Kaiparowits Basin, decades of microvertebrate studies were conducted in this and other areas of southern Utah by Richard Cifelli, Jeffrey Eaton, and their colleagues, who were studying the region's mammalian fauna before the project's initiation (Eaton and Cifelli, 1988; Eaton, Munk, and Hardman, 1998; Eaton, 1999; Eaton, Cifelli, et al., 1999; Eaton, Diem, et al., 1999). Concomitant fieldwork by J. Howard Hutchison and colleagues (Hutchison et al., 1997) involved surface reconnaissance in addition to microvertebrate screening and a greater focus on identifying the lower vertebrate fauna, including dinosaurs. Such early studies provided the first insight into the dinosaurian fauna of the region and the first (and in some cases only) comprehensive faunal lists for the formations reviewed here (Tables 22.1–22.5). These foundational contributions notwithstanding, the dinosaurian identifications generated by these studies are restricted almost entirely to surface float and isolated teeth, which recent discoveries indicate to be of limited taxonomic utility for theropods.

Table 22.1. Historical survey of the theropod fauna of the Iron Springs Formation derived from the literature

Taxon	Reference
Theropoda	
Theropoda indet.	Eaton, 1999
Paraves	
?Dromeosauridae	
?Dromeosauridae indet.	Eaton, 1999
Troodontidae	
Troodontidae indet.	Eaton, 1999

Table 22.2. Historical survey of the theropod fauna of the Dakota Formation derived from the literature

Taxon	Reference
Theropoda	
cf. *Richardoestesia* sp.	Eaton, Cifelli, et al., 1999
cf. *Paranychodon* sp.	Eaton, Cifelli, et al., 1999
Tyrannosauridae	
Tyrannosauridae indet.	Eaton, Cifelli, et al., 1999
Paraves	
Dromeosauridae	
Velociraptorinae indet.	Eaton, Cifelli, et al., 1999
Dromeosaurinae indet.	Eaton, Cifelli, et al., 1999
Troodontidae	
cf. *Troodon* sp.	Eaton, Cifelli, et al., 1999

Table 22.3. Historical survey of the theropod fauna of the Straight Cliffs Formation derived from the literature

Taxon	Reference
Theropoda	
Theropoda indet.	Eaton and Cifelli, 1988
cf. *Richardoestesia* sp.	Eaton, Cifelli, et al., 1999
cf. *Paranychodon* sp.	Parrish, 1999
Tyrannosauridae	
Tyrannosauridae indet.	Parrish, 1999
cf. *Aublysodon* sp.	Eaton, Cifelli, et al., 1999; Parrish, 1999
Paraves	
Dromeosauridae	
Velociraptorinae indet.	Eaton, Cifelli, et al., 1999; Parrish, 1999
Dromeosaurinae indet.	Eaton, Cifelli, et al., 1999; Parrish, 1999
Troodontidae	
Troodontidae indet.	Eaton, Cifelli, et al., 1999

Table 22.4. Historical survey of the theropod fauna of the Wahweap Formation derived from the literature[a]

Taxon	Reference
Theropoda	
Theropoda indet.	Eaton, Munk, and Hardman, 1998
cf. *Richardoestesia* sp.	Kirkland, 2001
cf. *Paranychodon* sp.	Kirkland, 2001
Tyrannosauridae	
Tyrannosauridae indet.[b]	Eaton, Cifelli, et al., 1999; Parrish, 1999
?Tyrannosauridae indet.	Eaton, Diem, et al., 1999
Tyrannosaurinae indet.	Kirkland, 2001
Aublysodontinae indet.	Kirkland, 2001
cf. *Aublysodon* sp.	Parrish, 1999
?*Aublysodon* sp.[b]	Eaton, Diem, et al., 1999
Paraves	
Dromeosauridae	
Dromeosauridae indet.[c]	Eaton, Diem, et al., 1999
Velociraptorinae indet.	Kirkland, 2001; Eaton, Cifelli, et al., 1999; Parrish, 1999
Dromeosaurinae indet.	Kirkland, 2001; Eaton, Cifelli, et al., 1999; Parrish, 1999
Troodontidae	
Troodontidae indet.	Kirkland, 2001
Troodon sp.	Eaton, Cifelli, et al., 1999
cf. *Troodon* sp.	Parrish, 1999

[a] Edward G. Sable, U.S. Geological Survey (pers. comm., 1994, in Eaton et al., 1999).

[b] Equivalent to Wahweap Formation.

[c] Equivalent to Wahweap Formation or younger.

Here we provide a review on the theropod fauna of the Upper Cretaceous formations in southern Utah with a focus on macrovertebrate materials recovered during the 2001–2009 field seasons by KBP crews and isolated teeth collected by the KBP and the RAM. KBP fieldwork is currently concentrated in the middle and upper Campanian Wahweap and Kaiparowits formations. As a result, the Wahweap and Kaiparowits formations have undergone the greatest recent advances in elucidating dinosaur diversity and comprise the bulk of this review. However, we also summarize the Late Cretaceous theropod fauna of southern Utah's Iron Springs, Dakota, and Straight Cliffs formations based almost exclusively on the aforementioned microvertebrate collections. Summaries of the chronostratigraphic relationships and radiometric ages of these formations and equivalent formations to the north and south on Laramidia are presented elsewhere in this volume. Finally, we discuss the surprising discovery of a dinosaur skeleton from the Upper Cretaceous marine Tropic Shale of

Table 22.5. Historical survey of the theropod fauna of the Kaiparowits Formation derived from the literature

Taxon	Reference
Theropoda	
cf. *Richardoestesia* sp.	Parrish, 1999
cf. *Paranychodon* sp.	Parrish, 1999
Tyrannosauridae	
Tyrannosauridae indet.	Eaton, Cifelli, et al., 1999; Parrish, 1999
cf. *Albertosaurus* sp.	Hutchison et al., 1997
Ornithomimidae	
Ornithomimus velox	Eaton, Cifelli, et al., 1999
Ornithomimus sp.	Hutchison et al., 1997
Paraves	
Dromeosauridae	
Dromeosauridae indet.	Hutchison et al., 1997
Velociraptorinae indet.	Eaton, Cifelli, et al., 1999; Parrish, 1999
Dromeosaurinae indet.	Parrish, 1999
Troodontidae	
Troodontidae indet.	Hutchison et al., 1997; Parrish, 1999
Troodon sp.	Eaton, Cifelli, et al., 1999
Aves	
Avisaurus sp.	Hutchison, 1993

southern Utah, *Nothronychus graffami*, a member of a rare group of predominantly plant-eating theropods known as therizinosaurians.

A preliminary report representing an earlier, less comprehensive version of this chapter and restricted to the theropod fauna of the Kaiparowits Formation was published as part of a symposium on Grand Staircase–Escalante National Monument in 2006 (Zanno et al., 2010). This work greatly expands upon that review and includes recent discoveries as well as updated research results for nearly all theropod subclades, specimen reevaluations, and new tooth morphotypes.

Institution Abbreviations BYU, Brigham Young University, Provo, Utah, U.S.A.; CMN, Canadian Museum of Nature, Ottawa, Ontario, Canada; MNA, Museum of Northern Arizona, Flagstaff, Arizona, U.S.A.; RAM, Raymond M. Alf Museum of Paleontology, Claremont, California, U.S.A.; UCMP, University of California Museum of Paleontology, Berkeley, California, U.S.A.; UMNH, Natural History Museum of Utah (formerly Utah Museum of Natural History), Salt Lake City, Utah, U.S.A.; YPM, Yale Peabody Museum, New Haven, Connecticut, U.S.A.

THEROPOD DIVERSITY IN THE KAIPAROWITS BASIN

Tyrannosauridae

Tyrannosaurids (tyrant reptiles) are a group of large-bodied theropods bearing massive skulls, diminutive arms, and stereoscopic vision (Holtz, 2004; Stevens, 2006). These highly specialized dinosaurs presumably functioned, alongside the alligatoroid *Deinosuchus*, as top predators within Late Cretaceous terrestrial ecosystems of the Western Interior Basin. At ~6300 kg, the Maastrichtian-aged *Tyrannosaurus rex* falls among the top body masses known for any terrestrial carnivore (Christiansen and Fariña, 2004; Therrien and Henderson, 2007); however, tyrannosaurids from preceding Campanian Age Western Interior Basin deposits were typically much smaller bodied (~2500 kg). Emerging evidence from China and Europe suggests that the group was more morphologically and ecologically diverse than previously appreciated. This evidence has complicated our understanding of the evolution of the tyrannosaurid body plan (Hutt et al., 2001; Xu et al., 2004, 2006; Li et al., 2009; Rauhut, Milner, and Moore-Fay, 2009; Sereno et al., 2009). Although small-bodied, Early Cretaceous tyrannosauroids possessed protofeathers (Xu et al., 2004), integument may have varied across the clade or during ontogeny.

Late Campanian tyrannosaurids have been well represented in northern Western Interior Basin formations for

22.1. Reconstruction of the holotype skull of *Teratophoneus curriei* (BYU 8120) from the Kaiparowits Formation. Reconstruction and original photograph by Rob Gaston of Gaston Design, Inc., Fruita, Colorado.

more than a century (Osborn, 1905; Lambe, 1914; Currie, 2003); however, tyrannosaurid species inhabiting southern Western Interior Basin ecosystems during this interval have remained enigmatic. Before the initiation of the KBP, the only diagnostic tyrannosaurid material recovered from the Kaiparowits Basin consisted of a partial associated skull and postcranial skeleton (BYU 8120/9396, BYU 8120/9397, BYU 8120/9398, BYU 826/9402, and BYU 13719, hereafter collectively referred to as BYU 8120) collected from the Kaiparowits Formation by BYU masters student Sam Webb in 1981 (Fig. 22.1). This material was recently described and named *Teratophoneus curriei* (Carr et al., 2011), three decades after its initial discovery (Carr, 2005). Exact locality data for BYU 8120 are not available; however, a contemporary newspaper article describing the location of the site places the specimen in the middle member of the Kaiparowits Formation in The Blues outcrop area northeast of Henrieville, Utah (Webb, 1981).

The skull of *Teratophoneus curriei* includes the left maxilla and lacrimal, the right jugal, frontal, and squamosal, both occipitals, the left prootic, the basioccipital, the right basisphenoid, both quadrates, the left articular, and the left dentary (Carr et al., 2011; Fig. 1). Postcranial materials assigned to this species include a single cervical vertebra (C3), a single midcaudal centrum, the left scapula, coracoid, humerus, and ulna, and the left femur (Carr et al., 2011; Loewen et al., in prep.). Although many of the elements of BYU 8120 are incomplete, this specimen provides the first clear evidence of a unique tyrannosaurid taxon from the Kaiparowits Formation, and its unusually short, deep skull sheds new light on

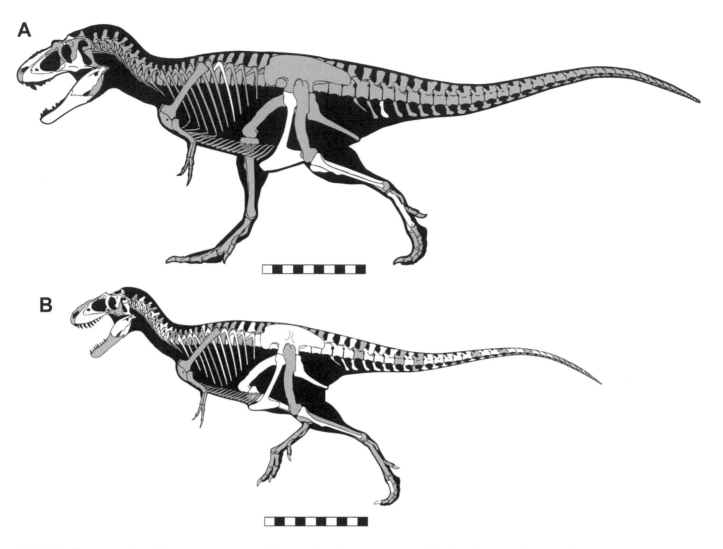

22.2. Skeletal reconstructions of the two new tyrannosaurid genera from the Late Cretaceous of Utah with recovered materials shown in white. (A), new species from the Wahweap Formation based entirely on UMNH VP 20200 and (B), *Teratophoneus curriei* from the Kaiparowits Formation based on UMNH VP 16690, with supplementary elements from the type and other specimens. Adapted with permission from artwork by Scott Hartman.

the ecomorphological diversity in tyrannosaurids during this time period (Carr et al., 2011).

Recent fieldwork in the Wahweap and Kaiparowits formations by the UMNH and RAM has produced numerous isolated tyrannosaurid elements as well as eight associated specimens that vastly augment our knowledge of *Teratophoneus* and southern Campanian tyrannosaurid diversity more generally.

The most complete Wahweap specimen (UMNH VP 20200) was discovered in the lower part of the Middle Member in 2009 (Fig. 22.2A). The specimen preserves portions of the skull including the right maxilla, both co-ossified nasals, the right frontal, the left laterosphenoid, the left jugal, the right quadrate, the left palatine, the left dentary, the left splenial, the left surangular, and the left prearticular. Postcranial elements include a single caudal chevron; both pubes; the left tibia, fibula, and metatarsals II, III, and IV (Loewen et al., in prep.). UMNH VP 20200 was recovered from sediments

dated to approximately 80 Ma (Jinnah et al., 2009), making it the oldest definitive tyrannosaurid yet known from the Western Interior Basin. As such, the specimen promises to provide valuable information on the origin and early evolution of Tyrannosauridae on the North American continent.

In contrast to the paucity of materials known from the Wahweap Formation, tyrannosaurid remains are abundant in the Kaiparowits. An exceptionally well-preserved associated subadult skeleton referable to *Teratophoneus curriei* (Loewen et al., in prep.) was discovered in 2004 (UMNH VP 16690; Fig. 22.2B). Not only does UMNH VP 16690 vastly expand our understanding of the anatomy of *Teratophoneus*, it also represents one of the most complete and phylogenetically informative tyrannosaurid individuals thus far collected from the southern Western Interior Basin formations.

The subadult specimen of *Teratophoneus* (UMNH VP 16690) is approximately 65% complete and preserves most

22.3. Reconstruction of the skull of a juvenile *Teratophoneus curriei* (UMNH VP 16690) from the Kaiparowits Formation. Reconstruction and original photograph by Rob Gaston of Gaston Design, Inc.

of the skull, axial column, pelvis, and a portion of the right leg (Fig. 22.2B). Recovered cranial materials include the following: the left maxilla; both lacrimals and postorbitals; the right squamosal; the left quadratojugal and quadrate; both frontals, parietals, angulars, and surangulars; a complete braincase; the left prearticular and articular, and multiple teeth (Fig. 22.3). In total, the skull lacks only the premaxillae, right maxilla, nasals, jugals, left squamosal, palatines, vomers, dentaries, splenials, and right prearticular and articular. Preserved axial materials include the atlas, seven cervical vertebrae and ribs, eight dorsal vertebrae and 14 dorsal ribs, two sacral vertebrae, 34 caudal vertebrae, and 19 chevrons. The appendicular skeleton is known from parts of both ilia, both pubes, both ischia, the right femur, tibia, and fibula, a single pedal phalanx, and an ungual (Wiersma and Loewen, this volume, Chapter 27).

A number of associated but shattered elements from the rostral portion of the skull, in addition to several shattered teeth, suggest that the facial skeleton of UMNH VP 16690 may have been trampled before burial (Wiersma, Loewen, and Getty, 2009; Wiersma and Loewen, this volume, Chapter 27). The reconstructed body size of this individual (approximately 6 m) (Fig. 22.2B), together with partial neurocentral fusion in preserved dorsal and sacral vertebrae, is suggestive of a subadult age for the animal at the time of death (Brochu, 1996).

The UMNH and RAM have collected other associated, less complete tyrannosaurid individuals and isolated tyrannosaurid elements from the Kaiparowits Formation during the course of the KBP. Specimens include: a partial subadult

dentary (RAM 8395), associated hind limb materials including a left femur, partial right tibia, complete right astragalus and calcaneum, partial right metatarsal III, and complete pedal PIII-1 and PIII-2 (RAM 9132); associated juvenile cranial material, including fused parietals, a partial unfused frontal, and partial dentary (UMNH VP 12586); partial limb elements and teeth (UMNH VP 16161); fragmentary limb elements, a pedal phalanx, and an ungual (UMNH VP 16692); associated limb and skull fragments, including a partial dentary, pedal phalanx, and ungual (UMNH VP 16693); a tooth, caudal vertebrae, left femur, tibia, fibula, metatarsal III, single pedal phalanx, and ungual of a large adult individual (UMNH VP 11302); isolated fused parietals (UMNH VP 16225); an isolated humerus (UMNH VP 12223); an isolated lacrimal from a large adult; and an isolated jugal from a large adult (UMNH VP 16691). The taxonomic identity of these specimens is currently under review (Loewen et al., in prep.).

Recent phylogenies (Brusatte et al., 2010; Carr et al., 2011) recover *Teratophoneus* as a derived tyrannosaurine tyrannosaurid just basal to a clade composed of the Campanian–Maastrichtian taxa *Daspletosaurus*, *Tyrannosaurus*, and *Tarbosaurus*. These authors also recover the southern laramidian taxon *Bistahieversor* (Carr and Williamson, 2000, 2010) from the late Campanian of New Mexico as a derived tyrannosauroid outside of Tyrannosauridae proper.

Research on UMNH VP 16690 supports the distinction between *Teratophoneus*, *Bistahieverser sealeyi*, and the Wahweap tyrannosaurid (UMNH VP 20200), documenting the presence of at least three distinct tyrannosauroid taxa from the southern Western Interior Basin. Moreover, preliminary phylogenetic analyses suggest that these three southern tyrannosauroid species form a clade to the exclusion of all other Campanian tyrannosaurids known from northern formations (i.e., *Gorgosaurus*, *Albertosaurus*, *Daspletosaurus*, *Daspletosaurus* n. sp. from the Dinosaur Park Formation and *Daspletosaurus* n. sp. from the Two Medicine Formation), lending critical information to the geographic distribution of theropod taxa across the Late Cretaceous Western Interior Basin (Loewen et al., in prep.).

The abundance of tyrannosaurid material collected during the relatively brief time span of the KBP challenges previous statements that the upper Campanian formations of New Mexico exceed those of Utah with regard to tyrannosaurid preservation (Carr and Williamson, 2000), and highlights the importance of fossil-bearing strata of southern Utah in understanding dinosaur evolution in the Western Interior Basin. Study of the diagnostic tyrannosaurid material recovered from the Kaiparowits Formation will permit a more comprehensive understanding of tyrannosauroid diversity, biogeography, and evolution during the Late Campanian.

Ornithomimidae

Ornithomimids (ostrich mimics) were medium-bodied (200–700 kg; Christiansen and Fariña, 2004), lightly built dinosaurs, estimated to be among the fastest theropods (Coombs, 1978; Thulborn, 1982; Christiansen, 1998). They are generally regarded as close relatives of tyrannosauroids, falling within Coelurosauria but outside of Maniraptora, the more derived subclade that includes modern birds. Advanced members of the group possess toothless, keratinous covered beaks, elongate necks, and gracile limbs. The discovery of *Sinornithomimus* in a mass-death assemblage prompted speculation that ornithomimids were social animals, congregating in large groups (Kobayashi et al., 1999; Varricchio et al., 2008). The diet of ornithomimids has been a matter of some debate; hypotheses have been put forth ranging from myrmecophagy (Russell, 1972) to filter feeding (Norell, Clark, and Makovicky, 2001). However, most recent studies make a strong argument for herbivory (Kobayashi et al., 1999; Barrett, 2005; Zanno, Gillette, et al., 2009; Zanno and Makovicky, 2010), based in part on the recovery of stomach stones in several remarkably preserved specimens (Kobayashi et al., 1999; Ji et al., 2003; Wings, 2007).

Owing to a lack of teeth, ornithomimids are not generally represented in faunal lists generated on the basis of microvertebrate studies. Thus, no ornithomimids have been reported from the Iron Springs, Dakota, Straight Cliffs, or Wahweap formations (Tables 22.1–22.4). However, their absence from these formations is almost certainly the result of collection and/or taphonomic biases because ornithomimid remains have been recovered from both older and younger formations within the Colorado Plateau (Marsh, 1890; Ostrom, 1970). Ornithomimid skeletal remains form an abundant constituent of the theropod materials recovered from the Kaiparowits Formation, and along with those of tyrannosaurids, they are among the most commonly recovered.

Despite their relative abundance, little progress has been made in identifying the ornithomimid remains from the Kaiparowits Formation. Thirty years ago, an ornithomimid specimen consisting of a nearly complete hind limb, fragmentary pelvis, and partial axial column (MNA PI.1762A) was collected from the Kaiparowits Formation by the Museum of Northern Arizona and subsequently referred to the late Maastrichtian taxon *Ornithomimus velox* by DeCourten and Russell (1985).

The holotype of *O. velox* (YPM 542) is fragmentary, comprising a distal tibia with astragalus, incomplete left metatarsus and second pedal digit, together with potentially associated manual elements (YPM 548). Its taxonomic status is currently uncertain. Russell (1972) noted that the two supposedly diagnostic characteristics of *O. velox* provided by

previous authors (shortness of the metatarsus and metatarsal II longer than metatarsal IV) were derived from a reconstruction of the incompletely preserved metatarsus (Marsh, 1890). At the time, the length and proportion of the metatarsals of *O. velox* could not be determined from the type specimen; therefore, these traits could not be evaluated with confidence (Russell, 1972). Moreover, Makovicky, Kobayashi, and Currie (2004) note only a single manual character as potentially diagnostic for *O. velox*–metacarpal I being the longest in the metacarpus. However, this trait is shared with *Ornithomimus edmontonicus*, and therefore, it would only be diagnostic to species level if these taxa were to prove synonymous. Although it is currently unclear whether *O. velox* can be regarded as a valid taxon, work on the status of this enigmatic taxon is forthcoming. Additional preparation of the holotype metatarsus (YPM 542) is permitting accurate measurements of its proportion, and detailed morphometric analyses of the manus (YPM 548) (Neabore et al., 2007; Lavender et al., 2010) is offering valuable insights that may help resolve these issues.

Even if *O. velox* is upheld as a valid taxon by additional research, the referral of the Kaiparowits specimen MNA PI.1762A to this species is dubious. DeCourten and Russell's (1985) justification for the referral of MNA PI.1762A to *O. velox* lies in pedal ungual morphology (which they identify as similar in both specimens), relative proportions of the pes, and temporal equivalence (at the time, palynological evidence supported a Lancian age for the Kaiparowits Formation [Lohrengel, 1969], suggesting that the Kaiparowits and the Denver Formation of Colorado, from which the type specimen of *O. velox* is described [Marsh, 1890] are coeval).

The specific ratio used by DeCourten and Russell (1985) to assign the Kaiparowits specimen to *O. velox* (ratio of the length of the second pedal ungual to the basal phalanx of digit II) is given by the authors as 0.61–0.64 in "pre-Lancian" North American taxa (*Ornithomimus edmontonicus*, sensu Makovicky, Kobayashi, and Currie, 2004, and *Struthiomimus altus*), 0.78 in MNA PI.1762A, and 0.88 in the holotype of *O. velox*. Given that the proportions of MNA PI.1762A are in fact closer to other North American ornithomimid species than to *O. velox* (as given), we do not see the justification for this association. Moreover, although proportional characteristics have been proposed as diagnostic for individual ornithomimid taxa (Russell, 1972), recent studies (Kobayashi, Makovicky, and Currie, 2006) have challenged the validity of most of these differentiations. Kobayashi, Makovicky, and Currie (2006) cite characteristics of the skull, forelimb, and caudal vertebrae as diagnostic for ornithomimids; however, they do not identify any diagnostic features of the pes among North American taxa. Finally, although Longrich (2008b) outlines differences between the pedal unguals of several species of

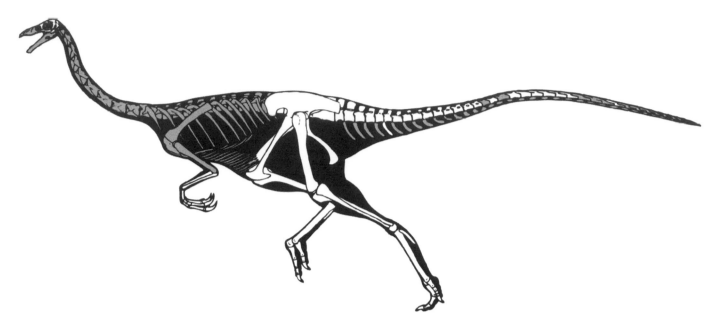

22.4. Skeletal reconstruction of *Ornithomimus,* showing ornithomimid elements recovered from the Kaiparowits Formation in white (multiple specimens and individuals represented). Adapted with permission from artwork by Greg Paul.

North American ornithomimids that may ultimately prove useful in discerning the evolutionary relationships of MNA PI.1762A, it is unclear how large of a sample was considered in the study and whether these distinctions can be discretely quantified by gap coding (Archie, 1985). Ultimately, additional study of Western Interior Basin ornithomimid materials is needed to confirm or refute these various distinctions.

Recent fieldwork conducted by the KBP has added significant morphological data to bear upon the identity of Kaiparowits ornithomimid materials (Fig. 22.4), including the contentious specimen MNA PI.1762A. Newly recovered

elements include associated caudal vertebrae, metatarsal fragments, and phalanges (UMNH VP 12223), two isolated tibiae (UMNH VP 9553 and UMNH VP 16698), an isolated articulated foot with an associated limb bone (UMNH VP 19467); an associated partial rear skeleton including caudal vertebrae, a partial pelvis, and partial foot (UMNH VP 20188); a badly damaged skull (UMNH VP unnumbered), as well as the first articulated forelimb material from the formation (UMNH VP 16385). The forelimb materials (UMNH VP 16385) consist of an incomplete and partially crushed manus, carpus, and antebrachium (Fig. 22.5). Additional material recently collected

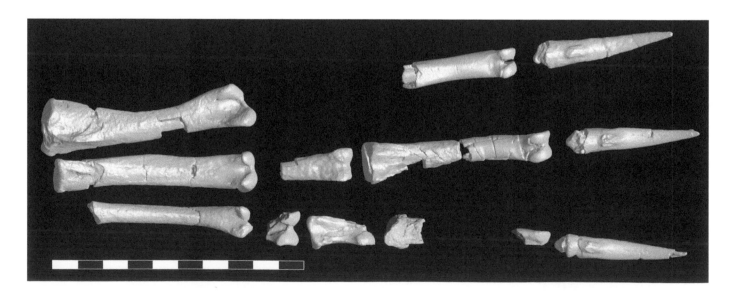

22.5. Three-dimensional surface scan of the left ornithomimid manus (UMNH VP 16385) from the Kaiparowits Formation in dorsal view, showing subequal metacarpals I and II and straight ungual morphology. Placement and identity of phalangeal fragments in digit III is approximate. Scale bar = 10 mm.

Zanno et al.

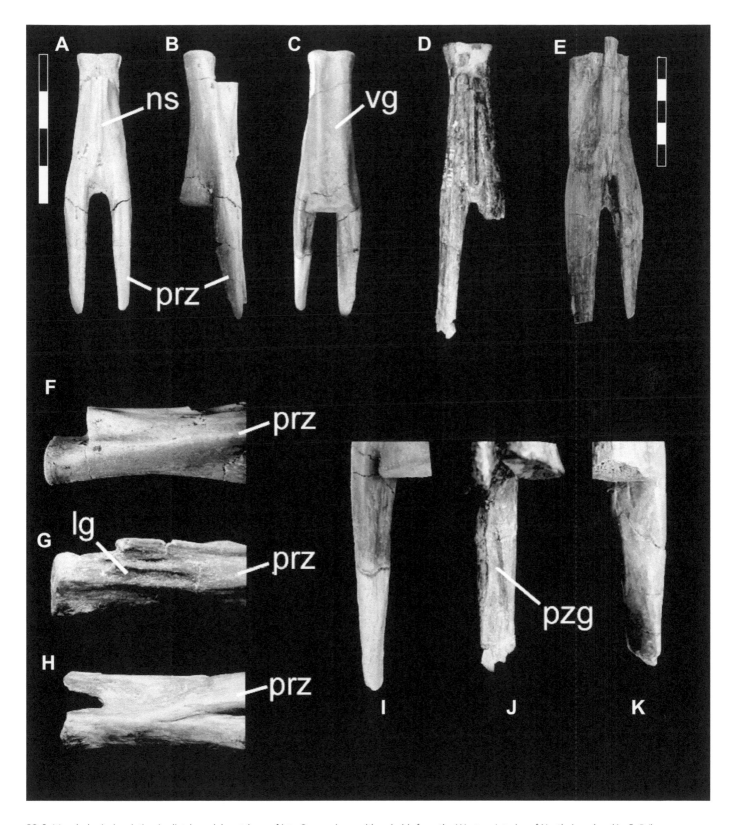

22.6. Morphological variation in distal caudal vertebrae of late Campanian ornithomimids from the Western Interior of North America. (A–C, F, I) Ornithomimidae incertae cedis from the Kaiparowits Formation (UMNH VP 16260); (D, G) and (J) CMN 12228 *Dromiceiomimus brevetertius*, Kobayashi, Makovicky, and Currie, 2006; (*Ornithomimus edmontonicus* sensu Makovicky, Kobayashi, and Currie, 2004); and (E, H, K) *Struthiomimus altus* (CMN 2102/8902). Kaiparowits ornithomimid shown in (A) dorsal, (B) right lateral, and (C) ventral views. Late Campanian ornithomimid caudals in right lateral views (F–H) showing the absence of lateral groove for the prezygapophyses in all but *Dromiceiomimus* (G). Prezygapophyses of ornithomimid caudal vertebrae in ventral views (I–K) showing presence of a ventral groove in *Dromiceiomimus* (J). lg, lateral groove on the centrum caused by articulation with the prezygapophyses; ns, neural spine remnant; prz, prezygapophysis; pzg, ventral groove on the prezygapophysis; vg, ventral groove on centrum. Upper left scale bar = 4 mm (A–C). Upper right scale bar = 5 mm (E). All other views not to scale.

by the Raymond M. Alf Museum (RAM 6794) includes the left pelvis (complete ilium, partial pubis and ischium), left hind limb (including partial femur, partial tibia, fibula, metatarsals II–V, and incomplete set of phalanges), partial right hind limb (complete femur, tibia, with partial metatarsals and incomplete set of phalanges), 15 caudal vertebrae, and several incomplete ribs. RAM 6794 thus provides useful comparative material to MNA PI.1762A.

Preliminary examination of the forelimb materials (UMNH VP 16385) reveals similarities to O. edmontonicus in the relative size of metacarpal I and in ungual morphology. However, isolated caudal vertebrae collected by the UMNH (UMNH VP 16260; Fig. 22.6A–C, F, I) lack the diagnostic, deeply grooved articulation between pre- and postzygapophyses (Fig. 22.6F–H), as well as the prezygapophyseal ventral groove (Fig. 22.6I–K) of Dromiceiomimus brevetertius (Kobayashi, Makovicky, and Currie, 2006) (O. edmontonicus sensu Makovicky, Kobayashi, and Currie, 2004). A more comprehensive investigation of ornithomimid materials from the Kaiparowits Formation, including reevaluation of MNA PI.1762A, is currently being undertaken by researchers at the UMNH, RAM, and the College of the Holy Cross (Zanno, Sampson et al., 2005; Zanno, Loewen, et al., 2009; Neabore et al., 2007; Lavender et al., 2010) and is expected to provide additional insights regarding the taxonomic and systematic relationships of North American ornithomimids.

Therizinosauria

Therizinosaurians (scythe reptiles) are regarded as one of the most enigmatic theropod subclades owing to their remarkably bizarre anatomy. Derived members are characterized by a keratinous beak, a potbelly, and over 1-m-long scythe-shaped claws on the hands. With some species weighing in at over 6 tons, therizinosaurians achieve some of the largest body sizes among maniraptoran theropods (Zanno and Makovicky, in prep.) and are generally considered chiefly herbivorous, partly as a result of their enlarged gut and small, leaf-shaped teeth (Paul, 1984; Weishampel and Norman, 1989; Russell, 1997; Zhang et al., 2001; Kirkland et al., 2005; Zanno, Gillette, et al., 2009; Zanno and Makovicky, 2010). At least early members of the clade were feathered (Xu, Tang, and Wang, 1999).

For the first half century after their discovery, therizinosaurians were known exclusively from Asia. However, recent years have seen the discovery of three therizinosaurian species in western North America (Kirkland and Wolfe, 2001; Kirkland et al., 2005; Zanno, Gillette, et al., 2009). In addition, although still controversial, several isolated elements from Upper Cetaceous formations of the northern Western Interior Basin have been referred to the clade including two isolated frontals, an ungual, an astragalus, and a cervical vertebra (Russell, 1984; Currie, 1987b, 1992; Ryan and Russell, 2001).

Only a single discovery marks the presence of therizinosaurians in the Late Cretaceous of Utah – the recently described species Nothronychus graffami, from the lower Turonian portion of the Tropic Shale (Gillette et al., 2001; Zanno, Gillette, et al., 2009) – although description of its sister species Nothronychus mckinleyi from middle Turonian sediments in nearby New Mexico (Kirkland and Wolfe, 2001) confirmed the existence of the group in the Late Cretaceous of North America nearly a decade earlier. Nothronychus graffami is a large-bodied member of the derived subclade Therizinosauridae and is represented by the most complete skeleton yet discovered for the clade (Fig. 22.7). Recent phylogenetic studies indicate that this taxon is more closely related to Late Cretaceous taxa from Asia than to the Early Cretaceous North American therizinosaurian Falcarius utahensis, supporting the idea of faunal interchange between western North America and Asia in the late Early Cretaceous (Russell, 1993; Cifelli et al., 1997; Kirkland et al., 1997, 1999; Zanno and Makovicky, 2011). As therizinosaurids were not unequivocally recognized as North American inhabitants until relatively recently, and given that their morphology (especially the teeth) is highly convergent with other clades of herbivorous dinosaurs, it is possible that therizinosaurid remains have been collected in other Upper Cretaceous formations in the Western Interior Basin and that future investigation may reveal a more widespread presence than currently recognized.

Oviraptorosauria

Oviraptorosaurians (egg thieves) are a predominantly endentulous group of feathered maniraptoran dinosaurs that are often adorned with a cranial crest, keratinous beak, and powerful arms bearing formidable claws.. Members of the group are predominantly small (3 kg) to medium bodied (250 kg); however, an exemplary Late Cretaceous species from China (Gigantoraptor, Xu et al., 2007) is estimated at over 3000 kg (Zanno and Makovicky, in prep.).

Oviraptorosaurians are widely known in Asia, where upward of a dozen taxa have been named, many from remarkably complete skeletons. In contrast, only a few North American members of the clade have been described (generally referred to as caenagnathids, although this taxonomy is currently contentious; Zanno and Sampson, 2005), and almost all are known from isolated and fragmentary materials. An exception to this is a new Maastrichtian-age species from South Dakota, which is represented by two partial skeletons awaiting description (Triebold, Nuss, and Nuss, 2000).

Zanno et al.

22.7. Skeletal reconstruction of the therizinosaurid *Nothronychus graffami* based on the holotype (UMNH VP 16420) from the Upper Cretaceous Tropic Shale, southern Utah. Modified from Zanno, Gillette, et al. (2009).

Although the dietary preference of advanced members of the clade is presently unclear, primitive members of the group possess incisiform front teeth and stomach stones indicative of an herbivorous component to the diet (Ji et al., 1998; Xu et al., 2002; Barrett, 2005; Zanno, Gillette, et al., 2009; Longrich, Currie, and Dong, 2010; Zanno and Makovicky, 2010). Several other rarely elucidated aspects of dinosaur paleobiology are known for oviraptorosaurians, including details about egg laying (Sato et al., 2005; Erickson et al., 2007) and brooding behavior (Norell et al., 1994, 1995). Although these dinosaurs are remarkably similar to birds in anatomy and behavior, current phylogenetic analyses demonstrate that these similarities are the result of convergence rather than ancestry (Sues, 1997; Makovicky and Sues, 1998; Norell, Makovicky, and Currie, 2001; Rauhut, 2003; Lu et al., 2004).

Because derived oviraptorosaurians lack teeth, their remains are not commonly recovered in microvertebrate assemblages. This bias may explain their absence from prior faunal surveys of Upper Cretaceous formations of southern Utah (Tables 22.1–22.5). Thus, as is the case with toothless ornithomimids, oviraptorosaurians are not yet recognized in the Iron Springs, Dakota, Straight Cliffs, or Wahweap formations, although they too are known from both earlier and later formations in the Western Interior (Sues, 1997; Makovicky and Sues, 1998).

Definitive oviraptorosaurian remains were not known from the prolific Kaiparowits Formation until UMNH crews recovered the first diagnostic skeletal material in 2002. A nearly complete left manus (missing only the second ungual), carpus, and distal antebrachium (UMNH VP 12765) of a new oviraptorosaurian were recovered in articulation within a remnant of channel sandstone (Fig. 22.8). As is often the case with Kaiparowits specimens (Lund et al., 2008), exceptional soft tissue preservation makes allusion to the keratinous sheath of one ungual. Additional elements—including fragmentary metatarsals and pedal phalanges, and a partial, articulated pedal digit with ungual—were salvaged from the surrounding hillside. The specimen, dubbed *Hagryphus giganteus*, represents the first dinosaur taxon to be named from Grand Staircase–Escalante National Monument and is notably larger than its named northern cousins *Chirostenotes* and *Elmisaurus* (Fig. 22.9), with an estimated body size increase of 30–40% (Zanno and Sampson, 2005) and an estimated mass of ~180 kg (Zanno and Makovicky, in prep.). Proportionally gigantic isolated unguals of oviraptorosaurians are also known from Alberta (P. Currie, pers. comm., 2003) and the undescribed skeleton of a new oviraptorosaurian from South Dakota (Triebold, Nuss, and Nuss, 2000) represents another relatively large-bodied species akin to *Hagryphus*.

Hagryphus is the first North American oviraptorosaurian described from south of Montana and South Dakota and represents the southernmost limit yet identified for this enigmatic group of theropods within the Western Interior Basin. Although the holotype represents the only published account of oviraptorosaurian materials from the formation, the recent recognition of a distal second right metatarsal (RAM 12433) closely comparing to an oviraptorosaurian specimen CMN 8538 (formerly *Macrophalangia*) from Alberta, Canada, suggests that other fragmentary oviraptorosaurian materials may have been collected that await identification.

Dromaeosauridae

Dromaeosaurid theropods (running reptiles) are swift small- to medium-bodied predators (0.6–350 kg; Turner et al., 2007) distinctive in possessing lightly built skeletons, an enlarged sickle claw on the second digit of the foot, and stiff tails reinforced by dramatically elongated bony struts. Dromaeosauridae is one of the more diverse maniraptoran clades, and its members are considered to be among the closest extinct cousins to birds. Several small-bodied, feathered species have been found within exceptionally prolific Lower Cretaceous lake beds in China (Xu, Wang, and Wu, 1999; Xu, Zhou, and Wang, 2000)—including the miniature "four-winged" dinosaur *Microraptor*—and the majority of dromaeosaurid species are known from Asia. Yet the group has a relatively diverse North American representation, with at least five documented species, including the largest dromaeosaurid known to date, *Utahraptor*, recovered from Lower Cretaceous beds in central Utah (Kirkland, Gaston, and Burge, 1993).

Dromaeosaurids are known to have been widespread across the Upper Cretaceous Western Interior Basin (Norell and Makovicky, 2004). Their serrated, blade-like teeth are commonly recovered in microvertebrate localities within Upper Cretaceous beds of southern Utah. At least one species is known from the Iron Springs Formation on the basis of isolated teeth (Eaton, 1999; Table 22.1), whereas at least two dromaeosaurid tooth morphotypes representing a minimum of two species are known from the Dakota, Straight Cliffs, and Wahweap formations (Tables 22.2–22.5). These teeth

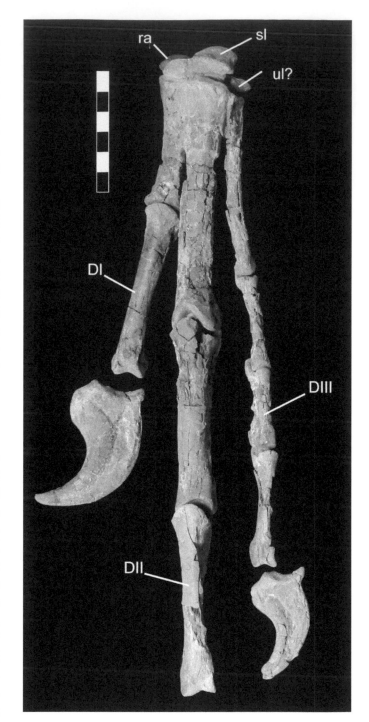

22.8. Holotype manus of the oviraptorosaurian *Hagryphus giganteus* (UMNH VP 12765) in dorsal view. DI, digit I; DII, digit II; DIII, digit III. Scale bar = 5 cm.

have commonly been referred to the subclades Dromaeosaurinae and Velociraptorinae (Eaton, 1999; Eaton, Cifelli, et al., 1999; Eaton, Diem, et al., 1999; Parrish, 1999; Kirkland, 2001). However, recent studies of dromaeosaurid evolutionary relationships vary, and many no longer recover these two major divisions (Hwang et al., 2002; Senter et al., 2004; Norell et al., 2006; Senter, 2007; Xu et al., 2009). Furthermore, it is

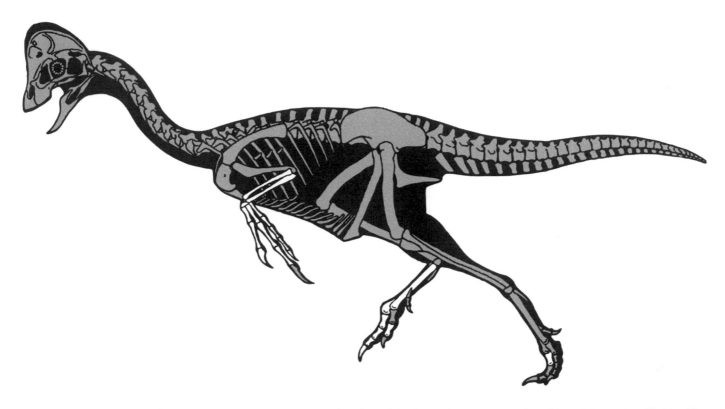

22.9. Skeletal reconstruction of *Hagryphus giganteus,* showing recovered portions of the skeleton (UMNH VP 12765) in white. Missing sections filled in with skeletal materials recovered from the Hell Creek Formation and adapted with permission from artwork by Scott Hartman.

not clear that these divisions are reflected with consistency in tooth morphology. For instance, Currie and Varricchio (2004:table 4.2) demonstrate that the dimensions and denticle counts of the teeth of the Late Cretaceous North American "velociraptorine" *Saurornitholestes* fall within the range of variation observed in the dromaeosaurid *Atrociraptor* from the Horseshoe Canyon Formation, which is often recovered as closer to *Dromaeosaurus* (the internal specifier for Dromaeosaurinae) than *Velociraptor* (the internal specifier for Velociraptorinae) (Makovicky, Apesteguia, and Agnolin, 2005; Senter, 2007; Zhang et al., 2008; Xu et al., 2009). They also note that the maxillary teeth of the North American dromaeosaurids *Bambiraptor* (Burnham et al., 2000), *Deinonychus* (Ostrom, 1969), *Saurornitholestes* (Currie, Rigby, and Sloan, 1990), *Atrociraptor,* and the Asian species *Velociraptor* (Barsbold and Osmólska, 1999) are all "closely comparable in terms of tooth shape, carina position, denticles size, and denticles shape" (Currie and Varricchio, 2004:122), yet current phylogenetic analyses recover these taxa as spread across the dromaeosaurid family tree (Hu et al., 2009) suggesting caution in the taxonomic assignment of isolated teeth of this morphotype beyond Dromaeosauridae.

Collection of teeth from microvertebrate localities suggested the presence of "dromaeosaurine" and "velociraptorine" dromaeosaurids in the Kaiparowits Formation over a decade ago (Hutchison et al., 1997; Table 22.5). Subsequent collection and detailed examination of UMNH and RAM collections has verified the existence of at least two dromaeosaurid species in the formation on the basis of isolated teeth recovered as surface float or within burial sites of herbivorous dinosaurs (Fig. 22.10A–D). In earlier publications (Zanno, Sampson et al., 2005; Gates et al., 2010), we provisionally referred these to cf. *Dromaeosaurus* and cf. *Saurornitholestes*

Table 22.6. Revised theropod fauna of the Wahweap and Kaiparowits formations, this study

Wahweap Formation	Kaiparowits Formation
Theropoda	Theropoda
Theropoda indet.	Theropoda indet.
Tyrannosauridae	Tyrannosauridae
Genus et sp. nov.	Genus et sp. nov.
Paraves	Ornithomimidae
Dromeosauridae	Ornithomimidae indet.
Morphotype A	Oviraptorosauria
Morphotype B	*Hagryphus giganteus*
Troodontidae	Paraves
Troodontidae indet.	Dromeosauridae indet.
	Morphotype A
	Morphotype B
	Troodontidae
	Talos sampsoni
	Aves
	? cf. *Avisaurus*

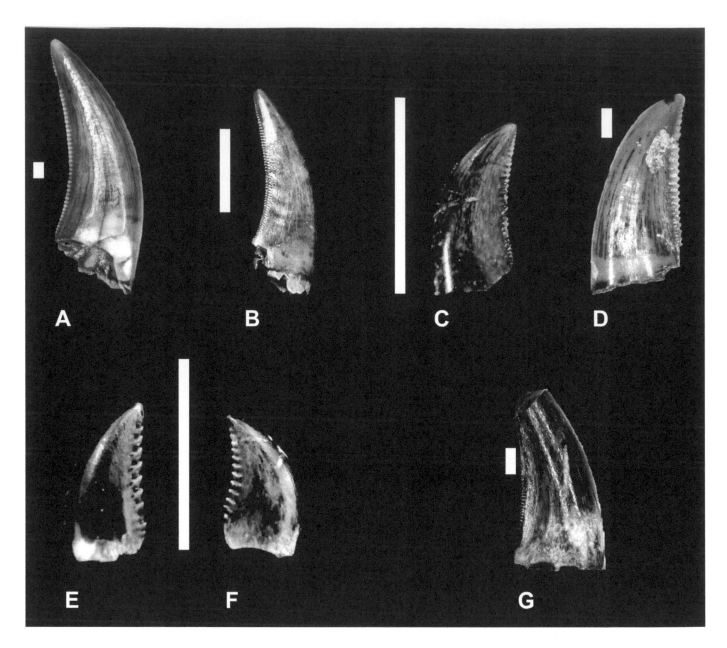

22.10. Paravian theropod teeth from the Kaiparowits Formation. (A) UMNH VP 16306 Dromaeosauridae morphotype A, (B) RAM 9136 Dromaeosauridae morphotype A, (C) UMNH VP 11803 Dromaeosauridae morphotype B, (D) RAM 12084 Dromaeosauridae morphotype B, (E) UMNH VP 12507 Troodontidae maxillary tooth, (F) UMNH VP 12507 Troodontidae dentary tooth, and (G) RAM 12083 Theropoda incertae sedis (cf. "*Richardoestesia*"). Scale bar = 1 cm.

(sensu Sankey, 2001; Sankey et al., 2002) based on comparisons with teeth from the approximately coeval Dinosaur Park and Aguja formations. Here we adopt a more general nomenclature that considers new information regarding the limited paleogeographical ranges of Western Interior Basin theropods by restricting the taxonomic identify of these teeth to Dromaeosauridae indet. morphotype A (*Dromaeosaurus* type) and morphotype B (*Saurornitholestes/Atrociraptor/Bambiraptor* type) (Table 22.6). Dromaeosauridae postcranial materials remain rare in the Kaiparowits Formation and include an isolated pedal phalanx PII-I (UMNH VP 12494) and isolated unguals.

Troodontidae

Troodontids (wounding teeth) are a group of feathered maniraptoran dinosaurs, notable for exhibiting some of the smallest body sizes (0.1 kg; Therrien and Henderson, 2007; Turner et al., 2007) and the largest relative brain sizes (Currie, 2005) within Dinosauria. The group is known almost exclusively from Asia, and until recently, only two genera were recognized in North America: *Troodon* (sensu Currie, 1987a, 2005) and *Pectinodon* (Carpenter, 1982; Longrich, 2008a), although other potentially valid troodontid species have been described and or subsumed into existing taxa (e.g.,

Zanno et al.

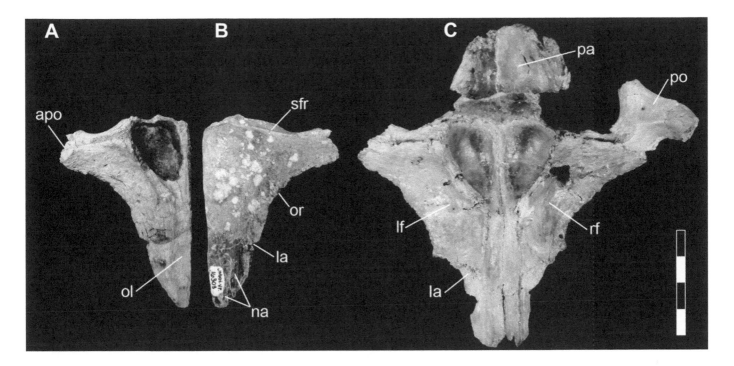

22.11. Morphology of troodontid frontals from the late Campanian of the Western Interior of North America. (A–B) UMNH VP 16303 troodontid incertae cedis from the Kaiparowits Formation of Utah in (A) ventral and (B) dorsal views. (C) CMN 12340, *Troodon formosus* cranium from the Dinosaur Park Formation, Alberta, Canada, in ventral view. apo, articular surface for the postorbital; la, articular surface for the lacrimal; lf, left frontal; na, articular surface for the nasal; ol, olfactory lobe; or, orbital margin; pa; parietal; po; postorbital; rf, right frontal; sfr, ridge on the rostral margin of the supratemporal fenestra. Scale bar = 4 cm.

Stenonychosaurus, Koparion). As a result of their distinctive teeth and the possible presence of seeds preserved as gut contents in the primitive species *Jinfengopteryx* (Xu and Norell, 2006), several authors have proposed an omnivorous or herbivorous diet for at least some members of the clade (Holtz, Brinkman, and Chandler, 1998; Zanno, Gillette, et al., 2009; Zanno and Makovicy, 2010). However, other species possess blade-like teeth, suggesting that dietary diversity, including carnivory, likely characterized the group (Currie and Dong, 2001; Zanno and Makovicy, 2010).

Several troodontid specimens preserve remarkable evidence of behavior. Perhaps the most intriguing troodontid is the miniature Chinese species *Mei long*. The subadult holotype was discovered preserved in a distinctive avian-like sleeping posture with its head rotated back toward the tail and tucked under its wing-like forelimb (Xu and Norell, 2004). Although not as widely touted, the holotype specimen of another Asian species, *Sinornithoides youngi* (Currie and Dong, 2001), is also preserved in this posture. Troodontid nests, eggs, and embryos are known from Upper Cretaceous formations in the Western Interior Basin, and studies on clutch volume, reproductive physiology, and egg-laying behavior (Varricchio et al., 1997; Varricchio, Horner, and Jackson, 2002; Varricchio and Jackson, 2004) have yielded important information on the reproductive biology of these rare theropods.

Troodontid teeth, which are widely recognized as diagnostic for the clade (Currie, 1987a; Makovicky and Norell, 2004), have been collected from the Iron Springs, Dakota, and Straight Cliffs formations and exclusively referred to Troodontidae (Eaton, 1999; Eaton, Cifelli, et al., 1999; Tables 22.1–22.3), whereas troodontid teeth from the Wahweap Formation have been identified as Troodontidae indet. (Kirkland, 2001), cf. *Troodon* (Parrish, 1999), and *Troodon* sp. (Eaton, Cifelli, et al., 1999; Table 22.4).

The presence of troodontid theropods in the Kaiparowits Formation was originally documented on the basis of isolated teeth (Hutchison et al., 1997; Eaton, Cifelli, et al., 1999). Numerous troodontid teeth have been collected by the UMNH and RAM over the past decade (Fig. 22.10E, F), adding to those collected during earlier microvertebrate surveys (Hutchison et al., 1997; Eaton, Cifelli, et al., 1999). Several authors referred these teeth to *Troodon* sp. (Hutchison et al., 1997; Eaton, Cifelli, et al., 1999; Table 22.5). However, troodontid cranial and postcranial materials have been recovered from the Kaiparowits Formation recently that shed doubt on this referral.

During the field season of 2005, an isolated left frontal of a troodontid (UMNH VP 16303) was recovered (Zanno, 2007; Fig. 22.11A, B), and 3 years later, in 2008, a partially articulated, partial postcranial skeleton representing a subadult troodontid individual (UMNH VP 19479) was found (Zanno,

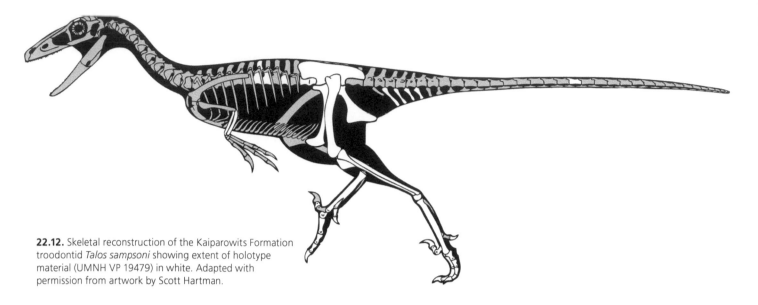

22.12. Skeletal reconstruction of the Kaiparowits Formation troodontid *Talos sampsoni* showing extent of holotype material (UMNH VP 19479) in white. Adapted with permission from artwork by Scott Hartman.

Varricchio, et al., 2009; Zanno et al., 2011). The frontal shares several features with *Troodon formosus* (CMN 12340) from the contemporaneous Dinosaur Park Formation in Alberta, such as an elongate, triangular morphology, an extensive orbital rim, a prominent ridge defining the rostral limit of the supratemporal fenestra, and a large, laterally extensive postorbital process (Fig. 22.11B, C). However, UMNH VP 16303 also varies from CMN 12340 and other troodontid frontals recovered from coeval formations across the Western Interior Basin (particularly the Two Medicine Formation) in the shape of the orbital margin, supratemporal ridge, and orbital bulbs.

The partial postcranial skeleton (UMNH VP 19479; Fig. 22.12) also differs from specimens currently considered part of the *T. formosus* paradigm and was recently made the holotype for the new genus and species *Talos sampsoni* (Zanno et al., 2011). In particular UMNH VP 19479 is significantly more gracile in metatarsal morphology (Fig. 22.13) when compared to materials from *T. formosus* individuals of smaller and larger body size, which indicates that its slender nature is not solely attributable to ontogenetic variation (Zanno et al., 2011). The specimen also exhibits autapomorphic features of the foot and ankle.

Additional work is currently underway to determine whether UMNH VP 16303 is distinguishable from other Western Interior Basin troodontid frontals and to investigate possible variation between the teeth of *T. formosus* and troodontid teeth from the Kaiparowits formation to determine whether late Campanian Western Interior Basin troodontid teeth are diagnostic beyond the level of Troodontidae.

Finally, a poorly preserved, partial paravian skeleton collected by Howard Hutchison of the University of California at Berkeley's Museum of Paleontology in 1994 was identified by him as a dromaeosaurid, and this identification was retained in earlier KBP publications (Zanno, Sampson et al., 2005; Zanno et al., 2010). The specimen, UCMP 149171, consists of a proximal tibia, fragmentary metatarsals, pedal phalanges, and pedal unguals, as well as some fragmentary skull material, including the basioccipital, fused parietals, and portions of the squamosals. The tibial morphology of UCMP 149171 suggests that these materials represent another troodontid specimen (Zanno et al., 2011). Further study is needed to determine how these specimens, along with additional isolated materials such as isolated caudal vertebrae, compare to troodontid materials across the Western Interior Basin.

Aves

Abundant evidence exists in support of the hypothesis that birds are the direct descendants of maniraptoran theropods, and thus represent a specialized clade of living dinosaurs (Padian and Chiappe, 1998). Just as living birds comprise one of the most diverse vertebrate groups in our modern ecosystems, the group is known to have had a strong representation during the Cretaceous. Over 30 genera of Cretaceous birds, most preserving evidence of feathers, are known from the Early Cretaceous of China alone (Zhou and Li, 2010, and references therein). Many of these taxa retain primitive morphology such as teeth, clawed fingers on the hand, and a long tail, such as the famous Jurassic bird *Archaeopteryx*.

To date, avian remains have not been documented in the Iron Springs, Dakota, Straight Cliffs, and Wahweap formations (Tables 22.1–22.4). A number of fragmentary avian skeletal elements have been collected during the KBP. A single paravian tooth (RAM 11906) possessing a constricted junction

Zanno et al.

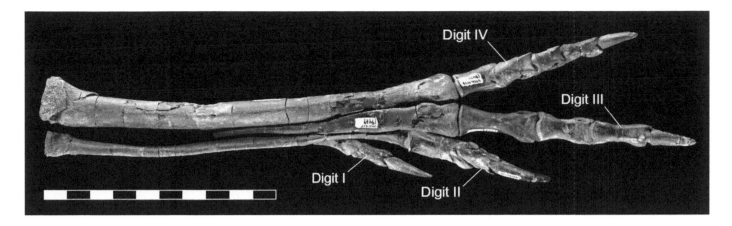

22.13. Left foot of the holotype (UMNH VP 19479) of *Talos sampsoni* in dorsal view, illustrating its gracile proportions, arctometatarsalian condition, subequal third and fourth metatarsals, and well-developed raptorial digit II. Phalanges of digits I and II inverted from right foot. Scale bar = 10 mm.

between root and crown was collected from the Kaiparowits Formation by the RAM and may be referable to Aves. A proximal portion of a right coracoid referable to Enantiornithines (RAM 14306) is also known.

Currently, the only named avian taxon known from the Kaiparowits Formation is *Avisaurus* (Hutchison, 1993). Two species of *Avisaurus* are known from Upper Cretaceous formations in Montana, *A. archibaldi* (Brett-Surman and Paul, 1985) from the Hell Creek Formation (formerly also known from the Lecho Formation in Argentina, now separated into a new species *Soroavisaurus australis*; Chiappe, 1993) and *A. gloriae* (Varricchio and Chiappe, 1995) from the Upper Two Medicine Formation. Both are known solely from the tarsometatarsus. By comparison, the Kaiparowits specimen represents the most complete enantiornithine bird fossil known from North America to date, preserving a large portion of the skeleton including a partial axial column with pygostyle, a well-developed pectoral girdle and forelimb with a robust keel, U-shaped furcula, and papillae remigiales, and a robust tarsometatarsus with highly recurved unguals (Hutchison, 1993). Preliminary study indicates that this specimen represents a new species of *Avisaurus*, but after publication of an abstract describing the find (Hutchison, 1993), no subsequent research has been undertaken to name the specimen.

Theropoda indet.

The San Diego Museum of Natural History and Jeff Eaton recovered a theropod specimen consisting of limb material, including metatarsals (UMNH VP 19477) from the John Henry Member of the Straight Cliffs Formation, Cedar Canyon (J. Kirkland, pers. comm., 2010). However, UMNH VP 19477 was not fully prepared when we wrote this chapter and is not yet assignable beyond Coelurosauria.

Teeth similar to the enigmatic theropod tooth morphotypes *"Richardoestesia"* and *"Paronychodon"* are reported throughout the Dakota, Straight Cliffs, Wahweap, and Kaiparowits formations (Hutchison et al., 1997; Eaton, Cifelli, et al., 1999; Eaton, Diem, et al., 1999; Parrish, 1999; Kirkland, 2001; Tables 22.2–22.5). The KBP and RAM crews also collected teeth sharing features with the morphotype *"Richardoestesia"* from the Kaiparowits Formation (UMNH VP 12684; RAM 12083; Fig. 22.10G). These teeth exhibit minute denticles on the distal carinae (~0.14–0.15 mm wide as in *"R. gilmori"* sensu Currie, Rigby, and Sloan, 1990), slight recurvature, and either minute, low-relief denticles on mesial carinae (UMNH VP 12684) or no denticles on the mesial carinae (RAM 12083). However, tooth morphotypes similar to *"Paronychodon"* have not yet been confirmed in UMNH or RAM collections.

The taxonomy of isolated teeth is problematic. Teeth falling within the parameters of the *"Richardoestesia"* and *"Paronychodon"* morphotypes from multiple Upper Cretaceous formations across the Western Interior Basin (including those of southern Utah) may represent multiple theropod taxa and these taxa may or may not represent natural evolutionary groupings. Furthermore, some disparate tooth types referred to these two taxa and found within the same formation may belong to the same species (Longrich, 2008a). At minimum, given the distinct taxonomic composition of the geographically intervening Kaiparowits Formation, teeth referred to *"Richardoestesia"* and *"Paronychodon"* from the Kaiparowits Formation and Aguja Formation of Texas (Sankey, 2001) probably represent taxa distinct from those in Upper Cretaceous formations of Alberta, Canada. Here we refer these isolated teeth to the suprageneric level, as we have done in other instances in this chapter, considering them theropod indet. (Table 22.6).

DISCUSSION

Isolated Tooth Taxonomy and Fragmentary Western Interior Basin Remains

Although proposals have been put forth supporting the taxonomic utility of tooth morphology in small theropods (Currie, Rigby, and Sloan, 1990; Fiorillo and Currie, 1994; Sankey, 2001; Sankey et al., 2002; Samman et al., 2005; Smith, 2005; Larson, 2008; Longrich, 2008a), these studies have focused either on the intraformational identification of isolated theropod teeth, which are generally referred to an existing species of the same formation on the basis of similarity, or intraspecific variation in tooth morphology (ontogenetic or tooth positional studies). Thus far, broad-scale studies of interspecific, interformational diagnostic utility have not been considered. Indeed, such a study would require a sufficient sample of dentigerous small theropod skeletons of closely related species across the Western Interior Basin, and although our knowledge of Western Interior Basin theropods is fast improving, we still lack the accompanying skeletal evidence to conduct such a query for most theropod clades. Data arguing against taxonomic utility of isolated theropod teeth are found in Farlow et al. (1991) and Currie and Varricchio (2004), who demonstrate significant overlap in morphological parameters of isolated dromaeosaurid teeth from different genera, and detailed studies by Samman et al. (2005), which fail to identify isolated tyrannosaurid teeth from the same formation even to genus level.

Indeed, the recent recovery of diagnostic troodontid skeletal material from the Kaiparowits Formation (i.e., *Talos sampsoni*) indicates that previous referral of isolated troodontid teeth from that formation to *Troodon* sp. (Eaton, Cifelli, et al., 1999) is dubious. The same may be true for the referral of Wahweap troodontid teeth to *Troodon* sp. (Eaton, Cifelli, et al., 1999) and the problematic taxon *Aublysodon* sp. (Eaton, Diem, et al., 1999). Whether or not the Kaiparowits and/or Wahweap troodontid teeth show significant morphological variation as to merit distinction from *T. formosus* without associated skeletal material is currently under study. This single example underscores the larger potential problem of assigning isolated teeth to genus or species over broad geographic areas within the Western Interior Basin, and we caution against the referral of any isolated teeth and/or undiagnostic skeletal remains to existing genera or species (Eaton, Cifelli, et al., 1999; Eaton, Diem, et al., 1999; Lucas, Heckert, and Sullivan, 2000; Sullivan and Lucas, 2000, 2006; Sankey, 2001; Sullivan, 2006) without more substantive morphological evidence. In the past, this practice was born out of the assumption that theropod diversity across the

Western Interior Basin was low and that many taxa probably had pan-Laramidian geographic ranges. Now that it has been demonstrated that coeval formations in the upper Campanian of the Western Interior Basin exhibit unique dinosaur faunas with unexpectedly limited geographic ranges (Smith et al., 2003, 2004; Sampson et al., 2004, 2009, 2010; Gates and Evans, 2005; Zanno, Gates, et al., 2005; Zanno, Sampson, et al., 2005; Lucas et al., 2006; Gates and Sampson, 2007; Zanno, 2007; Wagner and Lehman, 2009; Zanno, Loewen, et al., 2009; Gates et al., 2010; Sampson and Loewen, 2010; Sampson et al., this volume, Chapter 28), it is more reasonable to assume that teeth or fragmentary skeletal remains from Western Interior Basin formations with poorly known faunas are more likely to represent new species than be referable to existing taxa.

Biogeography and Diversity Patterns in the Western Interior Basin

Nearly all Late Cretaceous theropod clades known to have inhabited North America can now be documented in Upper Cretaceous formations of southern Utah, including tyrannosaurids (except Iron Springs), ornithomimids (Kaiparowits), therizinosaurids (Tropic Shale), oviraptorosaurians (Kaiparowits), dromaeosaurids, troodontids, and avians (Kaiparowits).

Thus far, comprehensive, quantitative biogeographical studies of dinosaurs across the Western Interior Basin have been conducted only for coeval upper Campanian formations (Gates et al., 2010, 2012). As the theropod fauna of earlier Upper Cretaceous formations are poorly known at the present time, we focus on identifying biogeographical patterns of theropods inhabiting the Kaiparowits and coeval Western Interior Basin ecosystems here.

Recent radiometric dates derived from several bentonite horizons within the Kaiparowits Formation establish the formation as coeval with fossiliferous portions of the upper Campanian Dinosaur Park, upper Judith River, and upper Two Medicine formations (Roberts, Deino, and Chan, 2005). As such, the ecological diversity preserved in the Kaiparowits offers essential insight into the phylogeny and biogeographic patterns of theropod dinosaurs within the Western Interior Basin during this interval.

Results of the KBP demonstrate that all major groups of theropod dinosaurs known from northern upper Campanian formations were present in the Kaiparowits ecosystem (Zanno, Sampson et al., 2005). However, despite the apparent homogeneity among theropod subclades within upper Campanian Western Interior Basin formations, taxa that can be identified to the species level appear to be endemic to restricted geographic areas. Over the last ten years, enough diagnostic material has been collected to verify that the Kaiparowits tyrannosaurids, troodontids, and oviraptorosaurs are local endemics. The identity of Kaiparowits ornithomimid materials are, as of yet, unconfirmed; however, our preliminary work suggests they may be referable to a new taxon. Dromaeosaurid materials remain too fragmentary to assess taxonomically. Thus, only a single theropod taxon from the Kaiparowits Formation is presently referred to a previously named Western Interior Basin genus—*Avisaurus*—and no formal study of this specimen has been conducted to confirm this assignment. Additional study of dromaeosaurid and ornithomimid materials may identify regionally distributed genera or species inhabiting the Kaiparowits; however, the currently confirmed theropod fauna of the Kaiparowits Formation was endemic to a small geographic area.

Paleoenviromental interpretations of upper Campanian Western Interior Basin formations suggest they span a range of habitats, from wet alluvial to arid coastal plain settings (Eberth, 1990; Rogers, 1990; Roberts, 2007). Yet this substantial environmental variation appears to have had little effect upon the subclade composition of the theropod fauna (although they may be affecting local endemicity and speciation patterns). Rather than determining the diversity of theropods, paleoenvironmental conditions may have impacted the relative abundance of these clades within Western Interior Basin formations, especially if depositional regimes were sampling different primary habitat or driving selective taphonomic processes. Preliminary evidence seems to support this hypothesis, as tyrannosaurids (*Gorgosaurus*), and paravians (*Saurornitholestes* and *Troodon*) are reported as the most abundant theropods in the Dinosaur Park Formation (Currie, 1987b, 2005), whereas tyrannosaurids and ornithomimids are the most commonly recovered theropods in the Kaiparowits Formation. Ultimately, although we find these patterns of interest, a larger sample size and greater taphonomic control is needed to determine whether these differences are reflections of variation in regional ecology or simply the result of skewed sampling or preservational biases.

CONCLUSIONS

The newly recovered theropod fauna of Upper Cretaceous formations of southern Utah is dramatically augmenting our understanding of the taxonomy, biogeography, and phylogeny of theropod taxa from the Western Interior Basin. Recent discoveries include: (1) two new tyrannosaurid genera, including *Teratophoneus curriei* and an unnamed Wahweap taxon, which represents the oldest North American member of the clade discovered to date; (2) the large-bodied oviraptorosaurian *Hagryphus giganteus* (Zanno and Sampson, 2005); (3) the troodontid *Talos sampsoni*, which provides the first substantial evidence for troodontid diversity in the Western Interior Basin during the Campanian (Zanno et al., 2011); and (4) the most complete therizinosaurid skeleton recovered to date, *Nothronychus graffami*. Research in progress on the first diagnostic ornithomimid forelimb material confirms a close relationship between the Kaiparowits ornithomimid and the genus *Ornithomimus*, although study is ongoing. Reinvestigation of ornithomimid specimens from the Kaiparowits Formation raises considerable questions about the referral of these materials to the Maastrichtian species *Ornithomimus velox* and suggests that they represent a new taxon. Finally, although isolated teeth substantiate the presence of at least two dromaeosaurids in the Kaiparowits Formation, new paleobiogeographical evidence suggests that the previous referral of these teeth to *Dromaeosaurus* and *Saurornitholestes* (Zanno, Sampson et al., 2005; Gates et al., 2010) is dubious. The presence of *Troodon* sp. in the Wahweap and Kaiparowits formations (Eaton, Cifelli, et al., 1999) is not supported by skeletal data. Finally, we find considerable problems with the utility of isolated theropod tooth "taxa" and caution against the use of these morphotypes in paleoecological and paleobiogeographical studies of Western Interior Basin theropods.

Although dinosaur diversity data for other Upper Cretaceous formations of southern Utah are predominantly limited to microvertebrate samples, our increased understanding of the Kaiparowits theropod fauna permits broad paleobiogeographical comparisons with other coeval upper Campanian formations in the Western Interior Basin. Such comparisons demonstrate a surprising amount of homogeneity in theropod taxa at the subclade level, particularly given the perceived variation in paleoenvironments. However, the data also establish a high degree of endemicity at the species level. As a result, we hypothesize that unique paleoenvironmental signatures may be better expressed through relative abundance rather than presence/absence data for theropod clades within the Western Interior Basin.

Finally, in stark contrast to previous assumptions of faunal homogeneity across the Western Interior Basin, the latest faunal data from the Kaiparowits Formation indicate geographically restricted theropod species ranges for most groups during at least the late Campanian. This pattern provides a new evolutionary context for the preliminary identification of isolated theropod materials and raises doubts regarding the referral of fragmentary theropod materials to existing genera or species from other Western Interior Basin formations without substantial morphological evidence.

ACKNOWLEDGMENTS

We thank Grand Staircase–Escalante National Monument and the Bureau of Land Management for their continued support throughout the duration of this project. Special thanks go to A. Titus (Grand Staircase–Escalante National Monument) for his ongoing efforts both in and out of the field. Without dozens of dedicated students, faculty, and museum volunteers from the UMNH, the University of Utah, the Raymond M. Alf Museum of Paleontology, and other collaborating institutions, this project could never have achieved the success we document here. Additional thanks go to M. Getty, E. Lund, J. Gentry, E. Roberts, J. Wiersma, D. Lofgren, M. Stokes, and T. Gates for their substantial contributions in the field and in the lab. We are grateful to H. Hutchison and the University of California at Berkeley's Museum of Paleontology for access to specimens. Skeletal drawings were modified from artwork by S. Hartman and G. Paul. Funding was provided by Grand Staircase–Escalante National Monument, UMNH, Discovery Channel, University of Utah Graduate Research Fellowships (L.E.Z., M.A.L.), National Science Foundation (NSF 0745454, Kaiparowits Basin Project), J. Caldwell-Meeker Postdoctoral Fellowship at the Field Museum (L.E.Z.), and by the Sherman Fairchild Foundation and the Fisher Foundation for Holy Cross student fieldwork.

REFERENCES CITED

Archie, J. W. 1985. Methods for coding variable morphological features for numerical taxonomic analysis. Systematic Zoology 34:326–345.

Barrett, P. M. 2005. The diet of ostrich dinosaurs (Theropoda: Ornithomimosauria). Palaeontology 48:347–358.

Barsbold, R., and H. Osmólska. 1999. The skull of Velociraptor (Theropoda) from the Late Cretaceous of Mongolia. Acta Palaeontologica Polonica 44:189–219.

Brett-Surman, M. K., and G. S. Paul. 1985. A new family of bird-like dinosaurs linking Laurasia and Gondwanaland. Journal of Vertebrate Paleontology 5:133–138.

Brochu, C. A. 1996. Closure of the neurocentral sutures during crocodilian ontogeny: implications for maturity assessment in fossil archosaurs. Journal of Vertebrate Paleontology 16:49–62.

Brusatte, S. L., M. A. Norell, T. D. Carr, G. M. Erickson, J. R. Hutchinson, A. M. Balanoff, G. S. Bever, J. N. Choiniere, P. J. Makovicky, and X. Xu. 2010. Tyrannosaur paleobiology: new research on ancient exemplar organisms. Science 329:1481–1485.

Burnham, D. A., K. L. Derstler, P. J. Currie, R. T. Bakker, Z.-H. Zhou, and J. H. Ostrom. 2000. Remarkable new birdlike dinosaur (Theropoda: Maniraptora) from the Upper Cretaceous of Montana. University of Kansas Paleontological Contributions 13:1–14.

Carpenter, K. 1982. Baby dinosaurs from the Late Cretaceous Lance and Hell Creek formations, and a description of a new species of theropods. University of Wyoming, Contributions to Geology 20:123–134.

Carr, T. D. 2005. Phylogeny of Tyrannosauroidea (Dinosauria: Coelurosauria) with special reference to North American forms. Ph.D. dissertation, University of Toronto, Toronto, Ontario, Canada.

Carr, T. D., and T. E. Williamson. 2000. A review of Tyrannosauridae (Dinosauria: Coelurosauria) from New Mexico; pp. 113–145 in S. G. Lucas and A. B. Heckert (eds.), Dinosaurs of New Mexico. New Mexico Museum of Natural History and Science Bulletin 17.

Carr, T. D., and T. E. Williamson. 2010. Bistahieversor sealeyi, gen. et sp. nov., a new tyrannosauroid from New Mexico and the origin of deep snouts in Tyrannosauroidea. Journal of Vertebrate Paleontology 31:1–16.

Carr, T. D., T. E. Williamson, B. B. Britt, and K. Stadtman. 2011. Evidence for high taxonomic and morphologic tyrannosauroid diversity in the Late Cretaceous (late Campanian) of the American Southwest and a new short-skulled tyrannosauroid from the Kaiparowits Formation of Utah. Naturwissenschaften 98:241–246.

Chiappe, L. M. 1993. Enantiornithine (Aves) tarsometatarsi from the Cretaceous Lecho Formation of northwestern Argentina. American Museum Novitates 3083:1–27.

Christiansen, P. 1998. Strength indicator values of theropod long bones with comments on limb proportions and cursorial potential. Gaia 15:241–255.

Christiansen, P., and R. A. Fariña. 2004. Mass prediction in theropod dinosaurs. Historical Biology 16:85–92.

Cifelli, R. L., J. I. Kirkland, A. Weil, A. L. Deino, B. Kowallis. 1997. High-precision ^{40}Ar/^{39}Ar geochronology and the advent of North America's Late Cretaceous terrestrial fauna. Proceedings of the National Academy of Sciences of the United States of America 94:111673–11167.

Coombs, W. P., Jr. 1978. Theoretical aspects of cursorial adaptations in dinosaurs. Quarterly Review of Biology 53:393–418.

Currie, P. J. 1987a. Bird-like characteristis of the jaws and teeth of troodontid theropods (Dinosauria: Saurischia). Journal of Vertebrate Paleontology 7:72–81.

Currie, P. J. 1987b. Theropods of the Judith River Formation of Dinosaur Provincial Park, Alberta, Canada; pp. 152–160 in P. J. Currie and E. H. Koster (eds.), Fourth Symposium on Mesozoic Terrestrial Ecosystems, Short Papers. Occasional Papers of the Royal Tyrrell Museum of Paleontology, Drumheller, Alberta, Canada.

Currie, P. J. 1992. Saurischian dinosaurs of the Late Cretaceous of Asia and North America; pp. 237–249 in N. J. Mateer and P. J. Chen (eds.), Aspects of Nonmarine Cretaceous Geology. Ocean Press, Beijing, China.

Currie, P. J. 2003. Cranial anatomy of tyrannosaurid dinosaurs from the Late Cretaceous of Alberta, Canada. Acta Palaeontologica Polonica 48:191–226.

Currie, P. J. 2005. Theropods including birds; pp. 367–397 in P. J. Currie and E. B. Koppelhus (eds.), Dinosaur Provincial Park: A Spectacular Ecosystem Revealed. Indiana University Press, Bloomington, Indiana.

Currie, P. J., and Z.-H. Dong. 2001. New information on Cretaceous troodontids (Dinosauria, Theropoda) from the People's Republic of China. Canadian Journal of Earth Science 38:1753–1766.

Currie, P. J., and D. J. Varricchio. 2004. A new dromaeosaurid from the Horseshoe Canyon Formation (Upper Cretaceous) of Alberta, Canada; pp. 112–132 in P. J. Currie, E. B. Koppelhus, M. A. Shugar, and J. L. Wright (eds.), Feathered Dragons. Indiana University Press, Bloomington, Indiana.

Currie, P. J., K. Rigby Jr., and R. E. Sloan. 1990. Theropod teeth from the Judith River Formation of southern Alberta, Canada; pp. 107–125 in K. Carpenter and P. J. Currie (eds.), Dinosaur Systematics: Approaches and Perspectives. Cambridge University Press, New York.

DeCourten, F. L., and D. A. Russell. 1985. A specimen of Ornithomimus velox (Theropoda: Ornithomimidae) from the terminal Cretaceous Kaiparowits Formation of Southern Utah. Journal of Paleontology 59:1091–1099.

Eaton, J. G. 1999. Vertebrate paleontology of the Iron Springs Formation, Upper Cretaceous, southwestern Utah; pp. 339–344 in D. D. Gillette (ed.), Vertebrate Paleontology in Utah. Utah Geological Survey Miscellaneous Publication 99–1.

Eaton, J. G., and R. L. Cifelli. 1988. Preliminary report on Late Cretaceous mammals of the Kaiparowits Plateau, Southern Utah. Contributions to Geology, University of Wyoming 26:45–55.

Eaton, J. G., H. Munk, and M. A. Hardman. 1998. A new vertebrate fossil locality within the Wahweap Formation (Upper Cretaceous) of Bryce Canyon National Park and its bearing on the presence of the Kaiparowits Formation on the Paunsaugunt Plateau; pp. 36–40 in V. L.

Santucci and L. McClelland (eds.), National Park Service Geologic Resources Division Technical Report NPS/NRGRD/GRDTR-98/01. National Park Service Paleontological Research Volume 3.

Eaton, J. G., R. L. Cifelli, J. H. Hutchison, J. I. Kirkland, and J. M. Parrish. 1999. Cretaceous vertebrate faunas from the Kaiparowits Plateau, south-central Utah; pp. 345–353 in D. D. Gillette (ed.), Vertebrate Paleontology in Utah. Utah Geological Survey Miscellaneous Publication 99–1.

Eaton, J. G., S. Diem, J. D. Archibald, C. Schierup, and H. Munk. 1999. Vertebrate Paleontology of the Upper Cretaceous Rocks of the Markagunt Plateau, southwestern Utah; pp. 323–334 in D. D. Gillette (ed.), Vertebrate Paleontology in Utah. Utah Geological Survey Miscellaneous Publication 99–1.

Eberth, D. A. 1990. Stratigraphy and sedimentology of vertebrate microfossil sites in the uppermost Judith River Formation (Campanian), Dinosaur Provincial Park, Alberta, Canada. Palaeogeography, Palaeoclimatology, Palaeoecology 78:1–36.

Erickson, G. M, K. Curry Rogers, D. J. Varricchio, M. A. Norell, and X. Xu. 2007. Growth patterns in brooding dinosaurs reveals the timing of sexual maturity in non-avian dinosaurs and the genesis of the avian condition. Bio Letters 3:558–561.

Farlow, J. O., D. L. Brinkman, W. L. Abler, and P. J. Currie. 1991. Size, shape, and serration density of theropod dinosaur lateral teeth. Modern Geology 16:161–198.

Fiorillo, A. R., and P. J. Currie. 1994. Theropod teeth from the Judith River Formation (Upper Cretaceous) of south-central Montana. Journal of Vertebrate Paleontology 14:74–80.

Gates, T. A., and D. C. Evans. 2005. Biogeography of Campanian hadrosaurid dinosaurs from western North America; pp. 33–39 in D. R. Braman, F. Therrien, E. B. Koppelhus, and W. Taylor (eds.), Dinosaur Park Symposium: Short Papers, Abstracts, and Program. Tyrrell Museum of Paleontology, Drumheller, Alberta, Canada.

Gates, T. A., and S. D. Sampson. 2007. A new species of Gryposaurus (Dinosauria: Hadrosauridae) from the Late Campanian Kaiparowits Formation of southern Utah. Zoological Journal of the Linnean Society 51:351–376.

Gates, T. A., A. Prieto-Márquez, and L. E. Zanno. 2012. Mountain building triggered Late Cretaceous North American megaherbivore dinosaur radiation. PLoS One 7(8):e42135.

Gates, T. A., S. D. Sampson, L. E. Zanno, E. M. Roberts, J. G. Eaton, R. L. Nydam, J. H. Hutchison, J. A. Smith, M. A. Loewen, and M. A. Getty. 2010. The fauna of the Late Cretaceous Kaiparowits Formation: testing previous ideas about late Campanian vertebrate biogeography in the Western Interior Basin. Paleogeography, Paleoclimatology, and Paleoecology 291:371–387.

Gillette, D. D., L. B. Albright, A. L. Titus, and M. Graffam. 2001. Discovery and paleogeographic implications of a therizinosaurid dinosaur from the Turonian (Late Cretaceous) of Southern Utah. Journal of Vertebrate Paleontology 21:54A.

Holtz, T. R., Jr. 2004. Tyrannosauroidea; pp. 111–136 in D. B. Weishampel, P. Dodson, and H. Osmólska (eds.), The Dinosauria. University of California Press, Berkeley, California.

Holtz, T. R., Jr., D. L. Brinkman, and C. L. Chandler. 1998. Denticle morphometrics and a possible omnivorous feeding habit for the theropod dinosaur Troodon; 159–166 in B. P. Pérez-Moreno, T. R. Holtz Jr., J. L. Sanz, and J. J. Moratalla (eds.), Aspects of Theropod Paleobiology. Gaia 15.

Hu, D., L. Hou, L. Zhang, and X. Xu. 2009. A pre-Archaeopteryx troodontid theropod from China with long feathers on the metatarsus. Nature 61:640–643.

Hutchison, J. H. 1993. Avisaurus; a "dinosaur" grows wings. Journal of Vertebrate Paleontology 13:43A.

Hutchison, J. H., J. G. Eaton, P. A. Holroyd, and M. B. Goodwin. 1997. Larger vertebrates of the Kaiparowits Formation (Campanian) in the Grand Staircase–Escalante National Monument and adjacent areas; pp. 391–398 in L. M. Hill (ed.), Learning from the Land: Grand Staircase–Escalante National Monument Science Symposium Proceedings. U.S. Department of the Interior, Bureau of Land Management.

Hutt, S., D. W. Naish, D. M. Martill, M. J. Barker, P. Newberry. 2001. A preliminary account of a new tyrannosauroid theropod from the Wessex Formation (Early Cretaceous) of southern England. Cretaceous Research 22:227–242.

Hwang, S. H., M. A. Norell, J. Qiang, and K. Gao. 2002. New specimens of Microraptor zhaoianus (Theropoda: Dromaeosauridae) from Northeastern China. American Museum Novitates 3381:1–31.

Ji, Q., P. J. Currie, M. A. Norell, and S.-A. Ji. 1998. Two feathered dinosaurs from northeastern China. Nature 393:753–761.

Ji, Q., M. A. Norell, P. J. Makovicky, K.-Q. Gao, S. Ji., and C. Yuan. 2003. An early ostrich dinosaur and implications for ornithomimosaur phylogeny. American Museum Novitates 3420:1–19.

Jinnah, Z. A., E. M. Roberts, A. L. Deino, J. S. Larsen, P. K. Link, and C. M. Fanning. 2009. New 40Ar-39Ar and detrital zircon U-Pb ages for the Upper Cretaceous Wahweap and Kaiparowits formations on the Kaiparowits Plateau, Utah: implications for regional correlation, provenance, and biostratigraphy. Cretaceous Research 30:287–299.

Kirkland, J. I. 2001. The Quest for New Dinosaurs at Grand Staircase–Escalante National Monument. Utah Geological Survey Survey Notes 33.

Kirkland, J. I., and D. G. Wolfe. 2001. First definitive Therizinosaurid (Dinosauria; Theropoda) from North America. Journal of Vertebrate Paleontology 21:410–414.

Kirkland, J. I., R. Gaston, and D. Burge. 1993. A large dromaeosaur (Theropoda) from the Lower Cretaceous of Eastern Utah. Hunteria 2:1–16.

Kirkland, J. I., L. E. Zanno, S. D. Sampson, J. M. Clark, and D. D. DeBlieux. 2005. A primitive therizinosauroid dinosaur from the Early Cretaceous of Utah. Nature 435:84–87.

Kirkland J. I., R. L. Cifelli, B. B. Britt, D. L. Burge, F. L. DeCourten, J. G. Eaton, and J. M. Parrish. 1999. Distribution of vertebrate faunas in the Cedar Mountain Formation, east-central Utah;

pp. 201–218 in D. D. Gillette (ed.), Vertebrate Paleontology in Utah. Utah Geological Survey Miscellaneous Publication 99–1.

Kirkland J. I., B. Britt, D. L. Burge, K. Carpenter, R. Cifelli, F. DeCourten, J. Eaton, S. Hasiotis, and T. Lawson. 1997. Lower to middle Cretaceous dinosaur faunas: a key to understanding 35 million years of tectonics, evolution, and biogeography. Brigham Young University Geology Studies 42:69–103.

Kobayashi, Y., P. J. Makovicky, and P. J. Currie. 2006. Ornithomimids (Theropoda: Dinosauria) from the Late Cretaceous of Alberta, Canada. Journal of Vertebrate Paleontology 26:86A.

Kobayashi, Y., J. C. Lu, Z. M. Dong, R. Barsbold, Y. Azuma, and Y. Tomida. 1999. Herbivorous diet in an ornithomimid dinosaur. Nature 402:480–481.

Lambe, L. 1914. On a new genus and species of carnivorous dinosaur from the Belly River Formation of Alberta, with a description of the skull of Stephanosaurus marginatus from the same horizon. Ottawa Naturalist 28:13–20.

Larson, D. W. 2008. Diversity and variation of theropod dinosaur teeth from the uppermost Santonian Milk River Formation (Upper Cretaceous), Alberta: a quantitative method supporting identification of the oldest dinosaur tooth assemblage in Canada. Canadian Journal of Earth Sciences 45:1455–1468.

Lavender, Z., A. Drake, M. Loewen, L. Zanno, and L. Claessens. 2010. Three-dimensional geometric morphometric analysis and univariant measurement analysis on an undescribed Ornithomimus manus. Journal of Vertebrate Paleontology 30:120A.

Li, D., M. A. Norell, K.-Q. Gao, N. D. Smith, P. J. Makovicky. 2009. A longirostrine tyrannosauroid from the Early Cretaceous of China. Proceedings of the Royal Society B: Biological Sciences 277:183–190.

Lohrengel, C. F., II. 1969. Palynology of the Kaiparowits Formation, Garfield County, Utah. Brigham Young University Geology Studies 16:61–80.

Longrich, N. 2008a. Small theropod teeth from the Lance Formation of Wyoming, U.S.A.; pp. 135–158 in J. T. Sankey and S. Baszio (eds.), Vertebrate Microfossil Assemblages: Their Role in Paleoecology and Paleobiogeography. Indiana University Press, Bloomington, Indiana.

Longrich, N. 2008b. A new, large ornithomimid from the Cretaceous Dinosaur Park Formation of Alberta, Canada: implications for the study of dissociated dinosaur remains. Palaeontology 51:983–997.

Longrich, N. R., P. J. Currie, and Z.-M. Dong. 2010. A new oviraptorid (Dinosauria: Theropoda) from the Upper Cretaceous of Bayan Mandahu, Inner Mongolia. Palaeontology 53:945–960.

Lu, J., Y. Tomida, Y. Azuma, Z. M. Dong, and Y. N. Lee. 2004. New oviraptorid dinosaur (Dinosauria: Oviraptorosauria) from the Nemegt Formation of Southwestern Mongolia. Bulletin of the National Science Museum Series C 30:95–130.

Lucas, S. G., A. B. Heckert, R. M. Sullivan. 2000. Cretaceous dinosaurs of New Mexico; pp. 83–92 in S. G. Lucas and A. B. Heckert (eds.), Dinosaurs of New Mexico. New Mexico Museum of Natural History and Science Bulletin 17.

Lucas, S. G., J. A. Spielmann, R. M. Sullivan, A. P. Hunt, and T. A. Gates. 2006. *Anasazisaurus,* a hadrosaurian dinosaur from the Upper Cretaceous of New Mexico; pp. 293–297 in S. G. Lucas and R. M. Sullivan (eds.), Late Cretaceous Vertebrates of the Western Interior. New Mexico Museum of Natural History and Science Bulletin 35.

Lund, E. K., M. A. Loewen, M. A. Getty, S. D. Sampson and E. M. Roberts. 2008. Preservation of dinosaur integumentary impressions in the Upper Cretaceous Kaiparowits Formation, Grand Staircase–Escalante National Monument, southern Utah. Journal of Vertebrate Paleontology 28:108A.

Makovicky, P. J., and M. A. Norell. 2004. Troodontidae; pp. 184–195 in D. B. Weishampel, P. Dodson, and H. Osmólska (eds.), The Dinosauria. University of California Press, Berkeley, California.

Makovicky, P., and H. D. Sues. 1998. Anatomy and phylogenetic relationships of the theropod dinosaur *Microvenator celer* from the Lower Cretaceous of Montana. American Museum Novitates 3240:1–27.

Makovicky, P. J., S. Apesteguia, and F. L. Agnolin. 2005. The earliest dromaeosaurid theropod from South America. Nature 437:1007–1011.

Makovicky, P. J., Y. Kobayashi, and P. J. Currie. 2004. Ornithomimosauria; pp. 137–150 in D. B. Weishampel, P. Dodson, and H. Osmólska (eds.), The Dinosauria. University of California Press, Berkeley, California.

Marsh, O. C. 1890. Description of new dinosaurian reptiles. American Journal of Science 39:81–86.

Neabore, S., M. A. Loewen, L. E. Zanno, M. A. Getty, and L. Claessens. 2007. Three-dimensional scanning and analysis of the first diagnostic ornithomimid forelimb material from the Late Cretaceous Kaiparowits Formation. Journal of Vertebrate Paleontology 27:123A.

Norell, M. A., and P. J. Makovicky. 2004. Dromaeosauridae; pp. 196–209 in D. B. Weishampel, P. Dodson, and H. Osmólska (eds.), The Dinosauria. University of California Press, Berkeley, California.

Norell, M. A., J. M. Clark, and P. J. Makovicky. 2001. Phylogenetic relationships among coelurosaurian theropods; pp. 49–67 in J. Gauthier and L. F. Gall (eds.), New Perspectives on the Origin and Early Evolution of Birds: Proceedings of the International Symposium in Honor of John H. Ostrom. New Haven, Connecticut.

Norell, M. A., P. J. Makovicky, and P. J. Currie. 2001. The beaks of ostrich dinosaurs. Nature 412:871–874.

Norell, M. A., J. M. Clark, L. M. Chiappe, and D. Dashzeveg. 1995. A nesting dinosaur. Nature 378:774–776.

Norell, M. A., J. M. Clark, A. H. Turner, P. J. Makovicky, R. Barsbold, and T. Rowe. 2006. A new dromaeosaurid theropod from Ukhaa Tolgod (Ömnögov, Mongolia). American Museum Novitates 3545:1–25.

Norell, M. A., J. M. Clark, D. Dashzeveg, R. Barsbold, L. M. Chiappe, A. R. Davidson, M. C. McKenna, A. Perle, and M. J. Novacek. 1994. A theropod dinosaur embryo and the affinities of the Flaming Cliffs dinosaur eggs. Science 266:779–781.

Osborn, H. F. 1905. *Tyrannosaurus* and other Cretaceous carnivorous dinosaurs. Bulletin of the American Museum of Natural History 21:259–265.

Ostrom, J. H. 1969. Osteology of *Deinonychus antirrhopus,* an unusual theropod from the lower Cretaceous of Montana. Bulletin Peabody Museum of Natural History 30:1–165.

Ostrom, J. H. 1970. Stratigraphy and Paleontology of the Cloverly Formation (Lower Cretaceous) of the Bighorn Basin Area, Wyoming and Montana. Peabody Museum of Natural History Bulletin 35.

Padian, K., and L. M. Chiappe. 1998. The origin and early evolution of birds. Biological Reviews 73:1–42.

Parrish, J. M. 1999. Dinosaur teeth from the Upper Cretaceous (Turonian–Judithian) of southern Utah; pp. 319–321 in D. D. Gillette (ed.), Vertebrate Paleontology in Utah. Utah Geological Survey Miscellaneous Publication 99-1.

Paul, G. S. 1984. The segnosaurian dinosaurs: relics of the Prosauropod–Ornithischian transition? Journal of Vertebrate Paleontology 4:507–515.

Rauhut, O. W. M. 2003. The Interrelationships and Evolution of Basal Theropod Dinosaurs. Special Papers in Paleontology 69. Palaeontological Association, London.

Rauhut, O. W. M., A. C. Milner, and S. Moore-Fay. 2009. Osteology and phylogenetic position of the theropod dinosaur *Proceratosaurus bradleyi* from the Middle Jurassic of England. Zoological Journal of the Linnean Society 158:155–195.

Roberts, E. M. 2007. Facies architecture and depositional environments of the Upper Cretaceous Kaiparowits Formation, southern Utah. Sedimentary Geology 197:207–233.

Roberts, E. M., A. L. Deino, and M. A. Chan. 2005. ^{40}Ar/^{39}Zr age of the Kaiparowits Formation, southern Utah, and correlation of contemporaneous Campanian strata and vertebrate faunas along the margin of the Western Interior Basin. Cretaceous Research 26:307–318.

Rogers, R. 1990. Taphonomy of three dinosaur bone beds in the Upper Cretaceous Two Medicine Formation of Northwestern Montana: evidence for drought-related mortality. Palaios 5:394–413.

Russell, D. A. 1972. Ostrich dinosaurs from the Late Cretaceous of Western Canada. Canadian Journal of Earth Sciences 9:375–402.

Russell, D. A. 1984. A checklist of the families and genera of North American dinosaurs. National Museum of Natural Science, National Museums of Canada, Syllogeus 53:1–35.

Russell, D. A. 1993. The role of Central Asia in dinosaurian biogeography. Canadian Journal of Earth Sciences 30:2002–2012.

Russell, D. A. 1997. Therizinosauria; pp. 729–730 in P. J. Currie and K. Padian (eds.), Encyclopedia of Dinosaurs. Academic Press, San Diego, California.

Ryan, M. J., and A. P. Russell. 2001. Dinosaurs of Alberta (exclusive of Aves); pp. 279–297 in D. H. Tanke and K. Carpenter (eds.), Mesozoic Vertebrate Life. Indiana University Press, Bloomington, Indiana.

Samman, T., G. L. Powell, P. J. Currie, and L. V. Hills. 2005. Morphometry of the teeth of western North American tyrannosaurids and its applicability to quantitative classification. Acta Palaeontologica Polonica 50:757–776.

Sampson, S. D., and M. A. Loewen. 2010. Unraveling a radiation: a review of the diversity, stratigraphic distribution, biogeography, and evolution of horned dinosaurs (Ornithischia: Ceratopsidae); pp. 405–467 in M. J. Ryan, B. J. Chinnery-Allgeier, and D. A. Eberth (eds.), New Perspectives on Horned Dinosaurs. Indiana University Press, Bloomington, Indiana.

Sampson, S. D., M. A. Loewen, E. M. Roberts, J. A. Smith, L. E. Zanno, and T. A. Gates. 2004. Provincialism in Late Cretaceous terrestrial faunas: new evidence from the Campanian Kaiparowits Formation, Utah. Journal of Vertebrate Paleontology 24:108A.

Sampson, S. D., M. A. Loewen, A. A. Farke, E. M. Roberts, C. A. Forster, J. A. Smith, and A. L. Titus. 2010. New horned dinosaurs from Utah provide evidence for intracontinental dinosaur endemism. PLoS One 5(9):e12292.

Sampson, S. D., E. M. Roberts, M. A. Loewen, M. A. Getty, T. A. Gates, L. E. Zanno, and A. L. Titus. 2009. Dinosaurs and other macrovertebrates from the Upper Cretaceous (Campanian) deposits of Grand Staircase–Escalante National Monument. Advances in Western Interior Late Cretaceous Paleontology and Geology (St. George) Abstracts with Program, 42A.

Sankey, J. T. 2001. Late Campanian southern dinosaurs, Aguja Formation, Big Bend, Texas. Journal of Paleontology 75:208–215.

Sankey, J. T., D. B. Brinkman, M. Guenther, and P. J. Currie. 2002. Small theropod and bird teeth from the Late Cretaceous (Late Campanian) Judith River Group, Alberta. Journal of Paleontology 76:751–763.

Sato, T., Y. N. Cheng, X. C. Wu, D. K. Zelenitsky, and Y. F. Hsiao. 2005. A pair of shelled eggs inside a female dinosaur. Science 308:375.

Senter, P. 2007. A new look at Coelurosauria (Dinosauria: Theropoda). Journal of Systematic Paleontology 5:429–463.

Senter, P., R. Barsbold, B. B. Brooks, and D. A. Burnham. 2004. Systematics and evolution of Dromaeosauridae (Dinosauria, Theropoda). Bulletin of Gunma Museum of Natural History 8:1–20.

Sereno, P. C., L. Tan, S. L. Brusatte, H. J. Kriegstein, X. Zhao, K. Cloward. 2009. Tyrannosaurid skeletal design first evolved at small body size. Science 326:418–422.

Smith, J. A., S. D. Sampson, E. Roberts, M. Getty, and M. Loewen. 2004. A new chasmosaurine ceratopsian from the Upper Cretaceous Kaiparowits Formation, Grand Staircase–Escalante National Monument, Utah. Journal of Vertebrate Paleontology 24:114A.

Smith, J. A., S. D. Sampson, T. A. Gates, E. M. Roberts, M. A. Getty, and L. E. Zanno. 2003. Fossil vertebrates from the Kaiparowits Fm., Grand Staircase–Escalante National Monument: an important window into the Late Cretaceous of Utah. Journal of Vertebrate Paleontology 23:98A.

Smith, J. B. 2005. Heterodonty in *Tyrannosaurus rex:* implications for the taxonomic and systematic utility of theropod dentitions. Journal of Vertebrate Paleontology 25:865–887.

Stevens, K. A. 2006. Binocular vision in theropod dinosaurs. Journal of Vertebrate Paleontology 26:321–330.

Sues, H. D. 1997. On Chirostenotes, a Late Cretaceous oviraptorosaur (Dinosauria: Theropoda) from Western North America. Journal of Vertebrate Paleontology 17:698–716.

Sullivan, R. M. 2006. Saurornitholestes robustus, n. sp. (Theropoda: Dromaeosauridae) from the Upper Cretaceous Kirtland Formation (De-Na-Zin Member), San Juan Basin, New Mexico; pp. 253–256 in S. G. Lucas and R. M. Sullivan, eds, Late Cretaceous Vertebrates of the Western Interior. New Mexico Museum of Natural History and Science Bulletin 35.

Sullivan, R. M., and S. G. Lucas. 2000. First occurrence of Saurornitholestes (Theropoda: Dromaeosauridae) from the Upper Cretaceous of New Mexico; pp. 105–108 in S. G. Lucas and A. B. Heckert (eds.), Dinosaurs of New Mexico. New Mexico Museum of Natural History and Science Bulletin 35.

Sullivan, R. M., and S. G. Lucas. 2006. The Pachycephalosaurid dinosaur Stegoceras validum from the Upper Cretaceous Fruitland Formation, San Juan Basin, New Mexico; pp. 329–330 in S. G. Lucas and R. M. Sullivan (eds.), Late Cretaceous Vertebrates of the Western Interior. New Mexico Museum of Natural History and Science Bulletin 35.

Therrien, F., and D. M. Henderson. 2007. My theropod is bigger than yours . . . or not: estimating body size from skull length in theropods. Journal of Vertebrate Paleontology 27:108–115.

Thulborn, R. A. 1982. Speeds and gaits of dinosaurs. Palaeogeography, Palaeoclimatology, Palaeoecology 38:227–256.

Triebold, M., F. Nuss, and C. Nuss. 2000. Initial report of a new North American oviraptor. Graves Museum of Archaeology and Natural History, Publications in Paleontology 2:1–25.

Turner, A., D. Pol, J. A., Clarke, G. M. Erickson, and M. A. Norell. 2007. A basal dromaeosaurid and size evolution preceding avian flight. Science 317:1378–1381.

Varricchio, D. J., and L. M. Chiappe. 1995. A new enantiornithine bird from the Upper Cretaceous Two Medicine Formation of Montana. Journal of Vertebrate Paleontology 15:201–204.

Varricchio, D. J., and F. D. Jackson. 2004. Two eggs sunny-side up: reproductive physiology in the dinosaur Troodon formosus; pp. 215–233 in P. J. Currie, E. B. Koppelhus, M. A. Shugar, and J. A. Wright (eds.), Feathered Dragons: Studies on the Transition from Dinosaurs to Birds. Indiana University Press, Bloomington, Indiana.

Varricchio, D. J., J. R. Horner, and F. D. Jackson. 2002. Embryos and eggs for the Cretaceous theropod dinosaur Troodon formosus. Journal of Vertebrate Paleontology 22:564–576.

Varricchio, D. J., F. Jackson, J. J. Borkowski, and J. R. Horner. 1997. Nest and egg clutches of the dinosaur Troodon formosus and the evolution of avian reproductive traits. Nature 385:247–250.

Varricchio, D. J., P. C. Sereno, X. Zhao, L. Tan, J. A. Wilson, and G. H. Lyon. 2008. Mud-trapped herd captures evidence of distinctive dinosaur sociality. Acta Palaeontologica Polonica 53:567–578.

Wagner, J. R., and T. M. Lehman. 2009. An enigmatic new lambeosaurine hadrosaur (Reptilia: Dinosauria) from the Upper Shale Member of the Campanian Aguja Formation of Trans-Pecos Texas. Journal of Vertebrate Paleontology 29:605–611.

Webb, L. 1981. Piecing together the past means a lot of boning up. Deseret News, Salt Lake City, Utah, September 3, 1981.

Weishampel, D. B., and D. B. Norman. 1989. Vertebrate herbivory in the Mesozoic; jaws, plants, and evolutionary metrics; pp. 87–100 in J. O. Farlow (ed.), Paleobiology of the Dinosaurs. Geological Society of America Special Paper 238.

Wiersma, J., M. A. Loewen, and M. A. Getty. 2009. Taphonomy of a subadult tyrannosaur from the Upper Campanian Kaiparowits Formation of Utah. Journal of Vertebrate Paleontology 29:201A.

Wings, O. 2007. A review of gastrolith function with implications for fossil vertebrates and a revised classification. Acta Palaeontologica Polonica 52:1–16.

Xu, X., and M. A. Norell. 2004. A new troodontid dinosaur from China with avian-like sleeping posture. Nature 431:838–841.

Xu, X., and M. A. Norell. 2006. Non-avian dinosaur fossils from the Lower Cretaceous Jehol Group of western Liaoning, China. Geological Journal 41:413–437.

Xu, X., Z.-L. Tang, and X.-L.Wang. 1999. A therizinosauroid dinosaur with integumentary structures from China. Nature 399:350–354.

Xu, X., X. L. Wang, and X. C. Wu. 1999. A dromaeosaurid dinosaur with a filamentous integument from the Yixian Formation of China. Nature 401:262–266.

Xu, X., Z. Zhou, and X. L. Wang. 2000. The smallest known non-avian theropod dinosaur. Nature 408:705–708.

Xu, X., Y.-N. Cheng, X.-L. Wang, and C.-H. Chang. 2002. An unusual oviraptorosaurian dinosaur from China. Nature 419:291–293.

Xu, X., Q. Tan, J. Wang, X. Zhao, and L. Tan. 2007. A gigantic bird-like dinosaur from the Late Cretaceous of China. Nature 447:844–847.

Xu, X., M. A. Norell, X. Kuang, X. Want, Q. Zhao, and C. Jia. 2004. Basal tyrannosauroid from China and evidence for protofeathers in tyrannosauroids. Nature 431:680–684.

Xu X., J. M. Clark, C. A. Forster, M. A. Norell, G. M. Erickson, D. A. Eberth, C. Jia, and Q. Zhao. 2006. A basal tyrannosauroid dinosaur from the Late Jurassic of China. Nature 439:715–717.

Xu, X., Q. Zhao, M. Norell, C. Sullivan, D. Hone, G. Erickson, X. Wang, F. Han, and Y. Guo. 2009. A new feathered maniraptoran dinosaur fossil that fills a morphological gap in avian origin. Chinese Science Bulletin 54:430–435.

Zanno, L. E. 2007. A new troodontid theropod from the late Campanian Kaiparowits Formation, southern Utah. Journal of Vertebrate Paleontology 27:170A.

Zanno, L. E., and P. J. Makovicky. 2010. Herbivorous ecomorphology and specialization patterns in theropod dinosaur evolution. Proceedings of the National Academy of Sciences of the United States of America 108:232–237.

Zanno, L. E., and P. J. Makovicky. 2011. On the earliest record of Cretaceous tyrannosauroids in western North America: implications for an Early Cretaceous Laurasian interchange event. Historical Biology 23:317–325.

Zanno, L. E., and S. D. Sampson. 2005. A new oviraptorosaur (Theropoda: Maniraptora) from the Late Cretaceous (Campanian) of Utah. Journal of Vertebrate Paleontology 25:897–904.

Zanno, L. E., D. D. Gillette, L. B. Albright, and A. L. Titus. 2009. A new North American therizinosaurid and the role of herbivory in "predatory" dinosaur evolution. Proceedings of the Royal Society B: Biological Sciences 276:3505–3511.

Zanno, L. E., M. A. Loewen, L. P. A. M. Claessens, and S. D. Sampson. 2009. Late Cretaceous theropod dinosaurs of southern Utah: progress and perspectives. Advances in Western Interior Late Cretaceous Paleontology and Geology (St. George), p. 50A.

Zanno, L. E., S. D. Sampson, E. M. Roberts, and T. A. Gates. 2005. Late Campanian theropod diversity across the Western Interior Basin. Journal of Vertebrate Paleontology 25:133–134A.

Zanno, L. E., T. A. Gates, S. D. Sampson, J. A. Smith, and M. A. Getty. 2005. Dinosaur diversity and biogeographical implications of the Kaiparowits Formation (Late Campanian), Grand Staircase–Escalante National Monument, southern Utah. Geological Society of America Abstracts with Programs 37:115A.

Zanno, L. E., D. J. Varricchio, P. M. O'Connor, A. L. Titus, and M. J. Knell. 2011. A new troodontid Theropod, Talos sampsoni gen. et sp. nov., from the Upper Cretaceous Western Interior Basin of North America. PLoS One 6: e24487.

Zanno, L. E., D. J. Varricchio, A. L. Titus, N. Wilkins, and M. J. Knell. 2009. A new troodontid (Theropoda: Paraves) specimen from the upper Campanian Kaiparowits Formation, Southern Utah: estimating the taxonomic diversity of North American Troodontidae. Journal of Vertebrate Paleontology 29:205A.

Zanno, L. E., J. P. Wiersma, M. A. Loewen, S. D. Sampson, and M. A. Getty. 2010. A preliminary report on the theropod dinosaur fauna of the late Campanian Kaiparowits Formation, Grand Staircase–Escalante National Monument, Utah; pp. 173–186 in M. Eaton (ed.), Learning from the Land, Grand Staircase–Escalante National Monument Science Symposium Proceedings. Grand Staircase–Escalante Partners, Kanab, Utah.

Zhang, F., Z. Zhou, X. Xu, X. Wang, and C. Sullivan. 2008. A bizarre Jurassic maniraptoran from China with elongate ribbon-like feathers. Nature 455:1105–1108.

Zhang, X.-H., X. Xing, X.-J. Zhao, P. C. Sereno, X.-W. Kuang, and L. Tan. 2001. A long-necked therizinosauroid dinosaur from the Upper Cretaceous Iren Dabasu Formation of Nei Mongol, People's Republic of China. Vertebrata PalAsiatica 10:282–290.

Zhou, Z., and F. Z. Z. Li. 2010. A new Lower Cretaceous bird from China and tooth reduction in early avian evolution. Proceedings of the Royal Society B Biological Sciences 277:219–227.

A Trackmaker for *Saurexallopus*: Ichnological Evidence for Oviraptorosaurian Tracks from the Upper Cretaceous of Western North America

23

Gerard Gierlinski and Martin Lockley

IDENTIFYING THE TRACKMAKER OF *SAUREXALLOPUS* HAS been an intriguing dilemma since footprints of this kind were first described in 1997. The large bird-like tracks of *Saurexallopus* are known from the Upper Cretaceous of North America. At this time, the only body fossils of animals with tetradactyl bird-like feet large enough to produce these distinctive ichnites are oviraptosaurs.

INTRODUCTION

Saurexallopus lovei was originally described from the Harebell Formation (Maastrichtian) of Wyoming by Harris et al. (1996) on the basis of a small assemblage of tracks preserved as natural casts (Fig. 23.1A, B). These tracks were first named incorrectly as *Exallopus*, a name preoccupied by a polychaete genus. Thus, Harris (1997) later renamed the tracks as *Saurexallopus*. More recently, a well-preserved representative of this ichnogenus named as *Saurexallopus zerbsti* was reported from the Maastrichtian Lance Formation of Wyoming by Lockley, Nadon, and Currie (2004) (Fig. 23.1C), and less well-preserved tracks cf. *Saurexallopus* ichnosp. indet. were also identified in the Maastrichtian Laramie Formation of Colorado (Fig. 23.1D).

The affinity of these tracks was discussed by previous authors without any conclusion. Harris et al. (1996) discussed and rejected such possible *Saurexallopus* trackmakers as psittacosaurids, protoceratopsians, pachycephalosaurs, hypsilophodontids, therizinosauroids, and several other nonavian theropods. Clearly the *Saurexallopus* trackmaker was a medium-sized gracile species, with very slender toes. Considering only the *Saurexallopus* footprint morphology, with a foot length (FL) which varies between 22 and 35 cm long, a hip height (h) of 110 to 175 cm is inferred from the formula h = 5 × FL. These dimensions in turn suggest a medium-sized bipedal trackmaker comparable in size to an emu or ostrich. Although an avian trackmaker seems plausible, no such large birds are known from body fossils in the Late Cretaceous of North America. In this regard, the traces of digits II–IV are reminiscent of tracks attributable to gracile coelurosaurian

dinosaurs such as ornithomimisaurs. However, tracks attributed to ornithomimisaurs (Lockley et al., in press) lack any trace of a hallux, as is consistent with foot skeleton evidence. Thus, we must consider other possible trackmakers.

Institutional Abbreviations CMN, Canadian Museum of Nature, Ottawa; CU-MWC, University of Colorado and Museum of Western Colorado joint collection, Colorado; DMNH, Denver Museum of Natural History, Denver, Colorado; HGM, Hunan Geological Museum, Changsha, China; MUZ.PIG, Geological Museum of the Polish Geological Institute, Warsaw.

DESCRIPTION

Saurexallopus tracks (Fig. 23.1) are functionally tetradactyl tracks with slender, widely divaricated toes and prominent hallux extended medially to posteromedially. Digits II and IV are subequal in length. Digit I is the shortest, whereas digit III is the longest one, typical for theropod tracks.

We can observe some degree of morphological variations among *Saurexallopus* footprints. As noted by Lockley, Nadon, and Currie (2004), two existing ichnospecies of *Saurexallopus* display a significantly different hallux configuration. *Saurexallopus lovei* Harris, 1997, shows a straight hallux directed medially, while *S. zerbsti* Lockley, Nadon, and Currie, 2004, possesses a more reversed and curved hallux, extended posteromedially. Average divarication between lateral digits (I and IV) of *Saurexallopus lovei* equals 111.4°. The footprint length of *S. lovei* varies between 28.4 and 31.5 cm, while width varies between 23 and 31.5 cm. *Saurexallopus zerbsti* is 35 cm long and 30 cm wide, and divarication between lateral toes is ~175.0°. Both ichnospecies from Wyoming clearly show characteristic triangular, elongate metatarsophalangeal pads, located directly behind the third toe. A distinctive feature of *Saurexallopus* morphology is the anterior position of the hypex between digits II and III relative the digits III and IV. This is the reverse of the pattern seen in many theropod dinosaurs and birds (Lockley, 2007:fig. 14).

526

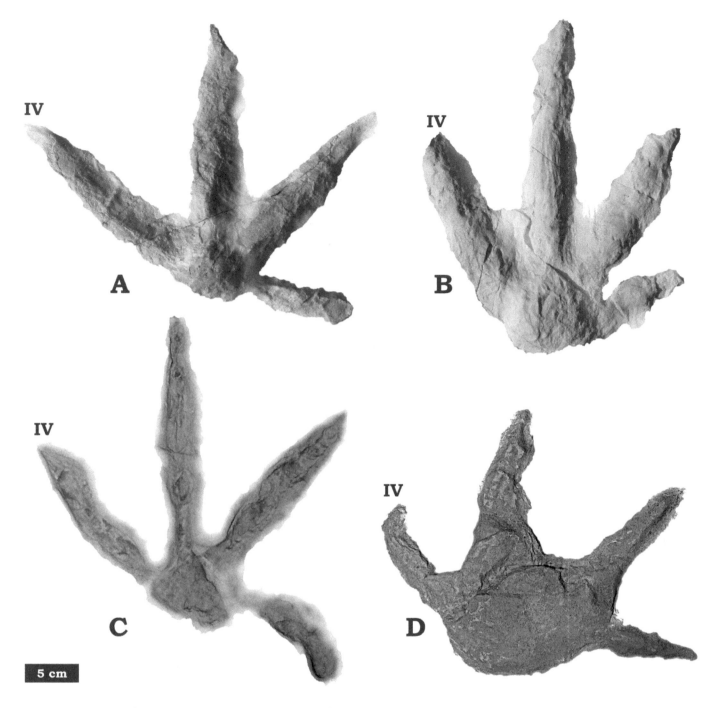

23.1. Various specimens of *Saurexallopus* from the Upper Cretaceous of western North America. (A) Holotype of *Saurexallopus lovei* Harris, 1997 (DMNH 5989), from the Harebell Formation of Wyoming; (B) plesiotype of *Saurexallopus lovei* (DMNH 5993) from the Harebell Formation of Wyoming; (C) holotype of *Saurexallopus zerbsti* Lockley, Nadon, and Currie, 2004 (specimen CU-MWC 224.2), from the Lance Formation of Wyoming; (D) *Saurexallopus* sp. (CU-MWC 220.24) from the Laramie Formation of Colorado.

The Colorado specimen of *Saurexallopus* (Fig. 23.1D) has broader proximal pad than these two ichnospecies from Wyoming and this proximal metatarsophalangeal pad is not so clearly separated, but this difference is evidently caused by extramorphologic factors attributable to its preservation. This footprint is 22 cm long and 27 cm wide. The angle between the lateral toes equals 137°. The hallux is directed medially as in *S. lovei*.

DISCUSSION

The discrete, centrally located, triangular metatarsophalangeal pad of *Saurexallopus* suggests an arctometatarsalian arrangement of metatarsals. In such a configuration, the metatarsals II, III, and IV are closely packed together and the proximal phalanx of digit III (phalanx III-1) is often relatively long. Thus, a triangular space between the distal ends of

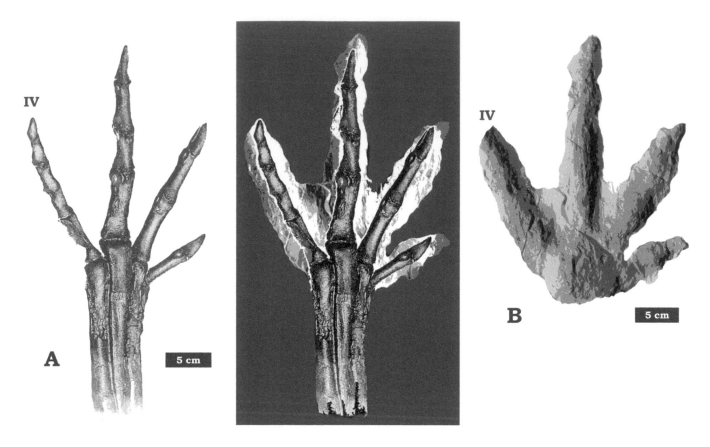

23.2. Articulated *Chirostenotes* pes specimen CMN 8538 (A) superimposed onto (B) *Saurexallopus* track specimen DMNH 5993.

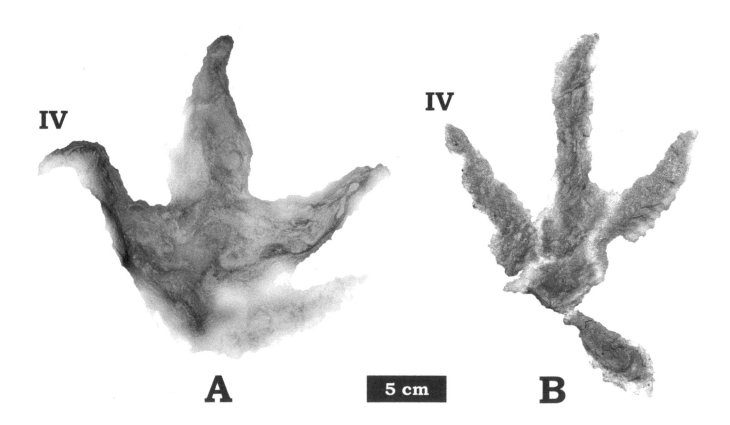

23.3. *Saurexallopus*-like footprints from the Upper Cretaceous of Eurasia. (A) *Xiangxipus chenxiensis* Zeng, 1982 (HGM HV003–4 and replica CU214.35), from the Hunan province of China; (B) cf. *Saurexallopus* (Muz.PIG 1704.II.4) from the Roztocze Hills of Poland.

Gierlinski and Lockley

metatarsals II, III, and IV was compressed laterally (parasagittally), and the pad beneath the metatarsophalangeal joint of digit IV and II was located relatively farther posteriorly, while the elongation of phalanx III1 has replaced a pad beneath the joint between phalanx III1 and III2 anteriorly.

The distinctive features of *Saurexallopus* including the reversed hallux and the swelled proximal pad behind digit III reflect strictly avian modifications of the pedal morphology. The reversed hallux and the swollen proximal pad below digit III seem to have evolved to extend the traction surface axially and medially from the third digit to the hallux. This adaptation observed in the feet of many modern birds evidently enhances their grasping ability. Thus, it seems that the *Saurexallopus* trackmaker could have been an avian theropod. Although there are few such large Maastrichtian tetradactyl avialians that could fit those footprints, the oviraptorosaurs (*Chirostenotes* Gilmore, 1924; and *Hagryphus* Zanno and Sampson, 2005) are plausible trackmakers because they have a sufficiently well-developed first toe that could have been able to produce *Saurexallopus* tracks (Fig. 23.2).

Additionally, we predict that *Saurexallopus* should occur in Asia, where oviraptorosaurs are common (Lockley and Gierlinski, 2009). For instance, a morphology close to *Saurexallopus* is exhibited by *Xiangxipus chenxiensis* Zeng, 1982 (Fig. 23.3A), from the Upper Cretaceous of Hunan province (China). The *Saurexallopus*-like form (Fig. 23.3B) is also reported from the Late Cretaceous track assemblage of Roztocze Hills (Poland), which is generally characterized by an ichnofauna with Asiatic characteristics (Gierlinski, 2009).

CONCLUSIONS

The distinctive Upper Cretaceous track *Saurexallopus*, distinguished by a large hallux, probably represents a medium-sized, gracile, maniraptoran theropod. The trackmaker was emu to ostrich sized, but an ornithomimisaur trackmaker is ruled out by the evidence of a large hallux. The most likely trackmaker candidate is here inferred to have been an oviraptorosaur.

REFERENCES CITED

Gierlinski, G. D. 2009. A preliminary report on new dinosaur tracks from the Triassic, Jurassic and Cretaceous of Poland; pp. 75–90 in C. A. Salas (ed.), Actas de las IV Jornadas Internacionales sobre Paleontologia de Dinosaurios y su Entorno. Colectivo Arqueológico-Paleontológico de Salas de los Infantes, Burgos, Spain.

Gilmore, C. W. 1924. Contributions of vertebrate paleontology. Geological Survey of Canada 38:1–89.

Harris, J. D. 1997. Four-toed theropod footprints and a paleomagnetic age from the Whetstone Falls Member of the Harebell Formation (upper Cretaceous: Maastrichtian), northwestern Wyoming: a correction. Cretaceous Research 18:139.

Harris, J. D., K. R. Johnson, J. Hicks, and L. Tauxe. 1996. Four-toed theropod footprints and a paleomagnetic age from the Whetstone Falls

Member of the Harebell Formation (upper Cretaceous: Maastrichtian), northwestern Wyoming. Cretaceous Research 17:381–401.

Lockley, M. G. 2007. The morphodynamics of dinosaurs, other archosaurs and their trackways: holistic insights into relationships between feet, limbs and the whole body; pp. 27–51 in R. Bromley and R. Melchor (eds.), Ichnology at the Crossroads: A Multidimensional Approach to the Science of Organism–Substrate Interactions. Society of Economic Paleontologists and Mineralogists Special Publication 88.

Lockley, M. G., and G. D. Gierlinski. 2009. A probable trackmaker for *Saurexallopus*. Advances in Western Interior Late Cretaceous Paleontology and Geology (St. George) Abstracts with Program, p. 33.

Lockley, M. G., G. Nadon, and P. J. Currie. 2004. A diverse dinosaur–bird footprint

assemblage from the Lance Formation, Upper Cretaceous, eastern Wyoming: implications for ichnotaxonomy. Ichnos 11:229–249.

Lockley, M. G., K. Cart, J. Martin, and A. R. C. Milner. In press. New Theropod Tracksites from the Upper Cretaceous Mesaverde Group, Western Colorado: Implications for Ornithomimosaur Track Morphology. New Mexico Museum of Natural History and Science Bulletin 53.

Zanno, L. E., and S. D. Sampson. 2005. A new oviraptorosaur (Theropoda, Maniraptora) from the Late Cretaceous (Campanian) of Utah. Journal of Vertebrate Paleontology 25:897–904.

Zeng, X. Y. 1982. Dinosaur footprints found in red beds of Yuan Ma Basin, west of Hunan, Xiangxi. Hunan Geology 1:57–58. [Chinese with English abstract]

First Report of Probable Therizinosaur (cf. *Macropodosaurus*) Tracks from North America, with Notes on the Neglected Vertebrate Ichnofauna of the Ferron Sandstone (Late Cretaceous) of Central Utah

24

Gerard Gierlinski and Martin Lockley

ALTHOUGH KNOWN FOR SOME YEARS, VERTEBRATE tracks from the Ferron Sandstone in the San Rafael Swell (Emery County, Utah) have never been described in detail. A new find attributed to cf. *Macropodosaurus*, one of the least known dinosaur tracks, is the first report of the ichnogenus from North America. *Macropodosaurus* is otherwise known only from the type locality in central Asia and a recently reported site in Poland. Recent studies suggest that the distinctive tetradactyl tracks may represent the aberrant theropodan group known as the Therizinosauroidea. We also reinterpret the so-called Moore Trackway as that of a quadrupedal rather than a bipedal ornithopod.

INTRODUCTION

The purpose of this chapter is to report on two dinosaur tracksites in the Ferron Sandstone Coal Cliffs region of Emery County, Utah (Fig. 24.1). The first site, alongside 803 Road about 6 km east-southeast of Moore, was previously reported by DeCourten (1998) and referred to by Jones (2001) as the Moore Trackway at the Moore Tracksite. Cassingham (2005) referred to the locality as the Moore Cutoff Road site. Although previously described by Jones (2001), the tracks need interpretation in detail and herein are attributed to a quadrupedal ornithopod with an unusual gait.

The second site, which yields cf. *Macropodosaurus*, is in Muddy Creek Canyon ~6 km south southeast of the town of Emery, and was only recently discovered by one of us (G.G.) in 2008. This find is significant because *Macropodosaurus* has not been previously reported from North America, and it adds to the known Late Cretaceous vertebrate ichnodiversity of the region.

It is outside the scope of this short report to describe the geology of these two sites in detail. The Mesozoic vertebrate ichnology of parts of Utah is well known. For example, in Grand County to the east, many vertebrate tracksites have been documented in the Moab region (Lockley and Hunt, 1995). The situation is somewhat different in the San Rafael Swell region of Emery County. Vertebrate tracksites are known at a few scattered locations, including a famous Early Jurassic (Navajo Formation) theropod site in Buckhorn Draw reported in various field guides (Massey and Wilson, 2006). Likewise, Mickelson et al. (2004) reported an important pterosaur tracksite ~16 km east of Ferron. Various tracksites are also known from the Upper Cretaceous Mesaverde Group in the Price-Castle Dale region (Robinson, 1991; Lockley and Hunt, 1995; Lockley, 1999). The most accessible collections of Upper Cretaceous tracks are those held in the College of Eastern Utah Museum in Price. These specimens, some of which have been known since the 1930s (Strevell, 1932), mostly originate from coal mines (Parker and Rowley, 1989).

Although scattered sites are known from the Lower Cretaceous Cedar Mountain Formation of eastern and central Utah (Lockley et al., 1999, 2004), there is a significant ichnological gap between this unit and the Mesaverde Group (Wolfe, 2006). The present report is a preliminary step toward filling in this gap. Undoubtedly, there is great potential for further vertebrate trackway discoveries in this region.

Institutional Abbreviations CU, Dinosaur Tracks Museum, University of Colorado, Denver, Colorado; MUZ.PIG, Geological Museum of the Polish Geological Institute, Warsaw.

MOORE TRACKSITE AT THE 803 ROAD LOCALITY

As shown in Fig. 24.1, this locality consists of a large fallen block close to the north side of the road (803) that connects the Interstate 70 corridor with the small, near-abandoned settlement of Moore. The site is well known because it is adjacent to spectacular petroglyphs (Cassingham, 2005) and is just east of the large water gap that cuts through the Coal Cliffs to the west. The exact point of origin of the fallen block is not known, but it must have originated from the

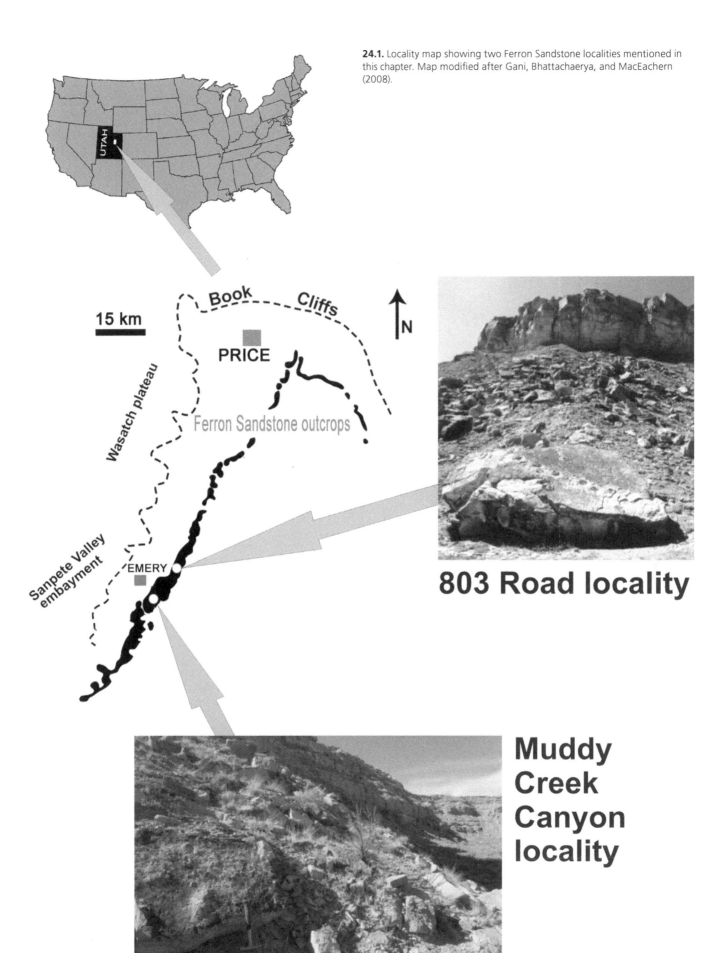

24.1. Locality map showing two Ferron Sandstone localities mentioned in this chapter. Map modified after Gani, Bhattachaerya, and MacEachern (2008).

15 km

Book Cliffs

N

PRICE

Wasatch plateau

Ferron Sandstone outcrops

Sanpete Valley embayment

EMERY

803 Road locality

Muddy Creek Canyon locality

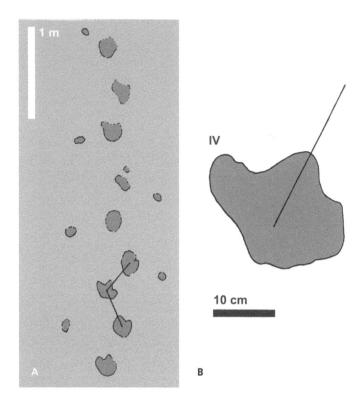

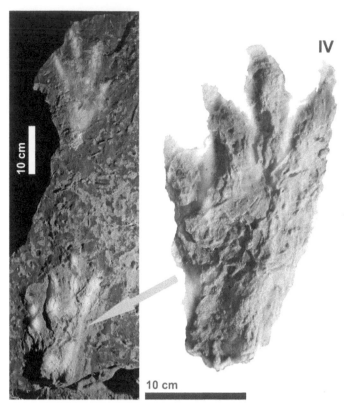

24.2. Maps of the ornithopod trackway from the Moore Tracksite (803 Road locality). (A) Revised interpretation of the trackway as representing a quadrupedal ornithopod with a very wide straddle of manus tracks. (B) Sketch map interpretation of Jones (2001:fig. 2) indicating a bipedal trackmaker.

24.3. Replica (CU 219.3) of a natural cast of cf. *Macropodosaurus*, from the Muddy Creek Canyon locality, shown in relation to its origin on the surface of the track-bearing block at the outcrop.

cliff-forming sandstone sequence at the top of the shale slope (Fig. 24.1). Jones (2001) described the tracks, briefly inferring that the block originated from the very top of the Ferron Sandstone escarpment. Herein we offer our interpretations.

The tracks appear to comprise a single trackway consisting of nine poorly preserved natural casts interpreted as tridactyl pes impressions (Fig. 24.2). In addition, there appear to be at least six smaller casts, interpreted as manus impressions, five of which occur in a regular alternating sequence on either side of the pes sequence (Fig. 24.2A). Thus, we agree with Jones (2001) that nine pes tracks are visible, although he illustrated only eight (Fig. 24.2B). However, we disagree with his conclusion that that there is no evidence of foreprints.

One pes is sufficiently well preserved to yield length and width measurement of 20 cm and 25 cm, respectively (Fig. 24.2A). The pes is rotated inward at ~25° relative to the trackway midline with a step of ~45 cm and a pace angulation of ~115°. By contrast, because the presumed manus tracks are so far from the midline, the steps are much longer, averaging 1.10 m (n = 4) with correspondingly lower pace angulation values (53°, n = 3). These measurements are somewhat at variance with those presented by Jones (2001), who noted that length and width varied from 18 to 40 cm and from

18 to 27 cm, respectively. These measurements cover much extramorphological or preservational variability in the series of tracks that Jones variously described as domed, dished, and indistinguishable, depending on the state of preservation. He reported a consistent stride length of 97 cm, although he referred to pace rather than stride.

On the basis of the shape and length–width ratio of the pes (0.80), the small size of the manus, and the location of manus traces well outside the pes, we interpret these tracks as those of an ornithopod. Ornithopod trackways with manus tracks situated well outside the pes are known from several localities (Lockley and Wright, 2001), with the most striking example of a wide straddle of the forelimbs being the examples reported by Norman (1980) and Wright (1999) from the Wealden of England. Thus, our trackmaker attribution is more confident than that of Jones (2001), who considered it more than likely an ornithopod dinosaur. We can definitively demonstrate that the trackmaker was quadrupedal, not bipedal.

MUDDY CREEK CANYON LOCALITY

The tetradactyl track herein assigned to cf. *Macropodosaurus* is specimen CU 219.3 (Fig. 24.3) from Muddy Creek Canyon

Gierlinski and Lockley

(GPS coordinates: N 38°52.705'; W 111°12.325'). This track is the better of two that are natural casts on a fallen block. Although the precise stratigraphic horizon of origin of the tracks has not been determined, the Ferron Sandstone in this area consists of a thin lower and thick upper member, with the latter comprising seven transgressive-regressive sequences (Gani, Bhattachaerya, and MacEachern, 2008) of mid- to late Turonian age. Regionally, the Ferron Sandstone represents a large northward-prograding delta lobe consisting of various prodelta, delta front, shoreface, delta plain, and fluvial deposits (Chidsey, 2001; Gani, Bhattachaerya, and MacEachern, 2008).

Typically for theropod tracks, the trackway is very narrow, and the pedal prints are directed forward. The step is relatively short, and the ratio of pace length to the pes length equals 1:1.8. The footprints are ~30 cm long and ~15 cm wide (Fig. 24.3) with four slender, tapering digits that appear slightly curved inward distally (toward digit I). The I–II hypex, or point of separation between digits I and II, occurs at about half total track length, and the II–III and III–IV hypicies occur at about two thirds of the total track length measured from the heel. Toe traces are slightly wider (2.0–2.5 cm) at the hypicies than distally. The track also displays a subcircular to slightly oval heel pad comprising about 40% the length of the track. Digit III is the longest, and slightly longer than digits II and IV. Digit I is the shortest, comprising about 65% of total track length. Digit divarication angles are relatively low, with a total of 35–42° between digits I and IV.

DISCUSSION

The ichnogenus *Macropodosaurus* (Fig. 24.4) was first reported from the Cenomanian of Tajikistan by Zhakharov (1964), who considered its affinities uncertain. The tracks are tetradactyl with a length of ~50 cm, a width of 29 cm, and a stride of 150 cm. Haubold (1971) illustrated the ichnogenus and attributed it to a large theropod by placing it within Infraorder Carnosauria Huene, 1920. Haubold (1984) reiterated this labeling by placing the ichnogenus in the unranked taxon Carnosauria. Thulborn (1990:61) referred to Zhakarov's original report without further comment except to state that the tracks are of "uncertain affinities." More recently, however, Sennikov (2006) proposed that the trackmaker was probably a therizinosauroid. We consider this interpretation to be plausible.

Gierlinski (2009) reported a probable *Macropodosaurus* track from the Maastrichtian of Poland. The specimen (MUZ.PIG 1704.II.3) is 35 cm long and 20.5 cm wide, and therefore is about 60% length of the Tajikistan specimen and slightly larger (116%) than the Muddy Creek specimen. Gierlinski (2009) compared the specimen with the purported

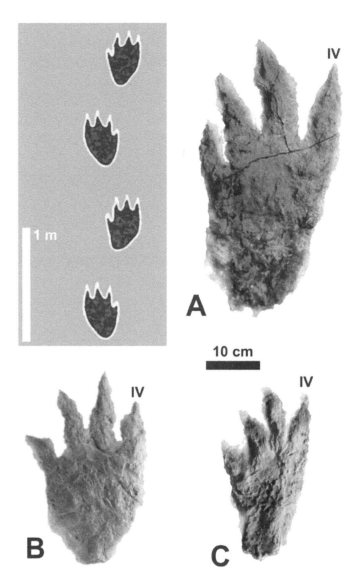

24.4. Comparison of known *Macropodosaurus* specimens. (A) Type *Macropodosaurus* trackway and individual footprint from the Cenomanian of Tajikistan after Sennikov (2006). (B) Maastrichtian-aged specimen Muz. PIG 1704.II.3 from Poland after Gierlinski (2009). (C) Turonian-aged Muddy Creek specimen CU 219-3 from Utah. All individual footprints at same scale.

prosauropod ichnogenus *Otozoum* from the Lower Jurassic, but noted that despite this convergence there is no evidence of Late Cretaceous prosauropods. In addition, *Otozoum* differs from *Macropodosaurus* in having well-developed digital pads. Nevertheless, the general similarity in foot morphology between the two ichnogenera suggests a degree of convergence between the footprints attributed to two completely different saurischian trackmakers (i.e., an early Jurassic prosauropod and a Late Cretaceous therizinosaur).

Sennikov (2006) revaluated *Macropodosaurus*, making a detailed case for therizinosaur affinity by illustrating the tracks with the foot skeletons of *Therizinosaurus* and the related genus *Erlikosaurus* superimposed. As a result of the small size of the sample and its scattered distribution in space and time, we have illustrated all the specimens together at the same scale for ease of comparison (Fig. 24.4).

According to Clark, Maryan'ska, and Barsbold (2004), the Therizinosauroidea, previously known as Segnosauria, consists of about a dozen genera, all of which are Cretaceous in age. With the exception of *Therizinosaurus* (Maleev, 1954), all these forms were described between 1979 and the present. Kirkland and Wolfe (2001) were the first to report a representative genus (*Nothronychus*) in North America. This genus occurs in the middle Turonian Moreno Hill Formation of New Mexico and the lower Turonian portion of the Tropic Shale of the Kaiparowits Basin of Utah and is therefore roughly the same age as the Ferron Sandstone track. This chronostratigraphic coincidence offers a measure of support for the interpretation of these tracks as therizinosauroid. The Ferron Sandstone track, however, is 40% smaller than the pedal skeleton of the Utah *Nothronychus* (*N. graffami*), which was described by Zanno et al. (2009).

Macropodosaurus is still poorly known as an ichnogenus and is morphologically convergent with other tracks such as *Otozoum*, which has in the past been interpreted as crocodylian in origin (Haubold, 1971). Crocodylian tracks and crocodylian skeletal remains are relatively common in the Cretaceous of western North America. Thus, it is possible that crocodylian tracks could be confused with *Macropodosaurus*. Several arguments, however, suggest that the Ferron track is not crocodylian. Most reports of Cretaceous crocodylian tracks pertain to incomplete swim tracks, which show only distal toe impressions (Kukihara, 2006; Lockley et al., 2010). Only one clear example of a walking crocodylian trackway (ichnogenus *Crocodylopodus*) is known from the Cretaceous (Fuentes Vidarte and Meijide Calvo, 1999), and in walking trackways, one expects to find a wide trackway, with short steps, associated manus impressions, interdigital web traces, and often a tail drag trace. None of these features is evident in the Muddy Creek Canyon track, which appears to be the trackway of a biped.

CONCLUSIONS

1. Vertebrate footprints from the Turonian-aged Ferron Sandstone have not previously been reported in any abundance or detail.

2. A reinterpretation of the Moore Trackway (beside 803 Road) suggests that it is ornithopod with an unusually wide straddle of the front limbs and not a biped, as previously implied by Jones (2001).

3. A newly discovered tetradactyl track from the Muddy Creek locality is best assigned to cf. *Macropodosaurus* and is likely attributable to a biped. A therizinosauroid foot is the best fit for this track on the basis of comparisons to skeletal remains.

4. This is only the third report of ichnogenus *Macropodosaurus* and the first in North America.

5. Although *Macropodosaurus* is convergent with other archosaur tracks, such as those of crocodylian or even prosauropod affinity, it is not convergent with any other known Cretaceous biped.

REFERENCES CITED

Cassingham, R. 2005. Explore the Roads Less Traveled in the San Rafael Swell: The Scenic Heart of Castle Country. Self Guided Driving Tours in the San Rafael Swell, Utah, Volume 2. Moab, Utah, Wayoutideas; and Castle Dale, Utah, Emery County Travel Bureau, 28 pp.

Chidsey, T. C., Jr. 2001. Geological and Petrophysical Characterization of the Ferron Sandstone for 3-D Simulation of a Fluvial–Deltaic Reservoir. Utah Geological Survey Report.

Clark, J. M., T. Maryańska, and R. Barsbold. 2004. Therizinosauroidea; pp. 151–164 in D. B. Weishampel, P. Dodson, and H. Osmólska (eds.), The Dinosauria (2nd edition). University of California Press, Berkeley, California.

DeCourten, F. 1998. Dinosaurs of Utah. University of Utah Press, Salt Lake City, Utah.

Fuentes Vidarte, C., and M. Meijide Calvo. 1999. Primeras Huellas de Cocodrilo en el Weald de Cameros (Sria, España) Nueva Familia Crocodilopodidae: Nuevo icnogenero: *Crocodylopodus* Nueva icnoespecie: *C. meijidei*. Actas de las jornadas internacionales sobre paleontologia de dinosairios y su entorno. Sala de los infantes (Burgos, España). Collectivo

Arqueologico-Paleontologico de Salas, pp. 329–338.

Gani, M., J. P. Bhattachaerya, and J. A. MacEachern. 2008. Using ichnology to determine relative influence of waves, storms, tides and rivers in deltaic deposits: examples from Cretaceous Western Interior Seaway, U.S.A. Society for Sedimentary Geology (SEPM) Applied Ichnology Short Course Notes 52:1–18.

Gierlinski, G. 2009. Preliminary report on new dinosaur tracksites in the Triassic, Jurassic and Cretaceous of Poland. Actas de Las IV Jornadas Internacionales sobre Paleontologia de Dinosaurios y su Entorno. Salas de los Infantes, Burgos, Spain, pp. 75–90.

Haubold, H. 1971. Ichnia amphibiorum et reptiliorum fossilium. Handbuch der Paläoherpetologie-Encyclopedia of Paleoherpetology, Volume 18. Gustav Fischer Verlag, Stuttgart.

Haubold, H. 1984. Saurierfahrten (2nd edition). A. Ziemsen Verlag, Wittenberg Lutherstadt.

Huene, F. v. 1920. Bemerkungen zur Systematik und Stammesgeschichte einiger Reptilien. Zeitschrift für Induktive Abstammungs und Vererbungslehre 22:209–212.

Jones, R. 2001. Dinosaur trackway from the Ferron Sandstone Member of the Mancos Shale Formation (Upper Cretaceous) of Central Utah; pp. 48–51 in V. L. Santucci and L. McClelland (eds.), Proceedings of the 6th Fossil Resource Conference. Geological Resources Division Technical Report NPS/NRGRD/GRDTR-01/01.

Kirkland, J. I., and D. G. Wolfe. 2001. First definitive therizonosaurid (Dinosauria, Theropoda) from North America. Journal of Vertebrate Paleontology 21:410–414.

Kukihara, R. 2006. Fossil footprint discoveries at John Martin Reservoir, Bent County, Colorado: new insights into the paleoecology of the Cretaceous dinosaur freeway. M.S. thesis, University of Colorado, Denver, Colorado.

Lockley, M. G. 1999. Pterosaur and bird tracks from a new locality in the Late Cretaceous of Utah; pp. 355–359 in D. D. Gillette (ed.), Vertebrate Paleontology in Utah. Utah Geological Survey Miscellaneous Publication 99–1.

Lockley, M. G., and A. P. Hunt. 1995. Dinosaur Tracks and Other Fossil Footprints of the Western United States. Columbia University Press, New York.

Lockley, M. G., and J. L. Wright. 2001. The trackways of large quadrupedal ornithopods from the Cretaceous: a review; pp. 428–442 in K. Carpenter and D. Tanke (eds.), Mesozoic Vertebrate Life: New Research Inspired by the Paleontology of Philip J. Currie. Indiana University Press, Bloomington, Indiana.

Lockley, M. G., D. White, J. Kirkland, and V. Santucci. 2004. Dinosaur tracks from the Cedar Mountain Formation (Lower Cretaceous), Arches National Park, Utah. Ichnos 11:285–293.

Lockley, M. G., J. I. Kirkland, F. L. DeCourten, B. B. Britt, and S. T. Hasiotis. 1999. Dinosaur tracks from the Cedar Mountain Formation of Eastern Utah: a preliminary report; pp. 253–257 in D. D. Gillette (ed.), Vertebrate Paleontology in Utah. Utah Geological Survey Miscellaneous Publication 99-1.

Lockley, M. G., D. Fanelli, K. Honda, K. Houck, and N. A. Mathews. 2010. Crocodile water ways and dinosaur freeways: implications of multiple swim track assemblages from the Cretaceous Dakota Group, Golden area, Colorado. New Mexico Museum of Natural History and Science Bulletin 51:137–156.

Maleev, E. A. 1954. [New turtle-like reptile in Mongolia]. Priroda 3:106–108. [Russian]

Massey, P., and J. Wilson. 2006. Four Wheel Drive Adventures Utah: The Ultimate Guide to the Utah Back Country for Anyone with a Sport Utility Vehicle. Adler Publishing, Castle Rock, Utah, 539 pp.

Mickelson, D., M. G. Lockley, J. Bishop, and J. Kirkland. 2004. A new pterosaur tracksite from the Jurassic Summerville Formation, near Ferron Utah. Ichnos 11:125–142.

Norman, D. B. 1980. On the ornithsichian dinosaur *Iguanodon bernisartensis* of Bernisart (Belgium). Memoirs de L'institut Royal des Sceinces Naturelle de Belgique 178:1–105.

Parker, L. R., and R. L. Rowley Jr. 1989. Dinosaur footprints from a coal mine in east-central Utah; pp. 361–366 in D. D. Gillette and M. G. Lockley (eds.), Dinosaur Tracks and Traces. Cambridge University Press, Cambridge.

Robinson, S. F. 1991. Bird and frog tracks from the late Cretaceous Black Hawk Formation in east central Utah. Utah Geological Association Publication 19:325–334.

Sennikov, A. G. 2006. [Reading segnosaur tracks]. Priroda 5:58–67 [Russian].

Strevell, C. N. 1932. Dinosauropeds. Deseret New Press, Salt Lake City, Utah, 15 pp.

Thulborn, T. 1990. Dinosaur Tracks. Chapman Hall, London.

Wolfe, D. G. 2006. Theropod dinosaur tracks from the Upper Cretaceous (Turonian) Moreno Hill Formation of New Mexico; pp. 115–117 in S. G. Lucas and R. M. Sullivan (eds.), Late Cretaceous Vertebrates from the Western Interior. New Mexico Museum of Natural History and Science Bulletin 35.

Wright, J. 1999. Ichnological evidence for the use of the forelimb in iguanodontid locomotion. Special Papers in Palaeontology 60:209–219.

Zanno, L. E., D. D. Gilette, L. B. Albright, and A. L. Titus. 2009. A new North American therizinosaurid and the role of herbivory in "predatory" dinosaur evolution. Proceedings of the Royal Society B 276:3505–3511.

Zhakharov, S. A. 1964. [On footprints of a Cenomanian dinosaur whose tracks were found in the Shirkent River Valley]; pp. 31–35 in V. M. Reiman (ed.), [Paleontology of Tadzikistan]. V. M. Akademiya Nauk Tadzhikskoi SSR, Dushanbe. [Russian with English summary]

Fossil Vertebrates from the Tropic Shale (Upper Cretaceous), Southern Utah

25

L. Barry Albright III, David D. Gillette, and Alan L. Titus

THE UPPER CENOMANIAN–UPPER TURONIAN TROPIC Shale of southern Utah contains an abundant and diverse vertebrate fauna. A decade of collecting, mostly by crews from the Museum of Northern Arizona, has secured specimens representing a dinosaur, at least five different short-necked plesiosaur genera, two genera of turtles, and a normal marine chondrichthyan–osteichthyan assemblage typical for the U.S. Western Interior. Most of the material is early and middle Turonian in age, reflecting farthest offshore conditions. Larger vertebrates are less common in the Cenomanian part of the formation, probably because of nearer-shore conditions. Chondrichthyans are represented mostly by teeth and dermal ossicles. The genera *Cretoxyrhina*, *Cretolamna*, *Squalicorax*, *Scapanorhyncus*, *Ptychotrygon*, and *Ptychodus* are most common. Osteichthyans are represented by both fragments and complete articulated skeletons. The osteichthyan assemblage is dominated by ichthyodectids such as *Xiphanctinus*, but remains largely unsampled and unstudied. Two genera of chelonians are known: the marine form *Desmatochelys*, and the nearshore genus *Naomichelys*. The plesiosaurs are represented by the relict pliosaurid *Brachauchenius* and four genera of polycotylids, *Palmulasaurus*, *Trinacromerum*, *Eopolycotylus*, and *Dolichorhynchops*. The lone dinosaur recovered is a remarkably complete specimen of the bizarre therizinosaur theropod, *Nothronychus*. Long-necked plesiosaurs remain unknown from the Tropic Shale. It is hypothesized that the turbid nature of the muddy western shore region of the Cretaceous Interior Seaway might have excluded those taxa.

INTRODUCTION

This chapter summarizes the vertebrate fauna of the Tropic Shale in Garfield and Kane Counties, southern Utah. Most of what is reported herein is based on work conducted primarily by the Museum of Northern Arizona (MNA) since 1999, but also builds upon earlier studies by MNA, Northern Arizona University, and the Utah Geological Survey. Work by MNA was done as part of a large-scale program designed by Grand Staircase–Escalante National Monument to study in detail the biodiversity of all Late Cretaceous units exposed

within the monument, with an emphasis on the richly fossiliferous Straight Cliffs, Wahweap, and Kaiparowits formations, as well as the Tropic Shale. Because vast exposures of Tropic Shale also occur on adjacent lands within Glen Canyon National Recreation Area and a tract administered by the Utah School and Institutional Trust Lands Administration, this chapter, like many others in this volume, is not restricted to work conducted solely within Grand Staircase–Escalante National Monument.

Before the monument's designation in 1996, the Tropic Shale was known primarily for its exceptional record of marine invertebrates, from which were developed detailed molluscan biostratigraphies across the Cenomanian–Turonian stage boundary (Peterson, 1969; Cobban and Hook, 1984; Elder, 1991; Kirkland, 1991, 1996; Olesen, 1991; Dyman et al., 2002). In fact, no Tropic Shale vertebrate fauna was illustrated until Gillette, Hayden, and Titus (1999) recorded an articulated partial large plesiosaur skeleton from near Orderville, Utah (west of Grand Staircase–Escalante National Monument). However, Lucas's (1994) description of plesiosaur remains and the report of Elliott, Irby, and Hutchinson (1997) of a marine turtle from equivalent strata (Mancos Shale) at Black Mesa, Arizona, approximately 100 km southeast of the Kaiparowits Basin, hinted that the Tropic Shale also would contain a significant marine vertebrate fauna.

Since our initial investigation in 1999, knowledge of the Tropic Shale vertebrate record has improved dramatically. Taxa cataloged and/or described include chondrichthyans, osteichthyans, plesiosaurs, turtles, and even a dinosaur. Five taxa of plesiosaurs (Albright, Gillette, and Titus, 2007a, 2007b; Schmeisser McKean, 2012) and the dinosaur (Zanno et al., 2009; see also Zanno et al., this volume, Chapter 22) have been described in detail.

The overall goal of this ongoing research has been to search for and recover vertebrate fossils from the Tropic Shale in order to more fully characterize the level of vertebrate biodiversity that existed approximately 93 million years ago along the western margin of the Cretaceous Western Interior Seaway—a time not well represented in the terrestrial record of North America.

Institutional Abbreviations MNA, Museum of Northern Arizona; UMNH, Natural History Museum of Utah (formerly Utah Museum of Natural History), Salt Lake City, Utah.

STRATIGRAPHIC AND TEMPORAL CONTEXT

All specimens discussed are from the Tropic Shale, a mostly drab gray sequence of open marine sediments named by Gregory and Moore (1931). The type section is near the town of Tropic, Utah; however, the formation crops out widely around the margins of the Kaiparowits Plateau in southern Utah (Fig. 25.1), although exposures are also found around the margins of the Paunsaugunt and Markagunt plateaus. The formation was deposited along the western margin of the Cretaceous Western Interior Seaway during the Greenhorn Cyclothem and spans the late Cenomanian *Sciponoceras gracile* through middle Turonian *Prionocyclis hyatti* North American ammonoid biozones (Cobban and Hook, 1984), or from about 94 to 90.5 Ma (Ogg, Agterberg, and Gradstein, 2004). Although Prokoph et al. (2001) redefined the Cenomanian–Turonian boundary at ~96.4 Ma, we follow the geomagnetic polarity timescale of Ogg, Agterberg, and Gradstein. (2004), which incorporates the ~93.5 Ma boundary date of Obradovich (1993). The Tropic Shale lies stratigraphically between the underlying Dakota Formation and overlying Straight Cliffs Formation and is differentiated from both, which are also partially marine, buy its mud-dominated character. It is equivalent to the Mancos Shale exposed at Black Mesa, Arizona (Elder, 1991; Kirkland, 1991, 1996; Leckie et al., 1991, 1998; Olesen, 1991), and the Tununk Member of the Mancos Shale in central Utah (Hintze, 1988).

The Tropic Shale thins westward from the area around Big Water, Utah, to where its onshore equivalent (in part), the Iron Springs Formation, crops out near Cedar City, Utah (Eaton, 1999). Cobban and Hook (1984) placed the paleoshoreline during latest Cenomanian to early Turonian time between Cedar City and St. George, or about 175 km (105 miles) west of Big Water (also see Cobban et al., 1994). This western shoreline provides a reasonable approximation of the greatest distance offshore that several of our vertebrate sites would have been located during maximum transgression of the Greenhorn Seaway, which occurred during the late early Turonian. Maximum transgression occurred in the Tropic Shale during the lower part of the *Mammites nodosoides* ammonoid biozone (Eaton et al., 1987), although Olesen (1991:165) concluded on the basis of foraminiferal ecology of the Mancos Shale at Black Mesa that "the timing of eustatic highstand in the Greenhorn Seaway [perhaps] should be moved into the overlying *C. woollgari* ammonite zone" (earliest middle Turonian). Olesen (1991) further determined

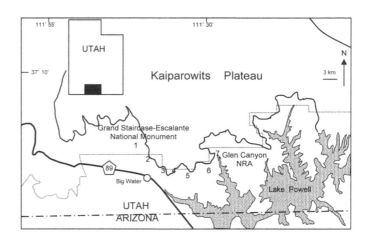

25.1. Area of study, primarily along southern border of Kaiparowits Plateau. Finely dotted line separates Grand Staircase–Escalante National Monument from Glen Canyon National Recreation Area. Although not delineated in the figure, there is a small section of Utah State Institutional and Trust Land immediately north of Big Water and south of the Monument boundary. Numbers refer to approximate locations where stratigraphic columns of Fig. 25.2 were measured (from Albright, Gillette, and Titus, 2007b, with permission from *Journal of Vertebrate Paleontology*).

that average seaway depth in this region during maximum transgression was less than 100 m.

Several workers have noted that the Tropic Shale conformably overlies and intertongues with the underlying Dakota Formation (Eaton et al., 1987; Gustason, 1989; am Ende, 1991) and that the contact is diachronous, ranging in age between the *Vascoceras diartianum* and *Euomphaloceras septemseriatum* ammonoid biozones (Titus, Roberts, and Albright, this volume, Chapter 2). Typically, where the basalmost Tropic Shale beds are within the *E. septemseriatum* ammonoid biozone, the underlying Dakota Formation is an indurated, bioturbated sandstone that is locally rich in the large oyster *Exogyra olisponensis*. The lower few meters of the Tropic Shale can locally contain eight or more distinct concretionary horizons, each of which contains a distinctive suite of biochronologically significant molluscan taxa. These are accompanied by several prominent, benchforming bentonites, many of which have been radioisotopically dated (Obradovich, 1993). Together, these provide a basis for detailed, bed-by-bed correlation with much of the Western Interior where deposits of the Greenhorn Seaway are exposed (Kauffman, Powell, and Hattin, 1969; Hattin, 1971, 1975, 1985; Elder, 1985, 1991; Elder and Kirkland, 1985; Eaton et al., 1987; Kirkland, 1991; Pratt, Kauffman, and Zelt, 1985; and references within). All of our finds, therefore, can be readily placed within a robust, high-resolution biochronostratigraphic framework, making regional faunal comparisons relatively straightforward.

Figure 25.2 shows a series of stratigraphic sections measured at or near sites from which important vertebrate

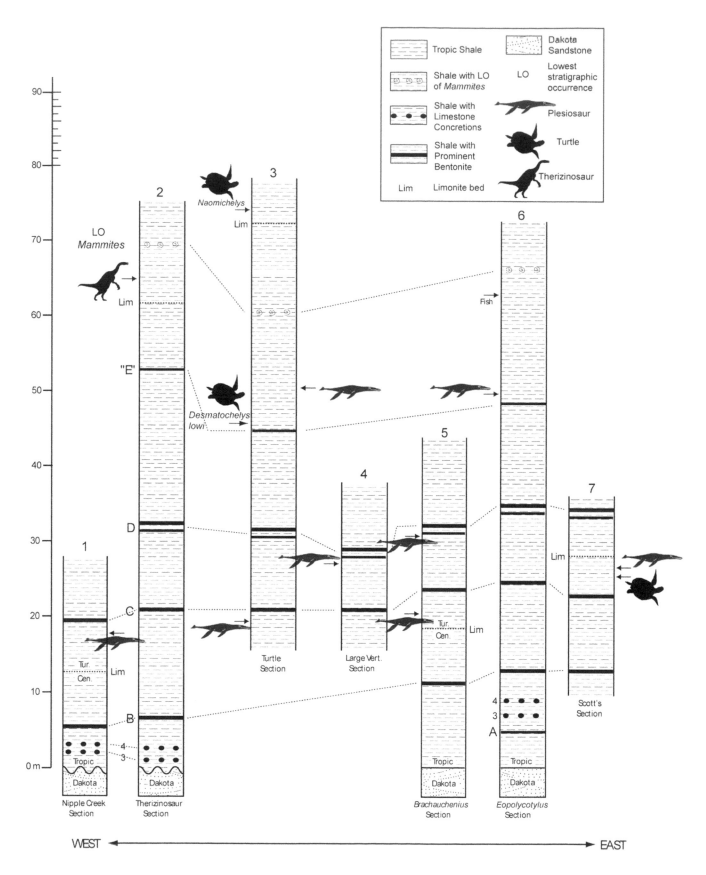

25.2. Generalized stratigraphic columns of lower Tropic Shale from which selected vertebrate remains (plesiosaurs, turtles, and the therizinosaur) were recovered. Arrows denote stratigraphic level of accompanying vertebrate taxon; capital letters A–E denote prominent bentonite marker beds; numbers 3 and 4 denote limestone concretionary beds; numbers above each column refer to corresponding localities on index map (Fig. 25.1); Cenomanian–Turonian boundary in columns 1 and 5 based on in situ biostratigraphic analysis; distance from west to east approximately 40 km (from Albright, Gillette, and Titus, 2007b, with permission from *Journal of Vertebrate Paleontology*).

Albright, Gillette, and Titus

remains were recovered from the Big Water region. Precise stratigraphic documentation of these finds is particularly significant because of the interval of time spanned by deposition of the Tropic Shale. The lower part of this unit, which is the most fossiliferous part for vertebrates, conformably spans the boundary between the Late Cretaceous Cenomanian and Turonian marine stages. This boundary represents a major, globally recognized, invertebrate and planktonic marine extinction event associated with Oceanic Anoxic Event II (Leckie et al., 1991, 1998; Prokoph et al., 2001; Keller and Pardo, 2004; and references therein). Important with respect to this event, all of our vertebrate discoveries have been found in direct association with biochronologically significant invertebrate taxa, and within stratigraphic proximity to radioisotopically dated volcanic ash horizons. Therefore, the Tropic Shale, with its high-resolution temporal control and its abundance of skeletal material in a single, broadly exposed stratigraphic section, provides a unique opportunity for accurately discerning patterns of biotic diversification and faunal turnover across and shortly following the Cenomanian–Turonian boundary event.

SYSTEMATIC PALEONTOLOGY

Entries in this section provide an abbreviated systematic paleontology of relevant taxa. Synonymies can be found in technical references. No new taxa are described, and we have not proposed any taxonomic changes. For most Chondrichthyes and Osteichthyes, precise stratigraphic positions have not been established; accordingly, ages should be considered latest Cenomanian to late early Turonian unless otherwise indicated. Ages for most Reptilia have been established. Ages for chondrichthyan teeth directly associated with plesiosaurs can be discerned from individual records for those taxa. Locality data and archives for each cataloged MNA specimen are on file at the Museum of Northern Arizona.

CHONDRICHTHYES Huxley, 1880
ELASMOBRANCHII Bonaparte, 1838
LAMNIFORMES Berg, 1958
MITSUKURINIDAE Jordan, 1898
SCAPANORHYNCHUS Woodward, 1889
*SCAPANORHYNCHUS
RAPHIODON* (Agassiz, 1844)

Referred Specimens MNA V10109, one tooth with complete crown, partial root; MNA V10082, one crown missing tip; MNA V10096, one partial crown missing base (Fig. 25.3A).

Description The slender, recurved crowns of these small teeth have fine striations on the labial surface that do not extend to the blade, like the anterior teeth of this species described by Cicimurri (2001) and Williamson, Kirkland, and Lucas (1993).

Discussion MNA V10109 was found associated with the plesiosaur MNA V10046 (cf. *Dolichorhynchops* sp.), suggesting that *S. raphiodon* may have been scavenging the carcass. According to Williamson, Kirkland, and Lucas (1993) this species occurs in the Upper Cenomanian–Middle Turonian Mancos Shale at Black Mesa, and ranges from Aptian to Maastrichtian.

ANACORACIDAE Casier, 1947
SQUALICORAX Whitley, 1939
SQUALICORAX CURVATUS Williston, 1900

Referred Specimens MNA V9974, V10106, V10108, V10056, V10062–V10064, V10067, V9461, V1009, one tooth each (Fig. 25.3B–F).

Description Identification of these teeth as *Squalicorax curvatus* follows Cicimurri (2001). These small teeth have distally inclined primary cusps and a low distal blade that intersect in a sharp angle. The edge of the primary cusp is finely serrated, almost straight in profile, and more than twice the length of the serrated distal blade. Labial and lingual surfaces of the crown are both convex, and the root is thick. The low root has a pair of short, divergent lobes. Teeth of this species can be distinguished from those of *S. falcatus* by their small size and morphology of the cusp and root (Cicimurri, 2001); teeth of *S. falcatus* are larger, the labial face of the crown is flat, and the lingual face is convex (both are convex in *S. curvatus*).

Discussion Williamson, Kirkland, and Lucas (1993) did not recognize this species from Black Mesa and concluded that recognition of species of *Squalicorax* is not possible from smaller teeth, like those recovered from the Tropic Shale. Our identification accepts the differences that Cicimurri (2001) recognized, but we agree that species in this genus are difficult to define. Therefore, the recognition of *S. curvatus* in the Tropic Shale fauna and the absence of this species from correlative beds at Black Mesa (Williamson, Kirkland, and Lucas, 1993) may be subjective. Williamson, Kirkland, and Lucas (1993) recognized *S. falcatus* in the Black Mesa fauna, which has larger teeth with much sharper apical angles than *S. curvatus*. On the other hand, if this difference is real, it might indicate separate ecological conditions, presumably deeper water and farther offshore for *S. falcatus* versus shallower water and nearshore for *S. curvatus*.

Like the specimens of *S. raphiodon* noted above, as well as those of *Cretolamna appendiculata* below, MNA V10106 and V10108 were found associated with plesiosaur MNA 10046, again suggesting that *S. curvatus* was also scavenging the carcass. MNA V10056, V10062–V10064, V10067,

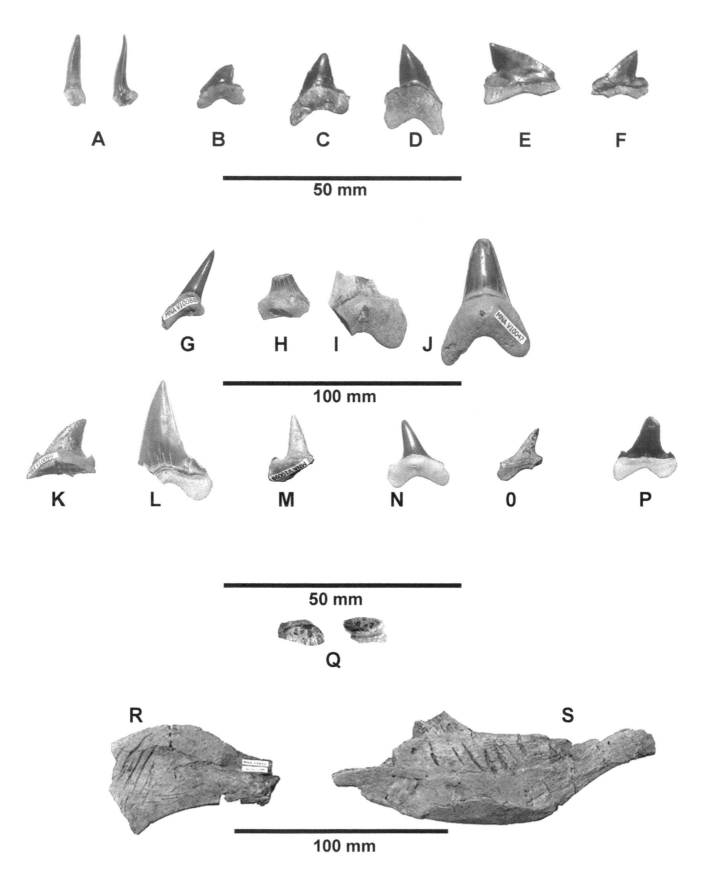

25.3. Chondrichthyans and examples of chondrichthyan predation from the Tropic Shale. (A) *Scapanorhynchus raphiodon,* MNA V10082. (B–F) *Squalicorax curvatus.* (B) MNA V9461; (C) MNA V10108; (D) MNA V10067; (E) MNA V10095; (F) MNA V10106; (G–J) *Cretoxyrhina mantelli.* (G) MNA V10088; (H–I) MNA V9490; (J) MNA V10047; (K–P) *Cretolamna appendiculata:* (K–L) MNA V10099; (M) MNA V10098; (N) MNA V10050; (O) MNAV10105; (P) MNA V9955; (Q) *Ptychotrygon* sp., MNA V10097 in occlusal and lateral views; (R–S) Tooth gouges on cranial elements of the plesiosaur *Brachauchenius lucasi:* (R) MNA V9433; (S) MNA V9432.

Albright, Gillette, and Titus

and V9461 were also found associated with plesiosaur skeletons (MNA V9445 and V9442), each possibly shed during scavenging.

CRETOXYRHINIDAE Gluckman, 1958
CRETOXYRHINA Gluckman, 1958
CRETOXYRHINA MANTELLI (Agassiz, 1843)

Referred Specimens MNA V9490, five associated teeth, all incomplete; MNA V10047, large, complete tooth; MNA V10088, partial tooth with complete crown (Fig. 25.3G–J).

Description MNA V10047 is large (height 34 mm) for the species with a very robust, rather than slender, crown and a massive root. Two of the teeth in MNA V9490 are at least as large as MNA V10047, but they are partial root and crown and cannot be measured for comparison. MNA V10088 is smaller, with a slender, recurved crown. As summarized by Williamson, Kirkland, and Lucas (1993), teeth of *Cretoxyrhina mantelli* are generally tall and narrow, and blades are smooth, with a flat labial surface and markedly convex lingual surface; lateral teeth tend to be slanted posteriorly (MNA V10088, Fig. 25.3G), and the root angle varies from obtuse on lateral teeth (Fig. 25.3G–I) to acute on anterior teeth (MNA V10047, Fig. 25.3J); lateral teeth may also have incipient lateral cusps (MNA V9490, Fig. 25.3I). Cutting edges are smooth and sharp, and the root is massive and often asymmetrical (Cicimurri, 2001), as in MNA V9490 (Fig. 25.3H–I), if missing root portions are projected. Teeth of *C. mantelli* are more robust than those of *Cretolamna* and generally lack the pronounced lateral cusplets and longitudinal ridges on the primary cusp like those of *Cretodus* (Cicimurri, 2001).

Discussion Teeth of this shark are relatively uncommon in the Tropic Shale, perhaps because this is a deep-water shark (Williamson, Kirkland, and Lucas, 1993). This shark is common in the Greenhorn Formation in South Dakota (Cicimurri, 2001) and widespread in the Cretaceous Western Interior Seaway, with a global geologic range of Cenomanian to Turonian (Williamson, Kirkland, and Lucas, 1993). Cicimurri (2001) reported occasional large teeth up to 3 cm in height, but most are much smaller.

CRETOLAMNA Gluckman, 1958
CRETOLAMNA APPENDICULATA (Agassiz, 1835)

Referred Specimens MNA V9955, one tooth missing tip of crown; MNA V10050, three associated, incomplete teeth; MNA V10105 and V10098, one tooth each; MNA V9464, six teeth; MNA V10099, two teeth (Fig. 25.3K–P).

Description MNA V10050 includes one tooth with a very small lateral cusp and two teeth lacking cusps. MNA V9464 includes one crownless root and five teeth with crowns, two

with lateral cusplets; MNA V10105 was found in the excavation of the plesiosaur MNA V10046 (cf. *Dolichorhynchops* sp.), perhaps as a result of scavenging of that carcass. One of the two teeth of MNA V10099 is large (approximately 33 mm height) with a prominent lateral cusplet with secondary ornamentation, and four deep enamel folds on the basal region of the labial surface. According to Cicimurri (2001), teeth of this species are generally small, not exceeding 2.5 cm height, indicating that MNA V10099 is unusually large (Fig. 25.3L). Teeth of *C. appendiculata* are tall, with weakly sigmoid crowns, a flat labial crown, a convex lingual crown, prominent lateral cusplets, a sharp cutting edge, a lingual boss, and bilobed roots that are relatively short; lateral teeth have a lower crown that is short and recurved as in MNA V10099 (Fig. 25.3K). Williamson, Kirkland, and Lucas (1993) described labial folds on some teeth in this species, as seen on MNA V10099 (Fig. 25.3L).

Discussion Williamson, Kirkland, and Lucas (1993) discussed the subtle differences that distinguish *Cretolamna appendiculata* from *Cretodus arcuata*, which we have not identified in the Tropic Shale fauna; *C. appendiculata* teeth are less robust, and the rearward curvature of lateral teeth is not as pronounced. Cicimurri (2001) distinguished this species from other cretoxyrhinids by the possession of lateral cusplets and from other similar genera in crown ornamentation. The geologic range of *C. appendiculata* is Albian to Maastrichtian in North America, and into the Eocene of Europe (Cappetta, 1987).

SCLERORHYNCHOIDEI Cappetta, 1980
Family uncertain
cf. PTYCHOTRYGON Jaekel, 1894
cf. PTYCHOTRYGON sp.

Referred Specimens MNA V10097, partial tooth (Fig. 25.3Q).

Discussion This incomplete tooth is not sufficiently preserved for confident identification to genus. It resembles *Ptychotrygon* from correlative beds at Black Mesa described by Williamson, Kirkland, and Lucas (1993). These authors recognized two species, *P. triangularis* and *P. rubyae*. *Ptychotrygon rubyae* is known only from the holotype and paratype (MNA V6404, single isolated tooth; and MNA V6403, four isolated teeth, respectively).

CTENACANTHIFORMES Cappetta, 1988
HYBODONTOIDEA Zangerl, 1981
PTYCHODONTIDAE Jaekel, 1898
PTYCHODUS Agassiz, 1835

Isolated and associated sets of pavement teeth of *Ptychodus* are often found in the blowout depressions common throughout the Tropic Shale, where the finer, powdery clays have been winnowed away leaving larger-sized particles. These deflation accumulations generally consist of individual teeth without organization or articulation, and cannot be considered with certainty as having originated from single individuals, nor from any specific stratigraphic level. In one case, however, several vertebrae and teeth were recovered in apparent association, suggestive of a single individual.

The Tropic Shale *Ptychodus* fauna is similar in composition to corresponding faunas of the Greenhorn Formation in western South Dakota reported by Cicimurri (2001) and at Black Mesa in northeastern Arizona by Williamson, Kirkland, and Lucas (1993), which is not surprising because all three are roughly the same age. Evidently all *Ptychodus* species in the Greenhorn Sea were widespread. Although it is tempting to conclude that less abundant species occupied different habitats than the abundant species, we consider the differences in abundance in museum collections to be the result of sampling bias.

Systematic paleontology follows Williamson, Lucas, and Kirkland (1991), Williamson, Kirkland, and Lucas (1993), Cicimurri (2001), and Shimada, Rigsby, and Kim (2009). The Tropic Shale fauna includes numerous isolated teeth, associated teeth in small numbers, and one large set of associated teeth and vertebrae. *Ptychodus* teeth occur throughout the lower and middle Tropic Shale with no apparent change at or across the Cenomanian–Turonian boundary. Identification to species is based entirely on tooth morphology.

PTYCHODUS DECURRENS Agassiz, 1843

Referred Specimens MNA V9434, 66 teeth and fragments; MNA V9435, associated dentition found disarticulated, 287 whole teeth, 137 fragmentary teeth, six whole vertebrae (more than half preserved), 12 fragmentary vertebrae (less than half preserved), and 97 small fragments of vertebrae; MNA V9492 (part), one tooth, and three additional teeth with anatomy that cannot be distinguished with confidence from *P. mammilaris*; MNA V9938, V9956, V9963, V9970, V9979, V9981, V9994, V9988, V9989, V9997, V9999, V10048, and V10125, one tooth each; MNA V9990 and V9992, three teeth each (Fig. 25.4A).

Description and Discussion Teeth of *Ptychodus decurrens* (Williamson, Kirkland, and Lucas, 1993; Cicimurri, 2001) can be quite large, but within a single dentition can range from large central teeth to small lateral teeth. Teeth of this species tend to be square or quadrilateral, with broad, low crowns that extend to the margin of the tooth and overhang the root. Enamel ornamentation consists of prominent transverse ridges that are perpendicular to the margin, rather than concentric, and branch marginally into increasingly finer ridges. Teeth of this species are low crowned, and the central cusp is not prominent.

The dentition of MNA V9435 includes very large central teeth (nearly 40 mm crown width), to very small marginal teeth (approximately 5 mm in crown width). In occlusal aspect most of the crowns are rectangular, but a few of the smaller, marginal teeth are nearly square. The central cusps are low on teeth of all sizes. Lateral teeth are slightly asymmetrical. The crown edges are convex anteriorly and correspondingly concave on the posterior margin for contact with adjoining teeth. Additional fossettes on the sides of some teeth mark the positions of adjoining teeth, especially those with diagonal placement. Crowns are weakly convex with a low central cusp that forms a central, rounded apex. The margins of the crowns are likewise weakly convex, taper toward the edge of the crown, and overhang the roots. Marginal areas surrounding the cusps are prominent in the medial teeth; marginal areas of lateral teeth become reduced and in most of the smaller teeth are absent. Roots are blocky, transversely elongate, weakly concave transversely, and weakly concave on anterior and posterior walls, producing a weak figure eight coronal cross section.

The crown ornamentation on larger teeth includes nine to 12 weakly arcuate, transverse ridges on the central cusp; the first two or three anterior ridges are divided into several segments, and the posteriormost ridge or two are segmented. On some teeth, the segmentation of crown ridges produces one or two rows of irregularly spaced tubercles. Ridges bifurcate at the junction with the marginal platform, and may bifurcate several times before the ridges reach the edge of the platform in an anastomosing pattern; with each bifurcation the ridges become smaller and the furrows between them narrower. Smaller teeth have narrower marginal platforms, but otherwise the ornamentation resembles that of larger teeth. The bifurcating ridges extend to the edge of the crown at right angles.

The four teeth in MNA V9492 have transverse ridges that break up toward the margins with faint bifurcations, but with granularity like that of *P. mammillaris*. Eight additional teeth in this set are referred to *P. mammillaris*, indicating a mixed accumulation of at least two individuals from different taxa.

25.4. Teeth of *Ptychodus*. (A) *Ptychodus decurrens*, MNA V9435; (B) *Ptychodus* sp., cf. *P. mammillaris*, MNA V9965; (C) *Ptychodus whipplei*, MNA V9973; (D) *Ptychodus occidentalis*, MNA V9960; (E) *Ptychodus anonymus*, MNA V9937; (F) *Ptychodus* species indeterminate, MNA V9982. Scale = 50 mm (A); scale = 25 mm (B–F). Left column, occlusal; center column, anterior; right column, lateral.

Albright, Gillette, and Titus

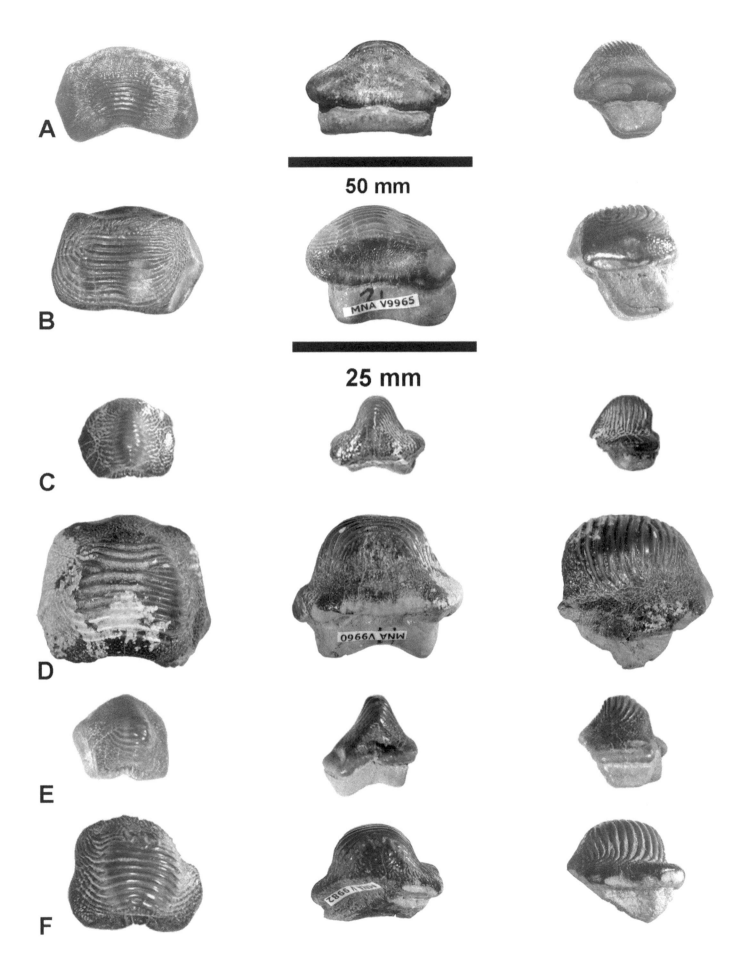

50 mm

25 mm

Three teeth in MNA v9990 were found in association with three high-crowned teeth identified as *P. whipplei*; this association appears to include two taxa.

The recognition of prominent marginal areas in medial teeth in this species differs from the general description of *P. decurrens* teeth by Shimada, Rigsby, and Kim (2009), who stated that *P. decurrens* teeth have marginal areas restricted to the anterior and posterior edges (implying little or no marginal development laterally and medially). *Ptychodus decurrens* teeth resemble those of *P. occidentalis*, but they have lower cusps and more restricted margins surrounding the cusp (Shimada, Rigsby, and Kim, 2009).

The definition of relative crown (or cusp) height has been largely a qualitative character (low cusp, high cusp, low crown, high crown). In all *Ptychodus* teeth we recovered from the Tropic Shale, the ratio of crown height to root height is relatively consistent for each species. We define a tooth as low crowned when the crown height is less than twice the height of the root (ratio < 2:1), as in *P. decurrens*; high crowned is when the crown height is greater than twice the height of the root (ratio > 2:1), as in *P. occidentalis* and cf. *P. mammillaris*, and generally much higher in *P. occidentalis*.

In the dentition of MNA v9435, large, central teeth have margins surrounding the central cusp that are roughly the same length and breadth. Smaller teeth have relatively larger cusps and corresponding reduction in marginal development, in some teeth so reduced that the margin is difficult to distinguish. This condition differs from the general statement by Shimada, Rigsby, and Kim (2009) that teeth of *P. decurrens* have restricted margins, which for MNA v9435 holds only for the smaller teeth. The low crown and weakly concave root distinguish these teeth from *P. occidentalis*. The marginal ridges are oriented at right angles to the edge of the platform, and show no tendency toward the concentric arrangement as in *P. anonymus* (Shimada, Rigsby, and Kim, 2009) and *P. mammillaris* (Williamson, Kirkland, and Lucas, 1993). Teeth of *P. whipplei* have steep-sided cusps and high crowns (Williamson, Lucas, and Kirkland, 1991; Williamson, Kirkland, and Lucas, 1993), rather than the low crowned and weakly convex cusps of *P. decurrens*.

Most of the large central teeth have pronounced, concave attrition fossettes on the margins of the crowns, rather than on the central apex, as observed for *P. decurrens* in the Black Mesa fauna by Williamson, Kirkland, and Lucas (1993). Those authors suggested that this might be an indication of a grinding rather than crushing function. Williamson, Kirkland, and Lucas (1993) found that this species is most abundant in deep-water deposits, and rare in shallow-water deposits, at Black Mesa.

The vertebral centra are compressed from preservation, primarily in the axial dimension, with anteroposterior length approximately 5 mm as preserved. Several are compressed obliquely, permitting an estimation of length in life of approximately 15 mm. Most have broken and/or incomplete edges. Transverse diameters range from approximately 42–50 mm for those specimens that can be measured.

PTYCHODUS sp. cf. *P. MAMMILLARIS* Agassiz, 1835

Referred Specimens MNA V10102, 73 associated teeth; MNA v9492 (part), eight teeth; MNA v9958, two teeth; MNA v9976, four teeth; MNA v9939, V9954, V9957, V9959, V9964, V9965, V9978, V9980, and V9984, one tooth each (Fig. 25.4B).

Description and Discussion Teeth of MNA V10102 were recovered in association, but there is no assurance they came from a single individual. Transverse diameters range from 30.9 mm for the largest tooth to less than 10 mm for the smallest teeth. In these and all others assigned to cf. *P. mammillaris*, the transverse dimension of the crown is only slightly greater than the anterior–posterior dimension. All are low crowned. In MNA V10102 the overall dental anatomy resembles that of *P. decurrens*, with a low apex and similar crown morphology, although the junction of the transverse ridges of the central cusp with the margin platforms is more abrupt. Typically 9–11 complete transverse ridges extend from side to side, and one or two anterior and posterior ridges are interrupted at the apex and are therefore incomplete. However, a distinctive difference in these teeth compared to those of *P. decurrens* is the configuration of the transverse ridges at their lateral extremity: on the former they are concentric on one or both margins. Asymmetrical teeth have well-developed concentric organization on one side of the apex; transverse ridges of the opposite side have only weakly developed concentric ridges, or they terminate with weakly expressed bifurcation that leads to granular texture on the margins. In a few highly asymmetrical teeth, the bifurcation extends to the margin. Most of the granular texture is weakly elongated and arranged in concentric lines paralleling the margins, rather than intersecting the margins as in *P. decurrens*. Roots resemble those of *P. decurrens* except that the transverse concavity is more pronounced in most of these teeth. The eight teeth in MNA v9492 have transverse ridges that become concentric or nearly concentric on one side of the crown, as in MNA V10102.

All teeth assigned to cf. *P. mammillaris* in the Tropic Shale fauna differ in having low crowns relative to the high, steep-sided crowns of *P. anonymus*. These teeth resemble *Ptychodus* teeth from northeastern Arizona that Williamson, Kirkland, and Lucas (1993) identified as *Ptychodus* sp. cf. *P. mammillaris*.

Albright, Gillette, and Titus

PTYCHODUS WHIPPLEI Marcou, 1858

Referred Specimens MNA V9990, three teeth; MNA V9967, V9968, V9971, V9983, V9986, V9987, V9992, V9997, V10070, V10077, one tooth each; MNA V9973, two teeth (one unusually large for this species, one small and asymmetrical) (Fig. 25.4C).

Description and Discussion Specimens are generally small and high-crowned, approximately 7–11 mm wide, with steeply inclined cusps and 7–9 transverse ridges that terminate at the junction of the cusp with the marginal shelf. The ornamentation bifurcates no more than once and becomes granular at the edges (Williamson, Kirkland, and Lucas, 1993). In occlusal aspect they are approximately square with a rounded anterior border. The anterior and posterior concavities on the crown range from weak to pronounced. Roots are broadly concave transversely, and have 1–3 nutrient foramina. Three teeth in MNA V9990 were found in association with three low-crowned teeth identified as *P. decurrens*; this association appears to include two taxa.

Williamson, Kirkland, and Lucas (1993) and Cicimurri (2001) summarized the known geologic occurrence of *P. whipplei* as common in the Turonian and extending into the Coniacian in the Cretaceous Western Interior Seaway, but neither listed this species for the Cenomanian. Our site records do not indicate exact stratigraphic positions for the MNA specimens. Future work and continued collecting will be required to determine if a faunal change in this genus occurred in the Tropic Shale across the Cenomanian–Turonian boundary.

PTYCHODUS OCCIDENTALIS Leidy, 1868

Referred Specimens MNA V9960, V9961, V9966, and V9972, one tooth each (Fig. 25.4D).

Description MNA V9960, V9966 and V9972 are low-crowned teeth with a steep-sided central cusp, distinctive marginal area medially and laterally, and slightly reduced marginal area anteriorly and posteriorly. MNA V9960 and V9966 are symmetrical, probably medial teeth; MNA V9972 is asymmetrical, with a lower crown height than the medial teeth. In anterior profile the roots are markedly concave. Crown ornamentation includes 10 transverse ridges that terminate at the junction of the cusp with the margin and disintegrate into a granular texture; the posterior depression on the crown is pronounced, and the corresponding projection on the anterior edge is rounded. In overall proportions the crowns in occlusal aspect are roughly equidimensional. In most respects MNA V9966 resembles the upper right second lateral tooth of *Ptychodus occidentalis* from the Greenhorn Limestone of Nebraska illustrated by Shimada, Rigsby, and

Kim (2009:fig. 5c), except that MNA V9966 is somewhat larger and the root is heart-shaped. MNA V9972 resembles the upper right second lateral tooth of the same *Ptychodus occidentalis* specimen from Nebraska with respect to shape, proportions, and ornamentation, but the crown height is less robust and more closely resembles the lower right first lateral tooth (Shimada, Rigsby, and Kim, 2009:figs. C and F, respectively). MNA V9960 resembles the upper right first lateral tooth of that specimen (Shimada, Rigsby, and Kim, 2009:fig. 5b) except that the crown profile is more rounded.

Discussion These teeth confirm the existence of *P. occidentalis* in the Tropic Shale fauna. They are distinguished from *P. decurrens* by a square rather than rectangular crown, high rather than low crown height, and a marginal area that is prominent laterally and medially, as well as anteriorly and posteriorly. Although Shimada, Rigsby, and Kim (2009) observed these characteristics of the marginal area for *P. decurrens* and *P. occidentalis* from the Greenhorn Limestone, the general statement that all teeth of the former lack medial and lateral margins (Shimada, Rigsby, and Kim, 2009:340) must be tempered with the recognition that medial teeth of that species have prominent lateral and medial marginal areas and that the marginal areas become more constricted, and are nearly absent, in smaller teeth. Williamson, Kirkland, and Lucas (1993) did not recognize this species in the Mancos Shale at Black Mesa.

PTYCHODUS ANONYMUS Williston, 1900

Referred Specimens MNA V10085, 11 teeth, and one fragmentary tooth; MNA V9455 and V9468, two teeth each; MNA V9937 and V9962, one tooth each (Fig. 25.4E).

Description and Discussion MNA V9962 and V9937 have high crowns, with ornamentation consistent with the anatomy of this species described and illustrated from South Dakota (Cicimurri, 2001:fig. 6a, b). They resemble other teeth in the Tropic Shale fauna that we identify as cf. *P. mammillaris*. The crown ornamentation curves laterally to become concentric on both sides of the cusp, and the ornamentation on the shelf is largely granular, with a faint indication of orientation or alignment parallel to the margin of the crown.

Teeth in the set MNA V10085 are all small and the crowns, in occlusal view, are roughly equidimensional. Cusps are steep sided, and several have a triangular profile in anterior aspect. Groove ornamentation consists of up to 12 transverse ridges that curve anteriorly at the junction with the shelf, but become discontinuous; ornamentation from the cusp outward consists of fragmented, discontinuous ridges that parallel the margin of the tooth crown. Several of the smaller teeth in this set are marginals, with asymmetrical crowns and reduced number of ornamentation ridges. Roots of three teeth are smooth and boxlike, smoothly concave, with two nutrient foramina; roots of the remaining nine are concave and deeply excavated by vascular grooves, quite unlike all other teeth we have observed from the Tropic Shale. We nevertheless conclude that these teeth all belong to the same taxon, recognizing the variation in root anatomy.

The steep-sided cusp and high crown distinguish these teeth from others assigned to cf. *P. mammillaris*. Williamson, Kirkland, and Lucas (1993) included *P. anonymus* in the synonymy of *P. mammilaris*.

PTYCHODUS sp. indet.

Referred Specimens MNA V9982, one tooth (Fig. 25.4F).

Description MNA V9982 is a complete lateral tooth, approximately 15 mm wide, with an asymmetric crown and tapered, rather than blocky, root. The crown includes a prominent, steep-sided cusp with distinct lateral margins and restricted anterior and posterior margins. The eight transverse ridges extend onto the margin of the crown without bifurcation into incomplete ridges or granular texture. The ridges curve anteriorly on the margins and converge on the anterolateral and anteromedial edges of the tooth.

Discussion Teeth of *P. whipplei* have a steep cusp, but transverse ridges become granular toward the margins of the teeth and do not extend to the edge of the crown. Central cusps of *P. occidentalis* and *P. anonymus* are prominent, but not as steeply inclined, and transverse ridges break up on the margins into a granular texture. Central cusps of *P. decurrens* are low and less pronounced, and crown ridges bifurcate into an anastomosing pattern. Ridges on the crowns of *P. mammillaris* are concentric, or nearly so, on the margins and parallel to the edge of the tooth rather than intersecting the edge. Identification of this tooth is problematical; without additional teeth for assessment of variation, species identification is unwarranted; but this tooth might represent another species not otherwise recognized in the fauna.

A mixed set of *Ptychodus* teeth (MNA V9466) not listed above consists of 47 teeth from at least three taxa (*P. decurrens*, *P. anonymus*, and *P. occidentalis*). Enamel surfaces of about one-third of these teeth are encrusted with colonial coral with very small corallites. In every case, enamel has been partially or completely dissolved, although the roots seem to be free of encrustation. Other teeth in this set seem to have partial loss of enamel, suggesting that these teeth have passed through the digestive tract of a larger animal. These teeth represent at least three individual prey that were passed through the digestive system of a predator and subsequently deposited as a mixed set of teeth on the sea bottom, whereupon they became a substrate for encrusting corals.

OSTEICHTHYES Huxley, 1880
ACTINOPTERYGII Cope, 1887
PYCNODONTIFORMES Berg, 1937
PYCNODONTOIDEI Nursall, 1996
Genus and species undetermined

Referred Specimens MNA V10076, left and right ?premaxillae with dentition (Fig. 25.5A).

Description This specimen is insufficiently preserved for generic identification. The teeth are crudely aligned in two rows parallel with the margin of the jaw. They are all circular to oval, and none are expanded into broad plates as found in some pycnodonts.

Discussion Although pycnodont fish are common elsewhere in Cretaceous marine sediments, this is the only specimen yet recovered from the Tropic Shale. Perhaps these fish were rare in this part of the Cretaceous Western Interior Seaway, but considering the generally small size of pycnodonts, it might also be an artifact of collection bias.

ICHTHYODECTIFORMES
Bardack and Sprinkle, 1969
ICHTHYODECTIDAE Crook, 1892

The Tropic Shale fauna contains at least three genera of this family. We follow the lower level classification of Bardack (1965) and Bardack and Sprinkle (1969).

GILLICUS Hay, 1898
GILLICUS ARCUATUS (Cope, 1875)

Referred Specimens MNA V10081, nearly complete articulated skeleton (Fig. 25.5B).

Discussion This specimen was found eroding (tail first) from a gentle slope near MNA V9445, the holotype skeleton of *Eopolycotylus rankini*, and at the same horizon. Part of the caudal fin was recovered as fragments on the surface. The remainder of the skeleton was completely articulated, including the skull. The delicate fins did not survive excavation as articulated elements, but photographs documented their position and shape. The skull and jaws are partially prepared, the remainder of the skeleton fully prepared. Teeth are not exposed, and no scales were recovered. As preserved, the specimen measures 138 cm from the tip of the snout to the base of the tail.

ICHTHYODECTES Cope, 1870
ICHTHYODECTES CTENODON Cope, 1870

Referred Specimens MNA V9467, two dentaries with crownless teeth, six vertebrae, skull fragments (Fig. 25.5C).

Albright, Gillette, and Titus

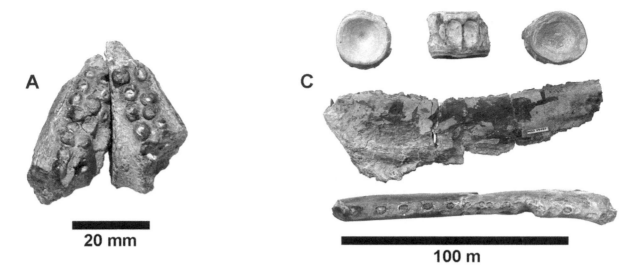

25.5. Osteichthyes from the Tropic Shale. (A) Pycnodont, gen. et sp. indet., MNA V10076; (B) *Gillicus arcuatus,* MNA V10081, right side; (C) *Ichthyodectes ctenodon,* MNA V9487, three vertebrae and two dentaries.

Discussion Identification of MNA v9467 is based on Bardack's (1965) description of the skull of this species. The two dentaries are elongated compared to the dentary of *Xiphactinus*. The dentary and teeth resemble corresponding elements described by Bardarck, including the angle of the symphysis with respect to the ventral margin of the bone, the orientation and size of teeth and alveoli, and dimensions.

ICHTHYODECTES sp., cf. *I. CTENODON* Cope, 1870

Referred Specimens MNA v9483, fragmentary lower jaw with crownless teeth.

Discussion Provisional assignment of this specimen to *Ichthyodectes ctenodon* is based on the anatomy of the jaw fragment and the size and spacing of the teeth. Bardack (1965) recognized only one species of *Ichthyodectes* in North America.

XIPHACTINUS Leidy, 1870
XIPHACTINUS sp., cf. *X. AUDAX* Leidy, 1870

Referred Specimens MNA v9482, partial skull, jaws, teeth, vertebrae; MNA v9484, partial skull, jaws, vertebrae, fins; MNA V10121, partial skull, partial pectoral fin, vertebrae; MNA V10000, fragmentary skull and vertebrae; MNA v9462, skull fragments with partial dentary and one fragmentary vertebra (Fig. 25.6).

Discussion MNA V9484 includes a complete premaxilla with three complete teeth, a complete maxilla with full dentition (but most crown tips missing), isolated skull and jaw elements, fragmentary left and right dentaries, several isolated teeth, left and right pectoral fins in articulation, six complete vertebrae, and bone fragments (Fig. 25.6). The skull elements of MNA V9482 are in partial articulation, but slightly crushed. MNA V10121 is less complete but well preserved. MNA V10000 includes partial jaw fragments with teeth, permitting assignment to this genus. MNA V9462 consists mainly of broken skull fragments, with a partial dentary in two parts containing at least nine teeth. All four specimens, representing medium-sized adults, are uncrushed and nearly three-dimensionally preserved.

Bardack (1965) and Bardack and Sprinkle (1969) reviewed the anatomy and evolution of fish in the family Chirocentridae, which includes only one modern genus, the wolf herrings (*Chirocentrus*), but which had an extensive geographic and geologic record from the Middle Jurassic to Late Cretaceous (Oxfordian to Campanian). Two common Late Cretaceous genera are *Gillicus* and *Xiphactinus*, both known from well-preserved specimens from the Green Horn Limestone and other roughly correlative formations in North America. Despite the differences in dentition, these two taxa are closely associated evolutionarily, and often intimately associated as fossils, with the recovery of several large *Xiphactinus* with *Gillicus* in the ribcage, evidently the large predator's last meal. *Xiphactinus* specimens are commonly large, reaching documented lengths of 6 m or more. Tentative identification of the Tropic Shale specimens to *X. audax* is consistent with its known geologic distribution in North America, but Bardack (1965) recognized several other species that differ in relatively minor details—and thus our "cf." designation.

REPTILIA

The reptilian fauna of the Tropic Shale includes short-neck plesiosaurs, turtles, and the exotic occurrence of a terrestrial dinosaur. Interestingly, remains of elasmosaurs, mosasaurs, and pterosaurs that occur elsewhere in sediments of the Cretaceous Western Interior Seaway have not yet been discovered.

TESTUDINATA Oppel, 1811
PROTOSTEGIDAE Cope, 1873
DESMATOCHELYS Williston, 1894
DESMATOCHELYS LOWI Williston, 1894

Referred Specimens MNA V9446, partial skeleton, including cervical and caudal vertebrae, ilium, ischium, right femur, left and right tibia, fibula, paddle elements including ungual phalanges, costal ribs, and large parts of the carapace and plastron (Fig. 25.7).

Discussion Comparisons with the specimen of *Desmatochelys lowi* noted by Elliott, Irby, and Hutchinson (1997) from exposures of Mancos Shale at Black Mesa, together with the descriptions and illustrations of this species by Zangerl and Sloan (1960), indicate that MNA V9446 is likely the same taxon. The documented occurrence of *Desmatochelys lowi* in Turonian strata of Nebraska, South Dakota, Kansas, Arizona, and Japan (Elliott, Irby, and Hutchinson, 1997; Hirayama, 1997) is compatible with this assignment.

PROTOSTEGIDAE
Genus et sp. indet.

Referred Specimens MNA V9458, partial carapace and plastron, vertebrae, pelvis, phalanges.

Currently under study, this specimen was provisionally identified as a possible new genus of protostegid by W. Joyce (pers. comm. to D.D.G., 2010).

Family Incertae sedis
NAOMICHELYS Hay, 1908
NAOMICHELYS sp.

Referred Specimens MNA V9461, fragmentary carapace and plastron, one limb fragment (Fig. 25.8).

Discussion Identification of this specimen as belonging to the genus
Naomichelys is based on the distinctive pustulate texture of the shell. Although the tubercles that result in this distinctive texture are slightly larger on the plastron than on the carapace (Fig. 25.8), the poor condition of the specimen precludes determination to species. MNA V9461 was recovered from sediments near the middle of the stratigraphic section of Tropic Shale in the Big Water area, approximately 14–15

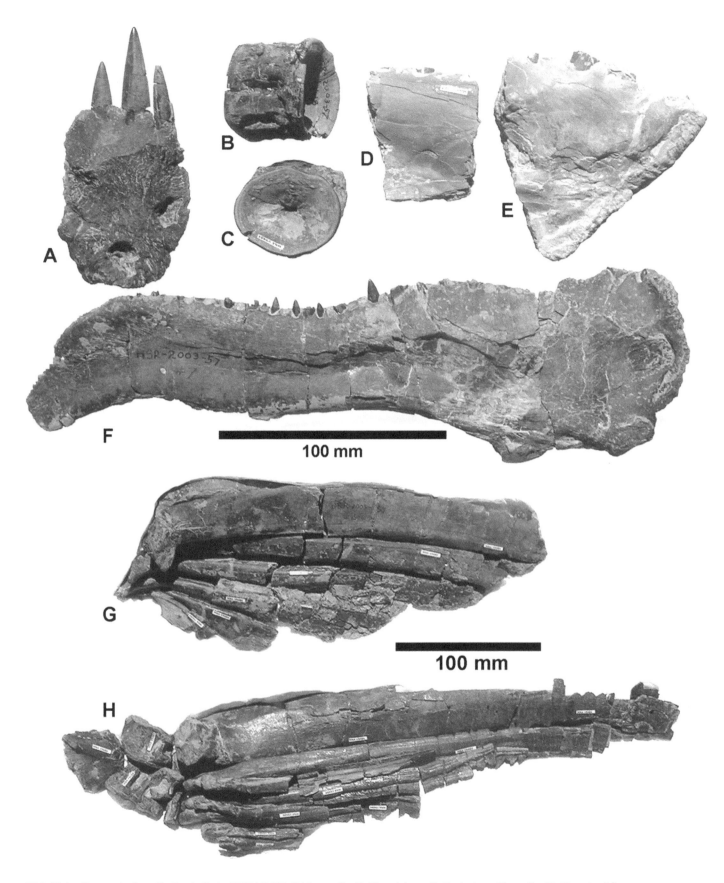

25.6. *Xiphactinus audax* from the Tropic Shale, MNA V9484. (A) Premaxilla; (B, C), vertebrae; (D, E), dentary; (F) maxilla; (G, H) pectoral fins.

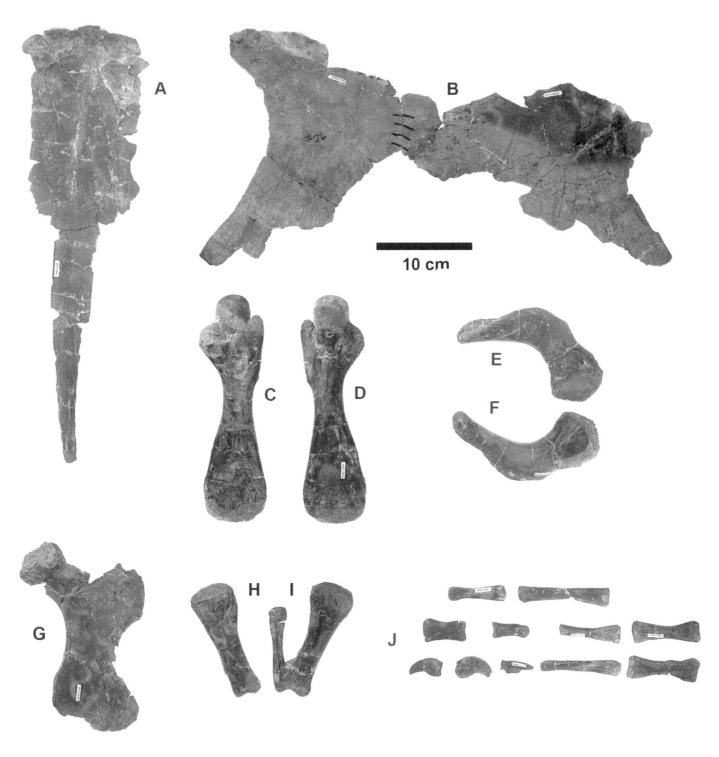

25.7. *Desmatochelys lowi* from the Tropic Shale. (A) Pleural with rib; (B) hypoplastron; (C, D) dorsal and ventral view of right femur; (E, F) dorsal and ventral view of ilium; (G) ischium; (H) tibia; (I) tibia and fibula; (J) phalanges.

m above the lowest appearance of the ammonoid *Mammites* (Fig. 25.2), which marks the boundary between the lower and middle Turonian. Typically considered an inhabitant of estuarine environments, the occurrence of this specimen in marine shale kilometers from the paleoshoreline suggests the carcass floated out to significantly deeper waters offshore. *Naomichelys* is also known from southern Utah in the terrestrial equivalent of the Tropic Shale, the Iron Springs Formation, and from the overlying Straight Cliffs and Wahweap formations (Eaton, 1999).

Albright, Gillette, and Titus

25.8. *Naomichelys* sp. from the Tropic Shale. (A) Fragment from margin of carapace; (B) fragment from margin of plastron.

SAUROPTERYGIA Owen, 1860
PLESIOSAURIA de Blainville, 1835
PLIOSAURIDAE Seeley, 1847
BRACHAUCHENIUS Williston, 1903
BRACHAUCHENIUS LUCASI Williston, 1903

Referred Specimens MNA V9433, nearly complete skull (cranium and mandible) with teeth, 12 cervical vertebrae, 22 dorsal vertebrae, ribs, partial scapula, left coracoid, partial right coracoid, right ilium, partial pubis, partial ischium; MNA V9432, partial cranium, partial mandible, teeth, and vertebrae of juvenile (Fig. 25.9). Note: The tooth of *B. lucasi* shown in fig. 9C–D of Albright, Gillette, and Titus (2007a) was incorrectly attributed to MNA V9433; in fact, it is part of MNA V9432.

Discussion These specimens, both from Glen Canyon National Recreation Area, represent the only material of *Brachauchenius* yet found in the Tropic Shale and the first record of this taxon along the western reaches of the Cretaceous Western Interior Seaway. Both were found only meters above the Cenomanian–Turonian boundary (MNA V9432 about 7 m above MNA V9433; Fig. 25.2), and about 300 m apart. MNA V9433 includes a nearly complete, but disarticulated skull, plus several vertebrae, ribs, and parts of the pectoral and pelvic girdles (Fig. 25.9). The second specimen, MNA V9432, is represented by a partial cranium and

mandible of a juvenile. In the description of these specimens, Albright, Gillette, and Titus (2007a) inadvertently concluded that the associated pelvic and pectoral elements were the first known for this taxon. After publication, however, it came to their attention that a nearly complete, articulated skeleton of *Brachauchenius* from early Cretaceous deposits in Columbia had been discovered previously, which included the pectoral and much of the pelvic girdles (Hampe, 2005).

The Tropic Shale occurrences of *Brachauchenius* in the lower Turonian stratigraphic levels place these specimens about midway through the temporal range of the genus, which Schumacher and Everhart (2005) noted was from the middle Cenomanian to middle Turonian.

POLYCOTYLIDAE Williston, 1908
PALMULASAURINAE Albright,
Gillette, and Titus, 2007c
PALMULASAURUS Albright,
Gillette, and Titus, 2007c
PALMULASAURUS QUADRATUS
(Albright, Gillette, and Titus, 2007b)

Referred Specimens Holotype (MNA V9442), anterior part of rostrum, anterior part of mandible, fragment of articular, fragment of occipital condyle, partial right coracoid, partial right scapula, complete right forelimb, right and left

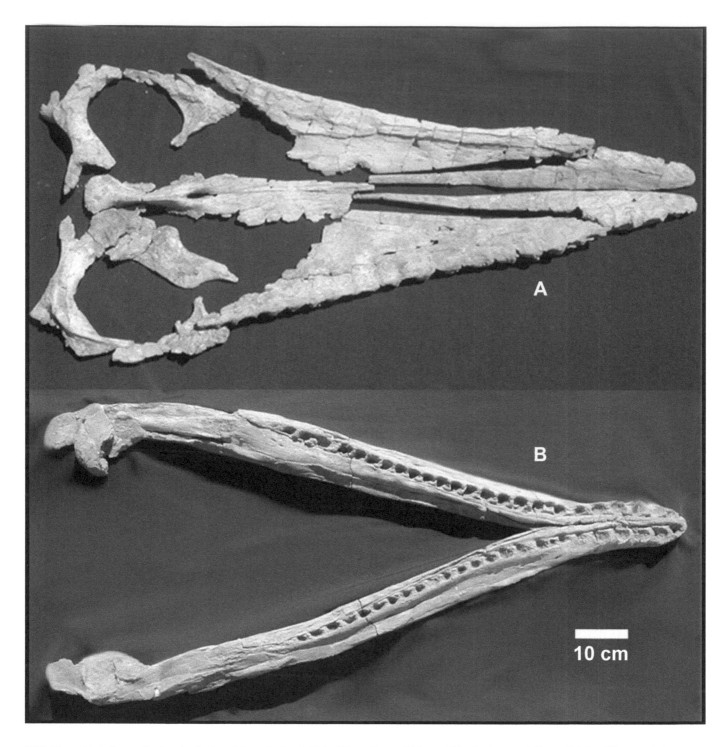

25.9. Pliosaurid plesiosaur, *Brachauchenius lucasi,* from the Tropic Shale. (A) cranium and (B) mandible of MNA V9433 (from Albright, Gillette, and Titus, 2007a, with permission from *Journal of Vertebrate Paleontology*).

ilium, right and left pubis, right and left ischium, complete right hind limb, one cervical vertebra, 19 dorsal vertebrae, three sacral vertebrae, 11 caudal vertebrae, several ribs (Fig. 25.10A–C).

Discussion Particularly diagnostic of this taxon are the nearly equidimensional radius and ulna in the forelimb and tibia and fibula in the hind limb, in contrast to all other known polycotylids in which these elements are relatively

25.10. (facing) Polycotylid plesiosaurs from the Tropic Shale. (A–C) *Palmulasaurus quadratus;* (A) right forelimb; (B) right hind limb; (C) left and right pubis, ischium, and ilium; (D, E) *Eopolycotylus rankini;* (D) right hind limb; (E) left and right ilium (from Albright, Gillette, and Titus, 2007b, with permission from *Journal of Vertebrate Paleontology*).

Albright, Gillette, and Titus

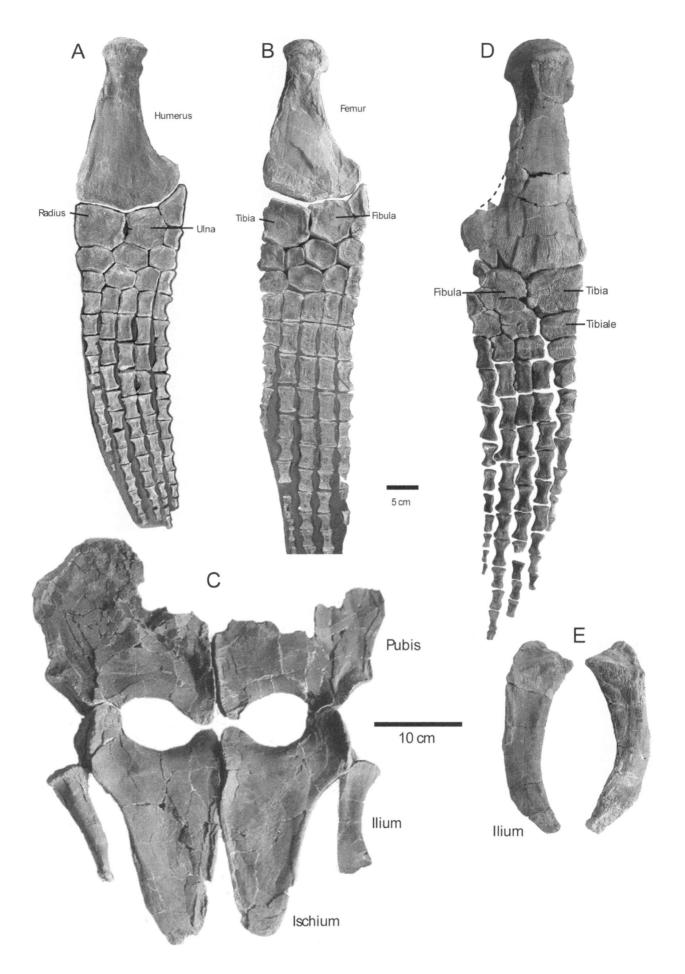

A: Humerus, Radius, Ulna

B: Femur, Tibia, Fibula

C: Pubis, Ilium, Ischium

D: Fibula, Tibia, Tibiale

E: Ilium

5 cm

10 cm

Fossil Vertebrates from the Tropic Shale (Upper Cretaceous)

short and broad. On the basis of this diagnostic morphology, Albright, Gillette, and Titus (2007b) included two additional specimens into this group one from the late Cenomanian of Japan (Sato and Storrs, 2000) and one from similarly aged deposits of South Dakota (Schumacher, 2007) and erected a new subfamily of the Polycotylidae, the Palmulasaurinae, to further emphasize their difference from all other known polycotylids. The latter were included in another new subfamily, the Polycotylinae. On the basis of new material of previously undescribed specimens and on a detailed reanalysis of a specimen that Adams (1997) referred to *"Trinacromerum bonneri"* O'Keefe (2008), concluded that the subfamilial split of the Polycotylidae may not be warranted. Citing the large amount of missing data for the Palmulasaurinae as the major reason for his conservative assessment, it will be interesting to revisit the subfamilial hypothesis upon inclusion of data from the much more complete, putative "palmulasaurine" from South Dakota, *Pahasapasaurus haasi* Schumacher, 2007.

POLYCOTYLINAE Albright,
Gillette, and Titus, 2007b
EOPOLYCOTYLUS Albright,
Gillette, and Titus, 2007b
EOPOLYCOTYLUS RANKINI
Albright, Gillette, and Titus, 2007b

Referred Specimens Holotype, MNA V9445 anterior part of rostrum, anterior part of mandible, right and left clavicle with fused interclavicle, right and left scapula, right and left coracoid, right humerus, right and left ilium, right and left pubis, right and left ischium, right and left femur, most of right rear paddle, most of left rear paddle, three cervical vertebrae, 29 dorsal vertebrae (including three sacrals), one caudal vertebra, and several ribs (Fig. 25.10D, E).

Discussion This specimen includes a significant portion of the axial and appendicular skeleton, with some skull material, of a polycotylid much larger than *Palmulasaurus*. The more traditional morphology of the radius/ulna and tibia/fibula seen in Fig. 25.10D clearly shows this taxon as a member of the "Polycotylinae" of Albright, Gillette, and Titus (2007b), but the unique morphology of the humerus indicates that it belongs to a taxon other than *Trinacromerum bentonianum*. Additionally, the morphologies of the ilia (Fig. 25.10E) and stout, relatively coarsely striated teeth are identical to that of the Campanian-aged *Polycotylus latippinus*. But *P. latippinus* has a highly diagnostic humerus and femur, both of which flare broadly at their distal ends, entirely unlike matching elements of the Tropic Shale species.

TRINACROMERUM Cragin, 1888
TRINACROMERUM
?BENTONIANUM (Cragin, 1888)

Referred Specimens UMNH accession number 96.10, "partial skeleton from the tail and trunk region, consisting of two sections of vertebrae in articulation, one with approximately 18 vertebrae, the other with approximately 30 vertebrae; isolated vertebrae, two teeth, a paddle bone, and additional unidentified elements," from upper part of the *Sciponoceras gracile* ammonoid biozone (upper Cenomanian) near Orderville, Kane County, Utah (Gillette, Hayden, and Titus, 1999:270); MNA V10044, two cervical vertebrae, from Glen Canyon National Recreation Area, early Turonian; MNA V10045, right femur with proximal paddle elements, from Glen Canyon National Recreation Area, early Turonian.

Discussion Though not conclusive, the cylinder-like shape of the vertebrae with little lateral or ventral constriction supports tentative referral of the Muddy Creek specimen and MNA V10044 to a large specimen of *T. bentonianum*. Referral of MNA V10045 to *Trinacromerum* is based primarily on the short and wide aspect of the tibia and fibula, and the femur is identified as such by the lack of sigmoidal curvature typically seen in plesiosaur humeri.

DOLICHORHYNCHOPS Williston, 1903
DOLICHORHYNCHOPS TROPICENSIS
Schmeisser McKean, 2012

Referred Specimen MNA V10046, nearly complete skeleton, with associated gastroliths.

Discussion MNA V10046 is the best plesiosaur specimen yet recovered from the Tropic Shale (Schmeisser McKean, 2012). The nearly complete skeleton, from early Turonian strata, includes the skull, jaws, and teeth plus 51 vertebrae, left and right pectoral elements and forelimbs, left and right pelvic elements and hind limbs, associated metapodials and phalanges, ribs, and 289 gastroliths.

Schmeisser (2006, 2008) and Schmeisser and Gillette (2009) tentatively identified this specimen as a new species of cf. *Dolichorhynchops* sp., noting the gracile, finely striated teeth, in addition to several other characters of the cranium and mandible. The Tropic species of *Dolichorhynchops* forms the basis of an 9 million year range extension for the genus – from the late Santonian at about 84 Ma downward to the early Turonian at about 93 Ma. This young adult specimen is small, with an estimated length of approximately 3 m. Three other species of *Dolichorhynchops* are currently recognized: *D. osborni*, *D. bonneri*, and *D. herschelensis* (Sato, 2005; O'Keefe, 2008).

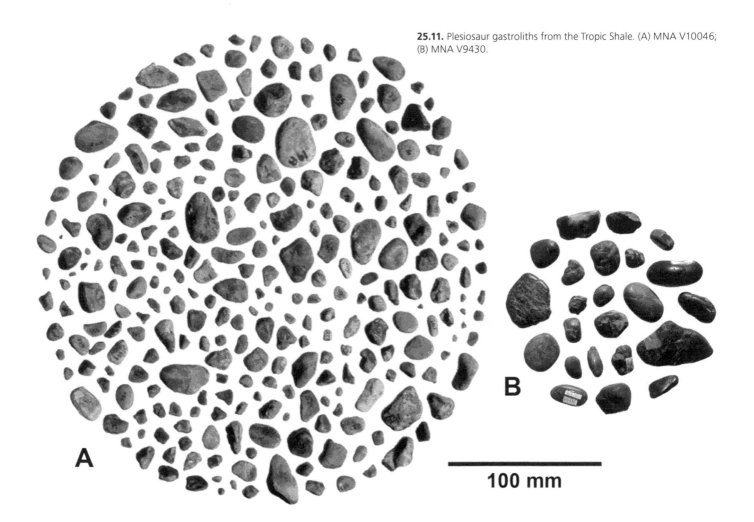

25.11. Plesiosaur gastroliths from the Tropic Shale. (A) MNA V10046; (B) MNA V9430.

100 mm

Schmeisser and Gillette (2009) described a large cluster of exotic stones directly and unambiguously associated with MNA V10046, which they concluded were gastroliths (Fig. 25.11). Another smaller cluster, but with larger stones, found isolated and not associated with a skeleton, is also tentatively identified as plesiosaur gastroliths (Fig. 25.11B). All of these stones are well rounded, highly polished, and composed mainly of black chert. Whittle and Everhart (2000) described a similar instance of probable plesiosaur gastroliths from the Upper Cretaceous Smoky Hill Chalk of Kansas and from the Lower Cretaceous Kiowa Shale of Kansas (Everhart, 2000).

ARCHOSAURIA Cope, 1869
THEROPODA Marsh, 1881
MANIRAPTORA Gauthier, 1986
THERIZINOSAURIDAE Maleev, 1954
NOTHRONYCHUS Kirkland and Wolfe, 2001
NOTHRONYCHUS GRAFFAMI Zanno et al., 2009

Referred Specimen Holotype (Fig. 25.12), UMNH VP 16420; relatively complete skeleton, missing skull and jaws, about

half the dorsal and cervical vertebrae, some of the ribs, and some phalanges (Zanno et al., 2009).

Discussion UMNH VP 16420 currently represents the most complete Late Cretaceous therizinosaur yet known. It was discovered and excavated by MNA field crews in 2000 and 2001 from a site near Big Water on land owned and managed by the School and Institutional Trust Lands Administration, State of Utah. At nearly the same time, J. Kirkland of the Utah Geological Survey and D. Wolfe of the Mesa Southwest Museum were publishing finds of therizinosaur bones they had recovered from slightly younger (middle Turonian), but terrestrial rather than marine sediments of western New Mexico (Kirkland and Wolfe, 2001). They named this first unequivocal North American therizinosaur *Nothronychus mckinleyi*. Before these two discoveries, therizinosaurs had not been recognized with certainty in North America.

The Tropic Shale therizinosaur was noted in several abstracts (Albright et al., 2001; Gillette et al., 2001a, 2001b; Gillette and Albright, 2003), but formal description was finally provided by Zanno et al. (2009), who concluded that the Tropic Shale specimen belonged to the same genus as that from New Mexico but that it represented a second species,

25.12. Mounted skeleton of *Nothronychus graffami* on exhibit at MNA from 2007 to 2009, produced by Gaston Design, Fruita, Colorado. Skull and cervical vertebrae modeled from *Erlikosaurus andrewsi*.

N. graffami. The story of the discovery of the Tropic Shale therizinosaur is described by Gillette (2007) in a popular-style account in an issue of MNA's *Plateau* series specifically devoted to this find. This issue accompanied a special exhibit at MNA, the center piece of which was a cast of the articulated skeleton (Fig. 25.12).

Among derived members of this family, UMNH VP 16420 is the most complete skeleton known, and it provides a unique foundation for interpretation of the anatomy and taxonomy of other therizinosaurs, and other theropods, known only from fragmentary remains (Zanno et al., 2009; see also Zanno et al., this volume, Chapter 22). This theropod lineage is unusual for its highly specialized feeding habits, and was almost certainly herbivorous. Its enormous gut, widely spaced, massive hips, stout rear legs, and gangly arms are unique in the world of theropod dinosaurs, which were otherwise primarily carnivorous.

The occurrence of a terrestrial dinosaur skeleton in marine sediments that were over 100 km from the shoreline is best explained by the floating and subsequent sinking of a carcass away from the coast where it apparently died – the "bloat and float" scenario. However, the nearly complete and partially articulated state of the specimen is hard to reconcile with the large marine predators, including both sharks and plesiosaurs, obviously inhabiting the same coastal waters. It should be noted that caudal vertebrae to nearly the tip of the tail, as well as the ultimate phalanges of both arms and one leg, were recovered. Essentially all that was missing were the head and most of the cervical vertebrae – a truly remarkable and exceptional circumstance of large vertebrate preservation. Figure 25.12 shows a reconstruction of *N. graffami*, with the skull and cervical vertebrae cast from the morphologically similar Late Cretaceous Mongolian therizinosaur, *Erlikosaurus andrewsi*.

DISCUSSION

One of the questions we set out to answer in this study was whether large marine reptiles of the Cretaceous Western Interior Seaway were affected by the globally widespread conditions of oceanic anoxia that occurred across the Cenomanian–Turonian (C-T) stage boundary between about 94.2 and 93.5 million years ago (Oceanic Anoxic Event II). As Voigt et al. (2008:66) noted, "The Cenomanian–Turonian Boundary Event is reflected by one of the most extreme carbon cycle perturbations in Earth's history." This event occurred at the

Albright, Gillette, and Titus

height of Late Cretaceous atmospheric greenhouse conditions and was accompanied by "a rapid and profound warming of surface and shallow waters . . . , fast rising sea level . . . , as well as the widespread occurrence of sediments indicating oxygen deficiency in the oceans." Raup and Sepkoski (1986) additionally noted that 24% of all invertebrate marine genera went extinct (but see discussion by Kerr, 2001, of so-called Lazarus taxa with respect to the c-t extinction event). These included deep water faunas as the oxygen minimum zone expanded onto shelf areas.

In contrast, the evidence from the Tropic Shale indicates that the oceanographic events across the c-t boundary had no significant, discernible impact on the large vertebrate fauna, particularly the plesiosaurs, of the Cretaceous Western Interior Seaway (Albright, Gillette, and Titus, 2007b). The growing faunal list for chondrichthyans, osteichthyans, and chelonians reported here seems to support the same conclusion. No marine vertebrates, plesiosaurs or otherwise, were found at, or immediately adjacent to, the c-t boundary (Fig. 25.2), easily located in the field by examination of molluscan faunas (especially an influx of *Mytiloides hattini*) in concert with a locally persistent limonite horizon that closely approximates the stage boundary and an epibole of the benthic oyster *Pycnodonte*.. Surprisingly, it appears that plesiosaur diversity was greater during the early Turonian than at any other time during the Late Cretaceous. This pattern somewhat resembles that noted by Eaton et al. (1997) for brackish-water species, which suffered little if any extinction across the c-t boundary, and for terrestrial vertebrate species, including mammals and dinosaurs, that showed an increase in diversity. Later in the early Turonian, however, during maximum transgression, Eaton et al. (1997) found significant extinction of freshwater aquatic species, such as fish and turtles.

The only published records of plesiosaurs from the Tropic Shale (and equivalent Mancos Shale at Black Mesa) prior to our work were those of Lucas (1994) and Gillette, Hayden, and Titus (1999). The remains of at least five taxa have now been recovered, dramatically altering our understanding of large reptile diversity in the Cretaceous Western Interior Seaway during the interval shortly following Oceanic Anoxic Event II and immediately preceding maximum transgression of the Greenhorn Cyclothem (Leckie et al., 1991, 1998; Prokoph et al., 2001; Keller and Pardo, 2004; Voigt et al., 2008; and references within). Both *Brachauchenius* and *T. bentonianum* are known from sediments deposited in the Seaway before and after the c-t boundary, thus indicating that they suffered few, if any, negative consequences as a result of this global oceanographic event. The early Turonian *Palmulasaurus* also seems to have crossed the boundary unscathed,

as members of the same "subfamily" from Japan and South Dakota (Sato and Storrs, 2000; Schumacher, 2007) occur in upper Cenomanian strata deposited during Oceanic Anoxic Event II. The appearance of *Eopolycotylus rankini* and *Dolichorhynchops tropicensis* further emphasizes the apparently minimal degree of disruption imposed upon the large vertebrate fauna of the Cretaceous Western Interior Seaway over this major oceanographic perturbation.

Myers and Lieberman (2011) examined the temporal and geographic distribution of selected taxa of invertebrates and vertebrates of the eastern and southern reaches of the Cretaceous Western Interior Seaway and concluded that the driving forces that controlled species distribution and range size were abiotic factors such as climate and sea-level changes. Our data on the vertebrates reported here are consistent with their conclusions.

Another question we sought to answer from our studies of the Tropic Shale concerns the apparent absence of other reptiles that are relatively common in Late Cretaceous marine strata deposited in what were the central and eastern regions of the Cretaceous Western Interior Seaway (Carpenter, 1999), especially elasmosaurs (long-necked plesiosaurs) and pterosaurs. To date, all plesiosaurs recovered from the Tropic Shale except *Brachauchenius* belong to the large-headed, short-necked ("pliosauromorph" morphology of O'Keefe, 2001) family, Polycotylidae, although *Brachauchenius*, a member of the Pliosauridae, was also a large-headed, short-necked form. The absence of pterosaurs in the Tropic Shale is less problematic considering the small size and delicate nature of their bones, and in the way in which the Tropic Shale rapidly erodes.

Although highly speculative, perhaps the absence of elasmosaurs in the Tropic Shale fauna, and along the western margin of the Cretaceous Western Interior Seaway in general, was a function of turbidity. As depicted in Fig. 25.13, a reconstruction of North America and the Cretaceous Western Interior Seaway during early Turonian time, the western margin of the seaway was analogous to the present day Atlantic Seaboard, with a mountain range not far inland from a coastal plain. The relative proximity of the coastline to a sediment source, the highland regions to the west, may have resulted in more turbid coastal waters with reduced visibility than the eastern margin of the seaway. There, coastal waters may have been less turbid with higher visibility as they lapped along a low-relief coastal region far distant from the Appalachian Mountains to the east, as deposits of Cretaceous chalk in Kansas and Nebraska attest.

In marine environments where visibility is relatively good, capturing prey (e.g., schooling fish) would be more difficult, requiring greater stealth, camouflage, or some other

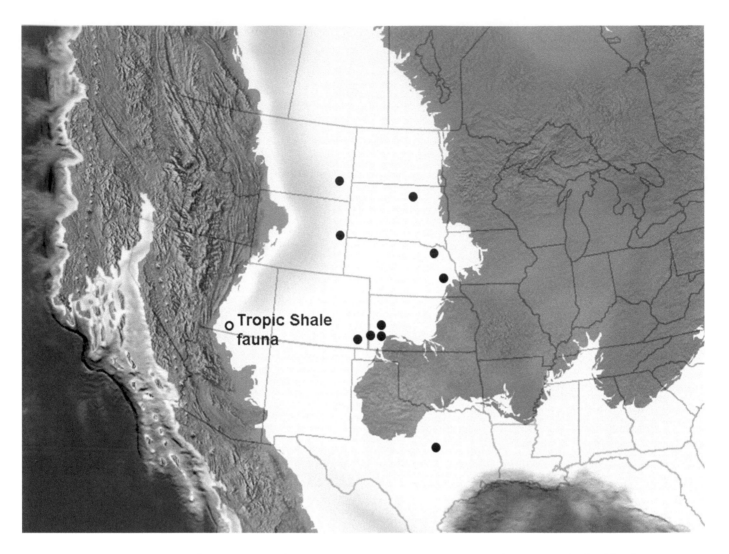

25.13. Paleogeographic reconstruction of North America and Cretaceous Western Interior Seaway approximately 93 mya showing localities of elasmosaurs denoted by closed circles (Carpenter, 1999) and of Tropic Shale plesiosaurs by open circle (map courtesy R. Blakey, Department of Geology, Northern Arizona University).

adaptation, than in more turbid conditions, where visibility would be limited. This supports Bakker's (1993) hypothesis that the highly elongated neck of elasmosaurids was an adaptation for striking prey from a distance, somewhat snake-like or heron-like, in relatively clear water.

Such an adaptation would have no benefit in turbid waters where the predator did not require the element of surprise. Both forms could, and did, inhabit clear-water environments, such as the central and eastern regions of the Cretaceous Western Interior Seaway, but so far, only short-necked forms have been found on the western margin. Obviously, the hypothesis would be falsified by the discovery of an elasmosaur in the Tropic Shale, the most likely candidate of which would be the Early Turonian *Libonectes morgani* (Carpenter, 1999).

The Tropic Shale fauna also includes a relatively diverse turtle fauna consisting of one genus with brackish-water affinities (*Naomichelys* sp.) and at least two species of large marine turtles, plus a relatively rich ichthyofauna with bottom-feeding and open-water pelagic forms like those of contemporary formations on the opposite shores of the seaway.

In addition to this new view of marine reptile diversity along the western margin of the Cretaceous Western Interior Seaway during early Turonian time, the unexpected, but highly fortuitous discovery of a therizinosaurid dinosaur provides a glimpse into this still poorly known terrestrial macrovertebrate fauna of North America. The discoveries of *Nothronychus graffami* in the Tropic Shale and *N. mckinleyi* from the Zuni Basin in western New Mexico provide tantalizing evidence that additional surprises will inevitably follow. That these bizarre new dinosaurs only recently surfaced after over 150 years of extensive prospecting across the American West by some of the world's great museums forces the realization that North America has yet to reveal all its secrets.

Albright, Gillette, and Titus

Research efforts such as those supported by Grand Staircase–Escalante National Monument for the last decade are already proving that the great dinosaur rush of the late 19th and early 20th centuries can no longer be considered a long-past golden era of paleontology. With each new field season, spectacular new discoveries are being made with surprising regularity, to which this volume attests. Those, in turn, prompt continual revision of previously held biological, ecological, and environmental theories, which, supported by ever more accurate and sophisticated data (e.g., isotopic, chronostratigraphic, geochronologic), are dramatically altering and clarifying our view of life in North America during the Late Cretaceous.

ACKNOWLEDGMENTS

We are particularly grateful to the Bureau of Land Management, through Grand Staircase–Escalante National Monument, and to the National Park Service, through Glen Canyon National Recreation Area, for providing financial support through Federal Assistance Agreements that helped fund a large part of the fieldwork and research reported in this chapter. The National Park Service, Bureau of Land Management, and the Utah Geological Survey are additionally thanked for granting the necessary permits required to conduct research on land they administer. We also want to acknowledge Northern Arizona University and the Museum of Northern Arizona for additional research funds. Donations to the MNA Geology Department from D. Jones, F. and C. Thomas, F. Peterson, J. and C. Ossenfort, and G. and B. Congreve kept the research on track and supported student involvement.

Field teams over the last decade include people too numerous to mention (but see acknowledgments in Albright, Gillette, and Titus, 2007a, and Gillette, 2007), although a few must be singled out. Foremost among these are M. Graffam and D. Rankin, who brought many of the discoveries noted above to our attention, and whose tireless commitment to all phases of fieldwork is directly responsible for much of the research reported herein. Our gratitude is extended to other members of the Big Water community, as well, for their interest in the scientific activities we conducted in their backyard, and for their help and hospitality. We also collectively thank the participants of MNA's 2000 Ventures Program. Student involvement included paid and volunteer work by undergraduates and graduate students from Northern Arizona University, Saint Johns University–College of Saint Benedict's, Loyola Marymount University; and the University of Puget Sound. The involvement of students in our Tropic Shale research has resulted in two master of science theses from Northern Arizona University (McCormick, 2006; Schmeisser, 2006) and two undergraduate research projects (Congreve, 2005; Johnson, Gillette, and Horton, 2006); it also stimulated two major MNA exhibits that brought this research to the general public: "Plesiosaur–Terror of the Cretaceous Sea" in 2003–2004, and "Therizinosaur–Mystery of the Sickle Claw Dinosaur" in 2007–2009 (Gillette, 2007). We additionally thank all who helped with those exhibits.

REFERENCES CITED

Adams, D. A. 1997. *Trinacromerum bonneri,* new species, last and fastest pliosaur of the Western Interior Seaway. Texas Journal of Science 49:179–198.

Agassiz, L. 1833–1844 [1833, 1835, 1839, 1843, 1844]. Recherches sur les poisons fossiles. 5 volumes. Imprimerie de Patitpierre, Neuchâtel, Switzerland.

Albright, L. B., III, D. D. Gillette, and A. L. Titus. 2007a. Plesiosaurs from the Upper Cretaceous (Cenomanian–Turonian) Tropic Shale of southern Utah, part 1: new records of the pliosaur *Brachauchenius lucasi.* Journal of Vertebrate Paleontology 27:31–40.

Albright, L. B., III, D. D. Gillette, and A. L. Titus. 2007b. Plesiosaurs from the Upper Cretaceous (Cenomanian–Turonian) Tropic Shale of southern Utah, part 2: Polycotylidae. Journal of Vertebrate Paleontology 27:41–58.

Albright, L. B., III, D. D. Gillette, and A. L. Titus. 2007c. Plesiosaurs from the Upper Cretaceous (Cenomanian–Turonian) Tropic Shale of southern Utah, part 2: Polycotylidae; replacement names for the preoccupied genus *Palmula* and the subfamily Palmulainae. Journal of Vertebrate Paleontology 27:1051.

Albright, L. B., D. D. Gillette, A. L. Titus, and M. H. Graffam. 2001. New pliosaur records from the Cretaceous (Turonian) Tropic Shale of southern Utah. Journal of Vertebrate Paleontology 21:27A.

am Ende, B. A. 1991. Depositional environments, palynology, and age of the Dakota Formation, south-central Utah; pp. 65–83 in J. D. Nations and J. G. Eaton (eds.), Stratigraphy, Depositional Environments, and Sedimentary Tectonics of the Western Margin, Cretaceous Western Interior Seaway. Geological Society of America Special Paper 260.

Bakker, R. T. 1993. Plesiosaur extinction cycles–events that mark the beginning, middle, and end of the Cretaceous; pp. 641–663 in W. G. E. Caldwell and E. G. Kauffman (eds.), Evolution of the Western Interior Basin. Geological Association of Canada Special Paper 39.

Bardack, D. 1965. Anatomy and Evolution of Chirocentrid Fishes. University of Kansas Paleontology Contributions 10.

Bardack, D., and G. Sprinkle. 1969. Morphology and relationships of saurocephalid fishes. Fieldiana Geology 16:297–340.

Bell, G. L., Jr., K. R. Barnes, and M. Polcyn. 2007. Chronostratigraphic distribution of mosasauroids from the Big Bend region of west Texas; p. 8 in M. J. Everhart (ed.), Abstract Booklet, Second Mosasaur Meeting, Sternberg Museum, Hays, Kansas, May 3–6, 2007.

Berg, L. S. 1937. A classification of fish-like vertebrates. Bulletin de l'Académie des Sciences de l'URSS 4:1277–1280.

Berg, L. S. 1958. System der Rezenten und Fossilen Fischartigen und Fische. Hochschulbücher für Biologie, Berlin.

Bonaparte, C. L. 1838. Inconografia della fauna italica, per le quattro classi degli animali vertebrati. 3 volumes, 1834–1841.

Buchy, M.-C., and K. T. Smith. 2007. Basal mosasauroids from the Turonian of Vallecillo, Nuevo León, Mexico; p. 8 in M. J. Everhart (ed.), Abstract Booklet, Second Mosasaur Meeting, Sternberg Museum, Hays, Kansas, May 3–6, 2007.

Cappetta, H. 1980. Les sélaciens du Crétacé Supérieur du Liban. II: Batoides. Palaeontographica Abteilung A 168:149–229.

Cappetta, H. 1987. Chondrichthyes II; Mesozoic and Cenozoic Elasmobranchii. Handbook of

Paleoichthyology 3B. Gustav Fischer Verlag, Stuggtart.

Cappetta, H. 1988. Les Torpédiniformes (Neoselachii, Batomorphii) des phosphates du Maroc. Observations sur la denture des genres actuels. Tertiary Research 10:21–52.

Carpenter, K. 1999. Revision of North American elasmosaurs from the Cretaceous of the Western Interior. Paludicola 2:148–173.

Casier, E. 1947. Constitution et évolution de la racine dentaire des Euselachii, II. Etude comparative des types. Bulletin du Musée Royal d'Histoire Naturelle de Belgique 23:1–32.

Cicimurri, D. J. 2001. Cretaceous elasmobranches of the Greenhorn Formation (Middle Cenomanian–Middle Turonian), Western South Dakota; pp. 27–43 in V. L. Santucci and L. McClelland (eds.), Proceedings of the Sixth Fossil Resource Conference. Geologic Resources Division Technical Report NPS/NRGRD/GRDTR-01/01.

Cobban, W. A., and S. C. Hook. 1984. Mid-Cretaceous molluscan biostratigraphy and paleogeography of southwestern part of the Western Interior, United States; pp. 257–271 in G. E. G. Westermann (ed.), Jurassic–Cretaceous Biochronology and Paleogeography of North America. Geological Association of Canada Special Paper 27.

Cobban, W. A., E. A. Merewether, T. D. Fouch, and J. D. Obradovich. 1994. Some Cretaceous shorelines in the western interior of the United States; pp. 393–414 in M. V. Caputo, J. A. Peterson, and K. J. Franczyk (eds.), Mesozoic Systems of the Rocky Mountain Region, U.S.A. U.S. Geological Survey, Denver, Colorado.

Congreve, C. R. 2005. High density population of the oyster Pycnodonte newberryi (Stanton) in the Tropic Shale (Cenomanian–Turonian) of the Kaiparowits Basin, southern Utah. Annual Meeting of the Geological Society of America, Abstracts with Programs 37(7):134.

Cope, E. D. 1869. Synopsis of the extinct Batrachia, Reptilia and Aves of North America. Transactions of the American Philosophical Society, n.s., 14:1–252.

Cope, E. D. 1870. On the Saurodontidae. American Philosophical Society, Proceedings 11:529–538.

Cope, E. D. 1873. On two new species of Saurodontidae. Proceedings of the Academy of Natural Science, Philadelphia 1873:337–339.

Cope, E. D. 1875. The Vertebrata of the Cretaceous formations of the West. Report of the U.S. Geological Survey Territory 2:1–303.

Cope, E. D. 1887. Zittel's manual of paleontology. American Naturalist 21:1014–1019.

Cragin, F. 1888. Preliminary description of a new or little known saurian from the Benton of Kansas. American Geologist 2:404–407.

Crook, A. R. 1892. Über einige fossile Knochenfische aus der mittleren Kreide von Kansas. Palaeontographica 39:107–124.

de Blainville, H. D. 1835. Description de quelques espèces de reptiles de la Californie, précédée de l'analyse d'un système general d'Erpetologie et d'Amphibiologie. Nouvelles Annales du Muséum (National) d'History Naturaelle, Paris 4:233–296.

Dyman, T. S., W. A. Cobban, L. E. Davis, R. L. Eves, G. L. Pollock, J. D. Obradovich, A. L. Titus, K. I. Takahashi, T. C. Hester, and D. Cantu. 2002.

Upper Cretaceous marine and brackish water strata at Grand Staircase–Escalante National Monument, Utah; pp. 1–36 in W. R. Lund (ed.), Field Guide to Geological Excursions in Southwestern Utah and Adjacent Areas of Arizona and Nevada. Field Trip Guide for the Geological Society of America Rocky Mountain Section, Cedar City, Utah, May 7–9, 2002.

Eaton, J. G. 1999. Vertebrate paleontology of the Iron Springs Formation, Upper Cretaceous, southwestern Utah; pp. 339–343 in D. D. Gillette (ed.), Vertebrate Paleontology in Utah. Utah Geological Survey Miscellaneous Publication 99–1.

Eaton, J. G., J. I. Kirkland, J. H. Hutchison, R. Denton, R. C. O'Neill, and J. M. Parrish. 1997. Nonmarine extinctions across the Cenomanian–Turonian boundary, southwestern Utah, with a comparison to the Cretaceous–Tertiary extinction event. Geological Society of America Bulletin 109:560–567.

Eaton, J. G., J. I. Kirkland, E. R. Gustason, J. D. Nations, K. J. Franczyk, T. A. Ryer, and D. A. Carr. 1987. Stratigraphy, correlation, and tectonic setting of Late Cretaceous rocks in the Kaiparowits and Black Mesa basins; pp. 113–125 in G. H. Davis and E. M. VandenDolder (eds.), Geologic Diversity of Arizona and its Margins: Excursions to Choice Areas. Geological Society of America, 100th Annual Meeting Fieldtrip Guidebook, Arizona Bureau of Geology and Mineral Technology, Special Paper 5.

Elder, W. P. 1985. Biotic patterns across the Cenomanian–Turonian extinction boundary near Pueblo, Colorado; pp. 157–169 in L. M. Pratt, E. G. Kauffman, and F. B. Zelt (eds.), Fine-grained deposits and biofacies of the Cretaceous Western Interior Seaway: Evidence of Cyclic Sedimentary Processes. Society of Economic Paleontologists and Mineralogists Midyear Field Trip Guidebook 4, 1985 Midyear Meeting.

Elder, W. P. 1991. Molluscan paleoecology and sedimentation patterns of the Cenomanian–Turonian extinction interval in the southern Colorado Plateau; pp. 113–137 in J. D. Nations and J. G. Eaton (eds.), Stratigraphy, Depositional Environments, and Sedimentary Tectonics of the Western Margin, Cretaceous Western Interior Seaway. Geological Society of America Special Paper 260.

Elder, W. P., and J. I. Kirkland. 1985. Stratigraphy and depositional environments of the Bridge Creek Limestone Member of the Greenhorn Limestone at Rock Canyon anticline near Pueblo, Colorado; pp. 122–134 in L. M. Pratt, E. G. Kauffman, and F. B. Zelt (eds.), Fine-Grained Deposits and Biofacies of the Cretaceous Western Interior Seaway: Evidence of Cyclic Sedimentary Processes. Society of Economic Paleontologists and Mineralogists Midyear Field Trip Guidebook 4, 1985 Midyear Meeting.

Elliott, D. K., G. V. Irby, and J. H. Hutchison. 1997. Desmatochelys lowi, a marine turtle from the Upper Cretaceous; pp. 243–257 in J. M. Callaway and E. L. Nicholls (eds.), Ancient Marine Reptiles. Academic Press, San Diego, California.

Everhart, M. J. 2000. Gastroliths associated with plesiosaur remains in the Sharon Spring

Member of the Pierre Shale (Late Cretaceous), western Kansas. Transactions of the Kansas Academy of Science 103(1–2):64–75.

Gauthier, J. A. 1986. Saurischian monophyly and the origin of birds; pp. 1–55 in K. Padian (ed.), The Origin of Birds and the Evolution of Flight. Memoirs of the California Academy of Sciences. California Academy of Sciences, San Francisco, California 8.

Gillette, D. D. 2007. Therizinosaur: mystery of the sickle-claw dinosaur. Plateau 4:1–72.

Gillette, D. D., and L. B. Albright. 2003. Further discoveries of Cenomanian–Turonian (Early Late Cretaceous) plesiosaurs from the Tropic Shale, southern Utah. Journal of Vertebrate Paleontology 23 (3, Supplement):55A.

Gillette, D. D., M. C. Hayden, and A. L. Titus. 1999. Occurrence and biostratigraphic framework of a plesiosaur from the Upper Cretaceous Tropic Shale of southwestern Utah; pp. 269–274 in D. D. Gillette (ed.), Vertebrate Paleontology in Utah. Utah Geological Survey Miscellaneous Publication 99–1.

Gillette, D. D., L. B. Albright, A. L. Titus, and M. H. Graffam. 2001a. A Late Cretaceous (Early Turonian) therizinosaurid dinosaur (Therizinosauridae, Theropoda) from the Tropic Shale of southern Utah, U.S.A. Paleobios 21 (2, Supplement):56.

Gillette, D. D., L. B. Albright, A. L. Titus, and M. H. Graffam. 2001b. Discovery and paleogeographic implications of a therizinosaurid dinosaur from the Turonian (late Cretaceous) of southern Utah, U.S.A. Journal of Vertebrate Paleontology (3, Supplement):54A.

Gluckman, L. S. 1958. [On the rate of lamnoid shark evolution.] Doklady Akademia Nauk SSSR 123:568–571. [Russian]

Gregory, H. E., and R. C. Moore. 1931. The Kaiparowits Region, a Geographic and Geologic Reconnaissance of Parts of Utah and Arizona. U.S. Geological Survey Professional Paper 164, 161 pp.

Gustason, E. R. 1989. Stratigraphy and sedimentology of the middle Cretaceous (Albian–Cenomanian) Dakota Formation, southwestern Utah. Ph.D. dissertation, University of Colorado, Boulder, Colorado, 376 pp.

Hampe, O. 2005. Considerations on a Braachauchenius skeleton (Pliosauroidea) from the lower Paja Formation (late Barremian) of Villa de Layva area (Columbia). Mitteilungen aus dem Museum für Naturkunde in Berlin, Geowissenschaftliche Reihe 8:37–51.

Hattin, D. E. 1971. Widespread, synchronously deposited beds in the Greenhorn Limestone (Upper Cretaceous) of Kansas and southern Colorado. American Association of Petroleum Geologists Bulletin 55:110–119.

Hattin, D. E. 1975. Stratigraphy and depositional environment of the Greenhorn Limestone (Upper Cretaceous) of Kansas. Kansas Geological Survey Bulletin 209:1–128.

Hattin, D. E. 1985. Distribution and significance of widespread, time-parallel pelagic limestone beds in Greenhorn Limestone (Upper Cretaceous) of the central Great Plains and southern Rocky Mountains; pp. 28–37 in L. M. Pratt, E. G. Kauffman, and F. B. Zelt (eds.), Fine-Grained Deposits and Biofacies of the Cretaceous Western Interior Seaway: Evidence of Cyclic Sedimentary Processes. Society of Economic Paleontologists and Mineralogists Midyear Field Trip Guidebook 4, 1985 Midyear Meeting.

Hay, O. P. 1898. Notes on species of *Ichthyodectes,* including the new species *I. cruentus,* and on the related and herein established genus *Gillicus.* American Journal of Science, ser. 4, 6:225–232.

Hay, O. P. 1908. The Fossil Turtles of North America. Carnegie Institution of Washington, Washington, D.C., 568 pp.

Hintze, L. F. 1988. Geologic history of Utah. Brigham Young University Geology Studies Special Publication 7:1–202.

Hirayama, R. 1997. Distribution and diversity of Cretaceous chelonioids; pp. 225–241 in J. M. Callaway and E. L. Nicholls (eds.), Ancient Marine Reptiles. Academic Press, San Diego, California.

Huxley, T. H. 1880. On the applications of the laws of evolution to the arrangement of the Vertebrata and more particularly of the Mammalia. Proceedings of the Zoological Society of London, 649–662.

Jaekel, O. 1894. Die eocanen selachies vom Monte Bolca. Ein Beitrag zur Morphogenied er Wirbelthiere 4:1–176.

Jaekel, O. 1898. Ueber *Hybodus.* Sitzungsberichte der Gesellshaft naturforschender Freunde zu Berlin 1898:135–146.

Johnson, B. W., Gillette, D. D., and T. W. Horton. 2006. Stable isotope stratigraphy of a therizinosaur-bearing section of the Tropic Shale near Big Water, Utah. 102nd Annual Meeting of the Cordilleran Section, GSA, 81st Annual Meeting of the Pacific Section, AAPG, and the Western Regional Meeting of the Alaska Section. GSA Abstracts with Program 38(5):9.

Jordan, D. S. 1898. Description of a species of fish (*Mitsukurina owstoni*) from Japan. The type of a distinct family of lamnoid sharks. California Academy of Science, Proceedings 1:199–204.

Kauffman, E. G., J. E. Powell, and D. E. Hattin. 1969. Cenomanian–Turonian facies across the Raton Basin. Mountain Geologist 6:93–118.

Keller, G., and A. Pardo. 2004. Age and paleoenvironment of the Cenomanian–Turonian global stratotype section and point at Pueblo, Colorado. Marine Micropaleontology 51:95–128.

Kerr, R. A. 2001. Mass extinctions face downsizing, extinction. Science 293:1037.

Kirkland, J. I. 1991. Lithostratigraphic and biostratigraphic framework for the Mancos Shale (Late Cenomanian to Middle Turonian) at Black Mesa, northeastern Arizona; pp. 85–112 in J. D. Nations and J. G. Eaton (eds.), Stratigraphy, Depositional Environments, and Sedimentary Tectonics of the Western Margin, Cretaceous Western Interior Seaway. Geological Society of America Special Paper 260.

Kirkland, J. I. 1996. Paleontology of the Greenhorn cyclothem (Cretaceous: Late Cenomanian–Middle Turonian) at Black Mesa, northeastern Arizona. New Mexico Museum of Natural History and Science Bulletin 9:1–132.

Kirkland, J. I., and D. G. Wolfe. 2001. The first definitive Therizinosaurid (Dinosauria; Theropoda) from North America. Journal of Vertebrate Paleontology 21:410–414.

Leckie, R. M., M. G. Schmidt, D. Finkelstein, and R. Yuretich. 1991. Paleoceanographic and paleoclimatic interpretations of the Mancos Shale (Upper Cretaceous), Black Mesa Basin, Arizona; pp. 139–152 in J. D. Nations and J. G. Eaton (eds.), Stratigraphy, Depositional Environments, and Sedimentary Tectonics of the Western Margin, Cretaceous Western Interior Seaway. Geological Society of America Special Paper 260.

Leckie, R. M., R. F. Yuretich, O. L. O. West, D. Finkelstein, and M. Schmidt. 1998. Paleoceanography of the southwestern Western Interior Sea during the time of the Cenomanian–Turonian boundary (Late Cretaceous); pp. 101–126 in W. E. Dean and M. A. Arthur (eds.), Stratigraphy and Paleoenvironments of the Cretaceous Western Interior Seaway, U.S.A. SEPM Concepts in Sedimentology and Paleontology 6.

Leidy, J. 1868. Notice of American species of *Ptychodus.* Proceedings of the Academy of Natural Sciences of Philadelphia 20:205–208.

Leidy, J. 1870. Remarks on ichthyodorulites and on certain fossil Mammalia. Academy of Natural Sciences, Proceedings 21:12–13.

Lucas, S. G. 1994. Late Cretaceous plesiosaurs (Euryapsida: Plesiosauroidea) from the Black Mesa Basin, Arizona, U.S.A. Journal of the Arizona–Nevada Academy of Science 28:41–45.

Maleev, E. A. 1954. New turtle-like reptile in Mongolia. Priroda 3:106–108. [Russian]

Marcou, J. 1858. Geology of North America. Zurcher and Furrer, Zurich, 144 pp.

Marsh, O. C. 1881. Principal characters of American Jurassic dinosaurs, part V. American Journal of Science 21:417–423.

McCormick, K. A. 2006. Depositional environments and taphonomy of vertebrate sties in the upper Cretaceous Tropic Shale, southern Utah. M.S. thesis, Northern Arizona University, Flagstaff, Arizona, 90 pp.

Mueller, B. D., S. Chatterjee, W. Cornell, D. Watkins, and G. L. Gurtler. 2007. The first mosasaur (Reptilia: Squamata) from the Llano Estacado of Northwest Texas; p. 17 in M. J. Everhart (ed.), Abstract Booklet, Second Mosasaur Meeting, Sternberg Museum. Hays, Kansas, May 3–6, 2007.

Myers, C. E., and B. S. Lieberman. 2011. Sharks that pass in the night: using geographical information systems to investigate competition in the Cretaceous Western Interior Seaway. Proceedings of the Royal Society B 278:681–689.

Nursall, J. R. 1996. The phylogeny of pycnodont fishes; pp. 125–152 in G. Arratia and G. Viohl (eds.), Mesozoic Fishes–Systematics and Palaeocology. Verlag Dr. Friedrich Pfeil, Munich.

Obradovich, J. D. 1993. Cretaceous time scale; pp. 379–396 in W. E. G. Caldwell and E. G. Kauffman (eds.), Evolution of the Western Interior Basin. Geological Association of Canada Special Paper 39.

Ogg, J. G., F. P. Agterberg, and F. M. Gradstein. 2004. The Cretaceous Period; pp. 344–383 in F. M. Gradstein, J. G. Ogg, and A. G. Smith (eds.), A Geologic Time Scale, 2004. Cambridge University Press, Cambridge.

O'Keefe, F. R. 2001. A cladistic analysis and taxonomic revision of the Plesiosauria (Reptilia:

Sauropterygia). Acta Zoologica Fennica 213:1–63.

O'Keefe, F. R. 2008. Cranial anatomy and taxonomy of *Dolichorhynchops bonneri* new combination, a polycothylid (Sauropterygia:Plesiosauria) from the Pierre Shale of Wyoming and South Dakota. Journal of Vertebrate Paleontology 28:664–676.

Olesen, J. 1991. Foraminiferal biostratigraphy and paleoecology of the Mancos Shale (Upper Cretaceous), southwestern Black Mesa, Arizona; pp. 153–166 in J. D. Nations and J. G. Eaton (eds.), Stratigraphy, Depositional Environments, and Sedimentary Tectonics of the Western Margin, Cretaceous Western Interior Seaway. Geological Society of America Special Paper 260.

Oppel, M. 1811. Die Ordnung, Familien und Gattungen der Reptilien, als Prodrom einer Natureschichte derselben. Lindauer, Munich, 87 pp.

Owen, R. 1860. On the orders of fossil and Recent Reptilia, and their distribution through time. Report of the British Association for the Advancement of Science 29:153–166.

Peterson, F. 1969. Cretaceous sedimentation and tectonism in the Kaiparowits region, Utah. U.S. Geological Survey Open-File Report 69–202, 259 pp.

Pratt, L. M., E. G. Kauffman, and F. B. Zelt (eds.). 1985. Fine-grained deposits and biofacies of the Cretaceous Western Interior Seaway: evidence of cyclic sedimentary processes. Society of Economic Paleontologists and Mineralogists Midyear Field Trip Guidebook 4, 1985 Midyear Meeting, 249 pp.

Prokoph, A., M. Villeneuve, F. P. Agterberg, and V. Rachold. 2001. Geochronology and calibration of global Milankovitch cyclicity at the Cenomanian–Turonian boundary. Geology 29:523–526.

Raup, D. M., and J. J. Sepkoski. 1986. Periodic extinction of families and genera. Science 231:833–836.

Ross, M. R. 2009. Charting the Late Cretaceous Seas: mosasaur richness and morphological diversification. Journal of Vertebrate Paleontology 29:409–416.

Sato, T. 2005. A new polycotylid plesiosaur (Reptilia: Sauropterygia) from the Upper Cretaceous Bearpaw Formation in Saskatchewan, Canada. Journal of Plaeontology 79:969–980.

Sato, T., and G. Storrs. 2000. An early polycotylid plesiosaur (Reptilia: Sauropterygia) from the Cretaceous of Hokkaido, Japan. Journal of Paleontology 74:907–914.

Schmeisser, R. L. 2006. Description of a new species of plesiosaur from the Upper Cretaceous Tropic Shale, southern Utah. M.S. thesis, Northern Arizona University, Flagstaff, Arizona, 199 pp.

Schmeisser, R. L. 2008. A new species of polycotylid plesiosaur from the early Turonian of Utah: extending the stratigraphic range of North American *Dolichorhynchops.* Journal of Vertebrate Paleontology 28:137A.

Schmeisser, R. L., and D. D. Gillette. 2009. Unusual occurrence of gastroliths in a polycotylid plesiosaur from the Upper Cretaceous Tropic Shale, southern Utah. Palaios 24:453–459.

Schmeisser McKean, R. L. 2012. A new species of polycotylid plesiosaur (Reptilia: Sauropterygia) from the Lower Turonian of Utah: extending the stratigraphic range of *Dolichorhynchops*. Cretaceous Research 34:184–199.

Schumacher, B. A. 2007. A new polycotylid plesiosaur (Reptilia; Sauropterygia) from the Greenhorn Limestone (Upper Cretaceous; lower upper Cenomanian), Black Hills, South Dakota; pp. 133–146 in J. E. Martin and D. C. Parris (eds.), The Geology and Paleontology of the Late Cretaceous Marine Deposits of the Dakotas. Geological Society of America Special Paper 427.

Schumacher, B., and M. J. Everhart. 2005. A stratigraphic and taxonomic review of pleisosaurs from the old "Fort Benton Group" of central Kansas: a new assessment of old records. Paludicola 5:33–54.

Seeley, H. 1847. Note on some genetic modifications of the plesiosaurian pectoral arch. Journal of the Geological Society of London 30:436–449

Shimada, K., C. Rigsby, and S. Kim. 2009. Partial skull of Late Cretaceous durophagous shark, *Ptychodus occidentalis* (Elasmobranchii: Ptycodontidae), from Nebraska, U.S.A. Journal of Vertebrate Paleontology 29:336–349.

Voigt, S., J. Erbacher, J. Mutterlose, W. Weiss, T. Westerhold, F. Wiese, M. Wilmsen, and T. Wonik. 2008. The Cenomanian–Turonian of the Wunstorf section (North Germany): global stratigraphic reference section and new orbital time scale for Oceanic Anoxic Event 2. Newsletters on Stratigraphy 43:65–89.

VonLoh, J. P., and G. L. Bell. 1998. New records of Turonian mosasauroids from the western United States; pp. 15–28 in J. E. Martin, J. W. Hoganson, and R. C. Benton (eds.), Proceedings of the Fifth Conference on Fossil Resources, October 1998, Dakoterra 5.

Whitley, G. P. 1939. Taxonomic notes on sharks and rays. Australian Journal of Zoology 9:227–262.

Whittle, C. H., and M. J. Everhart. 2000. Apparent and implied evolutionary trends in lithophagic vertebrates from New Mexico and elsewhere; pp. 75–82 in S. G. Lucas and A. B. Heckert (eds.), Dinosaurs of New Mexico. New Mexico Museum of Natural History and Science Bulletin 17.

Williamson, T. E., J. I. Kirkland, and S. G. Lucas. 1993. Selachians from the Greenhorn Cyclothem ("Middle" Cretaceous: Cenomanian–Turonian), Black Mesa, Arizona, and the paleogeographic distribution of Late Cretaceous selachians. Journal of Paleontology 67:447–474.

Williamson, T. E., S. G. Lucas, and J. I. Kirkland. 1991. The Cretaceous elasmobranch *Ptychodus decurrens* from North America. Geobios 24:1–4.

Williston, S. W. 1894. A new turtle from the Benton Cretaceous. Kansas University Quarterly 3:5–18.

Williston, S. W. 1900. Cretaceous fishes, selachians and pycnodonts. University Geological Survey of Kansas 6:237–256.

Williston, S. W. 1903. North American plesiosaurs, part 1. Field Columbian Museum, Geological Series 73(2):1–77.

Williston, S. W. 1908. North American plesiosaurs, *Trinacromerum*. Journal of Geology 16:715–736.

Woodward, A. S. 1889. Catalogue of the Fossil Fishes in the British Museum (Natural History). British Museum (Natural History), London, 475 pp.

Zangerl, R. 1981. Chondrichthyes I–Paleozoic Elasmobranchii, 3a; pp. 1–115 in H.-P. Shultze (ed.), Handbook of Paleoichthyology. Gustav Fischer Verlag, Stuttgart.

Zangerl, R., and R. E. Sloan. 1960. A new specimen of *Desmatochelys lowi* Williston. A primitive cheloniid sea turtle from the Cretaceous of South Dakota. Fieldiana Geology 14:7–40.

Zanno, L. E., D. D. Gillette, L. B. Albright, and A. L. Titus. 2009. A new North American therizinosaurid and the role of herbivory in "predatory" dinosaur evolution. Proceedings of the Royal Society B, Biological Sciences 276:3505–3512.

Paleontological Overview and Taphonomy of the Middle Campanian Wahweap Formation in Grand Staircase–Escalante National Monument

26

Donald D. DeBlieux, James I. Kirkland, Terry A. Gates, Jeffrey G. Eaton, Michael A. Getty, Scott D. Sampson, Mark A. Loewen, and Martha C. Hayden

THE WAHWEAP FORMATION PRESERVES THE MOST DI-verse middle Campanian terrestrial fauna in North America, based largely on information gained by the study of microvertebrate fossils collected by wet screen washing. These studies have documented a minimum of five freshwater shark species, three freshwater ray species, eight bony fish species, 11 amphibian species, 10 turtle species, four lizard taxa, five crocodilian taxa, 15 dinosaur taxa, and 33 mammal species. However, many of the turtles, crocodilians, and dinosaurs require more complete skeletal material for specific identification.

Teams from the Utah Geological Survey, University of Utah Museum of Natural History, and the Grand Staircase–Escalante National Monument have been conducting a paleontological inventory of the Wahweap Formation in the Kaiparowits Basin. In addition to providing data on the distribution of paleontological resources, this study has identified and recovered specimens that are adding to our knowledge of terrestrial faunas for the poorly known middle Campanian. Both trionychid and baenid turtle shells have been recovered and are presently under study. Cranial remains of a new species of long-horned centrosaurine ceratopsian have been described and are providing significant information about the evolution of the ceratopsid dinosaurs. A number of associated hadrosaurid skeletons have been identified in the field and several bonebeds have been excavated. A partial hadrosaurine hadrosaur skull has been identified as a new species congeneric with a taxon from Montana. An isolated hadrosaur maxilla is the only lambeosaurine material known from the Wahweap Formation and likely represents a new species. An isolated, but distinctive skull roof of a juvenile pachycephalosaur has been collected. Additionally, carnivorous dinosaur dental remains and ankylosaur scutes have been identified at a number of sites, but nothing diagnostic has yet been found.

Patterns of fossil distribution in the Wahweap Formation are beginning to emerge as a result of our work. A taphonomic analysis indicates that most vertebrate fossils are found in sandstone channel lag deposits and represent time-averaged accumulations of terrestrial and aquatic vertebrates. Mudstone sites, some of which preserve associated skeletons, are present but rare, and difficult to find because of rapid loss of bone on the surface as a result of weathering and the activity of plants. A number of dinosaur tracks and tracksites have been found. Most tracks occur as natural casts in thin sandstone units that likely formed as crevasse splay deposits filled in tracks made on the underlying mud. The majority of tracks recognized are tridactyl, likely belonging to hadrosaurs. Comparison of the Wahweap vertebrate fauna with those of the Aquilan and Judithian North American Land Mammal Ages indicates that a unique intermediate fauna is present.

INTRODUCTION

Between 2001 and 2010, teams from the Utah Geological Survey (UGS), Utah Museum of Natural History (UMNH), and Grand Staircase–Escalante National Monument conducted paleontological inventory of the middle Campanian Wahweap Formation on the Kaiparowits Plateau in Grand Staircase–Escalante National Monument. The UGS focused most of their investigations in the lower part of the formation (lower and middle members) along the southern and southeastern portion of the outcrop belt using the Smoky Mountain (BLM 230) and Heads of the Creeks (BLM 231) roads to provide access (Fig. 26.1). The UMNH worked primarily in the upper part of the formation (upper and capping sandstone members). The following is a summary of observations meant to characterize the overall paleontological makeup of the Wahweap Formation.

GEOLOGY

The geology of the Wahweap Formation is summarized in Chapters 2 and 4 of this volume. In general it can be characterized as being composed of drab mudstones, claystone,

26.1. A view of the Wahweap Formation along the Heads of the Creeks Road in Wesses Canyon, Grand Staircase–Escalante National Monument. The majority of rock in this picture is that of the middle member, with the upper member forming the top of the buttes in the distance.

siltstones, sandstones, and conglomerates, deposited in stream, levee, overbank, and lacustrine facies on a coastal plain near the edge of the Western Interior Seaway. It ranges between 360–460 m in thickness on the Kaiparowits Plateau and lies between the underlying Straight Cliffs Formation and overlying Kaiparowits Formation (Gregory and Moore, 1931).

Eaton (1991) redescribed in detail a reference section of the Wahweap Formation at Reynolds Point and divided it into four informal members: the lower, middle, upper, and capping sandstone. These members can be distinguished in the field by the relative ratio of sandstone to mudstone. The lower member is primarily composed of mudstone but is characterized by several laterally extensive amalgamated sandstone bodies thought to be deposited in a low-accommodation-space setting. The middle member is also mud dominated with sandstone units but lacks any continuous sandstone bodies and was deposited in a high-accommodation-space

setting. The upper member resembles the lower member, with several laterally extensive sandstone bodies (Eaton, 1991; Pollock, 1999; Jinnah et al., 2009). The capping sandstone member consists almost entirely of pure quartz sandstone and is thought to have been deposited as braided streams with sediments supplied from the west instead of from the southwest as in the lower three more feldspathic members (Goldstrand, 1992; Pollock, 1999; Lawton, Pollock, and Robison, 2003). The rocks of the Wahweap Formation are thought to have been deposited in a wet, seasonal climate (Jinnah et al., 2009, this volume, Chapter 4). For a more detailed review of Wahweap stratigraphy, depositional environments and age, see Jinnah, this volume, Chapter 4.

Eaton (1991) revised Peterson's (1969) lower contact of the Wahweap to coincide with the top of the broad bench of conglomeratic sandstone used by Gregory and Moore (1931). Thus, many of the fossil sites from the basal Wahweap Formation plot on existing maps in the Drip Tank Member

DeBlieux et al.

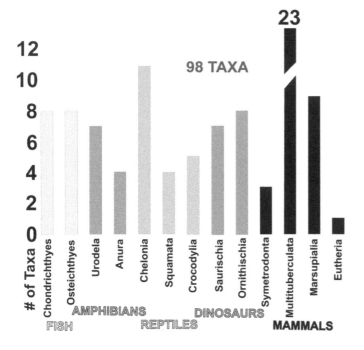

26.2. Wahweap fauna from Eaton and Cifelli (1988), Cifelli (1990b, 1990d), Eaton et al. (1999), Eaton (2002), and Parrish (1999). There were 57 identified taxa as of 1999.

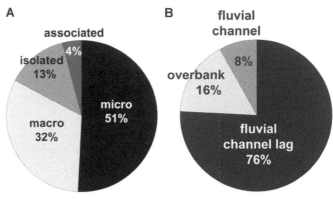

26.3. Relative frequencies of (A) taphonomic modes and (B) facies for 300 vertebrate localities in the Wahweap Formation of Grand Staircase–Escalante National Monument.

of the Straight Cliffs Formation (Doelling and Graham, 1972; Sargent and Hansen, 1982; Doelling and Willis, 2008). Similarly, Eaton (1991) restricted the definition of the upper contact of the capping sandstone member to that of Gregory and Moore (1931) and Peterson (1969). Subsequent studies of Wahweap Formation stratigraphy have relied on Eaton's (1991) described sections and nomenclature (Pollock, 1999; Lawton, Pollock, and Robison, 2003; Jinnah et al., 2009).

Pollock (1999) conducted the first stratigraphic research focused primarily on the Wahweap Formation. He documented that the lower three members of the Wahweap Formation are subfeldspathic and have a source in central Arizona, and that current data indicate sediment transport to the north–northeast parallel to the thrust belt, whereas the capping sandstone member is a quartz arenite with current data indicating sediment transport across the tectonic grain to the east and east–southeast. Comparisons with the Tarantula Mesa Sandstone in the Henry basin to the east (Smith, 1983; Eaton, 1990; Jinnah, this volume, Chapter 4) indicate that it makes sense stratigraphically to raise the capping sandstone member to the rank of formation and call it the Tarantula Mesa Sandstone, particularly because this lithologic unit is mapped as a discrete unit on the Smoky Mountain 30' by 60' geologic quadrangle map (Doelling and Willis, 2008).

Recent work on the geology of the Wahweap and Kaiparowits Formations, initiated since we began our project, has refined our understanding of the depositional environment and age of the Wahweap. The use of detrital zircons

and the radiometric dating of bentonites indicate the age of deposition of the Wahweap Formation at between 81 and 76 million years on the basis of ages in and just above the Wahweap Formation (Roberts, Deino, and Chan, 2005; Roberts, 2007; Jinnah et al., 2009), placing the Wahweap Formation in the middle Campanian. This means that the Wahweap Formation was deposited penecontemporaneously with several other fossil-bearing formations in western North America, including the lower Two Medicine Formation of Montana, the Foremost and Oldman Formations of Alberta, and the lower Judith River Formation of Montana (the upper part of which forms the basis of the type Judithian North American land mammal ages, NALMA).

PREVIOUS PALEONTOLOGICAL WORK

An extensive microvertebrate assemblage was collected from the Wahweap Formation (and other Late Cretaceous formations on the Kaiparowits Plateau) by Rich Cifelli at the Museum of Northern Arizona (MNA) and Jeff Eaton at the University of Colorado (UCM) by bulk sampling and wet screen washing. This research, which focuses on fossil mammals, has been continued by Cifelli, now working at the Oklahoma Museum of Natural History (OMNH) at the University of Oklahoma, and Eaton at Weber State University, with the UMNH as repository. These pioneering efforts resulted in the recognition of the most diverse pre-Judithian Campanian terrestrial fauna known in North America (Fig. 26.2; Eaton and Cifelli, 1988; Cifelli, 1990a, 1990b, 1990c, 1990d; Eaton, 2002). Extensive collections of Wahweap fossil mammals and their associated microvertebrate fauna are now housed at MNA, UCM, OMNH, and UMNH. These collections are an important scientific resource that continues to be studied and are producing a great deal of paleontological information, as can be seen by this volume.

26.4. Excavation at the Tibbet Springs Quarry (UMNH VP Loc. 1212). (A) Removing rock overburden with gas-powered jackhammer; (B) working with hand tools; (C) a hadrosaur tibia in place; (D) a partial hadrosaur dentary and tooth; (E) Jim Kirkland working the site; (F) using a gas-powered rock saw to pedestal a hadrosaur tibia.

These studies documented productive microvertebrate localities in the Wahweap Formation, including in the lower and middle members (Cifelli, 1990b) as well as in the upper member (Eaton, 1991), but not in the capping sandstone member. A number of isolated theropod teeth have been recovered from sandstone channel lag microsites. These teeth indicate the presence of dromaeosaurine and velociraptorine dromaeosaurids, in addition to troodontids

and tyrannosaurids (Parrish, 1999). The teeth of nodosaurid and ankylosaurid ankylosaurs are known from the Wahweap Formation (Eaton et al., 1999). We have collected solid osteoderms that are diagnostic of nodosaurids from a number of sites, but no other skeletal material has yet been discovered.

Dinosaur fossils from middle Campanian–aged rocks in North America are relatively rare in comparison with those

DeBlieux et al.

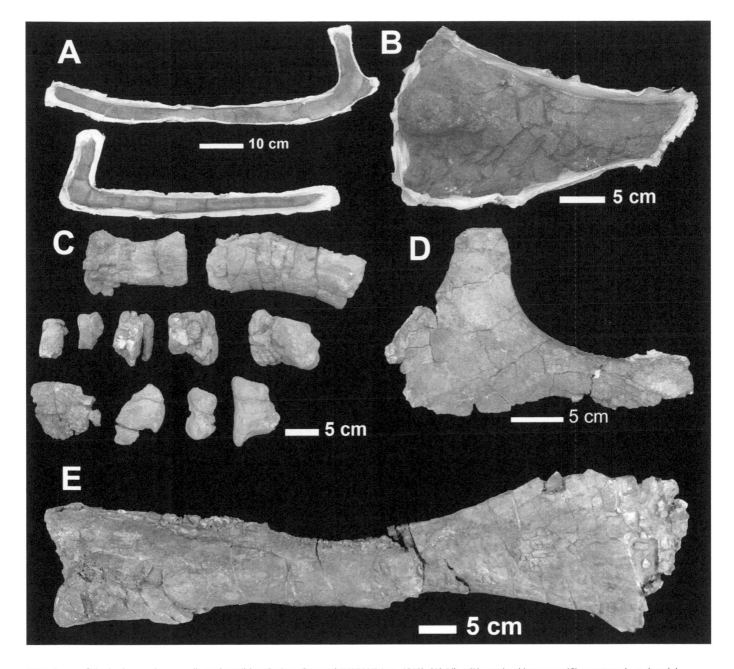

26.5. Some of the hadrosaur bones collected at Tibbet Springs Quarry (UMNH VP Loc. 1212). (A) Ribs; (B) proximal humerus; (C) metatarsals and pedal phalanges; (D) sternal; (E) tibia.

of the late Campanian, and any data from the Wahweap is significant. A leptoceratopsid-grade basal neoceratopsian *Cerasinops hodgkissi* has recently been described from the lower Two Medicine Formation in Montana (Chinnery and Horner, 2007) from rocks of approximately 80 Ma (Foreman et al., 2008). Ryan (2007) named the basal centrosaurine *Albertaceratops nesmoi* from a locality in the upper Oldman Formation of Alberta, Canada, that is middle Campanian in age. The hadrosaurine hadrosaurid *Gryposaurus latidens* (Horner, 1992) is known from the middle of the Two Medicine Formation at approximately 77 Ma (Foreman et al., 2008), and a new hadrosaurine hadrosaurid is known from the lower Two Medicine Formation (Gates et al., this volume, Chapter 19).

MACROVERTEBRATE TAPHONOMY

Our project has sought to assess the taphonomic and sedimentologic attributes of paleontological localities in the Wahweap Formation in order to better understand the distribution of sites and the environments in which these fossils were preserved. The analysis of Wahweap Formation taphofacies presented here is contrasted with the detailed studies of the taphonomy of the overlying Kaiparowits Formation by

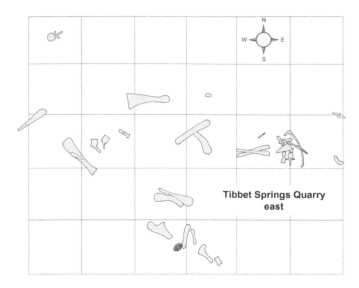

26.6. Quarry map of Tibbet Springs Quarry east (UMNH VP Loc. 1212). Each square represents 1 m².

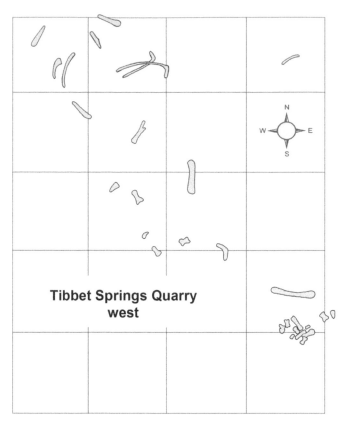

Tibbet Springs Quarry west

26.7. Quarry map of Tibbet Springs Quarry west (UMNH VP Loc. 1212). Each square represents 1 m².

Roberts, Chan, and Sampson (2003), Roberts, Deino, and Chan (2005), Roberts (2007), Getty et al. (2010), and Roberts et al. (this volume). The facies architecture and depositional environments that make up the Wahweap Formation are similar to those found in the overlying Kaiparowits Formation, although found in different proportions, and we see a somewhat similar pattern of fossil distribution and preservation. We analyzed data collected by us and others that are included in the Utah Paleontological Locality Database, which presently has 572 paleontological localities listed for the Wahweap Formation in Grand Staircase–Escalante National Monument. Of these, 444 (77%) are vertebrate bone sites, 89 (16%) are plant sites only, 21 (5%) are vertebrate tracksites, and 18 (4%) are invertebrate only. Of these sites, we analyzed data from 300 vertebrate localities that included sufficient taxonomic, sedimentologic, and taphonomic information.

Getty et al. (2010, based on Eberth and Currie, 2005) identified five primary taphonomic modes of vertebrate fossil preservation in the Kaiparowits Formation: mode A, articulated macrosites; mode B, associated macrosites; mode C, isolated macrosites; mode D, macrosfossil bonebeds; and mode E, vertebrate microfossil assemblages.

Articulated macrosites (mode A) have not yet been discovered in the Wahweap Formation. One locality, UMNH Loc. 1212 (Tibbet Springs Quarry), contains closely associated hadrosaur bones that almost fit the criterion of being articulated. A close association of pedal bones was collected but they had been dispersed enough to no longer be in anatomical position. This is also one of the only sites known in the formation that can be considered a multitaxic macrofossil bonebed (mode D). This is in contrast to many Late

Cretaceous Western Interior Basin formations that contain numerous localities with articulated remains of single or multiple individuals.

Associated macrosites (mode B) are also rare in the formation, and only 14 localities fall into this category and thus account for only 4% of vertebrate sites (Fig. 26.3). This is despite many years of extensive prospecting for associated vertebrate remains. Several of these associated macrosites are discussed below.

Isolated macrosites (mode C) consisting of single elements that are not found with other associated elements make up 13% of vertebrate localities. These are typically single skeletal elements preserved in fluvial channel sandstones, sandy mudstones, or overbank mudstones. Isolated postcranial skeletal elements in sandstone are usually not collected because they are typically not diagnostic beyond family level and difficult to collect. Isolated elements in mudstones have been exposed to erosion and are generally too fragmentary to be diagnostic.

Vertebrate microfossil assemblages (mode E) make up the vast majority of vertebrate localities in the Wahweap Formation. These accumulations of bones and teeth are most commonly found in intraformational conglomerate, where they

DeBlieux et al.

26.8. Overview of Jim's Hadrosaur Site (UMNH VP Loc. 1209) showing the approximate location of the contact between the lower and middle members.

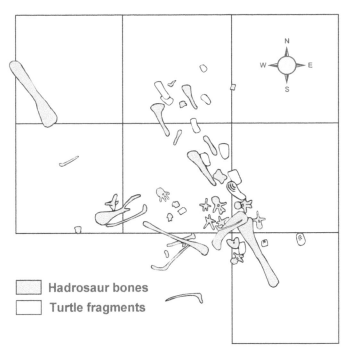

☐ **Hadrosaur bones**
☐ **Turtle fragments**

26.9. Quarry map of Jim's Hadrosaur Site (UMNH VP Loc. 1209). Each square represents 1 m².

occur along with mudstone rip-up and overbank carbonate clasts deposited as channel lags. These sites are typified by teeth, scales, scutes, rounded water-worn bone, wood fragments, and larger, more complete (but isolated) skeletal elements. Most of these sites are considered microsites because of the large number of small elements (<10 cm), but larger elements are present in many of these deposits. Microsites account for more than 50% of vertebrate sites in the formation (Fig. 26.3). Moreover, most of the sites which are considered macrosites (32% of the total) are also found in fluvial channel lag deposits that also contain small elements, so this distinction is somewhat arbitrary. Sandstone-hosted vertebrate localities account for 84% of all localities. Wahweap Formation microsites typically contain a large proportion of aquatic and semiaquatic vertebrates including fish (gar scales and teeth most common), turtles (scutes), crocodiles (teeth and scutes) and rare coprolites that likely represent time-averaged accumulations of skeletal material reworked from floodplain concentrations (Rogers and Kidwell, 2007; Roberts et al., this volume). Aquatic invertebrates (unionid bivalves, gastropods, and crabs) are associated with a little over 10% of these sites. Turtle and crocodile fossils were found in 54% of dinosaur bone sites with turtles being more common than crocodiles. Fish fossils were identified in 20% of dinosaur bone localities and invertebrates found in 12%. The large number of fossils of aquatic and semiaquatic vertebrates in Wahweap Formation rocks fits well with sedimentological evidence that the Wahweap Formation was deposited in a seasonally wet fluvial environment (Jinnah et al., 2009; Jinnah, this volume). Many groups of Wahweap fossil vertebrates such as mammals, amphibians, reptiles, and fish are known almost exclusively from channel lag microsites. In many cases, these fossils are incorporated into well-cemented rock at the base of channel sandstones where collection and processing is difficult. In places where this sandstone has eroded and broken down, this material can be collected for wet screen processing.

Microsites are also common in thin sandstone and conglomerate layers (lenses) that likely represent material washed onto the floodplain from channel concentrations during flooding events (splays). There are also microsites in which small bones and teeth have been concentrated in sandy mudstones; although these sites are more difficult to locate, they are more easily collected and processed. Microsites in carbonaceous mudstones indicative of overbank ponds, oxbow lakes, and wetlands are present but rare. The predominance of fluvial microsites in the Wahweap Formation is similar to that of many Late Cretaceous formations of the Western Interior Basin but is in contrast to the overlying Kaiparowits Formation in which pond microsites are more common than fluvial microsites (Roberts et al., this volume). Some of the larger vertebrate fossils that we have discovered and collected, such as the ceratopsian skulls described below, were found in channel lag deposits, representing large skeletal elements deposited in stream and river channels.

Our project spent a great deal of time prospecting for sites in mudstones, especially in the middle member. A number

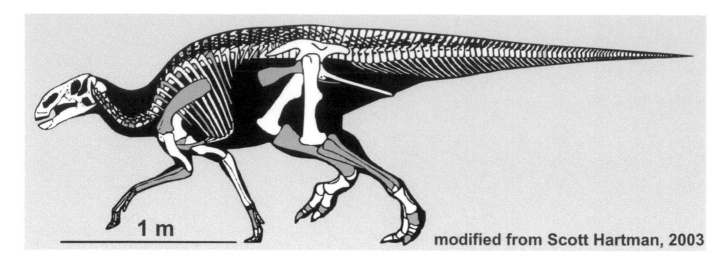

26.10. Skeleton of a hadrosaur showing the bones (grey) recovered from Jim's Hadrosaur Site (UMNH VP Loc. 1209).

of dinosaur macrosites were found in mudstones during our work yet these accounted for only 16% of recorded sites. Most of these sites contained only one or a few isolated bones, and no articulated specimens were found. At many sites, especially those on flats and gentle slopes, modern plant roots were found to have infiltrated and destroyed much of the bone. Wahweap fossil bone tends to be poorly mineralized, such that it erodes into small fragments soon after being exposed to weathering on the surface. In order to be found, the bone must be on a slope steep enough to prevent plants from taking root but not so steep that the bone fragments erode down the slope as soon as they are exposed. Only a couple mudstone sites preserve associated skeletons with a large number of elements. The amalgamated sandstone bodies in the lower and upper members indicate extensive reworking of floodplain sediments and help explain the lack of articulated and associated vertebrate skeletons in these members. The increased sedimentation rate in the middle member allowed the preservation of more extensive mudstone sediments. Several hypotheses might explain the relative lack, as compared with other Western Interior Basin formations, of associated skeletons in these rocks. Longer periods of subaerial exposure in a wet, humid environment and pedogenesis may have contributed to the weathering and breakdown of bone. Conversely, our examination of middle the member in the southeastern part of the outcrop belt indicates that, although there is evidence of some paleosol formation (layered siderite, blocky fracturing, root casts, and other paleosol features), mature paleosols are rare, and many of the fine-grained rocks represent lateral-accretion sediments, and that there continued to be extensive reworking of floodplain

sediments. Additional sedimentological study is needed to test these hypotheses and to evaluate the regional differences in various parts of the flood basin.

To date, thousands of acres have been prospected, yet this area represents only a small percentage of the available outcrop, so a great deal of potential for further discoveries remains. It also should be noted that much of our prospecting work took place during a prolonged period of relatively low rainfall in southern Utah. It is possible that during wetter periods more bone would be exposed on the surface, making it easier to find.

Here we examine several of the more notable vertebrate localities that we have excavated in the Wahweap Formation and give a preliminary taphonomic assessment.

UMNH VP Loc. 1212 *(Tibbet Springs Quarry)*

Although bonebed sites are rare in the Wahweap Formation, several have been identified and excavated. One of these is the Tibbet Springs Quarry, a multitaxic bonebed in the lower member within Tibbet Canyon. This site was discovered in 1984 by a team from NAU and UCM exploring the area for vertebrate microsites. At the initiation of our project the site was relocated, put into stratigraphic context, and in situ dinosaur bone (hadrosaur vertebra) was identified in what is now the east quarry area. This site is in a major tabular sandstone very low in the lower member, about 10 m above the contact with the Drip Tank Member. As noted above, this stratigraphic position would be placed in the Drip Tank Member on existing geologic maps. We began excavation of this site in 2001, with additional work in 2003 (Fig. 26.4). Roughly

26.11. (facing) Some of the axial elements recovered from Jim's Hadrosaur Site (UMNH VP Loc. 1209). (A–D) cervical vertebrae; (E–H) dorsal vertebrae; (I) scapula; (J) coracoid; (K) sternals; (L) ribs; (M) pubis.

DeBlieux et al.

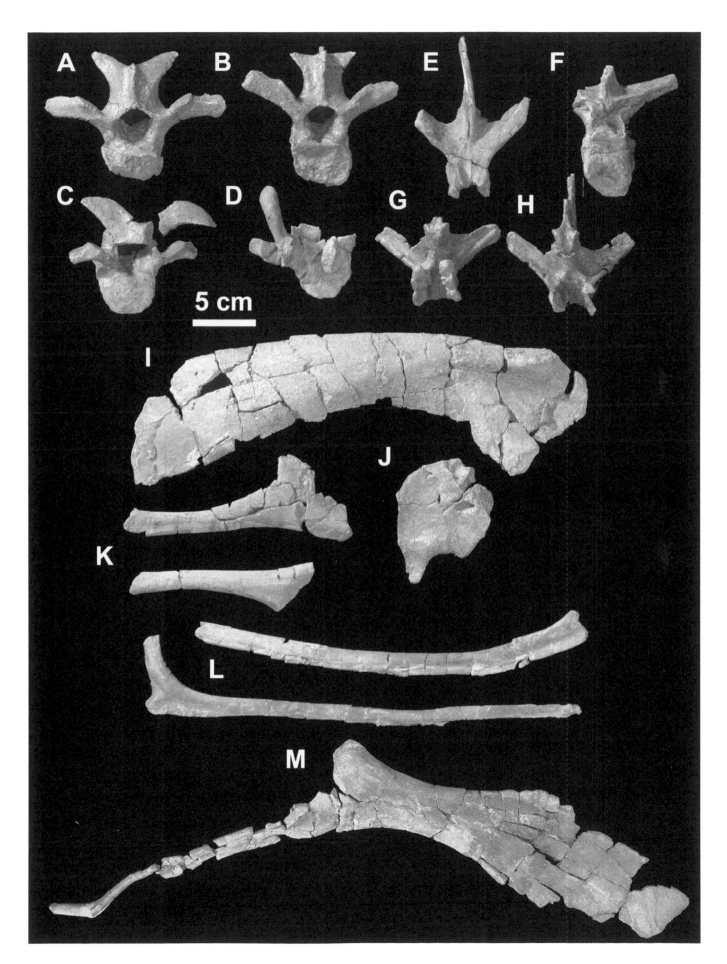

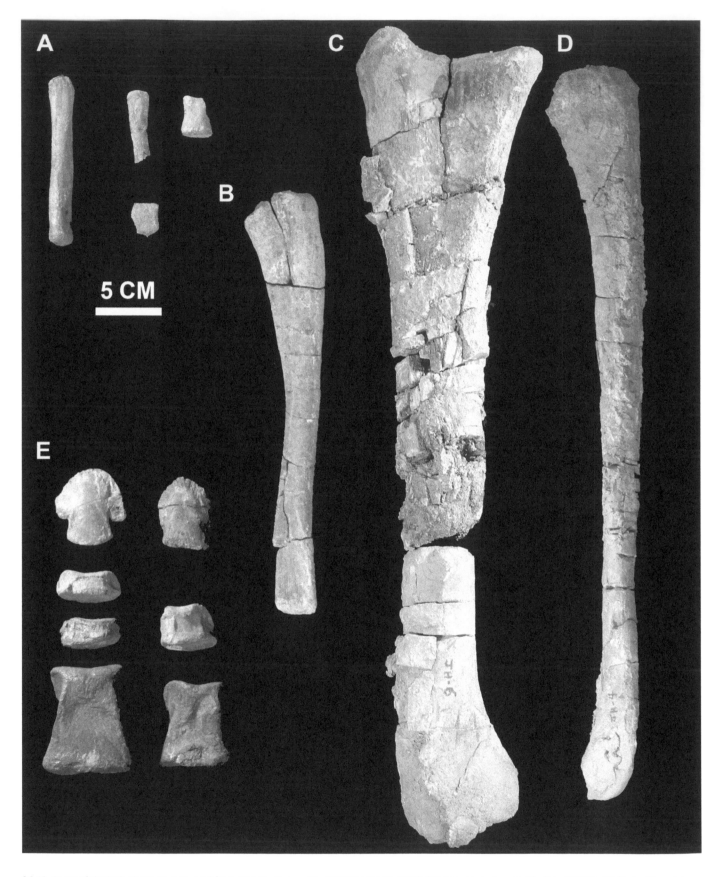

26.12. Some of the limb elements recovered from Jim's Hadrosaur Site (UMNH VP Loc. 1209). (A) manual phalanges; (B) ulna; (C) tibia; (D) fibula; (E) pedal phalanges.

DeBlieux et al.

26.13. Turtle scutes (UMNH VP 20213) from Jim's Hadrosaur Site (UMNH VP Loc. 1209).

30 bones belonging to hadrosaurs have been excavated and prepared (Fig. 26.5). In addition to hadrosaur bones, materials include a weathered flat element that may be part of a ceratopsian frill. Also, a large baenid turtle (UMNH VP 20645) was discovered in the sandstone ledge on the side of the ridge, at a stratigraphic horizon that is essentially at the same level as the main site and about 50 m northeast of the east quarry area (UMNH VP Loc. 562). Two primary areas of concentrated bone were excavated, an east area of 30 m² and a west area of 25 m² separated by approximately 45 m (Figs. 26.6, 26.7). Some of the skeletal elements are associated though none of them can be considered articulated. A skin impression was found in the east area associated with a hadrosaur limb bone and is the only skin impression yet found in the Wahweap Formation. Skin impressions are commonly associated with articulated hadrosaur bone localities found in sandstones in the overlying Kaiparowits Formation, indicating rapid burial of animal carcasses in fluvial channels (Lund et al., 2009). The bones at this site, like many in the formation, have been degraded by the activity of modern plant roots making it difficult to assess the number of bone modification features that would indicate a period of postmortem subaerial exposure. Overall, the bone preservation is excellent; however, a number of elements are broken and damaged, which may have resulted from a period of exposure before being washed into a fluvial channel. The close association of some elements suggests that they were minimally transported and/or that soft tissue may have been present, keeping some of the elements together. The presence of skin impressions is consistent with the latter. Whereas excavation at this quarry is difficult because of the hard sandstone matrix, it is one of the only extensive bonebeds yet located in the Wahweap Formation and additional excavation is warranted.

UMNH VP Loc. 1209 (Jim's Hadrosaur)

In 2004, several tiny scraps of bones were found on a small hill near the base of the middle member. Preliminary surface excavation showed that the site contained the associated remains of a juvenile hadrosaur in a sandy mudstone. Excavation of the site, known as Jim's Hadrosaur (Fig. 26.8), in the spring of 2005 resulted in the collection of over 30 elements of the postcranial skeleton. Elements recovered include large portions of the forelimb, hind limb, and girdles, as well as cervical, dorsal, and caudal vertebrae (Figs. 26.9, 26.10, 26.11, 26.12). Unfortunately, no portions of the skull were recovered that might allow for more specific taxonomic identification of this animal. In addition to the juvenile hadrosaur remains, portions of the carapace and plastron of an unidentified turtle of possible eucryptodire affinities were found mixed together with the hadrosaur bones (Fig. 26.13). A tooth of the extinct bowfin, *Melvius*, several freshwater crab claws (Fig. 26.14C), and poorly preserved unionid bivalves were also collected from the site. The fossils were contained in a muddy sandstone unit that also preserved abundant sections of compressed carbonized log fragments, commonly over a meter in length. Some log sections preserve evidence of burning (Sweeney et al., 2009), whereas others preserve insect burrowing (Fig. 26.14A). These insect traces look similar to *Asthenopodichnium*, a wood-boring trace fossil that has recently been recognized from the upper and capping sandstone members in Grand Staircase–Escalante National Monument (Moran et al., 2010). Smaller compressional plant fragments include mainly *Araucaria*, with rare taxodiaceous conifers and angiosperms (Fig. 26.14D–F). Bones and logs appear to be current oriented. We hypothesize that the juvenile hadrosaur carcass and the turtle carapace were washed down a small stream and were caught against a logjam. This site represents one of the few localities in the Wahweap Formation that contain an associated dinosaur skeleton, but illustrates the potential for further discoveries in this formation.

UMNH VP Loc. 275 (Bucky's Hadrosaur)

The largest bonebed found to date in the Wahweap Formation is Bucky's Hadrosaur Site (UMNH VP Loc. 275). This site is located in the northeastern portion of the Wahweap outcrop belt near Camp Flat and was excavated over several years by crews from the UMNH. Like Jim's Hadrosaur, it is in mudstone and of the middle member, only a few meters above the contact with the lower member. Deposited in a

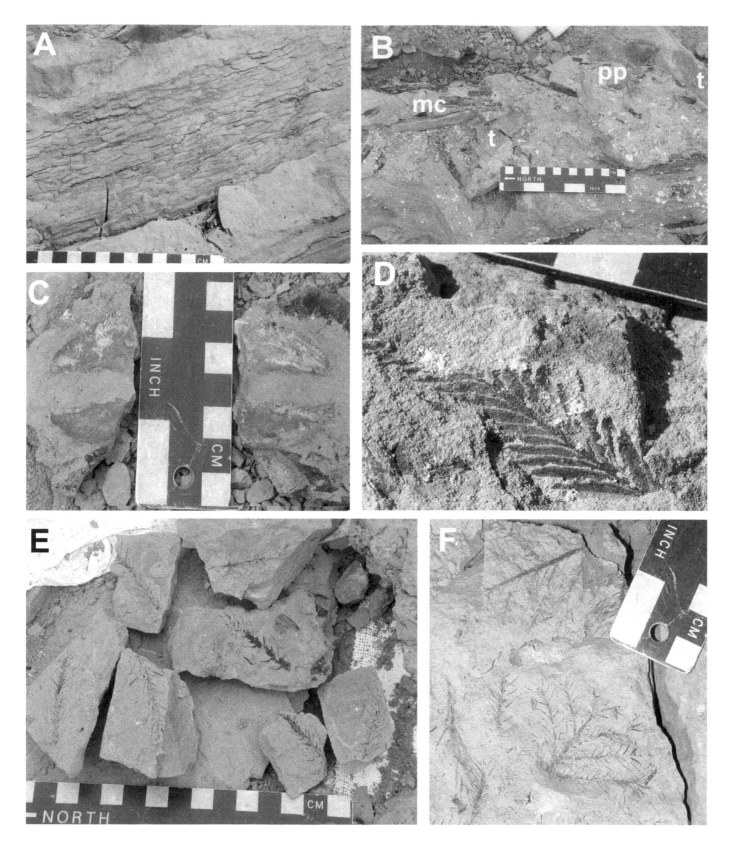

26.14. Fossils found at Jim's Hadrosaur Site (UMNH VP Loc. 1209). (A) compressed log section with possible evidence of borrowing and burning prior to burial. (B) in situ bones and carbonized plant fragments; (C) crab claws; (D) plant fossil belonging to the genus *Sequoia* (scale in centimeters); (E–F) plant fossils belonging to the genus *Araucaria*. Abbreviations: mc, metacarpal; pp, pedal phalanx; t, turtle scute Scale bar is in centimeters and inches.

DeBlieux et al.

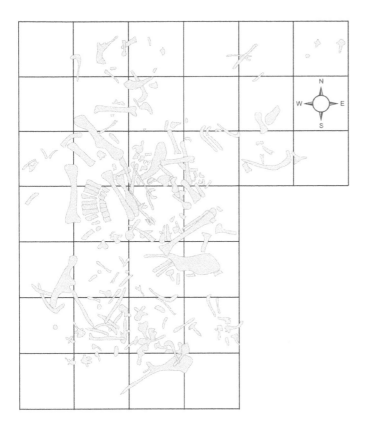

26.15. Quarry map of Bucky's Hadrosaur Site (UMNH VP Loc. 275). Each square represents 1 m².

26.16. Hadrosaur bones in situ at UMNH VP Loc. 1201. (A) proximal humerus; (B) coronoid process of the dentary; (C) distal half of the scapular blade. Scale bar = 10 cm.

back-swamp environment, two hadrosaur individuals, an adult and juvenile, were completely disarticulated over an area of more than 18.5 m² (Fig. 26.15); additional fossils found within the site include theropod teeth, fish bones, a turtle pelvis, and large freshwater crab claws. The site also preserves abundant plant remains consisting of numerous crisscrossing tree stems and clusters of leaf hash and several examples of taxodiaceous conifer needles. The excavation has thus far revealed approximately 70–80% of the adult postcranium and numerous elements of a much smaller juvenile specimen (Jinnah, Getty and Gates, 2009). The only skull material collected from this locality to date consists of a juvenile jugal, maxilla, and dentary. This site is extremely significant in that it has yielded the most complete dinosaur known from the Wahweap Formation, in addition to providing key insights into the paleoenvironment and taphonomy.

UMNH VP Loc. 1210 (Wesses Cove Hadrosaur)

A hadrosaurid proximal humerus, scapula, and dentary (Fig. 26.16, 26.17) were excavated from a site in the middle member in Wesses Cove. The three elements were closely associated within a 1 m² area. The humerus was exposed on the surface and all but proximal end had eroded away. This site is contained in a silty mudstone that may represent one of the few associated localities in an overbank facies. The scapula from this site bears theropod tooth marks indicating that the animal was partially scavenged.

Several other sites have yielded notable hadrosaurid material. A hadrosaur jugal (UMNH VP 19465) tentatively identified as cf. *Brachylophosaurus* (Gates et al., this volume) was collected from an unconsolidated channel lag in the middle member at a site in Wesses Canyon (UMNH VP Loc. 1099). Large hadrosaur bones, including a partial pubis and two femora, have also been discovered at a nearby site in Wesses Canyon (UMNH VP Loc. 1211).

UMNH VP Loc. 1276 (Cattleguard Site)

This site, near the base of the upper member in the vicinity of The Gut, contains elements of at least two hadrosaur individuals. These are primarily postcranial remains of a large and small individual of what appears to be the same taxon of Hadrosaur. Over sixty disarticulated elements were recovered, the majority of which are fragmentary ribs and vertebrae that appear to have been broken prior to deposition. Several well-preserved limb elements were recovered including a femur, tibia, fibula, partial ilium, and astragalus. The fossils are found in a dark gray carbonaceous mudstone that likely represents a pond environment, based on the large number of aquatic invertebrates (gastropods and bivalves) found in this site.

A site discovered within a channel sandstone unit of the upper member on Death Ridge includes partial limb bones and an isolated partial maxilla (UMNH VP 9548) tentatively identified as cf. *Brachylophosaurus*. For a more detailed discussion of the hadrosaur fossils from the Wahweap, see Gates et al. (this volume).

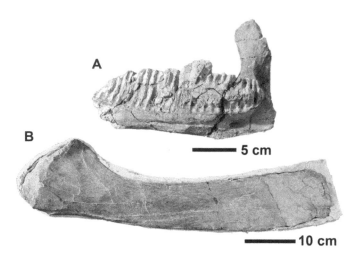

26.17. Hadrosaur bones from UMNH VP Loc. 1210. (A) Dentary (UMNH VP 20216); (B) scapula (UMNH VP 20215).

UMNH VP Loc. 1092 (Last Chance Ceratopsian)

One of the most successful aspects of our project has been the discovery of a number of scientifically significant ceratopsian dinosaur skulls. Along with hadrosaurs, ceratopsians are the most common large dinosaur fossils in the Wahweap Formation.

The single most spectacular and significant fossil found is the nearly complete skull of a new basal centrosaurine ceratopsid, *Diabloceratops eatoni* (UMNH VP 16699; Kirkland and DeBlieux, 2010). This isolated skull is one of the oldest and is the first diagnosable centrosaurine recovered south of Montana (Fig. 26.18).

In 2002, D.D. discovered the skull weathering out of a sandstone ledge 50 m above the base of the middle member at Last Chance Creek on the north side of Reynolds Point in the eastern Wahweap outcrop belt. The skull was contained in a tabular sandstone channel unit along with vertebrate microfossils and abundant clay rip-up clasts (Fig. 26.19). A rib and a splenial bone where also associated with this specimen. This is a classic channel lag deposit that fortuitously contained several large elements that had been washed into the channel.

UMNH VP Loc. 1092 (Nipple Butte Ceratopsian)

A partial centrosaurine skull was discovered by Joshua A. Smith in 1989 and recovered from the lower member of the Wahweap Formation. This was the first ceratopsid identified from the formation; it is not represented by enough critical elements to be confidently diagnosed, but has been tentatively placed in the genus *Diabloceratops* (Kirkland and DeBlieux,

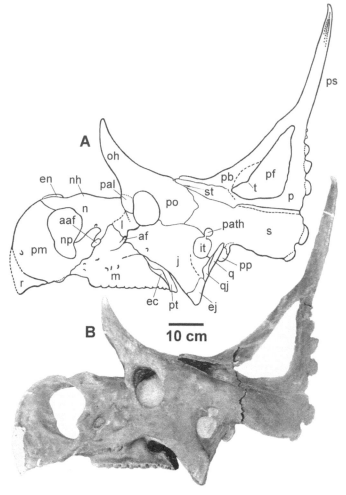

26.18. Left lateral view of *Diabloceratops eatoni* holotype (UMNH VP 16699). (A) Diagrammatic representation of skull; (B) skull. aaf, accessory antorbital fenestra; af, antorbital fenestra; ec, ectopterygoid; ej, epijugals; en, epinasal; j, jugal; l, lacrimal; it, infratemporal fenestra; m, maxilla; n, nasal; nh, nasal horn; np, caudal nasal process; oh, orbital horn; p, parietal; pal, palpebral; path, pathology; pb, parietal bar; pf, parietal fenestra; po, postorbital; pp, paroccipital process; ps, parietal spine; pt, pterygoid; q, quadrate; qj, quadratojugal; r, rostral; s, squamosal; st, supratemporal fenestra; t, triangular process on parietal.

2010). This skull was found in cross-bedded sandstone representing a classic channel lag facies that also included numerous vertebrate microfossils and clay rip-up clasts.

UMNH VP Loc. 1141 (Pilot Knoll Ceratopsian)

A centrosaurine frill was discovered by D.D. close to the base of the lower member on a mesa close to Pilot Knoll near the top of a major tabular sandstone channel unit (Fig. 26.20). The skull was excavated by the UMNH and is currently being prepared. It has a pair of large horns on the back of the frill that are much broader than those of *Diabloceratops* and resemble those of *Albertaceratops* (Ryan, 2007). Other features of the frill, however, share similarities with *Diabloceratops*.

26.19. Collection and preparation of the *Diabloceratops* skull. (A) Sawing out the block containing the skull; (B) the block during sawing; (C) sawing; (D) the block being transported by helicopter; (E) the block on a truck ready for transport to Salt Lake City; (F) beginning preparation of the skull block in the UGS preparation lab; (G) the ventral side of the skull during preparation; (H) the right side of the skull during preparation.

UMNH VP Loc. 146 (Death Ridge)

A partial centrosaurine frill was collected from a channel lag horizon in the upper member near Death Ridge (UMNH VP 4549, Fig. 26.21). This specimen consists of the rear portion of the parietal bar. A number of crescentic sockets are present for the attachment of epiparietals; the epiparietals were not yet fused to the parietal and have been lost. It does not appear that there would have been any large horns attached to the back of the parietal as in *Diabloceratops*.

UMNH VP Loc. 1268 (Star Seep Ceratopsian)

A partial associated ceratopsid skull was excavated from a mudstone site in the upper member of the Wahweap Formation near Star Seep by the UMNH and Ohio University (O'Connor et al., 2009). The skull is about 50% complete, including most of the diagnostic elements, and was found

broken up over an area of about 3 m². This material is currently being prepared and, like the other Wahweap ceratopsians, also pertains to Centrosaurinae. No chasmosaurine material has been discovered to date from the Wahweap Formation.

UMNH VP Loc. 1269 (Nipple Butte Tyrannosaur)

A partial skull and skeleton of a new species of tyrannosaur (Loewen et al., 2010) have been excavated by the UMNH from a site near Nipple Butte discovered by Scott Richardson in 2009. The specimen was found in a fine-grained mudstone unit and was disarticulated over a roughly 6 m² area (Fig. 26.22). The study of this skeleton should provide important information about this group because little is known from this time period and geographic location. A more detailed analysis of this specimen's taxonomy is given in Zanno et al. (this volume).

26.20. Pilot Knoll (UMNH VP Loc. 1141) partial centrosaurine ceratopsid skull in situ. View shows a cross section of the rear portion of the skull with the frill extending into the rock. Scale bar = 10 cm.

Pachycephalosaur

An isolated pachycephalosaur frontoparietal dome (UMNH VP 11939) was discovered by Jodi Vincent eroding out of a channel lag in the middle member in Clints Cove (UMNH VP Loc. 261). This is the only pachycephalosaur fossil known from the Wahweap Formation and, although it is only a single isolated element, it nevertheless provides a good deal of information (Evans et al., this volume, Chapter 20).

OTHER VERTEBRATES

Fish

Microsites in the Wahweap contain a number of fish, including four freshwater shark species, three freshwater ray species, and seven bony fish species (Eaton et al., 1999; Kirkland et al., this volume). Gar scales and teeth are common in channel lag sites. A number of mudstone microsites containing gar scales have also been identified. A well-preserved amiid jaw (UMNH VP 19464) was discovered in a field jacket containing a hadrosaur jugal from a site in Wesses Canyon (UMNH VP Loc. 1099; Fig. 26.23).

Turtles

Seven genera of turtles from at least six families have been identified from Wahweap microsites (Eaton et al., 1999). Turtle shell fragments, scutes, and the occasional complete shells are relatively abundant in Wahweap channel lag sites. Several turtles have been found in mudstone sites. A large, well-preserved baenid turtle shell including the carapace, plastron, and portions of both pectoral limbs (UMNH VP 20643) was collected from a mudstone site (UMNH Loc. 1266) in the middle member (Figs. 26.24, 26.25). The close association of elements indicates that the remains were deposited in a quiet-water setting.

Crocodyliforms

Crocodyliform fossils are present in the Wahweap primarily as isolated teeth and osteoderms recovered from sandstone channel lags. The osteoderms of a very large goniopholid (UMNH VP 9647) were recovered from the upper member on Death Ridge. The osteoderms are typical of goniopholid crocodyliforms in that they lack a defined keel, and in this case, are of very large size. Some of the largest scutes measure 9 cm in mediolateral length and 15 cm in anteroposterior width (Fig. 26.26). Rough speculation would place a size limit of this crocodilian at over 11 m in length. A small alligatoroid,

26.21. Partial centrosaurine ceratopsian frill (UMNH VP 4549) from the upper member on Death Ridge (UMNH VP Loc. 146).

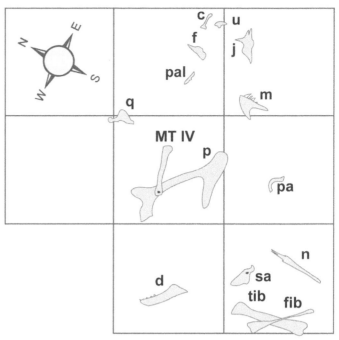

26.22. Quarry map of the Nipple Butte Tyrannosaur site (UMNH VP Loc. 1269). c, chevron; d, dentary; f, frontal; fib, fibula; j, jugal; m, maxilla; MT IV, fourth metatarsal; n, nasal; p, pubis; pa, parietal; pal, palatine; q, quadrate; sa, surangular; tib, tibia; u, ungual phalanx. Each square represents 1 m².

and *Deinosuchus* are also known to be present (Hutchison et al., 2009; Irmis et al., this volume, Chapter 17).

OTHER FOSSILS

Plants and invertebrates

Petrified log and wood sites are quite common in the Wahweap Formation (Fig. 26.27). Our survey work indicates that the lower interval of the Wahweap Formation, especially the mudstone portions of the lower member just above the Drip Tank Member of the Straight Cliffs Formation, contains abundant fossil logs. Fossil logs are also found at the base of many sandstone channel units and are concentrated on channel margins associated with crevasse splay sandstones (Eaton, 1991). Several sites have been located that preserve leaf and plant impressions. Many occur in hard sandstones and contain poorly preserved leaves and leaf fragments. Mudstone sites containing plant fossils are common but are found only by digging. As a result, many plant macrofossil sites are associated with vertebrate excavations such as Jim's Hadrosaur Site (Fig. 26.14) and Bucky's Hadrosaur Site.

The fossils of aquatic invertebrates are quite common in the Wahweap Formation and they are found in a number of facies representing fluvial channels, channel lags, and overbank ponds. The most common invertebrates are unionid bivalves, snails, and crabs (Figs. 26.28, 26.29). Mollusk shell conglomerates are found in sandstone units preserving unionid bivalves sometimes preserving crab claws and water-worn vertebrate bone. Aquatic invertebrates are

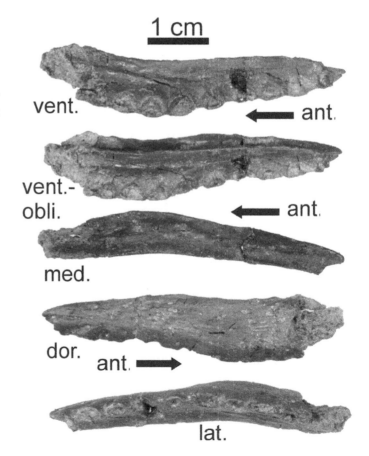

26.23. Amiid fish maxilla (UMNH VP 19464) from UMNH VP Loc. 1099 in the middle member. ant., anterior; dor., dorsal; lat., lateral; med., medial; vent., ventral; obli., oblique.

found in carbonaceous mudstones deposited in ponds and lakes. Preliminary findings suggest that these sites are more common in the middle and higher parts of the formation, which are generally less accessible than the lower part of the formation, where the main roads are located. Of scientific significance is the consistent occurrence of large crab claws in association with the mollusk sites. We do not know of this association being reported in the literature and is only known elsewhere in Utah from the Kaiparowits Plateau in the John Henry Member of the Straight Cliffs Formation, and from the Henry basin in the Masuk Formation.

Tracks and Track Facies

An interesting discovery in 2003 was the recognition of multiple track-bearing horizons within the Wahweap Formation. During a preliminary paleontological survey of the Grand Staircase–Escalante National Monument, Alden Hamblin located several tracks and tracksites in the Wahweap Formation (Hamblin and Foster, 2000). J.I.K. subsequently discovered an unmistakable track cast of a large three-toed dinosaur (Fig. 26.30), leading him to a thin sandstone unit composed almost entirely of dinosaur track casts. These track casts were formed when a thin layer of sand, possibly from a crevasse splay, was deposited over a mud layer trampled by hadrosaurs. The sandstone filled in the tracks and formed casts that replicated the form of the footprint. The replication was so precise that scratch marks from the scales on the skin of the dinosaur were preserved along with details of the foot anatomy.

In prospecting for fossils, paleontologists use a search image to help them pick out fossils from the background of rocks, plants, and other features in the environment that are not fossils. One of the keys to successful prospecting is the refinement of the search image needed to find fossils in a particular formation. For example, in the Wahweap Formation weathered bone often takes on an orange patina and this orange color becomes part of the search image used to pick out pieces of bone. During our previous prospecting we were likely looking past any tracks, which occur in thin, dark brown sandstone horizons, in our search for bone. Examination of the new tracksite by our prospecting team enabled us to refine our search image to look for track casts in the thin sandstone lenses that are common in the transition from the lower member to the middle member, a stratigraphic level that is well exposed in the Wesses Cove and Wesses

Canyon area. This resulted in the identification of at least 10 additional sites (Figs. 26.31–33) that preserve tracks, showing that tracks are much more common than previously thought, and adding an exciting new dimension to our study of the dinosaur fauna of the Wahweap Formation. Whereas most of the tracks appear to have been made by large hadrosaurid dinosaurs, several smaller tracks may belong to quadrupedal dinosaurs, possibly ceratopsians (Fig. 26.31D, F), and small tridactyl tracks may pertain to small ornithischians or theropods (Fig. 26.33). We observed a very similar mode of track preservation in the Kaiparowits Formation.

FAUNAL AGE

NALMAS are commonly used for Tertiary biochronology and have been established in the Mesozoic of North America only for the latter part of the Late Cretaceous. Initial work on the mammal fauna of the Wahweap showed similarities with the fauna of the Milk River Formation in Canada (Cifelli, 1990a, 1990b, 1990d; Eaton, 2002). The Milk River Formation fauna forms the basis of the Aquilan NALMA and was thought to be early Campanian in age. Thus, an early Campanian or late Santonian age was indicated for the Wahweap and the fauna was considered to be Aquilan. It is now known that the type Aquilan Milk River Formation is Santonian and earliest Campanian in age (Payenberg et al., 2002; Eaton, 2006). In addition, fossil mammals with affinities to those of the Milk River have been described from the Santonian-age John Henry Member of the Straight Cliffs Formation of Grand Staircase–Escalante National Monument. As noted above, the Wahweap Formation is firmly established as being middle Campanian in age (Roberts, Deino, and Chan, 2005; Roberts, 2007; Jinnah et al., 2009). The Judithian NALMA that comes after the Aquilan is based on a fauna from sites located near the top of the Judith River Formation in Montana; these localities are younger than the sites from which Wahweap mammals are found (Russell, 1975; Lillegraven and McKenna, 1986; Cifelli et al., 2004). Russell (1975) recognized a gap between the Aquilan and the overlying Judithian, as well as a gap between the Judithian and the overlying Edmontonian, where marine strata separate the sites preserving the terrestrial vertebrate faunas. The Kaiparowits Formation, which overlies the Wahweap Formation, has a mammalian fauna that is considered to be Judithian. Comparison of the Wahweap fauna with those from the Aquilan and Judithian shows that there are similarities to both, not surprising given

26.24. Collection of Don's Turtle (UMNH VP Loc.1266) in September 2003. (A) The site as it looked at discovery; (B) cleaning the turtle for jacketing; (C) the turtle ready for jacketing; (D) jacketing; (E) the turtle humerus in situ; (F) pedestalling the jacket before flipping; (G) team beginning the half-mile transport of the jacket to the road; (H) site after collection of the turtle showing the minimal impact of collecting in the mudstone exposures of the Wahweap Formation. After one or two rainstorms, any sign of disturbance at this site will have all but disappeared.

DeBlieux et al.

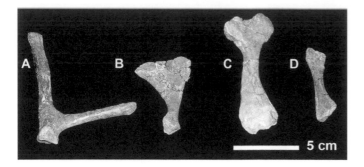

26.25. Associated turtle limb elements (UMNH VP 20643) from UMNH VP Loc. 1266. (A) Coracoid; (B) scapula; (C) humerus; (D) ulna.

26.26. Large goniopholid crocodyliform osteoderm (UMNH VP 9647) from Death Ridge.

26.27. Large petrified log in a sandstone ledge near the base of the lower member near Pilot Knoll.

that evidence suggests the Wahweap Formation is younger than rocks that contain the faunas of the Aquilan and much of it is older than the rocks that contain Judithian faunas. For example, despite lacking triconodontid mammals, the Wahweap Formation contains spalacotheriid symmedrodont mammals known from the Aquilan but not present in the Judithian (Russell, 1975). On the other hand, the Wahweap Formation contains multituberculate, marsupial, and eutherian mammals having affinities to Judithian faunas (Cifelli, 1990a, 1990d; Eaton et al., 1999; Eaton, 2006) but does not include the species that are used as the basis for the Judithian. In short, it appears that the Wahweap does not fit precisely in either the Aquilan or the Judithian. Thus, the Wahweap could be included in an expanded Aquilan, considered to be an intermediate fauna between the Aquilan and Judithian, or possibly made into its own Wahweapian NALMA.

Continued study of other elements of the microvertebrate fauna, such as fish, amphibians, and reptiles will also add information which can aid in biochronology. Of note is a generic-level turnover in the common freshwater myledaphid ray

lineage (Peng, Russell, and Brinkman, 2001; Kirkland et al., this volume). *Myledaphus bipartitus* occurs in the upper Judith River Formation of Montana, Dinosaur Park Formation of Alberta, and Kaiparowits Formation of Utah. Peng, Russell, and Brinkman (2001) recognized a new, morphologically less-ornamented ray in the underlying Foremost Formation and an absence of *Myledaphus bipartitus*. Similarly, within the Wahweap Formation, a less-ornamented species of ray is present instead of *Myledaphus bipartitus*, although the Wahweap and Foremost taxa are different species. The Aquilan ray teeth from the Milk River Formation and John Henry Member represent a third conspecific taxon, strengthening those biostratigraphic ties (Kirkland et al., this volume). A noteworthy observation is that, even though the mammalian fauna of the Foremost Formation is still under study, it does not include symmetrodonts (Brinkman, pers. comm., 2010). The mammalian faunas in the correlative basal Judith River Formation just to the south in northern Montana likewise do not include symmetrodonts (Montellano, 1992) and are

DeBlieux et al.

26.28. Crab claw from the middle member near Wesses Cove.

26.29. Unionid clam shells from the lower member near Drip Tank Canyon. Scale bar = 10 cm.

26.30. Jim Kirkland in front of the spectacular hadrosaur track cast he discovered during spring 2003. We collected this track and it is currently on display at the Grand Staircase–Escalante National Monument visitor center in Big Water, Utah.

considered Judithian as defined by the first occurrence of the multituberculates *Mesodma primaeva*, *Meniscoessus major*, *Meniscoessus intermedius*, and *Cimolomys clarki*, the marsupial *Turgidodon*, and the eutherian *Gypsonictops* (Cifelli et al., 2004).

Perhaps the Wahweapean NALMA could be defined on the occurrence of *Diabloceratops*, the ray *Crystomylus cifellii* (Kirkland et al., this volume, Chapter 9), together with a spectrum of its distinctive mammals (e.g. ?*Cimexomys gregoryi*, *Cimolomys milliensis*, *Symmetrodontoides foxi*, *Zygiocuspis goldingi*, *Varalphadon crebreforme*, *Iugomortiferum thoringtoni*).

Macrovertebrates, such as dinosaurs, may be more useful in biostratigraphy as more are found and studied. Another complicating factor that will have to be addressed is the variations in the faunas that have been proposed to result from difference in latitude (Lehman, 1997, 2001; Gates et al., 2010).

CONCLUSIONS

A paleontological inventory of the middle Campanian Wahweap Formation by teams from the UGS, UMNH, and Grand Staircase–Escalante National Monument has resulted in the identification of hundreds of new fossil sites. The vast majority of vertebrate fossils are found in sandstone-hosted channel lag microsites representing time-averaged accumulations of aquatic and terrestrial vertebrates. Larger vertebrate skeletal elements, including skulls, are sometimes found in these sites. Fossil sites in mudstone facies deposited in overbank settings such as alluvial plains, ponds, lakes, and swamps are present but rare, and difficult to find because of rapid loss of bone on the surface as a result of weathering and the activity of plants. Mudstone sites typically contain isolated elements but some have been found and excavated which preserve associated skeletons. In contrast to many Late Cretaceous terrestrial formations in the Western Interior Basin, no articulated skeletons have yet been discovered in the Wahweap Formation. A number of scientifically significant dinosaur sites have been discovered and excavated. As in other North American terrestrial vertebrate faunas of Campanian age, hadrosaurs and ceratopsians are the most common dinosaurs

26.31. Examples of some of the tracks discovered during 2003 in the Wahweap Formation. (A) Ornithopod track discovered by Alan Titus (Ka 1068t) consisting of contorted sediments on a sandstone surface (the paintbrush is 18 cm long); (B) ornithopod track cast on the underside of a thin sandstone block (Ka 1106t); (C) partial ornithopod track on the underside of a fallen block (Ka 1131t) with central and part of a side toe outlined. Arrow points to infilled cylindrical structure that is possibly from a tail drag; (D) small track casts from Ka 1127t. These are indistinct but may represent tracks from a quadrupedal dinosaur such as a ceratopsian; (E) ornithopod track from Ka 1127t. Arrow points to plant impressions; (F) indistinct track cast from Ka 1126t showing scratch marks.

found in the Wahweap Formation. Several hadrosaur bone-beds have been excavated, with the majority belonging to hadrosaurids. We now have cranial material belonging to at least four different specimens of ceratopsid dinosaur. The centrosaurine ceratopsid *Diabloceratops eatoni* is the most basal centrosaurine known to date. Pachycephalosaur and ankylosaur fossils have been found but are quite rare. The teeth of theropods are common in lag deposits but theropod

DeBlieux et al.

26.32. Trackside Ka1159t from the lower member south of Wesses Cove. (A) Ornithopod track cast fallen from sandstone cliff; (B) view of the cliff of interbedded sandstone and mudstone that contains the dinosaur track casts; (C, D) track casts in situ in the cliff. GPS unit is 14.5 cm in length.

bone sites are rare. An associated partial tyrannosaur skeleton is currently being prepared and studied.

A number of dinosaur tracks and tracksites have been found. Most tracks occur as natural casts in thin sandstone units that formed as crevasse splay deposits filled in tracks made on the underlying mud. The majority of tracks recognized are tridactyl, likely belonging to hadrosaurs. Comparison of the Wahweap vertebrate fauna with those of the Aquilan and Judithian NALMAS indicates that a unique intermediate fauna is present.

ACKNOWLEDGMENTS

We thank the staff of the Grand Staircase–Escalante National Monument and the Bureau of Land Management, especially A. Titus, M. Eaton, S. Foss, and D. Hunsaker for issuing permits and facilitating our work.

Help in the field was provided by V. Backus, B. and L. Baldazzi, J. DeBlieux, W. Elkington, S. Fischer, J. Gentry, J. Golden, J. Green, D. and S. Hughes, M. Madsen, B. Johnson, T. Mellenthin, D. Mickelson, A. Milner, S. Mosconi, G. Muller, P. Policelli, J. Smith, S. and S. Stevenson, J. Vincent, D. and T. Wade, D. Willcots, B. and A. Yensen, D. Zivkovic, Zion Helitack, and the UGS.

For help with compiling the faunal list, we thank D. Brinkman, R. Nydam, and C. Redman.

For helpful reviews and comments, we thank M. Hylland, M. Lowe, and S. Madsen. Funding has been provided by the Grand Staircase–Escalante National Monument, UMNH, and UGS.

Chinnery, B. J., and J. R. Horner. 2007. A new neoceratopsian dinosaur linking North American and Asian taxa. Journal of Vertebrate Paleontology 27:625–641.

Cifelli, R. L. 1990a. Cretaceous mammals of southern Utah. I. Marsupials from the Kaiparowits Formation (Judithian). Journal of Vertebrate Paleontology 10:295–319.

Cifelli, R. L. 1990b. Cretaceous mammals of southern Utah. II. Marsupials and marsupial-like mammals from the Wahweap Formation (early Campanian). Journal of Vertebrate Paleontology 10:320–331.

Cifelli, R. L. 1990c. Cretaceous mammals of southern Utah. III. Therian mammals from the Turonian (early Late Cretaceous). Journal of Vertebrate Paleontology 10:332–345.

Cifelli, R. L. 1990d. Cretaceous mammals of southern Utah. IV. Eutherian mammals from the Wahweap (Aquilan) and Kaiparowits (Judithian)

Formations. Journal of Vertebrate Paleontology 10:346–360.

Cifelli, R. L., J. J. Eberle, D. L. Lofgren, J. A. Lillegraven, and W. A. Clemens. 2004. Mammalian biochronology of the latest Cretaceous; pp. 21–42 in M. O. Woodburne (ed.), Late Cretaceous and Cenozoic Mammals of North America: Biostratigraphy and Geochronology. Columbia University Press, New York.

Doelling, H. H., and R. L. Graham. 1972. Southwestern Utah Coal Fields–Alton, Kaiparowits Plateau and Kolob-Harmony. Utah Geological and Minerological Survey Monograph Series 1.

Doelling, H. H., and G. C. Willis. 2008. Geological map of the Smoky Mountain 30'×60' quadrangle, Kane and San Juan counties, Utah, and Coconino County, Arizona. Utah Geological Survey Map 213DM.

Eaton, J. G. 1990. Stratigraphic revision of Campanian (Upper Cretaceous) rocks in the Henry Basin, Utah. Mountain Geologist 27:27–38.

Eaton, J. G. 1991. Biostratigraphic framework for Upper Cretaceous rocks of the Kaiparowits Plateau; pp. 47–63 in J. D. Nations, and J. G. Eaton (eds.), Stratigraphy, Depositional Environments, and Sedimentary Tectonics of the Western Margin, Cretaceous Western Interior Seaway. Geological Society of America Special Paper 260.

Eaton, J. G. 2002. Multituberculate mammals from the Wahweap (Campanian, Aquilan) and Kaiparowits (Campanian, Judithian) formations, within and near the Grand Staircase–Escalante National Monument, southern Utah. Utah Geological Survey Miscellaneous Publication 02–4.

Eaton, J. G. 2006. Santontian (Late Cretaceous) Mammals from the John Henry member of the Straight Cliffs Formation, Grand Staircase–Escalante National Monument, Utah. Journal of Vertebrate Paleontology 26:446–460.

Eaton, J. G, and R. L. Cifelli. 1988. Preliminary report on Late Cretaceous mammals of the Kaiparowits Plateau, southern Utah. Contributions to Geology, University of Wyoming 26:45–55.

Eaton, J. G., R. L. Cifelli, J. H. Hutchison, J. I., Kirkland, and J. M. Parrish. 1999. Cretaceous vertebrate faunas of the Kaiparowits Basin (Cenomanian–Campanian), southern Utah; pp. 345–353 in D. D. Gillette (ed.), Vertebrate Paleontology in Utah. Utah Geological Survey Miscellaneous Publication 99-1.

Eberth, D. A., and P. J. Currie. 2005. Vertebrate taphonomy and taphonomic trends; pp. 453–477 in P. J. Currie and E. B. Koppelhus (eds.) Dinosaur Provincial Park: A Spectacular Ancient Ecosystem Revealed. Indiana University Press, Bloomington, Indiana.

Foreman, B. Z., R. R. Rogers, A. L. Deino, K. R. Wirth, and J. T. Thole. 2008. Geochemical characterization of bentonite beds in the Two Medicine Formation (Campanian, Montana), including a new ^{40}Ar-^{39}Ar age. Cretaceous Research 29:373–385.

Gates, T. A., S. D. Sampson, L. E. Zanno, E. M. Roberts, J. G. Eaton, R. L. Nydam, J. H. Hutchison, J. A. Smith, M. A. Loewen, and M. A. Getty. 2010. Biogeography of terrestrial and freshwater vertebrates from the Late Cretaceous (Campanian) western interior of North America. Paleogeography 291:371–387.

Getty, M. A., M. A. Loewen, E. M. Roberts, A. L. Titus, and S. D. Sampson. 2010. Taphonomy of horned dinosaurs (Ornithischia: Ceratopsidae) from the Late Campanian Kaiparowits Formation, Grand Staircase–Escalante National Monument, Utah; pp. 478–494 in M. J. Ryan, B. J. Chinnery-Allgeier, and D. A. Eberth (eds.), New Perspectives on Horned Dinosaurs: The Royal Tyrell Museum Ceratopsian Symposium. Indiana University Press, Bloomington, Indiana.

Goldstrand, P. M. 1992. Evolution of Late Cretaceous and early Tertiary basins of southwest Utah based on clastic petrology. Journal of Sedimentary Petrology 62:495–507.

26.33. Small tridactyl track cast from the middle member near Wesses Canyon.

DeBlieux et al.

Gregory, H. E., and R. C. Moore. 1931. The Kaiparowits Region; Geographic and Geologic Reconnaissance of Parts of Utah and Arizona. U.S. Geological Survey Profesional Paper 164.

Hamblin, A. H., and J. R. Foster. 2000. Ancient animal footprints and traces in the Grand Staircase–Escalante National Monument; pp. 557–568 in D. A. Sprinkel, T. C. Chidsey Jr., and P. B. Anderson (eds.), Geology of Utah's Parks and Monuments. Utah Geological Association Publication 28.

Horner, J. R. 1992. Cranial Morphology of *Prosaurolophus* (Ornithischia: Hadrosauridae) with Descriptions of Two New Hadrosaurid Species and an Evaluation of Hadrosaurid Phylogenetic Relationships. Museum of the Rockies Occasional Paper 2.

Hutchison, J. H., R. B. Irmis, J. W. Sertich, S. J. Nesbitt, A. L. Titus, and L. P. Claessens. 2009. Late Cretaceous Crocodyliforms from Grand Staircase–Escalante National Monument, Utah. Advances in Late Cretaceous Paleontology and Geology, Abstracts with Program, p. 27.

Jinnah, Z. A., M. A., Getty, and T. A. Gates. 2009a. Taphonomy of a hadrosaur bonebed from the Upper Cretaceous Wahweap Formation, Grand Staircase–Escalante National Monument, Utah. Advances in Western Interior Late Cretaceous Paleontology and Geology, Abstracts with Program, p. 28.

Jinnah, Z. A., E. M. Roberts, A. L. Deino, J. S. Larsen, P. K. Link, and C. M. Fanning. 2009. New 40Ar-39Ar and detrital zircon U-Pb ages for the Upper Cretaceous Wahweap and Kaiparowits formations on the Kaiparowits Plateau, Utah: implications for regional correlation, provenance, and biostratigraphy. Cretaceous Research 30:287–299.

Kirkland, J. I., and D. D. DeBlieux. 2010. New basal centrosaurine ceratopsian skulls from the Wahweap Formation (Middle Campanian), Grand Staircase–Escalante National Monument, Southern Utah; pp. 117–140 in M. J. Ryan, B. J. Chinnery-Allgeier, and D. A. Eberth (eds.), New Perspectives on Horned Dinosaurs: The Royal Tyrell Museum Ceratopsian Symposium. Indiana University Press, Bloomington, Indiana.

Lawton, T. F., S. L. Pollock, and R. A. J. Robison. 2003. Integrating sandstone petrology and nonmarine sequence stratigraphy: application to the Late Cretaceous fluvial systems of southwestern Utah, U.S.A. Journal of Sedimentary Research 73:389–406.

Lehman, T. M. 1997. Late Campanian dinosaur biogeography in the western interior of North America; pp. 223–240 in D. Wolberg, and E. Stump (eds.), Dinofest International Proceedings. Philadelphia Academy of Sciences, Philadelphia, Pennsylvania.

Lehman, T. M. 2001. Late Cretaceous dinosaur provinciality; pp. 310–328 in D. H. Tanke and K. Carpenter (eds.), Mesozoic Vertebrate Life. Indiana University Press, Bloomington, Indiana.

Lillegraven, J. A., and M. C. McKenna. 1986. Fossil mammals from the "Mesaverde" Formation (Late Cretaceous, Judithian) of the Bighorn and Wind River basins, Wyoming, with definitions of Late Cretaceous North American land mammal "ages." American Museum Novitiates 2804:1–68.

Loewen, M. A., J. W. Sertich, R. B. Irmis, and S. D. Sampson. 2010. Tyrannosaurid evolution and intracontinental endemism in Laramidia: new evidence from the Campanian Wahweap Formation of Utah. Journal of Vertebrate Paleontology 30:123A.

Lund, E. K., M. A. Loewen, E. M. Roberts, S. D. Sampson, M. A. Getty. 2009. Dinosaur skin impressions in the Upper Cretaceous (Late Campanian) Kaiparowits Formation, Southern Utah: insights into preservation mode. Advances in Western Interior Late Cretaceous Paleontology and Geology, St. George, Utah, Abstracts with Program, p. 32.

Montellano, M. 1992. Mammalian fauna of the Judith River Formation (Late Cretaceous, Judithian), north central Montana. University of California Publications in Geological Sciences 136:1–115.

Moran, K., H. L. Hilbert-Wolf, K. Golder, H. F. Malenda, C. J. Smith, L. P. Storm, E. L. Simpson, M. C. Wizevich, and S. E. Tindall. 2010. Attributes of the wood-boring trace fossil *Asthenopodichnium* in the Late Cretaceous Wahweap Formation, Utah, U.S.A. Palaeogeography, Palaeoclimatology, Palaeoecology 297:662–669.

O'Connor, P. M., E. M. Roberts, L. Tapanila, A. Deino, T. Hieronymus, and Z. Jinnah. 2009. Reconnaissance paleontology in the Straight Cliffs and Wahweap formations (Turonian–Campanian), Grand Staircase–Escalante National Monument, Utah. Advances in Western Interior Late Cretaceous Paleontology and Geology, Abstracts with Program, p. 40.

Parrish, J. M. 1999. Dinosaur teeth from the Upper Cretaceous (Turonian–Judithian) of southern Utah; pp. 319–321 in D. Gillette (ed.), Vertebrate Paleontology in Utah. Utah Geological Survey Miscellaneous Publication 99-1.

Payenberg, T. H. D., D. R. Braman, D. W. Davis, and A. D. Miall. 2002. Litho- and chronostratigraphic relationships of the Milk River Formation in southern Alberta and Eagle Formation in Montana utilizing stratigraphy, U-Pb geochronology, and palynology. Canadian Journal of Earth Sciences 39:1553–1577.

Peng, J., A. P. Russell, and D. B. Brinkman. 2001. Vertebrate Microsite Assemblages (Exclusive of Mammals) from the Foremost and Oldman Formations of the Judith River Group (Campanian) of Southeastern Alberta: An Illustrated Guide. Provincial Museum of Alberta Natural History Occasional Paper 25.

Peterson, F. 1969. Cretaceous Sedimentation and Tectonism in the Southeastern Kaiparowits Region, Utah. U.S. Geological Survey Open-File Report 69–202.

Pollock, S. L. 1999. Provenance, geometry, lithofacies, and age of the Upper Cretaceous Wahweap Formation, Cordilleran Foreland Basin, southern Utah. M.Sc. thesis, New Mexico State University, Las Cruces, New Mexico.

Roberts, E. M. 2007. Facies architecture and depositional environments of the Upper Cretaceous Kaiparowits Formation, southern Utah. Sedimentary Geology 197:207–233.

Roberts, E. M., M. A. Chan, S. D. Sampson. 2003. Taphonomic analysis of the Late Cretaceous Kaiparowits Formation in the Grand Staircase–Escalante National Monument, southern Utah. Geolocical Society of America Abstracts with Programs 35:591A

Roberts, E. M., A. L. Deino, and M. A. Chan. 2005. 40Ar/39Ar age of the Kaiparowits Formation, southern Utah, and correlation of contemporaneous Campanian strata and vertebrate faunas along the margin of the Western Interior Basin. Cretaceous Research 26:307–318.

Rogers, R. R., and S. M. Kidwell. 2007. A conceptual framework for the genesis and analysis of vertebrate skeletal concentrations; pp. 1–63 in R. R. Rogers, D. A. Eberth, and A. R. Fiorillo (eds.), Bonebeds: Genesis, Analysis, and Paleobiological Significance. University of Chicago Press, Chicago, Illinois.

Russell, L. S. 1975. Mammalian faunal succession in the Cretaceous system of western North America; pp. 137–161 in W. G. E. Caldwell (ed.), The Cretaceous System in the Western Interior of North America. Geological Association of Canada Special Paper 13.

Ryan, M. J. 2007. A new basal centrosaurine ceratopsid from the Oldman Formation, southeastern Alberta. Journal of Paleontology 81:376–396.

Sargent, K. A., and D. E. Hansen. 1982. Bedrock geological map of the Kaiparowits coal-basin area, Utah. U.S. Geological Survey Miscellaneous Investigations Series Map I-1033-I.

Smith, C. 1983. Geology, depositional environments, and coal resources of the Mt. Pennell 2 NW quadrangle, Garfield County, Utah. Brigham Young University Geological Studies 30:145–169.

Sweeney, I. J., K. Chin, J. C. Hower, D. A. Budd, and D. G. Wolfe. 2009. Fossil wood from the middle Cretaceous Moreno Hill Formation: unique expressions of wood minerilization and implications for the processes of wood preservation. International Journal of Coal Geology 79:1–7.

27

Taphonomy of a Subadult *Teratophoneus curriei* (Tyrannosauridae) from the Upper Campanian Kaiparowits Formation of Utah

Jelle P. Wiersma and Mark A. Loewen

THE UPPER CAMPANIAN KAIPAROWITS FORMATION OF Utah has recently produced an associated subadult skeleton of the tyrannosaurid dinosaur *Teratophoneus curriei*. The approximately 65% complete skeleton includes most of the skull; representative elements from the entire axial column; a complete pelvis; and the right femur, tibia, and fibula. The quarry also has produced leaves, logs, abundant aquatic mollusks, a partial hatchling hadrosaurid, elements from a larger subadult hadrosaurid, a crocodilian, a bird, and a lizard, all preserved in a silty claystone, resulting in a multitaxic monodominant bonebed assemblage. We interpret the sedimentology of the quarry as the result of deposition in shallow standing water in a floodplain depression following an overbank flood. All of the elements lack any indication of surface weathering, insect modification, scavenging, or predation. The skull, vertebrae, and pelvis exhibit breakage penecontemporaneous with burial. These elements exhibit greenstick fractures perpendicular to the bedding plane. Recovered elements exhibit exceptional preservation with an absence subaerial weathering or predation marks. We suggest a taphonomic history of death and burial of the animal in standing water followed by an overbank flood event. This resulted in burial and disarticulation of the skeleton in fine-grained sediments, followed by partial trampling of the skull and skeleton in shallow standing water or waterlogged sediment. This interpretation contrasts with interpretations of most other vertebrate localities in fine-grained sediments within the Kaiparowits Formation, which exhibit multiple examples of predation and insect modification.

INTRODUCTION

As described previously in this volume (see Chapter 1), the Utah Museum of Natural History (UMNH) has focused a large portion of each field season since 2000 surveying for and excavating vertebrate fossils in the Cretaceous aged rocks of Grand Staircase–Escalante National Monument. During the 2004–2006 field seasons, UMNH crews excavated 133 elements of an associated and semiarticulated subadult tyrannosaur (UMNH VP 16690) from the upper Campanian Kaiparowits Formation. This specimen represents the single most complete tyrannosaur recovered to date from the Upper Cretaceous of Utah. Approximately 65% of the theropod skeleton was recovered from the quarry (informally known as "Kaiparowits, Jelle's Tyrannosaur Quarry" and labeled with the field number prefix KJT).

Comparisons of UMNH VP 16690 with other tyrannosaurid material from the Kaiparowits Formation including BYU 8120 reveals an overlap of most of the autapomorphies of *Teratophoneus curriei* (Carr et al., 2011). The type specimen of *Teratophoneus curriei* is a single individual with specimen numbers BYU 8120/9396; BYU 8120/937; BYU 826/9720; BYU 9398; and BYU 13719. This specimen is collectively refered to as BYU 8120. The maxilla has a maxillary fenestra situated far caudal to the rostral margin of the antorbital fossa. There is complete overlap of the frontal by the parietal on the caudal margin; and the braincase exhibits transversely oriented paroccipital processes that extend nearly directly laterally, instead of caudolaterally. Comparison with other tyrannosaurid materials suggests an affinity with short-faced southern Laramidian tyrannosaurids such as Bistahieversor sealeyi from the Kirtland Formation of New Mexico and an undescribed tyrannosaurid from the Wahweap Formation of Utah (Loewen et al., 2010). In contrast, UMNH VP 16690 differs widely from roughly coeval longirostran tyrannosaurids from the north of Laramidia such as *Gorgosaurus libratus*, *Albertosaurus sarcophagus*, and *Daspletosaurus torosus*.

The unique vertebrate assemblage and sedimentology observed at KJT led us to conduct a more thorough taphonomic analysis of the site. Summarized below are the results of this fieldwork, conducted between 2004 and 2009.

METHODS

The fossil remains of the KJT quarry including UMNH VP 16690 and other paleontologic specimens were excavated over the course of three field seasons spread out over the summer and fall of 2004, 2005, and 2006, totaling 15 weeks of excavation. During this time, the authors directed and participated in the excavation of specimen UMNH VP 16690. The KJT quarry is accessible to the Horse Mountain road by a 2-km hike and descent into a canyon on Horse Mountain east of the Cockscomb on the Kaiparowits Plateau. The quarry excavation was approximately 10 m long by 4 m wide. Above the bonebed, a total of over 50 m³ of overburden was removed using picks, shovels, and a gas-powered jackhammer. Detailed excavation, mapping, and collection of UMNH VP 16690 and other paleontological data were conducted by conventional methods. All in situ elements were mapped on a 1 m² grid and tied to a baseline. Elevation measurements were taken from the baseline on the majority of the elements in order to reconstruct the spatial distribution of skeletal elements in the quarry. Paleocurrent or orientation analysis was performed by measuring the orientation of the major long axes of bones. These measurements were later plotted on a rose diagram. The specimens were stabilized with polyvinyl acetate dissolved in acetone. They were removed from the quarry either as small jackets using medical plaster bandages or as larger standard burlap-and-plaster jackets with paper towels used as a separating agent.

A 3-m-thick stratigraphic section was measured with a Jacob staff and Brunton compass, and detailed lithologic descriptions were recorded from 1 m below the bonebed to 2 m above it. Particular interest was devoted to the sedimentology and sedimentary structures of silty mudstones and sandstone lenses situated in between mudstones, as these often provide important clues to hydraulic processes.

The assessment of the taxonomic identity of UMNH VP 16690 was based on firsthand observations of BYU 8120. Additionally, UMNH VP 16690 was compared to all other North American tyrannosauroids (and many other tyrannosauroids from around the world), many of which have been discussed at length in the literature (Brochu, 2003; Currie, 2003; Xu et al., 2004, 2006; Brusatte et al., 2009, 2010; Li et al., 2009; Sereno et al., 2009; Carr and Williamson, 2010; Carr et al., 2011). Taphonomic analysis was performed with particular attention to bone modification, such as greenstick fractures, predation, and insect modification, and was compared to taphonomic modes observed from previous vertebrate localities within the Kaiparowits Formation (Roberts, Rogers, and Foreman, 2007).

Institutional Abbreviations BYU, Brigham Young University Museum of Paleontology, Provo, Utah; UMNH VP, Natural History Museum of Utah (formerly Utah Museum of Natural History), Salt Lake City, Utah.

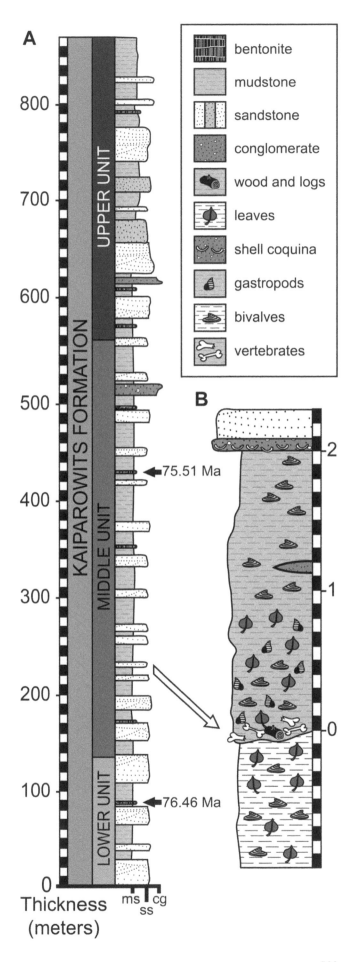

27.1. (A) Stratigraphic position of the KJT quarry in the Kaiparowits Formation. (B) Detailed stratigraphic section through the quarry.

Stratigraphic Data

The Kaiparowits Formation is approximately 850 m thick (Fig. 27.1) and is further subdivided into three informal members: the upper, middle, and lower members (Roberts, 2007, this volume, Chapter 6). The tyrannosaur quarry (KJT) from which UMNH VP 16690 was recovered is situated in the lower middle member, at approximately 230 m above the base of the Kaiparowits Formation (Fig. 27.1). Lithologies in the middle member are mostly silty mudstone and sandstone. Sedimentary structures and fossil content suggest these strata were deposited in a wide range of fluvial, lacustrine, and palustrine depositional environments in a relatively humid climate regime (Roberts, 2007). Chronostratigraphic ages have been obtained on the basis of $^{40}Ar/^{39}Ar$ dating of sanidine crystals present in bentonite ash beds and produced ages of 75.96 ± 0.14 Ma, 80 m above the basal contact and 74.21 ± 0.18 Ma, approximately 70 m below the upper contact of the formation (Roberts, Deino, and Chan, 2005), placing the Kaiparowits Formation and the KJT quarry firmly in the Upper Campanian. The Kaiparowits Formation correlates to other packages of strata across the Western Interior Basin, ranging from Alberta to New Mexico (Goodwin and Deino, 1989; Hunt and Lucas, 1992; Eberth and Hamblin, 1993; Rogers, Swisher, and Horner, 1993; Rogers, 1994; Fassett and Steiner, 1997; Sullivan and Lucas, 2006).

The bone-bearing bed at KJT consists mostly of fine-grained silty mudstones resting on a scoured 0.2–0.6-m-thick layer of silty mudstone. The scoured base of the bonebed shows up to 0.4 m of relief over the extent of the quarry floor. Lithologies above and below the bone layer are dominated by a grayish-blue silty mudstone with fine- to medium-sized grains of silt. Other than the basal scour below the bone layer, sedimentary structures in the silty mudstone facies are confined to occasional parallel laminae. There are no discrete burrows preserved, but the parallel laminae are obscured in a pattern consistent with bioturbation. The entire silty mudstone interval contains disarticulated, articulated, and fragmented mollusks. Fair amounts of vegetation in the form of coalified wood, other plant fragments, and angiosperm leaves are present in the bonebed above the scour. Overall, leaf material is found throughout the section from 1 m below the bonebed scour to 1 m above. Coalified wood is restricted to the bonebed unit itself. Bivalve shells are mostly restricted to below the top of the bonebed whereas gastropods are restricted to the bonebed unit above the scoured surface. The presence of nonmarine mollusk shells and shell fragments within the KJT bonebed are common, and their

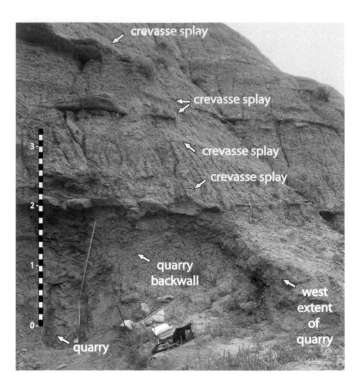

27.2. Facies relationships above the KJT quarry showing multiple sandstones interpreted as crevasse splays.

occurrence extends for several meters above and below the site. However, bivalve shells appear to form the dominant invertebrate fauna within the sediments just below and at the bonebed horizon, whereas gastropods become gradually more abundant in the sediments above the site.

The measured interval is bounded above and below by laterally discontinuous lenses of laminar and low angle cross-bedded sandstones (Figs. 27.1, 27.2). Laterally, there are tabular sandstones that grade into the mudstone facies of the quarry and extend laterally up to several hundreds of meters. These buff-colored, well-sorted, subangular, fine-grained tabular quartz arenite crevasse splays are concordantly interbedded within the silty mudstone outcrops, sometimes containing small mud rip-up clasts indicating high-velocity currents, possibly from high-water flooding stages. The lenses display sedimentary structures in the form of scour bases and reactivation surfaces at the transition between silty mudstones and sandstone, followed by near-horizontal planar laminations situated in the lower half of the sandstone layer and low-angle, unidirectional cross-bedding and occasional type 2 drift climbing ripple lamination in the upper half of the layer (Allen, 1973). Cross-bedded angles vary from 9° to 16° from the horizontal. These tabular sandstone outcrops vary from 0.1 to 0.6 m in thickness. Fossils are present in these sandstones but are heavily reworked and fragmentary, predominantly reduced to millimeter-sized angular mollusk shell fragments.

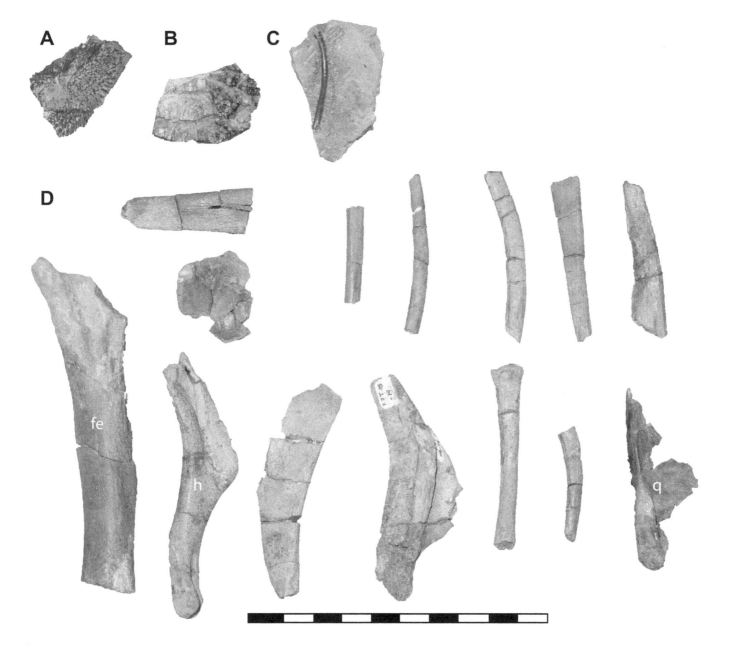

27.3. (A–D) Other taxa from the KJT quarry; (A) fish osteoderm; (B) crocodilian osteoderm; (C) bird bone; (D) partially complete juvenile hadrosaur. fe, femur; h, humerus; q, quadrate. Scale bar in total = 10 cm.

Paleontologic and Taphonomic Data

The bonebed contains unidentifiable plant material, angiosperm leaves, and sections of carbonized logs and branches. There were five logs and branches about 1 m long in the quarry. Some of these logs include short, broken secondary branches in places and an external surface impression of rotten wood. Tertiary branches appear stripped from the secondary branches, and no leaves are associated with or near the preserved logs. No bark impressions were preserved in the quarry. The logs were impossible to collect intact during excavation of the quarry, as they were preserved as friable coal. Invertebrate mollusks including up to 2-cm-long bivalves and small gastropods are abundant and frequently found within the bonebed mudstones. The subadult tyrannosaur specimen (UMNH VP 16690) is the dominant taxon in the bonebed; however, other vertebrate remains are present, including a quadrate, humerus, fibula, and vertebrae of a single hatchling hadrosaurid; a dentary from a subadult hadrosaurid; and a large, flat, weathered bone that may be part of a lambeosaurine scapular blade; together with a fish cranial element, a turtle osteoderm, a crocodilian osteoderm, an avian ulna, and a single lizard vertebra (Fig. 27.3).

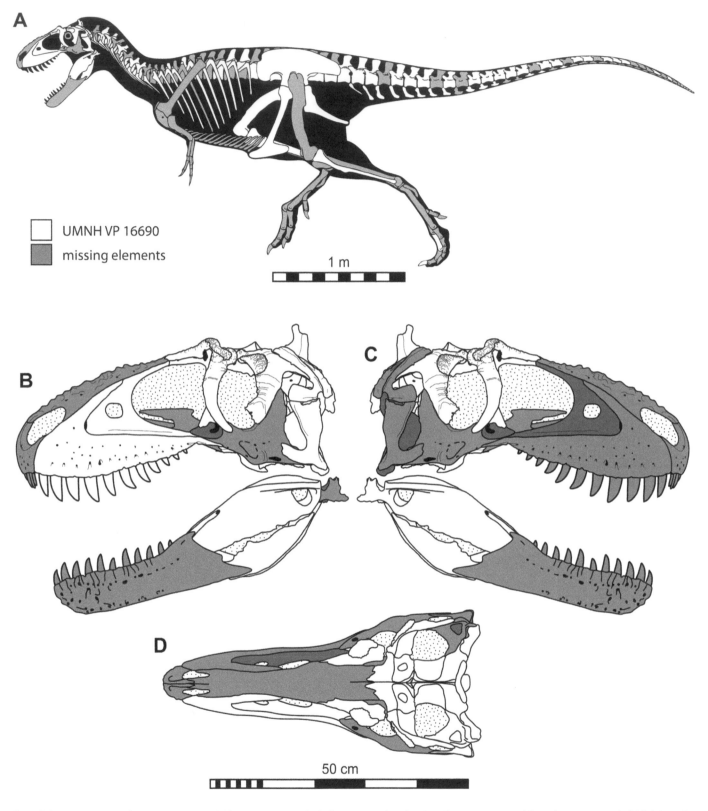

27.4. (A) Reconstruction of *Teratophoneus curriei* (UMNH VP 20200); skull reconstruction showing elements recovered from the KJT quarry in (B) left lateral, (C) right lateral, and (D) dorsal view. Skeletal reconstruction modified after artwork by S. Hartman.

Legend:
- □ UMNH VP 16690
- ▩ missing elements

1 m

50 cm

Wiersma and Loewen

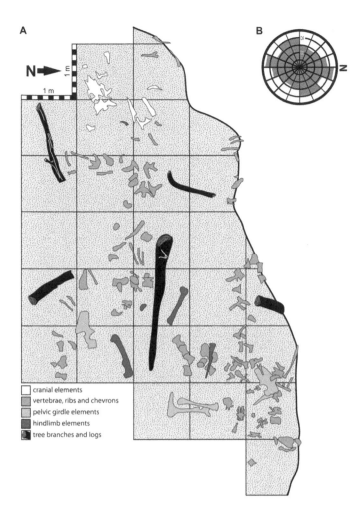

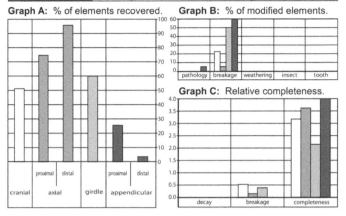

SKELETAL CLASS		ELEMENTS REPRESENTED	MAPPED ELEMENTS AND EXPECTED
CRANIAL		skull elements	31 of 58
AXIAL	proximal	cervical, dorsal & sacral verts	20 of 27
	distal	caudal verts and chevrons	50 of 62
GIRDLE		pelvic and shoulder elements	6 of 10
APPENDICULAR	proximal	limb elements	3 of 12
	distal	metapodials and phalanges	2 of 52

Graph A: % of elements recovered.   **Graph B:** % of modified elements.

Graph C: Relative completeness.

27.6. Tyrannosaurid elements analyzed from UMNH VP 20200 and taphonomic assessments. (Graph A) Percentage of tyrannosaurid elements recovered assuming a complete skeleton. These elements include 50% of the skull (with 76% represented by at least one side) and 71 % of all postcranial elements, with the exception gastralia. (Graph B) Percentage of elements showing modification. Taphonomic modification is limited to breakage. Tooth modification by carnivores is absent on all elements, as are insect modification, wet rot, and subaerial weathering. (Graph C) Amount of decay, breakage, and relative completeness of elements on a 0 to 4 scale. This scores breakage (average number of breaks) on a scale of 0 to 4, averaged for all of the elements of each skeletal class. Bone decay was absent. Most elements are more than 50% complete.

27.5. (A) Quarry map of the KJT showing the relationships of tyrannosaurid elements and logs in the quarry. (B) Rose diagram illustrating the near-random orientations of the tyrannosaurid elements in the quarry.

The most abundant vertebrate material in the quarry is tyrannosaurid, representing a single associated individual (UMNH VP 16690). The specimen includes most of the skull and axial column, the pelvis and part of the right hind limb. (Fig. 27.4). The skull includes the left maxilla, lacrimals, postorbitals, right squamosal, left quadratojugal, left quadrate, frontals, parietals, braincase, ectopterygoids, pterygoids, angulars, surangulars, left prearticular, left articular, and multiple teeth. The braincase is articulated; however, the rest of the skull is disarticulated but closely associated (Fig. 27.5). The skull is lacking the premaxillae, right maxilla, nasals, jugals, left squamosal, palatines, vomer, dentaries, splenials and the right prearticular and articular. Several associated rostral elements, including several teeth, are shattered, exhibiting greenstick fractures, suggesting that a large portion of the facial skeleton may have been trampled prior to burial (Wiersma, Loewen, and Getty, 2009).

Disarticulated axial elements of UMNH VP 16690 include the following: the atlas, seven cervical vertebrae and ribs,

eight dorsal vertebrae and 14 dorsal ribs, two sacral vertebrae, 34 caudal vertebrae, and 19 chevrons. The appendicular skeleton is represented by the parts of both ilia, both pubes, both ischia, the right femur, tibia, fibula, a single pedal phalanx, and an ungual. In total, the elements recovered make up 65% of the skeleton based on the number of expected elements in a complete skeleton. Full body size reconstruction of the skeleton indicates that it is approximately 6 m in length, a result consistent with a small subadult individual, further supported by the partial fusion of the neurocentral sutures from preserved cervical, dorsal, sacral, and anterior caudal vertebrae (Brochu, 1996).

A variety of both long and short elements appear absent from close association within the core concentration of the bonebed. A significant portion of the appendicular skeleton is missing, including forelimbs and shoulder girdles, the left leg, and the majority of the pes. Missing elements from the axial skeleton include several cervical, dorsal, and caudal vertebrae, some dorsal and cervical ribs, and the gastralia. The size of elements appears to have no correlation to

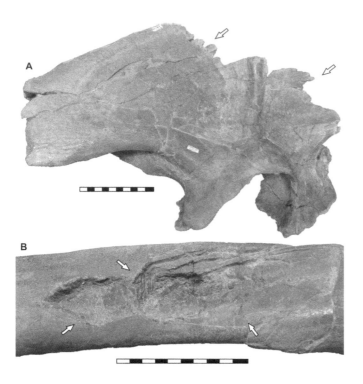

27.7. Breakage and crushing of appendicular elements. (A) Right ilium of UMNH VP 20200 showing breakage of the anterior end. (B) Crushing of the midshaft of the right tibia; distal end is to the right. Scale bars = 10 cm.

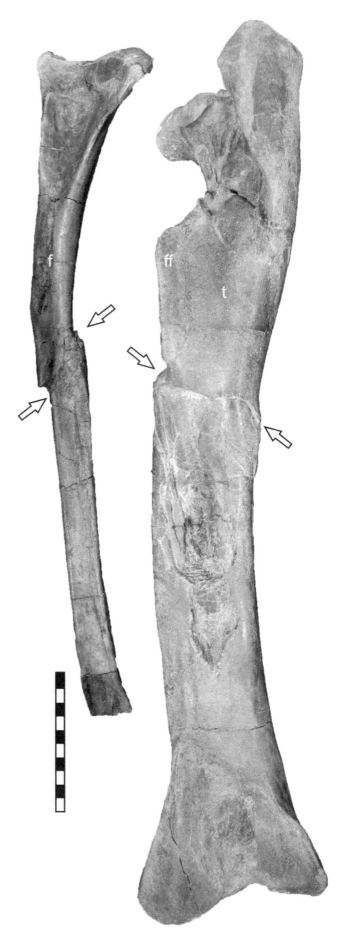

preservation; tiny skull bones and cervical ribs are preserved along with large limb bones such as the femur and tibia.

All of the elements of UMNH VP 16690 lack any indication of insect modification, scavenging, or predation (Fig. 27.6). The bones also lack any indication of surface weathering, root traces, wet rot, or mechanical abrasion. The front of the skull, some vertebrae, and the ilia all exhibit greenstick breakage consistent with fresh bone fractures penecontemporaneous with burial (Fig. 27.7). Some of the vertebrae and limb elements have been postdepositionally plastically deformed. Some pathologies occur in the form of premortem partially healed greenstick fractures of the upper shaft of the right tibia, just below the fibular flange, and the upper shaft of the right fibula (Fig. 27.8). These bones were associated in the quarry but were found over 1 m apart. Rearticulation of the tibia and fibula shows the parallel continuation of the fractures, suggesting breakage of both elements in a single event prior to disarticulation. We interpret these breaks as partially healed fractures in the lower right hind limb. There is a halo of healed bone associated with this break. This implies that the animal broke its leg during life. A break like

27.8. Pathology of UMNH VP 20200. The right fibula and tibia are broken and rehealed at the same place near the midshaft. Arrows show the line of breakage and rehealing. The fibula is in medial view and the tibia is in anterior view. f, fibula; ff, fibular flange of the tibia; t, tibia. Scale bar = 10 cm.

Wiersma and Loewen

this could have led to possible starvation or dehydration, a possible cause of death.

Recovered elements show stratigraphic relief up to 0.4 m between individual elements. Such relief might be attributed to differences in weight of individual bones, with heavier elements having sunk deeper into the mud compared to much lighter elements. For example, large, heavy bones such as the femur and the ilia were located at lower elevations compared to the vertebrae, ribs, and chevrons, which are much lighter in weight and were located at higher elevations. The vertebrate accumulations from KJT are considered a multitaxic monodominant bonebed (Eberth, Rogers, and Fiorillo, 2007) in which tyrannosaur material comprises more than 80% of the elements recovered and represents the dominant taxon.

DISCUSSION

Sedimentologic Interpretations

Sedimentologic indicators, including scoured siltstone below the bone layer, laterally equivalent tabular sandstone, and burial in fine-grained silty claystone and mudstone, suggest that the site represents a shallow standing body of water in a floodplain environment. We attribute its origin to a flooding episode related to the deposition of the lateral tabular sandstone bodies interpreted as crevasse splays.

Sedimentary structures are largely absent in the fine-grained silty mudstone deposits, suggesting a low-energy, palustrine, or perennial freshwater pond environmental setting. Decreasing bivalve abundance coincides with replacement by gastropods. Gastropod faunas are strong indicators for long-standing, low-energy environments such as perennial ponds and lacustrine waters. The blue-gray color indicates organic-rich, perpetually saturated conditions with minimal oxidation. At some point, the surface was briefly exposed by the disappearance of the lake and scours, and reactivation surfaces were formed by subsequent flooding episodes. Periodic increase of energy levels is furthermore supported by the presence of <1 cm unionid and gastropod fragments within the silty mudstone beds. The presence of such small shell fragments is possibly the result of in situ reworking of dead unionid shells that have been transported and reworked during flooding stages. Furthermore, some modern unionid mollusks prefer habitation in high-energy environments such as channel systems. Further work is needed to fully describe the range of preferred environments of unionid mollusks in the Kaiparowits Formation. The frequent encounter of broken unionid fragments from the silty mudstone beds within and below the bonebed suggest that bivalves were transported from their high-energy environments during flooding stages and the formation of crevasse splays in which they

became fragmented, possibly underwent some reworking, and eventually became deposited along the scour surface of the crevasse splay flood event.

Near-horizontal laminations in the bounding sandstone beds resulted from high-energy flow regimes, drowning the floodplains during high-water flooding stages and transporting bed loads of suspended sands from nearby channels with velocities high enough to prevent the formation of current ripples. Subsequent decreasing energy levels and lower current regimens resulted in the formation of unidirectional cross-beds within the upper half of the sandstone outcrops.

We interpret the depositional environment at the approximate time of bonebed formation as a fluvial channel system in combination with floodplain depressions in which periodic pulses of high-water-level floods. These floods resulted in breaching of levees, producing crevasse splays proximal to the channels and relatively long-term standing ponds within floodplain depressions more distal to the channel breaches. At present, the spatial relationships of the canyon in which the quarry is located make it impossible to trace the scoured quarry floor to a specific crevasse splay sandstone laterally. During flooding in fluvial systems, a lot of pelagic and suspended sediment flows around in the water and slowly settles far distal from the terminal end of the crevasse splay, producing a fine-grained sedimentary facies in the same horizon as the crevasse splay. The specific crevasse splay to which we attribute the scoured surface of the quarry may be lateral to the quarry in the subsurface to the north or south of the quarry. Cycles of floodplain mudstones, sandstone channels, and tabular sandstones repeat many times lateral to, below, and above the level of the quarry; this pattern of fluvial sedimentation represents the dominant pattern of sedimentation in this part of the middle unit of the Kaiparowits Formation (Roberts, 2007).

The depositional setting appears to be largely overlapping with the organic-poor mudrocks facies association within the lower Horseshoe Canyon Formation described by Straight and Eberth (2007). They suggest seasonally dry, inland floodplains with slow sedimentation rates relative to the organic-rich facies association. However, the sediments in which the KJT bonebed was located suggest rapid sedimentation rates in which only incipient pedogenesis might have occurred (Titus, pers. comm., 2011; Roberts, pers. comm., 2011).

Paleontologic and Taphonomic Interpretations

The subadult tyrannosaur skeleton is associated with elements of a hatchling hadrosaur, bird, fish, turtle, and crocodile, together making up a monodominant multitaxic bonebed (Eberth, Rogers, and Fiorillo, 2007). Many of the tyrannosaur bones show some form of modification in the form of plastic deformation, which we interpret as compaction deformation that occurred after burial.

In contrast to the majority of vertebrate localities within mudstones in the Kaiparowits Formation, there are no signs of predation in the form of tooth marks, insect borings, or shed teeth from other predatory animals commonly found in floodplain environments. The absence of root traces and surface weathering, along with an absence of predation, suggest that the animal was buried rapidly and did not sit out on the floodplain exposed to these modifying agents. Although scavenging of any sort appears to be absent, some form of interaction appears to have occurred between the carcass and other large animals because a large portion of the anterior skull, together with the distal end of the tibia and fibula show signs of greenstick predepositional breaks. These breaks are consistent with the bones being stepped on by a large animal, likely another dinosaur.

The lack of mechanical abrasion and any strong current orientation in elements suggests that the animal was not deposited by fluvial action as sedimentary particles under high-energy transport. Winnowing is not a factor in the deposition of UMNH VP 16690 when considering that small cranial elements and fragments of skull bones, thin cervical ribs, the atlas, toe bones, and distal caudal vertebrae and chevrons are preserved. With a major portion of the appendicular skeleton missing, the question arises of what processes might have caused small and large elements to be lost, yet allow for similar elements to remain preserved. These missing elements of the skeleton require further discussion. We believe that the loss of gastralia and the forelimbs can be explained by the expansion of the gastric gases in the rotting carcass, which eventually ruptured and might have contributed to the loss of these elements.

Winnowing by fluvial currents might explain the missing arms, leg, and feet but would not be consistent with the presence of the smaller skull and tail elements preserved in the quarry. Most of the anterior part of the skull is absent, and winnowing would not explain the removal of the premaxillae, right maxilla, jugals, and nasals without removing the disarticulated postorbitals and quadratojugals, which are smaller and more easily removed, than some of the missing elements. Certain elements—in particular those belonging to the pectoral region—often become removed from vertebrate carcasses, even in low-energy environments (Weigelt, 1989), and might partially explain the missing elements in UMNH VP 16690. Some elements possibly eroded away on the outcrop prior to the discovery of the specimen, and others might still remain in the back wall of the quarry.

The braincase of the skull was partially articulated with the parietals and frontals. Both postorbitals were associated, and the right postorbital and left squamosal were associated but displaced from life position. The left postorbital, lacrimal, and maxilla exhibit greenstick bone fractures and are crushed. We interpret this to be consistent with the skull lying upright on the bottom of the pond, listing to the right, having been stepped on just forward of the orbit, fracturing the rostrum of the skull. The left maxilla is completely crushed and was found lying parallel to the bedding plane. Multiple loose rooted teeth (more than are missing from the left maxilla) were found in the sediments around the skull, suggesting that the complete rostrum of the skull was disarticulated in place and elements dispersed from their original position in the quarry. Disturbance as a result of trampling may have destroyed these elements or shifted them out of the quarry (beyond the quarry back wall or beyond the eroded surface of discovery) and prevented their preservation or discovery. Element distribution shows that topographic differences extend as much as 0.4 m in elevation between individual elements in the quarry. Low energy environments with irregular topography are not uncommon in hosting vertebrate remains and might be the result of a number of external factors such as frequent flooding, erosion, or possible trampling by other large animals after carcasses are deposited (Weigelt, 1989; Wiersma, Loewen, and Getty, 2009).

Comparison with other macrovertebrate multitaxic bonebeds preserved in fine-grained sediments in the Kaiparowits Formation (Getty et al., 2003, 2010) reveals significant differences between the taphonomic mode and preservation of UMNH VP 16690. These vertebrate accumulations commonly exhibit multiple examples of wet rot, root traces, abrasion, and predation activity by carnivorous vertebrates and insects. The presence of such conditions suggest extensive subaerial exposure prior to burial and resulted in far less preserved accumulations of vertebrate bonebeds. We consider several possible taphonomic scenarios consistent with the detailed stratigraphy and sedimentology of the site, including the overall package of sediments and facies lateral to the quarry. The mudstone of the bonebed is consistent with low-energy lacustrine deposition. The entire package of sediments that underlie, overlay, and are lateral to this lacustrine mudstone suggests a combination of overbank fluvial and lacustrine–palustrine depositional environments. The sedimentologic and stratigraphic data suggest a floodplain environment adjacent to a channel that experienced periodic flooding events, producing crevasse splays and standing water in floodplain depressions.

CONCLUSIONS

We suggest a taphonomic history of death of a subadult *Teratophoneus* (UMNH VP 16690), possibly as the result of the pathologically partially rehealed fractured shafts of the right tibia and fibula, causing reduced mobility in the hind limb and/or reduced prey acquisition ability from which the animal eventually died on the floodplain. Bloating of the skeleton by decomposition and production of gas in the abdominal cavity may have allowed it to be transported during an overbank flooding event to its discovery location. Parts of the carcass, including the gastral basket, the forelimbs, parts of the skull, pelvis, and one hind limb, may have been lost to decomposition or scavenging before final settling in shallow standing water or waterlogged sediments. The abdominal cavity finally ruptured and degassed, and the carcass sank in standing water to the bottom of a shallow floodplain pond. This resulted in burial and disarticulation of the skeleton in fine-grained sediments, which was followed by partial trampling of the skull and vertebral column in shallow standing water or waterlogged sediment.

This is supported by taphonomic data including lack of current orientation in the fine-grained sediments in which the carcass came to rest, loose association of parts of the body, the presence of nonmarine gastropod faunas, the concentration of plant material, and the lack of fluvial sorting or abrasion, which are all consistent with a shallow floodplain pond. Missing elements may have been redistributed by light lacustrine swash while still being surrounded by rotten flesh. Some of the elements may have also eroded away before discovery in 2004. This interpretation contrasts with interpretations of most other vertebrate localities in fine-grained sediments within the Kaiparowits Formation, which exhibit multiple examples of predation and insect modification suggesting extensive subaerial exposure.

ACKNOWLEDGMENTS

The authors wish to thank Grand Staircase–Escalante National Monument (GSEMN) and the Bureau of Land Management (BLM) for their continued support throughout the duration of this project. All specimens were collected under permits from the BLM. Special thanks go to A. Titus (GSENM) for his ongoing efforts both in and out of the field. Without dozens of dedicated students, faculty, and museum volunteers from the Natural History Museum of Utah (NHMU), formerly the Utah Museum of Natural History (UMNH), the University of Utah, Utah Friends of Paleontology, the Web Schools, and collaborating institutions, this project could never have achieved the success we document here. Additional thanks go to E. Lund, J. Sertich, R. Irmis, and E. Roberts for their substantial contributions in the field and in the lab. M. Getty, E. Roberts, J. Sertich, and R. Irmis commented on an early version of this contribution. Skeletal drawings were modified from artwork by S. Hartman. Funding was provided by GSENM, NHMU, UMNH, and the National Science Foundation (NSF 0745454, Kaiparowits Basin Project).

REFERENCES CITED

Allen, J. R. L. 1973. A classification of climbing-ripple cross-lamination. Journal of the Geological Society 129:537–541.

Brochu, C. A. 1996. Closure of the neurocentral sutures during crocodilian ontogeny: implications for maturity assessment in fossil archosaurs. Journal of Vertebrate Paleontology 16:49–62.

Brochu, C. A. 2003. Osteology of Tyrannosaurus rex: insights from a nearly complete skeleton and high-resolution computed tomographic analysis of the skull. Journal of Vertebrate Paleontology 22(4, Supplement):1–138.

Brusatte, S. L., T. D. Carr, G. M. Erickson, G. S. Bever, and M. A. Norell. 2009. A long snouted, multi-horned tyrannosaurid from the Late Cretaceous of Mongolia. Proceedings of the National Academy of Sciences of the United States of America 106:17261–17266.

Brusatte, S. L., M. A. Norell, T. D. Carr, G. M. Erickson, J. R. Hutchinson, A. M. Balanoff, G. S. Bever, J. N. Choiniere, P, J. Makovicky, and X. Xing. 2010. Tyrannosaur paleobiology: new research on ancient exemplar organisms. Science 329:1487–1485.

Carr, T. D., and T. E. Williamson. 2010. Bistahieversor sealeyi, gen. et sp. nov., a new tyrannosauroid from New Mexico and the origin of deep snouts in Tyrannosauroidea. Journal of Vertebrate Paleontology 30:1–16.

Carr, T. D., T. E. Williamson, B. B. Britt, and K. Stadtman. 2011. Evidence for high taxonomic and morphologic tyrannosauroid diversity in the Late Cretaceous (Late Campanian) of the American Southwest and a new short-skulled tyrannosaurid from the Kaiparowits Formation of Utah. Naturwissenschaften 98:241–246.

Currie, P. J. 2003. Cranial anatomy of tyrannosaurid dinosaurs from the Late Cretaceous of Alberta, Canada. Acta Palaeontologica Polonica 48:191–226.

Eberth, D. A., and A. P. Hamblin. 1993. Tectonic, stratigraphic and sedimentologic significance of a regional discontinuity in the Upper Judith River Group (Belly River Wedge) of southern Alberta, Saskatchewan, and northern Montana. Canadian Journal of Earth Sciences 30:174–200.

Eberth, D. A., R. R. Rogers, and A. R. Fiorillo. 2007. A practical approach to the study of bonebeds; pp. 265–331 in R. R. Rogers, D. A. Eberth, A. R. Fiorillo (eds.), Bonebeds: Genesis, Analysis, and Paleobiological Significance. University of Chicago Press, Chicago, Illinois.

Fassett, J. E., and M. B. Steiner. 1997. Precise age of C33N–C32R magnetic polarity reversal, San Juan Basin, New Mexico and Colorado; pp. 239–247 in O. J. Anderson, B. S. Kues, and S. G. Lucas (eds.), Mesozoic Paleontology and Geology of the Four Corners Region. New Mexico Geological Society Guidebook 48.

Getty, M. A., M. A. Loewen, E. M. Roberts, A. L. Titus, and S. D. Sampson. 2010. Taphonomy of horned dinosaurs (Ornithischia: Ceratopsidae) from the Late Campanian Kaiparowits Formation, Grand Staircase–Escalante National Monument, Utah; pp. 478–494 in M. J. Ryan, B. J. Chinnery-Allgeier, and D. A. Eberth (eds.), New Perspectives on Horned Dinosaurs: The Royal Tyrrell Museum Ceratopsian Symposium. Indiana University Press, Bloomington, Indiana.

Getty, M., E. M. Roberts, M. A. Loewen, J. Smith, T. A. Gates, and S. D. Sampson. 2003. Taphonomy of a chasmosaurine ceratopsian skeleton from the Campanian Kaiparowits Formation, Grand Staircase–Escalante National Monument, Utah. Journal of Vertebrate Paleontology 23:54A.

Goodwin, M. B., and A. L. Deino. 1989. The first radiometric ages from the Judith River Formation (Upper Cretaceous), Hill County, Montana. Canadian Journal of Earth Sciences 26:1384–1391.

Hunt, A. P., and S. G. Lucas. 1992. Stratigraphy, paleontology and age of the Fruitland and Kirtland formations (Upper Cretaceous), San

Juan Basin, New Mexico; pp. 217–239 in S. G. Lucas, B. S. Kues, T. E. Williamson, and A. P. Hunt (eds.), New Mexico Geological Society Guidebook, 43rd Field Conference, San Juan Basin IV.

Li, D., M. A. Norell, K. Gao, N. D. Smith, and P. J. Makovicky. 2009. A longirostrine tyrannosauroid from the Early Cretaceous of China. Proceedings of the Royal Society Series B 277:183–190.

Loewen, M. A., J. J. W. Sertich, R. B. Irmis, and S. D. Sampson 2010. Tyrannosaurid evolution and intracontinental endemism in Laramidia: new evidenece from the Campanian Wahweap Formation of Utah. Journal of Vertebrate Paleontology 30:123A.

Roberts, E. M., A. D. Deino, and M. A. Chan. 2005. ^{40}Ar/^{39}Ar age of the Kaiparowits Formation, southern Utah, and correlation of coeval strata and faunas along the margin of the Western Interior Basin. Cretaceous Research 26:307–318.

Roberts, E. M., R. R. Rogers, and B. Z. Foreman. 2007. Continental insect borings in dinosaur bone: examples from the Late Cretaceous of Madagascar and Utah. Journal of Paleontology 81:201–208.

Roberts, E. M. 2007. Facies architecture and depositional environments of the Upper Cretaceous Kaiparowits Formation, southern Utah. Sedimentary Geology 197:207–233.

Rogers, R. R. 1994. Nature and origin of through-going discontinuities in nonmarine foreland basin strata, Upper Cretaceous, Montana: implications for sequence analysis. Geology 22:1119–1123.

Rogers, R. R., C. C. Swisher III, and J. R. Horner. 1993. ^{40}Ar/^{39}Ar age correlation of the nonmarine Two Medicine Formation (Upper Cretaceous), northwestern Montana, U.S.A. Canadian Journal of Earth Sciences 30:1066–1075.

Sereno, P. C., L. Tan, S. L. Brusatte, H. J. Keiegstein, X. Zhao, and K. Cloward. 2009. Tyrannosaurid skeleton design first evolved at small body size. Science 326:418–422.

Straight, W. H., and D. A. Eberth. 2002. Testing the utility of vertebrate remains in recognizing patterns in fluvial deposits: an example from the Lower Horseshoe Canyon Formation, Alberta. Palaios 17:472–490.

Sullivan, R. M., and S. G. Lucas. 2006. The Kirtlandian Land Vertebrate "Age" –faunal composition, temporal position and biostratigraphic correlation in the nonmarine Upper Cretaceous of Western North America; pp. 7–29 in S. G. Lucas and R. M. Sullivan (eds.), Late Cretaceous Vertebrates from the Western Interior. New Mexico Museum of Natural History and Science Bulletin 35.

Weigelt, J. 1989. Translated by J. Schaefer. Recent Vertebrate Carcasses and Their Paleobiological Implications. University of Chicago Press, Chicago, Illinois.

Wiersma, J. P., M. A. Loewen, and M. A. Getty. 2009. Taphonomy of a subadult tyrannosaur from the Upper Campanian Kaiparowits Formation of Utah. Journal of Vertebrate Paleontology 29:201A.

Xu, X., M. A. Norell, X. Kuang, X. Wang, Q. Zhao, and C. Jia. 2004. Basal tyrannosauroids from China and evidence for protofeathers in tyrannosauroids. Nature 431:680–684.

Xu, X., J. M. Clark, C. A. Forster, M. A. Norell, G. M. Erickson, D. A. Eberth, C. Jia, and Q. Zhao. 2006. A basal tyrannosauroid dinosaur from the Late Jurassic of China. Nature 439:715–718.

A New Macrovertebrate Assemblage from the Late Cretaceous (Campanian) of Southern Utah

28

Scott D. Sampson, Mark A. Loewen, Eric M. Roberts, and Michael A. Getty

FOR MOST OF THE LATE CRETACEOUS, A SHALLOW EPEIRic sea subdivided North America into eastern and western landmasses—Appalachia and Laramidia, respectively. Whereas little is known of Appalachian faunas, Laramidia has yielded an abundant terrestrial fossil record, arguably the best continent-scale example for any Mesozoic time interval. To date, however, the bulk of these fossils have been recovered from the northern portion of Laramidia, in particular Alberta and Montana. The relatively poor fossil record from southern Laramidia has limited our ability to test several key biogeographic and evolutionary hypotheses.

In 2000, the Utah Museum of Natural History (now the Natural History Museum of Utah), University of Utah, launched an interdisciplinary project in Grand Staircase–Escalante National Monument aimed at exploring Late Cretaceous terrestrial and freshwater ecosystems preserved within the Kaiparowits Basin. Although emphasis has been placed on collection and study of macrovertebrates, from the outset the "Kaiparowits Basin Project" has targeted a spectrum of data sources, spanning sedimentology, stratigraphy, paleobotany, and ichnology, as well as vertebrate and invertebrate paleontology. This large-scale, interdisciplinary effort has thus far concentrated on two Campanian-aged geologic formations, the Kaiparowits and Wahweap, with spectacular results that include a previously unknown assemblage of dinosaurs and other macrovertebrates. Turtles, crocodylians, and dinosaurs all exhibit relatively high species diversity, with many endemic taxa. The fossiliferous Kaiparowits Formation alone has yielded remains of 14 turtle taxa, six crocodylian taxa, and 16 dinosaur taxa, with evidence that these communities inhabited a wet, largely swampy environment with vegetation ranging from cypress trees (lowest elevation) to dicot forests (better-drained settings) to forests of taxodiaceous and other gymnosperms (well-drained settings). These and other finds are stratigraphically constrained by multiple radiometric dates, allowing robust comparisons with coeval northern vertebrate assemblages. Considered in unison, five distinct lines of evidence—taxonomy, phylogeny, stratigraphy, paleoenvironment, and biogeography—provide strong support for the vertebrate provincialism hypothesis, which postulates the occurrence of latitudinally arrayed biotic "provinces" on Laramidia for at least a portion of the Late Cretaceous. Paleontologically, the Kaiparowits Formation has now been established as the best-known Campanian unit from southern Laramidia, exceeded in the north only by the Dinosaur Park Formation of Alberta.

INTRODUCTION AND BACKGROUND

Laramidia: A Late Cretaceous Island Continent

Late Cretaceous biotic evolution took place against a background of global-scale environmental changes. At the beginning of this interval, active tectonism triggered high levels of volcanism, an increase in global atmospheric temperatures (peaking in the Turonian), and an abrupt rise in global sea levels, which in turn resulting in flooding of low-lying regions on multiple continents (Haq, Hardenbol, and Vail, 1987; Barrera and Savin, 1999; Catuneanu, Sweet, and Miall, 2000). Beginning about 96 Ma, an epeiric sea inundating the central region of North America subdivided the continent into eastern and western landmasses (Fig. 28.1). For most of the Late Cretaceous (early Cenomanian through late Campanian), this shallow body of water, the Cretaceous Interior Seaway, isolated the eastern and western landmasses, referred to as Appalachia and Laramidia, respectively (Archibald, 1996). The dynamic margins of this seaway underwent large-scale transgressions and regressions, alternately contracting and expanding the areal extent of the adjacent landmasses. High sea levels also isolated Appalachia and Laramidia from surrounding continents (i.e., Asia, Europe, South America).

Whereas we know relatively little about the Late Cretaceous biotas of Appalachia, those from Laramidia are much better established, thanks to fossiliferous formations, geologic uplift, an abundance of arid and poorly vegetated "badlands," and more than a century of intense sampling and study. Today, we know more about Campanian faunas from Laramidia than from any other continent-scale region of Mesozoic age. Campanian fossil localities extend virtually

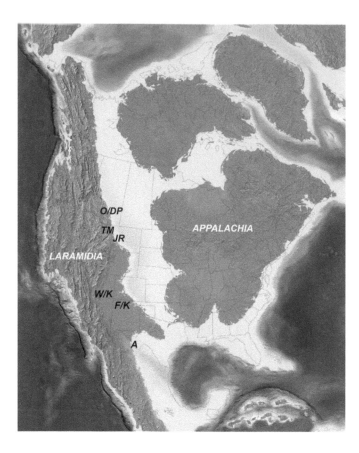

28.1. Late Campanian (75 Ma) North America, depicting the Western Interior Seaway subdividing the continent into Laramidia and Appalachia. Key Campanian-aged geologic formations are shown. A, Aguja Formation; DP/O, Dinosaur Park Formation; F/K, Fruitland and Kirtland formations; JR, Judith River Formation; TM, Two Medicine Formation; W/K, Wahweap and Kaiparowits formations. Paleogeography after Blakey (2009).

the entire latitudinal span of this landmass, from the North Slope of Alaska to Coahuila in central Mexico. The bulk of these localities occur in key regions of the Western Interior Basin, including Alberta (Oldman and Dinosaur Park formations), Montana (Judith River and Two Medicine formations), Utah (Wahweap and Kaiparowits formations), New Mexico (Kirtland and Fruitland formations), and Texas (Aguja Formation). The terrestrial ecosystems recorded in these units were juxtaposed between the rising Cordilleran Overthrust Belt to the west and the Western Interior Seaway to the east. Although variably sized as a result of seaway fluctuations, the total combined area of dinosaur-rich habitats on Campanian Laramidia encompassed approximately 4 million square kilometers (Lehman, 1987; Scotese, 2001), less than one-fifth the area of present-day North America.

To date, the biotic implications of Laramidia as an isolated, diminutive, and persistent landmass has not been fully considered. Some researchers have referred to this "island continent" simply as North America, or western North America. Alternatively, others have focused on subaerial connections

between Asia and western North America, variously referring to the conjoined landmass as Paleolaurasia (Russell, 1993), Asiamerica (Russell, 1995), or the Asian–American peninsula (Lehman, 1997, 2001). Yet current evidence indicates that Laramidia existed in isolation (at least as with regard to most nonvolant elements of terrestrial biotas) for about 20 million years—from at least the late Turonian (~70 Ma; Vandermark et al., 2009) until exposure of Beringia established subaerial links with Asia during the latest Campanian or early Maastrichtian (~70 Ma; Russell, 1995; Averianov and Archibald, 2003; Longrich and Currie, 2008; Pereda-Suberbiola, 2009). Shortly thereafter, during the mid-Maastrichtian (~68 Ma), retreat of the Western Interior Seaway reunited Appalachia and Laramidia as a single continent, North America, that has persisted to the present day. A connection between Laramidia and South America via the paleo-Panamanian land bridge was also established in the last 5 million years of the Cretaceous Period (Cifelli et al., 1997; D'Emic, Wilson, and Thompson, 2010; Prieto-Marquez and Salinas, 2010). The key point is that Laramidia appears to have been a distinct landmass for about 20 million years, about one third of the Cretaceous Period and a great majority of the Late Cretaceous. During that lengthy interval of isolation, and despite its diminutive size, Laramidia generated numerous endemic forms, including arguably the greatest Mesozoic florescence of dinosaurs. Thus, the establishment of land bridges with surrounding landmasses beginning in the late Campanian likely resulted in significant events of biotic exchange. In order better to understand the evolution of Mesozoic terrestrial ecosystems, we propose that investigators give greater consideration to the unique role of Laramidia, beginning with more widespread use of this name.

Latest Cretaceous dinosaurs from Laramidia and North America (i.e., after the retreat of the seaway) have been pivotal in the formation and assessment of many hypotheses relating to dinosaur physiology, behavior, rates of metabolism and growth, nesting and parental care, and herding and pack hunting (Farlow, Dodson, and Chinsamy, 1995, and references therein). With a robust and rapidly growing database, great potential now exists to begin using the Laramidian database to address fundamental and previously inaccessible questions about Mesozoic terrestrial ecosystems. For example, what was the tempo and mode of macroevolution? Did faunal turnover occur sporadically across vertebrate lineages or was it concentrated in pulses of changes that affected many lineages simultaneously? What were the ecological and evolutionary effects of latitude on biotic communities in the hothouse world of the Mesozoic?

In addition to its robust fossil record, Laramidia possesses another major advantage for resolving ecoevolutionary

Sampson et al.

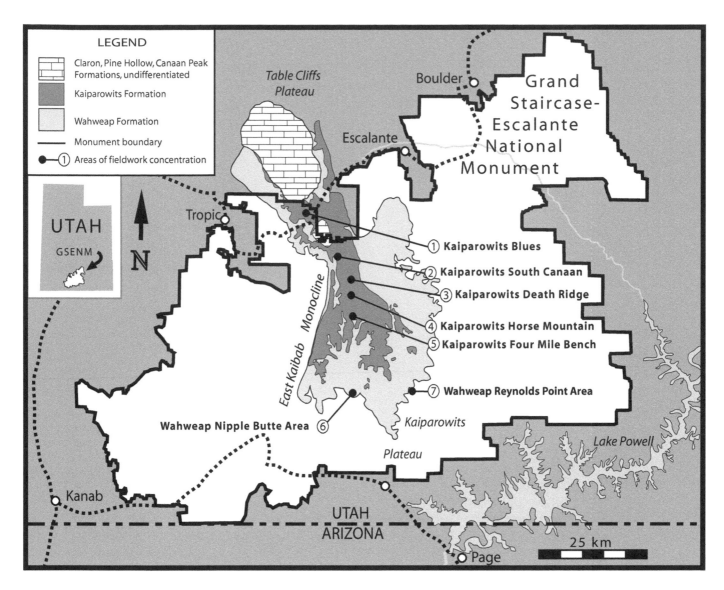

28.2. Map of Grand Staircase–Escalante National Monument (Grand Staircase–Escalante National Monument), Utah, showing the outcrop extent of the Wahweap and Kaiparowits formations, as well as areas of fieldwork concentration within these units. Inset map depicts the location of Grand Staircase–Escalante National Monument within Utah.

questions—its size. With floras and faunas largely constrained to a latitudinally arrayed series of small basins, testing of hypotheses is aided by the fact that terrestrial and freshwater biotas effectively had nowhere else to go; thus, for example, we can have greater confidence that the appearance and disappearance of biotic elements reflects evolutionary turnover rather than simply ecological displacement. Nevertheless, answering questions like those posed above requires detailed knowledge of floras and faunas from multiple formations of comparable age. Although the late Campanian of Laramidia is likely the best time and place to search for answers, at present only a handful of geologic formations preserve well-sampled terrestrial and freshwater biotas in a detailed stratigraphic framework (e.g., Dinosaur Park Formation, Alberta;

Currie and Koppelhus, 2005). Particularly problematic has been the relative dearth of fossil evidence from southern Laramidia, necessary for comparing northern and southern biotas.

To give perhaps the most pertinent example, it has been hypothesized that vertebrates during the Campanian and Maastrichtian were separated into latitudinally arrayed regional faunas (Russell, 1967; Sloan, 1969; Lehman, 1987, 1997, 2001). This finding is surprising, given the small areal extent of Laramidia and the large body sizes of many Laramidian dinosaurs. If confirmed, it would have profound ecological and evolutionary implications. Yet this hypothesis has been challenged on grounds of both temporal and geographic sampling (Sullivan, 2003; Sullivan and Lucas, 2006;

Vavrek and Larsson, 2010), with definitive testing hampered in particular by the lack of well-dated, well-sampled southern faunas. As will be detailed below, recent evidence presented by our working group (Gates et al., 2010; Sampson, Gates, et al., 2010; Sampson, Loewen, et al., 2010) is fully consistent with the presence of distinct northern and southern faunas during at least a portion of the late Campanian.

Grand Staircase–Escalante National Monument

Background Grand Staircase–Escalante National Monument encompasses a vast region (~1.9 million acres) of rugged terrain in southern Utah (Fig. 28.2). Grand Staircase–Escalante National Monument, the last major area within the contiguous United States to be mapped, was formally designated by Presidential Proclamation in 1996, established in large part to facilitate preservation and study of its diverse natural resources, both living and fossil. Several abundantly exposed, fossil-bearing Upper Cretaceous formations in particular the Dakota Formation, Tropic Shale, Straight Cliffs Formation, Wahweap Formation, and Kaiparowits Formation are concentrated in the central region of the monument known as the Kaiparowits Plateau. Considered together, these formations preserve one of the most continuous Cenomanian–Campanian terrestrial records anywhere in the world (Eaton and Cifelli, 1988). Extensive study of Grand Staircase–Escalante National Monument microvertebrates undertaken primarily during the last two decades of the 20th century documented the tremendous paleontological potential of these deposits. The bulk of this work concentrated on surface collection and screen washing of microfossils, with an emphasis on mammals (Cifelli, 1990a, 1990b, 1990c, 1990d; Eaton, 1991, 1999a, 1999b, 2002; Gillette and Hayden, 1997; Eaton et al., 1999). Although a smattering of dinosaur and other macrofossils were collected in the Monument during this interval particularly by crews from Brigham Young University and the University of California, Berkeley (Parrish and Eaton, 1991; Hutchison, 1993; Eaton et al., 1999) most of the recovered remains were fragmentary. Nevertheless, these finds demonstrated the considerable potential for recovering vertebrate macrofossils on the Kaiparowits Plateau of Grand Staircase–Escalante National Monument.

Kaiparowits Basin Project In 2000, the Utah Museum of Natural History (UMNH, recently renamed the Natural History Museum of Utah) at the University of Utah began conducting paleontological field surveys in Grand Staircase–Escalante National Monument under a 5-year assistance agreement (#JSA015003), which resulted in financial assistance from the Bureau of Land Management. This collaborative partnership rapidly developed into a larger, multidisciplinary project that was renewed under a second assistance agreement (#JSA071004). Now known as Kaiparowits Basin Project (KBP), this large-scale research effort has focused on collecting and researching terrestrial and freshwater vertebrate fossils from the Kaiparowits Basin of Grand Staircase–Escalante National Monument (Table 28.1). In particular, the project has thus far concentrated on macrofossils preserved within two Campanian-aged formations, the Wahweap and Kaiparowits, both exposed extensively on the Kaiparowits Plateau (Fig. 28.2).

The KBP has had four primary objectives: (1) collect vertebrate, invertebrate, trace, and paleobotanical fossils from the Kaiparowits and Wahweap formations to provide the first detailed window from Utah into this critical time period; (2) establish a high-precision temporal, depositional, taphonomic, and paleoenvironmental context for the richly fossiliferous Kaiparowits–Wahweap alluvial package in the Kaiparowits Basin; (3) use the paleontological and geological results to reconstruct as accurately as possible the succession of Late Cretaceous terrestrial ecosystems in the Kaiparowits Basin; and (4) place the above findings into a regional (i.e., Laramidian) framework and investigate large-scale patterns and processes relating to ecology, evolution, taphonomy and biogeography.

Table 28.1. Summary statistics for the first decade of the Kaiparowits Basin Project, 2000–2009

Work	Description	Value
Fieldwork	Days of fieldwork logged	724
	Person-hours of fieldwork logged	~36,700
	University student participants	69
	Undergraduate students	41
	Graduate students	28
	Nonstudent volunteers	79
	Total acres surveyed	~43,000
	Vertebrate localities discovered	783
	Major localities excavated	27
	Collected localities with associated dinosaur specimens	77
	Hadrosauridae (associated articulated skulls)	33 (15)
	Ceratopsia (associated articulated centrosaurine skulls)	18 (12)
	Ankylosauria (associated articulated skulls)	8 (1)
	Tyrannosauridae (associated articulated skulls)	7 (2)
	Ornithomimidae	4
	Maniraptora	6
	Associated turtle specimens collected (associated articulated skulls)	115 (3)
	Associated crocodylian specimens collected (associated articulated skulls)	26 (9)
	Total number of vertebrate specimens collected	~4200
Preparation laboratory work	Person-hours of volunteer preparation logged	~44,000
	Student participants	85
	Nonstudent volunteers	235

Sampson et al.

Table 28.2. Currently documented macrovertebrate diversity of the Upper Cretaceous (middle Campanian) Wahweap Formation, Grand Staircase–Escalante National Monument, Utah

Testudines	Alligatoroidea	Genus et sp. nov.
Cryptodira	*Deinosuchus* sp.	Thyreophora
Paracryptodira	Genus et sp. indet.	Ankylosauria
Pleurosternidae	Dinosauria	Nodosauridae
Compsemys sp. indet.	Ornithischia	Genus et sp. indet.
Baenidae	Ornithopoda	Ankylosauridae
Denazinamys nodosa	Hadrosauridae	Genus et sp. indet.
Eurcryptodira	Hadrosaurinae	Saurischia
Adocidae	*Acristavus gagslarsoni*	Theropoda
Adocus sp. indet.	Genus et sp. nov. B	Coelurosauria
Nanhsiungchelyidae	Lambeosaurinae	Ornithomimidae
Basilemys sp. indet.	Genus et sp. indet.	Genus et sp. indet.
Trionychidae	Basal Neornithischia	Tyrannosauridae
Cf. *Aspideretes* sp. indet.	Genus et sp. indet.	Genus et sp. nov.
Family incertae cedis	Marginocephalia	Maniraptora
Naomichelys sp. indet.	Ceratopsidae	Troodontidae
Crocodylia	Centrosaurinae	Genus et sp. nov.
Mesoeucrocodylia	*Diabloceratops eatoni*	Dromaeosauridae
Genus et sp. indet.	Genus et sp. nov. A	Genus et sp. indet. A
Neosuchia	Genus et sp. nov. B	Genus et sp. indet. B
Genus et sp. indet.	Pachephalosauridae	

Indet., taxa of uncertain identity; nov., taxa identified as new, yet currently undescribed.

Although initially spearheaded by teams from the University of Utah, from its inception the KBP has collaborated with other institutions and organizations in the discovery, collection, preparation, curation, and study of specimens. Key partners in this multidisciplinary effort have been Grand Staircase–Escalante National Monument (Kanab, Utah), the Bureau of Land Management (Salt Lake City, Utah), the Denver Museum of Nature and Science (Denver, Colorado), Utah Geological Survey (Salt Lake City, Utah), Weber State University (Ogden, Utah), the Raymond Alf Museum of Paleontology (Claremont, California), and the College of the Holy Cross (Worcester, Massachusetts).

Physical Environment The Kaiparowits Plateau, located south of the Aquarius Plateau, is bounded to the east by the East Kaibab Monocline and to the west by the Straight Cliffs (Fig. 28.2). The middle Campanian Wahweap Formation crops out extensively in the central and southern regions of the Kaiparowits Plateau within Grand Staircase–Escalante National Monument, whereas exposures of the late Campanian Kaiparowits Formation are concentrated in the northern portion of this central plateau. More specifically, the Kaiparowits Formation is exposed as extensive badlands between Powell Point and Horse Mountain (Fig. 28.2). Wahweap Formation exposures are much more extensive, extending from Canaan Peak in the north down to Tibbet Canyon in the south. Although the total area of exposed rock is relatively vast, both units tend to be cliff forming, greatly limiting the proportion of this area that can be effectively prospected or studied in detail. To date, KBP fieldwork has been concentrated primarily in the northern region of the Plateau in outcrops on and around Powell Point, Canaan

Peak, Horse Mountain, and Death Ridge. Vegetation covers much of Kaiparowits Plateau, with bedrock exposures often limited to steep and exposed slopes. Access to both formations is limited largely by the rugged nature of the terrain in which fossils are exposed and by the paucity of roads passing through these areas.

Field Methods Project fieldwork has entailed systematic surveys of Late Cretaceous rock outcrops ("badlands") within the boundaries of Grand Staircase–Escalante National Monument, utilizing small crews (<12 individuals). Whereas most early forays concentrated on rock exposures close to vehicle-traveled roads, substantial time and effort has been devoted to assessing fossil resources in more remote areas. The latter has been accomplished in part through overnight backpacking trips, and occasionally helicopter support is provided to establish remote backcountry camps. Surface collection was conducted in accordance with Grand Staircase–Escalante National Monument guidelines, ensuring that fossil localities are adequately assessed with minimum surficial disturbance. On the basis of the results of this work, potential quarry locations were identified and excavation permits submitted. Standard excavation methods were used at quarry localities, with heavier specimens removed via helicopter airlift.

Summary of Results Over the past decade, the KBP has resulted in identification of many hundreds of Late Cretaceous fossil localities in the Kaiparowits and Wahweap formations, several dozen of which have been excavated and prepared to reveal a previously unknown and surprisingly well-preserved dinosaur fauna (Tables 28.2, 28.3). During this time, the total area surveyed in both the Wahweap and

Testudines	Genus et sp. indet.	Pachycephalosauridae
Cryptodira	Neosuchia	Genus et sp. indet. A
Paracryptodira	Genus et sp. indet. A	Genus et sp. indet. B
Pleurosternidae	Genus et sp. indet. B	Thyreophora
Compsemys victa	Alligatoroidea	Ankylosauria
Baenidae	*Deinosuchus hatcheri*	Nodosauridae
Boremys grandis	Genus et sp. indet.	Genus et sp. indet.
Denazinemys nodosa	Globidonta	Ankylosauridae
Neurankylus sp. nov. A	*Brachychampsa* sp. nov.	Genus and species nov. A
Neurankylus sp. nov. B	Dinosauria	Genus and species nov. B
Plesiobaena sp. indet.	Ornithischia	Saurischia
Eurcryptodira	Ornithopoda	Theropoda
Adocidae	Hadrosauridae	Coelurosauria
Adocus sp. indet.	Hadrosaurinae	Ornithomimidae
Chelydridae	*Gryposaurus monumentensis*	Genus and species nov.
Genus et sp. indet.	*Gryposaurus* sp. indet.	Tyrannosauridae
Kinosternidae	Lambeosaurinae	*Teratophoneus curriei*
Genus et sp. indet.	*Parasaurolophus* sp. indet.	Maniraptora
Nanhsiungchelyidae	Basal Neornithischia	Oviraptorosauria
Basilemys nobilis	Genus et sp. indet.	*Hagryphus giganteus*
Trionychidae	Marginocephalia	Troodontidae
Helopanoplia sp. indet.	Ceratopsidae	*Talos sampsoni*
Aspideretoides sp. indet.	Centrosaurinae	Dromaeosauridae
Derrisemys sp. indet.	Genus and species nov.	Genus et sp. indet. A
Genus et sp. indet.	*Chasmosaurinae*	Genus et sp. indet. B
Crocodylia	*Kosmoceratops richardsoni*	
Mesoeucrocodylia	*Utahceratops gettyi*	

Indet., taxa of uncertain identity; nov., taxa identified as new, yet currently undescribed.

Kaiparowits formations is approximately 43,000 acres. Almost 800 vertebrate localities have been discovered and mapped by the UMNH alone, including 77 associated dinosaur specimens. Results to date have been abundant and spectacular, exceeding initial expectations in virtually all categories, with many discoveries representing new genera and/or species (Kirkland et al., 2002; Sampson et al., 2002, 2004; Smith et al., 2003; Titus et al., 2005; Zanno and Sampson, 2005; Gates and Sampson, 2007; Sampson, Loewen, et al., 2010). Moreover, preservation is frequently exceptional particularly with regard to the abundant dinosaur remains including articulated skeletons and integumentary impressions. Common nondinosaurian macrovertebrate taxa include a variety of crocodylians, turtles and fishes. Also common are trace fossils, encompassing not only vertebrate traces such as dinosaur tracks, but also remarkably preserved insect traces (Roberts and Tapanila, 2006; Roberts, Rogers, and Foreman, 2007). Additional paleontological results include collection of microvertebrates (most recovered from the surface or within macrovertebrate quarries), collection and preliminary identification of a diversity of plant morphotypes (Miller et al., this volume, Chapter 7), and collection of freshwater invertebrates (Roberts, Tapanila, and Mijal, 2008).

Since its inception, the KBP has sought to place newly discovered fossils into a well-constrained stratigraphic and paleoenvironmental context. As a result, geologic research has paralleled the paleontological work, providing key insights into these Late Cretaceous paleoenvironments. Highly significant is the discovery of multiple volcanic ash (bentonite) horizons at varying stratigraphic levels within the Wahweap and Kaiparowits formations, enabling radiometric dating controls to be placed on both formations and the fossils interred within them (Roberts, Deino, and Chan, 2005; Jinnah et al., 2009). Geologic work, combined with taphonomic and paleontological results, has also provided key insights into the paleoenvironmental context of the Wahweap and Kaiparowits formations (Miller et al., this volume, Chapter 7; Roberts et al., this volume, Chapter 6; Titus, Roberts, and Albright, this volume, Chapter 2; Jinnah and Roberts, 2011).

This chapter summarizes a decade of research into the geology and paleontology of the Kaiparowits and Wahweap formations, highlighting in particular macrovertebrate discoveries produced by the KBP. Several other chapters in this volume address specific aspects of the project in much greater detail (Gates et al., this volume, Chapter 19; Evans et al., this volume, Chapter 20; Hutchison et al., this volume, Chapter 13; Irmis et al., this volume, Chapter 17; Miller et al., this volume, Chapter 7; Loewen, Burns et al., this volume, Chapters 18; Loewen, Farke et al., this volume Chapter 21; Roberts et al., this volume, Chapter 6; Zanno et al., this volume, Chapter 22), and interested readers are referred to these contributions. The aim of the present contribution is to provide a broad overview of the project results and explore implications for our understanding of the diversity, biogeography, paleoecology, and evolution of Campanian faunas on Laramidia.

GEOLOGIC AND PALEOENVIRONMENTAL CONTEXT

Wahweap Formation

The 360–458-m-thick Wahweap Formation occurs between the underlying Straight Cliffs Formation and the overlying Kaiparowits Formation (Gregory and Moore, 1931; Pollock, 1999). A middle Campanian age spanning 81–76 Ma is indicated by a combination of detrital zircon analysis and radiometric dating (Roberts, Deino, and Chan, 2005; Roberts, 2007; Jinnah et al., 2009; Jinnah, this volume, Chapter 4; Roberts et al., this volume, Chapter 6). The Wahweap Formation consists of fluvial and estuarine sandstones and flood-basin mudstones deposited in a coastal plain setting adjacent to the western shoreline of the Western Interior Seaway. Four informal members are recognized (Eaton, 1991; Jinnah and Roberts, 2011): (1) a mud-dominated lower member with multiple laterally extensive sandstone bodies; (2) a mud-dominated middle member lacking laterally extensive sand bodies; (3) a sand-dominated upper member, though with abundant mud facies; and (4) a capping sandstone member dominated by major tabular sandstones. Whereas fluvial systems in the first three members appear to be composed predominantly of meandering rivers, that of the capping sandstone indicates the presence of braided river systems. The lowermost pair of members records progressive retreat of the seaway. Tidally influenced channels in the upper member indicate a marine incursion during this interval, likely associated with the Claggett highstand identified in the northern region of the Western Interior Basin (Jinnah and Roberts, 2011).

In a recent analysis of the sedimentology of the Wahweap Formation, Jinnah and Roberts (2011) described 10 facies associations divided into channel and flood-basin deposits, together with a series of paleoenvironmental interpretations. These authors regard the lower member as indicative of a waterlogged, densely vegetated, relatively nearshore coastal floodplain subject to frequent flooding. Similarly, the mud-dominated middle member appears to record a low-energy, waterlogged, and highly vegetated environment with abundant ponds and swamps. In contrast, geology and paleontology of the upper member are suggestive of a brackish-water, estuarine setting associated with marine highstand and a consequent reduction in accommodation space. Finally, the capping sandstone, marked by a sequence boundary at

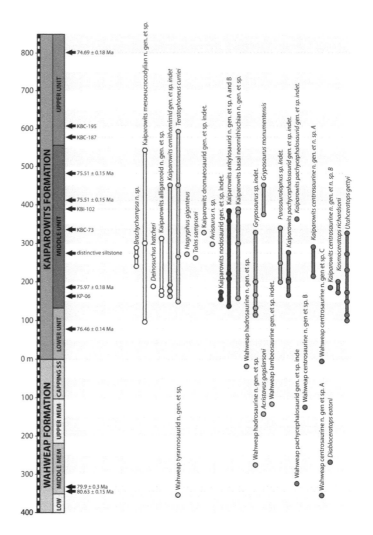

28.3. Stratigraphic distributions of known macrovertebrate (turtle, crocodile, and dinosaur) taxa from the Wahweap and Kaiparowits formations, Grand Staircase–Escalante National Monument, southern Utah. Dots indicate fossil occurrences of a given taxon; bars indicate inferred stratigraphic range. indet., taxa of uncertain identity; nov., taxa identified as new, yet currently undescribed. Scale on y-axis denotes meters of stratigraphic section.

its base, represents a major departure from the underlying members, recording dense braided river systems operating in low accommodation space.

Diagnostic vertebrate fossils are considerably less common in the Wahweap Formation than in the overlying Kaiparowits Formation (Fig. 28.3). Nevertheless, numerous localities have been recorded in the lower, middle, and upper members. The dearth of sites within the capping sandstone is likely the result both of the cliff-forming nature of this unit and the depositional environment. Although the bulk of Wahweap Formation fossils consist of fragmentary and eroded specimens within channel lag sandstones, multiple dinosaur bonebeds have been discovered, the majority of these preserved in overbank mudstones. A detailed study of the formation's flora, although planned, has yet to be conducted. On the basis of the abundance and quality of fossil localities discovered to date, it is likely that continued efforts in this unit will ultimately yield a robust macrovertebrate record.

Kaiparowits Formation

Much more is known of the geology and paleoenvironments of the Kaiparowits Formation because this unit has been more intensively studied than the Wahweap Formation. Roberts (2007) subdivided the Kaiparowits Formation into three informal units—lower, middle, and upper—largely on the basis of differing ratios of sandstone to mudstone. High-precision radiometric ($^{40}Ar/^{39}Ar$ and U-Pb) dating documents deposition over a 2 million year interval during the late Campanian (~76.6–74.5 Ma; Roberts, 2005b, 2007; Roberts et al., this volume, Chapter 6). These dates coincide closely with those of other fossiliferous formations further to the north—in particular the Dinosaur Park Formation, which has also been characterized as having been deposited in a wet, swampy coastal plain setting—allowing critical paleontologic comparisons between northern and southern Laramidia (the Campanian Hot Zone of Roberts et al., this volume, Chapter 6). Considered in conjunction with the remarkable thickness of the Kaiparowits Formation (860 m), these dates also indicate an extremely rapid sediment accumulation rate (Roberts, 2005b, 2007; Roberts et al., this volume, Chapter 6). Such rapid deposition of a richly fossiliferous unit generates exceptional potential for high-resolution ecological and evolutionary studies. Accordingly, detailed stratigraphic studies coupled with multiple, dated ash beds intercalated throughout the formation have allowed placement of the fossils into a high-resolution temporal sequence. The majority of diagnostic macrovertebrate remains have been recovered from the middle unit and the upper portion of the lower unit (Fig. 28.3). The paucity of specimens from the upper unit likely results at least in part from the cliff-forming, and thus relatively inaccessible, nature of these facies.

Roberts (2007) conducted a comprehensive review of the sedimentology of the Kaiparowits Formation, recognizing a total of 14 lithofacies and nine facies. Dominant facies include tabular and lenticular sandstone units typically exceeding 5 m in thickness, indicative of meandering and anastomosing rivers. Sandy and carbonaceous mudstones, particularly common in the middle unit, are interpreted as perennial ponds and lakes, as well as channel-fill oxbow ponds. Other, less common facies include inclined heterolithic sandstones, intraformational conglomerates, finely laminated calcareous siltstones, and minor tabular and lenticular sandstones (Roberts, 2007; see also Roberts et al., this volume, Chapter 6).

The taphonomy of the Kaiparowits Formation has also been the subject of rigorous study (Getty et al., 2003, 2010; Getty, Loewen, and Roberts, 2006; Roberts, 2005a, 2007; Roberts, Tapanila, and Mijal, 2008; Lund et al., 2009; Roberts et al., this volume, Chapter 6). Other than the aforementioned abundance of aquatic faunal and floral elements (Gates et al., 2010; Tapanila and Roberts, this volume, Chapter, 8; Roberts et al., this volume, Chapter 6; Miller et al., this volume, Chapter 7), perhaps the most striking taphonomic signal of the Kaiparowits Formation is the near absence of large-scale dinosaur bonebeds that characterize penecontemporaneous formations to the north (Two Medicine Formation, Judith River Formation, Dinosaur Park Formation). Instead, macrovertebrates are generally recovered either as isolated elements or single skeletons, partially to fully articulated. Preservation of these skeletons, particularly when within channel sandstones, tends to be exceptional, frequently associated with abundant skin impressions (Lund et al., 2009). In contrast, although exceptions do occur, specimens preserved in overbank mudstones are typically more poorly preserved, with high levels of bone modification (i.e., trample marks, tooth traces, insect traces, and wet rot–style weathering; Roberts, 2005a). The latter signal is likely linked to rapid decomposition in a wet, humid climate, resulting in greatly reduced potential for preservation in overbank settings.

The contention that the Kaiparowits Formation was deposited in a wet climate receives independent support from a diverse megaflora recently recovered from this unit (Miller et al., this volume, Chapter 7). Angiosperms comprise the bulk of morphotypes documented thus far, with platanoid dicots being most dominant. Also diverse are aquatic angiosperms, including multiple mega- and palynofloral species. Nonangiosperms include ferns, cycads, equisetopsids, and conifers. Fieldwork in 2009 and 2010 has revealed several new, highly productive plant macrofossil localities, establishing the Kaiparowits flora as the best-known Campanian example worldwide (Miller, pers. comm.). Miller and colleagues used physiognomic analysis of the dicot portion of the megafloral sample to estimate a mean annual temperature of 20.16 ± 2.33°C and a mean annual precipitation of 1.78 m.

A synthesis of paleobiogeographic, sedimentologic, paleontologic, and paleobotanical data (Roberts, 2007; Miller et al., this volume, Chapter 7; Roberts et al., this volume, Chapter 6) permits a first-order assessment of the physiography, paleoclimate, and paleoenvironments of the Kaiparowits Formation. This unit was deposited in a broad, crescent-shaped basin bounded by rising mountains to the west and north (Sevier thrust belts), as well as to the south (Mogollon Highlands), and by the Western Interior Seaway about 100 km to the east. The basin itself was flat, interrupted only by sporadic undulations in the topography. Dissecting the region was a series of meandering to anastomosing fluvial and distributary channels, with extensive floodplains characterized by back swamps, oxbow ponds, and lakes. The paleoclimate of this region during the late Campanian was subtropical and relatively wet, with temperature, humidity, and precipitation akin to the present day Gulf Coast of North America.

Persistent swamps and wetlands were dominated by cypress trees up to 30 m tall, accompanied by aquatic plants such as giant duckweed, water lettuce and other floating angiosperms. Aquatic and epiphytic ferns were present in these floodplain lowlands as well. Somewhat better-drained regions were dominated by dicot forests of 10–20 m tall trees, with occasional palmate and pinnate fronded palms alongside an understory that included ferns. Finally, well-drained soils more proximal to the uplands and/or distal to the channels and swamps were dominated by Cupressaceae and other members of Pinales up to 30 m in height, with an understory of cycads, small dicot trees/bushes, and perhaps ferns. Abundant leaf damage with numerous damage types is consistent with a diverse insect fauna. Vertebrates included a large diversity of fishes, amphibians, lizards, mammals, turtles, crocodylians, and dinosaurs. Included in the latter were large and small ornithischian herbivores alongside theropod carnivores and omnivores (avian and nonavian).

PALEONTOLOGIC RESULTS

Overview

Before 2000, paleontological work in the Wahweap and Kaiparowits formations was biased heavily toward collection of microvertebrates (Cifelli, 1990a, 1990b, 1990c, 1990d; Eaton, 1991, 1999a, 1999b, 2002; Eaton et al., 1999). Augmented by continued efforts during the past decade, this research has established a robust foundation for understanding the microvertebrate faunas that existed in the region of southern Utah during the middle and late Campanian. Indeed, the highly diverse assemblages of fishes, amphibians, squamates, and mammals are currently among the best known for the entire Late Cretaceous Western Interior Basin. Of these clades, the Kaiparowits mammals appear to exhibit the highest degree of endemism, although endemic taxa are present in all of them (Gates et al., 2010). Given that the KBP has predominantly targeted macrovertebrates, microvertebrate clades are not addressed further here (but see other volume contributions). The paleontologic results summarized below address three macrovertebrate clades—turtles, crocodylians, and dinosaurs—all of which are represented by numerous specimens collected by KBP teams during the past decade.

Turtles

Although often regarded as mesovertebrates rather than macrovertebrates, turtles are included in this discussion because numerous partial to complete specimens have been recovered by the KBP. Moreover, chelonians frequently occur as macroscale fossils that fall well within the size ranges of smaller crocodylians and dinosaurs. In addition to the specimens collected during the tenure of the KBP, numerous turtle specimens were recovered before 2000, the majority of these by Howard Hutchison (Hutchison et al., 1997, this volume, Chapter 13; Eaton et al., 1999).

With multiple lines of evidence pointing toward wet, swampy paleoenvironments for both the Wahweap and the Kaiparowits formations, it is perhaps not surprising that chelonian remains are some of the most common fossils found in these units. Although the bulk of specimens consist of carapace and plastron fragments, numerous partial to complete turtle shells (plastron and carapace) have been recovered, some with associated cranial and postcranial skeletal elements.

To date, the Wahweap Formation has yielded remains of seven turtle genera pertaining to six "families" (Table 28.2; Hutchison et al., 1997). For the most part, species-level diagnoses have not been possible as a result of the fragmentary nature of most of the materials; however, the collection is currently under study. Included within this diversity are pleurosternids (*Compsemys*), adocids (*Adocus*), trionychids (*Aspideretes*), baenids (*Baena* and *Boremys*), nanhsiungchelyids (*Basilemys*), and neurankylids (*Neurankylus*). Turtle remains in the Wahweap occur most commonly as fragmentary scutes and shells in channel lags, but several more complete specimens are known, some of them recovered from overbank mudstones.

The overlying Kaiparowits Formation has produced remains of no less than 14 distinct turtle taxa, making it the second most diverse chelonian assemblage known (Table 28.3; Hutchison et al., this volume, Chapter 13). Most of the represented clades (families, as well as multiple genera) mirror those present in the Wahweap Formation. Kaiparowits turtle diversity includes pleurosternids (*Compsemys*), baenids (*Neurankylus*, *Denzinemys*, *Boremys*, and *Plesiobaena*), adocids (*Adocus*), nanhsiungchelyids (*Basilemys*), and trionychids (*Helopanoplia*, *Aspideretoides*, and *Derrisemys*). Two additional clades, chelydrids and kinosternids, occur quite commonly as fragmentary specimens, and materials pertaining to these small-bodied taxa have not yet been diagnosed to genus. Two Kaiparowits turtle taxa—the eucryptodire *Basilemys nobilis* and an undescribed species of the paracryptodire *Neurankylus*—approach a meter in length, ranking this pair as the largest nonmarine turtles of the North American

Cretaceous (Hutchison et al., this volume, Chapter 13). Many of the Kaiparowits turtle fossils are exceptionally preserved, including distinct sulci and sutural surfaces along with additional skeletal elements. One largely complete *Adocus* specimen even preserves eggs within the carapace (Hutchison et al., this volume, Chapter 13; Knell et al., 2010). The majority of these turtles have been recovered from the lower and middle units of the formation, with the best-preserved specimens typically found at the base of fluvial channels. The most commonly represented turtle families (trionychids, baenids, and adocids) have all been linked to fluvial habitats (Hutchison et al., this volume, Chapter 13).

Crocodylians

Crocodylian remains are relatively common in the Wahweap and Kaiparowits formations (Tables 28.2, 28.3), providing further evidence of wet paleoenvironments. However, the diversity of this clade in the two target units remains incompletely understood because of the fragmentary basis of most taxa (Eaton et al., 1999; Wiersma, Hutchison, and Gates, 2004; Irmis et al., this volume, Chapter 17). Wahweap Formation crocodylians are known almost entirely from isolated teeth and osteoderms, with a minimum of three indeterminate taxa presently recognized: (1) an unidentified, large-bodied mesoeucrocodylian, either a goniopholid or pholidosaurid; and (2) the large alligatoroid *Deinosuchus*; and (3) a much smaller-bodied neosuchian, possibly an alligatoroid.

Crocodylians are significantly better known from the Kaiparowits Formation. In addition to the recovery of isolated elements – in particular, teeth, osteoderms, dentaries, and frontals – several taxa are known from partial associated to articulated skeletons. In all, the lower and middle units have produced remains of six distinct, apparently coeval taxa, making this the most diverse crocodylian assemblage from Laramidia, and one of the most diverse known from any time period. One of these taxa is a large-bodied mesoeucrocodylian – once again, either a goniopholid or pholidosaurid – known from teeth, osteoderms, and several fragmentary postcranial skeletons; it cannot yet be determined if this taxon is the same or different as the Wahweap Formation mesoeucrocodylian. Also present is an unidentified neosuchian, possibly an alligatoroid, based on very fragmentary remains. A third crocodylian taxon, known only from dental and postcranial fragments, is clearly a distinct taxon but further resolution is currently not possible. The remaining three taxa are all alligatoroids. A small-bodied taxon is currently founded on a lower jaw and two nearly complete, articulated skeletons. Another new, somewhat larger taxon is an as yet undescribed species of *Brachychampsa*, recognized from a

pair of associated specimens and isolated skull elements. Finally, the fourth alligatoroid is *Deinosuchus hatcheri*, by far the largest crocodylian in the assemblage. Indeed *Deinosuchus* is one of the largest known crocodylians from any place or time period, achieving body sizes up to 12 m (Titus et al., in prep.). The diverse morphologies of these apparently coeval Kaiparowits Formation taxa suggest equally diverse lifestyles (Wiersma, Hutchison, and Gates, 2004; Irmis et al., this volume, Chapter 17). Figure 28.3 shows the stratigraphic distribution of currently known crocodylian taxa from the two target formations.

Dinosaurs

Overview Arguably the most notable achievement of the KBP is recovery of a new, highly diverse assemblage of dinosaurs, with virtually every diagnosable species unknown elsewhere. A 1997 review of fossil vertebrates from the Kaiparowits Formation (Hutchison et al., 1997) listed the presence of eight distinct dinosaur taxa, all known from fragmentary remains, mostly teeth, that prohibited confident genus and species identifications. Since 2000, the KBP has documented the occurrence of at least 16 nonavian dinosaur taxa in this unit, a doubling of the previous estimate in one decade (Table 28.3). Over 70 partial to nearly complete dinosaur skeletons have now been recovered from the Wahweap and Kaiparowits formations, including remains of basal neornithischian ornithopods (i.e., "hypsilophodonts"), lambeosaurine and hadrosaurine hadrosaurids, centrosaurine and chasmosaurine ceratopsids, pachycephalosaurids, ankylosaurid and nodosaurids ankylosaurians, and tyrannosaurid and maniraptoran theropods (Tables 28.2, 28.3). Preservation can be exceptional, including nearly complete skulls (e.g., hadrosaurs, ceratopsids, tyrannosaurids) and integumentary impressions (hadrosaurs and ceratopsids). The stratigraphic distributions of dinosaur taxa currently known from the Wahweap and Kaiparowits formations are shown in Fig. 28.3.

Thyreophorans Thyreophoran, or armored, dinosaurs are represented in the Wahweap and Kaiparowits formations by representative specimens from the two major clades of ankylosaurians: Nodosauridae and Ankylosauridae (Loewen, Burns et al., this volume, Chapter 18). The Wahweap assemblage is fragmentary, based largely on isolated teeth and osteoderms that cannot be assigned to genus and species. Similarly, the Kaiparowits Formation nodosaurid, known only from teeth and a shoulder spike recovered from the middle unit, cannot yet be diagnosed. The ankylosaurid materials, in contrast, clearly pertain to a pair of new genera and species, based upon recently discovered, mostly complete skulls, one of which includes multiple postcranial elements and dermal armor, including a distal tail segment with a tail club (Loewen, Burns, et al., this volume, Chapter 18). These materials, also recovered from the middle unit, are currently under study.

Ornithopods Abundant remains of ornithopod dinosaurs, including small-bodied basal neornithischians (i.e., "hypsilophodonts") and large-bodied hadrosaurids, have been recovered from the two target formations. These discoveries are reviewed by Gates and colleagues (this volume, Chapter 19; see also Gates, 2004), and the following discussion is derived largely from their summary. Thus far, the Wahweap Formation has yielded only isolated teeth of basal neornithischians, which appear to compare closely with those recovered from the overlying Kaiparowits Formation.

With regard to Wahweap Formation hadrosaurids, lambeosaurines are represented by a single isolated maxilla recovered from the upper unit. This specimen resembles *Parasaurolophus* in certain characteristics but cannot yet be assigned with confidence to a particular genus or species. In contrast to the above clades, hadrosaurines are known from isolated specimens and bonebeds in the lower, middle, and upper members of the Wahweap Formation. Bonebed materials from the lower member cannot be assigned with confidence to a particular genus or species within Hadrosaurinae. In the middle member, a diagnostic partial skull is referable to the recently described genus *Acristavus* (also present in Montana; Gates et al., 2011; Gates et al., this volume, Chapter 19), closely allied with *Brachylophosaurus* and *Maiasaura*. A bonebed locality in the middle member has produced multiple individuals, juvenile and adult, of what is likely a second hadrosaurine taxon. Finally, a third possible hadrosaurine, this one perhaps referable to *Brachylophosaurus*, is known from a maxilla and partial postcranium from the upper member. In summary, hadrosaur diversity in the Wahweap Formation currently consists of at least one lambeosaurine and two hadrosaurines, with the likelihood that the sample contains a third distinct hadrosaurine taxon. Interestingly, none of the Wahweap hadrosaurines show close affinities with counterparts known from the overlying Kaiparowits Formation. Studies of this collection are underway.

Ornithopod fossils are relatively common in the Kaiparowits Formation. Basal neornithischians are represented by at least one new taxon (Gates et al., this volume, Chapter 19). This clade, informally known as "hypsilophodonts," was unrepresented in the formation prior to the onset of the KBP. At present, the new taxon is known from eight specimens, including partial skeletons with cranial and postcranial elements ranging from juvenile through adult. These materials have been recovered from the lower and middle units, with the most taxonomically informative specimens derived from

the latter. One of these specimens, collected from an overbank mudstone, includes a partial skull suggesting close affinities to northern Laramidian taxa *Orodromeus* and *Zephyrosaurus* (Gates et al., this volume, Chapter 19).

Hadrosaurid diversity in the Kaiparowits Formation presently consists of at least one species of the lambeosaurine *Parasaurolophus* and at least two species of the hadrosaurine *Gryposaurus*. The tube-crested *Parasaurolophus* is currently known from five partial skulls, a partial postcranium, and several bonebed elements, all recovered from the middle unit. Several of the specimens include partial crests, greatly elucidating the anatomy of these elaborate structures. The crest materials, although somewhat distinctive in shape and orientation, indicate close affinity with *P. cyrtocristatus* from the late Campanian of New Mexico. Studies currently underway will assess whether the Grand Staircase–Escalante National Monument materials are conspecific with *P. cyrtocristatus* or represent a distinct species.

The noncrested hadrosaurine *Gryposaurus*, with its distinctive nasal "hump," is not only the most commonly occurring hadrosaurid in the Kaiparowits Formation, but the most frequently found dinosaur. The current collection includes seven partial to nearly complete skulls, three of which include associated postcranial elements. The assemblage, which ranges from near the base of the lower unit to the uppermost facies of the middle unit, can be subdivided stratigraphically into two morphs interpreted as distinct species (Gates and Sampson, 2007). A more gracile form restricted to the lower unit bears close affinities with *G. notabilis* from the late Campanian of Alberta, but may represent a new species (Gates et al., this volume, Chapter 19). A second, hyperrobust form, *G. monumentensis*, occurs in the topmost facies of the lower unit and throughout most of the middle unit. In addition to its overall robustness, *G. monumentensis* possesses apomorphic skull and lower jaw structures that distinguish it from other species of *Gryposaurus* (Gates and Sampson, 2007). Both *Gryposaurus* species are known from nearly complete skulls.

Preservation of dinosaur specimens in the Kaiparowits Formation is often exceptional, particularly for articulated specimens recovered from channel sand bodies. In addition to relatively pristine bone quality, more than a dozen ornithischian dinosaur specimens found in this unit preserve skin impressions (Lund et al., 2009). Although integumentary impressions have been found in association with a basal neornithischian and a centrosaurine ceratopsid, the bulk of these specimens are preserved with skeletons of the hadrosaurine *Gryposaurus*. Previous occurrences of such skin impressions have typically been attributed to mummification in arid environments and referred to as "dinosaur mummies."

The Kaiparowits skin impressions, associated with skeletal remains entombed within channel sandstones, suggest an alternative taphonomic mode – rapid burial in fluvial settings (Lund et al., 2009).

Marginocephalians The Marginocephalia, or margin-headed dinosaurs, include the dome-headed Pachychephalo-sauria and the horned Ceratopsia (Sereno, 1986). Examples of both clades have been recovered from the two geologic units under consideration. With regard to pachycephalosaurs, the present sample is limited to isolated teeth and several fragmentary skull elements representing at least two taxa, one each from the Wahweap and Kaiparowits formations (Evans et al., this volume, Chapter 20). A single, isolated frontoparietal dome plus isolated teeth are known from the Wahweap Formation (Kirkland and DeBlieux, 2005; Evans et al., this volume, Chapter 20). The dome, recovered from the middle member, compares closely with that of *Stegoceras*, a small-bodied taxon also known from the late Campanian of Laramidia. However, certain features of this specimen, in particular a transversely narrow nasal buttress, suggest that it may represent a new taxon (Kirkland and DeBlieux, 2005; Evans et al., this volume, Chapter 20).

Pachycephalosaurs are currently represented in the Kaiparowits Formation by six fragmentary cranial specimens – four partial frontoparietal domes and two squamosals – and one relatively complete cranium; all recovered from the middle unit. Most of the Kaiparowits sample also suggests a close affinity with *Stegoceras*. One isolated squamosal in particular shares multiple similarities with *S. validum* (Evans et al., this volume, Chapter 20). However, as with the Wahweap example, this specimen possesses some unique features that could indicate a distinct species. Another frontoparietal dome from the Kaiparowits appears to exhibit affinities with *Hanssuesia*, but its fragmentary and weathered condition precludes confident assignment. A frontoparietal dome recovered in 2010 appears distinctive in some respects from the rest of the domes recovered in the formation, but a definitive diagnosis has not yet been made (Evans et al., this volume, Chapter 20). Finally, a well-preserved partial skull discovered in 2011 and currently under study is likely to provide significant insights. Additional work is needed to resolve the species diversity of pachycephalosaurs in the Kaiparowits Formation.

In contrast, horned dinosaurs are represented by numerous ceratopsid specimens, including nearly complete skulls with associated postcrania, attributed to six taxa – three each in the Wahweap and Kaiparowits formations. As with other late Campanian formations from Laramidia, ceratopsids are second only to hadrosaurids in abundance of specimens. The Wahweap Formation has yielded remains of three new genera of short-frilled ceratopsids, or centrosaurines. One of these taxa, found in lower member, is known from a partial parietal frill bearing a pair of large, laterally curving spikes (Loewen, Farke et al., this volume, Chapter 21). These spikes are extremely broad transversely and flattened in cross section. Occurring in sediments dating to approximately 81 Ma (Jinnah, pers. comm.), this taxon is the oldest known diagnosable ceratopsid.

Much better known is *Diabloceratops eatoni*, represented by an exquisitely preserved, nearly complete skull recovered from the middle unit (Kirkland and DeBlieux, 2010). *Diabloceratops* is remarkable for its mosaic of primitive and derived features. Symplesiomorphic characters include well-developed postorbital horn cores, also present in the basal (nonceratopsid) neoceratopsian *Zuniceratops*. The most notable autapomorphy of this taxon is the single pair of robust spikes projecting rearward from the caudal margin of the parietal. A phylogenetic analysis conducted by Kirkland and DeBlieux (2010) places *Diabloceratops* as the most basal centrosaurine. A third centrosaurine taxon, collected from the capping sandstone of the Wahweap Formation, consists of a partial parietal that appears distinct from either of the above pair of taxa in that it lacks any indication of large spikes (Loewen, Farke et al., this volume, Chapter 21). To date, ceratopsids pertaining to the long-frilled clade Chasmosaurinae are absent from the Wahweap Formation.

In contrast, chasmosaurine remains are relatively common finds in the Kaiparowits Formation, whereas centrosaurine fossils are comparatively rare. Two new chasmosaurine taxa have been identified from this formation, both represented by multiple specimens that span the upper portion of the lower unit and the lower portion of the middle unit (Sampson, Loewen, et al., 2010). The first, *Utahceratops gettyi*, is unique in possessing short, rounded, laterally projecting supraorbital horn cores. Several features, including a deep, transversely constricted median embayment on the caudal frill margin, suggest a sister relationship with *Pentaceratops sternbergii* from the late Campanian of New Mexico. The second chasmosaurine, *Kosmoceratops richardsoni*, is characterized by elongate, laterally projecting supraorbital horn cores and a short, broad frill with 10 pronounced hooks decorating the caudal margin. Among other features, the structure of the frill indicates that *Kosmoceratops* is closely allied to *Vagaceratops* (previously *Chasmosaurus irvinensis*) from the late Campanian of Alberta (Sampson, Loewen, et al., 2010). Although clearly a member of the long-frilled Chasmosaurinae on the basis of the possession of several key synapomorphies, *Kosmoceratops* resembles the typical centrosaurine condition in its frill, which is relatively short with extreme ornamentation. Indeed the frill of *Kosmoceratops* has the shortest length-to-breadth ratio of any ceratopsid, including all centrosaurines. With a total of 15 well-developed horns or horn-like processes, this animal also ranks as the most ornate-headed dinosaur known.

The third ceratopsid dinosaur known from the Kaiparowits Formation is an undescribed centrosaurine also recovered in the middle unit. It is based upon a mostly complete skull with partial postcranium, as well as two additional specimens: an isolated squamosal and several associated skull elements in a multitaxic bonebed. Like *Kosmoceratops*, the skull of this animal contrasts with the standard conformation of its clade, in this case bearing a long, low nasal horn core, hyperelongate supraorbital horn cores, and a relatively simple, unadorned frill. In recent years it has been established that the basal condition in centrosaurines includes elongate supraorbital horns (Ryan, 2007; Kirkland and DeBlieux, 2010; Sampson and Loewen, 2010). However, the supraorbital horns of the new Utah taxon are extremely elongate and unique in being directed strongly laterally and forward rather than upward.

The Wahweap and Kaiparowits formations of Utah are largely coeval with the Foremost, Oldman, and Dinosaur Park formations of southern Alberta (Roberts et al., this volume, Chapter 6). At present, the most diverse ceratopsid assemblage is known from the latter two formations, with two taxa known from the Oldman Formation (*Albertoceratops nesmoi*, *Centrosaurus brinkmani*), and seven taxa recognized from the Dinosaur Park Formation (*Centrosaurus apertus*, *Styracosaurus albertensis*, *Pachyrhinosaurus* sp., *Chasmosaurus kaiseni*, *Chasmosaurus belli*, *Mojoceratops perifania*, and *Vagaceratops irvinensis*) (Sampson and Loewen, 2010). The Wahweap–Kaiparowits sequence, with six recognized taxa, now ranks second in total diversity of ceratopsids (Loewen et al., this volume, Chapter 21). Given that no species-level taxa were known from the Kaiparowits Formation a decade ago, and that intensive fieldwork has been ongoing in southern Alberta for over a century, it seems likely that many more taxa await discovery in southern Laramidia, including within the boundaries of Grand Staircase–Escalante National Monument.

Theropods As one would anticipate based upon other Campanian-aged, Western Interior Basin dinosaur faunas, the new macrovertebrate assemblage from Grand Staircase–Escalante National Monument includes a rich diversity of theropods: at least five taxa in the Wahweap Formation and seven taxa in the Kaiparowits Formation. Represented clades consist of Tyrannosauridae (both formations), Ornithomimidae (both formations), Oviraptorosauria (Kaiparowits Formation only), Troodontidae (both formations), Dromaeosauridae (both formations), and Aviale (Kaiparowits Formation only). The present review is derived primarily from the summary presented in Zanno et al. (this volume, Chapter 22), augmented by studies addressing specific taxa (Hutchison, 1993; Zanno and Sampson, 2005; Neabore et al., 2007; Loewen et al., 2010). At present, with several notable exceptions described below, most theropod taxa from the target formations are known only from fragmentary materials.

Chief among the exceptions are the large-bodied tyrannosaurids. Before the inception of the KBP, tyrannosaurid remains from Wahweap and Kaiparowits formations consisted of a single, fragmentary partial skull from the latter unit (in the collections of Brigham Young University), together with isolated teeth and postcranial fragments. In 2009, the oldest known tyrannosaurid remains were discovered in the lower member of the Wahweap Formation, consisting of a well-preserved partial skull and postcranium of a remarkable new taxon. The skull of this medium-sized taxon is relatively robust, tall dorsoventrally and short rostrocaudally, with a transversely broad occiput and a low tooth count. The recently named *Teratophoneus curriei* (Carr et al., 2011)—based on multiple specimens, including the 65% complete skull and postcranium of a subadult individual—has been identified from the lower and middle units of the Kaiparowits Formation. The latter animal shares several derived features with the Wahweap tyrannosaurid, as well as with *Bistahaieversor* from the late Campanian of New Mexico. A phylogenetic analysis places all three of these southern, Campanian-aged tyrannosaurids in a previously unrecognized clade with *T. rex* and other (primarily Maastrichtian) taxa from North America and Asia (Loewen et al., 2010).

Ornithomimids, mostly small- to midsized cursorial theropods, are relatively common elements in Campanian and Maastrichtian terrestrial faunas from the Western Interior Basin. Although the diets of ornithomimids have been a matter of contention, in large part because the bulk of taxa are edentulous, recent studies tend to regard them as herbivorous (Barrett, 2005; Zanno et al., 2009). To date, no ornithomimid remains have been recovered from the Wahweap Formation, although this absence is likely the result of collection and/or preservation bias given that the clade is represented in older and younger geologic units in Utah (Zanno et al., this volume, Chapter 22). In contrast, the Kaiparowits Formation has produced several associated partial skeletons of ornithomimids, although as yet no skull materials. The first postcranial elements, collected about three decades ago by the Museum of Northern Arizona, were referred to *Ornithomimus velox* (DeCourten and Russell, 1985), an assignment that has subsequently been challenged (Zanno et al., this volume, Chapter 22). Over the past decade, UMNH and Raymond Alf Museum crews have discovered abundant isolated ornithomimid elements and several associated postcranial specimens from the lower and middle units of the Kaiparowits Formation. One of these specimens includes a partial pelvis, hind limbs, and caudal vertebral series. These materials are currently under study (Neabore et al., 2007; Zanno et al., this volume, Chapter 22), in part to determine if they represent a new taxon.

Oviraptorosaurs are a clade of small- to midsized maniraptoran theropods best known from Asia, but with several North American representatives known from fragmentary remains. Like ornithomimids, all but the most basal taxa are edentulous and may well have been omnivorous or even herbivorous (Ji et al., 1998; Xu et al., 2002; Barrett, 2005; Zanno et al., 2009). Although oviraptorosaurs are unknown from the Wahweap Formation, a single taxon, *Hagryphus giganteus*, has been recovered from the middle unit of the Kaiparowits Formation. *Hagryphus* is based on a single, articulated specimen consisting of a distal forelimb (including a mostly complete manus) and partial pes (Zanno and Sampson, 2005). The species epithet refers to the body size of this taxon, which is 30–40% larger than other Campanian-aged oviraptorosaurs from the Western Interior Basin (i.e., *Chirostenotes, Elmisaurus*). *Hagryphus* is the first oviraptorosaur described from southern Laramidia (Zanno and Sampson, 2005).

Troodontids are another clade of small-bodied maniraptorans predominantly known from Asia. As with the previous two groups, dietary habits have been a matter of debate, with some authors positing omnivory or herbivory (Holtz, 1998; Zanno et al., 2009). At present, two genera (*Troodon* and *Pectinodon*) are recognized from northern Laramidia. Evidence for this clade in the Wahweap Formation is limited to isolated teeth, which exhibit unusually large denticles (Eaton et al., 1999; Zanno et al., this volume, Chapter 22). These characteristic teeth have long been recognized in the Kaiparowits Formation as well. Additionally, recent work by crews from Grand Staircase–Escalante National Monument and the Raymond Alf Museum has yielded an isolated frontal and a subadult partial postcranium from the fossiliferous middle unit. The frontal compares closely with that of *Troodon formosus* from the Campanian of northern Laramidia (especially Alberta), yet it also possesses differences that may be taxonomically relevant. In contrast, the postcranial remains differ distinctly from those of *T. formosus*, particularly in being much more gracile overall, and these materials formed the basis for the new genus and species, *Talos sampsoni* (Zanno et al., 2011). An additional poorly preserved partial skull and postcranium from the Kaiparowits Formation is in the collection of the University of California Museum of Paleontology.

Dromaeosaurids are yet another clade of small-bodied, lightly built maniraptoran theropods best known from Asia. In contrast to ornithomimids, oviraptorosaurs, and

troodontids, dromaeosaurs are regarded as exclusively carnivorous, a contention based largely on their narrow-bladed, serrated teeth that closely resemble those of more basal, predatory theropods. North American dromaeosaurids include five recognized taxa, including the midsized *Utahraptor* from the Early Cretaceous of central Utah. Dromaeosaurids are represented in both the Wahweap and Kaiparowits Formations by teeth pertaining to two morphs, traditionally grouped into "dromaeosaurines" and "velociraptorines" (Hutchison et al., 1997; Zanno et al., this volume, Chapter 22). Zanno and colleagues (this volume, Chapter 22) assign the Kaiparowits examples to two distinct genera–Dromaeosauridae indet. morphotype A and morphotype B–and the same system is here applied to dromaeosaurids from the Wahweap Formation (Table 28.2).

Birds are now confidently regarded as the direct descendents of Jurassic maniraptoran theropods (Chiappe, 2004). Although little doubt exists that Mesozoic avifaunas were highly diverse (Chiappe and Dyke, 2006), the fossil record of Late Cretaceous birds is poorly established for Laramidia (and most other landmasses). At present, avian fossils are unknown from the Wahweap Formation. Bird remains from the Kaiparowits Formation consist mostly of teeth and fragmentary limb elements (Zanno et al., this volume, Chapter 22). The single striking exception is a partial postcranium–including vertebrae, pectoral girdle, forelimb, and tarsometatarsus–referable to an undescribed species of the enantiornithine *Avisaurus* (Hutchison, 1993). Two other species of *Avisaurus* are each known from an isolated tibiotarsus. Given the paucity of avian remains from the Late Cretaceous of North America, the Kaiparowits specimen is likely to prove highly significant once it is fully described.

DISCUSSION

The paleontological results summarized above from the Wahweap and Kaiparowits formations offer a direct glimpse into a remarkable, previously unknown assemblage of Campanian macrovertebrates, with important implications for Mesozoic terrestrial ecosystems. Granted, the KBP still has a long way to go before achieving its goal of providing a reasonably comprehensive census of vertebrate diversity in southern Utah during the mid- to late Campanian; at present, the bulk of taxa among turtles, crocodylians, and dinosaurs are still known only from fragmentary remains. Small-bodied taxa in particular are incompletely known, and it is very likely that additional, as yet unrepresented clades will be discovered. Despite greater than a century of fieldwork, the presence of microraptorine theropods was confirmed only recently in the Dinosaur Park Formation (Longrich and Currie, 2009). The occurrence of well-preserved remains of fragile taxa such as

Avisaurus (Archibald, 1996) within the Kaiparowits Formation indicates that additional small-bodied theropod taxa also await discovery in southern Utah. Nevertheless, building upon a robust understanding of microvertebrates (Eaton et al., 1999, Eaton and Cifelli, this volume, Chapter 14; Gates et al., 2010; Brinkman et al., this volume, Chapter 10; Gardner et al., this volume, Chapter 11; Nydam, this volume, Chapter 16), the past decade of intensive macrovertebrate research has established the Kaiparowits Formation as the standard-bearer for the Campanian of southern Laramidia, exceeded in the north only by the Dinosaur Park Formation of Alberta. As a direct result of this work, key patterns and processes have become evident, or at least strongly supported, and multiple new questions have been raised.

Ecological Implications

The most obvious primary finding pertains to latitudinal variation in vertebrate diversity on Laramidia and its ecological implications. In particular, the Kaiparowits Formation (paleolatitude ~ 37.5°N) can now be compared directly with the closely coeval Dinosaur Park Formation (~59°N; PLATES, 2009). With regard to the presence of major macrovertebrate clades, the emerging picture is one of cosmopolitanism on much of this "island continent." For example, among turtles, baenids, chelydrids, adocids, nanhsiungchelyids, and trionychids are all present in both the Dinosaur Park Formation (Brinkman, 2005) and the Kaiparowits Formation (Hutchison et al., this volume, Chapter 13). Taxonomic overlap among chelonians extends to the genus level in several instances (*Boremys, Neurankylus, Adocus, Aspideretoides*). Similarly, crocodylian diversity in this same pair of formations includes taxa of alligatoroid globidontans, with at least one genus (*Brachychampsa*) common to both units (Wu, 2005; Irmis et al., this volume, Chapter 17). Cladal diversity of dinosaurs recovered from the Wahweap and Kaiparowits formations also resembles closely that found elsewhere in the Campanian Western Interior Basin (Currie, 2005; Ryan and Evans, 2005). That is, ornithischian diversity consists of hadrosaurids, "hypsilophodont"-grade ornithopods, pachycephalosaurs, ceratopsids, and ankylosaurians, with hadrosaurids and ceratopsids by far the most dominant in terms of abundance. Similarly, theropod diversity includes tyrannosaurids in the top carnivore role, alongside a familiar variety of smaller-bodied forms: ornithomimids, oviraptorosaurs, troodontids, and dromaeosaurids.

The widespread occurrence of these major clades of macrovertebrates is unsurprising, given the diminutive size of Laramidia. Much more striking is the finding that numerous genera and species within these groups appear to be endemic to either the north or south (Gates et al., 2010).

Among turtles, taxa present in the Kaiparowits Formation that are absent in the north of Campanian Laramidia include: *Compsemys, Denazinemys nodosa, Plesiobaena* sp. nov., *Helopanoplia*, and *Derrisemys.*. Genus- and species-level resolution is poorly established for crocodylians of the Wahweap and Kaiparowits formations, so latitudinal variation in diversity is more difficult to assess for this group of archosaurs. Nevertheless, some taxa present in the Kaiparowits Formation (e.g., Alligatoroidea n. sp.) are unique to that unit, and there appears to be a greater diversity of nonalligatoroids in the Kaiparowits Formation that is absent further north (Irmis et al., this volume, Chapter 17). An exception to this endemic trend is the alligatoroid *Deinosuchus hatcheri*; although not yet known from the Dinosaur Park Formation, this giant crocodylian has been recovered as far north as Montana (Titus et al., in prep.). Most remarkable of all is the pattern observed among dinosaurs; of the dozen or so dinosaur taxa from the Wahweap and Kaiparowits formations that have been assigned to species, *all* are endemic to southern Laramidia (Gates et al., 2010). The same claim can be made for other geologic units in the southern portion of Laramidia (e.g., Kirtland and Fruitland formations, New Mexico; Aguja Formation, Texas). The corollary of this finding is that of that dozens of mid- to late Campanian dinosaurs found in multiple formations in Alberta and Montana (e.g., Dinosaur Park Formation, Two Medicine Formation, Judith River Formation), none has been conclusively identified in southern Utah.

Considered in conjunction with patterns observed among microvertebrates (Gates et al., 2010), these findings provide strong support for the presence of highly divergent faunas in the north and south of Laramidia during the late Campanian (Lehman, 1997, 2001). Previous criticisms of this hypothesis have focused on problems of temporal resolution and geographic sampling. Sullivan and Lucas (2006) argued that the perceived latitudinal differences in diversity resulted from diachronous faunas—that is, a time transgressive taxonomic distribution that merely generates the appearance of provincialism. The series of recent radiometric analyses summarized above (Roberts, 2005b; Roberts, Deino, and Chan, 2005; Jinnah et al., 2009; this volume, ChapterRoberts et al., this volume, Chapter 6) now establishes that key Campanian-aged formations in the north (Oldman and Dinosaur Park formations) and south (Wahweap and Kaiparowits formations) were deposited contemporaneously, allowing for direct comparisons of their constituent faunas.

Vavrek and Larsson (2010) conducted a detailed statistical analysis of the four best-sampled dinosaur faunas of Maastrichtian age, searching for biogeographic signals. They argued that the perceived latitudinal differences in Maastrichtian faunas were artifactual, the result of temporal and geographic sampling biases between and among geologic formations. Although this paper does not assess Campanian-aged faunas, the authors infer that their findings apply to Late Cretaceous dinosaurs generally, concluding that, "These results suggest that dinosaurs were not as restricted in their ranges as once thought, and that the fauna as a whole was largely homogenous" (Vavrek and Larsson, 2010:3–4). Although certainly intriguing, the results of this study do not refute the dinosaur provincialism hypothesis. First, Vavrek and Larsson used only a small, culled sample of four geologic formations in their analysis, resulting in a narrow latitudinal span (less than 15°) ranging from southern Alberta to Northern Wyoming. Second, some taxa are known only from the north or south (e.g., *Alamosaurus* in the south). So the presence of more widely distributed northern and southern dinosaur "provinces" in the Western Interior Basin during the Maastrichtian remains an open question.

More to the point with regard to the present discussion, late Campanian-aged geologic formations in the Western Interior Basin are now much better sampled, both in terms of temporal resolution and geographic sampling, than are units of Maastrichtian age. A recent quantitative biogeographic analysis of late Campanian vertebrates on Laramidia conducted by our working group (Gates et al., 2010) recovered strong evidence of highly divergent northern and southern faunas, with numerous taxa endemic to north or south. However, this comprehensive study, which incorporated recent KBP findings from the Kaiparowits Formation, could not resolve the nature of the interface between latitudinally separated faunas. This study listed two primary alternatives: 1) discrete "provinces" (at least two) separated by a zone (or zones) of faunal mixing; and 2) a continuous latitudinal cline or gradient, with no discrete zones of endemism. Further resolution of this issue would benefit greatly from sampling of late Campanian vertebrates between southern Utah and northern Montana. However, an additional test would be phylogenetic studies of key clades. The hypothesis of discrete zones of endemism would predict the occurrence of not just endemic species, but endemic clades, given that evolution would proceed semi-independently in the distinct provinces.

An obvious outstanding problem pertains to the nature of the barrier that restricted north–south dispersal on Laramidia. Current paleontological evidence points toward a barrier emplaced approximately at the latitude of northern Utah and Colorado (Lehman, 1987, 1997) between ~77 and 76 Ma (Sampson, Loewen, et al., 2010). Lacking any evidence of a physical obstacle to movement, and despite the fact that paleotemperature gradients were substantially lower than those of the present day, Lehman (1987, 1997, 2001) postulated climate as the critical factor. He noted the presence of

distinct palynological signatures north and south, and proposed that a latitudinal climate gradient resulted in distinctive floras as well as faunas. Today, evidence of a physical barrier is still lacking, although possible alternatives include: (1) an unidentified, east–west-trending mountain range such as the Uinta Range of Utah; (2) a major, persistent river system; and (3) maximum transgression of the Cretaceous Interior Seaway, flooding the coastal and alluvial plains; and (4) climatic differences. Of these four, the first option has recently been argued by Gates and coauthors (2012), who postulate that the onset of Laramide orogeny increased rates of speciation in some megaherbivore dinosaur clades by isolating populations. The second option seems implausible in that it is difficult to imagine a river system being sufficiently stable for deep time intervals to act as a barrier to floral and faunal dispersal. The remaining possiblities, transgression and climatic gradients, remain possible causal factors, and additional work is required to assess these alternatives.

The two marine cyclothems that effected Campanian terrestrial ecosystems were the Claggett and Bearpaw sea-level cycles, which varied significantly in timing of their origins (Kauffman, 1977). The base of the Claggett Cyclothem corresponds to a global eustatic sea-level rise that began ~80 million years ago. Peak transgression occurred during the *Baculites maclearni* ammonite biozone (~79.7 Ma; Hicks, Obradovich, and Tauxe, 1999), and its influences on terrestrial ecosystems have been well documented both to the north (Judith River–Two Medicine clastic wedge of Montana; Rogers, 1998) and to the south (Straight Cliffs–Wahweap clastic wedge; Jinnah and Roberts, 2011). In contrast, the Bearpaw cyclothem has been less easy to understand and correlate because of its time-transgressive nature across the Western Interior Basin, apparently initiating earlier in the south than in the northern Western Interior Basin (Roberts, 2007). The recent improvement in radiometric dating across the Western Interior Basin opens up excellent opportunities to reinvestigate the interplay between sea-level change, climate and vertebrate ecology and evolution in the Western Interior Basin. This problem should be a key area of future research.

Whatever the nature of the latitudinal barrier, the apparent lack of species-level taxa common to the north and south strongly indicates the presence of endemic dinosaur faunas for at least a portion of the late Campanian. Sampson, Loewen, et al. (2010) document the coeval presence of distinct chasmosaurine ceratopsids in the north and south of Laramidia, with *Chasmosaurus russelli* from the Dinosaur Park Formation occurring during the same interval as *Kosmoceratops richardsoni* and *Utahceratops gettyi* from the Kaiparowits Formation. An undescribed centrosaurine ceratopsid from the same interval within the Kaiparowits

Formation overlaps stratigraphically with *Centrosaurus apertus* of the Dinosaur Park Formation (Sampson and Loewen, 2010). Similarly, with regard to tyrannosaurids, *Teratophoneus curriei* from the middle unit of the Kaiparowits Formation overlaps stratigraphically with *Gorgosaurus* and *Daspletosaurus* in the Dinosaur Park Formation (Loewen et al., 2010).

Finally, Gates et al. (this volume, Chapter 19) document the concurrent presence of distinct hadrosaurid taxa in the Kaiparowits and Dinosaur Park formations. The hadrosaurine *Gryposaurus monumentensis*, from the middle unit of the Kaiparowits Formation, is temporally equivalent with *Prosaurolophus* from Alberta and Montana. The unidentified species of *Gryposaurus* from lower in section within the Kaiparowits Formation overlaps with *G. notabilis* and *G. incurvimanus* from the lower Dinosaur Park Formation. The lambeosaurine *Parasaurolophus* sp., which occurs in the same stratigraphic interval of the Kaiparowits Formation as *Gryposaurus* sp., overlaps stratigraphically with *Lambeosaurus lambei* and *Prosaurolophus maximus* in Alberta and *Hypacrosaurus stebingeri* in Montana. Other species of *Parasaurolophus* are found in different formations at putatively distinct stratigraphic levels: *P. cyrtocristatus* and *P. tubicen* from the Fruitland and Kirtland formations of New Mexico, respectively, and *P. walkeri* from the lower Dinosaur Park Formation of Alberta (Gates et al., this volume, Chapter 19). It must be noted, however, that current assertions of stratigraphic overlap between northern and southern forms are likely to be modified as the fossil record expands and our knowledge of chronostratigraphy becomes more refined.

The fact that no late Campanian dinosaur species can be confidently placed in both the north and south of Laramidia, despite the identification of dozens of taxa, is a striking observation. Before 2000, counterarguments to the provincialism hypothesis that were based on temporal and/or geographic sampling biases could not be excluded. With the new paleontologic and geochronologic data from the Kaiparowits Formation, these alternative hypotheses can now be effectively ruled out. Thus, we must account for largely distinct northern and southern macrovertebrate faunas for at least a million years of late Campanian time. This finding, arguably the only substantive example of intracontinental dinosaur biogeography for the entire Mesozoic Era, has major paleobiological implications. Current best estimates of late Campanian dinosaur faunas suggest the presence of at least 16 species at any given moment in ecological time. Of this number, approximately half qualify as giants, exceeding 1000 kg in average adult body mass, with most of these taxa exceeding 2000 kg.

The most comprehensive latitudinal comparison currently available spans a one million year interval of late Campanian time, between about 76.5 and 75.5 Ma, for which a reasonably complete fossil record is known for the Dinosaur Park Formation of southern Alberta and the Kaiparowits Formation of southern Utah. More specifically, on the basis of chronostratigraphic analyses, the fossiliferous middle unit of the Kaiparowits Formation is temporally equivalent to the lower half of the Dinosaur Park Formation (Roberts et al., this volume, Chapter 6). By far the most commonly found dinosaurs in the target intervals of both units are hadrosaurids, with at least three taxa occurring in the Dinosaur Park Formation, and two in the Kaiparowits Formation. The Dinosaur Park Formation includes two ceratopsids at any given time interval, whereas the middle portion of the Kaiparowits Formation (thus far the only portion of this unit that has been sampled) has three ceratopsids for certain: the chasmosaurines *Utahceratops gettyi* and *Kosmoceratops richardsoni*, as well as an undescribed genus and species of centrosaurine. With regard to ankylosaurians, the Dinosaur Park Formation has three representatives, whereas the Kaiparowits Formation has at least two taxa.

In the best sampled temporal window of the Campanian (~76.5–75.5 Ma), there appears to have been at least three ceratopsid taxa in the south and two in the north, two hadrosaurids in the south and three in the north, two ankylosaurids in the south and two in the north, and one tyrannosaurid in the south and two in the north, for a total of eight giant dinosaur taxa in the south and nine in the north. At present, given the minimal evidence of coeval dinosaur faunas further north or south on Laramidia, we cannot yet have any confidence that dinosaur faunas were homogenous even within the north or south of Laramidia during this interval. Indeed the bulk of dinosaur taxa are known only from one formation. This finding raises at least the possibility of additional, latitude-based faunal differentiation.

Studies of extant vertebrates demonstrate that large body sizes correlate closely with large individual home ranges and extensive species ranges, likely because of heightened dietary needs (Brown and Maurer, 1989; Brown, Marquet, and Taper, 1993; Marquet and Taper, 1998; Burness, Diamond, and Flannery, 2001). Thus, landmass size can restrict species diversity of megavertebrates, with bigger landmasses capable of supporting higher diversities (Burness, Diamond, and Flannery, 2001). For example, today the entire continent of Africa is home to six terrestrial species with a mean body mass in excess of 1000 kg: giraffe, black rhinoceros, white rhinoceros, hippopotamus (mostly freshwater rather than terrestrial), and two species of African elephant. All of these taxa are herbivorous; the continent's largest carnivore, the African lion, is typically less than 300 kg. How, then, could a landmass approximately the size of India plus Pakistan support (a minimum of) 13 species of giant herbivores and at least three species of giant carnivores, all weighing well in excess of 1000 kg? What does this finding tell us about the paleobiology of dinosaurs? Answering these questions is beyond the scope of this chapter, but the answers may well pertain to dinosaur physiology. One alternative is that most dinosaur clades possessed intermediate metabolic rates between those of typical ectotherms and endotherms (McNab, 2009; Sampson, 2009). An additional alternative is that Late Cretaceous floras offered a more abundant and/or more nutrient-rich resource for megaherbivores.

Although current evidence points strongly toward latitudinally arrayed dinosaur faunas on Laramidia during at least part of the late Campanian, it is still very possible that dinosaurs undertook large-scale migrations (contra Lehman, 2001; Bell and Snively, 2008; Sampson and Loewen, 2010). Wherever dinosaurs fell on the metabolic spectrum, a herd of hundreds (or even thousands) of ceratopsid or hadrosaurids clearly would have been forced to remain mobile in order to find sufficient food. Although current evidence indicates that dinosaurs did not migrate between the northern and southern regions of Laramidia, there would have had ample space – over 2000 miles, or 3200 km, in straight-line distance – for long distance migrations within these regions. By comparison, the longest annual terrestrial migrations today, those of North America caribou, are less than 400 miles (640 km) in straight-line distance, although an individual caribou may walk about 3000 miles (4800 km) during the course of a year (Fancy et al., 1989).

Evolutionary Implications

The results to date of the KBP also have important evolutionary implications. For example, the stratigraphic occurrences of dinosaurs in the Wahweap and Kaiparowits formations offer an unusual glimpse into faunal turnover. With regard to hadrosaurines, the Wahweap Formation has produced evidence only of *Acristavus* and other materials suggestive of close affiliations with *Brachylophosaurus*. In contrast, hadrosaurines from the Kaiparowits Formation all pertain to the distantly related genus *Gryposaurus* (Gates and Sampson, 2007; Gates et al., this volume, Chapter 19). Thus, rather than in situ evolution, this pattern of turnover indicates lineage replacement of hadrosaurines in at least the Utah region of Laramidia during the mid- to late Campanian (Gates et al., this volume, Chapter 19). A similar pattern has been documented in Alberta (Ryan and Evans, 2005); the mid-Campanian Oldman Formation has produced remains of only one hadrosaurine, *Brachylophosaurus*, whereas the overlying Dinosaur Park Formation has yielded evidence of *Gryposaurus* and *Prosaurolophus*. Within the Kaiparowits

Formation, *Gryposaurus* c.f. *notablis* is replaced by *G. monumentensis*. Although it is feasible that the latter pair of species represents ancestor and descendent forms, respectively, other alternatives are equally feasible and this hypothesis requires much additional testing.

Among ceratopsids, all three known taxa from the Wahweap Formation pertain to Centrosaurinae, whereas taxa of both centrosaurines and chasmosaurines (two chasmosaurines and one centrosaurine) occur in the overlying Kaiparowits Formation. Once again, a similar pattern has been documented in Alberta, with the Oldman Formation producing remains of at least three centrosaurine species and no chasmosaurines, in contrast to occurrence of both clades in the overlying Dinosaur Park Formation (Ryan and Evans, 2005). It is difficult to say at this point if the lack of chasmosaurines in the mid-Campanian is attributable to a genuine absence or to sampling bias, but the parallel is striking and may conceivably relate to biogeographic and/or evolutionary patterns.

The phylogenetic analysis of Chasmosaurinae that accompanied descriptions of *Kosmoceratops* and *Utahceratops* (Sampson, Loewen, et al., 2010) found evidence for distinct northern and southern clades during much of the late Campanian. *Chasmosaurus*, known only from Alberta's Dinosaur Provincial Park, was replaced near the end of this stage by *Vagaceratops*, apparently a descendant of a southern clade of chasmosaurines. Moreover, the results suggest that all more derived chasmosaurines from the latest Campanian and Maastrichtian (e.g., *Triceratops*, *Torosaurus*, and *Anchiceratops*) were also derived from a southern clade that includes *Pentaceratops* and *Utahceratops*. Intriguingly, the tyrannosaurid evidence exhibits a parallel pattern. For much of the late Campanian, *Daspletosaurus* and *Gorgosaurus* are the only known tyrannosaurids in northern Laramidia, whereas the new finds from Grand Staircase–Escalante National Monument suggest that the south was home to distinct taxa from a separate tyrannosaurid subclade. During the Maastrichtian, *Daspletosaurus* and *Gorgosaurus* are replaced in the north by *Tyrannosaurus*, which appears to be descended from the southern clade. If supported by additional analyses of these and other clades, this pattern suggests a northward displacement of provinces or biomes in the latest Campanian. Although highly speculative at this juncture, this displacement might feasibly have been associated with climate change linked to retreat of the Cretaceous Interior Seaway.

CONCLUSIONS

Recent work conducted in Grand Staircase–Escalante National Monument has revealed new Campanian-aged vertebrate assemblages highlighted by an array of previously unknown dinosaurs, many represented by exceptionally preserved materials. The discovery and analysis of macrofossils from the Kaiparowits and Wahweap formations within Grand Staircase–Escalante National Monument, together with an increased understanding of their geologic context, are: 1) illuminating the diversity and evolutionary history of vertebrates, invertebrates, and plants from the Kaiparowits Basin; 2) enabling reconstruction of the essential elements of several successive Late Cretaceous terrestrial ecosystems; and 3) permitting testing of key hypotheses relating to the tempo and mode of evolutionary change. Over the duration of this project, the Kaiparowits Formation in particular has become the best-documented Upper Cretaceous unit in the southern Western Interior Basin. Although the animals preserved belong to major groups that were present in coeval ecosystems further to the north (e.g., Montana, Alberta), there appears to have been a high degree of latitudinally based, species-level provinciality during this stage of the Late Cretaceous (Gates et al., 2010; Sampson, Loewen, et al., 2010). More specifically, the presence of north–south zonation of macrovertebrates suggests that, despite body sizes generally exceeding those of large-bodied mammals, late Campanian dinosaur species in the North American Western Interior possessed relatively small species ranges. Inhabiting a narrow, north–south-oriented belt of coastal and alluvial plains, these faunas may have been sensitive to latitudinal zonation of environments, despite the fact that paleotemperature gradients were markedly reduced relative to those of the present day, and changes in the sea level of the Western Interior Seaway. This in turn raises interesting questions relating to the physiology, ecology, and evolution of dinosaurs generally.

Finally, although the present discussion reviews work conducted to date, it should be noted that much of the research summarized herein is in progress. Several new taxa now under study will be described over the next few years, and more complete descriptions of existing taxa will be completed. Fieldwork also continues. With regard to the Kaiparowits Formation, most taxa are known only from fragmentary remains, and more complete specimens will undoubtedly lead to important insights. Moreover, sampling within this formation is currently biased heavily toward the middle unit, and it is possible that the lower and upper members will produce a number of new taxa. The Dinosaur Park Formation, by way of comparison, has yielded evidence of stratigraphically arrayed taxa, and perhaps a succession of faunas (Ryan and Evans, 2005). The KBP will continue efforts in the Wahweap Formation, which has generated several key specimens in recent years, and plans are in place to extend this work into the underlying Straight Cliffs Formation, also abundantly exposed in Grand Staircase–Escalante National Monument.

In short, an equivalent review conducted as little as five years from now will likely include numerous additional insights.

ACKNOWLEDGMENTS

Given its interdisciplinary framework and broad overarching goals, the great strength of the KBP has been productive collaborations. We sincerely thank our primary research collaborators: S. Bowring (MIT); D. Brinkman and J. Gardener (Royal Tyrrell Museum of Palaeontology); L. Claessens (College of the Holy Cross); A. Deino (Berkeley Geochronology Center); J. Eaton (Weber State University); D. Evans (University of Toronto); A. Farke and D. Lofgren (Raymond Alf Museum of Paleontology); H. Fricke (Colorado College); B. Gates and L. Zanno (Field Museum); H. Hutchison (University of California, Berkeley); R. Irmis (University of Utah); Z. Jinnah (University of the Witwatersrand); J. Kirkland and D. DeBlieux (U.S. Geological Survey); I. Miller and K. Johnson (Denver Museum of Nature & Science); R. Nydam (Arizona College of Osteopathic Medicine); P. O'Connor, co–principal investigator (Ohio University); and L. Tapanila (Idaho State University). For their unflagging support with permits, funding, field logistics, and actual fieldwork, we are grateful to Grand Staircase–Escalante National Monument and the Bureau of Land Management. Key individuals within these organizations include A. Titus (Grand Staircase–Escalante National Monument/BLM), M. Eaton (National Landscape Conservation System/BLM), and S. Foss (BLM). A. Titus in particular has been an outstanding partner in all phases of this project, including permits and funding, fossil prospection and excavation, preparation and research. For fieldwork, preparation, and curation support, sincere and hearty thanks go to the UMNH and Grand Staircase–Escalante National Monument paleontology volunteers. Without the ongoing efforts of these volunteers, including dozens of undergraduate and graduate students, there simply would be no KBP. For comments on earlier drafts of this chapter or portions thereof we thank D. Brinkman, B. Gates, R. Irmis, and L. Zanno. Funding for this work was provided by grants to S.D.S. from Grand Staircase–Escalante National Monument and the Bureau of Land Management, the National Science Foundation (EAR 0745454, 0819953), and the Discovery Channel.

Archibald, J. D. 1996. Dinosaur Extinction and the End of an Era. Columbia University Press, New York.

Averianov, A., and J. D. Archibald. 2003. Mammals from the Upper Cretaceous Aitym Formation, Kyzylkum Desert, Uzbekistan. Cretaceous Research 24:171–191.

Barrera, E., and S. M. Savin. 1999. Evolution of late Campanian–Maastrichtian marine climates and oceans; pp. 245–282 in E. Barrera, and C. C. Johnson (eds.), Evolution of the Cretaceous Ocean-Climate System. Geologic Society of America Special Paper 332.

Barrett, P. M. 2005. The diet of ostrich dinosaurs (Theropoda: Ornithomimosauria). Palaeontology 48:347–358.

Bell, P. R., and E. Snively. 2008. Polar dinosaurs on parade: a review of dinosaur migration. Alcheringa 32:271–284.

Blakey, R. C. 2009. Regional paleogeography. Northern Arizona University. Images and supporting materials available at http://jan.ucc.nau.edu/rcb7/regionaltext.html. Accessed January 2011.

Brinkman, D. B. 2005. Turtles: diversity, paleoecology, and distribution; pp. 202–220 in P. J. Currie and E. B. Koppelhus (eds.), Dinosaur Provincial Park: A Spectacular Ancient Ecosystem Revealed. Indiana University Press, Bloomington, Indiana.

Brown, J. H., and B. A. Maurer. 1989. Macroecology: the division of food and space among species on continents. Science 243:1145–1150.

Brown J. H., P. A. Marquet, and M. L. Taper. 1993. Evolution of body size: consequences of an energetic definition of fitness. American Naturalist 142:573–584.

Burness G. P., J. Diamond, and T. Flannery. 2001. Dinosaurs, dragons, and dwarves: the evolution of maximal body size. Proceedings of the National Academy of Sciences of the United States of America 98:14518–14523.

Carr, T., T. E. Williamson, B. B. Britt, and K. Stadtman. 2011. Evidence for high taxonomic and morphologic tyrannosauroid diversity in the Late Cretaceous (Late Campanian) of the American Southwest and a new short-skulled tyrannosaurid from the Kaiparowits Formation of Utah. Naturwissenschaften 98:241–246.

Catuneanu, O., A. R. Sweet, and A. D. Miall. 2000. Reciprocal stratigraphy of the Campanian–Paleocene Western Interior of North America. Sedimentary Geology 134:235–255.

Cifelli, R. L. 1990a. Cretaceous mammals of southern Utah. I. Marsupial mammals from the Kaiparowits Formation (Judithian). Journal of Vertebrate Paleontology 10:295–319.

Cifelli, R. L. 1990b. Cretaceous mammals of southern Utah. II. Marsupials and marsupial-like mammals from the Wahweap Formation (early Campanian). Journal of Vertebrate Paleontology 10:320–331.

Cifelli, R. L. 1990c. Cretaceous mammals of southern Utah. III. Therian mammals from the Turonian (early Late Cretaceous). Journal of Vertebrate Paleontology 10:332–345.

Cifelli, R. L. 1990d. Cretaceous mammals of southern Utah. IV. Eutherian mammals from the Wahweap (Aquilan) and Kaiparowits (Judithian) formations. Journal of Vertebrate Paleontology 10:346–360.

Cifelli, R. L., J. I. Kirkland, A. Weil, A. L. Deino, and B. J. Kowallis. 1997. High-precision ^{40}Ar/^{39}Ar geochronology and the advent of North America's Late Cretaceous terrestrial fauna. Proceedings of the National Academy of Sciences of the United States of America 94:11163–11167.

Chiappe, L. M. 2004. The closest relatives of birds. Ornitologia Neotropical 15(Supplement):1–16.

Chiappe, L. M., and G. J. Dyke. 2006. The early evolution of birds. Journal of the Paleontological Society of Korea 22:133–151.

Currie, P. J. 2005. Theropods, including birds; pp. 367–397 in P. J. Currie and E. B. Koppelhus (eds.), Dinosaur Provincial Park: A Spectacular Ancient Ecosystem Revealed. Indiana University Press, Bloomington, Indiana.

Currie, P. J., and E. B. Koppelhus (eds.). 2005. Dinosaur Provincial Park: A Spectacular Ancient Ecosystem Revealed. Indiana University Press, Bloomington, Indiana.

DeCourten, F. L., and D. A. Russell. 1985. A specimen of *Ornithomimus velox* (Theropoda, Ornithomimidae) from the terminal Cretaceous Kaiparowits Formation of southern Utah. Journal of Vertebrate Paleontology 59:1091–1099.

D'Emic, M. D., J. A. Wilson, and R. Thompson. 2010. The end of the sauropod hiatus in North America. Palaeogeography, Palaeoclimatology, Palaeoecology 297:486–490.

Eaton, J. G. 1991. Biostratigraphic framework for Upper Cretaceous rocks of the Kaiparowits Plateau, southern Utah; pp. 47–63 in J. D. Nations and J. G. Eaton (eds.), Stratigraphy, Depositional Environments, and Sedimentary Tectonics of the Western Margin, Cretaceous Western Interior Seaway. Geological Society of America Special Paper 260.

Eaton, J. G. 1999a. Vertebrate paleontology of the Paunsaugunt Plateau, Upper Cretaceous, southwestern Utah; pp. 335–338 in D. D. Gillette (ed.), Vertebrate Paleontology in Utah. Utah Geological Survey Miscellaneous Publication 99-1.

Eaton, J. G. 1999b. Vertebrate paleontology of the Iron Springs Formation, Upper Cretaceous, southwestern Utah; pp. 339–343 in D. D. Gillette (ed.), Vertebrate Paleontology in Utah. Utah Geological Survey Miscellaneous Publication 99-1.

Eaton, J. G. 2002. Multituberculate Mammals from the Wahweap (Campanian, Aquilan) and Kaiparowits (Campanian, Judithian) Formations, Within and Near the Grand Staircase–Escalante National Monument, Southern Utah. Utah Geological Survey Miscellaneous Publication 02-4.

Eaton, J. G., and R. L Cifelli. 1988. Preliminary report on Late Cretaceous mammals of the Kaiparowits Plateau, southern Utah. Contributions to Geology, University of Wyoming 26:45–56.

Eaton, J. G., and J. D. Nations. 1991. Introduction; tectonic setting along the margin of the Cretaceous Western Interior Seaway, southwestern Utah and northern Arizona; pp. 1–8 in J. D. Nations and J. G. Eaton (eds.), Stratigraphy, Depositional Environments, and Sedimentary Tectonics of the Western Margin, Cretaceous Western Interior Seaway. Geological Society of America Special Paper 260.

Eaton, J. G., R. L. Cifelli, J. H. Hutchison, J. I. Kirkland, and J. M. Parrish. 1999. Cretaceous vertebrate faunas from the Kaiparowits Plateau, south central Utah; pp. 345–353 in D. D. Gillette (ed.), Vertebrate Paleontology in Utah. Utah Geological Survey Miscellaneous Publication 99-1.

Fancy, S. G., L. F. Pank, K. R. Whitten, and W. L. Reglin. 1989. Seasonal movements of caribou in arctic Alaska as determined by satellite. Canadian Journal of Zoology 67:644–650

Farlow, J. O., P. Dodson, and A. Chinsamy. 1995. Dinosaur biology. Annual Review of Ecology and Systematics 26:445–471.

Gates, T. 2004. Hadrosaurian dinosaur diversity from the Upper Campanian Kaiparowits Formation, southern Utah. Journal of Vertebrate Paleontology 24:63A.

Gates, T. A., and S. D. Sampson. 2007. A new species of *Gryposaurus* (Dinosauria: Hadrosauridae) from the Upper Campanian Kaiparowits Formation of Utah. Zoological Journal of the Linnean Society 151:351–376.

Gates, T. A., J. R. Horner, R. R. Hanna, and C. R. Nelson. 2011. New unadorned hadrosaurine hadrosaurid (Dinosauria, Ornithopoda) from the Campanian of North America. Journal of Vertebrate Paleontology 31:798–811.

Gates, T. A., A. Prieto-Márquez, L. E. Zanno. 2012. Mountain building triggered Late Cretaceous North American megaherbivore dinosaur radiation. PLoS One 7(8):e42135.

Gates, T. A., S. D. Sampson, L. E. Zanno, E. M. Roberts, J. G. Eaton, R. L. Nydam, J. H. Hutchison, J. A. Smith, M. A. Loewen, and M. A. Getty. 2010. Biogeography of terrestrial and freshwater vertebrates from the Late Cretaceous (Campanian) Western Interior of North America. Palaeogeography, Palaeoclimatology, Palaeoecology 291:371–387.

Getty, M. A., M. A. Loewen, and E. M. Roberts. 2006. Collection and use of taphonomic data from vertebrate localities: lessons from six years of paleontological inventory and excavation in Grand Staircase–Escalante National Monument, Utah. Journal of Vertebrate Paleontology 26:67A.

Getty, M. A., M. A. Loewen, E. M. Roberts, A. L. Titus, and S. D. Sampson. 2010. Taphonomy of horned dinosaurs (Ornithischia: Ceratopsidae) from the late Campanian Kaiparowits Formation, Grand Staircase–Escalante National Monument, Utah; pp. 478–494 in M. J. Ryan, B. J. Chinnery-Allgeier, and D. A. Eberth (eds.), New Perspectives on Horned Dinosaurs. Indiana University Press, Bloomington, Indiana.

Getty, M. A., E. M. Roberts, M. A. Loewen, J. A. Smith, T. A. Gates, and S. D. Sampson. 2003. Taphonomy of a chasmosaurine ceratopsid skeleton from the Campanian Kaiparowits Formation, Grand Staircase–Escalante National Monument, Utah. Journal of Vertebrate Paleontology 23:54A–55A.

Sampson et al.

Gillette, D. D., and M. C. Hayden. 1997. A Preliminary Assessment of Paleontological Resources within the Grand Staircase–Escalante National Monument, Utah. Utah Geological Survey Circular C-96.

Gregory, H. E., and R. C. Moore. 1931. The Kaiparowits Region: Geographic and Geologic Reconnaissance of Parts of Utah and Arizona. U.S. Geological Survey Professional Paper 164.

Haq, B. U., J. Hardenbol, and P. R. Vail. 1987. Chronology of fluctuating sea levels since the Triassic. Science 235:1156–1167.

Hicks, J. F., J. D. Obradovich, and L. Tauxe. 1999. Magnetostratigraphy, isotopic age calibration and intercontinental correlation of the Red Bird section of the Pierre Shale, Niobrara County, Wyoming, U.S.A. Cretaceous Research 20:1–27.

Holtz, T. R., Jr. 1988. Denticle morphometrics and a possible omnivorous feeding habit for the theropod dinosaur *Troodon*. In B. P. Pérez-Moreno, T. R. Holtz Jr., J. L. Sanz, and J. J. Moratalla (eds.), Aspects of Theropod Paleobiology. Gaia 15:159–166.

Hutchison, J. H. 1993. *Avisaurus;* a "dinosaur" grows wings. Journal of Vertebrate Paleontology 13:43A.

Hutchison, J. H., J. G. Eaton, P. A. Holroyd, and M. B. Goodwin. 1997. Larger vertebrates of the Kaiparowits Formation (Campanian) in the Grand Staircase–Escalante National Monument and adjacent areas; pp. 391–398 in L. M. Hill (ed.), Learning from the Land: Grand Staircase–Escalante National Monument Science Symposium Proceedings. U.S. Department of the Interior, Bureau of Land Management.

Ji, Q., P. J. Currie, M. A. Norell, and S.-A. Ji. 1998. Two feathered dinosaurs from northeastern China. Nature 393:753–761.

Jinnah, Z. A., and E. M. Roberts. 2011. Facies associations, paleoenvironment, and base-level changes in the upper cretaceous Wahweap Formation, Utah, U.S.A. Journal of Sedimentary Research 81:266–283.

Jinnah, Z. A., E. M. Roberts, A. L. Deino, J. S. Larsen, P. K. Link, and C. M. Fanning. 2009. New 40Ar-39Ar and detrital zircon U-Pb ages for the Upper Cretaceous Wahweap and Kaiparowits formations on the Kaiparowits Plateau, Utah: implications for regional correlation, provenance, and biostratigraphy. Cretaceous Research 30:287–299.

Kauffman, E. G. 1977. Geological and biological overview: Western Interior Cretaceous Basin. Mountain Geologist 14:75–99.

Kirkland, J. I., and D. D. DeBlieux. 2005. Dinosaur remains from the Lower to Middle Campanian Wahweap Formation at Grand Staircase–Escalante National Monument, southern Utah. Journal of Vertebrate Paleontology 25:78A.

Kirkland, J. I., and D. D. DeBlieux. 2010. New centrosaurine ceratopsian skulls from the Wahweap Formation (middle Campanian), Grand Staircase–Escalante National Monument, southern Utah; pp. 117–140 in M. J. Ryan, B. J. Chinnery-Allgeier, and D. A. Eberth (eds.), New Perspectives on Horned Dinosaurs. Indiana University Press, Bloomington, Indiana.

Kirkland, J. I., D. D. DeBlieux, J. A. Smith, and S. D. Sampson. 2002. New ceratopsid remains from the lower Campanian Wahweap Formation, Grand Staircase–Escalante National Monument,

Utah. Journal of Vertebrate Paleontology 22:74A.

Knell, M. J., F. D. Jackson, A. L. Titus, and L. B. Albright III. 2010. A gravid fossil turtle from the Upper Cretaceous (Campanian) Kaiparowits Formation, southern Utah. Historical Biology 23:57–62.

Lehman, T. M. 1987. Late Maastrichtian paleoenvironments and dinosaur biogeography in the western interior of North America. Palaeogeography, Palaeoclimatology, Palaeoecology 60:189–217.

Lehman, T. M. 1997. Late Campanian dinosaur biogeography in the Western Interior of North America; pp. 223–240 in D. L. Wolberg, E. Stump, and G. D. Rosenberg (eds.), Dinofest International Symposium. Philadelphia Academy of Natural Sciences.

Lehman, T. M. 2001. Late Cretaceous dinosaur provinciality; pp. 310–328 in D. H. Tanke and K. Carpenter (eds.), Mesozoic Vertebrate Life. Indiana University Press, Bloomington, Indiana.

Loewen, M. A., J. W. Sertich, R. B. Irmis, and S. D. Sampson. 2010. Tyrannosaurid evolution and intracontinental endemism in Laramidia: new evidence from the Campanian Wahweap Formation of Utah. Journal of Vertebrate Paleontology 30:123A.

Longrich, N. R., and P. J. Currie. 2008. *Albertonykus borealis,* a new alvarezsaur (Dinosauria: Theropoda) from the early Maastrichtian of Alberta, Canada: implications for the systematics and ecology of Alvarezsauridae. Cretaceous Research 30:239–252.

Longrich, N. R., and P. J. Currie 2009. A microraptorine (Dinosauria–Dromaeosauridae) from the Late Cretaceous of North America. Proceedings of the National Academy of Sciences of the United States of America 106:5002–5007.

Lund, E. K., M. A. Loewen, E. M. Roberts, S. D. Sampson, and M. A. Getty. 2009. Dinosaur skin impressions in the Upper Cretaceous (Late Campanian) Kaiparowits Formation, Southern Utah: insights into preservation mode. Advances in Western Interior Late Cretaceous Paleontology and Geology, St. George, Utah, Abstracts with Program, p. 32. Bureau of Land Management.

Marquet, P. A., and M. L. Taper. 1998. On size and area: patterns of mammalian body size extremes across landmasses. Evolutionary Ecology 12:127–139.

McNab, B. K. 2009. Resources and energetics determined dinosaur maximal size. Proceedings of the National Academy of Sciences of the United States of America 106:12184–12188.

Neabore, S., M. A. Loewen, L. E. Zanno, M. A. Getty, and L. Claessens. 2007. Three-dimensional scanning and analysis of the first diagnostic ornithomimid forelimb material from the Late Cretaceous Kaiparowits Formation. Journal of Vertebrate Paleontology 27:123A.

Parrish, J. M., and J. G. Eaton. 1991. Diversity and evolution of dinosaurs in the Cretaceous of the Kaiparowits Plateau, Utah. Journal of Vertebrate Paleontology 11:50A.

Pereda-Suberbiola, X. 2009. Biogeographical affinities of Late Cretaceous continental tetrapods of Europe: a review. Bulletin de la Société Géologique de France 180:57–71.

PLATES. 2009. The PLATES Project; L. Lawver, I. Dalziel (principal investigators), and L. Gahagan (database and software manager). University of Texas Institute for Geophysics, Austin, Texas.

Pollock, S. L. 1999. Provenance, geometry, lithofacies, and age of the Upper Cretaceous Wahweap Formation, Cordilleran Foreland Basin, southern Utah. M.Sc. thesis, New Mexico State University, Las Cruces, New Mexico.

Prieto-Marquez, A., and G. Salinas. 2010. A re-evaluation of *Secernosaurus koerneri* and *Kritosaurus australis* (Dinosauria, Hadrosauridae) from the Late Cretaceous of Argentina. Journal of Vertebrate Paleontology 30:813–837.

Roberts, E. M. 2005a. Taphonomy of wet alluvial systems: a case example from the Cretaceous Western Interior Basin of Utah. Journal of Vertebrate Paleontology 25:105A.

Roberts, E. M. 2005b. Stratigraphic, taphonomic and paleoenvironmental analysis of the Upper Cretaceous Kaiparowits Formation, Grand Staircase–Escalante National Monument, Southern Utah. Ph.D. dissertation, University of Utah, Salt Lake City, Utah.

Roberts, E. M. 2007. Facies architecture and depositional environments of the Upper Cretaceous Kaiparowits Formation, southern Utah. Sedimentary Geology 197:207–233.

Roberts, E. M., and L. Tapanila. 2006. A new social insect nest trace from the Late Cretaceous Kaiparowits Formation of southern Utah. Journal of Paleontology 80:768–774.

Roberts, E. M., A. D. Deino, and M. A. Chan. 2005. 40Ar/39Ar age of the Kaiparowits Formation, southern Utah, and correlation of coeval strata and faunas along the margin of the Western Interior Basin. Cretaceous Research 26:307–318.

Roberts, E. M., R. R. Rogers, and B. Z. Foreman. 2007. Continental insect borings in dinosaur bone: examples from the Late Cretaceous of Madagascar and Utah. Journal of Paleontology 81:201–208.

Roberts, E. M., L. Tapanila, and B. Mijal. 2008. Taphonomy and sedimentology of storm-generated continental shell beds: a case example from the Cretaceous Western Interior Basin. Journal of Geology 116:462–479.

Rogers, R. R. 1998. Sequence analysis of the Upper Cretaceous Two Medicine and Judith River formations, Montana: nonmarine response to the Claggett and Bearpaw marine cycles. Journal of Sedimentary Research 68:615–631.

Russell, D. A. 1967. A census of dinosaur specimens collected in western Canada. National Museum of Canada Natural History Papers 36:1–13.

Russell, D. A. 1993. The role of central Asia in dinosaurian biogeography. Canadian Journal of Earth Sciences 30:2002–2012.

Russell, D. A. 1995. China and the lost worlds of the dinosaur era. Historical Biology 10:3–12.

Ryan, M. J. 2007. A new basal centrosaurine ceratopsid from the Oldman Formation, southeastern Alberta. Journal of Paleontology 81:376–396.

Ryan, M. J., and D. C. Evans. 2005. Ornithischian dinosaurs; pp. 312–348 in P. J. Currie and E. B. Koppelhus (eds.), Dinosaur Provincial Park:

A Spectacular Ancient Ecosystem Revealed. Indiana University Press, Bloomington, Indiana.

Sampson, S. D. 2009. Dinosaur Odyssey: Fossil Threads in the Web of Life. University of California Press, Berkeley, California.

Sampson, S. D., and M. A. Loewen. 2010. Unraveling a radiation: a review of the diversity, stratigraphic distribution, biogeography, and evolution of horned dinosaurs. (Ornithischia: Ceratopsidae); pp. 405–427 in M. J. Ryan, B. J. Chinnery-Allgeier, and D. A. Eberth (eds.), New Perspectives on Horned Dinosaurs. Indiana University Press, Bloomington, Indiana.

Sampson, S. D., M. A. Loewen, E. M. Roberts, J. A. Smith, L. E. Zanno, and T. A. Gates. 2004. Provincialism in Late Cretaceous terrestrial faunas: new evidence from the Campanian Kaiparowits Formation of Utah. Journal of Vertebrate Paleontology 24:108A.

Sampson, S. D., M. A. Loewen, T. A. Gates, M. A. Getty, and L. E. Zanno. 2002. New evidence of dinosaurs and other vertebrates from the Upper Cretaceous Wahweap and Kaiparowits Formations, Grand Staircase–Escalante National Monument. Geological Society of America, Rocky Mountain Section Meeting Abstracts with Progam.

Sampson, S. D., M. A. Loewen, A. A. Farke, E. M. Roberts, C. A. Forster, J. A. Smith, and A. L. Titus. 2010. New horned dinosaurs from Utah provide evidence for intracontinental dinosaur endemism. PLoS One 5(9):e12292.

Sampson, S. D., T. A. Gates, E. M. Roberts, M. A. Getty, L. E. Zanno, M. A. Loewen, J. A. Smith, E. K. Lund, J. Sertich, and A. L. Titus. 2010. Grand Staircase–Escalante National Monument: a new and critical window into the world of dinosaurs; pp. 161–179 in M. Eaton (ed.), Learning from the Land, Grand Staircase–Escalante National Monument Science Symposium Proceedings. Grand Staircase–Escalante Partners.

Scotese, C. R. 2001. Atlas of Earth History. Volume 1, Paleogeography. Paleomap Project, Arlington, Virginia.

Sereno, P. C. 1986. Phylogeny of the bird-hipped dinosaurs (order Ornithischia). National Geographic Research 2:234–256.

Sloan, R. E. 1969. Cretaceous and Paleocene terrestrial communities of western North America. North America Paleontological Convention, Chicago, Proceedings E:427–E453.

Smith, J. A., M. A. Getty, T. A. Gates, E. M. Roberts, and S. D. Sampson. 2003. Fossil vertebrates from the Kaiparowits Formation, Grand Staircase–Escalante National Monument: an important window into the Late Cretaceous of Utah. Journal of Vertebrate Paleontology 23:98A.

Sullivan, R. M. 2003. Revision of the dinosaur Stegoceras Lambe (Ornithischia, Pachycephalosauridae). Journal of Vertebrate Paleontology 23:181–207.

Sullivan, R. M., and S. G. Lucas. 2006. The Kirtlandian Land-Vertebrate "Age" – faunal composition, temporal position and biostratigraphic correlation in the nonmarine Upper Cretaceous of western North America; pp. 7–29 in S. G. Lucas and R. M. Sullivan (eds.), Late Cretaceous Vertebrates from the Western Interior. New Mexico Museum of Natural History and Science Bulletin 35.

Titus, A. L., J. D. Powell, E. M. Roberts, S. D. Sampson, S. L. Pollock, J. I. Kirkland, and L. B. Albright. 2005. Late Cretaceous stratigraphy, depositional environments, and macrovertebrate paleontology of the Kaiparowits Plateau, Grand Staircase–Escalante National Monument, Utah; pp. 101–128 in J. Pederson and C. M. Dehler (eds.), Interior Western United States. Geological Society of America Field Guide 6.

Vandermark, D., J. A. Tarduno, D. B. Brinkman, R. D. Cottrell, and S. Mason. 2009. New Late Cretaceous macrobaenid turtle with Asian affinities from the High Canadian Arctic: dispersal via ice-free polar routes. Geology 37:183–186.

Vavrek, M. J., and H. C. E. Larsson. 2010. Low beta diversity of Maastrichtian dinosaurs of North America. Proceedings of the National Academy of Sciences of the United States of America 107:8265–8268.

Wiersma, J., J. H. Hutchison, and T. A. Gates. 2004. Crocodilian diversity in the Upper Cretaceous Kaiparowits Formation (Upper Campanian), Utah. Journal of Vertebrate Paleontology 24:129A.

Wu, X.-C. 2005. Crocodylians; pp. 277–291 in P. J. Currie, and E. B. Koppelhus (eds.), Dinosaur Provincial Park: A Spectacular Ancient Ecosystem Revealed. Indiana University Press, Bloomington, Indiana.

Xu, X., Y.-N. Cheng, X.-L. Wang, and C.-H. Chang. 2002. An unusual oviraptorosaurian dinosaur from China. Nature 419:291–293.

Zanno, L. E., and S. D. Sampson. 2005. A new oviraptorosaur (Theropoda: Maniraptora) from the late Campanian of Utah and the status of the North American Oviraptorosauria. Journal of Vertebrate Paleontology 25:897–904.

Zanno, L. E., D. D. Gillette, L. B. Albright, and A. L. Titus. 2009. A new North American therizinosaurid and the role of herbivory in "predatory" dinosaur evolution. Proceedings of the Royal Society B: Biological Sciences 276:3505–3511.

Zanno L. E., D. J. Varricchio, P. M. O'Connor, A. L. Titus, and M. J. Knell. 2011. A new troodontid theropod, Talos sampsoni gen. et sp. nov., from the Upper Cretaceous Western Interior Basin of North America. PLoS One 6(9):e24487.

Index

ALAN L. TITUS is Monument Paleontologist at Grand Staircase–Escalante National Monument in Utah and Adjunct Curator, Natural History Museum of Utah.

MARK A. LOEWEN is Adjunct Assistant Professor, Department of Geology and Geophysics, University of Utah and Research Associate, Natural History Museum of Utah.

This book was designed by Jamison Cockerham and set in type by Tony Brewer at Indiana University Press, and printed by Sheridan Books, Inc.

The fonts are Electra, designed by William A. Dwiggins in 1935, Frutiger, designed by Adrian Frutiger in 1975, and Futura, designed by Paul Renner in 1927. All were published by Adobe Systems Incorporated.

Lightning Source UK Ltd.
Milton Keynes UK
UKHW050647220721
387536UK00002B/44